PHARMACEUTICAL MANUFACTURING ENCYCLOPEDIA

Third Edition

Volume 3

H through P

Volume 1:

Preface

Contents

A through B

Volume 2:

C through G

Volume 3:

H through P

Volume 4:

Q through Z

Raw Materials Index

Trade Names Index

PHARMACEUTICAL MANUFACTURING ENCYCLOPEDIA

Third Edition

Volume 3

H through P

William Andrew Publishing

Norwich, NY, U.S.A.

Cover by Brent Beckley

Library of Congress Cataloging-in-Publication Data

Pharmaceutical manufacturing encyclopedia. -- 3rd ed.
p. ; cm.
Rev. ed. of: Pharmaceutical manufacturing encyclopedia / by Marshall Sittig. c1988.
Includes bibliographical references and index.
ISBN-13: 978-0-8155-1526-5 (set : alk. paper)
ISBN-10: 0-8155-1526-X (set : alk paper)
1. Drugs--Synthesis--Dictionaries. 2. Pharmaceutical chemistry--Dictionaries. I. Sittig, Marshall. Pharmaceutical manufacturing encyclopedia. II. William Andrew Publishing.
[DNLM: 1. Pharmaceutical Preparations--Encyclopedias--English. 2. Chemistry, Pharmaceutical--Encyclopedias--English. QV 13 P536 2007]

RS402.5.S58 2007
615'.19--dc22

2006023472

Printed in India
This book is printed on acid-free paper.

10 9 8 7 6 5 4 3 2 1

Published by:
William Andrew Publishing
13 Eaton Avenue
Norwich, NY 13815
1-800-932-7045
www.williamandrew.com

Exclusively distributed in the Indian Subcontinent by Affiliated East-West Press Pvt. Ltd., G-1/16, Ansari Road, Darya Ganj, New Delhi 110 002

NOTICE

H

HALAZEPAM

Therapeutic Function: Tranquilizer

Chemical Name: 7-Chloro-1,3-dihydro-5-phenyl-1-(2,2,2-trifluoroethyl)-2H-1,4-benzodiazepine-2-one

Common Name: -

Structural Formula:

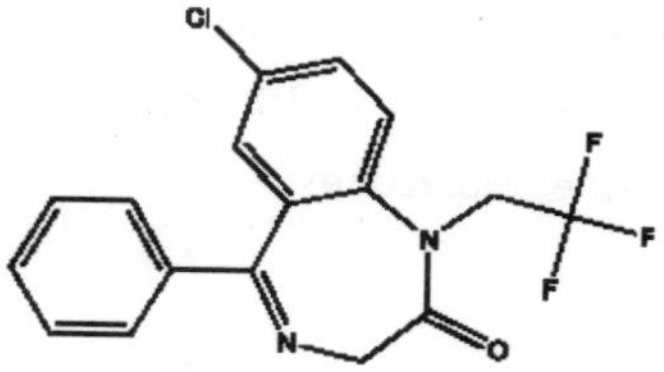

Chemical Abstracts Registry No.: 23092-77-3

Trade Name	Manufacturer	Country	Year Introduced
Paxipam	Schering	US	1981

Raw Materials

7-Chloro-1,3-dihydro-5-phenyl-2H-1,4-benzodiazepine-2-one
Sodium
Methanol
2,2,2-Trifluoroethyl iodide

Manufacturing Process

Prepare a solution of sodium methylate by dissolving 3.9 g of sodium metal in 500 ml of methanol. Add 39.0 g of 7-chloro-1,3-dihydro-5-phenyl-2H-1,4-benzodiazepine-2-one. Evaporate the reaction mixture to a residue and dissolve the residue in 170 ml of dimethylformamide. Add 30 g of 2,2,2-trifluoroethyl iodide and stir at room temperature for ½ hour, then heat to 60°C to 70°C for an additional 7 hours. Add 19 g of 2,2,2-trifluoroethyl iodide and resume the heating and stirring at 60°C to 70°C for an additional 16 hours. Filter off the solids and evaporate the filtrate to a residue in vacuo. Triturate the residue with water and extract with ethyl ether. Wash the ethereal extract with water, dry over anhydrous sodium sulfate and evaporate the solvent to a residue.

Extract the residue with ethyl ether and filter. Concentrate the ethereal extract to a residue. Dissolve the residue in benzene and chromatograph on 300 g of alumina contained in a glass column 1.5 inches in diameter to give the crude product. Elute with benzene. Crystallize this product from acetone-petroleum ether to obtain the product.

References

Merck Index 4472
DFU 3 (2) 109 (1978)
PDR p.1645
DOT 9 (6) 237 (1973), 11 (5) 191, 211 (1975) and 18 (8) 367 (1982)
I.N. p.476
REM p.1062
Topliss, J.G.; US Patents 3,429,874; Feb. 25, 1969 and 3,641,147; Feb. 8, 1972; both as signed to Schering Corp.

HALAZONE

Therapeutic Function: Antiseptic, Antiseptic

Chemical Name: Benzoic acid, 4-((dichloroamino)sulfonyl)-

Common Name: Alazone; Halazone

Structural Formula:

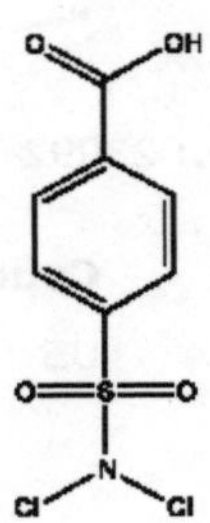

Chemical Abstracts Registry No.: 80-13-7

Trade Name	Manufacturer	Country	Year Introduced
Clordiapenser Disinf Alazone	Molteni Farmaceutici SpA.	-	-
Pantocid	Organika	-	-

Raw Materials

4-(Aminosulfonyl)benzoic acid
Sodium hypochlorite

Manufacturing Process

20 kg p-aminosulfonyl benzoic acid was mixed with 400 L 0.5 N NaOCl. After an acidification with 100 L 1 N acetic acid, a precipitate dropped out. It was filtered off, washed and dried. The obtained 4-((dichloroamino)sulfonyl) benzoic acid had MP: 203°C.

References

Claasz M., D.R. Patent No. 318,899; Feb. 18; 1920

HALCINONIDE

Therapeutic Function: Topical corticosteroid

Chemical Name: 21-Chloro-9α-fluoro-δ4-pregnene-11β,16α,17α-triol-3,20-dione16,17-acetonide

Common Name: -

Structural Formula:

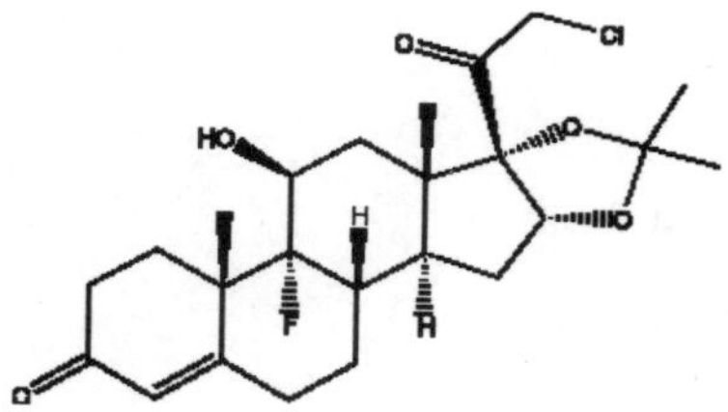

Chemical Abstracts Registry No.: 3093-35-4

Trade Name	Manufacturer	Country	Year Introduced
Halog	Squibb	US	1974
Halciderm	Squibb	UK	1974
Halciderm	Squibb	Italy	1976
Halog	Von Heyden	W. Germany	1977
Halog	Squibb	France	1979
Halciderm	Squibb	US	1980
Adcortin	Sankyo	Japan	1982
Beta Corton	Spirig	Switz.	-
Dihalog	Heyden	W. Germany	-
Halcimat	Heyden	W. Germany	-
Halcort	Fair	UK	-
Volog	Squibb	-	-

Raw Materials

16α-Hydroxy-9α-fluorohydrocortisone acetonide
Methanesulfonyl chloride
Lithium chloride

Manufacturing Process

(A) 16α-Hydroxy-9α-fluorohydrocortisone acetonide 27-mesylate: To a solution of 1.5 g of 16α-hydroxy-9α-fluorohydroortisone acetonide in 15 ml of dry pyridine is added at 0°C. 1.5 ml of methane-sulfonyl chloride. After standing in the refrigerator for 2,5 hours, excess methanesulfonyl chloride is destroyed by the addition of a small amount of ice, after which ice-water is added slowly to precipitate the reaction product. After ½ hour in the refrigerator the material is filtered off, washed thoroughly with water and dried in vacuo. The resulting crude material after recrystallization from acetone-hexane gives the pure 21-mesylate of the following properties: melting point about 225°C to 227°C (decomposition); $[\alpha]_D^{23}$+112° (c, 0.5 in chloroform).

(B) 21-Chloro-9α-fluoro-δ4-pregnene-11β,16α,17α-triol-3,20-dione16,17-acetonide : A solution of 200 mg of the acetonide 21-mesylate from part (A) and 900 mg of lithium chloride in 25 ml of dimethylformamide is kept at 100°C for 24 hours. The mixture is poured on ice, extracted with chloroform and the chloroform extract washed with water and dried over sodium sulfate. Evaporation of the solvent in vacuo furnishes the crystalline chloride, which after recrystallization from acetone-ethanol has a melting point about 276°C to 277°C.

References

Merck Index 4474
Kleeman and Engel p.454
PDR p.1745
OCDS Vol.2 p.187 (1980)
DOT 10 (11) 305 (1974)
I.N. p.477
REM p.972
Difazio, L.T. and Augustine, M.A.; US Patent 3,892,857; July 1, 1975; assigned to E.R. Squibb and Sons, Inc.

HALOPERIDOL

Therapeutic Function: Antidyskinetic, Antipsychotic

Chemical Name: 4-[4-(4-Chlorophenyl)-4-hydroxy-1-piperidinyl]-1-(4-fluorophenyl)-1-butanone

Common Name: -

Structural Formula:

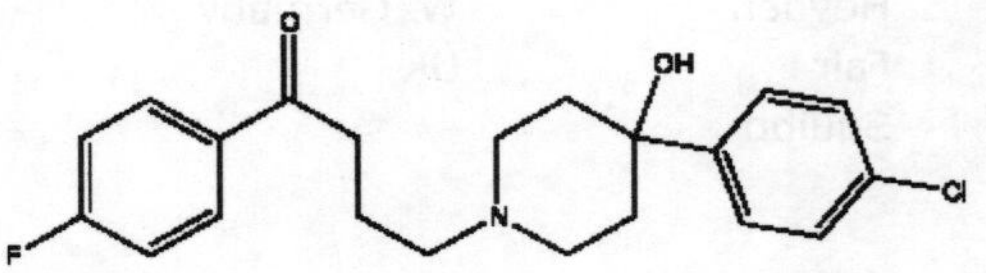

Chemical Abstracts Registry No.: 52-86-8

Trade Name	Manufacturer	Country	Year Introduced
Haldol	Janssen-Le Brun	France	1960
Haldol	McNeil	US	1967
Serenace	Searle	UK	1969
Fortunan	Steinhard	UK	1983
Bioperidolo	Firma	Italy	-
Brotopon	Pfizer Taito	Japan	-
Einalon S	Maruko	Japan	-
Eukystol	Merckle	W. Germany	-
Halidol	Abic	Israel	-
Halo Just	Horita	Japan	-
Haloperidol	Mohan	Japan	-
Halosten	Shionogi	Japan	-
Keselan	Sumitomo	Japan	-
Linton	Yoshitomi	Japan	-
Pacedol	Protea	Australia	-
Peluces	Isei	Japan	-
Peridor	Unipharm	Israel	-
Selezyme	Sawai	Japan	-
Serenace	Dainippon	Japan	-
Serenase	Lusofarmaco	Italy	-
Serenase	Orion	Finland	-
Sigaperidol	Siegfried	Switz.	-
Vesadol	Le Brun	France	-

Raw Materials

4-(4-Chlorophenyl)piperidin-4-ol hydrochloride
1,1-Dimethoxy-1-(4-fluorophenyl)-4-chlorobutane
Hydrogen chloride
Ammonia

Manufacturing Process

A stirred slurry of 120.0 parts 4-(4-chlorophenyl)-piperidin-4-ol hydrochloride and 40.0 parts of potassium iodide in 500 parts of water is warmed to a temperature of about 35°C under a nitrogen atmosphere. Then, 70.0 parts of potassium hydroxide is added. After further heating to about 55°C. 138.0 parts of 1,1 dimethoxy-1-(4-fluorophenyl)-4-chlorobutane is added. The temperature is then raised to about 102°C and heating continued for 3.5 hours. After cooling to about 75°C. 785 parts of toluene is added to the reaction mixture and stirred for about 5 minutes. An additional 320 parts of toluene is added and the water and organic layers separated. 102 parts of methanol is used to rinse the flask and added to the organic layer to provide a solution of 4-(4-chlorophenyl)-1-[4-(4-fluorophenyl)-4,4-dimethoxybutyl]-piperidin-4-ol. Then, 59 parts of concentrated hydrochloric acid is added to a stirred solution of the organic layer to precipitate a solid. The solid is filtered, rinsed twice with 550 parts by volume portions of a 10:9:1 acetone-toluene-methanol mixture, twice with 400 parts by volume portions of a 10:l acetone-methanol mixture, and air-dried. The dried solid is then dissolved in 1,950 parts of methanol with gentle heating on a steam bath. The resulting solution is filtered and 300 parts by volume of concentrated ammonium hydroxide is added. Heating is continued to reflux and maintained thereat for about 1 hour.

Then, 2,520 parts of water is added and the slurry stirred at about 75°C for 1.5 hours. After cooling to about 25°C. the solid is filtered, washed twice with 600 parts by volume portions of a 3:1 mixture of water-methanol, and air-dried. The resulting product, 4-[4-chlorophenyl)-4-hydroxypiperidino]-4'-fluorobutyrophenone, is obtained in 32.5% yield. This product melts at about 148.5°C to 150.5°C.

References

Merck Index 4480
Kleeman and Engel p.454
PDR p.1089
OCDS Vol.1 p.306 (1977)
DOT 9 (6) 234 (1973)
I.N. p.478
REM p.1088
Dryden, H.L. Jr. and Erickson, R.A.; US Patent 4,086,234; April 25, 1978; assigned to G.D. Searle and Co.

HALOPREDONE ACETATE

Therapeutic Function: Topical antiinflammatory

Chemical Name: 17,21-Bis(acetyloxy)-2-bromo-6,9-difluoro-11-hydroxypregna-1,4-diene-3,20-dione

Common Name: -

Structural Formula:

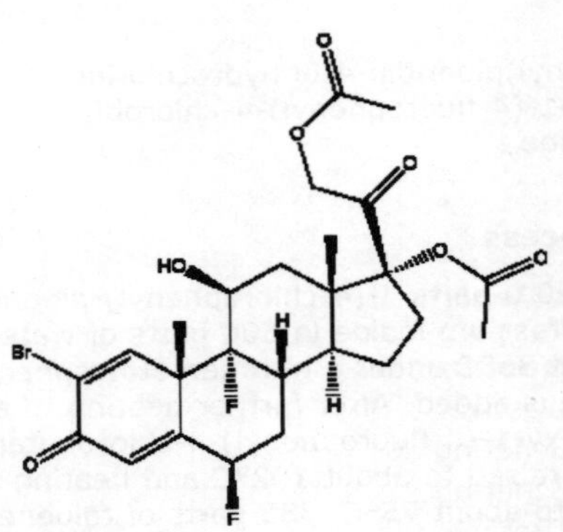

Chemical Abstracts Registry No.: 57781-14-3; 57781-15-4 (Base)

Trade Name	Manufacturer	Country	Year Introduced
Topicon	Pierrel	Italy	1983

Raw Materials

2-Bromo-6β-fluoro-17α,21 dihydroxy-9β,11β-oxido-pregna-1,4-diene-3,20-dione-17-21-diacetate
Hydrogen fluoride

Manufacturing Process

100 ml of a 70% hydrofluoric acid aqueous solution were cooled to -10°C in a polyethylene flask equipped with electromagnetic stirrer. 10 g of 2-bromo-6β-fluoro-17α,21-dihydroxy-9β, 11β-oxido-pregna-1,4-diene-3,20-dione-17,21-diacetate were added under stirring during 15 minutes. After 0.5 hour the reaction mixture was precipitated in water and ammonia. The solid was collected by filtration, washed with water and dried to a constant weight, giving about 9.5 g of 2-bromo-6β,9α-difluoro-11β,17α,21 -trihydroxypregna-1,4-diene-3,20-dione-17,21-diacetate.

References

Merck Index 4481
DFU 1 (11) 526 (1976)
Kleeman and Engel p.456
OCDS Vol.3 p.99 (1984)
I.N. p.478
Riva, M. and Toscano, L.; US Patent 4,272,446; June 9, 1981; assigned to Pierrel S.p.A. (Italy)

HALOPROGIN

Therapeutic Function: Antibacterial

Chemical Name: 3-Iodo-2-propynyl-2,4,5-trichlorophenyl ether

Common Name: 2,4,5-Trichlorophenyl-γ-iodopropargyl ether

Structural Formula:

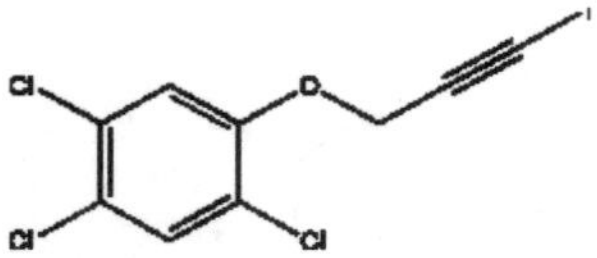

Chemical Abstracts Registry No.: 777-11-7

Trade Name	Manufacturer	Country	Year Introduced
Halotex	Westwood	US	1972
Mycanden	Schering	W. Germany	1975
Mycilan	Schering	France	1978
Mycilan	Theraplix	France	-
Polik	Meiji	Japan	-

Raw Materials

2,4,5-Trichlorophenyl propargyl ether
Copper chloride
Iodine

Manufacturing Process

4.7 grams of 2,4,5-trichlorophenyl propargyl ether (MP 64° to 65°C) are added to an aqueous solution of cupro-ammonium complex salt which has been prepared by warming a mixture of 4.0 grams of cuprous chloride, 11.0 grams of ammonium carbonate and 20 cc of water to 50°C. The resulting admixture is shaken vigorously. The cuprous acetylide deposited is filtered, washed with water and suspended in 100 cc of water, and the suspension is mixed under agitation with a solution of 5.0 grams of iodine and 5.0 grams of potassium iodide in 15 cc of water. The mixture is stirred for a period of 1 hour. The precipitate is filtered, washed with water and extracted with ether. After the drying of the ethereal extract, the solvent is distilled off. Recrystallization of the residue from n-hexane gives about 5.6 grams of 2,4,5-trichlorophenyi iodopropargyl ether, MP 114° to 115°C.

References

Merck Index 4483
Kleeman and Engel p.456
PDR p.1891
DOT 8 (8) 292 (1972)
I.N. p.478
REM p.1228
Ski, S.,Nomiya, B. and Ogawa, H.; US Patent 3,322,813; May 30, 1967; assigned to Meiji Seika Kaisha, Ltd., Japan

HALOTHANE

Therapeutic Function: Inhalation anesthetic

Chemical Name: 2-Bromo-2-chloro-1,1,1-trifluoroethane

Common Name: -

Structural Formula:

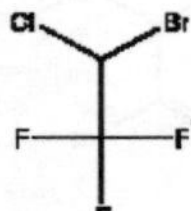

Chemical Abstracts Registry No.: 151-67-7

Trade Name	Manufacturer	Country	Year Introduced
Fluothane	Ayerst	US	1958
Fluopan	Propan-Lipworth	S. Africa	-
Fluothane	I.C.I.	UK	-
Halan	Arzneimittelwerk Dresden	E. Germany	-
Halothan Hoechst	Hoechst	W. Germany	-
Halovis	Vister	Italy	-
Narcotan	Spofa	Czechoslovakia	-
Rhodialothan	Rhodia Pharma	W. Germany	-
Somnothane	Hoechst	-	-

Raw Materials

1,1,1-Trifluoro-2-chloroethane
Bromine

Manufacturing Process

According to US Patent 2,849,502, the apparatus used consisted of a 2" x 24" silica tube packed with silica chips and enclosed in a vertical electric furnace. 1,1,1-trifluoro-2-chloroethane as vapor and bromine as liquid were introduced into a narrow tube passing down the inside of the reaction tube. The mixed reactants then passed up through the reaction tube which was maintained at a temperature of about 465°C. The reaction products were passed through a water-cooled condenser which condensed out most of the desired 1,1,1-trifluoro-2-bromo-2-chloroethane along with any high boiling by-products and unchanged bromine.

This condensate was washed with dilute caustic soda solution and dried over calcium chloride. The exit gases from this condenser were scrubbed with water and dilute caustic soda solution, dried and passed to a condenser cooled with a mixture of solid carbon dioxide and trichloroethylene which caused the unchanged 1,1,1-trifluoro-2-chloroethane to condense. This second condensate was then combined with the first and the mixture was fractionally distilled.

During a run of 2 hours 620 grams of 1,1,1-trifluoro-2-chloroethane and 630 grams of bromine were fed to the reactor and the product was worked up as described above. On fractional distillation there was obtained a first cut up to 50°C consisting of unchanged 1,1,1-trifluoro-2-chloroethane, then a middle cut between 50° and 52°C consisting of substantially pure 1,1,1-trifluoro-2-bromo-chloroethaneand a higher boiling residue that contained a further quantity of the desired product together with some 1,1,1-trifluoro-2,2-dibromo-2-chloroethane. On redistillation of the middle fraction pure 1,1,1-trifluoro-2-bromo-2-chloroethane was obtained with BP 50° to 50.5°C.

References

Merck Index 4486
Kleeman and Engel p.457
PDR p.620
I.N. p. 479
REM p.1042
Suckling, C.W. and Raventos, J.; US Patent 2,2349,502; August 26, 1958; assigned to Imperial Chemical Industries Limited, England
Suckling, C.W. and Raventos, J.; US Patent 2,921,098; January 12, 1960; assigned to Imperial Chemical Industries, Limited, England
Scherer, O. and Kuhn, H.; US Patent 2,959,624; November 8, 1960; assigned to Farbwerke Hoechst AG vormals Meister Lucius and Bruning, Germany
McGinty, R.L.; US Patent 3,082263; March 19, 1963; assigned to Imperial Chemical Industries Limited, England

HALQUINOL

Therapeutic Function: Antibacterial, Antidiarrheal, Antiseborreic

Chemical Name: 8-Quinolinol, 5,7-dichloro-, mixt. with 5-chloro-8-quinolinol and 7-chloro-8-quinolinol

Common Name: Chlorchinolinolum; Chloroxine; Chlorquinol; Cloroxinum; Halquinol(s)

Structural Formula:

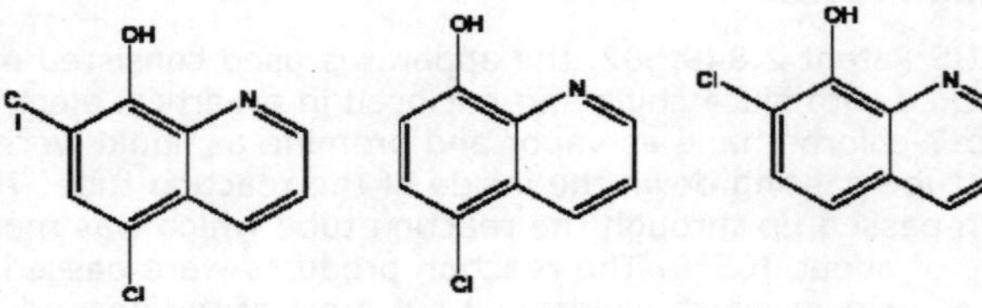

Chemical Abstracts Registry No.: 8067-69-4

Trade Name	Manufacturer	Country	Year Introduced
Capitrol	Westwood Squibb Pharmaceuticals Inc.	-	-
5,7-Dichloro-8-hydroxyquinoline, 99%	ABCR GmbH and Co. KG	-	-
Valpeda	Roche Products Ltd.	-	-

Raw Materials

8-Hydroxyquinoline
Sulfuryl chloride

Manufacturing Process

8-Quinolinol (mixture of 57-74% 5,7-dichloro-8-hydroxyquinoline, 23-40% 5-chloro-8-hydroxyquinoline and less than 3% 7-chloro-8-hydroxyquinoline) was synthesized by interaction of 8-hydroxyquinoline with sulfuryl chloride in glacial acetic acid at 50°C for 1.5 hours. The reaction mixture was cooled, pH was made 6.0 (by addition NH_4OH). The suspension obtained was centrifuged, washed with water and the mixture of chlorohydroxyquinolines was dried.

References

Brevet DInvention FR Patent No. 1,372,414; Aug. 2, 1961; Assigned to OLIN MATHIESON CHEMICAL CORPORATION

HEPARIN

Therapeutic Function: Anticoagulant

Chemical Name: Mucopolysaccaride polysulfuric acid ester

Common Name: -

Structural Formula: A highly acidic mucopolysaccharide formed of equal parts of sulfated D-glucosamine and D-glucuronic acid with sulfaminic bridges

Chemical Abstracts Registry No.: 9005-49-6

Trade Name	Manufacturer	Country	Year Introduced
Heparin	Upjohn	US	1942
Heprinar	Armour	US	1976
Chemyparin	S.I.T.	Italy	-
Clearane	Jamco	Italy	-
Disebrin	Tubi Lux Pharma	Italy	-
Embolex	Sandoz	US	-
Endoprin	Endo	US	-
Eparina	Vister	Italy	-
Eparinoral	Bruco	Italy	-
Eparinovis	Vis	Italy	-
Fioricet	Sandoz	US	-
Hamocura	Nordmark	W. Germany	-
Hepacort Plus	Rona Labs	UK	-
Hepa Gel	Spirig	Switz.	-
Heparin-Pos	Ursapharm	W. Germany	-
Heparin Sodium	Tokyo Tanabe	Japan	-
Heparinin	Sankyo	Japan	-
Hepathromb	Arzneimittelwerk Dresden	E. Germany	-
Hep-Lock	Elkins-Sinn	US	-
Hepsal	Weddell	UK	-
Liquaemin	Organon	US	-
Minihep	Leo	UK	-
Percase	Solac	France	-
Praecivenin	Pfleger	W. Germany	-
Pularin	Evans	UK	-
Thrombareduct	Azuchemie	W. Germany	-
Thromophob	Nordmark	W. Germany	-
Thrombo-Vetren	Promonta	W. Germany	-

Raw Materials

Beef intenstine
Chloroform
Water
Toluene

Manufacturing Process

5,000 pounds of beef intestine was introduced into a stainless steel reactor, jacketed with thermostated water and steam. 200 gallons of water and 10 gallons of chloroform were added. The mixture was agitated, the temperature was raised to 90°F and the agitation stopped. 5 gallons of toluene was added and the vessel closed. Autolysis was continued for 17 hours.

The extractant solution, consisting of 30 gallons of glacial acetic acid, 35 gallons of 30% aqueous ammonia, 50% sodium hydroxide to adjust the pH to 9.6 at 80°F and water to make 300 gallons, was added to the tissue. With agitation, the temperature was raised to 60°C and held there for 2 hours. Then steam was applied and the temperature was raised to boiling. 200 pounds of coarse filter aid (perlite) was added and the mixture filtered through a string discharge vacuum filter. The cake was washed with 200 gallons of hot water on the filter.

The filtrate was allowed to stand overnight and the fat skimmed off the top. After cooling to 100°F, the filtrate was transferred to a tank with thermostated water and the temperature set at 95° to 100°F. 24 gallons of pancreatic extract, prepared as described above, was added in 4-gallon increments every 12 hours for 3 days. The batch was brought to a boil and cooled to room temperature.

The batch was then filtered into a vessel and assayed for heparin content. 40,000,000 units were found in 1,000 gallons of filtrate. 20 kg of n-octylamine was added and 105 pounds of glacial acetic acid was added to bring the pH to 6.5. 20 gallons of methyl isobutyl ketone was added and the whole mixture was vigorously agitated for 1 hour. The mixture was then allowed to stand overnight. The clear, aqueous phase was drained off and discarded. The grayish-brown interphase was then removed, together with a small amount of the ketone phase, and transferred into a small kettle. The interphase volume was 7 gallons.

30 gallons of methanol was added and the mixture warmed to 120°F and then the pH was adjusted to 9.0. The mixture was then allowed to settle overnight. The solids were collected with vacuum and washed with 5 gallons of methanol. The cake was then suspended in 5 gallons of water and the heparin precipitated with 10 gallons of methanol. The solids were collected under vacuum. The dry weight of the cake was 1,000 grams and the total units were 38,000,000, according to US Patent 2,884,358.

References

Merck Index 4543
Kleeman and Engel p.458
PDR pp.872, 887, 1286, 1581, 1845, 1949
I.N. p.481
REM p.828
Bush, J.A., Freeman, L.D. and Hagerty, E.B.; US Patent 2,884,358; April 28, 1959; assigned to Southern California Gland Company
Nomine, G., Penasse, L. and Barthelemy, P.; US Patent 2,989,438; June 20, 1961; assigned to UCLAF, France
Toccaceli, N.; US Patent 3,016,331; January 9, 1962; assigned to Ormonoterapia Richter SpA, Italy

HEPRONICATE

Therapeutic Function: Vasodilator

Chemical Name: Nicotinic acid triester with 2-hexyl-2-(hydroxymethyl)-1,3-propanediol

Common Name: -

Structural Formula:

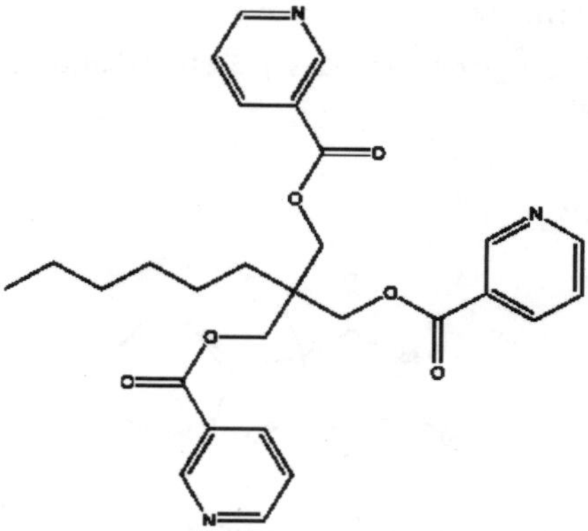

Chemical Abstracts Registry No.: 7237-81-2

Trade Name	Manufacturer	Country	Year Introduced
Megrin	Yoshitomi	Japan	1972

Raw Materials

2-Hexyl-2-(hydroxymethyl)-1,3-propanediol
Nicotinic acid
p-Toluenesulfonyl chloride

Manufacturing Process

In 50 ml of pyridine were dissolved 50 grams of nicotinic acid and 50 grams of p-toluene-sulfonyl chloride. While stirring, the mixture gradually became hot and colorless, and finally solidified. To the mixture was added dropwise a solution of 19 grams of 2-hexyl-2-(hydroxymethyl)-1,3-propanediol in 400 ml of pyridine at a temperature below 80°C. The mixture was heated at 115° to 125°C on an oil bath for 1 hour. After cooling, the mixture was poured into 300 ml of ice water, and extracted with toluene. The toluene layer was washed in sequence with water, aqueous sodium carbonate and water, dried over potassium carbonate, and then the toluene was distilled off. The oily residue was crystallized from ethanol to give 30 grams of 2-hexyl-2-(hydroxymethyl)-1,3-propanediol trinicotinate, melting at 94° to 96°C. The yield was 59.5%.

References

Merck Index 4545
Kleeman and Engel p.459
DOT 8 (8) 314 (1972)
I.N. p.482
Nakanishi, M., Kobayashi, R. and Arimura, K.; US Patent 3,384,642; May 21, 1968; as signed to Yoshitomi Pharmaceutical Industries, Ltd., Japan

HEPTABARBITAL

Therapeutic Function: Hypnotic, Sedative

Chemical Name: 5-(1-Cyclohepten-1-yl)-5-ethyl-2,4,6(1H,3H,5H)-pyrimidinetrione

Common Name: -

Structural Formula:

Chemical Abstracts Registry No.: 509-86-4

Trade Name	Manufacturer	Country	Year Introduced
Medomine	Ciba Geigy	France	1948
Medomin	Geigy	US	1955

Raw Materials

Ethanol
Sodium
Ethyl bromide
Urea
Cycloheptanone
Cyanoacetic acid methyl ester
Hydrogen chloride

Manufacturing Process

112 g of cycloheptanone (suberone) are mixed with 130 g of cyanoacetic acid methyl ester, 2 g of piperidine are added, and the mixture is heated on the water bath at 60°C for several hours until no more water separates from the reaction mixture. The water layer is removed, and the remainder is subjected to distillation in vacuo. The fraction distilling at 160°C to 175°C under a pressure of 20 mm is collected separately; it consists of cycloheptenyl-cyanoacetic acid methyl ester. The first fractions can be subjected to a fresh condensing reaction after addition of more piperidine.

The cycloheptenyl-cyanoacetic acid methyl ester so obtained is a colorless liquid boiling at 174°C under a pressure of 20 mm.

Into this compound, an ethyl radical is introduced at the same C-atom to which the cycloheptenyl radical is connected. This is done, for example, in the following way:

19.3 g of the above ester are added to a solution of 2.3 g of sodium in 40 cc of absolute ethyl alcohol. To this mixture, 13.0 g of ethyl bromide are gradually added while cooling, and the reaction mixture is heated under reflux on a water bath until it has become neutral. The mixture is then taken up in water, the aqueous layer is separated and the cycloheptenyl-ethyl-cyanoacetic acid methyl ester so formed distills at 169°C to 170°C under a pressure of 20 mm.

22.1 g of this latter substance are dissolved in a solution of 4.6 g of sodium in 100 cc of absolute ethyl alcohol. 12 g of urea are further added thereto, and the whole solution is heated to about 80°C for about eight hours. The alcohol is then distilled off in vacuo, the residue is dissolved in cold water, and from this solution, C-C-cycloheptenyl-ethyl barbituric acid is obtained by saponification with diluted hydrochloric acid. The crude product is recrystallized from diluted ethyl alcohol and forms colorless needles of faintly bitter taste and melting point 174°C.

The sodium salt of this acid may be prepared by dissolving 2.5 g of the acid in a solution of 0.23 g of sodium in 20 cc of ethyl alcohol, and the salt forms, after evaporating the alcohol, a colorless, water-soluble powder.

References

Merck Index 4546
Kleeman and Engel p.459
OCDS Vol.1 pp.269, 272 (1977)
I.N. p.482
Taub, W.; US Patent 2,501,551; March 21, 1950

HETACILLIN POTASSIUM

Therapeutic Function: Antibacterial

Chemical Name: 6-(2,2-Dimethyl-5-oxo-4-phenyl-1-imidazolidinyl)-3,3-dimethyl-7-oxo-4-thia-1-azabicyclo[3.2.0]heptane-2-carboxylic acid potassium salt

Common Name: Phenazacillin

Structural Formula:

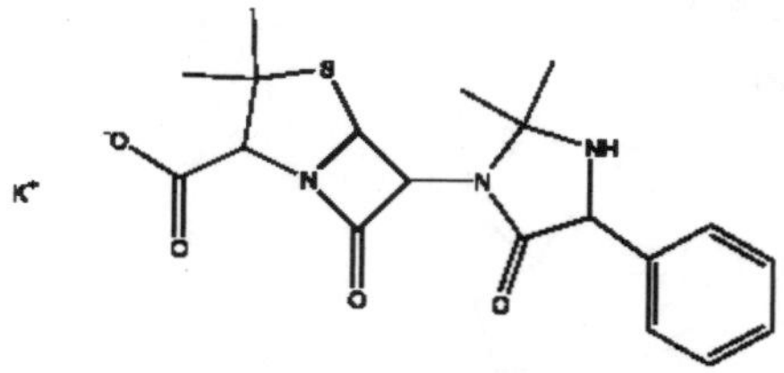

Chemical Abstracts Registry No.: 5321-32-4; 3511-16-8 (Base)

Trade Name	Manufacturer	Country	Year Introduced
Natacillin	Bristol Banyu	Japan	1970
Versapen	Bristol	Italy	1970
Versapen	Bristol	France	1970
Versapen	Bristol	US	1971
Hetabiotic	Bristol-Myers	-	-
Hetacin-K	Bristol	-	-
Penplenum	Bristol	W. Germany	-
Uropen	Bristol	-	-

Raw Materials

α-Aminobenzylpenicillin
Acetone

Manufacturing Process

To 100 grams of α-aminobenzylpenicillin slurried in 2,500 ml of acetone is added 200 ml of a 22% solution of potassium ethylhexanoate in dry n-butanol and the mixture is warmed to 45°C whereupon the acid dissolves. After the mixture is agitated for 1 hour at 40° to 45°C, the product begins to crystallize out. Agitation is continued for 4 hours at 45°C after which the product, the potassium salt of hetacillin, is collected by filtration, washed with 500 ml of dry acetone, dried for 17 hours at 40°C and found to weigh 70.0 grams.

References

Merck Index 4564
Kleeman and Engel p.460
OCDS Vol. 1 p.414 (1977)
DOT 3 (1) 12 (1967)
I.N. p.483
REM p. 1200
Johnson, D.A. and Panetta, C.A.; US Patent 3,198,804; August 3,1965; assigned to Bristol-Myers Company

HETERONIUM BROMIDE

Therapeutic Function: Anticholinergic

Chemical Name: Pyrrolidinium, 3-((hydroxyphenyl-2-thienylacetyl)oxy)-1,1-dimethyl-, bromide

Common Name: Heteronium bromide

Structural Formula:

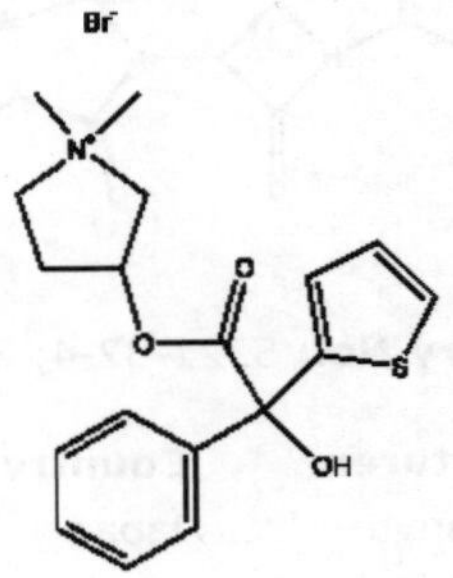

Chemical Abstracts Registry No.: 7247-57-6

Trade Name	Manufacturer	Country	Year Introduced
Hetrum bromide	Lilly	-	-
Heteronium bromide	Shanghai Lansheng Corporation	-	-

Raw Materials

1-Methyl-3-pyrrolidinol
Sodium methylate
Methyl bromide
Methyl phenyl-2-thienylglycolate
Hydrochloric acid
Potassium carbonate

Manufacturing Process

A solution of 5.0 g of 1-methyl-3-pyrrolidinol and 12.5 g of methyl phenyl-2-thienylglycolate in 200 ml of n-heptane is heated to reflux temperature, a trace of sodium methylate is added, and refluxing of the mixture is continued for 3 h. A Dean-Stark trap is incorporated in the reflux apparatus to separate and withdraw the methanol produced by the reaction. At the end of the reflux period, the reaction mixture is cooled and extracted two times with 100 ml portions of 2 N hydrochloric acid. The acid extracts are combined, basified with potassium carbonate, and extracted three times with 50 ml portions of benzene. The benzene extracts are combined, washed two times with 25 ml portions of water, and dried. The benzene is removed from the dried extract by evaporation at reduced pressure. The residue is 1-methyl-3-pyrrolidylphenyl-2-thienylglycolate in the form of a clear liquid.

1-Methyl-3-pyrrolidylphenyl-2-thienylglycolate is dissolved in ethyl acetate to a concentration of about 20% by weight. To the solution is added a small excess of an anhydrous ethanolic solution of hydrogen chloride, whereupon 1-methyl-3-pyrrolidylphenyl-2-thienylglycolate hydrochloride separates as a white crystalline solid and is removed by filtration.

5.0 g of 1-methyl-3-pyrrolidylphenyl-2-thienylglycolate hydrochloride are dissolved in 50 ml of water. The aqueous solution is basified by adding solid potassium carbonate with stirring, and the basified mixture is extracted three times with 25 ml portions of benzene. The benzene extracts are combined, dried with anhydrous magnesium sulfate, and evaporated to dryness. The residue is dissolved in methyl ethyl ketone. Into the resulting solution is bubbled an excess amount of methyl bromide, and the mixture is permitted to stand at room temperature, whereupon a white precipitate appears. The precipitate is removed by filtration, washed with methyl ethyl ketone, recrystallized from absolute ethanol, and dried. So the 1-methyl-3-pyrrolidylphenyl-2-thienylglycolate methobromide was obtained.

1-Methyl-3-pyrrolidylphenyl-2-thienylglycolate methobromide is readily separated into two diastereoisomeric fractions by fractional crystallization. The crude mixed 1-methyl-3-pyrrolidylphenyl-2-thienylglycolate methobromide is dissolved in the minimum quantity of methanol, decolorized with charcoal, diluted with 4 or 5 volumes of ethyl acetate, concentrated on a steam bath until crystals begin to appear, and cooled. The resulting crystals are the α-diastereoisomer of 1-methyl-3-pyrrolidylphenyl-2-thienylglycolate methobromide, melting point 210°-211°C.

References

Ainsworth C., Ryan C.W.; US Patent No. 3,138,614; June 23, 1964; Assigned: Eli Lilly and Company, Indianapolis, Ind., a corporation of Indiana

HEXACHLOROPHENE

Therapeutic Function: Topical antiinfective

Chemical Name: 2,2'-Methylenebis(3,4,6-trichlorophenol)

Common Name: -

Structural Formula:

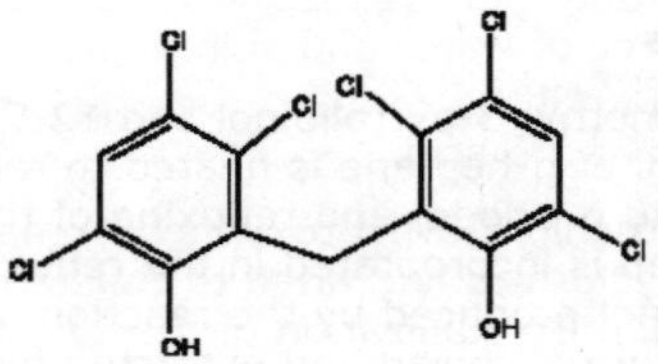

Chemical Abstracts Registry No.: 70-30-4

Trade Name	Manufacturer	Country	Year Introduced
Gamophen	Ethicon	US	1950
Phisohex	Winthrop	US	1954
Germa-Medica	Huntington	US	1979
Hexascrub	Prof. Disposables	US	1980
Pre-Op	Davis and Geck	US	1980
Turgex	Xttrium	US	1981
Coopaphene	McDougall and Robertson	UK	-
Dermadex	Alconox	US	-
Dermohex	Hartz	Canada	-
G-11	Givaudan	Switz.	-
Germibon	Gamir	Spain	-
Heksaden	Deva	Turkey	-
Hexal	Fischer	Israel	-
Solu-Heks	Mustafa Nevzat	Turkey	-
Soy-Dome	Dome	US	-
Ster-Zac	Hough	UK	-
Wescohex	West	US	-
Westasept	West	US	-

Raw Materials

2,4,5-Trichlorophenol
Paraformaldehyde

Manufacturing Process

A mixture of 198 grams of 2,4,5-trichlorophenol and 18.8 grams of paraformaldehyde was heated to 65°C and well stirred. 65 grams of oleum 20% was added dropwise and the addition was so regulated that the temperature increased, without the application of external heat, until it reached 135°C at the end of the acid addition, which took 10 to 15 minutes. The contents of the reaction vessel were stirred for 2 minutes more and then allowed to run into a solution of 100 grams of sodium hydroxide in 1,000 cc of water.

The reaction flask was washed with a solution of 25 grams of sodium hydroxide in 250 cc of water. The combined alkaline solutions were heated to boiling for 5 minutes. A small amount (6 grams) of alkali-insoluble material remained and was filtered off. Sulfuric acid (62% H_2SO_4 content) was then added at room temperature dropwise under stirring to the filtrate until a pH of 10.3 was reached. This required about 80 grams of the acid. The monosodium salt of bis-(3,5,6-trichloro-2-hydroxyphenyl) methane precipitated out of solution and was filtered and then washed with 200 cc of water. The salt was then suspended in 2,000 cc of water and sulfuric acid (62% H_2SO_4 content) was added under stirring until the contents were acid to Congo red paper. This required about 30 grams of the acid.

The resulting bis-(3,5,6-trichloro-2-hydroxyphenyl) methane was filtered, washed with water until acid-free and dried to constant weight at 100°C (170 grams, MP 154° to 158°C). Crystallization of the 170 grams of dried bis-(3,5,6-trichloro-2-hydroxyphenyl)methane from 300 grams toluene yielded a first crop amounting to 105 grams of substantially pure bis-(3,5,6-trichloro-2-hydroxyphenyl) methane, having a MP of 161° to 163°C (from US Patent 2,435,593).

References

Merck Index 4574
Kleeman and Engel p.461
PDR p.1926
I.N. p.484
REM p.1161
Gump, W.S.; US Patent 2,250,480; July 29, 1941; assigned to Burton T. Bush, Inc.
Luthy, M. and Gump, W.S.; US Patent 2,435,593; February 10, 1948; assigned to Burton T. Bush, Inc.
Gump, W.S., Luthy, M. and Krebs, H.G.; US Patent 2,812,365; November 5, 1957; as signed to The Givaudan Corporation

HEXAFLUORENIUM BROMIDE

Therapeutic Function: Muscle relaxant

Chemical Name: 1,6-Hexanediaminium, N,N'-di-9H-fluoren-9-yl-N,N,N',N'-tetramethyl-, dibromide

Common Name: Hexafluorenium bromide; Hexafluronium bromide

Structural Formula:

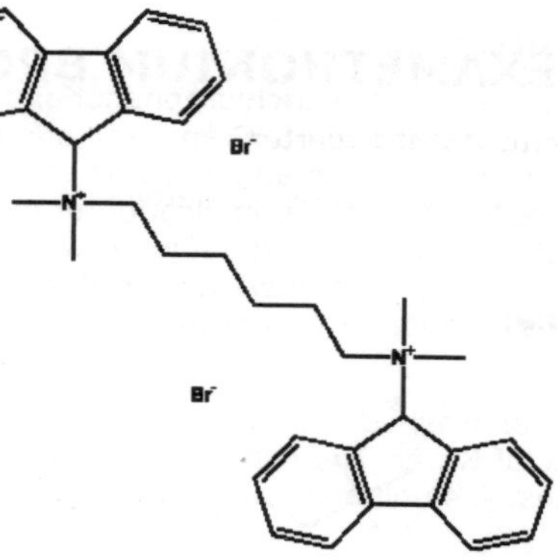

Chemical Abstracts Registry No.: 317-52-2

Trade Name	Manufacturer	Country	Year Introduced
Mylaxen	Mallinckrodt	-	-
Mylaxen	Wallace Labs	-	-
Hexafluorenium Bromide	Shanghai Lansheng Corporation	-	-
Myorelax	Neisler	-	-

Raw Materials

9-Bromofluorene
1,6-Bis(dimethylamino)hexane
2-Dimethylaminofluorene
1,6-Dibromohexane

Manufacturing Process

2 Methods of producing of hexamethylene bis(2-fluorenyldimethylammonium) dibromide:

1. To a solution of 13.2 g (0.054 mole) of 9-bromofluorene in 200 ml of acetonitrile was added 3.1 g (0.018 mole) of 1,6-bis(dimethylamino)hexane. Heat was evolved and solid began to precipitate immediately. After the initial reaction had subsided the mixture was refluxed for 3.5 h on a steam bath. The solid was collected and recrystallized three times from propanol to give 10.66 g (89% of the theoretical yield based on alkylenediamine) of hexamethylene bis(9-fluorenyldimethylammonium) dibromide, as white crystals melting at 188°-189°C.

2. A mixture of 4.18 g (0.02 mole) of 2-dimethylaminofluorene and 1.22 g (0.005 mole) of 1,6-dibromohexane was heated at 115°C for 24 h, after which time 10 ml of dry dioxane was added and the mixture refluxed for an additional 37 h. The solid was filtered off washed with hot acetonitrile and twice recrystallized from propanol to give 0.45 g of hexamethylene bis(2-fluorenyldimethylammonium) dibromide, melting at 220°-222°C.

References

Cavallito Ch. J. Gray A.P.; US Patent No. 2,783,237; Feb. 26, 1957; Assigned: Irwin, Neisler and Company, Decature, III., a corporation of Illinois

HEXAMETHONIUM BROMIDE

Therapeutic Function: Antihypertensive

Chemical Name: N,N,N,N',N',N'-Hexamethyl-1,6-hexanediaminium bromide

Common Name: -

Structural Formula:

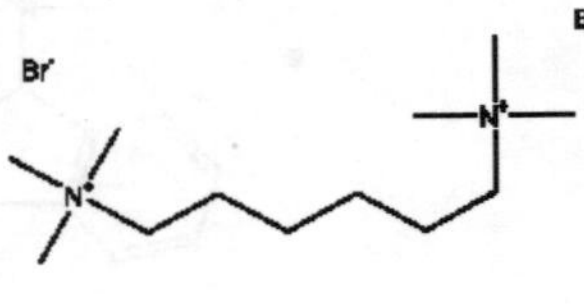

Chemical Abstracts Registry No.: 60-26-4 (Base)

Trade Name	Manufacturer	Country	Year Introduced
Bistrium	Squibb	US	1951
Hexanium	Adrian-Marinier	France	-
Methobromin	Yamanouchi	Japan	-
Vegolysen	May and Baker	-	-

Raw Materials

Hexamethylene diamine
Sodium hydroxide
Dimethyl sulfate
Hydrogen bromide

Manufacturing Process

Hexamethylene diamine (116 g), sodium carbonate (466 g), and water (800 ml) were heated to 60°C, and dimethyl sulfate (830 g) added with stirring over 1% hours keeping the temperature below 90°C. The reaction mixture was then stirred at 90°C for 2 hours, then cooled to 20°C, acetone (1,200 ml) added and the whole cooled to 0°C.

The solid formed was removed by filtration and washed with acetone (150 ml). Filtrate and washings were diluted with water to 4 liters and heated to 60°C under reflux. To this was added a solution prepared from embonic acid (388 g), sodium hydroxide (80 g) and water (5 liters), the whole refluxed for 10 minutes and thereafter allowed to cool overnight.

The resultant embonate (530 g) was filtered off, washed twice with a solution of acetone (75 ml) in water (425 ml), and dried at 100°C to give an amorphous yellow powder, MP 290°C to 291°C (with decomp.). 588 g of the embonate was dissolved in boiling water (4 liters).

Hydrobromic acid 50% w/w (325 g) diluted with water (2 liters) was added slowly at the boil and the precipitated embonic acid removed by filtering hot and washing twice with hot water (1 liter). The filtrate and washings were evaporated to dryness in a steam pan and the residue recrystallized from ethyl alcohol (1,200 ml), to yield the dibromide (320 g).

References

Merck Index 4582
Kleeman and Engel p.462
I.N. p.485
Barber, H.J.; US Patent 2,641,610; June 9, 1953; assigned to May and Baker, Ltd. (UK)

HEXCARBACHOLINE BROMIDE

Therapeutic Function: Muscle relaxant

Chemical Name: Choline bromide hexamethylenedicarbamate

Common Name: Carbolonium bromide; Hexabiscarbacholine; Hexcarbacholine bromide

Structural Formula:

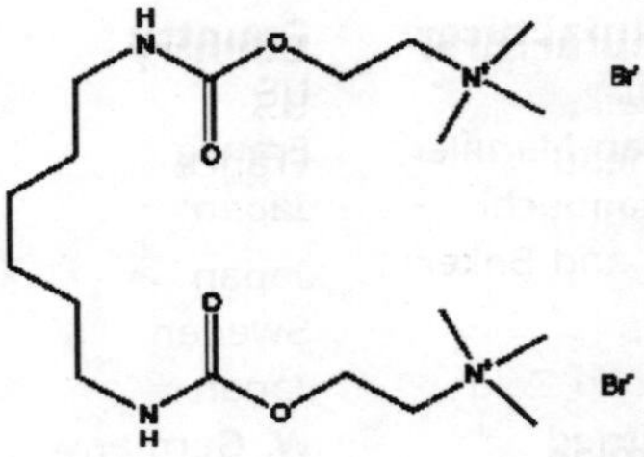

Chemical Abstracts Registry No.: 306-41-2

Trade Name	Manufacturer	Country	Year Introduced
Imbretil	Burroughs-Wellcome	-	-

Raw Materials

β-Dimethylaminoethyl ester hexamethylene biscarbamic acid
Methyl bromide

Manufacturing Process

42 parts β-dimethylaminoethyl ester hexamethylene biscarbamic acid in 350 parts of acetone was added to 100 parts methyl bromide. The reaction product was fallen after 24 hours at room temperature and was recrystallized from ethanol or a mixture of methanol and acetone to give 56.7 parts of choline bromide hexamethylenedicarbamate (hexacarbocholine bromide). Yield 87%. MP: 174°-176°C.

References

Schmidt O. et al.; DB Patent No. 1,021,842; June 23, 1954; Assigned to Osterreichishe Stikstoffwerke Aktiengesselshaft, Linz (Austria)

HEXESTROL

Therapeutic Function: Estrogen

Chemical Name: 4,4'-(1,2-Diethyl-1,2-ethanediyl)bisphenol

Common Name: Dihydrodiethylstilbestrol; Hexoestrol

Structural Formula:

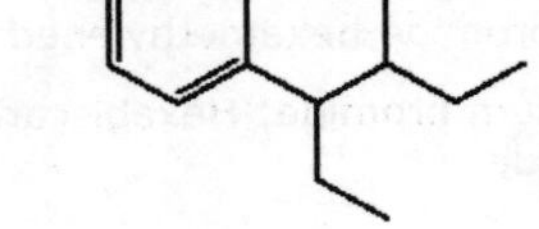

Chemical Abstracts Registry No.: 84-16-2

Trade Name	Manufacturer	Country	Year Introduced
Estra Plex	Rowell	US	1956
Cycloestrol	Bruneau	France	-
Estrene	Lepetit	-	-
Femirogen	Fuso	Japan	-
Folliplex	Recip	Sweden	-
Hexron	Teikoku Zoki	Japan	-
Hormoestrol	Siegfried	W. Germany	-
Syntex	Pharmacia	Sweden	-
Synthovo	Boots	UK	-

Raw Materials

Hydrogen chloride
Phosphorus (Red)
Hydrogen iodide
p-Hydroxypropiophenone
Sodium amalgam

Manufacturing Process

50 parts by weight of p-hydroxypropiophenone are dissolved in 200 parts by weight of a 12.5% solution of caustic soda and shaken with 350 parts by weight of 3% sodium amalgam. The sodium salt of the pinacol thereby precipitating is reacted with glacial acetic acid, whereby the free pinacol is obtained (MP 205°C to 210°C, after purification 215°C to 217°C). The yield amounts to 95% of the theoretical. The pinacol is suspended in ether and gaseous hydrogen chloride introduced, whereby water separates and the pinacolin formed is dissolved in the ether, from which it is obtained by evaporation as a viscous oil (diacetate of MP 91°C). The yield is quantitative.

40 parts by weight of pinacolin are dissolved in ethyl alcohol and gradually treated with 80 parts by weight of sodium under reflux. The solution is decomposed with water and the pinacolin alcohol formed extracted from the neutalized solution with ether. The pinacolin alcohol is a viscous oil which is characterized by a dibenzoate of MP 172°C. The yield is 95% of the theoretical.

30 parts by weight of pinacolin alcohol are dissolved in 25 parts by weight of glacial acetic acid and heated for 30 minutes to 135°C to 140°C after having added 20 parts by weight of hydroiodic acid (specific gravity = 1.94) and 5 parts by weight of red phosphorus. The whole is filtered, the solution poured into water, extracted with ether and the ether solution washed with bicarbonate. The oil remaining after distilling off the ether is taken up in chloroform, whereby hexoestrol [α,β-(p,p-dihydroxy-diphenyl)-α,β-diethyl-ethane] crystallizes out. MP after recrystallization from benzene: 185°C. Yield: 20%.

References

Merck Index 4593
DFU 8 (5) 413 (1983)
Kleeman and Engel p.466
OCDS Vol.1 p.102 (1977)
I.N. p.486

Wallis, E.S. and Bernstein, S.; US Patent 2,357,985; September 12, 1944; assigned to Research Corporation
Adler, E., Gie, G.J. and von Euler, H.; US Patent 2,421,401; June 3, 1947; assigned to Hoffmann-La Roche, Inc.

HEXETIDINE

Therapeutic Function: Antifungal

Chemical Name: 1,3-Bis(2-ethylhexyl)hexahydro-5-methyl-5-pyrimidinamine

Common Name: -

Structural Formula:

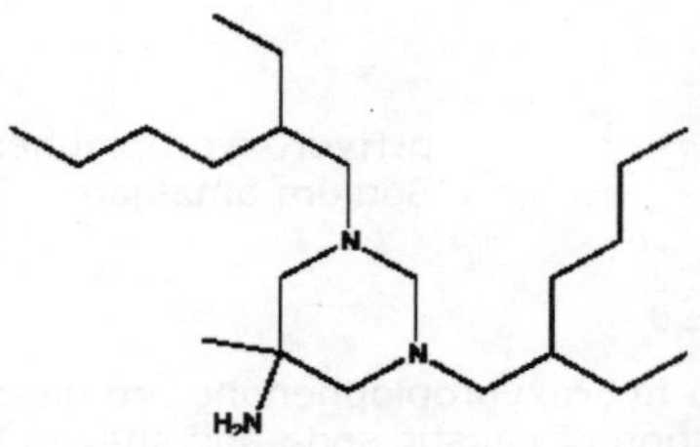

Chemical Abstracts Registry No.: 141-94-6

Trade Name	Manufacturer	Country	Year Introduced
Sterisil	Warner Lambert	US	1956
Oraseptic	Parke Davis	Italy	1960
Hextril	Substantia	France	1961
Hexoral	Goedecke	W. Germany	1967
Oraldene	Warner	France	1969
Oraldene	Warner	UK	1969
Bactidol	Warner-Chilcott	-	-
Bucosept	La Campana	Mexico	-
Collu-Hextril	Substantia	France	-
Drossadin	Drossapharm	Switz.	-
Glypesin	Stada	W. Germany	-
Sterisol	Warner-Chilcott	-	-

Raw Materials

Nitroethane
Hydrogen
Nickel Raney
Formaldehyde
2-Ethylhexylamine

Manufacturing Process

Nitroethane and formaldehyde are first reacted to give 2-methyl-2-nitro-1,3-propanediol. This is reacted with 2-ethylhexylamine and formaldehyde to give 5-nitro-1,3-bis(2-ethylhexyl)-5-methyl-hexahydropyrimidine.

To a hydrogenation apparatus containing 500 ml of methanol and 10 g of Raney nickel catalyst were continuously added over a period of one hour, 240 g of 5-nitro-1,3-bis(2-ethylhexyl)-5-methylhexahydropyrimidine. During the one-hour period, the resulting mixture was hydrogenated at approximately 1,000 pounds per square inch utilizing room temperature as the initial temperature and gradually increasing the temperature to about 70°C. At the end of the one-hour period, hydrogenation was stopped. The reaction mixture was first filtered to remove the catalyst and was then distilled at atmospheric pressure at a temperature of 70°C to remove methanol. 197.5 g of 5-amino-1,3-bis(2-ethylhexyl)-5-methylhexahydropyrimidine were collected.

References

Merck Index 4597
Kleeman and Engel p.463
I.N. p.487
Bell, W.O. and Neckar, A.E.; US Patent 3,054,797; September 18, 1962; assigned to Commercial Solvents Corp.

HEXOBENDINE

Therapeutic Function: Vasodilator

Chemical Name: 3,4,5-Trimethoxybenzoic acid 1,2-ethanediylbis-[(methylimino)-3,1-propanediyl]ester

Common Name: -

Structural Formula:

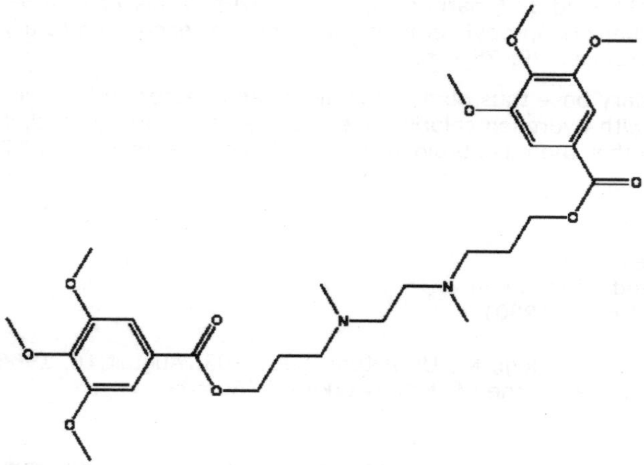

Chemical Abstracts Registry No.: 54-03-5; 50-62-4 (Dihydrochloride salt)

Trade Name	Manufacturer	Country	Year Introduced
Reoxyl	Hormonchemie	W. Germany	1966
Ustimon	Merck Clevenot	France	1969

Trade Name	Manufacturer	Country	Year Introduced
Flussicor	Farmalabor	Italy	1971
Andiamine	Polfa	Poland	-
Hityl	Biosedra	France	-
Instenon	Byk Gulden	W. Germany	-

Raw Materials

Methyl acrylate
Lithium aluminum hydride
N,N'-Dimethylethylenediamine
3,4,5-Trimethoxybenzoic acid chloride

Manufacturing Process

Methylacrylate and N,N'-dimethylethylenediamine are first reacted and that product reduced with lithium aluminum hydride to give a compound A.

To a solution of 13 parts of compound A and 12 parts by volume of absolute pyridine in 80 parts by volume of absolute dioxane there are added dropwise and under constant stirring 35 parts of 3,4,5-trimethoxybenzoyl chloride dissolved in 70 parts by volume of absolute dioxane in the course of 30 minutes. The mixture is stirred for a further 3 hours at a temperature of 100°C and the excess solvent is then evaporated in vacuo. The residue of the evaporation is treated with ethyl acetate and saturated sodium carbonate solution, whereafter the organic phase is separated, treated with water, dried with sodium sulfate and the solvent is removed in vacuo. The residue thus obtained is taken up in ether and separated from 4 parts of insoluble trimethoxybenzoic acid anhydride by filtration. After evaporation of the ether there are obtained 32.5 parts of N,N'-dimethyl-N,N'-bis-[3-(3,4,5-trimethoxybenzoxy)propyl]-ethylene diamine, corresponding to a yield of 86% of the theoretical. MP: 75°C to 77°C.

The di-tertiary base thus obtained is dissolved in ether and the solution is saturated with hydrogen chloride gas. After isolation and reprecipitation from methanol-ether there is obtained the dihydrochloride melting at 170°C to 174°C.

References

Merck Index 4600
Kleeman and Engel p.464
OCDS Vol.2 p.92 (1980)
I.N. p.487
Kraupp, O. and Schlogl, K.; US Patent 3,267,103; August 16, 1966; assigned to Oesterreichische Stickstoffwerke AG (Austria)

HEXOCYCLIUM METHYL SULFATE

Therapeutic Function: Spasmolytic

Chemical Name: 4-(2-Cyclohexyl-2-hydroxy-2-phenylethyl)-1,1-dimethylpiperazinium methyl sulfate

Common Name: -

Structural Formula:

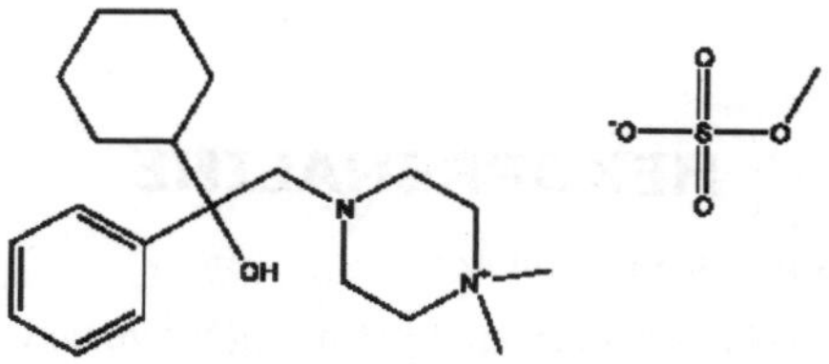

Chemical Abstracts Registry No.: 115-63-9

Trade Name	Manufacturer	Country	Year Introduced
Tral	Abbott	US	1957
Traline	Abbott	France	1959

Raw Materials

N-Phenacyl-N'-methylpiperazine
Cyclohexyl bromide
Magnesium
Dimethyl sulfate

Manufacturing Process

In a 2-liter, 3-necked, round-bottomed flask equipped with a stirrer, dropping funnel, and a condenser protected with a calcium chloride drying tube is placed 13.7 grams (0.57 mol) of magnesium turnings and the magnesium is covered with 200 cc of anhydrous ether. A crystal of iodine is added to the flask and 92.9 grams (0.57 mol) of cyclohexyl bromide dissolved in 300 cc of anhydrous ether is added dropwise with stirring while the reaction proceeds. After the addition of the cyclohexyl bromide is completed, the resulting mixture is stirred and heated on a steam bath for 3 hours. The mixture is cooled to room temperature and 49.5 grams (0.227 mol) of N-phenacyl-N'-methylpiperazine dissolved in 50 cc of anhydrous ether is added dropwise and the resulting mixture is stirred and re fluxed for about 16 hours.

The reaction mixture is cooled and 50 grams of ammonium chloride dissolved in 200 cc of water is added dropwise thereto with stirring. The decomposed Grignard complex is then filtered. Benzene is added to the ether filtrate and the solvents are removed therefrom on a steam bath. The residue is fractionated and the base, N-(β-cyclohexyl-β-hydroxy-β-phenyl-ethyl)N'-methylpiperazine, is obtained as a liquid having a boiling point of 196° to 203°C at a pressure of 4.0 mm.

To 3.8 grams of the base dissolved in 35 cc of ethyl alcohol is added 1.6 grams of dimethyl sulfate. The solution is allowed to stand at room temperature for about 12 hours. The salt formed is filtered, recrystallized from ethyl alcohol, and is found to have a melting point of 203° to 204°C.

References

Merck Index 4601
Kleeman and Engel p.465
PDR p.553
I.N. p.488

REM p.918
Weston, A.W.; US Patent 2,907,765; October 6, 1959; assigned to Abbott Laboratories

HEXOPRENALINE

Therapeutic Function: Bronchodilator

Chemical Name: 4,4'-[1,6-Hexanediylbis[imino(1-hydroxy-2,1-ethanediyl)]] bis-1,2-benzenediol

Common Name: N,N'-Bis[2-(3,4-dihydroxyphenyl)-2-hydroxyethyl] hexamethylenediamine

Structural Formula:

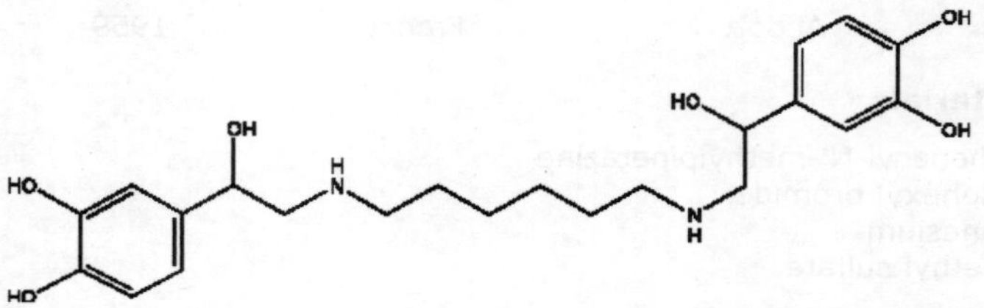

Chemical Abstracts Registry No.: 3215-70-1

Trade Name	Manufacturer	Country	Year Introduced
Etoscol	Byk Gulden	W. Germany	1973
Hexoprenaline	Morishita	Japan	1976
Leanol	Yoshitomi	Japan	1976
Bronalin	Byk Liprandi	Argentina	-
Gynipral	Chemie Linz	Austria	-
Ipradol	Chemie Linz	Austria	-
Prelin	Farmos	Finland	-

Raw Materials

Chloroaceto pyrocatechol
N,N'-Dibenzylhexamethylene diamine
Hydrogen

Manufacturing Process

The N,N'-dibenzyl-N,N'-bis-[2-(3',4'-dihydroxyphenyl)-2-oxoethyl]-hexamethylene-diamine dichlorohydrate-monohydrate used as the starting material was prepared as follows: 2 mols of chloroaceto pyrocatechin were dissolved in 2,000 cc of acetone and heated to boiling with 2 mols of N,N'-dibenzylhexamethylene-diamine for 12 hours, almost the theoretical quantity of N,N'-dibenzylhexamethylene-diamine dichlorohydrate being precipitated and removed by suction after cooling. Excess HCl was added to the filtrate, approximately 66% of the theoretically possible quantity of crude dichlorohydrate of the N,N'-dibenzyl-N,N'-bis-[2-(3',4'-dihydroxyphenyl)-2-

oxoethyll-hexamethylene-diamine being precipitated. The product was cleaned by recrystallization from water with the addition of animal charcoal. After drying the substance contained water of crystallization at ambient temperature, MP 206° to 209.5°C.

Five grams of N,N'-dibenzyl-N,N'-bis[2-(3',4'-dihydroxyphenyl)-2-oxoethyl]-hexamethylenediamine dichlorohydrate as a monohydrate were hydrogenated under considerable agitation by means of 2.0 grams of 10% palladium-carbon, with hydrogen in a mixture of 270 cc of methanol and 50 cc of water at 45°C and normal pressure. After about 4 hours the theoretical quantity of hydrogen (4 mols of hydrogen per 1 mol of substance) was absorbed for the splitting off of the two benzyl radicals and the reduction of the two carbonyl groups to carbinol groups, and the hydrogenation came to a stop.

After separation of the catalyst the product was concentrated until dry, the residue was triturated with acetone, the resulting crystallizate was removed by suction and washed with acetone. The yield of N,N'-bis-2-(3',4'-dihydroxyphenyl)-2-hydroxyethyll-hexamethylene-diamine dichlorohydrate was 3.3 grams, i.e., 92% of the theoretical value. A quantity of 2.8 grams having a melting point of 197.5° to 198°C was obtained by precipitation from a mixture of methanol-ether.

Free N,N'-bis-[2-(3',4'-dihydroxyphenyl)-2-hydroxyethyl]-hexamethylene-diamine can be separated from these salts by the addition of the equivalent quantity of caustic alkali solution. It has a melting point of 162° to 165°C and contains half a mol of water of crystallization.

N,N'-bis-[2-(3',4'-dihydroxyphenyl)-2-hydroxyethyl]-hexamethylene-diaminesulfate (MP 222° to 228°C) can be obtained by reacting the base with the equivalent quantity of sulfuric acid in an alcohol solution, followed by concentration and precipitation from water-alcohol solution.

References

Merck Index 4603
Kleeman and Engel p. 466
I.N. p. 488
Schmid, O., Lerchenthal, H.S.-M., Zolss, G., Gratz, R. and Wismayr, K.; US Patent 3,329,709; July 4, 1967; assigned to Oesterreichische Stickstoffwerke AG, Austria

HEXYLCAINE HYDROCHLORIDE

Therapeutic Function: Local anesthetic

Chemical Name: 1-Cyclohexylamino-2-propylbenzoate hydrochloride

Common Name: -

Structural Formula:

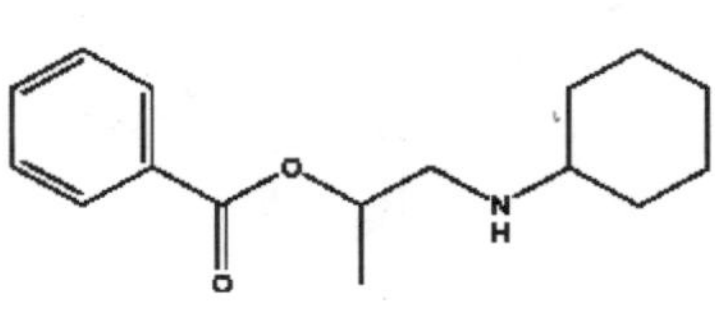

HCl

Chemical Abstracts Registry No.: 532-76-3; 532-77-4 (Base)

Trade Name	Manufacturer	Country	Year Introduced
Cyclaine	MSD	US	1952

Raw Materials

1-Cyclohexylamino-2-propanol
Benzoyl chloride
Hydrogen chloride

Manufacturing Process

A solution of 0.1 mol of 1-cyclohexylamino-2-propanol in 30 grams of chloroform was saturated with dry hydrogen chloride gas, with cooling. A solution of 0.1 mol of benzoyl chloride in 30 grams of chloroform was added and the solution was heated in a bath at 50° to 55°C for four days under a reflux condenser protected from atmospheric moisture. Then the solvent was removed by vacuum distillation while the mixture was warmed on a water bath. Benzene was then added to the syrupy residue and the reaction product crystallized out after the benzene was removed by vacuum distillation.

The crystallized solid residue was washed with anhydrous ether to remove any unreacted benzoyl chloride. The 1-cyclohexylamino-2-propyl benzoate hydrochloride obtained was purified by two recrystallizations from absolute alcohol. It melted at 177° to 178.5°C.

References

Merck Index 4605
Kleeman and Engel p. 467
OCDS Vol. 1 p. 12 (1977)
I.N. p. 488
REM p. 1056
Cope, A.C.; US Patent 2,486,374; November 1,1949; assigned to Sharp and Dohme, Inc.

HISTAPYRRODINE HYDROCHLORIDE

Therapeutic Function: Antihistaminic

Chemical Name: 1-(2-N-Benzylanilinoethyl)pyrrolidine, hydrochloride

Common Name: Histapyrrodinum

Structural Formula:

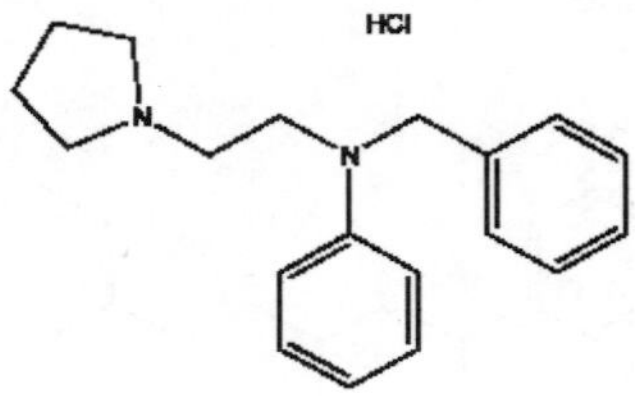

HCl

Chemical Abstracts Registry No.: 6113-17-3; 493-80-1 (Base)

Trade Name	Manufacturer	Country	Year Introduced
Calcistin	Galenus	-	-
Calcistin	Hestia Pharma	-	-
Domistan	Servier	-	-
Luvistin	Boehringer, Ing.	-	-

Raw Materials

Pyrrolidine	N-p-Chloroethyl-N-benzylaniline
Caustic soda	N-β-Chloroethylpyrrolidine hydrochloride
Copper	N-Benzylaniline
Aniline	Bensylchloride

Manufacturing Process

3 Methods of producing of N-(β-N-phenyl-N-benzylaminoethyl)pyrrolidine:

1. A mixture of 100 parts of N-p-chloroethyl-N-benzylaniline and 100 parts of freshly distilled pyrrolidine are heated to boiling under reflux cooling for about 10 h. Diluted aqueous caustic soda lye is added to the reaction mixture until it shows an alkaline reaction. The reaction product formed and unchanged pyrrolidine are extracted with ether. The extract thus obtained is a subjected to a distillation. After the ether has evaporated, there is produced N-(β-N-phenyl-N-benzylaminoethyl)pyrrolidine, which boils at from 198° to 205°C under 1 mm pressure.

2. 34 parts of N-β-chloroethylpyrrolidine hydrochloride are dissolved together with 110 parts of N-benzylaniline in 400 parts of alcohol, and the solution is heated to boiling under reflux cooling for 15 h, while adding 0.5 part of copper powder. After filtering off the copper powder and driving off the alcohol, dilute caustic soda solution is added until the solution shows an alkaline reaction, whereupon the oil separated is distilled in vacuum. After a forerun of excess N-benzylaniline, there passes over as a main fraction the N-(β-N-phenyl-N-benzylaminoethyl)pyrrolidine, which boils at between 198° and 205°C under 1 mm pressure (mercury gauge).

3. 34 parts of N-β-chloroethylpyrrolidine hydrochloride are added to a solution of 120 parts of aniline in 500 parts of alcohol and the mixture is heated to boiling under reflux cooling for 12 h. After rendering the mixture alkaline by adding caustic soda solution, the excess aniline is driven off with steam. The N-(β-N-phenylaminoethyl)pyrrolidine left is fractionated in vacuo (boiling point 160° to 165°C), 98 parts of this compound are dissolved in 250 parts of 10% aqueous caustic soda solution and admixed with 65 parts of bensylchloride while vigorously stirring at from 50° to 60°C. The temperature is raised to 80° to 90°C for 1 h and the product worked up as indicated in the preceding examples. There is obtained the N-(β-N-phenyl-N-benzylaminoethyl) pyrrolidine in a 95% yield.

In practice it is usually used as hydrochloride.

References

Hopff H. et al.; US Patent No. 2,623,880; December 30, 1952

HOMOFENAZINE

Therapeutic Function: Tranquilizer

Chemical Name: Hexahydro-4-[3-[2-(trifluoromethyl)phenothiazin-10-yl] propyl]-1H-1,4-diazepine-1-ethanol

Common Name: -

Structural Formula:

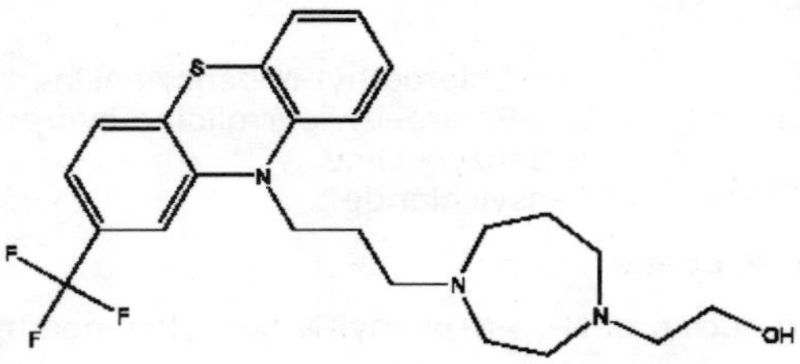

Chemical Abstracts Registry No.: 3833-998

Trade Name	Manufacturer	Country	Year Introduced
Pasaden	Homburg	Italy	1972
Oldagen	Purissimus	Argentina	-

Raw Materials

Sodium amide
3-Trifluoromethyl-phenothiazine
2-Chloroethanol
3-Bromopropylhomopiperazine

Manufacturing Process

35 parts of 3-trifluoromethyl-phenothiazine in 200 parts of toluene were reacted with 6.1 parts soda amide and then with 28.8 parts of 3-bromopropyl-homopiperazine. After a 2 hour reaction period the reaction mixture was washed with water twice and then extracted with dilute HCl, the resulting extract alkalized with excess K_2CO_3 and the precipitated base taken up in ether. After drying the ether extract and evaporation of the ether, the residue was distilled. 20.3 parts of 3-trifluoromethyl-10-(3'-homopiperazino)-propyl-phenothiazine having a boiling point of 225° to 230°C at 1 mm Hg pressure were obtained.

20 parts of 3-trifluoromethyl-10-(3'-homopiperazino)-propyl-phenothiazine in 100 parts of butanol were refluxed for 4 hours together with 5.5 parts of 2-chloroethanol and 11 parts potassium carbonate. The reaction mixture was diluted with 200 parts of ether, then washed three times with water and dried with potassium carbonate. After evaporation of the solvent the residue was distilled under a vacuum of 1 mm Hg. 17.5 parts of 3-trifluoro methyl-10-[3'-(4"-(2"-hydroxyethyl)-homopiperazino)-propyl]-phenothiazine distilled over at 230° to 240°C. The difumarate of this base had a melting point of 148°C.

References

Merck Index 4633

Kleeman and Engel p. 468
I.N. p. 492
Schuler, W.A., Beschke, H. and von Schlichtergroll, A.; US Patent 3,040,043; June 19, 1962; assigned to Deutsche Gold-und Silber-Scheideanstaltvormals Roessler, Germany

HYDRALAZINE HYDROCHLORIDE

Therapeutic Function: Antihypertensive

Chemical Name: 1(2H)-Phthalazinone hydrazone hydrochloride

Common Name: 1-Hydrazinophthalazine hydrochloride

Structural Formula:

Chemical Abstracts Registry No.: 304-20-1; 86-54-4 (Base)

Trade Name	Manufacturer	Country	Year Introduced
Apresoline HCl	Ciba	US	1952
Lopres	Tutag	US	1971
Aiselazine	Hotta	Japan	-
Alphpress	Unipharm	Israel	-
Anaspasmin	Vitacain	Japan	-
Aprelazine	Kaigai	Japan	-
Apresazide	Ciba	US	-
Aprezine	Kanto	Japan	-
Basedock D	Sawai	Japan	-
Deselazine	Kobayashi	Japan	-
Diucholin	Toyama	Japan	-
Dralzine	Lemmon	US	-
Homoton	Horii	Japan	-
Hydrapres	Rubio	Spain	-
Hydrapress	Isei	Japan	-
Hydroserpine	Zenith	US	-
Hypatol	Yamanouchi	Japan	-
Hyperazine	Seiko	Japan	-
Hypos	Nippon Shinyaku	Japan	-
Ipolina	Lafare	Italy	-
Lopress	Reid-Provident	US	-
Pressfall	Nissin	Japan	-
Prospectin	Maruishi	Japan	-
Ser-Ap-Es	Ciba	US	-

Trade Name	Manufacturer	Country	Year Introduced
Serpasil	Ciba	US	-
Solesorin	Hishiyama	Japan	-
Supres	Protea	Australia	-
Tetrasoline	Maruko	Japan	-
Unipres	Reid-Rowell	US	-

Raw Materials

Phthalazone
Hydrazine hydrate
Phosphorus oxychloride
Hydrogen chloride

Manufacturing Process

30 parts by weight of phthalazone are converted to 1-chlorophthalazine by the method described in Ber. d. deutsch. chem. Ges., vol 26, page 521 (1893). The freshly obtained yet moist chloro compound is heated on the water bath for two hours in a mixture of 100 parts by volume of ethyl alcohol and 90 parts by volume of hydrazine hydrate. Preferably after filtering, 1-hydrazine-phthalazine crystallizes out in yellow needles on cooling.

It is filtered with suction and washed with cold ethyl alcohol. The compound is crystallized from methyl alcohol, and melts, when rapidly heated, at 172° to 173°C. On warming in alcoholic or aqueous hydrochloric acid, the hydrochloride of MP 273°C (with decomposition) is obtained.

References

Merck Index 4661
Kleeman and Engel p. 468
PDR pp.789, 812, 830, 993, 1449, 1600, 1999
OCDS Vol. 1 p. 353 (1977)
I.N. p. 494
REM p.847 Hartmann, M. and Druey, J.; US Patent 2,484,029; October 11, 1949; assigned to Ciba Pharmaceutical Products, Inc.

HYDROCHLOROTHIAZIDE

Therapeutic Function: Diuretic

Chemical Name: 6-Chloro-3,4-dihydro-2H-1,2,4-benzothiadiazine-7-sulfonamide-1,1-dioxide

Common Name: Chlorosulthiadil

Structural Formula:

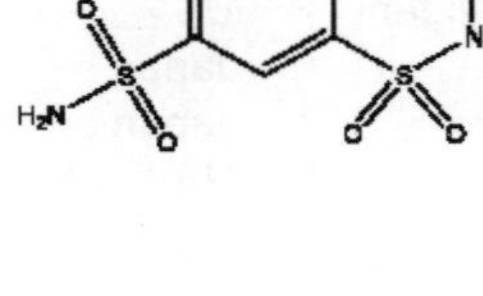

Chemical Abstracts Registry No.: 58-93-5

Trade Name	Manufacturer	Country	Year Introduced
Hydrodiuril	MSD	US	1959
Oretic	Abbott	US	1959
Esidrix	Ciba	US	1959
Esidrex	Ciba Geigy	France	1960
Thiuretic	Parke Davis	US	1974
Lexxor	Lemmon	US	1974
Aidactazide	Searle	US	-
Aldoril	MSD	US	-
Apresazide	Ciba	US	-
Apresoline	Ciba	US	-
Catiazida	Infale	Spain	-
Chemhydrazide	Chemo-Drug	Canada	-
Clothia	Iwaki	Japan	-
Chlorzide	Foy	US	-
Deidran	Pharma. Farm. Spec.	Italy	-
Delco-Retic	Delco	US	-
Dichlorosal	Teva	Israel	-
Dichlotride	Merck-Banyu	Japan	-
Didral	Caber	Italy	-
Dihydran	A.F.I.	Norway	-
Diidrotiazide	Omikron-Gagliardi	Italy	-
Direma	Distillers	UK	-
Dithiazid	Arcana	Austria	-
Diuchlor H	Medic	Canada	-
Diurogen	Gentili	Italy	-
Diursana H	Santos	Spain	-
Dixidrasi	Vaillant	Italy	-
Dyazide	SKF	US	-
Esoidrina	Bouty	Italy	-
Esimil	Ciba	US	-
HHR	Schein	US	-
Hidrosaluretil	Gayoso Wellcome	Spain	-
Hyclosid	Pharmacal	Finland	-
Hydoril	Cenci	US	-
Hydrazide	Powell	Canada	-
Hydrex	Orion	Finland	-
Hydrite	Verdun	Canada	-
Hydro-D	Halsey	US	-
Hydrodiuretex	Barlow Cote	Canada	-
Hydropres	MSD	US	-
Hydroserpine	Schein	US	-
Hydrozide	Elliott-Marion	Canada	-
Hytrid	Leiras	Finland	-

Trade Name	Manufacturer	Country	Year Introduced
Idrodiuvis	Vis	Italy	-
Inderide	Ayerst	US	-
Ivaugan	Voigt	W. Germany	-
Jen-Diril	Jenkins	US	-
Lopressor	Geigy	US	-
Loqua	Columbia	US	-
Manuril	I.C.N.	Canada	-
Maschitt	Showa	Japan	-
Maxzide	Lederle	US	-
Mikorten	Zensei	Japan	-
Moduretic	MSD	US	-
Natrimax	Trianon	Canada	-
Nefrol	Riva	Canada	-
Neo-Codema	Neo	Canada	-
Neo-Flumen	Serono	Italy	-
Neo-Minzil	Valeas	Italy	-
Neo-Saluretic	Lafare	Italy	-
Newtolide	Towa	Japan	-
Novodiurex	Oti	Italy	-
Novohydrazide	Novopharm	Canada	-
Pantemon	Tatsumi	Japan	-
Ro-Hydrazide	Robinson	US	-
Saldiuril	Bieffe	Italy	-
Ser-Ap-Es	Ciba	US	-
Serpasil	Ciba	US	-
Spironazide	Schein	US	-
Tenzide	Metro Med	US	-
Thiadril	Vangard	US	-
Thiaretic	Blue Line	US	-
Timolide	MSD	US	-
Unazid	Pliva	Yugoslavia	-
Unipres	Reid-Rowell	US	-
Urirex	Pharmador	S. Africa	-
Urodiazin	Apogepha	E. Germany	-
Urozide	I.C.N.	Canada	-
Zide	Reid-Provident	US	-

Raw Materials

5-Chloro-2,4-disulfamylaniline
Paraformaldehyde

Manufacturing Process

As described in US Patent 3,163,645, a mixture of 2.9 grams of 5-chloro-2,4-disulfamyl aniline in 15 ml of anhydrous diethyleneglycol dimethyl ether, 0.5 ml of an ethyl acetate solution containing 109.5 grams of hydrogen chloride per 1,000 ml and 0.33 grams (0.011 mol) of paraformaldehyde is heated to 80° to 90°C and maintained at that temperature for 1 hour. The resulting mixture is cooled to room temperature and concentrated to one-third of its volume under reduced pressure, diluted with water, then allowed to crystallize. The product is filtered off and recrystallized from water, to yield the desired 6-chloro-7-sulfamyl-3,4-dihydro-2H-[1,2,4]-benzothiadiazine-1,1-dioxide, MP 266° to 268°C, yield 1.4 grams. By replacing paraformaldehyde by 0.84 gram of 1,1 -dimethoxymethane and proceeding as above, the same compound is obtained.

As described in US Patent 3,025,292, the desired product may be made by hydrogenation of chlorothiazide. Three grams of 6-chloro-7-sulfamyl-1,2,4-benzothiadiazine-1,1-dioxide (chlorothiazide) is suspended in 100 ml of methanol. Then 1.0 gram of a 5% ruthenium on charcoal catalyst is added, and the mixture is reduced at room temperature and at an initial hydrogen pressure of 39 psig. The theoretical amount of hydrogen to form the 3,4-dihydro derivative is absorbed after a period of about 10 hours.

The reduction mixture then is heated to boiling and filtered hot to remove the catalyst. The catalyst is washed with a little methanol and the combined filtrate is concentrated to a volume of about 25 ml by evaporation on a steam bath. Upon cooling to room temperature, white crystals separate which are filtered, washed with water, and dried in vacuo at room temperature over phosphorus pentoxide overnight. The weight of 6-chloro-7-sulfamyl 3,4-dihydro-1,2,4-benzothiadiazine-1,1-dioxide obtained is 1.26 grams; MP 268.5° to 270°C. Dilution of the above filtrate with water to a volume of about 125 ml gives a second crop of product having the same melting point and weighing 1.22 grams, giving a combined yield of 83%. When the product is mixed with an authentic sample of 6-chloro-7-sulfamyl-3,4-dihydro-1,2,4-benzothiadiazine-1,1-dioxide, prepared by another method, the melting point is not depressed.

References

Merck Index 4683

Kleeman and Engel p. 469

PDR pp. 546, 625, 789, 812, 896, 1014, 1137, 1184, 1201, 1211, 1449, 1606, 1674, 1713, 1999

OCDS Vol. 1 p. 358 (1977)

DOT 16 (4) 141 (1980), (8) 266 (1980), 17 (5) 213 (1981), 19 (3) 172 (1983) and 19 (9) 496 (1983)

I.N. p.495

REM p. 939

Jones, W.H. and Novello, F.C.; US Patent 3,025,292; March 13, 1962; assigned to Merck and Co., Inc.

Downing, G.V., Jr.; US Patent 3,043,840; July 10, 1962; assigned to Merck and Co., Inc.

de Stevens, G. and Werner, L.H.; US Patent 3,163,645; December 29, 1964; assigned to Ciba Corporation

Irons, J.S. and Cook, T.M.; US Patent 3,164,588; January 5, 1965; assigned to Merck and Co., Inc.

HYDROCODONE

Therapeutic Function: Narcotic analgesic, Antitussive

Chemical Name: Morphinan-6-one, 4,5-epoxy-3-methoxy-17-methyl-, (5α)-

Common Name: Dihydrocodeinone; Hydrocodone; Hydrocon

Structural Formula:

Chemical Abstracts Registry No.: 125-29-1

Trade Name	Manufacturer	Country	Year Introduced
Dicodid	Knoll	Belgium	-
Dicodid	Knoll	Germany	-
Dicodid	Knoll	Switz.	-
Tucodil	Dilmen	Turkey	-
Vicodin	Knoll	USA	-

Raw Materials

Codeine
Sulfuric acid
Palladium on charcoal

Manufacturing Process

60 g of codeine about 94.5% (anhydrous codeine alkaloid basis) was dissolved in a solution made from 10 ml concentrated sulfuric acid and 390 ml water The mixture was refluxed for one hour with 25.0 g of 5% Pd on charcoal. The hot solution was immediately filtered and the catalyst was washed with 400 ml of dilute sulfuric acid of the same strength as was used in the rearrangement described above. To the combined cooled filtrate and wash, 750 ml if benzene was added, after which the mixture was cooled to 15°C, stirred, and made alkaline to pH 10 by addition of 80 ml of 40% NaOH. After shaking and separating the aqueous layer was extracted twice with 500 ml benzene. The combined benzene extracts are then extracted three times with 500 ml and twice with 400 ml portions of fresh 10% sodium bisulfite solution. Crude dihydrocodeinone was precipitated from bisulfite solution by eddition of 180 ml 40% NaOH at 15°C (to pH 10). The product was filtered, washed well and air-dried at room temperature. The melting point was about 184°C and yield about 35-38 g or 58-59%. The product was darken and the rest of Pd on charcoal and the original alkaloid were removed with column of Al_2O_3 (eluent - dry ethylene chloride) to give the dihyrocodeinone.

References

Baizer M.M. et al.; US Patent No. 2,544,291; Mar. 6, 1951
Harclerode W.H. et al.; US Patent No. 6,512,117 B1; Jan. 28, 2003; Assigned Abbott Laboratories, Abbot Park, 11 (US)

HYDROCORTAMATE HYDROCHLORIDE

Therapeutic Function: Corticosteroid

Chemical Name: Cortisol 21-ester with N,N-diethylglycine hydrochloride

Common Name: -

Structural Formula:

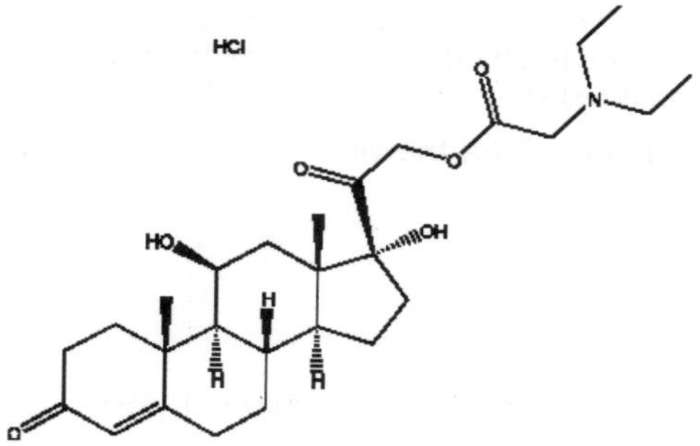

Chemical Abstracts Registry No.: 76-47-1 (Base)

Trade Name	Manufacturer	Country	Year Introduced
Magnacort	Pfizer	US	1956
Etacort	Angelini	Italy	-

Raw Materials

Hydrocortisone
Chloroacetic anhydride
Diethylamine

Manufacturing Process

1 g of hydrocortisone is introduced with stirring into 5 cc of anhydrous pyridine. After heating to 45°C and then cooling again to 0°C to 5°C there is slowly added dropwise a freshly prepared solution of 0.52 g (1 mol + 10%) of chloracetic anhydride in 4 cc of absolute ether, The reaction temperature should not exceed 10°C. During the whole time of reaction a stream of nitrogen is passed through the reaction mixture in order to achieve an exhaustive evaporation of the added ether. The batch is slowly allowed to come to room temperature, an operation requiring 4 to 5 hours, and then 0.1 cc of water is added for decomposition of the excess of anhydride. The reaction solution is introduced dropwise with stirrinq within 1 hour into 100 cc of water as a result of which the 21-chloracetate of hydrocortisone is deposited. After filtration with suction, washing is carried out with water, 5% hydrochloric acid, water, 2% sodium bicarbonate solution and water again. The substance is then dried in a vacuum desiccator. The white chloracetate thus obtained melts at 213°C to 214°C with decomposition. It is free from nitrogen and the yield amounts to 93.4% of the theoretical.

1 g of hydrocortisone-21-chloracetate is dissolved in 15 cc of anhydrous and peroxide-free tetrahydrofuran. The solution produced is treated with a solution

of 0.42 g of diethylamine in 15 cc of tetrahydrofuran. The reaction mixture is allowed to stand for 24 hours at room temperature. The separated diethylamine hydrochloride is filtered with suction and the filtrate evaporated under vacuum in a nitrogen atmosphere at 40°C. The residue is triturated with a little absolute ether and suction filtered. It is washed on the filter with a little ether and then with hexane. The 21-diethylaminoacetate of hydrocortisone melts at 150°C to 162°C. The base can be recrystallized from ethyl acetate but its melting point remains practically unchanged at 162°C to 163°C. The yield amounts to 72.5% of the theoretical. For conversion of the base into the hydrochloride it is suspended in ether and the suspension treated with ethereal hydrochloric acid. The hydrochloride is filtered with suction and recrystallized from ethanol; MP 222°C with decomposition.

With a starting quantity of 14g, the yield amounted to 85.4% of the theoretical.

References

Merck Index 4688
Hydrocortisone 775
I.N. p. 497
Schering A.G.; British Patent 879,208; October 4, 1961

HYDROCORTISONE

Therapeutic Function: Glucocorticoid

Chemical Name: 11β,17,21-Trihydroxypregn-4-ene-3,20-dione

Common Name: 17-Hydroxycorticosterone

Structural Formula:

Chemical Abstracts Registry No.: 50-23-7

Trade Name	Manufacturer	Country	Year Introduced
Hydrocortone	MSD	US	1952
Cortef	Upjohn	US	1953
Cortril	Pfizer	US	1954
Cortifan	Schering	US	1954
Otosone-F	Broemmel	US	1955
Cortispray	National	US	1956
Domolene-HC	Dome	US	1960
Texacort	Texas Pharm	US	1960

Trade Name	Manufacturer	Country	Year Introduced
Cortenema	Rowell	US	1966
Lubricort	Texas Pharm	US	1968
Proctocort	Rowell	US	1969
Hautosone	Merrell National	US	1970
Dermacort	Rowell	US	1972
Rectoid	Pharmacia	US	1977
Alphaderm	Norwich Eaton	US	1978
H-Cort	Pharm. Assoc.	US	1979
Dermoiate	Schering	US	1979
Clear-Aid	Squibb	US	1980
Hycort	Elder	US	1981
Prep-Cort	Whitehall	US	1981
Corizone-5	Thompson	US	1982
Flexicort	Westwood	US	1982
Aeroseb	Allergan	US	-
Ala-Cort	Del Ray	US	-
Algicortis	Vaillant	Italy	-
Aliersone	Mallard	US	-
Alphacortison	Norwich Eaton	US	-
Alphaderm	Norwich	US	-
Balneol-HC	Rowell	US	-
Barseb-HC	Barnes Hind	US	-
Bio-Cortex	Ries	US	-
Carmol HC	Syntex	US	-
Cleiton	Kodama	Japan	-
Cobadex	Cox	UK	-
Cortanal	Canada Pharmacal	Canada	-
Cort-Dome	Dome	US	-
Cortes	Taisho	Japan	-
Cortesal	Pharmacia	Sweden	-
Corticaine	Glaxo	US	-
Cortifair	Pharmafair	US	-
Cortiment	Ferring	Sweden	-
Cortiphate	Travenol	US	-
Cortisporin	Burroughs-Wellcome	US	-
Cortolotion	Kempthorne Prosser	New Zealand	-
Cortril	Pfizer	US	-
Cremesone	Dalin	US	-
Di-Hydrotic	Legere	US	-
Dioderm	Dermal	UK	-
Durel-Cort	Durel	US	-
Ecosone	Star	US	-
Efcortelan	Glaxo	UK	-
Egocort	Ego	Australia	-
Excerate	Foji Zoki	Japan	-

Trade Name	Manufacturer	Country	Year Introduced
FEP	Boots	US	-
Gyno-Cortisone	Lyocentre	France	-
HC-Cream	C and M Pharmacal	US	-
Heb-Cort	Barnes Hind	US	-
Hidroaltesona	Alter	Spain	-
Hycor	Sigma	Australia	-
Hycort	Douglas	US	-
Hycortole	Premo	US	-
Hydrocort	Ferring	W. Germany	-
Hydrocortex	Kenyon	US	-
Hydrofoam	U.S.V.	US	-
Hydrotisona	Roussel-Lutetia	Argentina	-
Hytone	Dermik	US	-
Idracemi	Farmigea	Italy	-
Lexocort	Lexington	US	-
Microcort	Alto	US	-
Milliderm	A.L.	Norway	-
Octicair	Pharmafair	US	-
Optef	Upjohn	-	-
Otic-HC	Hauck	US	-
Otobiotic	Schering	US	-
Otocort	Lemmon	US	-
Pedicort	Pedinol	US	-
Penecort	Herbert	US	-
Pyocidin	Berlex	US	-
Rectocort	Welcker-Lyster	Canada	-
Rectoid	Pharmacia	Sweden	-
Sigmacort	Sigma	Australia	-
Signef	Fellows-Testagar	US	-
Sterocort	Omega	Canada	-
Synacort	Syntex	US	-
Tega-Cort	Ortega	US	-
Vanoxide	Dermik	US	-
Vioform	Ciba	US	-
Viosol	Wallace	US	-
Vytone	Dermik	US	-

Raw Materials

Bacterium *Cunninghamella blakesleeana*
11-Desoxy-17-hydroxycorticosterone

Manufacturing Process

The following example from US Patent 2,602,769 illustrates the preparation of 17-hydroxycorticosterone (compound F) from 11-desoxy-17-hydroxycorticosterone (compound S). A medium was prepared from 0.5% peptone, 2% dextrose, 0.5% soybean meal, 0.5% KH_2PO_4, 0.5% sodium chloride and 0.3% yeast extract in tap water. To 200 ml of this sterilized medium was added an inoculum of the vegetative mycella of *Cunninghamella blakesleeana*. The spores had first been transferred from a sport slant to a broth medium and the broth medium was aerobically incubated at 24°C for 24 to 72 hours in a .reciprocating shaker until the development of vegetative growth. The inoculated medium containing added vegetative mycella of *Cunninghamella blakesleeana* was incubated for 48 hours at 24°C following which was added 66 mg of compound S, 11-desoxy-17-hydroxycorticosterone in solution in a minimum of ethanol, and incubation was maintained for 7 hours at 24°C. The beer containing steroid was diluted with 800 ml of acetone, shaken 1 hour on a reciprocating shaker and filtered. The cake was suspended in 500 ml of acetone, shaken another hour and again filtered. The filtrates were combined and the acetone was volatilized under reduced pressure at 50°C. Acetone was then added, if necessary, to bring the concentration to 20% acetone and this resulting aqueous acetone solution was extracted five times each with one-third volume of Skellysolve B petroleum ether to remove fatty materials. These extracts were back washed two times with one-tenth volume of 20% aqueous acetone and the washings were added to the main acetone extract.

The combined acetone extracts were extracted six times with one-fourth volume of ethylene dichloride and the ethylene dichloride extract was evaporated under vacuum to leave the steroid residue. This steroid residue was taken up in a minimum of methylene chloride and applied to the top of a column packed with 30 grams of silica which had been previously triturated with 21 ml of ethylene glycol. Then various developing mixtures, saturated with ethylene glycol, were passed over the column. Cuts were made as each steroid was eluted as determined by the lowering of the absorption of light at 240 nm on the automatic chromatographic fraction cutter.

Band Solvent Tube No. (60ml) Crude Solids (mg)

1 Cyclohexane 1-4 11

2 Cyclohexane-methylene chloride 3:1 5-13 6.4 compound S

3 Cyclohexane-methylene chloride 1:1 14-16 3.0

4 Cyclohexane-methylene chloride 2:3 17-23 6.0 compound E

5 Cyclohexane-methylene chloride 1:4 24-38 12.2 compound F

6 Methylene chloride 39-59 4.8

A 7.7 mg portion of band 5 was taken up in a minimum of acetone and refrigerated until crystals separated. This cold acetone mixture was centrifuged and the supernatant liquid removed by pipette. To the remaining crystals, a few drops of ice-cold ether-acetone, three to one mixture, were added, shaken, recentrifuged and the supernatant wash liquid removed by pipette. The ether-acetone wash was repeated. The resulting crystals were dried under vacuum yielding 3.3 mg of pure compound F, 17-hydroxycorticosterone.

References

Merck Index 4689
Kleeman and Engel p. 470
PDR pp. 671, 684, 739, 821, 833, 908, 928, 933, 1033, 1073, 1250, 1397, 1404, 1429, 1446, 1576, 1645, 1800, 1886
OCDS Vol. 1 p. 190 (1977)
DOT 12 (9) 343 (1976)

I.N. p. 497
REM p, 967
Murray, H.C. and Peterson, D.H.; US Patent 2,602,769; July 8, 1952; assigned to The Upjohn Company
Murray,H.C. and Peterson, D.H.; US Patent 2,649,400; August 18, 1953; assigned to The Upjohn Company
Murray,H.C. and Peterson, D.H.; US Patent 2,649,402; August 18, 1953; assigned to The Upjohn Company
Mann,K.M., Drake, H.A. and Rayman, D.E.; US Patent 2,794,816; June 4, 1957; assigned to The Upjohn Company

HYDROCORTISONE SODIUM PHOSPHATE

Therapeutic Function: Glucocorticoid

Chemical Name: 11β,17-Dihydroxy-21-(phosphonoxy)pregn-4-ene-3,20-dione disodium salt

Common Name: -

Structural Formula:

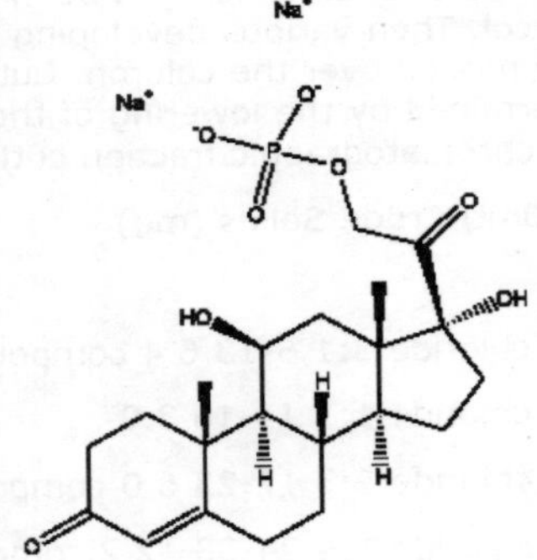

Chemical Abstracts Registry No.: 6000-74-4; 3863-59-0 (Phosphate base)

Trade Name	Manufacturer	Country	Year Introduced
Corphos	Tilden Yates	US	1959
Hydrocortone Phosphate	MSD	US	1960
Cortiphate	Travenol	US	1962
Ocu-Cort	Dome	US	1963
Actocortin	Cooper	W. Germany	-
Efcortesol	Glaxo	UK	-
Efcortesol	Glaxo	Italy	-
Flebocortid	Richter	-	-
Gleiton	Sankyo Zoki	Japan	-

Raw Materials

21-Iodo-11β:17α-dihydroxypregn-4-ene-3,20-dione

Phosphoric acid
Sodium hydroxide

Manufacturing Process

21-iodo-11β:17α-dihydroxypregn-4-ene-3,20-dione (5.0 g) in pure acetonitrile (125 ml) was mixed with a solution of 90% phosphoric acid (2.5 ml) and triethylamine (7.5 ml) in acetonitrile (125 ml) and boiled under reflux for 4 hours. The solvent was removed in vacuo and the residue, dissolved in ethanol (20 ml) and water (80 ml), was passed down a column of Zeo-Karb 225 (H^+form) (60 g) made up in 20% alcohol. Elution was continued with 20% alcohol (50 ml), 50% alcohol (50 ml) and alcohol (150 ml). The eluate was at first cloudy, but by the end of the elution it was clear and nonacid.

The eluate was titrated to pH 7 with 0.972N NaOH (63 ml). Removal of solvent left a gum, which was boiled with methanol (400 ml) for 20 minutes. The solid insoluble inorganic phosphate was filtered off and washed with methanol (200 ml). The slightly cloudy filtrate was filtered again, and evaporated to dryness in vacuo. The residual gum dissolved readily in water (40 ml) and on addition of acetone (600 ml) to the solution a mixture of sodium salts of hydrocortisone 21-phosphate separated as a white solid. This was collected after 2 days, washed with acetone and dried at 100°C/0.l mm/2 hr to constant weight. Yield 4.45 g.

References

Merck Index 4691
Kleeman and Engel p. 473
I.N. p. 498
REM p.968
Elks, J. and Phillips, G.H.; US Patent 2,936,313; May 10, 1960; assigned to Glaxo Laboratories, Ltd. (UK)

HYDROFLUMETHIAZIDE

Therapeutic Function: Diuretic, Antihypertensive

Chemical Name: 3,4-Dihydro-6-(trifluoromethyl)-2H-1,2,4-benzothiadiazine-7-sulfonamide 1,1-dioxide

Common Name: -

Structural Formula:

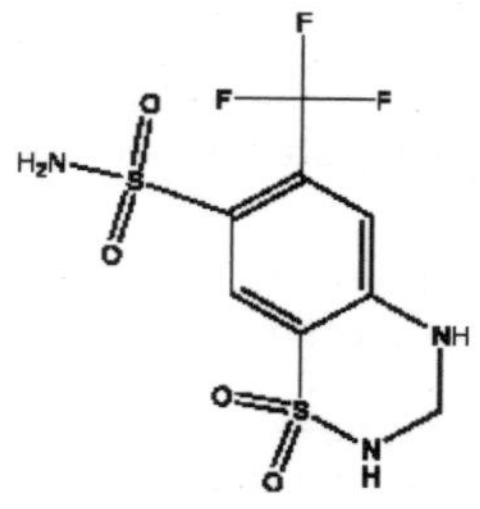

Chemical Abstracts Registry No.: 135-09-1

Trade Name	Manufacturer	Country	Year Introduced
Saluron	Bristol	US	1959
Leodrine	Leo	France	1960
Diucardin	Ayerst	US	1974
Di-Ademil	Squibb-Showa	Japan	-
Enjit	Meiji	Japan	-
Fluorodiuvis	Vis	Italy	-
Hydrenox	Boots	UK	-
Leodrine	Leo	France	-
Naclex	Glaxo	UK	-
Olmagran	Heyden	W. Germany	-
Plurine	Leo	France	-
Rivosil	Benvegna	Italy	-
Rotezon	Mitsui	Japan	-
Rontyl	Leo-Sankyo	Japan	-
Vergonil	Ferrosan	Denmark	-

Raw Materials

α,α,α-Trifluoro-m-toluidine
Chlorosulfonic acid
Ammonia
Paraformaldehyde

Manufacturing Process

(a) Preparation of 5-Trifluoromethylaniline-2,4-Disulfonyl Chloride: 113 ml of chlorosulfonic acid was cooled in an ice-bath, and to the acid was added dropwise while stirring 26.6 grams of α,α,α-trifluoro-m-toluidine. 105 grams of sodium chloride was added during 1 to 2 hours, whereafter the temperature of the reaction mixture was raised slowly to 150° to 160°C, which temperature was maintained for 3 hours. After cooling the mixture, ice-cooled water was added, whereby 5-trifluoromethylaniline-2,4-disulfonyl chloride separated out from the mixture.

(b) Preparation of 5-Trifluoromethyl-2,4-Disulfamylaniline: The 5-trifluoromethylaniline-2,4-disulfonyl chloride obtained in step (a) was taken up in ether and the ether solution dried with magnesium sulfate. The ether was removed from the solution by distillation, the residue was cooled to 0°C and 60 ml of ice-cooled, concentrated ammonia water was added while stirring. The solution was then heated for one hour on a steam bath and evaporated in vacuo to crystallization. The crystallized product was 5-trifluoromethyl-2,4-disulfamylaniline, which was filtered off, washed with water and dried in a vacuum exsiccator over phosphorus pentoxide. After recrystallization from a mixture of 30% ethanol and 70% water the compound had a MP of 247° to 248°C.

(e) Preparation of 6-Trifluoromethyl-7-Sulfamyl-3,4-Dihydro-1,2,4-Benzothiadiazine-1,1-Dioxide: 3.2 grams of 5-trifluoromethyl-2,4-disulfamylaniline was added to a solution of 0.33 gram of paraformaldehyde in 25 ml of methyl Cellosolve (2-methoxy ethanol) together with a catalytic amount of p-toluenesulfonic acid, and the mixture was boiled with reflux for 5 hours. The solvent was then distilled off in vacuo, and the residue triturated with 30 ml of ethyl acetate. 6-trifluoromethyl-7-sulfamyl-3,4-dihydro-1,2,4-benzothiadiazine-1,1-dioxide crystallized out. After recrystallization from methanol/water the substance had a MP of 272° to 273°C.

References

Merck Index 4695
Kleeman and Engel p. 474
PDR pp.617, 709, 1606, 1999
OCDS Vol. 1 p. 358 (1977)
I.N. D. 499
REM p. 939
Lund, F., Lyngby, K. and Godtfredsen, W.O.; US Patent 3,254,076; May 31, 1966; assigned to Lovens Kemiske Fabrik Ved A. Kongsted, Denmark

HYDROQUINONE

Therapeutic Function: Depigmentor

Chemical Name: 1,4-Benzenediol

Common Name: Quinol

Structural Formula:

HO–C_6H_4–OH

Chemical Abstracts Registry No.: 123-31-9

Trade Name	Manufacturer	Country	Year Introduced
Quinnone	Dermohr	US	1980
Melanek	Neutrogena	US	1981
Black and White	Plough	US	-
Eldopaque	Elder	US	-
Eldoquin	Elder	US	-
Phiaquin	Phial	Australia	-
Phiaquin	Robins	US	-
Solaquin	Elder	US	-

Raw Materials

Acetylene
Methanol

Manufacturing Process

Into a pressure reactor there was charged 100 ml of methanol and 1 g of diruthenium nonacarbonyl. The reactor was closed, cooled in solid carbon dioxide/acetone, and evacuated. Acetylene, to the extent of 1 mol (26 g), was metered into the cold reactor. Carbon monoxide was then pressured into this vessel at 835-980 atmospheres, during a period of 16.5 hours; while the reactor was maintained at 100°C to 150°C. The reactor was then cooled to room temperature and opened.

The reaction mixture was removed from the vessel and distilled at a pressure of 30-60 mm, and a bath temperature of 30°C to 50°C until the methanol had all been removed. The extremely viscous tarry residue remaining in the still pot was given a very crude distillation, the distillate boiling at 82°C to 132°C/2 mm. In an attempt to purify this distillate by a more careful distillation, 5.3 g of a liquid distilling from 53°C to 150°C/5 mm was collected. At this point, much solid sublimate was noted not only in this distillate but in the condenser of the still. 7 g of the solid sublimate was scraped out of the condenser of the still. Recrystallization of the sublimate from ethyl acetate containing a small amount of petroleum ether gave beautiful crystals melting at 175°C to 177°C (5 g). Infrared analysis confirmed that this compound was hydroquinone (9% conversion).

References

Merck Index 4719
PDR pp. 865, 1268
I.N. p. 499
REM p. 788
Howk, B.W. and Sauer, J.C.; US Patent 3,055,949; September 25, 1962; assigned to E.I. du Pont de Nemours and Co.

HYDROXOCOBALAMIN

Therapeutic Function: Hematopoietic vitamin

Chemical Name: Cobinamide hydroxide phosphate 3'-ester with 5,6-dimethyl-1-α-D-ribofuranosylbenzimidazole inner salt

Common Name: Vitamin B_{12a}

Structural Formula:

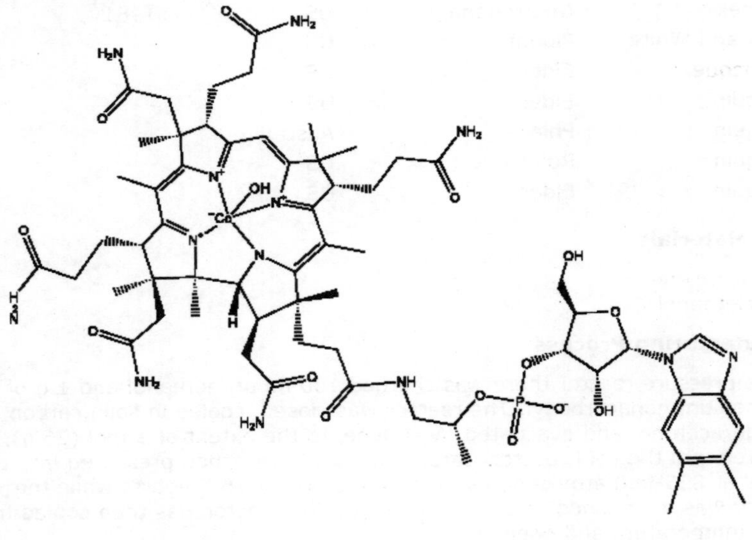

Chemical Abstracts Registry No.: 13422-51-0

Trade Name	Manufacturer	Country	Year Introduced
Alpha-Redisol	MSD	US	1962
Ducobee-Hy	Breon	US	1962
Rubramin-OH	Squibb	US	1963
Hycobal-12	Canfield	US	1964
Hydroxo B-12	Philips Roxane	US	1964
Neo-Vi-Twel	SMP	US	1964
Neo-Betalin 12	Lilly	US	1964
Sustwelve	Ascher	US	1964
Rubesol-LA	Central	US	1965
Sytobex-X	Parke Davis	US	1966
Acimexan	Cimex	Switz.	-
Anemisol	Tobishi	Japan	-
Aqua-B	Nippon Zoki	Japan	-
Aquo-Cytobion	Merck	W. Germany	-
Axlon	Albert Roussel	W. Germany	-
Behepan	Kabi Vitrum	Sweden	-
Berubi	Redel	W. Germany	-
Bistin	Yamanouchi	Japan	-
Bradiruba	Ibirn	Italy	-
Cobalidrina	Italsuisse	Italy	-
Cobalamin H	Otsuka	Japan	-
Cobalvit	Tosi-Novara	Italy	-
Colsamine	Kanyo	Japan	-
Docevita	Boizot	Spain	-
Dolevern	Seiko	Japan	-
Erycytol	Sanabo	Austria	-
Fravit B-12	Francia	Italy	-
Fresmin S	Takeda	Japan	-
Funacomin-F	Funai	Japan	-
Hicobala	Mitaka	Japan	-
Hicobalan	Maruko	Japan	-
Hydocobamin	Hishiyama	Japan	-
Hydocomin	Sanwa	Japan	-
Hydroxo 5000	Hepatrol	France	-
Hydroxomin	Tokyo Hosei	Japan	-
Idoxo 812	Ferrosan	Denmark	-
Idro-Apavit	Locatelli	Italy	-
Idrobamina	Tiber	Italy	-
Idrocobalmin	Panther-Osfa	Italy	-
Idrospes B12	Ausonia	Italy	-
Idrozima	Labif	Italy	-
Laseramin	Choseido	Japan	-
Longicobal	Farber-R.E.F.	Italy	-
Masblon H	Fuso	Japan	-

Trade Name	Manufacturer	Country	Year Introduced
Natur B12	Panthox and Burck	Italy	-
Nichicoba	Nichiiko	Japan	-
Novobedouze	Bouchara	France	-
OH-BIZ	Morishita	Japan	-
Oxobemin	Vitrum	Sweden	-
Rasedon	Sawai	Japan	-
Red	Neopharmed	Italy	-
Red-B	Kowa	Japan	-
Redisol H	Merck-Banyu	Japan	-
Rossobivit	Medici	Italy	-
Rubitard B12	Proter	Italy	-
Runova	Squibb-Sankyo	Japan	-
Solco H	Tobishi	Japan	-
Tsuerumin S	Mohan	Japan	-
Twelvmin	Mohan	Japan	-
Vigolatin	Kowa	Japan	-

Raw Materials

Cyanocobalamin
Hydrogen

Manufacturing Process

A solution containing 26.3 mg of vitamin B_{12} in 15 ml of water was shaken with 78 mg of platinum oxide catalyst and hydrogen gas under substantially atmospheric pressure at 25°C for 20 hours. Hydrogen was absorbed. During the absorption of hydrogen the color of the solution changed from red to brown. The solution was separated from the catalyst and evaporated to dryness in vacuo. The residue was then dissolved in 1 ml of water and then diluted with about 6 ml of acetone.

After standing for several hours a small amount of precipitate (about 2 to 3 mg) was formed and was then separated from the solution. This solution was diluted with an additional 2 ml of acetone and again allowed to stand for several hours. During this time about 4 to 5 mg of noncrystalline precipitate formed. This solid was separated from the solution and an additional 2 ml of acetone was added to the solution. On standing, vitamin B_{12a} began to crystallize in the form of red needles. After standing for 24 hours, the crystalline material was separated, yield 12 mg. By further dilution of the mother liquor with acetone additional crystalline precipitate formed (from US Patent 2,738,302).

References

Merck Index 4720
Kleeman and Engel p. 475
I.N. p. 500
REM pp. 1020, 1023
Kaczka, E.A., Wolf, D.E. and Folkers, K.; US Patent 2,738,301; March 13, 1956; assigned to Merck AND Co., Inc.
Kaczka, E.A., Wolf, D.E. and Folkers, K.;US Patent 2,738,302; March 13, 1956; assigned to Merck 81Co., Inc.

HYDROXYCHLOROQUINE SULFATE

Therapeutic Function: Antimalarial

Chemical Name: 2-[[4-[(7-Chloro-4-quinolinyl)amino]pentyl]ethylamino] ethanol sulfate

Common Name: -

Structural Formula:

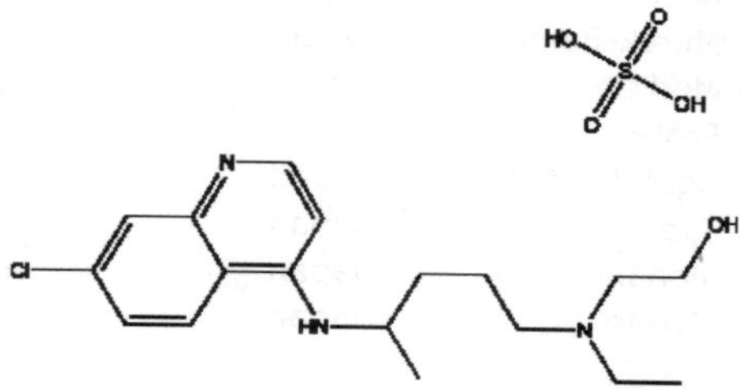

Chemical Abstracts Registry No.: 747-36-4; 118-42-3 (Base)

Trade Name	Manufacturer	Country	Year Introduced
Plaquenil	Winthrop	US	1956
Plaquenil	Winthrop	France	1960
Ercoquin	Erco	Denmark	-
Eroquin	Shionogi	Japan	-
Oxiklorin	Orion	Finland	-
Quensyl	Winthrop	W. Germany	-
Rhyumapirine S	Nichiiko	Japan	-
Toremonil	Iwaki	Japan	-

Raw Materials

Ammonia
Hydrogen
Sulfuric acid
Phosphoric acid
1-Chloro-4-pentanone
N-Ethyl-N-2-hydroxyethylamine
4,7-Dichloroquinoline

Manufacturing Process

A mixture of 323 grams of 1-chloro-4-pentanone, 480 grams of N-ethyl-N-2-hydroxyethylamine and 400 grams of sodium chloride (to aid in subsequent filtration) in 1.3 liters of xylene was heated with stirring on a steam bath for two hours and then refluxed for three hours. After standing overnight, the mixture was filtered and the filter cake washed with xylene. The filtrate was fractionally distilled, yielding 207.3 grams of a fraction distilling at 89° to 90°C at 0.35 mm; n_D^{25} = 1.4600. This fraction, 1-(N-ethyl-N-2-hydroxyethylamino)-4-pentanone, was used in the next step of the synthesis. A sample of the fraction was further purified by distillation through a column and gave an analytically pure sample of 1-(N-ethyl-N-2-hydroxyethylamino)-4-pentanone, boiling at 85° to 87°C at 0.4 mm.

The 1-(N-ethyl-N-2-hydroxyethylamino)-4-pentanone from above (284.2 grams) was dissolved in 300 grams of 28% ammoniacal methanol and reduced catalytically with Raney nickel (at an initial pressure of 1,000 pounds) at room temperature. After 24 hours the catalyst was filtered off and the product distilled in vacuo through a column, yielding 254 grams of a fraction distilling at 88.5° to 96°C at 0.3 mm and comprising mainly 5-(N-ethyl-N-2-hydroxyethylamino)-2-pentylamine. An analytical sample of this fraction distilled at 93°C at 0.6 mm.

A mixture of 90 grams of 4,7-dichloroquinoline, 90 grams of phenol, 1 gram of potassium iodide and 132 grams of 5-(N-ethyl-N-2-hydroxyethylamino)-2-pentylamine from above was heated with stirring for 13 hours at 125° to 130°C. Methanol (1.9 liters) was added and the the mixture was filtered with charcoal. The filtrate was treated with 270 cc of a solution of 100 grams of phosphoric acid in 300 cc of methanol. The walls of the flask containing the filtrate were scratched with a glass rod and the mixture was allowed to stand for two days. The solid was filtered off, washed with methanol and dried, yielding 101 grams of crude 7-chloro-4-[5-(N-ethyl-N-2-hydroxyethylamino)-2-pentyl]aminoquinoline diphosphate, MP 155° to 156°C.

Additional quinoline diphosphate was obtained as a gummy mass from the filtrate by concentrating the latter to about half its volume and adding acetone. The crude gummy diphosphate was dissolved in water, basified with ammonium hydroxide and the resulting liberated basic quinoline extracted with chloroform. After removal of the chloroform by distillation, the residue was dissolved in ether and crystallization was induced by scratching the walls of the flask with the glass rod. About 30 grams of the crude quinoline base, melting at 77° to 82°C, separated. Recrystallization of this material from ethylene dichloride or ethyl acetate yielded the purified 7-chloro-4-[5-(N-ethyl-N-2-hydroxyethylamino)2-pentyl] aminoquinoline, MP 89° to 91°C.

The base may then be dissolved in ethanol and precipitated as the sulfate by reaction with an equimolar quantity of sulfuric acid.

References

Merck Index 4729
Kleeman and Engel p. 476
PDR p. 1926
OCDS Vol. 1 p. 342 (1977)
I.N. p. 502
REM p. 1220
Surrey, A.R.; US Patent 2,546,658; March 27, 1951; assigned to Sterling Drug Inc.

HYDROXYDIONE SODIUM SUCCINATE

Therapeutic Function: Anesthetic

Chemical Name: 21-(3-Carboxy-1-oxopropoxy)-5β-pregnane-3,20-dione sodium salt

Common Name: -

Chemical Abstracts Registry No.: 53-10-1

Structural Formula:

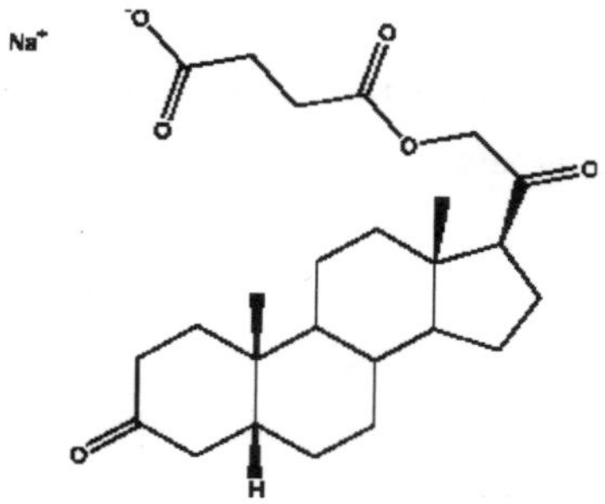

Trade Name	Manufacturer	Country	Year Introduced
Viadril	Pfizer	US	1957
Predion	V.N.I.Kh.F.I.	USSR	-

Raw Materials

Desoxycorticosterone
Hydrogen
Succinic anhydride

Manufacturing Process

A solution of 20 g of desoxycorticosterone in 190 ml of absolute ethanol was stirred in an atmosphere of hydrogen in the presence of 1.68 g of 25% palladium on calcium carbonate catalyst. After 20 hours, approximately 1 molar equivalent of hydrogen had been absorbed and hydrogen uptake had ceased. The catalyst was removed by filtration and the filtrate evaporated in vacuo to yield 20 g of nearly pure product, MP 135°C to 140°C. The crude product was demonstrated to be free of starting material by paper chromatography. A highly purified product was obtained by recrystallization from acetone-water with cooling in an ice bath, yield 14.5 g, MP 152°C to 154°C. The product was characterized by analysis and by absence of ultraviolet absorption.

A solution of 14 g of 21-hydroxypregnane-3,20-dione and of 14 g of recrystallized succinic anhydride in 140 ml of dry pyridine was allowed to stand at room temperature for 18 hours, then cooled in an ice bath and poured in a fine stream into 1.5 liters of ice water. Excess pyridine was neutralized with 3N hydrochloric acid and the solution further diluted with 2 liters of ice water. The precipitated product was filtered, washed with water and dried in vacuo at 50°C affording 18 g of solid MP 192°C to 195°C. Recrystallization of a small sample afforded analytically pure material, MP 200°C.

References

Merck Index 4734
I.N. p. 502
Laubach, G.D.; US Patent 2,708,651 ; May 17, 1955; assigned to Chas. Pfizer and Co., Inc.

HYDROXYPHENAMATE

Therapeutic Function: Tranquilizer

Chemical Name: 2-Phenyl-1,2-butanediol-1-carbamate

Common Name: Oxyfenamate

Structural Formula:

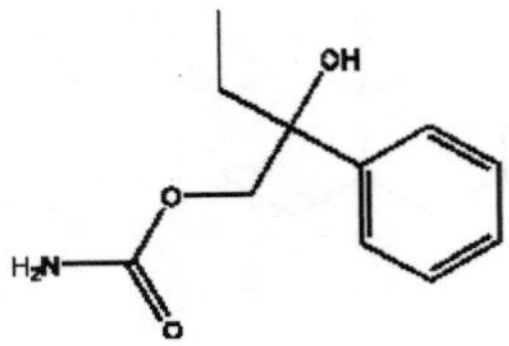

Chemical Abstracts Registry No.: 50-19-1

Trade Name	Manufacturer	Country	Year Introduced
Listica	Armour	US	1961
Listica	Armour-Montagu	France	1975

Raw Materials

2-Phenyl-1,2-butanediol
Ethyl chloroformate
Ammonia

Manufacturing Process

2-Phenyl-2-hydroxy-butyl carbamate was prepared by the following method:

49.81 g of 2-phenyl-1,2-butanediol and 25.01 g of pyridine were dissolved in 500 ml of benzene and cooled to 5°C. 34.01 g of ethyl chloroformate was added over a period of 36 hour at 4°C to 8°C. The reaction mixture was warmed to room temperature and stirred for 2 hours and then extracted with 100 cc each of the following:

Water, 15% hydrochloric acid, 10% sodium bicarbonate and finally water. The solvent was stripped off. The residual oil was mixed with 300 ml of 28% aqueous ammonia for 1 hour. The ammonia and water were vacuum distilled at a temperature of 40°C or less. Then 300 cc of carbon tetrachloride was added and the solution dried with sodium sulfate. The solution was cooled at 0°C and then filtered. The crystals were washed with cold carbon tetrachloride and vacuum dried. The yield was 57 g of dried product having a melting point of 55°C to 56.5°C.

References

Merck Index 4756
OCDS Vol. 1 p. 220 (1977)
I.N. p. 718
Sifferd, R.H. and Braitberg, L.D.; US Patent 3,066,164; November 27, 1962; assigned to Armour Pharmaceutical

HYDROXYPROGESTERONE

Therapeutic Function: Progestin

Chemical Name: Pregn-4-ene-3,20-dione, 17-hydroxy-

Common Name: Hydroxyprogesterone

Structural Formula:

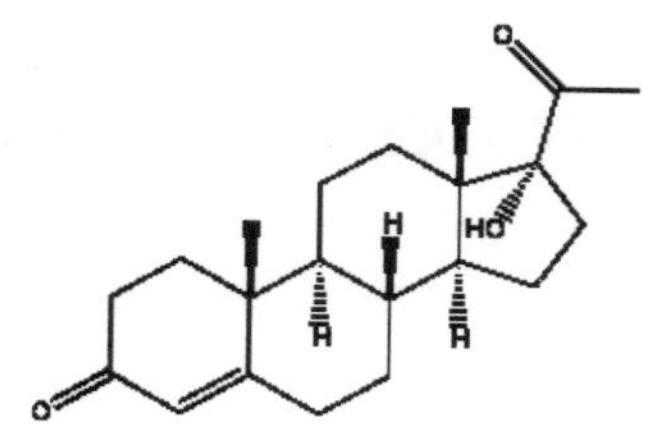

Chemical Abstracts Registry No.: 68-96-2

Trade Name	Manufacturer	Country	Year Introduced
Prodox	Upjohn	-	-
Fulterm	Micro Nova Pharmaceuticals Ltd.	-	-
Fetaron	Ochoa Laboratories (P) Ltd.	-	-

Raw Materials

Formic acid
Acetic anhydride
Potassium hydroxide
δ(5)-Pregnen-3β,17α-diol-20-one
4-Toluenesulfonic acid
Aluminum isopropylate

Manufacturing Process

A suspension of 90.0 g of δ^5-pregnen-3β,17α-diol-20-one in 2300 ml of 85% formic acid was shaken for 2 h at a temperature of 70C. During this time the compound partially dissolved and at the same time a new crystalline substance appeared in the solution. After cooling, the precipitate was filtered, thus giving 80.0 g of the 3-formate of δ^5-pregnen-3β,17α-diol-20-one having a melting point of 204°-207°C.

5.0 g of the 3-formate of δ^5-pregnen-3β,17α-diol-20-one suspended in 120 ml of acetic anhydride was treated with 1.5 g of p-toluenesulfonic acid and the mixture was stirred for 9 h at room temperature. It was poured into water and after 2 h standing, the precipitate was filtered and washed to neutral, thus yielding the 3-formate 17-acetate of δ^5-pregnen-3β,17α-diol-20-one in a yield of over 90%.

1.0 g of 3-formate 17-acetate of δ^5-pregnen-3β,17α-diol-20-one was dissolved in 30 ml of xylene and 10 ml of cyclohexanone and 4 ml of the solution were distilled in order to remove traces of moisture. 1.0 g of aluminum isopropylate was added to the hot solution and the mixture was refluxed for 45 min. After cooling to 90°C, water was added and the organic solvents were removed by steam distiliation. Salt was added to the aqueous suspension and the residue was filtered, dried and extracted with hot acetone. The acetone solution was evaporated to dryness and the residue was crystallized from chloroform-methanol, thus giving 610.0 mg of the 17-acetate of δ^4-pregnen-17α-ol-3,20-dione (17-acetoxy-progesterone) with a melting point of 239°-240°C.

Saponification of this compound with 1% methanolic potassium hydroxide yielded 80% of δ^4-pregnen-17α-ol-3,20-dione.

References

Ringold H.J. et al.; US Patent No. 2,802,839; August 13, 1957; Assigned: Syntex S.A., Mexico City, Mexico, a corporation of Mexico

HYDROXYPROGESTERONE CAPROATE

Therapeutic Function: Progestin

Chemical Name: 17-[(1-Oxohexyl)oxy]pregn-4-ene-3,20-dione

Common Name: -

Structural Formula:

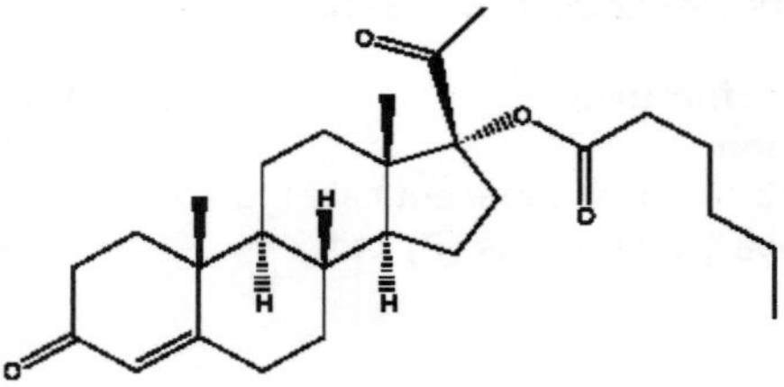

Chemical Abstracts Registry No.: 630-56-8

Trade Name	Manufacturer	Country	Year Introduced
Delalutin	Squibb	US	1956
Hyproval	Tutag	US	1976
Corluton Depot	I.E. Kimya Evi	Turkey	-
Caprogen Depot	Kanto	Japan	-
Depolut	Taro	Israel	-
Depot-Progen	Hokuriku	Japan	-
Hormofort	Kobanyai	Hungary	-
Idrogestene	Farmila	Italy	-
Kaprogest	Polfa	Poland	-
Lutopron	Cipla Limited	India	-
Pergestron	Dexter	Spain	-
Primolut-Depot	Schering	UK	-
Prodox	Legere	US	-
Proge	Mochida	Japan	-
Progestron-Depo	Galenika	Yugoslavia	-

Raw Materials

Hydrogen chloride	17α-Oxypregnene-(5)-ol-(3)-one-(20)-acetate-(3)
Cyclohexanone	Caproic acid anhydride

Manufacturing Process

40 grams of 17α-oxypregnene-(5)-ol-(3)-one-(20)-acetate-(3) is brought to reaction with 22 grams of p-toluolsulfonic acid and 850 cc of caproic acid anhydride under a nitrogen atmosphere for 5 days at room temperature or 2,5 days at 37°C. The excess anhydride is blown off with steam in the presence of 200 cc of pyridine and the distillation residue is extracted with ether and worked up as usual. The remaining oil is brought to crystallization with pentane and the raw 17α-oxypregnenolone-3-acetate-17-caproate is recrystallized from methanol. The crystals are needle-like and have a MP of 104° to 105°C. This substance is partially saponified by refluxing for 1 hour in 1,800 cc of methanol in the presence of 13 cc of concentrated hydrochloric acid. After evaporation of the methanol under vacuum, the dry residue is recrystallized from isopropyl ether or methanol (dense needles). The thus obtained 17α-oxypregnenolone-17-caproate melts at 145° to 146.5°C.

By oxidation in 100 cc of absolute toluol with 425 cc of cyclohexanone and 155 cc of a 20% aluminum isopropylate solution in absolute toluol and after repeated crystallizations from isopropyl ether or methanol, 24 grams of pure 17α-oxyprogesterone-17-caproate is obtained, MP 119° to 121°C (dense needles).

References

Merck Index 4761
Kleeman and Engel 479
PDR p. 1033
OCDS Vol. 1 pp. 176, 190 (1977)
DOT 19 (2) 112 (1983)
I.N. p. 505
REM p. 991
Kaspar, E., Pawlowski, K.H., Junkmann, K.and Schenck, M.; US Patent 2,753,360; July 3, 1956; assigned to Firma Schering AG, Germany

HYDROXYPROPYL CELLULOSE

Therapeutic Function: Topical protectant, Ophthalmic vehicle

Chemical Name: Cellulose 2-hydroxypropyl ether

Common Name: Hyprolose

Structural Formula:

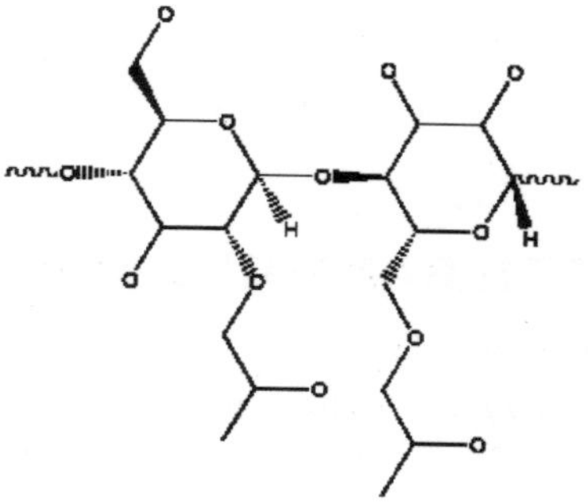

Chemical Abstracts Registry No.: 9004-64-2

Trade Name	Manufacturer	Country	Year Introduced
Lacrisert	MSD	US	1981

Raw Materials

Cotton linters
Propylene oxide
Sodium hydroxide
Acetic acid

Manufacturing Process

Charge:

	Parts
Purified cotton linters	1
Tertiary butanol	10
Water	1.4
Sodium hydroxide	0.1
Hexane	9.5
Propylene oxide	2.85

Procedure:

The tertiary butanol, water and sodium hydroxide were mixed and the mixture cooled to 20°C. The purified cotton linters were added to the mixture and aged at 20°C for one hour while stirring. Excess liquid was filtered off the resulting alkali cellulose so that the resulting alkali cellulose filter cake weighed 3.08 parts. This filter cake was broken up and slurried in the hexane, placed in a pressure vessel the pressure of which was increased to 100 psig with nitrogen, and then the pressure was vented to 5 psig. The propylene oxide was added to the pressure vessel and then the pressure was increased to 25 psig with nitrogen. The resulting charge was heated to 85°C in 30 minutes and then reacted at this temperature and 25 psig pressure for six hours. The charge was cooled to 30°C. the pressure vessel vented and 0.14 part of glacial acetic acid added. The excess hexane was filtered off from the resulting hydroxypropyl cellulose product, the product was purified by washing with hot water (85°C to 95°C) and then dried at 130°C using a two-roll drum drier.

References

Merck Index 4763
PDR p. 1191
DOT 18 (7) 338 (1982)
I.N. p. 509
REM p. 1298
Klug, E.D.; US Patent 3,278,521; October 11, 1966; assigned to Hercules, Inc.

HYDROXYSTILBAMIDINE ISETHIONATE

Therapeutic Function: Fungicide

Chemical Name: 2-Hydroxy-4,4'-stilbenedicarboxamidine di(β-hydroxyethanesulfonate)

Common Name: -

Structural Formula:

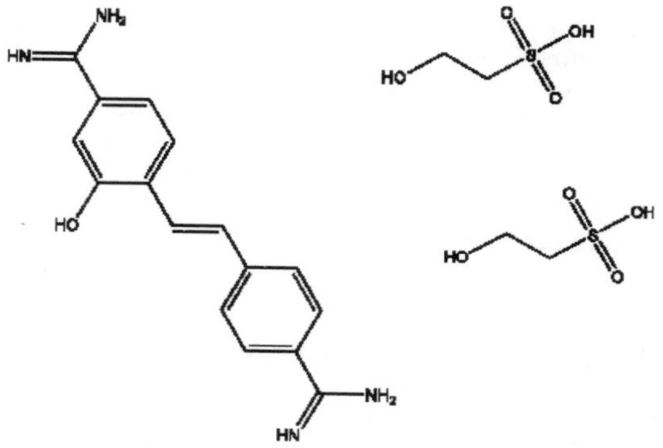

Chemical Abstracts Registry No.: 533-22-2; 495-99-8 (Base)

Trade Name	Manufacturer	Country	Year Introduced
Hydroxystilbamidin Isethionate	Merrell National	US	1954
Hydroxystilbamide	May and Baker	UK	-

Raw Materials

Ammonia
Sulfuric acid
Sodium nitrate
Isethionic acid
2-Nitro-p-tolunitrile
Stannous chloride
4-Cyanobenzaldehyde
Hydrogen chloride

Manufacturing Process

Preparation of 2-Nitro-4,4'-Dicyanostilbene: 10 grams of 2-nitro-p-tolunitrileand 8.1 grams of 4-cyano-benzaldehyde were heated to 170° to 180°C, 1.2 and 0.6 cc of piperidine were added at quarter-hour intervals, heating was continued for a further one and a quarter hours, the product cooled, triturated with glacial acetic acid and filtered. The residue was crystallized from glacial acetic acid as yellow needles, MP 290°C.

Preparation of 2-Amino-4,4'-Dicyanostilbene: 10.0 grams of 2-nitro-4,4'-dicyanostilbene thus prepared were suspended in 200 cc of glacial acetic acid and a hot solution of 50 grams of stannous chloride ($SnCl_2 \cdot 2H_2O$) in 50 cc of concentrated hydrochloric acid was quickly added. Rapid reaction occurred and the boiling was continued for a further 4 minutes, the reaction mixture was cooled, filtered, and the stannous chloride residue decomposed with 25% aqueous caustic soda solution. The liberated amine crystallized from glacial acetic acid as yellow needles, MP 232°C.

Preparation of 2-Hydroxy-4,4'-Dicyanostilbene: 10 grams of 2-amino-4,4'-dicyanostilbene thus prepared were dissolved in 400 cc of boiling glacial acetic acid and 200 cc of dilute sulfuric acid added; the solution was suddenly chilled and diazotized over one and a half hours at 5° to 10°C with sodium nitrate (3.0 grams/15 cc H_2O). The diazonium salt solution was decomposed by boiling for 15 minutes with 600 cc of 55% aqueous sulfuric acid solution; the solution was diluted, cooled and filtered. The residue crystallized from ethyl alcohol as lemon yellow prismatic needles, MP 296°C

Preparation of 2-Hydroxy -4,4'-Diamidinostilbene Dihydrochloride: 10 grams of 2-hydroxy-4,4'-dicyanostilbene were suspended in 250 cc of absolute ethyl alcohol and the mixture saturated with dry hydrogen chloride at 0°C. The whole was left for eight days at room temperature. The imino-ether hydrochloride formed was filtered off, washed with dry ether and dried in the air for a short time. It was then added to 250 cc of 10% ethyl alcoholic ammonia and the whole heated for 5 hours at 45°C. The 2-hydroxy-4,4'-diamidinostilbene dihydrochloride which separated was crystallized from 10% hydrochloric acid. It forms pale yellow needles, MP 357°C (decomposition).

Preparation of the Final Isethionate Product: The diisethionate may be produced by treating a solution of the dihydrochloride with alkali carbonate, separating and dissolving the resultant base in aqueous isethionic acid and precipitating the diisethionate with acetone. The product may be purified by dissolving in hot methyl alcohol containing a trace of water followed by precipitation by the cautious addition of acetone. The diiseihionate has a MP of 286°C.

References

Merck Index 4768
Kleeman and Engel p. 480
I.N. p. 506
REM p. 1230
Ewins, A.J.; US Patent 2,510,047; May 30, 1950; assigned to May and Baker Ltd., England

HYDROXYTRYPTOPHAN

Therapeutic Function: Central stimulant

Chemical Name: 5-Hydroxytryptophan

Common Name: -

Structural Formula:

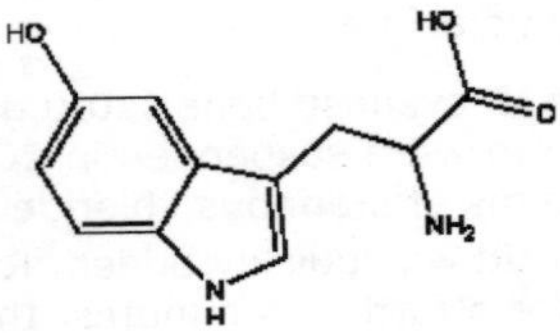

Chemical Abstracts Registry No.: 56-69-9

Trade Name	Manufacturer	Country	Year Introduced
Quietim	Nativelle	France	1973
Tript-Oh	Sigma Tau	Italy	1980
Levothym	Karlspharma	W. Germany	1980

Raw Materials

Acrolein	4-Benzyloxyaniline HCl
Sodium nitrite	Hydrochloric acid
Sodium hydroxide	Stannous chloride
Hydrogen	Diethylacetylamino malonate

Manufacturing Process

Preparation of 4-benzyloxyphenylhydrazine: 200 grams 4-benzyloxyaniline hydrochloride was suspended in a mixture of 264 ml concentrated hydrochloric acid, 528 ml water and 732 grams crushed ice. A solution of 62.4 grams sodium nitrite in 136 ml water was added below the surface of the stirred suspension at -10-(+/-)2°C during 10 minutes. After stirring for 1 hour at 0°C, the suspension was treated with acid-washed charcoal and filtered.

The filtrate was cooled and maintained at -8°C while a solution of 500 grams of stannous chloride in 760 ml concentrated hydrochloric acid was added with stirring. The mixture was stirred for 2 hours at -8°C and the 4-benzyloxyphenylhydrazine hydrochloride which separated was filtered off and washed with water. The product was crystallized by adding 800 ml hot water to a 3 liter solution in ethanol and had a MP of 185° to 189°C (yield 168.5 grams, 79%).

Preparation of Ethyl α-Acetylamino-α-Carbethoxy-β-(5-Benryloxy-Indolyl-3)-Propionate: 4-benzyloxyphenylhydrazine hydrochloride was converted to the corresponding base 2 to 3 hours before use: 28 grams of the hydrochloride was suspended in 500 ml chloroform and shaken with 55 ml 2N sodium hydroxide in 100 ml water. The chloroform was separated and the aqueous phase reextracted with chloroform (2 x 100 ml). After washing with 100 ml water, the chloroform solution was dried over sodium sulfate, filtered and evaporated at 30° to 35°C, leaving 4-benzyloxyphenylhydrazine as a friable buff-colored solid (23 grams, 97% from hydrochloride).

6.1 grams freshly distilled acrylic aldehyde (acrolein) in 9.7 ml chlorobenzene was added at 30°C over 30 minutes to a stirred suspension of 24.2 grams diethyl acetylaminomalonate in 37.5 ml chlorobenzene containing a catalytic amount (0.25 ml) of 50% w/v aqueous sodium hydroxide. After a further 30 minutes the resultant solution was warmed and 23 grams 4-benzyloxyphenylhydrazine was added at 45°C. The mixture was stirred and heated at 65° to 70°C for 1 hour to complete condensation, when a red solution was formed.

The resultant chlorobenzene solution was added to 440 ml N sulfuric acid and the suspension was refluxed with stirring for 6 hours. The product was extracted with chloroform (250 + 100 ml), and the chloroform solution washed with water (3 x 100 ml), separated and dried over sodium sulfate. After filtration and concentration at 40°C to 100 ml, 300 ml light petroleum (BP 40° to 60°C) was added to the warm chloroform-chlorobenzene solution. 33.1 grams ethyl α-acetylamino-α-carbethoxy-β-(5-benzyloxyindolyl-3)-propionate crystallized on cooling from the mixture. It was recrystallized by dissolving in 200 ml benzene and adding 100 ml light petroleum (BP 60° to 80°C) at the boiling point. After cooling, the buff crystals were collected, washed with cold benzene/light petroleum (1:1) mixture (50 ml), and dried at 55°C (yield 26.0 grams, 54%, MP 164° to 165°C).

Preparation of α-Acetylamino-α-Carboxy-β-(5-Benzyloxy-Indolyl-3)-Propionic Acid: 18 grams ethyl α-acetylamino-α-carbethoxy-β-(5-benzyloxy-indolyl-3)-

propionate was suspended in 85 ml water containing 8.5 grams sodium charcoal. The suspension was refluxed for 4 hours, decolorizing charcoal added, and the solution filtered hot through Hyflo Supercel.

After cooling in ice to 10°C, the solution was acidified with 24 ml concentrated hydrochloric acid. The solid which separated was filtered off, washed with water (3 x 30 ml) and dried in vacuo over silica gel, to give α-acetylamino-α-carboxy-(5-benzyloxy-indolyl-3)-propionic acid, MP 144° to 146°C (15.0 grams, 95%) sufficiently pure for use in the next stage.

Preparation of α-Acetylamino-β-(5-Benzyloxy-Indolyl-3)-Propionic Acid: 15 grams α-acetylamino-α-carboxy-β-(5-benzyloxy-indolyl-3)-propionic acid was suspended in 225 ml water and the suspension refluxed and stirred in a stream of nitrogen until evolution of carbon dioxide ceased (about 2 hours). After cooling somewhat, 120 ml ethyl alcohol was added and the suspension refluxed until the product dissolved. Charcoal was added to the solutior the mixture filtered hot, and the filter-cake washed with 50 ml hot 50% aqueous ethanol. α-Acetylamino-β-(5-benzyloxy-indolyl-3)-propionic acid, MP 164° to 166°C, which crystallized from the filtrate on cooling, was collected, washed with an ice-cold mixture of 15 ml ethanol and 45 ml water, and dried in vacuo over silica gel (yield 11.1 grams, 83%).

Preparation of 5-Benzyloxytryptophan: 11 grams α-acetylamino-β-(5-benzyloxy-indolyl-3)-propionic acid was suspended in a solution of 12 grams sodium hydroxide in 90 ml water and refluxed for 24 hours. Charcoal was added to the resultant solution and the mixture filtered hot. 150 ml 2N hydrochloric acid was added to the filtrate at 70°C and 5-benzyloxytryptophan crystallized on cooling. After washing with water and drying in vacuo over silica gel, the amino acid (6.9 grams, 71%) had MP (sealed evacuated tube) 232°C, with softening, finally melting at 237° to 238°C (decomposition). Charcoal was added to the filtrate, which was filtered hot and adjusted to pH 2. On cooling a second crop of 5-benzyloxytryptophan was obtained (2.2 grams, 23%), MP (sealed evacuated tube) 230°C, with softening, finally melting at 233° to 237°C (decomposition). The overall yield of 5-benzyloxytryptophan was 9.1 grams (94%).

Preparation of 5-Hydroxytryptophan: 0.4 gram palladium chloride and 1.7 grams acid-washed charcoal were suspended in 157 ml water and hydrogenated at room temperature and atmospheric pressure until no further hydrogen uptake occurred. A suspension of 14.2 grams 5-benzyloxytryptophan in 175 ml ethyl alcohol was added and the mixture hydrogenated under similar conditions. A hydrogen uptake slightly in excess of theory was obtained. The suspension was warmed for a few minutes on the steam bath and filtered hot. The filter-cake was washed with hot water (3 x 20 ml) and the filtrate evaporated to 20 ml under reduced pressure in a nitrogen atmosphere.

The resultant mass of colorless crystals was triturated with 250 ml ice-cold ethyl alcohol under hydrogen, filtered, and washed with cold ethyl alcohol (2 x 15 ml). The 5-hydroxytryptophan (6.9 grams, 69%) had MP (sealed evacuated tube) 288°C, with softening, finally melting at 249° to 247°C (decomposition). Concentration of the liquors under reduced pressure in a nitrogen atmosphere, and trituration as before, gave a second crop (0.9 gram, 9%). The combined crops (7.8 grams) were dissolved in 120 ml hot water, charcoal added, and the mixture filtered hot. The filtrate was concentrated in a nitrogen atmosphere under reduced pressure and ethyl alcohol added. The 5-hydroxytryptophan then crystallized as colorless microneedles (6.5 grams,

65%), had MP (sealed evacuated tube) 290°C, with slight softening, finally melting at 295° to 297°C (decomposition).

References

Merck Index 4771
DOT9 (6) 224 (1973). 10 (9) 323 and 10, 262 (1974)
REM p. 1083 Ash, A.S.F.; British Patent 845,034; August 17, 1960; Assigned to May and Baker Ltd., UK

HYDROXYUREA

Therapeutic Function: Cancer chemotherapy

Chemical Name: Hydroxycarbamide

Common Name: -

Structural Formula:

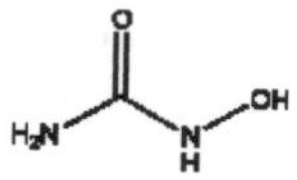

Chemical Abstracts Registry No.: 127-07-1

Trade Name	Manufacturer	Country	Year Introduced
Hydrea	Squibb	UK	1967
Hydrea	Squibb	US	1968
Litalir	Heyden	W. Germany	1968
Hydrea	Squibb	France	1969
Biosuppressin	Biogal	Hungary	-
Hidroks	Yurtoglu	Turkey	-
Onco-Carbide	Simes	Italy	-

Raw Materials

Hydroxylamine hydrochloride
Sodium cyanate

Manufacturing Process

The procedure may be illustrated by the following equations relating to the preparation of hydroxyurea from hydroxylamine hydrochloride:

(1) $R_4N^+Cl^- + NaNCO = R_4N^+NCO^- + NaCl$

(2) $R_4N^+NCO^- + H_2NOH\ HCl = R_4N^+Cl^- + HONH\text{-}CO\text{-}NH_2$

Equation (1) shows the simple conversion of a quaternary ammonium anion exchange resin from the chloride form to the cyanate form. Equation (2) shows the reaction of the resin in the cyanate form with hydroxylamine hydrochloride whereby hydroxyurea is formed and the anion Cl-is retained by the quaternary resin.

A 90 x 6 cm column was packed with 2 kg of granular Amberlite IRA-410 resin in the chloride form (a vinylpyridine/divinylbenzene copolymer quaternized with dimethyl sulfate and converted to chloride) and washed with 3 kg of a 10% aqueous solution of sodium cyanate. This changed the resin from the chloride to the cyanate form. Sodium chloride and excess sodium cyanate were then washed from the column with distilled water until the effluent failed to give a white precipitate with silver nitrate. The reaction of equation (2) was conducted by elutriating the column with a solution of 105 grams (1.5 mols) of hydroxylamine hydrochloride in 400 ml water at about 15°C.

A hot (50° to 70°C) reaction zone developed near the top of the column and about 30 minutes was required for this hot zone to descend the full length of the column. The reaction solution was followed in the column by 2.5 liters of distilled water. Collection of the product was begun when hydroxyurea could be detected in the effluent, as indicated by a black precipitate on warming a sample with a silver nitrate test solution. All the effluents were combined and vacuum evaporated at 35°C to give 90 grams of tan residue corresponding to 79% yield of crude product. After recrystallization from 100 ml of water heated to 75°C, the colorless product was dried in a vacuum desiccator over phosphorus pentoxide to give 60.6 grams (53% yield) of hydroxyurea, MP 133° to 136°C.

References

Merck Index 4772
Kleeman and Engel. p. 476
PDR p. 1746
I.N. p. 501
REM p. 1155
Graham, P.J.; US Patent 2,705,727; April 5, 1955; assigned to E.I. du Pont de Nemours and Company

HYDROXYZINE HYDROCHLORIDE

Therapeutic Function: Tranquilizer

Chemical Name: 2-[2-[4-[(4-Chlorophenyl)phenylmethyl]-1-piperazinyl] ethoxy]ethanol hydrochloride

Common Name: -

Structural Formula:

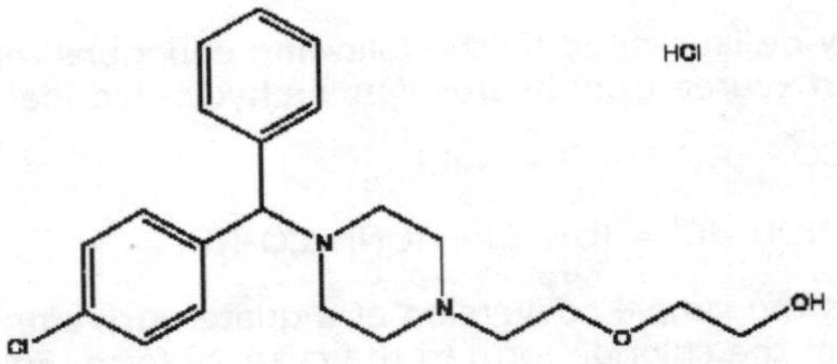

HCl

Cl N N O OH

Chemical Abstracts Registry No.: 2192-20-3; 68-88-2 (Base)

Trade Name	Manufacturer	Country	Year Introduced
Atarax	UCB	France	1956
Atarax	Roerig	US	1956
Vistaril	Pfizer	US	1958
Quiess	O'Neal Jones	US	1958
Hyzine	Hyrex	US	1980
Orgatrax	Organon	US	1980
Durrax	Dermik	US	1983
Alamon	Grelan	Japan	-
Arcanax	Arcana	Austria	-
Atazina	Panthox and Burck	Italy	-
Disron	Teikoku	Japan	-
Masmoran	Pfizer	W. Germany	-
Marax	Roerig	US	-
Neucalm	Legere	US	-
Neurozina	Farge	Italy	-
Theozine	Schein	US	-

Raw Materials

N-Mono-1-p-chlorobenzohydrylpiperazine
1-Chloro-2-(2-hydroxyethoxy)ethane
Sodium hydroxide
Hydrogen chloride

Manufacturing Process

A mixture of 0.1 mol of N-mono-1-p-chlorobenzohydrylpiperazine and 0.1 mol of 1-chloro-2-(2-hydroxy-ethoxy)-ethane is heated for 3 hours to 150°C. The mass is then taken up in 100 ml of benzene and 100 ml of a 10% aqueous solution of NaOH; decanting takes place, and the benzene solution is washed with water and the solvent is evaporated. Vacuum distilling of the residue yields 1-p-chlorobenzohydryl-4-[2-(2-hydroxy-ethoxy)-ethyl]-piperazine, BP 220°C/0.5 mm Hg.

The corresponding dihydrochloride is prepared by dissolving this base in about twice its weight of alcohol, by treating it with excess of gaseous HCl and by precipitating it with ether. The solvent is decanted and the residue, dissolved in a minimum of alcohol, crystallizes on the addition of ether, MP 193°C.

References

Merck Index 4773
Kleeman and Engel p. 480
PDR pp.832, 872, 993, 1033, 1288, 1416, 1520, 1528, 1606, 1989, 1999
OCDS Vol. 1 p. 59 (1977)
I.N.p.506
REM p. 1071
Morren, H.; US Patent 2,899,436; August 11, 1959; assigned to Union Chimique Belge Societe Anonyme, Belgium

HYMECROMONE

Therapeutic Function: Choleretic, Spasmolytic, Sunscreen agent

Chemical Name: Coumarin, 7-hydroxy-4-methyl-

Common Name: Himekromon; Hymecromone; Imecromone; Methylumbelliferone; Resocyanine

Structural Formula:

Chemical Abstracts Registry No.: 90-33-5

Trade Name	Manufacturer	Country	Year Introduced
Odeston	Merck	-	-
Cantabilin	Ilsan IIac	-	-
Isochol	Leciva	-	-
Cholestil	Pabianicke Zaklady Farmaceutyczne Polfa	-	-
Unichol	Laevosan	-	-
Bilcolic	Unifa	-	-
Biliton H	Berlin-Chemie	-	-
Cantabilin	Formenti	-	-
Cantabilin	Lipha	-	-
Cantabilin	Biofranco	-	-
Cantabilin	Lipha Sante/Aron-Med	-	-
Cantabilin	Remek	-	-
Cantabilina	Biofranco	-	-
Cantabiline	Lipha	-	-
Cantabiline	Remek	-	-
Cantabiline	Lipha Sante/Aron-Med	-	-
Cholonerton	Dolorgiet	-	-
Cholonerton	Pro. Med. CS Praha a.s.	-	-
Cholonerton	Sanova	-	-
Bilicanta	Boehringer Mannheim	-	-
Cholestil	Polfa	-	-
Cholirene	Uni-Pharma	-	-
Cholonerton	Sanova	-	-
Crodimon	Roussel	-	-
Himecol	Kissei	-	-
Himekromon	Zdravlje	-	-
Mendiaxon	Byk Gulden	-	-
Mendiaxon	Hemofarm	-	-

Raw Materials

Resorcin
3-Oxo-butyric acid ethyl ester
Sulfuric acid
Phosphorous pentaoxide

Manufacturing Process

Resorcin reacted with 3-oxo-butyric acid ethyl ester in the presence sulfuric acid and phosphorous pentaoxide and 4-methyl-7-hydroxycoumarine (hymecromone) was obtained.

References

Kleemann A., Engel J.; Pharmazeutische Wirkstoffe, GeorgThieme Verlag Stuttgart. New York, 1982

I

IBUPROFEN

Therapeutic Function: Antiinflammatory

Chemical Name: α-Methyl-4-(2-methylpropyl)benzene acetic acid

Common Name: 2-(4-Isobutylphenyl)propionic acid

Structural Formula:

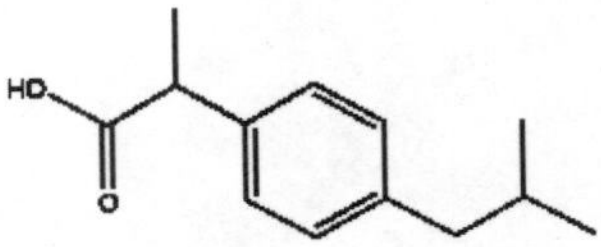

Chemical Abstracts Registry No.: 15687-27-1

Trade Name	Manufacturer	Country	Year Introduced
Brufen	Boots	UK	1969
Brufen	Kakenyaku Kako	Japan	1971
Brufen	Labaz	W. Germany	1971
Brufen	Formenti	Italy	1972
Brufen	Dacour	France	1972
Motrin	Upjohn	US	1974
Rufen	Boots	US	1981
Advil	Whitehall	US	-
Algofen	Ibirn	Italy	-
Andran	Takata	Japan	-
Anflagen	Ohta	Japan	-
Artofen	Lkapharm	Israel	-
Artril	Eczacibasi	Turkey	-
Artril 300	Farmasa	Brazil	-
Bluton	Morishita	Japan	-
Brufamic	Teigo	Japan	-
Buburone	Towa Yakuhin	Japan	-
Butylenin	Sanken	Japan	-

Trade Name	Manufacturer	Country	Year Introduced
Daiprophen	Daito	Japan	-
Donjust-B	Horita	Japan	-
Ebufac	D.D.S.A.	UK	-
Epinal	Mitsubishi Yuka	Japan	-
Epobron	Ono	Japan	-
Eputes	Kobayashi Kako	Japan	-
Focus	Angelini	Italy	-
IB-100	Hishiyama	Japan	-
Iborufen	Kyoritsu Yamagata	Japan	-
Ibucasen	Casen	Spain	-
Ibulav	A.L.	Norway	-
Ibumetin	Benzon	Denmark	-
Ibuprocin	Nisshin	Japan	-
Ibo-Slo	Lipha	UK	-
Inflam	Protea	Australia	-
Lamidon	Kowa	Japan	-
Landelun	Tsuruhara	Japan	-
Liptan	Kowa	Japan	-
Manypren	Zensei	Japan	-
Mono-Attritin	Atmos	W. Germany	-
Mynosedin	Toho Yakuhin	Japan	-
Napacetin	Toyama	Japan	-
Neobrufen	Liade	Spain	-
Nobfelon	Toho	Japan	-
Nobfen	Toho	Japan	-
Nobgen	Kanebo, Ltd.	Japan	-
Nurofen	Crookes	UK	-
Opturem	Kade	W. Germany	-
Paduden	Terapia	Rumania	-
Pantrop	Nippon Zoki	Japan	-
Rebugen	Dessy	Italy	-
Roidenin	Showa	Japan	-
Saren	Bracco	Italy	-
Sednafen	Taisho	Japan	-

Raw Materials

Sulfur
Sodium
Ethyl iodide
Ethanol
Isobutylbenzene
Sodium hydroxide
Acetyl chloride
Ethyl carbonate

Manufacturing Process

Isobutylbenzene is first acetylated to give isobutylacetophenone. 4-i-butylacetophenone (40 g), sulfur (11 g) and morpholine (30 ml) were refluxed for 16 hours, cooled, acetic acid (170 ml) and concentrated hydrochloric acid (280 ml) were added and the mixture was refluxed for a further 7 hours. The mixture was concentrated in vacuo to remove acetic acid and the concentrate was diluted with water.

The oil which separated was isolated with ether, the ethereal solution was extracted with aqueous sodium carbonate and this extract was acidified with hydrochloric acid. The oil was isolated with ether, evaporated to dryness and the residue was esterified by refluxing with ethanol (100 ml) and concentrated sulfuric acid (3 ml) for 5 hours. The excess alcohol was distilled off, the residue was diluted with water, and the oil which separated was isolated with ether. The ethereal solution was washed with sodium carbonate solution; then with water and was dried. The ether was evaporated off and the oil was distilled to give ethyl 4-i-butylphenylacetate.

Sodium ethoxide from sodium (3.67 g) in absolute alcohol (64 ml) was added over 20 minutes with stirring to a mixture of ethyl 4-i-butylphenylacetate (28.14 g) and ethyl carbonate (102 ml) at 100°C. The reaction flask was fitted with a Fenske column through which alcohol and then ethyl carbonate distilled. After 1 hour when the still head reached 124°C heating was discontinued. Glacial acetic acid (12 ml) and water (50 ml) was added to the stirred ice-cooled mixture and the ester isolated in ether, washed with sodium carbonate solution, water and distilled to give ethyl 4-i-butylphenylmalonate.

Ethyl 4-i-butylphenylmalonate (27.53 g) in absolute alcohol (25 ml) was added with stirring to a solution of sodium ethoxide From sodium (2.17 g) in absolute alcohol (75 ml). Ethyl iodide (15 ml) was added and the mixture refluxed for 2% hours, the alcohol distilled and the residue diluted with water, extracted with ether, washed with sodium bisulfite, water, and evaporated to dryness.

The residual oil was stirred and refluxed with sodium hydroxide (75 ml of 5 N), water (45 ml) and 95% ethanol (120 ml). Within a few minutes a sodium salt separated and after 1 hour the solid was collected, washed with ethanol, dissolved in hot water and acidified with dilute hydrochloric acid to give the methyl malonic acid which was collected and dried in vacuo MP 177° to 180°C (dec.).

The malonic acid (9 g) was heated to 210° to 220°C in an oil bath for 20 minutes until decarboxylation had ceased. The propionic acid was cooled and recrystallized from light petroleum (BP 60° to 80°C). Two further recrystallizations from the same solvent gave colorless prisms of 2-(4-isobutylphenyl)propionicacid MP 75° to 77.5°C. (The procedure was reported in US Patent 3,228,831.)

References

Merck Index 4797
Kleeman and Engel p. 482
PDR pp. 687, 728, 830, 1854, 1897
OCDS Vol.1 p.86 (1977) and 2, 218, 356 (1980)
DOT 5 (3) 101 (1969)
I.N. p. 510
REM p. 1117
Nicholson, J.S. and Adams, S.S.; US Patent 3,228,831; January 11, 1966; assigned to Boots Pure Drug Company Limited, England

Nicholson, J.S. and Adams, S.S.; US Patent 3,385,886; May 28, 1968; assigned to Boots Pure Drug Company Limited, England

IBUPROXAM

Therapeutic Function: Antiinflammatory

Chemical Name: N-Hydroxy-α-ethyl-4-(2-methylpropyl)benzene-acetamide

Common Name: -

Structural Formula:

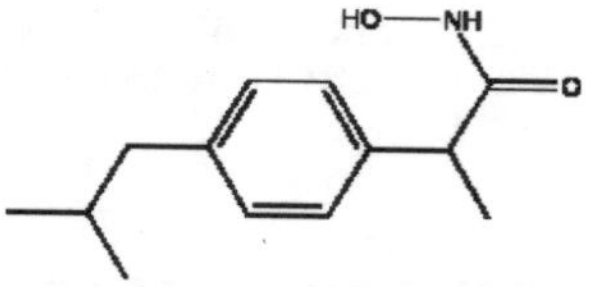

Chemical Abstracts Registry No.: 53648-05-8

Trade Name	Manufacturer	Country	Year Introduced
Ibudros	Manetti-Roberts	Italy	1978
Ibudros	Ferrer	Spain	-

Raw Materials

2-(4-Isobutylphenyl)propionic acid
Hydroxylamine hydrochloride
Ethanol
Potassium hydroxide

Manufacturing Process

In a 1,000 ml three-necked flask equipped with a stirrer, a dropping funnel and a silica gel guard pipe, 46.7 g hydroxylamine hydrochloride are dissolved cold in 480 ml methanol. Separately a solution of 56.1 g KOH in 280 ml methanol is prepared, heated to 30°C and admixed, dropwise under stirring to the hydroxylamine solution. All successive temperature increases during this admixture are prevented by cooling in an ice bath. After the whole KOH solution has been admixed, the mixture is left standing for 5 minutes so as to attain the complete precipitation of the KCl.

Separately, 72.02 g ethyl 2-(4-isobutylphenyl)-propionate, obtained by the esterification of 2-(4-isobutylphenyl)-propionic acid with ethanol and concentrated H_2SO_4, are solved with 100 ml methanol, this solution is introduced drop by drop into the reaction flask, and stirred and cooled for 5 hours on an ice bath. Thereafter it is suction filtered, the residue is washed with all together 50 ml methanol, the wash is added to the filtrate, thereafter the whole is evaporated in a water bath with a rotating evaporator at a reduced pressure, until 100-200 ml of a concentrated solution are obtained. This solution is poured into a 200 ml beaker into which are stirred approximately 1,000 ml 1.25N acetic acid. This mixture is left standing for 24 hours, thereafter suction filtered. The resulting filtrate is taken up with 100 ml petroleum ether at 40°C to 60°C, in order to solve any possible residue of unreacted starting ester, and refiltered. Approximately 50g of 2-(4-isobutylphenyl)-propiohydroxamic acid are obtained, having a melting point of 119°C to 121°C on Kofler's hot stage.

References

Merck Index 4798
DFU 2 (12) 808 (1977)
I.N. p. 511
Orzalesi, G. and Selleri, R.; US Patent 4,082,707; April 4, 1978; assigned to Societa Italo-Britannica L. Manetti-H. Roberts and Co. (Italy)

IBUTILIDE FUMARATE

Therapeutic Function: Antiarrhythmic

Chemical Name: Methanesulfonamide, N-(4-(4-(ethylheptylamino)-1-hydroxybutyl)phenyl)-, (2E)-2-butenedioate (2:1) (salt)

Common Name: Ibutilide fumarate

Structural Formula:

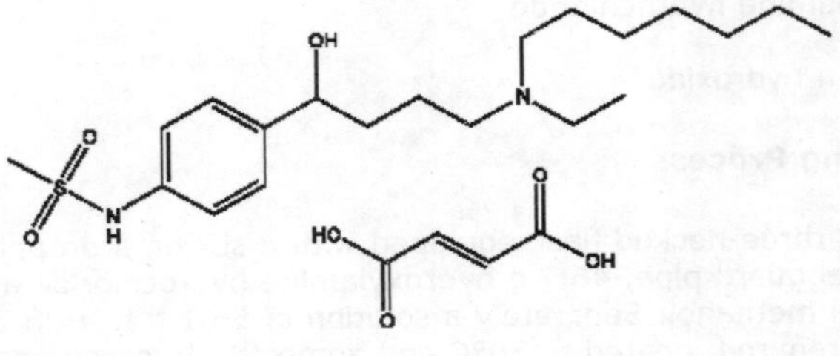

Chemical Abstracts Registry No.: 122647-32-9; 122647-31-8 (Base)

Trade Name	**Manufacturer**	**Country**	**Year Introduced**
Corvert	Pharmacia and Upjohn	USA	-

Raw Materials

Aniline
Methanesulfonyl chloride

Succinic anhydride
Fumaric acid
Ethylheptylamine
Lithium aluminum hydride
1-Hydroxybenzotriazole
N,N'-Dicyclohexylcarbodiimide
Sodium potassium tartrate

Manufacturing Process

A mechanically stirred solution of aniline (139.7 g, 1.5 mole) in pyridine (2 L), under N_2 is cooled in an ice bath. Methanesulfonyl chloride (171.8 g, 1.5 mole) is added dropwise to this solution while the temperature is maintained at 15°-20°C, which results in a red-orange color change in the reaction mixture. After the addition is complete the ice bath is removed and the reaction is allowed to continue at room temperature. The reaction is complete after 2.5 h. The reaction mixture is concentrated in vacuo and the residue is combined with 700 ml of water which results in crystallization of a dark red material.

This material is filtered and washed several times with water. The filtered material is dissolved in CH_2Cl_2, washed with brine, dried (Na_2SO_4), and concentrated in vacuo. The residue is dissolved in hot ethyl acetate, treated with Darco (decolorizing carbon) and crystallized to yield methanesulfonanilide which had a melting point: 93°-94°C.

A mechanically stirred suspension of aluminum chloride (88.0 g, 0.66 moles) and 150 ml of carbon disulfide under N_2 is cooled in an ice bath. Methanesulfonanilide (30.0 g, 0.175 mol) and succinic anhydride (17.5 g, 0.175 mol) are combined and added rapidly to the cooled reaction mixture. The ice bath is removed and the mixture is stirred at room temperature for 6 h. The reaction mixture is then heated to 55°C and allowed to continue for 18 h. The reaction mixture is separated into two layers the bottom of which solidifies.

The upper layer is decanted and the remaining solid layer is decomposed with ice. The resulting suspension is filtered and the solid is washed several times with methylene chloride and dissolved in a mixture of saturated sodium bicarbonate (500 ml) and water (500 ml). This solution is acidified (pH 2) with HCl and the resulting precipitate is collected by filtration, redissolved in $NaHCO_3$ and reprecipitated with HCl. The solid, 4-[(methylsulfonyl)amino]-γ-oxobenzenebutanoic acid, is collected by filtration. Melting point 198°-200°C.

A stirred solution of 4-[(methylsulfonyl)amino]-γ-oxobenzenebutanoic acid (12.0 g, 0.044 mol) in DMF (100 ml) under N_2 is cooled in an ice bath to 5°C and treated with 1-hydroxybenzotriazole (5.94 g, 0.044 mol) and N,N'-dicyclohexylcarbodiimide (9.08 g, 0.044 mol). After 1 hour, ethylheptylamine (6.3 g, 0.044 mol) is added, after an additional 30 min the ice bath is removed and the mixture is kept at room temperature for 18 h.

The reaction mixture is filtered over a Celite filter aid and the filtrate is concentrated under vacuum. The resulting material is dissolved in CH_2Cl_2, washed with dilute HCl, $NaHCO_3$ and concentrated. The residue is chromatographed over silica gel (1.25 kg) with 5% MeOH : 1% NH4OH : CH_2Cl_2. The N-ethyl-N-heptyl-γ-oxo-4-[(methylsulfonyl)amino] benzenebutanamide thus obtained is crystallized from EtOAc to yield 10.77 g, melting point 100°-102°C.

To a N_2 covered suspension of 0.29 g (7.57 mmol) of $LiAlH_4$ in 10 ml of THF cooled in an ice bath is added a solution of 1.0 g (2.52 mmol) of N-ethyl-N-heptyl-γ-oxo-4-[methylsulfonyl)amino]benzenebutanamide in 10 ml of THF over 6 min. The ice bath is then removed and the mixture heated at reflux for 27 h and then stirred at room temperature for 2 days. The mixture is cooled in an ice bath and there is added dropwise 10 ml of aqueous sodium potassium tartrate followed by EtOAc and H_2O to keep the mixture fluid.

The aqueous fraction is extracted once with EtOAc and the combined EtOAc fractions are washed in turn with H_2O and concentrated in vacuo. The residue is chromatographed on a 200 ml silica gel column (elution with 6% MeOH : CH_2Cl_2 containing 0.5% NH_4OH) and 9.7 ml fractions were collected and treated with Et2O and aqueous $NaHCO_3$. The organic layer is concentrated in vacuo to yield N-[4-[4-(ethylheptylamino)-1-hydroxybutyl]phenyl] methanesulfonamide.

Preparation of fumarate (WO Patent 01/07417). To dichloromethane solution of 4-[4-N-[(Ethylheptylamino)-1-hydroxybutyl]phenyl]methanesulfonamide is added hemimolar quantities of fumaric acid and heated to reflux until a clear solution was obtained. Upon cooling the fumarate of 4-[4-N-[(Ethylheptylamino)-1-hydroxybutyl]phenyl]methanesulfonamide was obtained.

References

Jackson B.H., Jr.; US Patent No. 5,155,268; Oct. 13, 1992; Assigned: The Upjohn Company (Kalamazoo, MI)

IDOXURIDINE

Therapeutic Function: Antiviral (ophthalmic)

Chemical Name: 2'-Deoxy-5-iodouridine

Common Name: -

Structural Formula:

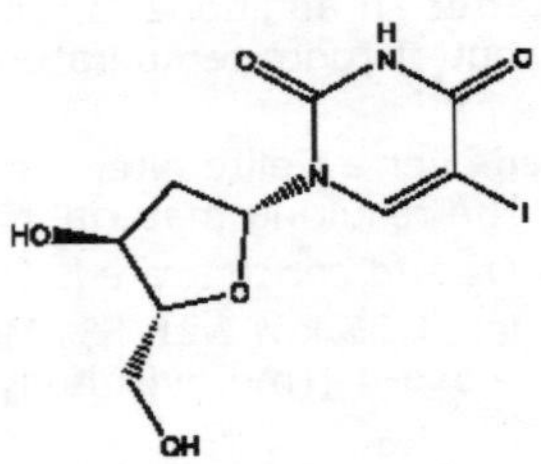

Chemical Abstracts Registry No.: 54-42-2

Trade Name	Manufacturer	Country	Year Introduced
Dendrid	Alcon	US	1963
Stoxil	SKF	US	1963
Herplex	Allergan	US	1963
Idoxene	Spodefeil	UK	1963
Idoviran	Chauvin-Blache	France	1963
Herpetil	Farmila	Italy	1963
Spectanefran	Pharm-Allergan	W. Germany	1964
Cheratil	Francia	Italy	-
Colircusi Virucida	Cusi	Spain	-
Dendrit	Smith and Nephew	UK	-
Gel "V"	P.O.S.	France	-
Herpid	W.B. Pharm.	UK	-
Herpidu	Dispersa	Switz.	-
IDU	Pliva	Yugoslavia	-
IDU Ophthalmic	Sumitomo	Japan	-
Iducher	Farmigea	Italy	-
Iduridin	Ferring	Sweden	-
Idustatin	Isnardi	Italy	-
Kerecid	SKF	UK	-
Oftan-Idurin	Star	Finland	-
Ophthalmadine	S.A.S.Sci.	UK	-
Synmiol	Winzer	W. Germany	-
Virexin	Vinas	Spain	-
Virunguent	Hermal	W. Germany	-
Virusan	Ikapharm	Israel	-
Vistaspectran	Allergan	W. Germany	-
Zostrum	W.B. Pharm.	UK	-

Raw Materials

5-Iodouracil
Acetic anhydride
Acetic acid
3,5-Di-p-toluyl-desoxy-D-ribofuranosyl chloride
Sodium hydroxide

Manufacturing Process

5 g of 5-iodo-uracil (obtained according to T.B. Johnson et al., J. Biol. Chem. 1905/6, 1, 310) in 15 cc of acetic anhydride are heated under reflux for 4,5 hours. The acetylated derivative crystallizes on cooling. The crystallized product is chilled for ½ hour then filtered with suction, washed with acetic anhydride and then with ether and dried. 4.5 g of 1-acetyl-5-iodo-uracil, MP 167°C, are thus obtained.

1.51 g of mercuric acetate are dissolved in 50 cc of methanol under reflux and 1.35 g of 1-acetyl-5-iodo-uracilare added. A white precipitate is soon formed. The reaction mixture is kept under reflux for % hour and then allowed to cool to room temperature. The precipitate is then filtered with suction, washed with methanol and dried.

2.1 g of monomercuric 5-iodo-uracil, MP 280°C, are thus obtained as a colorless powder, insoluble in water and the majority of the usual organic solvents, such as benzene, chloroform, alcohol, ether and acetone.

1.46 g of 5-iodo-uracil monomercuric derivative are introduced into 50 cc of chloroform and 20 to 30 cc of the solvent are distilled off under normal pressure to ensure good dehydration of the reaction medium. The mixture is cooled to room temperature and 2.59 g of 3,5-di-p-toluyl-desoxy-D-ribofuranosyl chloride added. The mixture is agitated for 6 hours with glass balls, filtered, rinsed with chloroform and the filtrate is successively washed with an aqueous sodium iodide solution, with water, with a saturated solution of sodium bicarbonate and again with water. The product is dried over sodium sulfate, filtered and evaporated to dryness.

The residue crystallizes in ether and yields about 600 mg of β-3',5'-di-p-toluyl-2'-desoxy-5-iodo-uridine which is recrystallized from toluene. The product is obtained as colorless crystals, soluble in chloroform and pyridine, sparingly soluble in acetone, benzene ether and alcohol, insoluble in water, MP 193°C.

206 mg of 3',5'-di-p-toluyl-2'-desoxy-5-iodo-uridineare heated at 80°C with 2.5 cc of caustic soda solution (0.4 N) for ½ hour. The solution obtained is cooled, filtered and then acidified with acetic acid. The desoxy-iodo-uridine and the p-toluic acid crystallize. Ether is added to dissolve the p-toluic acid, the mixture is chilled, filtered with suction, washed with water and ether, and dried. The residue is recrystallized from water and 100 mg of 5-iodo-2'-desoxy-uridine, are obtained.

References

Merck Index 4804
Kleeman and Engel p. 483
DOT 7 (5) 191 (1971) and 10 (10) 268 (1974)
I.N. p. 512
REM p. 1232
Roussel-Uclaf; British Patent 1,024,156; March 30, 1966

IFENPRODIL TARTRATE

Therapeutic Function: Vasodilator

Chemical Name: α-(4-Hydroxyphenyl)-β-methyl-4-(phenylmethyl)-1-piperidineethanol tartrate

Common Name: -

Structural Formula:

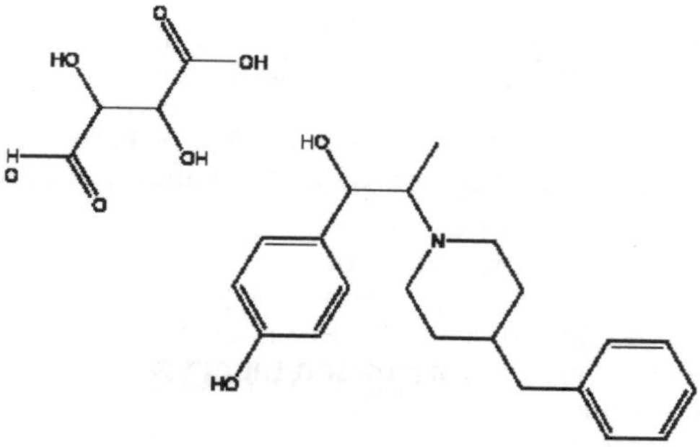

Chemical Abstracts Registry No.: 23210-58-4; 23210-56-2 (Base)

Trade Name	Manufacturer	Country	Year Introduced
Vadilex	Carriere	France	1972
Cerocral	Funai	Japan	1979
Angiotrofin	Montpellier	Argentina	-
Dilvax	Promeco	Argentina	-
Validex	Robert and Carriere	France	-

Raw Materials

Benzyl chloride	4-Benzylpiperidine
Hydrogen	4-Hydroxypropiophenone
Bromine	Tartaric acid

Manufacturing Process

The initial steps involve reacting benzyl chloride with 4-hydroxypropiophenone. The benzyloxypropiophene thus obtained is first brominated and then reacted with 4-benzylpiperidine to give 1-(p-benzyloxyphenyl)-2-(4-benzyl-piperidino)propan-1-one.

The neutral tartrate may be prepared directly by reduction of 1-(p-benzyloxyphenyl)-2-(4-benzyl-piperidino)propan-1-one. For the reduction, a mixture of 175 g of ketone (0.425 mol) and 32 g of tartaric acid (0.213 mol) is hydrogenated at 50°C under pressure of 50 kg/cm^2 in 440 ml of methanol in the presence of 12 g of palladium on charcoal.

The catalyst is filtered off at elevated temperature, and the filtrate is concentrated by evaporation under reduced pressure to a volume of 300 ml and added in a thin stream to 2.5 liters of diethyl ether with mechanical agitation. The precipitate is separated, washed with diethyl ether and dried in vacuo at 80° to 85°C for several hours. 325 g (96% yield) of the neutral tartrate of 1-(p-hydroxyphenyl)-2-(4-benzyl-piperidino)propan-1-ol are obtained.

References

Merck Index 4806
Kleeman and Engel p. 484
OCDS Vol. 2 p. 39 (1980)
I.N. p. 513
Carron. M.C.E.. Carron.C.L.C. and Bucher.B.P.; US Patent 3,509,164; April 28, 1970; assigned to societe Anonyme des Laboratoires Robert et Carriere, France

IFOSFAMIDE

Therapeutic Function: Antineoplastic

Chemical Name: N,3-Bis(2-chloroethyl)tetrahydro-2H-1,3,2-oxazaphosphorin-2-amine-2-oxide

Common Name: Isoendoxan

Structural Formula:

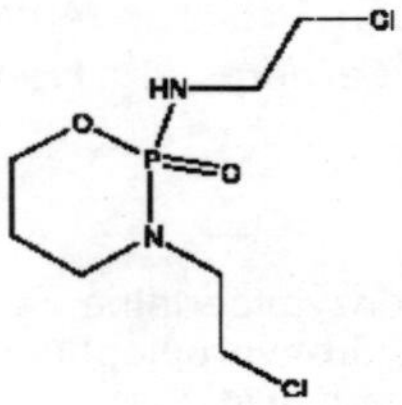

Chemical Abstracts Registry No.: 3778-73-2

Trade Name	Manufacturer	Country	Year Introduced
Holoxan	Lucien	France	1976
Holoxan	Asta	W. Germany	1977
Mitoxana	W.B. Pharm.	UK	1979
Holoxan	Asta-Werke	Switz.	1979
Holoxan	Schering	Italy	1981
Cyfos	Mead Johnson	-	-
Naxamide	Mead Johnson	-	-

Raw Materials

N-(2-Chloroethyl)amine HCl
N-(2-Chloroethyl)-N,O-propylene phosphoric acid ester amide HCl
Triethylamine

Manufacturing Process

127.6 g (1.1 mols) of N-(2-chloroethyl)-amine hydrochloride are suspended in a solution of 218 g (1 mol) of N-(2-chloroethyl)-N,O-propylene phosphoric acid ester amide monochloride in 600 cc of methylene dichloride, and 212 g of triethylamine are added thereto dropwise with stirring. The reaction mixture is heated to boiling by the reaction heat. After termination of the addition, the reaction mixture is heated to boiling for another 2 hours. Thereafter, it is cooled to room temperature and the precipitated triethylamine hydrochloride is separated by filtration with suction. The filtrate is extracted with about 60cc of dilute hydrochloric acid (pH 3), then twice with about 60 cc of water, thereafter with about 60 cc of dilute soda lye and finally twice with about 60 cc of water. After drying over anhydrous sodium sulfate, methylene dichloride is distilled off under normal pressure. The oily residue is dried in a vacuum and thereafter extracted in a perforator with 500 cc of anhydrous ether. The oily extract crystallizes upon inoculation and standing in an ice box. After standing for several hours, the precipitate is filtered off, washed with a small amount of cold ether and dried in a vacuum at room temperature. Yield: 185 g (71% of the theoretical). This material is also identified as 3-(2-chloroethyl)-2-(2-chloroethylamino)-tetrahydro-2H-1,3,2-oxazaphosphorin-2-oxide; generic name: ifosfamide. F.P.: 39°C to 41°C.

References

Merck Index 4807
Kleeman and Engel p. 485
OCDS Vol. 3 p. 151 (1984)
DOT 12 (11) 450 (1976) and 16 (5) 171 (1980)
I.N. p. 513
REM p. 1155 Arnold, H., Brock, N., Bourseaux, F. and Bekel, H.; US Patent 3,732,340; May 8, 1973; as signed to Asta-Werke A.G. Chemische Fabrik (W. Germany)

IMIPENEM

Therapeutic Function: Antibiotic

Chemical Name: 1-Azabicyclo[3.2.0]hept-2-ene-2-carboxylic acid, 6-(1-hydroxyethyl)-3-((2-((iminomethyl)amino)ethyl)thio)-7-oxo-, (5R-(5-α,6-α(R*)))-

Common Name: Gorillamicin; Imipemide; Imipenem

Chemical Abstracts Registry No.: 64221-86-9

Trade Name	Manufacturer	Country	Year Introduced
Primaxin	Merck and Co., Inc.	-	-

Structural Formula:

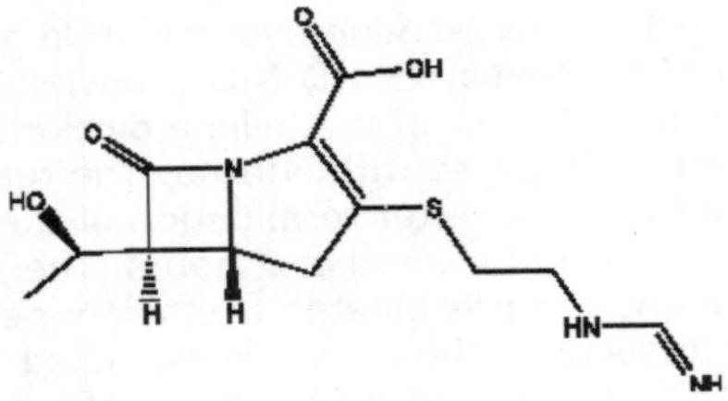

Raw Materials

Thienamycin
Methyl formimidate hydrochloride
6-(1)-Hydroxyethyl-1-azabicyclo[3.2.0]heptane-3,7-dione-2-carboxylate
N,S-Bistrimethylsilyl-N-formimidoylcysteamine

Manufacturing Process

Preparation of N-formimidoyl thienamycin:

Thienamycin (517 mg) is dissolved in pH 7 0.1 N phosphate buffer (25 ml) and cooled in an ice bath with magnetic stirring. The solution is adjusted to pH 8.5 using 2.5 N sodium hydroxide solution dispensed from an automatic burette. While maintaining a pH of 8.5, methyl formimidate hydrochloride (711 mg) is added portionwise over 2-3 minutes. After an additional 10 min, the pH of the solution is brought to 7.0 using 2.5 N hydrochloric acid. The solution is chromatographed on a column of XAD-2 resin (150 ml) which is eluted with water. The N-formimidoyl thienamycin derivative (imipenem) elutes in 1.5-2.0 column volumes (200-300 ml) and is lyophilized to a white solid (217 mg). UV (pH 7 0.1 N phosphate buffer); λ_{max}297 nm (8,590); ir (Nujol mull) 1767 Cm-1(β-lactam).

Another method preparation of imipenem:

6-(1)-Hydroxyethyl-1-azabicyclo(3.2.0)heptane-3,7-dione-2-carboxylate is converted to the diphenoxyphosphate enol ester and this in turn reacted with N,S-bistrimethylsilyl-N-formimidoylcysteamine (use of the bistrimethylsilylated reagent is necessary in order to avoid side reactions caused by cyclization reactions). As a result the $(PhO)_2OPO$-groups are converted to Me_3SiN = CHNH-groups. Removal of the protecting groups complete the synthesis of 1-azabicyclo[3.2.0]hept-2-ene-2-carboxylic acid, 6-(1-hydroxyethyl)-3-((2-((iminomethyl)amino)ethyl)thio)-7-oxo-, (5R-(5-alpha,6-alpha(R*)))-.

References

Merck Index, Monograph number: 4954, Twelfth edition, 1996, Editor: S. Budavari; Merck and Co., Inc.
Shinkai I. et al.; Tetrahedron Lett.; 1982, 23, 4903

IMIPRAMINE HYDROCHLORIDE

Therapeutic Function: Antidepressant

Chemical Name: 10,11-Dihydro-N,N-dimethyl-5H-dibenz[b,f]azepine-5-propanamine hydrochloride

Common Name: Imizin

Structural Formula:

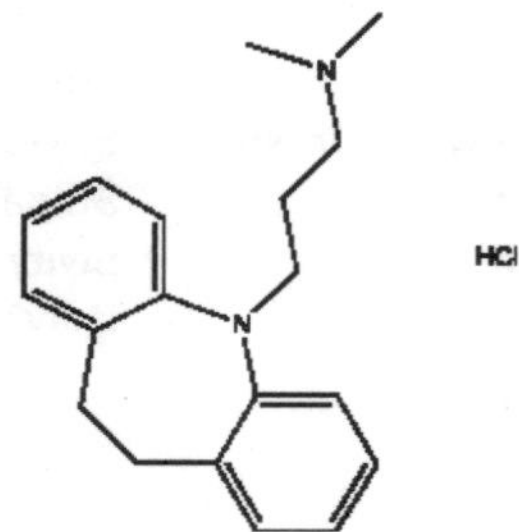

Chemical Abstracts Registry No.: 113-52-0; 50-49-7 (Base)

Trade Name	Manufacturer	Country	Year Introduced
Tofranil	Ciba Geigy	France	1959
Tofranil	Ciba Geigy	US	1959
Presamine	U.S.V. Pharm.	US	1971
SK-Pramine	SKF	US	1974
Janim ine	Abbott	US	1975
WDD Tab	Tutag	US	1979
Berkomine	Berk	UK	-
Censtim	Ohio Medical	US	-
Chemipramine	Chemo-Drug	Canada	-
Chemoreptin	Toho Iyaku	Japan	-
Chrytemin	Fujinaga	Japan	-
Depress	Toho	Japan	-
Deprinol	Dumex	Denmark	-
Dimipressin	Drugs, Ltd.	UK	-
Dynaprin	Monico	Italy	-
Eupramin	Pliva	Yugoslavia	-
Feinalmin	Sanko	Japan	-
I.A.-Pram	Inter-Alia Pharm.	UK	-
Imavate	Robins	US	-
Imidol	Yoshitomi	Japan	-
Imilanyle	Takata	Japan	-
Imipramine	Lederle	US	-
Imipranil	Medica	Finland	-
Imiprin	Protea	Australia	-

Trade Name	Manufacturer	Country	Year Introduced
Impranil	Barlow Cote	Canada	-
Impril	I.C.N.	-	-
Intalpran	Inter-Alia Pharm.	UK	-
Iprogen	Genethic	UK	-
Iramil	Knoll	W. Germany	-
Melipramin	EGYT	Hungary	-
Meripramin	Kanebo, Ltd.	Japan	-
Norpramine	Norton	UK	-
Novopramine	Novopharm	Canada	-
Primonil	Ikapharm	Israel	-
Prodepress	Medac	Australia	-
Pryleugan	Arzneimittelwerk Dresden	E. Germany	-
Psychof orin	Pharmachim	Bulgaria	-
Servipramine	Servipharm	Switz.	-
Surplix	Vis	Italy	-

Raw Materials

Iminodibenzyl
3-Dimethylamino n-propyl chloride
Sodium amide
Hydrogen chloride

Manufacturing Process

20 parts of imino dibenzyl are dissolved in 100 parts by volume of absolutely dry benzene. A suspension of 4 parts $NaNH_2$ in 50 parts by volume of absolute benzene are then added dropwise at 50° to 60°C after which the mixture is boiled for an hour under reflux. 13 parts of 3-dimethylamino n-propyl chloride are then added dropwise at 40° to 50°C and the mixture is boiled for 10 hours under reflux. After cooling, the benzene solution is thoroughly washed with water, whereupon the basic constituents are extracted with dilute hydrochloric acid.

The hydrochloric extract is then made alkaline and the separated base is extracted with ether. After drying, the solvent is evaporated and the residue is distilled in the high vacuum, whereby the N-(3-dimethylaminopropyl)-imino dibenzyl passes over at a temperature of 160°C under 0.1 mm pressure. The chlorohydrate with a melting point of 174° to 175°C is obtained therefrom with alcoholic hydrochloric acid.

References

Merck Index 4817
Kleeman and Engel p. 485
PDR pp. 527, 673, 901, 993, 1569, 1606, 1723
OCDS Vol. 1 p.401 (1977); 2, 420 (1980) and 3, 32 (1984)
I.N. p. 514
REM p. 1095

Haefliger, F. and Schindler, W.; US Patent 2,554,736; May 29, 1951; assigned to J.R. Geigy AG, Switzerland

IMPROSULFAN TOSYLATE

Therapeutic Function: Antitumor

Chemical Name: Bis-(3-methanesulfonyloxypropyl)amine

Common Name: -

Structural Formula:

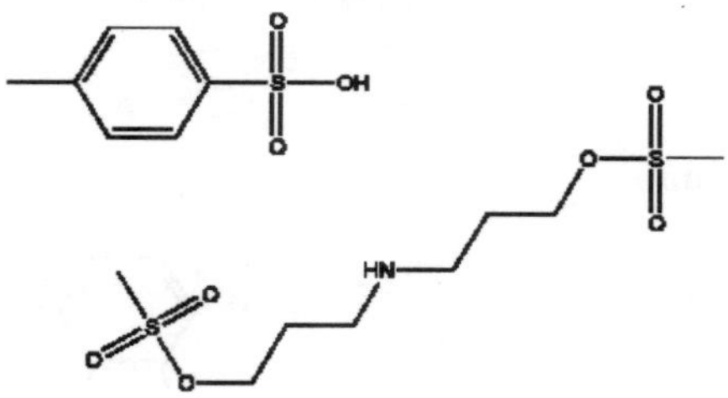

Chemical Abstracts Registry No.: 13425-98-4 (Base)

Trade Name	Manufacturer	Country	Year Introduced
Protecton	Yoshitomi	Japan	1980

Raw Materials

Bis-(3-Methylsulfonyloxypropyl)amine hydrochloride
Sodium carbonate
4-Toluenesulfonic acid

Manufacturing Process

A solution of 5 g of bis(3-methylsulfonyloxypropyl)amine hydrochloride in 20 ml of ice water is neutralized with 1N sodium carbonate solution. The resulting amine base is extracted with five 20 ml portions of chloroform. The combined extract is dried over anhydrous sodium sulfate, the solvent is distilled off under reduced pressure, and the residue is dissolved in 20 ml of ethanol. To the ethanol solution is added slowly with stirring under ice cooling a solution of 2.6 g of p-toluenesulfonic acid in 30 ml of ethanol. The white precipitate formed is collected by filtration and recrystallized from ethanol to give 5.0 g of white crystalline bis(3-methylsulfonyloxypropyl)amine p-toluenesulfonate melting at 115°C to 116°C.

References

Merck Index 4823
DFU 4 (2) 106 (1979)
DOT 16 (12) 422 (1980)
I.N. p. 515
Yoshitomi Pharmaceutical Industries, Ltd.; British Patent 1,272,497; April 26, 1972

INDALPINE

Therapeutic Function: Antidepressant

Chemical Name: 4-[2-(3-Indolyl)ethyl]piperidine

Common Name: -

Structural Formula:

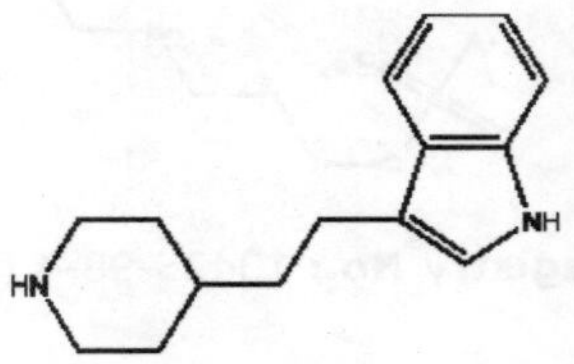

Chemical Abstracts Registry No.: 63758-79-2

Trade Name	Manufacturer	Country	Year Introduced
Upstene	Fournier	France	1983

Raw Materials

Bis(methoxy-2-ethoxy)sodium aluminum hydride
(Indolyl-3)(piperidinyl-4-methyl)ketone

Manufacturing Process

0.5 g of bis(methoxy-2ethoxy)sodium aluminum hydride in a 70% solution in toluene is added to a solution of 0.29 g of (indolyl-3)(piperidyl-4-methyl) ketone in 10 ml of toluene. The mixture is heated under refluxing conditions for 15 hours, then cooled to 0°C. 10 ml of an aqueous solution of 5N sodium hydroxide is added dropwise thereto, followed by stirring for 1 hour. The organic phase is decanted, washed with water, dried using potassium carbonate and evaporated under partial vacuum. 0.26 g of oil is obtained, which is purified by chromatography and hydrochloride formation. The product obtained is 0.1 g of (indolyl-3)-2-ethyl-4-piperidine hydrochloride which has a melting point of 167°C.

References

DFU 4 (12) 873 (1979)
DOT 19 (10) 584 (1983)
Champseix, A.A., Gueremy, C.G.A. and LeFur, G.R.; US Patent 4,064,255; December 20, 1977; assigned to Mar-Pha Societe D'Etudes et D'Exploitation De Marques

INDANAZOLINE

Therapeutic Function: Nasal decongestant

Chemical Name: 2-(4-Indanylamino)-2-imidazoline

Common Name: -

Structural Formula:

Chemical Abstracts Registry No.: 40507-78-6

Trade Name	Manufacturer	Country	Year Introduced
Farial	Nordmark-Werke	W. Germany	1980
Farial	Knoll	Switz.	1983

Raw Materials

N-4-Indanyl thiourea
Methyl iodide
Ethylene diamine

Manufacturing Process

38.5 g (0.1 mol) of N-4-indanyl thiourea are dissolved in 250 cc of methanol. 42,6 g (0.3 mol) of methyl iodide are added thereto and the mixture is refluxed for 2,5 hours. The mixture thereafter is cooled and the solvent is removed in a rotation evaporator in a vacuum. Thus, 57.5 g of N-4-indanyl-S-methylisothiuronium hydroiodide (86% of theoretical) are obtained. Melting point 144°C to 146°C.

33.49 (0.1 mol) of N-4-indanyl-S-methylisothiuronium hydroiodide are mixed with 9.0 g (0.15 mol) of anhydrous ethylenediamine. The mixture is slowly heated to 80°C and heating is continued until the termination of the formation of methylmercaptan (about 4 hours). After cooling the residue is dissolved in 2N hydrochloric acid and the solution is extracted with chloroform. The extract is discarded and the aqueous phase is rendered alkaline by the addition of 10% soda lye. The resulting solution is extracted with chloroform and the extract is washed with water, dried over anhydrous sodium sulfate and the solvent is removed. An oily residue is obtained which upon standing soon crystallizes.

The product is recrystallized from petroleum ether having a boiling range of 100°C to 140°C in the presence of activated carbon. Thus, 11.1 g of 2-(4-indanylamino)-2-imidazoline (55% of theoretical) are obtained as the free base. Melting point 109°C to 113°C.

References

Merck Index 4826
DFU 6 (7) 417 (1981)
DOT 17 (10) 413 (1981)
I.N. p. 516
May, H.J. and Berg, A.; US Patent 3,882,229; May 6, 1975; assigned to Nordmark-Werke GmbH

INDANOREX

Therapeutic Function: Anorexic

Chemical Name: 1H-Inden-2-ol, 2-(1-aminopropyl)-2,3-dihydro-

Common Name: Indanorex

Structural Formula:

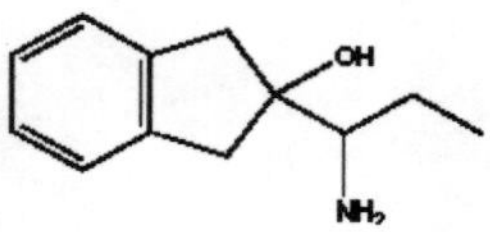

Chemical Abstracts Registry No.: 16112-96-2

Trade Name	Manufacturer	Country	Year Introduced
Indanorex	Shanghai Lansheng Corporation	-	-
Dietor	Logeais	-	-

Raw Materials

Pyridine	2-Cyano-2-hydroxyindane
Ethyl bromide	Trimethylchlorosilane
Magnesium	Sodium borohydride

Manufacturing Process

125 g (1.15 mol) trimethylchlorosilane was added to 159 g (1 mol) 2-cyano-2-hydroxyindane in 600 ml pyridine by stirring for 2 hours, whereupon the mixture was heated at 1 hour at 40°C for 1 hour. TA light precipitate was filtered off and washed with 1 liter of benzene. The filtrate was washed with water, dried over magnesium sulfate and concentrated in vacuum. The residue was dissolved in benzene 2 times and 2 times the solvent was removed in vacuum in order to any pyridine and water was present. Yield of silano-organic compound as a clear brown liquid was 230 g (100%); BP: 95°C/15 mm Hg.

10.5 g magnesium in 30 ml of dry ether was mixed with 55.5 g of ethyl bromide in 62 ml ether during about 1 hour at the temperature of boiling ether. After that the mixture was heated to 45°C in order to finish the synthesis of ethyl magnesium bromide. On cooling to 0°C it was stood for 2 hours and 50 g the above prepared silano-organic compound in 620 ml ether was added at the temperature about 5°C, whereupon in was stirred else 30 minutes at the ambient temperature and then was placed into ice bath. 82 ml of methanol was added to the prepared mixture during 1 hour. The temperature was kept about 18°C. On 30 minutes stirring the mixture was evaporated to 1/3 volume in vacuum at the temperature about 20°C. The residue was with ethanol diluted, the ethanol was evaporated, whereupon 100 ml methanol was added. 16.4 g sodium borohydride was added to the methanol solution for 1 hour at ice cooling, and stirred 2 hours at 3°C. It stood at ambient temperature overnight. Then the mixture was cooled to 10°C. 250 ml of hydrochloric acid (conc.) was added. After that it was heated at 45°C for 1 hour. The solvent was removed to dryness in vacuum. The pasty residue was in 250 ml of water dissolved and with chloroform washed. The 2-(1-aminopropyl)-2-indanol was precipitated by adding of 750 ml 10% sodium carbonate to pH 10. It was extracted with chloroform and dried over magnesium sulfate.

Chloroform was evaporated to give 13 g desired 2-(1-aminopropyl)-2-indanol. MP: 92°C. Yield 33%. IR spectrum and thin layer chromatography confirmed the structure of prepared compound.

References

Naillard J.G.; DB Patent No. 2,422,879; July 16, 1973; Laboratories Jacques Logeais, Issy-les-Moulineaux, Seine, Frankreich

INDAPAMIDE

Therapeutic Function: Diuretic

Chemical Name: 3-(Aminosulfonyl)-4-chloro-N-(2,3-dihydro-2-methyl-1H-indol-1-yl)-benzamide

Common Name: Metindamide

Structural Formula:

Chemical Abstracts Registry No.: 26807-65-8

Trade Name	Manufacturer	Country	Year Introduced
Natrilix	Pharmacodex	W. Germany	1976
Fludex	Eutherapie	France	1977
Natrilix	Servier	UK	1978
Natrilix	Servier	Australia	1983
Lozol	Revlon	US	1983
Arifon	Servier	France	-
Bajaten	Volpino	Argentina	-
Idamix	Gentili	Italy	-
Lozide	Servier	France	-
Nap-Sival	Promeco	Argentina	-
Noranat	Labinca	Argentina	-
Pressural	Polifarma	Italy	-
Tertensil	Servier	France	-

Raw Materials

3-Sulfamyl-4-chloro-benzoyl chloride
N-Amino-2-methyl indoline

Manufacturing Process

A total of 8.9 parts of 3-sulfamyl-4-chloro-benzoylchloride in a solution of 50 parts of anhydrous tetrahydrofuran are added portionwise in the course of 60 minutes, while stirring, to a solution of 5.2 parts of N-amino-2-methyl indoline and 3.5 parts of triethylamine in 150 parts of anhydrous tetrahydrofuran. The reaction mixture is left to stand 3 hours at room temperature, then the precipitated chiorhydrate of triethylamine is filtered off. The filtrate is evaporated under vacuum and the residue is crystallized from a solution of 60 parts of isopropanol in 75 parts of water. There are obtained 9 parts of N-(3-sulfamyl-4-chlorobenzamido)-2-methyl indoline, MP (K) 184° to 186°C, MP (MK) 160° to 162°C (isopropanol/water). [The melting points being determined on a Kofler heater plate under the microscope (MK) or on a Kofler Bank (K)].

References

Merck Index 4828
Kleeman and Engel p. 487
PDR p. 1816
OCDS Vol. 2 p. 349 (1980)
DOT 12 (8) 313 (1976) and 13 (1) 41 (1977)
I.N. p. 516
REM p. 944
Beregl, L., Hugon, P., Laubie, M.; US Patent 3,56591 1; February 23,1971; assigned to Science Union et Cie, Societe Francaise de Recherche Medicale, France

INDENOLOL

Therapeutic Function: Beta-adrenergic blocker

Chemical Name: 1-[1H-Inden-4(or 7)-yloxy]-3-[(1-methylethyl)amino]-2-propanol

Common Name: -

Structural Formula:

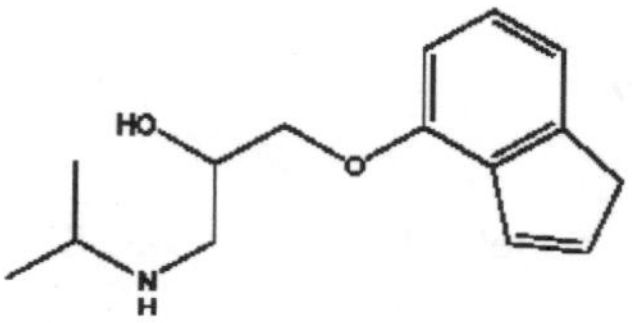

Chemical Abstracts Registry No.: 60607-68-3

Trade Name	Manufacturer	Country	Year Introduced
Pulsan	Yamanouchi	Japan	1979
Iambeta	Yamanouchi	Japan	-
Iambeta	Poli	Italy	-

Raw Materials

4-Hydroxyindene
Isopropylamine
Epichlorohydrin
Hydrogen chloride

Manufacturing Process

(a) A mixture of 0.9 g of 4-hydroxyindene, 2.0 g of 1,2-epoxy-3-chloropropane (epichlorohydrin), 2.7 g of potassium carbonate and 15 ml of acetone was refluxed at about 57°C for 24 hours. Acetone was removed by vacuum distillation, the residue was washed with 10 ml of water and then extracted with 20 ml of ether three times. The ether extract was dried with magnesium sulfate, filtered and subjected to column chromatography using a column (having an inside diameter of about 3 cm and a height of about 50 cm) packed with silica gel. The 5th to 7th fractions (volume of one fraction is 50 ml) recovered from the chromatographic column using chloroform as the effluent were combined together and concentrated to provide 0.6 g of 4-(2,3-epoxypropoxy)indene.

(6) A mixture of 0.42 g of 4-(2,3-epoxypropoxy)indene, 1.20 g of isopropylamine and 20 ml of methanol was stirred in a flask at room temperature for 2 hours. Methanol and unchanged isopropylamine were removed by vacuum distillation and the residue was recrystallized from a mixture of n-hexane and ether to yield 0.41 g of 4-(3-isopropylamino-2-hydroxypropoxy)indene having a melting point of 88°C to 89°C.

(c) To a solution of 0.41 g of 4-(3-isopropylamino-2-hydroxypropoxy)indene in 80 ml of absolute ether there was added dropwise a hydrochloric acid-ether mixture at 0°C with stirring. The precipitates thus formed were recovered by filtration and recrystallized from a mixture of ethanol and ether to provide 0.44 g of the hydrochloride of 4-(3-isopropylamino-2-hydroxypropoxy)indene. Melting point 147°C to 148°C.

References

Merck Index 4831
DFU 2 (11) 730 (1977)
Kleeman and Engel p. 487
DOT 16 (1) 24 (1980)
I.N. p. 516
Murakami, M., Murase, K., Niigata, K., Tachikawa, S. and Takenaka, T.; US Patent 4,045,482; August 30, 1977; assigned to Yamanouchi Pharmaceutical Co., Ltd. (Japan)

INDOMETHACIN

Therapeutic Function: Antiinflammatory

Chemical Name: 1-(p-Chlorobenzoyl)-5-methoxy-2-methylindole-3-acetic acid

Common Name: -

Structural Formula:

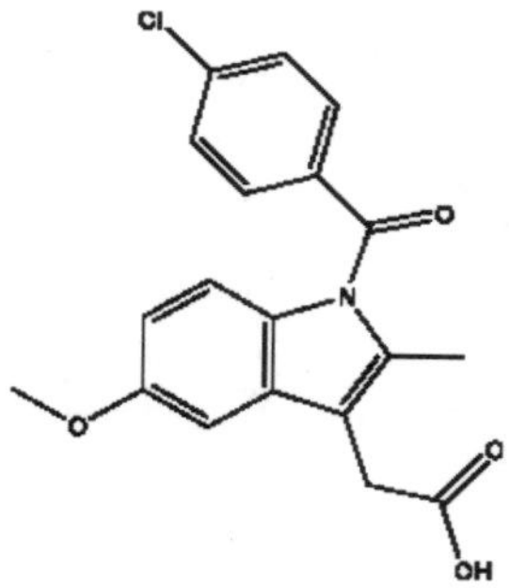

Chemical Abstracts Registry No.: 53-86-1

Trade Name	Manufacturer	Country	Year Introduced
Indocin	MSD	US	1965
Amuno	MSD	W. Germany	1965
Indocid	MSD-Chibret	France	1966
Indocid	MSD	UK	1966
Mefacen	Chiesi	Italy	1967
Algometacin	Biagini	Italy	-
Argun	Merckle	W. Germany	-
Arthrexin	Lennon	S. Africa	-
Artracin	D.D.S.A.	UK	-
Artrinova	Llorens	Spain	-
Artrivia	Lifasa	Spain	-
Artrobase	Libra	Italy	-
Artrocid	Schoum	Italy	-
Bonidon	Mepha	Switz.	-
Boutycin	Bouty	Italy	-
Calmocin	Mulda	Turkey	-
Cidalgon	Ecobi	Italy	-
Confortid	Dumex	Denmark	-
Durametacin	Durachemie	W. Germany	-
Endol	Deva	Turkey	-
Endomet	Dif-Dogu	Turkey	-
Endsetin	Nobel	Turkey	-
Imbrilon	Berk	UK	-
Imet	Firma	Italy	-
Indacin	Merck-Banyu	Japan	-
Inderapollon	Kaigai	Japan	-
Indetrit	Medica	Finland	-
Indium	Pharma Williams	Italy	-
Indo	Arcana	Austria	-
Indodur	Medica	Finland	-

Trade Name	Manufacturer	Country	Year Introduced
Indolag	Lagap	Switz.	-
Indolene	Italprofar	Italy	-
Indone RC	Sawai	Japan	-
Indomed	Teva	Israel	-
Indomet	Ratiopharm	W. Germany	-
Indomethine	Kowa	Japan	-
Indometin	Orion	Finland	-
Indorektal	Sanorania	W. Germany	-
Indoremed	Remed Econerica	W. Germany	-
Indo-Tablinen	Sanorania	W. Germany	-
Indotard	Benzon	Denmark	-
Indren	Spofa	Czechoslovakia	-
Inflazon	Taisho	Japan	-
Inmecin	Nippon Chemiphar	Japan	-
Inmetocin	Tobishi	Japan	-
Inmetsin	Farmos	Finland	-
Inteban	Sumitomo	Japan	-
Lausit	Showa	Japan	-
Metacen	Chiesi	Italy	-
Metartril	Ifisa	Italy	-
Methabid	Pharmador	S. Africa	-
Methazine	Sankyo	Japan	-
Metindol	Polfa	Poland	-
Mezolin	Meiji	Japan	-
Mobilan	Galen	US	-
Novomethacin	Novopharm	Canada	-
Osmogit	Merck-Frosst	Canada	-
Peralgon	S.A.R.M.	Italy	-
Ralicid	Waldheim	Austria	-
Rheumacin	Protea	Australia	-
Romacid	I.E. Kimya Evi	Turkey	-
Sadoreum	Mediolanum	Italy	-
Salinac	Nippon Kayaru	Japan	-
Takosashin S	Taiho	Japan	-
Tannex	Duncan Flockhart	UK	-
Zalbico	Toyo	Japan	-

Raw Materials

t-Butyl alcohol
Sodium hydride
p-Chlorobenzoyl chloride
Dicyclohexylcarbodiimide
2-Methyl-5-methoxy-3-indolyl acetic acid

Manufacturing Process

(A) 2-Methyl-5-Merhoxy-3-Indolylacetic Anhydride: Dicyclohexylcarbodiimide (10 g, 0.049 mol) is dissolved in a solution of 2-methyl-5-methoxy-3-

indolylacetic acid (22 g, 0.10 mol) in 200 ml of THF, and the solution is allowed to stand at room temperature for 2 hours. The precipitated urea is removed by filtration, and the filtrate is evaporated in vacuo to a residue and flushed with Skellysolve 6. The residual oily anhydride is used without purification in the next step.

(B) t-Butyl 2-Methyl-5-Merhoxy-3-Indolylacetate: t-Butyl alcohol (25 ml) and fused zinc chloride (0.3 g) are added to the anhydride from Part A. The solution is refluxed for 16 hours and excess alcohol is removed in vacuo. The residue is dissolved in ether, washed several times with saturated bicarbonate, water, and saturated salt solution. After drying over magnesium sulfate, the solution is treated with charcoal, evaporated, and flushed several times with Skellysolve B for complete removal of alcohol. The residual oily ester (18 g, 93%) is used without purification.

(C) t-Buryl 1-p-Chlorobenzoyl-2-Methyl-5-Mefhoxy-3-Indolylacetate: A stirred solution of ester (18 g, 0.065 mol) in dry DMF (450 ml) is cooled to 4°C in an ice bath, and sodium hydride (4.9 g, 0.098 mol, 50% susp.) is added in portions. After 15 minutes, p-chlorobenzoyl chloride (15 g, 0.085 mol) is added dropwise during 10 minutes, and the mixture is stirred for 9 hours without replenishing the ice bath. The mixture is then poured into one liter of 5% acetic acid, extracted with a mixture of ether and benzene, washed thoroughly with water, bicarbonate, saturated salt, dried over magnesium sulfate, treated with charcoal, and evaporated to a residue which partly crystallizes. This is shaken with ether, filtered and the filtrate is evaporated to a residue (17 g) which solidifies after being refrigerated overnight.

The crude product is boiled with 300 ml of Skellysolve 6, cooled to room temperature, decanted from some gummy material, treated with charcoal, concentrated to 100 ml, and allowed to crystallize. The product thus obtained (10 g) is recrystallized from 50 ml of methanol and gives 4.5 g of analytically pure material, MP 103° to 104°C.

(D) 1 -p-Chlorobenzoyl-2-Methyl-5-Methoxy-3-Indolylacetic Acid: A mixture of 1 g ester and 0.1 g powdered porous plate is heated in an oil bath at 210°C with magnetic stirring under a blanket of nitrogen for about 2 hours. No intensification of color (pale yellow) occurs during this period. After cooling under nitrogen, the product is dissolved in benzene and ether, filtered, and extracted with bicarbonate. The aqueous solution is filtered with suction to remove ether, neutralized with acetic acid, and then acidified weakly with dilute hydrochloric acid. The crude product (0.4 g, 47%) is recrystallized from aqueous ethanol and dried in vacuo at 65°C: MP 151°C.

References

Merck Index 4852
Kleeman and Engel p. 488
PDR pp.993, 1034, 1187, 1354, 1606, 1999
OCDS Vol. 1 p. 318 (1977); 2, 345 (1980) and 3, 165 (1984)
DOT 1 (4) 125 (1965); 18 (8) 373 (1982) and 19 (5) 286 (1983)
I.N. p. 517
REM p. 1118
Shen, T.-Y.; US Patent 3,161,654; December 15, 1964; assigned to Merck and Co., Inc.

INDOPROFEN

Therapeutic Function: Antiinflammatory

Chemical Name: 4-(1,3-Dihydro-1-oxo-2H-isoindol-2-yl)-α-methylbenzeneacetic acid

Common Name: -

Structural Formula:

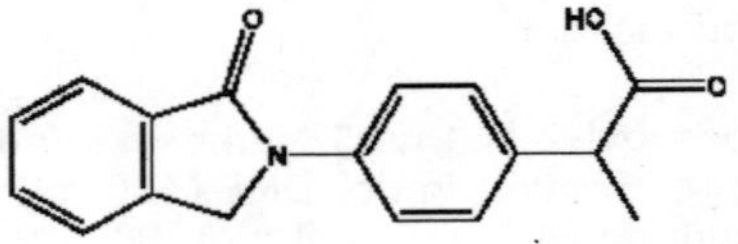

Chemical Abstracts Registry No.: 31842-01-0

Trade Name	Manufacturer	Country	Year Introduced
Flosint	Carlo Erba	Italy	1976
Flosin	Carlo Erba	W. Germany	1982
Flosin	Carlo Erba	Switz.	1982
Flosint	Carlo Erba	UK	1982
Fenint	Montedison	W. Germany	-
Praxis	Lisapharma	Italy	-

Raw Materials

Ethyl-α-(4-aminophenyl)propionate
Ethyl 2-chloromethyl benzoate
Potassium hydroxide

Manufacturing Process

The mixture of 7.9 g of ethyl α-(4-aminophenyl)propionate and 8.3 g of ethyl 2-chloromethylbenzoate is refluxed under nitrogen for one hour. The residue is recrystallized from hexane, to yield the ethyl α-[4-(1-oxo-isoindolino)-phenyl]-propionate of the formula

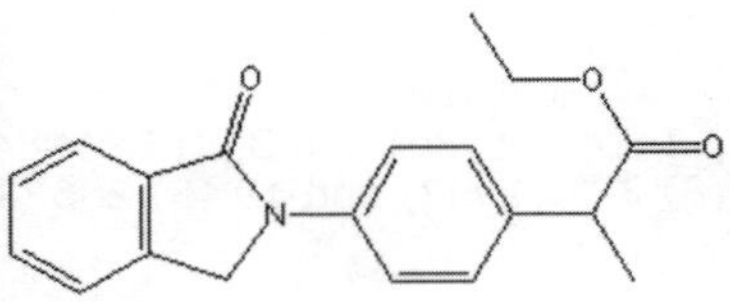

melting at 104° to 106°C. The mixture of 4.5 g thereof, 1.6 g of potassium hydroxide, 2 ml of water and 250 ml of ethanol is refluxed under nitrogen for 2 hours and evaporated under reduced pressure. The residue is taken up in water, the solution washed with chloroform, acidified with hydrochloric acid and extracted with ethyl acetate. The extract is dried, evaporated and the residue recrystallized from ethyl acetate, to yield the corresponding free acid melting at 208° to 210°C. (Procedure reported in US Patent 3,767,805.)

References

Merck Index 4853
DFU 1 (5) 242 (1976)
Kleeman and Engel p. 489
OCDS Vol. 3 p. 171 (1984)
DOT 13 (5) 200 (1977)
I.N. p. 517
Carney, R.W.J. and de Stevens, G.; US Patent 3,767,805; October 23, 1973; assigned to Ciba-Geigy Corporation
Carlo Erba, S.P.A., Italy; British Patent 1,344,663; January 23, 1974

INDORAMIN

Therapeutic Function: Antihypertensive

Chemical Name: N-[1-[2-(1H-Indol-3-yl)ethyl]-4-piperidinyl]benzamide

Common Name: -

Structural Formula:

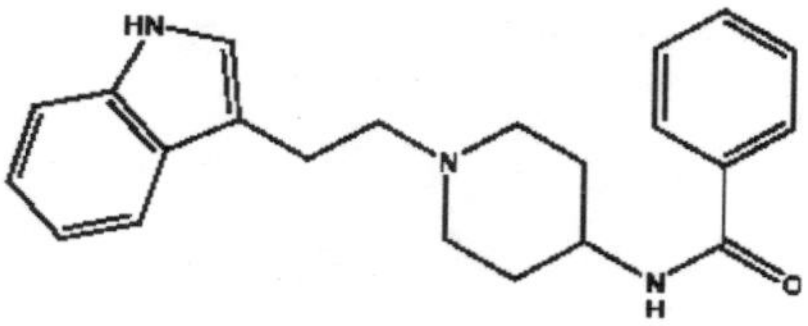

HN N N H O

Chemical Abstracts Registry No.: 26844-12-2

Trade Name	Manufacturer	Country	Year Introduced
Baratol	Wyeth	UK	1981
Wydora	Wyeth	W. Germany	1983

Raw Materials

4-Benzamido-1-(2-(3-indolyl)ethyl]pyridinium bromide
Hydrogen

Manufacturing Process

4-Benzamido-1-[2-(3-indolyl)ethyl] pyridinium bromide (3.0 g) was dissolved in 91% ethanol (300 ml) containing triethylamine (0.08 g) and freshly prepared W7 Raney nickel catalyst (ca 3 g) was added. The mixture was hydrogenated in an autoclave at 400 psi hydrogen pressure and 50°C for 4 hours. After filtering off the catalyst the filtrate was evaporated in vacuo and the residue was shaken with a mixture of chloroform and 2N sodium hydroxide solution. The resulting insoluble material was filtered off and dried to give 1.61 g of product, MP 203°C to 206°C. Recrystallization from ethanol gave the title compound as colorless needles (1.34 g), MP 208°C to 210°C.

References

Merck Index 4854
DFU 1 (10) 476 (1976)
OCDS Vol. 2 p. 344 (1980)
DOT 17 (10) 420 (1981)
I.N. p. 518
Archibald, J.L. and Jackson, J.L.; US Patent 3,527,761; September 8,1970; assigned to John Wyeth and Brother, Ltd. (UK)

INOSINE

Therapeutic Function: Cardiotonic

Chemical Name: 9-β-D-Ribofuranosylhypoxanthine

Common Name: Hypoxanthine riboside

Structural Formula:

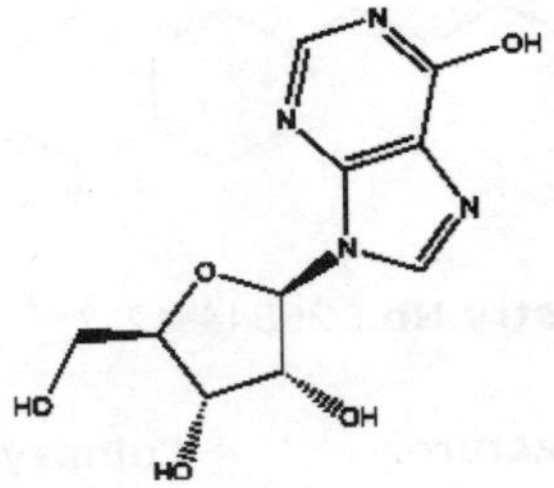

Chemical Abstracts Registry No.: 58-63-6

Trade Name	Manufacturer	Country	Year Introduced
Foreart	Guarnieri	Italy	1970
Oximain	Made	Spain	-
Ribonosine	Toyo Jozo	Japan	-
Salinite	Shinshin	Japan	-

Trade Name	Manufacturer	Country	Year Introduced
Tebertin	Berenguer-Beneyto	Spain	-
Trophicardyl	Innothera	France	-
Virusina	Dukron	Italy	-

Raw Materials

Adenosine
Barium nitrite
Sulfuric acid

Manufacturing Process

As described in US Patent 3,049,536, inosine may be prepared starting with adenosine.

The Deamination of Adenosine: 20 g of adenosine are dissolved in one liter of water by warming, and after cooling to room temperature 120 g of barium nitrite (monohydrate) are added to the solution. Under stirring there is added in time intervals of one hour 160 cc of 2 N sulfuric acid after each time interval. After the third addition, the reaction mass is allowed to stand for 3 hours at room temperature. The solution is then tested for barium, and if some barium is still present a slight excess of sulfuric acid is added. 300 cc of methanol is then added. In order to drive off the excess of nitrous acid, CO_2 is conducted through the solution until the solution is free of nitrous acid as determined by testing with potassium iodide-starch paper. The precipitated barium sulfate is separated by centrifugation. The residue is washed one time with about 500 cc of water. The total volume of the centrifugate is about 2.3 liters.

Isolation of Inosine by Ion Exchange Method: Half of the above clear centrifugate (1.15 liters) is treated with 250 cc of anion exchange (bicarbonate form) and stirred together therewith for 16 hours at room temperature. The pH value is increased thereby to about 4 to 5. The ion exchanger is filtered off under suction and washed 3 times, each time with 150 cc of water. The solution is brought to a pH value of 7 by means of normal sodium hydroxide (total volume of the solution about 1.55 liters), and concentrated to a volume of about 100 cc under vacuum.

The inosine is crystallized overnight in an ice box and the inosine is then filtered off by suction, washed with a small amount of ice water and dried at a temperature of 105°C. A first fraction of crude inosine consisting of 5.4 g having a purity of 99% is obtained. Further fractions of crude inosine are obtained from the mother liquid by concentration, the total amount constituting 3.2 g having a purity of 96 to 98%. The yield of crude inosine is 8.6 g which is equal to 86%.

Recrystallization of the Crude Inosine: 17.0 g of crude inosine are dissolved in 400 cc of 80% ethanol in a water bath, filtered while hot and brought to crystallization in an ice box. After standing overnight the crystalline material is filtered off under suction and washed with ice water. The pure inosine is dried in a drying chamber at a temperature of 105°C. The yield of pure inosine is 15.0 g which is equal to 75%. The yield can be further increased by working up the mother liquor of the crystallization as set forth above.

Alternatively, inosine may be made by fermentation as described in US Patent 3,111,459. 3 ml portions of a culture medium consisting of glucose (5 g/dl), ammonium chloride (0.4 g/dl), urea (0.4 g/dl), KH_2PO_4 (0.1 g/dl), $MgSO_4 \cdot 7H_2O$ (0.02 g/dl), Mn^{++} (2 ppm), Fe^{++} (2ppm), casein hydrolyzate (0.2 g/dl), yeast extract (0.2 g/dl), corn steep liquor (0.2 ml/dl), polypeptone (0.1 g/dl), meat extract (0.1 g/dl) and sodium ribonucleate (10 mg/dl) were poured into respective test tubes and each tube was sterilized at 115°C for 10 minutes. Thereafter separately sterilized calcium carbonate was added in the amount of 2 g/dl and then cells of Bacillus subtilis S26910 were inoculated into the above media and cultured with shaking at 30°C for 20 hours.

The resulting culture liquids were utilized for seeding, 20 ml of the medium having the composition described above were poured into a 500 ml shaking flask and sterilized at 115°C for 10 minutes and five drops of the above seed were added, and then cultured with shaking at 30°C for 65 hours. Thereafter 0.15 g/dl of inosine were accumulated.

The inosine-containing solution, which was obtained by separating the cells from the resulting fermentation liquid, was treated with both decolorizing resins and anion exchange resins by means of a conventional method and then acetone was added to crystallize the inosine. 1.47 g of the crude crystals of inosine were obtained from 3.5 liters of the culture liquid containing 1 g of inosine per liter.

References

Merck Index 4858
I.N. p. 519
Reiff, F., Huber, G. and Holle, K.; US Patent 3,049,536; August 14, 1962; assigned to Zellstoff Fabrik Waldhof, Germany
Motozakl, S., Tsunoda, T., Aoki, R., Okumura, S., Kondo, Y., Muramatsu, N., Momose, H. and Tamagawa, Y.;US Patent 3,111,459; November 19, 1963; assigned to Ajinomoto KK, Japan

INOSITOL

Therapeutic Function: Vitamin, Lipotropic

Chemical Name: Myo-Inositol

Common Name: Hexahydroxycyclohexane; Cyclohexitol

Chemical Abstracts Registry No.: 87-89-8

Raw Materials

Starch
Calcium hydroxide

Structural Formula:

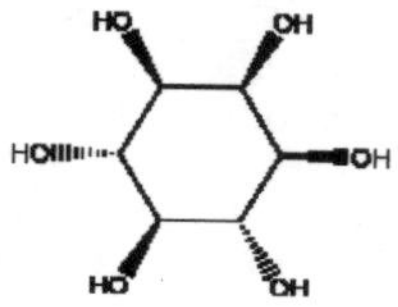

Trade Name	Manufacturer	Country	Year Introduced
Inositol	Comm. Solvents	US	1949
Amino-Ceru	Milex	US	-
Inosital	Biomedica Foscama	Italy	-
Inositine	Vis	Italy	-
Lipo-BC	Legere	US	-
Mega-B	Arco	US	-
Megadose	Arco	US	-

Manufacturing Process

Inactive inositol may be prepared from starch factory steep water which is the liquid in which corn is steeped to soften the covering of the corn kernel and to thoroughly soften the entire kernel. It contains approximately 1% sulfurous acid (H_2SO_3) in solution. A typical example of such treatment consists in adding to the acid steep water, lime $Ca(OH)_2$ or CaO to approximate neutrality, or to a pH of 6.0 to 8.0, at which range the insoluble "phytin" is precipitated. This precipitate of impure "phytin" or calcium phytate is removed by suitable means, as stated before, and may be mixed with (1) 1 to 10% acid solution; or (2) diluted with water; or (3) the solution may be made alkaline. This alkaline or neutral or acid mixture is placed in a suitable container in an autoclave or steam digester, and the steam turned on whereupon the reaction is allowed to proceed as long as desired. The autoclave in which the mixture has been placed may be heated by generating steam therein, by means of an electric heater, or by suitable heat from outside. A pressure of from 1 to 200 pounds steam for 1 to 18 hours may be used, the time required being correspondingly less for higher pressures. A suitable pressure is 80 pounds. The time expected for 80 pounds is three hours.

After hydrolysis or decomposition is complete, pressure is released, the autoclave cooled, the mixture removed, diluted, and made alkaline with $Ca(OH)_2$, $Ba(OH)_2$,etc., brought to boiling, thoroughly agitated with steam, the insoluble sludge allowed to settle, and the supernatant liquid removed by decantation, siphoning or filtration. The supernatant liquid is concentrated in an open vessel, or in vacuum, to remove the precipitating inorganic impurities as calcium carbonate ($CaCO_3$), magnesium carbonate ($MgCO_3$), etc. The liquid is concentrated until it becomes thick and syrupy. The concentrated solution is filtered, cooled, and agitated by a suitable mechanical means to precipitate i-inositol. The iI-inositol is removed by filtration, the mother liquor concentrated, and the process repeated until the solution becomes too thick to filter advantageously. A filter press may be employed to remove further quantities of i-inositol, or the thick residue may be diluted with a reagent in which i-inositol is insoluble; as, for example, acetic acid (CH_3COOH) and alcohol-acetic acid (C_2H_5OH, CH_3COOH, etc.). On cooling and stirring the solution, additional i-inositol, etc., results and can be removed by filtration or other mechanical means. The i-inositol may be recrystallized by dissolving the crude product in boiling water, and reprecipitated by cooling and stirring. The final crystallization from a hot water solution to which an equal volume of alcohol is added with cooling and stirring, gives a purer product.

References

Merck Index 4861
PDR pp. 581, 1033, 1263, 1734
I.N. p. 519
REM p. 1015
Bartow, E.and Walker, W.W.; US Patent 2,112,553; March 29, 1938
Elkin, M. and Meadows, C.M.; US Patent 2,414,365; January 14, 1947; assigned to American Cyanamid Co.

INOSITOL NIACINATE

Therapeutic Function: Vasodilator

Chemical Name: Myo-Inositol hexa-3-pyridine carboxylate

Common Name: Inositol hexanicotinate

Structural Formula:

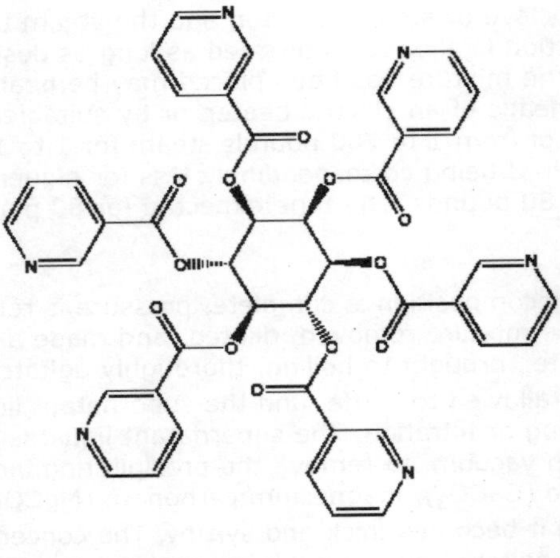

Chemical Abstracts Registry No.: 6556-11-2

Trade Name	Manufacturer	Country	Year Introduced
Hexanicotol	Philadelphia	US	1962
Dilexpal	Winthrop	France	1968
Bendigon	Bayer	W. Germany	-
Clevamin	Kowa	Japan	-
Cycnate	Toyo	Japan	-
Ebelin	Samva	Japan	-
Hammovenad	Bastian Werk	W. Germany	-
Hexalmin	Maruishi	Japan	-
Hexainosineat	Hishiyama	Japan	-
Hexanate	Nippon Chemiphar	Japan	-
Hexanicit	Yoshitomi	Japan	-
Hexate	Mohan	Japan	-
Hexatin	Kobayashi	Japan	-
Hexit	Toho	Japan	-
Inochinate	Nichiiko	Japan	-
Inosinit	Kanto	Japan	-
Kotanicit	Kotani	Japan	-
Mesonex	Tokyo Tanabe	Japan	-
Mesosit	Toyo Jozo	Japan	-
Nasky	Nikken	Japan	-
Neonitin	Chugai	Japan	-
Nicosamin	Toyama	Japan	-
Nicosinate	Toyo Ono	Japan	-
Nicosinit	Hokuriku	Japan	-
Nicotol	Maruko	Japan	-
Nicoxatin	Fuso	Japan	-
Romanit	Kowa	Japan	-
Salex	Iwaki	Japan	-
Sannecit	Sanko	Japan	-
Secotinen	Seiko	Japan	-
Shikioit	Shiri	Japan	-
Xatolone	Showa	Japan	-
Yonomol	Sawai	Japan	-

Raw Materials

Nicotinic acid
Phosphorus oxychloride
meso-Inositol

Manufacturing Process

100 g of nicotinic acid were suspended in 265 ml of distilled and dried pyridine without stirring. 68 g of phosphorus oxychloride were added dropwise to this mixture under continual stirring. The temperature of the reactants, initially at 20°C, was allowed to rise to about 60°C, and this temperature was maintained for a further 60 minutes. Thereafter 24.5 g of meso-inositol were

added gradually, the temperature being controlled so that it did not exceed about 80°C. The reactants were maintained at this temperature for from 2 to 3 hours, and thereafter the reaction mixture was poured into 500 ml of water. The pyridine salts formed during the reaction readily dissolved, and the meso-inositol hexanicotinate which had formed crystallized out. The ester was filtered off and washed with water and acetone or alcohol. Finally, the meso-inositol hexanicotinate was dried at 100°C.

The yield was 90%, the melting point of the product was 258°C to 260°C. and the chlorine content <0.01%.

References

Merck Index 4863
Kleeman and Engel p. 490
I.N. p. 519
A.B. Bofors; British Patent 1,053,689; January 4, 1967

INSULIN

Therapeutic Function: Antidiabetic

Chemical Name: Complex polypeptide hormone with molecular weight over 6,000

Common Name: -

Structural Formula: A protein that has the normal structure of the natural antidiabetic principle produced by the human pancreas

Chemical Abstracts Registry No.: 9004-10-8

Trade Name	Manufacturer	Country	Year Introduced
Humulin	Lilly	US	1982
Humulin	Lilly	UK	1982
Humulin	Lilly	Switz.	1983
Huminsulin	Lilly	W. Germany	1983
Velosulin	Leo	Switz.	1983
Monotard	Squibb	US	1983
Monotard	Nova	W. Germany	1983
Actrapid	Squibb	US	1983
Actrapid	Novo	W. Germany	1983
Basal-H	Hoechst	W. Germany	1983
Iletin	Lilly	US	-
Insulatard	Nordisk	US	-
Mixtard	Nordisk	US	-
Novolin	Squibb-Novo	US	-
Velosulin	Nordisk	US	-

Raw Materials

Beef pancreas glands
Ethanol

Manufacturing Process

40 pounds of frozen beef pancreas glands were hashed and extracted by stirring with 45,500 cc of 85% alcohol containing 925 cc of phosphoric acid. The acidity of the extraction mixture was pH 3.0 and the alcohol concentration approximately 65% after equilibrium was attained. The pancreatic meat solids removed were then reextracted by stirring in 45,000 cc of 65% alcohol. The pH of the combined filtrates was raised to pH 8.0 by addition of ammonium hydroxide to precipitate inert proteins and phsophoric acid salts. The solids were removed by filtration and sulfuric acid was then added to the filtrate to bring the pH to 3.5. The acidified extracts were then concentrated under reduced pressure to an alcohol concentration of 20%. Lipoidal material was removed by filtration and the filtrate concentrated under reduced pressure to the aqueous phase. Lipoidal material was then removed by filtration and the insulin containing filtrate biologically assayed for insulin activity. The biological assay showed the insulin recovered to be equivalent to 1425 I.U. for each pound of pancreas glands processed.

References

Merck Index 4866
PDR pp. 1054, 1270, 1777
DOT 19 (2) 111 and (5) 262 (1983)
REM p. 973
Maxwell, L.C.and Hinkel, W.P.; US Patent 2,695,861; November 30, 1954; assigned to Armour and Co.

INSULIN ISOPHANE

Therapeutic Function: Hypoglycemic

Chemical Name: See structure

Common Name: Isophane insulin injection

Structural Formula: Isophane insulin

Chemical Abstracts Registry No.: 53027-39-7

Trade Name	Manufacturer	Country	Year Introduced
NPH-Iletin	Lilly	US	1950
Protaphane	Novo	US	1981
Humulin-I	Lilly	UK	1982

Trade Name	Manufacturer	Country	Year Introduced
Insulatard	Leo	Switz.	1983
Novolin N	Squibb-Novo	US	-

Raw Materials

Zinc insulin
Selmiridine sulfate

Manufacturing Process

This is a crystalline product of insulin and an alkaline protein where the protein/insulin ratio is called the isophane ratio. This product gives a delayed and uniform insulin action with a reduction in the number of insulin doses necessary per day. Such a preparation may be made as follows: 1.6 g of zinc-insulin crystals containing 0.4% of zinc are dissolved in 400 ml of water, with the aid of 25 ml of 0.1 N hydrochloric acid. To this are added aqueous solutions of 3 ml of tricresol, 7.6 g of sodium chloride, and sufficient sodium phosphate buffer that the final concentration is 1/75 molar and the pH is 6.9.

Then 0.14 g of salmiridine sulfate dissolved in water is added, while shaking. Salmiridine is a protamine derived from the sperm of Salmo irideus Gibbons, or rainbow trout. Salmiridine-insulin (a protamine-insulin) containing zinc is promptly precipitated. Enough water is now added to make a total of one liter, and the whole is shaken again. After standing for about an hour, the precipitated salmiridine-insulin is found to have become crystalline.

This crystalline salmiridine-insulin can be removed if desired, as by filtration; but it is not necessary to do that, as the suspension of crystalline salmiridine-insulin may be preserved as thus prepared, and dispensed and used (in the same manner as known preparations of protamine insulin and protamine-zinc-insulin are used) in the original suspending medium in which it is formed.

References

PDR p.1778
REM p.974
Krayenbuhl, C.H. and Rosenberg, T,; US Patent 2,538,018; January 16, 1951; assigned to Nordisk Insulinlaboratorium, Denmark

INSULIN ZINC SUSPENSION

Therapeutic Function: Hypoglycemic

Chemical Name: Insulin zinc suspension

Common Name: -

Structural Formula: Sterile suspension, in a buffered water medium, of insulin modified by the addition of zinc chloride in a manner such that the solid phase of the suspension conisists of a mixture of crystals and amorphous material in a ratio of approximately 7:3

Chemical Abstracts Registry No.: 8049-62-5

Trade Name	Manufacturer	Country	Year Introduced
Lente Insulin	Squibb	US	1971
Iletin I	Lilly	US	-
Protamine	Lilly	US	-
Semilente	Squibb-Novo	US	-
Ultralente	Squibb-Novo	US	-

Raw Materials

Insulin
Zinc chloride

Manufacturing Process

First, a series of stock solutions are made.

Stock Solution 1: 2.18 g of recrystallized insulin are dissolved in 25 ml of 0.1 N hydrochloric acid, and distilled water to a volume of 125 ml is added.

Stock Solution 2: To 20 ml of an aqueous zinc chloride solution containing 1% zinc is added distilled water to a volume of 125 ml.

Stock Solution 3: 1.36 g of sodium acetate with 3 mols crystal water are dissolved in distilled water to a volume of 100 ml.

Then, 1.3 ml of glycerine are mixed with 0.5 ml of a 25% solution of methyl p-hydroxybenzoate in ethanol, and 50 ml of distilled water are added. To the produced mixture are, after sterile filtration, added 10 ml of the stock solution 1, 2.5 ml of the stock solution 2 and 10 ml of the stock solution 3, after which 3.0 ml of sterile 0.1 N sodium hydroxide are added, and the mixture is filled up with sterile distilled water to a volume of 100 ml. The insulin will be precipitated amorphously by the admixture of the sodium hydroxide, and the produced suspension acquires the pH value of 7. It will contain approximately 1 gamma zinc per insulin unit.

References

Merck Index 4869
PDR pp. 1055, 1777
REM p.975
Petersen, K., Schlichtkrull, J. and Halias-Moller, K.; US Patent 2,882,203; April 14, 1959 assigned to Novo Terapeutisk Laboratorium A/S, Denmark

INTERFERON

Therapeutic Function: Antineoplastic, Antiviral

Chemical Name: See structural Formula

Common Name: -

Structural Formula: Interferons (complex protein)

Chemical Abstracts Registry No.: 9008-11-1

Trade Name	Manufacturer	Country	Year Introduced
Fiblaferon	Bioferon	W. Germany	1983
Wellferon	Burroughs-Wellcome	-	-

Raw Materials

Semliki Forest arborvirus
Animal kidneys
Trypsin

Manufacturing Process

Semliki Forest arborvirus was grown in chick embryo tissue culture. The infectious tissue culture liquid was decanted and diluted with medium 199 to give a preparation containing between 10^6 and $10^{6.5}$ mouse ID_{50} of virus/ml.

Calf kidneys, dog kidneys and rhesus monkey kidneys were treated with trypsin to give suspensions of cells. The suspensions were centrifuged and the packed cells diluted with 400 volumes (calf cells) or 200 volumes (dog cells and rhesus monkey cells) of a growth medium consisting of 5% horse serum and 0.5% lactalbumen hydrolysate in Earle's saline, with 100 units/ml each of penicillin and streptomycin. These media were used separately to produce Semliki Forest/calf interferon, Semliki Forest/dog interferon and Semliki Forest/rhesus monkey interferon. The cellcontaining growth medium was dispensed into 500 ml medical flat bottles (70 ml in each). The cultures were incubated at 36°C. Confluent sheets of cells (monolayers) were formed in 5 to 6 days. The growth medium was then removed and the monolayers were washed with isotonic phosphate-buffered saline, pH 7.5.

Each bottle for interferon production received the arborvirus preparation in medium 199 (0.5 ml) and further medium 199 (50 ml); some bottles received only medium 199 (50 ml) and no virus and served as controls. The bottles were incubated for 3 to 5 days at 36°C.

The supernatants containing the interferons were decanted from monolayers, pooled, and tested for freedom from bacteria. Residual arborvirus was inactivated by acid and heat as follows. The liquid was brought to pH 2 by the addition of 0.3N hydrochloric acid in Earle's saline (minus sodium chloride and sodium bicarbonate), kept at 4°C for 24 hours, and then brought back to pH 7 by the addition of 0.3N sodium hydroxide in distilled water. The liquid was then heated at 56°C for 30 minutes.

At this stage the interferon preparations were assayed and submitted to safety tests for the absence of contaminating viruses.

Rhesus monkey kidney infected with Semliki Forest arborvirus gave interferon of titre 1.5 log interferon units/2 ml. (The interferon unit, determined in a volume of 2 ml, is the dilution of interferon which produced a half-maximal score for degree of cytopathic effect in virus-infected tissue culture tubes at the time when the control without interferon first showed the maximal score.)

Each interferon preparation was ultracentrifuged at 20,000 revolutions per minute for one hour to remove tissue debris and inactivated virus. The supernatant was dialyzed against distilled water (1:400) for 24 hours at 4°C. The material was then freeze-dried. Thedried product was reconstituted in one-tenth of the original volume in distilled water and dispensed into ampoules. Reconstituted solutions were assayed for interferon activity, examined for toxicity, and tested for sterility.

References

Merck Index 4870
DOT 18 (8) 393 (1982)
I.N. p. 520
REM p. 1233
Sellers,R.F.; British Patent 960,769; June 17, 1964; assigned to The Wellcome Foundation Ltd. (UK)

IOBENZAMIC ACID

Therapeutic Function: Diagnostic aid

Chemical Name: β-Alanine, N-(3-amino-2,4,6-triiodobenzoyl)-N-phenyl-

Common Name: Acide iobenzamique; Acidum iobenzamicum; Acidum jobenzamicum; Iobenzamic acid

Structural Formula:

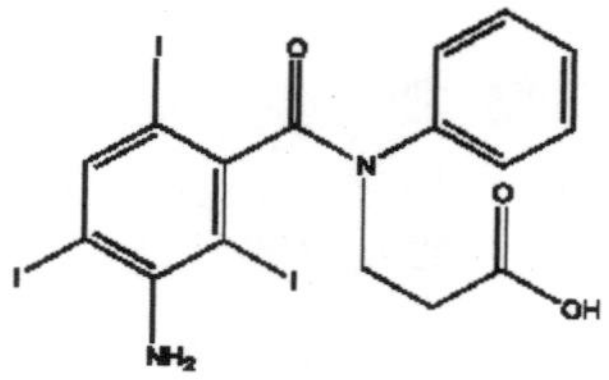

I O N O OH I I NH2

Chemical Abstracts Registry No.: 3115-05-7

Trade Name	Manufacturer	Country	Year Introduced
Bilibyk	Byk Gulden	-	-
Osbil	Upjohn	-	-
Osbil	M and B	-	-

Raw Materials

Methyl β-anilinopropionate
3-Amino-2,4,6-triiodobenzoyl chloride
Hydrogen chloride
Sodium hydroxide

Manufacturing Process

1013.2 g 3-amino-2,4,6-triiodobenzoyl chloride are dissolved in a minimum amount of hot dioxane and caused to flow into 700.6 g molten methyl β-anilinopropionate with stirring. After the ensuing exothermic reaction has gradually terminated the reaction mixture is heated on a steam bath for about 3 h and while boiling hot has then methanol and methanolic 3 N HCl added thereto, where after it is cooled. The precipitated solids are separated and washed with ether. 953.0 g methyl β-N-(3-amino-2,4,6-triiodobenzoyl)-phenylaminopropionate are obtained having a melting paint of 156°-157°C. Yield 74.2% of theory.

The methyl ester may be precipitated as the free acid by dissolving in dioxane, addition of 3.15 N methanolic sodium hydroxide solution, pouring the resulting reaction mixture in water, and acidulating the solution with 6 N HCl. After separation and drying, 553.6 g β-N-(3-amino-2,4,6-triiodobenzoyl)-phenylaminopropionic acid having a melting point of 133°-134.5°C are obtained from 571.0 g methyl ester. Overall yield 73.46% of theory.

References

Obendorf W.H.; US Patent No. 3,051,745; Aug. 28, 1962; Assigned: Osterreichische Stickstoffwerke Aktiengeselischaft, Linz, Austria

IOCARMIC ACID

Therapeutic Function: Diagnostic aid

Chemical Name: Benzoic acid, 3,3'-((1,6-dioxo-1,6-hexanediyl)diimino) bis(2,4,6-triiodo-5-((methylamino)carbonyl)-

Common Name: Acide iocarmique, Acidum iocarmicum, Acidum jocarmicum, Iocarmate meglumine, Iocarmic acid, Meglumine iocarmate

Structural Formula:

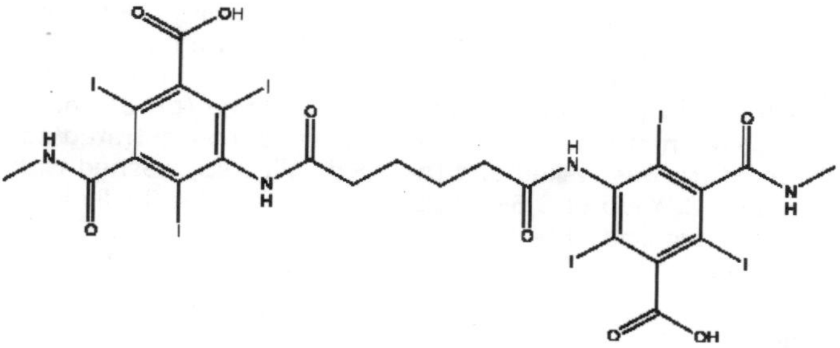

Chemical Abstracts Registry No.: 10397-75-8

Trade Name	Manufacturer	Country	Year Introduced
Myelotrast	Winthrop	-	-
Myelotrast	Guerbet	-	-
Iocarmic acid	Yick-Vic Chemicals and Pharmaceuticals (HK) Ltd.	-	-
Iocarmic acid	Shanghai Lansheng Corporation	-	-
Dimer-X	Byk Gulden	-	-

Raw Materials

Acetic acid	5-Amino-2,4,6-triiodo-N-methylisophthalamic acid
Adipoyl chloride	Dimethylacetamide
Hydrochloric acid	Sodium hydroxide

Manufacturing Process

5-Amino-2,4,6-triiodo-N-methylisophthalamic acid (228.0 g, 4 mole) was added to stirred, heated dimethylacetamide (400 ml). When the temperature reached 95°C, adipoyl chloride (27.5 g, 0.15 mole) was added all at once, followed by an equal amount added slowly over a period of 15 min (a total of 55.0 g). After addition of the adipoyl chloride the solution was stirred at about 95°C for another 15 min, then poured into 2 L of hot water. As the above mixture cooled to room temperature a gum separated. The mother liquor was discarded and the gum was dissolved in water (2 L) with sufficient sodium hydroxide to complete solution. The solution was acidified with hydrochloric and acetic acids, treated with decolorizing charcoal and filtered. The filtrate was then strongly acidified with hydrochloric acid, which caused the separation of an apparently amorphous granular solid. This was filtered off, digested 0.5 h with hot ethanol (500 ml) collected, washed with ethanol and dried at 110°C. Yield of crude 5,5-(adipoyldiimino)-bis[2,4,6-triiodo-N-methylisophthalamic acid].

The 5,5-(adipoyldiimino)-bis[2,4,6-triiodo-N-methylisophthalamic acid] was precipitated a second and third time from its sodium salt solution. The third precipitate was then dissolved in hot dimethylformamide (400 ml), and water (1.5 L) was slowly added. The mixture was digested and the hot mixture filtered, yielding a crystalline product which, after drying at 110°C, weighed 126.0 g (neutral equivalent, 724). This product was dissolved in dilute sodium hydroxide solution (1 L) and the solution was acidified (pH 5) and filtered into a hot stirred solution of hydrochloric acid (25 ml of concentrated acid in 75 ml water). The mixture was chilled and the solid collected, washed with water and dried at 110°C. Yield of 5,5-(adipoyldiimino)-bis[2,4,6-triiodo-N-methylisophthalamic acid] 114.0 g (45%). Melting point, 302°C (corrected), with decomposition.

References

Hoey G.B.; US Patent No. 3,290,366; Dec. 6, 1966; Assigned: Mallinckrodt Chemical Works, St. Louis, Mo., a corporation of Missouri

IOCETAMIC ACID

Therapeutic Function: Diagnostic aid

Chemical Name: Propanoic acid, 3-(acetyl(3-amino-2,4,6-triiodophenyl) amino)-2-methyl-

Common Name: Acide iocetamique; Acidum iocetamicum; Acidum jocetamicum; Iocetamic acid

Structural Formula:

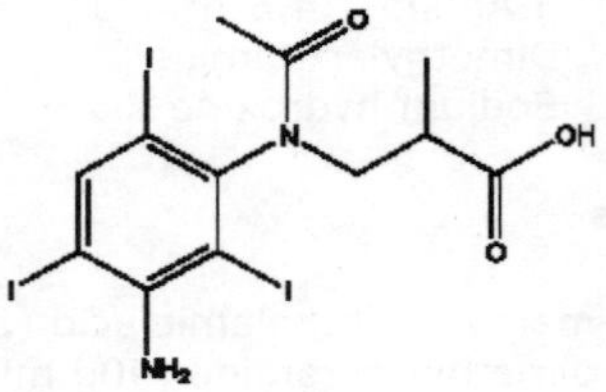

Chemical Abstracts Registry No.: 16034-77-8

Trade Name	Manufacturer	Country	Year Introduced
Cholebrine	Nicholas	-	-
Cholimil	Takeda	-	-

Raw Materials

m-Nitroaniline	Methacrylic acid
Pyridine	Sodium chloride

Acetic acid
Aceticanhydride
Iodine monochloride
Ammonia
Nickel Raney

Manufacturing Process

A mixture of equimolecular amounts of m-nitraniline (69.0 g), methacrylic acid (43.0 g), and pyridine (39.5 g) was heated to 125°C for 20 h, and was then poured into 500 ml water. A semi-crystalline product separated from the liquid. The supernatant liquid was decanted and 500 ml fresh water was added. The pH was adjusted to 7-7.5 by means of sodium hydroxide. The precipitate was filtered with suction and washed with water. When the filtrate was acidified with acetic acid, a precipitate of N-(3-nitrophenyl)-β-amino-isobutyric acid was formed. When filtered, washed with water and dried, it weighed 56.0 g and had a melting point of 130°-131°C (recryst. from alcohol). The yield was 50% based on nitraniline.

0.25 mol (56.0 g) N-(3-nitrophenyl)-β-aminoisobutyric acid, prepared as described above, 160 ml glacial acetic acid, and 40 ml acetic anhydride were heated 48 h to 50°C, and the reaction mixture was poured into 600 ml water. A crystalline precipitate of N-acetyl-N-(3-nitrophenyl)-β-aminoisobutyric acid formed gradually. When recovered, it weighed 56.0 g (84% yield). Melting point 146°-148°C.

100.0 g N-acetyl-N-(3-nitrophenyl)-β-amino-isobutyric acid were dissolved in 1 L water and 40 ml 25% aqueous ammonia, and the solution was hydrogenated in the presence of about 10.0 g Raney nickel at 20°C and about 450 p.s.i. until the pressure drop indicated the complete conversion of the NO_2 groups to NH_2. Thus N-acetyl-N-(3-aminophenyl)-β-amino-isobutyric acid was obtained.

The solution of N-acetyl-N-(3-aminophenyl)-β-amino-isobutyric acid was filtered after standing overnight, mixed with an equal volume of acetic acid, and there after with a solution of 275.0 g iodine monochloride and 200.0 g sodium chloride in 1 L water. The mixture was kept at 50°C with stirring for 48 h. Light brown crystals of crude N-acetyl-N-(2,4,6-triiodo-3-aminophenyl)-β-amino-isobutyric acid precipitated, were filtered off, washed with water, and dried. They weighed 195.0 g (84.5% yield based on N-acetyl-N 3-nitrophenyl-p-amino-isobutyric acid).

References

Korver J.A.; US Patent No. 3,661,975; May 9, 1972
GB Patent No. 1,116,586; Nov. 25, 1965; Assigned: Dagra N.V. a limited liability company incorporated, Diemen Nethelands

IODAMIDE

Therapeutic Function: Diagnostic aid (radiopaque medium)

Chemical Name: 3-(Acetylamino)-5-[(acetylamino)methyl]-2,4,6-triiodobenzoic acid

Common Name: Ametriodinic acid

Structural Formula:

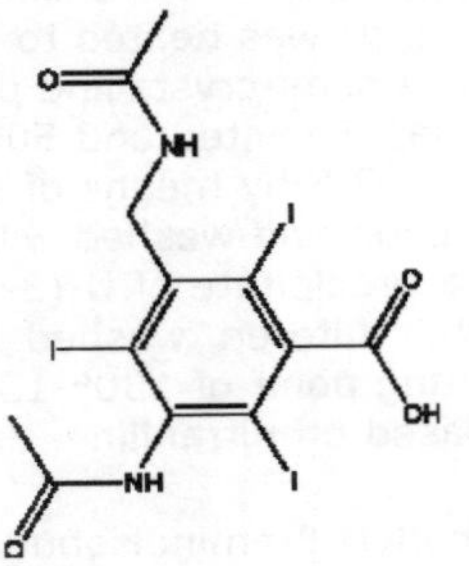

Chemical Abstracts Registry No.: 440-58-4

Trade Name	Manufacturer	Country	Year Introduced
Uromiro	Heyden	W. Germany	1965
Uromiro	Bracco	Italy	1970
Angiomiron	Schering	W. Germany	-
Contraxin	Takeda	Japan	-
Isteropac	Bracco	Italy	-
Opacist	Bracco	Italy	-

Raw Materials

3-Acetylaminomethyl-4-chloro-5-nitrobenzoic acid
Hydrogen
Potassium iodide dichloride
Acetic anhydride

Manufacturing Process

65.4 g (0.24 mol) 3-acetylaminomethyl-4-chloro-5-nitrobenzoic acid were dissolved in a mixture of 48 ml 10N sodium hydroxide and 1,800 ml water. 12 g of a 10% palladium catalyst on a carbon carrier were added, and the nitrobenzoic acid derivative was hydrogenated at slightly elevated temperature and at atmospheric pressure. The hydrogen was avidly absorbed. The nitro group was fully reduced to the corresponding amino radical within about 20 to 40 minutes, and 99 to 100% of the amount of chlorine ions to be theoretically expected was formed. Hydrogen absorption then stopped.

The catalyst was removed by filtration. The filtrate was diluted to about 18 liters, and was acidified with 15 ml concentrated hydrochloric acid. With vigorous stirring, 1,152 ml N $KICl_2$ solution were run into the diluted filtrate over a period of about 20 to 30 minutes. A solid precipitate was formed, and was filtered off after about six hours. The solid material was washed with water, with sodium bisulfite solution, and again with water. It was dissolved in aqueous ammonium hydroxide solution, the solution was filtered, and the filtrate was acidified with concentrated hydrochloric acid containing a small amount of sodium bisulfite. After a short time, the precipitate formed was filtered with suction, washed with water, and dried.

There were obtained 109 g 3-acetylaminomethyl-5-amino-2,4,6-triiodobenzoic acid which decomposes and melts at approximately 230°C. The equivalent weight was determined experimentally as being 591, as compared to a theoretical value of 586.

A suspension of 40 g 3-acetylaminomethyl-5-amino-2,4,6-triodobenzoic acid in 180 ml acetic anhydride were mixed with 0.4 ml concentrated sulfuric acid. An exothermic reaction was thereby initiated. Acetylation was completed by heating to 80°C for three hours. The reaction mixture was then evaporated to dryness in a vacuum at a temperature not exceeding 50°C. The residue was treated with a mixture of 30 ml concentrated aqueous ammonium hydroxide and 40 ml water, whereby the solid material dissolved with spontaneous heating. Within a few minutes, the ammonium salt of the acetylated product started precipitating. The precipitate and residual liquid were cooled externally with ice after about 15 minutes. The salt was separated from the liquid by filtration with suction, and was washed with ice cold saturated ammonium chloride solution.

The salt was dissolved in 300 ml water, and insoluble matter was removed from the solution by filtration. The free acid was precipitated from the filtrate at 50°C to 60°C by the addition of 40 ml 1:1 hydrochloric acid. The precipitate was filtered off after a few hours, washed with water, and dried. There were obtained 34 g 3-acetylaminomethyl-5-acetylamino-2,4,6-triiodobenzoic acid (79% of theoretical yield) having a melting point of 246°C to 248°C. The equivalent weight of this practically pure acid was found to be 631 as compared to the calculated value of 627.96.

When recrystallized from glacial acetic acid, the pure acid melts at 255°C to 257°C.

References

Merck Index 4878
Kleeman and Engel p. 493
I.N. p. 521
REM p. 1269
Felder, E. and Pitre, D.; US Patent 3,360,436: December 26, 1967; assigned to Eprova Ltd. (Switz.)

IODIPAMIDE

Therapeutic Function: Diagnostic aid (radiopaque medium)

Chemical Name: 3,3'-[(1,6-Dioxo-1,6-hexanediyl)diimino]bis[2,4,6-triodobenzoic acid]

Common Name: Adipodione

Structural Formula:

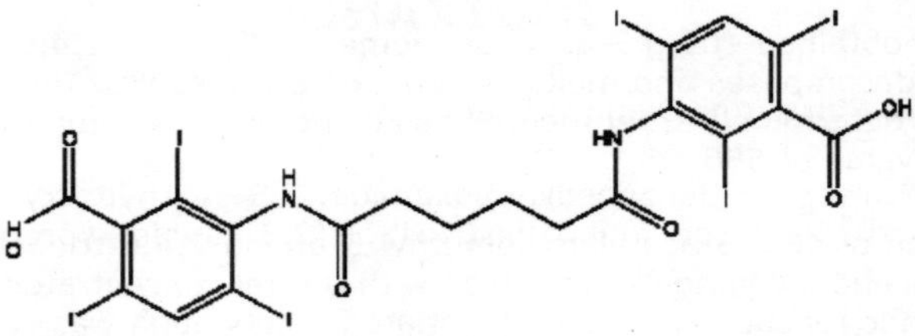

Chemical Abstracts Registry No.: 606-17-7

Trade Name	Manufacturer	Country	Year Introduced
Cholografin	Squibb	US	1954
Intralibix	Guerbet	France	1955
Biligrafin	Schering	W. Germany	-
Endocistobil	Bracco	Italy	-
Endografin	Schering	W. Germany	-
Radio-Selectan Biliare	S.E.P.P.S.	France	-
Transbilix	Guerbet	France	-
Ultrabil	Spofa	Czechoslovakia	-

Raw Materials

2,4,6-Triodo-3-amino benzoic acid
Adipic acid dichloride

Manufacturing Process

125 g of 2,4,6-triiodo-3-amino benzoic acid are dissolved in 250 cc of chlorobenzene and 15 g of adipic acid dichloride are added at a temperature between 110° and 130°C drop by drop to the solution. After evolution of hydrochloric acid (about 2 to 3 hours) has ceased, the precipitated crude adipic acid di-(3-carboxy-2,4,6-triiodo anilide) of the above formula is filtered hot with suction, washed with chlorobenzene, extracted by boiling with methanol and, for purification, dissolved in an amount of methanolic caustic soda solution required for neutralization, filtered with charcoal, and precipitated with dilute hydrochloric acid. Yield: 82.3 g, MP 306° to 308°C (with decomposition).

References

Merck Index 4890
Kleeman and Engel p. 16

I.N. p. 46
REM p. 1265
Priewe, H. and Rutkowski, R.; US Patent 2,776,241;January 1,1957;assigned to Schering AG, Germany

IODIXANOL

Therapeutic Function: Diagnostic aid

Chemical Name: 1,3-Benzenedicarboxamide, 5,5'-((2-hydroxy-1,3-propanediyl)bis(acetylimino))bis(N,N'-bis(2,3-dihydroxypropyl)-2,4,6-triiodo-

Common Name: Iodixanol

Structural Formula:

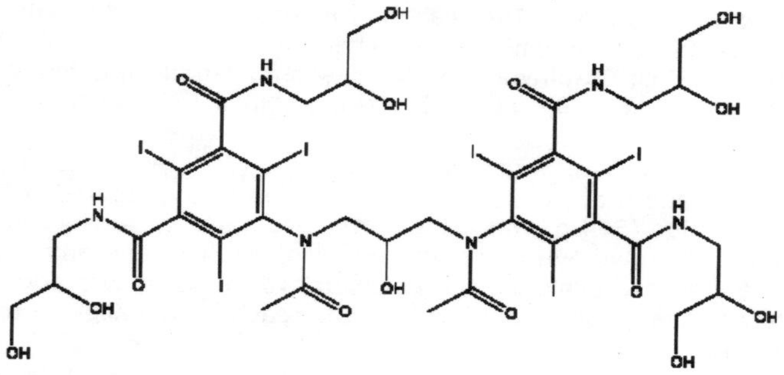

Chemical Abstracts Registry No.: 92339-11-2

Trade Name	Manufacturer	Country	Year Introduced
OptiPrep	Nycomed Pharma	Norway	-
OptiPrep	Axis-Shield PoC AS	Norway	-
Visipaque	Nycomed Ireland	Ireland	-

Raw Materials

Hydrochloric acid
$NaICl_2$
Sulfuric acid
2-Methoxyethanol
Epichlorohydrin
Dimethyl 5-nitroisophthalate
1-Amino-2,3-propandiol
Sodium hydroxide
Acetic anhydride

Manufacturing Process

Dimethyl 5-nitroisophthalate (215 g) and 1-amino-2,3-propandiol (196 g) were refluxed in methanol (500 ml). After twenty hours the solution was cooled and stored in a refrigerator overnight. The product 5-nitro-N,N'-bis(2,3-dihydroxypropyl)-isophthalamide was filtered and washed with methanol. Yield: 270 g (84%). M.p. 128-132°C.

5-Nitro-N,N'-bis(2,3-dihydroxypropyl)-isophthalamide (18.1 g) was suspended in water (250 ml), conc. hydrochloric acid (4.2 ml) and 10% PdO/charcoal (0.5 g) were added, and the mixture hydrogenated in a Parr apparatus for one day. After filtration the filtrate was heated at 80-90°C and 3.88 M $NaICl_2$ (42.5 ml) was added through a dropping funnel over 1 hour. The solution was heated for 2.5 hours. After cooling to 20°C 5-amino-2,4,6-triiodo-N,N'-bis(2,3-dihydroxypropyl)-isophthalamide crystallized out.

5-Amino-2,4,6-triiodo-N,N'-bis(2,3-dihydroxypropyl)-isophthalamide (110 g) was suspended in acetic anhydride (480 ml) and heated to 50°C. Concentrated sulfuric acid (3 ml) was then added. The starting material was dissolved after a few minutes, and the reaction mixture was heated at 60°C for 75 min. After cooling the residue dissolved in methanol (300 ml) with water (150 ml) the solution was heated to 50°C and the pH adjusted to about 10.5 by 10 N sodium hydroxide. After 4-5 hours the pH didn't decrease, and the hydrolysis was complete. The reaction mixture was cooled to 20°C and neutralized by adding hydrochloric acid. After stirring overnight 5-Acetamido-2,4,6-triiodo-N,N'-bis(2,3-dihydroxypropyl)-isophthalamide was filtered and washed with water. Yield: 94 g (80%). Melting point 275°C, dec.

2-Methoxyethanol (300 ml) and sodium hydroxide (20 g) was added to the reactor at 50°C, and 5-acetamido-N,N'-bis(2,3-dihydroxypropyl)-2,4,6-triiodoisophtalamide (304 g) was added after two hours of stirring. All solids were allowed to dissolve overnight before cooling to 30°C and adjustment to pH 12 with diluted hydrochloric acid. Epichlorohydrin (11 g) was added to the solution after further cooling to 15°C, and the reaction was allowed to proceed for 51 hours. As a result 1,3-bis(acetamido)-N,N'-bis[3,5-bis(2,3-dihydroxypropylaminocarbonyl)-2,4,6-triiodophenyl]-2-hydroxypropane was obtained.

References

Nordal V., Holtermann H.; US Patent No. 4,250,113; Feb. 10, 1981; Assigned to Nyegaard and Co. A/S, Oslo, Norway

Malthe-Sorenssen D., et al.; US Patent No. 6,232,499 B1; May 15, 2001; Assigned to Nycomed Imaging AS, Oslo (NO)

IODOALPHIONIC ACID

Therapeutic Function: Diagnostic aid (radiopaque medium)

Chemical Name: 4-Hydroxy-3,5-diiodo-α-phenylbenzenepropanoic acid

Common Name: Pheniodol

Structural Formula:

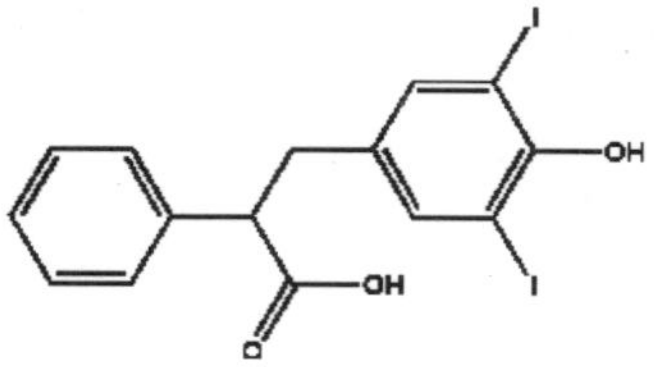

Chemical Abstracts Registry No.: 577-91-3

Trade Name	Manufacturer	Country	Year Introduced
Priodax	Schering	US	1943
Perfectochol	Lafayette	US	1952
Bilopsyl	Labaz	-	-
Choletrast	Burroughs-Wellcome	-	-

Raw Materials

Dextro-β-(4-hydroxyphenyl)-α-phenylpropionic acid
Iodine
Dimethylaminoethanol
Acetic acid

Manufacturing Process

Dextro-β-(4-hydroxyphenyl)-α-phenylpropionic acid (24 g) was dissolved in 630 ml of water containing 8.0 g of sodium hydroxide, and, with good stirring at 25°C, 51 g of iodine and 51 g of potassium iodide dissolved in 240 ml of water was added dropwise over a period of 30 minutes. During this period another 8 g of sodium hydroxide dissolved in 60 ml of water was added in order to keep the reaction mixture alkaline to phenolphthalein. Stirring was continued for 15 minutes longer. The resulting solution was made acid to Congo red with concentrated hydrochloric acid, and about 5 g of sodium bisulfite was added to partially decolorize the resulting slurry. The solid was collected by filtration and washed well with water.

The crude iodinated acid was then dissolved in 500 ml of 95% alcohol, 10 g of dimethylaminoethanol was added, the solution was decolorized with activated charcoal and filtered at 70°C. After keeping the filtrate for several hours at 5°C. the heavy crystalline precipitate which formed was collected by filtration and washed with acetone. The mother liquors were concentrated to 150 ml and cooled to give a second crop which was further purified by recrystallization from 50 ml of 95% alcohol. In this way a total of 36.0 g of dimethylaminoethanol salt of dextro-β-(3,5-diiodo-4-hydroxy)-α-phenylpropionic acid, MP 151° to 153°C, was obtained. The melting point of the dimethylaminoethanol salt of unresolved β-(3,5-diiodo-4-hydroxy)-α-phenylpropionic acid was 142° to 144°C.

The pure dimethylaminoethanol salt was dissolved in 400 ml of 50% acetic acid at 90°C and then cooled to 5°C. The solid which precipitated was collected by filtration, washed with water, cold 50% acetic acid and finally with low-boiling petroleum ether. After drying in vacuo there was obtained 24 g of hydrated dextro-β-(3,5-diiodo-4-hydroxy)-α-phenylpropionic acid, MP 80° to 85°C.

References

Merck Index 4893
I.N. p. 756
Tullar, B.F. and Hoppe, J.O.; US Patent 2,552,696; May 15, 1951; assigned to Sterling Drug Inc.

IODOXAMIC ACID

Therapeutic Function: Diagnostic aid

Chemical Name: Benzoic acid, 3,3'-((1,16-dioxo-4,7,10,13-tetraoxahexadecane-1,16-diyl)diimino)bis(2,4,6-triiodo-

Common Name: Acide iodoxamique, Acidum iodoxamicum, Iodoxamic acid

Structural Formula:

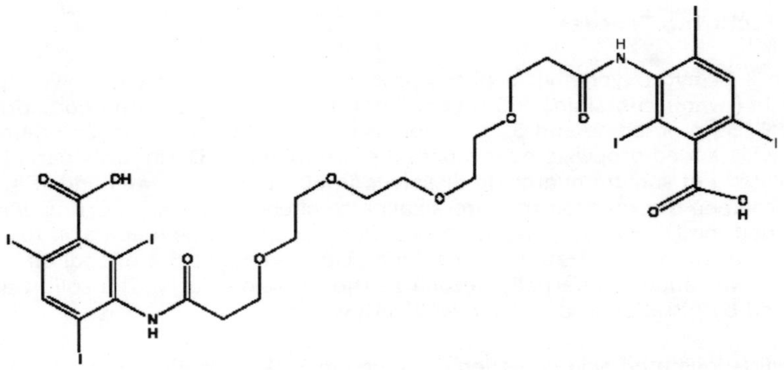

Chemical Abstracts Registry No.: 31127-82-9

Trade Name	Manufacturer	Country	Year Introduced
Endobil	Bracco	-	-
Endobil	Krka	-	-

Raw Materials

4,7,10,13-Tetraoxahexadecane-1,16-dinitrile

Thionyl chloride
3-Amino-2,4,6-triiodobenzoic acid
Dimethylacetamide

Manufacturing Process

148.5 g 4,7,10,13-tetraoxahexadecane-1,16-dinitrile (U.S. Patent No. 2,401,607) was added to a solution of 232 g (2.45 mol) concentrate sulfuric acid in 290 ml absolute ethanol at 15°C. The mixture was heated at reflux for 15 hours, cooled and poured into 1000 g ice and 250 g ammonium sulfate. It was extracted with methylene chloride, dried and a solvent was removed in vacuum. The residue was distilled to give 4,7,10,13-tetraoxahexadecane-1,16-dicarbonic acid dimethyl ester; BP: 190°-195°C/0.005 mm Hg.

1 mol above prepared diester was saponificated with equivalent of NaOH in water. The reaction mixture was heated for 90 minutes. On cooling it was extracted with ether and the water layer was evaporated to dryness. The residue was washed with acetone. The obtained disodium salt of 4,7,10,13-tetraoxahexadecane-1,16-dicarbonic acid (yield 100%; MP: 102°-104°C) was acidified with calculated quantity of HCl to give the dicarbonic acid. The solvent was evaporated to dryness. Acetone was added to the residue for removing a by-product (sodium chloride) by filtration. Acetone was evaporated and the residue was extracted with ether, dried and evaporated. The residual liquid was 4,7,10,13-tetraoxahexadecane-1,16-dicarbonic acid.

100 ml thionyl chloride was cautious added to 56 g above prepared diacid and heated at 40°-50°C and excess thionyl chloride was distilled in vacuum. The residue was the 4,7,10,13-tetraoxahexadecane-1,16-dicarbonic acid dichloride. The desired iodoxamic acid was prepared from above dichloride and 3-amino-2,4,6-trijode benzoic acid, in dimethylacetamide.

References

Felder and Pitre; D.B. Patent No. 1,937,211; July 22 1969; Milan, Italy

IOGLYCAMIC ACID

Therapeutic Function: Diagnostic aid (radiopaque medium)

Chemical Name: 3,3'-[Oxybis[(1-oxo-2,1-ethanediyl)imino]]bis[2,3,6-triiodobenzoic acid]

Common Name: -

Chemical Abstracts Registry No.: 2618-25-9

Structural Formula:

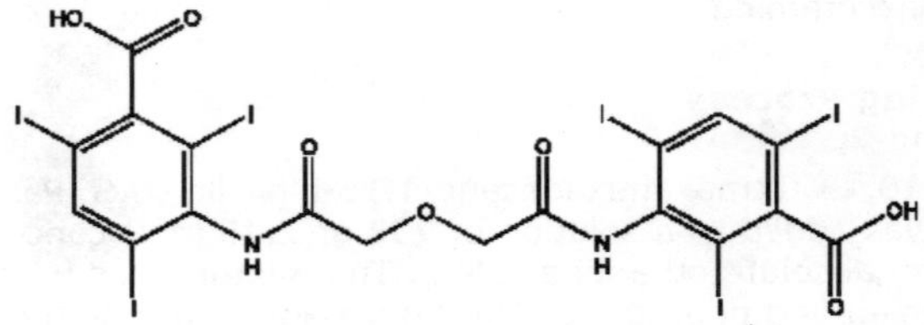

Trade Name	Manufacturer	Country	Year Introduced
Biligram	Schering	W. Germany	1971
BiIigram	Schering	UK	1972
Biligram	S.E.P.P.S.	France	1974
Bilivistan	Schering	Italy	-
Rayvist	Schering	W. Germany	-

Raw Materials

2,4,6-Triiodo-aminobenzoic acid
Diglycolic acid dichloride

Manufacturing Process

910 g of dry 2,4,6-triiodo amino benzoic acid are dissolved with stirring in 4,800 cc of dry, boiling chlorobenzene. A solution of 151.7 g diglycolic acid dichloride in 100 cc of dry chlorobenzene is slowly added to this solution and the mixture is further heated for 4 to 5 hours under reflux until development of hydrogen chloride has ceased. The resulting precipitate is filtered from the warm solution with suction and washed with chlorobenzene and then with ether. The microcrystalline, almost colorless crude product, 942 g, consists of the a-modification of diglycolic acid di-(3-carboxy-2,4,6-triiodo anilide).

The crude product is suspended, while stirring, in 2.5 liters of pure methanol and a solution of 73 g of pure sodium hydroxide in the same weight of water, diluted with 675 cc methanol, is slowly added to this suspension until the acid is dissolved and the pH of this solution reaches 9.0. The solution is allowed to stand at this pH for 15 minutes. The pH is then brought to 4.0 by addition of 10% acetic acid and 17 g of charcoal are stirred in. After 15 minutes the coal is filtered off and the clear filtrate is slowly added to a stirred solution of 415 cc of pure, concentrated hydrochloric acid in 4.15 liters of 50% methanol. After ½ hour of stirring and decanting after 1 hour, the precipitate is easily filtered off with suction, washed with little methanol and thoroughly with water, until the thixotropic residue is free of hydrochloric acid. In order to obtain a product of highest purity, this treatment is repeated two times. The resulting pure product, after drying in vacuo at 50°C still containing one molecule of methanol per two molecules of the acid (plus 4 molecules of water), must be suspended in boiling water and steamed out. The hot suspension is filtered with suction, the white microcrystalline residue is dried in vacuo at 50°C to give 860 g (83.5% of the theoretical yield) of the pure dihydrate of the diglycolic acid di-(3-carboxy-2,4,6-triiodo anilide), β-modification.

References

Merck Index 4912
Kleeman and Engel p. 494
I.N. p. 28
Priewe, H. and Rutkowski, R.; US Patent 2,853,424; September 23, 1958; assigned to Schering A.G. (W. Germany)

IOPAMIDOL

Therapeutic Function: Diagnostic aid (radiopaque medium)

Chemical Name: 5-(α-Hydroxypropionylamino)-2,4,6-triiodoisophthalic acid di-(1,3-dihydroxyisopropylamide)

Common Name: -

Structural Formula:

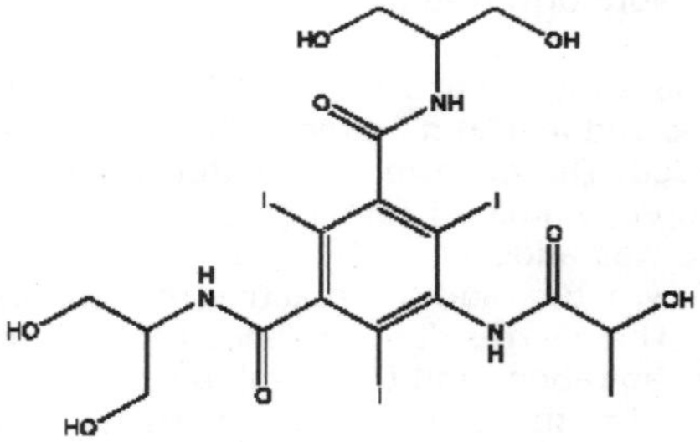

Chemical Abstracts Registry No.: 60166-93-0

Trade Name	Manufacturer	Country	Year Introduced
Iopamiro	Bracco	Italy	1981
Solutrast	Byk Gulden	W. Germany	1981
Niopam	Merck	UK	1982
Iopamiro	Astra	Sweden	1983
Isovue	Squibb	-	-

Raw Materials

5-Amino-2,4,6-triiodo-isophthalic acid
Thionyl chloride
DL-2-Acetoxypropionyl chloride
2-Amino-1,3-propanediol

Manufacturing Process

400 g (0.72 mol) 5-amino-2,4,6-triiodo-isophthalic acid was added to 200 ml thionyl chloride, the mixture was stirred at a boil for 6 hours, and the resulting solution was evaporated. The residue was dissolved in anhydrous ethyl acetate, and the solution was again evaporated to dryness. The solid material was dissolved in 4,000 ml ethyl acetate, and the solution was stirred into an ice-cold solution of 500 g sodium chloride and 200 g sodium bicarbonate in 2.5 liters water. The organic phase was separated from the aqueous solution, washed with aqueous sodium solution, dried by contact with anhydrous calcium chloride, and evaporated to dryness.

The residue of 420 g 5-amino-2,4,6-triiodo-isophthalyl chloride (97.5% yield) had a melting point above 300°C when recrystallized from toluene.

300 g (0.503 mol) 5-amino-2,4,6-triiodo-isophthalyl chloride was dissolved in 1,200 ml dimethylacetamide, and 187 g (126 mol) DL-2-acetoxypropionyl chloride was added dropwise to the solution with agitation. The mixture was permitted to stand overnight at ambient temperature and was then evaporated in a vacuum to approximately 400 ml. The oily residue was stirred into ice water to precipitate 353 g crystalline DL-5-(α-acetoxypropionylamino)-2,4,6-triiodo-isophthalyl chloride (98% yield) which was purified by suspension in warm chloroform free alcohol.

The purified intermediate melted at 210°C. 70.9 g (0.10 mol) of the intermediate was dissolved in 150 ml dimethylacetamide, and 15 g (0.08 mol) tributylamine was added. The mixture was heated to 50°C, and 56.6 g (0.62 mol) 1,3-dihydroxyisopropylamine (2-amino-1,3-propanediol) dissolved in 80 ml dimethylacetamide was added drop by drop. The reaction went to completion within a few hours, and the reaction mixture was evaporated to dryness in a vacuum. The oily residue was added to 350 ml methylene chloride with vigorous agitation, and the resulting precipitate was filtered off and purified by repeated suspension of warm methylene chloride.

Work-up of the reaction mixture yielded 56.5 g (73.5%) DL-5-α-hydroxypropionylamino-2,4,6-triiodo-isophthalic acid di-(1,3-dihydroxyisopropylamide) which was recrystallized from aqueous ethanol and melted with decomposition above 300°C.

References

Merck Index 4915
DFU 4 (12) 876 (1979)
I.N. p. 524
Felder, E., Vitale, R.S. and Pitre, D.E.; US Patent 4,001,323; January 4, 1977; assigned to Savac AG

IOPANOIC ACID

Therapeutic Function: Diagnostic aid (radiopaque medium)

Chemical Name: 3-Amino-α-ethyl-2,4,6-triiodobenzenepropanoic acid

Common Name: -

Structural Formula:

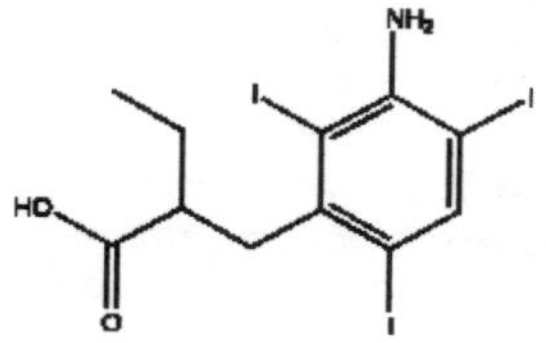

Chemical Abstracts Registry No.: 96-83-3

Trade Name	Manufacturer	Country	Year Introduced
Telepaque	Winthrop	US	1952
Telepaque	Winthrop	France	1955
Ace-Line	Maruishi	Japan	-
Biliopaco	Rovi	Spain	-
Chole-Contrast	Orion	Finland	-
Cistobil	Bracco	Italy	-
Colegraf	Estedi	Spain	-
Holevid	Krka	Yugoslavia	-
Leabar	Toyo	Japan	-
Molpaque	Tokyo Tanabe	Japan	-
Neocontrast	Bama-Geve	Spain	-
Polognost	Polfa	Poland	-
Teletrast	Astra	-	-

Raw Materials

m-Nitrobenzaldehyde
Butyric anhydride
Hydrogen
Iodine monochloride

Manufacturing Process

(A) Preparation of α-Ethyl-m-Nitrocinnamic Acid: This acid is prepared from 100 g of m-nitrobenzaldehyde, 210 g of butyric anhydride and 73 g of sodium butyrate. The crude α-ethyl-m-nitrocinnamicacid is crystallized from ethanol giving about 105 g, MP 140° to 142°C. From the filtrates there may be isolated a small amount of a stereoisomer, which when pure melts at 105° to 106°C.

(B) Preparation of m-Amino-α-Ethylhydrocinnamic Acid: A mixture of 50 g of α-ethyl-m-nitrocinnamic acid, 9.1 g of sodium hydroxide, 600 cc of water and 5 teaspoons of Raney nickel catalyst is shaken at 32°C in an atmosphere of hydrogen at an initial pressure of 450 psi until the calculated amount of hydrogen is absorbed. The filtered solution is acidified with hydrochloric acid, made basic with ammonium hydroxide and again acidified with acetic acid. Upon concentration of this solution, an oil separates which crystallizes upon standing, giving about 20 g, MP 60° to 68°C. Complete evaporation of the filtrate and extraction of the residue of inorganic salts with ether gives about 20 g of additional material, MP 54° to 59°C. Recrystallization of the combined product from benzene petroleum ether gives about 35 g of m-amino-α-ethylhydrocinnamic acid, MP 67° to 70°C.

(C) Preparation of β-(3-Amino-2,4,6-Triiodophenyll-α-Ethylpropionic Acid: A solution of 5.0 g of m-amino-α-ethylhydrocinnamic acid in 100 cc of water containing 5 cc of concentrated hydrochloric acid is added over a period of ½ hour to a stirred solution of 3.2 cc of iodine monochloride in 25 cc of water and 25 cc of concentrated hydrochloric acid heated to 60°C. After addition is complete, the heating is continued for one hour longer at 60° to 70°C. A black oil separates which gradually solidifies.

The mixture is then cooled and sodium bisulfite added to decolorize. Recrystallization of the product from methanol gives about 8 g, MP 147° to 150°C. The β-(3-amino-2,4,6-triiodophenyl)-α-ethylpropionic acid may be purified further by precipitation of the morpholine salt from ether solution and regeneration of the free amino acid by treatment of a methanol solution of the morpholine salt with sulfur dioxide. The pure amino acid has the MP 155° to 156.5°C.

References

Merck Index 4916
Kleeman and Engel p. 495
DOT 15 (7) 310 (1979)
I.N. p. 28
REM p. 1266
Archer, S.;US Patent 2,705,726; April 5, 1955; assigned to Sterling Drug Inc.

IOPHENDYLATE

Therapeutic Function: Diagnostic aid (radiopaque medium)

Chemical Name: Ethyl 10-(p-iodophenyl)undecylate

Common Name: -

Structural Formula:

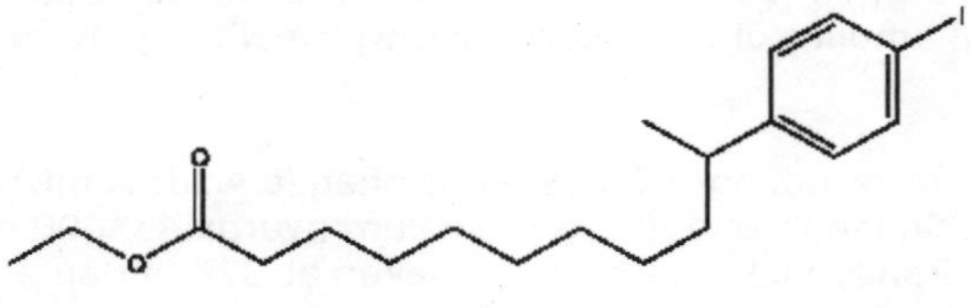

Chemical Abstracts Registry No.: 99-79-6

Trade Name	Manufacturer	Country	Year Introduced
Pantopaque	Lafayette	US	1944
Ethiodan	Allen and Hanburys	UK	-

Raw Materials

Ethyl undecylenate
Iodobenzene

Manufacturing Process

60 volumes of ethyl undecylenate is introduced gradually at 7° to 8°C during 35 minutes to a well-cooled mixture of 52.5 parts of aluminum chloride and 150 volumes of iodobenzene. The mixture is decomposed with cracked ice and dilute hydrochloric acid. The iodobenzene layer is washed with sodium bisulfite solution and with water, and then distilled. The composition of matter having the probable formula, ethyl 4-iodophenyl-undecylate, is a colorless liquid boiling at 196° to 198°C/1 mm, and of specific gravity of 1.26/20°C.

References

Merck Index 4917
Kleeman and Engel p. 494
REM p. 1267
Strain, W.H., Plati, J.T. and Warren, S.L.;US Patent 2,348,231; May 9, 1944; assigned to Noned Corporation and Eastman Kodak Company

IOPHENOIC ACID

Therapeutic Function: Diagnostic aid

Chemical Name: α-Ethyl-3-hydroxy-2,4,6-triiodobenzenepropanoic acid

Common Name: Iophenoxic acid; Teridax

Structural Formula:

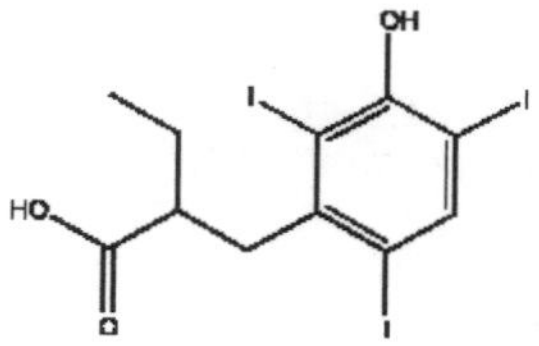

Chemical Abstracts Registry No.: 96-84-4

Trade Name	Manufacturer	Country	Year Introduced
Teridax	Schering	-	-

Raw Materials

Sodium propionate	m-Hydroxybenzaldehyde
Propionic anhydride	Hydrochloric acid
Potassium hydroxide	Acetic acid
Sodium amalgam	Iodine monochloride

Manufacturing Process

A mixture of m-hydroxy-benzaldehyde, fused sodium propionate and of butyric anhydride was stirred and refluxed at heating.

The mixture was then poured into water and acidified with hydrochloric acid. The organic material was extracted with chloroform, the chloroform was exaporated, and the residue stirred for one and 1.5 h with dilute potassium hydroxide solution. Acetic acid was added to make the solution almost neutral, but still slightly basic, the mixture was stirred with activated charcoal for about 15 min, filtered, and the filtrate acidified to Congo red with hydrochloric acid. A crystalline product was obtained upon cooling for several hours, and this was collected by filtration and recrystallized from water giving α-ethyl-m-hydroxycinnamic acid.

A solution of α-ethyl-m-hydroxycinnamic acid and potassium hydroxide in water was added to 3% sodium amalgam, and the mixture was stirred while heating on a steam bath for several hours. The mixture was then cooled, the mercury separated, and the reaction mixture was acidified and extracted with ether. The ether extracts were concentrated giving a residue containing α-ethyl-β-(m-hydroxyphenyl)propionic acid.

α-Ethyl-β-(m-hydroxyphenyl)propionic acid was dissolved in acetic acid. The solution was warmed on a steam bath, and water was added followed iodine monochloride. The mixture was stirred and heated for several hours, and water was then added to cause precipitation of the product. The semi-solid precipitate was triturated with a small amount of 95% alcohol, collected by filtration and washed with low boiling petroleum ether. Recrystallization from dilute alcohol, using charcoal for decolorization, gave α-ethyl-β-(2,4,6-triiodo-3-hydroxyphenyl)propionic acid.

References

Albany S.A.; US Patent No. 2,931,830; April 5, 1960; Assigned: Sterling Drug Inc., New York, N.Y, a corporation of Delaware

IOPRONIC ACID

Therapeutic Function: Diagnostic aid (radiopaque medium)

Chemical Name: 2-[[2-[3-(Acetylamino)-2,4,6-triiodophen-oxy]ethoxy] methyl]-butanoic acid

Common Name: -

Structural Formula:

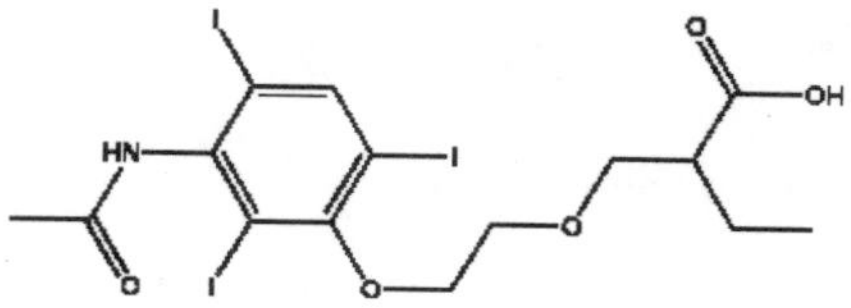

Chemical Abstracts Registry No.: 37723-78-7

Trade Name	Manufacturer	Country	Year Introduced
Bilimiru	Bracco	Italy	1974
Bilimiro	Byk Gulden	W. Germany	1980

Raw Materials

Sodium
Sodium hydroxide
3-(2-Iodoethoxy)-2-ethylpropionic acid ethyl ester
3-Acetylamino-2,4,6-triiodophenol
Hydrogen chloride

Manufacturing Process

A solution of 192 g 3-acetylamino-2,4,6-triiodophenol, sodium (0.35 mol) in 350 ml dimethylacetamide, was mixed with 107.5 g 3-(2-iodoethoxy)-2-ethylpropionic acid ethyl ester (0.35 mol) at 90°C with stirring over a period of about 20 to 30 minutes. Stirring was continued while the mixture was held at 95°C to 100°C for 16 hours. The solvent was then removed by distillation in a vacuum, and the residue was poured into 4,000 ml water. The solid precipitate formed was recovered and washed with water, dilute sodium carbonate solution, dilute sodium bisulfite solution, and again with much water. The ethyl ester was obtained in a yield of 220 g (90%). When recrystallized from 75% aqueous ethanol, it melted at 80°C to 86°C.

The ester (70 g, 0.1 mol) was saponified in a boiling mixture of 250 ml methanol and 250 ml water to which 100 ml N sodium hydroxide solution was added in small batches with stirring. The methanol was distilled from the saponification mixture, the residue was mixed with water and extracted with ethyl acetate. The aqueous phase was acidified with hydrochloric acid in the presence of sodium bisulfite.

The free acid gradually crystallized from the acidified solution in the amount of

42.4 g (63% yield). When recrystallized from 50% ethanol and from ethyl acetate, it melted at 130°C.

References

Merck Index 4919
I.N. p. 29
Felder, E. and Pitre, D.; US Patent 3,842,124; October 15, 1974;assigned to Bracco Industria Chimica, Societa per Azioni (Italy)

IOPYDOL

Therapeutic Function: Diagnostic aid

Chemical Name: 4(1H)-Pyridinone, 1-(2,3-dihydroxypropyl)-3,5-diiodo-

Common Name: Iopydol; Jopydolum

Structural Formula:

OH
OH
N
I
I
O

Chemical Abstracts Registry No.: 5579-92-0

Trade Name	Manufacturer	Country	Year Introduced
Hytrast Vial.	Guerbet	-	-
Hytrast	Byk Gulden	-	-

Raw Materials

4-Pyridone
Chloroiodide

Manufacturing Process

50 g 4-pyridone was dissolved in 300 ml 20% hydrochloric acid and slowly mixed with 180 g of chloroiodide in 200 ml 20% hydrochloric acid. Then the obtained dark solution was strong alkalized with sodium hydroxide. The solvent was removed to dryness and sodium salt diiod-4-pyridone was

obtained. It was dissolved in water, filtered and equivalent quantity of glacial acetic acid or hydrochloric acid was added to give 3,5-diiod-4-pyridone as a powder. 34.5 g 3,5-diiod-pyridone was dissolved in 100 ml 1 N NaOH by heating on a water bath and added to a 15 g of monochlorohydrine added. The mixture was heated on water bath before a dense pasty mass obtained. It was filtered off and treated with 1 N NaOH. The obtained 1-(2,3-dihydroxypropyl)-3,5-diiodo-4(1H)-pyridinone was recrystallized from diluted hydrochloric acid. MP: 161°C.

References

D.R. Patent No. 579,224; Dec. 30, 1930; I.G.Farbenindustrie Akt.-Ges. in Frankfurt a. M.

IOTHALAMATE SODIUM

Therapeutic Function: Diagnostic aid

Chemical Name: Benzoic acid, 3-(acetylamino)-2,4,6-triiodo-5-((methylamino)carbonyl)-, monosodium salt

Common Name: Iothalamate sodium; Jodtalaminatrium; Sodium isothalamate

Structural Formula:

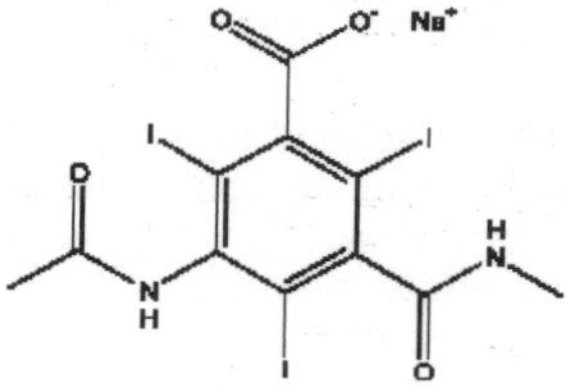

Chemical Abstracts Registry No.: 1225-20-3

Trade Name	Manufacturer	Country	Year Introduced
Angio-Conray	Mallinckrodt Inc.	USA	-
Iothalamate Sodium	Mallinckrodt Inc.	USA	-
Vascoray	Mallinckrodt Inc.	USA	-
Vascoray	Tyco Healthcare	Canada	-

Raw Materials

Methylamine
Palladium on charcoal
5-Nitroisophthalic acid, dimethyl ester
Iodine monochloride

Manufacturing Process

Normal aqueous sodium hydroxide (0.02 eq) was added at room temperature with rapid swirling to a solution of 5-nitroisophthalic acid, dimethyl ester, (4.8 g, 0.02 mole) in acetone-methanol (100 ml each). The clear solution immediately assumed a deep red-purple color which gradually lightened to a brown color over a 25-minute period. On standing overnight the solution lightened in color to a pale pink.

The solvent was evaporated, and the residue extracted with warm water (50 ml). The residue of unsaponified diester (0.23 g), 4.2%; m.p. 115°-117°C was filtered off, and the filtrate was acidified toprecipitate the crude monomethyl ester of 5-nitroisoftalic acid. Yield 3.4 g (75%). M.p. 170.5°-175.5°C.

The preparation was repeated on a larger scale with certain variations. Methanolic potassium hydroxide was substituted for the aqueous sodium hydroxide, and acetone was used as the solvent for the 5-nitroisophthalic acid, dimethyl ester. Yield, 78%. M.P. 175°-179°C (corrected).

Crude 5-nitroisophthalic acid, monomethyl ester (46.3 g, 0.21 mole) was dissolved in 35% aqueous methylamine solution (500 ml). On standing, the orange solution became blood red. The reaction mixture was evaporated overnight on the steam bath, the cool residue was treated with 50 ml of water and the solution was acidified with hydrochloric acid. A yellow precipitate of crude N-methyl-5-nitroisophthalamic acid was separated and dried (yield 41.5 g). This acid was redissolved in dilute ammonia solution and the resulting solution (pH 5.2) was treated with charcoal. Acidification of the treated solution yield a pale yellow product of neutral equivalent 213. A small portion (10 g) was recrystaliized from 1:1 water-ethanol (300 ml) to yield orange N-methyl-5-nitroisophthalamic acid. M.p. 251°-252.5°C.

Crude N-methyl-5-nitrosophthalamic acid (11.2 g, 0.05 mole) was reduced with hydrogen in a low pressure hydrogenator. The solvent was anhydrous methanol (250 ml) and the catalyst was 5% palladium on charcoal slurried in 10 ml of water. After the theoretical quantity of hydrogen for reduction of the nitro group had been absorbed the solution was filtered to remove the catalyst and the solvent was evaporated under reduced pressure, leaving a white residue of crude 5-amino-N-methylisophthaIamic acid. M.p. 227°-230°C (corrected).

The crude 5-amino-N-methylisophthalamic acid was dissolved in hydrochloric acid (100 ml concentrated acid and 100 ml of water) and this solution was diluted to 1 liter with water. Iodine monochloride (27.4 g of 95% ICI, 0.16 mole) in concentrated hydrochloric acid (30 ml) was added in one portion to the stirred solution maintained at 54°C. The solution was heated on a steam bath. After 2 hours the solution was diluted to 1.5 liters and after 3 hours titration of an aliquot indicated that 50% of the iodine monochloride had been consumed. Precipitation of a solid began after 33/4 hours of reaction (75°C). Intermittent heating and stirring was continued for 4 days, 10 g of 95% iodine monochloride was added during the third day. After 4 days, titration of an aliquot indicated that 96% of the theoretical quantity of iodine monochloride had been consumed. The precipitated solid was filtered off, washed with water and dried at 75°C under reduced pressure. Yield of 5-amino-2,4,6-triiodo-N-methylisophthalamic acid 20.6 g. M.p. 266-268°C (dec.).

1.95 molar $KICl_2$ a solution (1144 ml, 2.22 moles) was added during 0.5 hour to a stirred suspension of 5-amino-N-methylisophthalamic acid (196 g, 1.01 moles) in 2.5 liters of water. After three hours of additional stirring, a solution of sodium hydroxide (88 g, 2.2 moles of NaOH in 200 ml of water) was added. Then, additional 1.95 molar $KICl_2$ solution (522 ml, 1.01 mole) was added during 0.5 hour. The reaction mixture was stirred overnight after which the crude product was collected and purified by conversion first to the ammonium salt, then to the free acid. Yield of 5-amino-2,4,6-triiodo-N-methylisophthalamic acid, 310 g (53.6%).

Acetyl chloride (17 ml, 0.24 mole) was added in portions during 10 minutes to a stirred slurry of 5-amino-2,4,6-triiodo-N-methylisophthalamic acid (57.2 g, 0.1 mole) in dimethylacetamide (120 ml). Solution occurred in 0.5-1 hour and after a total of 1.5 hours 20 ml of water was added and the reaction mixture was evaporated to a thick slurry. The product was purified by twice dissolving it as a sodium salt and precipitating the free acid by the addition of mineral acid. The resulting nearly colorless 5-acetamido-2,4,6-triiodo-N-methylisophthalamic acid decomposed at about 285°C but did not melt below 300°C. Yield, 47 g (76.5%).

5-Acetamido-2,4,6-triiodo-N-methylisophthalamic acid was slurried in water and dissolved by the addition of an equivalent quantity of sodium hydroxide. The solution was evaporated to dryness to yield the sodium salt of 5-acetamido-2,4,6-triiodo-N-methylisophthalamic acid. Its solubility in water at 25°C is approximately 85 g per 100 ml of solution. The acute intravenous LD_{50} of this salt in male albino mice is approximately 19.2 g/kg.

References

Hoey G.B. et al.; US Patent No. 3,145,197; Aug. 18, 1964

IOTHALMATE MEGLUMINE

Therapeutic Function: Diagnostic aid (radiopaque medium)

Chemical Name: 3-(Acetylamino)-2,4,6-triiodo-5-[(methylamino)carbonyl]-benzoic acid

Common Name: -

Chemical Abstracts Registry No.: 13087-53-1; 2276-90-6 (Acid)

Raw Materials

5-Amino-2,4,6-triiodo-N-methylisophthalamic acid
Acetic anhydride
N-Methyl glucamine

Structural Formula:

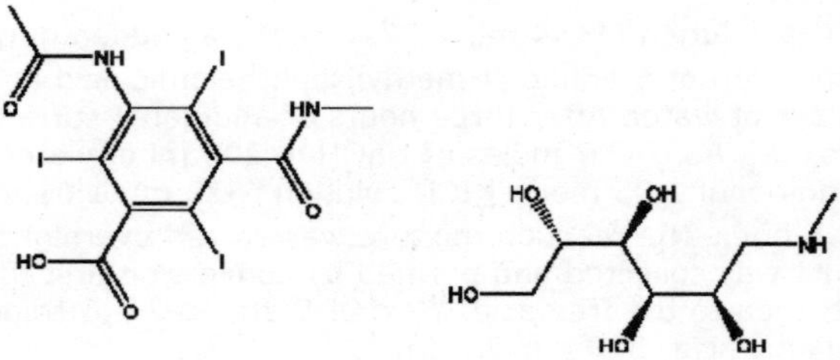

Trade Name	**Manufacturer**	**Country**	**Year Introduced**
Conray	Mallinckrodt Inc.	US	1962
Conray	Byk Gulden	W. Germany	1964
Contrix	Guerbet	France	1965
Angio-Conray	Daiichi	Japan	-
Cysto-Conray	Mallinckrodt Inc.	US	-
Gastro-Conray	May and Baker	UK	-
Sombril	Rovi	Spain	-
Vascoray	Mallinckrodt Inc.	US	-
Vascoray	Astra	Sweden	-

Manufacturing Process

Crude 5-amino-2,4,6-triiodo-N-methylisophthalamic acid (21.0 g) was dissolved in warm dimethylacetamide (40 ml) and acetic anhydride (30 ml) and concentrated sulfuric acid (2 drops) were added. This solution was heated on the steam bath for 2 hours, then heated at 110°C for 5 minutes, then cooled. Water and ammonium hydroxide were added to destroy the excess acetic anhydride, after which the mixture was evaporated to a volume of 50 ml. The cooled solution was acidified with concentrated hydrochloric acid and a tan solid was collected. The crude product was dissolved in 100 ml of water containing a slight excess of sodium hydroxide. The pH was adjusted to 4.5 with acetic acid, and the solution was treated with charcoal. The colorless solution was acidified with concentrated hydrochloric acid and cooled, and the precipitate was filtered off and dried under reduced pressure. The resulting 5-acetamido-2,4,6-triiodo-N-methylisophthalamic acid decomposes about 285°C and does not melt below 300°C.

5-acetamido-2,4,6-triiodo-N-methylisophthalamic acid was slurried in water and dissolved by the addition of an equivalent quantity of N-methylglucamine. The solution was evaporated to dryness to yield the meglumate salt of 5-acetamido-2,4,6-triiodo-N-methylisophthalamic acid.

References

Merck Index 4922
Kleeman and Engel p. 496
I.N. p. 29
REM p. 1269

Hoey, G.B.; US Patent 3,145,197; August 18,1964; assigned to Mallinckrodt Chemical Works

IOTHIOURACIL

Therapeutic Function: Thyroid inhibitor

Chemical Name: 2,3-Dihydro-5-iodo-2-thioxo-4(1H)-pyrimidinone

Common Name: Iodothiouracil

Structural Formula:

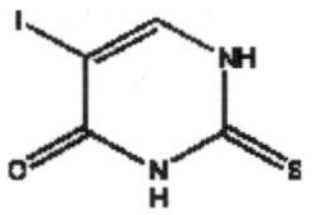

Chemical Abstracts Registry No.: 5984-97-4

Trade Name	Manufacturer	Country	Year Introduced
Itrumil	Ciba	US	1951

Raw Materials

5-Iodo-2-benzyl thiouracil
Acetic anhydride

Manufacturing Process

As an illustrative example 64.4 g of 5-iodo-2-benzyl thiouracil were deposited in the reaction vessel and dissolved by adding 400 cc of glacial acetic acid containing 10 cc of acetic anhydride and the reaction vessel was connected tightly with the reflux condenser. The second vessel or generator was charged with 95 cc of acetic anhydride and the vessel connected to a vessel such as a dropping funnel or equivalent containing 75 cc of a 50% solution of hydroiodic acid which was added slowly, as by dropwise addition, to the acetic anhydride in the generator. The mixture in the generator soon became hot and the hydrogen iodide which evolved passed continuously through the connecting conduit into the reaction flask just above the level of liquid therein. As the hydrogen iodide contacted the solution of the 2 benzyl derivative,a ring of the debenzylated product formed under the inlet conduct. This operation was continued until all of the hydroidic acid was added to the generator vessel. The hydrogen iodide remaining in the generator was driven over into the reaction vessel by heating the generator. It was ascertained that the reaction is complete when no more precipitate forms in the main reaction vessel. During the reaction vapors evolved were condensed in the condenser and returned to the reaction vessel as reflux. The upper end of the reflux is preferably connected with a vent leading to a drying chamber.

The reaction vessel was cooled and the precipitate separated by pouring or decanting off the supernatant liquor. The precipitate of the 5-iodo-2-thiouracil was then thoroughly washed, as, for example, on a Buchner funnel. The precipitate was then extracted twice with hot glacial acetic acid to remove unreacted material and then washed thoroughly by alternate washes with alcohol and water. The product was then further purified by dissolving it in warm dilute sodium hydroxide and after cooling was reprecipitated by careful acidulation with acetic acid. Utilizing this procedure 37 g of purified 5-iodo-2-thiouracil were obtained.

The supernatant liquid separated from the precipitate was concentrated in vacuo and 7.4 g of the unreacted 5-iodo-2-benzyl thiouracil were recovered. This obviously may be utilized for further debenzylation.

As pointed out previously, the 5-iodo-2-thiouracil is carefully dried, preferably in a vacuum over P_2O_5.

References

Merck Index 4924
OCDS Vol. 1 p. 265 (1977)
I.N. p. 573
Barrett, H.W.; US Patent 2,585,615; February 12, 1952; assigned to The Chemical Foundation

IOTROXIC ACID

Therapeutic Function: Diagnostic aid (radiopaque medium)

Chemical Name: 3,3'-[Oxybis(ethyleneoxymethylenecarbonylimino)]bis-[2,4,6-triiodo-benzoic acid]

Common Name: -

Structural Formula:

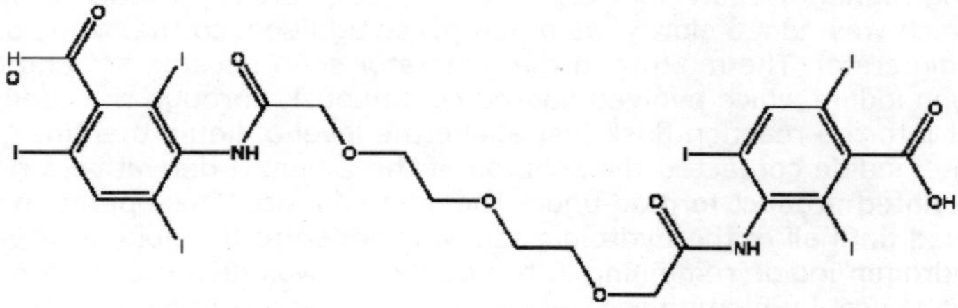

Chemical Abstracts Registry No.: 51022-74-3

Trade Name	Manufacturer	Country	Year Introduced
Biliscopin	Schering	W. Germany	1978
Biliscopin	Schering	Switz.	1981
Biliscopin	Nippon Schering	Japan	1982
Chologram	Schering	Italy	1982

Raw Materials

3-Amino-2,4,6-triiodobenzoic acid
3,6,9-Trioxaundecane diacid dichloride

Manufacturing Process

(a) Condensation in dimethylacetamide: To a suspension of 51.5 g of anhydrous 3-amino-2,4,6-triiodo-benzoic acid (0.1 mol) in 100 ml of dimethylacetamide were slowly added dropwise, while stirring, 15.5 g of 3,6,9-trioxaundecanediacid dichloride (0.06 mol), during which the temperature gradually rose to about 50°C and the whole passed into solution. After being stirred overnight, the solution was added dropwise to 1 liter of a 0.28 N solution of sodium hydroxide, and then 200 ml of 2 N hydrochloric acid were cautiously added. The precipitate was filtered off with suction, washed with water and dried. The yield was practically quantitative.

(b) Condensation in dioxan: 15.5 g of 3,6,9-trioxaundecane diacid dichloride were added dropwise at about 95°C to a solution of 51.5 g of anhydrous 3-amino-2,4,6-triiodo-benzoic acid in 52 ml of anhydrous dioxan. After further stirring and heating for 3 hours, the solution was cooled, stirred dropwise into 500 ml of a 0.4 N solution of sodium hydroxide, and further worked up as described in paragraph (a). The yield was practically quantitative.

(c) Purification: To the crude product obtained as described under paragraph (a) or (b) in 300 ml of methanol was slowly added a quantity (about 15 ml) of a 12 N solution of sodium hydroxide such that a test portion diluted with water had a pH-value of 8 to 9. After stirring the mixture overnight, the sodium salt of 3,6,9-trioxaundecane-1,11-dioyl-bis-(3-carboxy-2,4,6-triodo-anilide) which crystallized out was filtered off with suction, was hed with methanol and dried. Yield: 92 g (90% of the theoretical yield).

A solution of the salt in 900 ml of water was treated with active carbon, and concentrated hydrochloric acid was added until the pH-value was 1. The precipitate was filtered off with suction, washed with water, and dried at 50°C.

The yield of pure 3,6,9-trioxaundecane-1,11-dioyl-bis-(3-carboxy-2,4,6-triiodo-anilide) was 80 g (80% of the theoretical yield). The substance melted at 175°C with sintering.

References

Kleeman and Engel p. 497
DOT 15 (1) 48 (1979)
I.N. p. 30
Schering, A.G.; British Patent 1,501,507; February 15, 1978

IOXITALAMIC ACID

Therapeutic Function: Diagnostic aid

Chemical Name: 3-(Acetylamino)-5-(((2-hydroxyethyl)amino)carbonyl)-2,4,6-triiodobenzoic acid

Common Name: Acide ioxitalamique; Acidum ioxitalamicum; Acidum joxitalamicum; Ioxitalamic acid

Structural Formula:

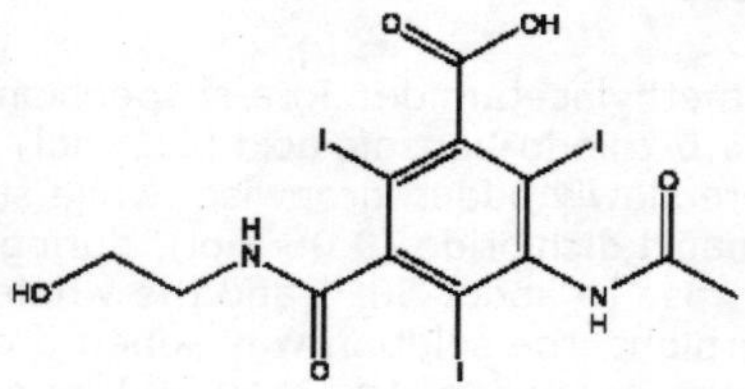

Chemical Abstracts Registry No.: 28179-44-4

Trade Name	Manufacturer	Country	Year Introduced
Oxilan	Cook Imaging Corporation	-	-
Telebrix	Laboratory Guerbet	-	-

Raw Materials

Acetic acid	3-Methoxycarboxyl-5-nitrobenzoic acid
Hydrochloric acid	Ammonium chloride
$NaICl_2$	Palladium oxide on charcoal
Thionyl chloride	Sodium bicarbonate
Sodium hydroxide	Acetic anhydride
Sulfuric acid	Ethanolamine
Triethylamine	Ammonia

Manufacturing Process

3-Methoxycarboxyl-5-nitrobenzoic acid (25 g) was hydrogenated in methanol (500 ml) using palladium oxide on charcoal (2.5 g 10%) at atmospheric pressure. When the exothermic reaction was completed the catalyst was fluttered off. After cooling the solution at -20°C for 2.5 h, 12.7 g of 3-amino-5-methoxycarbonylbenzoic acid was isolated. An additional 6.5 g of it was isolated by concentrating the mother liquor.

The 3-amino-5-methoxycarbonylbenzoic acid (12.0 g) was suspended in water (280 ml), dissolved by addition of concentrated hydrochloric acid (7.1 ml) and glacial acetic acid (28.5 ml). At 60°-70°C $NaICl_2$ solution (73 ml, 58.7 g ICl/100 ml) was added dropwise while stirring in the course of about 3 h. The reaction mixture was heated at 80°-90°C for additional 3 h while stirring.

After cooling to room temperature the mother liquor was decanted and the residue dissolved as ammonium salt in water (80 ml). The ammonium salt was precipitated by adding ammonium chloride (2.4 g) and cooling to 0°C. The ammonium salt was filtered off and dissolved in water (140 ml), charcoaled twice at 80°C and the acid was precipitated at room temperature by addition of hydrochloric acid and was filtered off. The crude product was dissolved in ethyl acetate (100 ml) and the solution was washed 3 times with hydrochloric acid (2 N). By evaporating the solvent, 19 g of 3-amino-5-methoxycarbonyl-2,4,6-triiodobenzoic acid was isolated. Melting point 170°-176°C.

A mixture of 3-amino-5-methoxycarbonyl-2,4,6-triiodobenzoic acid (198 g) and thionyl chloride (400 ml) was heated while stirring at 70°C for 16 h. The solid material dissolved slowly. Thionyl chloride was evaporated in vacuo, the residue dissolved in chloroform (1000 ml), the solution washed with water (80 ml each), twice with saturated sodium bicarbonate, then 5 times with 2 N sodium hydroxide solution and finally with water to neutral. The solution was dried with $CaCl_2$ filtered and evaporated to dryness. The 3-amino-5-methoxycarbonyl-2,4,6-triiodobenzoyl chloride was dried at 50°C in vacuo. Yield: 203.0 g. Melting point 55°-60°C.

To the 3-amino-5-methoxycarbonyl-2,4,6-triiodobenzoyl chloride (53.0 g) was added acetic anhydride (106 ml). After stirring at room temperature for 20 min then insoluble material was filtered off (3-4 g). To the filtrate was added concentrated sulfuric acid (0.3 ml) whereby a yellowish product started to precipitate. The temperature reached about 50°C. The 3-acetamido-5-methoxycarbonyl-2,4,6-triiodobenzoyl chloride was isolated after storing in refrigerator overnight. Yield: 39.0 g. Melting point 210°-215°C.

The 3-acetamido-5-methoxycarbonyl-2,4,6-triiodobenzoyl chloride was dissolved in a mixture of dioxan and dimethylformamide. In the course of 2 h this solution was added dropwise to a solution of ethanolamine and triethylamine in dioxan. The stirring was continued. A sticky precipitate was filtered off. The filtrate was evaporated to dryness in vacuo. The residue was triturated with aqueous sodium bicarbonate, filtered off and mixed with first fraction. The combined solids were then suspended in aqueous sodium bicarbonate filtered off washed with water and dried in vacuo to give methyl 5-acetamido-2,4,6-triiodo-(N-β-hydroxyethyl)-isophthalamate.

The methyl 5-acetamido-2,4,6-triiodo-(N-β-hydroxyethyl)-isophthalamate was mixed with fresh distilled ethanolamine and stirred. The excess ethanolamine was removed in vacuo at 50°-60°C. The residue was dissolved in water, and charcoaled at pH 5.5. The crude product was precipitated with hydrochloric acid (pH 0.5) and filtered after stirring at 0°C. 5-Acetamido-2,4,6-triiodo-(N-β-hydroxyethyl)isophthalamic acid was suspended in ethanol and dissolved by addition of concentrated ammonia. The ammonium salt started to precipitate in the course and was isolated after stirring. The salt was dissolved in water, filtered and the acid was precipitated with hydrochloric acid (pH 0.5). After stirring the product was filtered off and dried in vacuo.

References

Savesen S. et al.; US Patent No. 3,702,866; Nov. 14, 1972; Assigned: Nyegaard and Co.A/S, Nycoveien, Norway

IPRATROPIUM BROMIDE

Therapeutic Function: Bronchodilator

Chemical Name: 3-(3-Hydroxy-1-oxo-2-phenylpropoxy)-8-methyl-8-(1-methylethyl)-8-azoniabicyclo[3.2.1]octane bromide

Common Name: -

Structural Formula:

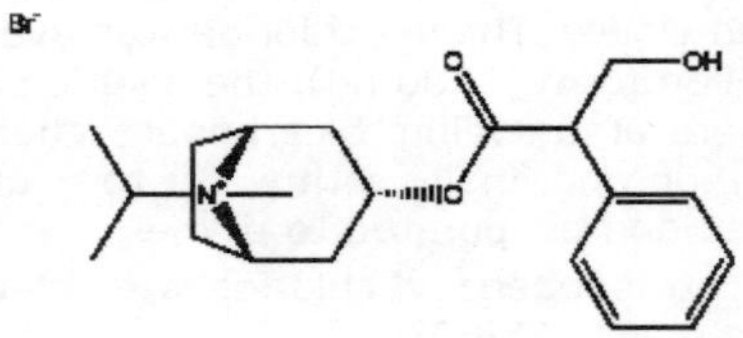

Chemical Abstracts Registry No.: 22254-24-6

Trade Name	Manufacturer	Country	Year Introduced
Atrovent	Boehringer Ingelheim	W. Germany	1975
Atrovent	Boehringer Ingelheim	UK	1977
Atrovent	De Angeli	Italy	1980
Breva	Valeas	Italy	1980
Atrovent	Teijin	Japan	1981
Atrovent	Boehringer Ingelheim	Canada	1982
Atem	Chiesi	Italy	-
Itrop	Boehringer Ingelheim	-	-
Vagos	Valeas	Italy	-

Raw Materials

N-Isopropyl-noratropine
Methyl bromide

Manufacturing Process

21 1.5 g (0.667 mol) of N-isopropyl-noratropine were dissolved at 60°C in 2.11 liters of absolute toluene in a 3-liter glass pressure tube. While the solution was still warm, 95 g (1 mol) of ice-cold methylbromide were added, and the pressure tube was sealed immediately thereafter. The reaction mixture was kept at 60°C for four days. After one hour of standing, the formation of crystals began. At the end of four days the crystals were separated by vacuum filtration at 60°C, washed with 600 cc of toluene at 60°C, and dried in vacuo in a drying cabinet at 100°C. Raw yield: 263.7 g (95.8% of theory). MP: 224°C to 225°C (decomp.). The raw product was refluxed with 2.5 liters of chloroform for 30 minutes, vacuum filtered while

hot, washed with 200 cc of chloroform, and dried in a vacuum drying cabinet at 100°C. Yield: 249 g (90.6% of theory). MP: 226°C to 228°C (decomp.). The purified product was recrystallized from 1.2 liters of n-propanol, washed with 200cc of n-propanol and dried in a vacuum drying cabinet at 100°C. Yield: 237 g (86.15% of theory). MP: 230°C to 232°C (decomp.). By evaporation of the mother liquor to 100 cc another 6.0 g of the pure product, MP 230°C to 231.5°C (decomp.), were obtained.

References

Merck Index 4929
Kleeman and Engel p. 498
OCDS Vol.3 p.160 (1984)
DOT 11 (12) 461 (1975) and 17 (7) 299 (1981)
I.N. p. 525
REM p.916 Zeile, K., Schulz, W., Banholzer, R. and Wick, H.; US Patent 3,505,337; April 7, 1970; assigned to Boehringer Ingelheim G.m.b.H. (W. Germany)

IPRONIAZID

Therapeutic Function: Antidepressant, Monoamine oxidase inhibitor

Chemical Name: 4-Pyridinecarboxylic acid 2-(1-methylethyl)hydrazide

Common Name: -

Structural Formula:

Chemical Abstracts Registry No.: 54-92-2

Trade Name	Manufacturer	Country	Year Introduced
Marsilid	Roche	US	1952
Marsilid	Roche	France	1960
Ellepibina	L.P.B.	Italy	-
Ipronid	A.F.I.	Norway	-
Rivivol	Zambeletti	Italy	-

Raw Materials

Isonicotinyl hydrazide
Acetone
Hydrogen

Manufacturing Process

A mixture of 40 g of isonicotinyl hydrazine and 600 cc of acetone was heated on a steam bath until solution was complete. Upon cooling the reaction mixture, 1-isonicotinyl-2-isopropylidene hydrazine precipitated in the form of white needles; MP 161°C to 161.5°C.

A solution of 20 g of 1-isonicotinyl-2-isopropylidene hydrazine in 150 cc of methanol was reduced with hydrogen at room temperature and 50 psi using 300 mg of platinum black as a catalyst.

References

Merck Index 4934
Kleeman and Engel p. 499
OCDS Vol. 1 p. 254 (1977)
I.N. p. 525
Fox, H.H.; US Patent 2,685,585; August 3, 1954; assigned to Hoffmann-La Roche, Inc.

IPRONIDAZOLE

Therapeutic Function: Antiprotozoal

Chemical Name: 1-Methyl-2-(1-methylethyl)-5-nitro-1H-imidazole

Common Name: 2-Isopropyl-1-methyl-5-nitroimidazole

Structural Formula:

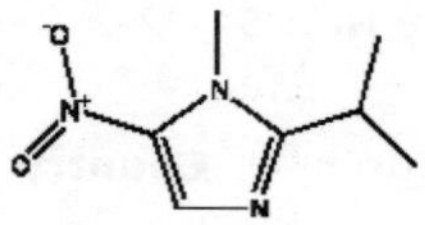

Chemical Abstracts Registry No.: 14885-29-1

Trade Name	Manufacturer	Country	Year Introduced
Ipropran	Roche	W. Germany	1981

Raw Materials

2-Isopropyl-4-nitroimidazole
Dimethyl sulfate

Manufacturing Process

2-Isopropyl-4 (or 5-nitroimidazole) (31 g = 0.2 mol), dioxane (70 g) and dimethylsulfate (28 g = 0.22 mol) were heated on a steam bath under reflux for 45 minutes. The solvent was removed in vacuo on a steam bath, the residue dissolved in 20 ml of water and the product precipitated by the gradual addition of 80 g of 25% sodium hydroxide solution at 0°C. A small additional amount was obtained by extraction of the mother liquor with methylene chloride. The product melted at 60°C.

The product was purified as follows. 60 g of product was dissolved in 3N aqueous hydrochloric acid, the solution was treated with charcoal and filtered. The filtrate was neutralized by the gradual addition of aqueous concentrated ammonia at 0°C to 5°C under stirring whereupon the product precipitated in white plates as the neutralization proceeded. The precipitate was filtered by suction, washed on the filter with 50 ml of ice cold water and dried at room temperature, MP 60°C.

The hydrochloride salt was formed by reacting the product, dissolved in isopropanol, with 25% ethanolic hydrochloric acid, whereupon the salt precipitated and was isolated. It has a melting point of 177°C to 182°C (dec). Similarly, the bisulfate salt was formed using 96% sulfuric acid. It has a MP of 151.5°C to 152.5°C.

References

Merck Index 4934
OCDS Vol. 2 p. 244 (1980)
I.N. p. 525
Hoffer, M. and Mitrovic, M.; US Patent 3,502,776; March 24, 1970; assigned to Hoffmann-La Roche Inc.

IRBESARTAN

Therapeutic Function: Antihypertensive

Chemical Name: 1,3-Diazaspiro[4.4]non-1-en-4-one, 2-butyl-3-((2'-(1H-tetrazol-5-yl)(1,1'-biphenyl)-4-yl)methyl)-

Common Name: Irbesartan

Chemical Abstracts Registry No.: 138402-11-6

Structural Formula:

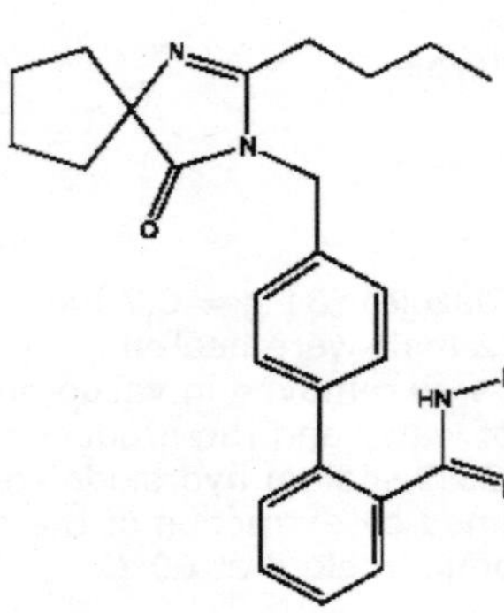

Trade Name	Manufacturer	Country	Year Introduced
Aprovel	Sanofi-Winthrop Industrie	France	-
Irbesartan	Bristol-Myers Squibb	-	-
Irbest	Biochem Pharma Industries	-	-
Irovel-H	Sun Pharmaceuticals Industries Ltd.	India	-
Irovel	Sun Pharmaceuticals Industries Ltd.	India	-
Xarb	Nicholas Piramal India Ltd. (Npil)	India	-
Xarb-H	Nicholas Piramal India Ltd. (Npil)	India	-

Raw Materials

Ethyl valerimidate	1-Aminocyclopentanecarboxylic acid
Sodium cyanide	Cyclopentanone
Triethylamine	Oxalic acid dihydrate
Valeryl chloride	4-Bromomethyl-2'-cyanobiphenyl
Tributyltin azide	Trityl chloride

Manufacturing Process

1. Synthesis of 2-n-Butyl-4-spirocyclopentane-2-imidazolin-5-one

Method 1:

The ethyl ester of 1-aminocyclopentanecarboxylic acid is prepared according to Adkins and Billica (J. Amer. Chem. Soc., 1948, 70, 3121).

Ethyl valerimidate hydrochloride is prepared according to Mac Elvain (J. Amer. Chem. Soc., 1942, 64, 1825-1827) and then freed from its hydrochloride by reaction with potassium carbonate and extraction with CH_2Cl_2.

The ethyl ester of 1-aminocyclopentanecarboxylic acid (1.57 g) and ethyl valerimidate (1.56 g) are dissolved in 12 ml of xylene containing 6 drops of acetic acid. After refluxing for 6.5 h, the reaction medium is concentrated under vacuum, the residue is chromatographed on silica gel using a chloroform/methanol/acetic acid mixture (94/4/2; v/v/v) as the eluent. The fraction containing the expected product is evaporated several times in the presence of xylene and then benzene in order to remove the acetic acid. 1.91 g of 2-n-butyl-4-spirocyclopentane-2-imidazolin-5-one are obtained in the form of a thick oil.

Method 2:

1.97 g of sodium cyanide are dissolved in 3.9 ml of water in a round-bottomed flask and a solution containing 2.33 g of ammonium chloride in 5.9 ml of water and 3.5 ml of 20% aqueous ammonia is added; finally, 3 g of cyclopentanone in 3.8 ml of methanol are added to the flask. After stirring for 1.5 h, the mixture is heated at 60°C for 45 min, heating is then stopped, stirring is continued for 45 min and the mixture is then cooled to 25°C. It is extracted several times with methylene chloride.

The 1-aminocyclopentanenitrile obtained is dissolved in 300 ml of acetone, and a solution of 2.25 g of oxalic acid dihydrate in 200 ml of acetone is added, with stirring. The precipitate of 1-aminocyclopentanenitrile formed is filtered off.

5.1 g of the oxalate obtained in the previous step are treated with 7.65 ml of concentrated sulfuric acid (d = 1.84) over 45 min, with stirring. The evolution of a gas is observed and the temperature rises to 100°C. The mixture is cooled to about 35°C and poured into a mixture of ice and concentrated aqueous ammonia (10 g/2.8 ml). The suspension formed is extracted with chloroform containing 5% of methanol. The 1-aminocyclopentanecarboxamide was obtained.

3 g of the compound prepared in the previous step are placed in 70 ml of anhydrous THF and 3.3 ml of triethylamine, and 3 ml of valeryl chloride in 10 ml of anhydrous THF are added, with stirring. A white suspension is formed. The intermediate which is formed, but not isolated, is 1-(N-valeryl)aminocyclopentanecarboxamide. 6 g of potassium hydroxide pellets, 7 ml of water and 16 ml of methanol are added. The mixture is refluxed for 2.5 h and 9 g of ammonium chloride are then added. After stirring for 15 min, the mixture is concentrated under vacuum. The residue of the 2-n-butyl-4-spirocyclopentane-2-imidazolin-5-one obtained is taken up in water and extracted with ethyl acetate.

2. Synthesis of 2-n-butyl-4-spirocyclopentane-1-[(2'-(tetrazol-5-yl)biphenyl-4-yl)-methyl]-2-imidazolin-5-one

A mixture containing 250 mg of sodium hydride (as an 80% dispersion in mineral oil) and 5 ml of DMF is prepared under a nitrogen atmosphere and a solution containing 0.97 g of 2-n-butyl-4-spirocyclopentane-2-imidazolin-5-one in 10 ml of DMF is added dropwise. The mixture is stirred for 30 min at 20°C and a solution of 1.5 g of 4-bromomethyl-2'-cyanobiphenyl in 10 ml of DMF is then added. After stirring for 1 h at 20°C, the DMF is evaporated off under reduced pressure, the residue is then taken up with ethyl acetate, filtered and evaporated. The residue of 1-[(2'-cyanobiphenyl-4-yl)methyl]-2-n-butyl-4-spirocyclopentane-2-imidazolin-5-one is purefied by chromatography.

1.56 g of the previous product, 2.6 g of tributyltin azide and 30 ml of xylene are refluxed for 66 h. The xylene is then evaporated off and the residue is dissolved in 20 ml of CH_2Cl_2 and 5 ml of THF with the addition of 0.8 ml of 10 N sodium hydroxide solution and, after stirring for 30 min, 2.5 g of trityl chloride, and the mixture is stirred for 26 h. After evaporation of the solvents, the residue is taken up in ethyl acetate in ethyl acetate, washed with water and then with a 3% solution of potassium bisulfate and water. It is dried and evaporated. The residue is chromatographed on alumina using a hexane/ethyl acetate mixture (9/1; v/v) as the eluent to give 1.97 g of the 2-n-butyl-4-spirocyclopentane-1-[(2'-(triphenylmethyltetrazol-5-yl)biphenyl-4-yl)methyl]-2-imidazolin-5-one. Melting point 150-152°C.

1.96 g of the product prepared in the previous step are dissolved in 10 ml of methanol and 10 ml of THF. After the reaction medium has been cooled to 5°C, 1.5 ml of 4 N hydrochloric acid are added and the mixture is stirred for 3 h at 20°C and 1 h at 30°C. After evaporation of the solvents, the residue is taken up in water and the pH is brought to 12 by the addition of 10 N sodium hydroxide solution. The aqueous phase is extracted with ether, toluene and ether again. The aqueous phase is acidified to pH 2 by the addition of 1 N hydrochloric acid and then extracted with ethyl acetate and the extract is evaporated. The aqueous phase is acidified to pH 2 by the addition of 1 N hydrochloric acid and then extracted with ethyl acetate and the extract is dried and evaporated. The white solid obtained is dried at 50°C under 0.05 mm of mercury to give 840 mg of the 2-n-butyl-4-spirocyclopentane-1-[(2'-(tetrazol-5-yl)biphenyl-4-yl)methyl]-2-imidazolin-5-one. Melting point 180-181°C.

References

Bernhart C. et al.; US Patent No. 5,270,317; Dec. 14, 1993; Assigned to Elf Sanofi, Paris, France

IRINOTECAN HYDROCHLORIDE

Therapeutic Function: Antineoplastic

Chemical Name: (1,4'-Bipiperidine)-1'-carboxylic acid, (4S)-3,4,12,14-tetrahydro-4,11-diethyl-4-hydroxy-3,4-dioxo-1H-pyrano(3',4':6,7) indolizino[1,2-b]quinolin-9-yl ester, monohydrochloride

Common Name: Camptetin hydrochloride; Irinotecan hydrochloride

Chemical Abstracts Registry No.: 100286-90-6

Structural Formula:

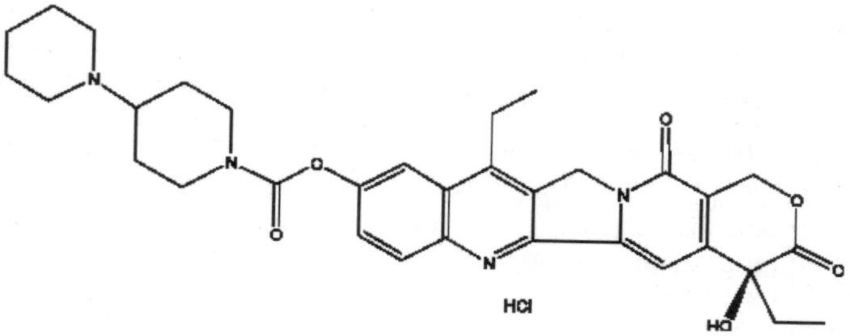

HCl

Trade Name	Manufacturer	Country	Year Introduced
Campto	Rhone-Poulenc Rorer	France	-
Camptosar	Pharmacia Canada Inc.	Canada	-
Irinotecan hydrochloride	Ohua Pharmaceutical Technology Co., Ltd.	China	-
Irinotel Inj.	Dabur Pharmaceuticals Ltd.	India	-

Raw Materials

Triethylamine
Phosgene dimer
4-Piperidinopiperidine
7-Ethyl-10-hydroxycamptothecin
1-Chlorocarbonyl-4-piperidinopiperidine

Manufacturing Process

7-Ethyl-10-[4-(1-piperidino)-1-piperidino]carbonyloxycamptothecin was synthesized by 2 methods.

Method 1.

7-Ethyl-10-hydroxycamptothecin (500 mg, 1.27 mmol) was suspended in dry dioxane (400 ml) and dissolved therein by adding triethylamine (2 ml) to the suspension under warming. This solution was stirred at room temperature while introducing thereinto phosgene prepared toties quoties by decomposing phosgene dimer (trichloromethoxychloroformate, 400 ml) in the presence of an active carbon catalyst. After 0.5 hours, consumption of the starting materials was confirmed and insoluble 10-chlorocarbonyloxy-7-ethylcamptothecin was removed by filtration.

10-Chlorocarbonyloxy-7-ethylcamptothecin (300 mg, 0.66 mmol) is suspended in dry dioxane (50 ml). To this suspension is added 4-piperidinopiperidine (330 mg, 1.96 mmol) as the amine, the reaction followed by the after-treatment was carried out whereby the 7-ethyl-10-[4-(1-piperidino)-1-piperidino]carbonyloxycamptothecin title compound (154 mg, 39.8%) was obtained.

Method 2.

7-Ethyl-10-hydroxycamptothecin (790 mg, 2.01 mmol) and 1-chlorocarbonyl-4-piperidinopiperidine (910 mg, 3.95 mmol) were dissolved in anhydrous pyridine (50 ml), and the mixture was stirred for 1 hour at 20°C. The reaction mixture was evaporated to dryness in vacuo, the residue was dissolved in $CHCl_3$ (200 ml). The solution was washed successively with a 7% aqueous solution of $NaHCO_3$ (200 ml), a saturated aqueous solution NaCl, and the $CHCl_3$ layer was filtered, and evaporated in vacuo. The residual material was decolorized by passing it through a short silica gel column. 7-Ethyl-10-[4-(1-piperidino)-1-piperidino]carbonyloxycamptothecin was obtained as a pale yellow mass, which was recrystallized from ethanol (ca. 60 ml) to give colorless needles (750 mg, 63.5% in yield).

To an ice-cooled suspension in distilled water (15 ml) of 7-ethyl-10-[1-(4-piperidino)piperidino]carbonyloxycamptothecin (1.00 g, 1.7 mmol) was added 0.1 N HCl (15.3 ml, 1.53 mmol), and the suspension was stirred vigorously for 5 minutes under cooling in an ice bath and filtered off. 7-Ethyl-10-[4-(1-piperidino)-1-piperidino]carbonyloxycamptothecin hydrochloride was obtained in yield 96%.

References

Miyasaka T. et al.; US Patent No. 4,604,463; August 5, 1986; Assigned to Kabushiki Kaisha Yakult Honsha, Tokyo, Japan

ISAXONINE PHOSPHATE

Therapeutic Function: Neurotropic

Chemical Name: N-(1-Methylethyl)-2-pyrimidinamine

Common Name: -

Structural Formula:

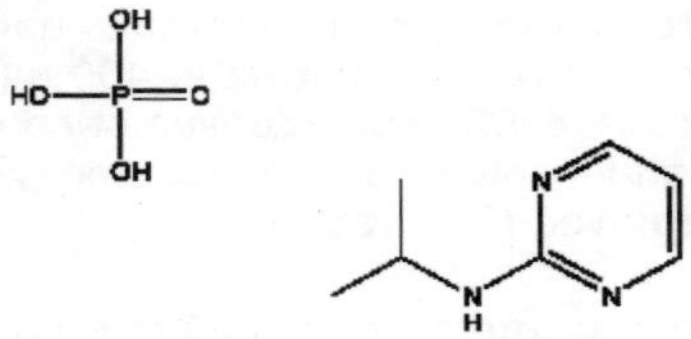

Chemical Abstracts Registry No.: 4214-72-6 (Base)

Trade Name	Manufacturer	Country	Year Introduced
Nerfactor	Ipsen	France	1981

Raw Materials

2-Isopropylamino pyrimidine
Phosphoric acid

Manufacturing Process

6 liters of ethanol and 685 g (5 mold of 2-isopropylamino pyrimidine were added to a 10 liter reactor and stirred. To the solution were added 600 g (5.2 mols) of phosphoric acid and the mixture was boiled under reflux for one hour. There was obtained a dark green solution which was treated with 30 g of carbon black. After separation and crystallization while stirring overnight, the crystallized product was separated, washed with ethanol and dried at 50°C. There was obtained 1,027 g (87% yield) of a white powder melting at 125°C. The analysis of the compound showed a good correspondence with the formula $C_7H_{14}O_4N_3P$.

References

Merck Index 4953
DFU 1 (5) 315 (1982)
Esanu, A.; US Patent 4,073,895; February 14, 1978; assigned to Societe D'Etudes de Produits Chimiques (France)

ISOAMINILE

Therapeutic Function: Antitussive

Chemical Name: α-[2-(Dimethylamino)propyl]-α-(1-methylethyl) benzeneacetonitrile

Common Name: -

Structural Formula:

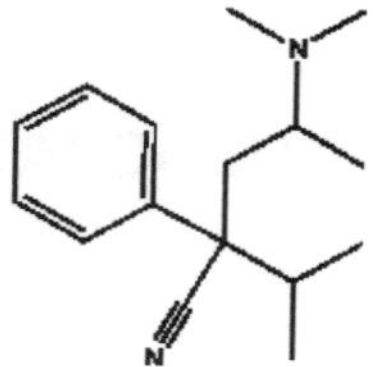

Chemical Abstracts Registry No.: 77-51-0

Trade Name	Manufacturer	Country	Year Introduced
Peracon	Toyo Jozo	Japan	1969
Dimyril	Fisons	UK	-

Trade Name	Manufacturer	Country	Year Introduced
Mucalan	Delagrange	France	-
Sedotosse	Panthox and Burck	Italy	-

Raw Materials

α-Isopropyl phenyl acetonitrile
Sodium amide
2-Dimethylamino-1-chloropropane

Manufacturing Process

140 cc of benzene and 24 g of α-isopropyl phenyl acetonitrile are added to 7.5 g of sodium amide. The mixture is stirred and refluxed for one hour. After cooling, 25 g of 2-dimethylamino-1-chloropropane, dissolved in 20 cc of benzene, are added and stirring and refluxing of the mixture is continued for 4 hours. After the reaction is completed, water is added to the reaction mixture. The benzene layer is separated from the aqueous layer and is extracted by means of 4N hydrochloric acid. The acid solution is rendered alkaline.

The separated oil is taken up in ether. After drying the ethereal solution over sodium sulfate and distilling off the ether, the resulting crude α-isopropyl-α-(β'-dimethylamino propyl) phenyl acetonitrile is purified by distillation in a vacuum. The compound boils at 138° to 146°C/3 mm, according to US Patent 2,934,557.

References

Merck Index 4956
Kleeman and Engel p. 499
OCDS Vol. 1 p.82 (1977)
I.N. p. 527
Stuhmer, W. and Funke, S.; US Patent 2,934,557; April 26, 1960; assigned to Kali-Chemie AG, Germany
Dickinson, H.M.N.; US Patent 3,074,996; January 22, 1963; assigned to Abbott Labs.

ISOBORNYL THIOCYANOACETATE

Therapeutic Function: Pediculicide

Chemical Name: Thiocyanatoacetic acid 1,7,7-trimethylbicyclo[2.2.1]-hept-2-yl ester

Common Name: -

Structural Formula:

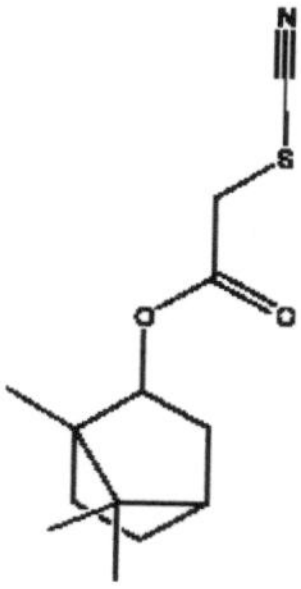

Chemical Abstracts Registry No.: 115-31-1

Trade Name	Manufacturer	Country	Year Introduced
Bornate	Wyeth	US	1946

Raw Materials

Camphene
Chloroacetic acid
Potassium thiocyanate

Manufacturing Process

200 g of camphene and 150 g of chloroacetic acid were heated 16 hours at 125°C, cooled to room temperature and the resulting product washed with water. In this way, 177 g of isobornyl monochloroacetate, analyzing 12.8%, by weight, chlorine was recovered. 174 g of the isobornyl monochloroacetate was dissolved in 300 cc of ethyl alcohol, 100 g of potassium thiocyanate added to this solution and the mixture refluxed for a period of 8 hours. 276 g of a product was recovered, which analyzed as follows: chlorine, 0.2% by wt. and sulfur, 10.9% by wt. This analysis shows the product to be principally isobornyl thiocyanoacetate.

References

Merck Index 4976
I.N. p. 527
Borglin, J.N.; US Patent 2,217,611; October 8, 1940; assigned to Hercules Powder Co.

ISOCARBOXAZIDE

Therapeutic Function: Antidepressant

Chemical Name: 5-Methyl-3-isoxazolecarboxylic acid 2-benzylhydrazide

Common Name: -

Structural Formula:

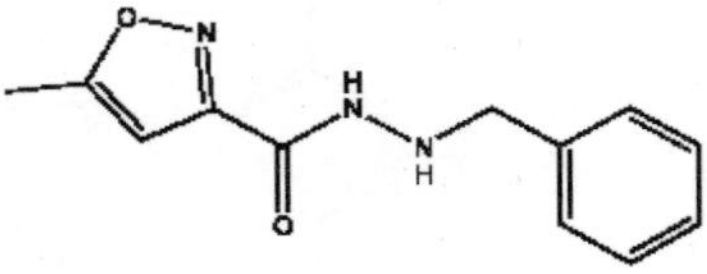

Chemical Abstracts Registry No.: 59-63-2

Trade Name	Manufacturer	Country	Year Introduced
Marplan	Roche	US	1959
Marplan	Roche	France	1961
Enerzer	Takeda	Japan	-

Raw Materials

5-Methyl-3-isoxazole carboxylic acid hydrazide
Benzaldehyde
Lithium aluminum hydride

Manufacturing Process

800 g of benzaldehyde was added to a hot solution (75°C) of 7 liters of ethanol containing 720 g of 5-methyl-2-isoxazole carboxylic acid hydrazide. The solution was stirred for ten minutes at which time the product began to crystallize. On cooling at 4°C for 14 hours, the solid was filtered off under vacuum and the solid filter cake was washed twice using 250 ml of ice cold ethanol for each washing. The lbenzylidene-2-(5-methyl-3-isoxazolylcarbonyl) hydrazine was recrystallized from ethanol, MP 199°C to 200°C.

115 g of 1-benzylidene-2-(5-methyl-3-isoxazolylcarbonyl)hydrazine was added portionwise over the period of an hour to 5 liters of anhydrous ether containing 18.5 g of lithium aluminum hydride. The reaction mixture was stirred for four hours and permitted to stand overnight. The excess lithium aluminum hydride was decomposed with 250 ml of ethyl acetate and 150 ml of water was added to decompose the complex. The solid was separated by filtration and the ether layer was concentrated to about 500 mi. 200 ml of benzene was added to dehydrate the solution. Concentration was continued until a solid remained. The 1-benzyl-2-(5-methyl-3-isoxazolylcarbonyl) hydrazine was recrystallized from methanol, MP 105°C to 106°C.

References

Merck Index 5003
Kleeman and Engel p. 500

PDR p. 1490
OCDS Vol. 1 p. 233 (1977) and 2,266 (1980)
I.N. p. 527
REM p. 1095
Gardner, T.S., Lee, J. and Wenis, E.; US Patent 2,908,688; October 13,1959; assigned to Hoffmann-La Roche, Inc.

ISOCONAZOLE NITRATE

Therapeutic Function: Antibacterial, Antifungal

Chemical Name: 1-[2,4-Dichloro-β-[(2,6-dichlorobenzyl)oxy]phenylethyl] imidazole nitrate

Common Name: -

Structural Formula:

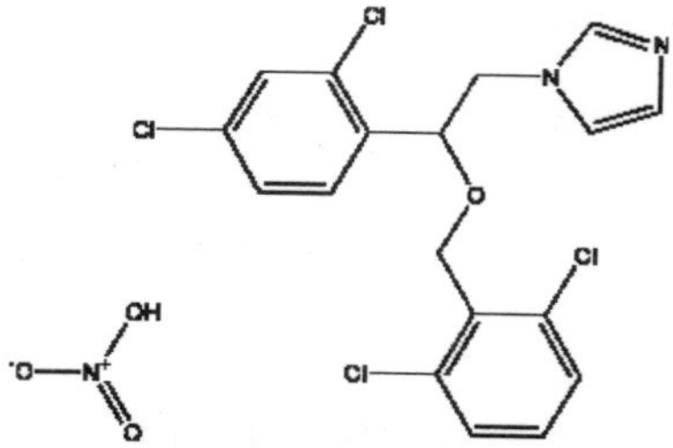

Chemical Abstracts Registry No.: 24168-96-5; 27523-40-6 (Base)

Trade Name	Manufacturer	Country	Year Introduced
Fazol	Fournier	France	1979
Travogen	Schering	W. Germany	1979
Travogen	Schering	Switz.	1980
Travogyn	Keymer	UK	1981
Adestan	Nihon Schering	Japan	1982
Travogen	Schering	Australia	-
Icaden	Schering	W. Germany	-
Gyno-Travogen	Schering	W. Germany	-

Raw Materials

α-(2,4-Dichlorophenyl)imidazole-1-ethanol
Sodium hydride
2,6-Dichlorotenzyl chloride

Manufacturing Process

To a stirred and refluxing solution of 40 parts of benzene and 35 parts of dimethylformamide (both solvents previously dried azeotropically) are added successively 1.6 parts of sodium hydride and 7.7 parts of α-(2,4-dichlorophenyl)imidazole-1-ethanol, (cooling on ice is necessary). After the addition is complete, stirring and refluxing is continued for 30 minutes. Then there are added 7.8 parts of 2.6-dichlorobenzyl chloride and the whole is stirred at reflux for another 3 hours. The reaction mixture is poured onto water and the product 1-[2,4-dichloro-b-(2,6-dichlorobenzyloxy)phenethyl] imidazole, is extracted with benzene. The extract is washed twice with water, dried, filtered and evaporated in vacuo. The base residue is dissolved in a mixture of acetone and diisopropyl ether and to this solution is added an excess of concentrated nitric acid solution. The precipitated nitrate salt is filtered off and recrystallized from a mixture of methanol and diisopropyl ether, yielding 1-[2,4-dichloro-b-(2,6-dichlorobenzyloxy)phenethyl]imidazole nitrate; melting point 179°C.

References

Merck Index 5007
DFU 4 (11) 814 (1979)
Kleeman and Engel p. 500
DOT 15 (12) 542 (1979) and 17 (9) 388 (1981)
I.N. p. 528
Godefroi, E.F. and Heeres, J.; US Patents 3,717,655; February 20, 1973 and 3,839,574; October 1,1974; both assigned to Janssen Pharmaceutica NV

ISOETHARINE

Therapeutic Function: Sympathomimetic, Bronchodilator

Chemical Name: 2-Benzenediol, 4-(1-hydroxy-2-((1-methylethyl)amino) butyl)-

Common Name: Etyprenalinum; Isoetharine

Structural Formula:

HO
N
H
OH
OH

Chemical Abstracts Registry No.: 530-08-5

Trade Name	Manufacturer	Country	Year Introduced
Dilabron	Sterling Winthrop	-	-
Isoetharine	Shanghai Lansheng Corporation	-	-

Raw Materials

3,4-Dibenzyl-hydroxy-butyrophenone
Bromine
Isopropylamine
Palladium on carbon

Manufacturing Process

36 g 3,4dibenzyl-hydroxy-butyrophenone was dissolved in 125 ml methylene chloride and 16 g bromine was added dropwise after 15 g calcium carbonate. An obtained precipitate was filtered off. The filtrate was distilled to dryness to give 1-(3,4-dibenzyloxy-phenyl)-2-bromo-butan-1-one. It was dissolved in a portion of ethanol and 15 g of isopropylamine was added. The mixture had stood for night at room temperature and thereafter ether was added. The isopropylamine hydrobromide had fallen. It was filtered off, the filtrate was shook with 200 ml of diluted hydrochloric acid. At that 4-(1-hydroxy-2-isopropylamino-butyl)-benzene-1,2-diol chlorohydrate was separated as an oil. The oil was diluted with the 5 volumes of ethanol and the calculated quantity hydrogen was passed through in a presence of palladium on coal catalyst. The catalyst was filtered off and a solvent was removed to dryness, the residue had crystallized by grinding with acetone. The obtained 4-(1-hydroxy-2-isopropylamino-butyl)-benzene-1,2-diol chlorohydrate was recrystallized from methanol. MP: 212°-213°C. Chlorohydrate may be changed into isoetharine with the calculated quantity of any base.

References

D.R. Patent No. 638,650; June 8, 1934; I.G.Farbenindustrie Akt.-Ges. in Frankfurt a.M.

ISOFLUPREDONE

Therapeutic Function: Glucocorticoid

Chemical Name: Pregna-1,4-diene-3,20-dione, 9-fluoro-11,17,21-trihydroxy-, (11β)-

Common Name: Deltafludrocortisone, 9-Fluorprednisolone, Isoflupredone

Chemical Abstracts Registry No.: 338-95-4

Structural Formula:

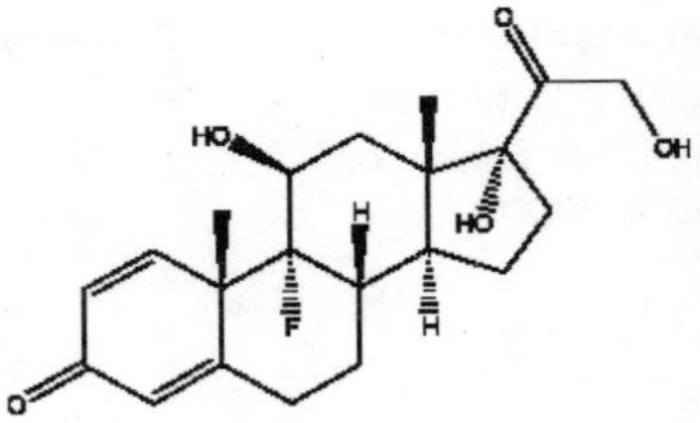

Trade Name	Manufacturer	Country	Year Introduced
Isoflupredon	Farmabios	-	-
Abicorten	Fatro	-	-

Raw Materials

Potassium dihydrogen phosphate
Disodium hydrogen phosphate
Corynebacterium simplex
9α-Fluoro-4-pregnen-11β,17α,21-triol-3,20-diene

Manufacturing Process

A 100 ml broth culture containing a 0.1% yeast extract concentration, 9.0 ml of 0.2 M KH_2PO_4 and 9.0 ml of 0.2 M Na_2HPO_4 contained in a 300 ml Erlenmeyer flask, is seeded with 1 ml of a 24-h broth culture of Corynehacierium simplex (A. T, C. C. 6946). The flask is incubated at 28°C for 24 h. A second 300 ml Erlenmeyer flask containing 150 mg of sterile 9α-fluoro-4-pregnen-11β,17α,21-triol-3,20-diene in 5.0 ml acetone is inoculated with the 24 h culture of *Corynebacterium simplex* (A. T. C. C. 6946). The culture-containing steroid solution is incubated for 48 h at 28° to 30°C.

The product is extracted with chloroform and isolated by evaporation to dryness. Recrytstallizstion of the residue affords 9α-fluoro-$\delta^{1,4}$-pregnadiene-11β,17α,21-triol-3.20-dione as a solid.

References

Belleville A.N.; US Patent No. 2,837,464; June 3, 1958; Assigned: Schering Corporation, Bloomfield, N.J., a corporation of New Jersey

ISOFLURANE

Therapeutic Function: Inhalation anesthetic

Chemical Name: 1-Chloro-2,2,2-trifluoroethyl difluoromethyl ether

Common Name: -

Structural Formula:

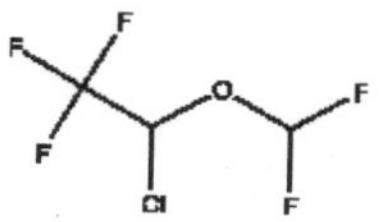

Chemical Abstracts Registry No.: 26675-46-7

Trade Name	Manufacturer	Country	Year Introduced
Forane	Ohio Medical	US	1980
Aerrane	Ohio Medical	Switz.	1983
Aerrane	Ohio Medical	UK	1983

Raw Materials

1-Chloro-2,2,2-trifluoroethyl dichloromethyl ether
Hydrogen fluoride

Manufacturing Process

A 1-liter 3-necked stainless steel flask was fitted with a copper "Dry Ice" cold finger condenser, a stainless steel stirring shaft and gland and a copper gas inlet tube. To the flask there was then added 50 g (0.23 mol) of $CF_3CHClOCHCl_2$ and 1.5 g of $SbCl_5$ · HF gas was then slowly bubbled through the stirred mixture which was maintained at 0°C. The reaction was run until 0.35 mol of HCl was collected, as indicated by the titration of the effluent gas which was dissolved in water. Following the fluorination 26 g of material were recovered and determined to be 90% pure by vapor phase chromatography. Fractional distillation using a 30 x 0.5 cm column packed with glass helices gave the pure product, BP 48°C to 48.5°C.

References

Merck Index 5021
DOT 16 (11) 374 (1980)
I.N. p. 528
REM p. 1042
Terrell, R.C.; US Patent 3,535,388; October 20, 1970; assigned to Air Reduction Co., Inc.

ISOFLUROPHATE

Therapeutic Function: Cholinergic (ophthalmic)

Chemical Name: Phosphorofluoridic acid bis(1-methylethyl)ester

Common Name: Fluostigmine

Structural Formula:

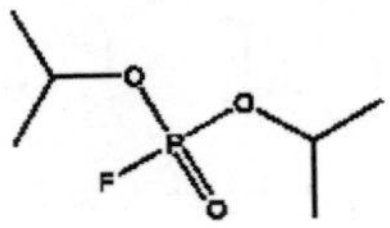

Chemical Abstracts Registry No.: 55-91-4

Trade Name	Manufacturer	Country	Year Introduced
Floropryl	MSD	US	1949
D.F.P.	Sumitomo	Japan	-
D.F.P.	Boots	UK	-
D.F.P.	Winzer	W. Germany	-
Diflupyl	Labaz	-	-
Fluopryl	MSD	-	-

Raw Materials

Isopropanol
Chlorine
Phosphorus trichloride
Sodium fluoride

Manufacturing Process

212 lb (3.54 lb-mols) of isopropanol containing less than 0.2 wt % of water was cooled with brine to -5°C in a jacketed reactor. 160 lb (1.16 lb-mols) of phosphorus trichloride was gradually added to the isopropanol with cooling and stirring during a period of 4 hours. The temperature of the reaction was not allowed to exceed 12°C and the system was maintained under slight negative pressure (about 700 mm) to remove undesirable vapors.

After completion of the addition, the mixture was stirred for ½ hour and then subjected to a pressure of 12 to 100 mm of mercury. Chlorine was then passed into the crude reaction product at a rate of 12 lb/hr, the temperature of the reaction being kept below 12°C by brine cooling. The end of the reaction was indicated by a temperature drop which occurred after a total of 122 lb of chlorine (1.72 lb-mols, 48% excess) was used.

To remove excess chlorine, hydrogen chloride and isopropyl chloride, the well-stirred mixture was subjected to a pressure of 12 to 100 mm of mercury for 2 hours. The temperature was gradually raised to 20°C during this time by passing steam into the jacket of the reactor. 10 gallons of benzene was then added and distilled off under reduced pressure, gradually raising the temperature of the reaction mixture to 30°C. The last traces of hydrogen chloride were removed by adding an additional 10 gallons of benzene which was distilled off under reduced pressure at reactor temperatures not exceeding 50°C. The total time required for the removal of the volatile acid components of the reaction mixture was 4 hours.

The mixture was then cooled to 20°C and 19 gallons of benzene was added. This was followed by the introduction of 123.5 lb (2.80 lb-mols) of dry powdered sodium fluoride (95% pure). The mixture was stirred and heated to the refluxing temperature in a period of 1 hour and held at this temperature (95° to 98°C) for 4 hours. The product obtained was cooled and filtered to yield a filter cake which was washed with three 5-gallon portions of benzene. The filtrate and washing were then combined and distilled under reduced pressure. There was obtained 158 lb (74% yield of theory based on PCl_3) of diisopropyl fluorophosphate, BP 62°C at 9 mm and 46°C at 5 mm.

References

Merck Index 5022
Kleeman and Engel p. 501
PDR p. 1179
I.N.p. 437
REM p. 899
Hardy, E.E. and Kosolapoff, G.M.; US Patent 2,409,039; October 8, 1946; assigned to Monsanto Chemical Company

ISOMETHEPTENE

Therapeutic Function: Muscle relaxant

Chemical Name: N,1,5-Trimethyl-4-hexenylamine

Common Name: Methyl isooctenylamine

Structural Formula:

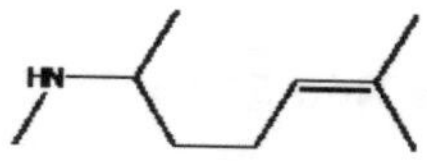

Chemical Abstracts Registry No.: 503-01-5

Trade Name	Manufacturer	Country	Year Introduced
Octinum	Knoll	US	1948
Cesal	Dainippon	Japan	-
Midrin	Camrick	US	-
Migralam	Bart	US	-

Raw Materials

Methyl heptenone
Methylamine

Manufacturing Process

Methyl heptenone dissolved in 75% alcohol is reduced with activated aluminum in the presence of methylamine to give isometheptene.

References

Merck Index 5031
Kleeman and Engel p. 502
PDR pp. 654,781
I.N. p. 529
REM p. 891
Klavehn, W. and Wolf, A.; US Patent 2,230,753; February 4, 1941; assigned to E.Bilhuber Corporation, Germany
Klavehn, W. and Wolf, A.; US Patent 2,230,754; February 4,1941; assigned to E. Bilhuber Corporation, Germany

ISONIAZID

Therapeutic Function: Antitubercular

Chemical Name: 4-Pyridinecarboxylic acid hydrazide

Common Name: Isonicotinic acid hydrazide

Structural Formula:

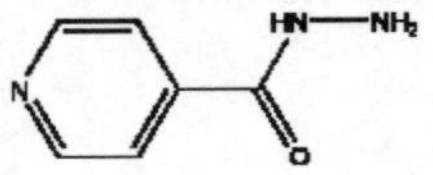

Chemical Abstracts Registry No.: 54-85-3

Trade Name	Manufacturer	Country	Year Introduced
Nyrazid	Squibb	US	1952
Niconyl	Parke Davis	US	1952
INH	Lilly	US	1952
Tisin	U.S.V. Pharm.	US	1952
Pyrizidin	Warner Lambert	US	1952
Cotinazin	Pfizer	US	1952
Tyvid	Merrell National	US	1952
Ditubin	Schering	US	1952

Trade Name	Manufacturer	Country	Year Introduced
Rimafon	Roche	US	1952
Armazide	Armour	US	1952
Anteben	Dainippon	Japan	-
Cedin	Lyssia	W. Germany	-
Cernidon	Gayoso Wellcome	Spain	-
Cin Vis	Vis	Italy	-
Dardex	Llorente	Spain	-
Diazid	Nippon Shinyaku	Japan	-
Dinacrin	Winthrop-Stearns	Philippines	-
Dow-Isoniazid	Dow	US	-
Eutizon	Pliva	Yugoslavia	-
Fimazid	Wassermann	Spain	-
Hidrafasa	Lifasa	Spain	-
Hidranic	Efeyn	Spain	-
Hidrazinda	Jorba	Spain	-
Hiperazida	Martin Santos	Spain	-
Hycozid	Takeda	Japan	-
Hydra	Otsura	Japan	-
Hyzyd	Mallinckrodt Inc.	US	-
Idrazil	Bracco	Italy	-
INH-Burgthal	Conzen	W. Germany	-
Iscotin	Daiichi	Japan	-
Isobicini	Maggioni	Italy	-
Iso-Dexter	Dexter	Spain	-
Isotamine	I.C.N.	Canada	-
Isozide	I.C.N.	Canada	-
Kridan	Cidan	Spain	-
Lefos	Bicsa	Spain	-
Lubacida	Alfar	Spain	-
Neoteben	Bayer	W. Germany	-
Neo-Tizide	Aesca	Austria	-
Niadrin	Enzo	US	-
Niazid	Sankyo	Japan	-
Nicazide	Wassermann	Italy	-
Niconyl	Parke Davis	US	-
Nicotibina	Zambeletti	Italy	-
Nicotbine	Abic	Israel	-
Nicotubin	Leiras	Finland	-
Nicozid	Piam	Italy	-
Nicozide	Premo	US	-
Niplen	Tanabe	Japan	-
Panazid	Panray	US	-
Pycazide	Smith and Nephew	UK	-
Pyrizidin	Nepera	US	-
Rifamate	Merrell Dow	US	-

Trade Name	Manufacturer	Country	Year Introduced
Rimifon	Roche	France	-
Sumifon	Sumitomo	Japan	-
TB-Phlogin	Heyl	W. Germany	-
Tebesium	Hefa-Frenon	W. Germany	-
Tebilon	Kwizda	Austria	-
Tibinide	Ferrosan	Denmark	-
Tibizina	Farmochimica	Italy	-
Tubanox	Morgens	Spain	-
Tuberon	Shionogi	Japan	-
Tubilysin	Orion	Finland	-
Zidaf imia	Santos	Spain	-
Zideluy	Miluy	Spain	-

Raw Materials

4-Cyanopyridine
Hydrazine hydrate

Manufacturing Process

4 parts of 4-cyanopyridine in 12 parts of water were reacted with 4 parts of hydrazine hydrate in the presence of 0.08 part of sodium hydroxide at 100°C under reflux for 7 hours. The product, after filtration and evaporation to dryness, was crystallized from ethanol. The yield of isonicotinyl hydrazide amounted to 3.27 parts which is 62% of the theoretical.

References

Merck Index 5032
Kleeman and Engel p. 503
PDR pp. 798, 830, 1237
OCDS Vol. 1 p. 254 (1977) and 2,266 (1980)
I.N. p. 529
REM p. 1214
Gasson, E.J.; US Patent 2,830,994; April 15, 1958; assigned to The Distillers Company Limited, Scotland
Fox, H.H.; US Patent 2,596,069; May 6, 1952; assigned to Hoffmann-La Roche Inc.

ISOPROPAMIDE IODIDE

Therapeutic Function: Spasmolytic

Chemical Name: γ-(Aminocarbonyl)-N-methyl-N,N-bis(1-methylethyl)-γ-phenylbenzenepropanaminium iodide

Common Name: Diisopropylaminodiphenylbutyramide methiodide

Structural Formula:

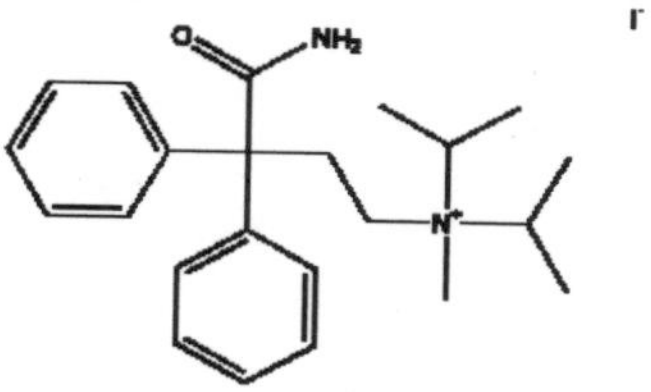

Chemical Abstracts Registry No.: 71-81-8

Trade Name	Manufacturer	Country	Year Introduced
Darbid	SKF	US	1957
Priamide	Delalande	France	1959
Combid	SKF	US	-
Dipramid	Valeas	Italy	-
Marygin M	Sumitomo	Japan	-
Ornade	SKF	US	-
Prochlor-Iso	Schein	US	-
Pro-Iso	Zenith	US	-
Tyrimide	SKF	UK	-

Raw Materials

γ-Diisopropylamino-α,α-iphenylbutyronitrile
Sulfuric acid
Methyl iodide

Manufacturing Process

γ-Diisopropylamino-α,α-diphenylbutyronitrile (60 g) was added in several portions to a mixture of sulfuric acid (150 ml) and water (15 ml) and the solution was heated 3% hours on the steam bath and then poured on ice and made basic with NH_4OH. The γ-diisopropylamino-α,α-diphenylbutyramide precipitated as a solid, which was taken up in methylene chloride from an aqueous slurry. The methylene chloride was separated and dried by filtering through anhydrous K_2CO_3. The solvent was removed by distillation, leaving the amide which was crystallized from Skellysolve B five times and found then to have MP 87.0° to 88.5°C.

γ-Diisopropylamino-α,α-diphenylbutyramide in propanol was refluxed 4 hours in the presence of excess methyl iodide. Upon dilution of the solution with ethyl acetate (100 ml per 50 ml isopropyl alcohol) and cooling γ-diisopropylamino-α,α-diphenylbutyramide methiodide precipitated, was collected by filtration and recrystallized (9.0 g) by dissolving in a hot mixture of 100 ml isopropyl alcohol and 10 ml methanol and then diluting with 90 ml Skellysolve B, to give 8.3 g recrystallized product, MP 182° to 184°C.

References

Merck Index 5051
Kleeman and Engel p. 504
PDR pp. 1606, 1706, 1711, 1999
I.N. p. 531
REM p. 916
Speeter, M.E.; US Patent 2,823,233; February 11, 1956; assigned to Bristol Laboratories Inc.

ISOPROTERENOL SULFATE

Therapeutic Function: Bronchodilator

Chemical Name: 4-[1-Hydroxy-2-[(1-methylethyl)amino]ethyl]-1,2-benzenediol sulfate

Common Name: Isoprenaline sulfate; Isopropylarterenol sulfate

Structural Formula:

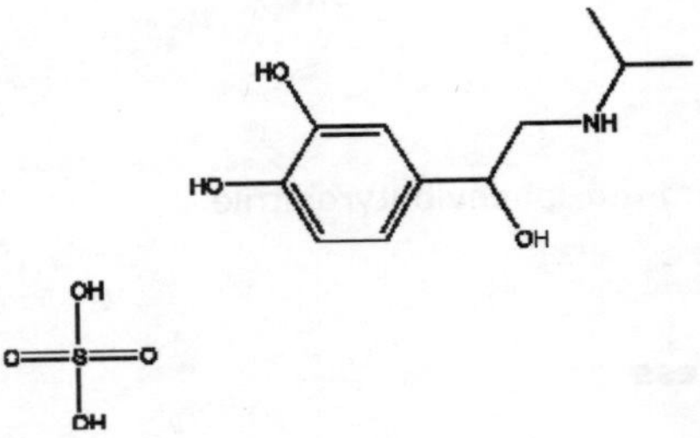

Chemical Abstracts Registry No.: 299-95-6; 7683-59-2 (Base)

Trade Name	Manufacturer	Country	Year Introduced
Isonorin	Smith, Miller and Patch	US	1949
Norisodrine	Abbott	US	1950
Medihaler-Iso	Riker	US	1956
Luf-Iso	Mallinckrodt Inc.	US	1974
Aleudrin	Lewis	UK	-
Aludrin	Boehringer Ingelheim	W. Germany	-
Asmadren	A.F.I.	Norway	-
Asthpul	Nippon Shoji	Japan	-

Trade Name	Manufacturer	Country	Year Introduced
Dyspnoesan	Noury Pharma	Netherlands	-
Ingelan	Boehringer Ingelheim	W. Germany	-
Isomenyl	Kaken	Japan	-
Meterdos-Iso	West-Silten	UK	-
Nebair	Warner-Chilcott	US	-
Novodrin	VEB Berlin Chemie	E. Germany	-
Prenomiser	Fisons	UK	-
Propynalin	Ferrosan	Denmark	-
Proternol	Nikken	Japan	-
Sedansol "Iso"	Nippon Zoki	Japan	-
Vapo-N-Iso	Fisons	US	-

Raw Materials

3,4-Dihydroxy-ω-chloroacetophenone
Hydrogen
Isopropylamine
Sulfuric acid

Manufacturing Process

As described in US Patent 2,308,232, 100 g 3,4-dihydroxy-ω-chloroacetophenone, 200 cc ethyl alcohol and 200 cc of about 50% aqueous isopropylamine solution are boiled during 3 hours on the water bath with the use of a reflux condenser, whereupon neutralizing with diluted sulfuric acid is carried out and the sulfate, obtained upon cooling, from alcohol of 50% is recrystallized; its MP is 245°C.

21 g 3,4-dihydroxy-ω-isopropylaminoacetophenone sulfate are hydrogenated with 50 cc methyl alcohol and 50 cc water, 0.5 g carbon and 3 cc palladium chloride solution of 2%. After 2 hours the hydrogen absorption comes to a standstill, after the theoretical quantity of hydrogen has been absorbed. After concentrating, the isopropylaminomethyl-(3,4-dihydroxyphenyl)carbinolsulfate crystallizes out. It has a MP of 180°C after refining.

References

Merck Index 5065
Kleeman and Engel p. 503
OCDS Vol. 1 p. 63 (1977); 2, 37, 107 (1980) and 3, 20 (1984)
I.N. p. 531
REM p.886
Scheuing, G. and Thoma, O.; US Patent 2,308,232; January 12, 1943
Delmar, G.S. and Macallum, E.N.; US Patent 2,715,141; August 9, 1955; assigned to Delmar Chemicals Limited, Canada

ISOSORBIDE DINITRATE

Therapeutic Function: Coronary vasodilator

Chemical Name: 1,4:3,6-Dianhydro-D-glucidol dinitrate

Common Name: Dinitrosorbide

Structural Formula:

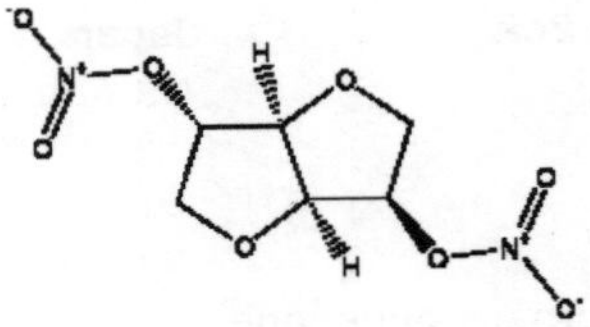

Chemical Abstracts Registry No.: 87-33-2

Trade Name	Manufacturer	Country	Year Introduced
Isordil	Ives	US	1959
Sorbitrate	Stuart	US	1968
Isordil	Ayerst	UK	1971
Sorquad	Tutag	US	1972
ISDN	Cooper	US	1975
Iso-Bid	Geriatric	US	1975
Isomotic	Alcon	US	1980
Dilatrate	Reed Carnrick	US	1981
Cardio-10	Nicholas	W. Germany	-
Cardis	Iwaki	Japan	-
Carvanil	Banyu	Japan	-
Cardopax	Erco	Denmark	-
Carvasin	Ayerst	Italy	-
Cedocard	Tillotts	UK	-
Cordil	Disco	Israel	-
Cornilat	Galenika	Yugoslavia	-
Corovliss	Boehringer Mannheim	W. Germany	-
Difutrat	Srbolek	Yugoslavia	-
Dilatrate	Reed Carnrick	US	-
Diretan	Ono	Japan	-
Duranitrate	Durachemie	W. Germany	-
Isobid	Geriatric	US	-
Isocardide	Sam-On	Israel	-
ISO-D	Dunhall	US	-
Isoket	Gebro	Austria	-
Isomack	Mack	W. Germany	-
Isopuren	Klinge	W. Germany	-

Trade Name	Manufacturer	Country	Year Introduced
Isordil	Wyeth	US	-
Isotrate	Hauck	US	-
Laserdil	Laser	US	-
Marrolingual	Pohl-Boskamp	W. Germany	-
Maycor	Parke Davis	W. Germany	-
Metonitron	Petazon	Switz.	-
Nitorol R	Eisai	Japan	-
Nitroret	Hishiyama	Japan	-
Nitrosit	Pharmacal	Finland	-
Nitrosorbide	Lusofarmaco	Italy	-
Nitro-Tablinen	Sanorania	W. Germany	-
Nosim	Richet	Argentina	-
Risordan	Theraplix	France	-
Soni-Slo	Lipha	UK	-
Sorbangil	Kabi Vitrum	Sweden	-
Sorbid	I.E. Kimya Evi	Turkey	-
Tinidil	Pliva	Yugoslavia	-
Vascardin	Nicholas	UK	-

Raw Materials

1,4:3,6-Dianhydro-D-glucitol
Nitric acid

Manufacturing Process

An aqueous syrup of 1,4:3,6-dianhydro-D-glucitol is slowly added to a cooled mixture of HNO_3 and H_2SO_4. After standing a few minutes the mixture is poured into cold water and the precipitated product is collected and recrystallized from ethanol.

References

Merck Index 5074
Kleeman and Engel p. 505
PDR pp.830, 905, 928, 993, 1442, 1606, 1784, 1951, 1999
I.N. p. 533
REM p. 853
Cordes, G., Munch, U. and Giesselmann, E.; US Patent 4,156,736; May 29, 1979; assigned to Sanol Schwarz-Monheim G.m.b.H. (W. Germany)

ISOTHIPENDYL HYDROCHLORIDE

Therapeutic Function: Antihistaminic

Chemical Name: 10-(2-Dimethylamino-2-methylethyl)-10H-pyrido[3,2-b][1,4]benzothiazine hydrochloride

Common Name: -

Structural Formula:

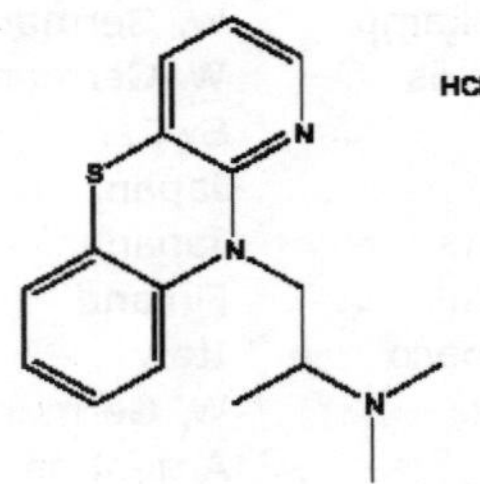

Chemical Abstracts Registry No.: 1225-60-1 ; 482-15-5 (Base)

Trade Name	Manufacturer	Country	Year Introduced
Theruhistin	Ayerst	US	1957
Andantol	Gerda	France	1957
Aczen NS	Kanebo, Ltd.	Japan	-
Adantol	Imidas	Brazil	-
Andanton	Lacer	Spain	-
Nilergex	I.C.I.	UK	-
Thiodantol	Teva	Israel	-

Raw Materials

Sulfur
Sodium amide
Hydrogen chloride
Phenylpyridylamine
Dimethylaminoisopropyl chloride

Manufacturing Process

85 parts of phenylpyridyl amine, 21 parts of powdered sulfur and 1.7 parts of iodine were heated to 275°C for two hours. Evolution of hydrogen sulfide began when the mixture reached a temperature of 250°C and became vigorous when it reached 275°C. Such evolution of hydrogen sulfide diminished after about one hour at 275°C. A light oil was distilled from the reaction mixture under vacuum (pressure = 2-3 mm Hg). This oil which contained phenylpyridyl amine in addition to the thiophenylpyridyl amine was then treated at boiling temperature with approximately the theoretical amount of 2-3 normal HCl until complete solution resulted with formation of the HCl salts of the amines. The solution was then treated with 1 to 2% (based upon the substance mixture) of active carbon and then filtered hot. The nitrate was then cooled to 0°C whereupon the thiophenylpyridyl amine hydrochloride crystallized out while the phenylpyridyl amine hydrochloride remained in solution. The thiophenylpyridyl amine hydrochloride was filtered off and suspended in water and the pH adjusted with half concentrated ammonia to 8. The thiophenylpyridyl amine set free was filtered off and dried. It was in the form of gold yellow needles and had a melting point of 114°C to 115°C.

40 parts of thiophenylpyridyl amine were dissolved in 200 parts of water free toluene. After the addition of 16 parts of soda amide, the mixture was refluxed for 1% hours. Thereafter, 28 parts of dimethylaminoisopropyl chloride in 30 parts of water free toluene were dropped in and the temperature maintained at 20°C to 25°C for 30 minutes. Thereafter, the mixture was heated at 60°C for 30 minutes and subsequently refluxed for 20 minutes. Water and hydrochloride acid were then added to the reaction mixture and this mixture rendered alkaline with NaOH and then the alkalized mixture shaken out with ether. The dimethylaminoisopropyl-N9-thiophenylpyridyl amine base thus obtained was vacuum distilled. It was then converted to hydrochloride salt. The monohydrochloride salt is almost white in color and melts at 213°C to 216°C. The yield was almost 100% of the theoretical.

References

Merck Index 5077
Kleeman and Engel p. 505
OCDS Vol. 1 p.430 (1977)
I.N.p. 534
Schuler, W.A. and Klebe, H.; US Patent 2,974,139; March 7, 1961; assigned to Degussa (W. Germany)

ISOTRETINOIN

Therapeutic Function: Antiacne, Keratolytic

Chemical Name: Retinoic acid, 13-cis-

Common Name: cis-Retinoic acid; Isotretinoin

Structural Formula:

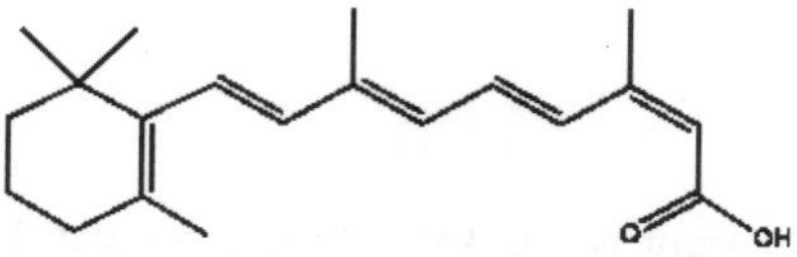

Chemical Abstracts Registry No.: 4759-48-2

Trade Name	Manufacturer	Country	Year Introduced
Accutane	Roche	-	-
Amnesteem	Bertek	-	-
Amnesteem	Mylan Laboratories Inc.	-	-
Amnesteem	Genpharm Inc.	-	-
Claravis	BARR	-	-
Isotrex	Stiefel	-	-

Trade Name	Manufacturer	Country	Year Introduced
Oratane	Douglas	-	-
Roaccutane	R.P. Scherer GmbH and Co. KG	Germany	-
Sotret	Ranbaxy	India	-
Tasmar	Hoffmann - La Roche Inc.	-	-

Raw Materials

Butyl lithium
Diisopropylamine
Methyl 3,3-dimethyl acrylate
β-Ionylidene acetaldehyde

Manufacturing Process

Under an atmosphere of nitrogen, a solution of n-butyl lithium in hexane (321 ml, 15%) was added to a solution of diisopropylamine (48.6 g, 0.48 mole) in tetrahydrofuran (1000 ml) at -30°C and the mixture was stirred for one hour. The reaction mixture was then cooled to -72°C and methyl 3,3-dimethyl acrylate (55 g, 0.48 mole) was added to it. Stirring was continued at -65° to -75°C for 30 min. To the resulting mixture, a solution of β-ionylidene acetaldehyde (100 g, 0.458 mole, 9-trans content: 80%) was added and the reaction mixture was stirred at -65° to -75°C for 1 h. The reaction mixture was then warmed to 40°C and stirred at this temperature for 3 h. Solvent was removed under vacuum and the reaction mixture was diluted with water (700 ml) and methanol (300 ml). Activated charcoal (4 g) was then added and the mixture was refluxed for 30 min. The heterogeneous mixture was filtered through hyflo and the hyflo bed was washed with methanol (300 ml) and water (150 ml). The aqueous methanolic layer was then extracted with hexanes (2 x 500 ml) and acidified with 10% sulfuric acid to pH 2.80.5. The desired product was then extracted with dichloromethane (2 x 500 ml). The combined dichloromethane layer was washed with water (2 x 300 ml) and concentrated in vacuo to afford the desired isotretinoin. Crystallization from methanol (200 ml) afforded isotretinoin (44 g) in greater than 99% HPLC purity.

References

Salman M., et al.; US Patent No. 6,441,226 B1; August 27, 2001; Assigned: Ranbaxy Laboratories Limited, New Delhi (IN)

ISOXICAM

Therapeutic Function: Antiinflammatory

Chemical Name: 4-Hydroxy-3-(5-methyl-3-isoxazolocarbamyl)-2-methyl-2H-1,2-benzothiazine 1,1-dioxide

Common Name: -

Structural Formula:

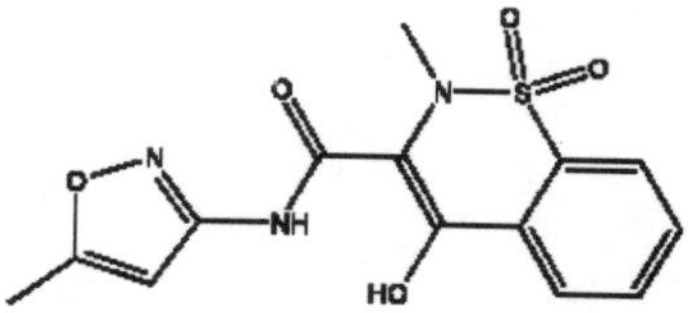

Chemical Abstracts Registry No.: 34552-84-6

Trade Name	Manufacturer	Country	Year Introduced
Pacyl	Warner Lambert	Switz.	1983
Pacyl	Adenylchemie	W. Germany	1983
Maxicam	Parke Davis	-	-

Raw Materials

3-Carbethoxy-4-hydroxy-2-methyl-2H-1,2-benzothiazine-1,1-dioxide
3-Amino-5-methyl-isoxazole

Manufacturing Process

A mixture of 40.5 g (0.15 mol) of 3-carbethoxy-4-hydroxy-2-methyl-2H-1,2-benzothiazine 1,1-dioxide, 20.6 g (0.21 mol) of 3-amino-5-methylisoxazole, and 2,500 ml of xylene was refluxed for 24 hours in a Soxhlet apparatus, the thimble of which contained 60 g of Linde type 4A molecular sieve. The mixture was cooled to 25°C and the resulting crystalline precipitate was collected and washed with ether to give 44 g of crude product. Recrystallization from 1,600 ml of 1,4-dioxan gave 34.7 g of material, MP 265°C to 271°C dec.

References

Merck Index 5085
DFU 1 (3) 123 (1976)
OCDS Vol. 2 p. 394 (1980)
DOT 19 (2) 119 (1983) and 19 (7) 414 (1983)
I.N. p. 534
Zinnes, H., Schwartz, M.L. and Shavel, J. Jr.; US Patent 3,787,324; January 22, 1974; assigned to Warner-Lambert Co.

ISOXSUPRINE HYDROCHLORIDE

Therapeutic Function: Vasodilator

Chemical Name: 4-Hydroxy-α-[1-[(1-methyl-2-phenoxyethyl)amipo]ethyl] benzenemethanol hydrochloride

Common Name: -

Structural Formula:

Chemical Abstracts Registry No.: 579-56-6; 395-28-8 (Base)

Trade Name	Manufacturer	Country	Year Introduced
Duvadilan	Duphar	France	1958
Vasodilan	Mead Johnson	US	1959
Cardilan	Ferrosan	Denmark	-
Defencin	Bristol	UK	-
Isokulin	Toho Iyaku	Japan	-
Isolait	Elder	US	-
Largiven	Bristol	Italy	-
Suprilent	Duphar	Belgium	-
Synzedrin	Teisan	Japan	-
Trophodilan	Duphar	France	-
Vahodilan	Morita	Japan	-
Vaxoprin	Guidotti	Italy	-
Vasodilene	Chiesi	Italy	-
Vasolan	Disco	Israel	-
Vasoplex	Frika	Austria	-
Vasosuprina	Lusofarmaco	Italy	-
Xuprin	Duphar	Belgium	-

Raw Materials

1-Phenoxy-2-aminopropane
1-(4'-Benzyloxyphenyl)-2-bromopropanone-1
Hydrogen

Manufacturing Process

To a solution of 30.7 g (0.203 mol) of 1-phenoxy-2-aminopropane in 150 ml of ethanol there was added 31.9 g (0.100 mol) of 1-(4'-benzyloxyphenyl)-2-bromopropanone-1. The mixture was heated to boiling temperature and the solution was then refluxed in a reflux condenser for 3 hours. Most of the ethanol was then distilled off in vacuo, Then to the residue there was added about 150 ml of diethyl ether. The hydrogen bromide salt of 1-phenoxy-2-aminopropane was filtered off and washed with diethyl ether.

The collected ethereal filtrates were acidified with 50 ml of 4 N hydrochloric acid and this solution was stirred vigorously. The hydrochloride of 1-(4'-benzyloxyphenyl)-2-(1'-methyl-2-phenoxy-ethylamino)propanone-1 precipitated out, was filtered off, washed with water and then with diethyl ether. Then this substance was dried in vacuo. The yield was 37.7 g, i.e., 89% of the theoretically possible yield, calculated on 1-(4'-benzyloxyphenyl)-2-bromine propanone-1. This substance had a light yellow color and melted at 197 to 198°C, while decomposing.

Then 21.89 g of the hydrochloride salt was dissolved in 600 ml of 80% aqueous ethanol. With the addition of a palladium carbon catalyst, this solution was hydrogenated at room temperature under a hydrogen pressure of about 1.1 atmospheres. After 2 mols hydrogen had been absorbed, the catalyst was filtered off and the filtrate was evaporated in vacuo until crystallization occurred. Then the crystals were dissolved by heating in the smallest possible quantity of water and after cooling, the crystallized substance was filtered off, washed with water and dried in vacuo. The yield was 6.80 g, i.e., 39% of the theoretically possible yield. The resultant product recrystallized from water melted at 203° to 204°C.

References

Merck Index 5086
Kleeman and Engel p. 506
PDR pp.830, 993, 1129, 1569, 1606, 1999
OCDS Vol. 1 p. 69 (1977)
I.N. p. 534
REM p. 892
Moed, H.D.; US Patent 3)256,836; October 2,1962;assigned to North American Philips Company

ITRACONAZOLE

Therapeutic Function: Antifungal

Chemical Name: 3H-1,2,4-Triazol-3-one, 4-(4-(4-(4-((2-(2,4-dichlorophenyl)-2-(1H-1,2,4-triazol-1-ylmethyl)-1,3-dioxolan-4-yl) methoxy)phenyl)-1-piperazinyl)phenyl)-2,4-dihydro-2-(1-methylpropyl)-

Common Name: Itraconazole

Chemical Abstracts Registry No.: 84625-61-6

Structural Formula:

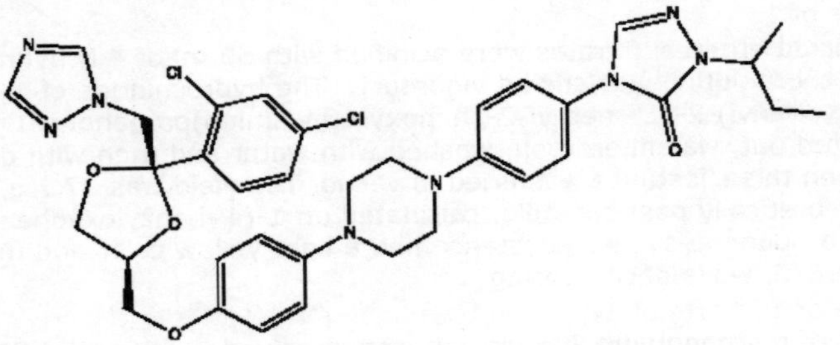

Trade Name	Manufacturer	Country	Year Introduced
Canadiol	Esteve	-	-
Canditral	Glenmark Pharmaceuticals Ltd.	India	-
Funit	Nobel	-	-
Itaspor	Intas Pharmaceuticals Pvt. Ltd.	India	-
Itraconazole	Janssen Pharmaceutica Inc.	USA	-
Itraconazole pellets	Chemo Iberica	Spain	-
Kanazol	Slaviamed	Yugoslavia	-
Micoral	Sintyal	-	-
Orungal	Janssen Pharmaceutica N.V.	Belgium	-
Sporacid	Dexa Medica	-	-
Sporanox	Farmasa/Neo Quimica	-	-
Sporanox	Janssen Pharmaceutica Inc.	Belgium	-
Sporanox	Ortho Biotech. Inc.	USA	-
Sporanox	Johnson and Johnson	India	-
Sporex	Toprak	-	-

Raw Materials

Hydrogen
Hydrobromic acid
Sodium hydride
1-(4-Methoxyphenyl)piperazine dihydrochloride
cis-[2-(2,4-Dichlorophenyl)-2-(1H-1,2,4-triazol-1-ylmethyl)-1,3-dioxolan-4-ylmethyl]methanesulfonate
4-Methyl-2-pentanone
1-Chloro-4-nitrobenzene
Potassium carbonate

Manufacturing Process

Synthesis of cis-4-{4-[4-{4-[2-(2,4-dichlorophenyl)-2-(1H-1,2,4-triazol-1-ylmethyl)-1,3-dioxolan-4-ylmethoxy]phenyl}-1-piperazinyl]phenyl}-2,4-dihydro-2-(methylpropyl)-3H-1,2,4-triazol-3-one is showed by the same procedure as for cis-4-{4-[4-{4-[2-(2,4-dichlorophenyl)-2-(1H-1,2,4-triazol-1-ylmethyl)-1,3- dioxolan-4-ylmethoxy]phenyl}-1-piperazinyl]phenyl}-2,4-

dihydro-2-propyl-3H-1,2,4-triazol-3-one described in the patent.

A mixture of 13.4 parts of 1-(4-methoxyphenyl)piperazine dihydrochloride, 7.9 parts of 1-chloro-4-nitrobenzene, 10 parts of potassium carbonate and 90 parts of N,N-dimethylformamide is stirred and refluxed overnight. The reaction mixture is diluted with water and the product is extracted twice with trichloromethane. The residue is triturated in 4-methyl-2-pentanone. The product is filtered off and crystallized from 1,4-dioxane, yielding 10.5 parts (67%) of 1-(4-methoxyphenyl)-4-(4-nitrophenyl)piperazine; melting point 195.1°C.

A mixture of 12 parts of 1-(4-methoxyphenyl)-4-(4-nitrophenyl)piperazine, 200 parts of methanol and 225 parts of tetrahydrofuran is hydrogenated at normal pressure and at 20°C with 2 parts of palladium-on-charcoal catalyst 10%. After the calculated amount of hydrogen is taken up, the catalyst is filtered off and washed with N,N-dimethylacetamide. Product is filtered off and crystallized from 1-butanol, yielding 8 parts (74%) of 4-[4-(4-methoxyphenyl)-1-piperazinyl]benzenamine; melting point 191.8°C.

A mixture of 30 parts of 4-[4-(4-methoxyphenyl)-1-piperazinyl]benzenamine and 300 parts of a hydrobromic acid solution 48% in water is stirred and refluxed for 10 days. The reaction mixture is evaporated and the residue is alkalized with sodium hydroxide. The mixture is filtered and the filtrate is acidified with acetic acid. The precipitated product is filtered off and crystallized from 1,4-dioxane, yielding 12 parts (44%) of 2,4-dihydro-4-{4-[4-(4-hydroxyphenyl)-1-piperazinyl]phenyl}-2-(1-methylpropyl)-3H-1,2,4-triazol-3-one.

To a stirred solution of 2,4-dihydro-4-{4-[4-(4-hydroxyphenyl)-1-piperazinyl] phenyl}-2-(1-methylpropyl)-3H-1,2,4-triazol-3-one in 100 parts of dimethyl sulfoxide are added 0.3 parts of sodium hydride dispersion 78% and the whole is stirred at 50°C till foaming has ceased. Then there are added 3.7 parts of cis-[2-(2,4-dichlorophenyl)-2-(1H-1,2,4-triazol-1-ylmethyl)-1,3-dioxolan-4-ylmethyl]methanesulfonate and stirring is continued for 3 hours at 100°C. The reaction mixture is cooled and poured onto water. The product is extracted with dichloromethane. The extracts are washed with a diluted sodium hydroxide solution and filtered. The residue is crystallized from 1-butanol. The product yield 4.3 parts (75%) of cis-4-{4-[4-{4-[2-(2,4-dichlorophenyl)-2-(1H-1,2,4-triazol-1-ylmethyl)-1,3-dioxolan-4-ylmethoxy] phenyl}-1-piperazinyl]phenyl}-2,4-dihydro-2-(methylpropyl)-3H-1,2,4-triazol-3-one.

References

Heeres, J., Backx, L.J.J.; US Patent No. 4,267,179; May 12,1981; Assigned to Janssen Pharmaceutica, N.V., Beerse, Belgium

IVERMECTIN

Therapeutic Function: Antiprotozoal

Chemical Name: A mixture of Ivermectin component B_{1a} and Ivermectin component B_{1b}

Common Name: Hyvermectin; Ivermectin

Structural Formula:

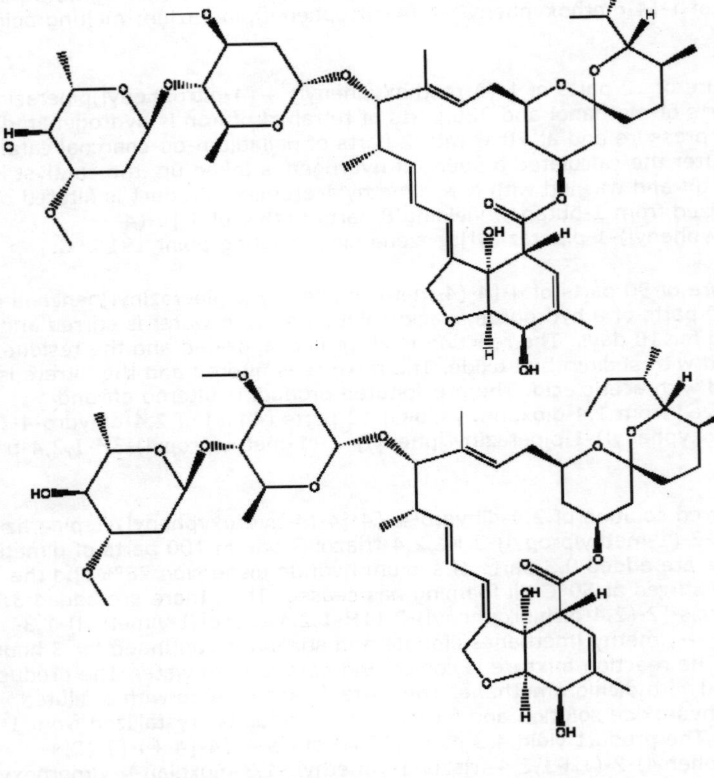

Chemical Abstracts Registry No.: 70288-86-7; 74564-75-3

Trade Name	Manufacturer	Country	Year Introduced
Equimectrin Paste	Merial Limited	USA	-
Eqvalan	Merial	UK	-
Heartgard	Merial Limited	USA	-
Iverhart	Virbac Corporation	-	-
Ivermectol	Ochoa Laboratories (P) Ltd.	-	-
Ivomec	Merial Limited	USA	-
Jetamec	Merial Limited	USA	-
Mectizan	Merck Sharp and Dohme	-	-
Merial	Merial Limited	USA	-

Trade Name	Manufacturer	Country	Year Introduced
Panomec	Merial Limited for Canada	USA	-
Paramax MCS	Coopers	-	-
Qualimec	Janssen	-	-
Stromectol	Merck and Company, Inc.	-	-
Stromectol, Mectizan Generic	Sintofarma/Cifarma	-	-
Zimecterin	Merial	-	-

Raw Materials

Avermectin B_1
Hydrogen
Hydrazine hydrate
Rhodium trichloride trihydrate
Triphenylphosphine
tris-(Hexylphenyl)-phosphine

Manufacturing Process

Avermectin is produced by biotechnological methods with the aid of Streptomyces avermitilis.

Preparation of Catalyst I

Rhodium trichloride trihydrate (1.00 g, 3.80 mmol) was dissolved in water (5.0 ml) with heating (70°C). A solution of triphenylphosphine (1.95 g, 7.43 mmol) in acetone (25.0 ml) was then added under a nitrogen atmosphere in the course of 20 min. After 10 min hydrazine hydrate (1.90 ml; 39.09 mmol) was added with stirring and the mixture was heated at reflux temperature for 3 hours, then kept at 45°C for a further 1 hour. The crystalline solid was filtered off under nitrogen and washed with a little acetone and then with diethyl ether. 1.05 g of an orange-coloured solid were obtained.

Hydrogenation with catalyst I

The catalyst (10 mg) was dissolved in toluene (25 ml) and added under argon to the solution of a mixture (1.1 g) of avermectin B_{1a} (96%) and avermectin B_{1b} (4%) and of 100 mg of triphenylphosphine in toluene (25 ml) in a stainless steel autoclave. This starting material was then hydrogenated at 88°C under a hydrogen pressure of 20 bar with stirring of the solution. After 10 hours, HPLC analysis revealed a content of 86% dihydro-avermectin B_{1a} and of 4 % dihydroavermectin B_{1b}, and also of 3% tetrahydroavermectin B_{1a}.

Preparation of Catalyst II

Under an atmosphere of argon, a mixture of 7.5 mg of rhodium trichloride, 30.0 mg of tris-(hexylphenyl)-phosphine, 3 ml of acetone and 15 ml of hydrazine hydrate is heated with stirring and reflux cooling for 4 hours.

Hydrogenation with catalyst II

The catalyst is added to a solution of 4.3 g of avermectin (B_{1a} and B_{1b} mixture) in 25 ml of a mixture of acetone and cyclohexane in a ratio of 2:1. After addition of 51.4 mg of tris-(mexylphenyl)phosphine, the hydrogenation is carried out in a steel autoclave at a hydrogen pressure 5 bar and at 88°C. After a hydrogenation time of 4 hours, 8.9% of starting material, 89.9% of ivermectin (B_{1a} and B_{1b} mixture), tetrahydroavermectin content <0.1% was obtained (according to HPLC analysis).

Removing of the catalyst system

The crude product after distillative removal of the solvent mixture, dissolved in a mixture of 35 ml of methanol and 20 ml of water and this solution is extracted with 25 ml of cyclohexane in a separating funnel. The phases are separated and concentrated under reduced pressure. The extraction is repeated twice in the same manner.

Result:

The crude product of the hydrogenation

690 ppm of Rh

The resulting product contains

after the 1st extraction	39 ppm of Rh
after the 2nd extraction	29 ppm of Rh
after the 3rd extraction	22 ppm of Rh

References

Arlt D., Bonse G., Reisewitz F.; US Patent No. 5,656,748; August 12, 1997; Assigned: Bayer AG, Leverkusen, Germany

Arlt D., Bonse G., Reisewitz F.; US Patent No. 6,072,052; June 6, 2000; Assigned: Bayer AG, Leverkusen, Germany

J

JOSAMYCIN

Therapeutic Function: Antibiotic

Chemical Name: Leucomycin V, 3-acetate-4β-(3-methylbutanoate)

Common Name: Josamycin; Leucomycin A3; Platenomycin A3; Turimycin A5; Yosamicina

Structural Formula:

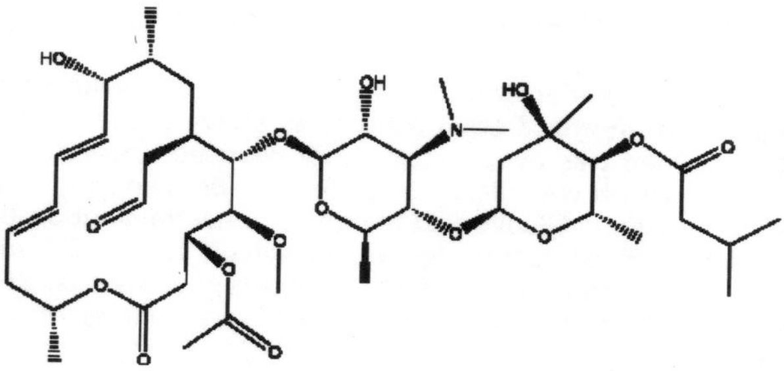

Chemical Abstracts Registry No.: 16846-24-5

Trade Name	Manufacturer	Country	Year Introduced
Josalid	Biochemie	-	-
Josalid	Schering	-	-
Jomybel	Sarva	-	-
Josamina	Novag	-	-
Josamycin	Yamanouchi Pharmaceutical Co., Ltd.	-	-
Josamycin	Shanghai Lansheng Corporation	-	-
Vilprafen	Heinrich Mack	-	-
Proxacin	Yamanouchi	-	-

Raw Materials

Soybean meal	*Streptomyces narbonensis var. josatny ceticus*
Starch	Dipotassium hydrogen phosphate
Glucose	Magnesium sulfate
Sodium chloride	Hydrochloric acid
Sodium hydroxide	

Manufacturing Process

100 ml of a culture medium consisting of water containing 1.5% soybean meal, 1% starch, 1% glucose, 0.3% sodium chloride, 0.1% dipotassium hydrogen phosphate, and 0.05% magnesium sulfate was placed in a 500 ml flask and sterilized for 20 min at 120°C. After cooling, the culture medium was inoculated with strain A 205-P_2 *Streptomyces narbonensis var. josatny ceticus*, and the strain was subjected to shaking culture at 27°-29°C and at 130 strokes per min and 8 cm amplitude. After 3 days of culture, the culture fluids in such 100 flasks were combined together and filtered to give 8700 ml of culture filtrate. The pH of the filtrate was 6.4 and showed an inhibition zone of 25 mm. to Bacillus subtilis (PCI 219 strain). The filtrate was extracted with 8700 ml of ethyl acetate. The extract (7300 ml) thus obtained was concentrated to 730 ml under vacuum at temperatures lower than 50°C, 360 ml of water added, and then concentrated hydrochloric acid added to ad just the pH to 2.0, whereby josamycin was transferred to the aqueous layer. After adjusting the pH of the aqueous layer to 7.5 by the addition of 0.1 N sodium hydroxide, josamycin was extracted with 180 ml of ethyl acetate.

Josamycin was then transferred to 90 ml of an aqueous solution at pH 2.0 and extracted again with 45 ml of ethyl acetate as above process. Ethyl acetate solution thus obtained was evaporated under reduced pressure to give a solidified product, which was dissolved in 5 ml of benzene to remove impurities and the product, solidified from the benzene solution by evaporating under reduced pressure, was dissolved in a small amount of ethyl acetate and subjected to an alumina chromatography. That is, Brockman alumina (Merck) was treated with hydrochloric acid, sufficiently rinsed with water, and activated by heating for 5 h at 150°C.

50.0 g of thus treated alumina was filled in a glass tube of 1.6 cm in diameter by using ethyl acetate. The above prepared ethyl acetate solution was added to the alumina column and the product was eluted with 200 ml of ethyl acetate. The eluate thus obtained was concentrated under reduced pressure and the solid product thus obtained was dissolved in 5 ml of benzene and 50 ml of n-hexane added to give 0.18 g of amorphous josamycin having a purity of above 90%.

References

Umezawa H., Osono T.; US Patent No. 3,636,197; Jan. 18, 1972; Assigned: Yamanouchi Pharmaceutical Co., Ltd., Tokyo, Japan

K

KANAMYCIN SULFATE

Therapeutic Function: Antibacterial

Chemical Name: O-3-Amino-3-deoxy-α-D-glucopyranosyl-(1-6)-O-[6-amino-6-deoxy-α-D-glucopyranosyl-(1-4)-2-deoxy-D-streptamine sulfate

Common Name: -

Structural Formula:

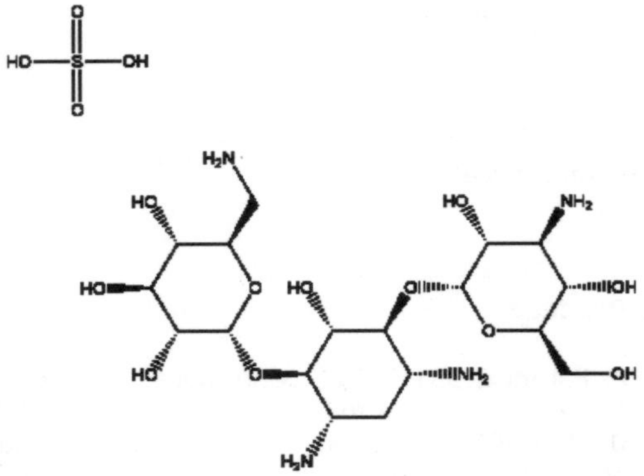

Chemical Abstracts Registry No.: 25389-94-0; 8063-07-8 (Base)

Trade Name	Manufacturer	Country	Year Introduced
Kantrex	Bristol	US	1958
Kanamycine	Bristol	France	1959
Kanabristol	Bristol	W. Germany	1969
Klebcil	Beecham	US	1979
Enterokanacin	Labif	Italy	-
Kamycine	Bristol	France	-
Kanabiol	Osfa	Italy	-
Kanabiot	Galepharma Iberica	Spain	-
Kanacet	Boniscontro-Gazzone	Italy	-

Trade Name	Manufacturer	Country	Year Introduced
Kanacillin	Banyu	Japan	-
Kanacyclin	Banyu	Japan	-
Kanacyn	Continental Pharma	Belgium	-
Kanafil	Farmila	Italy	-
Kanafuracin	Fujita	Japan	-
Kanahidro	Medical	Spain	-
Kanamicina Normon	Normon	Spain	-
Kanamycin	Ferosan	Denmark	-
Kanamytrex	Basotherm	W. Germany	-
Kanapiam	Piam	Italy	-
Kanaqua	Andromaco	Spain	-
Kanasig	Sigma	Australia	-
Kanatrol	Lusofarmaco	Italy	-
Kanescin	Torlan	Spain	-
Kano	Pierrel	Italy	-
Keimicina	Robin	Italy	-
Koptin	Chinoin	Mexico	-
Ophtalmokal ixan	Bristol	France	-
Orakanamicil	Merifarma	Italy	-
Otokal ixan	Bristol	France	-
Visiokan	S.I.F.I.	Italy	-

Raw Materials

Bacterium S. Kanamyceticus
Soybean meal
Dextrin

Manufacturing Process

As described in US Patent 2,931,798, Streptomyces kanamyceticus (K2-J) was first cultured in shake flasks in the following media: (a) 0.75% meat extract, 0.75% peptone, 0.3% NaCl, with 1.0% of starch, dextrin, maltose, glucose, lactose, sucrose or glycerol; or (b) 2.0% soybean meal, 0.05% KCl, 0.05% $MgSO_4 \cdot 7H_2O$, 0.5% NaCl, 0.2% $NaNO_3$, with 1.0% of starch, dextrin, maltose, glucose, lactose, sucrose or glycerol. The initial pH of all media was adjusted to 7.0. After 24 to 48 hours shaking in some cases the pH decreased to about 6.0 to 6.8, but from 72 to 120 hours the pH rose and became 7.5 to 8.6. The production of kanamycin was apparent after 48 hours and, depending on the media; the maximum production was found after 72 to 120 hours.

The yield was highest with starch or dextrin, intermediate and about the same with sucrose, glucose, maltose and lactose and poorest with glycerol. Kanamycin was produced by media containing soybean meal, peanut meal, cottonseed meal, corn steep liquor, peptone, yeast extract or meat extract, with or without sodium nitrate. Commercially available soybean meal was recognized to be one of the best nitrogen sources. The addition of corn steep liquor, peptone, yeast extract or nitrate to the soybean meal promoted the production of kanamycin.

The brownish white kanamycin (5 g) was dissolved in 50 ml of 60% aqueous methanol, insoluble material was removed and to the filtrate 40 ml of 60% aqueous methanol containing 2,000 mg of ammonium sulfate was added, and the precipitated kanamycin sulfate was collected, washed with 50 ml of 80% aqueous methanol, and dried. Thus, 4.5 g of kanamycin sulfate was obtained as a light brownish powder.

References

Merck Index 5118
Kleeman and Engel p. 508
PDR p. 698
I.N.p.539
REM p. 1181
Umezawa, H., Maeda, K. and Ueda, M.; US Patent 2,931,798; April 5, 1960
Extraction:
Johnson, D.A., Hardcastle, G.A., Jr. and Perron, Y.G.; US Patent 2,936,307; May 10, 1960; assigned to Bristol-Myers Company
Purification:
Johnson, D.A. and Harcastle, G.A., Jr.; US Patent 2,967,177; January 3, 1961; assigned to Bristol-Myers Company
Separation Process:
Rothrock, J.W. and Putter, I.; US Patent 3,032,547; May 1, 1962; assigned to Merck and Co.,Inc.

KAWAIN

Therapeutic Function: Anesthetic; Tranquilizer

Chemical Name: (R-(E))-5,6-Dihydro-4-methoxy-6-styryl-2H-pyran-2-one

Common Name: Cavain; Gonosan; Kavain; Kava pyrone; Kawain

Structural Formula:

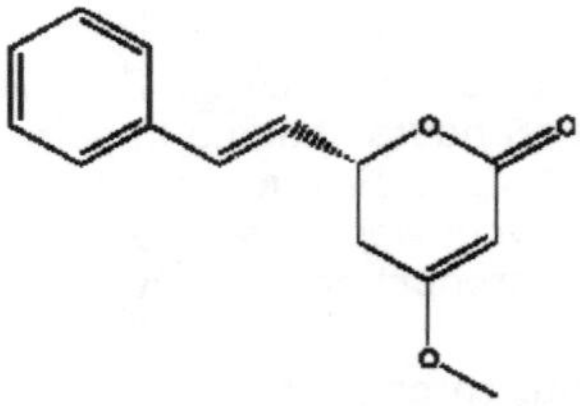

Chemical Abstracts Registry No.: 500-64-1

Trade Name	Manufacturer	Country	Year Introduced
Largon	Klinge	-	-

Raw Materials

Ethyl acetoacetate
Bromosuccinimide
Zinc

Manufacturing Process

To 1170 g ethyl acetoacetate at 100-110°C was added a little at time 1605 g bromosuccinimide. After cooling to the mixture was added 300 ml of carbon tetrachloride. From the mixture was isolated ethyl ester of bromoacetoacetic acid which was distilled at 105-125°C/18 mm; yield 67%.

By condensation of the mixture 1400 g ethyl ester of bromoacetoacetic acid, 700 g bromosuccineimide, 500 ml benzene and 350 mg zinc was prepared (R)-5,6-dihydro-4-methoxy-6-styryl-2H-pyran-2-one; melting point 157°C, yield 60-70%.

References

Fr. Brevet D'Invention 1,526,596; June 9, 1967; Assigned to Spezialchemie G.m.b.H and Co. Residant en Republique Federale d'Allemgane

KEBUZONE

Therapeutic Function: Antirheumatic

Chemical Name: 4-(3-Oxobutyl)-1,2-diphenyl-3,5-pyrazolidinedione

Common Name: Ketophenylbutazone

Structural Formula:

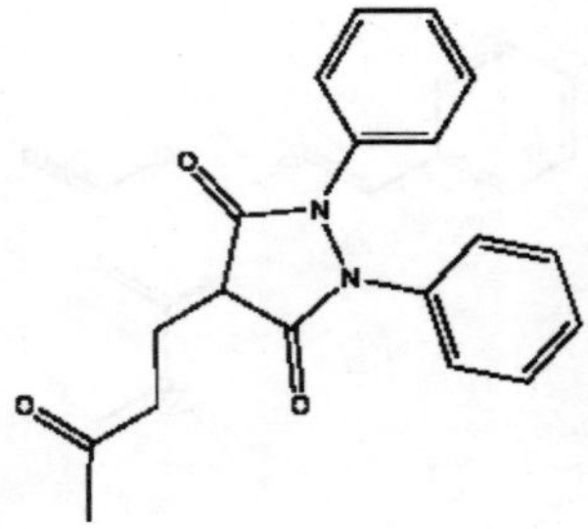

Chemical Abstracts Registry No.: 853-34-9

Trade Name	Manufacturer	Country	Year Introduced
Chebutan	Bioindustria	Italy	1961
Phloguron	Steiner	W. Germany	1976
Chetazolidine	Zeria	Japan	-
Chetopir	Sidus	Italy	-
Chetosol	Aristochimica	Italy	-
Copirene	Marxer	Italy	-
Ejor	Elea	Argentina	-
Hichillos	Kotani	Japan	-
Kebuzon	Steiner	W. Germany	-
Kentan-S	Sawai	Japan	-
Ketazon	Kyowa	Japan	-
Ketazone	Spofa	Czechoslovakia	-
Ketobutan	Santen	Japan	-
Ketobutane	Yamagata	Japan	-
Ketobutazone	Toho	Japan	-
Ketofen	Francia	Italy	-
Ketophezon	Kissei Pharmaceutical Co., Ltd.	Japan	-
Neo-Panalgyl	Italsuisse	Italy	-
Neuphenyl	Ohta	Japan	-
Pecnon	Sanken	Japan	-
Reumo Campil	Lopez-Brea	Spain	-
Vintop	Maruro	Japan	-

Raw Materials

Diethyl malonate
Hydrazobenzene
Sodium ethoxide
Ethylene glycol
Methyl vinyl ketone
Acetone

Manufacturing Process

(a) 3,3-ethylene dioxyburyl malonic acid diethyl ester: Diethylmalonate is reacted with methyl vinyl ketone and the resulting oxobutyl diethylmalonate is reacted with ethylene glycol.

(b) 1,2-diphenyl-4-(3',3'-ethylene dioxybutyl)3,5-dioxopyrazolidine: 274 parts of (3,3-ethylene dioxybutyl)-malonic acid diethyl ester are dissolved in 100 parts by volume of abs. benzene and 57 parts of sodium ethylate and 184 parts of hydrazobenzene are added. Heat is generated. The reaction mass is boiled for 15 hours under reflux. After cooling, it is poured into water, separated and the aqueous part is washed twice with benzene. The benzene solutions are washed three times with 2N sodium carbonate solution and the unified aqueous solutions are acidified with 2N hydrochloric acid. The 1,2-phenyl-4-(3',3'-ethylene dioxybutyl)-3,5-dioxopyrazolidine which precipitates can be recrystallized from alcohol. Melting point 165°C to 167°C.

(c) 1,2-diphenyl-4-(3'-oxobutyl)-3,5-dioxopyrazolidine: 36.6 parts of 1,2-diphenyl-4-(3',3'-ethylene dioxybutyl)-3,5-dioxopyrazolidine in 750 parts by volume of acetone are boiled under reflux for 18 hours with 0.35 part of p-toluene sulfonic acid. The solution is then filtered, 1,500 parts of water are added and the whole is allowed to stand for 24 hours at 5°C. The 1,2-diphenyl-4-(3'-oxobutyl)-3,5-dioxopyrazolidine which precipitates is filtered off under suction and washed with 50% acetone. Melting point from alcohol/water mixture: 115.5°C to 116.5°C. Sometimes a crystal form is obtained which melts at 127.5°C to 128.5°C.

References

Merck Index 5125
Kleeman and Engel p. 509
I.N. p. 540
Denss, R., Pfister, R.and Hafliger, F.; US Patent 2910,481; October 27, 1959; assiged to Geigy Chemical Corp.

KETAMINE HYDROCHLORIDE

Therapeutic Function: Anesthetic

Chemical Name: 2-(o-Chlorophenyl)-2-(methylamino)-cyclohexanone hydrochloride

Common Name: -

Structural Formula:

HCl Cl H N O

Chemical Abstracts Registry No.: 1867-66-9; 6740-88-1 (Base)

Trade Name	Manufacturer	Country	Year Introduced
Ketanest	Parke Davis	W. Germany	1969
Ketanest	Parke Davis	UK	1970
Ketalar	Parke Davis	US	1970
Ketalar	Sankyo	Japan	1970
Ketalar	Parke Davis	France	1970
Ketaject	Bristol	US	1970
Ketalar	Parke Davis	Italy	1972

Raw Materials

Cyclopentyl bromide
Methylamine
Bromine
o-Chlorobenzonitrile
Magnesium

Manufacturing Process

The 1-hydroxycyclopentyl-(o-chlorophenyl)-ketone N-methylimine used as an intermediate is prepared as follows. To the Grignard reagent prepared from 119.0 g of cyclopentyl bromide and 19.4 g of magnesium is added 55.2 g of o-chlorobenzonitrile. The reaction mixture is stirred for 3 days and thereafter hydrolyzed in the usual manner. From the hydrolysis there is obtained o-chlorophenylcyclopentylketone, BP 96° to 97°C (0.3 mm), n_D^{25}1.5452. To 21.0 g of the ketone is added 10.0 g of bromine in 80 ml of carbon tetrachloride.

1-Bromocyclopentyl-(o-chlorophenyl)-ketone, BP 111° to 114°C (0.1 mm) is isolated in the usual manner. Since it is unstable, it must be used immediately. The bromoketone (29.0 g) is dissolved in 50 ml of liquid methylamine. After one hour, the excess liquid methylamine is allowed to evaporate. The organic residue is dissolved in pentane, and upon evaporation of the solvent, 1-hydroxycyclopentyl-(o-chlorophenyl)-ketone N-methylimine, MP 62°C, is isolated.

1-Hydroxycyclopentyl-(o-chlorophenyl)-ketone N-methylimine (2.0 g) is dissolved in 15 ml of Decalin and refluxed for 2,5 hours. After evaporation of the Decalin under reduced pressure, the residue is extracted with dilute hydrochloric acid, the solution treated with decolorizing charcoal, and the resulting acidic solution is made basic. The liberated product, 2-methylamino-2-(o-chlorophenyl)-cyclohexanone, after crystallization from pentane-ether, has MP 92° to 93°C. The hydrochloride of this compound has MP 262° to 263°C.

References

Merck Index 5133
Kleeman and Engel p. 510
PDR p. 1356
OCDSVol.1 p.57 (1977) and 2, 16 (1980)
DOT 2 (4) 152 (1966); 6 (2) 42 (1970) and 2,16 (1980)
I.N. p. 542
REM p. 1045
Stevens, C.L.; US Patent 3,254,124; May 31, 1966; assigned to Parke, Davis and Company

KETAZOLAM

Therapeutic Function: Antianxiety

1996 Ketazolam

Chemical Name: 11-Chloro-8,12b-dihydro-2,8-dimethyl-12b-phenyl-4H-[1,3]-oxazino-[3,2-d][1,4]benzodiazepine-4,7-(6H)-dione

Common Name: -

Structural Formula:

Chemical Abstracts Registry No.: 27223-35-4

Trade Name	Manufacturer	Country	Year Introduced
Anxon	Beecham	UK	1980
Solatran	Beecham	Switz.	1980
Solatran	Beecham	W. Germany	1980
Unakalm	Upjohn	France	1981
Ansietin	Exa	Argentina	-
Contamex	Beecham-Wulfing	W. Germany	-
Loftran	Beecham	-	-

Raw Materials

2-(2-Amino-N-methylacetamido)-5-chlorobenzophenone
Diketene

Manufacturing Process

A solution of 0.7 g of 2-(2-amino-N-methylacetamido)-5-chlorobenzophenone in 10 ml of a 50% solution (by weight) of diketene in acetone is refluxed for 3 hours and then evaporated to give a brown oil. The oil is chromatographed on 200 g of silica gel using a 1:1 (by volume) mixture of ethyl acetate cyclohexane; 25 ml fractions are collected. Fractions 11-14 are combined, mixed with chloroform, evaporated and triturated with ether to give 0.337 g of 11-chloro-8,12b-dihydro-2,8-dimethyl-12b-phenyl-4H-[1,3]oxazino[3,2-d][1.4] benzodiazepine-4,7(6H)-dione as a pale yellow solid, MP 174°C to 176°C.

References

Merck Index 5134
DFU 1 (6) 293 (1976)

OCDS Vol. 1 p. 369 (1977)
DOT 16 (9) 293 (1980)
I.N. p. 542
Szmuszkoviez, J.; US Patent 3,575965; April 20, 1971; assigned to The Upjohn Co.

KETOCONAZOLE

Therapeutic Function: Antifungal

Chemical Name: 1-Acetyl-4-[4-[[2-(2,4-dichlorophenyl)-2(1H-imidazol-1-ylmethyl)-1,3-dioxolan-4-yl]methoxy]phenyl]piperazine

Common Name: -

Structural Formula:

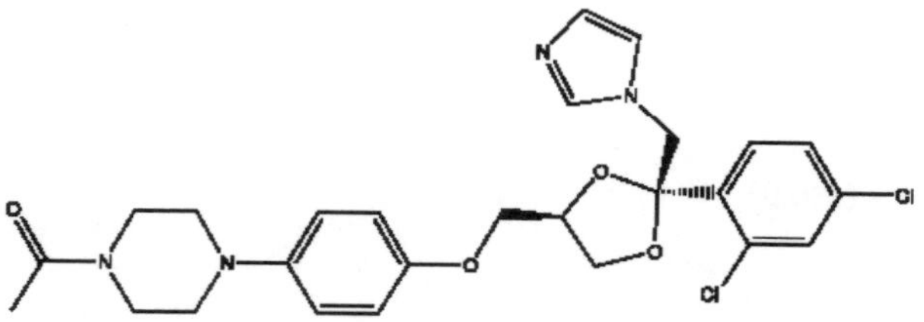

Chemical Abstracts Registry No.: 65277-42-1

Trade Name	Manufacturer	Country	Year Introduced
Nizoral	Janssen	US	1981
Nizoral	Janssen	W. Germany	1981
Nizoral	Janaen	Switz.	1981
Nizoral	Janssen	UK	1981
Nizoral	Janssen-Le Brun	France	1983
Nizoral	Janssen	Italy	1983
Ketazol	Exa	Argentina	-

Raw Materials

4-(1-Piperazinyl)phenol dihydrobromide
Acetic anhydride
cis-2-(2,4-Dichlorophenyl)-2-(1H-imidazol-1-ylmethyl)-1,3-dioxolan-4-yl methyl methane sulfonate

Manufacturing Process

(A) A mixture of 33.8 parts of 4-(4-piperazinyl)phenol dihydrobromide, 11.2 parts of acetic acid anhydride, 42 parts of potassium carbonate and 300 parts of 1,4-dioxane is stirred and refluxed for 3 days. The reaction mixture is filtered and the filtrate is evaporated. The solid residue is stirred in water and sodium hydrogen carbonate is added. The whole is stirred for 30 minutes. The precipitated product is filtered off and dissolved in a diluted hydrochloric acid solution. The solution is extracted with trichloromethane. The acid aqueous phase is separated and neutralized with ammonium hydroxide. The product is filtered off and crystallized from ethanol, yielding 5.7 parts of 1-acetyl-4-(4-hydroxyphenyl)piperazine; MP 181-183°C.

(B) A mixture of 2.4 parts of 1-acetyl-4-(4-hydroxyphenyl)piperazine, 0.4 part of sodium hydride dispersion 78%; 75 parts of dimethylsulfoxide and 22.5 parts of benzene is stirred for one hour at 40°C. Then there are added 4.2 parts of cis-2-(2,4-dichlorophenyl)-2-(1H-imidazol-1-ylmethyl)-1,3-dioxolan-4-ylmethyl methane sulfonate and stirring is continued overnight at 100°C. The reaction mixture is cooled and diluted with water. The product is extracted with 1,1'-oxybisethane. The extract is dried, filtered and evaporated. The residue is crystallized from 4-methyl-2-pentanone. The product is filtered off and dried, yielding 3.2 parts (59%) of cis-1-acetyl-4-[2-(2,4-dichlorophenyl)-2-(1H-imidazol-1-ylmethyl)-1,3-dioxolan-4-ylmethoxyl phenyl]piperazine; MP 146°C.

References

Merck Index 5139
DFU 4 (7) 496 (1979)
PDR p. 956
OCDS Vol. 3 p. 132 (1984)
DOT 17 (9) 377 (1981)
I.N. p. 542
REM p. 1229
Heeres, J., Backx, L.J.J. and Mostmans, J.H.; US Patent 4,144,346; March 13,1979; assigned to Janssen Pharmaceutica N.V. (Belgium)

KETOPROFEN

Therapeutic Function: Antiinflammatory

Chemical Name: m-Benzoylhydratropic acid

Common Name: 2-(3-Benzoyiphenyl)propionic acid

Structural Formula:

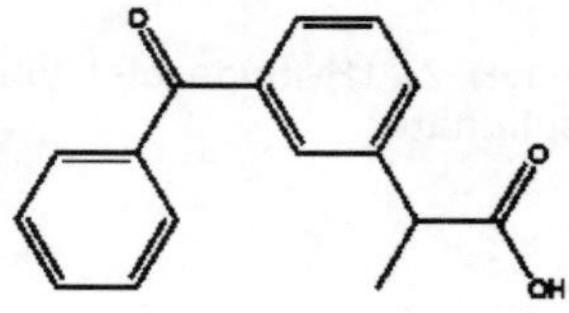

O
OH

Chemical Abstracts Registry No.: 22071-15-4

Trade Name	Manufacturer	Country	Year Introduced
Profenid	Specia	France	1973
Orudis	May and Baker	UK	1973
Alrheumin	Bayropharm	W. Germany	1975
Orudis	Farmitalia	Italy	1975
Keto	Sigurta	Italy	1976
Orudis	Hokuriku	Japan	1978
Capisten	Kissei Pharmaceutical Co., Ltd.	Japan	1978
Inflen	Ohta	Japan	1983
Zaditen	Sandoz	Japan	1983
Orudis	Leo Rhodia	Sweden	1983
Alrheumat	Bayer	UK	-
Arcental	Janovich	Spain	-
Dexal	Pulitzer	Italy	-
Fastum	Manetti-Roberts	Italy	-
Flexen	Italfarmaco	Italy	-
Helenil	Roux-Ocefa	Argentina	-
Iso-K	San Carlo	Italy	-
Kefenid	S.I.T.	Italy	-
Ketalgin	I.B.P.	Italy	-
Ketofen	Nobel	Turkey	-
Keton	Ilsan	Turkey	-
Ketonal	Lek	Yugoslavia	-
Ketopron	Biosintetica	Brazil	-
Ketoprosil	Liberman	Spain	-
Ketoval	Valles Mestre	Spain	-
Kevadon	Lemonier	Argentina	-
Knavon	Belupo Ltd.	Yugoslavia	-
Lertus	Exa	Argentina	-
Meprofen	A.G.I.P.S.	Italy	-
Niflam	Alkaloid	Yugoslavia	-
Profenid	Specia	France	-
Remauric	Lifepharma	Spain	-
Romin	Fako	Turkey	-
Salient	Biomedica Foscama	Italy	-
Sinketol	Italchemie	Italy	-
Wasserprofen	Wassermann	Spain	-

Raw Materials

Ethanol
(3-Benzoylphenyl)acetonitrile
Sulfuric acid
Sodium
Methyl iodide

Manufacturing Process

In an initial step, the sodium derivative of ethyl (3-benzoylphenyl) cyanoacetate is prepared as follows: (3-benzoylphenyl)acetonitrile (170 9) is dissolved in ethyl carbonate (900 g). There is added, over a period of 2 hours, a sodium ethoxide solution [prepared from sodium (17.7 g) and anhydrous ethanol (400 cc)], the reaction mixture being heated at about 105° to 115°C and ethanol being continuously distilled. A product precipitates. Toluene (500 cc) is added, and then, after distillation of 50 cc of toluene, the product is allowed to cool. Diethyl ether (600 cc) is added and the mixture is stirred for 1 hour. The crystals which form are filtered off and washed with diethyl ether (600 cc) to give the sodium derivative of ethyl (3-benzoylphenyl)cyanoacetate (131 g).

Then, ethyl methyl(3-benzoylphenyl)cyanoacetate employed as an intermediate material is prepared as follows: The sodium derivative of ethyl (3-benzoylphenyl)cyanoacetate (131 g) is dissolved in anhydrous ethanol (2 liters). Methyl iodide (236 g) is added and the mixture is heated under reflux for 22 hours, and then concentrated to dryness under reduced pressure (10 mm Hg). The residue is taken up in methylene chloride (900 cc) and water (500 cc) and acidified with 4N hydrochloric acid (10 cc). The methylene chloride solution is decanted, washed with water (400 cc) and dried over anhydrous sodium sulfate. The methylene chloride solution is filtered through a column containing alumina (1,500 g). Elution is effected with methylene chloride (6 liters), and the solvent is evaporated under reduced pressure (10 mm Hg) to give ethyl methyl(3-benzoylphenyl)cyanoacetate (48 g) in the form of an oil.

In the final production preparation, a mixture of ethyl methyl(3-benzoylphenyl)cyanoacetate (48 g), concentrated sulfuric acid (125 cc) and water (125 cc) is heated under reflux under nitrogen for 4 hours, and water (180 cc) is then added. The reaction mixture is extracted with diethyl ether (300 cc) and the ethereal solution is extracted with N sodium hydroxide (300 cc). The alkaline solution is treated with decolorizing charcoal (2 g) and then acidified with concentrated hydrochloric acid (40 cc). An oil separates out, which is extracted with methylene chloride (450 cc), washed with water (100 cc) and dried over anhydrous sodium sulfate. The product is concentrated to dryness under reduced pressure (20 mm Hg) to give a brown oil (33.8 g).

This oil is dissolved in benzene (100 cc) and chromatographed through silica (430 g). After elution with ethyl acetate, there is collected a fraction of 21 liters, which is concentrated to dryness under reduced pressure (20 mm Hg). The crystalline residue (32.5 g) is recrystallized from acetonitrile (100 cc) and a product (16.4 g), MP 94°C, is obtained. On recrystallization from a mixture of benzene (60 cc) and petroleum ether (200 cc), there is finally obtained 2-(3-benzoylphenyl)propionic acid (13.5 g), MP 94°C.

References

Merck Index 5142
Kleeman and Engel p. 511
OCDS Vol. 2 p. 64 (1980)
DOT 9 (11) 469 (1973) and 19 (3) 160 (1983)
I.N. p. 543

Farge, D., Messer, M.N. and Moutonnier, C.; USPatent 3,641,127; February 8,1972; assigned to RhonePoulenc S.A., France

KETOROLAC TROMETHAMINE

Therapeutic Function: Analgesic, Antiinflammatory

Chemical Name: 1H-Pyrrolizine-1-carboxylic acid, 2,3-dihydro-5-benzoyl-, (+-)-, compd. with 2-amino-2-(hydroxymethyl)-1,3-propanediol (1:1)

Common Name: Ketorolac tromethamine; Ketorolac trometamol; Trometamol keterolac

Structural Formula:

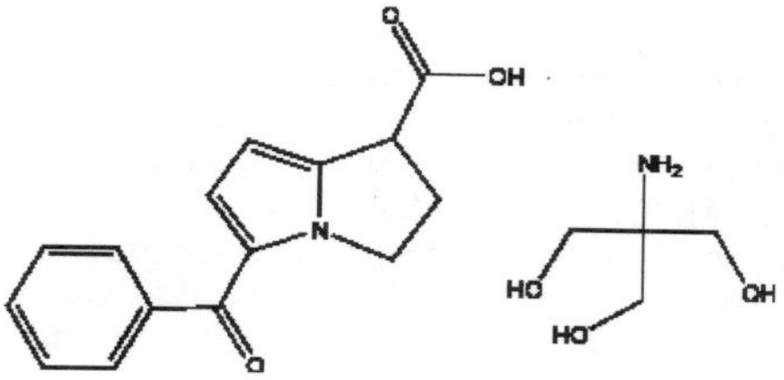

Chemical Abstracts Registry No.: 74103-07-4

Trade Name	Manufacturer	Country	Year Introduced
Acular	Allergan	India	-
Apo-Ketorolac Ophthalmic Solution	Apotex Inc.	-	-
Ketorolac Tromethamine Injection USP	Sabex Inc.	Canada	-
Toradol	Syntex	Switz.	-
Toradol	Roche	-	-
Toradol	Apotex Inc.	Canada	-
Toradol	Novopharm	Canada	-
Toradol	Nu-Pharm Inc.	Canada	-
Toradol	Ratiopharm	Canada	-

Raw Materials

Bromine
Methylaniline
Triethylamine
Butyl diglyme
Pyrrole
Diglyme
Sodium hydroxide
4-Chlorobutanoyl chloride
Phosphorus tribromide
Methyl magnesium chloride
ALIQUAT (phase transfer catalyst)
Phosphorus oxychloride
Benzoyl chloride

Manufacturing Process

Preparation of 2-bromo-4-chloro-N-methyl-N-phenylbutanamide.

4-Chlorobutanoyl chloride (62 g, 440 mmol) and phosphorus tribromide (3 g) were added to a distillation flask, and heated to 90°C. Bromine (77.5 g, 485 mmol) was added over eight hours, with the solution being allowed to decolorize between additions. After the addition was complete, and the solution decolorized, a vacuum was slowly applied, and the acid gases and phosphorus tribromide scrubbed. Unreacted starting material was distilled at 98-100°C/22 mm Hg, and the temperature slowly increased to 105°C, where a mixture of 2-bromo-4-chlorobutanoyl chloride and 2-bromo-4-chlorobutanoyl bromide began to distill. Pure 2-bromo-4-chlorobutanoyl bromide distilled at approximately 108°C. The combined yield of 2-bromo-4-chlorobutanoyl chloride and 2-bromo-4-chlorobutanoyl bromide was 100.5 g, with a bromide/chloride ratio of approximately 6:1. The mixture of 2-bromo-4-chlorobutanoyl bromide and chloride is directly usable in the preparation of the butanamide, if desired, or may be separated and either component used.

2-Bromo-4-chlorobutanoyl bromide (300 mmol) was added to a solution of N-methylaniline (320 mmol) and triethylamine (330 mmol) in toluene (340 mL). The reaction was exothermic, and the mixture was cooled to maintain the temperature at about 40°C. After the addition was complete, the resulting mixture was stirred for 30 minutes, 150 mL water was added, and the mixture was stirred further. The aqueous and organic phases were separated, and the organic phase was washed with 5% hydrochloric acid and with water. The toluene was evaporated completely under vacuum to yield 86.3 g 2-bromo-4-chloro-N-methyl-N-phenylbutanamide (98% yield, approximately 95-96% pure).

A solution of methylmagnesium chloride in butyl diglyme (4.0 L, 2.8 M, 11.2 mol, 2.8 equivalents with respect to 2-bromo-4-chloro-N-methyl-N-phenylbutanamide) was added to a 12 L 4-necked round bottom flask fitted with a mechanical stirrer and two 1 L addition funnels, under a nitrogen atmosphere. 2-Bromo-4-chloro-N-methyl-N-phenyl-butanamide (3.98 mol) was added to the first addition flask, and pyrrole (3.04 equivalents with respect to 2-bromo-4-chloro-N-methyl-N-phenyl-butanamide) was added to the second. The pyrrole was slowly added to the methylmagnesium chloride/butyl diglyme solution at 45-50°C over 3 hours. The resulting viscous mixture was cooled to 25°C and stirred for 30 min. 2-Bromo-4-chloro-N-methyl-N-phenylbutanamide was added to the resulting mixture over a period of 2 hours at 25-30°C, and the resulting solution was stirred for another 3 hours.

The dark colored reaction mixture was transferred into 5.76 mol 2 N hydrochloric acid with rapid stirring for 1 hour. The aqueous phase was removed, and 0.8 L 15 weight % ammonium chloride in water was added to the organic phase. The resulting mixture was stirred at 35-40°C for 10 min, the aqueous phase then removed, and hexanes (2.4 L) added. The resulting suspension was cooled to -20°C and maintained at that temperature for a few minutes. The precipitate was filtered in a 300 mL sintered glass funnel and washed with hexanes (1 L). Drying of the solid under vacuum at 25-30°C yielded 4-chloro-N-methyl-N-phenyl-2-(2-pyrrolyl)butanamide (81% yield).

A solution of 4-chloro-N-methyl-N-phenyl-2-(2-pyrrolyl)butanamide in toluene was added dropwise at 85°C over 40 min to 1 hour to a stirred suspension of ALIQUAT 336 (phase transfer catalyst, 2 mol % with respect to pyrrolylbutanamide) and granular sodium hydroxide (3 equivalents) in toluene (50 mL). After the addition was complete, the suspension was stirred under a nitrogen atmosphere at a temperature of 85°C for 30 min, then cooled to 35°C. Cooled water (200 mL) was rapidly added to the mixture and stirred for 15 min at 25°C. The solution was rinsed with water and the layers were separated. The organic layer was washed with water, then distilled under atmospheric pressure to recover the toluene and water. The resultant solution was cooled to 50°C and allowed to crystallize after the addition of hexane and a seed crystal. The suspension was cooled to 5°C and stirred for 15 minutes. The resultant precipitate was filtered, washed with 100 mL of hexane, and dried under vacuum at 25°C to yield approximately 38 g (63%) N-methyl-N-phenyl-2,3-dihydro-1H-pyrrolizine-1-carboxamide. This solid was recrystallized from toluene to yield colorless crystals of N-methyl-N-phenyl-2,3-dihydro-1H-pyrrolizine-1-carboxamide, melting point 112-112.5°C.

Benzoyl chloride (4.3 mol) was added dropwise to a rapidly stirring mixture of piperidine (4.3 mol), sodium hydroxide (4.7 mol), toluene (1 L), and water (1.7 L) over a period of 70 min. After the addition was complete, the mixture was stirred at 25°C for one hour. The organic and aqueous phases were separated, and the organic phase was washed with 2 N hydrochloric acid, concentrated by rotary evaporation, and distilled under vacuum to yield benzoylpiperidine as a colorless liquid which crystallized on standing (95% yield, boiling point 169-171°C).

N-Methyl-N-phenyl-2,3-dihydro-1H-pyrrolizine-1-carboxamide (480 mmol) and toluene (100 mL) were added to a mixture of benzoylpiperidine (1.05 eq.) and phosphorus oxychloride (0.96 eq.), which had been stirred at 25°C for 1 hour. An additional 100 mL toluene was added. The suspension was heated to at 40-45°C for 4 hours. The resulting syrup was transferred into a rapidly stirring solution of sodium hydroxide (4.5 mol), piperidine (1.0 mL), and water (650 mL) at 25-35°C and the mixture was stirred for 1 hour. A mixture of toluene (100 mL), water (50 mL), and sodium hydroxide (12 g, 300 mmol) was added to the reaction flask, and the reaction mixture was stirred at 25°C for 1 hour. The suspension was then heated to 75°C and the layers were separated. The organic layer was cooled to 60°C and hexane (100 mL) was slowly added, and the solution slowly stirred and cooled to -15°C. The precipitate was filtered, washed with toluene/hexane (2:1) and then with hexane, and dried under vacuum at 25°C to yield 5-benzoyl-N-methyl-N-phenyl-2,3-dihydro-1H-pyrrolizine-1-carboxamide (83.5% yield).

Preparation of ketorolac tromethamine.

A mixture of 34.4 g (100 mmol) 5-benzoyl-N-methyl-N-phenyl-2,3-dihydro-1H-pyrrolizine-1-carboxamide, 25 g sodium hydroxide in 25 mL water, and 80 mL methanol was refluxed for 5 hours. The mixture was cooled to room temperature, stirred under nitrogen for sixteen hours, and then diluted with 80 mL of water. The mixture was extracted with toluene, and the aqueous and organic phases were separated. The aqueous phase was acidified with 6 N hydrochloric acid. The resulting precipitate was extracted with dichloromethane. The combined extract was treated with activated clay decolorizing agent (4.5 g) for 30 minutes, filtered, and concentrated by atmospheric distillation. Hexane was added and the mixture allowed to cool to 0-5°C. The product, 5-benzoyl-2,3-dihydro-1H-pyrrolizine-1-carboxylic acid (ketorolac) was collected by filtration, washed with 100 mL of hexane/dichloromethane (7:3), and dried at 60°C under vacuum, to yield ketorolac (83.4% yield), melting point 152-162°C. Ketorolac (25 g) and 11.9 g tromethamine were dissolved in 175 mL methanol. The solution was filtered and the filter washed with 40 mL methanol. The resulting solution was concentrated by vacuum distillation. Ethylacetate was added to precipitate the ketorolac tromethamine; and the solution was cooled to room temperature for two hours, cooled further to 0°C, and filtered. The precipitate was washed with ethyl acetate/methanol (4:1) and dried under vacuum at a temperature of 65°C, to yield ketorolac tromethamine (95% yield).

References

Yarrington P.J. et al.; US Patent No. 6,197,976; 03.06.2001; Assigned to Syntex (U.S.A.) LLC

KETOTIFEN

Therapeutic Function: Anti-asthmatic, Antihistaminic

Chemical Name: 4-(1-Methyl-4-piperidylidene)-4H-benzo[4,5]cyclohepta [1,2-b]-thiophen-10(9H)-one

Common Name: -

Structural Formula:

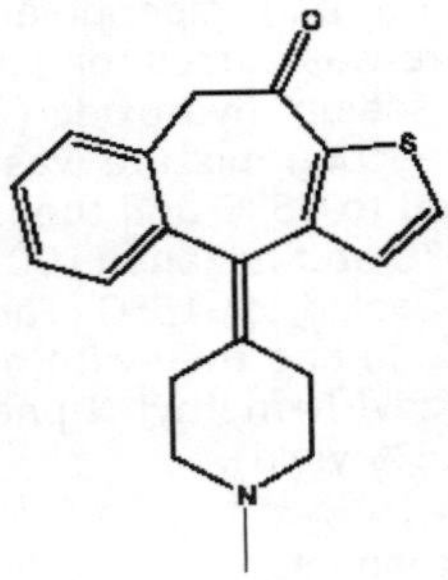

Chemical Abstracts Registry No.: 34580-13-7

Raw Materials

4-Chloro-1-methylpiperidine
Magnesium
10-Methoxy-4H-benzo[4,5]cyclohepta[1,2-b]thiophen-4-one
Hydrogen chloride

Trade Name	Manufacturer	Country	Year Introduced
Zaditen	Wander	Switz.	1978
Zaditen	Sandoz	W. Germany	1979
Zaditen	Sandoz	UK	1979
Zaditen	Sandoz	France	1980
Zaditen	Sandoz	Italy	1982
Zaditen	Sandoz	Japan	1983
Totifen	Chiesi	Italy	1983
Zasten	Sandoz	-	-

Manufacturing Process

3.07 g of iodine-activated magnesium shavings are covered with a layer of 25 cc of tetrahydrofuran, and approximately 1/10 of a solution of 17.7 g of 4-chloro-1-methylpiperidine base in 70 cc of absolute tetrahydrofuran is added. The Grignard reaction is initiated by the addition of a few drops of 1,2-dibromoethane. The remaining 4-chloro-1-methylpiperidine solution is then added dropwise to the magnesium at such a rate that the reaction mixture boils continuously at reflux without external heating. Boiling at reflux is then continued for 1 hour. 15.3 g of 10-methoxy-4H-benzo[4,5]cyclohepta[1,2-b]thiophen-4-one are subsequently added portionwise at 20°C, within 40 minutes, with slight cooling. After stirring at 20°C for 1,5 hours, the reaction solution is poured on a mixture of 180 g of ice and 20 g of ammonium chloride. The free base is extracted with chloroform.

The chloroform solution is concentrated and the residue recrystallized from 270 cc of absolute ethanol. The pure 10-methoxy-4-(1-methyl-4-piperidyl)-4H-benzo[4,5]cyclohepta[1,2-b]thiophen-4-ol base, having a melting point of 194°C to 196°C, is obtained in this manner. Microanalysis corresponds with the formula $C_{20}H_{23}NO_2S$.

A mixture of 3.4 g of 10-methoxy-4-(1-methyl-4-piperidyl)-4H-benzo[4,5] cyclohepta [1,2-b]thiophen-4-ol base and 40 cc of 3N hydrochloric acid is kept in a boiling water bath at 95°C to 100°C for 1 hour. The mixture is subsequently made alkaline with concentrated caustic soda solution at 20°C while cooling, and the free base is extracted with chloroform. The chloroform solution is concentrated, and the residue is recrystallized from ethanol/water 1:1. The pure 4-(1-methyl-4-piperidylidene)-4H-benzo[4,5]cyclohepta [1,2-b] thiophen-10(9H)-one base, having a melting point of 152°C to 153°C, is obtained in this manner.

References

Merck Index 5144
DFU 2 (2) 108 (1977)
Kleeman and Engel p. 512
OCDS Vol. 3 p. 239 (1984)
DOT 14 (8) 370 (1978)
I.N. p. 543
Bourquin, J.P., Schwarb, G. and Waldvogel, E.; US Patents 3,682,930; Aug. 8, 1972; 3,770,728; Nov. 6, 1973 and 3,960,894; June 1, 1976; all assigned to Sandoz, Ltd.

L

LABETALOL HYDROCHLORIDE

Therapeutic Function: Alpha-adrenergic blocker, Beta-adrenergic blocker

Chemical Name: 2-Hydroxy-5-[1-hydroxy-2-[(1-methyl-3-phenylpropyl) amino]ethyl]benzamide hydrochloride

Common Name: Ibidomide

Structural Formula:

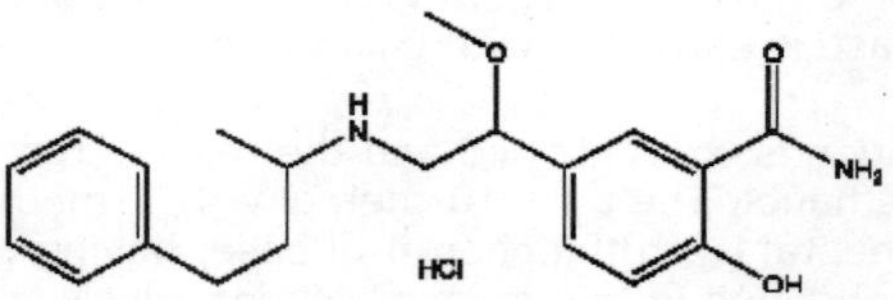

Chemical Abstracts Registry No.: 36894-69-6; 32780-64-6 (Hydrochloride salt)

Trade Name	Manufacturer	Country	Year Introduced
Trandate	Allen and Hanburys	UK	1977
Trandate	Glaxo	W. Germany	1977
Labetalol	Duncan	Italy	1978
Trandate	Glaxo	Switz.	1979
Trandate	Glaxo	France	1980
Trandate	Glaxo	Japan	1983
Abetol	C.T.	Italy	-
Labelol	Elea	Argentina	-
Lamitol	Pliva	Yugoslavia	-
Lolum	Farmochimica	Italy	-
Mitalolo	Ellem	Italy	-
Normodyne	Schering	US	-
Presdate	Alfa Farm.	Italy	-

Raw Materials

5-Bromoacetylsalicylamide
N-Benzyl-N-(1-methyl-3-phenylpropyl)amine
Hydrogen

Manufacturing Process

(a) 5-Bromoacetylsalicylamide (2.6 g), N-benzyl-N-(1-methyl-3-phenylpropyl) amine (4.8 g) and methyl ethyl ketone (50 ml) were heated at reflux for 40 minutes. The solvent was removed and the residue was treated with benzene. The secondary amine hydrobromide was filtered off and discarded, and the filtrate was evaporated to dryness. The residue was treated with an excess of ethanolic hydrogen chloride when 5-[N-benzyl-N-(1-methyl-3-phenylpropyl)-glycyl]-salicylamide hydrochloride (1.15 g) crystallized out, MP 139°C to 141°C.

(b) 5-[N-benzyl-N-(1-methyl-3-phenylpropyl)glycyl]-salicylamide hydrochloride (0.75 g), 10% mixture of PdO and PtO on carbon catalyst (0.1 g) and ethanol (20 ml) were shaken at room temperature and pressure with hydrogen until uptake ceased. The catalyst was filtered off and the filtrate evaporated to dryness. The residue was crystallized from ethanol to give 5-[1-hydroxy-2-(1-methyl-3-phenylpropyl)aminoethyl]salicylamide hydrochloride as a white solid (0.40 g), MP 188°C.

References

Merck Index 5166
DFU 1 (3) 125 (1976)
Kleeman and Engel p. 513
PDRpp.913, 1638
OCDS Vol. 3 p. 24 (1984) and 18 (8) 378 (1982)
DOT 13 (11) 493 (1977)
I.N. p. 547
REMp. 904
Lunts, L.H.C. and Collin, D.T.; US Patent 4,012,444; March 15, 1977; assigned to Allen and Hanburys Ltd. (UK)

LACTULOSE

Therapeutic Function: Laxative

Chemical Name: 4-O-β-D-Galactopyranosyl-D-fructose

Common Name: -

Chemical Abstracts Registry No.: 4618-18-2

Structural Formula:

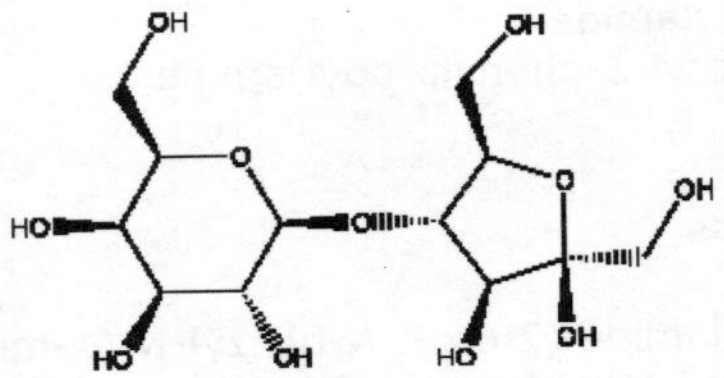

Trade Name	**Manufacturer**	**Country**	**Year Introduced**
Duphalac	Philips-Duphar	UK	1969
Bifiteral	Philips-Duphar	W. Germany	1971
Duphalac	Duphar	France	1972
Duphalac	Duphar	Italy	1973
Gatinar	Duphar	UK	1973
Lactulose	Nikken	Japan	1973
Cephulac	Merrell Dow	US	1976
Duphalac	Philips Roxane	US	1977
Chronulac	Merrell Dow	US	1979
Dia-Colon	Piam	Italy	-
Epalfen	Zambon	Italy	-
Laevilac	Wander	W. Germany	-
Laevolac	Laevosan	Austria	-
Monilac	Chugai	Japan	-

Raw Materials

Lactose
Sodium aluminate

Manufacturing Process

105 g of lactose monohydrate were dissolved in 500 ml of water. 48 g of $NaAlO_2$ was dissolved in 100 ml of water and was then added to the lactose solution. The mixture was then diluted to one liter to provide a pH of 11.5. The reactant concentrations of 48 g of sodium aluminate and 105 g of lactose are equivalent to a mol ratio of two mols of aluminate to one mol of lactose. The mixture was then heated to 50°C and 100 ml aliquots were removed at periodic intervals to determine the level of conversion. The reaction was terminated after three hours by adding sufficient 30% HCl to lower the pH to 4.2. The pH was then raised to neutrality, i.e., 6.5 to 7.0, with ammonium hydroxide so as to completely precipitate insoluble aluminum hydroxide. The precipitate was then removed by vacuum filtration and the filtrate was analyzed for the presence of ketose sugar by chromatographic analysis. The chromatographic analysis of the filtrate confirmed that the main component of the filtrate was lactulose and not the monosaccharide ketose sugar, fructose.

References

Merck Index 5184
Kleeman and Engel p. 513
PDR p. 1224
I.N. p. 548
REM p. 814
Guth, J.H. and Tumerman, L.; US Patent 3,546,206; December 8, 1970; assigned to Kraftco Corp.

LAMIVUDINE

Therapeutic Function: Antiviral

Chemical Name: 2(1H)-Pyrimidinone, 4-amino-1-((2R,5S)-2-(hydroxymethyl)-1,3-oxathiolan-5-yl)-

Common Name: Lamivudine

Structural Formula:

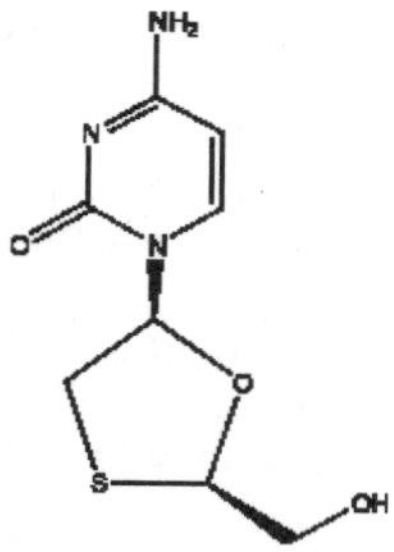

Chemical Abstracts Registry No.: 134678-17-4

Trade Name	Manufacturer	Country	Year Introduced
Epivir (3TC)	GlaxoSmithKline	USA	-
Hepitec	Glaxo Smithkline	-	-
Heptovir	GlaxoSmithKline	Canada	-
Ladiwin	Cadila Healthcare	India	-
Ladiwin	Zydus Biogen	India	-
Lamda	Le Sante	India	-
Lamidac	Zydus Alidac	India	-
Lamivir	Cipla Limited	India	-
Lamivir-Hbv	Cipla Limited	India	-
Lamivudine	GlaxoSmithKline	USA	-
Zeffix	Glaxo Wellcome	UK	-

Raw Materials

Thiobenzoic acid
Potassium t-butoxide
1-Benzoyl glycerol
Cytosine
Trimethylsilyl chloride
4-Toluenesulfonic acid
Bromoacetaldehyde diethyl acetal
Sodium periodate
Hexamethyldisilazane

Manufacturing Process

To a solution of potassium t-butoxide (0.11 mol) in 100 ml DMF was added thiobenzoic acid (0.11 mol) and the solution partially evaporated in vacuo, benzene added in two consecutive portions and evaporated in vacuo each time. To the residual DMF solution was added bromoacetaldehyde diethyl acetal (0.1 mol) and the mixture stirred at 120°C for 15 h. After cooling, it was poured onto water (500 ml), the product extracted with ether, the extract washed with aqueous $NaHCO_3$ followed by water, then dried and the solvent removed in vacuo. The residue was distilled in vacuo to give 17.2 g of pure 2-thiobenzoyl acetaldehyde diethyl acetal, boiling point 131-133°C/0.07 mm.

The 2-thiobenzoyl acetaldehyde diethyl acetal (17.2 g) was dissolved in 100 ml THF followed by the addition of 6 g NaOH in 20 ml H_2O. The mixture was refluxed under N_2 for 15 h, then cooled and diluted with water (200 ml) and the product extracted with ether (3 x 200 ml). The extract was dried, the solvent removed in vacuo and the residue distilled to yield 7.1 g of mercaptoacetaldehyde diethylacetal.

50 g of the 1-benzoyl glycerol in a mixture of 500 ml of CH_2Cl_2 and 25 ml of H_2O was treated portionwise with 80 g of NaIO4 under vigorous stirring at room temperature. After addition, stirring was continued for 2 h after which time 100 g of $MgSO_4$ was added and stirring continued for 30 min. The mixture was filtered, the filtrate evaporated in vacuo and the residue distilled to yield 26 g of pure benzoyloxyacetaldehyde, boiling point 92-94°C/0.25 mm.

2-Benzoyloxymethyl-5-ethoxy-1,3-oxathiolane:

The mercaptoacetaldehyde diethylacetal (7 g) was mixed in 100 ml of toluene with 7 g of the above benzoyloxyacetaldehyde, a few crystals of p-toluenesulfonic acid added and the mixture place in an oil-bath at 120°C under N_2. The formed ethanol was allowed to distill over, the mixture kept at 120°C for 30 min longer than cooled and washed with aqueous $NaHCO_3$, dried and evaporated in vacuo. The residue was distilled in vacuo to yield 9.8 g of 2-benzoyloxymethyl-5-ethoxy-1,3-oxathiolane as a mixture of cis- and trans-isomers, boiling point 140-143°C/0.1 mm.

Cis- and trans-2-benzoyloxymethyl-5-cytosin-1'-yl-1,3-oxathiolane:

A mixture of 2.7 g of cytosine, 30 ml of hexamethyldisilazane (HMDS) and 0.3 ml of trimethylsilyl chloride (TMSCl) was heated under reflux under dry N_2 untila clear solution resulted (3 L) and the excess reagents evaporated in vacuo. The remaining volatiles were removed under high vacuum, the solid residue taken up in 250 ml of dichlorethane and 5 g of the 2-

benzoyloxymethyl-5-ethoxy-1,3-oxathiolane in 50 ml of dichloroethane added under dry argon followed by 4.7 ml of trimethylsilyl triflate. After 3 days of heating under reflux under argon, it was cooled and poured onto 300 ml of saturated aqueous $NaHCO_3$. The organic layer was collected, the aqueous phase extracted with $CH_2Cl_2$and the combined extracts washed with water, dried and evaporated in vacuo. The residue was purified by chromatography on silica gel using CH_2Cl_2-CH_3OH 9:1 as the eluant to give 2.5 g of a pure mixture of cis- and trans-2-benzoyloxymethyl-5-cytosin-1'-yl-1,3-oxathiolane in a 1:1 ratio. These were separated as the N-acetyl derivatives.

The preceding mixture of cis- and trans-2-benzoyloxymethyl-5-cytosin-1'-yl-1,3-oxathiolane (2.5 g) in 100 ml of dry pyridine containing 0.1 g of 4-dimethylaminopyridine (DMAP) was treated with acetic anhydride (7 ml) at room temperature and after 16 h, the mixture was poured onto cold water followed by extraction with CH_2Cl_2. The extract was washed with water, dried, and evaporated in vacuo. Toluene was added to the residue, then evaporated in vacuo and the residual oil purified by chromatography on silica gel using EtOAc-CH_3OH 99:1 as the eluant to yield 1.35 g of pure trans-2-benzoyloxymethyl-5-(N^4-acetyl-cytosin-1'-yl)-1,3-oxathiolaneas the fast moving product and 1.20 g of pure cis-2-benzoyloxymethyl-5-cytosin-1'-yl-1,3-oxathiolan as the slow moving component, melting point 158-160°C.

Cis- and trans-isomers of 2-hydroxymethyl-5-(cytosin-1'-yl)-1,3-oxathiolane was obtained by action of methanolic ammonia at 24°C.

References

Delleau B., Nguyen-Ba N.; US Patent No. 5,047,407; 09.10.1991; Assigned to IAF BioChem International, Inc.

LAMOTRIGINE

Therapeutic Function: Anticonvulsant

Chemical Name: 1,2,4-Triazine-3,5-diamine, 6-(2,3-dichlorophenyl)-

Common Name: Lamotrigine

Structural Formula:

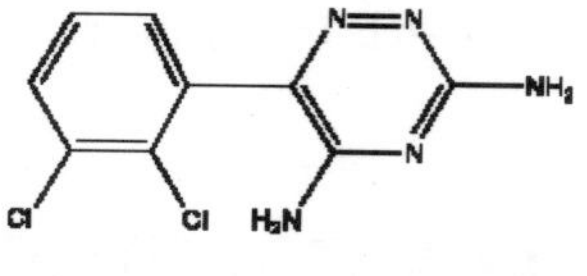

N=N NH2 N Cl Cl H2N

Chemical Abstracts Registry No.: 84057-84-1

Trade Name	Manufacturer	Country	Year Introduced
Lamepil	Innova (IPCA)	India	-
Lametec	Protech Biosystems	India	-
Lamictal	Glaxo Wellcome	-	-
Lamidus-Dt	Zydus Neurosciences	India	-
Lamitor	Torrent	India	-
Lamitor-Dt	Torrent Pharmaceuticals Ltd.	India	-
Lamotrigine	GlaxoSmithKline	-	-
Lysin	Pfizer	-	-
Vero-Lamotrigine	Okasa Pharma	Japan	-

Raw Materials

2,3-Dichlorophenylglyoxylamide
Aminoguanidine hydrochloride
Concentrated hydrochloric acid
Ethanol

Manufacturing Process

A mixture of 2,3-dichlorophenylglyoxylamide (54.5 g, 0.25 mol), aminoguanidine hydrochloride (33.15 g, 0.30 mol), ethanol (1 liter) and concentrated hydrochloric acid (4 ml) were heated under reflux for 6 hours at pH 1.5. The resulting solution was evaporated to dryness, the solid was dissolved in water (2 L; resulting pH 2.5) and the solution was basified to pH 13 by the addition of 50% aqueous sodium hydroxide (45 ml) at <15°C. The mixture was filtered, the solid washed with 0.88 N ammonia solution and dried to give (E)-2-(2',3'-dichlorophenyl)-2-(guanidinylimino)acetamide (59.5 g, 87%) m.p. 231-233°C. Recrystallisation of this product (2.2 g) from n-propanol (60 ml) afforded pure material (1.83 g, 83%), m.p. 238-239°C (decomp.).

(E)-2-(2',3'-dichlorophenyl)-2-(guanidinylimino)acetamide (0.3 g) was dissolved in ethanol (10 ml) and was irradiated by exposure to sunlight. After 5 days 6-(2,3-dichlorophenyl)-1,2,4-triazine-3,5-diamine (Lamotrigine) was detected by TLC in the liquor material. Melting point of lamotrigine 218°C.

References

Winter R.G., Sawler D.A., Germain A.; US Patent No. 5,912,345; 06.15.1999; Assigned to Glaxo Wellcome Inc. (Research Triangle Park, NC)
Winter R.G. et al.; US Patent No. 5,047,407; June 15, 1999; Assigned to Glaxo Wellcome Inc. (Research Triangle Park, NC)

LANATOSIDE C

Therapeutic Function: Cardiotonic

Chemical Name: Card-20(22)-enolide, 3-((O-beta-D-glucopyranosyl-(1-4)-O-3-O-acetyl-2,6-dideoxy-beta-D-ribo-hexopyranosyl-(1-4)-O-2,6-dideoxy-beta-D-ribo-hexopyranosyl-(1-4)-2,6-dideoxy-beta-D-ribo-hexopyranosyl)oxy)-12,14-dihydroxy-, (3beta,5beta,12beta)-

Common Name: Celanidum; Glycoside C from Digitalis lanata; Lanatoside C

Structural Formula:

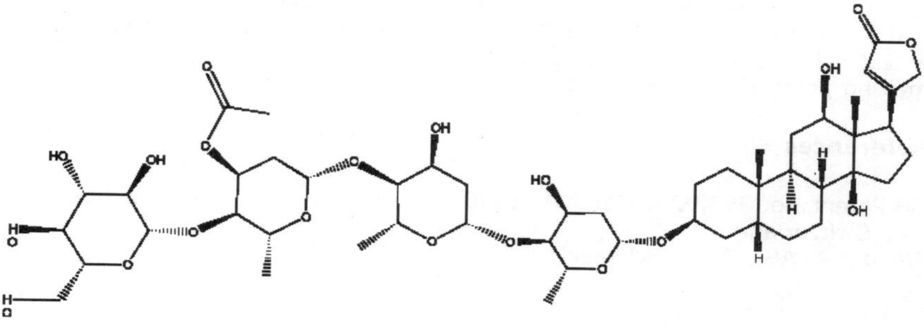

Chemical Abstracts Registry No.: 17575-22-3

Trade Name	Manufacturer	Country	Year Introduced
Lanatozid C	Biofarm	-	-
Cedigalan	Zdravle	-	-

Raw Materials

Dry leaves of digitalis lanata
Lead hydroxide
Tannin

Manufacturing Process

2000 parts of dry leaves of digitalis lanata are finely ground with 500 parts of sodium chloride, they are then wetted with 1000 parts of water and extracted with 30,000 parts of chloroform. The filtered extract is completely evaporated in vacuum at a low temperature and to the remaining residue are added 1000 parts of dry ether and the whole mixture is left under the ether until the thick viscous mass has been transformed into a hard body. The ether is then poured away and the residue is digested with 1000 parts of ether for about 2 hours under a reflux condenser. After cooling down the mixture, it is filtered, and the residue obtained, which is now in form of a brittle mass, which is then dried in vacuum in order to completely eliminate the remaining ether present, and pulverized. The pulverized mass is advantageously subjected once more to the treatment with ether. The yellow greenish powder thus obtained is then dissolved in 1000 parts of methyl alcohol and to the solution, so obtained a fine suspension of 30 parts lead hydroxide in 1000 parts water is added with stirring. The solution obtained is neutralized, stirred for about 2 hours, and filtered, and the clear yellowish filtrate is preferably treated again with a small quantity of tannin precipitating substance. The clear filtrate thus obtained is concentrated in vacuum at a low temperature to about 200 parts, whereby the difficultly soluble portion of the glucoside mixture precipitates. The solution is then filtered. The precipitate is dissolved in a small quantity of methyl alcohol and is treated with a small quantity of water whereby the new product begins to precipitate in a crystalline form. By repeated crystallization from methyl alcohol, without addition of water, the glucoside may be obtained in the form of a perfectly pure compound; it does not change its properties even on further recrystallization. The glucoside, freshly crystallized from methyl alcohol and dried in vacuo, had MP: 248°C with decomposition, when heated rapidly. At 230°-235°C, the substance begins to sinter and becomes quite soft; the melting point, therefore, is not well defined.

References

GB Patent No. 357,926; March 1, 1930; Chemiche Fabrik Sandoz of Basle, Switzerland

Wander A. AG., Bern (Schweiz); S.P. No. 245219; Oct. 31, 1946

LANSOPRAZOLE

Therapeutic Function: Antiulcer

Chemical Name: 1H-Benzimidazole, 2-(((3-methyl-4-(2,2,2-trifluoroethoxy)-2-pyridinyl)methyl)sulfinyl)-

Common Name: Lansoprazole

Structural Formula:

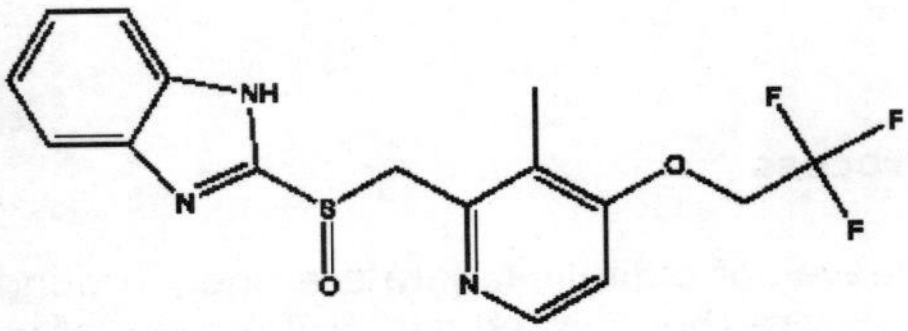

Chemical Abstracts Registry No.: 103577-45-3

Trade Name	Manufacturer	Country	Year Introduced
Acilanz	Themis Pharmaceuticals Ltd.	India	-
Lams OD	Recon Healthcare Ltd.	India	-
Lancid	Brown and Burk Pharmaceuticals Ltd.	India	-
Lansoptol	Krka	Slovenia	-
Lansoprazole	Chemo Iberica	Spain	-
Lansoprazole	Wyeth Pharmaceuticals	-	-
Lanzap	Dr. Reddy's Laboratories Ltd.	India	-

Raw Materials

Diethyl azodicarboxylate
Triphenylphosphine
Sodium hypochlorite
2-Hydroxymethyl-3-methyl-4-(2,2,2-trifluoroethoxy)pyridine
2-Mercaptobenzimidazole
Tetramethyl-1-piperidinyloxy free radical
Tetrabutylammonium chloride

Manufacturing Process

Preparation of 2-[3-methyl-4-(2,2,2-trifluoroethoxy)-2-pyridyl]methylthio-1H-benzimidazole:

A mixture of 6.63 g of 2-hydroxymethyl-3-methyl-4-(2,2,2-trifluoroethoxy) pyridine (30 mmol), 4.5 g of 2-mercaptobenzimidazol (30 mmol) and 8.67 g of triphenylphosphine (33 mmol) was dissolved in 100 ml of tetrahydrofuran, 5.75 g of diethyl azodicarboxylate (33 mmol) dissolved in 30 ml of tetrahydrofuran was added dropwise thereto at room temperature, and stirred for 1 hour. The reaction mixture was concentrated under a reduced pressure, the resulting residue was combined with 100 ml of ethylacetate, and extracted twice with 50 ml portions of 1 N HCl. The aqueous layer was then washed with 50 ml of diethylether; neutralized with 1 N NaOH to adjust the pH to 7. The resulting precipitates were filtrated, washed with water, and dried, to obtain 10.06 g of 2-[3-methyl-4-(2,2,2-trifluoroethoxy)-2-pyridyl]methylthio-1H-benzimidazole as a white solid (yield: 95%), m.p.142-144°C.

4.46 g of 2-[3-methyl-4-(2,2,2-trifluoroethoxy)-2-pyridyl]methylthio-1H-benzimidazole (12 mmol) and 18.74 mg of tetramethyl-1-piperidinyloxy free radical (1 mol %, used as a catalyst) were dissolved in 40 ml of tetrahydrofuran, and combined with 166.76 mg of tetrabutylammonium chloride (5 mol %) dissolved in 20 ml of distilled water. The resulting mixture was cooled to 0°C and 13.6 ml of NaOCl (12%, 2.2 equivalent) dissolved in 20 ml of distilled water was added thereto over 2 hours at 0°C, stirred for 10 min, and then for additional 10 min at 20°C. Then, the reaction mixture was extracted with 40 ml of ethylacetate and the organic layer was washed with sat. $NaHCO_3$ (30 ml) and then with sat. brine (30 ml), dried over anhydrous $MgSO_4$, and the solvent was removed therefrom. The resulting crude product as recrystallized from acetone/hexane, to obtain 3.99 g of 2-[3-methyl-4-(2,2,2-trifluoroethoxy)-2-pyridyl]methylsulphinyl-1H-benzimidazol (lansoprazole) as a white-light brown solid (yield: 90%), melting point 164-165°C (decomposition).

References

Moon Y.-H. et al.; US Patent No. 6,423,846; 07.23.2002; Assigned to Hanmi Pharm. Co., Ltd. (KR)

LATANOPROST

Therapeutic Function: Antiglaucoma

Chemical Name: 5-Heptenoic acid, 7-((1R,2R,3R,5S)-3,5-dihydroxy-2-((3R)-3-hydroxy-5-phenylpentyl)cyclopentyl)-, 1-methylethyl ester, (5Z)-

Common Name: Latanoprost

Structural Formula:

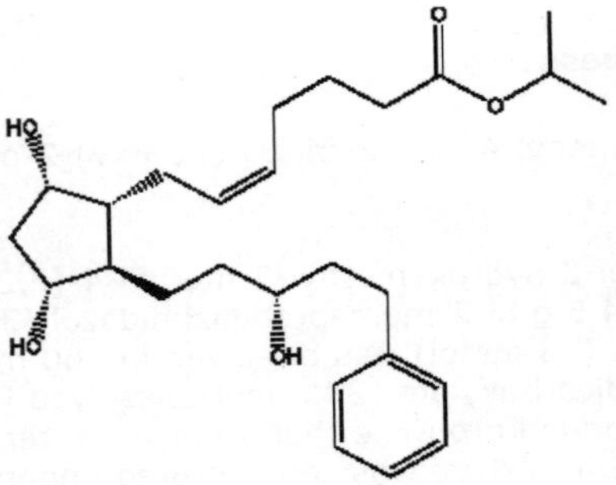

Chemical Abstracts Registry No.: 130209-82-4

Trade Name	Manufacturer	Country	Year Introduced
Latanoprost	Pharmacia and Upjohn	USA	-
Latanoprost	Milmet Pharma Ltd.	India	-
Xalatan	Pharmacia and Upjohn	USA	-
Xalatan	Pharmacia India (P) Ltd.	India	-

Raw Materials

Lithium chloride
Triethylamine
2-Iodopropane
Cesium carbonate
Dimethyl-(2-oxo-4-phenylbutyl)phosphonate
(1S,5R,6R,7R)-6-Formyl-7-(benzyloxy)-2-oxabicyclo[3.3.0]octan-3-one
Platinum on carbon
Chlorodiisopinocampheylborane, (-)-
Potassium hydroxide
tris(Hydroxymethyl)aminomethane

Manufacturing Process

Lithium chloride (2.6 g) is dissolved in THF (170 mL). Dimethyl-(2-oxo-4-phenylbutyl)phosphonate (7.87 g) and triethylamine (4.3 mL) are added. The mixture is stirred and cooled to -10°C. A solution of the Corey aldehyde benzoate, (1S,5R,6R,7R)-6-formyl-7-(benzyloxy)-2-oxabicyclo[3.3.0]octan-3-one (8.42 g) in THF (75 mL) is added to the reaction mixture over three hours. The resulting mixture is stirred for 18 hours at -10°C. At the end of this time, methyl t-butyl ether (MTBE) (100 mL) is added and the mixture warmed to 0-20°C. Sodium bisulfite (38%, 100 mL) is added and the two-phase mixture was stirred for 10 min. The phases are separated and the organic phase is washed with saturated aqueous sodium bicarbonate solution (100 mL). The organic phase is separated and concentrated under reduced pressure to a volume of <100 mL. Ethyl acetate (200 mL) is added and the mixture is concentrated to a volume of 50 mL. MTBE (100 mL) is added and the mixture is allowed to cool to 20-25°C for 1 hour. The mixture is then cooled to -20°C for 2 hours. The solids were filtered, washed with MTBE and dried on a nitrogen to give [3aR-[3aα,4α(E),5β,6aα]]-5-(benzoyloxy) hexahydro-4-(3-oxo-5-phenyl-1-pentenyl)-2H-cyclopenta[b]furan-2-one, m.p. 117-118°C.

[3aR-[3aα,4α(E),5β,6aα]]-5-(Benzoyloxy)hexahydro-4-(3-oxo-5-phenyl-1-pentenyl)-2H-cyclopenta[b]furan-2-one(10.0 g, 0.0247 mole) in THF (100 mL) is cooled to -38 to -42°C and is added a solution of (-)-chlorodiisopinocampheylborane (2 M in hexane; 43 mL) is added at <-35°C. When the addition is complete, the mixture is stirred at -38 to -42°C for 18 hours. At this time acetone (12.7 mL) is added and the mixture is allowed to warm to 20-25°C and stirred for two hours. MTBE (100 mL) is added and then a solution of sodium bicarbonate (10 g) in water (150 mL) is added. The two phase mixture is stirred for 15 min. The phases are separated and the organic phase is washed with water (100 mL) and concentrated in vacuum. MTBE (300 mL) is added and the mixture then concentrated. Acetonitrile (100 mL) is added and the mixture is again concentrated. Acetonitrile (150 mL) and heptane (100 mL) are added. The two-phase mixture is stirred for 5 min and then allowed to settle. The phases are separated. The acetonitrile phase is extracted with heptane. The acetonitrile phase is concentrated. A portion of the concentrate is purified by chromatography (silica gel, heptane/ethyl acetate, 1/1) to give [3aR-[3aα,4a(1E,3S),5β,6aα]]-5-(benzoyloxy)hexahydro-4-(3 -hydroxy-5-phenyl-1-pentenyl)-2H-cyclopenta[b]furan-2-one, m.p. 78-81°C.

[3aR-[3aα,4a(1E,3S),5β,6aα]]-5-(Benzoyloxy)hexahydro-4-(3-hydroxy-5-phenyl-1-pentenyl)-2H-cyclopenta[b]furan-2-oneis dissolved in THF (125 mL). Platinum on carbon catalyst (5%, 1 g) and triethylamine (3.4 mL) are added. The mixture is purged with nitrogen and then and the mixture is stirred vigorously under 5 psi hydrogen at 20°C. When the reaction was complete as measured by HPLC, the reaction is purged with nitrogen. The mixture is filtered over celite. The filtrate is concentrated under reduced pressure to give the crude product. A portion of the product is purified by chromatography (silica gel, heptane/ethyl acetate, 1/1) to give [3aR-[3aα,4a(1E,3S),5β,6aα]]-5-(benzoyloxy)hexahydro-4-(3-hydroxy-5-phenyl-1-pentyl)-2H-cyclopenta[b]furan-2-one, m.p. 68-70°C.

A mixture of potassium hydroxide in methanol (300 ml) and water (5 mL) is added to [3aR-[3aα,4a(1E,3S),5β,6aα]]-5-(benzoyloxy)hexahydro-4-(3 -hydroxy-5-phenyl-1-pentyl)-2H-cyclopenta[b]furan-2-one. The mixture is stirred and heated in an 80°C for 2 hours. When the reaction is complete, the mixture is concentrated under reduced pressure. Water (100 mL) and MTBE (100 mL) are added and the mixture stirred at 20-25°C for 15 min. The phases are allowed to separate. The product is in the aqueous phase and the organic phase is removed. The pH of the aqueous phase is adjusted to 1 to 1.5 by the addition of hydrochloric acid (3 N, about 60 mL are required). The solution is stirred at 20-25°C. After 30 min, MTBE (100 mL) is added and the mixture stirred for 12 hours. The phases are separated and the aqueous phase extracted once with MTBE (50 mL). The MTBE phases are combined and washed with sodium carbonate (1 N, 50 mL). The MTBE mixture is stirred with a solution of potassium hydroxide (2.8 g, 42.5 mmole) in water (100 mL) for 30 min. The phases are separated and the aqueous phase is added to a slurry of citric acid monohydrate (8.90 g) and ethyl acetate (100 mL). The mixture is stirred for 15 min and the phases are separated. The aqueous phase is extracted with ethyl acetate. The combined organic phases are dried over anhydrous sodium sulfate (8.90 g) for 15 min. The ethyl acetate extract is concentrated under reduced pressure to a volume of 100 mL. Ethyl acetate (200 mL) is added and the mixture is again concentrated to a volume of 100 mL. The resulting slurry is stirred at 0-5°C for 30 min. The solids are filtered and washed with heptane/ethyl acetate (1/1, 35 mL), then dried on a nitrogen to give 2-[(1R,2R,3R,5S)-3,5-dihydroxy-2-[(3R)-3-hydroxy-5-phenylpentyl]cyclopentyl]acetic acid.

2-[(1R,2R,3R,5S)-3,5-Dihydroxy-2-[(3R)-3-hydroxy-5-phenylpentyl] cyclopentyl]aceticacid (4.80 g) and toluene (100 mL) are stirred and the slurry heated to reflux for 30 min. Then the toluene is slowly distilled at atmospheric pressure to remove water. After about 1 hour of distillation, all acid is dissolved. The solution is then distilled to a volume of about 50 mL. The mixture is then cooled to about 80°C and ethyl acetate (25 mL) is added. The mixture is then cooled to 30C and heptane (20 mL) is added. The mixture is seeded with a small amount of [3aR-[3aα,4α(R),5β,6aα]]-hexahydro-5-hydroxy-4-(3-hydroxy-5-phenylpentyl)-2H-cyclopenta[b]furan-2-one. The mixture is stirred at about 30°C for 10 min, during which time crystallization occurred. Heptane (30 mL) is added over 15 min. The slurry is cooled to 20-25°C and stirred for 1 hour. The product is filtered and dried under nitrogen to give [3aR-[3aα,4α(R),5β,6aα]]-hexahydro-5-hydroxy-4-(3-hydroxy-5-phenylpentyl)-2H-cyclopenta[b]furan-2-one, m.p. 69-71°C.

[3aR-[3aα,4α(R),5β,6aα]]-Hexahydro-5-hydroxy-4-(3-hydroxy-5-phenylpentyl)-2H-cyclopenta[b]furan-2-one(1.0 g, 3.3 mmoles) is dissolved in methylene chloride (3 mL) and the mixture is placed in a sealable pressure tube. Add 1.0 mL of a mixture of trichloracetic acid (0.27 g) in methylene chloride (10 mL) followed by ethyl vinyl ether (6.3 mL). The pressure tube is closed and heated to 45°C for about 8 hours. At this time, triethylamine (0.12 mL) is added and the mixture is stirred for 10 min. The mixture is then concentrated under reduced pressure to give [3aR-[3aα,4α(R),5β,6aα]]-hexahydro-5-hydroxy-4-(3-hydroxy-5-phenylpentyl)-2H-cyclopenta[b]furan-2-one.

[3aR-[3aα,4α(R),5β,6aα]]-Hexahydro-5-hydroxy-4-(3-hydroxy-5-phenylpentyl)-2H-cyclopenta[b]furan-2-oneis dissolved in THF (14 mL) and the mixture cooled to -40°C. Using a syringe pump, diisobutyl aluminum hydride DIBAL (1.0 M, 3.78 mL in toluene) is added over 15 min. The mixture is stirred for 15 min after the completion of the addition, then ethyl acetate (0.38 mL) is added. The mixture is poured into a solution of potassium sodium tartarate (10 g in 30 mL of water) and warmed to 20-25°C. The two phase mixture is heated to 45°C for 1 hour and then cooled. The phases are separated and the organic phase is concentrated to give (3aR,4R,5R,6aS)-5-(1-ethoxyethoxy)-4-[(3R)-3-(1-ethoxyethoxy)-5-phenylpentyl]hexahydro-2H-cyclopenta[b]furan-2-ol.

(3aR,4R,5R,6aS)-5-(1-Ethoxyethoxy)-4-[(3R)-3-(1-ethoxyethoxy)-5-phenylpentyl]hexahydro-2H-cyclopenta[b]furan-2-olis dissolved in dry THF (10 mL) and added to a mixture containing potassium 5-(triphenylphosphoranylidene)pentaonate (prepared from 4-carboxybutyltriphenylphosphonium bromide and potassium t-butoxide solution in THF at 0°C, the resulting solution is then cooled to -10°C) at -10 to -5°C. The resulting mixture is stirred for about 3 hours at -5°C. Water (30 mL; 0°C) is added over 10 min, then ethyl acetate (20 mL) and aqueous tris (hydroxymethyl)aminomethane solution (10 mL) is added. The phases are separated and the organic phase is washed with aqueous tris(hydroxymethyl) aminomethane solution. The aqueous phases are combined and washed once with ethyl acetate, MTBE is added to the combined aqueous phases. The mixture is acidified to pH 3 with aqueous phosphoric acid (40%). The organic phase is separated and concentrated under reduced pressure to 20 mL. Solids (5-diphenylphosphinopentanoic acid) crystallized. MTBE (50 mL) is added and the slurry concentrated under reduced pressure to a volume of 20 mL. The solid is filtered and washed with MTBE (100 mL). The filtrate is concentrated under reduced pressure to give 7-[(1R,2R,3R,5S)-3-(1-ethoxyethoxy)-5-hydroxy-2-[(3R)-3-(1-ethoxyethoxy)-5- phenylpentyl]cyclopentyl-5-heptenoic acid.

7-[(1R,2R,3R,5S)-3-(1-Ethoxyethoxy)-5-hydroxy-2-[(3R)-3-(1-ethoxyethoxy)-5-phenylpentyl]cyclopentyl-5-heptenoic acid is dissolved in THF (30 mL). Water (15 mL) and phosphoric acid (85 wt %; 0.67 mL) are added and the mixture is heated to reflux for about 2 hours. The mixture is cooled and MTBE (30 mL) is added. The phases are separated. The organic phase is washed once with saline (100 mL). The organic phase is concentrated under reduced pressure. MTBE (3 times 50 mL) is added and concentrated under reduced pressure to give (5Z)-7-[(1R,2R,3R,5S)-3,5-dihydroxy-2-[(3R)-3-hydroxy-5-phenylpentyl]cyclopentyl]-5-heptenoic acid (Latanoprost Acid).

Latanoprost acid is dissolved in DMF (10 mL) and added to a slurry of cesium carbonate (1.6 g) in DMF (10 mL). 2-Iodopropane (0.49 mL) is added and the slurry is heated to 45°C for about 6 hours. When the reaction is complete, MTBE (40 mL) and water (50 mL) are added and the mixture is stirred for 15 min. The phases are separated and the aqueous phase is washed with MTBE (20 mL). The organic phases are combined and concentrated. The concentrate is chromatographed (silica gel) eluting with MTBE. The appropriate fractions are pooled and concentrated to give (5Z)-(9CI)-7-[(1R,2R,3R,5S)-3,5-Dihydroxy-2-[(3R)-3-hydroxy-5-phenylpentyl]cyclopentyl]-5-heptenoic acid 1-methylethyl ester (Latanoprost).

References

Henegar K.E.; US Patent No. 6,689,901; Feb. 10, 2004; Assigned to Pharmacia and Upjohn Company (Kalamazoo, MI)

LEFLUNOMIDE

Therapeutic Function: Immunosuppressive, Antiarthritic

Chemical Name: 4-Isoxazolecarboxamide, 5-methyl-N-(4-(trifluoromethyl) phenyl)

Common Name: Leflunomide

Structural Formula:

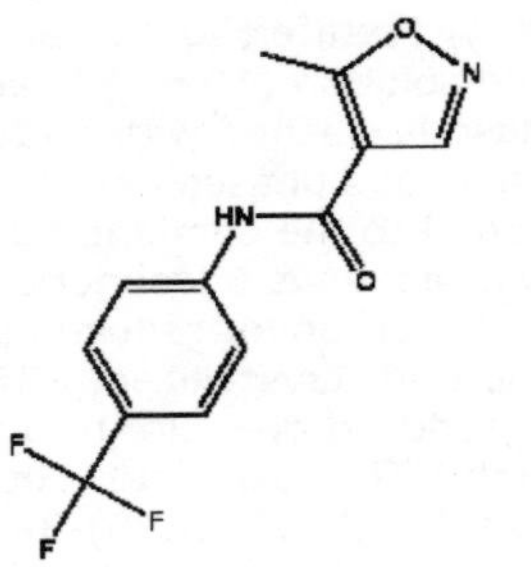

Chemical Abstracts Registry No.: 75706-12-6

Trade Name	**Manufacturer**	**Country**	**Year Introduced**
Arava	Aventis Pharma Deutschland	Germany	-
Arava	Aventis Pharmaceuticals	USA	-
Arava	Hoechst Marion Roussel	Germany	-
Leflunomide	Torrent	-	-

Raw Materials

Diketene
Acetic anhydride
Sodium hydroxide
5-Methylisoxazole-4-carboxylic acid chloride
Orthoformic acid triethyl ester
Hydroxylamine hydrochloride
4-Trifluoromethylaniline
Trifluoromethylaniline

Manufacturing Process

In US Patent No. 4,284,786 is described two methods of preparation of 5-methylisoxazole-4-carboxylic-(4-trifluoromethyl)-anilide.

The method 1

A mixture of 0.55 mole of diketene (46.3 g) and 30 ml of acetonitrile is added dropwise, at 75°C, to a solution of 0.5 mole of 4-trifluoromethylaniline (30.6 g) in 150 ml of acetonitrile. The mixture is heated to boiling under reflux for 2.5 hours. When it has cooled to room temperature, the crystals which are precipitated are filtered off, washed with cold ethanol and dried. This gives 79.1 g (64.5% of theory) of crystalline acetoacetic acid-4-trifluoromethylanilide, melting point (after recrystallization from ethanol) 155°C.

The acetonitrile phase is evaporated to dryness under reduced pressure. The crystalline residue (42.1 g) is recrystallized from 80 ml of ethanol. This gives a further 24.1 g (19.7% of theory) of crystals. Melting point (after recrystallization from ethanol) 155°C. Total yield: 84.2% of theory.

0.75 mole of acetoacetic acid 4-trifluoromethylanilide (183.9 g) is boiled under reflux for 1.5 hours with 0.83 mole of orthoformic acid triethyl ester (123 g) and 2.25 mole of acetic anhydride (229.7 g). After the mixture has cooled to room temperature, the crystals which have precipitated are filtered off and washed first with a small amount of acetic anhydride and then with petroleum ether. This gives 116.1 g (51.4% of theory) of crystalline 2-ethoxymethyleneacetoacetic acid 4-trifluoromethylanilide, melting point (after recrystallization from toluene) 124-125°C.

The combined filtrates are concentrated under reduced pressure. The crystals of the crystal paste which thereupon remains are filtered off, washed first with a small amount of acetic anhydride and then with petroleum ether and dried. A further 56.1 g (24.8% of theory) of crystals are thus obtained. Melting point (after recrystallization from toluene) 124-125°C. Total yield: 76.2% of theory.

A solution 0.1 mole of 2-ethoxymethyleneacetoacetic acid 4-trifluoromethylanilide (30.1 g) in 60 ml of ethanol is added dropwise at 5-10°C to the mixture of 0.11 mole of hydroxylamine hydrochloride (7.65 g) in 50 ml of water and 0.11 mole of sodium hydroxide (4.4 g) in 10 ml of water. The mixture is heated under reflux for 15 min. The crystals which are precipitated after cooling are filtered off, washed with water and dried. 19.6 g (72.6% of theory) of crystalline 5-methylisoxazole-4-carboxylic acid 4-trifluoromethyl-anilide are thus obtained, melting point (after recrystallization from toluene) 166.5°C.

The method 2

0.1 mole of 5-methylisoxazole-4-carboxylic acid chloride (14.6 g) and 20 ml of a 5 N potassium hydroxide solution are added dropwise to 0.1 mole of trifluoromethylaniline (16.1 g), suspended in 150 ml of water, in such a way that the pH of the reaction mixture does not rise above 5. The mixture is subsequently shaken with 150 ml of methylene chloride. The methylene chloride phase is washed with water and, after drying with sodium sulfate is, evaporated to dryness under reduced pressure. This gives 24.4 g (90.2% of theory) of a crystalline 5-methylisoxazole-4-carboxylic acid 4-trifluoromethyl-anilide, melting point (after recrystallization from toluene) 166.5°C.

References

Kammerer, F.-J., Schleyerbach R.; US Patent No. 4,284,786; August 18, 1981; Assigned: Hoechst Aktiengesellschaft (Frankfurt am Main, DE)

LETOSTEINE

Therapeutic Function: Mucolytic

Chemical Name: 4-Carboxythiazolidinyl-2-ethylmercapto-acetic acid ethyl ester

Common Name: -

Structural Formula:

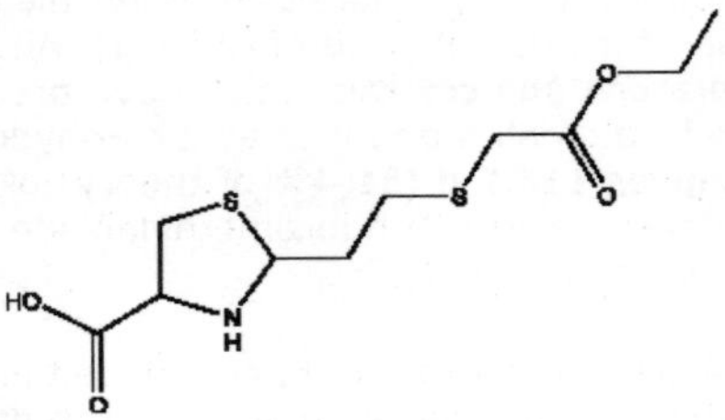

Chemical Abstracts Registry No.: 53943-88-7

Trade Name	Manufacturer	Country	Year Introduced
Viscotiol	Carlo Erba	France	1979
Viscotiol	Carlo Erba	Switz.	1980
Viscotiol	I.S.F.	Italy	1981

Raw Materials

Acrolein
Thioglycolic acid
Cysteine hydrochloride

Manufacturing Process

In an Erlenmeyer flask placed in an ice bath, and under a well-ventilated hood, a solution of 0.1 mol of acrolein in 100 ml of ether was introduced, With the aid of a bromine ampoule, 0.1 mol (= 11 ml) of the ethyl ester of thioglycolic acid containing 0.5 ml of triethylamine was added drop by drop.

One hour after completion of the addition, there was added 0.1 mol (15.6 g) of chlorhydrate of cysteine in alcoholic solution. The chlorhydrate of the expected derivative, which appeared in the form of a thick oil, was precipitated by addition of 0.1 mol (10 g) of potassium acetate in aqueous solution. The abundant precipitate obtained was filtered and washed in water and ether. The product was recrystallized in a minimum of absolute alcohol.

References

DFU 4 (10) 729 (1979)
Kleeman and Engel p. 516
DOT 16 (4) 109 (1980)
I.N. p. 553
Chodkiewicz, M.X.; US Patent 4,032,534; June 28, 1977; assigned to Ferlus-Chimie SA

LETROZOLE

Therapeutic Function: Antineoplastic

Chemical Name: Benzonitrile, 4,4'-(1H-1,2,4-triazol-1-ylmethylene)bis-

Common Name: Letrozole

Structural Formula:

Chemical Abstracts Registry No.: 112809-51-5

Trade Name	Manufacturer	Country	Year Introduced
Femara	Novartis Pharmaceuticals	-	-
Letrozole	Novartis Pharmaceuticals	-	-
Lets	Samarth Pharma Pvt. Ltd.	India	-
Letzole	VHB Life Sciences	India	-

Raw Materials

4-Bromomethylbenzonitrile
1H-1,2,4-Triazole
tert-BuOK
p-Fluorobenzinitrile

Manufacturing Process

From 4-bromomethylbenzonitrile and 1H-[1,2,4]triazole was obtained 4-[1,2,4]triazol-1-ylmethylbenzonitrile. Treatment of that with strong base (tert-BuOK) results in formation of the anion by removal of the relatively acidic benzyl proton. This anion was condensed with p-fluorobenzinitrile to give benzhydryl tetrazole (Letrozole).

References

Lang M. et al.; J. Ster; Biochem. Mol. Biol. 1993, V.44, P. 421
The Medical Letter, 1998, V 20, P.53-54
Org. Chem. Drug. Synth., V 6, P.86

LEUCOVORIN CALCIUM

Therapeutic Function: Antidote (folic acid antagonists), Antianemic

Chemical Name: Glutamic acid, N-(p-(((2-amino-5-formyl-5,6,7,8-tetrahydro-3-hydroxy-6-pteridinyl)methyl)amino)benzoyl)-, calcium salt (1:1), L-

Common Name: Calcio folinato; Calcium folinate; Leucovorin calcium

Structural Formula:

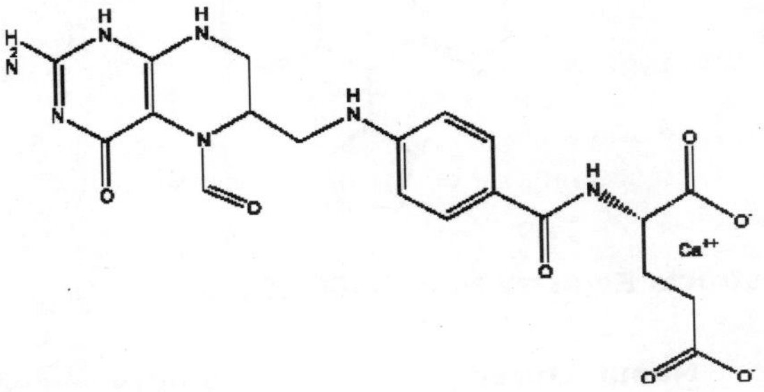

Chemical Abstracts Registry No.: 1492-18-8; 58-05-9 (Base)

Trade Name	Manufacturer	Country	Year Introduced
Leucovorin calcium	AstraZeneca	-	-
Leucovorin calcium	ROX	-	-
Wellcovorin	Immunex Corporation	-	-
Wellcovorin	Glaxo Wellcome	-	-

Raw Materials

5,10-Methenyl-5,6,7,8-tetrahydrofolic acid, chloride hydrochloride dihydrate
N,N-Diethylethanolamine

Manufacturing Process

5,10-Methenyl-5,6,7,8-tetrahydrofolic acid, chloride hydrochloride dihydrate (20 g) was added in one portion to 100 ml water at 60°C followed by N,N-diethylethanolamine (14.9 g) which adjusted the pH to 6. The mixture was maintained at reflux for 5 hours and the pH kept between 5.7 and 6.2 by addition of N,N-diethylethanolamine. The mixture was cooled, synthetic magnesium silicate (15 g) added and slurried, and filtered through celite and diluted with 40 ml SD3A (95% ethanol with 5% methanol). The filtrate was kept at -5°C for 16 hours, aqueous calcium chloride (4.0 g) was added dropwise to the cold filtrate, and the precipitate filtered, washed with SD3A (100 ml) and with ethyl acetate (100 ml) and dried under reduced pressure.

Yield of calcium leucovorin 91%.

References

Shive W.; US Patent No. 2,741,608; 04.10.1956; Assigned to Research Corporation

Wisowaty J.C., Swaringen R.A., Yeowell D.A.; US Patent No. 4,500,711; 02.19.1985; Assigned to Burroughs Wellcome Co. (Research Triangle Park, NC)

LEUPROLIDE ACETATE

Therapeutic Function: Antineoplastic

Chemical Name: 1-9-Luteinizing hormone-releasing factor (pig), 6-D-leucine-9-(N-ethyl-L-prolinamide)-L-deglycinamide, monoacetate

Common Name: Leuprolide acetate; Leuprorelin acetate

Structural Formula:

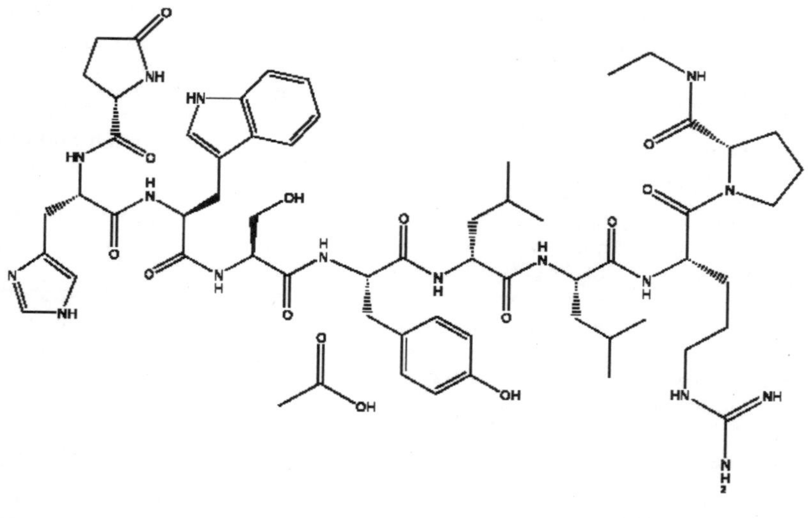

Chemical Abstracts Registry No.: 74381-53-6; 53714-56-0 (Base)

Trade Name	Manufacturer	Country	Year Introduced
Eligard	Atrix Laboratories, Inc.	-	-
Lupride	Inca (Sun)	India	-
Lupride Depot	Inca (Sun)	India	-
Lupron	TAP Pharmaceuticals	-	-
Viadur	ALZA Corp.	-	-

Raw Materials

Boc-Arg(Tos)	Chloromethylated divinylbenzene-styrene copolymer
Boc-Leu	Hydrogen fluoride
Boc-Tyr(Cl_2Bzl)	Boc-Ser(Bzl)
Boc-Trp	Boc-His(DNP)
pGlu	Dicyclohexylcarbodiimide
Boc-D-Leu	N-Ethyl dinitroaniline

Manufacturing Process

5-Oxo-L-prolyl-L-histidyl-L-tryptophanyl-L-seryl-L-tyrosyl-D-leucyl-L-leucyl-L-arginyl-L-prolylethylamideacetate was prepared by using of Boc strategy on a 2%-crosslinking chloromethylated divinylbenzene-styrene copolymer in a the Merrifield automatic sintesizer apparatus. 4.6 g of this resin/aminoacid material is used for the synthesis of the desired nonapeptide. Each N-blocked aminoacid is added in a three-fold access and allowed to couple to them, existing aminoacid-resin ester in the usual coupling cycle. Ordinarily the solvent used for the coupling reaction is dichloromethane or, when the solubility of the blocked aminoacid is low, a mixture of dichloromethane and DMF. Coupling is effected by the addition of a solution of dicyclohexylcarbodiimide in dichloromethane at a 2.9 fold excess. The sequence used for deprotection, neutralization and coupling of the next aminoacid is done in a fully automatic system. In this manner, the peptide is assembled using in turn Boc-Arg(Tos), Boc-Leu, Boc-D-Leu, Boc-Tyr(Cl_2Bzl), Boc-Ser(Bzl), Boc-Trp, Boc-His(DNP), and pGlu wherein all aminoacids are in the L-form except in the leucine so designated. A 250 mg sample of the above is placed in a hydrogen fluoride reaction with 250 mg vessel of anisole and about 5 ml of anhydrous hydrogen fluoride is distilled into it. After 1 hour at 0°C, the hydrogen fluoride is removed with a stream of dry nitrogen and the residue is taken up in 1% acetic acid. This solution is extracted with ether, and the aqueous phase applied to a 1 time 30 cm column of a highly basic ion exchange resin (marketed by Bio-Rad as AGl resin) in the acetate form. The product is eluted with 0.1 N acetic acid and localized using thin-layer chromatography ($CHCl_3$/MeOH/32% HOAc: 120/90/40, silica gel G, Cl_2/tolidine). The product bearing solution is lyophilized, rechromatographed on a Sephadex G-25 (marketed by Pharmacia of Uppsala, Sweden) column. The product eluted is collected and lyophilized to yield a fluffy white solid. An aminoacid analysis shows the expected ratio of all desired aminoacids assembled in the above fashion.

References

Adjei A.L., Johnson E.S., Kesterson J.W.; US Patent No. 4,897,256; 01.30.1990; Assignd to Abbott Laboratories (Abbott Park, IL)

Gendrich R.L., Rippel R.H., Seely J.H.; US Patent No. 4,005,063; 01.25.1977; Assigned: Abbott Laboratories (North Chicago, IL)

LEVALBUTEROL HYDROCHLORIDE

Therapeutic Function: Bronchodilator

Chemical Name: 1,3-Benzenedimethanol, α^1-(((1,1-dimethylethyl)amino) methyl)-4-hydroxy-, hydrochloride, (α^1R)-

Common Name: Levalbuterol hydrochloride; Levosalbutamol hydrochloride; (R)-Albuterol hydrochloride; (R)-Salbutamol hydrochloride

Structural Formula:

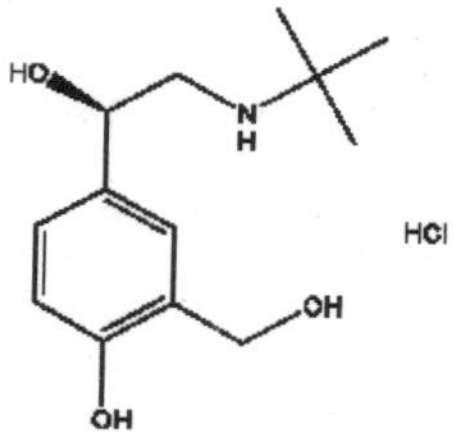

Chemical Abstracts Registry No.: 50293-90-8; 34391-04-3 (Base)

Trade Name	Manufacturer	Country	Year Introduced
Xopenex	Sepracor Inc.	USA	-

Raw Materials

Methyl 5-acetylsalicylate
t-Butylamine
Palladium on carbon
Hydrobromic acid
Borane-dimethyl sulfide

Manufacturing Process

Preparation of 5-glyoxyloyl-salicylic acid methyl ester hydrate using aqueous HBr

To a 3-neck flask immersed in an oil bath containing a solution of 40 g (0.206 mole) methyl 5-acetylsalicylate in 6 ml methylene chloride is charged with 82 ml of isopropanol. The solution is distilled to remove excess methylene chloride. When the internal temperature reaches 77°C, 126 ml (1.77 mole or 8.6 equivalents) of DMSO is added to the reaction mixture and the temperature of the mixture is increased to a temperature of 85° to 90°C. Then 33 ml (0.29 mole or 1.4 equivalents) of HBr (aqueous, 48%) is added to the mixture over a period of 20 minutes (exothermic), and the bath temperature is maintained at 95° to 100°C. As the addition of HBr nears completion distillation is initiated and dimethysulfide and isopropanol are distilled off. The mixture is stirred and the volume of the distillate monitored. After distillation of 82 ml of solvent, 20 ml of isopropanol is added slowly to maintain a steady rate of distillation. After the reaction completed as determined by high performance liquid chromatography (HPLC), the reaction mixture is quenched with 70 ml of 2.4 N H_2SO_4, the temperature of the reaction mixture is allowed to drop to 75°C and residual isopropanol is distilled off under vacuum. After a total of 165 ml distillate is collected, the title compound begins to precipitate. A mixture of 30 ml of acetonitrile and 70 ml of water is added slowly at 75°C with stirring. After 30 minutes of stirring, the reaction mixture is cooled to 15°C over a period of 90 minutes to complete the precipitation. The reaction mixture is filtered and the cake is washed with three 300 ml portions of water. The cake is dried in a draft oven at 50°C for 16 hours to give 39.5 g of the title compound (85% yield).

Preparation of albuterol from 5-glyoxyloyl-salicylic acid methyl ester

To a solution of 5-glyoxyloylsalicylic acid methyl ester hydrate (50 g, 0.221 mol) in ethylene glycol diethyl ether, 440 mL is added tertiary butylamine (16.2 g, 0.221 mol) at room temperature. The resulting light orange solution is stirred for 5 min until a clear solution is formed. The clear solution is then heated to reflux. Water and DME are distilled off azeotropically. After a total of 200 ml of distillate are collected, the solution is cooled to 25°C. The reaction mixture is slowly added to a solution containing 49 mL (0.49 mol) of 10.0 M borane-dimethyl sulfide in 220 mL of ethylene glycol diethyl ether (DME) at 70°C. The resulting reaction mixture is further refluxed for 2.5 hrs. After the reaction is completed as monitored by HPLC, excess DME is removed via vacuum distillation. The residue containing complexes of boron and arylethanolamine is subsequently cooled to 0°C. Quenching of the residue with 300 mL methanol gives the methylborate of arylethanolamine. The borate is then removed by azeotropic distillation as trimethylborate, leaving behind the desired arylethanolamine in the reaction mixture. An additional 300 ml of methanol and acetic acid (85 mL) are added to ensure the complete removal of trimethylborate via vacuum distillation to near dryness. The residue containing the boron-free arylethanolamine is cooled to 25°C and concentrated sulfuric acid (10.4 g, 0.221 mole) in water (64 mL) is added following by 570 ml of isopropyl alcohol. Albuterol sulfate is precipitated out as a white solid. After the reaction mixture is stirred at room temperature for 12 hrs and 0°C for 30 min the albuterol sulfate is filtered, washed with isopropyl alcohol (two 50 mL portions) and dried at 50°C for 12 hrs to give 49.75 g of the title compound (78% yield) as racemate.

The optically pure albuterol may be prepared by resolving a mixture of enantiomers methyl benzoate albuterol precursors which prepared by procedures well known to persons skilled in the art. The starting material 4-benzyl albuterol is commercially available from Cipla (Bombay, India).

(-)-D-Dibenzoyltartaric acid (D-DBTA) (32.2 g, 90 mmol, 1.0 eq) is added to a hot solution of racemic 4-benzyl albuterol (29.6 g, 90 mmol, 1.0 eq) in 180 mL of anhydrous denatured ethanol (type 3A, denatured with 5 vol % 2-propanol). The resulting solution is refluxed for 15 min and cooled to room temperature over 40 min and seeded with 99% ee (R)-4-benzyl albuterol D-DBTA salt. The mixture is cooled to 5°-10°C and stirred for 1 hour. The white solid is collected by filtration and dried at 40°C and 28 inches of Hg for 1 hour to give (R)-4-benzyl albuterol D-DBTA salt (31.8 g, 50% yield, 83.6% ee). The solid is redissolved in 240 mL of ethanol at 55°-60°C and the solution is cooled to room temperature and stirred at room temperature for 2 hours and at 0°-5°C for 1 hour. The resulting solid is collected by filtration and dried at

40°C and 28 inches of Hg for 2 hours as (R)-4-benzyl albuterol D-DBTA salt (22.9 g, 37.1% yield, 99.3% ee). The salt (22.9 g) is then treated with 204 mL of 5 wt % aq. Na_2CO_3 solution in 570 mL of ethyl acetate. The solid is worked-up, and recrystallization from 30 mL of ethyl acetate and 30 mL of n-heptane gives optically pure (R)-4-benzyl albuterol free base as a white powder (10.1 g, 34.1% yield from racemic compound 99.6% ee and 99.8% purity).

A mixture of (R)-4-benzyl albuterol as a free base (3.2 g, 9.73 mmol) and 10% Pd/C (0.64 g) in 24 mL of ethanol (denatured with 5 vol % 2-propanol) is shaken on a Parr-hydrogenator under 50 psi of hydrogen at room temperature for 3 hours. The catalyst is removed by filtration and the filtrate is concentrated to ca. 9 mL in volume containing crude (R)-albuterol and treated with anhydrous HCl in ether (1.0 M, 9.5 mL, 0.98 eq) at 0°-5°C. After 30 min at room temperature, 9 mL of methyl t-butyl ether (MTBE) is added, the resulting mixture is stirred at room temperature for 30 min and at 0°-5°C for 2 hours. The white solid (R)-albuterol hydrochloride is collected by filtration and recrystallized from 25 mL of ethanol and 12.5 mL of MTBE to give pure (R)-albuterol hydrochloride (2.17 g, 80.9% yield, 99.6% purity), white powder.

References

Tann C. et al.; US Patent No. 5,283,359; Feb. 1, 1994; Assigned to Schering Corp., Kenilworth, N.J.

Gao Y. et al.; US Patent No. 5,545,745; Aug. 13, 1996; Assigned to Sepracor, Inc. (Marlborough, MA)

LEVAMISOLE HYDROCHLORIDE

Therapeutic Function: Antiinflammatory

Chemical Name: L-2,3,5,6-Tetrahydro-6-phenylimidazo[2,1-b]thiazole hydrochloride

Common Name: L-Tetramisole hydrochloride

Structural Formula:

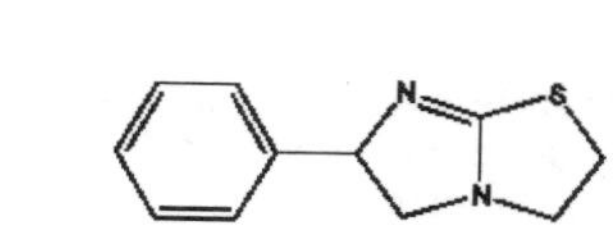

HCl

Chemical Abstracts Registry No.: 16695-80-5; 14769-73-4 (Base)

Trade Name	Manufacturer	Country	Year Introduced
Solaskil	Specia	France	1971
Ergamisol	Janssen	Italy	1978
Ascaryl	Abic	Israel	-
Meglum	Bago	Argentina	-
Niratic-Pur-On	Vet. Med. Handel	W. Germany	-
Tramisol	Lederle	US	-
Vermisol	Andreu	Spain	-

Raw Materials

1,2-Dibromoethane
Potassium hydroxide
Hydrogen chloride
DL-2-Thio-1-phenyl-imidazolidine
d-10-Gamphorsulfonic acid
Sodium hydroxide

Manufacturing Process

To a stirred and refluxed suspension of 17 parts of 1,2-dibromoethane, 7.8 parts of sodium hydrogen carbonate and 50 parts of 2-propanol is added a mixture of 3.4 parts of dl-2-thio-1-phenyl-imidazolidine, 9 parts of a 20% potassium hydroxide solution in 40 parts of 2-propanol over a period of about 1 hour. After the addition is complete, the whole is stirred and refluxed for an additional 3 hours. The reaction mixture is evaporated. To the residue are added 18 parts of a 15% potassium hydroxide solution. The whole is extracted with toluene. The extract is dried and evaporated. The oily residue is dissolved in acetone and gaseous hydrogen chloride is introduced into the solution. The precipitated solid salt is filtered off and recrystallized from 2-propanol, yielding dl-2,3,5,6-tetrahydro-6-phenyl-imidazo[2,1-b]thiazole hydrochloride; melting point 264°C to 266°C.

dl-6-phenyl-2,3,5,6-tetrahydroimidazo[2,1-b]thiazole hydrochloride, 188 g (0.785 mol), is suspended in a mixture of 500 ml of water and 500 ml of methylene chloride. The suspension is stirred mechanically while 20% sodium hydroxide solution is added until the solution is basic. Ice is added from time to time to keep the temperature below the boiling point of the methylene chloride. The methylene chloride layer is separated, washed with water, dried over potassium carbonate and evaporated. The oily residue crystallizes with the evolution of the heat when poured into a beaker containing 100 ml of ether. The free base is washed with ether. The yield of dl-6-phenyl-2,3,5,6-tetrahydroimidazo[2,-b]thiazole is 151.4 g (0.746 mol), 94%. The product has a melting point of 90°C.

A solution of 204.3 g (1 mol) of dl-6-phenyl-2,3,5,6-tetrahydroimidazo[2,1-b]thiazole and 232.3 g (1 mol) of d-10-camphorsulfonic acid in 1,750 ml of chloroform is allowed to crystallize overnight at -28°C. The solvate is recovered by filtration and washed with ice cold chloroform (400 ml). The solvate is dried (decomposed) under nitrogen 7 hours and then in air overnight. The yield of d(+)6-phenyl-2,3,5,6-tetrahydroimidazo[2,1-b]thiazole d-10-camphorsulfonate is 202.5 g (0.464 mol) 92.8%, melting point 139°C to 140°C $[\alpha]_D^{25}$+ 82.6 (C = 16, H_2O).

A solution of 150 g (0.344 mol) of d(+)6-phenyl-2,3,5,6-tetrahydroimidazo

[2,1-b]thiazole, d-10-camphorsulfonate in water is treated with 15.5 g (0.378 mol) of 98% sodium hydroxide and the liberated base extracted with chloroform. The chloroform solution is washed with water followed by sodium chloride solution and dried over magnesium sulfate. Evaporation of the solvent left 72.1 g of residue which crystallized shortly. The free base hereby obtained has a melting point of 60°C to 61.5°C and an optical rotation $[\alpha]_D^{25}$+ 85.1 (C = 10, $CHCl_3$).

The free base d(+)6-phenyl-2,3,5.6-tetrahydroimidazo[2.1-b]thiazole is dissolved in 112 ml of acetone and 178 ml of isopropanolic hydrogen chloride is added all at once. The hydrochloride crystallizes at once. After cooling to below 0°C, the salt is recovered by filtration and washed with acetone. The product weighs 75.2 g (0.312 mol), 91%, from the camphorsulfonate, melting point 227°C to 227.5°C $[\alpha]_D^{25}$+ 123.1 (C = 15, H_2O).

References

Merck Index 9055
DFU 4 (6) 420 (1979)
Kleeman and Engel p. 517
DOT 8 (6) 225 (1972) and 16 (10) 327, 359 (1980)
I.N. p. 554
REM p. 1156
Raeymaekers, A.H.M., Thienpont, D.C.I.C. and Demoen, P.J.A.W.; US Patents 3,274,209; September 20, 1966 and 3,364,112; January 16,1968; both assigned to Janssen Pharmaceutica NV
Bullock, M.W.; US Patent 3,463,786; August 26, 1969; assigned to American Cyanamid Co. Dewar, R.A., Maier, V.E. and Ingram, M.A.; US Patent 3,579,530; May 18, 1971; assigned to Imperial Chemical Industries of Australia and New Zealand Ltd.
Dewilde, F. and Frot, G.G.; US Patent 3,646,051; February 29, 1972; assigned to Rhone-Poulenc SA

LEVETIRACETAM

Therapeutic Function: Antiepileptic, Nootropic

Chemical Name: 1-Pyrrolidineacetamide, α-ethyl-2-oxo-, (αS)-

Common Name: Levetiracetam

Structural Formula:

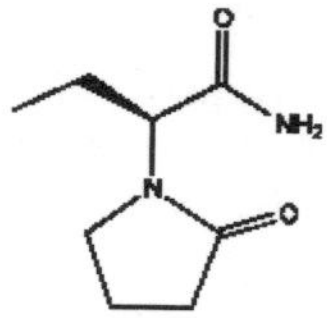

Chemical Abstracts Registry No.: 102767-28-2

Trade Name	Manufacturer	Country	Year Introduced
Keppra	UCB Pharma	Belgium	-
Levetiracetam	UCB Pharma	Belgium	-

Raw Materials

(+/-)-α-Ethyl-2-oxo-1-pyrrolidineacetic acid
(R)-(+)-α-Methyl-benzylamine
Triethylamine
Ethyl chloroformate

Manufacturing Process

(a) Preparation of the (R)-α-methyl-benzylamine salt of (S)-α-ethyl-2-oxo-1-pyrrolidineacetic acid

8.7 kg (50.8 moles) of racemic ()-α-ethyl-2-oxo-1-pyrrolidineacetic acid are suspended in 21.5 liters of anhydrous benzene in a 50 liter reactor. To this suspension is added gradually a solution containing 3.08 kg (25.45 moles) of (R)-(+)-α-methyl-benzylamine and 2.575 kg (25.49 moles) of triethylamine in 2.4 liters of anhydrous benzene. This mixture is then heated to reflux temperature until complete dissolution. It is then cooled and allowed to crystallize for a few hours. 5.73 kg of the (R)-α-methyl-benzylamine salt of (S)-α-ethyl-2-oxo-1-pyrrolidineacetic acid are thus obtained. Melting point: 148°-151°C. Yield: 77.1%.

This salt may be purified by heating under reflux in 48.3 liters of benzene for 4 hours. The mixture is cooled and filtered to obtain 5.040 kg of the desired salt. Melting point: 152°-153.5°C. Yield: 67.85%.

(b) Preparation of (S)-α-ethyl-2-oxo-1-pyrrolidineacetic acid

5.04 kg of the salt obtained in (a) above are dissolved in 9 liters of water. 710 g of a 30% sodium hydroxide solution are added slowly so that the pH of the solution reaches 12.6 and the temperature does not exceed 25°C. The solution is stirred for a further 20 minutes and the α-methylbenzylamine liberated is extracted with a total volume of 18 liters of benzene. The aqueous phase is then acidified to a pH of 1.1 by adding 3.2 liters of 6 N hydrochloric acid. The precipitate formed is filtered off, washed with water and dried. The filtrate is extracted repeatedly with a total volume of 50 liters of dichloromethane. The organic phase is dried over sodium sulfate and filtered and evaporated to dryness under reduced pressure. The residue obtained after the evaporation and the precipitate isolated previously, are dissolved together in 14 liters of hot dichloromethane. The dichloromethane is distilled and replaced at the distillation rate, by 14 liters of toluene from which the product crystallizes. The mixture is cooled to ambient temperature and the crystals are filtered off to obtain 2.78 kg of (S)-α-ethyl-2-oxo-1-pyrrolidineacetic acid. Melting point: 125.9°C. $[\alpha]_D^{20}$ = -26.4° (c = 1, acetone). Yield: 94.5%.

(c) Preparation of (S)-α-ethyl-2-oxo-1-pyrrolidineacetamide

34.2 g (0.2 mole) of (S)-α-ethyl-2-oxo-1-pyrrolidineacetic acid are suspended in 225 ml of dichloromethane cooled to -30°C. 24.3 g (0.24 mole) of triethylamine are added dropwise over 15 minutes. The reaction mixture is then cooled to -40°C and 24.3 g (0.224 mole) of ethyl chloroformate are added over 12 minutes. Thereafter, a stream of ammonia is passed through the mixture for 4 ½ hours. The reaction mixture is then allowed to return to ambient temperature and the ammonium salts formed are removed by filtration and washed with dichloromethane. The solvent is distilled off under reduced pressure. The solid residue thus obtained is dispersed in 55 ml toluene and the dispersion is stirred for 30 minutes and then filtered. The product is recrystallized from 280 ml of ethyl acetate in the presence of 9 g of 0.4 nm molecular sieve in powder form 24.6 g of (S)-α-ethyl-2-oxo-1-pyrrolidineacetamide are obtained. Melting point: 115°-118°C. $[\alpha]_D^{25}$ = -89.7° (c = 1, acetone). Yield: 72.3%.

References

Gobert et al.; US Patent No. 4,696,943; Sep. 29, 1987; Assigned to USB Societe Anonyme, Brusseles, Belgium

LEVOBUNOLOL HYDROCHLORIDE

Therapeutic Function: Beta-adrenergic blocker

Chemical Name: 1(2H)-Naphthalenone, 3,4-dihydro-5-((2S)-3-(tert-butylamino)-2-hydroxypropoxy)-, hydrochloride

Common Name: Levobunolol hydrochloride

Structural Formula:

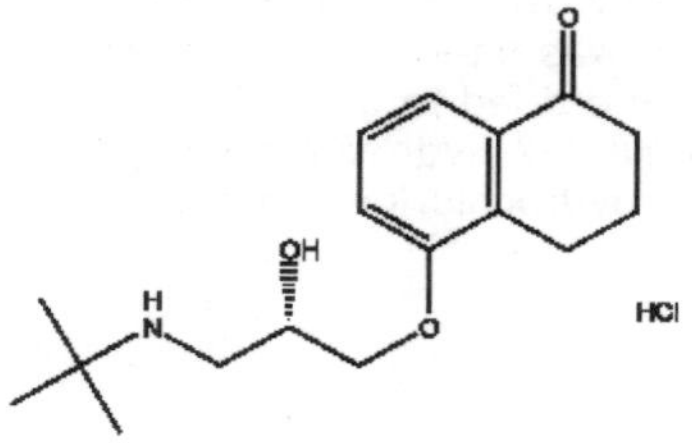

Chemical Abstracts Registry No.: 27912-14-7; 47141-42-4 (Base)

Trade Name	Manufacturer	Country	Year Introduced
Betagan	Allergan	India	-
Betagan Liquifilm	Allergan	Australia	-
Betagen	Ferraz	-	-
Sab-Levobunolol	Sabex Inc.	-	-

Raw Materials

Potassium hydroxide
t-Butylamine
5-Hydroxy-3,4-dihydro-1(2H)-naphthalenone
Tetrabutylammonium bromide
(R)-(-)-Epichlorhydrine

Manufacturing Process

9.62 g (59 mmoles) 5-hydroxy-3,4-dihydro-1(2H)-naphthalenone, 67 ml toluene, 0.36 g (1.1 mmoles) tetra-n-butylammonium bromide, 4.51 g (68 mmoles) 85% potassium hydroxide and 20 ml (254 mmoles) (R)-(-)-epichlorhydrine were placed in an appropriate flask fitted with efficient mechanical stirring, and the mixture was heated under reflux for two hours. The mixture was allowed to cool to 30°C, 50 ml toluene and 50 ml water were added and the mixture was vigorously stirred. The organic phase was removed and the aqueous phase extracted with 25 ml toluene. The combined organic phases were concentrated at reduced pressure, 31 ml (300 mmoles) tert-butylamine, 45 ml ethanol and 3 ml deionized water were added, and the solution was heated under reflux for one hour. The mixture was allowed to cool to 40°C and the volatile products were distilled at reduced pressure. Toluene (9 ml) was added to the residue and volatiles were distilled at reduced pressure. (S)-5-(2,3-Epoxypropoxy)-3,4-dihydro-1(2H)-naphthalenone with an optical purity greater than 95% was obtained. Toluene (75 ml) was added to the product, and then, 10 ml of 35% (w/v) hydrochloric acid and 110 ml water, and the mixture was stirred for fifteen minutes. The organic phase was decanted and the aqueous one was extracted with 50 ml toluene. The aqueous phase was basified by addition of a solution of 5.1 g sodium hydroxide in 150 ml water and extracted twice with toluene (100 and 50 ml, respectively). The combined organic extracts were dried with anhydrous sodium sulfate, decolorized with active charcoal and filtered.

To the above toluenic solution containing levobunolol as free base, 16 ml ethanol and the stoichiometric amount of hydrogen chloride were added. The stirred mixture was cooled below 10°C and kept at this temperature for one hour. The precipitated solid was filtered, washed with toluene, recrystallized twice from 43 ml ethanol and dried to give 10.0 g (51% yield) of (-)-3,4-dihydro-5-(3-(tert-butylamino)-2-hydroxypropoxy)-1(2H)-naphthalenone hydrochloride (levobunolol hydrochloride) having a rotary power at 25°C below -19°.

References

Camps G.P. et al.; CA Patent No. 2,119,052, Dec. 25, 1994; Medichem SA (ES)
Stampa D.D.C. et al.; US Patent No. 5,426,227; June 20, 1995; Assigned to Medichem S.A.

LEVOBUPIVACAINE HYDROCHLORIDE

Therapeutic Function: Local anesthetic

Chemical Name: 2-Piperidinecarboxamide, 1-butyl-N-(2,6-dimethylphenyl)-, monohydrochloride, (2S)-

Common Name: Levobupivacaine hydrochloride

Structural Formula:

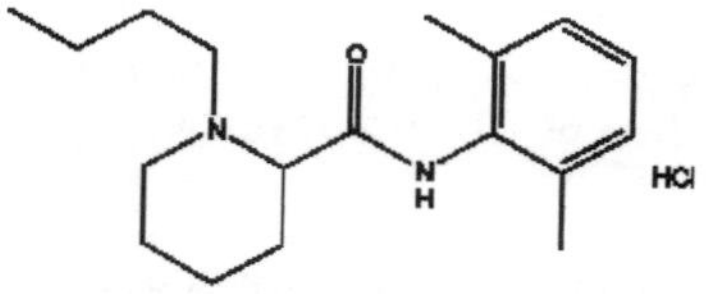

Chemical Abstracts Registry No.: 27262-48-2; 27262-47-1 (Base)

Trade Name	Manufacturer	Country	Year Introduced
Chirocaine	Abbott Laboratories	-	-
Chirochaine	Darwin Discovery	-	-

Raw Materials

Pipecolic acid
Amberlite
2,6-Dimethylaniline
Tartaric acid, L-
Phosphorus pentachloride
n-Butyl bromide

Manufacturing Process

Synthesis of L-pipecolic acid 2,6-xylidide (Patent US 4,695,576)

130 g of pipecolic acid and 158.6 g of Laevo (+)-tartaric acid are dissolved under stirring in 2 L 95% ethyl alcohol and 125 ml water at 80°C. The solution is allowed to cool to room temperature and after two days the crystallized D-pipecolic-tartrate is separated. The L-pipecolic-tartrate remains in solution. The filtrate is evaporated and dissolved in 5% acetic acid. Finally the solution is treated with Amberlite IR 45* in an ion exchanger. The eluate thus obtained is evaporated and the resulting crystalline residue is dried with potassium hydroxide in vacuo. The product obtained consists of L-pipecolic acid $[\alpha]_D^{24}$ = -26.2°(C = 5, H_2O).

4 g of phosphorus pentachloride was added to a suspension of 4 g of L-pipecolic acid hydrochloride in 40 ml acetylchloride. The initial reaction is effected at a temperature of about 35°C under stirring for 2 hours. The chlorination is completed by adding during a time period of about 10 minutes an additional two grams of phosphorus pentachloride and stirring over a further period of 4 hours while maintaining the suspension at a temperature of about 35°C. The resulting L-pipecolic acid chloride hydrochloride is filtered and washed with toluene and acetone. The crystalline residue is then dried in vacuo, m.p. 155°C.

A mixture of 2.7 ml 2,6-dimethylaniline, 4 ml acetone, and 4 ml Nmethylpyrrolidone is gradually added under stirring for 2 hours at 70°C to a suspension of 4 g of L-pipecolic acid chloride hydrochloride. This yields a crystalline product, which is filtered, washed with acetone and dried. This crystalline product is then dissolved in water and the base is precipitated by the addition of ammonia. The base is then extracted by the use of toluene and is recovered by evaporation. The base is recrystallized from a mixture of hexane and ethanol to yield L-pipecolic acid 2,6-xylidide. The melting point of this compound is 129-130°C.

Preparation of L-N-n-butylpipecilic acid 2,6-xylidide may de carried out by analogy with the preparation of L-N-n-propylpipecolic acid 2,6-xylidide (Patent US 5,777,124).

n-Butylbromide and potassium carbonate are added to a solution of L-pipecolic acid 2,6-xylidide dissolved in isopropyl alcohol. Thereafter, 5 ml of water is added to the mixture and the reaction is carried out for 4 hours at 72°C.

To complete the reaction, a further 0.8 ml n-butylbromide are added under continuous stirring and heating for 4 hours. The residue is treated with a mixture of 250 ml toluene and an equal amount of water at 50°C. The toluene layer is separated and washed three times with 100 ml warm water (40°C). A 175 ml portion of the toluene is removed by evaporation and the remainder is stored at +5°C for 6 hours to achieve crude crystalline L-N-n-butylpipecilic acid 2,6-xylidide. The crystalline product is separated by filtration, washed with some cooled toluene and dried at 70°C. Recrystallization may be carried from toluene. This product is dissolved in 100 ml ethanol and neutralized with concentrated hydrochloric acid. Ethanol is removed by evaporation and the hydrochloride product obtained is vacuum dried. Finally the latter is recrystallized from isopropyl alcohol.

References

Ekenstam B.T., Bovin Ch.; US Patent No. 4,695,576; Sep. 22, 1987; Assigned to Astra Lake Medel, Aktiebolag, Sweden

Zavareh H.Sh., Frampton G.A.Ch.; US Patent No. 5,777,124; Jul. 7, 1998; Assigned to Chiroscience Limited, Camdridge, United Kingdom

LEVOCABASTINE HYDROCHLORIDE

Therapeutic Function: Antihistaminic

Chemical Name: 4-Piperidinecarboxylic acid, 1-(cis-4-cyano-4-(4-fluorophenyl)cyclohexyl)-3-methyl-4-phenyl-, monohydrochloride, (3S,4R)-

Common Name: Levocabastine hydrochloride

Structural Formula:

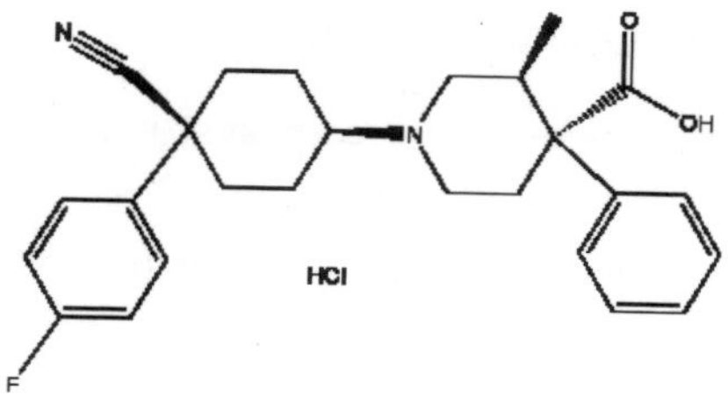

HCl

Chemical Abstracts Registry No.: 79547-78-7; 79516-68-0 (Base)

Trade Name	Manufacturer	Country	Year Introduced
Livostin	Janssen-Cilag	Belgium	-
Livostin	Novartis	-	-

Raw Materials

Ethyl acrylate
Palladium on charocal
3-Methyl-5-phenylpiperidine-4-carboxylic acid benzyl ester
p-Fluorophenylacetonitrile
Sodium methoxide

Manufacturing Process

4-Cyano-4-(4-fluorophenyl)-heptanedioic acid diethyl ester is obtained by addition of ethyl acrylate to the anion from p-fluorophenylacetonitrile. By base catalyzed cyclization of these diester (sodium methoxide, 60°C, xylene) is synthesized an intermediate that after decarboethoxylation gives 1-(4-fluorophenyl)-4-oxycyclohexanecarbonitrile. By condensation of 3-methyl-5-phenylpiperidine-4-carboxylic acid benzyl ester and 1-(4-fluorophenyl)-4-oxycyclohexanecarbonitrile under reductive hydrogenation conditions (palladium-on-charcoal catalyst, 50°C, in ethanol) is prepared benzyl ester 4-piperidinecarboxylic acid, 1-(4-cyano-4-(4-fluorophenyl)cyclohexyl)-3-methyl-4-phenyl-, (3S-(1(cis),3α,4β))-. The benzyl protecting group is then removed by hydrogenation method and 4-piperidinecarboxylic acid, 1-(4-cyano-4-(4-fluorophenyl)cyclohexyl)-3-methyl-4-phenyl-, (3S-(1(cis),3α,4β))- obtained is transformed into 4-piperidinecarboxylic acid, 1-(4-cyano-4-(4-fluorophenyl) cyclohexyl)-3-methyl-4-phenyl-, hydrochloride, (3S-(1(cis), 3α,4β))-(Levocabastine hydrochloride).

References

Merck Index
Stokbroekx R. A. et al.; Patent U.S. 4,369,184, January 18, 1983; Assigned to Janssen Pharmaceutica N.V. (Beerse, BE)
Stokbroekx R. A. et al.; Drug Dev. Res., 1986, 8, 87
G. Vanden Bussche, Drugs of the Future, 1986, 11, 841

LEVOCARNITINE

Therapeutic Function: Appetite stimulant

Chemical Name: 1-Propanaminium, 3-carboxy-2-hydroxy-N,N,N-trimethyl-, inner salt, (2R)-

Common Name: Carnitine; L-Carnitine; Levocarnitine; Vitamin B_T

Structural Formula:

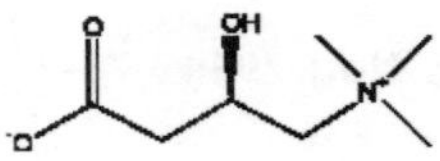

Chemical Abstracts Registry No.: 541-15-1

Trade Name	Manufacturer	Country	Year Introduced
Carnitene	Sigma Tau Industrie Farmaceutiche Riunite	Italy	-
Carnitor	Sigma Tau	Italy	-
Levocarnitine	Shire Pharmaceuticals	-	-

Raw Materials

Trimethylamine
Potassium iodide
Ethyl 4-chloro-3-oxobutyrate
Dichloro(p-cymene)ruthenium(II) dimer
Tetramethylammonium iodide
(+)-[2,2',5,5'-Tetramethyl-3,3'-bis(diphenylphosphino)]-4,4'-bithiophene

Manufacturing Process

a) Preparation of $[RuI_2p\text{-cymene}]_2$

Two g of $[RuCl_2p\text{-cymene}]_2$ and 50 ml of methylene chloride are placed under nitrogen in a flask; 66 mg of tetramethylammonium iodide and subsequently an aqueous solution (50 ml) containing 10.2 mg of KI are added to the solution. The mixture is left under vigorous stirring and in an inert atmosphere for approximately 15 hours at ambient temperature. The phases are separated. The aqueous phase is extracted with 2 times 40 ml of CH_2Cl_2. The gathered organic phases are washed with 3 x 40 ml of H_2O, dried on Na_2SO_4 and filtered on fume silica (dicalite). A red-brown solution is obtained which is vacuum-dried. 3.07 g of $[RuI_2p\text{-cymene}]_2$ are obtained.

b) Preparation of {[Ru (p-cymene) I (+)-TMBTP] I}

155 mg of $[RuI_2p\text{-cymene}]_2$ and 204 mg of (+)-TMBTP are placed under nitrogen in a flask, and the mixture of 80 ml of CH_2Cl_2 fvand 30 ml of MeOH degassed with nitrogen is added. The mixture is left at reflux under stirring for

1.5 h; it is then cooled and concentrated at reduced pressure. The dark red solid consisting of {[Ru (p-cymene) I (+)-TMBTP] I} is used as such in the enantioselective hydrogenation processes.

c) Preparation of ethyl (+)-(R)-4-chloro-hydroxybutyrate

14 kg of ethyl 4-chloro-3-oxobutyrate (titre 88%) and 6.2 g of {[Ru (p-cymene) I (+)-TMBTP] I}, are placed under argon in a 200-liter reactor, in 143 L of ethyl alcohol. The mixture is heated at 116°C and pressurized with hydrogen at 5-6 bar. Temperature rises up to 124°C and the reaction goes to completion within about 1 hour. The mixture is cooled, concentrated at reduced pressure, and the residue, analyzed with gaschromatography, has a 81% titre of ethyl (+)-(R)-4-chloro-3-hydroxybutyrate, with an e.e. of 96.7% reaction yield: 94%.

d) Preparation of L-carnitine

400 g of crude ethyl (+)-(R)-4-chloro-3-hydroxybutyrate, above prepared and 1 L of 45% trimethylamine in H_2O are placed in a 2-liter reactor. The reaction mixture is heated to 80°C and kept at this temperature for 15 h. After cooling and removing the excess of trimethylamine under nitrogen flow, the aqueous solution is extracted with 1.9 L of methylene chloride and analyzed with HPLC. L-carnitine is obtained with 75% yield.

References

Tiniti M. et al.; US Patent No. 6,566,552 B2; May 20, 2003; Assigned to Sigma-Tau Industrie Farmaceutiche Riunite S.p.A., Roma

LEVODOPA

Therapeutic Function: Antiparkinsonian

Chemical Name: 3-Hydroxy-L-tyrosine

Common Name: β-(3,4-Dihydroxyphenyl)-α-alanine; 2-Amino-3-(3,4-dihydroxyphenyl)propanoic acid

Structural Formula:

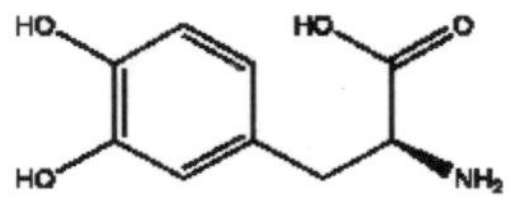

HO HO O HO NH_2

Chemical Abstracts Registry No.: 59-92-7

Trade Name	Manufacturer	Country	Year Introduced
Larodopa	Roche	US	1970
Dopar	Norwich Eaton	US	1970
Dopaidan	De Angeli	Italy	1970
Larodopa	Roche	W. Germany	1970
Larodopa	Roche	UK	1970
Larodopa	Roche	France	1970
Larodopa	Roche	Italy	1970
Brocadopa	Brocades	UK	1970
Levodopa	SKF	US	1971
Bendopa	I.C.N.	US	1971
Larodopa	Roche	Japan	1972
Biodopa	DDR Pharm	US	-
Ceredopa	Merckle	W. Germany	-
Cidandopa	Cidan	Spain	-
Dehdopa	De Angeli	Brazil	-
Dopacin	I.C.N.	Brazil	-
Dopaflex	EGYT	Hungary	-
Dopaidan	De Angeli	Italy	-
Dopalfher	Fher	Spain	-
Doparkin	Farmos	Finland	-
Doparkine	Armstrong	Argentina	-
Doparl	Kyowa	Japan	-
Dopasol	Daiichi	Japan	-
Dopason	Yurtoglu	Turkey	-
Dopaston	Sankyo	Japan	-
Eldopar	Weifa	Norway	-
Eldopatec	Labatec	Switz.	-
Eurodopa	Castejon	Spain	-
Levopa	Arco	Switz.	-
Maipedopa	Maipe	Spain	-
Medidopa	Medica	Finland	-
Novedopa	Torlan	Spain	-
Parkidopa	Farmos	Finland	-
Parmedin	Kwizda	Austria	-
Prodopa	Faulding	Australia	-
Syndopa	Sankyo	Japan	-
Weldopa	Smith and Nephew	UK	-

Raw Materials

Velvet beans
Acetic acid

Manufacturing Process

A charge of 1,000 g of ground velvet beans was extracted with 9 liters of 1% aqueous acetic acid at room temperature over a 20-hour period with occasional stirring during the first 4 hours. The liquor was decanted and the bean pulp slurry was vacuum filtered through a cake of acid-washed diatomaceous earth in a Buechner funnel. The decanted liquor was combined with the filtrate and concentrated under vacuum and a nitrogen atmosphere to a volume of 900 ml. After treating with acid-washed activated carbon, the concentrate was then filtered through acid-washed diatomaceous earth.

After concentrating the filtrate to approximately 400 ml, solids started crystallizing out at which time the filtrate was cooled by refrigerating at 5°C for several hours. Filtration gave 18.7 g of L-Dopa, MP 284° to 286°C (dec.); $[\alpha]_D$ 8.81° (1% solution in aqueous 4% HCl). The infrared spectrum and paper chromatography indicated very good L-Dopa according to US Patent 3,253,023.

Various synthetic routes are also described by Kleeman and Engel.

References

Merck Index 5298
Kleeman and Engel p. 520
PDR pp. 1210, 1489
DOT 9 (6) 247 (1973) and 10 (9) 317, 332 (1974)
I.N. p. 555
REM p, 930
Wysong, D.V.; US Patent 3,253,023; May 24, 1966; assigned to The Dow Chemical Company
Krieger, K.H., Lago, J. and Wantuck, J.A.; US Patent 3,405,159; October 8, 1968; assigned to Merck and Co.,Inc.

LEVOFLOXACIN

Therapeutic Function: Antibacterial

Chemical Name: 7H-Pyrido(1,2,3-de)-1,4-benzoxazine-6-carboxylic acid, 2,3-dihydro-9-fluoro-3-methyl-10-(4-methyl-1-piperazinyl)-7-oxo-, (S)-

Common Name: Levofloxacin; (S)-Ofloxacin

Structural Formula:

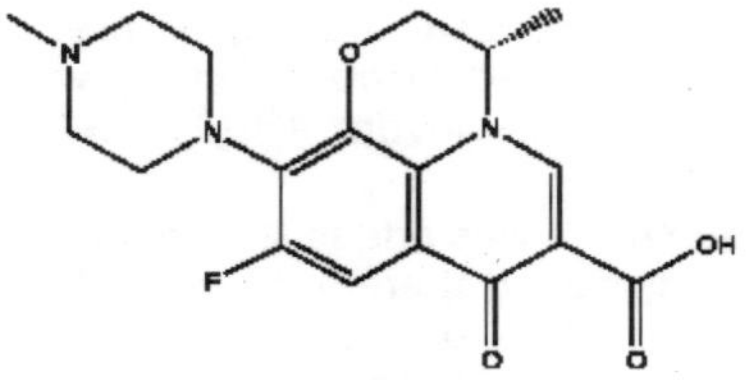

Chemical Abstracts Registry No.: 100986-85-4

Trade Name	Manufacturer	Country	Year Introduced
Fynal-500	Mankind Pharma Pvt. Ltd.	India	-
Levaquin	Hoechst Marion Roussel	-	-
Levoday	Recon Healthcare Ltd.	India	-
Levoff	Biochem Pharma Industries	-	-
Levoflox	Protech Biosystems	India	-
Levofloxacin	Hoechst Marion Roussel	-	-
Levo-Fq Inj.	Venus Remedies Limited	India	-
Levox	Claris Life Sciences	India	-
Lf	Finecure Pharma	India	-
Lofel	East African (I) Remedies Pvt. Ltd.	-	-
Lotor	Emcure Pharmaceuticals Ltd.	India	-
Loxof	Rexcel Pharmaceuticals	India	-
Qure	Aristo Pharmaceutical Ltd.	India	-
Tavanic	Aventis Pasteur	-	-
Tavanic	Hoechst Marion Roussel	Germany	-
T-Livo	Taurus Laboratories Pvt Ltd.	India	-
Voxin	Panacea Biotec Ltd.	India	-

Raw Materials

2,3-Difluoro-6-nitrophenol
Nickel Raney
Thionyl chloride
Acetic anhydride
N-Methylpiperazine
1-Acetoxy-3-chloro-2-propane
Lipoprotein lipase
3,5-Dinitrobenzoyl chloride
Diethyl ethoxymethylenemalonate
Boron trifluoride ethyl etherate

Manufacturing Process

()-3-Acetoxymethyl-7,8-difluoro-2,3-dihydro-4H-[1,4]benzoxazine (m.p. 73-74°C) was synthesized by hydrogenation of a compound prepared from 2,3-difluoro-6-nitrophenol, 1-acetoxy-3-chloro-2-propane and potassium iodide. The hydrogenation was carried out on Raney nickel. The resulting compound was dissolved in THF, and 3,5-dinitrobenzoyl chloride and pyridine were added thereto, followed by heating at 60°C for 3 hours. The mixture was concentrated, and the concentrate was dissolved in ethyl acetate, washed successively with diluted hydrochloric acid, an aqueous solution of sodium bicarbonate and water, dried over anhydrous sodium sulfate and concentrated. Addition of n-hexane to the concentrate caused precipitation of yellow crystals of a racemate. The yield of 3,5-dinitrobenzoyl derivative of the ()-3-acetoxymethyl-7,8-difluoro-2,3-dihydro-4H-[1,4]benzoxazine 3.93 g.

To 2.0 ml of Amberlite XAD 7 was added 2.0 ml of a 0.05 M phosphoric acid buffer (pH 7.0) having dissolved therein 20 mg of lipoprotein lipase, and the system was allowed to stand at room temperature for 18 hours to thereby adsorb the enzyme onto the resin. The resin was filtered. A solution of 250 mg of 3,5-dinitrobenzoyl derivative of ()-3-acetoxymethyl-7,8-difluoro-2,3-dihydro-4H-[1,4]benzoxazine as a substrate in 25 ml of a mixed solvent of benzene and n-hexane (4:1 by volume) was added to the resin, followed by allowing to react at 37°C for 4 hours. It was obtained 117 mg of a 3,5-dinitrobenzoyl derivative of the (-)-3-acetoxymethyl-7,8-difluoro-2,3-dihydro-4H-[1,4]benzoxazine and 65 mg of a derivative of the (-)-3-acetoxymethyl-7,8-difluoro-2,3-dihydro-4H-[1,4]benzoxazine.

In 135 ml THF was dissolved 3.03 g of a 3,5-dinitrobenzoyl derivative of (-)-3-acetoxymethyl-7,8-difluoro-2,3-dihydro-4H-[1,4]benzoxazine, and 135 ml of ethanol and 30 ml of 1.0 N potassium hydroxide were added to the solution. After 30 min 3 ml of acetic acid was added thereto for neutralization. The mixture was concentrated. The solid was subjected to column chromatography using 40 g of silica gel and eluted with chloroform/methanol to obtain 1.17 g of (-)-7,8-difluoro-2,3-dihydro-3-hydroxymethyl-4H-[1,4]benzoxazine; $[\alpha]_D^{22}$ = -14.1° (c = 1.80, $CHCl_3$).

To 1.17 g of (-)-7,8-difluoro-2,3-dihydro-3-hydroxymethyl-4H-[1,4] benzoxazine was added 2.77 g of thionyl chloride in pyridine. The reaction mixture was concentrated and the concentrate was subjected to column chromatography using 40 g of silica gel and eluted with chloroform to obtain 1.18 g of the reaction product as a colorless oily product. This product was dissolved in 30 ml of dimethyl sulfoxide, and 0.41 g of sodium borohydride was added thereto, followed by heating at 80-90°C for 1 hour. The reaction mixture was dissolved in 500 ml of benzene, washed with water to remove the dimethyl sulfoxide, dried over anhydrous sodium sulfate and concentrated under reduced pressure. The concentrate was subjected to column chromatography using 40 g of silica gel and eluted with benzene to obtain 0.80 g of (-)-7,8-difluoro-2,3-dihydro-3-methyl-4H-[1,4]benzoxazine as a colorless oily product; $[\alpha]_D^{25}$ = -9.6° (c = 2.17, $CHCl_3$). Optical Purity: >99% e.e.

To 1.13 g of (-)-7,8-difluoro-2,3-dihydro-3-methyl-4H-[1,4]benzoxazine was added 1.58 g of diethyl ethoxymethylenemalonate, and the mixture was stirred at 130-140°C for 70 min. The reaction mixture was subjected to column chromatography using 50 g of silica gel and eluted with chloroform to obtain 2.47 g of diethyl [(-)-7,8-difluoro-3-methyl-2,3-dihydro-4H-[1,4] benzoxazin-4-yl]methylenemalonate. This product was dissolved in 5 ml of acetic anhydride, and 10 ml of a mixture of acetic anhydride and concentrated sulfuric acid (2/1 by volume) with stirring under ice-cooling, followed by stirring at 50-60°C for 40 min. To the reaction mixture were added ice and an aqueous solution of sodium bicarbonate, and the product was extracted three times with 150 ml portions of chloroform. The combined extract was washed with water, dried over anhydrous sodium sulfate and concentrated. The precipitate was washed with a small amount of diethyl ether to yield 1.32 g of (-)-ethyl 9,10-difluoro-3-methyl-7-oxo-2,3-dihydro-7H-pyrido[1,2,3-de][1,4] benzoxazine-6-carboxylate.

In 12 ml of acetic acid was dissolved 1.20 g of the resulting compound, and 25 ml of concentrated hydrochloric acid was added, followed by refluxing at 120-130°C for 90 min. Upon allowing the reaction mixture to stand at room temperature, colorless crystals were precipitated, which were collected by filtration and washed successively with a small amount of water, ethanol and diethyl ether to obtain 0.96 g of (-)-9,10-difluoro-3-methyl-7-oxo-2,3-dihydro-7H-pyrido[1,2,3-de][1,4]benzoxazine-6-carboxylic acid.

In 30 ml of diethyl ether was suspended 324 mg of the resulting compound, and a large excess of boron trifluoride ethyl etherate was added thereto, followed by stirring at room temperature for 30 min to form a chelate compound. The product was collected by filtration and washed with a small amount of diethyl ether to obtain 373 mg of a powder. The powder was dissolved in 7 ml of dimethyl sulfoxide, and 136 mg of N-methylpiperazine and 228 mg of triethylamine were added thereto, followed by stirring at room temperature for 17 hours. The reaction mixture was concentrated to dryness under reduced pressure, and to the solid were added 15 ml of 95% methanol and 0.31 ml of triethylamine. The resulting mixture was refluxed for 3 hours. The reaction mixture was concentrated under reduced pressure, and the residue was filtered and washed successively with a small amount of ethanol and diethyl ether to obtain 350 mg of a white powder. Recrystallization from a mixed solvent of ethanol and thick aqueous ammonia gave 230 mg of S-(-)-ofloxacin (Levofloxacin).

Melting Point: 225-227°C (with decomposition); $[\alpha]_D^{23}$ = -76.9° (c = 0.39, 0.05 N NaOH).

References

Hayakawa I. et al.; US Patent No. 5,053,407; Oct. 1, 1991; Assigned to Daiichi Pharmaceutical Co., Ltd.

LEVOTHYROXINE SODIUM

Therapeutic Function: Thyroid hormone

Chemical Name: L-3,3',5,5'-Tetraiodothyronine sodium salt

Common Name: -

Structural Formula:

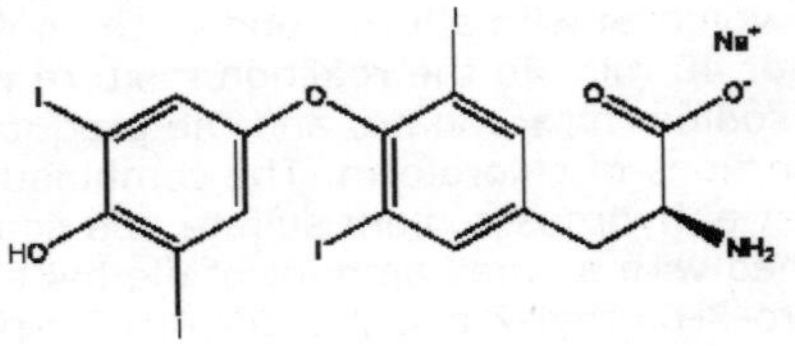

Chemical Abstracts Registry No.: 55-03-8; 51-48-9 (Base)

Trade Name	Manufacturer	Country	Year Introduced
Synthroid	Flint	US	1953
Letter	Armour	US	1965
Eltroxin	Glaxo	UK	-
Euthyrox	Merck	W. Germany	-
Eutirox	Bracco	Italy	-

Trade Name	Manufacturer	Country	Year Introduced
Levaxin	Nyegaard	Norway	-
Levothyrox	Merck Clevenot	France	-
Levotiron	Abdi Ibrahim	Turkey	-
Ro-Thyroxine	Robinson	US	-
Syntaroid	Travenol	US	-
Thevier	Glaxo	W. Germany	-
Thyradin-S	Teikoku Zoki	Japan	-
Thyraplex	Erco	Denmark	-
Thyrex	Sanabo	Austria	-

Raw Materials

Manganese sulfate
Sodium hydroxide
Hydrochloric acid
N-Acetyl-L-diiodotyrosinamide
Acetic acid

Manufacturing Process

A 9.30 g portion of N-acetyl-L-diiodotyrosinamide was suspended in 100 ml of 0.05M boric acid (H_3BO_3) and 100 ml of 95% ethanol, and the solid was dissolved by adjusting the pH to 10.5 with 2N sodium hydroxide (NaOH). A 15% (by weight) portion of manganese sulfate monohydrate was added and the solution heated at 44°C under conditions of oxygenation while being agitated mechanically. After approximately 24 hours of incubation, the precipitated product was collected and separated from the catalyst, providing the amide of N-acetyl-L-thyroxine in 30.6% yield. On hydrolysis (removal of both amide functions), achieved by refluxing in glacial acetic acid-hydrochloric acid (approximately 2:1), L-thyroxine is obtained. It was isolated as the sodium salt, containing approximately 5 molecules of water of hydration.

References

Merck Index 5303
Kleeman and Engel p. 525
PDR p. 993
OCDS Vol. 1 p. 97 (1977)
I.N. p. 558
REM p. 980
Anthony, P.Z. and Ginger, L.G.; US Patent 2,889,364; June 2, 1959;assigned to Baxter Laboratories, Inc.

LIDOCAINE

Therapeutic Function: Local anesthetic, Antiarrhythmic

Chemical Name: 2-(Diethylamino)-N-(2,6-dimethylphenyl)acetamide

Common Name: Lignocaine

Structural Formula:

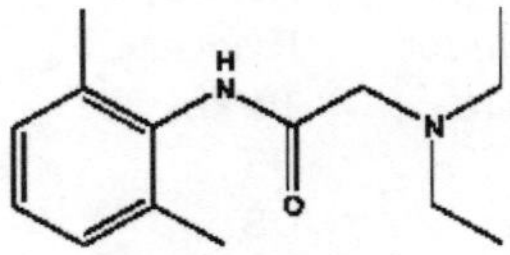

Chemical Abstracts Registry No.: 137-58-6; 73-78-9 (Hydrochloride salt)

Trade Name	Manufacturer	Country	Year Introduced
Xylocaine	Astra	US	1949
Anestacon	Contal	US	1967
Octocaine	Novocol	US	1980
Clinicaine	Johnson and Johnson	US	1982
Anestacain	Farmos	Finland	-
Anestecidan	Cidan	Spain	-
Baylocaine	Bay	US	-
Cidancaina	Cidan	Spain	-
Cito-Optadren	Fischer	Switz.	-
Dolicaine	Reid-Provident	US	-
Dulicaine	Dulcis	Monte Carlo	-
Duncaine	Duncan Flockhart	UK	-
Esracain	Hillel	Israel	-
Leotesin-N	Showa	Japan	-
Lida-Mantal	Dome	US	-
Lidocain	Bristol	US	-
Lidocard	Orion	Finland	-
Lidocaton	Pharmaton	Switz.	-
Lidocor	Gebro	Austria	-
Lido Pen	Survival Tech.	US	-
Lignane	Propan-Lipworth	S. Africa	-
Neo-Novutox	Braun	W. Germany	-
Ortoderm ina	Tiber	Italy	-
Qualigens	Qualipharma	Switz.	-
Rapidocaine	Sintetica	Switz.	-
Sedodent	Belupo Ltd.	Yugoslavia	-
Xylanaest	Gebro	Austria	-
Xylesin	Amino	Switz.	-
Xylestesin	Espe	W. Germany	-
Xylocard	Hassle	Sweden	-
Xylocitin	Jenapharm	E. Germany	-
Xyloneural	Gebro	Austria	-
Xylonor	Septodont	France	-
Xylotox	Willows-Francis	UK	-

Raw Materials

2,6-Xylidine
Chloroacetyl chloride
Diethylamine

Manufacturing Process

One mol of 2,6-xylidine is dissolved in 800 ml glacial acetic acid. The mixture is cooled to 10°C, after which 1.1 mol chloracetyl chloride is added at one time. The mixture is stirred vigorously during a few moments after which 1,000 ml half-saturated sodium acetate solution, or other buffering or alkalizing substance, is added at one time. The reaction mixture is shaken during half an hour. The precipitate formed which consists of ω-chloro-2,6-dimethyl-acetanilide is filtered off, washed with water and dried. The product is sufficiently pure for further treatment. The yield amounts to 70 to 80% of the theoretical amount.

One mole of the chloracetyl xylidide thus prepared and 2.5 to 3 mols diethyl amine are dissolved in 1,000 ml dry benzene. The mixture is refluxed for 4 to 5 hours. The separated diethyl amine hydrochloride is filtered off. The benzene solution is shaken out two times with 3N hydrochloric acid, the first time with 800 ml and the second time with 400 ml acid. To the combined acid extracts is added an approximately 30% solution of sodium hydroxide until the precipitate does not increase.

The precipitate, which sometimes is an oil, is taken up in ether. The ether solution is dried with anhydrous potassium carbonate after which the ether is driven off. The remaining crude substance is purified by vacuum distillation. During the distillation practically the entire quantity of the substance is carried over within a temperature interval of 1° to 2°C. The yield approaches the theoretical amount. MP 68° to 69°C. BP 180° to 182°C at 4 mm Hg; 159° to 160°C at 2 mm Hg. (Procedure is from US Patent 2,441,498.)

References

Merck Index 5310
DFU 8 (12) 1021 (1983)
Kleeman and Engel p. 526
PDR pp. 607, 888, 1569
OCDS Vol. 1 p. 16 (1977); 2, 95, 449 (1980) 813, 40 (1984)
I.N. p. 559
REM p. 1051
Lofgren, N.M. and Lundqvist, B.J.; US Patent 2,441,498; May 11, 1948; assigned to AB Astra, Sweden
Brown, C.L.M. and Poole, A.; US Patent 2,797,241; June 25, 1957

LIDOFLAZINE

Therapeutic Function: Coronary vasodilator

Chemical Name: 4-[4,4-Bis(4-fluorophenyl)butyl]-N-(2,6-dimethylphenyl)-1-piperazineacetamide

Common Name: -

Structural Formula:

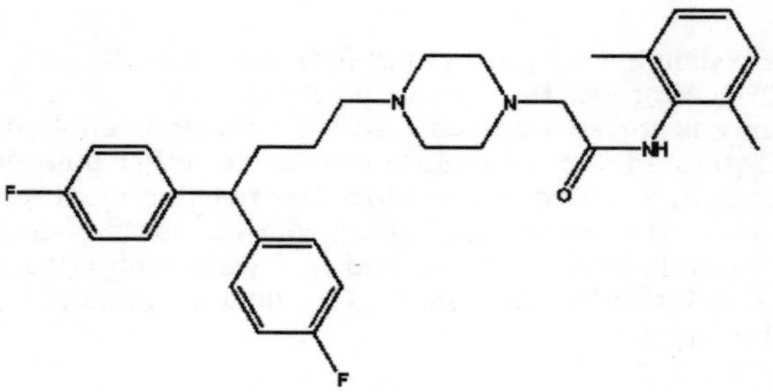

Chemical Abstracts Registry No.: 3416-26-0

Trade Name	Manufacturer	Country	Year Introduced
Clinium	Janssen	W. Germany	1969
Corflazine	Cassenne	France	1972
Clinium	Janssen	Italy	1974
Clinium	Janssen	UK	1980
Anginin	Yurtoglu	Turkey	-
Clavidene	Corvi	Italy	-
Clinium	McNeil	US	-
Klinium	Esteve	Spain	-
Klintab	Eczacibasi	Turkey	-

Raw Materials

1-[4,4-(Di-p-fluorophenyl)butyl]piperazine
N-(2-Chloroacetyl)-2,6-dimethylaniline

Manufacturing Process

A mixture of 6.6 parts 1-[4,4-di-(4-fluoro-phenyl)butyl]-piperazine, 4.33 parts N-(2-chloro-acetyl)-2,6-dimethyl-aniline, 3.2 parts sodium carbonate, a few crystals of potassium iodide in 200 parts 4-methyl-2-pentanone is stirred and refluxed for 70 hours. After cooling there are added 70 parts water. The organic layer is separated, dried over potassium carbonate, filtered and evaporated. The oily residue is dissolved in 80 parts diisopropylether and the solution is filtered hot. After cooling the filtrate at 0°C. the formed solid is filtered off and recystallired from 80 parts ether, yielding 1-[4,4-di-(4-fluoro-phenyl)butyl]-4-[(2,6-dimethylanilino-carbonyl)-methyl]-piperazine; MP 159°C to 161°C.

References

Merck Index 5311
Kleeman and Engel p. 526
OCDS Vol. 1 p. 279 (1977)
DOT 2 (4) 118 (1966) and 6 (1) 21 (1970)
I.N. p. 560
Hermans, H.K.F. and Schaper, W.K.A.; US Patent 3,267,164; August 16, 1966; assigned to Janssen Pharmaceutica N.V. (Belgium)

LINCOMYCIN

Therapeutic Function: Antibacterial

Chemical Name: Methyl 6,8-dideoxy-6-(1-methyl-4-propyl-2-pyrrolidinecarboxamido)-1-thio-D-erythro-D-galacto-octopyranoside

Common Name: Lincoinensin

Structural Formula:

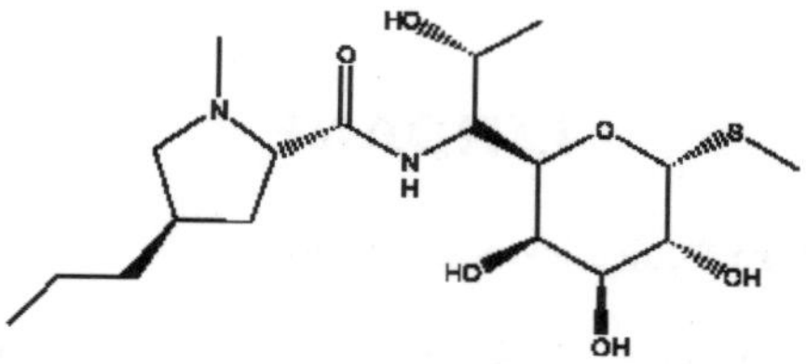

Chemical Abstracts Registry No.: 154-21-2; 859-18-7 (Hydrochloride salt)

Trade Name	Manufacturer	Country	Year Introduced
Lincocin	Upjohn	UK	1964
Lincocin	Upjohn	US	1965
Lincocine	Upjohn	France	1966
Albiotic	Upjohn	W. Germany	1966
Lincocin	Upjohn	Italy	1966
Cillimicina	Albert Pharma	Spain	-
Cillimycin	Hoechst	W. Germany	-
Lincolcina	Atral	Portugal	-
Mycivin	Boots	UK	-

Raw Materials

Bacterium Streptomyces lincolnensis
Nutrient medium

Manufacturing Process

As described in US Patent 3,086,912, the process comprises cultivating Streptomyces lincolnensis var. lincolnensis in an aqueous nutrient medium containing a source of assimilable carbohydrate and assimilable nitrogen under aerobic conditions until substantial activity is imparted to the medium by production of lincolnensin and isolating the lincolnensin so produced.

References

Merck Index 5328
Kleeman and Engel p. 527
PDR p. 1847
DOT 2 (2) 62 (1966)
I.N. p. 561
REM p. 1212
Bergy, M.E., Herr, R.R. and Mason, D.J.; US Patent 3,086,912; April 23, 1963;assigned to The Upjohn Company
Bergy, ME, Herr, R.R. and Mason, D.J.; US Patent 3,155,580; November 3, 1964; assigned to The Upjohn Company
Argoudelis, A.D., Bannister, B., Hoeksema, H., Kagan, F. and Magerlein, B.J.; US Patent 3,380,992; April 30, 1968;assigned to The Upjohn Company
Jariwala, S.L.; US Patent 4,091204; May 23, 1978; assigned to The Upjohn Company

LINDANE

Therapeutic Function: Pediculicide, Scabicide

Chemical Name: 1α,2α,3β,4α,5α,6β-Hexachlorocyclohexane

Common Name: γ-BHC

Structural Formula:

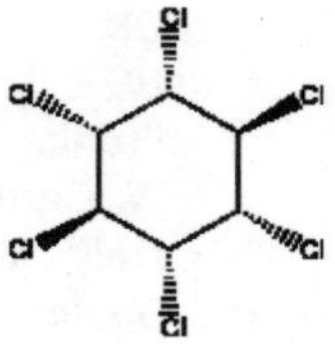

Chemical Abstracts Registry No.: 58-89-9

Trade Name	Manufacturer	Country	Year Introduced
Kwell	Reed Carnrick	US	1952
Gamene	Barnes Hind	US	1975
Escabiol	Stiefel	US	1979
Scabene	Stiefel	US	1981

Trade Name	Manufacturer	Country	Year Introduced
Bicide	Fischer	Israel	-
Gambex	Continental Ethicals	S. Africa	-
HCH-Salbe	VEB Leipziger Arz.	E. Germany	-
Jacutin	Hermal	W. Germany	-
Malice Shampoo	Restan	S. Africa	-
Quellada	Stafford-Miller	UK	-

Raw Materials

Benzene
Chlorine

Manufacturing Process

Chlorine gas was gradually passed into 660 parts of benzene contained in a lead-lined reaction vessel until 890 parts of the gas had been absorbed. The mixture was stirred continuously and the temperature maintained at 15°C to 20°C.

The supply of chlorine was then interrupted and the precipitated solid filtered off and dried. In weight, it was found to be equivalent to 900 parts. The mother liquid was then mixed with 330 parts of benzene and the mixture again treated with 890 parts of chlorine in the manner described.

After filtering the reaction mixture resulting from the second chlorination, the filtrate was again mixed with a smaller quantity of benzene and again chlorinated in a similar manner. In this way, a continuous process for the preparation of benzene hexachloride resulted.

That benzene hexachloride isomer mixture is then the raw material for lindane production. The production of lindane per se is not a chemical synthesis operation but a physical separation process. It is possible to influence the gamma isomer content of benzene hexachloride to an extent during the synthesis process. Basically, however, one is faced with the problem of separating a 99%-plus purity gamma isomer from a crude product containing perhaps 12 to 15% of the gamma isomer. The separation and concentration process is done by a carefully controlled solvent extraction and crystallization process. One such process is described by R.D. Donaldson et al. Another description of hexachlorocyclohexane isomer separation is given by R.H. Kimball.

References

Merck Index 5329
PDR pp. 1444, 1606, 1779
I.N. p. 561
REM pp. 1239, 1253
Donaldson, R.D. et al; US Patent 2,767,223; October 16, 1956; assigned to Allied Chemical and Dye Corp.
Kimball, R.H.; US Patent 2,767,224; October 16, 1956; assigned to Hooker Electrochemical Co.

Hay,J.K. and Webster, K.C.; US Patent 2,502,258; March 28, 1950; assigned to Imperial Chemical Industries, Ltd.
Hardie, T.; US Patent 2,218,148; October 15, 1940; assigned to Imperial Chemical Industries, Ltd.

LINEZOLID

Therapeutic Function: Antibacterial

Chemical Name: Acetamide, N-(((5S)-3-(3-fluoro-4-(4-morpholinyl)phenyl)-2-oxo-5-oxazolidinyl)methyl)-

Common Name: Linezolid

Structural Formula:

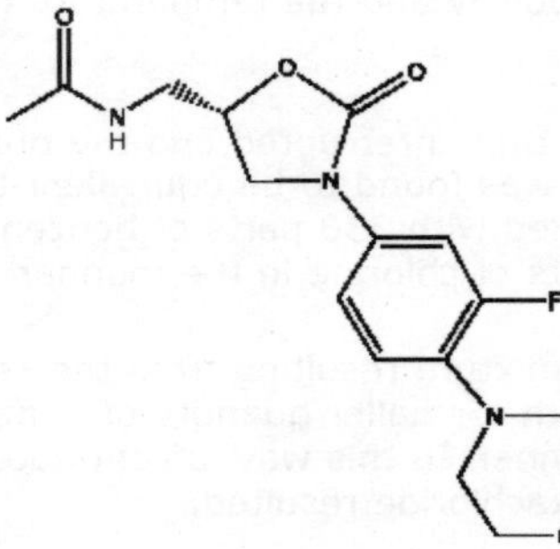

Chemical Abstracts Registry No.: 165800-03-3

Trade Name	Manufacturer	Country	Year Introduced
Linosept	Eros Pharma Ltd. (A div. of Microlabs)	India	-
Linospan	Cipla Limited	India	-
Zivox	Pharmacia and Upjohn	USA	-
Zyvox	Pharmacia and Upjohn	USA	-
Zyvoxam	Pharmacia and Upjohn	USA	-

Raw Materials

Butyl lithium	S-(+)-3-Chloro-1,2-propanediol
t-Amyl alcohol	Methanesulfonyl chloride
Citric acid	Potassium t-butoxide
Triethylamine	4-Nitrobenzenesulfonyl chloride
Salicylaldehyde	Hydrochloric acid
Morpholine	Diisopropylethylamine
Sodium azide	Benzyl chloroformate

(R)-Glycidyl butyrate Palladium on carbon
N-Carbobenzoxy-3-fluoro-4-morpholinylaniline

Manufacturing Process

The 1st method of synthesis (Patent US 5,837,870)

N-Carbobenzoxy-3-fluoro-4-morpholinylaniline (3.00 mmol, 1.000 eq) and tetrahydrofuran (3.5 ml) were agitated and cooled. The lithium t-amylate mixture [prepared in THF at 25°C from t-amyl alcohol (0.66 ml, 6.03 mmol, 2.00 eq) and butyl lithium (1.8 ml, 2.5 M in hexanes, 4.55 mmol, 1.5 eq)] is then added to the carbamate mixture at less than 8°C and rinsed in with THF (1 ml).

Tetrahydrofuran (3.2 ml) and S-(+)-3-chloro-1,2-propanediol (0.299 ml, 3.58 mmol, 1.19 eq) are mixed. The mixture of THF (3.2 ml) and S-(+)-3-chloro-1,2-propanediol (0.299 ml, 3.58 mmol, 1.19 eq) is cooled to -16°C and potassium t-butoxide (3.2 ml, 1.0 M) in THF (3.2 mmol, 1.07 eq) is added at less than -10°C. The resulting slurry is stirred at -14-0°C for 1 hour. Then added to the lithium anion mixture while maintaining both mixtures at 0°C, then rinsed in with THF (2 ml). The resultant slurry is stirred at 20-23°C for 2 hour and then cooled to 6°C and a mixture of citric acid monohydrate (0.4459 g, 2.122 mmol, 0.705 eq) in water (10 ml) is added. The resultant liquid phases are separated and the lower aqueous phase is washed with ethyl acetate (12 ml). The organic layers are combined and solvent is removed under reduced pressure until a net weight of 9.73 g remains. Heptane (10 ml) and water (5 ml) are added and solvent is removed 4-nitrobenzenesulfonyl chloride y reduced pressure until a total volume of 5 ml remains. The precipitated product is collected by vacuum filtration and washed with water (7 ml). The solids are dried in a stream of nitrogen to give (R)-[N-3-(3-fluoro-4-(4-morpholinylphenyl)-2-oxo-5-oxazolidinyl]methanol.

To a slurry of (R)-[N-3-(3-fluoro-4-(4-morpholinylphenyl)-2-oxo-5-oxazolidinyl]methanol (43.0 g, 145 mmol) and triethylamine (36 g, 355 mmol) in methylene chloride (450 ml) at 0°C is added a mixture of 4-nitrobenzenesulfonyl chloride (32 g, 145 mmol) in methylene chloride (55 ml). The mixture is stirred in a 0°C bath for 30 min and then quenched with hydrochloric extracted again with methylene chloride (200 ml). The combined organic extracts are then concentrated column chromatographed (silica gel, methanol/methylene chloride 1-2/98-99, about 8 L). The appropriate fractions are combined and concentrated to give the (R)-[N-3-(3-fluoro-4-(4-morpholinylphenyl)-2-oxo-5-oxazolidinyl]methanol 4-nitrobenzenesulfonate ester.

A mixture of (R)-[N-3-(3-fluoro-4-(4-morpholinylphenyl)-2-oxo-5-oxazolidinyl]methanol 4-nitrobenzenesulfonate ester, isopropanol (149 ml), acetonitrile (245 ml), salicylaldehyde (13.7 ml, 129 mmol) and aqueous ammonia (30%, 257 ml, 4.02 mol), is heated to 40°C and stirred at 39-42°C for 24 hours. The mixture is then cooled to -22°C and the precipitate collected by vacuum filtration, washed with water (10 ml) and dried to give the (S)-[N-3-(3-fluoro-4-(4-morpholinylphenyl)-2-oxo-5-oxazolidinyl]methylamine salicylaldehyde imine.

(S)-[N-3-(3-Fluoro-4-(4-morpholinylphenyl)-2-oxo-5-oxazolidinyl] methylaminesalicylaldehyde imine (1.0068 g, 2.521 mmol) is slurried in water (10 ml) and 37% aqueous hydrochloric acid (0.417 ml, 5.04 mmol) and stirred at 20-25°C for 15 hours. Toluene (10 ml) is added and the phases separated; then, the organic phase is washed with hydrochloric acid (1 M, 5 ml) and the combined aqueous phases are washed with toluene (10 ml). The toluene wash is back-extracted with hydrochloric acid (1 M, 5 ml). The combined aqueous phases are then adjusted to pH 13.0 with aqueous sodium hydroxide (50%, 1.83 g, 22.9 mmol). To the resultant slurry is then added methylene chloride (10 ml) and sodium chloride (1 g) and the phases separated. The aqueous phase is then washed with methylene chloride (10 ml). To the combined organic phases is then added acetic anhydride (0.472 ml, 5.00 mmol) while maintaining 24-27°C. The mixture is stirred 40 min, then water is added (5 ml). The phases are separated and the aqueous phase is washed with methylene chloride (5 ml). The combined organic phases are concentrated and ethyl acetate (25 ml) is added. The mixture is warmed to 70°C and then the resultant mixture is slowly cooled to -25°C. The precipitate is collected by vacuum filtration, washed with ethyl acetate (5 ml) and dried to give the (S)-[[N-3-(3-fluoro-4-(4-morpholinylphenyl)-2-oxo-5-oxazolidinyl] methyl]acetamide, HPLC major component (99.93 area % at 254 nm detection) retention time = 0.97 min, column Zorbax RX-C8, mobile phase 650 ml acetonitrile, 1.85 ml triethylamine, 1.30 ml acetic acid and sufficient water to make 1000 ml; flow rate = 3 ml/min.

The 2 th method of synthesis (Patent U.S. 5,688,792)

A solution of 19.9 g of morpholine, 14.8 g of diisopropylethylamine and 28.7 g of 3,4-difluoronitrobenzene in 100 mL of ethylacetate was refluxed under nitrogen for 4 hours. The mixture was allowed to cool to room temperature overnight, then 100 mL of ethyl acetate, 150 mL of methylene chloride, and 150 mL of water were added, and the aqueous layer extracted with methylene chloride and ethyl acetate. The combined organic layers were dried (Na_2SO_4) to give a yellow solid. This was recrystallized from acetone-water to give 3-fluoro-4-morpholinyl-nitrobenzene as a yellow solid, m.p. = 112-113°C.

To a suspension of 36.56 g of 3-fluoro-4-morpholinyl-nitrobenzene and 48.84 g of ammonium formate in 110 mL of tetrahydrofuran and 440 mL of methanol under nitrogen was added 0.524 g of 10% palladium on carbon. After stirring the mixture for 3 hours, the mixture was filtered through diatomaceous earth, and the filter pad was washed with ethyl acetate. The filtrate was concentrated to a volume of about 450 mL and then 200 mL of water was added. This was extracted with 300 mL of ethyl acetate, then the organic layer was washed with of water and then with brine, dried ($MgSO_4$), and concentrated to give a brown solid of 3-fluoro-4-morpholinyl-aniline.

To a solution of 28.91 g of 3-fluoro-4-morpholinyl-aniline and 27.88 g of sodium bicarbonate in 500 mL of acetone and 250 mL of water at 0°C was added 28.68 g of benzyl chloroformate. After stirring the mixture for 1.5 hours, the mixture was poured onto 1 L of ice and water, and the ice allowed to melt. The precipitated solid was collected by filtration and washed with of water, and then dried in a vacuum oven at 75°C to give a gray-purple solid. This was recrystallized from acetone-water to give a cream-colored solid of N-carbobenzyloxy-3-fluoro-4-morpholinylaniline, m.p. = 123-124°C.

To a solution of 39.01 g of N-carbobenzyloxy-3-fluoro-4-morpholinylaniline in

550 mL of THF at -78°C under nitrogen was added 77 mL of 1.6 M n-butyl lithium/hexane via syringe over 30 min, and the mixture stirred for an additional 40 min. At that time, a solution of 18.32 g of (R)-glycidyl butyrate in 30 mL of THF was added over 30 min, and after 1.5 hours, the flask was removed from the dry ice bath, and allowed to come to ambient temperature. After stirring the mixture overnight, 20 ml of saturated aqueous ammonium chloride was added, followed by 500 mL of water, and the aqueous layer extracted with ethyl acetate. The combined organic layers were washed with 200 mL of brine and dried ($MgSO_4$) to give a light purple solid. This is triturated with 1200 mL of 1:1 ethyl acetate/hexanes (v:v), then recrystallized from ethyl acetate/hexanes to give a white solid of (R)-N-[[3-(3-fluoro-4-morpholinyl)phenyl]-2-oxo-5-oxazolidinyl]methanol, m.p. 110-113°C.

To a solution of 13.28 g of (R)-N-[[3-(3-fluoro-4-morpholinyl]phenyl)-2-oxo-5-oxazolidinyl]methanol and 8.71 g of triethylamine in 100 mL of methylene chloride at 0°C under nitrogen was added 7.4 g of methanesulfonyl chloride over 4 min. The mixture was allowed to stir at 0°C for 30 min, then allowed to warm to ambient temperature. A white solid of (R)-N-[[3-[3-fluoro-4-morpholinyl]phenyl]-2-oxo-5-oxazolidinyl]methane sulfonate was precipitated and dried in a vacuum, m.p. = 183-184°C.

To a solution of (R)-N-[[3-(3-fluoro-4-morpholinyl)phenyl]-2-oxo-5-oxazolidinyl]methane sulfonate (9.05 g) in 200 mL of DMF was added 6.367 g of sodium azide, and the mixture heated at 85°C overnight. The mixture was cooled and poured into 500 mL of water and 150 mL of ethyl acetate. The aqueous layer was extracted with ethyl acetate, and the combined organic layers were dried ($MgSO_4$), and concentrated in vacuum. The brown oil of (R)-N-[[3-(3-fluoro-4-morpholinyl)phenyl)-2-oxo-5-oxazolidinyl]methyl]azide, containing some DMF, was utilized without further purification.

A flask containing the crude (R)-N-[[3-[3-fluoro-4-morpholinyl]phenyl]-2-oxo-5-oxazolidinyl]methyl]azide (24.2 mmol) in 500 mL of ethyl acetate was evacuated and filled with nitrogen (three times). Then 0.602 g of 10% palladium/carbon was added and the flask again evacuated and filled with nitrogen (three times), then with hydrogen from a balloon (four times). The mixture was stirred for 17 hours, then a fresh balloon of hydrogen was attached. After a period of 5 hours, the flask was evacuated and filled with nitrogen (three times), and 16 mL of pyridine and 10 mL of acetic anhydride were added. After a period of 2.5 hours, the mixture was filtered over diatomaceous earth, washing the pad with ethyl acetate, and the filtrate concentrated in vacuo to give a brown gummy solid. The residue was purified by chromatography (silica gel column, eluting with a gradient of 2-10% methanol/ethyl acetate (v/v); the combined proper fractions gave an off white solid, which was triturated with ethyl acetate and dried to give an off white solid of (S)-N-[[3-(3-fluoro-4-morpholinyl)phenyl)-2-oxo-5-oxazolidinyl] methyl]acetamide, m.p. 181.5-182.5°C.

References

Barbachyn M.R. et al.; US Patent No. 5,688,792; Nov. 18, 1997; Assigned to Pharmacia and Upjohn Company (Kalamazoo, MI)

Pearlman B.A. et al.; US Patent No. 5,837,870; Nov.17, 1998; Assigned to Pharmacia and Upjohn Company (Kalamazoo, MI)

Lobray B.D. et al.; Tetrahedron Lett., 1999, 40(26), 4855

LIOTHYRONINE

Therapeutic Function: Thyroid hormone

Chemical Name: O-(4-Hydroxy-3-iodophenyl)-3,5-diiodo-L-tyrosine

Common Name: 3,5,3'-Triiodothyronine; L-3-[4-(4-Hydroxy-3-iodophenoxy)-3,5-diiodophenyl]alanine

Structural Formula:

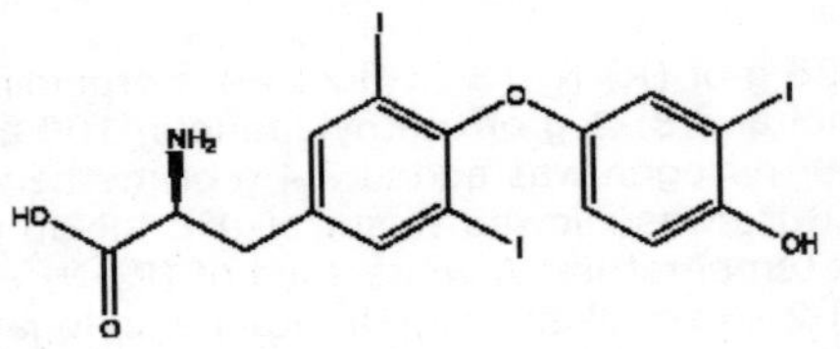

Chemical Abstracts Registry No.: 6893-02-3; 55-06-1 (Sodium Salt)

Trade Name	Manufacturer	Country	Year Introduced
Cytomel	SKF	US	1956
Cynomel	Merrell	France	1961
Cytobin	Norden	US	-
Cytomine	Darby	US	-
Ro-Thyronine	Robinson	US	-
Tertroxin	Glaxo	UK	-
Thybon	Hoechst	W. Germany	-
Thyronamin	Takeda	Japan	-
Thyronine	Taisho	Japan	-
Tiromel	Abdi Ibrahim	Turkey	-
Ti-Tre	Glaxo	Italy	-
Trijodthyronin	Nyegaard	Norway	-
Trithyron	Millot	France	-

Raw Materials

L-Diiodothyronine
Iodine

Manufacturing Process

The 3,5-diiodo compound used as a starting material is a known material and may be prepared by the method in British Patents 643,089 and 671,070 and in the Journal of the Chemical Society, London, 1949, page 3424.

Synthesis: L-diiodo thyronine (1.05 g) is dissolved in ammonia (specific gravity 0.880) (40 ml) and methanol (40 ml) and iodinated slowly with shaking with N-iodine in KI solution at room temperature. After iodination, most of the ammonia and methanol are removed by evaporation under diminished pressure, water is added to the original volume, the solution is heated to 60°C and brought to pH 4 with hydrochloric acid. A crystalline precipitate is obtained which after cooling to room temperature is collected and washed with water. At this stage, the crude triiodo thyronine is contaminated with thyroxine and a little unchanged diiodo thyronine.

Purification: The crude precipitate is dissolved in boiling 2N HCl (300 ml) and filtered from the relatively insoluble thyroxine hydrochloride. The hot filtrate is brought to pH 4 with 5N NaOH and triiodo thyronine again separates; after chilling at 0° to 4°C it is collected, washed with water and dried. The yield of triiodo thyronine is 70 to 75% of the theoretical. This triiodo thyronine still contains some thyroxine (about 10%).

The final purification consists of chromatographic separation of thyroxine and triiodo thyronine on a kieselguhr column using 20% chloroform in n-butanol equilibrated with 0.5N NaOH as the developing solvent. 80 to 100 mg triiodo thyronine is purified during each run on a 50 g kieselguhr column. Pure L-triiodo thyronine has MP 236° to 237°C (dec.) and $[\alpha]_D^{29.5°}$ = +21.5 in a 4.75% solution in a mixture of 1 part of N HCl and 2 parts of ethanol. Liothyronine is commonly used as the sodium salt.

References

Merck Index 5337
Kleeman and Engel p. 527
PDR pp. 1606, 1709
OCDS Vol. 1 p. 97 (1977)
I.N. p. 562
REM p. 980
Pitt-Rivers, R. and Gross,J.; US Patent 2,823,164; February 11,1958; assigned to National Research Development Corporation, England
Platt, J.T. and Wenner, W.; US Patent 2,784,222; March 5, 1957; assigned to Hoffmann-La Roche Inc.
Razdan, R.K. and Wetherill, L.A.; US Patent 2,993,928; July 25, 1961;assigned to Glaxo Laboratories, Ltd.

LISINOPRIL

Therapeutic Function: Antihypertensive

Chemical Name: L-Proline, 1-(N^2)-(1-carboxy-3-phenylpropyl)-L-lysyl)-,(1S)-

Common Name: Lisinopril

Chemical Abstracts Registry No.: 76547-98-3

Structural Formula:

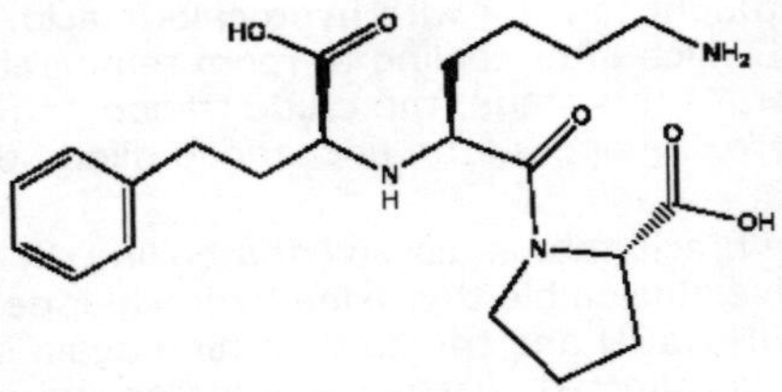

Trade Name	Manufacturer	Country	Year Introduced
Acebitor 5	Biddle Sawyer	India	-
Acemin	AstraZeneca	-	-
Acemin	ICI	-	-
Acetan	Merck Sharp and Dohme	-	-
Acetan	Kwizda	-	-
Acinopril	Nicholas Piramal India Ltd. (Npil)	India	-
Adicanil	Pharmathen	Greece	-
Biopril	Biochem Pharma Industries	India	-
Cipril	Cipla Limited	India	-
Cipril-H	Cipla Limited	India	-
Doxapril	Labs. Bago S. A.	Argentina	-
Ecalisin	Eurolabor	-	-
E.S.	Stadmed Private Limited	India	-
Hipril	Carsyon (Div. Of Microlabs)	India	-
Irumed	Belupo Ltd.	Croatia	-
Landolaxin	Faran Abee	Greece	-
L.P.L.	Concept Pharmaceuticals Ltd.	India	-
Lesopril	Themis Pharmaceuticals Ltd.	India	-
Linoril	Stadmed Private Limited	India	-
Linvas	Zydus Cadila	India	-
Lipril	Lupin Laboratories Ltd.	India	-
Lisinace	Alkem Laboratories Ltd.	India	-
Lisir	Kramer	India	-
Lislo	SPPL (Sarabhai Piramal Pharmaceuticals Ltd.)	India	-
Lisoril	IPCA laboratories Ltd.	India	-
Lisoril-5 Ht	IPCA laboratories Ltd.	India	-
Listril	Torrent Pharmaceuticals Ltd.	India	-
Listril Plus	Torrent Pharmaceuticals Ltd.	India	-
Nafodryl	Costas G. Xydias and Co.	Greece	-
Nivant	German Remedies Limited	Germany	-
Normopril	Otsira Genetica (A Div. Of Aristo Pharma Ltd.)	India	-
Odace	Zydus Medica	India	-

Trade Name	Manufacturer	Country	Year Introduced
Presokin	Chemopharma	Chile	-
Presokin	Sanitas	-	-
Prevace	RPG Life Sciences Ltd.	India	-
Sedotensil	Sanofi Winthrop	France	-
Sedotensil	Ramon	-	-
Tensopril	Syncro	-	-
Tensopril	IVAX Arg.	-	-
Tensopril	Teva	Israel	-
Veroxil	Anfarm	-	-
Zestril	AstraZeneca	Chile	-
Zestril	ICI India Limited	India	-

Raw Materials

Cyanoborohydride
XAD-2 resin
LH-20
2-Oxo-4-phenylbutyric acid
Sodium cyanoborohydride
t-BOC-L-lysyl-L-proline

Manufacturing Process

2-Oxo-4-phenylbutyric acid and t-BOC-L-lysyl-L-proline are condensed in the presence of sodium cyanoborohydride. Essentially all of the t-BOC protecting group is cleaved when the product is absorbed on strong acid ion exchange resin. The crude N-(1-carboxy-3-phenylpropyl)-L-lysyl-L-proline is eluted from the resin with 10% ammonia, freeze dried, and purified by gel filtration chromatography (LH-20). A minute peak for t-BOC protons in the NMR spectrum disappears when the product is treated with ethyl acetate that is 4 N in hydrogen chloride gas. The NMR spectrum of the resulting HCl salt of the product is consistent with structure. The mass spectrum shows a molecular ion at 693 m/e for the tetrasilylated species. Chromatography on XAD-2 resin using 3.5% acetonitrile in 0.1 molar ammonium hydroxide affords N-α-((1S)-1-carboxy-3-phenylpropyl)-L-lysyl-L-proline. The last peptide can be produced if 2-oxo-4-phenylbutyric acid and N-t-Boc-L-lysyl-L-proline are condensed in the presence of sodium cyanoborohydride. The product is absorbed on strong acid ion exchange resin, and eluted with 2% pyridine in water. Product-rich cuts are stripped to a glass and treated with 4 N HCl in ethylacetate to remove the t-Boc protecting group. The resulting hydrochloride salt is converted to the free base by absorbing on strong acid ion exchange resin and eluting with 2% pyridine in water. Freeze drying of product-rich cuts affords N-α-(1-carboxy-3-phenylpropyl)-L-lysyl-L-proline as a white fluffy solid. The NMR spectrum is consistent with structure. The mass spectrum shows a molecular ion at 549 for the disilylated species. Chromatography affords the desired isomer.

References

Harris E.E. et al.; US Patent No. 4,374,829; Feb. 22, 1983; Assigned: Merck and Co., Inc. (Rahway, NJ)

LODOXAMIDE

Therapeutic Function: Anti-asthmatic, Antiallergic

Chemical Name: 2,2'-((2-Chloro-5-cyano-1,3-phenylene)diimino)bis(2-oxo-acetic acid)

Common Name: Lodoxamide

Structural Formula:

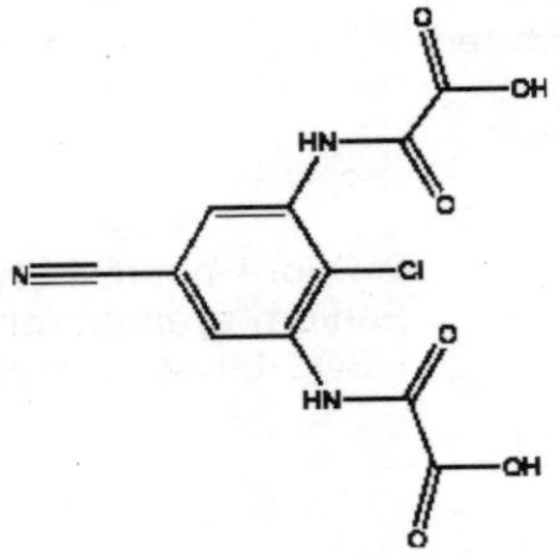

Chemical Abstracts Registry No.: 53882-12-5

Trade Name	Manufacturer	Country	Year Introduced
Lodoxamide	Alcon	USA	-

Raw Materials

Stannous chloride dihydrate
4-Chloro-3,5-dinitrobenzonitrile
Ethyloxalyl chloride
Concentrated hydrochloric acid

Manufacturing Process

To a solution of 1.56 mole of stannous chloride dihydrate in 860 ml of concentrated hydrochloric acid is added 0.2195 mole of 4-chloro-3,5-dinitrobenzonitrile. The mixture is stirred at room temperature for 2 hours and cooled to 0°C in an ice-salt bath. A cold solution of 50% sodium hydroxide is added to the mixture until strongly basic. During the addition the temperature is kept below 30°C.The precipitate is removed by filtration and extracted three times with 400 ml of ethyl acetate. The extracts are combined and added to the aqueous filtrate. The phases are shaken well for ten minutes and separated. The organic phase is evaporated to dryness in vacuo. The solid residue is recrystallized from ethanol-water. There is obtained 25.0 g (68%) of 4-chloro-3,5-diaminobenzonitrile, melting point 169-170°C.

To a solution of 0.34 mole of 4-chloro-3,5-diaminobenzonitrile in 160 ml of dry DMF is added 0.82 mole of triethylamine. The solution is cooled to 5°C and there is added 0.82 mole of ethyloxalyl chloride dropwise, keeping the temperature less than 15°C. The mixture is stirred for 1 hour and warmed to room temperature. The mixture is stirred at room temperature for 24 hours. The precipitate is removed by filtration and washed two times with ethyl acetate. The filtrate and washes are combined and the ethyl acetate distilled off in vacuo. The DMF solution is poured into 3 L of water. The semi-solid residue is removed by filtration. The residue is recrystallized from ethanol. There is obtained 72.4 g (58%) of diethyl N,N'-(2-chloro-5-cyano-m-phenylene)dioxamate, melting point 177-179°C.

A solution of 0.197 mole of diethyl (N,N'-(2-chloro-5-cyano-m-phenylene) dioxamate in 750 ml of methylene chloride is extracted with 465 ml of 1 N sodium hydroxide. The aqueous phase is separated and stirred for 20 min at room temperature. The solution is acidified with dilute hydrochloric acid. The precipitate is removed by filtration and washed with water. There is obtained 59.1 g (96%) of N,N'-(2-chloro-5-cyano-m-phenylene)dioxamic acid, melting point 212°C (dec.).

References

Hall Ch. M., Wright J.B.; US Patent No. 3,993,679; 11.23.1976; Assigned to The Upjohn Company

Aoki K.R. et al.; US Patent No. 5,457,126; Oct. 10, 1995; Assigned: Alcon Laboratories, Inc. (Fort Worth, TX)

LOFEPRAMINE HYDROCHLORIDE

Therapeutic Function: Antidepressant

Chemical Name: Ethanone, 1-(4-chlorophenyl)-2-((3-(10,11-dihydro-5H-dibenz[b,f]azepin-5-yl)propyl)methylamino)-, hydrochloride

Common Name: -

Structural Formula:

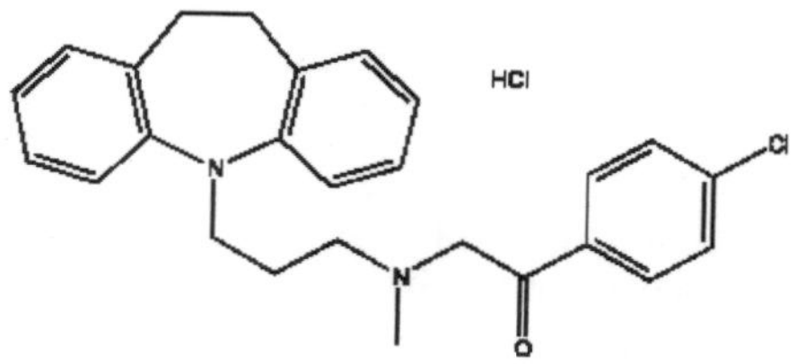

HCl

Chemical Abstracts Registry No.: 26786-32-3; 23047-25-8 (Base)

Trade Name	Manufacturer	Country	Year Introduced
Gamonil	E. Merck	-	-
Tymelyt	Leo	-	-

Trade Name	Manufacturer	Country	Year Introduced
Tymelyt	Lundbeck	-	-
Amplit	Daiichi	-	-
Gamanil	Merck KGaA	-	-
Timelit	UCB	-	-
Timelit	Montefarmaco	-	-
Lofepramine hydrochloride	Shanghai Lansheng Corporation	-	-
Deftan	Merck	-	-
Deprimyl	Merck Portuguesa	-	-
Emdalen	Merck	-	-
Lomont	Rosemont	-	-

Raw Materials

10,11-Dihydro-5H-dibenzo[b,f]azepine
1-(4-Chlorophenyl)-2-[(3-chloropropyl)methylamino]ethanone
Sodium amide

Manufacturing Process

9.8 parts of 10,11-dihydro-5H-dibenzo[b,f]azepine are dissolved in 10 parts of dry toluene and 3.1 parts of sodium amide are added and the mixture is refluxed and stirred for four hours. A solution of 13.5 parts of 1-(4-chlorophenyl)-2-[(3-chloropropyl)methylamino]ethanone in 20 parts of dry toluene is added dropwise and the mixture is stirred and refluxed for eight hours.

After cooling to room temperature water is carefully added to the reaction mixture and the toluene solution is extracted with water to which hydrochloric acid is added so that the aqueous phase obtains the pH-value of 5. The aqueous extract is discarded and the toluene phase is evaporated to dryness in vacuum. The residue is dissolved in 50 parts of methanol. Hydrogen gas is introduced to give the crystalline hydrochloride 1-(4-chlorophenyl)-2-((3-(10,11-dihydro-5H-dibenzo[b,f]azepin-5-yl)propyl)methylamino)ethanone; MP: 154°-156°C. The hydrochloride may be removed by adding an equivalent of any base (triethyl amine, sodium hydroxide and so on).

References

Eriksoo E. et al.; US Patent No. 3,637,660; Jan. 25, 1972; Assigned to Aktiebolaget Leo, Halsingborg, Sweden

LOFEXIDINE HYDROCHLORIDE

Therapeutic Function: Antihypertensive

Chemical Name: 2-[1-(2,6-Dichlorophenoxy)ethyl]-2-imidazoline hydrochloride

Common Name: -

Structural Formula:

HCl

Cl O N N H Cl

Chemical Abstracts Registry No.: 21498-08-8

Trade Name	Manufacturer	Country	Year Introduced
Lofetensin	Nattermann	W. Germany	1981

Raw Materials

α-2,6-Dichlorophenoxypropionitrile
Hydrogen chloride
Ethanol
Ethylenediamine

Manufacturing Process

10.4 ml of absolute ethanol are added to 57.5g of α-2,6-dichlorophenoxypropionitrile, followed by the introduction of 100 ml of chloroform dried over phosphorus pentoxide; 10.4 g of carefully dried hydrogen chloride being slowly introduced with stirring and cooling with ice/common salt. Most of the chloroform and excess hydrogen chloride is then removed by filtration in vacuo at room temperature, and dry ether added to the residue until the imido acid ester hydrochloride is quantitatively precipitated. The α-dichlorophenoxypropionimido acid ethyl ester hydrochloride can be obtained analytically pure in the form of white, strongly hygroscopic crystals by repeated dissolution in a little absolute ethanol in the absence of heat, and precipitation with ether.

The crude α-(2,6-dichlorophenoxy)propionamido acid ethyl ester hydrochloride is added in portions to a stirred, ice-cooled solution of 29.5 g of anhydrous ethylenediamine in 200 ml of absolute ethanol in such a way that the temperature does not exceed 0°C to 5°C. The cooling bath is then removed and the reaction mixture heated for 1 hour on a water bath to approximately 70°C.

After cooling, unreacted ethylenediamine is neutralized in a cooling mixture with the absolute ethanolic hydrochloric acid, filtered off from any components that are insoluble in ethanol and approximately two-thirds of the solvent filtered off under suction in a water jet pump vacuum. Residual quantities of ethylenediamine dihydrochloride are precipitated in fractions by the careful addition of ethyl methyl ketone, after which the imidazoline hydrochloride is separated off by the addition of dry ether. Following repeated recrystallization from ethanol ether, 2-[α-(2,6-dichlorophenoxy)ethyl]-δ^2-imidazoline hydrochloride is obtained in the form of small white crystals melting at 221°C to 223°C.

References

Merck Index 5388
DFU 3 (8) 592 (1978)
DOT 19 (9) 496 (1983)
I.N. p. 566
Baganz, H. and May, H.J.; US Patent 3,966,757; June 29, 1976; assigned to A. Natterman and Cie GmbH

LOFLAZEPATE ETHYL

Therapeutic Function: Tranquilizer

Chemical Name: 7-Chloro-5-(2-fluorophenyl)-2,3-dihydro-2-oxo-1H-1,4-benzodiazepine-3-carboxylic acid ethyl ester

Common Name: -

Structural Formula:

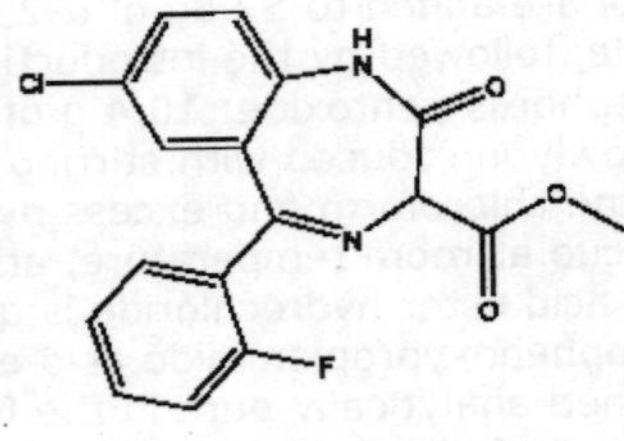

Chemical Abstracts Registry No.: 29177-84-2

Trade Name	Manufacturer	Country	Year Introduced
Victan	Clin Midy	France	1982

Raw Materials

2-Methylimidazole HCl
2-Amino-5-chloro-2'-fluoro-benzophenone
Ethyl aminomalonate hydrochloride

Manufacturing Process

(A) 1-(2-Amino-5-chlorophenyl)-1-(2-fluorophenyl)-2-aza-buty1-en-4-ol: A mixture of 40 g of 2-methylimidazole hydrochloride and of 90 g of 2-amino-5-chloro-2'-fluoro-benzophenone in 240 ml of ethanolamine is heated at 135°C for 2 hours. After cooling, the reaction mixture is poured into an aqueous sodium bicarbonate solution. The mixture is extracted with ether, the organic phase is washed repeatedly with water and is dried over sodium sulfate, and the solvent is evaporated to dryness. The residual oil is chromatographed on a silica column, elution being carried out with a 50/50 mixture of cyclohexane and ethyl acetate.

88 g of the expected amine are thus isolated. Melting point: 105°C to 110°C.

(B)1-(2-Amino-5-chlorophenyl)-1-(2-fluorophenyl)-3,3-bis-(ethoxycarbonyl)-2-aza-prop-1-ene: A mixture of 88 g of the product obtained above, 300 g of ethyl aminomalonate hydrochloride and 60 ml of acetic acid in 2.3 liters of absolute ethanol is heated to the reflux temperature for 6 hours. The alcohol and the acetic acid are evaporated in vacuo and the residue is taken up in ether. The solution is washed with a dilute sodium bicarbonate solution and then with water and is dried over sodium sulfate. The solvent is evaporated and the residue is then chromatographed on a silica column, using a 90/10 mixture of chloroform and ethyl acetate for the elution. An oil (64g) is thus obtained, and is used, without further treatment, for the cyclization.

A sample recrystallized from isopropyl ether has a melting point of 119°C.

(C) Compound of Code No. CM 6912: 25 g of the imine obtained under (B), dissolved in 400 ml of acetic acid, are heated at the reflux temperature for 1 hour. After evaporating the solvent in vacuo, the residue is taken up in methylene chloride. The solution is washed with a dilute sodium bicarbonate solution and then with water. After evaporating the solvent, the residue is chromatographed on silica, elution being carried out with an 80/20 mixture of ether and ethyl acetate. 9 g of benzodiazepine are thus obtained. Melting point: 196°C.

References

Merck Index 3766
DFU 6 (12) 772 (1981)
DOT 19 (1) 24 (1983)
I.N. p. 566
Demarne, H. and Hallot, A.; British Patent 1,538,165; January 17, 1979; assigned to C.M. Industries (France)

LOMEFLOXACIN HYDROCHLORIDE

Therapeutic Function: Antibacterial

Chemical Name: 3-Quinolinecarboxylic acid, 1,4-dihydro-6,8-difluoro-1-ethyl-7-(3-methyl-1-piperazinyl)-4-oxo-, hydrochloride

Common Name: Lomefloxacin hydrochloride

Structural Formula:

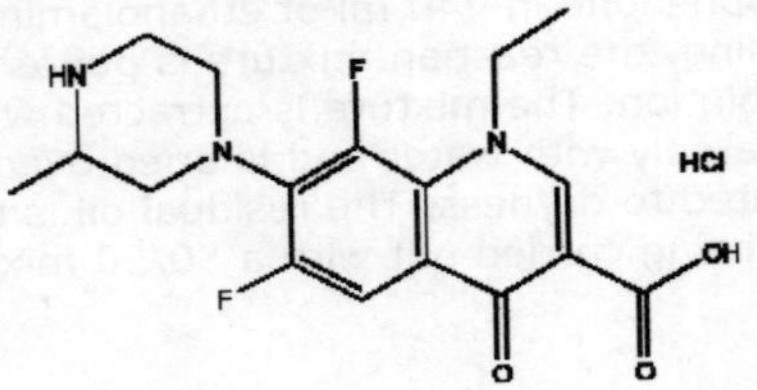

Chemical Abstracts Registry No.: 98079-52-8; 98079-51-7 (Base)

Trade Name	Manufacturer	Country	Year Introduced
Foxil	Sarabhai Chemicals	India	-
Liexina	Leti		-
Lomaday	Dr. Reddy's Laboratories Ltd.	India	-
Lomefloxacine hydrochloride	Searle Chemical Inc.	USA	-

Raw Materials

1-Ethyl-6,7,8-trifluoro-1,4-dihydro-4-oxoquinoline-3-carboxylic acid
2-Methylpiperazine
Pyridine
Sodium bicarbonate

Manufacturing Process

A mixture of 1.00 g of 1-ethyl-6,7,8-trifluoro-1,4-dihydro-4-oxoquinoline-3-carboxylic acid, 1.10 g of 2-methylpiperazine and 10 ml of pyridine was heated for 15 minutes under reflux. The reaction mixture was evaporated and methanol was added to the residue. The precipitate was filtered and recrystallized from ethanol to give 0.36 g of the 1-ethyl-6,8-difluoro-1,4-dihydro-7-(3-methyl-1-piperazinyl)-4-oxoquinoline-3-carboxylic acid as colorless needles, melting point 239.0-240.5°C.

By the usual manner the hydrochloride was prepared and recrystallized from water as colorless needles, melting point 290-300°C (decomp.).

References

Yasuo Itoh et al.; US Patent No. 4,528,287; 07.09.1985; Assigned to Hokuriku Pharmaceutical Co., Ltd.

LOMIFYLLINE

Therapeutic Function: Vasodilator

Chemical Name: Purin-2,6-dione, 1,2,3,6-tetrahydro-1,3-dimethyl-7-(5-oxohexyl)-

Common Name: Lomifylline

Structural Formula:

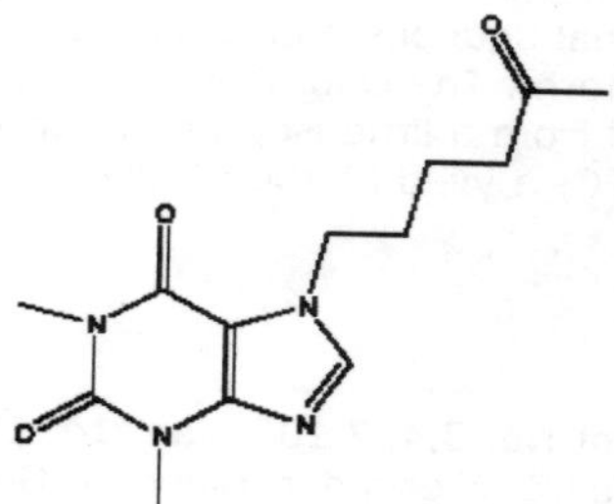

Chemical Abstracts Registry No.: 10226-54-7

Trade Name	Manufacturer	Country	Year Introduced
Lomifylline	Yick-Vic Chemicals and Pharmaceuticals (HK) Ltd.	-	-
Lomifylline	Shanghai Lansheng Corporation	-	-

Raw Materials

1,3-Dibromopropane
Ethyl acetoacetate
Hydrobromic acid
Theophylline sodium

Manufacturing Process

A mixture of 560 g of potassium carbonate, 700 ml of ethanol (96%), 404 g of 1,3-dibromopropane and 260 g of ethyl acetoacetate was heated with stirring to go 60°C. After the reaction had subsided, the reaction mixture was refluxed for 5 hours. Then the bulk of the alcohol was distilled off under ordinary pressure and the residue was mixed with 1.5 L of water. The resulting oily layer was separated, and the aqueous phase was extracted with benzene and the benzene layer was combined with the oil. After drying with sodium sulfate the benzene was distilled off and the residue was fractionally distilled 250 g (73% of theory) of 2-methyl-3-carbethoxy-5,6-dihydropyrane of boiling point 105°-108°C were obtained.

140 ml of 63% hydrobromic acid were slowly added at room temperature to 128 g of 2-methyl-3-carbethoxy-5,6-dihydropyrane, and much carbon dioxide

was evolved. After standing for 1 to 2 days at room temperature the mixture was diluted with an equal volume of iced water; the layer of dark colored oil formed was separated, the aqueous phase was extracted with chloroform, and the extract was combined with the oil and washed with a saturated solution of sodium bicarbonate. The solution was dried with sodium sulfate, the chloroform was distilled off under normal pressure, and the residue was fractionally distilled in vacuo. 109 g (81% of theory) of 1-bromohexanone-5 of boiling point 94°-98°C/12 mm Hg were obtained.

A solution of 10.0 g of 1-bromohexanone-5 in 100 ml of ethanol was gradually mixed at the boil with vigorous stirring with 11.3 g of the sodium salt of theophylline in 100 ml of water. After 3 hours refluxing the alcohol was distilled off, and the residual aqueous phase was cooled and made alkaline and extracted with chloroform. The chloroform solution was evaporated and the residue re-crystallized from a little isopropanol to yield 7-(5-oxohexyl) theophylline. MP: 75°-76°C; a yield of about 80% (calculated on the reacted theophylline).

References

Mohler W. et al.; US Patent No. 3,422,107; Jan. 14, 1969; Assigned to Chemische Werke Albert, Wiesbaden-Biebrich, Germany, a corporation of Germany

LONAPALENE

Therapeutic Function: Antipsoriatic

Chemical Name: 1,4-Naphthalenediol, 6-chloro-2,3-dimethoxy-, diacetate

Common Name: Lonapalene

Structural Formula:

O
O
O
O
Cl
O
O

Chemical Abstracts Registry No.: 91431-42-4

Trade Name	Manufacturer	Country	Year Introduced
Lonapalene	Syntex	-	-

Raw Materials

Sodium nitrate
6-Amino-2,3-dimethoxy-1,4-naphthoquinone
Hydrochloric acid
Copper chloride

Manufacturing Process

A solution of sodium nitrate (0.69 g, 10 mmol) in water (5 ml) was added at 0-5°C to a solution of 6-amino-2,3-dimethoxy-1,4-naphthoquinone (1.17 g, 5 mmol) in 5:1 acetic acid:water (25 ml) containing concentrated hydrochloric acid (1.7 ml). A further quantity of sodium nitrite (0.69 g) was then added to the reaction mixture after cooling to -5°C, followed by a solution of cuprous chloride (0.6 g) in concentrated hydrochloric acid (5 ml). The mixture was allowed to warm to room temperature and solid cuprous chloride was added portionwise until the mixture assumed a green color. Water was then added to the reaction mixture and the precipitated yellow solid filtered off, washed with water and recrystallized from methanol:water (2:1) giving 1.01 g of 6-chloro-2,3-dimethoxy-1,4-naphthoquinone, melting point 93-94°C (from ether-petrolium ether).

References

Jones Gordon H., Venuti Michael C., Young John M.; US Patent No. 4,466,981; August 21, 1984; Assigned to Syntex (U.S.A.) Inc. (Palo Alto, CA)
Jones G. H. et al.; Eur. pat. Appl. 107,512
Tetrahedron Letters 28, 4507 (1987)

LONAZOLAC

Therapeutic Function: Antiinflammatory

Chemical Name: 3-(4-Chlorophenyl)-1-phenyl-1H-pyrazole-4-acetic acid

Common Name: -

Chemical Abstracts Registry No.: 53808-88-1

Trade Name	Manufacturer	Country	Year Introduced
Irriten	Tosse	W. Germany	1981
Irritren	Byk Gulden	Switz.	1982

Structural Formula:

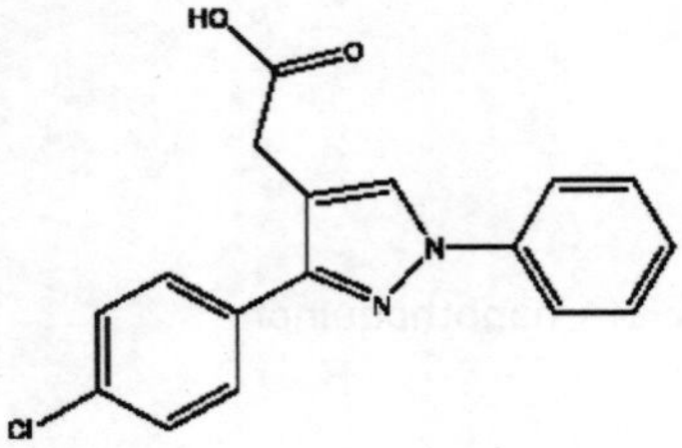

Raw Materials

1-Phenyl-3-(p-chlorophenyl)-pyrazol-4-acetonitrile
Hydrogen chloride

Manufacturing Process

17.6 g 1-phenyl-3-(p-chlorophenyl)-pyrazol-4-acetonitrile and 180 ml 25% aqueous hydrochloric acid were mixed and heated to the boiling temperature under reflux for 6 hours. To the mixture was then added dropwise concentrated aqueous sodium hydroxide until the pH of the mixture reached a value in the range from 3 to 5. The free pyrazol-4-acetic acid precipitated thereby was filtered off, redissolved in dilute aqueous sodium hydroxide, the solution cleared by treatment with activated carbon, and the pyrazol-4-acetic acid precipitated by acidifying the solution by the addition of dilute mineral acid, sulfuric acid. The filtered acid was crystallized from a mixture of ethanol and water. 17.1 g 1-phenyl-3-(p-chlorophenyl)pyrazol-4-acetic acid, melting at 148°C to 150°C, were obtained, representing a yield of 91%.

References

Merck Index 5392
DFU 7 (2) 110 (1982)
DOT 18 (4) 184 (1982)
I.N. p. 567
Rainer, G.; US Patent 4,146,721; March 27, 1979; assgned to Byk Gulden Lomberg Chemische Fabrik G.m.b.H. (W. Germany)

LOPERAMIDE HYDROCHLORIDE

Therapeutic Function: Antidiarrheal

Chemical Name: 4-(4-Chlorophenyl)-4-hydroxy-N,N-dimethyl-α,α-diphenyl-1-piperidinebutanamide hydrochloride

Common Name: -

Structural Formula:

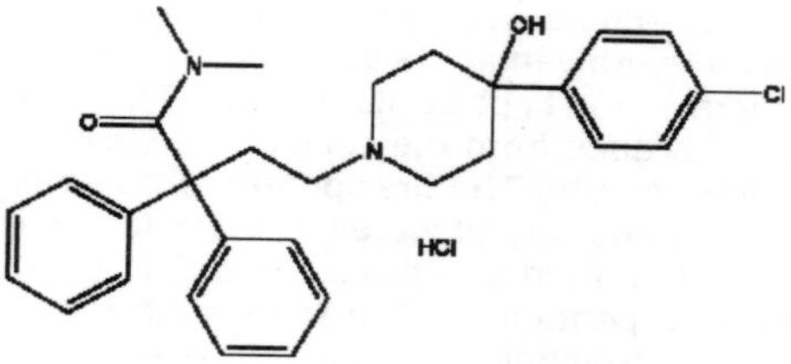

Chemical Abstracts Registry No.: 34552-83-5; 53179-11-6 (Base)

Trade Name	Manufacturer	Country	Year Introduced
Imodium	Janssen	UK	1975
Imodium	Janssen-Le Brun	France	1976
Imodium	Janssen	W. Germany	1976
Imodium	Ortho	US	1977
Dissenten	S.P.A.	Italy	1978
Imodium	Janssen	Italy	1979
Lopemid	Gentili	Italy	1979
Imodium	Januen	Switz.	1981
Imodium	Dainippon	Japan	-
Blox	Biomedica Foscama	Italy	-
Brek	Irbi	Italy	-
Fortasec	Esteve	Spain	-
Lopermid	Drifen	Turkey	-
Loperyl	Zambeletti	Italy	-
Regulane	Finadiet	Argentina	-
Seldiar	Krka	Yugoslavia	-
Tebloc	Dukron	Italy	-

Raw Materials

Hydrogen bromide	2-Oxo-3,3-diphenyl-tetrahydrofuran
Thionyl chloride	4-(p-Chlorophenyl)-4-piperidinol
Dimethylamine	Hydrogen chloride

Manufacturing Process

23.6 parts of 2-oxo-3,3-diphenyl-tetrahydrofuranare melted at 100°C in an oil-bath and gaseous hydrogen bromide is introduced into it during 3 hours. The reaction mixture is cooled and triturated in benzene. The product is filtered off, washed with petroleum ether and dried in an exsiccator, yielding 4-bromo-2,2-diphenylbutyric acid; MP 127.5%.

To a stirred suspension of 16 parts of 4-bromo-2,2-diphenylbutyric acid in 150 parts of chloroform are added dropwise 16 parts of thionyl chloride and the whole is stirred and refluxed for 2 hours. The reaction mixture is evaporated, yielding 4-bromo-2,2-diphenyl-butyrylchloride as a residue.

60 parts of 4-bromo-2,2-diphenylbutyrylchloride are dissolved in 400 parts of toluene and gaseous dimethylamine is introduced slowly into the solution while cooling (temperature is kept at about 0°C). The introduction is ceased when dimethylamine escapes from the cooler, and stirring is continued for 2 hours at ordinary temperature. The precipitated product is filtered off and dissolved in a minimum quantity of water. The product is extracted with chloroform. The extract is dried and evaporated. The residue solidifies on triturating in 4-methyl-2-pentanone. The solid is filtered off and dried, yielding dimethyl -(tetrahydro-3,3-diphenyl-2-furylidene)ammonium bromide; MP 169° to 171.5°C.

A mixture of 6.33 parts of 4-(p-chlorophenyl)-4-piperidinol, 8 parts of sodium carbonate, 0.2 part of potassium iodide and 240 parts of 4-methyl-2-pentanone is distilled azeotropically. Then there are added 12.12 parts of dimethyl-(tetrahydro-3,3-diphenyl-2-furylidene)ammonium bromide (from the preceding step) and the whole is stirred and refluxed for about 15 hours. The reaction mixture is filtered hot and the filtrate is evaporated.

The oily residue is dissolved in 2-propanol and to this solution is added an excess of 2-propanol previously saturated with gaseous hydrogen chloride. The whole is evaporated and the oily residue is warmed in diluted hydrochloric acid solution. Upon the addition of toluene, the salt is precipitated. It is filtered off, boiled in acetone, and filtered off again after cooling, yielding 4-(p-chlorophenyl)-4-hydroxy-N,N-dimethyl-α,α-diphenylpiperidine-1-butyramide hydrochloride; MP 222.1°C.

References

Merck Index 5396
Kleeman and Engel p. 530
PDR p. 953
OCDS Vol. 2 p. 334 (1980)
DOT 10 (6) 220 (1974)
I.N. p. 567
REM p. 814
Janssen, P.A.J., Niemegeers, C.J.E.J., Stokbroekx, R.A. and Vandenberk, J.; US Patent 3,714,159; January 30, 1973; and US Patent 3,884,916; May 20, 1975; both assigned to Janssen Pharmaceutica, NV, Belgium

LOPINAVIR

Therapeutic Function: Antiviral

Chemical Name: 1(2H)-Pyrimidineacetamide, N-((1S,3S,4S)-4-(((2,6-dimethylphenoxy)acetyl)amino)-3-hydroxy-5-phenyl-1-(phenylmethyl) pentyl)tetrahyrdo-α-1-methylethyl)-2-oxo-, (αS)-

Common Name: Lopinavir

Structural Formula:

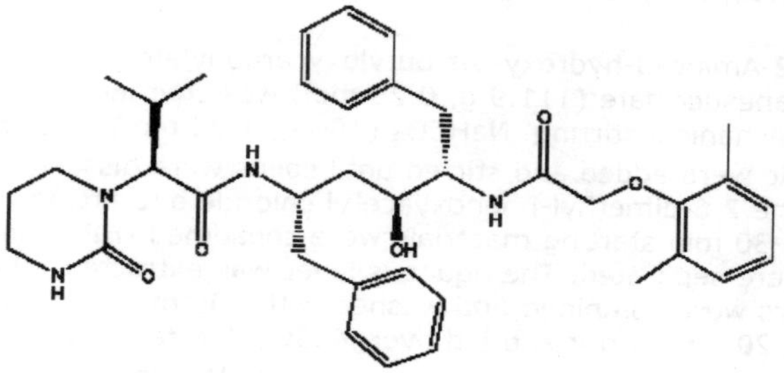

Chemical Abstracts Registry No.: 192725-17-0

Trade Name	**Manufacturer**	**Country**	**Year Introduced**
ABT 378	Abbott Laboratories	-	-
Aluviran	Abbott Laboratories	-	-
Kaletra	Abbott	-	-
Lopinavir	Abbott Laboratories	-	-

Raw Materials

Chloroacetic acid	2,6-Dimethylphenol
Oxalyl chloride	Sodium hydroxide
Trifluoroacetic acid	Valine methyl ester, (S)-, hydrochloride
Hydrogen chloride	Benzyl chloroformate
Nickel Raney	Palladium hydroxide
Lithium hydroxide	1-Ethyl-3-(3-dimethylaminopropyl)carbodiimide

2-Amino-3-hydroxy-5-t-butyloxycarbonylamino-1,6-diphenylhexanesuccinate, (2S,3S,5S)-

Manufacturing Process

2,6-Dimethylphenol (102.8 g, 0.842 mol) and chloroacetic acid (159.6 g, 1.68 mol) in 1000 ml of H_2O was added to a 3-L 3-necked round bottom flask with mechanical stirring and a water-cooled condenser. A solution of NaOH (134.9 g, 3.37 mol) dissolved in 500 ml of water was slowly added to the above mixture and heat to reflux. After 2 hours, additional chloroacetic acid (79.4 g, 0.84 mol) and NaOH solution (67.2 g, 1.68 mol in in 200 ml water) was added to the reaction mixture. After 19 hours, additional chloroacetic acid (39.8 g, 0.42 mol) and NaOH solution (33.6 g, 0.84 mol in in 100 ml water) was added to the reaction mixture and refluxing was continued until starting phenol was consumed. The reaction flask was cooled in and ice-water bath and acidified to pH 1 with conc. HCl, causing a precipitate to form. The resulting slurry was stirred in the ice bath for 1 hour then filtered. The solid was dissolved in hot water and cooled to crystallize 2,6-dimethylphenoxyacetic acid as white plates, m.p. 136-137°C, yield 78.8 g, 52%.

Oxalyl chloride (36.3 ml, 0.42 mol) was added to a slurry of 2,6-dimethylphenoxyacetic acid (50 g, 0.28 mol) in 500 ml toluene followed by addition of 5 drops of DMF and stirred at 20C for 30 min, then at 55°C for 1.5

hours. The toluene was removed in vacuum to afford 2,6-dimethylphenoxyacetyl chloride as an oil, yield 55 g, 100%.

(2S,3S,5S)-2-Amino-3-hydroxy-5-t-butyloxycarbonylamino-1,6-diphenylhexanesuccinate (111.9 g, 0.25 mol) was charged to a 2-L 3-necked flask with mechanical stirring. $NaHCO_3$ (106 g, 1.26 mol), 600 ml H_2O and 600 ml EtOAc were added and stirred until solids were dissolved (15 min). A solution of the 2,6-dimethyl-phenoxyacetyl chloride and EtOAc (100 ml) was added. After 30 min starting materials were consumed (HPLC analysis) and the layers were separated. The aqueous layer was extracted with EtOAc, the organic layers were combined and washed with 200 ml of 1 M NaOH, 200 ml of 10% HCl, 200 ml of brine, dried over $MgSO_4$, filtered and concentrated to provide (2S,3S,5S)-2-(2,6-dimethylphenoxyacetyl)amino-3-hydroxy-5-(t-butyloxycarbonylamino)-1,6-diphenylhexane as a white solid.

(2S,3S,5S)-2-(2,6-Dimethylphenoxyacetyl)amino-3-hydroxy-5-(t-butyloxycarbonylamino)-1,6-diphenylhexane(175.1 g, 0.32 mol) and 500 ml CH_2Cl_2 were mixed with CF_3CO_2H (249 ml, 3.2 mol) was added and stirred 20-25 min, then the reaction mixture was poured into a separatory funnel containing 1000 ml of water and 200 ml of CH_2Cl_2. The resulting mixture was shaken carefully and the layers were separated. The organic layer was washed again with 500 ml of water, then with $NaHCO_3$ andfinally with brine. The organic solution was dried over $MgSO_4$, filtered and concentrated to an oil. 300 ml of diethyl ether was added to the crude product and shaken to dissolve. Within minutes solid began to crystallize and the mixture became thick. Enough diethyl ether was added to make the mixture stirrable and the mixture was stirred at room temperature for 1 hour. The solid was filtered and dried to give (2S,3S,5S)-2-(2,6-dimethylphenoxyacetyl)amino-3-hydroxy-5-amino-1,6-diphenylhexane as a white needles, yield 115 g, 81%.

To a 12 L 3-neck round bottom flask was added isopropyl acetate (6.5 L). The solvent was cooled to 0°C in an ice-water bath and 3-amino-1-propanol (1.14 kg, 15.1 mol) was added in one portion. To this stirring solution, benzyl chloroformate (1.20 kg, 7.03 mol) was added dropwise over 2 hours while maintaining the internal temperature of the flask between 10-15°C. After the addition was complete, the reaction mixture was allowed to stir for an additional 0.3 hour after which time water (3.5 L) was added in one portion. The solution was then partitioned and washed with an additional 2 times 3.5 L of water. The organic layer was dried over potassium carbonate and concentrated to give a solid that was dissolved in excess isopropyl acetate and precipitated from solution by adding the compound to heptane. The solid was filtered under nitrogen to yield 1.20 kg (82%) of N-carbonylbenzyloxy-3-aminopropanol as a colorless solid.

A mixture of 335 mL of DMSO and 9 L of methylene chloride were chilled to -48°C. 313 mL of oxalyl chloride was added over 25 min at temperature below -40°C. At -48°C added 500 g of N-carbonylbenzyloxy-3-amino-1-propanol dissolved in 1 L of methylene chloride. 1325 mL of triethylamine was added at such a rate that the temperature remained below -40°C. After stirring an additional 15 min, the mixture was allowed to warm to -30°C, then added 2.5 L of 20% aqueous potassium dihydrogen phosphate. Stirred for one hour, then separated the layers, washed the organic layer with brine, and dried with magnesium sulfate. The resulting N-carbonylbenzyloxy-3-aminopropanal was kept in solution at -20°C.

To a 5 L 3-neck round bottom flask was added the crude carbonylbenzyloxy-3-aminopropanal (115 g, 0.555 mol) followed by addition of water (400 mL) and methanol (1600 mL). The reaction mixture was maintained at 25°C throughout the course of the reaction. After the solution became homogeneous. (S)-Valine methyl ester hydrochloride (90.2 g, 0.538 mol) was added in one portion followed by rapid addition of sodium acetate trihydrate (151 g, 1.11 mol) and sodium cycanoborohydride (73.2 g, 1.17 mol). The reaction mixture was allowed to stir at room temperature for 0.5 hour and was concentrated in vacuo. To this solution, saturated aq sodium bicarbonate (400 mL) was added and the mixture was extracted with isopropyl acetate (1 L). The organic layer was washed with water, dried over sodium sulfate, and concentrated to yield 150 g of crude product, which was dissolved in isopropyl acetate (300 mL) and heptane (2400 mL). Dry HCl was bubbled in and an oily solid precipitated out of solution. The liquid was decanted away from and the solid was dissolved in dichloromethane (3 L). The solution was washed with water (600 mL) and saturated aq sodium bicarbonate (600 mL) and dried over sodium sulfate. It was concentrated in vacuo to yield 105 g (59%) of N-(N-(benzyloxycarbonyl-3-amino)-propyl)valine methyl ester as a light yellow oil.

To a 3 L flask was added N-(N-(benzyloxycarbonyl-3-amino)-propyl)valine methyl ester (120 g, 0.372 mol) and methanol (1 L). This solution was allowed to stir in the presence of Raney Nickel (180 g) for 1 h. After removal of Raney Nickel by filtration, $Pd(OH)_2$ (24 g) was added and the solution was allowed to stir under 60 psi of a hydrogen atmosphere for 12 h. The solution was purged with nitrogen and repressurized with 60 psi of hydrogen for an additional 1 h. The solution was filtered and concentrated to give 63 g of N-(3-amino)-propyl)valine methyl ester as an an oil (90%). To this oil toluene (120 mL) was added and the solution was again concentrated in vacuo to give the desired product.

To a 5 L 3-neck round bottom flask with stir bar was added the crude N-(3-amino)-propyl)valine methyl ester (150 g, 0.8 mol) and dichloromethane (3.2 L). Carbonyldiimidazole (232 g, 1.44 mol) was added slowly in portions over 25 min. The solution was allowed to stir at ambient temperature for 40 hours. Water (200 mL) was added over 1 h with stirring until no more gas evolution occurred. A solution of 35% HCl was slowly added to the stirring solution until the solution became acidic. The solution was then partitioned and was washed with water. The organic layer was dried over sodium sulfate and was concentrated to yield 126 g (74%) of 2S-(1-tetrahydro-pyrimid-2-onyl)-3-methyl butanoic acid methyl ester as a colorless solid.

To a 12 L 3-neck round bottom flask with stir bar was added 2S-(1-tetrahydro-pyrimid-2-onyl)-3-methyl butanoic acid methyl ester (126 g, 0.588 mol), water (1.3 L), and THF (3.9 L). The solution was cooled to 0°C in an ice-water bath and lithium hydroxide monohydrate (74 g, 1.76 mol) was added in one portion with rapid stirring. The solution was allowed to stir at 0°C for 14 hours. It was then acidified to pH 11 by slow addition of 50% aq. phosphoric acid and the THF was removed in vacuo. The aqueous phase was washed with isopropyl acetate (2 L) and was subsequently acidified to pH by slow addition of 35% aq. HCl. The aqueous layer was then extracted with ethyl acetate. The combined organic layers were concentrated to give the desired product (105 g) as a white solid. The compound was then purified by addition of isopropyl acetate (500 mL) and ethanol (15 mL) and bringing the solution to a boil with rapid stirring until 50 mL of solvent had evaporated. The solution was cooled to 0°C and filtered to give 92 g (75%) of pure 2S-(1-tetrahydro-pyrimid-2-onyl)-3-methyl butanoic acid methyl ester.

The mixture of (2S,3S,5S)-2-(2,6-dimethylphenoxyacetyl)amino-3-hydroxy-5-amino-1,6-diphenylhexane (100 g, 0.22 mol), 2S-(1-tetrahydro-pyrimid-2-onyl)-3-methyl butanoic acid methyl ester (44.8 g, 0.22 mol) and 750 ml DMF was cooled in an ice/water bath. N-Hydroxybenzotriazole (90.9 g, 0.67 mol), 1-ethyl-3-[3-dimethylaminopropyl]carbodiimide (86 g, 0.45 mol) and triethylamine (62.5 ml, 0.45 mol) were added and the ice bath was removed, allowing the reaction mixture to stir with warming to room temperature for 5 hours. The mixture was diluted with 1000 ml of IPAC and quenched with 1000 ml of water. The mixture was shaken and separated, the aq. layer was extracted IPAC, the organics were washed with 10% HCl, solution of $NaHCO_3$ with 100 ml hexanes, then washed 500 ml water, and brine, dried over $MgSO_4$, filtered and concentrated to provide. (2S,3S,5S)-2-(2,6-dimethylphenoxyacetyl)amino-3-hydroxy-5-(2S-(1-tetrahydro-pyrimid-2-onyl)-3-methylbutanoyl)amino-1,6-diphenylhexane as a white foam.

References

Chemburkar S.R. et al.; US Patent No. 6,372,905; Apr.16, 2002; Assigned to Abbott Laboratories, Abbot Park,IL (US)

Sham H.L.; US Patent No. 5,914,332; Jun.22, 1999; Assigned: Abbott Laboratories (Abbott Park, IL)

LOPRAZOLAM

Therapeutic Function: Tranquilizer

Chemical Name: 8-Nitro-1,2-dihydro-2-(N-methyl-piperazin-1-yl)methylene-6-(o-chlorophenyl)-1H,4H-imidazo-[1,2-a][1,4]-benzodiazepin-1-one methanesulfonate

Common Name: -

Structural Formula:

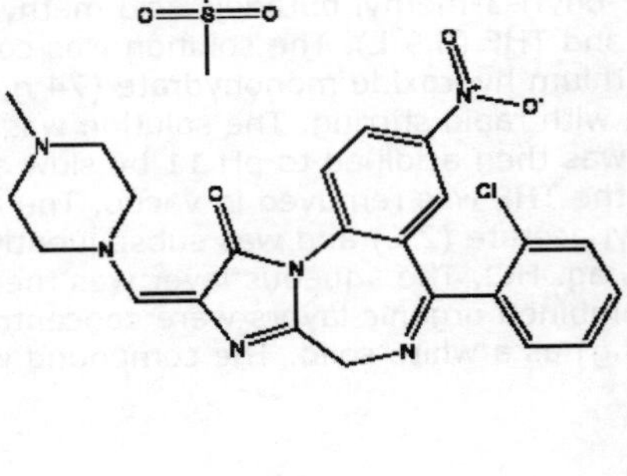

Chemical Abstracts Registry No.: 61197-93-1

Trade Name	Manufacturer	Country	Year Introduced
Avlane	J.A.S.M.	France	1981
Dormonoct	Roussel	UK	1983

Raw Materials

8-Nitro-1,2-dihydro-2-(N-methylpiperazin-1-yl)methylene-6-(o-chlorophenyl)-1H,4H-imidazo[1,2-a][1,4]benzodiazepin-1-one
Methanesulfonic acid

Manufacturing Process

1.1 g of methanesulfonic acid were added dropwise to a mixture of 4.6 g of 8-nitro-1,2-dihydro-2-(N-methylpiperazin-1-yl)methylene-6-(o-chlorophenyl)-1H,4H-imidazo-[1,2-a][1,4]benzodiazepin-1-one in 100 ml of anhydrous methylene chloride and 5 ml of methanol. Dry ether was slowly added until crystals formed on scratching and the solution was allowed to crystallize with further ether being added to complete the crystallization. The pale yellow solid was filtered off, washed with ether and crystallized from methylene chloride-methanol to obtain 5.4 g of 8-nitro-1,2-dihydro-2-(N-methylpiperazin-1-yl)methylene-6-(o-chlorophenyl)-1H,4H-imidazo-[1,2-a][1,4]-benzodiazepin-1-one methanesulfonate melting at 205°C to 210°C.

References

Merck Index 5399
DFU 5 (3) 144 (1980) (As Ru-31, 158) and 5 (12) 635 (1980)
Taylor, F.B. and Harrison, D.R.; US Patent 4,044,142; August 23, 1977; assigned to Roussel Uclaf.

LORACARBEF

Therapeutic Function: Antibiotic

Chemical Name: 1-Azabicyclo[4.2.0]oct-2-ene-2-carboxylic acid, 7-(((2R)-aminophenylacetyl)amino)-3-chloro-8-oxo-, (6R,7S)-

Common Name: Loracarbef

Chemical Abstracts Registry No.: 76470-66-1

Trade Name	Manufacturer	Country	Year Introduced
Lorabid	Eli Lilly and Company	-	-
Lorabid	Monarch. Pharm.	-	-
Lorax	Lilly	-	-

Structural Formula:

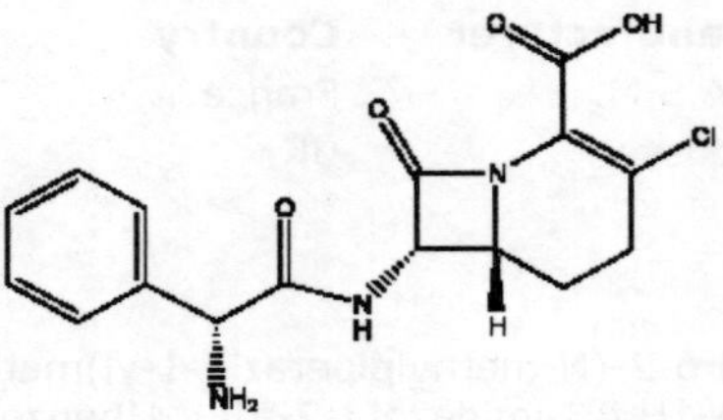

Raw Materials

(+/-)-cis-7-Amino-3-chloro-1-azabicyclo[4,2,0]oct-2-en-8-on-2-carboxylic acid trifluoroacetate
D-Phenylglycine methylester hydrochloride
Potassium phosphate buffer

Manufacturing Process

Loracarbef was obtained by biochemical method.

a) Cultivation of a microorganism having an ability of optically selective acylation

As a seed strain, Pseudomonas melanogenum ATCC 17808 [Biological properties are described in Journal of the Agricultural Chemical Society of Japan 37, 71 (1963)] is used.

As the seed medium, an aqueous solution containing 1% polypepton, 1% yeast extract, 0.5% meat extract, 0.5% sodium glutamate and 0.25% sodium chloride and adjusted to a pH of 7.0 with 5 N NaOH is used. One loopful of the seed strain is inoculated into 10 ml the seed medium and culturing is carried out at of 30°C for 24 hours. The whole amount of the seed medium is put into 300 ml of the culture medium in a 2 L Erlenmeyer flask and culturing is carried out at a temperature of 30°C. The composition of the culture medium is the same as that of the seed medium.

b) Preparation of cell suspension

After culturing for 24 hours, cell bodies are recovered from the culture broth by centrifugation and washed 2 times with 50 ml of 0.9% saline solution. The cells are suspended in a concentration of 20 mg/ml by dry weight in 1/30 M phosphate buffer (pH 6.5).

c) Preparation of a substrate solution

200 mg of the trifluoroacetate of ()-cis-7-amino-3-chloro-1-azabicyclo[4,2,0] oct-2-en-8-on-2-carboxylic acid (obtained by the method described in JPUPA No. 87791/80) and 800 mg of the hydrochloride of D-phenylglycine methylester are added in 9 ml of 1/30 M potassium phosphate buffer (pH

6.5). 5 N KOH is added in a small portion and the mixture is again adjusted to a pH of 6.5 to dissolve two starting compounds. Finally, deionized water is added to make 10 ml of a solution.

d) Enzyme reaction

In this step, 10 ml of the disrupted cell suspension is added to 10 ml of the substrate solution and enzyme reaction is carried out at a temperature of 30°C for 2 hours. The reaction is monitored by high speed liquid chromatography. Elution is carried out with 7% methanol - 0.2 M KH_2PO_4 solution. Reaction reaches maximum in a yield of 90% to the starting compound in 2 hours.

After the completion of reaction, cell bodies are removed from the reaction solution by centrifugation. The supernatant is concentrated under reduced pressure and charged on a column with 100 ml of Diaion HP-10. After adding 200 ml of deionized water, elution is carried out with 25% aqueous methanol solution. Then, the fractions containing the desired compound are concentrated under reduced pressure to make a 5 ml of concentrate. The concentrate is charged on a column packed with 130 ml of Sephadex-LH20 and elution is carried out with a solvent of water and methanol (50:50). The desired product is eluted in 55 ml to 75 ml of fractions. The fractions are concentrated under reduced pressure and lyophilized to obtain 78 mg (6R,7S)-7-(R)-phenylglycinamido-3-chloro-1-azabicyclo[4,2,0]oct-2-en-8-on-2-carboxylic acid of a white powder $[\alpha]_D^{21}$ = -75.8° (c = 0.4, H_2O), melting point 300°C or more (browning).

References

Hashimoto Yu. et al.; US Patent No. 4,335,211; 06.15.1982, Assigned to Kyowa Hakko Kogyo Co.

LORATADINE

Therapeutic Function: Antihistaminic, Antiallergic

Chemical Name: 1-Piperidinecarboxylic acid, 4-(8-chloro-5,6-dihydro-11H-benzo[5,6]cyclohepta[1,2-b]pyridin-11-ylidene)-, ethyl ester

Common Name: Loratadine

Chemical Abstracts Registry No.: 79794-75-5

Raw Materials

Pyridine	Titanium tetrachloride
Zinc	Ethyl 4-oxopiperidine-1-carboxylate
Tetrahydrofuran	8-Chloro-5,6-dihydrobenzo[5,6]cyclohepta[1,2-b]pyridin-11-one

Structural Formula:

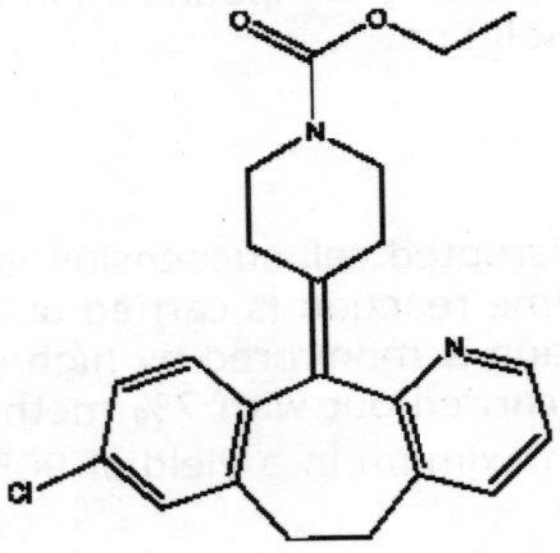

Trade Name	**Manufacturer**	**Country**	**Year Introduced**
Alledryl	Prater	-	-
Alergaliv	Sigma/Neo Quimica	-	-
Claratyne	Schering-Plough Pty	-	-
Claritin	Schering-Plough	-	-
Claritin 24 hour	Schering Canada Inc.	-	-
Difmedol	Faran	-	-
Histadin	Nodel	-	-
Loranil	Libbs	-	-
Lorin	Stadmed Private Limited	India	-
Lorinol	Micro Labs	India	-
Loronet	Klar Sehen Pvt. Ltd.	India	-
Noxin	Bussie	-	-
Ponderal	Biogen	-	-
Versal	Nycomed	-	-

Manufacturing Process

Preparation of Loratadine

In a two-liter vessel provided with a thermometer, a reflux condenser and nitrogen atmosphere, dry tetrahydrofuran (343 ml) was placed, and cooled between 0 and -5°C. Titanium tetrachloride (28.5 ml, 49.5 g, 0.255 mol) was slowly added with stirring (17 min.), keeping the temperature in the above indicated range, a yellow suspension being formed. After the addition was finished, stirring was continued for 10 min. Then, zinc dust (34.5 g, 0.524 mol) was added with stirring in approximately 15 min. keeping the temperature in the above cited range, and after addition was finished, stirring was continued at this temperature for 20 min., a blue suspension being formed. Then, pyridine (17 ml, 0.21 mol) was added with stirring, keeping the temperature in the above range, and then, a solution of 8-chloro-5,6-dihydrobenzo[5,6]cyclohepta[1,2-b]pyridin-11-one (30.0 g, 0.123 mol) and ethyl 4-oxopiperidine-1-carboxylate (25.2 g, 0.147 mol) in anhydrous tetrahydrofuran (96 ml) was added in about 20 min., with stirring and keeping the temperature in the above cited range. The, thus obtained, dark brown mixture was stirred for 3 h keeping the temperature in the above cited range, then was allowed to heat to room temperature and kept at this temperature for 2 h and then heated to 40°C for 17 h. The tetrahydrofuran was distilled off from the reaction mixture to give a black resin that was dissolved in dichloromethane (300 ml) and acidified by addition of isopropanol/HCl 7.2 N (97 ml). The mixture was stirred for 10 min, and the phases were separated, being the aqueous one extracted with dichloromethane (150 ml). The combined organic phases were washed 6 times with a mixture of water (125 ml) and 35% aqueous HCl (7.5 ml). Then, the organic phase was basified to pH 7.5-8.0 by addition of 30% aqueous NH_3. The mixture was stirred for 10 min and the phases were separated, and then washed 3 times with water (250 ml). The organic phase was dried with anhydrous sodium sulfate, filtered and the solvent eliminated in vacuo to give a residue (47.47 g) that was treated with acetonitrile (97 ml). The solid was filtered and crystallized from the same solvent to give pure Loratadine, m.p. 132-133°C (18.8 g, 40% yield).

References

Stampa A. et al.; US Patent No. 6,084,100; 007.04.2000; Assigned to Medichem, S.A.

LORAZEPAM

Therapeutic Function: Tranquilizer

Chemical Name: 7-Chloro-5-(o-chlorophenyl)-1,3-dihydro-3-hydroxy-2H-1,4-benzodiazepin-2-one

Common Name: -

Structural Formula:

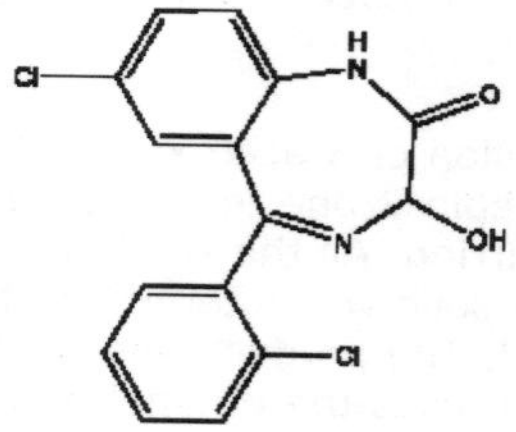

Chemical Abstracts Registry No.: 846-49-1

Trade Name	Manufacturer	Country	Year Introduced
Tavor	Wyeth	Italy	1972
Tavor	Wyeth	W. Germany	1972
Ativan	Wyeth	UK	1973
Temesta	Wyeth Byla	France	1973
Ativan	Wyeth	US	1977
Wwax	Wellcome	Japan	1978

Trade Name	Manufacturer	Country	Year Introduced
Bonton	Unipharm	Israel	-
Control	Sigurta	Italy	-
Emotion	Alpes	Argentina	-
Emotival	Armstrong	Argentina	-
Idalprem	Prem	Spain	-
Lorans	Schiapparelli	Italy	-
Lorivan	Disco	Israel	-
Lorsilan	Belupo Ltd.	Yugoslavia	-
Orfidal	Orfi	Spain	-
Piralone	Ferrer	Spain	-
Placidia	Fedal	Spain	-
Pro Dorm	Schurholz	W. Germany	-
Ouait	Jamco	Italy	-
Securit	Marxer	Italy	-
Sedarkey	Cuatrecasas-Darkey	Spain	-
Sedatival	Raffo	Argentina	-
Sedicepan	Septa	Spain	-
Sidenar	Syncro	Argentina	-

Raw Materials

Chloroacetyl chloride
Acetic anhydride
Methyl amine
2-Amino-2',5-dichlorobenzophenone
Hydroxylamine
Sodium hydroxide

Manufacturing Process

The starting material was 2-amino-2',5-dichlorobenzophenone which was reacted with hydroxylamine and then with chloroacetyl chloride. The intermediate thus obtained is reacted with methylamine and then with acetic anhydride.

To a slightly warm suspension of 3-acetoxy-7-chloro-5-(o-chlorophenyl)-1,3-dihydro-2H-1,4-benzodiazepin-2-one thus obtained was added 4N sodium hydroxide solution with stirring. All the solid dissolved and soon a thick white solid precipitated out. The solid was filtered, washed well with water and recrystallized from ethanol. The product was isolated as a solvate with 1 mol of ethanol. When heated it loses the ethanol of solvation and melts at 166°C to 168°C.

References

Merck Index 5400
Kleeman and Engel p. 530
PDR p. 1938
OCDS Vol. 1 p. 368 (1977)
DOT 7 (6) 210 (1971) and 9 (6) 238 (1973)
I.N. p.568
REM p. 1063

Bell, S.C. British Patent 1,057,492; February 1, 1967; assigned to American Home Products Corporation

Bell, S.C. US Patent 3,176,009; March 30, 1965; assigned to American Home Products Corp.

Bell, S.C.; US Patent 3,296,249; January 3, 1967; assigned to American Home Products Corp.

LORCAINIDE HYDROCHLORIDE

Therapeutic Function: Antiarrhythmic

Chemical Name: N-(p-Chlorophenyl)-N-(1-isopropylpiperidin-4-yl) phenylacetamide hydrochloride

Common Name: Isocainide hydrochloride; Socalnide hydrochloride

Structural Formula:

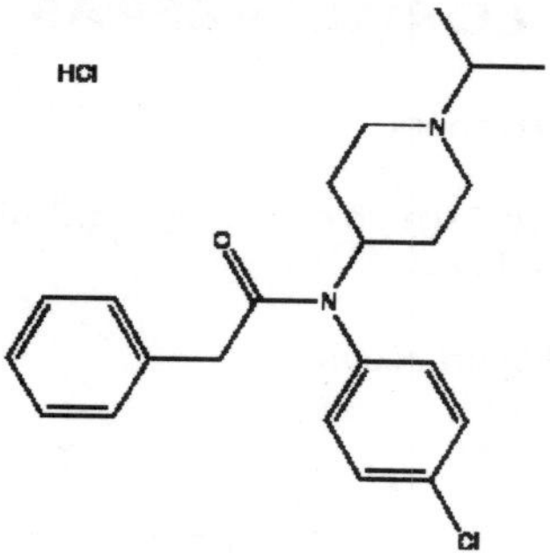

Chemical Abstracts Registry No.: 59729-31-6 (Base)

Trade Name	Manufacturer	Country	Year Introduced
Remivox	Janssen	W. Germany	1980

Raw Materials

N-(4-Chlorophenyl)-N-(piperidinyl)benzeneacetamide
2-Bromopropane
Hydrogen chloride

Manufacturing Process

To a stirred suspension of 5 parts of N-(4-chlorophenyl)-N-(4-piperidinyl) benzeneacetamide, 5 parts of sodium carbonate, a few crystals of potassium iodide in 200 parts of butanol is added dropwise 4 parts of 2-bromopropane at room temperature. After the addition is complete, the whole is stirred and refluxed for 20 hours. Then the second portion of 4 parts of 2-bromopropane is added and stirring and refluxing is continued for another 19 hours. The reaction mixture is cooled, filtered and the filtrate is evaporated. From the oily free base, the hydrochloride salt is prepared in the conventional manner in 1,1'-oxybisethane and 2-propanone. The precipitated solid salt is filtered off and crystallized from a mixture of 2-propanone and 2-propanol, yielding 2 parts of N-(4-chlorophenyl)-N-[1-(1-methylethyl)-4-piperidinyl] benzeneacetamide hydrochloride; melting point 263°C.

References

Merck Index 5401
DFU 3 (7) 518 (1978)
OCDS Vol. 3 p. 40 (1984)
DOT 18 (1) 17 and (10) 548 (1982)
I.N. p. 568
Sanczuk, S. and Hermans, H.K.F.; US Patent 4,196,210; April 1, 1980; assigned to Janssen Pharmaceutica NV

LORMETAZEPAM

Therapeutic Function: Hypnotic

Chemical Name: 7-Chloro-5-(o-chlorophenyl)-1,3-dihydro-3-hydroxy-1-methyl-2H-1,4-benzodiazepin-2-one

Common Name: N-Methyllorazepam

Structural Formula:

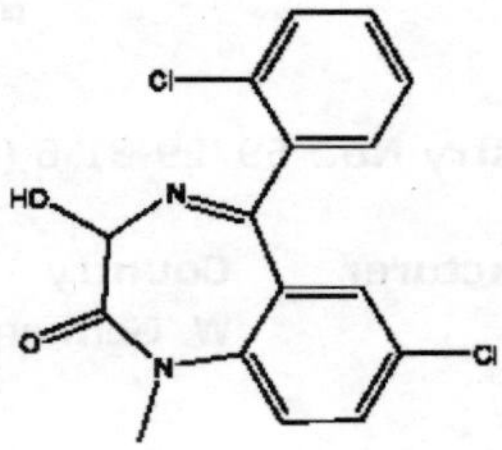

Cl HO N O N Cl

Chemical Abstracts Registry No.: 848-75-9

Trade Name	Manufacturer	Country	Year Introduced
Loramet	Wyeth	W. Germany	1980
Noctamid	Schering	W. Germany	1980
Loramet	Wyeth	Switz.	1981
Noctamid	Schering	UK	1981
Noctamid	Schering	France	1981

Trade Name	Manufacturer	Country	Year Introduced
Loramet	Wyeth	UK	1983
Loramid	Wyeth	W. Germany	-
Minias	Farmades	Italy	-
Pronoctan	Schering	-	-

Raw Materials

3-Acetoxy-7-chloro-1,3-dihydro-5-(o-chlorophenyl)-2H-1,4-benzodiazepin-2-one
Sodium hydroxide

Manufacturing Process

To a suspension of 3.4 g of 3-acetoxy-7-chloro-1,3-dihydro-5-(o-chlorophenyl)-2H-1,4-benzodiazepin-2-one in 80 ml of alcohol was added 6 ml of 4 N sodium hydroxide. After complete solution had taken place a solid precipitated that redissolved upon the addition of 80 ml of water. The solution was acidified with acetic acid to give white crystals. After recrystallization from alcohol the compound melted at 192°C to 194°C.

References

Merck Index 5403
DFU 5 (10) 495 (1980)
Kleeman and Engel p. 531
OCDS Vol. 3 p. 196 (1984)
DOT 17 (4) 137 (1981)
I.N. p. 569
American Home Products Co.; British Patent 1,022,642; March 16, 1966

LOSARTAN POTASSIUM

Therapeutic Function: Antihypertensive

Chemical Name: 1H-Imidazole-5-methanol, 2-butyl-4-chloro-1-((2'-(1H-tetrazol-5-yl)(1,1'-biphenyl)-4-yl)methyl)-, monopotassium salt

Common Name: Losartan potassium

Trade Name	Manufacturer	Country	Year Introduced
Alsartan	Aristo Pharmaceutical Ltd.	India	-
Cozaar	Merck Frosst	Canada	-
Czar	Argus	India	-
Losacar	Cadila	India	-
Losartan Potassium	Merck Sharp and Dohme	UK	-
Paxon	Gador	-	-
Redupress	Ache	-	-

Structural Formula:

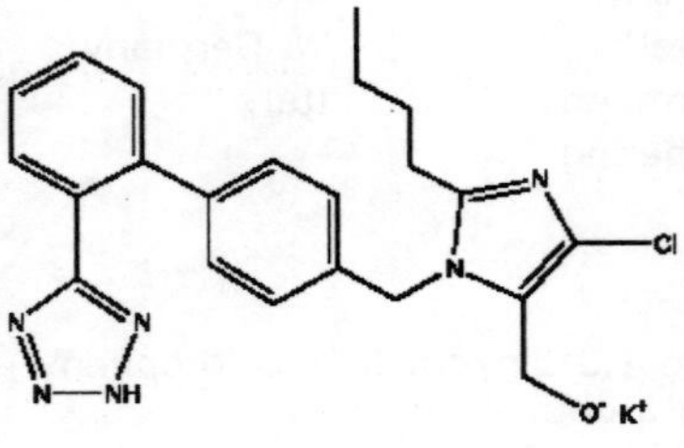

Chemical Abstracts Registry No.: 124750-99-8; 114798-26-4 (Base)

Raw Materials

Potassium hydroxide	Concentrated hydrochloric acid
Thionyl chloride	Ammonium hydroxide
Bromine	2-n-Butyl-4-chloro-5-(hydroxymethyl)-imidazole
Sodium azide	Methyl 4'-methylbiphenyl-2-carboxylate

Manufacturing Process

2-Butyl-4-chloro-1-(2'-(tetrazol-5-yl)biphenyl-4-ylmethyl)-1H-imidazole-5-methanolpotassium was synthesized in 5 stages.

1. Methyl 4'-methylbiphenyl-2-carboxylate (44.2 mmol), 0.5 N KOH in methanol (133 mmol), and water (50 mL) were mixed and refluxed under nitrogen. After 5 hours, the solvent was removed in vacuo and water (200 mL) and ethyl acetate (200 mL) added. The aqueous layer was acidified with concentrated hydrochloric acid to a pH of 3 and the layers were separated. The aqueous phase was extracted with ethyl acetate, the organic layers collected, dried ($MgSO_4$) and the solvent removed in vacuo to yield 8.71 g of a 4'-methylbiphenyl-2-carboxylic acid, melting point 140.0-145.0°C.

2. 4'-Methylbiphenyl-2-carboxylic acid (41 mmol) and thionyl chloride (411 mmol) were mixed and refluxed for 2 hours. The excess thionyl chloride was removed in vacuo and the residue was taken up in toluene. The toluene was removed by rotary evaporation. The crude acid chloride was then added slowly to cold (0°C) concentrated NH_4OH (50 mL) so that the temperature was kept below 15°C. After 15 minutes of stirring, water (100 mL) was added and solids precipitated. These were collected, washed with water and dried under high vacuum over P_2O_5 to yield 7.45 g of a white solid, melting point 126.0-128.5°C. The above product amide (35 mmol) and thionyl chloride (353 mmol) were mixed and refluxed for 3 hours. The thionyl chloride was removed using the same procedure as described above. The residue was washed with a little hexane to yield 6.64 g of 4'-methyl-2-cyanobiphenyl, melting point 44.0-47.0°C.

3. 4'-Methyl-2-cyanobiphenyl (5.59 g) was brominated using benzoyl peroxide as an initiator. The product was recrystallized from ether to yield 4.7 g of 4'-bromomethyl-2-cyanobiphenyl, melting point 114.5-120.0°C.

4. 4'-Bromomethyl-2-cyanobiphenyl (4.6 g) was alkylated onto 2-n-butyl-4-chloro-5-(hydroxymethyl)-imidazole. For separation of the product was used a flash chromatography in 1:1 hexane/ethyl acetate over silica gel. The regioisomeric products yielded 2.53 g of the faster eluting isomer. Recrystallization from acetonitrile yielded 1.57 g of analytically pure 2-n-butyl-4-chloro-1-[2'-cyanobiphenyl-4-yl)methyl]-5-(hydroxymethyl)-imidazole, melting point 153.5 -155.5°C.

5. 2-n-Butyl-4-chloro-1-[(2'-cyanobiphenyl-4-yl)-methyl]-5-(hydroxymethyl)-imidazole (10 mmole), sodium azide (10 mmol), and ammonium chloride (30 mmol) were mixed in DMF (150 mL) under N_2 at 100°C for 2 days, after which the temperature was raised to 120°C for 6 days. The reaction was cooled and 3 more equivalents each of ammonium chloride and sodium azide were added. The reaction was again heated for 5 days at 120°C. The reaction was cooled, the inorganic salts filtered, and the filtrate solvent removed in vacuo. Water (200 mL) and ethyl acetate (200 mL) were added to the residue and the layers were separated. The aqueous layer was extracted with ethyl acetate, the organic layers were collected, dried ($MgSO_4$) and the solvent removed in vacuo, to yield a dark yellow oil. The product was purified by flash chromatography in 100% ethyl acetate to 100% ethanol over silica gel to yield 5.60 g of a light yellow 2-n-butyl-4-chloro-5-hydroxymethyl-1-[(2'-(1H-tetrazol-5-yl)biphenyl-4-yl)methyl]imidazole. Recrystallization from acetonitrile yielded 4.36 g of light yellow crystals which still melted broadly. The crystals were taken up in 100 mL of hot acetonitrile. The solid that did not dissolve was filtered off to yield 1.04 g of product as a light yellow solid, melting point of 2-n-butyl-4-chloro-5-hydroxymethyl-1-[(2'-(1H-tetrazol-5-yl)biphenyl-4-yl)methyl]imidazole 183.5-184.5°C.

2-n-Butyl-4-chloro-5-hydroxymethyl-1-[(2'-(1H-tetrazol-5-yl)biphenyl-4-yl)methyl]imidazole may be converted to potassium salt.

References

Carini D. J., Duncia J. J., Wong, Pancras C. B.; US Patent No. 5,138,069; 08.11.1992; Assigned to E. I. Du Pont de Nemours and Company

LOTEPREDNOL ETABONATE

Therapeutic Function: Glucocorticoid

Chemical Name: Androsta-1,4-diene-17-carboxylic acid, 17-((ethoxycarbonyl)oxy)-11-hydroxy-3-oxo-, chloromethyl ester, (11β,17α)-

Common Name: Loteprednol etabonate

Trade Name	Manufacturer	Country	Year Introduced
Alrex	Bausch and Lomb	USA	-
Alrex	Pharmos	-	-
Lotemax	Bausch and Lomb	USA	-
Lenoxin	Pharmos	-	-

Structural Formula:

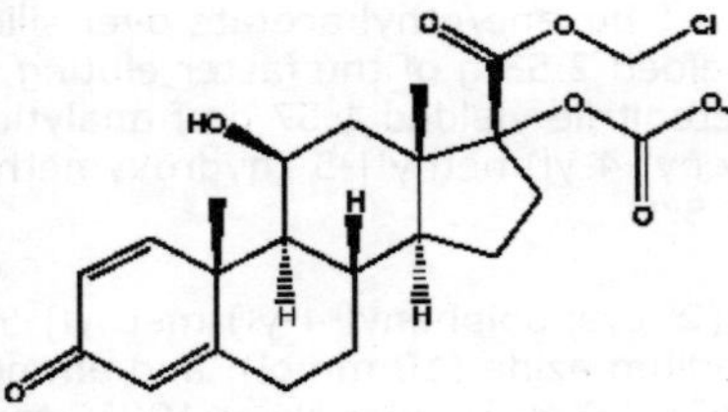

Chemical Abstracts Registry No.: 82034-46-6

Raw Materials

Hydrocortisone
Ethyl chloroformate
Sodium metaperiodate
Sodium hydroxide

Manufacturing Process

To a solution of hydrocortisone (15 g, 0.04 mol) in 120 ml of THF and 30 ml of methanol at room temperature is added a warm solution of sodium metaperiodate (25.7 g, 0.12 mol) in 100 ml of water. The reaction mixture is stirred at room temperature for 2 hours, then is concentrated under reduced pressure to remove the tetrahydrofuran and methanol. The solid is triturated with 50 ml of water, separated by filtration, washed with water and dried in vacuo at 50°C for 3 hours. The product, 11β,17α-dihydroxyandrost-4-en-3-one-17β-carboxylic acid (i.e., cortienic acid), is obtained in approximately 96% yield (13.76 g); melting point 231-234°C.

To a cold solution of 11β,17α-dihydroxyandrost-4-en-3-one-17β-carboxylic acid (5% weight/volume; 1 mol) and triethylamine (4 mol) in dichloromethane is added a 50% (weight/volume) solution of ethyl chloroformate (3.9 mol) in dichloromethane. The reaction mixture is allowed to warm to room temperature over a 2 hour period. The triethylamine hydrochloride precipitate which forms is removed by filtration and the filtration is washed successively with 3% sodium bicarbonate, 1% hydrochloric acid and water. The organic layer is separated, dried with magnesium sulfate, and filtered. The filtrate is concentrated in vacuo to a foam.

The foam is used in the next step below or chromatographed and crystallized for analysis. The product 17α-ethoxycarbonyloxy-11β-hydroxyandrost-4-en-3-one-17β-carboxylic acid, melting at 192-195°C C after chromatography and crystallization.

17α-Ethoxycarbonyloxy-11β-hydroxyandrost-4-en-3-one-17β-carboxylic acid is combined with an equivalent amount of 1 N sodium hydroxide in methanol and that solution is diluted to 100 times the original volume with ethyl ether. The suspension which results is refrigerated for 1 hour. Then, the crystals which form are removed by filtration, dried in an evacuated desiccator, and dissolved in hexamethylphosphoramide (10% weight/volume). A portion of the resultant solution containing 1 mole of the acid salt, i.e. of sodium 17αethoxycarbonyloxy-11β-hydroxyandrost-4-en-3-one-17β-carboxylate, is combined with 4 moles of chloromethyl iodide. The reaction mixture is maintained at room temperature for 3 hours, then is diluted to 10 times the original volume with ethyl acetate. The diluted reaction mixture is washed successively with 5% sodium thiosulfate, 3% sodium bicarbonate, and water. The organic layer is separated, dried with magnesium sulfate and filtered. The filtrate is concentrated in vacuo to a foam. The foam is purified by crystallization from ethyl ether or tetrahydrofuran/hexane. There is thus obtained chloromethyl-17α-ethoxycarbonyloxy-11β-hydroxyandrost-4-en-3-one-17β-carboxylate, melting at 197-200°C after crystallization.

References

Bodor N.S.; US Patent No. 4,996,335; Feb. 26, 1991; Assigned to Bodor; Nicholas S. (Gainesville, FL)

LOVASTATIN

Therapeutic Function: Antihyperlipidemic

Chemical Name: Butanoic acid, 2-methyl-, (1S,3R,7S,8S,8aR)-1,2,3,7,8,8a-hexahydro-3,7-dimethyl-8-(2-((2R,4R)-tetrahydro-4-hydroxy-6-oxo-2H-pyran-2-yl)ethyl)-1-naphthalenyl ester, (2S)-

Common Name: Lovastatin; Mevinolin; Monacolin K

Structural Formula:

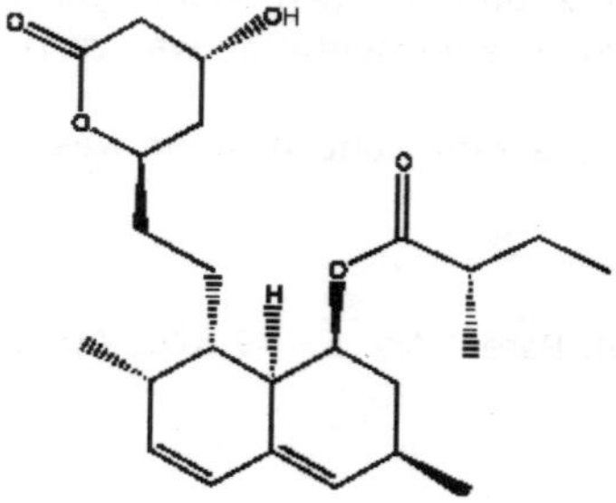

Chemical Abstracts Registry No.: 75330-75-5; 71949-96-7; 74133-25-8

Trade Name	Manufacturer	Country	Year Introduced
Aztatin	Sun Pharmaceuticals Industries Ltd.	India	-
Choletar	Krka	Slovenia	-
Favolip	SPPL (Sarabhai Piramal Pharmaceuticals Ltd.)	India	-

Trade Name	Manufacturer	Country	Year Introduced
Lovastatin	Ranbaxy	India	-
Lovasterol	Polpharma	Poland	-
Mevacor	Aetna Inc.	-	-
Mevacor	Merck and Company, Inc.	Germany	-
Medostatin	Medochemie Ltd.	Cyprus	-
Rovacor	Ranbaxy	India	-

Raw Materials

Coniothyrium fuckelii ATCC 74227
Nutrient medium

Manufacturing Process

1) *Coniothyrium fuckelii* ATCC 74227 was grown in a sterilizable fermentation apparatus with a volume of 15 L. The apparatus was equipped with an agitator, aerator, pH control system, dissolved oxygen control system, and a pump and feed system designed to allow the sterile addition of glucose solutions. The pH was controlled by the automatic addition of ammonium hydroxide or phosphoric acid to maintain the pH of the culture medium constant at 5.0. Periodically, the fermentation broth was sampled, measured for glucose concentration and an addition of glucose was made manually to maintain a concentration of glucose at approximately 2-5 g/L. After 192 hours of growth under these conditions, the concentration of biomass reached 65 g/L and the concentration of Lovastatin reached 102 mg/L.

2) A further medium for the growth of *Coniothyrium fuckelii* ATCC 74227, has the following composition: Glucose 12%, Peptone 1%, $(NH_4)_2SO_4$ 0.4%, $MgSO_4 \cdot 7H_2O$ 0.05%, P 2000 0.1% (Antifoam agent), L-isoleucine 0.2-1.5%, L-aspartic acid 0.2-1.5%. The fermentation was carried out as before.

With this medium, the lovastatin concentration was 430 mg/L.

References

Gerson D.F., Xiao X.; US Patent No. 5,409,820; Apr.,25, 1995; Assigned to Apoptex, Inc.

LOXAPINE

Therapeutic Function: Tranquilizer

Chemical Name: 2-Chloro-11-(4-methyl-1-piperazinyl)-dibenz[b,f][1,4] oxazepine

Common Name: Oxilapine

Structural Formula:

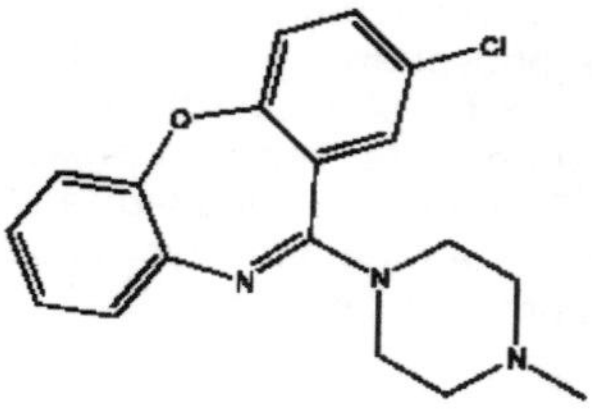

Chemical Abstracts Registry No.: 1977-10-2

Trade Name	Manufacturer	Country	Year Introduced
Loxitane	Lederle	US	1976
Loxapac	Lederle	France	1980
Loxapac	Cyanamid	Italy	1981
Daxolin	Dome	US	-

Raw Materials

o-(p-Chlorophenoxy)aniline
1-Methylpiperazine
Ethyl chloroformate
Phosphorus oxychloride

Manufacturing Process

One route is described in US Patent 3,412,193 as follows. To a mixture of o-(p-chlorophenoxy)aniline hydrochloride (prepared from 32 g of the base) in 50 ml of pyridine is added gradually while heating under reflux, 25 ml of ethyl chloroformate. After the addition is completed, the mixture is heated under reflux for one hour longer, and then evaporated under reduced pressure to an oily residue. The residue is taken up in 300 ml of water, and extracted with ether (approximately 200 ml).

The ether extract is separated, dried over sodium sulfate, and evaporated to an oily residue (40 g) which contains ethyl o-(p-chlorophenoxy)carbanilate and is used without further purification. The crude ethyl o-(p-chlorophenoxy) carbanilate is dissolved in 20 ml of benzene, and 20 ml of 1I-methylpiperazine and a small amount of sodium methylate (approximately 25 to 50 mg) are added. Benzene is then removed by slow distillation; and the mixture is heated overnight under reflux (approximately 16 hours).

Evaporation under reduced pressure then gives a solid residue which is dissolved in 400 ml of ether with heating. Concentration to half-volume under reduced pressure produces a precipitate which is collected, washed with petroleum ether and dried (36 g). A second crop of product is isolated from the filtrate. This product is dissolved in 200 ml of chloroform and treated with an excess of anhydrous hydrogen chloride. The resulting precipitate is collected and dried at 50°C (in vacuo), and 4-methyl-2'-(p-chlorophenoxy)-1-

piperazinecarboxanilide hydrochloride, MP 210° to 213°C, is thereby obtained.

A mixture of 4-methyl-2'-(p-chlorophenoxy)-1-piperazinecarboxanilide hydrochloride (6 g), 50 ml of phosphorus oxychloride and 10 g of phosphorus pentoxide is heated under reflux for about 24 hours, and then concentrated to a gummy residue by evaporation under reduced pressure. This residue is taken up in 150 ml of ether, 200 g of ice is added, and the mixture is made basic with concentrated aqueous ammonium hydroxide. The ether layer is separated, dried over potassium hydroxide pellets and evaporated to a solid residue (approximately 4 g).

This crude product is dissolved in 100 ml of dilute hydrochloric acid, the acid solution is extracted with ether, and the aqueous layer is made basic with sodium hydroxide solution (3N) in the presence of ether (approximately 250 ml). The ether layer is separated, dried over potassium hydroxide and evaporated to a white solid. Additional purification by repeating the formation of the hydrochloric acid salt and reprecipitation of the base is carried out. When purified in this manner, followed by drying at 80°C in vacuo over phosphorus pentoxide, 2-chloro-11-(4-methyl-1-piperazinyl)dibenz[b,f][1,4]oxazepine, MP 109° to 111°C, is obtained.

References

Merck Index 5404
Kleeman and Engel p. 532
PDR p. 1012
OCDS Vol. 2 p. 427 (1980)
DOT 14 (6) 248 (1978)
I.N. p. 569
REM p. 1089
Coppola, J.A.; US Patent 3,412,193; November 19, 1968;assigned to American Cyanamid Company
Schmutz, J., Hunziker, F. and Kunzle, F.M.; US Patent 3,546,226; December 8, 1970

LYMECYCLINE

Therapeutic Function: Antibiotic

Chemical Name: Lysine, N^6-((4-(dimethylamino)-1,4,4a,5,5a,6,11,12a-octahydro-3,6,10,12,12a-pentahydroxy-6-methyl-1,11-dioxo-2-naphthacenecarboxamido)methyl)-, (+)-

Common Name: Limeciclina; Lymecycline

Trade Name	Manufacturer	Country	Year Introduced
Armyl	Armour Pharm.	-	-
Tetralysal	Farmitalia	-	-
Tetralysal	Galderma International	-	-
Tetradin	Granelli	-	-

Structural Formula:

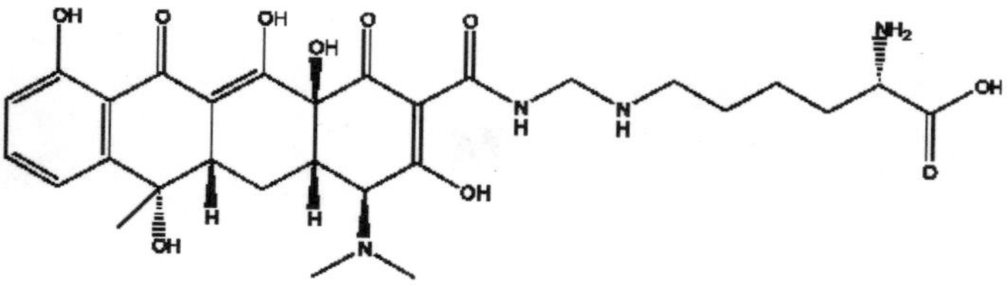

Chemical Abstracts Registry No.: 992-21-2

Raw Materials

L-Lysine hydrochloride
Formaldehyde
Tetracycline

Manufacturing Process

Amido-N-(lysinomethyl)tetracycline hydrochloride:

To 18.3 g of L-lysine hydrochloride dissolved in 100 ml of water is added 10 ml of 37% aqueous solution of formaldehyde. To the resultant mixture is added 44.0 g of anhydrous tetracycline dissolved in 500 ml of tetrahydrofuran. After thorough mixing the product forms over a period of about 15 min as an oily layer which after separation from the aqueous phase is added dropwise to 3 L of stirred isopropyl alcohol. The product after recovery by filtration, is reslurried with acetone, filtered and dried at 65°C at reduced pressure.

The product thus obtained has a bioassay of 500 mcg/mg (K. pneumonlae oxytetracyoline assay).

References

Blackwood R.K. et al.; US Patent No. 3,042,716; July 3, 1962; Assigned: Chas. Pfizer and Co., Inc., New York, N.Y., a corporation of Delaware

M

MAFENIDE ACETATE

Therapeutic Function: Antibacterial

Chemical Name: α-Acetylamino-p-toluenesulfonamide

Common Name: -

Structural Formula:

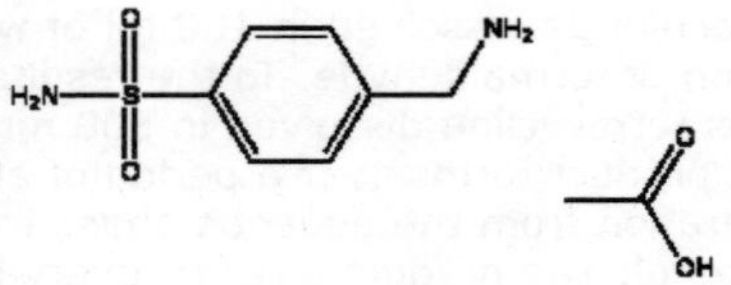

Chemical Abstracts Registry No.: 13009-99-9; 138-39-6 (Base)

Trade Name	Manufacturer	Country	Year Introduced
Sulfamylon	Winthrop	US	1949
Napaltan	Winthrop	W. Germany	1969
Sulfamylon	Winthrop	UK	1970
Mafatate	Torii	Japan	1980
Mafylon	Winthrop	-	-

Raw Materials

Acetylbenzylamine
Chlorosulfonic acid
Ammonia

Manufacturing Process

For the preparation of mafenide 50 g of acetylbenzylamine are introduced while stirring into 150 cc of chlorosulfonic acid, whereby the temperature is kept below 40°C by external cooling. After several hours' storing at ordinary temperature the mixture is heated for 1 hour in the boiling water-bath and after cooling, poured on to ice. Thereupon the 4-acetylaminoethyl-benzenesulfonic acid chloride precipitates at first in an oily form, but solidifies after short stirring to crystals. The product sucked off and washed with cold water is introduced into a 10% aqueous ammonia solution. Thereby dissolution takes place while heating and after a short time the 4-acetylaminomethyl-benzenesulfonic acid amide precipitates in a crystalline form. After heating to 70°C for 30 minutes the solution is cooled, filtered with suction and washed out. The product is obtained when recrystallized from water or dilute alcohol in colorless crystals melting at 177%. It is readily soluble in warm water, extremely readily soluble in dilute sodium hydroxide solution.

References

Merck Index 5466
Kleeman and Engel p. 534
PDR p. 1929
OCDS Vol. 2 p. 114 (1980)
DOT 5 (4) 132 (1969)
I.N. p. 574
REM p. 1162
Klarer, J.; US Patent 2,288,531; June 30, 1942; assigned to Winthrop Chemical Co., Inc.

MAGALDRATE

Therapeutic Function: Antacid

Chemical Name: Tetrakis(hydroxymagnesium)decahydroxydialuminate dihydrate

Common Name: Magnesium aluminate hydrate; Monalium hydrate

Structural Formula: $[Mg(OH)]_4[(OH)_4Al(OH)(HO)Al(OH)_4]2H_2O$

Trade Name	Manufacturer	Country	Year Introduced
Riopan	Ayerst	US	1960
Riopan	Byk Gulden	W. Germany	1981
Dynese	Galen	UK	1983
Bismag-Lac	Much	W. Germany	-

Chemical Abstracts Registry No.: 1317-26-6

Raw Materials

Aluminum chloride
Sodium hydroxide
Magnesium sulfate

Manufacturing Process

1 kg aluminum chloride hydrate was dissolved in 2 kg water and reacted with a solution of 1.2 kg sodium hydroxide in 2.5 kg water, under constant stirring. The resultant sodium aluminate solution was cooled to about 20°C and, with thorough stirring, it was reacted with 3.5 kg of a magnesium sulfate solution produced by dissolving 1 kg of magnesium sulfate anhydride in 2.5 kg water. The magnesium sulfate solution was introduced in a plurality of thin jets through several shower heads to avoid localized differences of concentration as much as possible. After all the magnesium sulfate was added, stirring was continued for about ½ hour.

A colorless, colloidal precipitate was formed and stirred thoroughly for about 15 minutes, whereupon it was filtered by suction. The raw product thus obtained was washed with water until it contained only about 0.5% water-soluble salts. After drying for 12 hours in a vacuum apparatus at 60°C and under a pressure of 12 mm Hg, the product had the form of hard pieces. The pieces were comminuted to powder in a ball mill and the powder was passed through a sieve (3,600 meshes per cm^2). The small residue on the sieve was again pulverized and passed through the same sieve. The yield was 870 g, or 99% of theoretical, calculated on the assumed formula $[Mg(OH)]_4[(HO)_4Al(OH)(HO)Al(OH)_4] \cdot 2H_2O$ with a molecular weight of 425.

References

Merck Index 5467
PDR p. 650
I.N. p. 574
REM p. 795
Hallmann, G.; US Patent 2,923,660; February 2,1960; assigned to Byk-Gulden Lomberg Chemische Fabrik GmbH, Germany

MALATHION

Therapeutic Function: Pediculicide

Chemical Name: Diethyl(dimethoxyphosphinothioyl)thiobutanedioate

Common Name: Mercaptothion (South Africa); Maldison (Australia and New Zealand); Carbofos (USSR)

Chemical Abstracts Registry No.: 121-75-5

Trade Name	Manufacturer	Country	Year Introduced
Prioderm	Purdue Frederick	US	1982
Organoderm	Mundipharma	W. Germany	1982
Derbac	Benque	UK	-
Lusap	Interdelta	Switz.	-
Taskil	Tasman Vaccine	UK	-

Structural Formula:

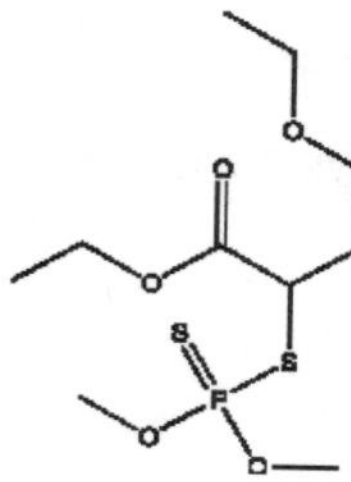

Raw Materials

O,O-Dimethyl phosphorodithioic acid
Diethyl maleate

Manufacturing Process

The feed materials for malathion manufacture are O,O-dimethyl phosphorodithioic acid and diethyl maleate or fumarate which react according to the equation:

An antipolymerization agent such as hydroquinone may be added to the reaction mixture to inhibit the polymerization of the maleate or fumarate compound under the reaction conditions. This reaction is preferably carried out at a temperature within the range of 20°C to 150°C. This reaction is preferably carried out at atmospheric pressure. Reaction time of 16 to 24 hours have been specified for this reaction by J.T. Cassaday. The reaction is preferably carried out in a solvent such as the low molecular weight aliphatic monohydric alcohols, ketones, aliphatic esters, aromatic hydrocarbons or trialkyl phosphates.

The reaction may be accelerated by using an aliphatic tertiary amine catalyst, usually within the range of 0.2 to 2.0% based on the total weight of the reactants. A stirred, jacketed reactor of conventional design may be used. After cooling, the reaction mixture may be taken up in benzene. It is then washed with 10% Na_2CO_3 and with water. The organic layer is dried over anhydrous Na_2SO_4, filtered and concentrated in vacuo to give the final product as residue.

References

Merck Index 5522
I.N. p. 575
REM p. 1240
Cassaday, J.T.; US Patent 2,578,652; December 18, 1951; assigned to American Cyanamid Co.
Backlund, G.R., Martino, J.F. and Divine, R.D.; US Patent 3,463,841; August 26, 1969; assigned to American Cyanamid Co.
Usui, M.; US Patent 2,962,521; November 29, 1960; assigned to Sumitomo Chemical Co.

MALTOSE

Therapeutic Function: Sugar supplement

Chemical Name: 4-O-α-Glucopyranosyl-D-glucose

Common Name: -

Structural Formula:

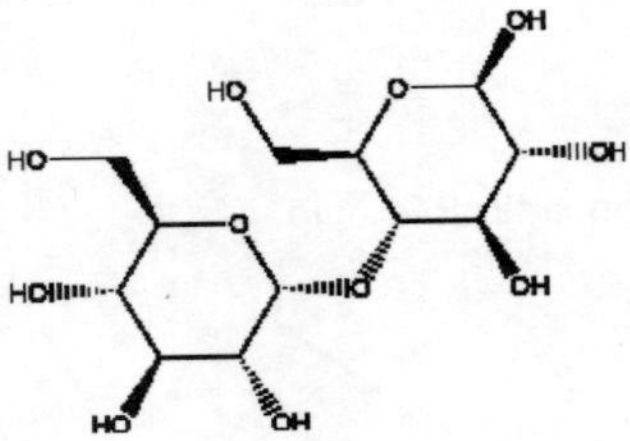

Chemical Abstracts Registry No.: 69-79-4

Trade Name	Manufacturer	Country	Year Introduced
Maltos-10	Otsuka	Japan	1974

Raw Materials

Starch
Water

Manufacturing Process

The process of manufacturing a maltose product from a suitably purified starch source includes preparing an aqueous starchy suspension, adjusting the acidity thereof to from 4.6 to 6.0 pH, liquefying the suspension by heating in the presence of a diastatic agent, diastatically saccharifying the liquefied mixture, filtering, and concentrating the liquid to a syrup.

References

Merck Index 5536
DOT 10 (11) 308 (1974)
REM p. 1029
Gore, H.C.; US Patent 1,657,079; January 24, 1928; assigned to The Fleischmann Co.

MANGAFODIPIR TRISODIUM

Therapeutic Function: Diagnostic aid

Chemical Name: Manganate(6-), ((N,N'-1,2-ethanediylbis(N-((3-hydroxy-2-methyl-5-((phosphonooxy)-methyl)-4-pyridinyl)methyl)glycinato)) (8-))-, trisodium trihydrogen, (OC-6-13)-

Common Name: Mangafodipir trisodium

Structural Formula:

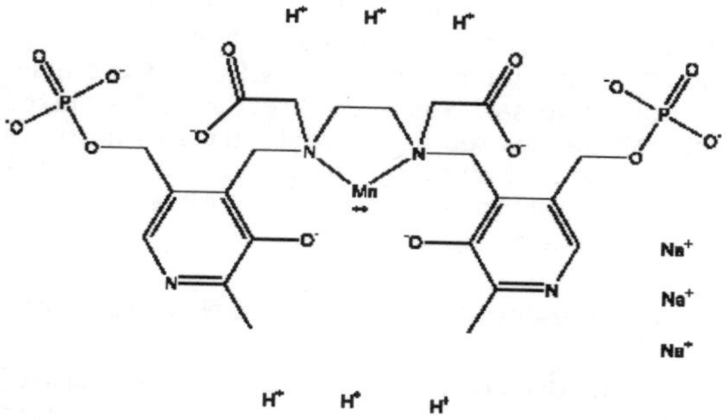

Chemical Abstracts Registry No.: 140678-14-4

Trade Name	Manufacturer	Country	Year Introduced
Teslascan	Nycomed Amersham Health Inc.	-	-

Raw Materials

Pyridoxal-5-phosphate
Platinum on carbon
Sodium hydroxide
Manganese dichloride tetrahydrate
1,2-Diaminoethane
Hydrogen
Bromoacetic acid

Manufacturing Process

Sodium N,N'-bis(pyridoxal-5-phosphate)ethylenediimine

A 265.2 g (1 mole) of pyridoxal-5-phosphate was slurried in 1 L of methanol, and 400 mL of 5 M NaOH was added. When the solution was homogeneous, 34.2 mL of 1,2-diaminoethane was added rapidly with vigorous stirring. The imine product sodium N,N'-bis(pyridoxal-5-phosphate)ethylenediimine or sodium 5-(N-(3-hydroxy-2-methyl-5-phosphonomethyl-4-pyridyl) methylideneaminoethyleneiminomethyl)-2-hydroxy-3-methyl-5-pyridylmethylphosphate was stirred for 1 hr, 400 mL of diethyl ether was added, and the slurry was filtered. The filtrate was washed with 600 mL of ethanol and dried at 60°C in vacuo. A 290 g of the bis-imine with a melting point of 215-220°C (decomp.) was isolated (90% yield, based on the tetra-sodium salt).

N,N'-bis(Pyridoxal-5-phosphate)ethylenediamine

To the diimine obtained was added 1.5 L of deionized water and 1.5 L of methanol. The yellow solution formed was stirred while sparging with nitrogen. Then 13 g of 5% Pt on carbon was added, and the apparatus was purged with hydrogen. The reaction was allowed to proceed for 5 hr with continuous addition of hydrogen. HPLC analysis showed complete reduction to the amine. The reaction mixture was sparged with nitrogen for 15 min and then filtered through Celite. The filtrate was concentrated in vacuo at 60°C to about 500 mL. The solution, containing N,N'-bis(pyridoxal-5-phosphate) ethylenediamine or 5-(N-(3-hydroxy-2-methyl-5-phosphonomethyl-4-pyridyl)methylaminoethyleneaminomethyl)-2-hydroxy-3-methyl-5-pyridylmethylphosphoric acid salt was used directly for the next step. If desired the diamine can be isolated as off-white crystals by the addition of 200 ml of 97% formic acid and allowing the product to crystallize at room temperature overnight. The diamine is isolated by filtration and washed with 150 mL of cold deionized water.

N,N'-Bis-(pyridoxal-5-phosphate)ethylenediamine-N,N'-diacetic acid synthesis

The diamine obtained was dissolved in a 100 g (2.5 mole) of NaOH, and 130 g (0.9 mole) of bromoacetic acid was dissolved in 180 mL of deionized water. Each solution was charged to an addition funnel. NaOH solution was added to the diamine solution to bring the pH to 11. The temperature was raised to 42°C, and bromoacetic acid and NaOH solution were added concurrently to maintain the pH at 11. The progress of the reaction was checked by HPLC. A 675 g of cation exchange resin (AMBERLITE IRC-50) was added, and the mixture was placed in a refrigerator for 14 hr. The pH had dropped to 6.5. The resin was removed by filtration, and the filtrate treated with 260 g of cation exchange resin (DOWEX 50W-X8). The pH dropped to about 4. The resin was removed by filtration, and the solution was concentrated in vacuo at 60°C to a viscous oil. The oil was dried in vacuo for 48 hr to yield a resinous solid containing N,N'-bis-(pyridoxal-5-phosphate)ethylenediamine-N,N'-diacetic acid or N,N'-bis(3-hydroxy-2-methyl-5-phosphonomethyl-4-pyridylmethyl) ethylenediamine-N,N'-diacetic acid (DPDP). The resinous solid obtained was dissolved in 600 mL of 88% formic acid, then 1.5 L of methanol and 2.2 L of ethanol was added, and the mixture was cooled to 0°C for 2 hr. The solvent mixture was decanted from the resulting gum. The gum was dissolved in 800 mL of deionized water which was then concentrated in vacuo to about 600-650 mL. Seed crystals were added, and the solution was allowed to stand overnight. The product was isolated by filtration, washed with 400 mL of cold water, 250 mL of ethanol, and then dried in vacuo to yield 65 g of DPDP in 85-90% purity by HPLC.

The 65 g of product was then dissolved in 75 mL of 88% formic acid containing 5 mL of deionized water with heating to 60°C. Cold water was added to a total volume of 1 L, and the solution was allowed to stand at 25°C for 16 hr to crystallize. The product was isolated by filtration, washed with 200 mL cold water, and dried in vacuo at 60°C to yield 55 g of DPDP in 93-95% purity by HPLC. A second recrystallization, using the same procedure yields 50 g of DPDP in 96-98% purity by HPLC, melting point 174-180°C (decomp.).

Sodium salt of Mn(DPDP)

A 4.16 g (6.25 mmole) portion of DPDP was dissolved in 15 mL of rigorously degassed water by the addition of 1.0 g (25 mmoles) of NaOH. A 1.25 g (6.25 mmole) quantity of manganese dichloride tetrahydrate was added, and the solution immediately turned yellow. After stirring for 30 min, 0.25 g (6.25 mmole) of solid NaOH was added to bring the pH up to 6.5. Then degassed water was added to bring the volume of the solution to 25 mL. The clear yellow solution was sterilized by being filtered through a 0.2 micron filter to yield the sodium salt of a manganese chelate complex of N,N'-bis-(pyridoxal-5-phosphate)ethylenediamine-N,N'-diacetic acid or N,N'-bis(3-hydroxy-2-methyl-5-phosphonomethyl-4-pyridylmethyl)ethylenediamine-N,N'-diacetic acid.

References

Rocklage S.M., Quay S.C.; US Patent No. 4,933,456; Jun. 12, 1990; Assigned to Salutar, Inc.

Rocklage S.M., Quay S.C.; US Patent No. 4,935,518; Jun. 19, 1990; Assigned to Salutar, Inc.

Towart R. et al.; US Patent No. 6,258,828; Jul 10. 2001; Assigned to Nycomed Imaging AS (Oslo, NO)

MANNITOL

Therapeutic Function: Diuretic, Diagnostic aid (kidney function)

Chemical Name: D-Mannitol

Common Name: -

Structural Formula:

HO OH OH HO HO OH

Chemical Abstracts Registry No.: 69-65-8

Trade Name	Manufacturer	Country	Year Introduced
Mannitol	MSD	US	1946
Osmitrol	Travenol	US	1964
Mannitol I.V.	Abbott	US	1968
Eufusol	Knoll	W. Germany	-

Trade Name	Manufacturer	Country	Year Introduced
Isotol	Baxter	Italy	-
Manit	Pliva	Yugoslavia	-
Mannidex	Pharmacia	Sweden	-
Osmofundin	Braun	W. Germany	-
Osmosol	Farmer Hill	Australia	-
Rectisol	McGaw	US	-

Raw Materials

Glucose
Hydrogen

Manufacturing Process

250 g of glucose is dissolved in distilled water to give a solution of 48% concentration. This solution is heated to 65°C and barium hydroxide added in quantity sufficient to make the concentration of the barium hydroxide 0.2 mol/liter. The solution is agitated and maintained at 65°C for 6 hours after the addition of the barium hydroxide. It is then cooled and neutralized to a pH of 6.8 with sulfuric acid. The precipitated barium sulfate is filtered out. A quantity of activated supported nickel catalyst containing 5 g of nickel is added.

The slurry is introduced into a 3-liter rocking autoclave, and hydrogen admitted to a pressure of 1,500 psi. The autoclave is heated to a temperature of 150°C in one hour and held at this temperature for 2.5 hours more. Pressure rises to about 1,800 psi and then declines to about 1,600 during the hydrogenation. The autoclave is then cooled, emptied, and the catalyst filtered from the product. The filtrate is then concentrated under vacuum on a hot water bath to remove a part of the water.

The concentrate is taken up in warm aqueous methanol so adjusted that the composition of the solvent is 90% methanol/10% water, and the weight of the solvent is 3 times the weight of the solids in the concentrate. This solution is cooled to 20°C and held overnight. The mannitol which crystallizes is filtered out. The filtrate is concentrated on a water bath under vacuum to remove methanol and adjusted to a water percentage of 16%. The resulting syrup is viscous, noncrystallizing and nongelling, and analysis shows a PN (Pyridine Number) of 32 and essentially no reducing sugar, according to US Patent 2,749,371.

References

Merck Index 5569
I.N. p. 576
REM p. 935
Kasehagen, L.; US Patent 2,642,462; June 16, 1953; assigned to Atlas Powder Company
Kasehagen, L.; US Patent 2,749,371; June 5, 1956; assigned to Atlas Powder Company
Kasehagen, L. and Luskin, M.M.; US Patent 2,759,024; August 14, 1956; assigned to Atlas Powder Company

MAPROTILINE HYDROCHLORIDE

Therapeutic Function: Antidepressant

Chemical Name: 9,10-Ethanoanthracene-9(10H)-propylamine, N-methyl-, hydrochloride

Common Name: Maprotiline hydrochloride

Structural Formula:

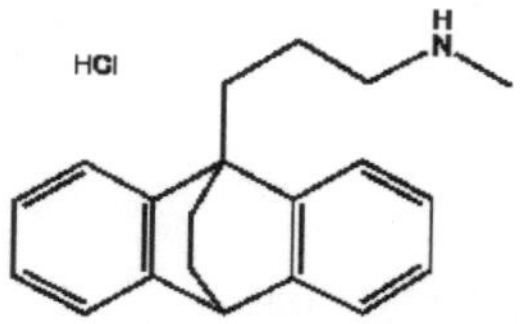

Chemical Abstracts Registry No.: 10347-81-6; 10262-69-8 (Base)

Trade Name	Manufacturer	Country	Year Introduced
Maprotiline hydrochloride	Mylan	-	-
Ludiomil	Novartis AE (ex. CIBA-GEIGY)	-	-
Retinyl	Kleva	-	-

Raw Materials

Thionyl chloride
Methylamine
Ethylene
3-(9-Anthryl)propionic acid
Lithium aluminum hydride

Manufacturing Process

9-(3-Hydroxypropyl)anthracene was prepared by reduction of 3-(9-anthryl) propionic acid with $LiAlH_4$. By action of thionylchloride and then methylamine the 9-(3-hydroxypropyl)anthracene was converted to 9-(3-methylaminopropyl) anthracene. By addition of ethylene to 9-(3-methylaminopropyl)anthracene (at 150°C, a pressure of ethylene 50 atm, 24 hours) was obtained 3-(9,10-dihydro-9,10-ethanoanthracene-9-yl)-N-methylpropylamine. Hydrochloride 3-(9,10-dihydro-9,10-ethanoanthracene-9-yl)-N-methylpropylamine may be prepared by action hydrochloric acid.

References

Merck Index, Monograph number: 5792, Twelfth edition, 1996, Editor: S. Budavari; Merck and Co., Inc.

Rev.: Scoginis, Clin. Chem. 1980, 26, 805-815

Boissier et al.; "Synthesis and Pharmacological Properties of New 9,10-Dihydro-9,10-ethanoanthracene Derivatives", J. Med. Chem. 10:86-91 (Jan. 1967)

MAZATICOL HYDROCHLORIDE

Therapeutic Function: Antiparkinsonian

Chemical Name: 6,6,9-Trimethyl-9-azabicyclo[3.3.1]non-3β-yl-di-2-thienylglycolate hydrochloride

Common Name:-

Structural Formula:

HCl

Chemical Abstracts Registry No.: 38738-59-9; 42024-98-6 (Base)

Trade Name	Manufacturer	Country	Year Introduced
Pentona	Tanabe	Japan	1978

Raw Materials

6,6,9-Trimethyl-9-azabicyclo[3.3.1]nonan-3α-ol
Methyl α,α-di(2-thienyl)glycolate

Manufacturing Process

A mixture of 1.0 g of 6,6,9-trimethyl-9-azabicyclo[3.3.1]nonan-3β-ol, methyl α,α-di-(2-thienyl)-glycolate and 30 mg of metallic sodium is heated at 80°C to 90°C for about 2 hours under reduced pressure. After cooling, ether is added to the reaction mixture. The mixture is extracted with 10% hydrochloric acid. The aqueous layer is alkalified with sodium carbonate and reextracted with ethyl acetate. The extract is washed with water, dried and concentrated to dryness. The residue thus obtained is treated with hydrogen chloride by conventional manner. 2.0 g of the α,α-di-(2-thienyl)glycolate of 6,6,9-trimethyl-9-azabicyclo[3.3.1]nonan-3β-ol hydrochloride are obtained. Yield 83%.

References

Kleeman and Engel p. 535
DOT 13 (2) 72 (1977)
I.N. p. 579
Yoneda, N., Ishihara.T., Kobayashi, T., Kondo, Y., Okumura, K., Kojima, M. and Nose, T.; US Patent 3,673,195; June 27, 1972; assigned to Tanabe Swiyaku Co.

MAZINDOL

Therapeutic Function: Antiobesity

Chemical Name: 5-(4-Chlorophenyl)-2,5-dihydro-3H-imidazo[2,1-a]isoindol-5-ol

Common Name: -

Structural Formula:

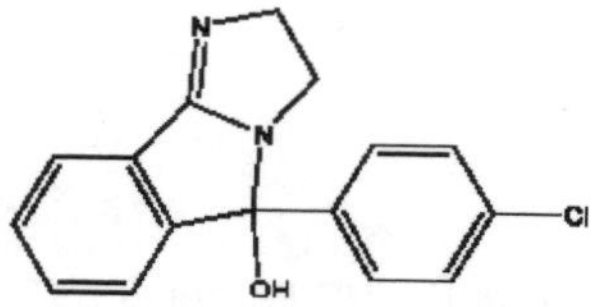

Chemical Abstracts Registry No.: 22232-71-9

Trade Name	Manufacturer	Country	Year Introduced
Sanorex	Sandoz	US	1973
Teronac	Wander	UK	1974
Teronac	Wander	W. Germany	1976
Mazildene	Farmochimica	Italy	1979
Mazanor	Wyeth	US	1980
Degonan	Spofa	Czechoslovakia	-
Magrilan	Sintyal	Argentina	-

Raw Materials

3-(p-Chlorophenyl)phthalimidine
Epichlorohydrin
Ethylene imine

Manufacturing Process

Step 1: 1-(p-Chlorophenyl)-3-Ethoxy-11H-Isoindole - Crystalline triethyloxonium borontetrafluoride (21 g) (prepared from 23 g of borontrifluoride etherate and 11 g of epichlorohydrin) is dissolved in 100 ml of absolute methylenechloride. 3-(p-Chlorophenyl) phthalimidine (21 g) is added and the reaction mixture is stirred overnight at room temperature. The resulting solution is poured onto 50 ml of saturated sodium carbonate, extracted with 500 ml of ether and dried. Upon evaporation of the solvent there is obtained crude material which is recrystallized from methylene chloride/hexane (1:1) to yield 1-(p-chlorophenyl)-3-ethoxy-1H-isoindole; MP 102° to 103°C.

Step 2: 5-(p-Chlorophenyl)-5-Hydroxy-2,3-Dihydro-5H-Imidazo[2,1-a]Isoindole - 1-(p-Chlorophenyl)-3-ethoxy-1H-isoindole (1 g), 2 g of ethyleneimine hydrotetrafluoroborate moistened with methylene chloride (containing approximately 0.66 g of dry salt) is refluxed in 25 ml of absolute toluene for 2 hours in an atmosphere of nitrogen. The resulting mixture is poured into 2 N sodium carbonate solution (25 ml) and extracted with ether. The ether solution is contacted with air for 6 days at room temperature to give the desired product. The crude material is recrystallized from acetone/hexane (1:1) to give 5-(p-chlorophenyl)-5-hydroxy-2,3-dihydro-5H-imidazo[2,1-a]isoindole; MP 198° to 199°C.

References

Merck Index 5585
Kleeman and Engel p. 535
PDR pp. 1595, 1958
OCDS Vol. 2 p. 462 (1980)
DOT 10 (1) 24 (1974)
I.N. p. 579
REM p. 892
Houlihan, W.J. and Eberle, M.K.; US Patent 3,597,445; August 3, 1971; assigned to Sandoz-Wander, Inc.
Sulkowski, T.S.; US Patent 3,763,178; October 2, 1973; assigned to American Home Products Corp.

MEBENDAZOLE

Therapeutic Function: Anthelmintic

Chemical Name: (5-Benzoyl-1H-benzimidazol-2-yl)carbamic acid methyl ester

Common Name: Methyl-5-benzoyl-2-benzimidazole carbamate

Structural Formula:

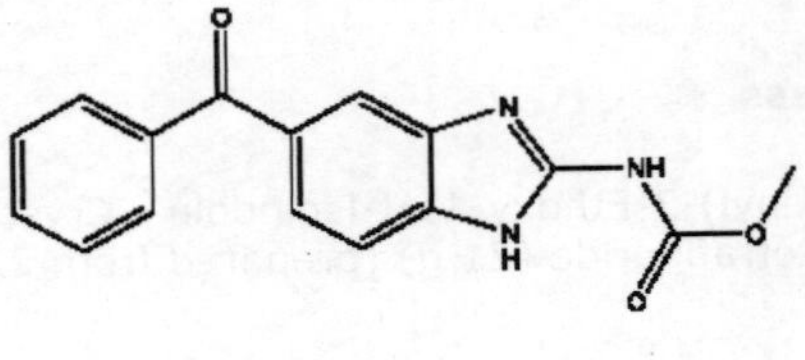

Chemical Abstracts Registry No.: 31431-39-7

Trade Name	Manufacturer	Country	Year Introduced
Vermox	Ortho	US	1975
Vermox	Janssen	UK	1976
Vermox	Janssen	W. Germany	1976
Vermox	Janssen	Italy	1978
Vermox	Janssen	Sweden	1983
Lomper	Esteve	Spain	-
Mebutar	Andromaco	Argentina	-
Panfugan	Byk Procienx	Brazil	-
Sirben	Andromaco	Brazil	-
Sufil	Cusi	Spain	-
Vermirax	Biosintetica	Brazil	-
Verpanil	Krka	Yugoslavia	-

Raw Materials

Ammonia
Hydrogen
Methyl chloroformate
4-Chloro-3-nitrobenzophenone
S-Methyl isothiourea sulfate

Manufacturing Process

A mixture of 5.2 parts of 4-chloro-3-nitrobenzophenone, 5 parts of ammonia, 72 parts of methanol and 13 parts of sulfolane is heated overnight at 125°C in a sealed tube. The reaction mixture is evaporated in vacuo. The semisolid residue is boiled in 100 parts of a diluted hydrochloric acid solution. After cooling, the precipitated product is filtered off and dissolved in chloroform. The chloroform phase is dried and evaporated. The residue is crystallized from toluene, yielding 4-amino-3-nitrobenzophenone; MP 141°C.

A mixture of 9.6 parts of 4-amino-3-nitrobenzophenone, 160 parts of methanol, 8 parts of concentrated hydrochloric acid and 1 part of palladium-on-charcoal catalyst 10% is hydrogenated at normal pressure and at room temperature. After the calculated amount of hydrogen is taken up, hydrogenation is stopped. The catalyst is filtered off and the solvent is evaporated. The solid residue is triturated in 2-propanol. The latter is partly evaporated and the solid product is filtered off, washed with 2-propanol and dried, yielding 3,4-diaminobenzophenone hydrochloride; MP 207°C.

7.8 parts of S-methylisothiourea sulfate are stirred in 10 parts of water in an ice bath and there are added 4.5 parts of methyl chloroformate. While keeping the temperature below 20°C, there are added dropwise, in the course of 10 minutes, 17 parts of sodium hydroxide solution 25% (pH 8±), followed by the addition of 5.6 parts of acetic acid (pH 5). To this mixture is added at 20°C a suspension of 7 parts of 3,4-diaminobenzophenone hydrochloride in 100 parts of water, followed by the addition of 2.3 parts of sodium acetate.

The whole is slowly heated to 85°C and stirred at this temperature for 45 minutes. The reaction mixture is cooled and the precipitated product is filtered off. It is washed successively with water and ethanol, dried and crystallized from a mixture of acetic acid and methanol, yielding methyl N-[5(6)-benzoyl-2-benzimidazolyl]carbamate; MP 288.5°C.

References

Merck Index 5589
Kleeman and Engel p. 536
PDR p. 960
OCDS Vol. 2 p. 353 (1980)
DOT 7 (5) 195 (1971); 9 (7) 299 (1973); 16 (10) 350 (1980) and 17 (6) 262 (1981)
I.N. p. 580
REM p. 1235
Van Gelder, J.L.H., Roevens, L.F.C. and Raeymaekers, A.H.M.; US Patent 3,657,267; April 18, 1972; assigned to Januen Pharmaceutica, NV, Belgium

MEBEVERINE HYDROCHLORIDE

Therapeutic Function: Spasmolytic

Chemical Name: 3,4-Dimethoxybenzoic acid 4-[ethyl-[2-(4-methoxyphenyl)-1-methylethyl]-amino]butyl ester hydrochloride

Common Name: -

Structural Formula:

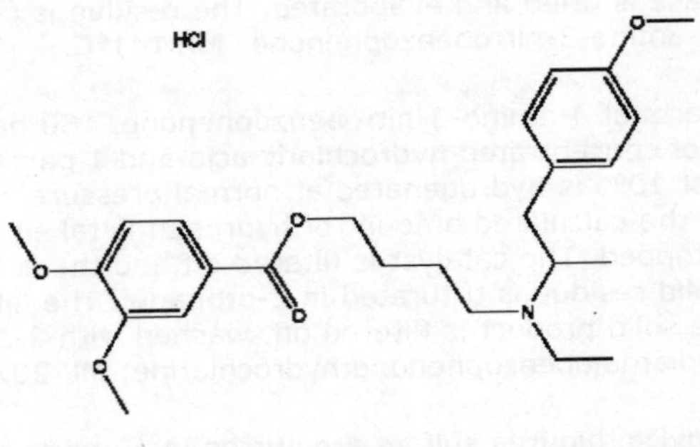

HCl

Chemical Abstracts Registry No.: 2753-45-9; 3625-06-7 (Base)

Trade Name	Manufacturer	Country	Year Introduced
Duspatalin	Duphar	France	1965
Colofac	Duphar	UK	1967
Duspatal	I.S.M.	Italy	1970
Duspatal	Thomae	W. Germany	1977
Duspatalin	Duphar	Switz.	1981

Raw Materials

Sodium	Tetramethylene dichloride
Ethanol	p-Methoxyphenyl acetone
Sodium iodide	3,4-Dimethoxybenzoic acid
Ethylamine	Hydrogen

Manufacturing Process

(A) Sodium-3,4-Dimethoxybenzoate: A solution of 91 g of 3,4-dimethoxybenzoic acid in 500 ml of boiling, absolute alcohol was added quickly to a solution of 11.5 g of sodium in 300 ml of absolute alcohol; after cooling to room temperature the resulting precipitate was filtered off and washed with 2 x 50 ml of absolute alcohol and 4 x 200 ml of ether and dried in air to constant weight; yield 92.5 g, MP about 265°C. The filtrate was bulked with the alcohol and ether washings, left to stand overnight, and a further precipitate then filtered off, washed with 3 x 100 ml of ether, and dried in air to constant weight. Yield 22.5 g, MP about 265°C. Total yield therefore 115 g (=113%).

(B) 4'-Chlorobutyl-3,4-Dimethoxybenzoate: 92 g of the sodium salt described under (A) (it contains at the most 81.5 g of sodium 3,4-dimethoxybenzoate) was boiled in 900 ml of tetramethylene dichloride for 90 hours; after cooling the mixture was filtered and the residue washed with 3 x 50 ml of ether. The filtrate was evaporated to dryness in vacuo and the residue (102 g) was distilled in vacuo. Fraction 1: 50° to 55°C/0.5 mm; 19 g (probably tetramethylene dichloride). Fraction 2: 175° to 184°C/0.5 mm; 77.5 g (=71%); Cl= 12.6% (calculated 13.0%). Remark: The second fraction partially solidified or became more viscous on standing, and even during the distillation.

(C) 4'-Iodobutyl-3,4-Dimethoxybenzoate: 32.5 g of 4'-chlorobutyl-3,4-dimethoxybenzoate and 19.5 g of sodium iodide (10% excess) were boiled in 150 ml of methyl ethyl ketone for 2.5 hours; after cooling and filtering off the sodium chloride produced, the reaction was found not to be entirely completed; boiling was then continued for another two hours; the reaction mixture was cooled, and the solid filtered off and washed with 2 x 100 ml of ether.

The filtrate was evaporated to dryness in vacuo and the residue was dissolved in 300 ml of ether and 100 ml of water; the layers were separated and the water layer was once again extracted with 100 ml of ether; then the ether layers were boiled and washed again with a solution of 3.5 g of sodium thiosulfate in 100 ml of water. The ether layer was dried over sodium sulfate. Finally the solution was filtered and the ether was evaporated; the residue was an almost colorless oil, which partially solidified or became more viscous after being left to stand for some time. Yield: 40 g (=92%), I=34.2% (calculated 34.9%).

(D) 4'-[N-Ethyl-1''-Methyl-2''-(4'''-Methoxyphenyl)Ethylamino]Butyl-3,4-Dimethoxybenzoate Hydrochloride: 10.3 g of 4'-iodobutyl-3,4-dimethoxybenzoate and 11.0 g of N-ethyl-p-methoxyphenylisopropylamine (obtained by catalytic reduction of an alcoholic solution of an excess quantity (60%) of p-methoxy-phenyl-acetone, to which was added a 33% (weight-for-weight) aqueous solution of ethylamine, with Pt as a catalyst), were boiled in 200 ml of methyl ethyl ketone for 20 hours, cooled and the iodine ion was determined; the reaction was found to be complete. Then the methyl ethyl ketone was evaporated in vacuo and the residue was dissolved in 300 ml of water and 30 ml of ether; the layers were separated and the water layer was extracted twice more with 20 ml portions of ether.

References

Merck Index 5590
Kleeman and Engel p. 537
OCDS Vol. 2 p. 54 (1980)
DOT 3 (4) 143 (1967)
I.N. p. 580
Phillips' Gloeilampenfabrieken; British Patent 1,009,082; November 3, 1965

MEBUTAMATE

Therapeutic Function: Antihypertensive

Chemical Name: 2-Methyl-2-(1-methylpropyl)-1,3-propandiol dicarbamate

Common Name: Dicamoylmethane

Structural Formula:

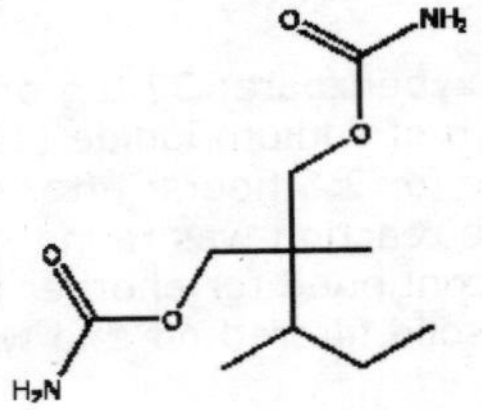

Chemical Abstracts Registry No.: 64-55-1

Trade Name	Manufacturer	Country	Year Introduced
Capla	Wallace	US	1961
Axiten	Zambon	Italy	-
Butatensin	Benvegna	Italy	-
Carbuten	Kalopharma	Italy	-
Dormate	Wallace	US	-
Ipotensivo	Vita	Italy	-
Mebutina	Formenti	Italy	-
No-Press	Janus	Italy	-
Prean	Chemil	Italy	-
Preminex	Dumex	Denmark	-

Trade Name	Manufacturer	Country	Year Introduced
Sigmafon	Lafare	Italy	-
Vallene	Simes	Italy	-

Raw Materials

Diethyl-sec-butyl methyl malonate
Lithium aluminum hydride
Ethyl urethane

Manufacturing Process

The following example illustrates the preparation of 2-methyl-2-sec-butyl-1,3-propanediol:

92 g of diethyl-sec-butyl methyl malonate were reduced in the usual manner using 22.8 g of lithium aluminum hydride in a suitable volume of anhydrous ethyl ether. The mixture was treated with 10% sulfuric acid and the ether soluble components extracted. The ether solution was dried, using a suitable drying agent, and the residue obtained by the removal of the ether was purified by distilling under reduced pressure. This material was further purified by redistillation. Approximately 46 g of 2-methyl-2-sec-butyl-1,3-propanediol were obtained as a clear colorless liquid, boiling point 92°C to 97°C (0.1 mm pressure).

The following example describes the preparation of 2-methyl-2-sec-butyl-1,3-propanediol dicarbamate using the urethane exchange method:

14.6 g of 2-methyl-2-sec-butyl-1,3-propanediol and 18.7 g ethyl urethane are dissolved in about 100 ml anhydrous toluene. 3 g of aluminum isopropylate are added and the mixture distilled to remove the ethyl alcohol formed in the condensation of ethyl urethane and the diol. The alcohol distills in the form of an azeotrope with toluene. Distillation is continued until the theoretical quantity of ethanol has been removed. The toluene is distilled from the mixture under reduced pressure and the residue dissolved in hot aqueous isopropanol solution. The hot solution is filtered and allowed to cool, whereupon approximately 14 g of product separates. The purified product represents a yield of about 60% of theoretical and melts at 77°C to 79°C.

References

Merck Index 5594
Kleeman and Engel p. 538
OCDS Vol. 1 p. 218 (1977)
I.N. p. 581
Berger, F.M. and Ludwig, B.J.; US Patent 2,878,280; March 17, 1959; assigned to Carter Products, Inc.

MECAMYLAMINE HYDROCHLORIDE

Therapeutic Function: Antihypertensive

Chemical Name: N,2,3,3-Tetramethylbicyclo[2.2.1]heptan-2-amine hydrochloride

Common Name: Dimecamin hydrochloride

Structural Formula:

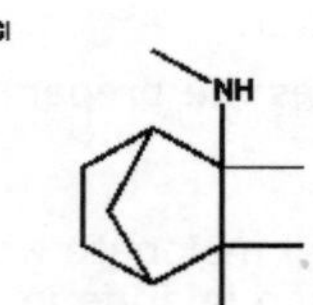

Chemical Abstracts Registry No.: 826-39-1; 60-40-2 (Base)

Trade Name	Manufacturer	Country	Year Introduced
Inversine	MSD	US	1956
Mevasine	MSD	W. Germany	-
Prexion	I.T.I.	Italy	-

Raw Materials

dl-Camphene
Sodium cyanide
Lithium aluminum hydride
Sulfuric acid

Manufacturing Process

Preparation of 2-(N-Formylamino)Isocamphane: Into a 5-liter 3-necked round bottom flask equipped with stirrer, dropping funnel and thermometer, was added 325 ml of glacial acetic acid. Then, portionwise, a total of 133 g of sodium cyanide (granular, 2.6 mols) was added with stirring while holding the temperature at 15°C. To the thick white slurry was added dropwise a previously prepared cold mixture of 325 ml glacial acetic acid and 360 ml concentrated sulfuric acid.

After addition of a few milliliters at 15°C, the thick slurry thins slowly and the remainder of the sulfuric-glacial acetic acid mixture was added at 0° to 2°C. A total of about 2 hours was required for the addition. After addition, stirring was continued for 15 minutes, Then dropwise, over an hour, a solution of 178 g (1.3 mold of dl-camphene in 50 ml of glacial acetic acid was added while keeping the temperature at about 0°C (±3°C).

Stirring was continued for two hours at 0°C during which time a slight pinkish-yellow color developed in the reaction mixture. The cooling bath was removed and the temperature allowed to rise to 15° to 20°C in about 2 to 3 hours. The ice bath was then re placed and while holding the temperature at about 20°C, the mixture was gradually diluted with 3 liters of water while stirring vigorously. After an hour or two of good agitation at room temperature, the oily product was extracted with 2 x 500 ml and 1 x 200 ml of chloroform and the combined extracts washed with 2 x 500 ml of water. The chloroform extract was then rendered neutral by stirring with 500 ml water and gradually adding solid sodium bicarbonate to the mixture until the aqueous phase had a pH of about 7; required, approximately 88 g of $NaHCO_3$.

After separation the chloroform layer was washed with 2 x 500 ml water, dried over calcium chloride, and after filtration the solvent was removed in vacuo on the steam bath. A solid somewhat sticky residue of 231.2 g was obtained. After removal of last traces of chloroform by repeated swishing with petroleum ether, the cake was finally refluxed with about 500 ml petroleum ether (BP 30° to 60°C) until a thick crystalline slurry was obtained. After refrigeration for a day, the white crystalline mass was filtered by suction, washed with petroleum ether (2 x 125 ml), then n-heptane (2 x 125 ml) and again with petroleum ether (2 x 125 ml). After air drying at room temperature to constant weight, 180.6 g of the dl-2-(N-formylamino)isocamphane melting at 160° to 165°C was obtained.

The combined petroleum ether and n-heptane washes were concentrated under diminished pressure and the residual oil dissolved in a minimum amount of hot petroleum ether (about 75 ml). The resulting solution was placed in the refrigerator for two days. The precipitated dl-2-(N-formylamino) isocamphane was then recovered by filtration and washed with petroleum ether and n-heptane as described above. Obtained, 12.6 g of product having a MP of 158° to 164°C.

The dl-2-(N-formylamino)isocamphane (193 g) was dissolved in 1.9 liters n-heptane by heating on a steam bath, After clarifying the solution by filtration, the clear filtrate was allowed to stand at room temperature until crystallization was complete. The crystalline product is filtered by suction, washed with a little cold n-heptane and air dried. The dl-2-(N-formylamino)isocamphane melted at 169° to 174°C.

Preparation of 2-(N-Methylamino)Isocamphane: To 4.23 liters of anhydrous ether in a 12-liter 3-necked flask fitted with a stirrer, reflux condenser and dropping funnel was quickly added 78 g (2.05 mols) of lithium aluminum hydride. The mixture was gently refluxed with stirring until all hydride had dissolved which required several hours.

A solution of 168 g (0.92 mol) of dl-2-(N-formylamino)isocamphane, prepared as described above, in 1.81 liters of anhydrous ether was then added during a period of about one hour with stirring. After addition, the mixture was refluxed for about 6 hours after which it was cooled slightly and 347 ml of water added with stirring, hydrogen gas being evolved during the addition, Stirring was continued until the precipitate changed to a powder, which was filtered by suction and washed with ether (a total of about 2 liters).

The combined filtrate and washes were concentrated to 1.6 liters and the concentrate containing the dl-2-(N-methylamino)isocamphane washed once with about 350 cc water, and then dried over anhydrous sodium sulfate. The dried ether concentrate was then cooled in an ice bath and with stirring a cold saturated ethereal-hydrogen chloride solution was added slowly until acid to Congo red; required, about 440 ml anhydrous ether saturated (at 0°C) with HCl gas, After precipitation was complete, the white crystalline dl-2-(N-methylamino)isocamphane hydrochloride was filtered, and washed with anhydrous ether (about 1 liter) until the washes were neutral. The dl-2-(N-methylamino)isocamphane hydrochloride was air dried at room temperature. Obtained, 156.5 g of product melting with decomposition at 249°C.

References

Merck Index 5595
Kleeman and Engal p. 538
I.N. p. 581
REM p.849
Pfister, K., Ill and Stein, G.A.; US Patent 2,831,027; April 15, 1958; assigned to Merck and Co., Inc.

MECILLINAM

Therapeutic Function: Antibacterial

Chemical Name: 6-[[(Hexahydro-1H-azepin-1-yl)methylene]amino]-3,3-dimethyl-7-oxo-4-thia-1-azabicyclo[3.2.0]heptane-2-carboxylic acid

Common Name: Amdinocillin

Structural Formula:

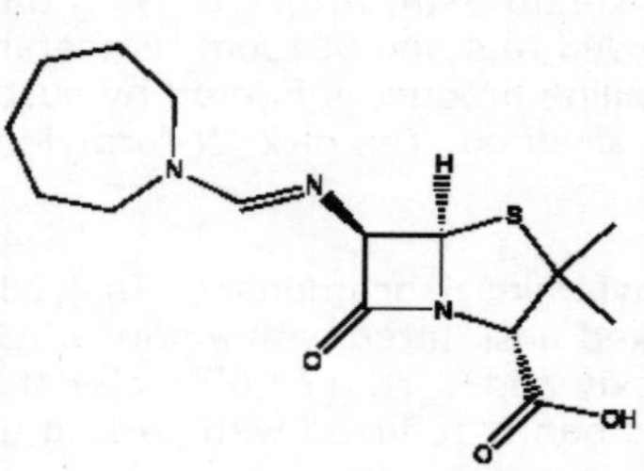

Chemical Abstracts Registry No.: 32887-01-7

Trade Name	Manufacturer	Country	Year Introduced
Selexidin	Leo	UK	1979
Celfuron	Roche	-	-

Raw Materials

Chloral
Hexamethyleneimine

Oxalyl chloride
Trimethylsilyl 6-aminopenicillinate

Manufacturing Process

The starting material N-formylhexamethyleneimine was prepared from hexamethyleneimine and chloral.

12.7 g of N-formylhexamethyleneimine were dissolved in 250 ml of dry ether. While stirring and cooling, 8.5 ml of oxalyl chloride in 50 ml of dry ether were added dropwise, whereafter the mixture was stirred overnight at room temperature. The precipitated amide chloride was filtered off and washed with dry ether, and was placed in an exsiccator.

A solution of the amide chloride (4.6 g) in dry, alcohol-free chloroform (20 ml) was added slowly to a solution of trimethylsilyl 6-amino-penicillanate (7.2 g) and triethylamine (3.5 ml) in dry, alcohol-free chloroform (50 ml) with stirring and cooling to -70°C. The temperature was raised to 0°C during 1.5 hours. The solution was evaporated to dryness in vacuo and the residue was triturated with dry ether (200 ml). The precipitate was filtered off and washed with dry ether. The filtrate was diluted with ether (200 ml). 2-Butanol (2.8 ml) was added dropwise with stirring and cooling to 0°C. The stirring was continued for 1/4 hour at 0°C, whereupon the precipitate was filtered off, washed with ether and dried. It was a white, amorphous powder, soluble in water.

References

Merck Index 390
Kleeman and Engel p. 539
OCDS Vol. 3 p. 208 (1984)
DOT 11 (11), 489 (1975) and 16 (6) 193 (1980)
I.N. p. 582
REM p. 1201
Lund, F.J.; British Patent 1,293,590; October 18, 1972; and US Patent 3,957,764; May 18, 1976; both assigned to Lovens Kemiske Fabrik Produktionsakties Lab (Denmark)

MECLIZINE HYDROCHLORIDE

Therapeutic Function: Antinauseant

Chemical Name: 1-[(4-Chlorophenyl)phenylmethyl]-4-[(3-methylphenyl) methyl]piperazine hydrochloride

Common Name: Meclozin; Histamethizine

Chemical Abstracts Registry No.: 1104-22-9; 569-65-3 (Base)

Structural Formula:

Trade Name	**Manufacturer**	**Country**	**Year Introduced**
Antivert	Roerig	US	1957
Ru-Vert M	Reid-Provident	US	1983
Ancolan	Duncan Flockhart	UK	-
Bonamine	Pfizer	W. Germany	-
Calmonal	Heyden	W. Germany	-
Chiclida	Torpens	Spain	-
Diadril	Pliva	Yugoslavia	-
Duremesan	Streuli	Switz.	-
Itinerol	Galenika	Switz.	-
Mecazine	Barlow Cote	Canada	-
Navicalur	Delagrange	France	-
Peremesin	Heyden	W. Germany	-
Postafen	U.C.B.	W. Germany	-
Supermesin	M.P.Q.	Spain	-
Suprimal	A.C.F.	Netherlands	-
Taizer	Pfizer Taito	Japan	-
V-Cline	Vangard	US	-
Veritab	Vista	US	-
Vertizine	Merchant	US	-

Raw Materials

1-p-Chlorobenzhydryl-4-benzyl-piperazine
Hydrogen
Sodium amide
m-Methyl benzyl chloride

Manufacturing Process

32.3 g of 1-p-chlorobenzhydryl-4-benzyl-piperazine, dissolved in 300 cm3 of alcohol are heated in an autoclave vessel, in the presence of Raney nickel, under a pressure of 100 kg H_2, at about 150°C for 6 hours. The catalyst is filtered, the solvent is evaporated and the residue is fractionated under a high vacuum. p-Chlorobenzylhydryl-piperazine (BP 180° to 185°C/1 mm Hg) is isolated with a yield of 75%. Then finely ground $NaNH_2$ is added. The mixture is heated under reflux for 1 hour, the mass is cooled and a molar equivalent of m-methyl benzyl chloride is added.

The solvent is evaporated and the residue is dissolved in chloroform. This solution is washed with a saturated solution of K_2CO_3 and dried on K_2CO_3. The solvent is evaporated and the residue is distilled under high vacuum. The product of the condensation distills near 230°C at 2 mm Hg pressure and the corresponding dihydrochloride melts at 217° to 224°C.

References

Merck Index 5598
Kleeman and Engel p. 540
PDR pp. 993, 1403, 1449, 1520, 1606, 1999
OCDS Vol. 1 p. 59 (1977)
I.N. p. 583
REM p. 808
Morren, H.; US Patent 2,709,169; May 24, 1955; assigned to Union Chimique Belge Societe Anonyme, Belgium

MECLOFENAMIC ACID

Therapeutic Function: Antiinflammatory

Chemical Name: N-(2,6-Dichloro-3-methylphenyl)anthranilic acid

Common Name:-

Structural Formula:

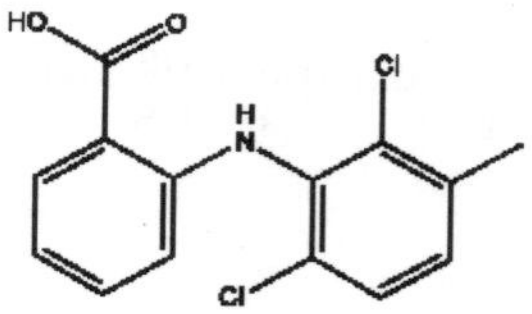

Chemical Abstracts Registry No.: 644-62-2; 6385-02-0 (Sodium Salt)

Trade Name	Manufacturer	Country	Year Introduced
Meclomen	Warner Lambert	US	1980
Meclomen	Parke Davis	Switz.	1982
Arquel	Parke Davis	-	--

Raw Materials

Potassium o-bromobenzoate
2,6-Dichloro-3-methylaniline
N-Ethylmorpholine

Manufacturing Process

A mixture consisting of 22.7 g potassium o-bromobenzoate, 16.6 g 2,6-dichloro-3-methylaniline, 12 ml N-ethylmorpholine, 60 ml diethylene glycol dimethyl ether, and 1.0 g anhydrous cupric bromide is heated in a nitrogen atmosphere at 145°C to 155°C for 2 hours. The reaction mixture is diluted with 60 ml diethylene glycol dimethyl ether and acidified with 25 ml concentrated hydrochloric acid. The acidic mixture is diluted with 100 ml of water and the liquid phase decanted from the insoluble oil. The insoluble oil is stirred with methanol and the crystalline N-(2,6-dichloro-3-methylphenyl) anthranilic acid which separates is collected and washed with methanol. The product, after recrystallization from acetone-water mixture, melts at 248°C to 250°C.

References

Merck Index 5600
DFU 3 (4) 307 (1978)
Kleeman and Engel p. 539
PDR p. 1366
OCDS Vol. 1 p. 110 (1977) and 2,88 (1980)
DOT 17 (6) 250 (1981)
I.N. p. 31
REM p. 1118
Scherrer, R.A. and Short, F.W.; US Patent 3,313,848; April 11, 1967; assigned to Parke-Davis and Co.

MEDAZEPAM

Therapeutic Function: Tranquilizer

Chemical Name: 7-Chloro-2,3-dihydro-1-methyl-5-phenyl-1H-1,4-benzodiazepine

Common Name: -

Structural Formula:

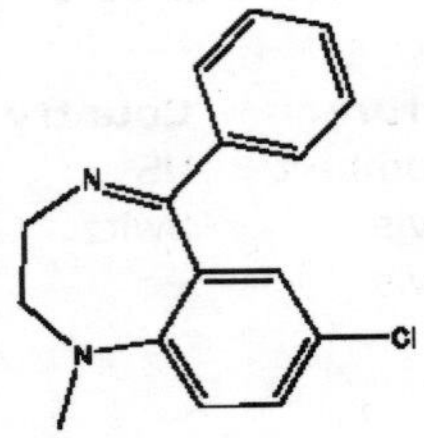

Chemical Abstracts Registry No.: 2898-12-6

Trade Name	Manufacturer	Country	Year Introduced
Nobrium	Roche	Italy	1969
Nobrium	Roche	W. Germany	1969
Nobrium	Roche	France	1970
Lesmit	Shionogi	Japan	1971
Nobrium	Roche	Japan	1971
Nobrium	Roche	UK	1971
Azepamid	Taiyo	Japan	-
Becamedic	Nemi	Argentina	-
Benson	Farber-R.E.F.	Italy	-
Cerase	Torii	Japan	-
Diepin	Biosintetica	Brazil	-
Enobrin	I.E. Kimya Evi	Turkey	-
Esmail	Richter	Mexico	-
Glorium	Teva	Israel	-
Kobazepam	Nihon Yakuhin	Japan	-
Lerisum	Poli	Italy	-
Medaurin	Isis	Yugoslavia	-
Megasedan	Andreu	Spain	-
Metonas	Kanto	Japan	-
Mezepan	Hosbon	Brazil	-
Narsis	Sumitomo	Japan	-
Nivelton	Lemonier	Argentina	-
Nobraskin	Fako	Turkey	-
Nobral	Nobel	Turkey	-
Pazital	Andromaco	Spain	-
Psiquium	Sintofarma	Brazil	-
Rudotel	Arzneimittelwerk Dresden	E. Germany	-
Sedepam	Sawai	Japan	-
Serenium	Richter	Brazil	-
Tranqulax	Hokuriku	Japan	-
Vegatar	Orion	Finland	-

Raw Materials

Calcium carbonate
Sodium hydroxide
Oxalic acid
Acetic anhydride
5-Chloro-N-methylanthranilic acid
Bromoethylamine hydrobromide
Phosphorus oxychloride
Bromobenzene magnesium

Manufacturing Process

(A) Preparation of 4-Acetyl-7-Chloro-1,2,3,4-Tetrahydro-1-Methyl-5H-1,4-Benzodiazepin-5-one: A mixture of 68.5 g (0.37 mol) of 5-chloro-N-methylanthranilic acid, 51 g (0.51 mol) of calcium carbonate, 76 g (0.37 mol) of bromoethylamine hydrobromide and 2.5 liters of water was stirred and heated under reflux for 3 hours. A solution of 23.4 g (0.26 mol) of anhydrous oxalic acid in 250 ml of water was slowly added to the refluxing mixture. The precipitated calcium oxalate was filtered off, and the filtrate adjusted to pH 7 with concentrated ammonium hydroxide. The filtrate was then concentrated to dryness in vacuo and the residue heated on the steam bath with 400 ml of 6 N ethanolic hydrogen chloride until the residue was crystalline. Filtration gave 122 g of N-(aminoethyl)-5-chloro-N-methylanthranilic acid hydrochloride as a solid.

A mixture of 100 g of this solid and 1 liter of acetic anhydride was stirred and heated under reflux for 1.5 hours and then allowed to stand for 18 hours at room temperature. The excess acetic anhydride was removed in vacuo, and the residue was treated with one liter of water and ice and sufficient sodium bicarbonate to make neutral. The solid was collected, sucked dry on the filter, and triturated with hot ethanol. The ethanol solution on cooling gave 30.8 g of 4-acetyl-7-chloro-1,2,3,4-tetrahydro-1-methyl-5H-1,4-benzodiazepin-5-one.

(B) Preparation of 7-Chloro-1,2,3,4-Tetrahydro-1-Methyl-5H-1,4-Benzodiazepin-5-one: A mixture of 25.25 g (0.1 mol) of 4-acetyl-7-chloro-1,2,3,4-tetrahydro-1-methyl-5H-1,4-benzodiazepin-5-one, 33.3 ml (0.1 mol) of 3 N sodium hydroxide and 350 ml of ethanol was heated under reflux for 15 minutes and then concentrated to dryness in vacuo. The residue was treated with 500 ml of water, collected and washed with ethanol to give 20.2 g of 7-chloro-1,2,3,4-tetrahydro-1-methyl-5H-1,4-benzodiazepin-5-one.

(C) Preparation of 7-Chloro-2,3-Dihydro-1-Methyl-5-Phenyl-1H-1,4-Benzodiazepine: A mixture of 4.7 g (22.6 mol) of 7-chloro-1,2,3,4-tetrahydro-1-methyl-5H-1,4-benzodiazepin-5-one and 100 ml of phosphorus oxychloride was heated in an oil bath at 100°C for 15 minutes. The solution was concentrated to dryness in vacuo. The residue was partitioned between methylene chloride and cold saturated sodium bicarbonate solution. The methylene chloride phase was dried over sodium sulfate and sodium bicarbonate, filtered, diluted with benzene and concentrated in vacuo to produce crude 5,7-dichloro-2,3-dihydro-1-methyl-1H-1,4-benzodiazepine.

The residue was dissolved in 75 ml of tetrahydrofuran, treated with charcoal, and sodium sulfate and filtered. This solution was added to a solution in 250 ml of tetrahydrofuran of phenyl magnesium bromide prepared from 17.7 ml (0.17 mol) of bromobenzene. This mixture was stirred and heated under reflux for 1 hour. It was then cooled and diluted with 400 ml of ether and sufficient 3 N hydrochloric acid to make it acidic. The aqueous phase was separated, adjusted to pH 8 with 3 N sodium hydroxide and extracted 3 times with 200 ml of ether. The ether extracts were combined, washed with water and dried over sodium sulfate. The residue left on removal of the ether in vacuo was crystallized from petroleum ether to give 3.3 g of 7-chloro-2,3-dihydro-1-methyl-5-phenyl-1H-1,4-benzodiazepine, according to US Patent 3,624,703.

A variety of alternative routes are outlined by Kleeman and Engel.

References

Merck Index 5609
Kleeman and Engel p. 542
OCDS Vol. 1 p. 368 (1977)
DOT 5 (4) 150 (1969) and 9 (6) 238 (1973)
I.N. p. 584

Reeder, E. and Sternbach, L.H.; US Patent 3,109,843; November 5, 1963; assigned to Hoffmann-LaRoche Inc.
Archer, G.A. and Sternbach, L.H.; US Patent 3,131,178; April 28, 1964; assigned to Hoffmann-LaRoche Inc.
Reeder, E. and Sternbach, L.H.; US Patent 3,141,890; July 21, 1964; assigned to Hoffmann-LaRoche Inc.
Reeder, E. and Sternbach, L.H.; US Patent 3,144,439; August 11, 1964; assigned to Hoffmann-LaRoche Inc.
Field, G.F., Sternbach, L.H. and Zally, W.J.; US Patent 3,624,073; November 30, 1971; assigned to Hoffmann-LaRoche Inc.

MEDIBAZINE DIHYDROCHLORIDE

Therapeutic Function: Coronary vasodilator

Chemical Name: Piperazine, 1-(1,3-benzodioxol-5-ylmethyl)-4-(diphenylmethyl)-, dihydrochloride

Common Name: Medibazine dihydrochloride

Structural Formula:

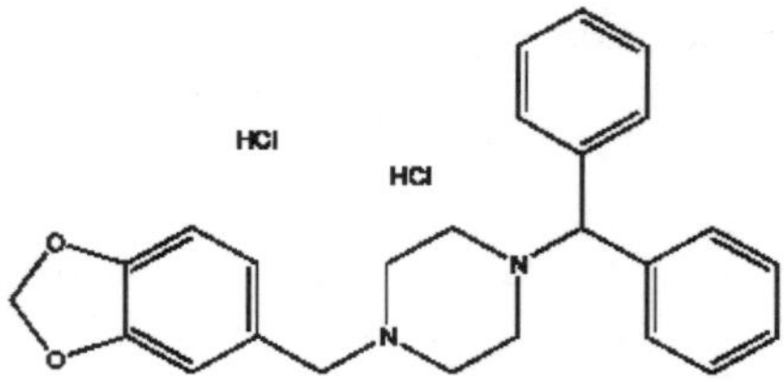

HCl HCl

Chemical Abstracts Registry No.: 96588-03-3; 53-31-6 (Base)

Trade Name	Manufacturer	Country	Year Introduced
Vialibran	Servier	-	-

Raw Materials

Piperonyl-1-piperazine
Methanesulfonic acid
Benzhydryl chloride

Manufacturing Process

To a solution of 32 g piperonyl piperazine in 100 ml anhydrous toluene, 10 g sodium carbonate are added and 35.2 g benzhydryl chloride are added dropwise. The mixture is then heated to reflux for 7 hours with vigorous agitation. Then the mixture is cooled, the salt that has formed is filtered out and 100 ml water is added. The organic layer is extracted with several batches of 10% methane sulfonic acid. The acid extracts are combined and washed with ether then alkalized with sodium carbonate. The mixture is extracted with several batches of chloroform and the combined chloroform solutions are washed several times with water. After drying and solvent evaporation, the crude base of 1-diphenylmethyl-4-piperonyl-piperazineis isolated and the hydrochloride thereof is formed in acetone. After recrystallization 22.5 g of the dihydrochloride are finally obtained. M.P.: 228°C, from methanol.

References

Regnier G. et al.; US Patent No. 3,119,826; Jan. 28, 1964; Assigned to Societe en nom collectif dite: Science Union et Compagnie-Societe Francaise de Recherche Medicale, Suresnes, Seine, France, a French society

MEDIGOXIN

Therapeutic Function: Cardiotonic

Chemical Name: 3β,12β,14β-Trihydroxy-5β-card-20(22)-enolide-3-(4'''-o-methyltridigitoxoside)

Common Name: β-Methyldigoxin

Structural Formula:

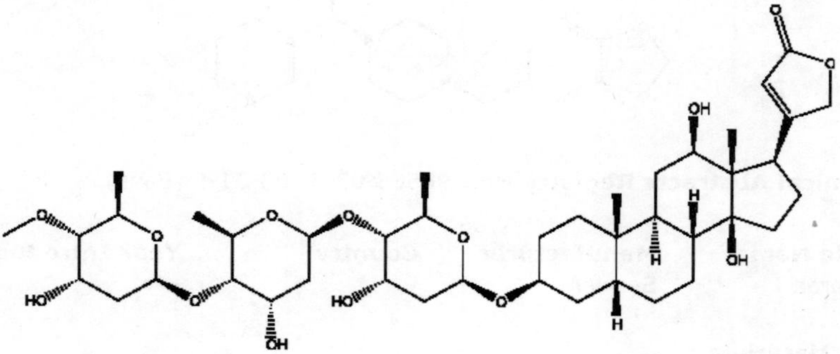

Chemical Abstracts Registry No.: 30685-43-9

Trade Name	Manufacturer	Country	Year Introduced
Lanitop	Boehringer Mannheim	W. Germany	1972
Lanitop	Boehringer Mannheim	Italy	1973
Lanitop	Roussel	UK	1976
Lanirapid	Yamanouchi	Japan	1979
Cardiolan	Tosi-Novara	Italy	-
Digicor	Lek	Yugoslavia	-
Intensain-Lanitop	Boehringer Mannheim	W. Germany	-

Raw Materials

Digoxin
Methyl mesylate

Manufacturing Process

Digoxin (10 g) is dissolved in a mixture of dimethylformamide (80 ml) and dioxane (80 ml) and then strontium hydroxide (3.5 g) and aluminum oxide (10 g, activity 1-2 according to Brockmann) are added. To this suspension methyl mesylate (9.3 g), dissolved in dioxane (80 ml) is added dropwise within one hour in the presence of an inert gas and under stirring. After the addition of the methylating agent is completed, the reaction mixture is stirred for further 5 hours, then chloroform (160 ml) is added, the precipitate is filtered off, washed with chloroform (100 ml), pyridine (40 ml) is added to the filtrate, which is then concentrated in vacuo to an oily residue. The latter is diluted with chloroform (300 ml) and extracted four times with distilled water (40 ml portions). The combined chloroform extracts are dried with anhydrous sodium sulfate and then concentrated in vacuo to a dry residue. Therefrom β-methyldigoxin is eluted on a SiO_2 column with a chloroformlethanol mixture (93:7). After recrystallization from ethyl acetate, saturated with water, the yield of β-methyldigoxin is 6.7 g; MP 225°C to 229°C. IR spectrum is identical with the spectrum of standard methyldigoxin.

References

Merck Index 3148
Kleeman and Engel p. 544
DOT 12 (8) 319, 323 (1976)
I.N.p.627
Pelan, E., Milohnoja, M. and Pezdirc, M.; US Patent 4,145,528; March 20, 1979; assigned to L.E.K. Tovarna Farmacevtskih in Kemicnih Izdelkov (Yugoslavia)

MEDROGESTONE

Therapeutic Function: Progestin

Chemical Name: 6,17-Dimethylpregna-4,6-diene-3,20-dione

Common Name: 6,17α-Dimethyl-6-dehydroprogesterone

Chemical Abstracts Registry No.: 977-79-7

Trade Name	Manufacturer	Country	Year Introduced
Colpro	Ayerst	Italy	1970
Colprone	Auclair	France	1972
Prothil	Kali-Ghemie	W. Germany	1975
Colpron	Arcana	Austria	-

Structural Formula:

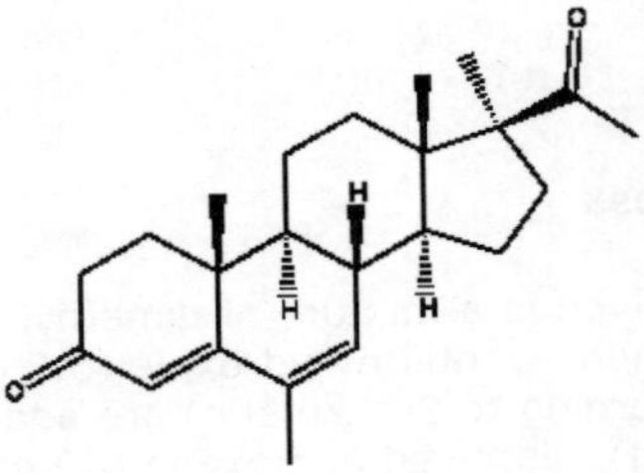

Raw Materials

Chromic acid	17α-Methyl-17β-carbomethoxyandrost-5-ene-3β-ol
Acetic anhydride	N-Bromosuccinimide
Hydrogen peroxide	Methyl magnesium bromide

Manufacturing Process

The manufacturing process as described in US Patent 3,170,936 uses the readily available methyl 3β-hydroxy-17α-methyl-δ^5-etienate (I), described by Plattner in Helv. Chim. Acta, vol. 31, p 603 (1948), as the starting material. The etienic acid ester (I) may also be called 17α-methyl-17β-carbomethoxyandrost-5-ene-3β-ol.

3β,5α,6β-Trihydroxy-17α-Methyl-17β-Carbomethoxyandrostane (II): 5 g of 17α-Methyl-17β-carbomethoxyandrost-5-ene-3β-ol (I) is dissolved in formic acid (50 ml) and heated on the steam bath for 10 minutes. The solution is cooled to room temperature and a crystalline solid precipitates. This is stirred, 30% hydrogen peroxide (5 ml) is added, and the reaction mixture is left at room temperature for 2 hours. The clear solution is poured into water 1300 ml) and the solid which precipitates is filtered.

It is dissolved in hot methanol and heated on the steam bath with 10% methanolic potassium hydroxide solution (15.8 ml) for 10 minutes. Then more potassium hydroxide solution (2 ml) is added, the solution is cooled and on dilution with water a solid (II), MP 245° to 255°C, is obtained. A second crop is obtained from the mother liquors. Several recrystallizations from acetone yield an analytical sample, MP 262° to 265°C, $[\alpha]_D^{24}$ is -2.1°.

3β-Acetoxy-5α-Hydroxy-17α-Methyl-17β-Carbomethoxyandrostane-6-one (IIIb): 3β,5α,6βp-Trihydroxy-17α-methyl-17β-carbomethoxyandrostane (II, 5.2 g) is dissolved in methanol (105 ml) to which ether (105 ml) and water (84 ml) are added. Then N-bromosuccinimide (5.2 g) is added with stirring and the clear solution is left in the refrigerator for 3 hours. The ether is removed under reduced pressure at room temperature and a crystalline solid (IIIa) separates, MP 268° to 272°C.

The above substance is dissolved in pyridine (15 ml) and acetic anhydride (7.5 ml), and heated on the steam bath for ½ hour. The product (IIIb) crystallizes from aqueous ethanol in leaflets, MP 237° to 239°C. An analytical sample has MP 241° to 243°C.

3β,5α,6β-Trihydroxy-6α,17α-Dimethyl-17β-Carbomethoxyandrostane (IV): 3β-Acetoxy-5α-hydroxy-17α-methyl-17β-carbomethoxyandrostan-6-one (III, 1.004 g) is dissolved in dry benzene (25 ml) and methyl magnesium bromide solution in ether (3 M, 10 ml) is added. The reaction mixture is diluted with dry tetrahydrofuran (25 ml) and allowed to stand at room temperature for 20 hours, Excess Grignard reagent is quenched by adding a saturated solution of ammonium chloride. The organic layer is separated and the aqueous layer is extracted with ethyl acetate.

After washing the combined extracts with ammonium chloride solution and water and working up in the usual way a white solid (IV) is obtained which after one recrystallization from aqueous methanol has MP 242° to 243°C. The infrared spectrum of this compound indicates the presence of a carbomethoxy group (1,730 cm^{-1}) and disappearance of the 6-keto group together with the presence of an ester group (1,727 cm^{-1}). This substance is used without further purification for the next step.

3β,5α,6β-Trihydroxy-6α,17α-Dimethylpregnan-20-one (V): Crude 3β,5α,6β-trihydroxy-6α,17α-dimethyl-17β-carbomethoxyandrostane (IV, 773 mg) is dissolved in dry benzene (25 ml) and tetrahydrofuran (freshly distilled over lithium aluminum hydride, 25 ml). To the stirred solution under dry N_2 there is added methyl magnesium bromide solution in ether (3 M, 10 ml) over a period of 10 minutes. Then the ether and tetrahydrofuran are almost all distilled and the resulting solution is refluxed for 3 hours (solid precipitates during the reaction). The reaction mixture is cooled and worked up in the same way as in the previous experiment leaving a white solid (V) with an infrared spectrum which indicates the presence of a 20-ketone group (1,690 cm^{-1}), a sample of which is recrystallized to MP 238° to 240°C.

Analysis confirmed the empirical formula $C_{23}H_{38}O_4H_2O$: Required: C, 69.60%; H, 10.17%. Found: C, 69.90%; H, 10.15%.

Alternatively, 25.0 g of either 3β,5α-dihydroxy-17α-methy-17β-carbomethoxyandrostan-6-one (IIIa) or 25.0 g of its 3β-acetate (IIIb), are dissolved in dry tetrahydrofuran (1,250 ml, freshly distilled over lithium aluminum hydride) and dry benzene (2,000 ml) is added. Methyl magnesium bromide in ether solution (3 M, 750 ml) is added to the stirred solution and the resulting mixture is stirred at room temperature for 16 hours. An additional quantity of methyl magnesium bromide solution in ether (2 M, 375 ml) is added, and 1,250 ml of the solvent mixture are distilled off. The resulting mixture is refluxed for 5 hours and worked up as described above, yielding compound (V) as a colorless oil.

5α,6β-Dihydroxy-6α,17α-Dimethylpregnane-3,20-dione (VI): Crude 3β,5α,6β-trihydroxy-6α,17α-Dimethylpregnan-20-one (V, 650 mg) is dissolved in acetone (freshly distilled over potassium permanganate, 150 ml) and cooled in an ice-water bath with stirring. Then excess chromic acid solution (8 N) is added and stirring is continued at room temperature for 4 minutes. The reaction mixture is poured into water and extracted with ethyl acetate. The combined extracts are washed with dilute sodium bicarbonate solution and water and then dried over magnesium sulfate. Removal of the solvent leaves a white solid (VI). This crude product is used for the next step. Its IR spectrum shows a strong band at 1,705 cm^{-1}. A sample is recrystallized to MP 243° to 245°C (dec.).

6,17α-Dimethyl-4,6-Pregnadiene-3,20-dione (VII): 5α,6β-Dihydroxy-6α,17α-dimethylpregnane-3,20-dione (VI, 553 mg) is dissolved in absolute ethanol (60 ml) and two drops of concentrated hydrochloric acid are added. This solution is heated on a steam bath for 45 minutes, cooled, diluted with water and extracted with ether. The combined extracts are washed with dilute sodium bicarbonate solution and water and subsequently dried over magnesium sulfate. After the solvent has been removed a syrup remains and the UV spectrum of this substance indicates the presence of a $\delta^{4,6}$-ketone.Elution of this material over alumina (Woelm, Grade III, 25 g) with 1:1 hexane-benzene gives a crystalline substance, MP 138° to 141°C which, after one recrystallization from ether, has an infrared spectrum identical to that of an authentic sample of 6,17α-dimethyl-4,6-pregnadiene-3,20-dione (VII).

References

Merck Index 5613
Kleeman and Engel p. 545
OCDS Vol. 1 p. 182 (1977)
I.N. p. 586
Deghenghi, R.; US Patent 3,133,913; May 19, 1964; assigned to American Home Products Corporation
Morand, P.F. and Deghenghi, R.; US Patent 3,170,936; February 23, 1965; assigned to American Home Products Corporation
Deghenghi, R.; US Patent 3,210,387; October 5, 1965; assigned to American Home Products Corporation

MEDROXYPROGESTERONE ACETATE

Therapeutic Function: Progestin

Chemical Name: 17-Acetoxy-6α-methyl-pregn-4-ene-3,20-dione

Common Name: 6α-Methyl-17α-acetoxyprogesterone

Structural Formula:

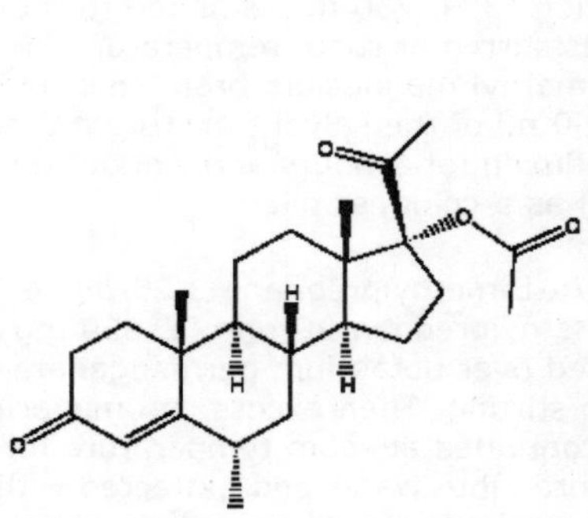

Chemical Abstracts Registry No.: 71-58-9; 520-85-4 (Base)

Trade Name	Manufacturer	Country	Year Introduced
Provera	Upjohn	US	1959
Farlutal	Carlo Erba	France	1962
Provest	Upjohn	US	1964
Amen	Carnrick	US	1975
Unison	Reid-Provident	US	1978
Mepred	Savage	US	1978
Curretab	Reid-Provident	US	1979
Farlutal	Carlo Erba	UK	1982
Depcorlutin	O'Neal, Jones and Feldman	US	-
Depo-Clinovir	Upjohn	W. Germany	-
Depo-Progevera	Alter	Spain	-
Depo-Provera	Upjohn	US	-
Gestapuran	Lovens	Denmark	-
Hysron	Kyowa	Japan	-
Luteocrin	Richter	Italy	-
Luteodione	Panther-Osfa	Italy	-
Luteos	Ion	Italy	-
Lutopolar	Farmos	Finland	-
Lutoral	Midy	Italy	-
Metilgestene	Farmila	Italy	-
Nadigest	Streuli	Switz.	-
Oragest	Ikapharm	Israel	-
Petogen	Petersen	S. Africa	-
P-Medrate	Tutag	US	-
Progevera	Alter	Spain	-
Sodelut G	Sodex	Switz.	-

Raw Materials

Sulfuric acid
Ethylene glycol
Peracetic acid
17α-Hydroxyprogesterone
Methyl magnesium bromide
Acetic anhydride

Manufacturing Process

Preparation of 17α-Hydroxyprogesterone 3,20-Bis-(Ethylene Ketal): A solution was prepared containing 50.0 g of 17α-hydroxyprogesterone in 1,000 ml of benzene, 100 ml of ethylene glycol and 2.5 g of p-toluenesulfonic acid monohydrate. This mixture was refluxed for a period of 17 hours using a calcium carbide water-trap to remove the water formed in the reaction. After this period of reflux 6.5 ml of pyridine was added to the solution, and the mixture cooled to room temperature.

The lower glycol layer was separated and washed with benzene. The benzene layer and the benzene washings were combined and the combined solution was divided into two equal portions, one of which was used for the isolation of 17α-hydroxyprogesterone 3,20-bis-(ethylene ketal) as follows. The benzene solution was washed with 5% sodium carbonate solution, water and saturated sodium chloride solution. After being dried over anhydrous magnesium sulfate the solution was concentrated to dryness at reduced pressure, The residue was recrystallized by taking up in hot methylene chloride, adding acetone and boiling to remove the methylene chloride until a final volume of about 200 ml was reached.

The solution was then refrigerated overnight and 17.8 g of crystals were removed by filtration. A second crop was obtained yielding 3.7 g of compound. The total yield of 17α-hydroxyprogesterone 3,20-bis-(ethylene ketal) was 20.3 g (64.3% of theory). Recrystallization of the crude 17α-hydroxyprogesterone 3,20-bis-(ethylene ketal) from methanol gave the pure bisketal of MP 209° to 211°C.

Preparation of 5α,6α-Oxido-17α-Hydroxyallopregnane-3,20-dione 3,20-Bis-(Ethylene Ketal): A solution was prepared by heating 19.96 g (0.0477 mol) of 17α-hydroxyprogesterone 3,20-bis-(ethylene ketal) and 500 ml of benzene. After the solution was effected the flask was cooled to 5°C and a mixture of 3.68 g (0.0449 mol) of sodium acetate and 174 ml of 40% peracetic acid was added with stirring. The reaction mixture was stirred in the ice bath for 3 hours. The lower peracid layer was separated, diluted with water and extracted twice with benzene.

The upper layer was neutralized by the addition of cold 10% sodium hydroxide solution while stirring in an ice bath. The rate of addition of the sodium hydroxide was regulated to keep the temperature below 10°C. The benzene extracts from the peracid layer were combined and washed with cold 10% sodium hydroxide solution and with saturated sodium chloride solution. All the aqueous layers were washed again with the same portion of benzene. The combined benzene layers were dried over anhydrous magnesium sulfate and concentrated to dryness at reduced pressure.

The residue was recrystallized from acetone using methylene chloride to aid in solution. The crystalline material was removed by filtration and was recrystallized from methylene chloride-acetone to yield a total of 8 g of 5α,6α-oxido-17α-hydroxyallopregnane-3,20-dione 3,20-bis-(ethylene ketal) of MP 211° to 215°C. For analytical purposes, another recrystallization from methylene chloride-acetone gave pure 5α,6α-oxido-17α-hydroxyallopregnane-3,20-dione 3,20-bis-(ethylene ketal) of MP 216° to 218.5°C.

Preparation of 5α,17α-Dihydroxy-6β-Methylallopregnane-3,20-dione 3,20-bis-(Ethylene Ketal): To a solution of 91.6 g of 5α,6α-oxido-17α-hydroxyallopregnane-3,20-dione 3,20-bis-(ethylene ketal) in 3,500 ml of freshly distilled tetrahydrofuran was added 1,170 ml of commercial 3 molar methyl magnesium bromide in ether solution. The reaction mixture was boiled to remove 1,800 ml of solvent by distillation and thereafter 1,000 ml of freshly distilled tetrahydrofuran was added.

Boiling was continued under reflux for a period of 16 hours. The solution was then concentrated to about one-half its original volume by distillation and was poured slowly with vigorous stirring into a large volume of ice water containing 340 g of ammonium chloride. The aqueous solution was saturated with sodium chloride and extracted with benzene. The benzene extract was washed with saturated brine, and both aqueous layers were washed again with the same portions of benzene.

The combined benzene layers were dried over anhydrous sodium carbonate and the solvent was removed at reduced pressure to give 90.5 g of crude crystalline 5α,17α-dihydroxy-6β-methylallopregnane-3,20-dione 3,20-bis-(ethylene ketal). Half of the residue, 45.2 g, was recrystallized from acetone and some methylene chloride to give 34.4 g of 5α,17α-dihydroxy-6β-methylallopregnane-3,20-dione 3,20-bis-(ethylene ketal). A sample recrystallized from acetone and methylene chloride for analysis melted at 160° to 163°C.

Preparation of 5α,17α-Dihydroxy-6β-Methylallopregnane-3,20-dione: A solution was prepared containing 38.9 g of 5α,7α-dihydroxy-6β-methylallopregnane-3,20-dione 3,20-bis-(ethylene ketal) in 389 ml of boiling acetone. Thereto was added 39 ml of 1 N sulfuric acid in portions under swirling and seeding with product. Boiling was continued for a period of 2 minutes and the mixture was allowed to stand at room temperature. Thereafter the mixture was diluted with 1,500 ml of water, chilled and filtered.

The precipitate was washed with water, dilute ammonium hydroxide and water, and dried in a vacuum oven overnight. The yield was 31.2 g which was recrystallized by dissolving in 1,200 ml of dimethylformamide, heating to 150°C, cooling slightly, and adding 12 ml of hot water. The recrystallized 5α,17α-dihydroxy-6β-methylallopregnane-3,20-dione thus obtained was 28.75 g of MP 270° to 275.5°C. After an additional recrystallization from aqueous dimethylformamide, the MP was 274° to 279°C.

Preparation of 6α-Methyl-17α-Hydroxyprogesterone: A suspension was made by introducing 2 g of 5α,17α-dihydroxy-6β-methylallopregnane-3,20-dione into 200 ml of chloroform. The suspension was chilled in an ice bath with stirring, and thereupon hydrogen chloride was bubbled through the reaction mixture for 80 minutes with continuous cooling and stirring. After bubbling in nitrogen for a period of 15 minutes the solution was washed with water, 1 N sodium bicarbonate solution and again with water.

The aqueous layers were rewashed with one portion of chloroform, and the washings combined with the remainder of the chloroform solution. After drying over anhydrous magnesium sulfate, the chloroform solution was concentrated to dryness, then taken up in a small volume of methylene chloride, treated with Magnesol anhydrous magnesium silicate and filtered. Acetone was added to the solution and the solution was boiled to remove the methylene chloride. After the solution was concentrated to a volume of about 15 ml it was chilled and the crystals were collected through filtration. The 1.37 g of crystals so obtained were recrystallized from acetone to give pure 6α-methyl-17α-hydroxyprogesterone of MP 220° to 223.5°C.

Preparation of 6α-Methyl-17-Hydroxyprogesterone 17-Acetate: 1 g of 6α-methyl-17α-hydroxyprogesterone was dissolved in a mixture of 10 ml of acetic acid and 2 ml of acetic anhydride by heating. After solution was effected the mixture was cooled to 15°C, and 0.3 g of p-toluenesulfonic acid was added. After allowing the mixture to stand for a period of 2.5 hours at room temperature, the pink solution was poured into ice water to give an amorphous solid which was recovered by filtration.

The precipitate was washed carefully with water and was then dissolved in 10 ml of methanol and 1.5 ml of methylene chloride. The solution was concentrated to 10 ml, diluted with 0.5 ml of 10% sodium hydroxide, boiled for one minute and cooled. The product, which crystallized on cooling, was recrystallized to give flakes of 6α-methyl-17α-hydroxyprogesterone 17-acetate, having a MP 205° to 209°C, according to US Patent 3,147,290.

References

Merck Index 5614
Kleeman and Engel p. 546
PDR pp. 777, 1447, 1839, 1858
OCDS Vol. 1 pp. 180, 186 (1977) and 2, 165 (1980)
DOT 4 (1) 14 (1968)
I.N. p. 586
REM p. 992
Miramontes, L.E., Romero, M.A. and Farjat, F.A.; US Patent 3,000,914; September 19, 1961; assigned to G.D. Searle and Co.
de Ruggieri, P. and Ferrari, C.; US Patent 3,043,832; July 10, 1962; assigned to Ormonoterapia Richter S.p.A., Italy
Camerino, B., Modelli, R., Patelli, B., Sala, G. and Baldratti, G.; US Patent 3,061,616; October 30, 1962; assigned to Societa Farmaceutici Italia, Italy
Patchett, A.A. and Hoffman, F.G.; US Patent 3,084,174; April 2, 1963; assigned to Merck and Co., Inc.
Beyler, R.E.; US Patent 3,105,840; October 1, 1963; assigned to Merck and Co.
Spero, G.B.; US Patent 3,147,290; September 1, 1964; assigned to The Upjohn Company

MEDRYLAMINE

Therapeutic Function: Antihistaminic

Chemical Name: 2-(p-Methoxy-alpha-phenylbenzyloxy)-N,N-dimethylethylamine

Common Name: Medrylamine

Structural Formula:

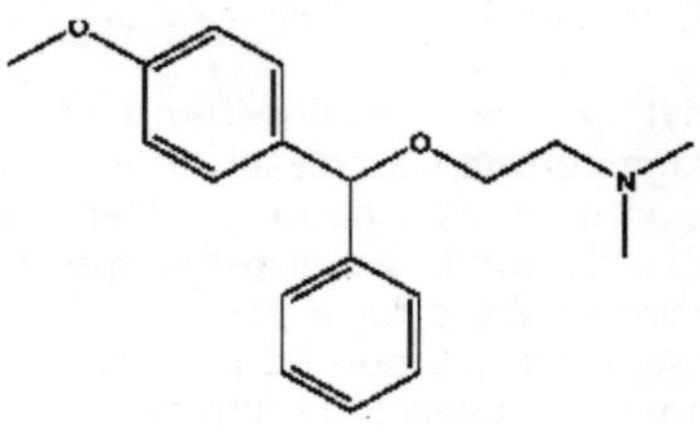

Chemical Abstracts Registry No.: 524-99-2

Trade Name	Manufacturer	Country	Year Introduced
Medrylamine	Shanghai Lansheng Corporation	-	-

Raw Materials

4-Methoxybenzhydryl chloride
2-Dimethylaminoethanol

Manufacturing Process

A 116 parts by weight of phenyl-p-methoxy-chloromethane (4-methoxybenzhydryl chloride) were added to 500 parts by weight of toluene and to the resulting mixture 65 parts by weight of 2-dimethylaminoethanol was added. Then the entire mixture was refluxed for two hours. The refluxed mixture was cooled and 250 parts by weight of a 10% solution of sodium hydroxide were added. This alkaline mixture was then steam distilled until the distillate was only weakly alkaline, for example a pH 7.5-8.

The residue of this steam distillation was then mixed with 200 parts by weight of benzene and washed with water until the wash waters were practically neutral. The benzene solution was then evaporated to dryness until the resulting mixture was of constant weight (145 parts). This product was oily and was then dissolved in 3000 parts by weight of dry ether; and treated while being agitated with the theoretically equivalent amount of dry hydrochloric acid dissolved in ether. This product is an oily product, which solidified after standing overnight in an ice box. The ether solution was decanted off and the solidified residue was dissolved in 1500 parts by weight of dioxane, and then precipitated with 3000 parts by weight of ether while being continuously agitated. The resulting solid product was the hydrochloride of 2-(p-methoxy-α-phenylbenzyloxy)-N,N-dimethylethylamine 100 parts by weight of the desired product were obtained corresponding to a yield of 62% and having a melting point of 141°C.

References

Morren H.; US Patent No. 2,668,856; Feb. 9, 1954; Assigned to Union Chemique Belge, S. A., Brussels, Belgium, a corporation of Belgium

MEDRYSONE

Therapeutic Function: Glucocorticoid

Chemical Name: 11β-Hydroxy-6α-methylpregn-4-ene-3,20-dione

Common Name: Hydroxymesterone; 6α-Methyl-11β-hydroxyprogesterone

Structural Formula:

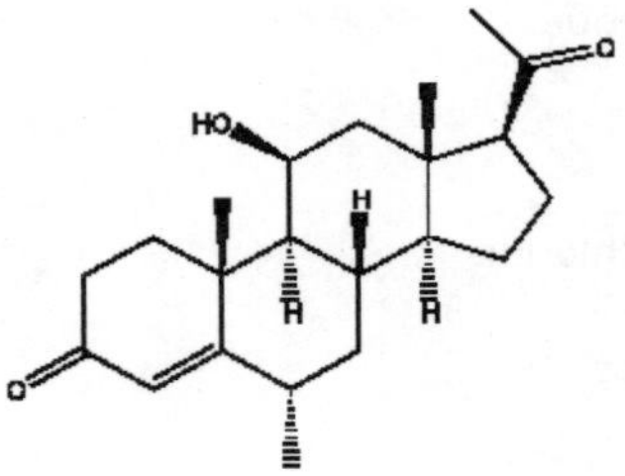

Chemical Abstracts Registry No.: 2668-66-8

Trade Name	Manufacturer	Country	Year Introduced
HMS	Allergan	US	1970
Visudrisone	Italseber	Italy	1970
Spectamedryn	Pharm-Allergan	W. Germany	1975
Medrysone Faure	Faure	France	1976
Ipoflogin	Tubi Lux	Italy	-
Medrifar	Farmila	Italy	-
Medritonic	Llorens	Spain	-
Medroptil	Farmigea	Italy	-
Ophthocortin	Winzer	W. Germany	-
Sedestrol	Poen	Argentina	-

Raw Materials

11-Keto-6β-methylprogesterone
Ethylene glycol
Lithium aluminum hydride

Manufacturing Process

Preparation of 11-Keto-6β-Methylprogesterone 3,20-bis-(Ethylene Ketal): A mixture of 5 g of 11-keto-6β-methylprogesterone [Spero et al, A Am. Chem. Soc., 78, 6213 (1956)], 503 ml of benzene, 26 ml of ethylene glycol, and 0.152 g of p-toluenesulfonic acid monohydrate was stirred and heated under reflux for 22 hours while water was removed by means of a water trap. The reaction mixture was then cooled to 30°C, 0.4 ml of pyridine was added, and stirring was continued for 10 minutes.

The reaction mixture was then shaken with 110 ml of water and the organic and aqueous layers separated. The organic layer was dried over sodium sulfate and evaporated under diminished pressure giving a residue. The thus obtained residue was recrystallized from methanol giving 2.68 g of 11-keto-6β-methyl progesterone 3,20-bis-(ethylene ketal) having a MP of 168° to 175°C.

Preparation of 11β-Hydroxy-6α-Methylprogesterone: A mixture of 2.68 g of

11-keto-6β-methylprogesterone 3,20-bis-(ethylene ketal), 161 ml of tetrahydrofuran (previously distilled from lithium aluminum hydride), 1.34 g of lithium aluminum hydride and 14.5 ml of absolute ether was stirred and refluxed under nitrogen for 1.5 hours, then 27 ml of water was added cautiously, to decompose excess hydride. The resulting mixture was filtered and the filter cake was washed with 135 ml of ether. The combined filtrate and wash was shaken with 135 ml of water and separated. The aqueous layer was washed with four 55-ml portions of ether, then the organic layer and the washes were combined, washed once with water, and evaporated to dryness under diminished pressure leaving a tan residue.

The thus-obtained residue was dissolved in a mixture of 268 ml of methanol and 26.8 mi of 3 N aqueous sulfuric acid and heated under reflux for 40 minutes, with a color change from yellow to green. The reaction mixture was then cooled, neutralized by addition of 127 ml of 5% sodium bicarbonate solution, and concentrated under reduced pressure until almost all the methanol was removed. The resulting solid was removed by filtration, washed with water, dried, and twice crystallized from ethyl acetate to give 1.1 g of 11β-hydroxy-6α-methylprogesterone having a MP of 155° to 158°C, according to US Patent 2,864,837.

References

Merck Index 5616
Kleeman and Engel p. 548
OCDS Vol. 2 p. 200 (1980)
DOT 6 (5) 184 (1970)
I.N. p. 587
REM p. 972
Sebek, O.K., Spero, G.B. and Thompson, J.L.; US Patent 2,864,837; assigned to The Upjohn Company
Spero, G.B. and Thompson, J.L.; US Patent 2,968,655; January 17,1961; assigned to The Upjohn Company

MEFENAMIC ACID

Therapeutic Function: Analgesic

Chemical Name: 2-[2,3-Dimethylphenyl)amino]benzoic acid

Common Name: N-(2,3-Xylyl)anthranilic acid

Structural Formula:

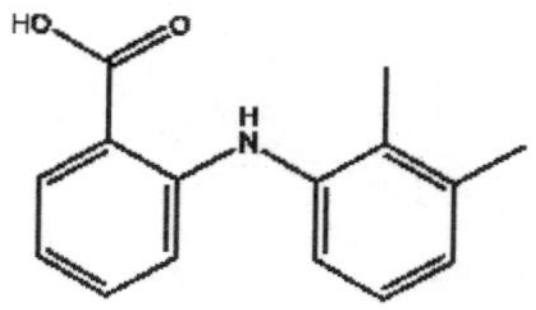

Chemical Abstracts Registry No.: 61-68-7

Trade Name	Manufacturer	Country	Year Introduced
Ponstan	Parke Davis	UK	1963
Ponalar	Parke Davis	W. Germany	1964
Ponstyl	Parke Davis	France	1967
Ponstel	Parke Davis	US	1967
Bafameritin	Hishiyama	Japan	-
Bonabol	Sawai	Japan	-
Fenamin	Yurtoglu	Turkey	-
Lysalgo	Schiapparelli	Italy	-
Mefacit	Polfa	Poland	-
Mefedolo	Ion	Italy	-
Parkemed	Parke Davis	W. Germany	-
Rolan	Nobel	Turkey	-
Spantac	Uji	Japan	-
Vialidin	Italfarmaco	Italy	-

Raw Materials

Potassium o-bromobenzoate
2,3-Dimethylaniline

Manufacturing Process

A mixture of 800 g of potassium o-bromo-benzoate, 1,500 ml of bis-(2-methoxyethyl)ether, 355 g of N-ethyl-morpholine, 375 g of 2,3-dimethylaniline, and 30 g of cupric acetate is heated gradually with stirring to 140°C over a period of 90 minutes. The hot reaction mixture is then acidified with 260 mi of concentrated hydrochloric acid and the acidified mixture divided into 2 equal portions. One liter of water is added to each portion and the mixtures allowed to cool. The N-(2,3-dimethylphenyl)anthranilic acid which separates upon cooling is collected by filtration and recrystallized from bis(2-methoxyethyl)ether; MP 229° to 230°C (corr.).

References

Merck Index 5617
Kleeman and Engel p. 548
PDR p. 1383
OCDS Vol. 1 p. 110 (1977) and 2, 280 (1980)
DOT 1 (2) 59 (1965)
I.N. p. 31
REM p. 1118
Scherrer, R.A.; US Patent 3,138,636; June 23, 1964; assigned to Parke, Davis and Company

MEFENOREX HYDROCHLORIDE

Therapeutic Function: Anorexic

Chemical Name: N-(3-Chloropropyl)-α-methylphenylethylamine hydrochloride

Common Name: -

Structural Formula:

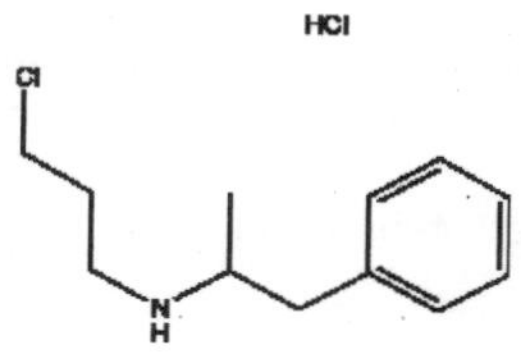

Chemical Abstracts Registry No.: 5586-87-8; 17243-57-1 (Base)

Trade Name	Manufacturer	Country	Year Introduced
Pondinil	Roche	France	1970
Rondimen	Homburg	W. Germany	1976
Anexate	Roche	US	-
Doracil	Gador	Argentina	-

Raw Materials

β-Chloropropionaldehyde
1-Phenyl-2-aminopropane
Hydrogen

Manufacturing Process

9.5 parts of β-chloropropionaldehyde were added slowly, at a temperature of 0°C to a solution of 31.5 parts of 1-phenyl-2-aminopropane in 150 parts of methanol. Thereafter, 0.2 part of platinum oxide was added to the reaction mixture following which the mixture was reacted with hydrogen, in a shaking vessel, until the theoretical quantity of hydrogen had been taken up. When the hydrogenation reaction was completed, the catalyst was removed by filtration and the filtrate neutralized with hydrochloric acid. Subsequently, the filtrate was evaporated to dryness and recrystallized from isopropyl alcohol. The thus-obtained N-(3-chloropropyl)-α-methylphenethylamine hydrochloride melted at 128°C to 130°C.

References

Merck Index 5618
Kleeman and Engel p. 549

OCDS Vol. 2 p. 47 (1980)
DOT 6 (4) 133 (1970)
I.N. p. 587
Schuler, W.A., Schlichtegroll, A.V., Beschke, H. and Klingler, K.H.; US Patent 3,485,926; December 23, 1969; assigned to Hoffmann-LaRoche, Inc.

MEFLOQUINE

Therapeutic Function: Antimalarial

Chemical Name: 4-Quinolinemethanol, 2,8-bis(trifluoromethyl)-α-2-piperidinyl-, (R*,S*)-, (+-)-

Common Name: Meflochina; Mefloquine

Structural Formula:

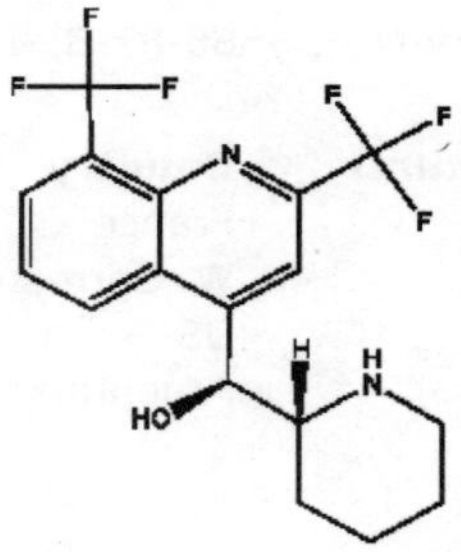

Chemical Abstracts Registry No.: 53230-10-7

Trade Name	Manufacturer	Country	Year Introduced
Lariam	Roche Pharmaceuticals	Switz.	-
Mephaquin	Mepha	-	-

Raw Materials

1,1'-Carbonyldiimidazole
2-Bromopyridine
Butyl lithium
2-Pyridylacetonitrile
2,8-Bis(trifluoromethyl)quinoline-4-carboxylic acid
4-Chloro-2,8-bis(trifluoromethyl)quinoline
N,O-Dimethylhydroxylamine hydrochloride
Benzyltriethylammonium chloride
Sodium hydroxide

Manufacturing Process

The first method of synthesis of 2,8-bis(trifluoromethyl)-4-quinolinyl-2-pyridinylmethanone

N-Methoxy-N-methyl-2,8-bis(trifluoromethyl)-quinoline-4-carboxamide was prepared using synthetic methodology reported by Thiesen et al (J. Org. Chem. 1988, 53, 2374). To a suspension of 12.5 g (40.4 mmol) 2,8-bis(trifluoromethyl)quinoline-4-carboxylic acid (was prepared by the method of Hickmann et al. (U.S. Patent No. 4,327,215)) in 200 ml CH_2Cl_2 was added 1,1'-carbonyldiimidazole (7.3 g, 45 mmol) and N,O-dimethylhydroxylamine hydrochloride (4.25 g, 45 mmol). The resulting deep red solution was stirred overnight, then poured into dilute hydrochloric acid (0.25 M, 200 ml). The organic phase was separated, and washed with dilute sodium hydroxide and brine, and dried ($MgSO_4$). The solvents was evaporated to leave a viscous brown oil, which was filtered through a pad of silica gel using ethyl acetate-hexane (1:1) as eluent to give N-methoxy-N-methyl-2,8-bis(trifluoromethyl)-quinoline-4-carboxamide as a yellowish oil, 14.3 g (98%), which solidified on standing. This material was broken up under hexane to afford the product as a solid, melting point 93-95°C. Analysis of this material by HPLC showed it to be >99.8% pure.

To a solution of the N-methoxy-N-methyl-2,8-bis(trifluoromethyl)-quinoline-4-carboxamide amide (10 g, 28.4 mmol) in anhydrous ether (100 ml) was added a solution of 2-pyridyl lithium (Pinder et al (J. Med. Chem. 1968, 11, 267)) [formed by addition of 2-bromopyridine (3.3 ml, 34.6 mmol) to a solution of butyl lithium (29.7 ml of a commercial 1.6 M solution, diluted with an equal quantity of ether) at -78°C] at -78°C. Analysis of the reaction by TLC after 10 min showed that no starting material remained. The reaction was allowed to warm to room temperature, then poured into aqueous ammonium acetate, and extracted with ether, the combined organic layers washed with brine and dried ($MgSO_4$). Filtration through a pad of silica gel using ethyl acetate-hexane (1:1) afforded 9.0 g (84%) of the crude 2,8-bis(trifluoromethyl)-4-quinolinyl-2-pyridinylmethanone. This was recrystallised from isopropyl alcohol to give the product as colourless needles, identical to that described in the literature (Hickmann et al.; Pinder et al.; Ohnmacht et al.; and Adam et al. (Tetrahedron 1991, 36, 7609)).

The second method of synthesis of 2,8-bis(trifluoromethyl)-4-quinolinyl-2-pyridinylmethanone

In a round bottom flask (100 ml) were placed 4-chloro-2,8-bis(trifluoromethyl)quinoline (0.0385 mole, 11.52 g), 2-pyridylacetonitrile (0.0423 mole, 5.0 g), benzyltriethylammonium chloride (0.26 g, 3 mole %), THF (35 ml) and aq NaOH (20 N, 9.63 ml, 0.192 moles). On stirring the colour of the solution became cherry red. The reaction temperature was increased to 5-0°C and stirred for further 1 hour. Monitoring of the reaction mixture by thin layer chromatography (TLC) or gas liquid chromatography (GLC) indicated complete consumption of 4-chloroquinoline to give nitrile. The reaction temperature was lowered to 20-25°C followed by addition of 30% H_2O_2 (13 ml, 0.1154 moles). TLC and GLC monitoring indicated complete conversion of nitrile compound to 2,8-bis(trifluoromethyl)-4-quinolinyl-2-pyridinylmethanone. Reaction mixture was cooled to 0-5°C and neutralized by ortho-phosphoric acid (85% aq, 4.5 ml). THF was distilled off, followed by addition of water (30 ml) and extraction with toluene. The crude product was crystallized from isopropanol to obtain 2,8-bis(trifluoromethyl)-4-quinolinyl]-2-pyridinylmethanone. Yield = 13.17 g (92%), melting point 123°C.

References

Chawla H.P.S. et al.; US Patent No. 6,500,955; Dec. 31, 2002; Assegned to National Institute of Pharmaceutical Education and Research

Fletcher A., Szhepard R., US Patent No. 6,664,397; Dec. 16, 2002; Assigned to Vernalis Research Limited

MEFRUSIDE

Therapeutic Function: Diuretic

Chemical Name: 4-Chloro-N'-methyl-N'-[tetrahydro-2-methyl-2-furanyl) methyl]-1,3-benzene-disulfonamide

Common Name: -

Structural Formula:

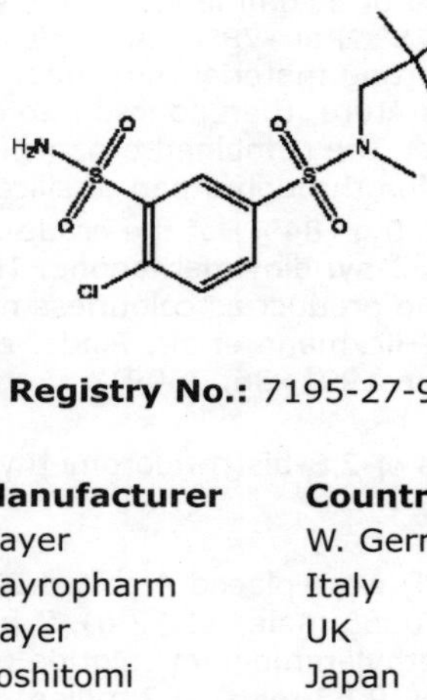

Chemical Abstracts Registry No.: 7195-27-9

Trade Name	Manufacturer	Country	Year Introduced
Baycaron	Bayer	W. Germany	1967
Mefrusal	Bayropharm	Italy	1969
Baycaron	Bayer	UK	1971
Baycaron	Yoshitomi	Japan	1975
Bendigon	Bayer	W. Germany	-
Caprinol	Bayer	W. Germany	-
Sali-Presinol	Bayer	W. Germany	-

Raw Materials

Hydrogen
α-Methyl-α-cyanotetrahydrofuran
Dimethyl sulfate
4-Chloro-3-sulfamyl benzene sulfochloride

Manufacturing Process

By hydrogenation of α-methyl-α-cyanotetrahydrofuran with Raney nickel as catalyst, α-methyl-α-tetrahydrofurfuryl amine is obtained (BP 48°C/12 mm Hg) which is alkylated by dimethyl sulfate to give α-methyl-α-tetrahydrofurfurylmethylamine (BP 70°C/40 mm Hg). The amine is then reacted with 4-chloro-3-sulfamyl benzene sulfochloride in the presence of an acid acceptor. The mixture is stirred overnight, the solvent (acetone or pyridine) is driven off under vacuum and the residue is recrystallized from alcohol.

References

Merck Index 5621
Kleeman and Engel p. 550
OCDS Vol. 1 p. 134 (1977)
I.N. p. 588
Horstmann, H., Wollweber, H. and Meng, K.; British Patent 1,031,916; June 2, 1966; assigned to Farbenfabriken Bayer AG, Germany Horstmann, H., Wollweber, H. and Meng, K.; US Patent 3,356,692; December 5, 1967; assigned to Farbenfabriken Bayer AG

MEGESTROL ACETATE

Therapeutic Function: Cancer chemotherapy

Chemical Name: 17α-Hydroxy-6-methylpregna-4,6-diene-3,20-dione acetate

Common Name: -

Structural Formula:

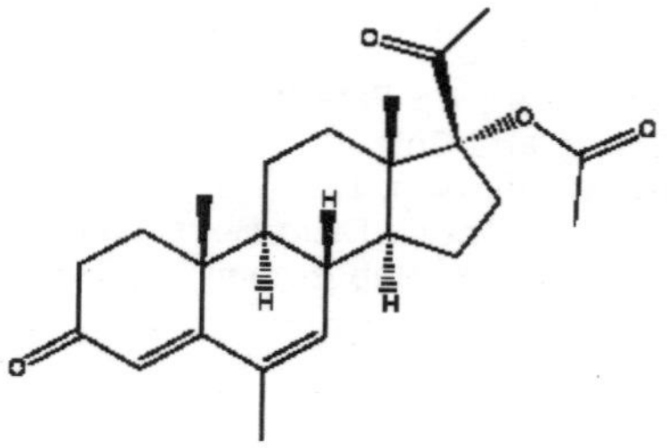

Chemical Abstracts Registry No.: 595-33-5

Trade Name	Manufacturer	Country	Year Introduced
Megestat	Bristol	W. Germany	1964
Megace	Bristol	UK	1967
Megace	Mead Johnson	US	1982
Pallace	Bristol	US	1982

Trade Name	Manufacturer	Country	Year Introduced
Megestat	Bristol	Switz.	1983
Megeron	Neofarma	Finland	-
Minigest	Novo	-	-
Niagestin	Novo	-	-
Ovarid	Glaxo	-	-
Volplan	B.D.H.	UK	-

Raw Materials

17α-Acetoxy-3β-hydroxy-6-methylpregn-5-ene-20-one
Aluminum-t-butoxide
p-Benzoquinone

Manufacturing Process

The following preparation is given in US Patent 3,356,573. 17α-Acetoxy-3β-hydroxy-6-methylpregn-5-ene-20-one (1 g), aluminum tert-butoxide (1 g) and p-benzoquinone (6 g) were dissolved in dry benzene (100 ml) and the mixture was heated under reflux for 30 minutes. The reaction mixture was cooled and washed with potassium hydroxide solution until the benzene layer was colorless. The benzene was washed with water, dried and evaporated to dryness under reduced pressure. The residue crystallized from aqueous methanol to give 17α-acetoxy-6-methylpregna-4,6-diene-3,20-dione, needles, MP 214° to 216°C.

References

Merck Index 5623
Kleeman and Engel p. 550
PDR p. 721
OCDS Vol. 1 p. 180 (1977)
DOT4 (1) 17 (1968)
I.N. p. 588
REM p. 993
Dodson, R.M. and Sollman, P.B.; US Patent 2,891,079; June 16, 1959; assigned to G.D. Searle and Co.
Kirk, D.N., Petrow, V. and Williamson, D.M.; US Patent 3,356,573; December 5, 1967; assigned to The British Drug Houses Limited, England
Cross, A.D.; US Patent 3,400,137; September 3, 1968; assigned to Syntex Corporation, Panama

MELITRACEN

Therapeutic Function: Antidepressant

Chemical Name: 3-(10,10-Dimethyl-9(10H)-anthracenylidene)-N,N-dimethyl-1-propanamine

Common Name: -

Structural Formula:

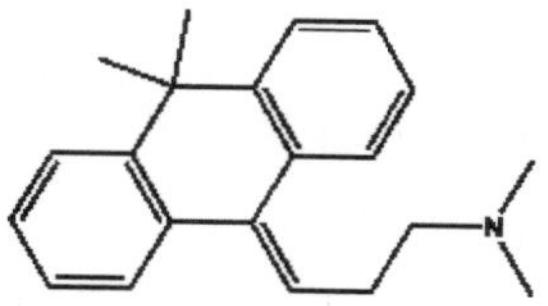

Chemical Abstracts Registry No.: 5118-29-6; 10563-70-9 (Hydrochloride salt)

Trade Name	Manufacturer	Country	Year Introduced
Trausabun	Lusofarma	W. Germany	1965
Meixeran	Lusofarma	Italy	1975
Dixeran	Lundbeck	-	-
Thymeol	Takeda	Japan	-

Raw Materials

Magnesium
Sulfuric acid
Hydrogen chloride
2-o-Benzoylphenylpropanol-2
Dimethylaminopropyl chloride

Manufacturing Process

24 g of 2-o-benzoylphenylpropanol-2 (MP 116°C) were dissolved in 250 ml of anhydrous ether and the resulting solution was added dropwise while stirring to a suspension of 0.22 mol of dimethylaminopropylmagnesium chloride in 100 ml of ether. The reaction mixture was refluxed for one hour on a steam bath, and water and dilute hydrochloric acid were added until the reaction was pH 4-5. The aqueous phase was separated and 60 ml of concentrated aqueous ammonia were added. The mixture was extracted with ether, and the ether phase was separated, dried and evaporated in a steam bath. The residue was dissolved in hot petroleum ether and the solution left standing to cool for some time, whereupon 4-dimethylamino-1-phenyl-1-[2-(2-hydroxy-2-propyl)phenyl]-butanol-1 crystallized out as white crystals which were sucked off. After drying they melted at 88°C to 90°C.

10 g of this compound were cautiously dissolved in 50 ml of concentrated sulfuric acid under cooling and the mixture was kept at room temperature for 24 hours, whereupon the reaction mixture was poured into 200 g of finely crushed ice, and concentrated aqueous ammonia was added to about pH 9, whereupon the oil which separated out was extracted with ether. The ether phase was separated, dried and the ether evaporated on a steam bath. The residue was dissolved in 20 ml of acetone and the solution neutralized with a solution of dry hydrogen chloride in ether. The white crystals of 9-γ-dimethylaminopropylidene-10,10-dimethyl-9,10-dihydroanthracene hydrochloride which separated out was filtered off and dried. Yield 9 g. MP 245°C to 247°C.

References

Merck Index 5642
Kleeman and Engel p. 552
OCDS Vol. 2 p. 220 (1980)
I.N. p. 589
Holm, T.O.; US Patent 3,190,893; June 22, 1965; assigned to Kefalas A/S (Denmark)

MELOXICAM

Therapeutic Function: Antiinflammatory

Chemical Name: 2H-1,2-Benzothiazine-3-carboxamide, 4-hydroxy-2-methyl-N-(5-methyl-2-thiazolyl)-, 1,1-dioxide

Common Name: Meloxicam; Mesoxicam

Structural Formula:

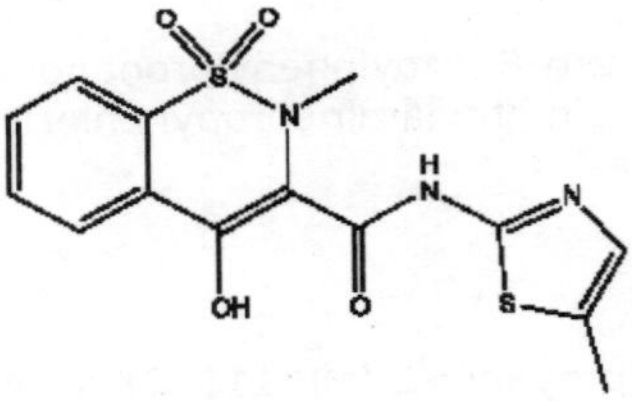

Chemical Abstracts Registry No.: 71125-38-7

Trade Name	Manufacturer	Country	Year Introduced
Novo-Meloxicam	Novopharm	-	-
Mobic	Boehringer Ingelheim	-	-
Meloxicam	AroKor Holdings Inc.	-	-
Meloxicam	NANJING PHARMA CHEMICAL PLANT	-	-
Meloxicam	SMS Pharmaceuticals Limited	-	-
Meloxicam	Hangzhou Verychem Science and Technology Co., Ltd.	-	-
Meloxicam	Technodrugs and Intermediates (P) Ltd.	-	-
Mobicox	Boehringer Ingelheim	-	-
Flexidol	Raffo	-	-
Flogoten	Montpellier	-	-
Miogesil	Glaxco-Wellcome	-	-

Trade Name	Manufacturer	Country	Year Introduced
Skudal	Schering-Plough	-	-
Bronax	Roemmers	-	-
Dominadol	Craveri	-	-
Tenaron	Sandoz	-	-
Telaroid	Cetus	-	-
Flexium	Sidus	-	-
Loxitenk	Biotenk	-	-
Meloxid	Klonal	-	-
Merapiran	Finadiet	-	-
Leutrol	Istituto De Angeli Ph. Spa	-	-
Leutrol	Abbott	-	-

Raw Materials

1,1-Dioxide of methyl 4-hydroxy-2-methyl-2H-1,2-benzothiazine-3-carboxylate
2-Amino-5-methylthiazole

Manufacturing Process

A mixture of 26.9 g (0.1 mol) of the 1,1-dioxide of methyl 4-hydroxy-2-methyl-2H-1,2-benzothiazine-3-carboxylate and 12.5 g (0.11 mol) of 2-amino-5-methylthiazole was refluxed in 4 liters of xylene for 24 hours in a nitrogen atmosphere. The methanol formed by the reaction was removed by means of a 4-A-molecular sieve mounted in a Soxhlet-extractor. The hot reaction solution was filtered. Upon cooling and standing overnight, the crude product separated out of the filtrate in the form of crystals (32.0 g, 91% of theory). After recrystallization from ethylene chloride 26.0 g (74% of theory) of 4-hydroxy-2-methyl-N-(5-methyl-2-thiazolyl)-2H-1,2-benzothiazine-3-carboxamide-1,1-dioxide were obtained; M.P.: 254°C (decomp.).

References

Trummlitz G., Engel W., Seeger E., Engelhardt G.; US Patent No. 4,233,299; November 11, 1980; Assigned to Boehringer Ingelheim GbH (DE)

MELPERONE

Therapeutic Function: Neuroleptic

Chemical Name: 1-(4-Fluorophenyl)-4-(4-methyl-1-piperidinyl)-1-butanone

Common Name: Flubuperone; Methylperone

Structural Formula:

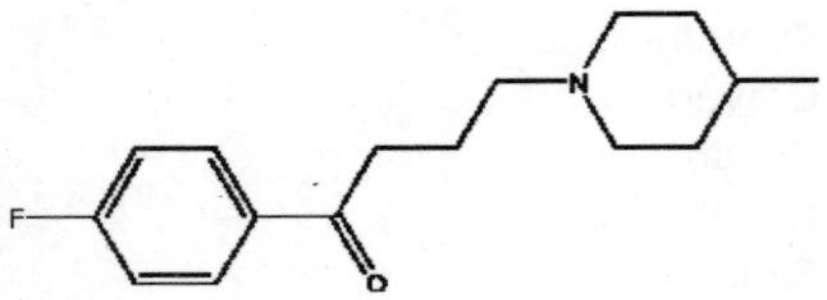

Chemical Abstracts Registry No.: 3575-80-2; 1622-79-3 (Hydrochloride salt)

Trade Name	Manufacturer	Country	Year Introduced
Eunerpan	Nordmark	W. Germany	1965
Buronil	Ferrosan	Sweden	-

Raw Materials

γ-Chloro-p-fluorobutyrophenone
4-Methylpiperidine

Manufacturing Process

A solution or dispersion consisting of 20.1 g (0.1 mol) of γ-chloro-p-fluorobutyrophenone, 19.8 g (0.2 mol) of 4-methylpiperidine and 0.1 g of potassium iodide in 150 ml toluene is heated in a sealed glass tube for 15 hours at 100°C to 110°C. The potassium iodide and the 4-methylpiperidine hydrochloride formed in the reaction are separated by filtration and the solvent removed from the filtrate by evaporation in vacuum on a steam bath. The residue is distilled and the fraction obtained at 120°C to 125°C and at a pressure lower than 0.1 mm Hg is collected. The base is dissolved in ether and the 4-fluoro-γ-(4-methylpiperidino)-butyrophenone precipitated as the hydrochloride. The reaction product is purified by recrystallization in ethanol/ether.

Yield 22.0 g (73% of theory). MP 209°C to 211°C.

References

Merck Index 5645
Kleeman and Engel p. 552
I.N. p. 590
Hernestam, S.E.H., Sterner, N.O.B. and Lassen, J.; US Patent 3,816,433; June 11, 1974; assigned to A.B. Ferrosan (Sweden)

MELPHALAN

Therapeutic Function: Cancer chemotherapy

Chemical Name: 4-[Bis(2-chloroethyl)amino]-L-phenylalanine

Common Name: Alanine nitrogen mustard; L-Sarcolysine

Structural Formula:

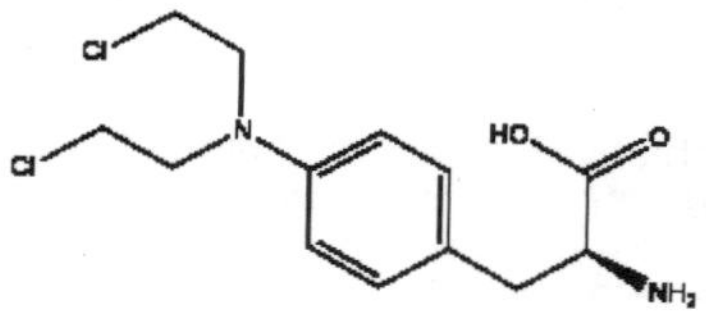

Chemical Abstracts Registry No.: 148-82-3

Trade Name	Manufacturer	Country	Year Introduced
Alkeran	Burroughs-Wellcome	US	1964
Alkeran	Wellcome	UK	1964
Alkeran	Wellcome	W. Germany	1965
Alkeran	Wellcome	France	1966
Alkeran	Wellcome	Italy	1968
Alkeran	Wellcome	Japan	1979

Raw Materials

Sodium carbonate
Acetic anhydride
Cinchonidine
Hydrogen
Phosphorus oxychloride
Diethyl sodium phthalidomalonate
p-Nitrobenzoyl chloride
Hydrogen chloride
Ethylene oxide

Manufacturing Process

Diethyl sodium phthalimidomalonate (Barger and Weichselbaum, Organic Syntheses, 1943, Coll. Vol. II, 384) (6.52 g) was dissolved in boiling methyl ethyl ketone (80 ml) and a solution of p-nitrobenzyl chloride (3.44 g; 1.0 mol) in the same solvent (20 ml) was added, Sodium iodide (ca 0.5 g) dissolved in hot methyl ethyl ketone (10 ml) was introduced, and produced an immediate precipitation. The mixture was refluxed for 1.5 hours, cooled, filtered, evaporated under vacuum and the residual gum crystallized from ethanol. The di-ethyl-p-nitrobenzyl-phthalimidomalonate formed colorless prisms (88%), MP 103° to 105°C, sharpening to 104° to 105°C on recrystallizing from ethanol.

Diethyl-p-nitrobenzyl-phthalimidomalonate (70 g) and sodium carbonate (70 g) in water (700 ml) were refluxed overnight with mechanical stirring (to avoid bumping). The clear brown solution was acidified with hydrochloric acid and refluxing and stirring were continued for a further 40 minutes. The mixture was cooled and the colorless precipitate (31 g) collected. A second crop (18.5 g) was obtained on evaporation of the mother liquors. Crystallization from aqueous ethanol gave the compound N-carboxybenzoyl-p-nitro-DL-phenylalanine as small needles, MP 198° to 200°C.

The N-carboxybenzoyl compound (2.7 g) was refluxed for 30 minutes with acetic anhydride (10 ml), the mixture taken to dryness (vacuum) and the residue heated with water. The cooled gummy product became granular on rubbing and crystallized from methyl ethyl ketone-petrol or aqueous ethanol in almost colorless needles, MP 184° to 186°C, of p-nitro-N-phthaloyl-DL-phenylalanine.

A solution of p-nitro-N-phthaloyl-DL-phenylalanine (1.0 g) in methanol (25 ml) and a solution of cinchonidine (0.865 g) in methanol (30 ml) were mixed. Crystallization soon set in. The mixture was left overnight, and the colorless needles (0.97 g), MP 209° to 210°C, collected. After two recrystallizations from methanol the cinchonidine salt of the D-acid had MP 211°C.

Evaporation of the mother liquors from the original cinchonidine experiment gave a gum which crystallized readily from aqueous ethanol in almost colorless needles (0.73 g), MP 191° to 192.5°C. Two recrystallizations from aqueous ethanol gave the cinchonidine salt of the L-acid, MP 192.5° to 194°C. To the salt (2.9 g) in warm ethanol (50 ml) was added water (50 ml) and a slight excess (ca 10 ml) of N aqueous sodium hydroxide. The mixture was diluted with water, cooled, filtered from the precipitated base and the filtrate acidified with hydrochloric acid. Refluxing with 2 N ethanolic hydrogen chloride yielded p-nitro-N-phthaloyl-L-phenylalanine ethyl ester, according to US Patent 3,032,585.

Then, as described in US Patent 3,032,584, ethyl N-phthaloyl p-nitrophenylalaninate (9.0 g) was hydrogenated in a mixture of ethyl acetate (120 g) and methanol (80 g) with a palladium-calcium carbonate (1% Pd) catalyst (1.4 g). When gas uptake was complete, the filtrate from the hydrogenation mixture was evaporated under reduced pressure. The residual gum was taken up in ether, the solution filtered, and a slight excess of a dry ethereal hydrogen chloride solution added slowly with stirring. The gummy precipitate became granular on rubbing and the ether-washed product was crystallized from ethyl acetate-acetone [1st crop, 2.8 g, MP 188° to 192°C (decomp.); 2nd crop, 3.9 g, MP 189° to 192°C (decomp.)] . Part of the first batch was recrystallized from ethyl acetate and gave very slightly tinted needles, MP 188° to 190°C (decomp.) of ethyl N-phthaloyl p-aminophenylalaninate hydrochloride.

The free base was obtained from the hydrochloride by adding a slight excess of dilute ammonium hydroxide to the aqueous solution, and crystallizing the product from aqueous methanol. A further recrystallization with charcoal treatment gave almost colorless needles, MP 110° to 112°C of ethyl N-phthaloyl p-aminophenylalaninate.

Ethyl N-phthaloyl p-aminophenylalaninate (3.15 g) (unrecrystallized) was suspended in water (50 g) and glacial acetic acid (30 g) added. To the clear solution, ethylene oxide (8.0 g) was added, the mixture allowed to stand for 17 hours, and then poured into water (350 g). The solution was neutralized with sodium hydrogen carbonate and the liberated gum extracted with ether. The ethereal solution was dried (magnesium sulfate) and evaporated. The residual gum (3.95 g) was dissolved in benzene (50 g) and the solution dried azeotropically by distilling off some of the solvent. Freshly distilled phosphorus oxychloride (8 g) was added and the mixture heated under reflux for 30 minutes.

The solvent was evaporated off under reduced pressure, and the residual gum refluxed with concentrated hydrochloric acid (50 g) for 6 hours. The solution was allowed to cool overnight. It was filtered from the phthalic acid crystals, and freeze-dried, and to the pink residue was added acetone (160 g) and ethyl acetate (50 g). The mixture was left in the cold room overnight and the clear pink supernatant liquid poured off. The pink gummy hydrochloride remaining in the flask was dissolved in water (20 g), saturated sodium acetate solution added until precipitation was complete, and the product collected and dried in a desiccator. The crude p-bis-(2-chloroethyl)-aminophenylalanine (3.6 g) was crystallized from methanol giving colorless needles, MP 172° to 174°C (decomp.) of p-bis-(2-chloroethyl)-aminophenylalanine.

References

Merck Index 5646
Kleeman and Engel p. 552
PDR p. 733
OCDS Vol. 2 p. 120 (1980)
I.N. p. 590
REM p. 1151
Bergel, F. and Stock, J.A.; US Patent 3,032,584; May 1, 1962; assigned to National Research Development Corporation, England
Bergel, F. and Stock, J.A.; US Patent 3,032,585; May 1, 1962; assigned to National Research Development Corporation, England

MEMANTINE

Therapeutic Function: Spasmolytic

Chemical Name: 3,5-Dimethyltricyclo[3.3.1.$1^{3,7}$]decanol-1-amine

Common Name: -

Structural Formula:

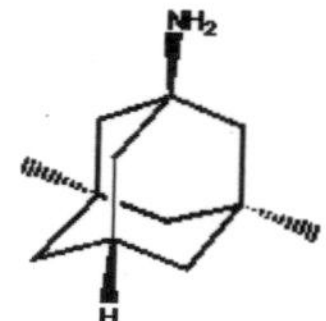

Chemical Abstracts Registry No.: 19982-08-2

Trade Name	Manufacturer	Country	Year Introduced
Akatinol	Merz	W. Germany	1983

Raw Materials

Acetonitrile	1,3-Dimethyladamantane
Bromine	Sodium hydroxide
Sulfuric acid	Hydrogen chloride

Manufacturing Process

A mixture of 24 g of 1,3-dimethyladamantane and 80 ml of bromine was refluxed for 6 hours. The reaction product mixture was cooled, taken up in about 200 ml of chloroform, and poured onto ice. The excess bromine was removed by adding sodium hydrosulfite. The chloroform layer was separated from the aqueous layer, dried, concentrated in vacuo, and distilled at reduced pressure to yield 30.5 g of product having a boiling point of about 118°C at 5-6 mm; n_D^{25} = 1.5169-1.5182. The product was identified by nuclear magnetic resonance (NMR) and elemental analyses as 1-bromo-3,5-dimethyladamantane.

A mixture of 20 g of 1-bromo-3,5-dimethyladamantane, 75 ml of acetonitrile, and 150 ml of concentrated sulfuric acid was allowed to react overnight at ambient room temperature. The red reaction product mixture was poured over crushed ice, and the white solid which precipitated was taken up in benzene and the benzene solution dried over sodium hydroxide pellets. The benzene solution was filtered from the drying agent and evaporated to dryness in vacuo to yield 18.2 g of product having a melting point of about 97°C and identified by infrared spectrum as 1-scetamido-3,5-dimethyladamantane.

A mixture of 18 g of 1-acetamido-3,5-dimethyladamantane, 38 g of sodium hydroxide, and 300 ml of diethylene glycol was refluxed for a period of 6 hours. The reaction product mixture was cooled and poured onto about 2,000 ml of crushed ice. The basic solution thus obtained was extracted five times with 250 ml portions of benzene and the aqueous layer was discarded. The combined benzene extracts were dried over sodium hydroxide and the dried benzene solution concentrated in vacuo to give a crude oil weighing 14 g and having n_D^{25} = 1.4941, A 4 g sample of the crude oil was dissolved in ether and the solution saturated with anhydrous hydrogen chloride. The solid which precipitated was filtered off and recrystallized from a mixture of alcohol and ether to yield product weighing 3.5 g and melting at 258°C.

It was identified by analysis as 1-amino-3,5-dimethyladamantane hydrochloride.

References

Merck Index A-7
DFU 1 (9) 427 (1976)
DOT 19 (6) 303 (1983)
I.N. p. 590
Mills, J. and Krumkalns, E.; US Patent 3,391,142; July 2, 1968; assigned to Eli Lilly and Co.

MENADIOL SODIUM DIPHOSPHATE

Therapeutic Function: Prothrombogenic vitamin

Chemical Name: 2-Methyl-1,4-naphthalenediol diphosphoric acid ester tetrasodium salt

Common Name:-

Structural Formula:

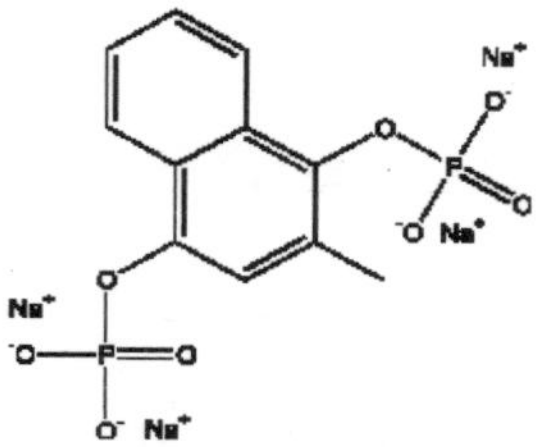

Chemical Abstracts Registry No.: 131-13-5; 84-98-0 (Phosphate)

Trade Name	Manufacturer	Country	Year Introduced
Synkayvite	Roche	US	1941
Analogue	Upjohn	US	1951
Kappadione	Lilly	US	1956
Carbocaina	Pierrel	Italy	-
Katij	Takeda	Japan	-
Thylokay	Squibb	-	-

Raw Materials

2-Methyl-1,4-naphthohydroquinone
Phosphorus oxychloride
Sodium hydroxide

Manufacturing Process

2,000 g 2-methyl-1,4-naphthohydroquinone diphosphoryl chloride (from the quinone and $POCl_3$) are dissolved in 2 liters ether and decomposed with 2 liters distilled water. The mixture is transferred to a separatory funnel and the aqueous layer separated from the ether layer, the latter being discarded. The aqueous layer is extracted with a further 2 liters of ether and again separated and discarded. The aqueous solution of the 2-methyl-1,4-naphthohydroquinone diphosphoric acid is extracted with successive portions of isobutyl carbinol in 500 cc quantities until the aqueous layer becomes almost colorless, after which this latter is discarded. The isobutyl carbinol solution is then concentrated to remove water and hydrochloric acid, and the crystalline residue neutralized with sodium hydroxide solution. The resulting solution of the sodium salt of 2-methyl-1,4-naphthohydroquinone diphosphoric ester is extracted with two successive portions of 1 liter acetone each and the latter discarded. Methanol and acetone are then added, filtered, and the product brought to crystallization by heating. Crystals of the sodium salt of 2-methyl-1,4-naphthohydroquinone diphosphoric acid ester are sucked off. The substance contains much moisture of crystallization and is dried in vacuum until it contains 21-22% moisture of crystallization as determined by drying at 145°C at 2 mm vacuum.

References

Merck Index 5649
Kleeman and Engel p. 553
PDR p. 1502
I.N. p. 591
REM p. 1010
Solmssen, U.V.; US Patent 2,345,690; April 4, 1944; assigned to Hoffmann-LaRoche, Inc.

MENADIONE

Therapeutic Function: Prothrombogenic vitamin

Chemical Name: 1,4-Naphthalenedione, 2-methyl-

Common Name: Menadione; Menaphthone; Menaquinone; Methylnaphthochinon(um); Vitamin K_3

Structural Formula:

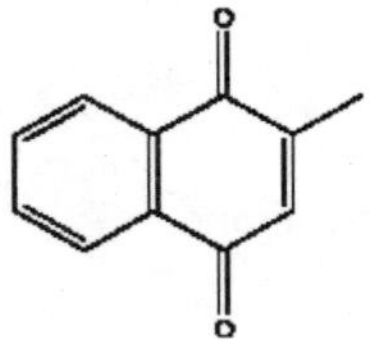

Chemical Abstracts Registry No.: 58-27-5

Trade Name	Manufacturer	Country	Year Introduced
Kappaxin	Sterling Winthrop	-	-
Kayquinone	Abbott	-	-
Thyloquinone	Bristol-Myers Squibb	-	-
K-Vitamin	Medica	-	-
Bilkaby	Bailly-Speab	-	-
Kapavit	Chropi	-	-
Kavitamin	Galenika	-	-
Vikaman	Disperga	-	-
Vit. K3	Agepha	-	-

Raw Materials

β-Methylnaphthalene
Sodium dichromate
Sulfuric acid

Manufacturing Process

Dissolve 100 g of β-methylnaphthalene in 500 g of carbon tetrachloride. Dissolve 500 g of commercial sodium dichromate in 175 g of hot water. Pour these two solutions into a 3-liter 3-necked flask, equipped with an efficient stirrer, a reflux condenser, and a dropping funnel. The flask should be put into a water bath held at 50°C. The contents of the flask agitated as violently as possible at adding (through the dropping funnel) 896 g of 77% (by weight) sulfuric acid. The rate of dropping depends on the efficiency of the reflux condenser. If the reaction tends to get out of hand due to overheating, cold water should be run into the water bath. After the addition of acid has been completed, keep the water bath at 70°C, for 1% to 2 hours. Then stop the agitation, cool the mixture and decant therefrom as much of the carbon tetrachloride layer as possible. Pour water into the flask; add 100 g more of carbon tetrachloride and stir for an additional ten minutes. The carbon tetrachloride layer will now settle to the bottom. The acid layer can be decanted and discarded, or worked up for those components desired. A complete separation of the acid from the carbon tetrachloride is effected by means of a separatory funnel. All of the carbon tetrachloride solutions are pooled and filtered to clarity through filter paper. The carbon tetrachloride may be distilled off from the quinone in a vacuum, using a water bath heated to 50°C.

The quinone is concentrated to a point where crystallization begins. Thereupon the concentrated solution is transferred to a beaker and allowed to crystallize at room temperature. Further crops of crystals are obtained by allowing the mother liquor to cool in an icebox, or by reducing it still further. If the β-methylnaphthalene starting material was pure, the 2-methyl-1,4-naphthoquinone obtained will have a melting point of 100°-104°C, without further purification. Vacuum sublimation has been found most effective to produce quinones of a very high purity. There is a rapid method of oxygenation of β-methylnaphthalene: 3 g of β-methylnaphthalene is dissolved in 180 grams of carbon tetrachloride. To this is added 15 grams of commercial sodium dichromate dissolved in 6 grams of hot water. The above solutions are placed in a 1-liter 3-necked flask equipped with an efficient stirrer, a dropping funnel, and a distilling condenser of large bore. The mixture is agitated energetically, and 25 grams of 77% sulfuric acid added through the dropping funnel. The time of addition need not be over 30 seconds. The flask is not subjected to exterior cooling, as the distilling carbon tetrachloride should hold the heat to the proper temperature, which should not exceed 85°C. After five minutes the reaction may be discontinued and the product handled similarly to the method outlined in above. The yield of 2-methyl-1,4-naphthoquinone will be around 80% of theoretical by the use of this rapid method.

References

Hyman J. et al; US Patent No. 2,402,226; June 18, 1946; Assigned to Velsicol Corporation, Chicago, Ill., a corporation of Illinois

MENADIONE SODIUM BISULFITE

Therapeutic Function: Prothrombogenic vitamin

Chemical Name: 2-Naphthalenesulfonic acid, 1,2,3,4-tetrahydro-2-methyl-1,4-dioxo-, sodium salt

Common Name: Menachinonum natrium bisulfurosum; Menadione sodium bisulfite; Menadionnatriumbisulfit; Menaphthone sodium bisulphite; Vikasol(um)

Structural Formula:

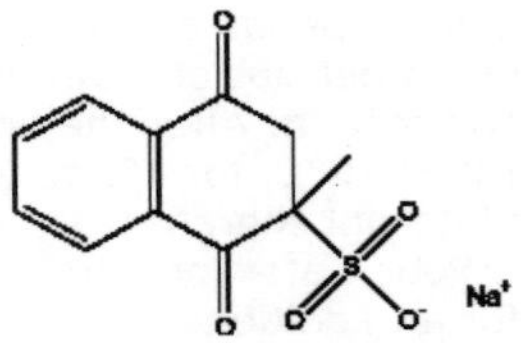

Chemical Abstracts Registry No.: 130-37-0

Trade Name	Manufacturer	Country	Year Introduced
Kavitol	Lannacher Heilmittel	-	-
Kaergona	Ibys	-	-
Libavit	Liba	-	-
Vitaminum	Polfa Warszawa	-	-

Raw Materials

2-Methyl-1,4-naphthoquinone
Sodium bisulfite

Manufacturing Process

The 2-naphthalenesulfonic acid, 1,2,3,4-tetrahydro-2-methyl-1,4-dioxo-, sodium salt, trihydrate can be prepared by mixing the 2-methyl-1,4-naphthoquinone with the bisulphite salt in the presence of water. Ordinarily gentle warming of the aqueous mixture is preferred to facilitate solution. The mixture of 2-methyl-1,4-naphthoquinone (250 mg; 1 molar equivalent); sodium bisulphite (149 mg; 1 molar equivalent); distilled water (250 ml) or 2-methyl-1,4-naphthoquinone (250 mg; 1 molar equivalent); potassium bisulphate (349 mg; 2 molar equivalent); distilled water 250 ml may be used. These examples representing preferred ratios of ingredients are merely illustrative and are not to be interpreted as limiting.

The bisulphite addition compounds have been found to be stable in sunlight and also to be heat stable. Tests, for example, carried out in ampoules have shown aqueous solutions of the compounds not to be decomposed after exposure to a month's sunlight, while other tests have shown the solutions of such compounds to retain their original potency (a) when stored in an oven at 60°C for 15 days or (b) when sterilized at 15 pounds for 0.5 hour in an autoclave at about 122°C. These properties emphasize the radical differences between the stable salts and the properties of 2-methyl-1,4-naphthoquinone, the characteristic instability of which is illustrated by its sensitivity, i. e., decomposition, when exposed to light.

The bisulphite addition compounds have a vitamin K activity equal to that of the 2-methyl-1,4-naphthoquinone contained in the molecule. The compounds, although suitable for oral administration, are particularly adaptable in aqueous solution for parenteral administration in the treatment of hemorrhagic conditions.

References

Moore M. B. et al; US Patent No. 2,367,302; Jan. 16, 1945; Assigned to Abbott Laboratories, North Chicago, Illinois

MENBUTONE

Therapeutic Function: Choleretic

Chemical Name: 4-Methoxy-6-oxo-1-naphthalene butanoic acid

Common Name: Methonaphthone

Structural Formula:

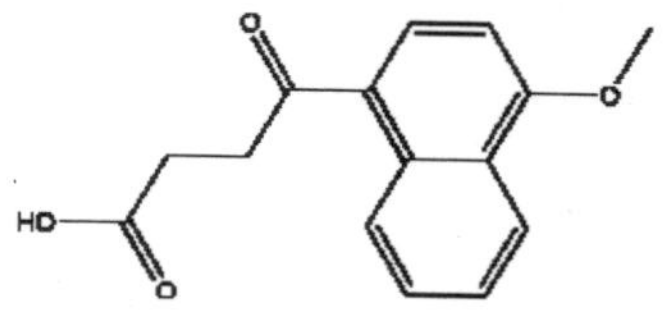

HO O O O

Chemical Abstracts Registry No.: 3562-99-0

Trade Name	Manufacturer	Country	Year Introduced
Hepalande	Delalande	W. Germany	1977
Sintobilina	A.F.I.	Italy	-

Raw Materials

α-Methoxynaphthalene
Succinic anhydride
Aluminum chloride
Hydrogen chloride
Sodium carbonate

Manufacturing Process

395 parts of (α-methoxynaphthalene and 265 parts of succinic anhydride are dissolved in 8,000 parts of dry benzene at room temperature. The resulting solution is stirred and 710 parts of anhydrous aluminum chloride are added over a period of twenty minutes. During the addition the temperature of the reaction mixture rises to about 60°C to 70°C. After the addition the reaction mixture is stirred for fifteen or twenty minutes at 60°C to 70°C and then refluxed for one hour. The hot reaction mixture is then poured onto a mixture of 5,000 parts of ice and 900 parts of concentrated hydrochloric acid. The benzene is removed by steam distillation and the hot aqueous residue is filtered to remove the insoluble β-(1-methoxy-4-naphthoyl)-propionic acid. The residue of the latter is dried and then dissolved in 16,000 parts of hot water containing 300 parts of sodium carbonate. The hot solution is treated with activated charcoal, filtered while hot, chilled and acidified. The residue of purified acid is collected on a filter, washed with water, and dried at 65°C. A yield of 552 parts of purified β-(1-methoxy)-4-naphthoyl)propionic acid, melting at 172°C to 173°C is obtained.

References

Merck Index 5656
I.N.p. 592
Burtner, R.R.; US Patent 2,623,065; December 23, 1952; assigned to G.D. Searle and Co.

MEPAZINE

Therapeutic Function: Tranquilizer

Chemical Name: 10-[(1-Methyl-3-piperidinyl)methyl]-10H-phenothiazine

Common Name: Mepasin; Pecazine

Structural Formula:

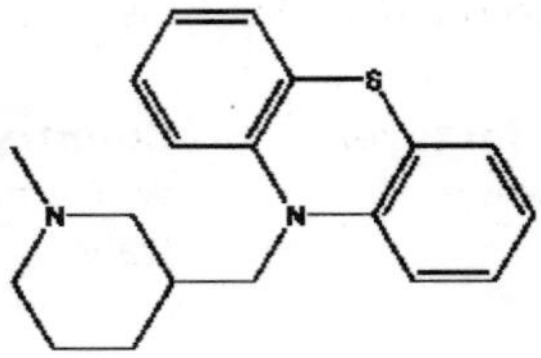

Chemical Abstracts Registry No.: 60-89-9; 2975-36-2 (Hydrochloride salt)

Trade Name	Manufacturer	Country	Year Introduced
Pacatal	Warner Lambert	US	1957
Pacatal	Promonta	W. Germany	-

Trade Name	Manufacturer	Country	Year Introduced
Lacumin	Lundbeck	-	-
Ravenil	Caber	Italy	-

Raw Materials

1-Methyl-3-bromomethylpiperidine
Phenothiazine
Sodium amide
Acetic acid

Manufacturing Process

A 500 cc flask equipped with a mechanical stirrer, reflux condenser and a soda-lime tube was filled with 230 cc of absolute xylene, 27.5 g of 1-methyl-3-bromomethylpiperidine, 53.3 g of phenothiazine and 14.2 g of finely powdered sodium amide, and the solution was heated under reflux for 6 hours. After cooling water was added and the batch was extracted with ether. As the hydrochloric acid salt of the obtained phenothiazine derivative is difficultly soluble in water, the further processing was carried out by way of the acetate. The etheric solution was extracted several times in a separating funnel with dilute acetic acid. The combined aqueous extracts were basified, extracted with ether, dried with potassium carbonate and, after removal of the ether, distilled in vacuo.

Yield = 64%; boiling point 230°C to 235°C at 4 mm; melting point of hydrochloride is 180°C to 181°C.

References

Merck Index 5672
Kleeman and Engel p. 689
I.N. p. 735
Schuler, W.A.; US Patent 2,784,185; March 5, 1957; assigned to Chemische Fabrik Promonta GmbH

MEPENZOLATE BROMIDE

Therapeutic Function: Spasmolytic

Chemical Name: 3-[(Hydroxydiphenylacetyl)oxy]-1,1-dimethylpiperidinium bromide

Common Name: N-Methyl-3-piperidylbenzilate methobromide

Structural Formula:

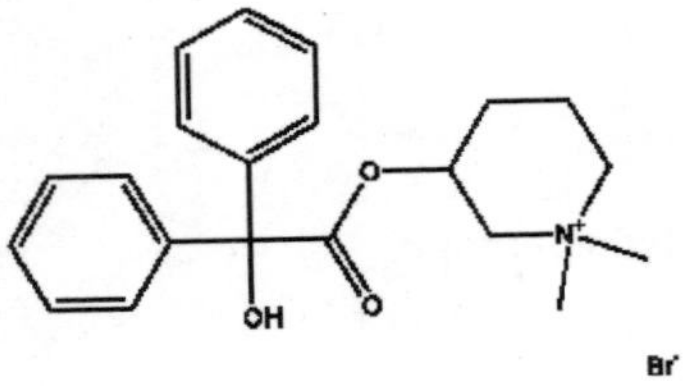

Chemical Abstracts Registry No.: 76-90-4

Trade Name	Manufacturer	Country	Year Introduced
Cantil	Merrell National	US	1956
Cantilon	Draco	Sweden	-
Colibantil	Tosi-Novara	Italy	-
Colum	Jamco	Italy	-
Eftoron	Maruko	Japan	-
Gastropodil	Fabo	Italy	-
Sachicoron	Zensei	Japan	-
Tendalin	Nihon Yakuhin	Japan	-
Tralanta	Sawai	Japan	-
Trancolon	Fujisawa	Japan	-

Raw Materials

N-Methyl-3-chloropiperidine
Benzilic acid
Methyl bromide

Manufacturing Process

A mixture containing 8 g (0.06 mol) of N-methyl-3-chloro-piperidine and 13.6 g (0.06 mol) of benzilic acid in 50 cc of anhydrous isopropyl alcohol was refluxed for 3 days; the isopropyl alcohol was removed by distillation in vacuo, the residue treated with dilute aqueous hydrochloric acid and the aqueous acid mixture extracted repeatedly with ether. The aqueous phase was separated, made strongly alkaline with 20% aqueous sodium hydroxide and extracted with ether. The ether extracts were dried with potassium carbonate and distilled; the product was collected at 175° to 176°C (0.03 mm), yield 11.5 g (59 %).The ester base thus prepared was then dissolved in 75 cc of isopropyl alcohol and 3.4 g (0.037 mol) methyl bromide added. The reaction mixture was allowed to stand at 30°C for 2 days and the product isolated by filtration, yield, 13 g (87%), MP 228° to 229°C dec.

References

Merck Index 5673
Kleeman and Engel p. 555

PDR p. 1223
I.N. p. 593
REM p. 916
Biel, J.H.; US Patent 2,918,408; December 22, 1959; assigned to Lakeside Laboratories, Inc.

MEPERIDINE HYDROCHLORIDE

Therapeutic Function: Narcotic analgesic

Chemical Name: 1-Methyl-4-phenyl-4-piperidinecarboxylic acid ethyl ester hydrochloride

Common Name: Isonipecaine hydrochloride; Pethidine hydrochloride

Structural Formula:

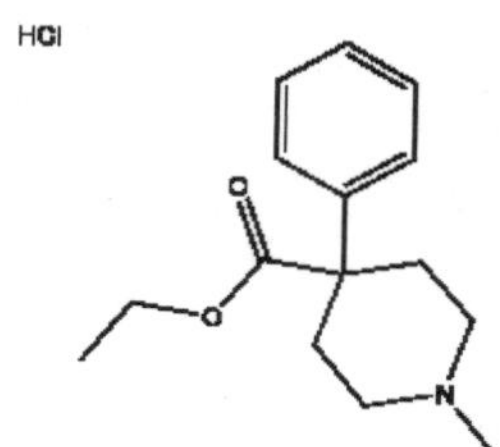

Chemical Abstracts Registry No.: 50-13-5; 57-42-1 (Base)

Trade Name	Manufacturer	Country	Year Introduced
Dolosal	Specia	France	1943
Dolantin	Hoechst	W. Germany	1943
Demerol	Winthrop	US	1944
Algil	Maggioni	Italy	-
Alodan	Gerot	Austria	-
Centralgin	Amino	Switz.	-
Demer-Idine	Sabex	Canada	-
Dolanquifa	Uquifa	Spain	-
Dolcontral	Arzneimittelwerk Dresden	E. Germany	-
Dolestine	Teva	Israel	-
Doloneurin	O.P.G.	Netherlands	-
Dolopethin	Gattiker	Switz.	-
Medfina	Carlo Erba	-	-
Pethidine Roche	Roche	UK	-
Supplosal	Specia	France	-

Raw Materials

Thionyl chloride
Sodium amide
Sulfuric acid
Hydrogen chloride
Diethanol methylamine
Benzyl cyanide
Ethanol

Manufacturing Process

80 parts of finely pulverized sodium amide are added in portions each of about ½ of the entire quantity, while stirring and cooling in a suitable manner, to a mixture of 756 parts of methyl-di(β-chloroethyl)-amine (prepared from di-ethanol-methylamine by means of thionyl chloride), 117 parts of benzyl cyanide and 600 parts of toluene. The reaction sets in at once at room temperature. The temperature is maintained between 30° and 40°C; when self-heating no longer occurs a further portion of the sodium amide is introduced. During the reaction heat is liberated and gaseous ammonia escapes.

The mixture is then slowly heated to the boiling point of toluene and kept boiling for one hour under reflux. After the mixture has been allowed to cool the sodium chloride which precipitates is separated by extraction with water. The solution of toluene is then extracted with dilute hydrochloric acid. From the hydrochloric acid extract the basic substance is separated in the form of an oil by means of caustic soda solution and is introduced into ether. The ethereal solution is dried with the aid of potassium carbonate and then distilled.

Under a pressure of 4.5 ml the 1-methyl-4-phenyl-piperidine-4-carboxylic acid nitrile passes over at a temperature of about 148°C in the form of a colorless oil; under a pressure of 6 ml it passes over at about 158°C. After having been allowed to cool the distillate solidifies completely to form a crystalline mass. Its solidification point is at 53°C; the yield amounts to about 135 parts, that is, about 2/3 of the theoretical yield. When recrystallized from isopropyl alcohol the hydrochloride of the nitrile forms colorless crystals, readily soluble in water and melting at 221° to 222°C.

The nitrile may best be saponified with methyl alcoholic potash while heating to 190° to 200°C with application of pressure. After the methyl alcohol has evaporated the salt is introduced into water and by the addition of dilute mineral acid until the alkaline reaction to phenolphthalein has just disappeared, the amphoteric 1-methyl-4-phenyl-piperidine-4-carboxylic acid is precipitated while hot in the form of a colorless, coarsely crystalline powder. When dried on the water bath the acid still contains 1 mol of crystal water which is lost only at a raised temperature. The acid melts at 299°C. Reaction with ethanol yields the ester melting at 30°C and subsequent reaction with HCl gives the hydrochloride melting at 187° to 188°C.

References

Merck Index 5674
Kleeman and Engel p. 707
PDR pp. 872, 1908, 1959, 1989
OCDS Vol. 1 p. 300 (1977); 2, 328 (1980) and 3, 116 (1984)

I.N. p. 750
REM p. 1108
Eisleb, O.; US Patent 2,167,351; July 25, 1939; assigned to Winthrop Chemical Company, Inc.

MEPHENESIN

Therapeutic Function: Muscle relaxant

Chemical Name: 3-(2-Methylphenoxy)-1,2-propanediol

Common Name: o-Cresyl glycerol ether; Glyceryl o-tolyl ether; Cresoxypropanediol; Cresoxydiol

Structural Formula:

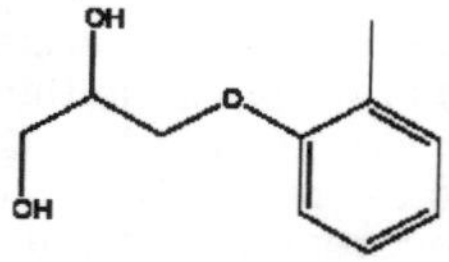

Chemical Abstracts Registry No.: 59-47-2

Trade Name	Manufacturer	Country	Year Introduced
Tolserol	Squibb	US	1948
Oranixon	Organon	US	1949
Avosyl	Schenley	US	-
Curaresin	Kyoto	Japan	-
Decontractyl	Robert and Carriere	France	-
Glyotol	US Standard	US	-
Myanesin	B.D.H.	UK	-
Myanol	Chugai	Japan	-
Myocuran	Deutsches Hydrierwerk	E. Germany	-
Myoserol	Sankyo	Japan	-
Myoxane	Ascher	US	-
Noctynol	Moore	UK	-
Prolax	Cole	US	-
Relaxar	Bouty	Italy	-
Rhex	Hobein	W. Germany	-
Spasmolyn	Heun	US	-
Tolosate	Brewer	US	-
Tolulox	Miller	US	-
Tolyspaz	Chicago Pharmacal	US	-

Raw Materials

3-Cresol
Glycerol

Manufacturing Process

Into an iron or copper reaction vessel having an efficient stirring device and furnished with a refluxing column and condenser, were charged 330 lb of high quality meta-cresol and 150 lb of glycerol, together with 25 lb of sodium acetate to serve as the catalyst in the reaction. The reaction mixture, of this composition, was then heated to 250°C. The water of the reaction distilled off during the heating as the ether formation proceeded, this removal of water from the reaction chamber being promoted by the presence of the excess of phenol, some of which also continued to distill over. Towards the end of the reaction, after about 12 hours, when about 60% of the glycerol had been converted, at which point the reaction slowed down and the distillate was mainly cresol, the batch was cooled and 50 gallons of water were added to it along with 150 lb of xylene. As the result of these additions and the cooling down of the material the batch stratified into an aqueous layer containing unreacted glycerol, polyglycerols and sodium acetate, and a nonaqueous layer containing the ethers that had been formed in the reaction, together with unreacted cresol which remained in the reaction chamber, dissolved in the xylene that had been added to the batch. The aqueous layer was then separated and the water content removed therefrom by evaporation to a degree suitable for the recovery of the glycerol and sodium acetate contents of the layer, for their reuse in the process in a succeeding batch therein. The separated nonaqueous layer containing the ethers was distilled to recover the xylene and cresol contents respectively as the early fractions of the layer thus subjected to distillation. The cresol thus recovered, together with the cresol recovered from the distillate obtained during the heating of the reaction mixture, was returned to the process for reuse in a succeeding batch. Redistillation of the ether mixture recovered is usually necessary and desirable, particularly from the point of view of removing last traces of cresol therefrom. The yield of mixed ethers in this example was about 200 lb, in the relative proportions stated of about 70 parts of monoether to 30 of diether.

References

Merck Index 5675
Kleeman and Engel p. 556
OCDS Vol. 1 p. 118 (1977)
I.N. p. 593
Carroll, M.F. and A. Boake Roberts and Co., Ltd.; British Patent 589,821; July 1, 1947

MEPHENESIN CARBAMATE

Therapeutic Function: Muscle relaxant

Chemical Name: 3-(2-Methylphenoxy)-1,2-propanediol 1-carbamate

Common Name: -

Structural Formula:

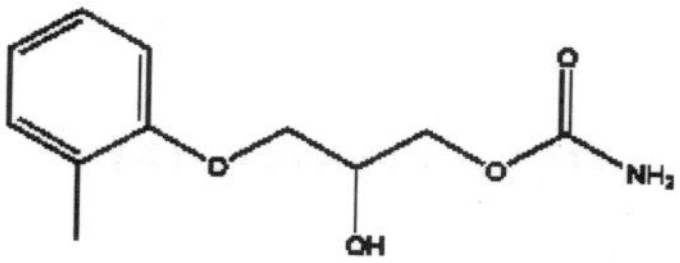

Chemical Abstracts Registry No.: 533-06-2

Trade Name	Manufacturer	Country	Year Introduced
Tolseram	Squibb	US	1954
Kinavosyl	Schenley	US	-

Raw Materials

3-o-Toloxy-1,2-propanediol
Phosgene
Ammonia

Manufacturing Process

A solution of 32 g (0.30 mol) phosgene in 200 ml benzene is added dropwise at 30°C to a stirred solution of 53.5 g (0.32 mol) 3-o-toloxy-1,2-propanediol in 400 ml benzene. The mixture is stirred for an hour after the addition is completed, and a solution of 39 g of dimethylaniline in 100 ml benzene is then added, and stirring continued for a half-hour. Ice water (about one-third volume) is then added, and the benzene layer formed is separated and stirred with 500 mi concentrated ammonia at 5°C for six hours. The precipitated solid (weighing about 55 g) is recovered and recrystallized from water. The product thus obtained in a yield of about 53 g is 3-(o-toloxy)-2-hydroxypropyl carbamate: it is a crystalline solid melting at about 93°C, and having a lower water-solubility and higher oil-solubility than 3-o-toloxy-1,2-propanediol.

References

Merck Index 5676
Kleeman and Engel p. 556
OCDS Vol. 1 p. 118 (1977)
I.N. p. 593
Lott, W.A. and Pribyl, E.; US Patent 2,609,386; September 2, 1952; assigned to E.R. Squibb and Sons

MEPHENOXALONE

Therapeutic Function: Tranquilizer

Chemical Name: 5-[(o-Methoxyphenoxy)methyl]-2-oxazolidinone

Common Name: Methoxadone

Structural Formula:

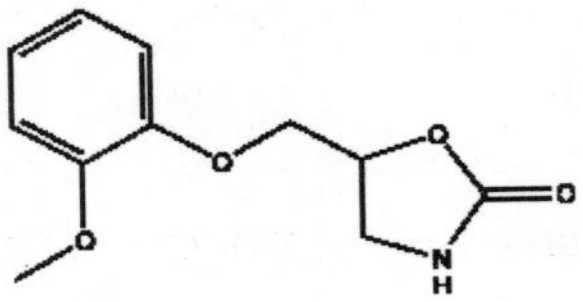

Chemical Abstracts Registry No.: 70-07-5

Trade Name	Manufacturer	Country	Year Introduced
Trepidone	Lederle	US	1961
Tranpoise	Robins	US	1962
Lenetran	Lakeside	US	1962
Xerene	Martinet	France	1964
Control-Om	O.M.	Switz.	-
Dorsiflex	Syntex-Medical	Switz.	-
Placidex	Toraude	-	-
Riself	Gibipharma	Italy	-

Raw Materials

3-o-Methoxyphenoxy-2-hydroxy-1-propylcarbamate
Urea

Manufacturing Process

A mixture of 24.1 g (0.10 mol) of 3-o-methoxyphenoxy-2-hydroxy-1-propyl carbamate and 6.0 g (0.10 mol) of urea was heated rapidly to the temperature range of 180°C to 200°C, and maintained there for five hours. The reaction melt was poured into 50% ethyl alcohol, from which the product crystallized as a white solid. The crude yield was 18.3 g (82%); melting point 131.5% to 137°C. Crystallization from water and 95% alcohol gave 9.0 g (40.3 %) of pure 5-o-methoxyphenoxymethyl-2-oxazolidone; melting point 141°C to 143°C. This melting point was not depressed when the material was mixed with an authentic sample. In additional runs acetone was used instead of ethyl alcohol with equivalent results.

It was found that when the heating time was reduced to three hours and a reaction temperature of 190°C to 200°C was maintained, equivalent yields (40 to 50%) were obtained, but that the yields were appreciably lowered when the heating time was further reduced to two hours. It was also found that when the temperature was lowered to the range of 170°C to 180°C the yield was significantly lowered.

When the material was isolated by extraction with chloroform and distillation, the yield of pure material was 58.5%.

References

Merck Index 5679
OCDS Vol. 1 p. 119 (1977)
I.N. p. 593
Lunsford, C.D.; US Patent 2,895,960; July 21, 1959; assigned to A.H. Robins Co., Inc.

MEPHENTERMINE

Therapeutic Function: Adrenergic (vasopressor)

Chemical Name: N,α,α-Trimethylbenzene ethanamine

Common Name: -

Structural Formula:

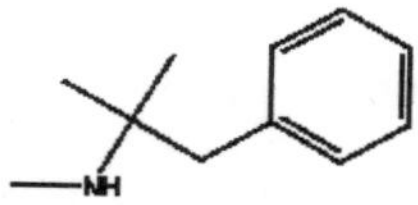

Chemical Abstracts Registry No.: 100-92-5

Trade Name	Manufacturer	Country	Year Introduced
Wyamine	Wyeth	US	1947

Raw Materials

2-(N-Methylamino)-2-methyl-1-phenyl-1-propanol
Thionyl chloride
Hydrogen

Manufacturing Process

0.5 g of 2-(N-methylamino)-2-methyl-1-phenyl-1-propanol was treated with 1 cc of thionyl chloride at room temperature. A vigorous reaction set in. The gummy material was stirred with a small amount of petroleum ether and allowed to stand overnight. The brown crystalline solid after washing with petroleum ether was recrystallized from a small amount of absolute alcohol with addition of charcoal followed by filtration. On dilution with several volumes of ether and refrigeration white granular crystals of 1-chloro-2-(N-methamino)-2-methyl-1-phenyl propane hydrochloride were deposited.

250 mg of 1-chloro-2-(N-methylamino)-2-methyl-1-phenyl propane hydrochloride was dissolved in 2 cc of warm methanol and hydrogenated in the presence of 250 mg of palladium barium carbonate catalyst with provision for the absorption of the carbon-dioxide formed. When the theoretical amount of hydrogen had been taken up the mixture was filtered to remove the catalyst, concentrated to small volume and extracted with ether. After separating the ether the residue was further concentrated yielding a white crystalline solid. This solid on solution in water, strongly alkalizing, extraction with ether and removal of the ether yielded 2-(N-methylamino)-2-methyl-1-phenyl propane identified as the picrate by melting point 155°C to 156°C and mixed melting point 154.0°C to 154.5°C, with an authentic sample melting at 150°C to 153°C.

References

Merck Index 5680
OCDS Vol. 1 p. 72 (1977)
I.N. p. 593
REM p. 887 Bruce, W.F., Szabo, J.L. and Tubis, S.; US Patent 2,597,445; May 28, 1952; assigned to Wyeth, Inc.

MEPHENYTOIN

Therapeutic Function: Anticonvulsant, Antiepileptic

Chemical Name: 2,4-Imidazolidinedione, 5-ethyl-3-methyl-5-phenyl-

Common Name: Mefenetoin; Mephenetoinum; Methantoinum; Mephenytoin; Methoin; Methylphenetoin; Methylphenylaethylhydatoinum

Structural Formula:

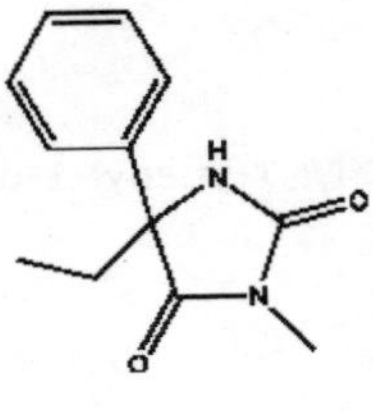

Chemical Abstracts Registry No.: 50-12-4

Trade Name	Manufacturer	Country	Year Introduced
Mesantoin	Sandoz	-	-
Sedantoinal	Sandoz	-	-
Gerot-Epilan	Gerot	-	-
Sacerno	EGYT	-	-

Raw Materials

Sodium
5-Phenylcyanacetamide
Bromine
Sodium bisulfite

Manufacturing Process

23 parts sodium was dissolved in 300 parts of ethanol and added to 160 parts of 5-phenylcyanacetamide in 750 parts of ethanol. A mixture was cooled straight away and a sodium salt of amide precipitated as a white powder. 200 parts of ethyl iodide was added to this mixture and heated for 1.5 hours. The ethanol was distilled off, water was added to the residue and rapidly hardened oil precipitated. After recrystallization from ethanol, 5-ethyl-5-phenylacetamide afforded; MP: 116°C.

100 parts of sodium hydroxide was solved in 500 parts of water and added to 83 parts of bromine by cooling. 5-Ethyl-5-phenylacetamide was added to above prepared mixture. It dissolved quickly, whereupon all mass was heated some time, cooled and stood at room temperature some hours. Then a solution of sodium bisulfite was added before the formed precipitate dissolved. The reaction mixture was filtered, the filtrate was acidified to give rapidly hardened oil. After recrystallization from ethanol 5-ethyl-3-methyl-5-phenyl-hydantoin was yielded as the bright needles; MP: 201°-202°C.

References

Chemishe Fabrik von Heyden Akt-Ges. in Radedeul b. Dresden; D.R. Patent No. 308,508; May 1914

MEPICYCLINE

Therapeutic Function: Antimicrobial

Chemical Name: 4-(Dimethylamino)-1,4,4a,5,5a,6,11,12a-octahydro-3,6,10,12,12a-pentahydroxy-N-[[4-(2-hydroxyethyl)-1-piperazinyl]methyl]-6-methyl-1,11-dioxo-2-naphthacenecarboxamide

Common Name: N-[[4-(2-Hydroxyethyl)-1-piperazinyl]methyl]tetracycline; Pipacycline

Trade Name	Manufacturer	Country	Year Introduced
Sieromicin	Sierochimica	Italy	1962
Ambra-Vena	Lepetit	-	-
Boniciclina	Boniscontro-Gazzone	Italy	-
Tetrasolvina	N.C.S.N.	Italy	-
Valtomicina	Midy	-	-

Structural Formula:

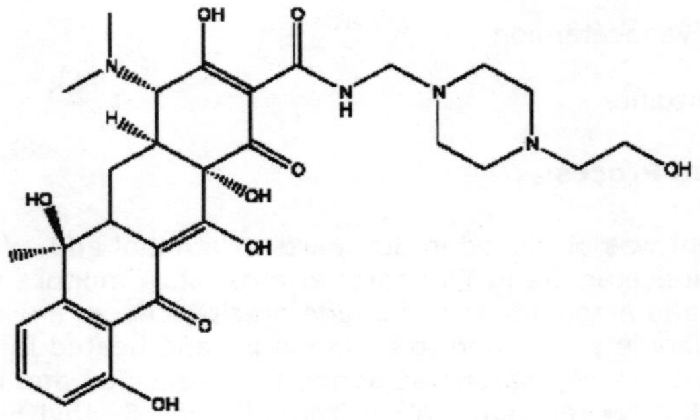

Raw Materials

N-(β-Hydroxyethyl)diethylene diamine
Paraformaldehyde
Tetracycline

Manufacturing Process

1.55 g p-formaldehyde were added to a solution of 7 g N-(β-hydroxyethyl)-diethylene diamine in 150 cc isopropanol and the whole was heated to 60°C for 30 minutes, to obtain complete dissolution; after cooling the solution to 40°C, 22.2 g of anhydrous tetracycline base were added as a fine powder and the reaction was allowed to proceed for 3 hours with agitation and while passing through a current of dry nitrogen; the solution was then filtered on a Buchner funnel and the filter cake was washed twice with 20 cc isopropanol; the crystalline cake was resuspended in 100 cc anhydrous ether, again filtered and washed 3 times with 50 cc anhydrous ether; finally, it was dried in vacuo and 28.6 g of product were obtained, namely a yield of 98%.

The characteristics of this product are as follows. It is a pale yellow, nonodorous, slightly bitter, crystalline powder, very soluble in water (>1.5 g/cc), soluble in methanol and formamide, slightly soluble in ethanol and isopropanol, insoluble in ether, benzene and chloroform; MP 162° to 163°C with decomposition (uncorrected).

References

Merck Index 7325
I.N. p. 775
Gradnik, B., Pedrazzoli, A. and Cipelletti, G.; US Patent 3,149,114; September 15, 1964; assigned to Societe d'Etudes de Recherches et d'Applications Scientifiques et Medicales, France

MEPINDOLOL

Therapeutic Function: Beta-adrenergic blocker

Chemical Name: 4-(2-Hydroxy-3-isopropylaminopropoxyl)-2-methylindole

Common Name: -

Structural Formula:

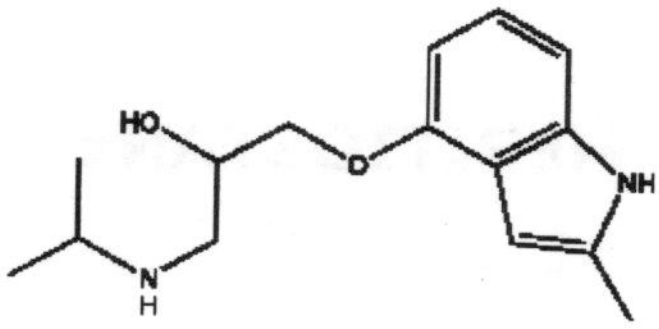

Chemical Abstracts Registry No.: 56396-94-2 (Sulfate salt)

Trade Name	Manufacturer	Country	Year Introduced
Corindolan	Schering	W. Germany	1980

Raw Materials

4-Benzyloxy-2-dimethylamino-methylindole
Hydrogen
Epichlorohydrin
Isopropylamine

Manufacturing Process

The 4-hydroxy-2-methylindole (MP 112°C to 115°C from benzene/ethyl acetate), used as starting material, may be obtained by hydrogenation of 4-benzyloxy-2-dimethylamino-methylindole (MP 117°C to 120°C from benzene) in the presence of a palladium catalyst (5% on aluminum oxide).

11.6 g of 4-hydroxy-2-methylindole are added to a solution of 3.1 g of sodium hydroxide in 150 cc of water, and then 12.4 cc of epichlorhydrin are added while stirring and in an atmosphere of nitrogen. The reaction mixture is further stirred at room temperature for 24 hours, is extracted 4 times with methylene chloride, and the combined organic layers which have been dried over magnesium sulfate are concentrated by evaporation at reduced pressure. The resulting residue is taken up in 150 cc of dioxane and 50 cc of isopropylamine, and the mixture is heated to the boil for 6 hours. The reaction mixture is evaporated to dryness at reduced pressure, the residue is shaken 4 times between ethyl acetate and a 1 N aqueous tartaric acid solution, and a 5 N caustic soda solution is then added to the combined tartaric acid phases until an alkaline reaction is obtained. The alkaline solution is then shaken out 6 times with methylene chloride, the combined extracts are dried over magnesium sulfate, and the solvent is evaporated in a vacuum. The oily viscous residue may be crystallized from ethyl acetate. The title compound has a MP of 95°C to 97°C.

References

Merck Index 5684
DFU 3 (5) 381 (1978)
DOT 17 (10) 426 (1981) and 18 (10) 551 (1982)
I.N. p. 594
Troxler, F. and Hofmann, A.; British Patent 1,260,907; January 19, 1972; assigned to Sandoz, Ltd.

MEPITIOSTANE

Therapeutic Function: Antiestrogen

Chemical Name: 17β-(1-Ethoxycyclopentyl)oxy-2α,3α-epithio-5α-androstane

Common Name: -

Structural Formula:

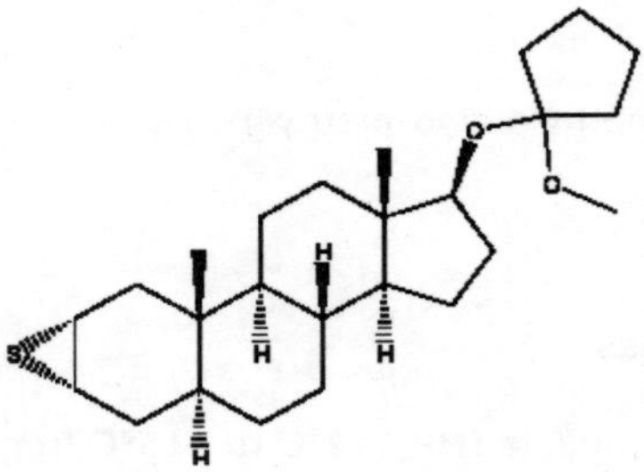

Chemical Abstracts Registry No.: 21362-69-6

Trade Name	Manufacturer	Country	Year Introduced
Thioderon	Shionogi	Japan	1979

Raw Materials

2α,3α-Epithio-5α-androstan-17β-ol
Methoxycyclopentene

Manufacturing Process

A mixture of 1.759 g of 2α,3α-epithio-5α-androstan-17β-ol, 2.3 ml of 1-methoxycyclopentene, 20 mg of pyridine salt of p-toluenesulfonic acid and 20 ml of t-butanol is stirred for 4 hours at room temperature. The reaction mixture is poured into an aqueous solution of sodium carbonate and the whole extracted with dichloromethane. The extract is dried over anhydrous sodium sulfate and evaporated to remove solvent. Purification of the residue by chromatography over alumina gives 1.487 g of 17β-(1-methoxycyclopentyl) oxy-2α,3α-epithio-5α-androstane. Yield 68.2%. MP 98°C to 101°C.

References

Merck Index 5687
DFU 3 (4) 311 (1978)
Kleeman and Engel p. 557
I.N. p. 594
Komeno, T.; US Patent 3,567,713; March 2, 1971; assigned to Shionogi and Co.

MEPIVACAINE

Therapeutic Function: Local anesthetic

Chemical Name: N-(2,6-Dimethylphenyl)-1-methyl-2-piperidinecarboxamide

Common Name: N-Methylpipecolic acid 2,6-dimethylanilide

Structural Formula:

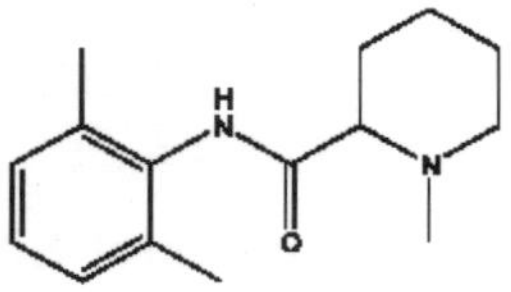

Chemical Abstracts Registry No.: 96-88-8; 16452-56-5 (Hydrochloride salt)

Trade Name	Manufacturer	Country	Year Introduced
Carboraine	Winthrop	US	1960
Chlorocain	Pharmac. Mfg.	UK	-
Isocaine	Novocol	US	-
Meaverin	Woelm Pharma	W. Germany	-
Mepivastesin	Espe	W. Germany	-
Scandicain	Astra	Sweden	-
Tevacaine	Teva	Israel	-

Raw Materials

Ethyl bromide

Magnesium
N-Methylpipecolic acid ethyl ester
2,6-Dimethylaniline

Manufacturing Process

Ethyl magnesium bromide is prepared in the usual way by reacting 185 parts by weight of ethyl bromide in 800 parts of anhydrous ether with 37 parts by weight of magnesium turnings. Under vigorous stirring 121 parts of 2,6-dimethyl aniline are added at a rate depending on the vigor of the gas evaporation. When the evolution of gas has ceased, 85 parts by weight of methylpipecolic acid ethyl ester are added to the 2,6-dimethyl aniline magnesium bromide slurry. The mixture is refluxed for ½ hour with continued stirring, after which it is cooled down. Dilute hydrochloric acid is added carefully in order to dissolve and hydrolyze the magnesium compound formed.

The pH is adjusted to 5.5 and the water phase separated and extracted with additional ether in order to remove the surplus dimethyl aniline. After addition of an excess of ammonia to the solution, the reaction product, N-methylpipecolic acid 2,6-dimethyl anilide, is recovered by extraction with isoamyl alcohol. The isoamyl alcohol solution is evaporated to dryness, the product dissolved in dilute hydrochloric acid, treated with charcoal and reprecipitated with NaOH. N-methylpipecolic acid 2,6-dimethyl anilide is obtained in crystalline form.

References

Merck Index 5688
Kleeman and Engel p. 558
PDR pp. 824, 1906
OCDS Vol. 1 p. 17 (1977)
I.N. p. 594
REM p. 1052
af Ekenstam, B.T. and Egner, B.P.H.; US Patent 2,799,679; July 16, 1957; assigned to AB Bofors, Sweden
Pettersson, B.G.; US Patent 4,110,331; August 29, 1978; assigned to AB Bofors

MEPREDNISONE

Therapeutic Function: Glucocorticoid

Chemical Name: 17,21-Dihydroxy-16β-methylpregna-1,4-diene-3,11,20-trione

Common Name: 16β-Methylprednisone

Chemical Abstracts Registry No.: 1247-42-3

Structural Formula:

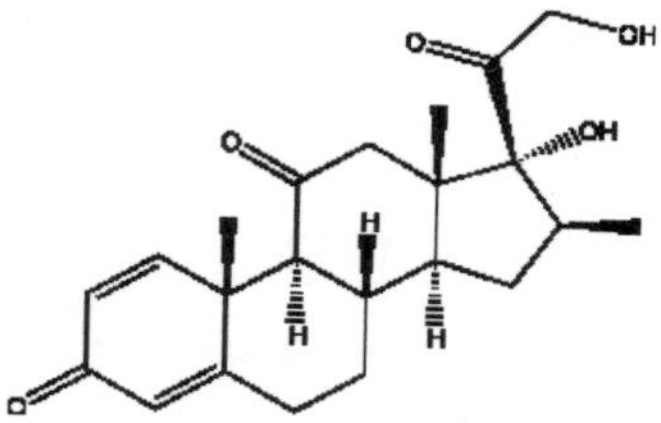

Trade Name	Manufacturer	Country	Year Introduced
Betapar	Parke Davis	US	1970
Betalone	Lepetit	France	-
Betapred	Schering	US	-
Corti-Bi	Sidus	Italy	-

Raw Materials

16β-Methylprednisone-21-acetate
Potassium bicarbonate
Bacterium *Bacillus sphaericus var. fusifermis*
Nutrient broth

Manufacturing Process

16β-Methylprednisone 21-acetate (0.5 g), when hydrolyzed by means of aqueous alcoholic potassium bicarbonate yields 16β-ethylprednisone. An alternative method of the preparation of the compound of this example is as follows. Bacillus *sphaericus var. fusifermis* (A.T.C.C. 7055) is incubated on a nutrient agar (composed of Bacto-beef extract, 3 g; Bacto-peptone, 5 9; sodium chloride, 8 9; agar, 15 g; and tap water, 1 liter) for 24 hours at 28°C.

To 100 ml of a sterile nutrient broth (composed of Bacto-beef extract, 3 9; Bacto-peptone, 5 9; per liter of tap water) in a 300 ml flask is added one loopful of the incubated culture and the broth mixture is further incubated for 24 hours at 28°C on a shaking machine. The broth culture so obtained is employed as an inoculum (1%). Into each of ten flasks containing 100 ml of sterile nutrient broth is added 1 ml of the inoculum. The flasks are agitated on a rotary shaker for 8 hours at 28°C at 240 strokes per minute. After this growth period, a solution of 25 mg of 16β-methylcortisonein 0.5 ml of methanol is aseptically added to each flask which in turn is reshaken and incubated for an additional 24 hours. The final pH is 7.8.

The contents of the flasks are then combined and extracted 3 times with two liters of chloroform per extraction. The combined chloroform extracts are evaporated to dryness yielding 310 mg of crude product. The crude steroid is purified by chromatography on a chromatographic system described by G.M. Shull, Abstracts of Papers of the 126th Meeting of the American Chemical Society, December 12-17, 1954, page 9a, paper No. 24. Chromatographic evaluation shows a quantitative conversion of the starting material to the diene when an authentic sample of the 16β-methylprednisone is used as a control. Alternatively, the crude product is recrystallized from acetone affording 225 mg of 16β-methylprednisone.

References

Merck Index 5689
Kleeman and Engel p. 558
I.N. p. 595
Rausser, R. and Oliveto, E.P.; US Patent 3,164,618; January 5, 1965; assigned to Schering Corporation

MEPROBAMATE

Therapeutic Function: Tranquilizer

Chemical Name: 2-Methyl-2-propyl-1,3-propanediol dicarbamate

Common Name: Procalmadiol; Procalmidol

Structural Formula:

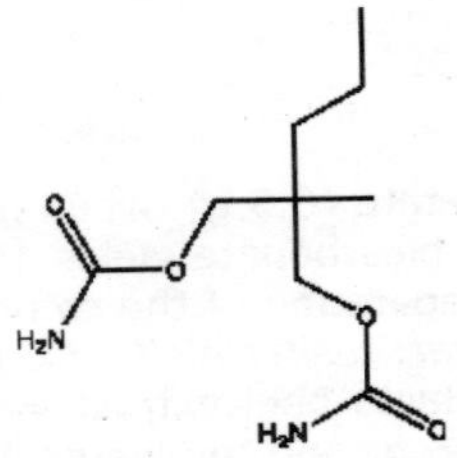

Chemical Abstracts Registry No.: 57-53-4

Trade Name	Manufacturer	Country	Year Introduced
Equanil	Wyeth	US	1955
Miltown	Wallace	US	1955
Meprotabs	Wallace	US	1957
Meprospan	Wallace	US	1958
Viobamate	Rowell	US	1963
Meprocon	Consol. Midland	US	1964
Canquil	Canfield	US	1964
Klort	Lemmon	US	1964
Equanil	Clin Midy	France	1967
SK-Bamate	SKF	US	1971
Amepromamat	Arcana	Austria	-
Amosene	Ferndale	US	-

Trade Name	Manufacturer	Country	Year Introduced
Aneural	Wyeth	W. Germany	-
Ansietan	Italfarmaco	Italy	-
Ansiowas	Wassermann	Spain	-
Artolon	Roter	Netherlands	-
Atraxin	Daiichi	Japan	-
Carb-A-Med	Chemieprodukte	Austria	-
Coprobate	Coastal	US	-
Cyrpon	Tropon	W. Germany	-
Dabrobamat	Dabrowski	W. Germany	-
Dapaz	Alter	Spain	-
Deprol	Wallace	US	-
Dormabrol	Kwizda	Austria	-
Dystoid	Makara	W. Germany	-
Ecuanil	Orfi	Spain	-
Edenal	Wassermann	Italy	-
Epikur	Agepha	Austria	-
Equagesic	Wyeth	US	-
Erina	Sumitomo	Japan	-
Gene-Bamate	Franca	Canada	-
Harmonin	Yoshitomi	Japan	-
Kesso-Bamate	McKesson	US	-
Lan-Dol	Bio-Chimique	Canada	-
Marbate	Mardale	US	-
Meditran	Medic	Canada	-
Mepavlon	I.C.I.	UK	-
Meprate	DDSA	UK	-
Mepriam	Lennon	US	-
Mepro	Rekah	Israel	-
Meproban	Draco	Sweden	-
Meprocon CMC	Consol. Midland	US	-
Meprodil	Streuli	Switz.	-
Meprodiol	Pirri	Italy	-
Meprol	Lokman	Turkey	-
Mepron	Choseido	Japan	-
Mepron	Hamilton	Australia	-
Mepronel	Heather Drug	US	-
Meprosa	Chemipharma	W. Germany	-
Meprotil	Brunner-Tillman	US	-
Meriprobate	Meriot	Canada	-
Microbamat	Werfft	Austria	-
Midixin	Reid-Provident	US	-
Miltaun	Mack	W. Germany	-
Misedant	Lemmon	US	-
M.P. Trantabs	Martin-Phillips	US	-
My-Trans	Heather Drug	US	-
Neo-Tran	Neo	Canada	-
Nervonus	Orion	Finland	-

Trade Name	Manufacturer	Country	Year Introduced
Neuramate	Halsey	US	-
Novamato	Torlan	Spain	-
Novomepro	Novopharm	Canada	-
Oasil	Simes	Italy	-
Paxin	Pierrel	Italy	-
Pensive	Norbrook	UK	-
Perequil	Lepetit	Italy	-
PMB Ayerst	Ayerst	US	-
Probasan	I.C.N.	Canada	-
Quietidon	Pharma. Farm. Spec.	Italy	-
Relaksin	Deva	Turkey	-
Restanil	Kabi	W. Germany	-
Sedanyl	Washington	Italy	-
Selene	Biomedica Foscama	Italy	-
Sopanil	Sopar	Belgium	-
Sowell	Cophar	Switz.	-
Stensolo	Salfa	Italy	-
TCM	Zenith	US	-
Trankilin	Biofarma	Turkey	-
Tranlisant	Vita	Canada	-
Trelmar	Elliott-Marion	Canada	-
Urbilat	Hor-Fer-Vit	W. Germany	-
Wescomep	Saunders	Canada	-
Xalogen	Ono	Japan	-

Raw Materials

2-Methyl-2-propyl-1,3-propanediol
Phosgene
Ammonia

Manufacturing Process

A solution containing 52.8 parts of 2-methyl-2-n-propyl-1,3-propanediol and 128 parts of acetone is added with stirring to 112 parts of liquid phosgene at such a rate that the temperature of the reaction is maintained at -5° to 0°C. The reaction is stirred one hour at about 0°C then cooled to -15°C. A cooled 30% solution of 32 parts of sodium hydroxide is added with stirring to the reaction at such a rate that the temperature is maintained at -15° to -5°C. The mixture is stirred for an additional ½ hour at about 0°C then cooled to -20°C. 180 parts of cooled ammonium hydroxide solution (28.6% NH_3) are added while cooling and with stirring at such a rate that the temperature rises slowly to 20°C and stirring is continued for an additional ½ hour. The mixture is poured with agitation into 1,700 parts of ice water. The solid which separates is removed by filtration and dried. Recrystallization from water gives 55 parts (63% of theoretical yield) of 2-methyl-2-n-propyl-1,3-propanediol dicarbamate, MP 104° to 105°C.

References

Merck Index 5690
Kleeman and Engel p. 559
PDR pp. 634, 830, 1024, 1606, 1723, 1874, 1880, 1947, 1949
OCDS Vol. 1 p. 218 (1977) and 2, 21 (1980)
I.N. p. 595
REM p. 1072
Berger, F.M. and Ludwig, B.J.; US Patent 2,724,720; November 22, 1955; assigned to Carter Products, Inc.

MEPROSCILLARIN

Therapeutic Function: Cardiotonic

Chemical Name: 14-β-Bufa-4,20,22-trienolide, 3-β-((6-deoxy-4-O-methyl-α-L-mannopyranosyl)oxy)-14-hydroxy-

Common Name: Meproscillarin; Methylproscilleridin; Rambufaside

Structural Formula:

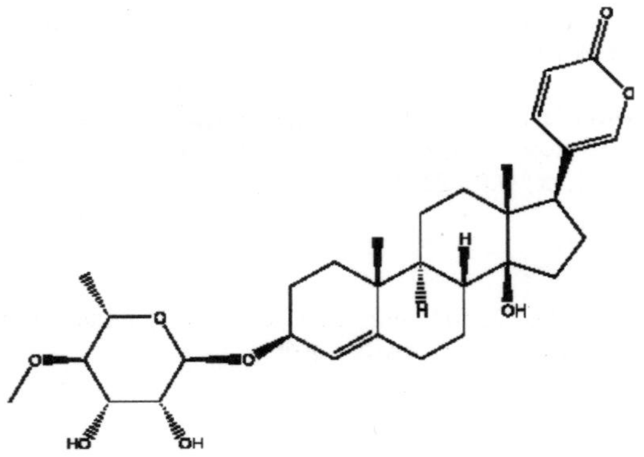

Chemical Abstracts Registry No.: 33396-37-1

Trade Name	Manufacturer	Country	Year Introduced
Clift	Knoll	-	-
Clift	Abbott Laboratories	-	-
Talusin	Lek D.D.	-	-

Raw Materials

Proscillaridin

Triethyl orto-formiate
4-Toluenesulfonic acid
Methyl iodide
Sodium hydride

Manufacturing Process

100 g proscillaridin (from *Scilla maritima L.*, enzymatic hydrolysis) was dissolved in 500 ml dry tetrahydrofuran, mixed with 100 ml of triethyl orto-formiate and 50 mg p-tholuene sulfonic acid and stirred for 15 minutes at 20°C.

It was put into a separating funnel and shook with 1 L ethyl acetate and 200 ml of 5% sodium hydroxide. The an organic layer was separated, with 2-3 L water washed (portions 400-500 ml), dried over sodium sulfate and distilled in vacuum to dryness at about 60°C. 121.2 g crude proscillaridin-2,3-ethyl orto-formiate yielded. It was dissolved in 1 L dimethylformamide, mixed with 200 ml methyl iodide and stirred with 20 g 55-60% suspension of sodium hydride at 20°C for 1 hour. 14 L ethyl acetate was added, 5 times with 1-2 L water shook and the organic layer was distilled to 1/4 of volume. The solution of proscillaridin-2,3-ethyl ortho-formiate-4-methyl ester obtained (about 1 L) was mixed with 2 L 0.002 N HCl and stood for 2 hours at 20°C. Then it was neutralized with 0.1 N sodium hydroxide and distilled in vacuum to about 1 L. The solution was shook with 2 L chloroform and 1 L water, organic layer was separated, water layer was 2 times was extracted with still 1 L chloroform and the pooled organic phase dried over sodium sulfate. Then the solvent was removed and 147 g of obtained product was purified by chromatography on silica gel in system chloroform/acetone 4:1.

63 g of crude amorphic 4-O-methylproscillaridin was isolated and recrystallized from methylene chloride/ethyl acetate to give 49.3 g (53% yield) the desired product; MP: 213°-217°C.

References

Kubinyi H.; D.B. Patent No. 2,301,382; Jan. 12, 1973; Assigned to Knoll AG, Chemische Fabriken, 6700 Ludwigshafen

MEPRYLCAINE HYDROCHLORIDE

Therapeutic Function: Local anesthetic

Chemical Name: 1-Propanol, 2-methyl-2-(propylamino)-, benzoate, hydrochloride

Common Name: Meprylcaine hydrochloride

Structural Formula:

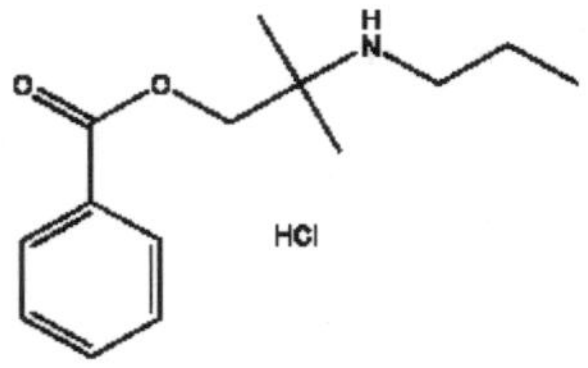

Chemical Abstracts Registry No.: 956-03-6; 495-70-5 (Base)

Trade Name	Manufacturer	Country	Year Introduced
Oracaine	ALK-Abello	-	-
Meprylcaine hydrochloride	Shanghai Lansheng Corporation	-	-

Raw Materials

N-(1,1-Dimethyl-2-hydroxyethyl)propylamine
Benzoyl chloride

Manufacturing Process

78 g of N-(1,1-dimethyl-2-hydroxy-ethyl)propylamine is added to a solution of 30 g of sodium hydroxide in 700 ml of water. To this is added 300 ml of ether and this is followed by the dropwise addition of 70 ml of benzoyl chloride benzoyl chloride while stirring and cooling. The benzoyl chloride is added at such a rate that the temperature does no rise above 30°C. When the addition is complete, the stirring is continued for another 30 minutes. The aqueous layer is removed and the ether layer is washed with water, dried and evaporated to leave an yellow oil 1-propanol, 2-methyl-2-(propylamino)-, benzoate. This is treated slowly with 45 ml of concentrated hydrochloric acid, during which addition a vigorous exothermic reaction ensues. When the reaction mixture is cooled, it solidifies to a pasty solid which is allowed to dry. This is dissolved in boiling isopropanol and allowed to cool when white crystals of the hydrochloride are formed. The resultant slurry is filtered and the hydrochloride of 1-propanol, 2-methyl-2-(propylamino)-, benzoate recrystallized from isopropanol to give white crystals; MP: 150°-151°C. It is useful in base and salt forms as such as a topical local anesthetic being administrable in oil or alcohol solution.

In practice it is usually used as hydrochloride.

References

Reasenberg J.R.; US Patent No. 2,767,207; Oct. 16, 1956; Assigned to Mizzy Inc., a corporation of New York

MEPTAZINOL

Therapeutic Function: Analgesic

Chemical Name: 3-Ethyl-3-(m-hydroxyphenyl)-1-methylhexahydro-1H-azepine

Common Name: -

Structural Formula:

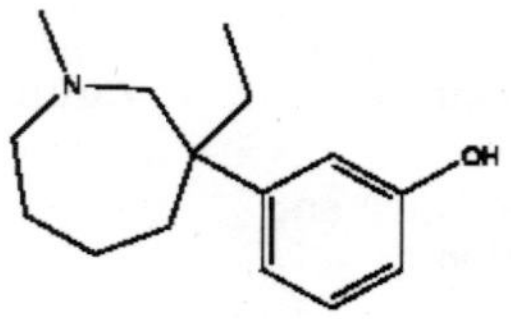

Chemical Abstracts Registry No.: 54340-58-8

Trade Name	Manufacturer	Country	Year Introduced
Meptid	Wyeth	UK	1983

Raw Materials

Formaldehyde
Sodium amide
Hydrogen
Hydrogen bromide
2-(m-Methoxyphenyl)butyronitrile
Ethyl-4-iodobutyrate
Lithium aluminum hydride

Manufacturing Process

2-(m-Methoxyphenyl)butyronitrile in dry ether was added to a stirred suspension of sodium amide in liquid ammonia. The mixture was stirred for 30 minutes then ethyl-4-iodobutyrate (99.25 g, 0.4 mol) in dry ether (200 ml) was added dropwise. The mixture was stirred at the temperature of refluxing liquid ammonia for 5 hours. Ammonium chloride (10 g) was added and the mixture allowed to warm to room temperature. Water (300 ml) was added, the organic layer separated, washed with water, 2 N sulfuric acid and water. After drying over magnesium sulfate and removing the ether, the product was distilled yielding ethyl 5-cyano-5-(mmethoxyphenyl)heptanoate.

That material was hydrogenated in cyclohexane using a Raney nickel catalyst. The product after distillation was recrystallized from ethyl acetate affording 10.0 g of 6-ethyl-(m-methoxyphenyl)hexahydro-2H-azepin-2one, MP 87°C to 88°C.

The azepinone (9.1 g) in dry tetrahydrofuran (50 ml) and ether (50 ml) was added dropwise to a stirred suspension of aluminum lithium hydride (7.5 g) in dry ether (50 ml). After heating under reflux for 3 hours the reaction mixture was worked up and distilled yielding 7.66 g of a compound which was a colorless oil, BP 108°C to 110°C/0.01 mm.

That product was then heated under reflux with 50% hydrobromic acid for 1.5 hours. The reaction mixture was evaporated to dryness and reevaporated with three portions of propan-2-ol. The oil obtained was dissolved in propan-2-ol and diluted with ether. 3-Ethyl-3-(m-hydroxyphenyl)hexahydro-1H-azepine was obtained. That material in turn was reductively methylated by hydrogenation in the presence of formaldehyde in absolute ethanol solution to give 3-ethyl-3-(m-methoxyphenyl)-1-methylhexahydro-1H-azepine.

The methoxy group was converted to a hydroxy group by refluxing with 80% HBr giving meptazinol hydrobromide.

References

Merck Index A-8
DFU 1 (2) 68 (1976)
DOT 19 (7) 415 (1983)
I.N. p. 597
Cavalla, J.F. and White, A.C.; British Patent 1,285,025; August 9, 1972; assigned to John Wyeth & Brother Ltd.
Cavalla, J.F. and White, A.C.; US Patent 3,729,465; April 24, 1973; assigned to John Wyeth & Brother Ltd.
Cavalla, J.F. and White, A.C.; US Patent 4,197,241; April 8, 1980; assigned to John Wyeth & Brother Ltd.

MEQUITAZINE

Therapeutic Function: Antihistaminic

Chemical Name: 10-(1-Azabicyclo[2.2.2]oct-3-yl-methyl)-10H-phenothiazine

Common Name: -

Structural Formula:

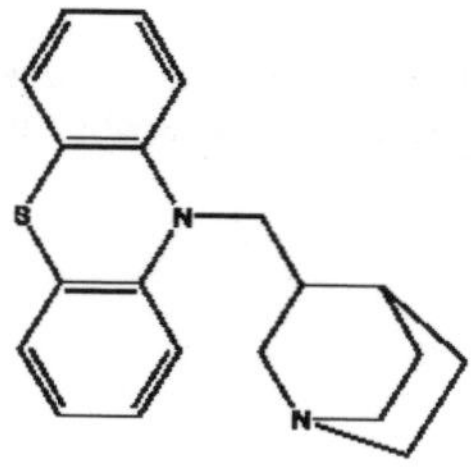

Chemical Abstracts Registry No.: 29216-28-2

Trade Name	Manufacturer	Country	Year Introduced
Primalan	Berk	UK	1976
Primalan	Spret-Mauchant	France	1976
Metaplexan	Bad. Arzneimittel	W. Germany	1977
Nipolazin	Nippon Shoji	Japan	1983
Zesulan	Toyo Jozo	Japan	1983
Instotal	IMA	Argentina	-
Mircol	Pharmuka	Belgium	-
Vigigan	Spret-Mauchant	France	-

Raw Materials

Phenothiazine
Sodium amide
3-Chloromethyl quinuclidine HCl

Manufacturing Process

30 g of phenothiazine were added, all at once, to a suspension of 6 g of sodium amide in 240 ml of anhydrous xylene. The mixture was agitated and heated to reflux. When evolution of ammonia ceased (5 hours), 15 g of 3-chloromethyl-quinuclidine hydrochloride were added portionwise over a period of 50 minutes and reflux was then maintained for 22 hours. After cooling to room temperature, 250 ml of distilled water and 250 ml of ethyl acetate were added to the reaction mixture. The aqueous phase was decanted and extracted twice with a total of 250 ml of methyl acetate. The combined organic extracts were extracted three times with a total of 750 ml of a 10% aqueous solution of tartaric acid. The combined acid solutions were treated with 5 g of animal charcoal, filtered and rendered alkaline on an ice bath with 96 ml of 10 N aqueous caustic soda. The oil which separated was extracted three times with a total of 1.500 ml of ethyl acetate. The combined organic extracts were washed to neutrality by washing twice with a total of 1 liter of distilled water, dried over anhydrous magnesium sulfate and evaporated under reduced pressure on a water bath at 45°C. 17 g of oil were obtained which was purified by chromatography on an inert alumina column. 13.3 g of crystallized product were obtained. 10-(3-Quinuclidinyl-methyl)-phenothiazine having a MP of 130°C to 131°C was obtained by recrystallization in boiling acetonitrile.

The 3-chloromethyl-quinuclidine hydrochloride used as starting material in this process can be obtained as described by Grob and coll., Helv. Chim. Acta, 37 (1954),1689.

References

Merck Index 5694
Kleeman and Engel p. 562
DOT 15 (4)199 (1979)
I.N. p. 597
Gueremy, C., Labey, R., Wirth, D. and Auclair, M.; US Patent 3,987,042; October 19, 1976

MERALLURIDE

Therapeutic Function: Diuretic

Chemical Name: [3-[[[(3-Carboxy-1-oxopropyl)amino]carbonyl]amino]-2-methoxypropyl]-hydroxymercury mixture with 3,7-dihydro-1,3-dimethyl-1H-purine-2,6-dione

Common Name: [3-[3-(3-Carboxypropionyl)ureido]-2-methoxypropyl] hydroxymercury mixture with theophylline

Structural Formula:

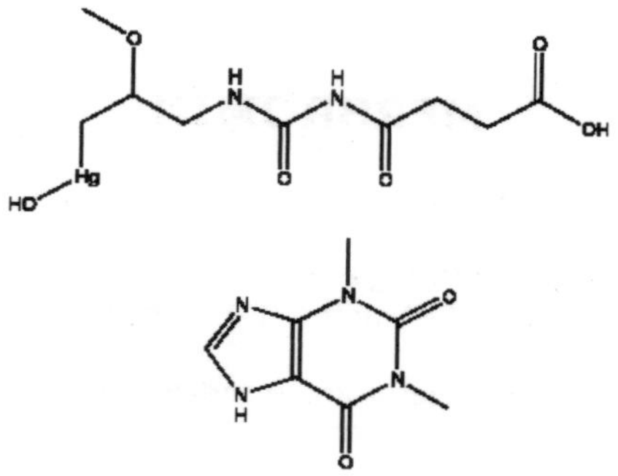

Chemical Abstracts Registry No.: 8069-64-5

Trade Name	Manufacturer	Country	Year Introduced
Mercuhydrin	Merrell National	US	1943
Mercardac	Parke Davis	US	-
Mercadon	Parke Davis	US	-

Raw Materials

Allyl carbamide
Succinic anhydride
Mercury acetate
Theophylline

Manufacturing Process

First, to produce the mercury component, a pulverized mixture of 50 g of allylcarbamide and 50 g of succinic anhydride is heated for 30 minutes at 110°C. After cooling the fused mass is ground with 50 cc of cold water and the crystalline mass after quick filtering from the liquid is recrystallized from hot water. The white crystalline needles having a MP of 142° to 144°C are allyl-succinyl-carbamide. In order to produce a mercury compound thereof a mixture of 20 g of the allyl-succinyl-carbamide and 30 g of mercury acetate is shaken for 3 hours with methanol. The scarcely soluble precipitate of the mercury compound after filtration is washed with methanol and with water and dried in vacuum. The white powder melts at 185° to 186°C under decomposition. Then, condensation with an equimolar proportion of theophylline yields meralluride.

References

Merck Index 5696
OCDS Vol. 1 p, 224 (1977)
I.N. p. 598
Geiger, E., Vargha, L. and Richter, L.; US Patent 2,208,941; July 23, 1940; assigned to Chemical Works of Gedeon Richter Ltd., Hungary

MERBROMIN

Therapeutic Function: Antiseptic

Chemical Name: Mercury, (2',7'-dibromo-3',6'-dihydroxy-3-oxospiro (isobenzofuran-1(3H),9'-(9H)xanthen)-4'-yl)hydroxy-, disodium salt

Common Name: Merbromin; Merbromine sodique; Mercuresceine; Mercurobromfluorescein; Mercurochrome; Mercurocromo

Structural Formula:

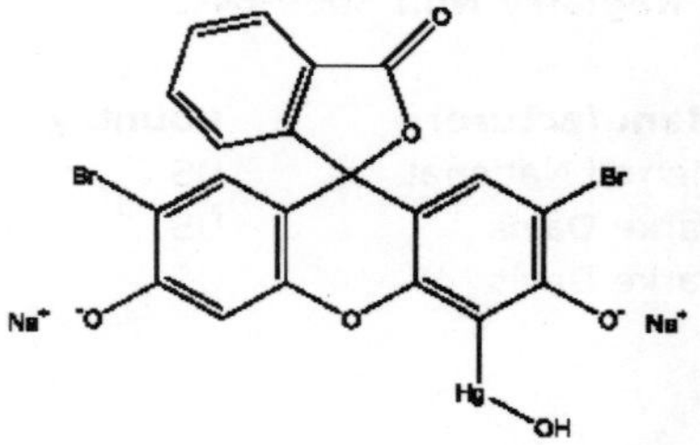

Chemical Abstracts Registry No.: 129-16-8

Trade Name	Manufacturer	Country	Year Introduced
Mercurin	Monik	-	-
Mersol	Merkez	-	-
Pharmadose mercuresceine	Gilbert	-	-

Raw Materials

2,7-Dibromofluorescein
Mercury (II) oxide

Manufacturing Process

49 g 2,7-dibromofluorescein are dissolved in a solution of 8 g of sodium hydroxide in 50 ml of water, and diluted to 200 ml 12.5 ml of glacial acetic acid are added to this solution with stirring. A homogeneous pasty precipitate results with vigorously stirring. A filtered solution of about 22.5 g of mercuric oxide in 25 ml of glacial acid and 50 ml water, diluted after solution to 100 ml, is then added to the suspended precipitate, and the whole diluted to about 500 ml. The mixture is boiled until a small portion of filtered solution gives no test for mercury when treated with ammonium sulfide, the approximate time required for this operation being about 4.5-6 hours. As the boiling continues the precipitate become darker in color and more granular. It is washed, preferably by centrifuging, to remove acetic acid and sodium acetate, and dried at about 110°C. By close adherence to above conditions an almost quantitative yield may be secured. The product may be regarded as consisting essentially of 2,7-dibromo-4-hydroxymercuryfluorescein, resulting from substantially complete hydrolysis of an acetoxy-mercury compound, which probably formed as an intermediate. It is red powder, which is insoluble in the usual solvents but dissolves in two equivalents of sodium hydroxide yielding a deep cherry-red solution. The solution has the tendency to decomposition on long standing.

References

White E. C.; US Patent No. 1,535,003; April 21, 1925

MERCAPTOMERIN SODIUM

Therapeutic Function: Diuretic

Chemical Name: [3-[[(3-Carboxy-2,2,3-trimethylcyclopentyl)carbonyl] amino]-2-methoxypropyl](mercaptoacetato-S)mercury disodium salt

Common Name: -

Structural Formula:

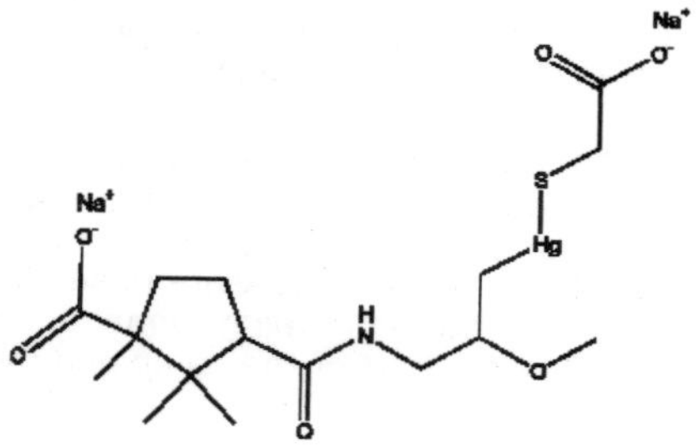

Chemical Abstracts Registry No.: 21259-76-7

Trade Name	Manufacturer	Country	Year Introduced
Thiomerin	Wyeth	US	1949
Diucardyn	Ayerst	-	-
Thio-Novurit	Chinoin	Hungary	-

Raw Materials

dl-N-Allylcamphoramic acid
Mercury acetate
Sodium methylate
Thioglycolic acid

Manufacturing Process

(A) Preparation of dl-N-(γ-Chloromercuri-β-Methoxy)-Propylcamphoramic Acid: A suspension of 31.9 g (= 0.10 M) of mercuric acetate in 25 ml of methanol is stirred for 30 minutes at room temperature in a 4-necked flask equipped with stirrer, dropping funnel, drying tube and thermometer. To this suspension is added dropwise and with stirring, a solution of 23.9 g (= 0.10M) of dl-N-allylcamphoramic acid in 65 ml of methanol over a period of 30 minutes. The temperature of the reaction mixture should not rise over 30°C. The stirring is continued for one hour. The reaction mixture is allowed to stand at room temperature overnight in the dark to complete the reaction. A solution of 5.9 g (= 0.10M) of sodium chloride in 25 ml of water is added and the stirring is continued for four hours. The small amount of gray precipitate produced is removed by centrifuging. The colorless, clear supernatant is concentrated to about half of its original volume and then dropped into 300 ml of water with stirring.

The white precipitate which forms is filtered and dried at 80°C, yielding 45 g of chloromercuri acid (= 89% of the theory), MP 106° to 109°C (decomp.). This compound is finally obtained in analytically pure form and with a constant melting point by two recrystallizations from acetone-water giving a MP of 131° to 132°C with decomposition.

(B) Preparation of the Chloromercuri Acid Sodium Salt Solution: 50.6 g (= 0.100 M) of the chloromercuri acid (dried over $CaCl_2$, at 0.1 mm and room temperature overnight) is dissolved in 100 ml of warm methanol. To this solution 6.0 g (= 0.111 M) of sodium methylate is added in small portions with constant stirring, so that the temperature of the solution does not rise over 30°C. The solution is centrifuged, and the glass is rinsed with 10 ml of methanol. The final pH of the combined solutions is 8.5.

(C) Preparation of the Disodium Thioglycolate Solution: The following steps are carried out under nitrogen. To 9.2 g (= 0.100 M) of freshly distilled thioglycolic acid (BP at 2 mm, 84° to 85°C) in 100 ml of methanol in a flask is added 12.0 g (= 0.222 M) of sodium methylate in small portions with stirring. The turbid solution is poured into a dropping funnel and the flask is rinsed with 20 ml of methanol. The final pH of the combined methanolic solutions is 11, according to US Patent 2,834,795.

To 50 cc of a carefully purified aqueous solution of the sodium salt of N(γ-chloromercuri-β-methoxy-propyl)-d-α-camphoramic acid containing 40 mg of mercury per cc is added 10 cc of a solution containing 1.14 g (1 mol equivalent) of sodium thioglycolate and the mixture is then evaporated to dryness at room temperature and reduced pressure in the presence of a desiccant. The product is an amorphous white powder which decomposes at 156° to 158°C (uncorr.), and which was found on analysis to have a mercury content of 33.0%, according to US Patent 2,576,349.

References

Merck Index 5701
OCDS Vol. 1 p. 224 (1977)
I.N. p. 599
Lehman, R.A.; US Patent 2,576,349; November 27, 1951; assigned to Wyeth Incorporated
Wendt, G.R.; US Patent 2,834,795; May 13,1958; assigned to American Home Products Corporation

MERCAPTOPURINE

Therapeutic Function: Cancer chemotherapy

Chemical Name: 6-Purinethiol

Common Name: -

Structural Formula:

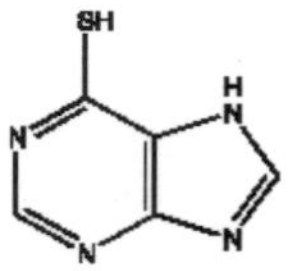

Chemical Abstracts Registry No.: 50-44-2

Trade Name	Manufacturer	Country	Year Introduced
Purinethol	Sandoz	France	1950
Purinethol	Burroughs-Wellcome	US	1953
Classen	Nippon Shoji	Japan	-
Ismipur	I.S.M.	Italy	-
Leukerin	Takeda	Japan	-
Mercaleukin	Arzneimittelwerk Dresden	E. Germany	-
Mern	Tanabe	Japan	-
6-MP	Dojin	Japan	-
Oncomercaptopurina	Simes	Belgium	-
Puri-Nethol	Burroughs-Wellcome	UK	-
Thioinosie	Morishita	Japan	-

Raw Materials

4-Amino-6-chloro-5-nitropyrimidine
Hydrogen sulfide
Formic acid
Sodium hydroxide

Manufacturing Process

7.5 g of 4-amino-6-chloro-5-nitropyrimidine was suspended in 200 ml of 1 N potassium hydrosulfide and heated on the steam bath for 2 hours while passing hydrogen sulfide through the reaction mixture. The reaction mixture was allowed to cool slowly, acidified with 10 N sulfuric acid and chilled. The precipitate consisted of 4,5-diamino-6-mercaptopyrimidine and sulfur. It was boiled with 300 ml of water, filtered hot and then chilled. The product precipitated as pale yellow needles (4.2 g); an additional 0.95 g was obtained by concentration of the mother liquors to 100 ml.

A mixture of 2 g of 4,5-diamino-6-mercaptopyrimidine and 10 ml of 98% formic acid was heated at 70°C for two hours and then evaporated to dryness on the steam bath to give as a residue, 7-amino-thiazolo (5,4-d) pyrimidine.

To 820 mg of 7-amino-thiazolo[5,4-d]pyrimidine was added 2.5 cc of 2 N sodium hydroxide. The water was removed under reduced pressure. The sodium salt was then heated at 240°C for one hour, during which time it melted, gave off water and resolidified. The sodium salt of 6-mercaptopurine was dissolved in 15 ml of water and acidified to pH 5 with acetic acid. Yellow crystals of 6-mercaptopurine hydrate precipitated, according to US Patent 2,933,498.

References

Merck Index 5702
Kleeman & Engel p. 563
PDR p. 759
I.N. p. 599
REM p. 1151
Hitchings, G.H. and Elion, G.B.; US Patent 2,721,866; October 25, 1955; assigned to Burroughs Wellcome & Co. (USA.) Inc.
Hitchings, G.H. and Elion, G.B.; US Patent 2,724,711; November 22, 1955; assigned to Burroughs Wellcome & Co. (USA.) Inc.
Hitchings, G.H. and Elion, G.B.; US Patent 2,933,498; April 19, 1960; assigned to Burroughs Wellcome & Co. (USA.) Inc.

MEROPENEM

Therapeutic Function: Antibiotic

Chemical Name: 1-Azabicyclo(3.2.0)hept-2-ene-2-carboxylic acid, 3-((5-(2-dimethylamino)carbonyl)-3-pyrrolidinyl)thio)-6-(1-hydroxyethyl)-4-methyl-7-oxo-, trihydrate, (4R- (3(3S*,5S*),4alpha,5beta,6beta(R*)))-

Common Name: Meropenem

Structural Formula:

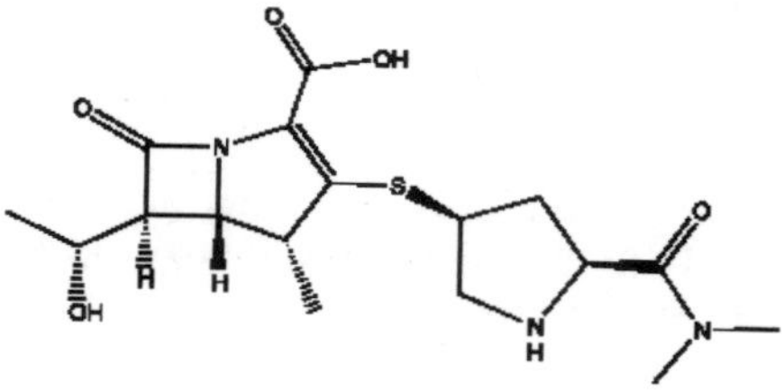

Chemical Abstracts Registry No.: 96036-03-2

Trade Name	Manufacturer	Country	Year Introduced
Merrem	AstraZeneca	-	-
Meronem	AstraZeneca	-	-
Meropenem	AstraZeneca	-	-
Merozen	AstraZeneca	-	-
Zeropenem	Hoechst Marion Roussel	-	-

Raw Materials

Ethyl chloroformate
Triethylamine
Thioacetic acid
Palladium on carbon
Zinc
Diphenyl chlorophosphate
Hydrogen
Diisopropylethylamine
(3R,4R)-4-Acetoxy-3-[(R)-1-(t-butyldimethylsilyloxy)ethyl]-2-azetidinone
Methanesulfonyl chloride
Diethylaluminium chloride
Orpholinopropanesulfonic acid buffer
Benzyl-α-bromopropionate
trans-1-(p-Nitrobenzyloxycarbonyl)-4-hydroxy-L-proline
Methanesulfonyl chloride
t-Butyldimethylsilyl chloride

Manufacturing Process

3.10 g of trans-1-(p-nitrobenzyloxycarbonyl)-4-hydroxy-L-proline and 1.10 g of triethylamine were dissolved in 40 ml of dried tetrahydrofuran, and a solution of 1.20 g of ethyl chloroformate in 10 ml of dried tetrahydrofuran was added dropwise thereto at -25-35°C. After stirring at the same temperature for 50 min, 10 ml of concentrated aqueous ammonia was added dropwise to the mixture at -25-40°C. The temperature was then gradually elevated to room temperature, and the reaction mixture was stirred for 1 hour, followed by concentration under reduced pressure. To the residue were added 20 ml of water and 50 ml of diethyl ether. After ice-cooling, the thus formed white crystals were separated by filtration, washed successively with cool water and cool diethyl ether, and dried under reduced pressure to yield trans-1-(p-nitrobenzyloxycarbonyl)-4-hydroxy-L-prolineamide. Melting point: 163.3-

164.0°C.

A solution of 1.89 g of methanesulfonyl chloride in 10 ml of dried tetrahydrofuran was added dropwise to a suspension of 2.32 g of trans-1-(p-nitrobenzyloxycarbonyl)-4-hydroxy-L-prolineamide and 1.67 g of triethylamine in 40 ml of dried tetrahydrofuran at room temperature. After stirring for 1 hour, the reaction mixture was concentrated under reduced pressure, and to the residue were added 30 ml of water and 30 ml of diethyl ether. After cooling, the resulting white crystals were separated by filtration, washed successively with cool water and cool diethyl ether and dried under reduced pressure to obtain trans-1-(p-nitrobenzyloxycarbonyl)-4-methanesulfonyloxy-L-prolineamide. Melting point: 149.5-151°C.

A solution of 642 mg of thioacetic acid in 14 ml of dried dimethylformamide was added to a suspension of 374 mg of 50% sodium hydride in 13 ml of dried dimethylformamide in a nitrogen stream, followed by stirring at room temperature for 25 minutes. To the mixture were added 975 mg of sodium iodide and then a solution of 2.52 g of trans-1-(p-nitrobenzyloxycarbonyl)-4-methanesulfonyloxy-L-prolineamide in 12 ml of dried dimethylformamide, and the resulting mixture was heated to 70°C for 6 hours while stirring. The reaction mixture was poured into a cool aqueous solution of sodium chloride and extracted with benzene. The extract was washed successively with a 10% aqueous solution of sodium sulfate and a sodium chloride aqueous solution, dried over sodium sulfate and distilled off to remove the solvent. The resulting crude crystals were washed with a warm mixed solvent of tetrahydrofuran and benzene to obtain (2S,4S)-1-(p-nitrobenzyloxycarbonyl)- 2-carbamoyl-4-acetylthio-L-prolineamide. Melting point: 168.5-169.5°C.

950 mg of (2S,4S)-1-(p-nitrobenzyloxycarbonyl)-2-carbamoyl-4-acetylthiopyrrolidine was dissolved in 95 ml of methanol, and 2.59 ml of a 1 N aqueous solution of sodium hydroxide was added thereto at room temperature in an argon stream, followed by stirring at that temperature for 15 min. The reaction mixture was neutralized with 2.59 ml of a 1 N aqueous solution of hydrochloric acid and distilled off under reduced pressure to remove the methanol. The thus precipitated crystals were filtered and washed with water to obtain (2S,4S)-1-(p-nitrobenzyloxycarbonyl)-2-carbamoyl-4-mercaptopyrrolidine. Melting point: 158-162°C.

To 1.33 g (20 mM) of activated zinc was added 20 ml of dried tetrahydrofuran, and 8.8 ml of a 15% n-hexane solution of diethylaluminium chloride was added thereto in a nitrogen stream under ice-cooling. A solution prepared by dissolving 1.49 g (5.2 mM) of (3R,4R)-4-acetoxy-3-[(R)-1-(t-butyldimethylsilyloxy)ethyl]-2-azetidinone and 3.73 g (15.3 mM) of benzyl-α-bromopropionate in 13.3 ml of dried tetrahydrofuran was added dropwise to the mixture over a period of 30 to 40 min, followed by stirring for 1 hours. Under ice-cooling, 2.8 ml of pyridine, 13.2 ml of water, 26.5 ml of ethyl acetate and 13.2 ml of a 1 N hydrochloric acid aqueous solution were successively added thereto, and the resulting mixture was filtered using Celite. The filtrate was washed with water, and the organic layer was dried over sodium sulfate and distilled off to remove the solvent. The resulting oily residue was subjected to silica gel column chromatography to obtain an isomeric mixture of 4-(1-benzyloxycarbonyl)ethyl-3-[(R)-1-(t-butyldimethylsilyloxy)ethyl]-2-azetidinone.

The isomeric mixture was separated into each compound by Lober column chromatography using silica gel and 1.5% isopropanol/n-hexane as an eluent to obtain the compound (1a) and the compound (1b) as oily substances.

200 mg of 4-(1-benzyloxycarbonyl)ethyl-3-[(R)-1-(t-butyl-dimethylsilyloxy) ethyl]-2-azetidinone (1a) was dissolved in 2 ml of dried dimethylformamide. 126 mg of triethylamine was added to the resulting solution, and then 151 mg of t-butyldimethylsilyl chloride was added thereto, followed by stirring at room temperature overnight. The reaction mixture was diluted with ethyl acetate, washed with water, dried over sodium sulfate and purified by silica gel chromatography to obtain 4-(1-benzyloxycarbonyl)ethyl-3-[(R)-1-(t-butyldimethylsilyloxy)ethyl]-1-(t-butyldimethylsilyl)-2-azetidinone (2a).

184 mg of (2a) was dissolved in 4 ml of methanol, and the resulting solution was stirred together with 20 mg of 10% palladium-on-carbon at an atmospheric pressure of hydrogen for 2 hours. The catalyst was removed by filtration, and the filtrate was concentrated under reduced pressure to obtain 4-(1-carboxy)ethyl-3-[(R)-1-(t-butyldimethylsilyloxy)ethyl]-1-(t-butyldimethylsilyl)-2-azetidinone (3a).

(4R,5R,6S,8R)-p-Nitrobenzyl-4-methyl-6-(1-hydroxyethyl)-1-azabicyclo [3.2.0]-hept-3,7-dione-2-carboxylate was obtained from 170 mg of 4-(1-carboxy)ethyl-3-[(R)-1-(t-butyldimethylsilyloxy)ethyl]-1-(t-butyldimethylsilyl)-2-azetidinone (3a) according to the method described in Japanese Patent Application OPI No. 26887/83, pages 64-65.

(a) 53 mg of (4R,5R,6S,8R)-p-nitrobenzyl-4-methyl-6-(1-hydroxyethyl)-1-azabicyclo[3,2,0]-hept-3,7-dione-2-carboxylate was dissolved in 5 ml of dry acetonitrile, and 57 mg of diisopropylethylamine and then 43 mg of diphenyl chlorophosphate were added thereto. After stirring for 2.5 hours, 57 mg of [2S,4S]-1-p-nitrobenzyloxycarbonyl-2-dimethylaminocarbonyl-4-mercaptopyrrolidine was added to the mixture, followed by stirring for 1 hour. The reaction solution was diluted with ethyl acetate, washed with water, dried over magnesium sulfate and the solvent was distilled off. The residue was purified by silica gel thin layer chromatography to obtain 35 mg of (4R,5S,6S,8R,2'S,4'S)-p-nitrobenzyl-3-[4-(1-p-nitrobenzyloxycarbonyl-2-dimethylaminecarbonyl)pyrrolidinylthio]-4-methyl-6-(1-hydroxyethyl)-1-azabicyclo[3,2,0]-hept-2-ene-7-one-2-carboxylate.

(b) 25 mg of (4R,5S,6S,8R,2'S,4'S)-p-nitrobenzyl-3-[4-(1-p-nitrobenzyloxycarbonyl-2-dimethylaminecarbonyl)pyrrolidinylthio]-4-methyl-6-(1-hydroxyethyl)-1-azabicyclo[3,2,0]hept-2-ene-7-one-2-carboxylate was dissolved in a mixture of 1.9 ml of tetrahydrofuran and 0.3 ml of ethanol, and the mixture was hydrogenated in a morpholinopropanesulfonic acid buffer solution (pH = 7.0, 1.9 ml) under atmospheric pressure of hydrogen for 3 hours at room temperature in the presence of 30 mg of 10% palladium-carbon, which had been activated in hydrogen atmosphere for 1 hour followed by washing with water. After filtering off the catalyst, tetrahydrofuran and ethanol were distilled off under reduced pressure, and the residual solution was washed with ethyl acetate. The aqueous layer was again distilled under reduced pressure to remove organic solvents, and the residual solution was subjected to polymer chromatography (CHP-20P) to obtain (4R,5S,6S,8R,2'S,4'S)-3-[4-(2-dimethylaminecarbonyl)pyrrolidinylthio]-4-methyl-6-(1-hydroxyethyl)-1-azabicyclo[3,2,0]hept-2-ene-7-one-2-carboxylic acid from the fraction eluted with water.

References

Merck Index, Monograph number: 5960, Twelfth edition, 1996, Editor: S. Budavari; Merck and Co., Inc.

Nishitani Y., Irie T.; US Patent No. 5,122,604; May 31, 1994; Assigned to: Shionogi Seiyaku Kabushiki Kaisha (Osaka, JP)

Sunagawa M.; US Patent No. 4,943,569; July 24, 1990; Assigned to Sumitomo Pharmaceuticals Co., Ltd. (Osaka, JP)

Sunagawa M. et al.; J. Antibiotics; 1990, 43, 519

MESALAMINE

Therapeutic Function: Antibacterial

Chemical Name: Salicylic acid, 5-amino-

Common Name: Acidum metaminosalicylicum; Fisalamine; Mesalamine; Mesalazine

Structural Formula:

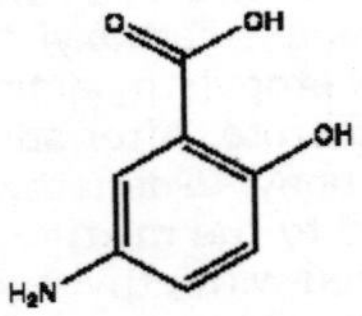

Chemical Abstracts Registry No.: 89-57-6

Trade Name	Manufacturer	Country	Year Introduced
Asacol	Proctor and Gamble	-	-
Asacol	Tillots Pharma AG Ziefen	-	-
Asalit	Merck	-	-
Mesacol	Sun Pharma	-	-
Mesalamine	Novopharm	-	-
Mesasal	Glaxol/Wellcome	-	-
Pentasa	Ferring	-	-
Salofalk	Minipharm	-	-
Walsa	Wallace Pharmaceuticals Ltd.	India	-

Raw Materials

5-Nitrosalicylic acid potassium salt
Hydrazine hydrate

Nickel Raney
Hydrogen

Manufacturing Process

Procedure A: To 5-nitrosalicylic acid potassium salt (55 g, 246 mmol) dissolved in water (200 mL) was added potassium hydroxide pellets to reach pH 11.5. To this solution 2 g of Raney nickel were added. The mixture was heated-up to reflux and hydrazine hydrate (40 mL, 80% in water, 64 mmol) was added dropwise during 3-4 hrs. The reflux was maintened until HPLC showed the disappearance of the starting material and the complete reduction of 5-nitrosalicylic acid (3-4 hrs). The hot mixture was filtered under nitrogen and the solution was collected. The solution was cooled to 40°C and the pH was adjusted to 2.3 by addition of 35% HCl aqueous solution. The precipitation of 5-aminosalicylic acid occurred. The solution was cooled at 0°C, and after standing at this temperature for 2 hr, the precipitate was filtered, washed with water, and dried at 60-70°C. 5-Aminosalicylic acid was obtained in 89% yield.

Procedure B: To 5-nitrosalicylic acid potassium salt (55 g, 246 mmol) dissolved in water (200 mL) was added potassium hydroxide pellets to reach pH 11.5. The solution was charged in a stainless steel autoclave and 2 g of Raney nickel are added. Hydrogen was introduced into the autoclave reaching a pressure of 8 atm. The mixture was heated-up to 100°C. The temperature was maintained until HPLC-test 5-aminosalicylic acid showed the disappearance of the starting material and the complete reduction of 5-aminosalicylic acid (6-8 hrs). Hydrogen was purged and replaced by nitrogen. The hot mixture was filtered under nitrogen, the filtrate was cooled to 40°C, and the pH was adjusted to 2.3 by addition of 35% HCl aqueous solution. The precipitation of the 5-aminosalicylic acid occurred. The solution was cooled at 0°C, and after standing at this temperature for 2 hr, the precipitate was filtered, washed with ion depleted water, and dried at 60-70°C.

References

Breviglieri G., Giacomo B., Contrini Sergio, Assanelli Cinzia, Eileen Campanab, Mauro Panunzio, Molecules 2001, 6, M260
Martelli, G., Spunta, G., Panunzio, M.; Tetrahedron Lett. 1998, 39, 6257-6260
Kennedy, J.F., Barker, S.A., Epton, J., Kennedy, G.R.; J. Chem. Soc. Perkin 1, 1973, 488-490

MESNA

Therapeutic Function: Mucolytic

Chemical Name: 2-Mercaptoethane sulfonic acid sodium salt

Common Name: -

Structural Formula:

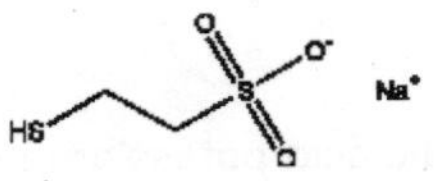

Chemical Abstracts Registry No.: 19767-45-4; 3375-50-6 (Base)

Trade Name	Manufacturer	Country	Year Introduced
Mistabronco	UCB	W. Germany	1973
Mistabron	Diethelm	Switz.	1978
Mucofluid	UCB Fraysse	France	1978
Mucofluid	UCB	Italy	1981
Uromitexan	W.B. Pharm.	UK	1983
Uromitexan	Asta	W. Germany	-

Raw Materials

β-S-Thiuronium ethanesulfonate
Ammonia

Manufacturing Process

2,100 g of β-S-thiuronium ethanesulfonate were placed in a solution of 2,100 cc of concentrated aqueous ammonia and 400 cc of water. The mixture was carefully warmed on a steam bath and an exothermic reaction ensured, at which point the β-S-thiuronium ethanesulfonate passed into solution. After standing for two hours at room temperature, the solution was concentrated until all of the excess ammonia had been removed.

The resultant clear solution from the ammonolysis reaction was processed through "Amberlite IR-120" ion exchange resin and converted into β-S-mercaptoethanesulfonic acid in 93.7% yield (based on β-S-thiuronium ethanesulfonate).

It is expedient not to heat the reaction mixture rapidly since this increases the loss of ammonia and effects an incomplete reaction. Heating the mixture too rapidly may retard the ammonolysis reaction entirely. The amount of ammonia used is considered to be a satisfactory minimum and larger quantities of ammonia are not found to have any beneficial effect on the reaction. It is also expedient to remove the excess ammonia before processing the guanidinium β-mercaptoethanesulfonate solution through the ion exchange resin since the resin will also remove the ammonia with the result that the capacity of the resin for the exchange of guanidinium ions will be reduced.

Although the preparation of β-mercaptoethanesulfonic acid through the ammonolysis reaction is the preferred method, it is also possible to prepare the sulfonic acid by the sodium hydroxide hydrolysis of β-S-thiuronium ethanesulfonate followed by the ion exchange treatment. The resulting acid, however, is generally not as satisfactory as that prepared by the ammonolysis reaction.

References

Merck Index 5754
Kleeman & Engel p. 563
DOT 8 (5) 180 (1972); 19 (10) 585 and (11) 608 (1983)
I.N. p. 601
Schramm, C.H. and Karlson, R.H.; US Patent 2,695,310; November 23, 1954; assigned to Lever Brothers Co.

MESORIDAZINE BESYLATE

Therapeutic Function: Tranquilizer

Chemical Name: 10-[2-(1-Methyl-2-piperidinyl)ethyl]-2-methylsulfinyl-10H-phenothiazine benzene sulfonate

Common Name: -

Structural Formula:

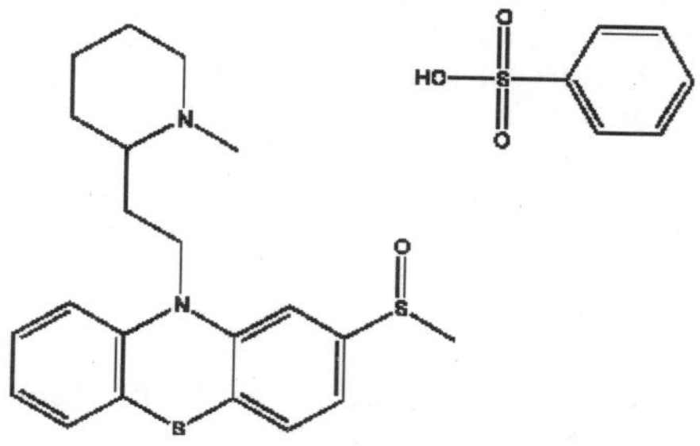

Chemical Abstracts Registry No.: 32672-69-8; 5588-33-0 (Base)

Trade Name	Manufacturer	Country	Year Introduced
Serentil	Sandoz	US	1970
Calodal	Heyden	Switz.	1980
Lidanil	Salvoxyl-Wander	France	-

Raw Materials

Acetic anhydride
Hydrogen peroxide
Sodium hydroxide
3-Methylmercaptophenothiazine
Potassium carbonate
2-(N-Methylpiperidyl-2')-1-chloroethane

Manufacturing Process

10.0 g of 3-methylmercapto phenothiazine and 17.5 cc of acetic acid anhydride are refluxed for 8 hours from an oil bath maintained at a temperature of 180°C. After concentration of the solution the residue is crystallized from ethanol. The pure 3-methylmercapto-10-acetyl phenothiazine melts at 89° to 91°C. For the purpose of oxidation 5.0 g of 3-methylmercapto-10-acetyl phenothiazine are dissolved in 50 cc of ethanol, refluxed from an oil bath maintained at 120°C and 1.6 cc of a 40% hydrogen peroxide solution are then added dropwise in the course of 30 minutes.

Heating is continued for another 5 hours and the reaction mixture is concentrated after 50 cc of water have been added. The residue is taken up in 40 cc of benzene and the benzene layer washed with 10 cc of water. After having been concentrated, the residue, crude 3-methylsulfinyl-10-acetyl phenothiazine, is dissolved in 55 cc of a 90% methanol solution for splitting off the acetyl group and, after 2.9 g of potassium carbonate have been added, it is boiled for 2 hours under reflux on an oil bath kept at a temperature of 120°C. After concentration, the residue is taken up in 50 cc of chloroform, the chloroform layer is washed with a total of 25 cc of water, dried over potassium carbonate, filtered and concentrated. After twice crystallizing the residue, each time from 50 cc of ethanol, analytically pure 3-methylsulfinyl phenothiazine (MP 193° to 195°C) is obtained.

A mixture of 10.0 g of 3-methylsulfinyl phenothiazine (MP 193° to 195°C), 6.1 g of finely powdered sodium hydroxide and 125 cc of toluene is boiled for 1 hour under reflux with a water separator on an oil bath kept at a temperature of 150°C, while the mixture is stirred. Without interrupting the boil a solution of 7.0 g of 2-(N-methyl-piperidyl-2')-1-chloroethane (BP 84°C/10 mm Hg) in 10 cc of toluene is added dropwise in the course of 1 hour, after which boiling is continued for another 3 hours. When the reaction mixture has cooled it is first washed with 25 cc of water three times and then extracted with 75 cc of a 15% aqueous tartaric acid solution. The tartaric acid extract is shaken out with 25 cc of benzene, 20 cc of concentrated caustic soda are added until the phenolphthalein reaction is alkaline, and the separated oily base is taken up in a total of 150 cc of benzene.

After having been washed with 50 cc of water the benzene layer is dried over potassium carbonate, filtered, allowed to stand over 10 g of alumina for about 1½ hours for partial decolorization, filtered again and concentrated under reduced pressure. The oily base which remains as a residue is directly converted into the tartrate. A solution cooled to 0°C, of 6.50 g of the free base in 100 cc of acetic acid ethyl ester is thoroughly shaken and poured into an ice cold solution of 2.66 g of tartaric acid in 410 cc of acetic acid ethyl ester. The precipitated, analytically pure, tartrate of 3-methylsulfinyl-10-[2'-N-methyl-piperidyl-2')-ethyl-l']-phenothiazine melts at 115° to 120°C (foam formation) and sinters above 80°C. The base is reacted with benzene sulfonic acid in a suitable solvent to give the besylate.

References

Merck Index 5755
Kleeman & Engel p. 564
PDR p. 681

OCDS Vol. 1 p. 389 (1977)
DOT 6 (6) 211 (1970) and 9 (6) 227 (1973)
I.N. p.601
REM p. 1089
Renz, J., Bourquin, J.-P. and Schwarb, G.; US Patent 3,084,161; April 2, 1963; assigned to Sandoz Ltd., Switzerland

MESTEROLONE

Therapeutic Function: Androgen

Chemical Name: 17β-Hydroxy-1α-methyl-5α-androstan-3-one

Common Name: -

Structural Formula:

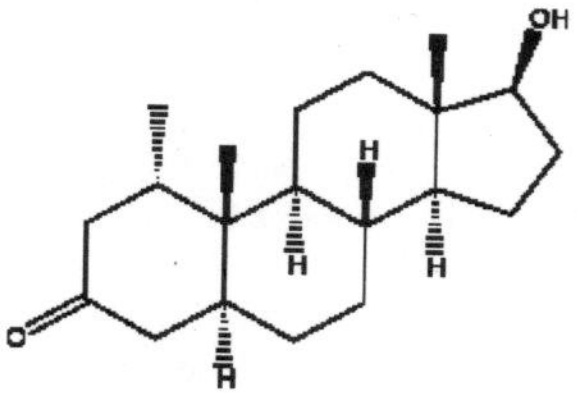

Chemical Abstracts Registry No.: 1424-00-6

Trade Name	Manufacturer	Country	Year Introduced
Proviron	Schering	W. Germany	1967
Proviron	Schering	Italy	1971
Pro-Viron	Schering	UK	1971
Proviron	S.E.P.P.S.	France	1975
Mestoran	Schering	W. Germany	-
Vistimon	Jenapharm	E. Germany	-

Raw Materials

1α-Methyl-androstan-17β-ol-3-one-17-acetate
Sodium hydroxide

Manufacturing Process

500 mg of 1α-methyl-androstan-17β-ol-3-one-17-acetate are heated under reflux for 90 minutes in a nitrogen atmosphere in 5 ml of 4% methanolic sodium hydroxide solution. The reaction mixture is then stirred into ice water,

the precipitated product filtered with suction and recrystallized from isopropyl ether. 1α-Methyl-androstan-17β-ol-3-one melts at 203.5° to 205°C.

References

Merck Index 5760
Kleeman & Engel p. 565
OCDS Vol. 1 p. 174 (1977)
I.N. p. 602
Schering AG, Germany; British Patent 977,082; December 2, 1964
Schering AG, Germany; British Patent 977,083; December 2, 1964
Wiechert, R.; US Patent 3,361,773; January 2, 1968; assigned to Schering A.G.

MESTRANOL

Therapeutic Function: Estrogen

Chemical Name: 3-Methoxy-19-nor-17α-pregna-1,3,5(10)-trien-20-yn-17-ol

Common Name: 17α-Ethynylestradiol 3-methyl ether

Structural Formula:

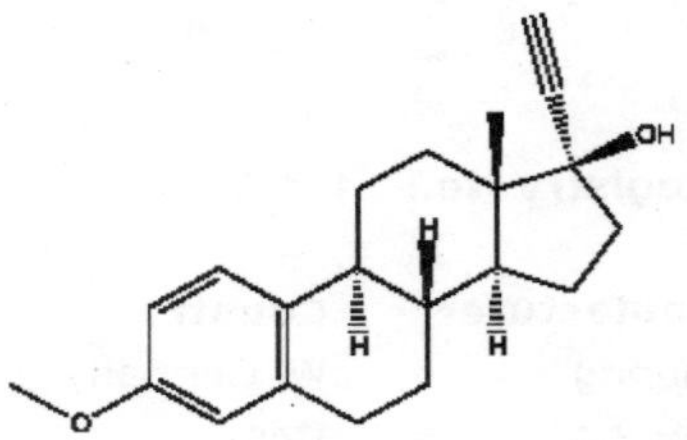

Chemical Abstracts Registry No.: 72-33-3

Trade Name	Manufacturer	Country	Year Introduced
Enovid	Searle	US	1957
Ortho-Novum	Ortho	US	1963
Enovid-E	Searle	US	1964
Norinyl	Syntex	US	1964
C-Quens	Lilly	US	1965
Ovulen	Searle	US	1966
Conceplan	Gruenenthal	W. Germany	-
Conovid	Searle	UK	-
Enavid	Dainippon	Japan	-
Estalor	Lilly	US	-
Gestamestrol	Hermal	W. Germany	-

Trade Name	Manufacturer	Country	Year Introduced
Lutedione	Teikoku Zoki	Japan	-
Lyndiol	Organon-Sankyo	Japan	-
Metrulen	Searle	UK	-
Noracycline	Ciba Geigy	France	-
Noriday	Syntex	US	-
Norinyl	Syntex	US	-
Norluten	Shionogi	Japan	-
Norquen	Syntex	US	-
Nuriphasic	Noury Pharma	W. Germany	-
Orgaluton	Organon	UK	-
O.V. 28	Biosedra	France	-
Ovanon	Organon	UK	-
Ovastol	Rendell	UK	-

Raw Materials

3-Methoxy-δ(^{1,3,5})-estratrien-17-one
Acetylene

Manufacturing Process

A stirred solution of 120 parts of 3-methoxy-$\delta^{1,3,5}$-estratrien-17-one in 2,600 parts of anhydrous toluene and 4,300 parts of anhydrous ether is saturated with a slow stream of acetylene. In the course of 30 minutes there is added a solution of 120 parts of potassium tert-amylate in 2,800 parts of anhydrous tert-pentanol. The passage of acetylene and stirring are continued for an additional 5 hours after which the reaction mixture is washed 5 times with 3,000-part portions of saturated ammonium chloride solution and then with water. It is then dried over anhydrous sodium sulfate and concentrated to dryness under vacuum. The residue is recrystallized from methanol. The 3-methoxy-17-ethynyl-$\delta^{1,3,5}$ estratrien-17-ol thus obtained melts at about 143° to 146°C. A further recrystallization from acetone yields crystals melting at about 150° to 151°C.

References

Merck Index 5762
Kleeman & Engel p. 566
PDR pp. 1297,1680,1793
OCDS Vol. 1 p. 162 (1977)
I.N. p. 602
REM p.989
Colton, F.B.; US Patent 2,666,769; January 19, 1954; assigned to G.D. Searle & Co.

MESULFEN

Therapeutic Function: Scabicide

Chemical Name: 2,7-Dimethylthianthrene

Common Name: Mesulfen; Mesulphen; Thianthol

Structural Formula:

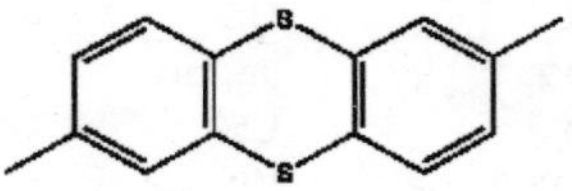

Chemical Abstracts Registry No.: 135-58-0

Trade Name	Manufacturer	Country	Year Introduced
Mesulfen	Synopharm GMBH and CO.KG	-	-
Odylen	Winthrop	-	-
Anacar	Teknofarma	-	-
Schwefelol	Solco	-	-
Sulfor	Takeda	-	-
Thiotal	Linz	-	-
Citemul S	Medopharm Arzneim	-	-

Raw Materials

Toluene
Sulfur
Aluminum chloride

Manufacturing Process

100 parts by weight of toluene, 330 parts of sulfur and 80 parts of aluminum chloride were heated to reflux on the oil bath before the formation of hydrogen sulfide and hydrogen chloride ended. The mixture was poured into water for removing the excess of aluminum chloride. A toluene layer was separated, dried and distilled in vacuum. When toluene was distilled off, about 500 parts by weight the residual yellow oil boiled at 150°-230°C/3 mm Hg was yielded. An objectionable odor was removed by shaking with sodium hydroxide. The residual sulfur (23-25%) was removed by washing with hydrogen peroxide. The residue (mesulfen) looked like fragile asphalt mass after distillation.

References

Weyland H. et al.; D.R. Patent No. 365,169; Sept. 4, 1919; Assigned to Farbenfabriken vorm. Friedr. Bayer and Co. in Leverkusen b. Koin a. Rh.

METAHEXAMIDE

Therapeutic Function: Oral hypoglycemic

Chemical Name: Benzenesulfonamide, 3-amino-N-((cyclohexylamino) carbonyl)-4-methyl-

Common Name: Metahexamide; Metahexanamide

Structural Formula:

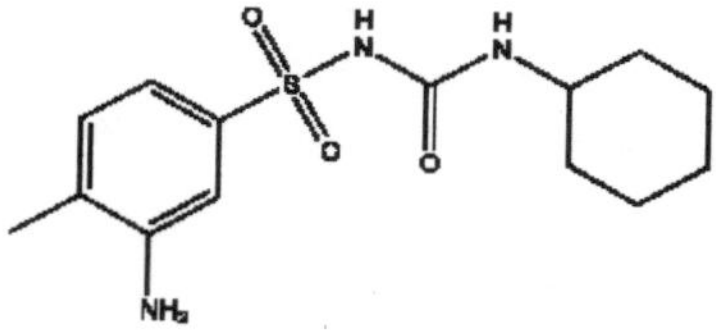

Chemical Abstracts Registry No.: 565-33-3

Trade Name	Manufacturer	Country	Year Introduced
Metahexamide	Shanghai Lansheng Corporation	-	-
Melanex	Upjohn	-	-

Raw Materials

Ethyl chloroformate
3-Acetylamino-p-toluene sulphonic acid amide
Cyclohexylamine

Manufacturing Process

39 g ethyl chloroformate are added dropwise to a mixture of 68.4 g 3-acetylamino-p-toluene sulphonic acid amide, 123 g potassium carbonate and 450 ml acetone for one hour while boiling under reflux. Refluxing is then continued for a further nine hours. The reaction mixture is cooled and mixed, while stirring, with a mixture of 450 ml water and 50 ml 2 N potassium hydroxide solution. Thereby two layers are formed. The upper layer, which consists of aqueous acetone, is separated. Acetone is distilled off in a vacuum. The pH-value of the resulting aqueous solution is adjusted to a pH of 8.8 by passing in gaseous carbon dioxide. Precipitated unchanged starting material is filtered off. The filtrate is rendered congo acid by the addition of dilute hydrochloric acid. The precipitated 3-acetylamino-p-toluene sulfnyl ethyl urethane is filtered off by suction, washed with water, and dried in a vacuum. The yield is 77%. The resulting compound melts at 183°-194°C.

54.3 g above prepared 3-acetylamino-p-toluene sulphonyl ethyl urethane are mixed with 37 ml dimethylformamide and 18 g cyclohexylamine. The resulting clear solution is heated at 70°C for 1.5 hours and at 110°C for 1.5 more hours. After cooling, the reaction mixture is poured into 500 ml water while stirring. The precipitated oily product crystallizes shortly. The crystals are filtered off by suction, washed with water and dried in a vacuum. Yield of 3-acetylamino-p-toluene sulphonyl cyclohexyl urea is 84%. MP: 174°C. The urea is saponified without further purification by heating it in 90 ml 5 N potassium hydroxide solution at 90°C for one hour. After dilution with 500 ml water the resulting reaction mixture is rendered acid (pH 6.5) by the addition of dilute hydrochloric acid. Thereby, 1-(3-amino-p-tolylsulfonyl)-3-cyclohexylurea separates in crystals, which are collected, washed with water, and dried. The yield is 86%. After recrystallization from ethanol the compound has MP: 151°-152°C.

References

Boehringer C.F., Soehne G.m.b.H., Germany; G.B. Patent No. 831,043; Feb. 26, 1957

METAMPICILLIN SODIUM

Therapeutic Function: Antibacterial

Chemical Name: 3,3-Dimethyl-6-[[(methyleneamino)phenylacetyl]amino]-7-oxo-4-thia-1-azabicyclo[3.2.0]heptane-2-carboxylic acid sodium salt

Common Name: -

Structural Formula:

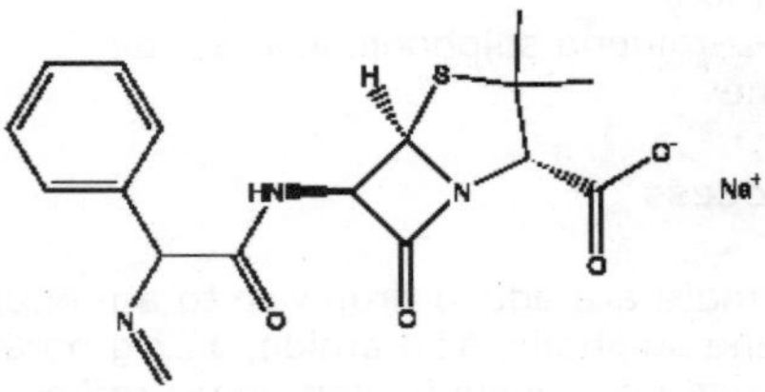

Chemical Abstracts Registry No.: 6489-61-8; 6489-97-0 (Base)

Trade Name	Manufacturer	Country	Year Introduced
Magnipen	Clin-Comar-Byla	Italy	1969
Magnipen	Clin Midy	France	1970
Actuapen	Larma	Spain	-
Ampilprats	Prats	Spain	-
Apliopenil	Miluy	Spain	-
Co-Metampicil	Sanchez-Covisa	Spain	-
Daniven	Aldon	Spain	-

Trade Name	Manufacturer	Country	Year Introduced
Fedacilina	Fedal	Spain	-
Janopen	Janovich	Spain	-
Madecilina	Made	Spain	-
Maipen	Maipe	Spain	-
Mempil	Kairon	Spain	-
Metabacter	Rubio	Spain	-
Metacidan	Cidan	Spain	-
Meta-Ferran	Ferran	Spain	-
Metakes	Kessler	Spain	-
Metambac	Wolner	Spain	-
Metampicef	Cecef	Spain	-
Metamplimedix	Medix	Spain	-
Metiskia	Iskia	Spain	-
Ocelina	Roux-Ocefa	Argentina	-
Pluriespec	Vir	Spain	-
Ruticina	Bernabo	Argentina	-
Tisquibron	Bryan	Spain	-
Venzoquimpe	Quimpe	Spain	-
Vigocina	Europa	Spain	-

Raw Materials

6-[D-(-)α-Aminophenylacetamido]penicillanic acid
Sodium bicarbonate
Formaldehyde

Manufacturing Process

0.01 mol of 6-[D(-)α-(aminophenylacetamido)]-penicillanic acid was suspended in 150 cc of water cooled to +5°C and treated with 0.01 mol of sodium bicarbonate.

The solution was treated with 0.01 mol of formaldehyde in aqueous solution, with agitation. The solution was then filtered to eliminate traces of insoluble product and the filtrate was lyophilized. Sodium 6-[D(-)-alpha-(methylene-amino-phenylacetamido)]-penicillanate was obtained.

References

Merck Index 5775
Kleeman & Engel p. 569
OCDS Vol. 1 p. 414 (1977)
DOT 6 (3) 85 (1970)
I.N. p. 604
Gradnick, B.; British Patent 1,081,093; August 31, 1967; assigned to Societe d'Etudes de Recherches et d'Applications Scientifiques et Medicales (E.R.A.S.M.E.) (France)

METAPRAMINE

Therapeutic Function: Antidepressant

Chemical Name: 10,11-Dihydro-5-methyl-10(methylamino)-5H-dibenz[b,f]azepine

Common Name: -

Structural Formula:

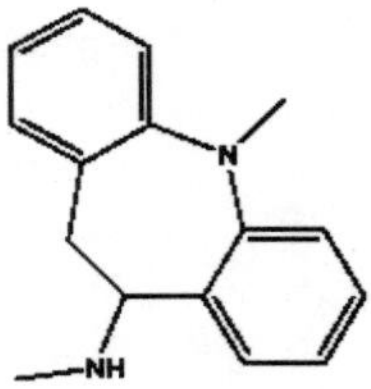

Chemical Abstracts Registry No.: 21730-16-5; 21737-55-3 (Hydrochloride salt)

Trade Name	Manufacturer	Country	Year Introduced
Timaxel	Specia	France	1983
Rodostene	Rhone Poulenc	France	-

Raw Materials

5-Methyl-dibenzo[b,f]azepine
Methylamine
Sodium hypochlorite

Manufacturing Process

5-Methyl-dibenzo[b,f]azepine (4.1 g), N-diethylaminoborane (1.7 g) and freshly distilled toluene (150 cc) are introduced into a 500 cc three-neck flask equipped with a dropping funnel and a condenser, and protected against moisture by a calcium chloride guard tube. The solution is heated under reflux (110°C) for 22 hours under a nitrogen atmosphere and then cooled. A 2 N aqueous sodium hydroxide solution (33 cc) is then run in followed by an 0.316 N aqueous methylchloramine solution (190 cc), the addition of which takes 9 minutes. The mixture is stirred for 1 hour and then decanted. The organic layer is washed with water until it has a pH of 6 and is then extracted with 2 N hydrochloric acid (5 times 50 cc), dried over sodium sulfate, filtered and evaporated. Recrystallization of the residue from petroleum ether yields some unconverted 5-methyl-dibenzo[b,f]azepine (2.17 g).

The aqueous acid solution is rendered alkaline by adding 2 N sodium hydroxide solution. After extracting with diethyl ether (3 times 100 cc), drying the extracts over potassium carbonate, treating them with decolorizing charcoal, filtering and evaporating the ether, a yellowish oil (0.9 g), identified as 5-methyl-10-methylamino-10,11-dihydrodibenzo[b,f]azepine, is obtained in a yield of 37.5%.

Methylchloramine can be prepared by adding an aqueous solution of sodium hypochlorite to an aqueous solution of methylamine in accordance with the process described by W.S. Metcalf, J. Chem. Soc.1942,148.

References

Merck Index 5781
DFU 6 (8) 479 (1981)
Kleeman& Engel p. 569
I.N. p. 605
Linares, H.; British Patent 1,323,219; July 11, 1973; assigned to Rhone-Poulenc SA
Fouche, J.C.L. and Gueremy, C.G.A.; US Patent 3,622,565; November 23, 1971; assigned to Rhone-Poulenc S.A.

METAPROTERENOL SULFATE

Therapeutic Function: Bronchodilator

Chemical Name: 5-[1-Hydroxy-2-[(1-methylethyl)amino]ethyl]-1,3-benzenediol sulfate

Common Name: Orciprenaline sulfate

Structural Formula:

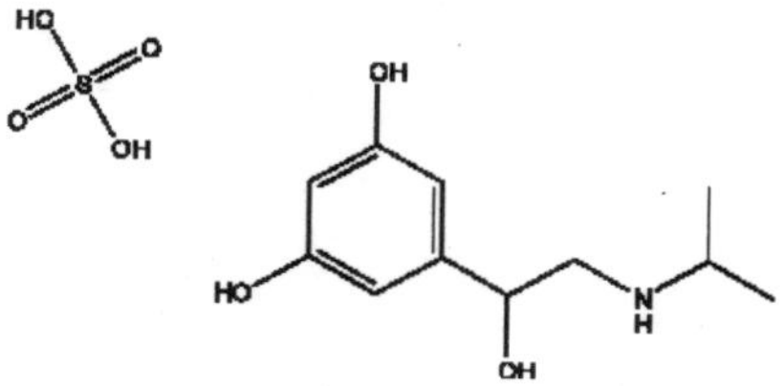

Chemical Abstracts Registry No.: 5874-97-5; 586-06-1 (Base)

Trade Name	Manufacturer	Country	Year Introduced
Alupent	Boehringer Ingelheim	W. Germany	1961
Dosalupent	Boehringer Ingelheim	Italy	1963
Alupent	Badrial	France	1966
Alupent	Boehringer Ingelheim	US	1973
Metaprel	Dorsey	US	1973
Alotec	Tanabe	Japan	-
Astmopent	Polfa	Poland	-

Trade Name	Manufacturer	Country	Year Introduced
Astop	Rafa	Israel	-
Lenasma	Ravasini	Italy	-
Novasmasol	Zambeletti	Italy	-

Raw Materials

3,5-Diacetoxyacetophenone
Isopropylamine
Bromine
Hydrogen

Manufacturing Process

In an initial operation, 3,5-diacetoxyacetophenone was reacted first with bromine and then with isopropylamine to give 1-(3,5-dihydroxyphenyl)-2-isopropylaminoethanone.

59 g of 1-(3,5-dihydroxy-phenyl)-2-isopropylaminoethanone (free base) were dissolved in 590 cc of methanol, and the solution was hydrogenated in the presence of about 80 g Raney nickel at room temperature and under a pressure of 5 atm. Hydrogen absorption was terminated after a few minutes. The catalyst was separated by vacuum filtration, and the filtrate, an ethanolic solution of 1-(3,5-dihydroxyphenyl)-1-hydroxy-2-isopropylaminoethane, was admixed with the calculated amount of an alcoholic 20% sulfuric acid solution. A crystalline precipitate formed which was filtered off and washed with alcohol. For purification, the product was dissolved in water and the solution was filtered through iron-free charcoal.

Thereafter, the filtrate was evaporated to dryness in vacuo and the residue was taken up in alcohol. The crystalline precipitate which separated out after some standing was separated by vacuum filtration and washed with alcohol. After recrystallization from 90% alcohol, 61 g (83.2% of theory) of 1-(3,5-dihydroxyphenyl)-1-hydroxy-2-isopropylamino-ethane sulfate, MP 202° to 203°C, was obtained.

References

Merck Index 5782
Kleeman & Engel p. 658
PDR pp. 674, 848
OCDSVol. 1 p.64 (1977)
I.N. p. 705
REM p. 887
Thoma, O. and Zeile, K.; US Patent 3,341,594; September 12, 1967; assigned to Boehringer Ingelheim G.m.b.H., Germany

METARAMINOL

Therapeutic Function: Hypertensive

Chemical Name: α-(1-Aminoethyl)-3-hydroxybenzenemethanol

Common Name: m-Hydroxynorephedrine; m-Hydroxypropadrine; Metaradrine

Structural Formula:

Chemical Abstracts Registry No.: 54-49-9

Trade Name	Manufacturer	Country	Year Introduced
Aramine	MSD	US	1952
Pressoral	Travenol	US	1963
Pressonex	Winthrop	US	1963
Aramine	MSD-Chibret	France	1963
Araminiurn	Sharp and Dohme	W. Germany	-
Ararninon	Merck-Banyu	Japan	-
Icopal B	Bayer	-	-
Levicor	Bioindustria	Italy	-
Metaraminol	Bristol	US	-

Raw Materials

m-Hydroxyphenylethyl ketone
Butyl nitrite
Hydrogen

Manufacturing Process

The hydrochloride of the m-hydroxyphenylpropanolamine may be prepared by dissolving or suspending 90 parts of m-hydroxyphenylethyl ketone, O = $C(C_6H_4\text{-}OH)\text{-}C_2H_5$, in about 400 parts of ether. Hydrogen chloride is slowly bubbled through the solution or suspension while agitating it and 61.8 g of butyl nitrite is added during the course of 60 to 90 minutes. During the addition of the butyl nitrite the suspended m-hydroxyphenylethyl ketone gradually dissolves. The mixture or solution is allowed to stand for at least an hour, but preferably overnight. It is then repeatedly extracted with dilute alkali until all alkali-soluble material is removed. The alkaline extract is slowly acidified and the precipitate which forms is crude m-hydroxyphenyl-α-oximinoethyl ketone. After recrystallization from water this melts at 138°C.

10.8 parts of the meta ketone is dissolved in about 125 parts of absolute alcohol containing 5.6 parts of hydrogen chloride. The solution is agitated with a catalyst such as the palladium catalyst above described in an atmsophere of hydrogen until no more hydrogen is absorbed. This requires from 60 to 90 minutes or more. When reduction is complete the catalyst is filtered off and the filtrate evaporated to dryness by being placed in a desiccator at ordinary temperature.

The residue is the hydrochloride of m-hydroxyphenyl-α-aminoethyl ketone. This is purified by recrystallization from absolute alcohol. It is then dissolved in 200 parts of water and agitated with a further quantity of the palladium catalyst in an atmosphere of hydrogen until saturated. The product thus recovered from the solution is the hydrochloride of m-hydroxyphenylpropanol amine. After recrystallization from absolute alcohol this melts at 177°C. The corresponding free base can be prepared from the hydrochloride by treatment with ammonia, according to US Patent 1,995,709.

Metaraminol is often used in the form of the bitartrate.

References

Merck Index 5783
Kleeman & Engel p. 570
PDR pp. 695, 1140
I.N. p. 605
REM p. 888
Bockmuhl, M., Ehrhart, G. and Stein, L.; US Patent 1,948,162; February 20, 1934; assigned to Winthrop Chemical Company, Inc.
Bockmuhl, M., Ehrhart, G. and Stein, L.; US Patent 1,951,302; March 13,1934; assigned to Winthrop Chemical Company, Inc.
Hartung, W.H.; US Patent 1,995,709; March 26,1935; assigned to Sharp & Dohme, Inc.

METAXALONE

Therapeutic Function: Muscle relaxant

Chemical Name: 5-(3,5-Dimethylphenoxymethyl)-2-oxazolidnone

Common Name: -

Structural Formula:

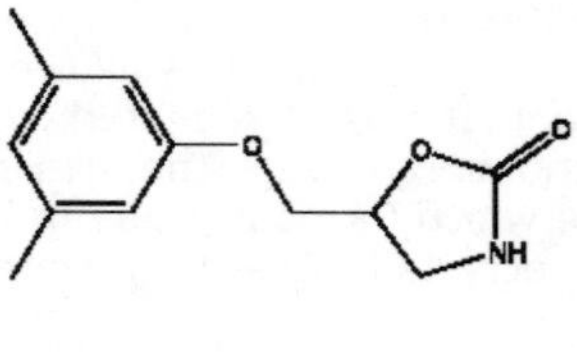

Chemical Abstracts Registry No.: 1665-48-1

Trade Name	Manufacturer	Country	Year Introduced
Skelaxin	Robins	US	1962

Raw Materials

Urea
3-(3',5'-Dimethylphenoxy)-1,2-propanediol

Manufacturing Process

Urea (118 g, 1.96 mols) was added to 192 g (0.98 mol) of 3-(3',5'-dimethylphenoxy)-1,2-propane-diol which had previously been heated to 150°C. The reaction mixture was then heated rapidly to 195° to 200°C and maintained at this temperature for 5 hours with constant stirring. The resulting mixture was partitioned between water and ethyl acetate and the ethyl acetate layer was dried over sodium sulfate and concentrated. The residue was distilled in vacuo and the fraction boiling at 220° to 225°C/1.5 mm was collected. Yield, 172 g (79%). The distillate was crystallized from dry ethyl acetate; MP, 121.5° to 123°C.

References

Merck Index 5785
Kleeman & Engel. p.571
PDR p. 783
OCDS Vol. 1 p. 119 (1977)
I.N.p. 606
REMp. 927
Lunsford, C.D.; US Patent 3,062,827; November 6, 1962; assigned to A.H. Robins Company, Inc.

METERGOLINE

Therapeutic Function: Analgesic

Chemical Name: [[(8β)-1,6-Dimethylergolin-8-yl]methyl]carbamic acid phenylmethyl ester

Common Name: Methyl-N-carbobenzoxy-dihydro-lysergamine

Chemical Abstracts Registry No.: 17692-51-2

Trade Name	Manufacturer	Country	Year Introduced
Liserdol	Farmitalia	Italy	1970

Raw Materials

1-Methyl-dihydro-lysergamine
Carbobenzoxy chloride

Structural Formula:

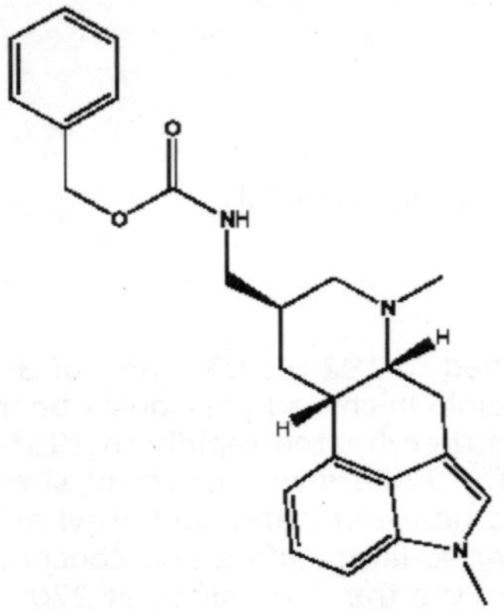

Manufacturing Process

16 g of 1-methyl-dihydro-lysergamine (the 10-position hydrogen has the α-configuration) are dissolved in 80 cc of anhydrous pyridine by mildly heating. To the solution, cooled to -10°C and stirred, 18 cc of 85% carbobenzoxy-chloride (in toluene) diluted in 36 cc of chloroform are added dropwise, rather rapidly. The mixture is kept at -10°C during the addition, and for 10 minutes afterwards. The cooling means is removed and the temperature is allowed to rise to room level in 10 minutes. The reaction mixture is diluted with 240 cc of chloroform and rapidly washed with 80 cc of 5% aqueous sodium hydroxide solution, with saturated aqueous sodium bicarbonate solution, and finally with water.

The chloroform solution is briefly dried over anhydrous sodium sulfate and evaporated to dryness in vacuo at 40°C. The oily residue is taken up in 160 cc of benzene and passed through a column containing 48 g of alumina. The column is then eluted with further 160 cc of benzene. The collected eluates are evaporated in vacuo at 40°C. The thick oily residue is mixed with a small amount of anhydrous diethyl ether. After some time a crystalline mass is obtained, which is collected and washed with a small amount of benzene and diethyl ether. 12 g of white crystals are obtained, melting at 146° to 148°C.

References

Merck Index 5790
I.N. p. 606
Camerino, B., Patelli, B. and Glaesser, A.; US Patent 3,238311; March 1, 1966; assigned to Societa Farmaceutici Italia, Italy

METFORMIN HYDROCHLORIDE

Therapeutic Function: Oral hypoglycemic

Chemical Name: Biguanide, 1,1-dimethyl-, hydrochloride

Common Name: Dimethylguanilguanidini chloridum; Metformin

Structural Formula:

Chemical Abstracts Registry No.: 1115-70-4; 15537-72-1; 657-24-9 (Base)

Trade Name	Manufacturer	Country	Year Introduced
Diabetex	Germania	-	-
Diabetex	Terrapharm	-	-
Diaformin	Alphapharm	-	-
Diaphage	UPM	-	-
D.B.I.	Montpellier	-	-
Glucomet	USV	India	-
Glucophage	Laboratoires Aron	France	-
Glucophage	Bristol-Meyers Squibb	USA	-
Metforal	Menarini	Italy	-

Raw Materials

Dimethylamine
Dicyanamide
Hydrogen chloride

Manufacturing Process

The boiling mixture of 1,000 L xylene, 450 kg dimethylamine and 840 kg dicyanamide was added 365 kg hydrogene chloride. Yield of biguanide, 1,1-dimethyl-, hydrochloride 1,588 kg (96%). Biguanide, 1,1-dimethyl-, hydrochloride may be recrystallysed from methanol.

References

Patent DE 1023757
Patent FR 2,322,860; Sep. 1975; Assigned to ARON S.A.R.L.

METHACYCLINE

Therapeutic Function: Antibiotic

Chemical Name: 4-Dimethylamino-1,4,4a,5,5a,6,11,12a-octahydro-3,5,10,12,12a-pentahydroxy-6-methylene-1,11-dioxo-2-naphthacenecarboxamide

Common Name: 6-Methylene-5-hydroxytetracycline

Structural Formula:

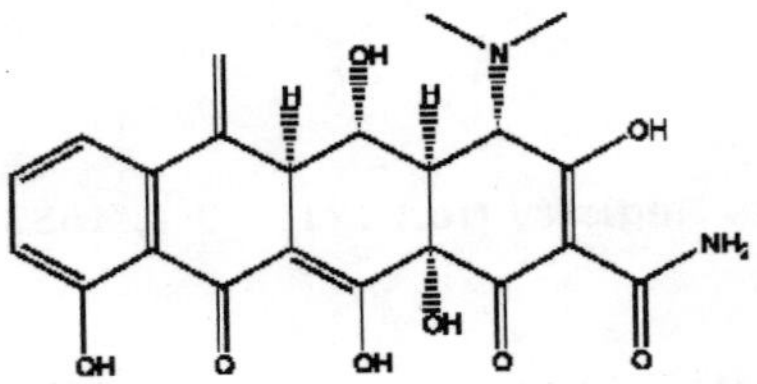

Chemical Abstracts Registry No.: 914-00-1; 3963-95-9 (Hydrochloride salt)

Trade Name	Manufacturer	Country	Year Introduced
Rondomycin	Pfizer	UK	1963
Megamycine	Creat	France	1966
Rondomycin	Wallace	US	1966
Adramycin	Janko	Japan	-
Apriclina	Lancet	Italy	-
Benciclina	Benvegna	Italy	-
Boscillina	Molteni	Italy	-
Brevicillina	Neopharmed	Italy	-
Ciclobiotic	Beta	Italy	-
Ciclum	Italsuisse	Italy	-
Duecap	Sam	Italy	-
Duplaciclina	Locatelli	Italy	-
Duramicina	Bergamon	Italy	-
Dynamicin	Medal	Italy	-
Esarondil	Terapeutico	Italy	-
Esquilin	Saita	Italy	-
Fitociclina	Ifisa	Italy	-
Franciclina	Francia	Italy	-
Francomicina	N.C.S.N.	Italy	-
Gammaciclina	Sthol	Italy	-
Globociclina	Importex	Italy	-
Idrossimicina	San Carlo	Italy	-
Isometa	Isom	Italy	-
Largomicina	Jamco	Italy	-
Medomycin	Medosan	Italy	-
Megamycine	C.R.E.A.T.	Italy	-
Metabiotic	Panther-Osfa	Italy	-
Metabioticon BG	Boniscontro-Gazzone	Italy	-

Trade Name	Manufacturer	Country	Year Introduced
Metac	Dima	Italy	-
Metacil	Ibirn	Italy	-
Metaclin	Medici	Italy	-
Metaclor	Esset	Italy	-
Metadomus	Medici Domus	Italy	-
Metagram	Zanardi	Italy	-
Metilenbiotic	Coli	Italy	-
Microcilina	Biotrading	Italy	-
Mit-Ciclina	Von Boch	Italy	-
Molciclina	Molteni	Italy	-
Optimicine	Biochemie	Austria	-
Ossirondil	Gazzini	Italy	-
Paveciclina	I.B.P.	Italy	-
Physiomycine	Roland-Marie	France	-
Piziacina	Farmochimica	Italy	-
Plurigram	Lafare	Italy	-
Prontomicina	Tosi-Novara	Italy	-
Quickmcina	Panthox and Burck	Italy	-
Radiomicin	Radiopharma	Italy	-
Rindex	Sidus	Italy	-
Rotilen	Arnelix	Italy	-
Sernamicina	Pharma Williams	Italy	-
Stafilon	A.G.I.P.S.	Italy	-
Tachiciclina	C.T.	Italy	-
Tetrabios	Ausonia	Italy	-
Tetranovo	Totalpharm	Italy	-
Tiberciclina	Tiber	Italy	-
Ticomicina	Benedetti	Italy	-
Treis-Ciclina	Ecobi	Italy	-
Valcin	Chemil	Italy	-
Vitabiotic	PHARMEX	Italy	-
Wassermicina	Wassermann	Italy	-
Yatrociclina	Italfarmaco	Italy	-
Zermicina	Pulitzer	Italy	-

Raw Materials

Oxytetracycline
Sulfur trioxide
Hydrogen fluoride

Manufacturing Process

To a stirred solution of 4.6 g (0.01 mol) of anhydrous oxytetracycline in 40 ml of dry tetrahydrofuran is added 3.5 g (0.021 mol) of pyridine-sulfur trioxide complex. After 16 hours of stirring at room temperature, the resulting suspension is filtered, and the solid is slurried with 25 ml of 2% hydrochloric acid for 10 minutes, filtered and thoroughly washed with methanol followed by ether. The pale yellow crystalline 5-oxytetracycline-6,12-hemiketal-12-sulfuric acid ester melts at 210°C.

500 mg 5-oxytetracycline-6,12-hemiketal-12-sulfuric acid ester, prepared as described, is added to 4 ml dry liquid hydrogen fluoride, and the mixture is stirred for 1.5 hours at ice bath temperature. The hydrogen fluoride is then evaporated in a stream of nitrogen and the resulting gummy solids are triturated with about 15 ml ether and filtered. The resulting solid hydrofluoride salt is further purified by suspending in water, adjusting the pH to about 4, and extracting the 6-methylene-5-oxytetracycline free base from the aqueous phase with ethyl acetate. The extract is separated and evaporated to dryness under reduced pressure. The resulting residue is triturated with ether and filtered, and the solid is recrystallized from methanol-acetone-ether-concentrated hydrochloric acid to obtain the product as a purified hydrochloride, according to US Patent 3,026,354.

References

Merck Index 5798
Kleeman & Engel p. 567
PDR p. 1881
OCDS Vol. 2 p. 227 (1980)
DOT 1 (1) 10 (1965)
I.N. p. 603
REM p. 1205
Blackwood, R.K., Rennhard, H.H., Beereboom, J.J. and Stephens, C.R., Jr.; US Patent 2,984,686; May 16, 1961; assigned to Chas. Pfizer & Co., Inc.
Biackwood, R.K.; US Patent 3,026,354; March20, 1962; assigned to Chas. Pfizer & Co., Inc.

METHADONE HYDROCHLORIDE

Therapeutic Function: Narcotic analgesic

Chemical Name: 6-Dimethylamino-4,4-diphenyl-3-heptanone hydrochloride

Common Name: Amidone hydrochloride

Structural Formula:

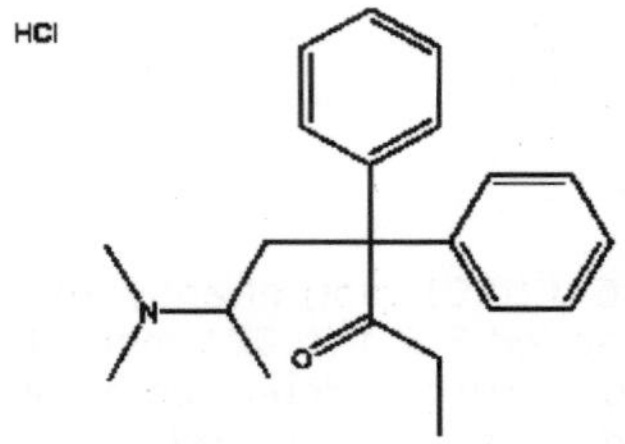

HCl

Chemical Abstracts Registry No.: 1095-90-5; 76-99-3 (Base)

Trade Name	Manufacturer	Country	Year Introduced
Dolophine	Lilly	US	1947
Adanon	Winthrop	US	1947
Westadone	Vitarine	US	1973
Adolan	Abic	Israel	-
Eptadone	Tosi	Italy	-
Heptadon	E.B.E.W.E.	Austria	-
Heptanal	Treupha	Switz.	-
Heptanon	Pliva	Yugoslavia	-
Ketalgin	Amino	Switz.	-
Mephenon	Spemsa	Italy	-
Optalgin	Dr. Wust	Switz.	-
Physeptone	Burroughs-Wellcome	UK	-

Raw Materials

Ethyl bromide
Diphenylacetonitrile
Hydrogen chloride
Magnesium
2-Chloro-1-dimethylaminopropane

Manufacturing Process

Diphenylacetonitrile is condensed with 2-chloro-1-dimethylaminopropane to give 4-(dimethylamino)-2,2-diphenyl valeronitrile. It is then reacted with ethyl magnesium bromide and then hydrolyzed using HCl to give methadone hydrochloride.

References

Merck Index 5799
Kleeman & Engel p. 573
PDR pp. 1048, 1061, 1571
OCDS Vol.1 pp.79, 289, 298 (1977) and 2, 328 (1980)
I.N. p. 607
REM p. 1109
Resolution of Optical Isomers:
Howe, E.E. and Tishler, M.; US Patent 2,644,010; June 30, 1953; assigned to Merck & Co., Inc.
Zaugg, H.E.; US Patent 2,983,757; May 9, 1961; assigned to Abbott Laboratories

METHALLENESTRIL

Therapeutic Function: Estrogen

Chemical Name: β-Ethyl-6-methoxy-α,α-dimethyl-2-naphthalenepropionic acid

Common Name: -

Structural Formula:

Chemical Abstracts Registry No.: 517-18-0

Trade Name	Manufacturer	Country	Year Introduced
Vallestril	Searle	US	1952
Cur-Men	Novapharma	Italy	-
Ercostrol	Erco	Denmark	-
Ercostrol	Green Cross	Japan	-

Raw Materials

Copper cyanide	2-Bromo-6-methoxynaphthalene
Ethyl bromide	Ethyl bromoisobutyrate
Magnesium	Potassium bisulfate
Hydrogen	Sodium hydroxide

Manufacturing Process

A first step involves the preparation of 2-cyano-6-methoxynaphthalene (cyanonerolin). 90 g of 2-bromo-6-methoxynaphthalene are heated with 60 g of cuprous cyanide in a metal bath at 240° to 250°C stirring for one hour. At the instant when the cuprous cyanide begins to react and dissolves, the mass turns brown, liquefies and heats up strongly. The molten mass is poured onto a cold surface, is pulverized and sifted. This powder is treated with dilute ammonia (1 liter of water to 300 cc of commercial ammonia solution). The solution is filtered on a Buchner filter and the precipitate that remains on the filter is washed with dilute ammonia and then with water.

After drying, the residue is treated in a Kumagawa extracting apparatus with boiling benzene. The benzene is evaporated and the residue is distilled in vacuo. About 50 g of cyanonerolin (BP = 205° to 208°C/14 mm) are obtained with a yield of about 70%. By recrystallization in 200 cc of methyl alcohol, 40 g of the product are obtained in absolutely pure state, in the shape of beautiful colorless needles (MP = 103°C with the Maquene block). By concentrating the mother liquor to half its original volume, a further 3.6 g of pure product are obtained.

The 2-cyano-6-methoxy-naphthaleneis in turn converted by successive reactions into: (a) β-ketonic ester, (b) ester-alcohol, (c) β-ethylene ester by dehydration, (d) saturated ester, and (e) [3-(6-methoxy-2-naphthyl)]2,2-dimethyl pentanoic acid which is the required product.

(A) Obtaining a β-Ketonic Ester by Reacting Ethyl Bromoisobutyrate with Cyanonerolin: 9 g of cyanonerolin are heated in a reflux apparatus for 40 minutes with 7 g of zinc and 19 g of ethyl bromoisobutyrate in the presence of 150 cc of anhydrous benzene. After cooling, the mixture is filtered to eliminate unreacted zinc and is hydrolyzed by stirring for one hour with dilute sulfuric acid (10 cc of sulfuric acid to 200 cc of water). The benzene layer is washed, dried and the solvent is eliminated. It is purified by recrystallization in methyl alcohol. 12.5 g of ketonic ester (MP = 72.5° to 73.5°C) are thus obtained in the form of large prismatic crystals.

(B) Obtaining an Ester-Alcohol by Reacting Magnesium Ethyl Bromide with the Previous Ketonic Ester: 10 g of the previous ester dissolved in 40 cc of anhydrous benzene are gradually poured while stirring into an iced solution of magnesium ethyl bromide prepared from 1.035 g of magnesium, 4.15 cc of ethyl bromide and 40 cc of anhydrous ether. After heating in a reflux apparatus for one-half hour, the mixture is poured into ice in the presence of ammonium chloride.

After washing the ether-benzene layer, the solvents are eliminated in vacuo and an ester-alcohol is thus obtained with a yield of 98%, in the form of a transparent resin. This resin, if treated with petroleum ether, yields 6.35 g of ester-alcohol in the form of fine needles (MP = 66.68°C) which are very soluble in the chief organic solvents and in petroleum ether.

(C) Conversion into Ethyl [3(6-Methoxy-2-Naphthyl)] 2,2-Dimethyl-3-Pentanoate by Dehydrating the Previous Ester-Alcohol: The semi-oily raw product of the previous reaction is dehydrated by heating with its own weight of potassium bisulfate to 180°C until boiling stops. After cooling, the magma is removed from the anhydrous ether in small portions. The ether is then evaporated and an ethylene ester is obtained in the form of an oil which slowly solidifies, with a yield of 98%. The product, after being purified by chromatography, melts at 48° to 51°C.

(D) Obtaining Ethyl [3-(6-Methoxy-2-Naphthyl)] 2,2-Dimethyl Pentanoate by Hydrogenation of the Previous Ethylene Ester: 3.5 g of the previous ethylene ester, purified by chromatography, are hydrogenated in the presence of 3.6 g of platinum in 30 cc of ether. The quantity of hydrogen fixed corresponds to the theoretical quantity calculated. After filtering, the ether is evaporated, 3.45 g of ester are thus obtained in the form of an oil which quickly solidifies. Purification is effected by chromatography.

(E) Obtaining [3-(6-Methoxy-2-Naphthyl)] 2,2-Dimethyl Pentanoic Acid: 2.5 g of the previous ester are saponified by means of 15 cc of soda lye and 25 cc of methyl glycol. The mixture is boiled for one hour, diluted with water and, after cooling, is treated twice with ether in order to eliminate the remaining neutral fractions. The aqueous layer is precipitated by means of 15 cc of acetic acid. 2.1 g of raw acid are obtained. After effecting two crystallizations in 10 parts of acetic acid mixed with 3 parts of water, fine needles are obtained which are grouped in rosettes and melt at 131.5° to 132.5°C.

References

Merck Index 5803
Kleeman & Engel p. 574
OCDS Vol. 1 p. 87 (1977)
I.N. p. 608
Horeau, A. and Jacques, J.; US Patent 2,547,123; April 3, 1951

METHAMPHETAMINE HYDROCHLORIDE

Therapeutic Function: Sympathomimetic, Central stimulant

Chemical Name: Benzeneethanamine, N,alpha-dimethyl-, hydrochloride, (S)-

Common Name: Desoxyephedrine hydrochloride; Metamfetamine hydrochloride; Metamphetamine hydrochloride; Methaphetamine hydrochloride; Methylamphetamine hydrochloride; Phenylmethylaminopropane hydrochloride

Structural Formula:

HCl

H
N

Chemical Abstracts Registry No.: 51-57-0 ; 537-46-2 (Base)

Trade Name	Manufacturer	Country	Year Introduced
Desoxyn	Abbott Laboratories	-	-
Amphedroxyn	Lilly	-	-
Destim	Central Pharm.	-	-
Drinalfa	Squibb	-	-
Gerobit	Gerot	-	-
Isophen	Knoll	-	-
Madrine	Langley	-	-
Methampex	Lemmon	-	-
Methampex	Teva	-	-
Methedrine	Burroughs-Wellcome	-	-
Pervitin	Temmler	-	-
Soxysympamine	Ferndale	-	-
Syndrox	McNeil	-	-
Tonedron	Grimault	-	-

Raw Materials

Ephedrine, (-)-
Phenylisopropylamine
Platinum on carbon
Hydrochloric acid

Manufacturing Process

2 Methods of prepearing of methamphetamine:

1. (-)-Ephedrin was reduced by hydrogenesation with hydrogen in the presence of Pt-C catalyst to give the (+)-N-α-dimethylphenethylamine (methamphetamine), melting point 172°-174°C.

2. Methamphetamine was obtained by the methylation of phenylisopropylamine.

To give methamphetamine hydrochloride the base methamphetamine was treated by eqimolar quantity of hydrochloric acid.

References

Haletsky A.M.; Pharmaceutical Chemistry, Medicina. L., 1966, 761p.
Emde H.; Helv. Chim. Acta 1929, v. 12, p. 365

METHANDROSTENOLONE

Therapeutic Function: Androgen, Anabolic

Chemical Name: 17β-Hydroxy-17-methylandrosta-1,4-dien-3-one

Common Name: Methandienone

Structural Formula:

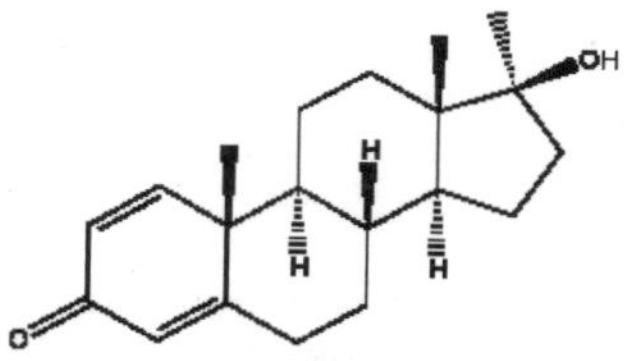

Chemical Abstracts Registry No.: 72-63-9

Trade Name	Manufacturer	Country	Year Introduced
Dianabol	Ciba	US	1960
Abirol	Takeda	Japan	-
Anabolin	Medica	Finland	-
Anoredan	Kodama	Japan	-
Encephan	Sato/Shinshin	-	-
Lanabolin	Labatec	Switz.	-
Metabolina	Guidi	Italy	-
Metanabol	Polfa	Poland	-
Metastenol	Farber-R.E.F.	Italy	-
Naposim	Terapia	Rumania	-
Nerobol	Galenika	Yugoslavia	-
Perbolin	Ion	Italy	-
Vanabol	Vitrum	Sweden	-

Raw Materials

Bacterium Didymella lycopersici
17β-Methyl testosterone
Selenium dioxide

Manufacturing Process

As described in US Patent 2,929,763, methandrostenolone may be made by a fermentation route. 2 g of sodium nitrate, 1 g of primary potassium orthophosphate, 0.5 g of magnesium sulfate heptahydrate, 0.5 g of potassium chloride, 50 g of glucose and 1 g of Difco yeast extract are dissolved in one liter of tap water, brought to pH 5 by addition of a sodium hydroxide solution and sterilized. The resulting nutrient solution is inoculated with 50 cc of a 4-day-old shaking culture of Didymella lycopersici and shaken for 48 hours at 27°C, whereby the culture becomes well developed.

To two liters of a culture so prepared there is added under sterile conditions a solution of 500 mg of 17α-methyl-testosterone in 15 cc of acetone. Shaking is carried out for 3 days at 27°C, the mycellium then filtered off with suction, washed with water and ethyl acetate and the combined filtrates extracted with ethyl acetate. The extraction residue obtained after evaporation of the solvent is dissolved in a little acetone. On addition of ether, the 1-dehydro-17α-methyl-testosterone is obtained in compact crystals. MP 163° to 164°C.

An alternative synthetic route is described in US Patent 2,900,398 as follows. A suspension of 30 g of 17α-methyl-testosterone and 10 g of selenium dioxide in 600 cc of tertiary amyl alcohol is treated with 60 g of magnesium powder and 6 cc of glacial acetic acid.

The mixture is refluxed for 24 hours with good stirring in an atmosphere of nitrogen, another 10 g of selenium dioxide being added after 10 hours. After some cooling, the suspension is filtered through some Hyflo and washed thoroughly with ethyl acetate. The resulting brown solution is evaporated in vacuo and the residue dissolved in ethyl acetate.

The ethyl acetate solution is then washed with water, dried and evaporated. To remove any selenium still present, the residue is dissolved in 200 cc of methanol and mixed with 100 g of iron powder and 2 g of active carbon. The mixture is heated for 30 minutes with stirring under reflux, then filtered with suction, washed with methanol and the solution evaporated in vacuo. The residue is then chromatographed on 900 g of aluminum oxide. The residues of the evaporated benzene and ether fractions are treated with active carbon in methanol or acetone, evaporated again, and the residue recrystallized from a mixture of acetone and ether. There are obtained 17.5 g of pure 1-dehydro-17α-methyl-testosterone which melts at 163° to 164°C.

References

Merck Index 5810
Kleeman & Engel p. 570
OCDS Vol. 1 p. 173 (1977)
I.N. p. 605

REM p.998
Wettstein, A., Hunger, A., Meystre, C. and Ehmann, L.; US Patent 2,900,398; August 18, 1959; assigned to Ciba Pharmaceutical Products, Inc.
Wettstein, A., Vischer, E. and Meystre, C.; US Patent 2,929,763; March 22, 1960; assigned to Ciba Pharmaceutical Products, Inc.

METHAPYRILENE HYDROCHLORIDE

Therapeutic Function: Antihistaminic

Chemical Name: N,N-Dimethyl-N'-2-pyridinyl-N'-(2-thienylmethyl)-1,2-ethanediamine hydrochloride

Common Name: Thenylpyramine hydrochloride

Structural Formula:

Chemical Abstracts Registry No.: 135-23-9; 91-80-5 (Base)

Trade Name	Manufacturer	Country	Year Introduced
Thenylene	Abbott	US	1947
Pyrathyn	Davis Sly	US	1947
Histadyl	Lilly	US	1948
Semikon	Beecham	US	1949
Lullamin	Reed Carnrick	US	1954
Dozar	Tutag	US	1956
Allergin	Myers-Carter	US	-
Allerest	Pharmacraft	US	-
Brexin	Savage	US	-
Citra	Boyle	US	-
Ephed-Organidin	Wallace	US	-
Excedrin P.M.	Bristol-Myers	US	-
Histadyl	Lilly	US	-
M.P.	Dymond	Canada	-
Sedanoct	Woalm-Pharma	W. Germany	-
Contac	Vonora	W. Germany	-
Co-Pyronil	Lilly	Italy	-

Raw Materials

2-Aminopyridine
2-Thenyl chloride
Hydrogen chloride
N,N-Dimethyl-β-chloroethylamine
Sodium amide

Manufacturing Process

To a slurry of sodamide in 200 cc of toluene representing 6.7 g of sodium was added at 30° to 40°C, 32.3 g (0.31 mol) of 2-aminopyridine. The mixture was heated to reflux temperature and was refluxed for 1½ hours. To the resulting mixture was added over a period of approximately one hour a solution of 32 g of freshly distilled N,N-dimethyl-β-chloroethylamine in 40 to 50 cc of dry toluene, The reaction mixture was then heated for 2 hours at reflux temperature. Thereafter, 200 cc of water was added and the toluene layer was separated and washed with water. The toluene was stripped from the mixture by distillation and the residue was distilled under reduced pressure. The distillate was refractionated and the portion distilled at 93° to 103°C/1 mm was recovered. Yield of N-(2-pyridyl)-N',N'-dimethyl-ethylenediamine, 60%.

A solution of 20 g (0.121 mol) of N-(2-pyridyl)-N',N'-dimethyl-ethylenediamine in 25 cc of toluene was added to a slurry of sodamide in 100 cc of toluene representing 2.8 g of sodium. The mixture was refluxed for one hour. To this mixture was added over a period of ½ hour a solution of 16 g (0.121 mol) of 2-thenyl chloride in 25 cc of toluene. The resulting reaction mixture was refluxed for 3 hours. Thereafter, water was added and the toluene layer was separated and washed with water.

The toluene was then stripped off by distillation and the residue was distilled under reduced pressure. The main fraction was redistilled. Yield of N-(2-pyridyl)-N-(2-thenyl)-N',N'-dimethyl-ethylenediamine was 69%; BP 130° to 140°C/0.4 mm. A portion of the product was dissolved in ether and an ether solution of hydrogen chloride was added. The monohydrochloride of N-(2-pyridyl)-N-(2-thenyl)-N',N'-dimethyl-ethylenediamine which separated was washed with ether and dried.

References

Merck Index 5819
Kleeman & Engel p. 575
OCDS Vol. 1 p. 54 (1977)
I.N. p. 609
Kyrides, L.P.; US Patent 2,581,868; January 8, 1952; assigned to Monsanto Chemical Company

METHAQUALONE

Therapeutic Function: Hypnotic

Chemical Name: 2-Methyl-3-o-tolyl-4(3H)-quinazolinone

Common Name: Metolquizolone; Ortonal

Structural Formula:

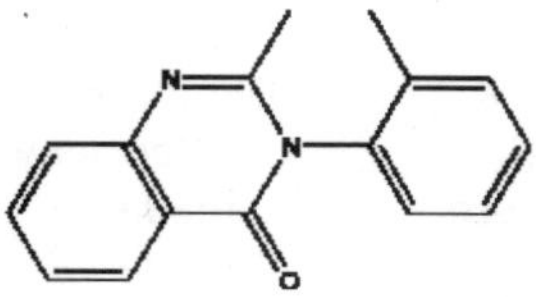

Chemical Abstracts Registry No.: 72-44-6; 340-56-7 (Hydrochloride salt)

Trade Name	Manufacturer	Country	Year Introduced
Quaalude	Lemmon	US	1965
Sopor	Amer. Crit. Care	US	1967
Somnafac	Cooper	US	1968
Parest	Lemmon	US	1969
Quaalude	Rorer	Italy	1969
Optimil	Wallace	US	1972
Aqualon	Arcana	Austria	-
Cateudyl	Cavor	Belgium	-
Citexal	Draco	Sweden	-
Divinoctal	I.S.H.	France	-
Dormigoa	Scheurich	W. Germany	-
Dormir	Langley	Australia	-
Dormutil	Isis-Chemie	E. Germany	-
Hyptor	Bio-Chimique	Canada	-
Hyminal	Eisai	Japan	-
Mandrax	I.S.H.	France	-
Mequelon	Merck-Frosst	Canada	-
Meroctan	Sanwa	Japan	-
Methadorm	Eri	Canada	-
Metasedil	Cooper	Switz.	-
Mollinox	Asperal	Belgium	-
Motolon	Chinoin	Hungary	-
Nene	Sankyo	Japan	-
Nobadorm	Streuli	Switz.	-
Normi-Nox	Herbrand	W. Germany	-
Normorest	Doitsu-Aoi	Japan	-
Noxybel	Probel	Belgium	-
Oblioser	Gamaprod.	Australia	-
Optinoxan	Robisch	W. Germany	-
Parmilene	Chiesi	Italy	-
Paxidorm	Wallace	US	-
Pexaqualone	Therapex	Canada	-
Pro-Dorm	Schurholz	W. Germany	-
Revonal	Merck	UK	-
Rouqualone	Rougier	Canada	-

Trade Name	Manufacturer	Country	Year Introduced
Sedalone	Pharbec	Canada	-
Sleepinal	Medichem	Australia	-
Somnium	Fargal	Italy	-
Sovelin	Weifa	Norway	-
Sovinal	N.D. and K.	Denmark	-
Spasmipront	Mack	W. Germany	-
Tiqualone	Barlow Cote	Canada	-
Tualone	I.C.N.	Canada	-

Raw Materials

Anthranilic acid
o-Toluidine
Acetic anhydride
Hydrogen chloride

Manufacturing Process

Anthranilic acid (1 part) is dissolved in acetic anhydride (2 parts) and the temperature raised progressively to 190° to 200°C while distillation takes place. The last traces of acetic acid are removed under vacuum and, after cooling to about 50° to 60°C, o-toluidine (1 part) is added in portions.

The temperature is then raised to 170° to 200°C when the excess water and o-toluidine is gradually distilled off, finally maintaining the temperature at 180° to 200°C for 2 hours. After cooling to about 100°C dilute hydrochloric acid (3 parts) is added and the mixture boiled and stirred. The solution is then neutralized with NaOH with stirring and the product which separates is recrystallized twice from alcohol after decolorizing with carbon. Yield: 70% of theoretical, LIP 114° to 115°C.

References

Merck Index 5820
Kleeman & Engel p. 576
OCDS Vol. 1 p.353 (1977)
DOT 9 (6) 245 (1973)
I.N. p.610
REM p. 1072
Laboratoires Toraude, France; British Patent 843,073; August 4, 1960

METHAZOLAMIDE

Therapeutic Function: Carbonic anhydrase inhibitor

Chemical Name: N-[5-(Aminosulfonyl)-3-methyl-1,3,4-thiadiazol-2(3H)-ylidene]acetamide

Common Name: -

Structural Formula:

Chemical Abstracts Registry No.: 554-57-4

Trade Name	Manufacturer	Country	Year Introduced
Neptazane	Lederle	US	1959
Neptazane	Theraplix	France	1961

Raw Materials

5-Acetylimino-4-methyl-2-benzylmercapto-δ2-1,3,4-thiadiazoline
Chlorine
Ammonia

Manufacturing Process

A suspension of 6 parts by weight of 5-acetylimino-4-methyl-2-benzylmercapto-δ2-1,3,4-thiadiazoline in 180 parts by volume of 33% aqueous acetic acid was chlorinated at 5°C for 30 minutes. The solid was filtered off, dried, and added portion-wise to 100 parts by volume of liquid ammonia. The ammonia was removed under a stream of dry nitrogen.

The residual solid was partially dissolved in 10 parts by volume of water, filtered, and acidified to give 5-acetylimino-4-methyl-δ2-1,3,4-thiadiazoline-2-sulfonamide. The product was purified by two recrystallizations from hot water.

References

Merck Index 5824
Kleeman & Engel p. 576
PDR p. 1021
OCDS Vol. 1 p. 250 (1977)
I.N. p. 610
REM p.936
Young, R.W., Wood, K.H. and Vaughan, J.R., Jr.; US Patent 2,783,241; February 26, 1957; assigned to American Cyanamid Company

METHDILAZINE HYDROCHLORIDE

Therapeutic Function: Antipruritic

Chemical Name: 10-[(1-Methyl-3-pyrrolidinyl)methyl]phenothiazine hydrochloride

Common Name: -

Structural Formula:

HCl

Chemical Abstracts Registry No.: 1229-35-2; 1982-37-2 (Base)

Trade Name	Manufacturer	Country	Year Introduced
Tacaryl	Westwood	US	1960
Dilosyn	Duncan Flockhart	UK	-
Disyncran	Allard	France	-
Tacryl	Pharmacia	Sweden	-

Raw Materials

1-Methyl-3-pyrrolidylmethyl chloride
Phenothiazine
Hydrogen chloride

Manufacturing Process

10.8 parts of 10-(1-methyl-3-pyrrolidylmethyl) phenothiazine (prepared from 1-methyl-3-pyrrolidylmethyl chloride by reaction with phenothiazine) in 80 parts of 99% isopropyl alcohol were treated with a solution of 1.33 parts of hydrogen chloride in 30 parts of the same solvent. The clear light yellow solution soon deposited white crystals of the acid addition salt. After cooling overnight at 0°C, the crystalline product was collected on a filter, washed with 99% isopropyl alcohol and anhydrous ether and then dried in a vacuum oven at 95°C. Yield 10.4 parts, MP 187.5° to 189°C.

References

Merck Index 5826
Kleeman & Engel p. 577
PDR p. 1895

OCDS Vol. 1 p. 387 (1977)
I.N. p. 611
REM p. 1129
Feldkamp, R.F. and Wu, Y.H.; US Patent 2,945,855; July 19, 1960; assigned to Mead Johnson & Company

METHENAMINE HIPPURATE

Therapeutic Function: Antibacterial (urinary)

Chemical Name: Hexamethylenetetramine hippurate

Common Name: -

Structural Formula:

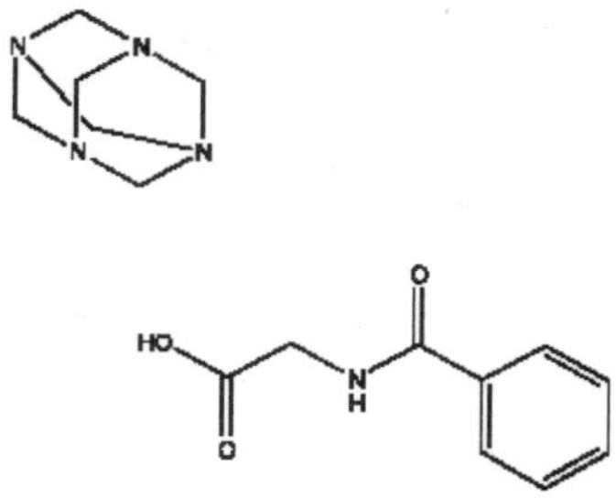

Chemical Abstracts Registry No.: 5714-73-8

Trade Name	Manufacturer	Country	Year Introduced
Hiprex	Merrell National	US	1967
Hiprex	Riker	UK	1971
Hiprex	Kettelhack Riker	W. Germany	1975
Hipeksal	Leiras	Finland	-
Hippuran	Orion	Finland	-
Lisogerm	Labofarma	Brazil	-
Urotractan	Klinge	W. Germany	-

Raw Materials

Hexamethylenetetramine
Hippuric acid

Manufacturing Process

179 g (1 mol) hippuric acid (benzoyl glycine) and 140 g (1 mol) hexamethylenetetramine were heated under reflux in 500 ml methanol. The small amount of water necessary to give a clear, homogeneous solution was

added to the resulting reaction mixture which was then evaporated to dryness. The residue soon crystallized, a procedure that could be greatly accelerated by seeding with crystals of hexamethylenetetramine hippurate from a previous preparation. The resulting solid product was broken up and pulverized. Hexamethylenetetramine hippurate is stable on exposure to air and is soluble in water and alcohol. It melts at 105° to 110°C.

References

Merck Index 5832
PDR pp. 1227, 1453
DOT 4 (3) 108 (1968)
I.N. p. 611
REM p. 1167
Galat, A.; US Patent 3,004,026; October 10, 1961

METHENOLONE ACETATE

Therapeutic Function: Anabolic

Chemical Name: 17β-Hydroxy-1β-methyl-5α-androst-l-ene-3-one acetate

Common Name: -

Structural Formula:

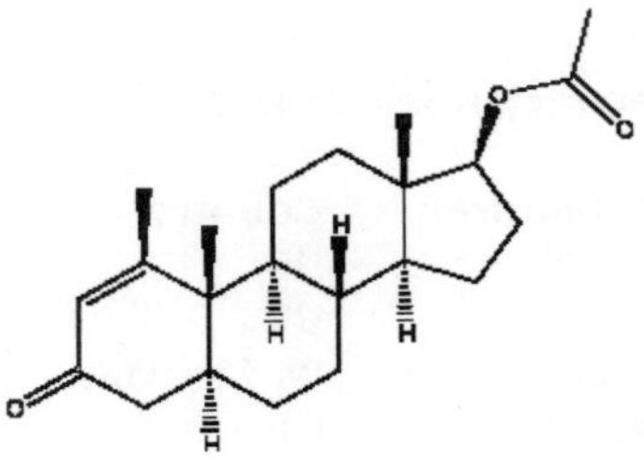

Chemical Abstracts Registry No.: 434-05-9; 153-00-4 (Base)

Trade Name	Manufacturer	Country	Year Introduced
Primobolan	Schering	W. Germany	1961
Dacomid	Schering	W. Germany	-
Fortabol	Schering	W. Germany	-
Neuro-Fortabol	Schering	W. Germany	-

Raw Materials

Methyl iodide	$\Delta^{(1,4,6)}$-Androstatrien-17β-ol-3-one-17-acetate
Magnesium	Hydrogen

Manufacturing Process

8.42 ml of methyl iodide are slowly added dropwise at room temperature with stirring in a nitrogen atmosphere to 3.067 g of magnesium turnings and 107 ml of absolute ether. After about 30 minutes, 185 ml of absolute tetrahydrofuran are slowly introduced and then liquid is distilled off until a boiling point of 62°C is reached. After cooling to room temperature, 613 mg of cuprous chloride are added and then 10 g of $\Delta^{1,4,6}$-androstatrien-17β-ol-3-one-17-acetate in 110 ml of tetrahydrofuran slowly introduced. After 30 minutes reaction time, the whole is cooled to 0C, the excess of Grignard reagent decomposed with saturated ammonium chloride solution, the product diluted with ether and the aqueous phase separated. The ethereal phase is washed consecutively with aqueous sodium thiosulfate solution, saturated ammonium chloride solution and water. It is dried over sodium sulfate and evaporated to dryness under vacuum. The residue is dissolved in 40 ml of pyridine and 20 ml of acetic anhydride and the solution kept for 16 hours at room temperature. It is then stirred into ice water and the precipitate filtered with suction, dried and recrystallized from isopropyl ether. 1α-Methyl-$\Delta^{4,6}$-androstadien-17β-ol-3-one-17-acetate is obtained. MP 156°C to 157°C; $[\alpha]_D^{25}$ = -33.8° (in $CHCl_3$; c = 0.9). Yield 65-70% of the theoretical.

4.67 g of 1α-methyl-$\Delta^{4,6}$-androstadien-17β-ol-3-one-17-acetate are dissolved in 273 ml of methanol and, after the addition of 350 mg of 10% palladium on calcium carbonate catalyst, hydrogenated until 1 mol equivalent of hydrogen has been taken up. After filtering off the catalyst, the solution is treated with 150 ml of 2N-hydrochloric acid and evaporated under vacuum to about 1/3 of the volume. The whole is then diluted with water and extracted with ether. The ethereal solution is washed with water until neutral, dried over sodium sulfate and evaporated. The crude product is heated on a steam bath for 90 minutes in 10 ml of pyridine and 10 ml of acetic anhydride. Extraction with ether is then carried out and the ethereal phase washed until neutral with water. The crude crystalline 1α-methyl-Δ^4-androsten-17β-ol-3-one-17-acetateobtained after drying and evaporation of the solution, melts at 122°C to 129°C. Yield 98% of the theoretical.

1α-Methyl-Δ^4-androsten-17β-ol-3-one-17-acetate when purified by recrystallization from isopropyl ether melts at 138°C to 139°C.

References

Merck Index 5839
Kleeman and Engel p. 571
OCDS Vol. 1 p. 175 (1977)
I.N. p. 606
Schering A.G.; British Patent 977,082; December 2, 1944

METHICILLIN SODIUM

Therapeutic Function: Antimicrobial

Chemical Name: 6-(2,6-Dimethoxybenzamido)-3,3-dimethyl-7-oxo-4-thia-1-azabicyclo[3.2.0]heptane-2-carboxylic acid sodium salt

Common Name: 2,6-Dimethoxyphenylpenicillin sodium salt

Structural Formula:

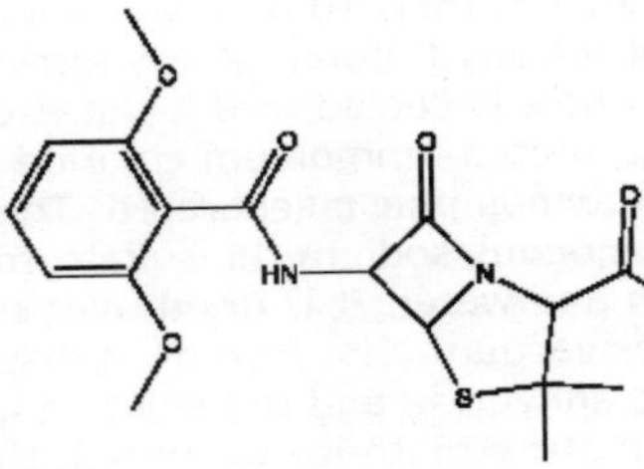

Chemical Abstracts Registry No.: 7246-14-2

Trade Name	Manufacturer	Country	Year Introduced
Celbenin	Beecham	UK	1960
Staphcillin	Bristol	US	1960
Dimocillin	Squibb	US	1961
Flabelline	Delagrange	France	1961
Celbenin	Beecham	US	1973
Azapen	Pfizer	US	1975
Baclyn	Sifrochimica	Italy	-
Celpillina	Farmitalia	Italy	-
Ellecillina	Ellea	Italy	-
Esapenil B.G.	Boniscontro-Gazzone	Italy	-
Metin	C.S.L.	Australia	-
Methocillin	Meiji	Japan	-
Penysol	Saita	Italy	-
Sintespen	Coli	Italy	-
Staficyn	Firma	Italy	-

Raw Materials

6-Aminopenicillanic acid
2,6-Dimethoxybenzoyl chloride

Manufacturing Process

To a stirred suspension of 6-aminopenicillanic acid (540 g) in dry alcohol-free chloroform (3.75 liters) was added dry triethylamine (697 ml), and the mixture stirred for 10 minutes at room temperature. It was then cooled in a bath of crushed ice while a solution of 2,6-dimethoxybenzoyl chloride (500 g) in dry alcohol-free chloroform (3.75 liters) was added in a steady stream over 20 minutes. When all the acid chloride had been added the cooling bath was removed and the mixture stirred for 1 hour at room temperature. The mixture was stirred vigorously and sufficient dilute hydrochloride acid (2.3 liters of 0.87 N) was added to give an aqueous layer of pH 2.5. The mixture was filtered, the layers separated, and only the chloroform layer was retained.

This was stirred vigorously while further dilute hydrochloric acid (0.69 liter of 0.87 N) was added to give an aqueous layer of pH 1. The layers were separated and again only the chloroform layer was retained. Then the chloroform layer was stirred vigorously while sufficient sodium bicarbonate solution (3.2 liters of 0.97 N) was added to give an aqueous layer of pH 6.7 to 7.0. The layers were separated and both were retained. The chloroform layer was stirred vigorously while sufficient sodium bicarbonate solution (50 ml of 0.97 N) was added to give an aqueous layer of pH 7.7, and again the layers were separated. The two bicarbonate extracts were combined, washed with ether (1 liter), and then concentrated at low temperature and pressure until the concentrate weighed 1,415 g.

The concentrate was treated with dry acetone (22 liters), the mixture well mixed, and then filtered to remove precipitated solid impurities. Further dry acetone (4 liters) was added to the filtrate, then the product started to crystallize slowly. Crystallization was allowed to proceed at a temperature between 0° and 3°C for 16 hours and then the product (563 g) was collected by filtration. Dry ether (7.5 liters) was added to the filtrate, and after several hours a second crop (203 g) of solid was collected. The two crops were combined to give sodium 2,6-dimethoxyphenylpenicillin monohydrate (766 g, 73%) as a white crystalline solid.

References

Merck Index 5842
Kleeman and Engel p. 591
PDR p. 713
OCDS Vol. 1 p.412 (1977)
I.N. p. 626
REM p. 1200
Doyle, F.P., Nayler, J.H.C. and Rolinson, G.N.; US Patent 2,951,839; September 6,1960

METHIMAZOLE

Therapeutic Function: Thyroid inhibitor

Chemical Name: 2H-Imidazole-2-thione, 1,3-dihydro-1-methyl-

Common Name: Mercazolyl(um); Methimazole; Methymazole; Thiamazole; Tiamazol

Chemical Abstracts Registry No.: 60-56-0

Structural Formula:

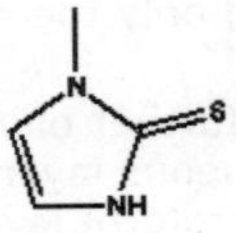

Trade Name	**Manufacturer**	**Country**	**Year Introduced**
Favistan	Asta	-	-
Favistan	Temmler Pharma	-	-
Tapazole	Lilly	-	-
Thiamazole	Marion Merrel Dow	-	-
Metisol	Polfa	-	-
Thyrozol	Merck	-	-

Raw Materials

Sulfuric acid	3,3-Diethoxypropylamine
Bromine	Potassium thiocyanate
Methylamine	Methylisothiocyanate
Hydrochloric acid	1,1-Diethoxyethane

Manufacturing Process

2 Methods of preparation of thiamazole:

1. To 2,2-diethoxyethylamine methylisothiocyanate was added and mixed after then 1-(2,2-diethoxy-ethyl)-3-methylthiourea was obtained.

The reaction of the 1-(2,2-diethoxyethyl)-3-methylthiourea with sulfuric acid yield thiamazole.

2. 1,1-Diethoxyethane was treated by bromine in the presence $CaCO_3$ and 2-bromo-1,1-diethoxyethane was obtained.

Then to the 2-bromo-1,1-diethoxyethane methylamine was added, mixed and reaction mixture was heated to 120°-130°C in autoclave. As the result (2,2-diethoxyethyl)methylamine was obtained.

(2,2-Diethoxyethyl)methylamine reacted with potassium thiocyanate in the presence of hydrochloric acid and give the thiamazole, yellow crystallic precipitate, melting point 144°-147°C.

References

Kleemann A., Engel J.; Pharmazeutische Wirkstoffe, GeorgThieme Verlag Stuttgart. New York, 1982

Chaletsky A.M. Pharmaceutical chemistry, Medicina, L., 1966, 761p.

METHIONINE

Therapeutic Function: Lipotropic

Chemical Name: 2-Amino-4-(methylthio)butyric acid

Common Name: -

Structural Formula:

Chemical Abstracts Registry No.: 63-68-3

Trade Name	Manufacturer	Country	Year Introduced
Meonine	Ives	US	1944
Lobamine	Opodex	France	1948
Oradash	Lambda	US	1955
Ammonil	Philips Roxane	US	1957
Dyprin	Lincoln	US	1958
Acimetion	Continental Pharma	Belgium	-
Amino-Serv	Milex	US	-
AminoPlex	Tyson	US	-
Antamon P.E.D.	Protea	S. Africa	-
Methnine	Medical Research	Australia	-
Monile	Cortunon	Canada	-
Ninol	Horner	Canada	-
Uracid	Wesley	US	-
Unanap	N. Amer. Pharm.	US	-
Urimeth	N. Amer. Pharm.	US	-

Raw Materials

Methyl mercaptan
Sodium cyanide
Sodium hydroxide
Acrolein
Ammonium chloride

Manufacturing Process

A 3-necked flask fitted with a stirrer, thermometer, gas inlet, dropping funnel, and brine-cooled reflux condenser was charged with 53 g (1.1 mol) methyl mercaptan and 0.35 g mercuric methyl mercaptide. After admitting 56 g (1.0

mol) of acrolein during the course of 15 minutes with an inside temperature of about 10°C, the temperature was allowed to rise spontaneously to 75°C, at which point an ice bath was applied. There was no indication of further reaction one hour after the addition of the acrolein. Distillation of the product gave 71 g (yield 68%) of β-methylmercaptopropionaldehyde, as described in US Patent 2,584,496.

Then as described in US Patent 2,732,400, β-methylmercaptopropionaldehyde (0.60 M) (56.5 g) is added to a stirred solution of sodium cyanide (0.66 M) (32.4 g) and ammonium chloride (0.63 M) (33.7 g) in water (140 ml). The temperature of the mixture rises to 49°C and is maintained at this point by heat evolution for about 5 minutes when it slowly begins to fall. Methanol (50 ml) is added and the mixture is stirred for 4 hours as the temperature falls to 28°C (room temperature).

After chilling to +12°C, additional methanol (35 ml) and a concentrated aqueous ammoniun hydroxide solution (1.4 M) (100 ml) are added and stirring is continued for 2 hours at a temperature maintained at from +5° to +15°C. The organic layer is separated and solvent is stripped from the aqueous layer at water aspirator pressure at a temperature below 40°C. The residue is extracted several times with chloroform and the chloroform extracts are combined with the separated oil. Chloroform is removed at water aspirator pressure at a temperature below 35°C to leave crude α-amino-γ-methylmercaptobutyronitrile (methionine nitrile) in 88% yield (68 g) as a clear, somewhat viscous oil.

The methionine nitrile (20 g) is dissolved in a solution prepared from 50 ml of aqueous 5 N sodium hydroxide solution and 65 ml of ethanol. The solution is then refluxed for 24 hours; ammonia is evolved. The solution is treated with activated carbon, filtered, acidified with glacial acetic acid (17 ml), chilled to -10°C and filtered to give crude product. This crude product is then slurried with a solution made up of 20 ml of water and 20 ml of methanol, filtered at -5° to +10°C and dried to give dl-methionine as white platelets.

References

Merck Index 5849
PDR pp. 1263,1807
I.N. p. 612
Pierson, E. and Tishler, M; US Patent 2,584,496; February 5, 1952; assigned to Merck and Co., Inc.
Weiss, M.J.; US Patent 2,732,400; January 24, 1956; assigned to American Cyanamid Company

METHITURAL

Therapeutic Function: Hypnotic, Sedative

Chemical Name: Dihydro-5-(1-methylbutyl)-5-[2-(methylthio)ethyl]-2-thioxo-4,6-(1H,5H)pyrimidinedione monosodium salt

Common Name: Methioturiate

Structural Formula:

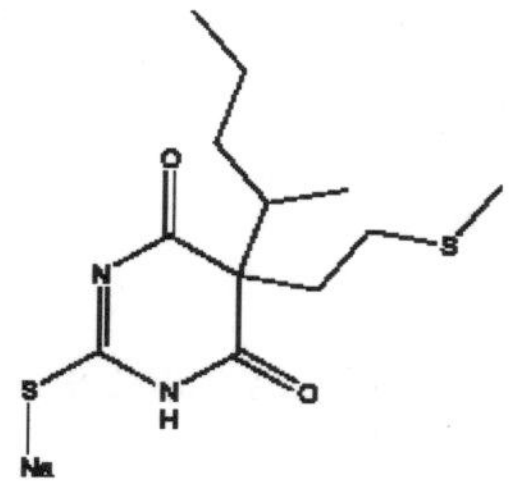

Chemical Abstracts Registry No.: 730-68-7

Trade Name	Manufacturer	Country	Year Introduced
Neraval	Schering	US	1956
Diogenal	Merck	-	-
Thiogenal	Merck	-	-

Raw Materials

Thiourea	Ethanol
Sodium	Sodium hydroxide
Sulfuric acid	β-Methyl-thioethyl-(1-methyl)-n-butyl-cyanoacetic acid ethyl ester

Manufacturing Process

A solution of 69 g of sodium in 1,380 cc of absolute alcohol is mixed with 257.4 g of β-methylthioethyl-(1-methyl)-n-butyl-cyano-acetic acid ethyl ester and 114 g of thiourea and the whole mass boiled under reflux with stirring for six hours. After concentration under vacuum the residue is taken up in 1.5 liters of water and shaken up thrice, each time with 300 cc of ether. The aqueous alcoholic layer is stripped, under vacuum, of the dissolved ether and mixed with 300 cc of 30% acetic acid under stirring and ice cooling. The precipitated material is sucked off, washed with water, dried and recrystallized from isopropyl alcohol. The thus obtained β-methyl-thioethyl-(1-methyl)-n-butyl-cyano-acetyl thiourea forms yellowish green crystals having a melting point of 229°C to 230C.

100 g of this product are boiled under reflux for three hours with 1 liter of 20% sulfuric acid. After cooling the mixture is taken up in ether, the ether solution washed with water, dried, filtered, concentrated and drawn off under vacuum. The residue is caused to crystallize by treatment with a mixture of 60 volume parts of methanol and 40 volume parts of petroleum benzene. The isolated crystals are recrystallized from the mentioned solvent mixture and yield thereby 5-β-methyl-thioethyl-5-(1-methyl)-n-butyl-2-thiobarbituric acid having a melting point of 79°C to 81°C.

20 g of the free acid are shaken up (in a machine) for one hour with 69.5 cc

n/l (normal) caustic soda. The filtered solution is concentrated under vacuum, the residue is taken up in absolute alcohol and again with drawn under vacuum. After two recrystallizations of the residue from isopropyl alcohol one obtains the readily water-soluble, analytically pure, sodium salt of the 5-β-methyl-thioethyl-5-(1 -methyl)-n-butyl-2-thiobarbituric acid.

References

Merck Index 5854
OCDS Vol. 1 p.275 (1977)
I.N. p. 612
Zima, O. and Von Werder, F.; US Patent 2,802,827; August 13,1957; assigned to Emanuel Merck (Germany)

METHIXENE HYDROCHLORIDE

Therapeutic Function: Spasmolytic

Chemical Name: 1-Methyl-3-(9H-thioxanthen-9-yl-methyl)piperidine hydrochloride

Common Name: -

Structural Formula:

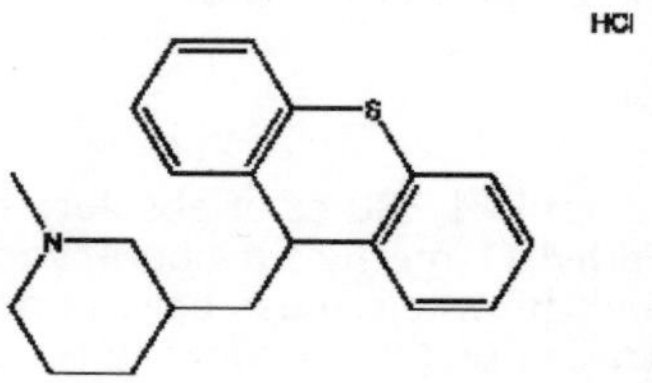

HCl

Chemical Abstracts Registry No.: 1553-34-0; 4969-02-2 (Base)

Trade Name	Manufacturer	Country	Year Introduced
Tremarit	Wander	W. Germany	1960
Tremaril	Wander	Italy	1962
Tremonil	Wander	UK	1963
Trest	Dorsey	US	1965
Atosil	Teikoku	Japan	-
Cholinfall	Tokyo Tanabe	Japan	-
Dalpan	Grelan	Japan	-
Inoball	Sawai	Japan	-
Methixart	Fuso	Japan	-
Methyloxan	Nippon Shoji	Japan	-

Trade Name	Manufacturer	Country	Year Introduced
Raunans	Kowa	Japan	-
Spasmenzyme	Salvoxyl-Wander	France	-
Thioperkin	Hokuriku	Japan	-

Raw Materials

Thioxanthene
Chlorobenzene
Sodium
N-Methyl-3-chloromethyl-piperidine
Hydrogen chloride

Manufacturing Process

To 4.9 g of finely pulverized sodium in 50 ml of absolute benzene add dropwise with stirring 12 g of chlorobenzene in 50 ml of absolute benzene. As soon as the exothermic reaction begins, maintain the temperature by cooling between 30° and 35°C, and continue stirring for 2 to 3 hours. To the resulting phenyl sodium add dropwise 19.8 g of thioxanthene in 120 ml of absolute benzene. The slightly exothermic reaction ceases after about 1 to 1½ hours.

To this newly formed 9-thioxanthyl sodium add dropwise, with stirring and cooling, 13.1 g of N-methyl-3-chloromethyl-piperidine in 30 to 40 ml of absolute benzene, then continue stirring at about 25°C for 1½ hours, and heat subsequently to 40°C for 1 hour. Decompose the resulting mixture by adding carefully a small amount of water, and then extract the newly formed base from the benzene solution by means of dilute hydrochloric acid. The aqueous hydrochloric solution is made alkaline by adding dilute sodium hydroxide, and the thioxanthene base is isolated by extraction with ether. This results in 22 g of a slightly yellow, viscous base of BP 171° to 175°C/0.07 mm.

The base is acidified with alcoholic hydrochloric acid. Alcohol-ether (1:2) is then added and the hydrochloride salt is crystallized as colorless flakes melting at 211° to 213°C.

References

Merck Index 5855
Kleeman and Engel p. 592
OCDS Vol. 1 p.400 (1977) and 2,413 (1980)
I.N. p. 628
REM p.919
Schmutz, J.; US Patent 2,905,590; September 22,1959; assigned to The Wander Company

METHOCARBAMOL

Therapeutic Function: Muscle relaxant

Chemical Name: 3-(o-Methoxyphenoxy)-1,2-propanediol-1-carbamate

Common Name: Guaiacol glyceryl ether carbamate

Structural Formula:

Chemical Abstracts Registry No.: 532-03-6

Trade Name	Manufacturer	Country	Year Introduced
Robaxin	Robins	US	1957
Lumirelax	Sarbach	France	1968
Robaxin	Brenner	W. Germany	1976
Carbametin	Uji	Japan	-
Carxin	Kanto	Japan	-
Delaxin	Ferndale	US	-
Methocabal	Zeria	Japan	-
Methocal	Daiko	Japan	-
Miowas	Wassermann	Italy	-
Myomethol	Abic	Israel	-
Parabaxin	Parmed	US	-
Relax	Ion	Italy	-
Robamol	Cenci	Italy	-
Robaxisal	Robins	US	-
Romethocarb	Robinson	US	-
Traumacut	Brenner	W. Germany	-
Tresortil	Gea	Denmark	-

Raw Materials

Guaiacol glyceryl ether
Phosgene
Ammonia

Manufacturing Process

The starting material for methocarbamol is 3-o-methoxyphenoxy-1,2-propanediol (guaiacol glyceryl ether) (see entry under Guaifenesin for its preparation). To a stirred suspension of 198.2 g (1.0 mol) of 3-o-methoxyphenoxy-1,2-propanediol in 1,000 ml of dry benzene contained in a 5-liter, 3-neck, round bottom flask equipped with a thermometer, dropping funnel and blade stirrer, was added dropwise (in 30 minutes) a solution of 98.9 g (1.0 mol) of phosgene in 400 ml of cold dry benzene. The mixture was stirred at 30°C until all solid material dissolved (about 3 hours was required) and stirring was continued for 30 minutes longer. To this mixture was added dropwise 79.1 g (1.0 mol) of dry pyridine, the temperature being held below 30°C by cooling. After addition of the pyridine, stirring at 30°C was continued for 30 minutes.

The mixture was cooled to 7°C, extracted with two 500-cc portions of ice water to remove pyridine hydrochloride, and the benzene solution of 3-o-methoxyphenoxy-2-hydroxypropyl chlorocarbonate was added to 500 ml of cold concentrated ammonium hydroxide. The mixture was vigorously stirred at 5°C for 6 hours, then the crude white precipitate of 3-o-methoxyphenoxy-2-hydroxypropyl carbamate was filtered off, dissolved in 1,500 ml of hot benzene and completely dried by codistillation of last traces of water with benzene, treated with decolorizing carbon and filtered while hot. On cooling 160 g of product crystallized as white needles melting at 88° to 90°C.

References

Merck Index 5856
Kleeman and Engel p. 578
PDR pp.830, 993, 1466, 1569, 1606, 1999
OCDS Vol. 1 p. 118 (1977)
I.N. p. 613
REM p. 927
Murphey, R.S.; US Patent 2,770,649; November 13,1956; assigned to A.H. Robins Company, Inc.

METHOHEXITAL SODIUM

Therapeutic Function: Anesthetic

Chemical Name: (+/-)-1-Methyl-5-(1-methyl-2-pentynyl)-5-(2-propenyl)-2,4,6-(1H,3H,5H)-pyrimidinetrione sodium salt

Common Name: Methohexitone

Structural Formula:

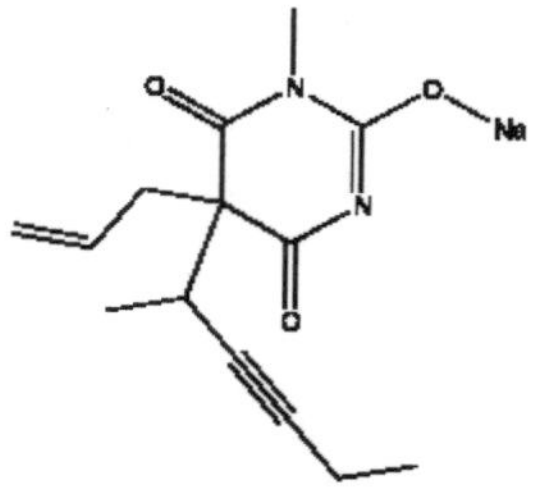

Chemical Abstracts Registry No.: 309-36-4

Trade Name	Manufacturer	Country	Year Introduced
Brevital	Lilly	US	1960
Brietal	Lilly	UK	1961
Brevimytal	Lilly	W. Germany	1963
Brietal	Lilly	Italy	1963

Raw Materials

Magnesium	Ethyl acetylene (1-butyne)
Sodium	Phosphorus tribromide
Allyl bromide	Ethyl bromide
Acetaldehyde	Diethyl malonate
Ethanol	Methyl urea

Manufacturing Process

Preparation of 3-Hexyne-2-ol: A solution of ethyl magnesium bromide was prepared by the reaction of 229 g of ethyl bromide and 48.6 g of magnesium in 750 ml of anhydrous ether. To the ether solution was then added with stirring a solution of 108 g of ethyl acetylene in 250 ml of cold anhydrous ether. The addition required approximately 3 hours, and the mixture was stirred and refluxed for a further period of 3½ hours. Thereafter there was added to the reaction mixture a solution of 88 g of freshly distilled acetaldehyde in 170 ml of anhydrous ether, over a period of about 45 minutes and at a temperature in the range of about -10° to 0°C.

The resulting reaction mixture was poured over about 1 kg of crushed ice, and neutralized with 10% aqueous hydrochloric acid. The organic phase of the resulting mixture was separated, and the aqueous phase was extracted 3 times with 250 ml portions of ether. The combined organic phase and ether washings were washed twice with water and dried over anhydrous potassium carbonate. The dried ether solution was fractionally distilled, and the 3-hexyne-2-ol formed in the reaction was collected as a fraction boiling at about 79° to 80°C at the pressure of 60 mm of mercury.

Preparation of 2-Bromo-3-Hexyne: A solution of 138 g of 3-hexyne-2-ol and 9 g of pyridine in 138 ml of anhydrous ether was treated with 175 g of phosphorus tribromide, added dropwise over a period of about 20 minutes at a temperature of about -10°C. The reaction mixture was permitted to come to room temperature while stirring for about 3 hours, and was then heated to refluxing for about 1 hour. After cooling, the reaction mixture was poured over about 50 g of crushed ice. A two-phase system formed, and the ether layer was separated, washed with dilute sodium bicarbonate solution, dried over anhydrous potassium carbonate and fractionally distilled. The 2-bromo-3-hexyne formed in the reaction was collected at 75°C at the pressure of 50 mm of mercury.

Preparation of Diethyl (1-Methyl-2-Pentynyl) Malonate: To a solution of 28.6 g of sodium in 430 ml of absolute ethanol were added 200 g of diethyl malonate. About half of the alcohol was removed by distillation in vacuo, and thereafter a solution of 200 g of 2bromo-3-hexyne in 100 ml of anhydrous ether was added slowly to the reaction mixture.

The heat of reaction brought about refluxing during the addition of the 2-bromo-3-hexyne, and when the addition was complete the reaction mixture was heated to refluxing for a further period of 30 minutes. A sufficient amount of water was then added to the reaction mixture to dissolve the sodium bromide which had formed, and the only organic layer was separated, washed with water and dried over anhydrous magnesium sulfate. The dried organic layer was then fractionally distilled under reduced pressure, and the diethyl (1-methyl-2-pentynyl) malonate formed in the reaction was collected at about 117° to 120°C at the pressure of 2 mm of mercury.

Preparation of Diethyl Allyl (1-Methyl-2-Pentynyl) Malonate: A solution of 12.1 g of sodium in 182 ml of absolute ethanol was prepared, and thereto were added 126.6 g of diethyl (1-methyl-2-pentynyl) malonate. Most of the ethanol was then distilled off under reduced pressure, and the residue was cooled and 63.5 g of allyl bromide were slowly added thereto. After completion of the addition, the mixture was refluxed for about 1 hour. The reaction mixture was cooled, treated with about 100 ml of water, and the oily organic layer which formed was removed, washed with water and dried over anhydrous magnesium sulfate. The dried oily organic material was fractionally distilled in vacuo, and diethyl allyl (1-methyl-2-pentynyl) malonate boiling at 105° to 107°C at the pressure of 1 mm of mercury was recovered.

Preparation of 1-Methyl-5-Allyl-5-(1-Methyl-2-Pentynyl) Barbituric Acid: A solution of 23.8 g of sodium in 360 ml of absolute alcohol was prepared and thereto were added 38.3 g of methyl urea and 96.8 g of diethyl allyl (1-methyl-2-pentynyl) malonate. The mixture was refluxed for about 20 hours, cooled, and the ethanol was removed by distillation in vacuo. The residue was dissolved in about 300 ml of water and the aqueous solution was washed with ether, and the washings were discarded. The aqueous solution was then acidified with acetic acid, and extracted with three 150 ml of portions of ether.

The combined ether extracts were washed with 5% aqueous sodium bicarbonate solution, dried over anhydrous sodium sulfate, and fractionally distilled in vacuo. The fraction boiling at about 145 to 150°C at the pressure of 0.5 mm of mercury, weighing 61 g and consisting of 1-methyl-5-allyl-5-(1-methyl-2-pentynyl) barbituric acid, was collected. The only distillate was substantially pure, and could be used as such in pharmaceutical preparation or a salt could be prepared therefrom according to the procedures disclosed hereinafter. On standing, the oil crystallized. The crystalline 1-methyl-5-allyl-5-(1-methyl-2-pentynyl) barbituric acid melted at about 60° to 64°C after recrystallization from dilute ethanol.

Preparation of Sodium 1-Methyl-5-Allyl-5-(1-Methyl-2-Pentynyl) Barbiturate: A solution of 61 g of 1-methyl-5-allyl-5-(1-methyl-2-pentynyl) barbituric acid in 100 ml of ether was extracted with 465 ml of 2% aqueous sodium hydroxide solution. The aqueous extract was washed with successive 75 ml and 50 ml portions of ether. The pH of the aqueous solution was adjusted to 11.7, using 5% aqueous sodium hydroxide solution. 5 g of decolorizing carbon were added to the solution with stirring; the mixture was permitted to stand for 20 minutes at room temperature, and the carbon was removed by filtration. A solution containing 4 g of sodium carbonate in 25 ml of water was added to the aqueous solution, and the mixture was filtered sterile through a porcelain filter candle of 02 porosity into sterile bottles. The aqueous solution was then dried from the frozen state, whereupon a sterile residue of sodium 1-methyl-

5-allyl-5-(1-methyl-2-pentynyl) barbiturate, weighing about 62 g was obtained.

References

Merck Index 5857
Kleeman and Engel p. 578
PDR p.1038
OCDS Vol. 1 p. 269 (1977)
I.N. p. 613
REM p.1046
Doran, W.J.; US Patent 2,872,448; February 3,1959; assigned to Eli Lilly and Company

METHOTREXATE

Therapeutic Function: Antineoplastic

Chemical Name: N-[4-[[(2,4-Diamino-6-pteridinyl)methyl]methylamino]-benzoyl]-L-glutamic acid

Common Name: Amethopterin

Structural Formula:

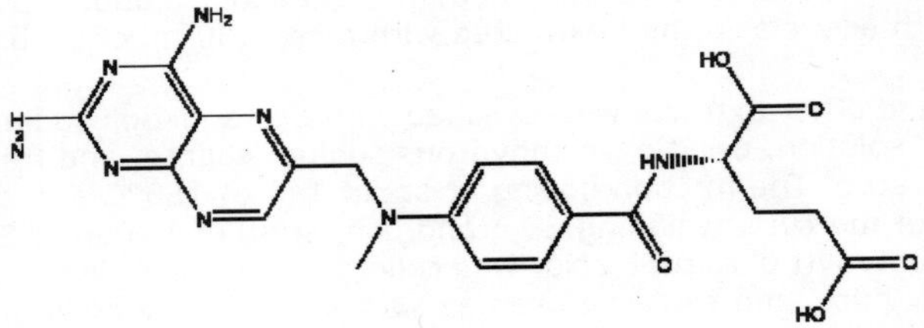

Chemical Abstracts Registry No.: 59-05-2

Trade Name	Manufacturer	Country	Year Introduced
Methotrexate	Lederle	US	1955
Mexate	Bristol	US	1979
Emtexate	Nordic	UK	1981
Folex	Adria	US	1983
Abitrexate	Abic	Israel	-
Emthexate	Pharmachemie	Netherlands	-
Ledertrexate	Lederle	France	-

Raw Materials

Diethyl-p-methylaminobenzoyl-L-glutamate

Aminomalononitrile tosylate
β-Bromopyruvaldoxime
Guanidine acetate

Manufacturing Process

5 g (15 mmol) of diethyl-p-methylaminobenzoyl-L-glutamate and 8.0 g of aminomalononitrile tosylate (65% by NMR assay, 20 mmol) were dissolved in warm ethanol (65 ml, with 15% water by volume). To this solution, cooled to 0°C, was added all at once and with vigorous stirring, 3.6 g of β-bromopyruvaldoxime (89% by NMR assay, 19 mmol). After 30 minutes the stirred mixture, which was allowed to warm slowly to room temperature, was neutralized with powdered $NaHCO_3$ to pH 6, stirring continued for four additional hours, and the resulting mixture filtered through Celite. The filtrate was evaporated under reduced pressure to a glasslike substance, which was taken up in 500 ml of chloroform. The resulting suspension was then filtered using Celite, and the filtrate was washed with water, dried with anhydrous $MgSO_4$, and evaporated to give an orange glasslike substance which was used directly in the next step.

To a 20% solution of titanium trichloride in water (39 mmol), stirred under nitrogen, was added a solution of 18 g (230 mmol) of ammonium acetate in 55 ml of water. Then, to this mixture, cooled to 10°C and stirred with an air-driven stirrer, was added over a period of 5 minutes a solution of the orange glassy substance above distilled in 60 ml of tetrahydrofuran. The mixture was vigorously stirred for 15 minutes while a rapid stream of nitrogen was passed through. After this time, 15 g of powdered sodium sulfite (120 mmol) was added to the mixture, which after several minutes turned from green to yellowish white. This mixture was stirred into 1 liter of chloroform, and the heavy yellow layer separated by use of a separatory funnel. This chloroform layer was washed with water, dried using anhydrous $MgSO_4$, and evaporated under reduced pressure to give a light orange glass, which was then chromatographed rapidly on a column made from 80 g of Baker silica gel, using 5% ethyl acetate in chloroform as the eluent.

The product obtained by evaporation of the eluate was recrystallized from ethanol-ether (1:10) to give a light yellow powder, MP 85 to 88°C. The yield was 4.4 g (63%).

A solution containing 4.8 g (10.2 mmol) of diethyl-N-[p-[[(2-amino-3-cyano-5-pyrazinyl)methyl] methylamino]benzoyl]glutamate and 5 g (42 mmol) of guanidine acetate in 40 ml of dimethylformamide was stirred under nitrogen at 120°C for six hours. The resulting solution was cooled to room temperature, filtered and evaporated to a glassy product using a rotary evaporator and a mechanical vacuum pump to insure a better vacuum. The residual glass was taken up in 500 ml of chloroform, the resulting suspension filtered using Celite, and the filtrate washed with water, dried using anhydrous $MgSO_4$, and evaporated to dryness. (The residual material was chromatographed rapidly on a column prepared from 250 g of Baker silica gel using, initially, 2% ethanol in chloroform, and then 5% ethanol in chloroform as eluents.) The material obtained by evaporation of the eluates was crystallized from ethanol-chloroform (4:1) to give small, pale yellow lustrous platelets, MP 142°C to 154°C; yield, 3.8 g (73%). Further crystallization of this material from ethanol-chloroform (4:1) raised the MP to 153°C to 155°C. The compound is completely racemic.

A sample of this product was hydrolyzed in a mixture of water and methanol in the presence of potassium hydroxide. Essentially pure methotrexate was thus obtained.

References

Merck Index 5861
Kleeman and Engel p. 579
PDR p. 1016
DOT 8 (11) 426 (1972) and 16 (5) 170 (1980)
I.N. p. 614
REM p. 1152
Wiecko, J.; US Patent 4,057,548; November 8,1977
Ellard, J.A.; US Patent 4,080,325; March 21,1978; assigned to US Dept. of Health, Education and Welfare

METHOTRIMEPRAZINE

Therapeutic Function: Analgesic

Chemical Name: 2-Methoxy-N,N,β-trimethyl-10H-phenothiazine-10-propanamine

Common Name: Levomepromazine

Structural Formula:

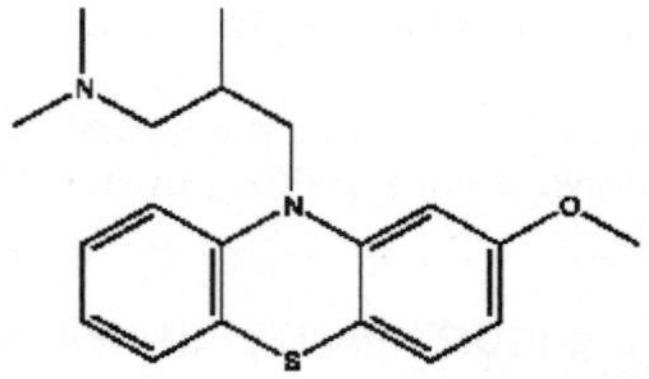

Chemical Abstracts Registry No.: 60-99-1; 1236-99-3 (Hydrochloride salt)

Trade Name	Manufacturer	Country	Year Introduced
Levoprome	Lederle	US	1966
Hirnamin	Shionogi	Japan	-
Levaru	Mohan	Japan	-
Levomezine	Toho	Japan	-
Levotomin	Shionogi	Japan	-
Nozinan	Farmalabor	Italy	-
Ronexine	Ikapharm	Israel	-

Trade Name	Manufacturer	Country	Year Introduced
Sinogan	Rhodia Iberica	Spain	-
Sofmin	Dainippon	Japan	-
Veractil	May and Baker	US	-

Raw Materials

3-Methoxyphenthiazine
Sodium amide
1-Dimethylamino-2-methyl-3-chloropropane

Manufacturing Process

95% sodamide (2.33 g) is added to a boiling solution of 3-methoxyphenthiazine (12 g) in anhydrous xylene (150 cc) and the mixture is heated with agitation under reflux for 1½ hours. A solution of 1-dimethylamino-2-methyl-3-chloropropane (8.2 g) in anhydrous xylene (90 cc) is then run in over a period of 45 minutes while the reaction temperature is maintained and heating under reflux is continued for 18 hours.

After cooling, the reaction mixture is agitated with a mixture of water (40 cc) and a normal solution of methanesulfonic acid (70 cc), the xylene layer is removed and the acid liquors are washed with ether (200 cc). The aqueous phase is then made alkaline with sodium hydroxide (d = 1.33; 10 cc) and the liberated base is extracted with ether. The ethereal solution is dried over anhydrous potassium carbonate and concentrated at normal pressure. On distillation of the residue under reduced pressure 3-(3-methoxy-10-phenthiazinyl)-2-methyl-1-dimethylaminopropane (11.3 g) is obtained, MP 103°C, BP 182° to 191°C/0.15 mm Hg. The hydrochloride prepared in isopropanol melts at about 90°C.

References

Merck Index 5862
Kleeman and Engel p. 522
DOT 3 (2) 62 (1967) and 9 (7) 227 (1971)
I.N. p. 556
REM p. 1113
Jacob, R.M. and Robert, J.G.; US Patent 2,837,518; June 3,1958; assigned to Societe des Usines Chimiques Rhone-Poulenc, France

METHOXAMINE HYDROCHLORIDE

Therapeutic Function: Hypertensive

Chemical Name: α-(1-Aminoethyl)-2,5-dimethoxybenzenemethanol hydrochloride

Common Name: -

Structural Formula:

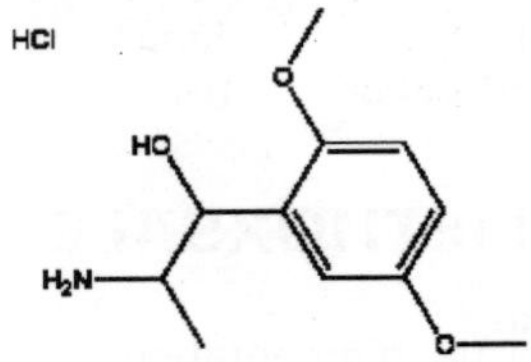

Chemical Abstracts Registry No.: 61-16-5; 390-28-3 (Base)

Trade Name	Manufacturer	Country	Year Introduced
Vasoxyl	Burroughs-Wellcome	US	1949
Idasal	Gayoso Wellcome	Spain	-
Mexan	Nippon Shinyaku	Japan	-
Vasylox	Burroughs-Wellcome	-	-

Raw Materials

2,5-Dimethoxypropiophenone
Methyl nitrite
Hydrogen

Manufacturing Process

2,5-Dimethoxypropiophenone is treated in absolute ether with methyl nitrite and hydrogen chloride. The hydrochloride of 2,5-dimethoxy-α-isonitrosopropiophenone crystallizes out of the solution. It is removed, the base is liberated and crystallized from benzene-heptane forming yellow leaflets that melt at about 97° to 98°C. This isonitrosoketone is dissolved in absolute alcohol containing an excess of hydrogen chloride and is hydrogenated with palladized charcoal, yielding β-(2,5-dimethoxyphenyl)-β-ketoisopropylamine hydrochloride, a salt that melts at about 176°C with decomposition.

12.3 g (1/20 mol) of β-(2,5-dimethoxyphenyl)-β-ketoisopropylamine hydrochloride (MP 176°C) is dissolved in 50 cc of water and hydrogenated with platinum oxide platinum black in the customary Adams-Burgess Parr apparatus. About 1/20 mol of hydrogen is absorbed, after which the solution is filtered off from the catalyst, evaporated to dryness in vacuo and recrystallized from absolute alcohol, absolute ether being added to decrease solubility. The hydrochloride is thus obtained in substantially theoretical yield. It crystallizes in plates and melts at 215°C.

References

Merck Index 5863
Kleeman and Engel p. 580
PDR p. 768
I.N. p. 614

REM p. 888
Baltzly, R., de Beer, E.J. and Buck, J.S.; US Patent 2,359,707; October 3,1944; assigned to Burroughs Wellcome and Co. (USA.) Inc.

METHOXSALEN

Therapeutic Function: Dermal pigmentation enhancer

Chemical Name: 9-Methoxy-7H-furo[3,2-g][1]benzopyran-7-one

Common Name: 8-Methoxypsoralen; Ammoidin; Xanthotoxin

Structural Formula:

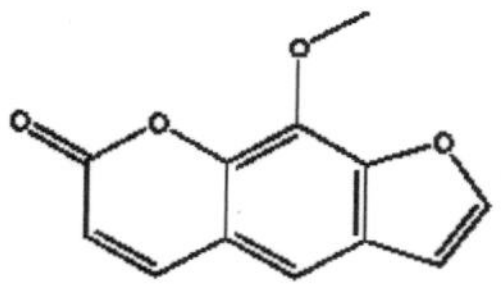

Chemical Abstracts Registry No.: 298-81-7

Trade Name	Manufacturer	Country	Year Introduced
Oxsoracen	Eider	US	1955
Meloxine	Upjohn	US	1958
Meladinine	Basotherm	W. Germany	-
Oxoralen	Farmochimica	Italy	-
Psoritin	Yurtoglu	Turkey	-
Puvalen	Star	Finland	-
Soloxsalen	I.C.N.	Canada	-

Raw Materials

8-Geranoxy psoralen
Sulfuric acid
Diazomethane

Manufacturing Process

It has been found that the compound 8-geranoxy psoralen is present in citrus oils, particularly lemon and lime oils. This compound can be isolated from the oil by a process which involves primarily absorption on an adsorbent material followed by elution with a suitable solvent.

(A) Cleavage of 8-Geranoxypsoralen: 275 mg of 8-geranoxypsoralen was dissolved with mechanical stirring in 4 ml glacial acetic acid. After 10 minutes, one drop of concentrated sulfuric acid was added to the solution. In 4 minutes thereafter a light tan precipitate began to form. Stirring was continued for 35 minutes and the reaction mixture was refrigerated for one hour and 20 minutes. The precipitate was then removed by suction filtration and washed on the filter with glacial acetic acid followed by ice-cold ethyl ether. The product, 8-hydroxypsoralen, weighed 115 mg, that is, 74% of theory.

(B) Methylation of 8-Hydroxypsoralen: 115 mg of 8-hydroxypsoralen was dissolved in 10 ml absolute methanol, an excess of diazomethane dissolved in ether was added and the mixture allowed to stand at room temperature with occasional stirring for 3 hours. The next day the reaction mixture was reduced in volume to 3 ml by evaporation on the steam bath and the concentrate was held in a refrigerator overnight. The next day, fine needles (80 mg) of 8-methoxypsoralen were filtered from the solution. The compound had a MP of 145 to 146°C and was obtained in a yield of 65% of theory.

There is also a wholly synthetic route to Methoxsalen as outlined by Kleeman and Engel.

References

Merck Index 5864
Kleeman and Engel p. 580
PDR p. 867
OCDS Vol. 1 p. 333 (1977)
I.N. p. 614
REM p. 788
Stanley, W.L. and Vannier, S.H.; US Patent 2,889,337; June 2,1959; assigned to the US Secretary of Agriculture
Glunz, L.J. and Dickson, D.E.; US Patent 4,129,575; December 12, 1978; assigned to Thomas C. Elder, Inc.
Liebman, A.A. and Liu, Y.-Y.; US Patent 4,147,703; April 3,1979; assigned to Hoffmann-LaRoche, Inc.

METHOXYFLURANE

Therapeutic Function: Inhalation anesthetic

Chemical Name: 2,2-Dichloro-1,1-difluoro-1-methoxyethane

Common Name: 1,1-Difluoro-2,2-dichloroethyl methyl ether

Structural Formula:

Cl Cl F O F

Chemical Abstracts Registry No.: 76-38-0

Trade Name	Manufacturer	Country	Year Introduced
Penthrane	Abbott	US	1962
Penthrane	Abbott	W. Germany	1962
Penthrane	Abbott	UK	1963
Anecotan	Spofa	Czechoslovakia	-
Methofane	Pitman-Moore	US	-

Raw Materials

1,1-Dichloro-2,2-difluoroethylene
Methanol

Manufacturing Process

Into a reactor equipped with agitator and temperature control jacket is charged approximately 100 lb (about 3 lb mols) of methanol, technical. This methanol is used in excess, and so it is both a reactant and a solvent in the synthesis.

Approximately 1 US gallon of ion exchange resin beads wet with methanol is then added to the methanol. This is in the hydroxide form with at least 0.7 milliequivalent OH^- per milliliter of wet beads. Approximately 190 lb of 1,1-dichloro-2,2-difluoroethylene (about 1.44 lb mols) is then added to the reactor and, within it, to the 100 lb of methanol through a sparge pipe while the beads are kept in suspension by agitation. Coolant is run through the jacket of the reactor during this addition because the reaction is exothermic. The temperature in the reaction medium is kept at 10° to 20°C, to prevent side reactions and to minimize losses of the dichlorodifluoroethylene, which boils at 17°C. Reaction time is affected by the rate of heat removal and the reaction normally takes from 4 to 8 hours, using the stated quantities and conditions. After the dichlorodifluoroethylene is added, the resin is checked for residual alkalinity. If the resin is alkaline to phenolphthalein, it is assumed to have been of sufficient capacity and is removed from the $CH_3OCF_2CHCl_2$-methanol mixture. If it is not alkaline to phenolphthalein, additional resin is added to insure complete reaction.

Essentially the same procedure can be carried out, employing as alkali any strongly alkaline substance, such as caustic soda in methanol solution. Control of the reaction rate may be accomplished by the rate of the addition of reactants and the amount of cooling applied to the reaction mixture. Agitation is employed to insure efficient contact of the reactants.

After removal of the resin catalyst, the excess methanol is extracted out of the mixture using three separate water washes, suitably of 25 gallons each. The water layer is decanted off, leaving product as an immiscible organic layer, after each wash. The 2,2-dichloro-1,1-difluoroethyl methyl ether containing intolerable unsaturated impurities may be purified and stabilized by a treatment with oxidizing agents such as air, oxygen, ozone, peroxy compounds, or other similar oxidizing agents, with subsequent removal of the decomposition or oxidation products and distilling if desired.

References

Merck Index 5869
Kleeman and Engel p. 581
PDR p. 547
I.N. p. 615
REM p. 1043
Larsen, E.R.; US Patent 3,264,356; August 2,1966; assigned to The Dow Chemical Company

METHSCOPOLAMINE BROMIDE

Therapeutic Function: Spasmolytic

Chemical Name: 7-(3-Hydroxy-1-oxo-2-phenylpropoxy)-9,9-dimethyl-3-oxa-9-azoniatricyclo-[3.3.1.02,4]nonane bromide

Common Name: Hyoscine methyl bromide

Structural Formula:

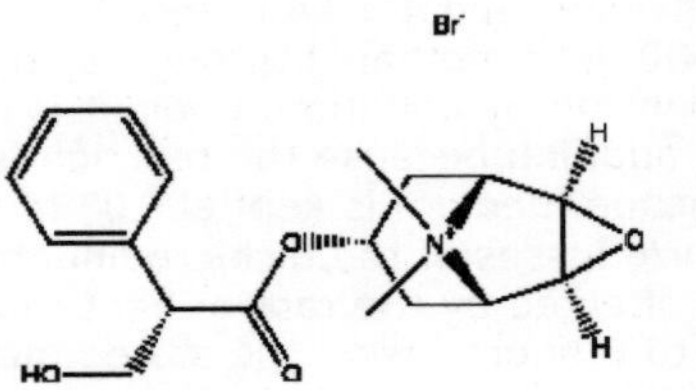

Chemical Abstracts Registry No.: 155-41-9

Trade Name	Manufacturer	Country	Year Introduced
Pamine	Upjohn	US	1953
Daipin	Daiichi Seiyaku	Japan	1972
Ace	Ono	Japan	-
Blocan	Estedi	Spain	-
Lescopine	Lincoln	US	-
Meporamin	Taiyo	Japan	-
Neo Avagal	Andrews	Australia	-
Parantin	Teva	Israel	-
Proscomide	Miller	US	-
Scopolate	Strasenburgh	US	-
Scordin	Ono	Japan	-
Skopyl	Farillon	UK	-

Raw Materials

Scopolamine hydrobromide trihydrate
Methyl bromide

Manufacturing Process

In a one-liter separatory funnel, 94 g (0.215 mol) of scopolamine hydrobromide trihydrate was dissolved in 250 ml of water, made alkaline by shaking with 40 g (1 mol) of sodium hydroxide in 150 ml of water, and the free base immediately extracted with ether. As scopolamine is somewhat soluble in water, the aqueous layer was saturated with potassium carbonate and again extracted with ether. The combined ether extracts were dried over anhydrous magnesium sulfate and the ether removed by distillation, leaving 65 g (0.214 mol; 100% yield) of nearly colorless oil. Then 100 g (1.05 mols) of cold methyl bromide was added to a chilled, 500-ml pressure flask containing the 65 g of scopolamine, the flask stoppered tightly with a clamp, and allowed to stand at room temperature for 96 hours.

The flask was cooled before opening, excess methyl bromide removed by filtration, and the white solid washed thoroughly with dry ether. The yield of crude scopolamine methyl bromide was 80g (94% yield; 93.5% over-all yield).

The salt was recrystallized from 550 ml of alcohol; first crop, 70 g, MP 212° to 214°C; second crop, 6 g, MP 195° to 200°C. The combined crops were again recrystallized from 500 ml of 3-A alcohol; MP 210° to 212°C. The third recrystallization from 600 ml of alcohol yielded 64 g, MP 214° to 216°C, a 75% yield based on scopolamine hydrobromide trihydrate starting material.

References

Merck Index 5881
Kleeman and Engel p. 582
PDR p. 1857
I.N. p. 508
REM p. 917
Visscher, F.E.; US Patent 2,753,288; July 3,1956; assigned to The Upjohn Company

METHSUXIMIDE

Therapeutic Function: Anticonvulsant

Chemical Name: 1,3-Dimethyl-3-phenyl-2,5-pyrrolidinedione

Common Name: Mesuximid

Structural Formula:

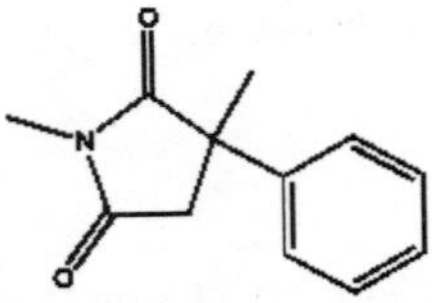

Chemical Abstracts Registry No.: 77-41-8

Trade Name	Manufacturer	Country	Year Introduced
Celontin	Parke Davis	US	1957
Petinutin	Parke Davis	W. Germany	-

Raw Materials

α-Phenyl-α-methylsuccinic acid
Methylamine

Manufacturing Process

100 g of α-phenyl-α-methylsuccinic acid and 110 g of 40% aqueous methyl amine are heated together at 200 to 250°C until no more distillate is obtained. Upon vacuum distillation of the residue, the N-methyl-α-phenyl-α-methylsuccinimide, of BP 121° to 122°C at 0.1 mm is obtained. After recrystallization from aqueous ethanol, this compound melts at 52° to 53°C.

References

Merck Index 5882
Kleeman and Engel p. 567
PDR p. 1320
OCDS Vol. 1 p. 228 (1977)
I.N. p. 602
REM p. 1079
Miller, C.A. and Long, L.M.; US Patent 2,643,257; June 23, 1953; assigned to Parke, Davis and Company

METHYLDOPA

Therapeutic Function: Antihypertensive

Chemical Name: 3-Hydroxy-α-methyl-L-tyrosine

Common Name: L-α-Methyl-3,4-dihydroxyphenylalanine

Structural Formula:

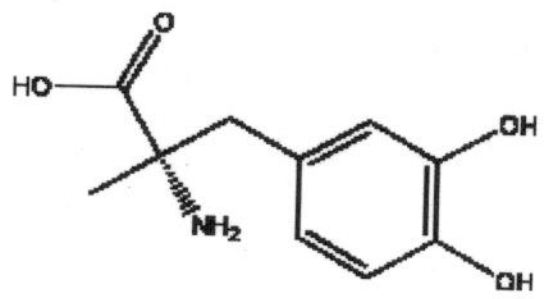

Chemical Abstracts Registry No.: 555-30-6

Trade Name	Manufacturer	Country	Year Introduced
Aldometil	MSD	W. Germany	1962
Aldomet	MSD	UK	1962
Aldomet	MSD	Italy	1962
Aldomet	MSD	US	1963
Aldomet	MSD-Chibret	France	1964
Adopal	Pharmacal	Finland	-
Aldomin	Teva	Israel	-
Aldoril	MSD	US	-
Alphamex	Protea	S. Africa	-
Becanta	Kissei Pharmaceutical Co., Ltd.	Japan	-
Caprinol	Bayer	W. Germany	-
Dansul	Nippon Yakko	Japan	-
Desens	Nissin	Japan	-
Dimal	Protea	Australia	-
Domecin	Sankyo	Japan	-
Dopamet	Berk	US	-
Dopamin	Hokuriku	Japan	-
Dopatec	Labatec	Switz.	-
Dopegyt	Gedeon Richter	Hungary	-
Equibar	Genekod	France	-
Grospisk	Toho Iyaku	Japan	-
Hydromet	MSD	France	-
Hyperten	Toho	Japan	-
Hypolag	Lagap	Switz.	-
Hy-Po-Tone	Lennon	S. Africa	-
Medimet	Medic	Canada	-
Medomet	D.D.S.A.	UK	-
Medopa	Kaigai	Japan	-
Medopal	A.L.	Norway	-
Medopren	Dietopharma	Italy	-
Metholes	Taisho	Japan	-
Methoplain	Kowa	Japan	-
Nichidopa	Nichiiko	Japan	-
Novomedopa	Novopharm	Canada	-
Polinal	Boehringer Yamanouchi	Japan	-
Sembrina	Boehringer Mannheim	Italy	-

Raw Materials

3-Hydroxy-4-methoxyphenylalanine
Hydrogen chloride

Manufacturing Process

The dl-α-methyl-3,4-dihydroxyphenylalanine may be made as described in US Patent 2,868,818. Five-tenths of a gram of 3-hydroxy-4-methoxyphenylalaninewas dissolved in 20 ml of concentrated hydrochloric acid, the solution saturated with hydrogen chloride and heated in a sealed tube at 150°C for 2 hours. The dark reaction mixture was concentrated to dryness in vacuo, excess acid removed by flushing several times with ethanol. On dissolving the dark residue in a minimum amount of water and adjusting the clarified solution to pH 6.5 with ammonium hydroxide the compound separated in fine crystals which were filtered, washed with alcohol and ether. The crystalline product had a MP of 299.5 to 300°C with decomposition.

Then, as described in US Patent 3,158,648, the optical isomers may be resolved as follows. 37 g of racemic α-methyl-3,4-dihydroxyphenylalanine are slurried at 35°C in 100 cc of 1.0 N hydrochloric acid. The excess solids are filtered leaving a saturated solution containing 34.6 g of racemic amino acid of which about 61% is present as the hydrochloride. The solution is then seeded at 35°C with 7 g of hydrated L-α-methyl-3,4-dihydroxyphenylalanine (6.2 g of anhydrous material). The mixture is then cooled to 20°C in 30 minutes and aged one hour at 20°C. The separated material is isolated by filtration, washed twice with 10 cc of cold water and dried in vacuo. The yield of product is 14.1 g of L-α-methyl-3,4-dihydroxyphenylalanine in the form of a sesquihydrate of 100% purity as determined by the rotation of the copper complex.

References

Merck Index 5928
Kleeman and Engel p. 583
PDR pp. 993, 1133
OCDS Vol. 1 p. 95 (1977)
DOT 10 (9) 323 (1974) and 19 (3) 170 (1983)
I.N. p. 618
REM p.846
Pfister, K., Ill and Stein, G.A.; US Patent 2,868,818; January 13, 1959; assigned to Merck and Co., Inc.
Jones, R.T., Krieger, K.H. and Lago, J.; US Patent 3,158,648; November 24,1964; assigned to Merck and Co., Inc.

METHYLERGONOVINE MALEATE

Therapeutic Function: Oxytocic

Chemical Name: 9,10-Didehydro-N-[1-(hydroxymethyl)propyl]-6-methylergoline-8-carboxamide maleate

Common Name: d-Lysergic acid dl-hydroxybutylamide-2; Methylergometrin maleate

Structural Formula:

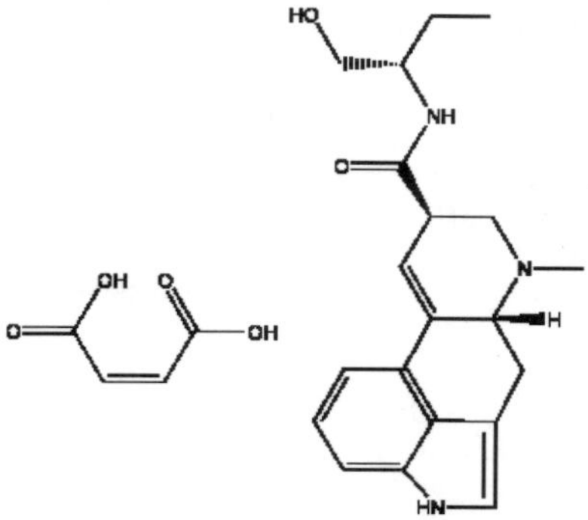

Chemical Abstracts Registry No.: 7054-07-1; 113-42-8 (Base)

Trade Name	Manufacturer	Country	Year Introduced
Methergine	Sandoz	US	1948
Methergin	Sandoz	France	1953
Ergotrate	Lilly	US	-
Levospan	Isei	Japan	-
Metenarin	Teikoku Zoki	Japan	-
Methylergobrevin	Arzneimittelwerk Dresden	E. Germany	-
Metiler	Adika	Turkey	-
Myomergin	Leiras	Finland	-
Ryegonovin	Morishita	Japan	-
Spametrin M	Sanzen	Japan	-
Takimetrin M	Nakataki	Japan	-
Uterin	Biofarma	Turkey	-

Raw Materials

d-Isolysergic acid azide
d-2-Aminobutanol-1

Manufacturing Process

To a freshly prepared solution of 2 parts of d-isolysergic acid azide in 300 parts of either is added an ethereal solution of 2 parts of d-2-aminobutanol-1 and the mixture is left to stand at room temperature during 12 hours. The yellowish clear solution is then washed several times with some water, dried over sodium sulfate and the ether evaporated in vacuo. The crystallized residue is treated with a small quantity of acetone and filtered. Yield: 2.2

parts of d-isolysergic acid-d-1-hydroxybutylamide-2. On recrystallization from some hot methanol the new compound is obtained in form of beautiful polygonal crystals that melt with some decomposition at 192° to 194°C (corr.).

1 part of the iso-compound is then dissolved in 10 parts of absolute ethanol and an alcoholic potassium hydroxide solution is added thereto. The mixture is left to stand at room temperature during 45 minutes. After this time equilibrium is reached between lysergic acid and the isolysergic acid forms, which can be checked by determination of the constancy of the optical rotation of the solution. When this point is reached, potassium hydroxide is transformed into potassium carbonate by bubbling through the solution a stream of carbon dioxide; the thick crystal paste of potassium carbonate is then diluted with 50 parts of ether, filtered and washed again with 50 parts of ether.

The alcoholic ethereal filtrate is then dried over calcined potassium carbonate and the solution evaporated, whereby 0.9 to 1 part of a mixture of d-lysergic acid-d-1-hydroxybutylamide-2 and of d-isolysergic acid-d-1-hydroxybutylamide-2 is obtained. In order to separate the isomers, the residue is dissolved in 15 parts of hot chloroform and filtered from the small quantity of inorganic salt, whereby on cooling down, the difficultly soluble chloroform compound of d-lysergic acid-d-1-hydroxybutylamide-2 crystallizes out. Yield: 0.4 part. This compound can be recrystallized from hot benzene, whereby crystals melting with some decomposition at 172°C (corr.) are obtained. It may then be reacted with maleic acid to give the maleate.

References

Merck Index 5943
Kleeman and Engel p. 584
PDR p. 1587
I.N. p. 619
REM p. 948
Stoll, A. and Hofmann, A.; US Patent 2,265,207; December 9,1941; assigned to Sandoz AG, Switzerland

METHYLHEXANEAMINE CARBONATE

Therapeutic Function: Nasal decongestant

Chemical Name: 4-Methyl-2-hexylamine carbonate

Common Name: -

Chemical Abstracts Registry No.: 105-41-9 (Base)

Trade Name	Manufacturer	Country	Year Introduced
Forthane	Lilly	US	1948

Structural Formula:

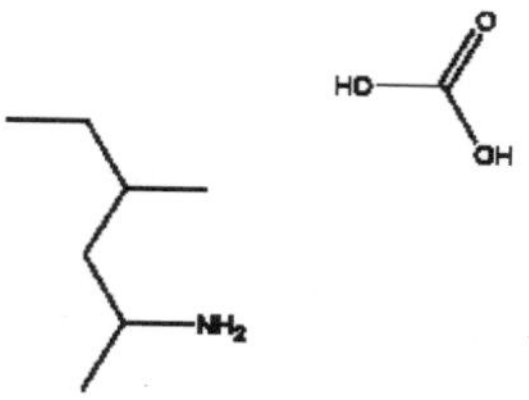

Raw Materials

4-Methylhexanone-2	Hydrogen
Hydroxylamine	Carbon dioxide

Manufacturing Process

One molecular equivalent of 4-methylhexanone-2 is reacted with slightly more than one molecular equivalent of hydroxylamine. Desirably, the hydroxylamine is prepared in the presence of the 4-methylhexanone-2 by reacting the hydrochloride or sulfate or other salt of the hydroxylamine with a suitable base, such as sodium carbonate or sodium hydroxide. Desirably, the reaction mixture is agitated for a few hours to insure the conversion of the 4-methylhexanone-2 to 4-methylhexanone-2 oxime.

The resulting 4-methylhexanone-2 oxime separates and is dried by any suitable means, such as with a dehydrating agent, for example, sodium sulfate or magnesium sulfate. After drying, 4-methylhexanone-2 oxime is reduced with hydrogen by means of a catalyst, such as Raney nickel, or by reaction of sodium and a primary alcohol, such as ethanol. The resulting 2-amino-4-methylhexane may be purified by distillation, as described in US Patent 2,350,318.

115 g (1 mol) of 2-amino-4-methylhexane and 9 g (0.5 mol) of water are placed in a tared 500 cc 3-necked flask which is equipped with a mechanical stirrer, a thermometer, and a gas delivery tube. The flask is surrounded by a cooling bath of ice and water. Dry carbon dioxide gas is introduced into the solution through the gas delivery tube, with constant stirring, until the increase in weight is approximately 22 g (0.5 mol). The temperature during this addition is maintained between 20° and 30°C. A viscous liquid results, and consists essentially of 2-amino-4-methylhexane carbonate. This also dissociates very slowly at room temperature to the free amine, carbon dioxide, and water; and is effective as an inhalant, according to US Patent 2,386,273.

References

Merck Index 5955
I.N. p. 620
Shonle, H.A. and Rohrmann, E.; US Patent 2,350,318; May 30,1944; assigned to Eli Lilly and Company
Shonle, H.A. and Rohrmann, E.; US Patent 2,386,273; October 9,1945; assigned to Eli Lilly and Company

METHYLMETHIONINSULFONIUM CHLORIDE

Therapeutic Function: Hepatoprotectant

Chemical Name: (3-Amino-3-carboxypropyl)dimethylsulphonium chloride

Common Name: Methiosulfonii chloridum; Methylmethioninesulfonium chloride; U-Vitamin; Vitamin U

Structural Formula:

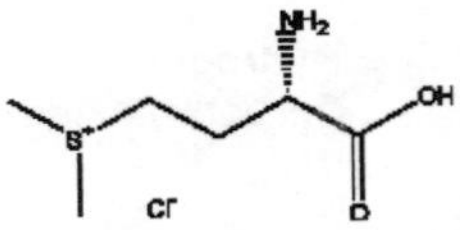

Chemical Abstracts Registry No.: 1115-84-0

Trade Name	Manufacturer	Country	Year Introduced
Cabagin-U	Kowa	-	-
Vitamin U	Medexport	-	-

Raw Materials

dl-Methionine
Methyl chloride

Manufacturing Process

298 g dl-methionine, 2000 ml water and 151.1 methyl chloride were put into autoclave (volume 3 L) made from V4-a steel. The mixture was heated at temperature 50°-55°C for 8 hours and pressure 12-13 atmospheres.

On cooling an excess of methyl chloride was evaporated, the slightly yellow residue was mixed with 0.2% activated coal and filtered. Water was evaporated in vacuum at 45°-55°C. 2 L methanol was poured into an almost colorless syrup obtained, cooled to -5°-(-10°)C. The methylmethioninesulfonium chloride crystallized.

It was filtered off to give 340 g (85.3%) of desired product 99.5 % purity (data of thin-layer chromatography).

References

Wagner H.; D.B. Patent No. 1,239,697; Feb. 20, 1963; Deutsche Gold- und Silber- Scheideanstalt vormals Roessler, Frankfurt/M.

METHYLOL RIBOFLAVIN

Therapeutic Function: Enzyme cofactor vitamin source

Chemical Name: Riboflavin monomethylol

Common Name: -

Structural Formula:

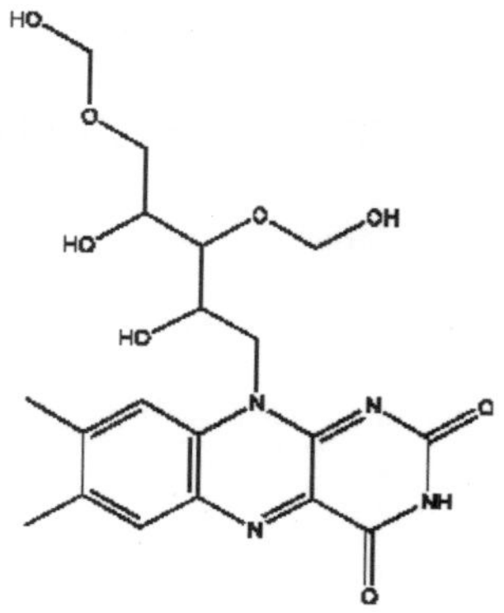

Chemical Abstracts Registry No.: 83-88-5 (Base)

Trade Name	Manufacturer	Country	Year Introduced
Hyflavin	Endo	US	1948

Raw Materials

Riboflavin
Formaldehyde

Manufacturing Process

100 g of riboflavin and 3 of potassium carbonate are suspended in 500 cc of the aqueous formaldehyde solution and the mixture is stirred at 30°C for 8 hours. At the end of this period, 5 cc of glacial acetic acid and 1 liter of methanol are added, with stirring. The solution is freed from undissolved material by filtration and the clear solution is poured slowly at about 20°C to 22°C with vigorous stirring into 8 liters of anhydrous acetone. The resultant precipitate is filtered off, washed repeatedly with anhydrous acetone and with ether, and then dried at room temperature and with vacuum. The resultant dried powder is dissolved in hot water at 95°C to give an aqueous solution of 20% by weight. This solution is kept in the dark at room temperature for 3 to 4 weeks, after which time a large amount of material crystallizes out of the solution. This crystallized material is removed by filtration and recrystallized from hot water. A small amount of dark red insoluble material is filtered from the hot solution. This recrystallization step is repeated four times. The

resultant end product is monomethylol riboflavin, which crystallized in small orange clusters. It has a melting point of 232°C to 234°C with decomposition, and it becomes dark when heated above 225°C.

References

Merck Index 5974
I.N. p. 621
Schoen, K. and Gordon, S.M.; US Patent 2,587,533; February 26,1952; assigned to Endo Products, Inc.

METHYLPENTYNOL

Therapeutic Function: Sedative

Chemical Name: 1-Pentyn-3-ol, 3-methyl-

Common Name: Meparfynol; Methylparafynol; Methylpentynol

Structural Formula:

Chemical Abstracts Registry No.: 77-75-8

Trade Name	Manufacturer	Country	Year Introduced
Citodorm	Haury	-	-
Dormiphen	Darck	-	-

Raw Materials

Sodium acetylene
Methyl ethyl ketone

Manufacturing Process

To 5 parts of sodium acetylene in absolute ether 6 parts of dry methyl ethyl ketone was slowly dropwise added with ice cooling and stirring. Than the reaction mixture was poured into excess of acetic acid by ice cooling and extracted with ether. The ether extract was washed with solution of potash for removing the diluted acetic acid and dried over potassium carbonate. The ether was distilled off and residual colorless oil methylpentynol had BP at 120°-121°C. Instead of sodium acetylene the solution of sodium acetylene in liquid ammonia may be successfully used.

References

Farbenfabriken vorm. Friedr. Bayer and CO. in Leverkusen b. Coln a. Rh.; D.R. Patent No. 285,770; Nov. 22, 1913

METHYLPHENIDATE HYDROCHLORIDE

Therapeutic Function: Psychostimulant

Chemical Name: α-Phenyl-2-piperidineacetic acid methyl ester hydrochloride

Common Name: -

Structural Formula:

HCl

NH

O

O

Chemical Abstracts Registry No.: 298-59-9; 113-45-1 (Base)

Trade Name	Manufacturer	Country	Year Introduced
Ritalin	Ciba	US	1958
Rubifen	Rubio	Spain	-

Raw Materials

Phenylacetonitrile
Methanol
Sodium amide
Hydrogen chloride
2-Chloropyridine
Hydrogen
Sulfuric acid

Manufacturing Process

As described in US Patent 2,507,631, 80 g of pulverized sodium amide are gradually added, while stirring and cooling, to a solution of 117 g of phenylacetonitrile and 113 g of 2-chloropyridine in 400 cc of absolute toluene. The mixture is then slowly heated to 110° to 120°C and maintained at this temperature for 1 hour. Water is added thereto after cooling, the toluene solution is shaken with dilute hydrochloric acid and the hydrochloric acid extracts are made alkaline with concentrated caustic soda solution. A solid mass is separated thereby which is taken up in acetic ester and distilled, α-phenyl-α-pyridyl-(2)-acetonitrile passing over at 150° to 155°C under 0.5 mm pressure. When recrystallized from ethyl acetate it melts at 88° to 89°C, the yield amounting to 135 g.

100 g of α-phenyl-α-pyridyl-(2)-acetonitrile are introduced into 400 cc of concentrated sulfuric acid, allowed to stand overnight at room temperature, poured into ice and rendered alkaline with sodium carbonate. α-Phenyl-α-pyridyl-(2)-acetamide is precipitated thereby which melts at 134°C after recrystallization from ethyl acetate.

100g of the resulting α-phenyl-α-pyridyl-(2)-acetamide, when dissolved in one liter of methyl alcohol and treated for 6 hours at water-bath temperature with hydrogen chloride, and after concentrating, diluting with water and rendering alkaline with sodium carbonate, yield 90 g of the α-phenyl-α-pyridyl-(2)-acetic acid methylester of MP 74° to 75°C (from alcohol of 50% strength).

The α-phenyl-α-piperidyl-(2)-acetic acid methylester of BP 135° to 137°C under 0.6 mm pressure is obtained in theoretical yield by hydrogenation of 50 g of α-phenyl-α-pyridyl(2)-acetic acid methylester in glacial acetic acid in the presence of 1 g of platinum catalyst at room temperature, while taking up 6 hydrogen atoms. Reaction with HCl gives the hydrochloride. Resolution of stereoisomers is described in US Patent 2,957,880.

References

Merck Index 5981
Kleeman and Engel p. 586
PDR p. 811
OCDS Vol. 1 p. 88 (1977)
I.N. p. 622
REM p. 1136
Hartmann, M. and Panizzon, L.; US Patent 2,507,631;May 16,1950; assigned to Ciba Pharmaceutical Products Inc.
Rornetsch, R.; US Patent 2,957,880; October 25,1960; assigned to Ciba Pharmaceutical Products Inc.

METHYLPHENOBARBITAL

Therapeutic Function: Anticonvulsant

Chemical Name: Barbituric acid, 5-ethyl-1-methyl-5-phenyl-

Common Name: Acidum methyl-phenyl-aethyl-barbiuricum; Enfenemal; Enphenemal; Mefoberbital; Mephobarbital; Methylfenobarbital; Methylphenobarbital; Methylphenobarbitone; Metilfenobarbitale

Chemical Abstracts Registry No.: 115-38-8

Trade Name	Manufacturer	Country	Year Introduced
Isonal	Roussel	-	-

Structural Formula:

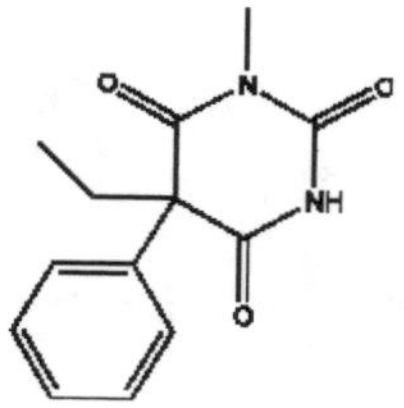

Raw Materials

Sodium
Phenyl ethyl malonic acid diethyl ester
Methylurea

Manufacturing Process

46 parts metallic sodium was dissolved in 1000 parts absolute alcohol. The obtained solution was mixed with 264 parts of phenyl ethyl malonic acid diethyl ester and 80 parts of monomethyl urea and heated for 8 hours at reflux. Alcohol was distilled off, the residue was dissolved in water and neutralized with diluted sulfuric acid. N-Methylethylphenylbarbituric acid precipitated as a powder. It was filtered off, washed to neutral and dissolved in 50 parts of boiling alcohol. On cooling the obtained methylphenobarbital precipitated as the colorless prisms. MP: 176.5°C. This compound may be also prepared by condensation of equivalents of phenyl malonic ester and monomethyl urea, which was dissolved in above described solution of sodium ethylate.

References

Taub L., Kropp W.; D.R. Patent No. 537,366; Oct. 15, 1931; I. G. Farbenindustrie Akt.-Ges. in Frankfurt a. M.

METHYLPREDNISOLONE

Therapeutic Function: Glucocorticoid

Chemical Name: 11β,17α,21-Trihydroxy-6α-methyl-l,4-pregnadiene-3,20-dione

Common Name: 1-Dehydro-6α-methylhydrocortisone

Structural Formula:

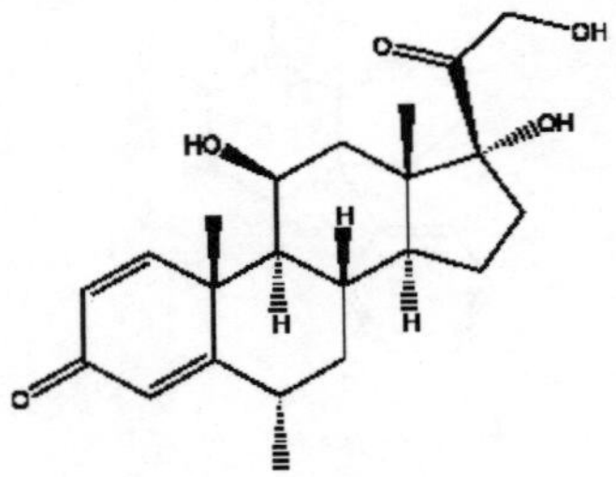

Chemical Abstracts Registry No.: 83-43-2

Trade Name	Manufacturer	Country	Year Introduced
Medrol	Upjohn	US	1957
Medrol	Upjohn	France	1959
A-Methapred	Abbott	US	1978
Solu-Medrol	Upjohn	Japan	1980
Caberdelta	Caber	Italy	-
Cortalfa	S.A.M.	Italy	-
Depo-Medrate	Upjohn	W. Germany	-
Emmetip	Magis	Italy	-
Esametone	Lisapharma	Italy	-
Eutisone	Eufarma	Italy	-
Firmacort	Firma	Italy	-
Horusona	Horus	Spain	-
Medesone	Fargal	Italy	-
Mega-Star	Ausonia	Italy	-
Metilbetasone	Coli	Italy	-
Metilcort	Gazzini	Italy	-
Metilprednilone	Guidi	Italy	-
Metilstendiolo	Panther-Osfa	Italy	-
Moderin	Alter	Spain	-
Nirypan	Jugoremedija	Yugoslavia	-
Nixolan	S.I.T.	Italy	-
Prednilen	Lenza	Italy	-
Prednol	Mustafa Nevzat	Turkey	-
Radiosone	Radiumpharma	Italy	-
Reactenol	Lafare	Italy	-
Sieropresol	Sierochimica	Italy	-
Summicort	Benvegna	Italy	-
Suprametil	Geistlich	Switz.	-
Urbason	Hoehst	Italy	-

Raw Materials

Bacterium *Septomyxa affinis*
Corn steep liquor
Glucose
64-α-Methylhydrocortisone

Manufacturing Process

The following process description is taken from US Patent 2,897,218. Six 100-ml portions of a medium in 250-ml Erlenmeyer flasks containing 1% glucose, 2% corn steep liquor (60% solids) and tap water was adjusted to a pH of 4.9. This medium was sterilized for 45 minutes at 15 psi pressure and inoculated with a one to two day growth of *Septomyxa affinis* ATCC 6737. The Erlenmeyer flasks were shaken at room temperature at about 24°C for a period of 3 days.

At the end of this period, this 600-ml volume was used as an inoculum for ten liters of the same glucose-corn steep liquor medium which in addition contained 10 ml of an antifoam (a mixture of lard oil and octadecanol). The fermentor was placed into the water bath, adjusted to 28°C, and the contents stirred (300 rpm) and aerated (0.5 liter air/10 liters beer). After 17 hours of incubation, when a good growth developed and the acidity rose to pH 6.7, 2 g of 6α-methylhydrocortisone plus 1 g of 3-ketobisnor-4-cholen-22-al, dissolved in 115 ml of dimethylformamide, was added and the incubation (conversion) carried out at the same temperature and aeration for 24 hours (final pH 7.9).

The mycelium (56 g dry weight) was filtered off and the steroidal material was extracted with methylene chloride, the methylene extracts evaporated to dryness, and the resulting residue chromatographed over a Florisil column. The column was packed with 200 g of Florisil and was developed with five 400-ml fractions each of methylene chloride, Skellysolve B-acetone mixtures of 9:1, 8:2, 7:3, 1:1, and methanol. The fraction eluted with Skellysolve B-acetone (7:3) weighed 1.545 g and on recrystallization from acetone gave, in three crops, 928 mg of product of MP 210° to 235°C. The sample prepared for analysis melted at 245° to 247°C.

References

Merck Index 5984
Kleeman and Engel p. 587
PDR pp. 1286,1606,1850
OCDS Vol. 1 p. 196 (1977)
I.N. p. 623
REM p.968
Sebek, O.K. and Spero, G.B.; US Patent 2,897,218; July 28, 1959; assigned to The Upjohn Company
Gould, D.H.; US Patent 3,053,832; September 11, 1962; assigned to Schering Corporation

METHYLSCOPOLAMINE NITRATE

Therapeutic Function: Anticholinergic, Spasmolytic

Chemical Name: 1αH,5αH-Tropanium, 6β,7β-epoxy-3α-hydroxy-8-methyl-, nitrate, (-)-

Common Name: Hyocsine methonitrate; Methylscopolamine nitrate; Scopolamine methylnitrate

Structural Formula:

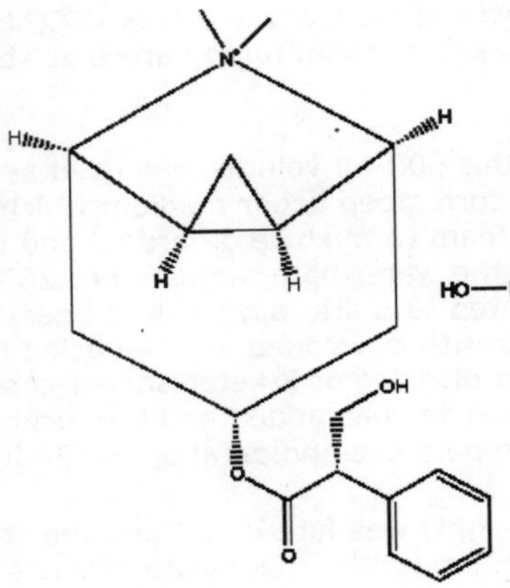

Chemical Abstracts Registry No.: 6106-46-3

Trade Name	Manufacturer	Country	Year Introduced
Transderm Scop	Novartis Consumer Health Inc.	-	-

Raw Materials

Scopolamine hydrobromide trihydrate
Methyl bromide
Silver nitrate

Manufacturing Process

In a one-liter separatory funnel, 94 g (0.215 mol) of scopolamine hydrobromide trihydrate was dissolved in 250 ml of water, made alkaline by shaking with 40 g (1 mol) of sodium hydroxide in 150 ml of water, and the free base immediately extracted with ether. As scopolamine is somewhat soluble in water, the aqueous layer was saturated with potassium carbonate and again extracted with ether. The combined ether extracts were dried over anhydrous magnesium sulfate and the ether removed by distillation, leaving 65 g (0.214 mol; 100% yield) of nearly colorless oil. Then 100 g (1.05 mol) of cold methyl bromide was added to a chilled, 500-ml pressure flask containing the 65 g of scopolamine, the flask stoppered tightly with a clamp, and allowed to stand at room temperature for 96 hours. The flask was cooled before opening, excess methyl bromide removed by distilling, and the white solid washed thoroughly with dry ether. The yield of crude N-methylscopolammonium bromide was 80 g (94 %t yield). The salt was recrystallized from 550 ml of alcohol; first crop, 70 g, MP: 212-214°C; second crop, 6 g, melting point 195-200°C. The combined crops were again recrystallized from 500 ml of alcohol; melting point 210-212°C. The third recrystallization from 600 ml of alcohol yielded 64 g, melting point 214-216°C, a 75 % yield based on scopolamine hydro-bromide trihydrate starting material. N-Methylscopolammonium bromide may be dissolved in water. An equivalent of solution $AgNO_3$ was added, a precipitated AgBr was filtered off. The filtrate was evaporated to dryness to give the desired N-methylscopolammonium nitrate.

References

Visscher F. E.; US Patent No. 2,753,288; July 3, 1956; Assigned to The Upjohn Company, Kalamazoo, Mich., a corporation of Michigan

METHYLTESTOSTERONE

Therapeutic Function: Androgen

Chemical Name: 17β-Hydroxy-17-methyl-androst-4-ene-3-one

Common Name: -

Structural Formula:

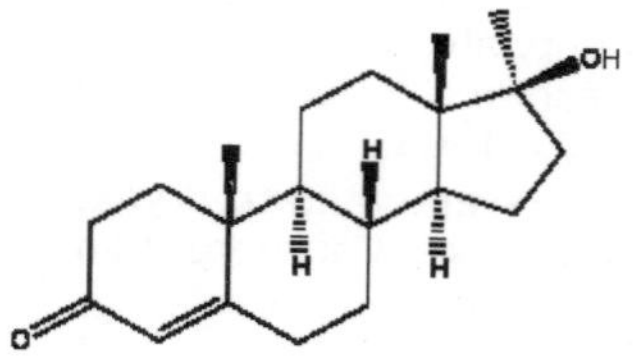

Chemical Abstracts Registry No.: 58-18-4

Trade Name	Manufacturer	Country	Year Introduced
Metandren	Ciba	US	1941
Oreton-M	Schering	US	1941
Neo-Hombreol	Organon	US	1941
Hormale	Key	US	1958
Android-S	Brown	US	-
Arcosterone	Arcum	US	-
Climaterine	Lucien	France	-

Trade Name	Manufacturer	Country	Year Introduced
Climatone	Paines and Byrne	UK	-
Dumone	Squibb	US	-
Estan	Schering	US	-
Gynosterone	Sam-On	Israel	-
Hormobin	Munir Sahin	Turkey	-
Malogen	Fellows-Testagar	US	-
Orchisterone	Negroni	Italy	-
Seksfort	Uranium	Turkey	-
Steronyl	Kay	US	-
Synandrets	Pfizer	US	-
Testipron	Kwizda	Austria	-
Testomet	Protea	Australia	-
Testora	Alcon	US	-
Testostelets	Barlow Cote	Canada	-
Testonic B	Sam-On	Israel	-
Testovis	Vister	Italy	-
Testred	I.C.N.	US	-
Virilon	Star	US	-

Raw Materials

17-Methyl-Δ(^{5,6})-androstenediol-(3,17)
Magnesium
Acetone
Methyl chloride

Manufacturing Process

0.6 g of 17-Methyl-$\Delta^{5,6}$-androstenediol-(3,17) is heated under reflux cooling during 20 hours in 50 cm^3 of benzene and 12 cm^3 of acetone with 3 g of tertiary chloromagnesium butylate, which may be prepared by conversion of acetone with methyl magnesium chloride. The magnesium is then removed by shaking out with dilute H_2SO_4; the benzene layer is washed with water, dried with sodium sulfate and then evaporated to dryness. Methyltestosterone (MP 160° to 162°C) is obtained in a yield of more than 75% of the theory, according to US Patent 2,384,335.

References

Merck Index 6000
Kleeman and Engel p. 588
PDR pp.645, 729, 802, 949, 1447, 1643, 1778
OCDS Vol. 1 p. 172 (1977)
I.N. p.625
REM p. 998
Miescher, K. and Wettstein, A.; US Patent 2,374,369; April 24, 1945; assigned to Ciba Pharmaceutical Products, Incorporated
Miescher, K. and Wettstein, A.; US Patent 2,374,370; April 24, 1945; assigned to Ciba Pharmaceutical Products, Incorporated

Oppenauer, R.; US Patent 2,384,335; September 4, 1946
Miescher, K.; US Patent 2,386,331; October 9, 1945; assigned to Ciba Pharmaceutical Products, Incorporated
Miescher, K.; US Patent 2,435,013; January 27,1948; assigned to Ciba Pharmaceutical Products, Incorporated

METHYPRYLON

Therapeutic Function: Sedative, Hypnotic

Chemical Name: 3,3-Diethyl-5-methyl-2,4-piperidinedione

Common Name: 2,4-Dioxo-3,3-diethyl-5-methylpiperidine

Structural Formula:

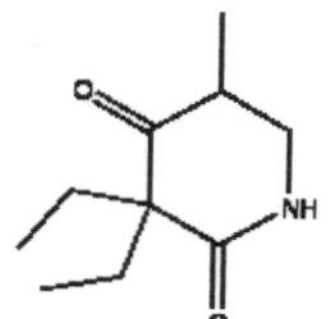

Chemical Abstracts Registry No.: 125-64-4

Trade Name	Manufacturer	Country	Year Introduced
Noludar	Roche	US	1955
Noctan	Yamanouchi	Japan	-
Nolurate	Roche	-	-

Raw Materials

2,4-Dioxo-3,3-diethyl-piperidine
Hydrogen
Sodium
Methyl formate

Manufacturing Process

24 parts by weight of powdered sodium are suspended in 100 parts by volume of absolute benzene and to this suspension is added a freshly prepared solution of 150 parts by weight of methyl formate and 165 parts by weight of 2,4-dioxo-3,3-diethyl-piperidine in 900 parts by volume of absolute benzene. By cooling with cold water, the temperature is maintained at 25° to 28°C. After being stirred for 12 hours 200 parts by volume of 0.6 N sodium hydroxide are added while cooling. The aqueous layer is separated and acidified to Congo red by means of 35% hydrochloric acid. The 2,4-dioxo-3,3-diethyl-5-oxymethylenepiperidine is precipitated in good yield as a solid. After having been recrystallized in chloroform/petroleum ether it melts at 140° to

141°C.

5 parts by weight of 2,4-dioxo-3,3-diethyl-5-oxymethylene-piperidine are hydrogenated in 25 parts by volume of methanol in the presence of about 2 parts by weight of Raney nickel at 120°C and under an elevated pressure of 100 atm. Once 2 mols of hydrogen are absorbed, the hydrogenation is interrupted, the solution is separated from the catalyst and concentrated and the residue is distilled in vacuo. The distillate, boiling between 178° and 185°C under a pressure of 16 mm, consists of 2,4-dioxo-3,3-diethyl-5-methyl-piperidine, which melts at 74° to 75°C.

The same compound is obtained when proceeding according to the following alternative procedure. A mixture of 39.4 parts by weight of 2,4-dioxo-3,3-diethyl-5-oxymethylenepiperidine and 27 parts by weight of dibutylamine are heated to 150°C in a closed vessel. The 2,4-dioxo-3,3-diethyl-5-dibutylamino-methylene-piperidine formed melts at 77°C after having been recrystallized in petroleum ether.

31 parts by weight of the latter compound are hydrogenated in 150 parts by volume of alcohol, containing 6 parts by weight of acetic acid, in the presence of 10 parts by weight of Raney nickel, at 120°C and under an elevated pressure of 100 atm. The catalyst is separated and the solution is distilled in vacuo. The 2,4-dioxo-3,3-diethyl-5-methyl-piperidine boils between 178° and 185°C under a pressure of 16 mm and melts at 74° to 75°C.

References

Merck Index 6010
Kleeman and Engel p. 590
PDR p. 1495
DOT 9 (6) 245 (1973)
I.N. p. 626
REM p. 1072
Frick, H.and Lutz, A.H.; US Patent 2,680,116; June 1, 1954; assigned to Hoffmann-LaRoche Inc.

METHYSERGIDE MALEATE

Therapeutic Function: Migraine therapy

Chemical Name: 9,10-Didehydro-N-[1-(hydroxymethyl)propyl]-1,6-dimethylergoline-8-carboxamide maleate

Common Name: 1-Methyl-d-lysergic acid butanolamide maleate

Chemical Abstracts Registry No.: 129-49-7; 361-37-5 (Base)

Structural Formula:

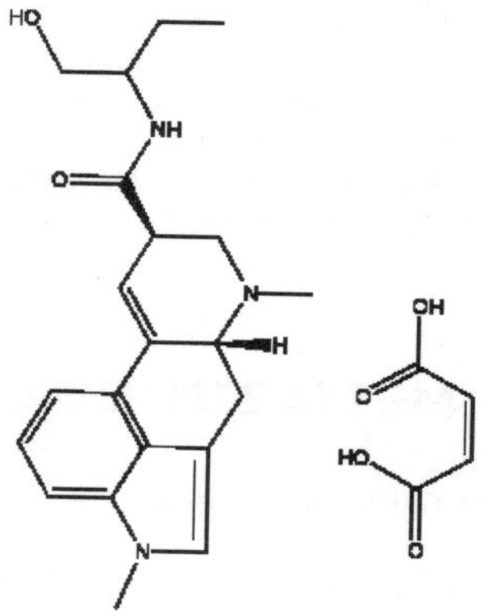

Trade Name	**Manufacturer**	**Country**	**Year Introduced**
Sansert	Sandoz	US	1962
Desernil	Sandoz	France	1962
Deseril	Sandoz	UK	1963

Raw Materials

Potassium
Methyl iodide
Lysergic acid-1'hydroxy-butylamide-2'
Ammonia
Maleic acid

Manufacturing Process

As described in US Patent 3,218,324, 0.9 part of potassium are dissolved in 500 parts by volume of liquid ammonia, then oxidized with ferric nitrate to potassium amide, after which 4.85 parts of lysergic acid-1'-hydroxy-butylamide-2' are dissolved in the obtained mixture. After 15 minutes there are added to the obtained yellow solution 4.1 parts of methyl iodide in 5 parts by volume of ether, the mixture being allowed to stand for 30 more minutes at -60°C. The liquid ammonia is thereupon evaporated and the dry residue is shaken out between water and chloroform. The mixture of bases which remains after the evaporation of the chloroform is chromatographed on a column of 250 parts of aluminum oxide, the desired 1-methyl-lysergic acid-1'-hydroxy-butylamide-2' being washed into the filtrate with chloroform and chloroform-0.2% ethanol. The 1-methyl-lysergic acid-1'hydroxy-butylamide-2' crystallizes from chloroform in the form of plates which melt at 194° to 196°C. Reaction with maleic acid gives the dimaleate, melting at 187° to 188°C.

References

Merck Index 6011
Kleeman and Engel p. 590

PDR p. 1596
OCDS Vol. 2 p. 477 (1980)
I.N. p. 626
REM pp. 949, 1113
Hofmann, A. and Troxler, F.; US Patent 3,113,133; December 3,1963; assigned to Sandoz Ltd., Switzerland
Hofmann, A. and Troxler, F.; US Patent 3,218,324; November 16,1965; assigned to Sandoz Ltd., Switzerland

METIAZINIC ACID

Therapeutic Function: Antiinflammatory

Chemical Name: 10-Methylphenothiazine-2-acetic acid

Common Name: -

Structural Formula:

Chemical Abstracts Registry No.: 13993-65-2

Trade Name	Manufacturer	Country	Year Introduced
Soripan	Specia	France	1970
Soripal	Torii	Japan	1977
Soripal	Farmalabor	Italy	1978
Ambrunate	Rhodia	Argentina	-
Metian	Horus	Spain	-
Novartril	Andromaco	Spain	-
Roimal	Nippon Rhodia	Japan	-
Soridermal	Specia	France	-

Raw Materials

10-Methyl-3-acetylphenthiazine
Morpholine
Sulfur
Potassium hydroxide

Manufacturing Process

10-Methyl-3-acetylphenthiazine is prepared in accordance with G. Cauquil and A. Casadevall, Bull. Soc.Chim., p 768 (1955). (10-Methyl-3-

phenthiazinyl)acetic acid (MP 146°C; 21.4 g) is prepared by Willgerodt's reaction (action of sulfur and morpholine, followed by hydrolysis) employing 10-methyl-3-acetylphenthiazine as starting material.

References

Merck Index 6013
Kleeman and Engel p. 591
I.N. p. 32
Farge, D., Jeanmart, C. and Messer, M.N.; US Patent 3,424,748; January 28, 1969; assigned to Rhone-Poulenc S.A., France

METICRANE

Therapeutic Function: Diuretic

Chemical Name: 2H-1-Benzothiopyran-7-sulfonamide, 3,4-dihydro-6-methyl-, 1,1-dioxide

Common Name: Meticrane

Structural Formula:

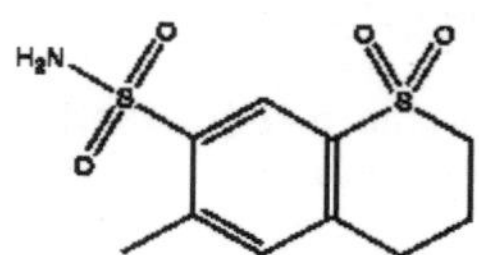

Chemical Abstracts Registry No.: 1084-65-7

Trade Name	Manufacturer	Country	Year Introduced
Fontilix	Diamant	-	-

Raw Materials

6-Methylthiachromane
Hydrogen peroxide
Chlorosulfonic acid
Ammonia

Manufacturing Process

64.5 g of 6-methylthiachromane were dissolved in 500 ml of acetic acid and 250 ml of 110-volume hydrogen peroxide were added. The solution was placed on a water bath for 1.5 hour; it was then diluted with iced water, the precipitate obtained was recovered, and this was washed and dried. There were obtained 59.6 g of 6-methylthiachromane-1,1-dioxide (yield: 77.5%; melting point 79°-81°C).

52 g of this product were added to 250 ml of chlorosulfonic acid and the mixture was placed on a water bath at a temperature of 70°-75°C for 2 hours. It was allowed to cool and poured onto crushed ice; the product was extracted by means of chloroform, the extracted solutions were washed and the chloroform was evaporated. There were obtained 67 g of crude 6-methyl-7-chloro-sulfonyl-thiachromane-1,1-dioxide (yield: 86%; melting point 158°-161°C).

47 g of this sulfochloride were introduced into 200 ml of liquid ammonia. The mixture was left to stand at ambient temperature until the ammonia evaporated. The residue was taken up in water and the solution was acidified. The precipitate formed was centrifuged, washed with water and dried. There were obtained 30.8 g of 6-methyl-7-sulfamido-thiachromane-1,1-dioxide, recrystallized from 2-methoxy ethanol (yield: 49%; MP: 236°-237°C).

References

Boissier J.R. et al.; US Patent No. 3,488,424; Jan. 6, 1970; Assigned to Societe Industrielle pour la Fabrication des Antibiotiques (S.I.F.A.), Paris, France, a French company

METOCLOPRAMIDE HYDROCHLORIDE

Therapeutic Function: Antiemetic

Chemical Name: 4-Amino-5-chloro-N-[(2-diethylamino)ethyl]-2-methoxybenzamide hydrochloride

Common Name: -

Structural Formula:

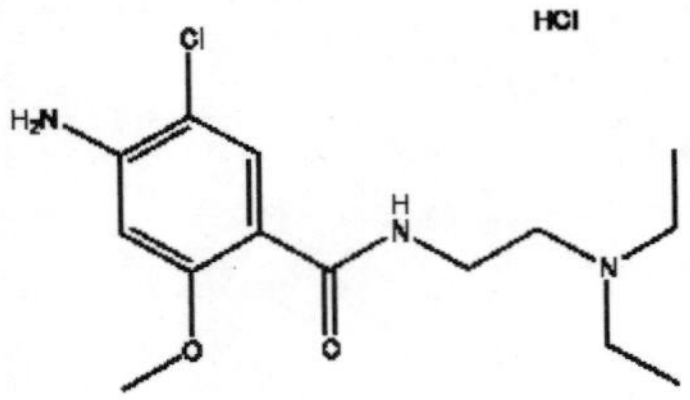

Chemical Abstracts Registry No.: 7232-21-5; 364-62-5 (Base)

Trade Name	Manufacturer	Country	Year Introduced
Primperan	Delagrange	France	1964
Paspertin	Kali-Chemie	W. Germany	1965
Maxolon	Beecham	UK	1967
Plasil	Richter	Italy	1967
Reglan	Robins	US	1979

Trade Name	Manufacturer	Country	Year Introduced
Metox	Steinhard	UK	1983
Ananda	Bonomelli-Hommel	Italy	-
Cerucal	Arzneimittelwerk Dresden	E. Germany	-
Clodil-Ion	Ion	Italy	-
Clopamon	Petersen	S. Africa	-
Clopan	Firma	Italy	-
Contromet	Script Intal	S. Africa	-
Digetres	Scalari	Italy	-
Donopon-GP	Sana	Japan	-
Elietin	Nippon Kayaku, Co.	Japan	-
Emesa	Mulda	Turkey	-
Emetisan	Phoenix	Argentina	-
Emperal	Neofarma	Finland	-
Gastronertron	Dolorgiet	W. Germany	-
Imperan	Bender	Austria	-
Kilozim	A.G.I.P.S.	Italy	-
Maxrean	Nordic	Canada	-
MCP-Ratiopharm	Ratiopharm	W. Germany	-
Meclopran	Lagap	Switz.	-
Metamide	Protea	Australia	-
Metoclol	Toyama	Japan	-
Metocobil	Beta	Italy	-
Metopram	Leiras	Finland	-
Metpamid	Sifar	Turkey	-
Moriperan	Morishita	Japan	-
Nadir	Oti	Italy	-
Netaf	Sintyal	Japan	-
Peraprin	Taiyo	Japan	-
Placitril	Sigurta	Italy	-
Pramiel	Nagase	Japan	-
Pramin	Rafa	Israel	-
Primperil	Lacefa	Argentina	-
Prindarl	Sawai	Japan	-
Prometin	Yamanouchi	Japan	-
Putoprin	Mohan	Japan	-
Quanto	Mediolanum	Italy	-
Randum	Scharper	Italy	-
Regastrol	Sarm	Italy	-
Reliveran	Finadiet	Argentina	-
Rimetin	Farmakhim	Bulgaria	-
Terperan	Teikoku Zoki	Japan	-
Viscal	Zoja	Italy	-

Raw Materials

o-Toluidine
Nitrous acid
Acetic anhydride
Nitric acid
Thionyl chloride
Chlorine
Potassium permanganate
N,N-Diethylene diamine
Hydrogen chloride
Dimethyl sulfate
Hydrogen

Manufacturing Process

The N-(diethylaminoethyl)-2-methoxy-4-aminobenzamide used as the starting material may be prepared from o-toluidine. The o-toluidine is initially nitrated with nitric acid to produce 4-nitro-o-toluidine. The 4-nitro-o-toluidine is then converted to 2-hydroxy-4-nitrotoluene by heating with nitrous acid. By reacting the resulting 2-hydroxy-4-nitrotoluene with dimethyl sulfate, 2-methoxy-4-nitrotoluene is formed. The 2-methoxy-4-nitrotoluene is oxidized with potassium permanganate to produce 2-methoxy-4-nitrobenzoic acid. The latter substituted benzoic acid is treated with thionyl chloride to form 2-methoxy-4-nitrobenzoyl chloride. A methyl ethyl ketone solution of the 2-methoxy-4-nitrobenzoyl chloride is added over a period of about 1½ hours to a methyl ethyl ketone solution containing an equal molecular quantity of N,N-diethylethylene diamine while stirring and maintaining the temperature between 0°C and 5°C. The N-(diethylaminoethyl)-2-methoxy-4-nitrobenzamide hydrochloride formed precipitates. It is filtered, washed twice with methyl ethyl ketone, dissolved in alcohol, and reduced catalytically in an absolute isopropyl alcohol solution to form N-(diethylaminoethyl)-2-methoxy-4-aminobenzamide. The base is obtained by precipitating with sodium hydroxide.

80 g (0.3mol) of N-(2-diethylaminoethyl)-2-methoxy-4-aminobenzamide are dissolved in small portions in 150 cc of acetic acid. The mixture is cooled and 45 g (0.45 mol) of acetic anhydride are added, and the solution obtained is heated for two hours on a water bath. After cooling, the solution is decanted into a round-bottomed flask with a stirrer, a thermometer and a tube for introducing the chlorine. It is stirred and the current of chlorine is passed through, the temperature being maintained between 20°C and 25°C. The stirring is continued for one hour after the completion of the absorption of the chlorine.

The mixture obtained is poured into 2 liters of water and the base is precipitated with 30% soda. The precipitated base is extracted with 400 cc of methylene chloride. After evaporation of the solvent, the N-(2-diethylaminoethyl)-2-methoxy-4-acetamino-5-chlorobenzamide formed crystallizes. The melting point is 86°C to 87°C and the yield is 95%.

To obtain the corresponding amino derivative, 109 g of base are heated under agitation in a round-bottomed flask with 300 cc of 35-36% concentrated hydrochloric acid and 600 cc of water. It is heated on a water bath until dissolution is complete, then maintained at boiling point for 90 minutes, cooled, diluted with 1 liter of water, and neutralized with about 350 cc of 30% soda. The N-(2-diethylaminoethyl)-2-methoxy-4-amino-5-chlorobenzamide formed crystallizes, is centrifuged and washed in water. Its melting point is 122°C and the yield is 74%.

To obtain the corresponding dihydrochloride, the base is dissolved in absolute alcohol (3 volumes) and to that solution is added 5 N alcoholic hydrochloric acid. The dihydrochloride precipitates, is centrifuged and washed with alcohol. It is a solid white material, having a melting point of 134°C to 135°C.

References

Merck Index 6019
Kleeman and Engel p. 593
PDR p. 1463
DOT 1 (2) 66 (1965); 16 (5) 159 (1980) and 19 (8) 476 (1983)
I.N. p. 629
REM p. 809
Thominet, M.L.; US Patent 3,177,252; April 6,1965; assigned to Soc. d'Etudes Scientifiques et Industrielles de l'Ile de France (France)

METOLAZONE

Therapeutic Function: Diuretic

Chemical Name: 7-Chloro-l,2,3,4-tetrahydro-2-methyl-3-(2-methylphenyl)-4-oxo-6-quinolinesulfonamide

Common Name: -

Structural Formula:

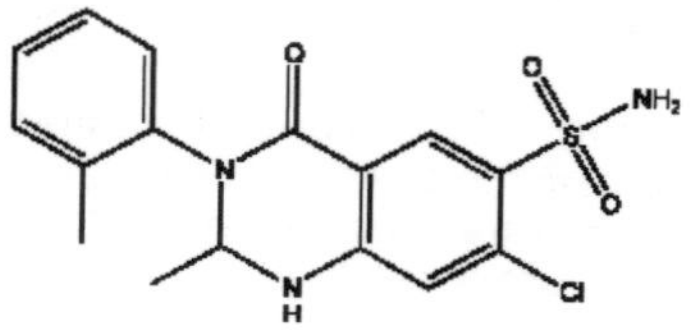

Chemical Abstracts Registry No.: 17560-51-9

Trade Name	Manufacturer	Country	Year Introduced
Zaroxolyn	Pennwalt	UK	1973
Zaroxolyn	Pennwalt	US	1974
Diulo	Searle	US	1978
Zaroxolyn	Searle	W. Germany	1978
Zaroxolyn	Sandoz	Switz.	1978
Zaroxolyn	I.S.F.	Italy	1981
Normeran	Sankyo	Japan	1982
Metenix	Hoechst	UK	-
Oldren	Roemmers	Argentina	-

Raw Materials

5-Chloro-2-methylaniline
o-Toluidine
Acetic anhydride
Phosphorus trichloride
Chlorosulfonic acid
Sodium borohydride
Ammonia

Manufacturing Process

Preparation of Intermediate Compound N-Acetyl-5-Chloro-2-Methylaniline: To a well-stirred mixture of 1,270 g (9 mols) of 5-chloro-2-methylaniline in 7.5 liters of water at 34°C was added all at once 1,710 ml (18 mols) of acetic anhydride. A solution was obtained and then almost immediately the product started to crystallize. The temperature rose to 60°C. The mixture was stirred until the temperature dropped to 30°C. The product was filtered and washed well with water. Yield 97% (1,640 g), MP 134° to 138°C. Product was air dried and then in vacuum over P_2O_5.

Preparation of Intermediate Compound 5-Chloro-2-Methyl-4-Sulfamylacetanilide: Into a 3-necked 3-liter flask fitted with stirrer and thermometer 540 ml of chlorosulfonic acid were placed and cooled in an ice bath to 20°C. 300 g of the acetanilide were added portionwise while stirring and maintaining temperature at 20°C. This addition takes approximately 20 minutes. Remove the ice bath and add 88 g of sodium chloride portionwise (approximately 1 tsp every 10 minutes), This addition takes approximately 1 hour. Some foaming takes place. Using heating mantle bring temperature up slowly (approximately ½ hour) to 75°C. Considerable foaming takes place and heating is continued another ½ hour until 92°C is reached. Foaming can be controlled by shutting off heat and with good stirring. Once the temperature of 92°C has been reached and foaming has subsided reaction can be left unattended. Keep reaction at 92°C for a total of 2½ hours.

Pour the hot reaction mixture onto 4 liters of crushed ice. Pour slowly and stir the ice mixture. What remains in the flask can be worked up by adding ice to it and swirling the contents. After approximately 3/4 of an hour, the solid is filtered and washed with approximately 600 ml water.

Break up cake into small pieces and add to 2.5 liters concentrated NH_4OH in 4 liter beaker. Stir. Solid goes into solution and then the sulfonamide precipitates out. Heat to 50°C and then turn off heat. After ½ hour cool in ice bath and filter. Wash cake with 600 ml water. Add cake to 2 liters 5% NaOH (130 ml 50% NaOH to 2 liters water). Filter and discard insolubles. While cooling filtrate add concentrated HCl until mixture is acid. Filter and wash cake until filtrate is neutral. Suck cake as dry as possible then air dry. Yield approximately 200 g (45%), MP 255° to 260°C.

Preparation of Intermediate Compound 4-Chloro-5-Sulfamyl-N-Acetylanthranilic Acid: To a hot solution (80°C) of 366 g (1.482 mols) of magnesium sulfate (Epsom salts) in 2.8 liters of water was added 130 g (0.495 mol) of powdered 5-chloro-2-methyl-4-sulfamylacetanilide. With stirring and maintaining the temperature at 83°C, 234 g (1.482 mols) of potassium permanganate was added portionwise over a period of 2 hours. The mixture was then kept at 85°C with stirring for an additional 3 hours. By this time the pink color of the permanganate had been discharged.

The mixture was cooled to 65°C and 250 g (2.0 mols) of sodium carbonate monohydrate was added. The warm reaction mixture was filtered and the cake washed with water. The filtrate was then slowly treated with concentrated hydrochloric acid until mixture tested acid. Product was then filtered, washed with water and dried. Yield 103 g (71.0%), MP 245° to 249°C (dec.).

Preparation of Intermediate Compound 2-Methyl-3-o-Tolyl-6-Sulfamyl-7-Chloro-4(3H)-Quinazolinone: Set up a 5-liter 3-necked flask fitted with a stirrer, condenser and a drying tube. To a stirred mixture of 100 g (0.342 mol) of powdered 4-chloro-5-sulfamyl-N-acetylanthranilic acid, 40.2 g (0.376 mol) of o-toluidine and 2.0 liters of dry toluene was added dropwise, over a period of 15 minutes, 21.7 ml (34.1 g) (0.248 mol) of phosphorus trichloride. The mixture was then refluxed for 10 hours. The solid turned somewhat gummy towards the latter part of the first hour. The mixture then became more free flowing as heating was continued. Let stand overnight. The yellow solid was filtered, washed with toluene and dried. The toluene filtrate was discarded. The dried solid was triturated with 1.5 liters of 10% sodium bicarbonate, filtered and the cake washed with water. The filtrate on acidification yielded 11.5 g of the starting acid. The damp product was dissolved in 4.5 liters of 95% ethanol and the solution treated with charcoal and filtered. On cooling filtrate yielded 69.5 g (55.5%) of the title compound, MP 271.5° to 274°C.

Preparation of the Final Compound 2-Methyl-3-o-Tolyl-6-Sulfamyl-7-Chloro-1,2,3,4-Tetrahydro-4(3H)-Quinazolinone: To 4 liters of dry diglyme in a 12-liter 3-necked flask fitted with a stirrer, thermometer and drying tube was added 5.34 g (0.04 mol) of aluminum chloride, while stirring. To the resulting solution was added 43.6 g (0.12 mol) of 2-methyl-3-o-tolyl-6-sulfamyl-7-chloro-4(3H)-quinazoline. A solution of 4.56 g (0.12 mol) of sodium borohydride in 1 liter of dry diglyme was added portionwise over a period of 1 hour while stirring the mixture. The mixture was then heated at 85°C, with stirring, for 1 hour.

After cooling the reaction mixture to 25°C in an ice bath 600 ml of water was added and then enough dilute hydrochloric acid (approximately 100 ml) to make the solution acid. The solvent was then removed under reduced pressure at 60° to 70°C. The very viscid residue solidified when triturated with water. The solid was filtered and washed with water. The solid was dissolved in approximately 400 ml 95% ethanol and the solution filtered through Celite. On cooling the solution yielded 30 g of colorless solid, MP 253° to 259°C. The filtrate was concentrated to 200 ml to yield another 4.6 g, MP 253° to 259°C.

The above product was then recrystallized from 900 ml of 95% ethanol after filtering the hot solution through Celite. Crystallization was initiated and the mixture agitated occasionally while being cooled in the refrigerator. Yield of product 29 g, MP 253° to 259°C. Concentration of the filtrate to 125 ml yielded another 7.5 g of product, MP 253° to 259°C. The product was recrystallized another time in the manner described above. Total yield, first and second crops, 28.8 g (66%), MP 250° to 255°C. Product was dried at 80°C in a vacuum, according to US Patent 3,360,518.

References

Merck Index 6024
Kleeman and Engel p. 594
PDR pp. 1401, 1668
OCDS Vol. 2 p. 384 (1980)
DOT 9 (12) 498 (1973)
I.N. p. 629
REM p. 940
Shetty, B.V.; US Patent 3,360,518; December 26, 1967; assigned to Wallace and Tiernan Inc.
Shetty, B.V.; US Patent 3,557,111; January 19, 1971

METOPIMAZINE

Therapeutic Function: Antiemetic

Chemical Name: 4-Piperidinecarboxamide, 1-(3-(2-(methylsulfonyl)-10H-phenothiazin-10-yl)propyl)-

Common Name: Metopimazine

Structural Formula:

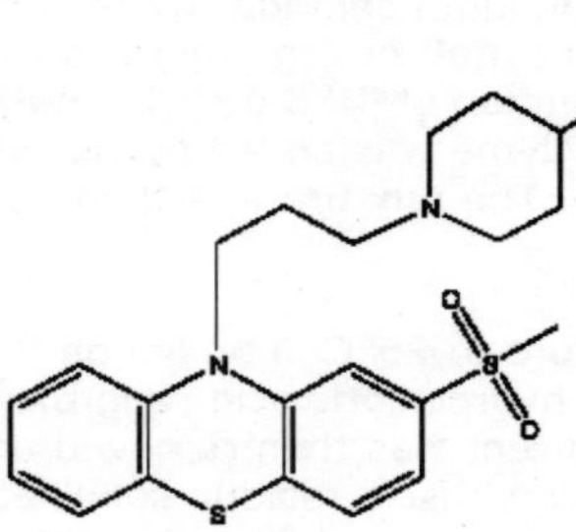

Chemical Abstracts Registry No.: 14008-44-7

Trade Name	Manufacturer	Country	Year Introduced
Nortrip	Rhodia	-	-

Raw Materials

2-Methylsulfonylphenothiazine
1-Bromo-3-chloropropane
Piperidine-4-carboxylic acid amide
Sodium
Ammonia

Manufacturing Process

2-Methylsulfonyl-10-(3-chloropropyl)phenothiazine was prepared by condensation of 1-bromo-3-chloropropane and 2-methylsulfonyl phenothiazine in liquid ammonia in presence of obtained in situ sodium amide.

10 g 2-methylsulfonyl-10-(3-chloropropyl)-phenothiazine, 4 g piperidine-4-carboxylic acid amide, 3.5 g dry sodium carbonate in 200 ml of ethanol was heated to reflux for 24 hours. Than 1.75 g sodium carbonate was added and the mixture was heated another 8 hours. After that the new 1.75 g portion of sodium carbonate was added and heated for 16 hours. The solvent was removed in vacuum (20 mm Hg). The residue was stirred with 50 ml water and 150 ml ethyl acetate. The organic layer was separated and extracted with 200 ml 1 N hydrochloric acid. The water layer was made alkaline with 4 N sodium hydroxide, extracted with ethyl acetate and dried over sodium sulfate. The solvent was removed in vacuum (20 mm Hg) to dryness. The obtained residue 2-methylsulfonyl-10-(3-(4-carbamoylpiperidino)propyl)phenothiazine was recrystallized from ethyl acetate. Yield 6 g; MP: 170°-171°C.

References

Jacob R. et al.; D.B. Patent No. 1,092,476; April 14, 1959; Societe des Usines Chimiques Rhone-Poulenc, Paris

METOPROLOL TARTRATE

Therapeutic Function: Beta-adrenergic blocker

Chemical Name: 1-[4-(2-Methoxyethyl)phenoxy]-3-[(1-methylethyl)amino]-2-propanol tartrate

Common Name: -

Structural Formula:

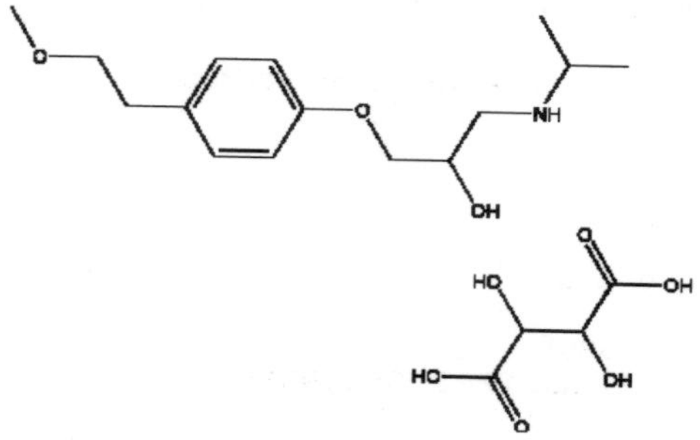

Chemical Abstracts Registry No.: 56392-17-7; 37350-58-6 (Base)

Trade Name	Manufacturer	Country	Year Introduced
Betaloc	Astra	UK	1975
Lopressor	Geigy	UK	1975
Beloc	Astra	W. Germany	1976
Lopressor	Ciba Geigy	W. Germany	1976
Lopressor	Ciba Geigy	Italy	1978
Selomen	Bracco	Italy	1978
Lopressor	Ciba Geigy	US	1978
Seloken	Searle	France	1980
Seloken	Fujisawa	Japan	1983
Lopresol	Takeda	Japan	1983
Lati 2	Unifa	Argentina	-
Neobloc	Unipharm	Israel	-
Prelis	Brunnengraber	W. Germany	-

Raw Materials

Isopropylamine
Tartaric acid
Epichlorohydrin
Sodium bicarbonate
p-(β-Methoxyethyl)phenol

Manufacturing Process

The starting material 1,2-epoxy-3-[p-(β-methoxyethyl)phenoxy]-propane was obtained from p-(β-methoxyethyl)-phenol which was reacted with epichlorohydrin whereafter the reaction product was distilled at 118°C to 128°C at a pressure of 0.35mm Hg.

1,2-Epoxy-3-[p-(β-methoxyethyl)-phenoxy]-propane (16.7g) was dissolved in 50 ml isopropanol and mixed with 20 ml isopropylamine. The mixture was heated in an autoclave on boiling water-bath overnight, whereafter it was evaporated and the remainder dissolved in 2 N HCI. The solution was extracted first with ether and thereafter with methylene chloride. After evaporating the methylene chloride phase, the hydrochloride of 1-isopropylamino-3-[p(β-methoxyethyl)-phenoxy] -propanol-2 was obtained which, after recrystallization from ethyl acetate, weighed 10.4 g. Melting point 83°C. Equivalent weight: found 304.0, calculated 303.8.

The hydrochloride is then converted to the tartrate.

References

Merck Index 6027
Kleeman and Engel p. 595
PDR p. 894
OCDS Vol. 2 p. 109 (1980)
DOT 11 (9) 360 (1975) and 17 (2) 65 (1981)
I.N. p. 630
REM p. 905
Brandstrom, A.E., Carlsson, P.A.E., Carlsson, S.A.I., Corrodi, H.R., Ek, L.and Ablad, B.A.H.; US Patent 3,873,600; March 25, 1975

METRIZAMIDE

Therapeutic Function: Diagnostic aid

Chemical Name: D-Glucose, 2-((3-(acetylamino)-5-(acetylmethylamino)-2,4,6-triiodobenzoyl)amino)-2-deoxy-

Common Name: Metrizamide

Structural Formula:

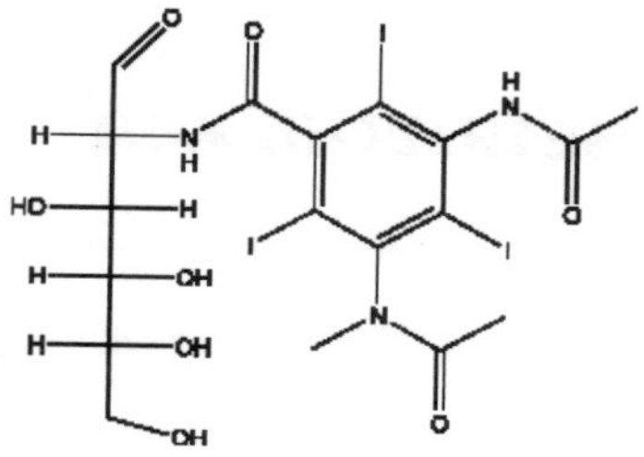

Chemical Abstracts Registry No.: 31112-62-6

Trade Name	Manufacturer	Country	Year Introduced
Amipaque	Winthrop laboratories	-	-

Raw Materials

3-(Acetylmethylamino)-2,4,6-triiodo-5-methylamino-benzoic acid
Thionyl chloride
Glucosamine

Manufacturing Process

1 mole 3-(acetyl-methyl-amino)-2,4,6-triiodo-5-methylamino-benzoic acid was suspended in thionyl chloride and reacted by stirring at 70°C for 16 hours. Excess thionyl chloride was distilled off in vacuum, the residue dissolved in chloroform, cooled in the ice bath, washed with iced water (3x100 ml), saturated sodium bicarbonate solution (3 x 100 ml), 2 N sodium carbonate solution (2 x 100 ml) and finally with water (3 x 100 ml). After drying with $CaCl_2$ the chloroform distilled off and residue dried in vacuum. Yield of 3-acetylamino-5-(acetyl-methyl-amino)-2,4,6-triiodo-benzoyl chloride: 66%; MP: 238°-240°C (the re-crystallization from tetrahydrofuran).

It (0.02 mole) was dissolved in dioxan (120 ml). To the solution was added (25 ml) and $NaHCO_3$ (0.0022 mol). Glucosamine (0.022 mol) was added in portions and reaction mixture left by stirring at room temperature for 24 hours. The solution was evaporated to dryness in vacuum, the residue dissolved in water (500 ml), filtered clear and run through an Amberlite IR

120 H^+ ion exchange column. The effluent was evaporated to dryness in vacuum resulting in a white crystalline residue. The crude 3-acetylamino-5-N-methyl-acetylamino-2,4,6-triiodobenzoyl glucosamine was crystallised from isopropanol (charcoal-treated when in solution), dissolved in water and charcoal-treated at 100°C for 20 min. The water was distilled off in vacuum and the white residue dried in vacuum at 70°C. MP: 190°-195°C.

References

Almen T.H.O.; US Patent No. 3,701,771; Oct. 31, 1972; Assigned to Nyegaard and Co. A/S; Norway

METRIZOIC ACID

Therapeutic Function: Diagnostic aid (radiopaque medium)

Chemical Name: 3-(Acetylamino)-5-(acetylmethylamino)-2,4,6-triiodobenzoic acid

Common Name: -

Structural Formula:

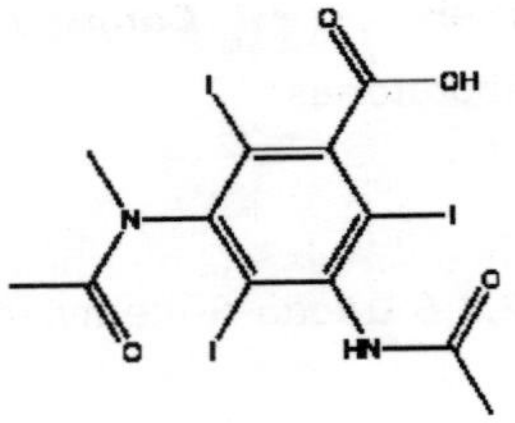

Chemical Abstracts Registry No.: 1949-45-7

Trade Name	Manufacturer	Country	Year Introduced
Isopaque	Winthrop	France	1973
Isopaque	Sterling	US	1975
Isopaque	Winthrop	Italy	1978
Ronpacon	Cilag Chemie	W. Germany	-

Raw Materials

Diatrizoic acid (diatrizoate)
Dimethyl sulfate

Manufacturing Process

3,5-Diacetamido-2,4,6-triiodobenzoic acid (diatrizoic acid) (see Diatrizoate entry for synthesis) (10 g) is suspended in water (10 ml), 5 N potassium hydroxide (4.3 equivalent) is added and the mixture cooled to about 15°C. Dimethyl sulfate (0.5 equivalent) dissolved in an equal volume of acetone is added drop by drop while stirring. After the reaction mixture has been stirred for about 1 hour hydrochloric acid (1:1) is added, with stirring to pH about 0.5. The precipitate is filtered, washed and suspended moist in 4 parts of water, concentrated ammonia is added to pH about 7 and the ammonium salt solution is isomerized at 90°C to 100°C for about one-half hour whereafter additional ammonia is added to pH about 9 followed by solid ammonium chloride (about 10% weight/volume) and the solution stirred overnight and the excess of 3,5-diacetamide-2,4,6-triiodobenzoic acid recovered as ammonium salt on the filter. The filtrate is precipitated by means of hydrochloric acid (1:1) at pH about 0.5 and the N-methyl-3,5-diacetamido-2,4,6-triiodobenzoic acid collected on a filter, washed and dried.

References

Merck Index 6032
Kleeman and Engel p. 597
I.N. p. 631
REM p. 1270
Holtermann, H., Haugen, L.G., Nordal, V. and Haavaldsen, J.L.; US Patent 3,178,473; April 13, 1965; assigned to Nyegaard and Co. A/S (Norway)

METRONIDAZOLE

Therapeutic Function: Antiprotozoal

Chemical Name: 2-Methyl-5-nitroimidazole-1-ethanol

Common Name: -

Structural Formula:

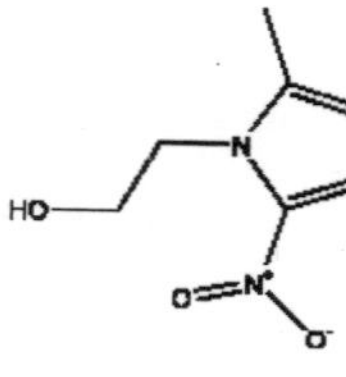

Chemical Abstracts Registry No.: 443-48-1

Trade Name	Manufacturer	Country	Year Introduced
Flagyl	Specia	France	1960
Flagyl	May and Baker	UK	1960
Flagyl	Rhone Poulenc	W. Germany	1961
Flagyl	Farmitalia	Italy	1962
Flagyl	Searle	US	1963

Trade Name	Manufacturer	Country	Year Introduced
Satric	Savage	US	1982
Metryl	Lemmon	US	1982
Metro IV	McGaw	US	1982
Protostat	Ortho	US	1983
Anaerobex	Gerot	Austria	-
Arilin	Wolff	W. Germany	-
Asuzol	Fuji	Japan	-
Clont	Bayer	W. Germany	-
Deflamon	SPA	Italy	-
Efloran	Krka	Yugoslavia	-
Elyzol	Dumex	Denmark	-
Entizol	Polfa	Poland	-
Flagemona	Phoenix	Argentina	-
Fossyol	Merckle	W. Germany	-
Gineflavir	Crosara	Italy	-
Klion	Kobanyai	Hungary	-
Kreucosan	Kreussler	W. Germany	-
Medazol	Belupo Ltd.	Yugoslavia	-
Meronidal	Kissei Pharmaceutical Co., Ltd.	Japan	-
Metrajil	Mulda	Turkey	-
Matrogil	Lkapharm	Israel	-
Metrolag	Lagap	Switz.	-
Monasin	Helvepharm	Switz.	-
Nalox	Omega	Argentina	-
Neo-Tric	Neo	Canada	-
Nida	Toyo Pharm.	Japan	-
Novonidazol	Novopharm	Canada	-
Orvagil	Galenika	Yugoslavia	-
Rathimed N	Pfleger	W. Germany	-
Rivozol	Rivopharm	Switz.	-
Rodogyl	Specia	France	-
Salandol	Sato	Japan	-
Sanatrichom	Godecke	W. Germany	-
Sawagyl	Sawai	Japan	-
Servizol	Servipharm	Switz.	-
Surimol	Labatec	Switz.	-
Takimetol	Nakataki	Japan	-
Tarozole	Taro	Israel	-
Tranoxa	Exa	Argentina	-
Trichazol	Will	Canada	-
Trichex	Gerot	Austria	-
Trichocide	Green Cross	Japan	-

Trade Name	Manufacturer	Country	Year Introduced
Tricho Cordes	Icthyol	W. Germany	-
TrichoGynaedron	Artesan	W. Germany	-
Trichomol	Gea	Denmark	-
Trichostop	Sigmapharm	Austria	-
Trichozole	Protea	Australia	-
Tricowas B	Wassermann	Spain	-
Trikamon	Elliott-Marion	Canada	-
Trikozol	Farmos	Finland	-
Trivazol	Vister	Italy	-
Vagilen	Farmigea	Italy	-
Vagimid	Apogepha	E. Germany	-
Vaginyl	D.D.S.A.	UK	-
Wagitran	Ono	Japan	-

Raw Materials

2-Methyl-5-nitroimidazole
Ethylene chlorohydrin

Manufacturing Process

2-Methyl-4(or 5)-nitroimidazole (127 g) is heated with ethylene chlorohydrin (795 g) for 18 hours at 128° to 130°C and the chlorohydrin (660 g) is then distilled under reduced pressure (30mm Hg). The residue is treated with water (300 cc) and filtered, and the filtrate is made alkaline by the addition of sodium hydroxide solution (d = 1.33, 100 cc). It is then extracted with chloroform (1,000 cc) and, after evaporation of the chloroform in vacuo, there is obtained a pasty mass (77 g) which is recrystallized from ethyl acetate (450 cc) in the presence of animal charcoal. There is thus obtained 1-(2-hydroxyethyl)-2-methyl-5-nitroimidazole (24 g) as a creamy white crystalline powder melting at 158° to 160°C.

References

Merck Index 6033
Kleeman and Engel p. 597
PDR pp. 830, 872, 876, 993, 1034, 1305, 1605, 1670, 1723, 1999
OCDS Vol. 1 p. 240 (1977)
DOT 13 (4) 147 (1977) 8117 (1) 34 (1981)
I.N. p. 632
REM p. 1222
Jacob, R.M., Regnier, G.L. and Crisan, C.; US Patent 2,944,061; July 5,1960; assigned to Societe des Usines Chimiques Rhone-Poulenc, France

METYRAPONE

Therapeutic Function: Diagnostic aid (pituitary function)

Chemical Name: 2-Methyl-1,2-di-3-pyridyl-1-propanone

Common Name: -

Structural Formula:

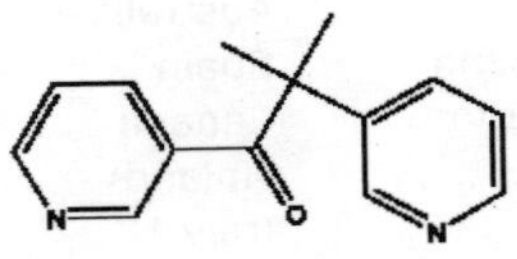

Chemical Abstracts Registry No.: 54-36-4

Trade Name	Manufacturer	Country	Year Introduced
Metopirone	Ciba	US	1961
Metopirone	Ciba	UK	1961
Metyrapone	Ciba	Switz.	1964
Metopiron	Ciba	W. Germany	1966

Raw Materials

3-Acetylpyridine
Sulfuric acid
Hydrogen
Hydroxylamine sulfate

Manufacturing Process

According to US Patent 2,966,493, the 2,3-bis-(3-pyridyl)-2,3-butanediol used as the starting material may be prepared as follows. A solution of 1,430 g of 3-acetyl-pyridine in 7,042 ml of a 1 N aqueous solution of potassium hydroxide is placed into a cathode chamber containing a mercury cathode with a surface of 353 cm^2 and is separated from an anode chamber by an Alundum membrane. As anode a platinum wire is used and the anolyte consists of a 1 N solution of aqueous potassium hydroxide which is replenished from time to time.

The electrolysis is carried out at a reference potential of -2.4 volts vs a standard calomel electrode. An initial current density of 0.0403 amp/cm^2 is obtained which drops to 0.0195 amp/cm^2 at the end of the reduction, which is carried on over a period of 1,682 minutes at 15° to 20°C. The catholyte is filtered, the solid material is washed with water and dried. 430 g of the 2,3-bis-(3-pyridyl)-butane-2,3-diol is recrystallized from water, MP 244° to 245°C.

A mixture of 3.43 g of 2,3-bis-(3-pyridyl)-2,3-butane-diol and 25 ml of concentrated sulfuric acid is heated to 76°C and kept at that temperature for 7½ hours. It is then poured on ice, neutralized with 50% aqueous solution of sodium hydroxide and the pH is adjusted to 8 with solid sodium carbonate. The aqueous solution is three times extracted with ethyl acetate, the separated organic layer dried over sodium sulfate and evaporated to dryness.

The residue is distilled and 1.86 g of viscous, colorless oil is obtained which is purified by distillation. BP 140° to 160°C/0.07 mm. The infrared spectrum shows the presence of a mixture of two compounds, one containing a conjugated, the other one an unconjugated carbonyl group, without the presence of a compound containing a hydroxyl group; thus the rearrangement has taken place.

The resulting mixture does not crystalize and is converted into a mixture of oximes by treatment of a solution of the mixture in 20 ml of ethanol with a solution of 1.8 g of hydroxylamine sulfate in 3 ml of water. 1.8 g of sodium acetate in 5 ml of water is added, and the mixture is refluxed for 5 hours, then extracted with ethyl acetate, and the ethyl acetate solution is washed with a saturated aqueous sodium chloride solution and dried over sodium sulfate. After evaporating the solvent, the residue is triturated with warm ether and 1.1 g of a crystalline oxime is obtained, MP 168° to 171°C.

0.1 g of the resulting oxime is dissolved in 5 ml of 2 N aqueous sulfuric acid and the mixture is refluxed for 3 hours and allowed to stand overnight. After being rendered basic by adding a concentrated aqueous solution of sodium hydroxide and adjusted to a pH of 8 with sodium carbonate, the mixture is extracted 3 times with ethyl acetate; the organic layer is washed with water, dried and evaporated. Upon distillation of the residue an oily product is obtained, BP 130° to 160°C/0.3 mm. Infrared analysis shows the presence of a uniform compound, containing a conjugated carbonyl group. The 2-methyl-1,2-bis-(3-pyridyl)-propane-1-one crystallizes upon standing at room temperature or by covering the oily distillate with pentane and cooling to -80°C and filtering the oily crystals. It melts after recrystallization from a mixture of ether, hexane and petroleum ether at 48° to 50°C.

References

Merck Index 6036
Kleernan and Engel p. 598
PDR p. 803
I.N. p. 633
REM p. 1276
Bencze, W.L. and Allen, M.J.; US Patent 2,923,710; February 2, 1960; assigned to Ciba Pharmaceutical Products, Inc.
Allen, M.J. and Bencze, W.L.; US Patent 2,966,493; December 27, 1960; assigned to Ciba Pharmaceutical Products, Inc.

METYROSINE

Therapeutic Function: Tyrosine hydroxylase inhibitor

Chemical Name: α-Methyl-L-tyrosine

Common Name: Metirosine

Structural Formula:

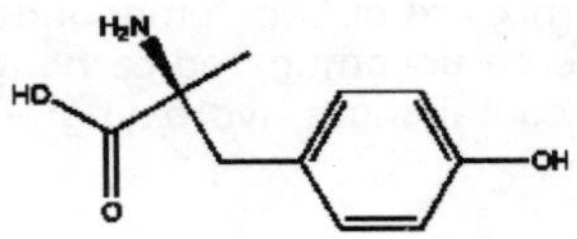

Chemical Abstracts Registry No.: 672-87-7

Trade Name	Manufacturer	Country	Year Introduced
Demser	MSD	US	1979

Raw Materials

Hydrogen
Sulfuric acid
α-Methyl-N-dichloroacetyl-p-nitrophenylalanine
Sodium nitrite
Hydrogen chloride

Manufacturing Process

50 g of α-methyl-N-dichloroacetyl-p-nitrophenylalanine was dissolved in 500 ml methanol, 300 mg of platinum oxide were added and the mixture reduced at 41 pounds of pressure; within an hour 14.5 pounds were used up (theory 12.4 pounds). After filtration of the catalyst, the red clear filtrate was concentrated in vacuo and the residual syrup flushed several times with ether. The crystalline residue thus obtained, after air drying, weighed 45.3 g (99.5%), MP unsharp at about 104°C to 108°C with decomposition. After two precipitations with ether from an alcoholic solution, the somewhat hygroscopic amine was dried over sulfuric acid for analysis.

10 g of the amine prepared above was dissolved in 5 ml of 50% sulfuric acid at room temperature; the viscous solution was then cooled in ice and a solution of sodium nitrite (2.4 g) in 10 ml water gradually added with agitation. A flocculent precipitate formed. After all the nitrite had been added, the mixture was aged in ice for an hour, after which it was allowed to warm up to room temperature. Nitrogen came off and the precipitate changed to a sticky oil. After heating on the steam bath until evolution of nitrogen ceased, the oil was extracted with ethyl acetate. After removal of the solvent in vacuo, 9.4 g of colored solid residue was obtained, which was refluxed with 150 ml hydrochloric acid (1:1) for 17 hours. The resulting dark solution; after Norite treatment and extraction with ethyl acetate, was concentrated in vacuo to dryness and the tan colored residue (7.4 g) sweetened with ethanol. Dissolution of the residue in minimum amount of ethanol and neutralization with diethylamine of the clarified solution, precipitated the α-methyl tyrosine, which was filtered, washed with ethanol (until free of chlorides) and ether. The crude amino acid melted at 309°C with decomposition. For further purification, it was dissolved in 250 ml of a saturated sulfur dioxide-water solution, and the solution, after Noriting, concentrated to about 80 ml, the tan colored solid filtered washed with ethanol and ether. Obtained 1.5 g of α-methyl tyrosine, MP 320°C dec.

References

Merck Index 6038
PDR p. 1167
DOT 16 (10) 346 (1980)
I.N. p. 628
REM p. 909
Pfister, K. III and Stein, G.A.; US Patent 2,868,818; January 13, 1959;assigned to Merck and Co., Inc.

MEXENONE

Therapeutic Function: Sunscreen agent

Chemical Name: (2-Hydroxy-4-methoxyphenyl)(4-methylphenyl)methanone

Common Name: 2-Hydroxy-4-methoxy-4'-methylbenzophenone

Structural Formula:

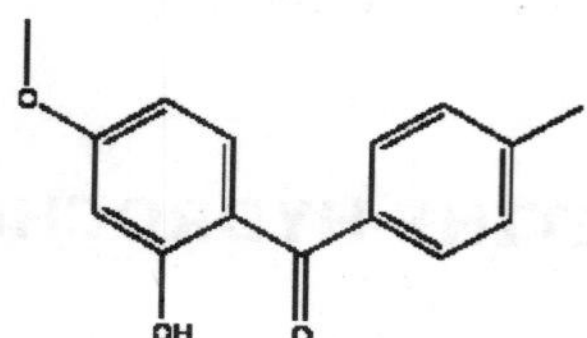

Chemical Abstracts Registry No.: 1641-17-4

Trade Name	Manufacturer	Country	Year Introduced
Uvistat-L	Ward Blenkinsop	UK	1960

Raw Materials

p-Toluoyl chloride
1,3-Dimethoxybenzene
Hydrogen chloride
Sodium hydroxide

Manufacturing Process

p-Toluoyl chloride is the starting material. To this is added chlorobenzene and 1,3-dimethoxybenzene. The reaction mixture is cooled to 12°C in an ice bath and aluminum chloride is added gradually, keeping the reaction below 30°C. The reaction is then gradually heated to 115°C with the evolution of hydrogen chloride gas. As the temperature increases, the reaction mixture becomes thicker. At 105°C, dimethyl formamide is added slowly. The reaction is heated at 115°C for a short time and is then poured into concentrated hydrochloric acid. The reaction mixture pours very easily and very cleanly. The acid mixture is heated with steam to dissolve all the material which had not hydrolyzed and the mixture is filtered. The red chlorobenzene layer is separated and washed twice with hot water.

To the chlorobenzene solution is then added sodium hydroxide dissolved in water and the chlorobenzene is removed by a steam distillation. After all of the chlorobenzene is removed, the precipitate which forms during the distillation is removed by filtration and discarded. The solution is cooled and acidified with hydrochloric acid, precipitating a tan solid. This is removed by filtration and washed acid-free. It is then treated with sodium bicarbonate solution to remove any acid present and is then washed with water to remove all traces of bicarbonate. After drying approximately a 75% yield of mexenone is obtained.

References

Merck Index 6045
Kleeman and Engel p. 598
OCDS Vol. 2 p. 175 (1980)
I.N. p. 633
Hardy. W.B. and Forster, W.S.; US Patent 2,773,903; December 11, 1956; assigned to American Cyanamid Company

MEXILETINE HYDROCHLORIDE

Therapeutic Function: Antiarrhythmic

Chemical Name: 1-(2,6-Dimethylphenoxy)-2-propanamine hydrochloride

Common Name: -

Structural Formula:

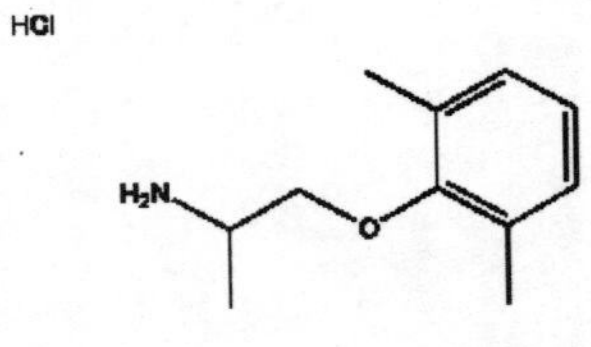

HCl

H_2N O

Chemical Abstracts Registry No.: 5370-01-4; 31828-71-4 (Base)

Trade Name	Manufacturer	Country	Year Introduced
Mexitil	Boehringer Ingelheim	US	1976
Mexitil	Boehringer Ingelheim	Switz.	1978
Mexitil	Boehringer Ingelheim	W. Germany	1979
Mexitil	Boehringer Ingelheim	France	1981
Mexitil	Boehringer Ingelheim	Italy	1982

Raw Materials

Dimethyl phenol
Hydrogen
Hydroxylamine
Chloroacetone
Sodium hydroxide

Manufacturing Process

The sodium salt of dimethyl phenol was reacted with chloroacetone and this product with hydroxylamine to give the starting material.

245 g of this 1-(2',6'-dimethyl-phenoxy)-propanone-(2)-oxime were dissolved in 1,300 cc of methanol, and the solution was hydrogenated at 5 atmospheres gauge and 60°C in the presence of Raney nickel. After the calculated amount of hydrogen had been absorbed, the catalyst was filtered off, the methanol was distilled out of the filtrate,and the residue, raw 1-(2',6'-dimethyl-phenoxy)-2-amino-propane, was dissolved in ethanol. The resulting solution was acidified with ethereal hydrochloric acid, the acidic solution was allowed to cool, and the precipitate formed thereby was collected by vacuum filtration. The filter cake was dissolved in ethanol and recystallized therefrom by addition of ether. 140.5 g (51.5% of theory) of a substance having a melting point of 203°C to 205°C were obtained, which was identified to be 1-(2',6'-dimethyl-phenoxy)-2-anino-propane hydrochloride.

References

Merck Index 6047
DFU 1 (4) 180 (1976)
Kleeman and Engel p. 598
DOT 12 (9) 361 (1976)
I.N. p.633
REM p.861
Koppe, H., Zeile, K., Kummer, W., Stahle, H. and Dannenberg, P.; US Patent 3,659,019; April 25,1972; assigned to Boehringer Ingelheim G.m.b.H. (W. Germany)

MEZLOCILLIN

Therapeutic Function: Antibiotic

Chemical Name: Sodium D(-)-α-[(3-methylsulfonyl-imidazolidin-2-on-1-yl)-carbonylamino]benzylpenicillin

Common Name: -

Chemical Abstracts Registry No.: 51481-65-3

Structural Formula:

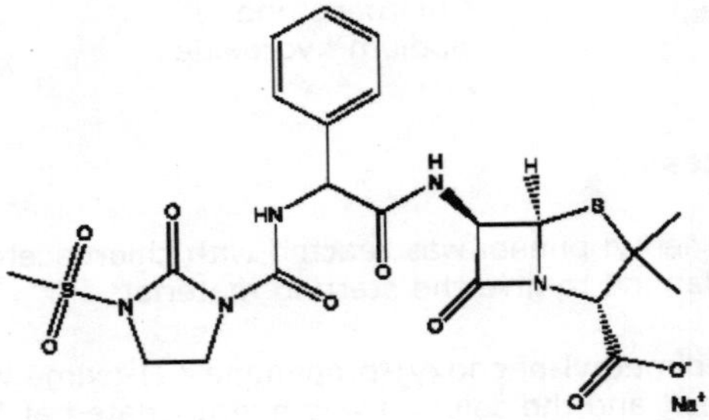

Trade Name	Manufacturer	Country	Year Introduced
Baypen	Bayer	W. Germany	1977
Baypen	Bayer	UK	1980
Baypen	Bayer	Switz.	1980
Baypen	Bayer	Italy	1981
Mezlin	Miles	US	1981
Baypen	Bayer Yakuhin	Japan	1982
Baypen	Bayer	France	1983
Baypen	Bayer	Sweden	1983
Baycipen	Bayer	-	-
Optocillin	Bayer	W. Germany	-

Raw Materials

Ampicillin
Methanesulfonyl chloride
2-Imidazolidone
Phosgene

Manufacturing Process

9.3 parts by weight of ampicillin were suspended in 80% strength aqueous tetrahydrofuran (140 parts by volume) and sufficient triethylamine (approximately 6.3 parts by volume) was added dropwise while stirring at 20°C, just to produce a clear solution and to give a pH value of between 7.5 and 8.2 (glass electrode). The mixture was cooled to 0°C and 5.1 parts by weight of 3-methylsulfonyl-imidazolidin-2-one-1-carbonyl chloride were added gradually in portions over the course of 30 minutes, while the mixture was stirred and kept at a pH value of between 7 and 8 by simultaneous addition of triethylamine.

The carbonyl chloride reactant was prepared by reacting 2-imidazolidone with methanesulfonyl chloride then that product with phosgene. The mixture was stirred for 10 minutes at 0°C and subsequently further stirred at room temperature until no further addition of triethylamine was necessary to maintain a pH value of 7 to 8. 150 parts by volume of water were added and the tetrahydrofuran was largely removed in a rotary evaporator at room temperature.

The residual aqueous solution was extracted once by shaking with ethyl acetate, covered with 250 parts by volume of fresh ethyl acetate and acidified to pH 1.5 to 2.0 with dilute hydrochloric acid while being cooled with ice. The organic phase was separated off, washed twice with 50 parts by volume of water at a time and dried for 1 hour over anhydrous $MgSO_4$ in a refrigerator. After filtration, about 45 parts by volume of a 1 molar solution of sodium 2-ethylhexanoate in ether containing methanol were added to the solution of the penicillin. The mixture was concentrated on a rotary evaporator until it had an oily consistency and was dissolved in a sufficient amount of methanol by vigorous shaking, and the solution was rapidly added dropwise, with vigorous stirring, to 500 parts by volume of ether which contained 10% of methanol.

The precipitate was allowed to settle for 30 minutes, the solution was decanted from the precipitate, and the latter was again suspended in ether, filtered off and washed with anhydrous ether. After drying over P_2O_5 in a vacuum desiccator, the sodium salt of the mezlocillin was obtained in the form of a white solid substance.

References

Merck Index 6049
DFU 2 (9) 200 (1977)
Kleeman and Engel p. 599
PDR p. 1254
OCDS Vol. 3 p. 206 (1984)
DOT 11 (11) 444 (1975) and 15 (2) 54 (1979)
I.N. p. 633
REM p. 1196
Konig, H.B., Schrock, W. and Metzger, K.G.; US Patents 3,972,869; August 3, 1976; 3,972,870; August 3, 1976; 3,974,141; August 10, 1976; 3,974,142; August 10, 1976; 3,975,375; August 17, 1976; 3,978,056; August 31, 1976; 3,983,105; September 28,1976; and

MIANSERIN

Therapeutic Function: Serotonin antagonist, Antihistaminic

Chemical Name: 1,2,3,4,10,14b-Hexahydro-2-methyldibenzo[c]pyrazino[1,2-a]azepine

Common Name: 2-Methyl-1,2,3,4,10,14b-hexahydro-2H-pyrazino-[1,2-f]morphanthridine

Chemical Abstracts Registry No.: 24219-97-4; 21535-47-7 (Hydrochloride salt)

Structural Formula:

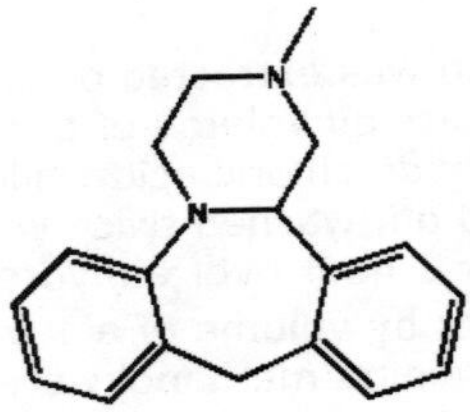

Trade Name	**Manufacturer**	**Country**	**Year Introduced**
Tolvin	Organon	W. Germany	1975
Bolvidon	Organon	UK	1976
Norval	Bencard	UK	1976
Lantanon	Ravasini	Italy	1976
Athymil	Organon	France	1979
Athmyl	Organon	Switz.	1980
Tetramide	Sankyo	Japan	1983

Raw Materials

2-Benzylaniline
Diethyloxalate
Chloroacetyl chloride
Lithium aluminum hydride
Polyphosphoric acid
Diborane
Methylamine

Manufacturing Process

(A) 25 g of 2-benzylaniline dissolved in 150 ml of benzene are cooled down in an ice bath to 8°C. To this solution are added 15 ml of pyridine and after that a solution of 15 ml of chloroacetyl chloride in 25 ml of benzene, maintaining the temperature of the reaction mixture at 10° to 15°C. After stirring for 1 hour at room temperature 25 ml of water are added and the mixture is shaken for 30 minutes. Next the mixture is sucked off and the benzene layer separated. Then the benzene layer is washed successively with 2 N HCl, a sodium carbonate solution and water. The extract dried on sodium sulfate is evaporated and the residue crystallized together with the crystals obtained already from benzene. Yield 18 g; MP 130° to 133°C.

(B) 40 g of N-chloroacetyl-2-benzylaniline are heated for 2 hours at 120°C together with 50 ml of phosphorus oxychloride and 320 g of polyphosphoric acid. Next the reaction mixture is poured on ice and extracted with benzene. The extract is washed and dried on sodium sulfate and the benzene distilled off. The product obtained (31g) yields after recrystallization 24 g of 6-chloromethyl-morphanthridine of MP 136° to 137°C.

(C) 10 g of 6-chloromethyl-morphanthridine are passed into 150 ml of a solution of methylamine in benzene (10%). After storage of the solution for 20 hours at 0° to 5°C the methylamine hydrochloride formed is sucked off and the filtrate evaporated to dryness. There remains as residue 11 g of crude

6-methylaminomethyl-morphanthridine.

(D) 11 g of crude 6-methylaminomethyl-morphanthridine are dissolved in 50 ml of absolute ether. While cooling in ice 2.7 g of lithium aluminumhydride, dissolved in 100 ml of absolute ether, are added. After boiling for 1 hour and cooling down in ice 11 ml of water are added slowly dropwise while stirring. After stirring for another 30 minutes at room temperature the mixture is sucked off and the filtrate evaporated to obtain 11 g of crude 5,6-dihydro-6-methylaminomethyl-morphanthridine in the form of a light yellow oil.

(E) 10 g of 5,6-dihydro-6-methylaminomethyl-morphanthridine are heated slowly, in 30 minutes, from 100° to 160°C with 7 g of pure diethyloxalate and after that from 160° to 180°C in 45 minutes. After cooling down the reaction mixture is stirred with benzene. The crystals are sucked off and yield after crystallization from dimethylformamide 9 g of 1,2-diketo-3(N)-methyl-2,3,4,4a-tetrahydro-1H-pyrazino-[1,2-f]-morphanthridine of MP 245° to 247°C.

(F) 9 g of the diketo-pyrazino-morphanthridine compound obtained above are reduced with diborane to give mianserin.

References

Merck Index 6050
Kleeman and Engel p. 599
OCDS Vol. 2 p. 451 (1980)
DOT 12 (1) 31 (1976)
I.N. p. 634
van der Burg,W.J. and Delobelle, J.; US Patent 3,534,041; October 13,1970; assigned to Organon Inc.

MIBEFRADIL HYDROCHLORIDE

Therapeutic Function: Coronary vasodilator

Chemical Name: Acetic acid, methoxy-, 2-(2-((3-(1H-benzimidazol-2-yl)propyl)methylamino)ethyl)-6-fluoro-1,2,3,4-tetrahydro-1-(1-methylethyl)-2-naphthalenyl ester, dihydrochloride, (1S-cis)-

Common Name: Mibefradil hydrochloride

Chemical Abstracts Registry No.: 116666-63-8; 116644-53-2 (Base)

Trade Name	Manufacturer	Country	Year Introduced
Posicor	Roche Pharmaceuticals	USA	-

Structural Formula:

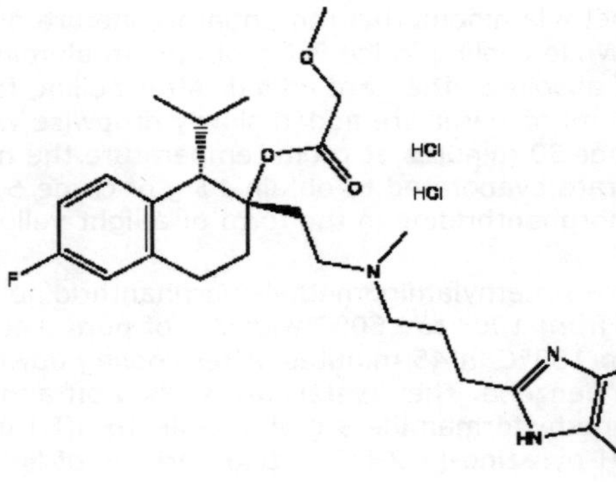

HCl
HCl

Raw Materials

Butyl lithium	[3-(1H-Benzimidazol-2-yl)propyl]methylamine
Diisopropylamine	Isopropenyl acetate
Lithium chloride	Potassium hydroxide
Methoxyacetyl chloride	
Sodium bis(2-methoxyethoxy)aluminum hydride	
(S)-6-Fluoro-1-isopropyl-3,4-dihydro-1H-naphthalen-2-one	

Manufacturing Process

To the solution of 5.35 g (28 mmol) [3-(1H-benzimidazol-2-yl)propyl]methylamine in 12.5 mL toluene was added by syringe 12.5 mL (11.42 g, 114 mmol) isopropenyl acetate. The reaction mixture was heated to reflux temperature, and stirred at that temperature for 1.75 hours, with reaction completion monitored by thin-layer chromatography (silica gel, eluting with 70% ethyl acetate/30% methanol). The product, N-[3-(1H-benzimidazol-2-yl)propyl]-N-methylacetamide, was obtained in quantitative yield.

Under a dry nitrogen atmosphere, a 2.5 molar solution of butyl lithium in hexane, 8.4 mL (21 mmol) was added by syringe to 20 mL pentane. The solution was cooled to 0°C and 2.75 mL (2.13 g, 21 mmol) diisopropylamine was added by syringe over six min. The solution was warmed to 25°C and stirred for three hours, then volatiles were removed in vacuo. THF, 20 mL, was added via syringe to the residue, and the resulting yellow solution cooled to 0°C. A solution of 2.42 g (10.5 mmol) N-[3-(1H-benzimidazol-2-yl)propyl]-N-methylacetamide in 10 mL THF was added by syringe over 9 min. The yellow solution was stirred for 15 min, then cooled to -78°C. (S)-6-Fluoro-1-isopropyl-3,4-dihydro-1H-naphthalen-2-one, 2.166 g, 87.2% pure (97.6:2.4 S:R), in 2 mL toluene was added by syringe over 12 min, and a further 2 mL toluene was used to complete the transfer. After stirring for two hours, the viscous yellow mixture was added to 50 mL water at less than 10°C. The suspension that formed was extracted with diethyl ether; and the extracts were dried over anhydrous magnesium sulfate, filtered, and concentrated in vacuo to afford 3.74 g of impure (1S,2S)-N-[3-(1H-benzimidazol-2-yl)propyl]-2-(6-fluoro-2-hydroxy-1-isopropyl-1,2,3,4-tetrahydronaphthalen-2-yl)-N-methylacetamide as a yellow foam. The foam was recrystallized from toluene, yield of a colorless solid 2.69 g, melting point 132-138°C. This material may be recrystallized a second time from toluene to remove residual (S)-6-fluoro-1-isopropyl-3,4-dihydro-1H-naphthalen-2-one if necessary.

(1S,2S)-N-[3-(1H-Benzimidazol-2-yl)propyl]-2-(6-fluoro-2-hydroxy-1-isopropyl-1,2,3,4-tetrahydronaphthalen-2-yl)-N-methylacetamidemay be synthesized by another method:

To the mixture 22.7 g (0.54 mol) dry lithium chloride and 100 mL THF at -15°C was added 160 mL 2 molar lithium diisopropylamide (0.32 mol) in heptane/THF/ethylbenzene was added. Then a solution of 36.6 g (0.16 mol) N-[3-(1H-benzimidazol-2-yl)-propyl]-N-methylacetamide in 140 mL toluene was added, the solution was stirred for 2 hours, and a further 155 mL toluene was added. (S)-6-Fluoro-1-isopropyl-3,4-dihydro-1H-naphthalen-2-one (29.9 g, 0.15 mol), in 15 mL toluene was added. After stirring at -10°C for 4 hours, the resulting solution was added to 200 mL ice water. The pH of the resulting mixture was adjusted to 7-8 by addition of a 71 g concentrated hydrochloric acid. The organic layer washed with water, then the solvents removed under reduced pressure to give 96 g of (1S,2S)-N-[3-(1H-benzimidazol-2-yl)propyl]-2-(6-fluoro-2-hydroxy-1-isopropyl-1,2,3,4-tetrahydronaphthalen-2-yl)-N-methylacetamide as a brown oil. The product was crystallysed from toluene, yield 45.3 g.

(1S,2S)-N-[3-(1H-Benzimidazol-2-yl)propyl]-2-(6-fluoro-2-hydroxy-1-isopropyl-1,2,3,4-tetrahydronaphthalen-2-yl)-N-methylacetamide,20.22 g (45.7 mmol), dissolved in 200 mL toluene at 40°C, was added by cannula over 40 min at 0°C to a suspension of sodium bis(2-methoxyethoxy)aluminum hydride in toluene, 40 mL (41.44 g suspension, 26.94 g sodium bis(2-methoxyethoxy)aluminum hydride, 133 mmol). The mixture was stirred at 0°C for 15 min, then at 35-40°C for 3 hours. The mixture was cooled to 25°C then added carefully to 70 g sodium hydroxide in 140 g ice. The resulting suspension was warmed to 25°C over 30 min, and the phases were separated. The aqueous phase was extracted with toluene; and the organic phase was washed twice with 10% aqueous sodium hydroxide, once with water, then once with saturated brine. The toluene phase was dried and concentrated in vacuo to afford 20.61 g of (1S,2S)-2-[2-{[3-(1H-benzimidazol-2-yl)propyl]methylmethylamino}ethyl]-6-fluoro-1-isopropyl-1,2,3,4-tetrahydronaphthalen-2-ol as a colorless foam.

To the mixture of 41.0 g (1S,2S)-2-[2-{[3-(1H-benzimidazol-2-yl)propyl]methylmethylamino}ethyl]-6-fluoro-1-isopropyl-1,2,3,4-tetrahydronaphthalen-2-ol, 240 mL water, and 240 mL toluene were added 22.4 g potassium hydroxide, and the mixture heated to 45-50°C for one hour. The resulting two-phase mixture was separated. To the organic phase was added 39.4 g (4.0 eq.) potassium carbonate sesquihydrate; then a solution of 21.0 g (17.7 mL, 3.25 eq.) methoxyacetyl chloride in 33 mL toluene was added over two hours at 25-30°C, and the resulting mixture stirred for an additional 30 min. Water, 200 mL, was added to quench the reaction. The organic phase, containing mibefradil as the free base was added an ethanol. To the stirred mixture of mibefradil and ethanol was added at 20°C a solution of 4.4 g of hydrogen chloride in 44.6 mL (35.0 g) ethanol. The mixture was heated to 50°C and 1.0 mL water was added, followed by a solution of 3.4 mL water in 332 mL methyl tert-butyl ether over one hour. The mixture was stirred for 3 hours. Mibefradil dihydrochloride crystals was seeded. A solution of 0.6 mL water in 65 mL methyl tert-butyl ether was added over one hour, and the mixture aged for a further 1.5 hours. The mixture was then cooled, and the resulting slurry of mibefradil dihydrochloride was filtered; yield 95%.

References

Harrington P. J.; US Patent No. 5,892,055; April 6, 1999; Assigned to Roche Colorado Corporation (Boulder, CO)

MICONAZOLE NITRATE

Therapeutic Function: Antifungal

Chemical Name: 1-[2,4-Dichloro-β-[(2,4-dichlorobenzyl)oxy]phenethyl] imidazole mononitrate

Common Name: -

Structural Formula:

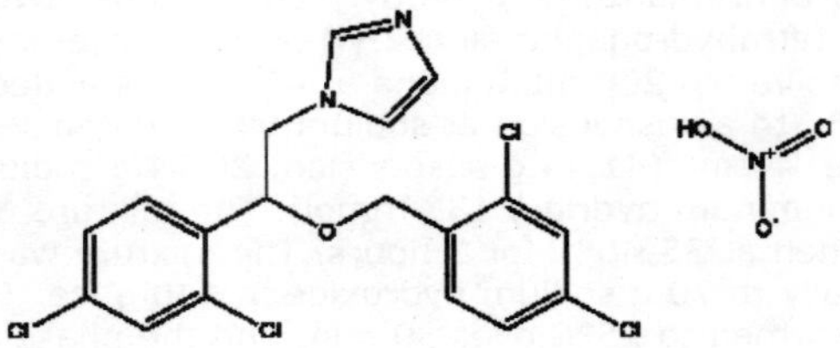

Chemical Abstracts Registry No.: 22832-87-7; 22916-47-8 (Base)

Trade Name	Manufacturer	Country	Year Introduced
Daktarin	Janssen	Italy	1974
Daktarin	Janssen	UK	1974
Daktar	Janssen	W. Germany	1974
Dermonistat	Ortho	UK	1974
Monistat	Ortho	US	1974
Daktarin	Le Brun	France	1975
Micatin	Johnson and Johnson	US	1976
Minostate	Janssen	US	1978
Andergin	Isom	Italy	1980
Frolid P	Mochida	Japan	1981
Aflorix	Gerardo Ramon	Argentina	-
Conofite	Pitman-Moore	US	-
Dektarin	Janssen	Italy	-
Deralbine	Andromaco	Argentina	-

Trade Name	Manufacturer	Country	Year Introduced
Epi-Monistat	Cilag	W. Germany	-
Florid	Mochida	Japan	-
Fungisdin	Esteve	Spain	-
Gyno-Daktarin	Le Brun	France	-
Gyno-Monistat	Cilag	W. Germany	-
Micatin	McNeil	US	-
Miconal	Ecobi	Italy	-
Micotef	Italfarmaco	Italy	-
Vodol	Andromaco	Brazil	-

Raw Materials

Imidazole	ω-Bromo-2,4-dichloroacetophenone
Sodium hydride	2,4-Dichlorobenzyl chloride
Nitric acid	Sodium borohydride

Manufacturing Process

Imidazole is reacted with ω-bromo-2,4-dichloroacetophenone and that product reduced with sodium borohydride.

A suspension of 10.3 parts of the α-(2,4-dichlorophenyl)imidazole-1-ethanol thus obtained and 2.1 parts of sodium hydride in 50 parts of dry tetrahydrofuran is stirred and refluxed for 2 hours. After this reaction time, the evolution of hydrogen is ceased. Then there are added successively 60 parts dimethylformamide and 8 parts of 2,4-dichlorobenzyl chloride and stirring and refluxing are continued for another 2 hours. The tetrahydrofuran is removed at atmospheric pressure. The dimethylformamide solution is poured onto water.

The product, 1-[2,4-dichloro-β-(2,4 -dichlorobenzyloxy)phenethyl]imidazole, is extracted with benzene. The extract is washed with water, dried, filtered and evaporated in vacuo. From the residual oily free base, the nitrate salt is prepared in the usual manner in 2-propanol by treatment with concentrated nitric acid, yielding, after recrystallization of the crude solid salt from a mixture of 2-propanol, methanol and diisopropyl ether, 1-[2,4-dichloro-β-dichlorobenzyloxy)phenethyl]imidazole nitrate; melting point 170.5°C.

References

Merck Index 6053
Kleeman and Engel p. 601
PDR pp.956, 1293
OCDS Vol. 2 p. 249 (1980)
DOT 7 ( ) 192 (1971) and 8 (6) 229 (1972)
I.N. p. 634
REM p. 1229
Godefroi, E.F. and Heeres, J.; US Patent 3,717,655; February 20,1973; assigned to Janssen Pharmaceutica NV
Godefroi, E.F. and Heeres, J.; US Patent 3,839,574; October 1,1974; assigned to Janssen Pharmaceutica NV

MICRONOMICIN

Therapeutic Function: Antibiotic

Chemical Name: O-2-Amino-2,3,4,6-tetradeoxy-6-(methylamino)-α-D-erythrohexopyranosyl(1-->4)-O-[3-deoxy-4-C-methyl-3-(methylamino)-β-L-arabinopyranosyl-(1-->6)-2-deoxy-D-streptamine

Common Name: 6'-N-Methylgentamicin C_{1a}; Sagamicin

Structural Formula:

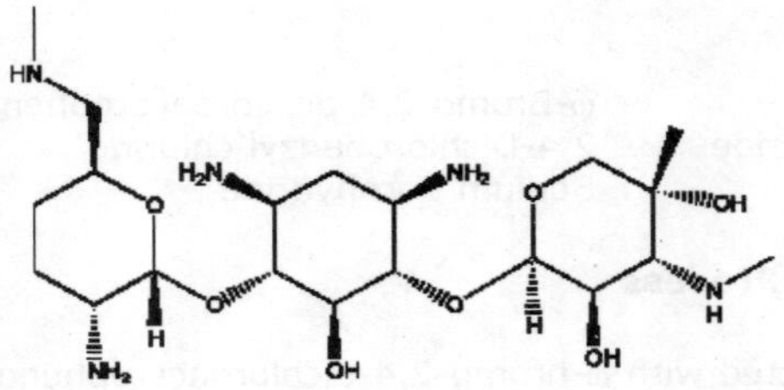

Chemical Abstracts Registry No.: 52093-21-7

Trade Name	Manufacturer	Country	Year Introduced
Sagamicin	Kyowa Hakko	Japan	1982

Raw Materials

Bacterium *Micromonospora sagamiensis*
Dextrin
Soybean meal

Manufacturing Process

A. Culturing of MK-65: In this example, *Micromonospora sagamiensis* MK-65 ATCC 21826 (FERM-P No. 1530) is used as the seed strain. One loopful of the seed strain is inoculated into 30 ml of a first seed medium in a 250 ml-Erlenmeyer flask. The first seed medium has the following composition:

	Percent
Dextrin	1
Glucose	1
Peptone	0.5
Yeast extract	0.5
$CaCO_3$(pH: 7.2 before sterilization)	0.1

Culturing is carried out with shaking at 30°C for 5 days. 30 ml of the seed culture is then inoculated into 300 ml of a second seed medium, of the same composition as the first seed medium, in a 2 liter-Erlenmeyer flask provided with baffles. The second seed culturing is carried out with shaking at 30°C for

2 days. Then 1.5 liters of the second seed culture (corresponding to the content of 5 flasks) is inoculated into 15 liters of a third seed medium of the same composition as set forth above, in a 30 liter-glass jar fermenter. Culturing in the jar fermenter is carried out with aeration (15 liters/minute) and stirring (350 rpm) at 30°C for 2 days. Then, 15 liters of the third seed culture is inoculated into 60 liters of a fourth seed medium of the same composition as set forth above, in a 300 liter-fermenter. Culturing in the fermenter is carried out with aeration (60 liters/minute) and stirring (150 rpm) at 30°C for 2 days. Finally, 60 liters of the fourth seed culture is inoculated into 600 liters of a fermentation medium having the following composition in a 1,000 liter-fermenter.

	Percent
Dextrin	5
Soybean meal	4
$CaCO_3$ (pH: 7.2 before sterilization)	0.7

Culturing in the fermenter is carried out with aeration (600 liters/minute) and stirring 150 rpm) at 35°C for 5 days.

B. Isolation of crude antibiotic: After the completion of fermentation, the culture liquor is adjusted to a pH of 2.0 with 12 N sulfuric acid and stirred for 30 minutes. Then, about 10 kg of a filter aid, Radiolite No. 600 (product of Showa Kagaku Kogyo Co., Ltd., Japan) is added thereto and the microbial cells are removed by filtration. The filtrate is adjusted to a pH of 8.0 with 6N sodium hydroxide and passed through a column packed with about 50 liters of a cation exchange resin, Amberlite IRC-50 (ammonia form). The active substance is adsorbed on the resin and the eluate is discarded. After washing the resin with water, the active substance is eluted out with 1N aqueous ammonia. The eluate is obtained in fractions and the activity of each of the fractions is determined against Bacillus subtilis No. 10707 by a paper disk method using an agar plate.

Active fractions are combined and concentrated in vacuo to about 5 liters. The concentrate is then adjusted to a pH of 8.0 with 6N sulfuric acid and passed through a column packed with 1 liter of an anion exchange resin, Dowex 1X2 (OH^-form). The column is washed with about 5 liters of water and the effluent and the washings containing active substance are combined and are concentrated to 1/15 by volume. The concentrate is adjusted to a pH of 10.5 with 6 N sodium hydroxide and 5 volumes of acetone is added thereto. The resultant precipitate is removed by filtration and the filtrate is concentrated to 500 ml. The concentrate is adjusted to a pH of 4.5 with 6 N sulfuric acid and 2.5 liters of methanol is added thereto. After cooling, a white precipitate is obtained. The precipitate is separated by filtration and washed with methanol. After drying in vacuo, about 300 g of white powder is obtained.

The white powder is a mixture of the sulfate of gentamicin C_{1a}, and the sulfate of XK-62-2, and exhibits an activity of 620 units/mg (the activity of 1 mg of pure product corresponds to 1,000 units).

C. Isolation and purification of XK-62-2: 100 g of the white powder obtained in the above step B are placed to form a thin, uniform layer on the upper part of a 5 cm x 150 cm column packed with about 3 kg of silica gel advancely suspended in a solvent of chloroform, isopropanol and 17% aqueous ammonia

(2:1:1 by volume). Thereafter, elution is carried out with the same solvent at a flow rate of about 250 ml/hour. The eluate is separated in 100 ml portions. The active fraction is subjected to paper chromatography to examine the components eluted. XK-62-2 is eluted in fraction Nos. 53-75 and gentamicin C1ais eluted in fraction Nos. 85-120. The fraction Nos. 53-75 are combined and concentrated under reduced pressure to sufficiently remove the solvent. The concentrate is then dissolved in a small amount of water. After freeze-drying the solution, about 38 g of a purified preparate of XK-62-2 (free base) is obtained. The preparate has an activity of 950 units/mg. Likewise, fraction Nos. 85-120 are combined and concentrated under reduced pressure to sufficiently remove the solvent. The concentrate is then dissolved in a small amount of water. After freeze-drying the solution, about 50 g of a purified preparate of gentamicin C_{1a} (free base) is obtained. The activity of the preparate is about 980 units/mg.

References

Merck Index A-9
DFU 4 (5) 360 (1979) (as sagamicin) and 6 (5) 332 (1980)
DOT 19 (4) 211 (1983)
I.N. p. 635
Nara, T., Takasawa, S.,Okachi, R., Kawamoto, I., Yamamoto, M., Sato, S., Sato, T. and Morikawa, A.; US Patent 4,045,298; August 30,1977; assigned to Abbott Laboratories

MIDAZOLAM MALEATE

Therapeutic Function: Anesthetic

Chemical Name: 8-Chloro-6-(2-fluorophenyl)-1-methyl-4H-imidazo-[1,5-a][1,4]benzodiazepine maleate

Common Name: -

Structural Formula:

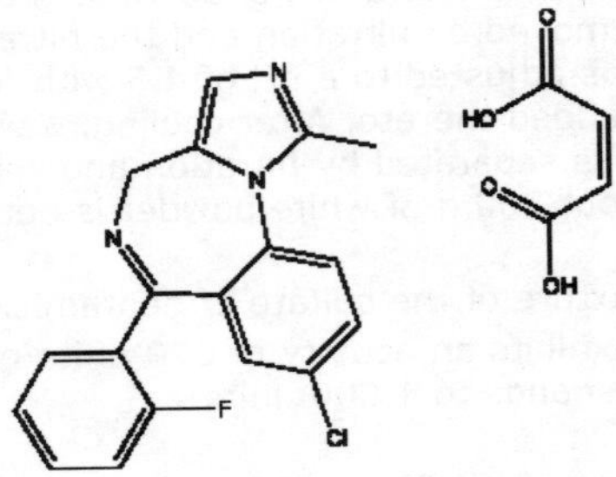

Chemical Abstracts Registry No.: 59467-70-8 (Base)

Trade Name	Manufacturer	Country	Year Introduced
Dormicum	Roche	Switz.	1982
Dormonid	Roche	-	-
Hypnovel	Roche	UK	-
Sorenor	Roche	-	-

Raw Materials

Acetic anhydride
Manganese dioxide
2-Aminomethyl-7-chloro-2,3-dihydro-5-(2-fluorophenyl)-1H-1,4-benzodiazepine
Polyphosphoric acid
Maleic acid

Manufacturing Process

Acetic anhydride (7 ml) was added to a solution of 6.16 g of crude 2-aminomethyl-7-chloro-2,3-dihydro-5-(2-fluorophenyl)-1H-1,4-benzodiazepine in 200 ml of methylene chloride. The solution was added to 200 ml of saturated aqueous sodium bicarbonate and the mixture was stirred for 20 minutes. The organic layer was separated, washed with sodium bicarbonate, dried over sodium sulfate and evaporated to leave resinous 2-acetylaminomethyl-7-chloro-2,3-dihydro-5-(2-fluorophenyl)-lH -l,4-benzodiazepine. This material was heated with 40 g of polyphosphoric acid at 150°C for 10 minutes. The cooled reaction mixture was dissolved in water, made alkaline with ammonia and ice and extracted with methylene chloride. The extracts were dried and evaporated and the residue was chromatographed over 120 g of silica gel using 20% methanol in methylene chloride. The clean fractions were combined and evaporated to yield resinous 8-chloro-3a,4-dihydro-6-(2-fluorophenyl)-1- methyl-4H-imidazo[1,5-a][1,4] -benzodiazepine.

A mixture of this material with 500 ml of toluene and 30 g of manganese dioxide was heated to reflux for 1½ hours. The manganese dioxide was separated by filtration over Celite. The filtrate was evaporated and the residue was crystallized from ether to yield 8-chloro-6-(2-fluorophenyl)-1-methyl-4H-imidazo[1,5-a][1,4]benzodiazepine, melting point 152°C to 154°C. The analytical sample was recrystallized from methylene chloride/hexane.

A warm solution of 6.5 g (0.02 mol) of 8-chloro-6-(2-fluorophenyl)-1-methyl-4H-imidazo[1,5-a] [1,4]-benzodiazepine in 30 ml of ethanol was combined with a warm solution of 2.6 g (0.022 mol) of maleic acid in 20 ml of ethanol. The mixture was diluted with 150 ml of ether and heated on the steam bath for 3 minutes. After cooling, the crystals were collected, washed with ether and dried in vacuo to yield 8-chloro-6-(2-fluorophenyl)-1-methyl-4H-imidazo[1.5-a] [1,4]-benzodiazepine maleate, melting point 148°C to 151°C.

References

Merck Index 6056
DFU 3 (11) 822 (1978)
OCDS Vol. 3 p. 197 (1984)
DOT 19 (2) 113; (4) 221 and (7) 378 (1983)
I.N.p. 635

F. Hoffmann-La Roche and Co.; British Patent 1,527,131; October 4,1978

MIDECAMYCIN

Therapeutic Function: Antibacterial

Chemical Name: Leucomycin V, 3,4B-dipropanoate

Common Name: Espinomycin

Structural Formula:

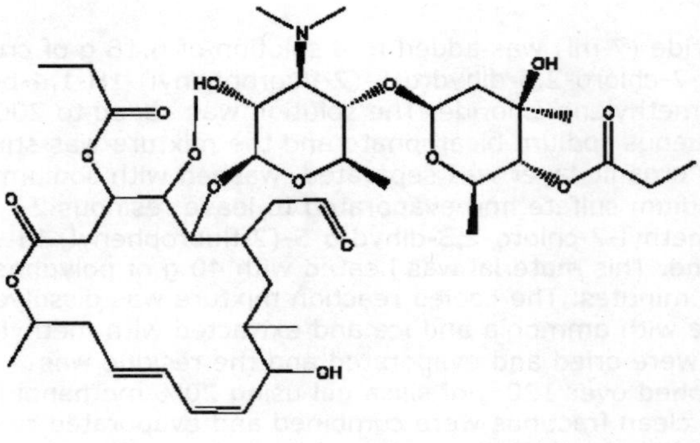

Chemical Abstracts Registry No.: 35457-80-8

Trade Name	Manufacturer	Country	Year Introduced
Medemycin	Meiji Seika	Japan	1974
Midecacine	Clin Midy	France	1978
Midecacine	Clin Midy	Switz.	1980
Midicacin	Midy	Italy	1981
Aboren	Promeco	Argentina	-
Macro-Dil	Roussel	-	-

Raw Materials

Bacterium *Streptomyces mycsrofaciens*
Starch
Vegetable protein

Manufacturing Process

The SF-837 strain, namely *Streptomyces mycarofaciens* identified as ATCC No. 21454 was inoculated to 60 liters of a liquid culture medium containing 2.5% saccharified starch, 4% soluble vegetable protein, 0.3% potassium chloride and 0.3% calcium carbonate at pH 7.0, and then stir-cultured in a jar-fermenter at 28°C for 35 hours under aeration. The resulting culture was filtered directly and the filter cake comprising the mycelium cake was washed with dilute hydrochloric acid.

The culture filtrate combined with the washing liquid was obtained at a total volume of 50 liters (potency 150 mcg/ml). The filtrate (pH 8) was then extracted with 25 liters of ethyl acetate and 22 liters of the ethyl acetate phase was concentrated to approximately 3 liters under reduced pressure. The concentrate was diluted with 1.5 liters of water, adjusted to pH 2.0 by addition of 5N hydrochloric acid and then shaken thoroughly. The aqueous phase was separated from the organic phase and this aqueous solution was adjusted to pH 8 by addition of 3N sodium hydroxide and then extracted with 800 rnl of ethyl acetate. The resulting ethyl acetate extract was then shaken similarly together with 500 ml of aqueous hydrochloric acid to transfer the active substances into the latter which was again extracted with 400 ml of ethyl ether at pH 8.The ether extract was dried with anhydrous sodium sulfate and concentrated under reduced pressure to give 16.5 g of light yellow colored powder.

12 g of this crude powder were dissolved in 200 ml of ethyl acetate and the solution was passed through a column of 600 ml of pulverized carbon which had been impregnated with ethyl acetate. The development was carried out using ethyl acetate as the solvent and the active fractions of eluate were collected to a total volume of 2,500 ml, which was then evaporated to dryness under reduced pressure to yield 5 g of a white colored powder. This powder was dissolved in 10 ml of benzene and the insoluble matters were filtered out. The filtered solution in benzene was then subjected to chromatographic isolation by passing through a column of 700 ml of silica gel which had been impregnated with benzene. The development of the active substances adsorbed on the silica gel was effected using a solvent system consisting of benzene-acetone (4:1), and the eluate was collected in fractions of each 20 ml. The active fractions No. 90-380 which gave a single spot in alumina thin layer chromatography and which could be recognized as containing the SF-837 substance purely in view of the Rf-value of the single spot were combined together to a total volume of 4,000 ml, and then concentrated under reduced pressure to yield 1.5 g of white colored powder of a melting point of 122°C to 124°C which was found by analysis to be the pure SF-837 substance free base.

References

Merck Index 6057
Kleeman and Engel p. 601
DOT 10 (2) 62 (1974)
I.N. p. 635
Tsuruoka,T., Shomura.T., Ezaki, N., Akita, E., Inoue, S., Fukatsu, S., Amano,S., Watanabe, H. and Niida, T.; US Patent 3,761,588; September 25,1973; assigned to Meiji Seika Kaisha, Ltd. (Japan)

MIDODRINE

Therapeutic Function: Peripheral vasotonic, Antihypotensive

Chemical Name: 2-Amino-N-[2-(2,5-dimethoxyphenyl)-2-hydroxyethyl]-acetamide

Common Name: -

Structural Formula:

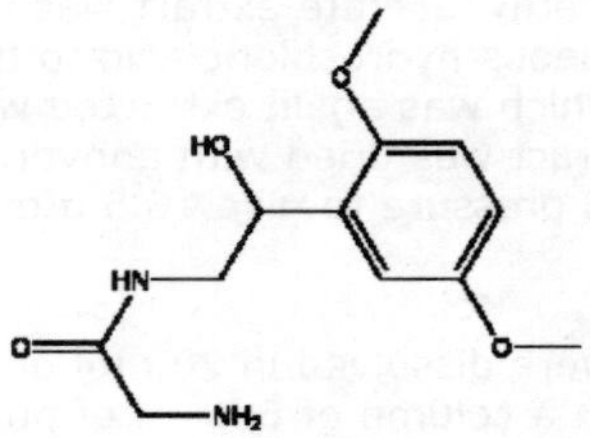

Chemical Abstracts Registry No.: 42794-76-3; 3092-17-9 (Hydrochloride salt)

Trade Name	Manufacturer	Country	Year Introduced
Gutron	Hormonchemie	W. Germany	1977
Gutron	Chemie Linz	Italy	1981
Alphamine	Centerchem	US	-

Raw Materials

Carbobenzoxyglycine
Isovaleric acid chloride
1-(2',5'-Dimethoxyphenyl)-2-aminoethanol-(1)
Hydrogen

Manufacturing Process

19.5 parts of carbobenzoxyglycine, 7.1 parts of triethylamine and 162 parts of dry toluene are mixed with 11.2 parts of isovaleric acid chloride at 0°C to form the mixed anhydride and the mixture is agitated for two hours at 0°C. 32.4 parts of 1-(2',5'-dimethoxyphenyl)-2-aminoethanol-(1) are then added, the mixture is agitated for four hours at a temperature between 0°C and +10°C and then left to stand overnight at that temperature. A thick crystal paste forms. The reaction product is dissolved in 450 parts of ethyl acetate and 200 parts of water. The ethyl acetate solution is separated, washed with hydrochloric acid, sodium bicarbonate solution and water, dried over sodium sulfate and inspissated. The inspissation residue is digested with 342 parts of xylene, the required product crystallizing out. 34.9 parts of 1-(2',5'-dimethoxyphenyl)-2-(N-carbobenzoxyglycineamido)-ethanol-(1) are obtained.

66.2 parts of 1-(2',5'-dimethoxyphenyl)-2-(N-carbobenzoxyglycineamido)-ethanol-(1) are hydrogenated in the presence of 6.6 parts of palladium carbon (10%) in 2,000 parts of glacial acetic acid. When no more hydrogen is absorbed (3 mols of hydrogen are used), hydrogenation stops. The catalyst is removed by suction and the equivalent quantity of hydrochloric acid in ethanol is added to the filtrate with agitation. During further agitation at room temperature 28.6 parts of crude 1-(2',5'-dimethoxyphenyl)-2-glycineamidoethanol-(1)hydrochloride crystallize, and are isolated and recrystallized from water-methanol for purification. 22.1 parts of pure product are obtained with a melting point of 192°C to 193°C.

An alternative synthesis route is described by Kleeman and Engel.

References

Merck Index 6058
Kleeman and Engel p. 602
DOT 18 (10 530 (1982)
I.N. p. 636 Wismayr, K., Schmid, O., Kilches, R. and Zolss, G.; US Patent 3,340,298; September 5, 1967; assigned to Oesterreichische Stickstoffwerke A.G. (Austria)

MIFEPRISTONE

Therapeutic Function: Antiprogesterone

Chemical Name: Estra-4,9-dien-3-one, 11-(4-(dimethylamino)phenyl)-17-hydroxy-17-(1-propynyl)-, (11β,17β)-

Common Name: Mifepristone

Structural Formula:

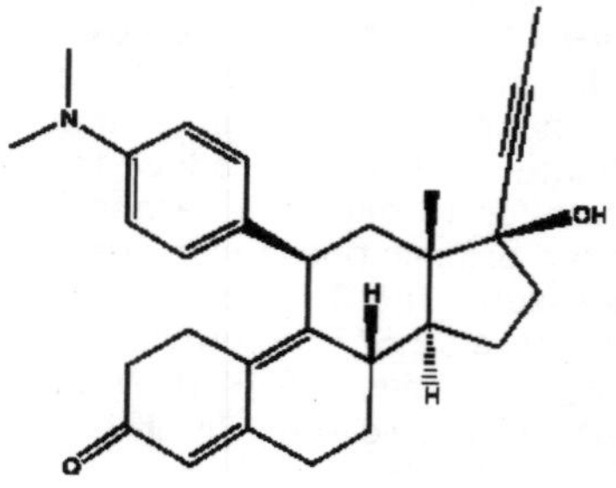

Chemical Abstracts Registry No.: 84371-65-3

Trade Name	Manufacturer	Country	Year Introduced
Mifegyne	HMR	-	-
Mifegyne	Exelgyn	-	-

Raw Materials

4-(N,N-Dimethylaminoethoxy)bromobenzene
1,2-Dibromoethane
Dimethylsulfide-cuprous bromide complex
3,3-[1,2-(Ethanediyl-bisoxy)]-5α,10α-epoxy-17α-prop-1-ynyl-δ(9(11))-estrene-17β-ol

Manufacturing Process

1st method of synthesis of mifepristone:

A solution of 24 g of 4-(N,N-dimethylaminoethoxy)bromobenzene was added dropwise over 45 min to magnesium in 90 ml of anhydrous tetrahydrofuran. 2 ml of 1,2-dibromoethane were added as catalyst. After the addition, the mixture was stirred at 25°C for one hour to obtain a solution of 0.7 M of 4-(N,N-dimethylaminoethoxy)-benzene magnesium bromide which was then added to a solution of 6.16 g of dimethylsulfide-cuprous bromide complex in 20 ml of tetrahydrofuran. The mixture was stirred at room temperature for 20 min and a solution of 3.7 g of 3,3-[1,2-(ethanediyl-bisoxy)]-5α,10α-epoxy-17α-prop-1-ynyl-δ(9(11))-estrene-17β-ol in 50 ml of tetrahydrofuran was added thereto dropwise over a few minutes. The mixture was stirred under an inert atmosphere for one hour and was then poured into a solution of 15 g of ammonium chloride in 20 ml of iced water. The mixture was extracted with ether and the organic phase was washed with aqueous saturated sodium chloride solution, was dried and evaporated to dryness under reduced pressure. The 18.3 g of oil were chromatographed over silica gel and eluted with chloroform to obtain 4.5 g of 3,3-[1,2-ethanediyl-bisoxy]-11β-[4-(N,N-dimethylaminoethoxy)phenyl]-17α-(prop-1-ynyl)-δ^9-estrene-5α,17β-diol with a specific rotation of $[\alpha]_D^{20}$ =-44(+/-)1.5° (c = 1% in chloroform).

9.5 ml of 2 N hydrochloric acid were added to a solution of 4.5 g of 3,3-[1,2-ethanediyl-bisoxy]-11β-[4-(N,N-dimethylaminoethoxy)phenyl]-17α-(prop-1-ynyl)-δ^9-estrene-5α,17β-diol in 20 ml of methanol and the solution was stirred at room temperature for 2 hours. 260 ml of ether and 110 ml of an aqueous saturated sodium bicarbonate solution were added to the mixture which was stirred at room temperature for 15 min. The decanted aqueous phase was extracted with ether and the organic phase was dried and evaporated to dryness under reduced pressure. The 3.3 g of residue were chromatographed over silica gel and eluted with a 92.5/7.5 methylene chloride-methanol mixture to obtain 1.8 g of amorphous 11β-[4-(N,N-dimethylaminoethoxy) phenyl]-17α-(prop-1-ynyl)-δ^{4,9}-estradiene-17β-ol-3-one with a specific rotation of $[\alpha]_D^{20}$ =+71° (c = 1% in chloroform).

2th method of synthesis of mifepristone (see scheme):

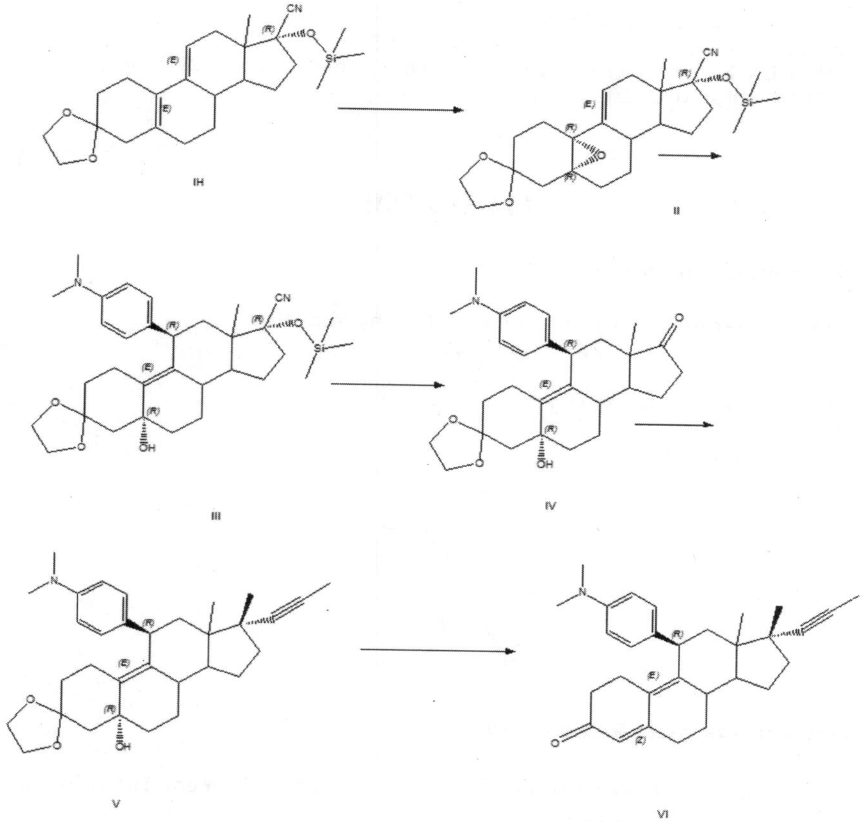

The oxidation of the diene I, which constitutes an intermediate for total synthesis of 19-nor steroids, with a reagent prepared from trifluoroacetic anhydride/hydrogen peroxide was obtained exclusively α-epoxide II. The condensation of II with the Grignard reagent from 4-bromo-N,N-dimethylaniline results in addition of the reagent at the 11β-position. This results in rearragement of the olefin to 9,10 and opening of the epoxide. The stereochemistry of the product obtained III is consistent with trans-opening of the oxirane, albeit at a remove of two carbon atoms. Mild hydrolysis removes the silyl cyanohydrin protecting group at the 17-position to give a ketone IV. Reaction of the ketone with propargyl lithium leads to V. Hydrolysis of that product under more strenuous condition results in removal of the acetal at 3; the resulting β-hydroxyketone then dehydrates to afford the 4,10(9)-dienone VI. Another name of VI is estra-4,9-dien-3-one, 11-(4-(dimethylamino) phenyl)-17-hydroxy-17-(1-propynyl)-, (11β,17β)- or mifepristone.

References

Merck Index, Monograph number: 6273, Twelfth edition, 1996, Editor: S. Budavari; Merck and Co., Inc.

Teutsch J.G. et al.; US Patent No. 4,386,085; May 31, 1983; Assigned to Roussell Uclaf, Paris, France

Velluz L. et al.; Compt. Rend., 257, 569 (1963)

Lednicer D.; Ed 'Chronicles of Drug Discovery', Vol. 3, p.1., ACS Books, Washington, DC, 1993, p.1

MIGLITOL

Therapeutic Function: Glucosidase inhibitor

Chemical Name: 3,4,5-Piperidinetriol, 1-(2-hydroxyethyl)-2-(hydroxymethyl)-, (2R-(2α,3β,4α,5β))-

Common Name: Glycet; Miglitol

Structural Formula:

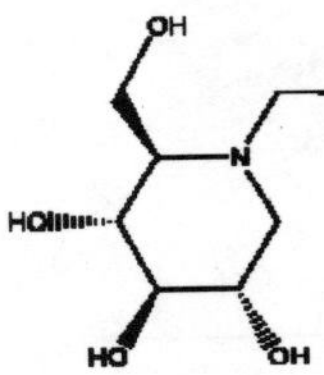

Chemical Abstracts Registry No.: 72432-03-2

Trade Name	Manufacturer	Country	Year Introduced
Miglitol	SMS Pharmaceuticals Limited	-	-
Miglitol	ZYF Pharm Chemical	-	-
Diastabol	Sanofi-Synthelabo	-	-
Diastabol	Sanofi Aventis	-	-
Diastabol	Bayer Pharma	-	-
Glyset	Pharmacia and Upjohn	-	-

Raw Materials

Triethylamine	6-Amino-6-desoxy-L-sorbose hydrochloride
Sodium borohydride	Dimethylaminoborane
Lewatit	1,5-Didesoxy-1,5-imino-D-glucitol of 6-amino-6-desoxy-L-sorbose hydrochloride

Manufacturing Process

50 g (0.23 mole) of 6-amino-6-desoxy-L-sorbose hydrochloride were dissolved

in 500 ml of distilled water, and the solution was added in the course of one hour to a solution of 11.2 g of dimethylaminoborane in 500 ml of distilled water, whilst stirring at a temperature of 50°C. The mixture was stirred for one hour at room temperature and one hour at 50°C, 5 ml of triethylamine were added to it, and it was then poured over a column containing 800 ml of strongly basic ion exchanger ("Lewatit" MP 500 OH--form). The exchanger was washed with distilled water, and the runnings collected were concentrated to a syrup on a rotary evaporator. The concentrated syrup was crystallised at 50°C on addition of a large amount of ethanol. The suspension of crystals was cooled and filtered off under suction, and the crystalline product was dried in a vacuum drying cabinet. Yield: 30 g, 80% of theory. MP: 192°-193°C.

25 g (0.115 mole) 1,5-didesoxy-1,5-imino-D-glucitol of 6-amino-6-desoxy-L-sorbose hydrochloride were dissolved in 200 ml of distilled water, and the solution was added at 5°C to a mixture of 4.8 g of $NaBH_4$, 250 ml of ethanol/water 1:1 and 16.2 ml of triethylamine, whilst stirring. The mixture was further stirred for one hour at room temperature and one hour at 50°C., and the reaction mixture was poured over a column containing 400 ml of strongly basic ion exchanger ("Lewatit" MP 500 OH-form). The exchanger was washed with distilled water, and the eluate collected was concentrated to a syrup in a rotary evaporator. Th syrup was taken up with 200 ml of distilled water, and the mixture was poured over a column containing 400 ml of acid ion exchanger ("Lewatit" S 100 H+-form). The column was rinsed with distilled water, and the product was eluted with 10% strength ammonia water. The runnings, rendered alkaline with ammonia, were collected and were concentrated to a syrup in a rotary evaporator. The syrup was crystallized, whilst warm, with a large amount of ethanol, and the suspension of crystals was cooled and is filtered off under suction, and the crystalline product was dried in a vacuum drying cabinet. Yield of 1,5-dideoxy-1,5-((2-hydroxyethyl)imino)-D-glucitol 14 g, 74% of theory. MP: 192°-193°C.

References

Koebernick W.; US Patent No. 4,611,058; September 9, 1986; Assigned to Bayer Aktiengesellschaft (Leverkusen, DE)

MILRINONE LACTATE

Therapeutic Function: Cardiotonic

Chemical Name: Propanoic acid, 2-hydroxy-, compd. with 1,6-dihydro-2-methyl-6-oxo(3,4'-bipyridine)-5-carbonitrile

Common Name: Milrinone lactate

Chemical Abstracts Registry No.: 100286-97-3; 78415-72-2 (Base)

Trade Name	Manufacturer	Country	Year Introduced
Primacor	Baxter Healthcare Corporation	-	-

Structural Formula:

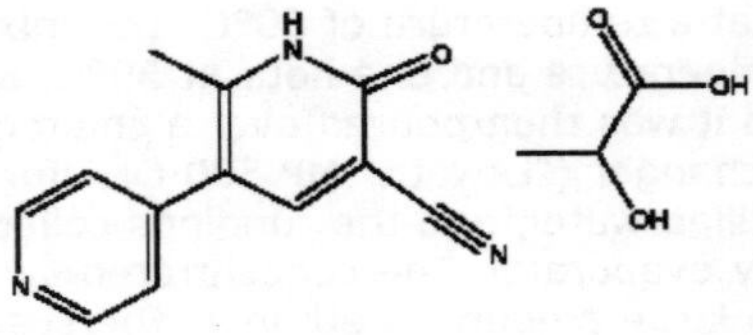

Raw Materials

α-Cyanoacetamide	1-(4-Pyridinyl)-2-propanone
Sodium methoxide	Dimethylformamide dimethyl acetal
Malononitrile	

Manufacturing Process

A mixture containing 20 g of 1-(4-pyridinyl)-2-propanone and 30 ml of hexamethylphosphoramide was diluted with 65 ml of dimethylformamide dimethyl acetal and the resulting mixture was refluxed for 30 min. TLC analysis showed a single spot, thereby indicating completion of the reaction (in another run, the reaction appeared to be complete after 30 min at room temperature). The mixture was evaporated under reduced pressure and a pressure, thereby resulting in a crystalline residue weighing 24 g. The residue was purified by continuous chromatographic extraction on alumina (about 150 g) using refluxing chloroform as eluant. After 90 min, the extract was heated in vacuo to remove the chloroform, thereby leaving, as a light yellow crystalline material, 23.2 g of 1-(4-pyridinyl)-2-(dimethylamino)ethenyl methyl ketone, alternatively named 4-dimethylamino-2-(4-pyridinyl)-3-buten-2-one.

To a mixture containing 23 g of 1-(4-pyridinyl)-2-(dimethylamino)ethenyl methyl ketone and 11 g of α-cyanoacetamide dissolved in 400 ml of dimethylformamide was added with stirring 14 g of sodium methoxide and the resulting reaction mixture was heated in an oil bath under gentle reflux for one hour. TLC analysis showed no starting material in the reaction mixture which was then concentrated in vacuo on a rotary evaporator to a volume of about 80 ml. The concentrate was treated with about 160 ml of acetonitrile and the resulting mixture was stirred on a rotary evaporator with warming until homogenous and then cooled. The crystalline product was collected, rinsed successively with acetonitrile and ether, and dried overnight at 55°C to yield 28 g of crystalline product, namely, sodium salt of 1,2-dihydro-6-methyl-2-oxo-5-(4-pyridinyl)nicotinonitrile, the presence of cyano being confirmed by IR analysis. An 8 g portion of said sodium salt was dissolved in 75 ml of hot water, the aqueous solution treated with decolorizing charcoal, filtered, the filtrate again treated with decolorizing charcoal and filtered, and the filtrate acidified with 6 N hydrochloric acid by dropwise addition to a pH of 3. The acidic mixture was diluted with ethanol and cooled. The crystalline product was collected, dried, recrystallized from dimethylformamide-water and dried to produce 3.75 g of 1,2-dihydro-6-methyl-2-oxo-5-(4-pyridinyl)nicotinonitrile, m.p. >300°C.

Another method of preparation of 1,2-dihydro-6-methyl-2-oxo-5-(4-

pyridinyl)nicotinonitrile (Patent US 4,413,127)

A 69.5 g portion of 1-ethoxy-2-(4-pyridinyl)ethenyl methyl ketone was dissolved in 300 ml of ethanol and to the solution was added 13.2 g of malononitrile. The resulting mixture was refluxed for 5 hours, crystals starting to separate after about 30 min of refluxing. The reaction mixture was allowed to cool to room temperature and, the precipitate of fine needles was filtered, washed with ethanol and dried in a vacuum at 90°C to yield 25.4 g of 1,2-dihydro-6-methyl-2-oxo-5-(4-pyridinyl)-nicotinonitrile, m.p. >300°C. Concentration of the mother liquor provided another 2.1 g of product, m.p. >300°C.

To a aqueous solution of 1,2-dihydro-6-methyl-2-oxo-5-(4-pyridinyl) nicotinonitrile was added one molar equivalent of lactic acid to prepare the monolactate of 1,2-dihydro-6-methyl-2-oxo-5-(4-pyridinyl)nicotinonitrile.

References

Lesher G. Y. et al.; US Patent No. 4,413,127; Feb. 2, 1982; Assigned to Sterling Drug Inc. (New York, NY)

Lesher G.Y. et al.; US Patent No. 4,313,951; Feb. 2, 1982; Assigneed to Sterling Drug Inc. (New York, NY)

MINAPRINE

Therapeutic Function: Antidepressant

Chemical Name: 3-(2-Morpholinoethylamino)-4-methyl-6-phenylpyridazine dihydrochloride

Common Name: -

Structural Formula:

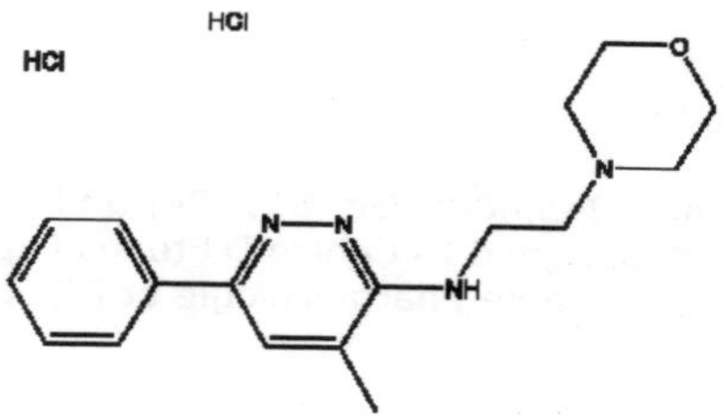

Chemical Abstracts Registry No.: 25905-77-5; 25953-17-7 (Dihydrochloride salt)

Trade Name	Manufacturer	Country	Year Introduced
Cantor	Clin Midy	France	1979
Kantor	Gador	Argentina	1983

Raw Materials

3-Chloro-4-methyl-6-phenylpyridazine
N-(2-Aminoethyl)morpholine
Hydrogen chloride

Manufacturing Process

(a) Preparation of the free base: A mixture comprising 0.1 mol (20.4 g) of 3-chloro-4-methyl-6-phenylpyridazine and 0.2 mol (26.2 g) of N-(2-aminoethyl)-morpholine in 100 ml of n-butanol, with a pinch of copper powder, was heated under reflux for 12 hours. At the end of this time, the hot solution was poured into 200 ml of cold water. The resulting mixture was filtered through a sintered glass filter and the precipitate washed with ether. The filtrate and the ether washings were placed in a separating funnel and extracted with two 150 ml portions of ether. The ethereal layer was then extracted with about 250 ml of N sulfuric acid.

The acid solution was made alkaline with a 10% aqueous solution of sodium carbonate, and left to crystallize overnight.

The solution was filtered, yielding the colorless needles which were recrystallized from isopropanol. The yield was 15 g (53%).

(b) Preparation of the hydrochloride: The base was dissolved in the smallest amount possible of anhydrous acetone. Double that volume of anhydrous ether was added, and a stream of hydrogen chloride gas was passed through the solution. The hydrochloride salt obtained was recrystallized from absolute alcohol. The yield after recrystallization was 17 g (90%).

References

Merck Index 6066
DFU 2 (12) 811 (1977)
Kleeman and Engel p. 602
I.N. p. 637
Laborit, H.; British Patent 1,345,880; Feb. 6, 1974; and US Patent 4,169,158; Sept. 25, 1979; both assigned to Centre D'Etudes Experimentales et Cliniques de Physiobiologie de Pharmacologie et D'Eutonologie (C.E.P.B.E.P.E.)

MINOCYCLINE

Therapeutic Function: Antibiotic

Chemical Name: 4,7-Bis(dimethylamino)-1,4,4a,5,5a,6,11,12a-octahydro-3,10,12,12a-tetrahydroxy-1,11-dioxo-2-naphthacenecarboxamide

Common Name: 7-Dimethylamino-6-demethyl-6-deoxytetracycline

Structural Formula:

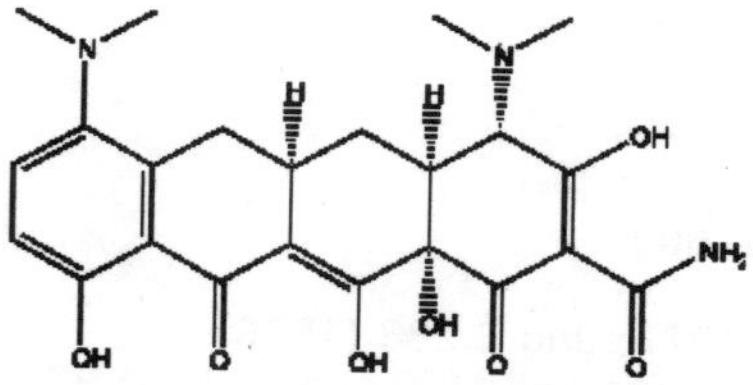

Chemical Abstracts Registry No.: 10118-90 8; 13614-98-7 (Hydrochloride salt)

Trade Name	Manufacturer	Country	Year Introduced
Minocin	Lederle	US	1971
Minomycin	Lederle	Japan	1971
Klinomycin	Lederle	W. Germany	1972
Minocin	Lederle	Italy	1972
Minomycin	Takeda	Japan	1972
Vectrin	Parke Davis	US	1973
Minocin	Lederle	UK	1973
Mynocine	Lederle	France	1973
Ultramycin	Parke Davis	-	-

Raw Materials

6-Demethyltetracycline
Dibenzyl azodicarboxylate
Hydrogen

Manufacturing Process

Preparation of 7-(N,N'-Dicarbobenzyloxyhydrazino)-6-Demethyltetracycline: A 1.0 g portion of 6-demethyltetracycline was dissolved in a mixture of 9.6 ml of tetrahydrofuran and 10.4 ml of methanesulfonic acid at -10°C. The mixture was allowed to warm to 0°C. A solution of 0.86 g of dibenzyl azodicarboxylate in 0.5 ml of tetrahydrofuran was added dropwise and the mixture was stirred for 2 hours while the temperature was maintained at 0°C. The reaction mixture was added to ether. The product was filtered off, washed with ether and then dried. The 7-(N,N'-dicarbobenzyloxyhydrazino)-6-demethyltetracycline was identified by paper chromatography.

Reductive Methylation of 7-(N,N'-Dicarbobenzyloxyhydrazino)-6-Demethyl-6-Deoxytetracycline to 7-Dimethylamino-6-Demethyl-6-Deoxytetracycline: A solution of 100 mg of 7(N,N'-dicarbobenzyloxyhydrazino)-6-demethyl-6-deoxytetracycline in 2.6 ml of methanol, 0.4 ml of 40% aqueous formaldehyde solution and 50 mg of 5% palladium on carbon catalyst was hydrogenated at room temperature and two atmospheres pressure. Uptake of the hydrogen was complete in 3 hours. The catalyst was filtered off and the solution was taken to dryness under reduced pressure. The residue was triturated with ether and then identified as 7-dimethylamino-6-demethyl-6-deoxytetracycline by comparison with an authentic sample, according to US Patent 3,483,251.

References

Merck Index 6068
Kleeman and Engel p. 603
PDR p. 1018
OCDS Vol. 1 p. 214 (1977) and 2,288 (1980)
DOT 5 (2) 75 (1969); 7 (5) 188 (1971) and 8 (3) 93 (1972)
I.N. p.637
REM p. 1206
Boothe, J.H. and Petisi, J.; US Patent 3,148,212; September 8,1964; assigned to American Cyanamid Company
Petisi, J. and Boothe, J.H.; US Patent 3,226,436; December 28,1965; assigned to American Cyanamid Company
Winterbottom, R., Bitha, P. and Kissman, H.M.; US Patent 3,345,410; October 3,1967; assigned to American Cyanamid Company
Zambrano, R.T.; US Patent 3,403,179; September 24,1968; assigned to American Cyanamid Company
Zambrano, R.T.; US Patent 3,483,251 ;December 9,1969; assigned to American Cyanamid Company

MINOXIDIL

Therapeutic Function: Antihypertensive

Chemical Name: 6-Amino-1,2-dihydro-1-hydroxy-2-imino-4-piperidinopyrimidine

Common Name: -

Structural Formula:

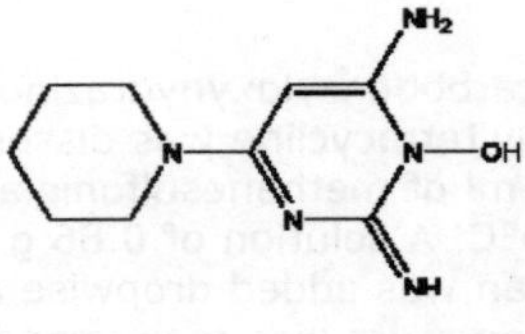

NH_2 N N–OH N NH

Chemical Abstracts Registry No.: 38304-91-5

Raw Materials

Barbituric acid	2,4,6-Trichloropyrimidine
Ammonia	m-Chloroperbenzoic acid
Piperidine	Phosphorus oxychloride

Trade Name	Manufacturer	Country	Year Introduced
Loniten	Upjohn	US	1979
Loniten	Upjohn	UK	1980
Loniten	Upjohn	Switz.	1981
Loniten	Upjohn	W. Germany	1982
Loniten	Upjohn	Italy	1983
Prexidil	Bioindustria	Italy	1983

Manufacturing Process

Barbituric acid is reacted with phosphorus oxychloride then with 2,4,6-trichloropyrimidine and that product with ammonia to give 4-chloro-2,6-diaminopyritnidine.

A 30 g (0.15 mol) quantity of 4-chloro-2,6-diaminopyrimidine is dissolved in 600 ml of hot 3A alcohol, the solution cooled to 0°C to 10°C and 41.8 g (0.24 mol) of m-chloroperbenzoic acid is added. The mixture is held at 0°C to 10°C for 4 hours and filtered. The solid is shaken for 2 hours in 0.24 mol of 10% sodium hydroxide and filtered. The solid is washed with water and dried to yield 193 g of crude product. This product is extracted for 1 hour with 900 ml of boiling acetonitrile to yield 14.8 g (44.7% yield) of 6-amino-4-chloro-1,2-dihydro-1-hydroxy-2-iminopyrimidine, melting point 193°C.

A mixture of 3.0 g (0.019 mol) of 6-amino-4-chloro-1,2-dihydro-1-hydroxy-2-iminopyrimidine and 35 ml of piperidine is refluxed for 1.5 hours, cooled and filtered. The solid is shaken for 20 minutes in a solution of 0.8 g of sodium hydroxide in 30 ml of water and filtered. The solid is washed with water and extracted with 800 ml of boiling acetonitrile and filtered to yield 3.5 g (89%) yield of 6-amino-4-chloro-1,2-dihydro-1-hydroxy-2-iminopyrimidine, melting point 248°C, decomposition at 259°C to 261°C.

References

Merck Index 6069
DFU 2 (6) 383 (1977)
Kleeman and Engel p. 604
PDR p. 1848
OCDS Vol. 1 p. 262 (1977)
DOT8 (7) 277 (1972) and 16 (9) 298 (1980)
I.N.p. 638
REM p. 848
Anthony, W.C. and Ursprung, J.J.; US Patents 3,382,247; May 7,1968 and 3,382,248; May 7,1968; both assigned to The Upjohn Co.
Anthony, W.C.; US Patent 3,644,364; February 22,1972; assigned to The Upjohn Co.

MIRTAZAPINE

Therapeutic Function: Antidepressant, Antihistaminic, Antidiuretic

Chemical Name: Pyrazino[2,1-a]pyrido[2,3-c](2)benzazepine, 1,2,3,4,10,14b-hexahydro-2-methyl-

Common Name: Azamiaserin; Mepirzapin; Mirtazapine

Structural Formula:

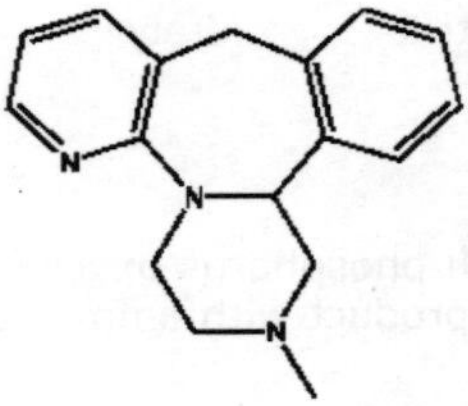

Chemical Abstracts Registry No.: 61337-67-5

Trade Name	Manufacturer	Country	Year Introduced
Mirazep	Syncro (A Div. of Microlabs)	India	-
Mirt	Panacea Biotec Ltd.	India	-
Mirtaz	Sun Pharmaceuticals Industries Ltd.	India	-
Remeron	Organon	Netherlands	-
Zispin	Triton	-	-

Raw Materials

Potassium fluoride
2-Chloronicotinonitrile
Sulfuric acid
1-Methyl-3-phenylpiperazine
Lithium aluminum hydride

Manufacturing Process

1) 1-(3-Cyanopyridyl-2)-2-phenyl-4-methylpiperazine

17.43 g (0.3 mol) potassium fluoride is added to a solution of 13.85 g (0.1 mol) 2-chloronicotinonitrile and 17.62 g (0.1 mol) 1-methyl-3-phenylpiperazine in 250 ml dry DMF and the suspension is heated at 140°C under a nitrogen atmosphere for 20 hours. After cooling, the reaction mixture is poured into 1,250 ml water. The aqueous phase is extracted four times with ethyl acetate the combined organic extracts are washed with 100 ml water. After drying, the extracts are evaporated. The crude oil may be used as such for the following step. The nitrile obtained may however also be purified by column chromatography on SiO_2, with hexane-acetone (95:5). In this way, 21.9 g (79%) pure 1-(3-cyanopyridyl-2)-4-methyl-2-phenylpiperazine is obtained; the substance crystallizes from petroleum ether; melting point 66.5-67.5°C.

2) 1-(3-Carboxypyridyl-2)-2-phenyl-4-methylpiperazine

The solution of 19.5 g (0.07 mol) 1-(3-cyanopyridyl-2)-4-methyl-2-phenylpiperazine in 390 ml of a solution of 25 g KOH/ 100 ml ethanol is heated at 100°C for 24 hours. After cooling, water (390 ml) is added. The alcohol is evaporated under vacuum and the cloudy solution remaining is extracted twice with 100 ml methylene chloride. The residual aqueous phase is cooled and the pH is adjusted to 7 with 2 N HCl, after which it is extracted with chloroform. After drying the chloroform extract, it is evaporated and 16.2 g 1-(3-carboxypyridyl-2)-4-methyl-2-phenylpiperazine is obtained as a colourless oil. Crystallization from ethanol gives a crystalline substance with a melting point of 161-162°C.

3) 1-(3-Hydroxymethylpyridyl-2)-2-phenyl-4-methylpiperazine

20.4 g (0.07 mol) 1-(3-carboxypyridyl-2)-2-phenyl-4-methylpiperazine is dissolved in 300 ml dry THF and gradually added to a boiling suspension of 20.4 g LiAlH4in 600 ml dry THF under a nitrogen atmosphere. The mixture is boiled for 4 hours, after which it is cooled in an ice-bath and decomposed by adding 81.6 ml water. The inorganic salts are filtered off. The filtrate is dried and solvent is removed by evaporation, giving a yield of 18.39 g (93%) 1-(3-hydroxymethylpyridyl-2)-2-phenyl-4-methylpiperazine. Recrystallization from ether gives a crystalline product (white needles) of melting point 124-126°C.

4)2-Methyl-1,2,3,4,10,14b-hexahydro-benzo[c]pyrazino-[1,2-a]-pyrido[2,3-c]azepine(Mirtazapine)

6.5 ml concentrated sulfuric acid is added dropwise at room temperature to 3.25 g of 1-(3-hydroxymethylpyridyl-2)-2-phenyl-4-methylpiperazine. During the addition, the temperature rises to 35°C. The whole is subsequently stirred for a few hours, after which 60 g ice is added and the mixture is made alkaline with concentrated ammonia (22 ml). The reaction mixture is then extracted with chloroform. The chloroform extracts are dried and concentrated. The crude reaction product crystallizes when ether is added, and the solid obtained is recrystallized from petroleum ether. Yield of 2-methyl-1,2,3,4,10,14b-hexahydro-benzo[c]pyrazino-[1,2-a]-pyrido[2,3-c]azepine 2.43 g; melting point: 114-116°C.

References

Maeda, et al.; US Patent No. 6,660,730; Dec. 9, 2003; Assigned to Sumika Fine Chemicals Co., Ltd.

MISOPROSTOL

Therapeutic Function: Antiulcer

Chemical Name: Prost-13-en-1-oic acid, 11,16-dihydroxy-16-methyl-9-oxo-, methyl ester, (11α,13E)-(+-)-

Common Name: Misoprostol

Structural Formula:

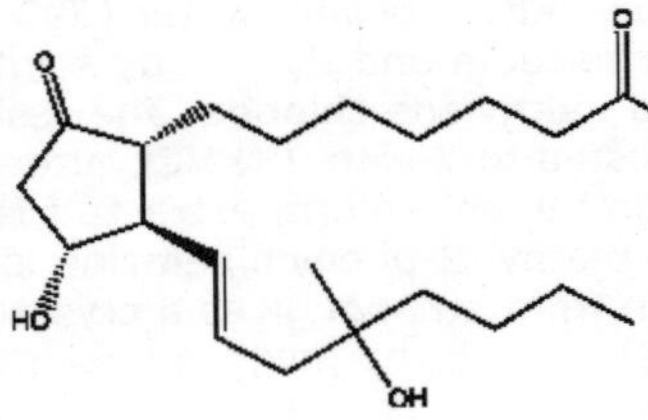

Chemical Abstracts Registry No.: 59122-46-2; 62015-39-8

Trade Name	Manufacturer	Country	Year Introduced
Cytotec	Searle, division of Monsanto plc.	UK	-
Misoprost	Cipla Limited	India	-
Misoprostol	IVAX	-	-
Misoprostol	G.D. Searle and Co.	USA	-

Raw Materials

(E)-Trimethyl[[1-methyl-1-[3-(tributylstannyl)-2-propenyl]pentyl]oxy]silane
Copper (I) iodide
Methyllithium
Methyl-5-oxo-3-[(triethylsilyl)oxy]-1-cyclopentene-1-heptanoate

Manufacturing Process

1 method of synthesis

To a 1000 ml dried flask under a nitrogen atmosphere was added 74.6 g of (E)-trimethyl-[[1-methyl-1-[3-(tributylstannyl)-2-propenyl]pentyl]oxy]silane, 125 ml anhydrous THF and 24.2 g of copper (I) iodide. The mixture was stirred at room temperature for 30 minutes and then it was cooled to -25 to -30°C. 98.8 ml of methyllithium (2.86 M) in DEM was added dropwise and the resultant solution was stirred at -15°C for 2 hours. Then the reaction mixture was cooled to -78°C and 25 g of methyl-5-oxo-3-[(triethylsilyl)oxy]-1-cyclopentene-1-heptanoate in 100 ml of THF was added rapidly. After stirring the mixture for 5 min at -78°C, it was quenched into a mixture of 750 ml of aqueous ammonium chloride solution and 200 ml of ammonium hydroxide. The resulting mixture was warmed to room temperature and stirred until a deep blue aqueous layer was obtained. Ethyl acetate (250 ml) was used for extraction. Then the combined organic layers were washed with brine and subsequently dried over magnesium sulfate. After a filtration and concentration under reduced pressure, an oil (105 g) was obtained. This oil containing the protected prostaglandin was subjected to acidic deprotection (cat. PPTS, acetone and water) and purification (chromatography on silica gel) to provide 15.8 g (60%) of misoprostol was identical.

2 method of synthesis

To a 300 ml dried flask under a nitrogen atmosphere was added 4.45 g of copper (I) iodide and 60 ml of anhydrous THF. The mixture was cooled to 0°C 35 ml of 1.4 M methyllithium in diethyl ether was added dropwise and the resultant solution was stirred at 0°C for 30 min. 13.7 g of (E)-trimethyl-[[1-methyl-1-[3-(tributylstannyl)-2-propenyl]pentyl]oxy]silane in 5 ml of THF was added and then the mixture was stirred at 0°C for 30 min. Then an additional 1.5 ml of 1.4 M methyllithium in diethyl ether was added and the mixture was stirred for another 30 min. The reaction mixture was cooled to -78°C and 10 g of methyl 5-oxo-3-[(triethylsilyl)oxy]-1-cyclopentene-1-heptanoate in 10 ml of THF was added rapidly. After stirring the mixture for 5 min at -78°C, it was quenched into 210 ml of basic aqueous ammonium chloride solution. The resulting mixture was warmed to room temperature and stirred until a deep blue aqueous layer was obtained. Ethyl acetate was used for extraction. Then the combined organic layers were washed with water (10 ml), then with brine (25 ml) and subsequently dried over magnesium sulfate. After a filtration and concentration under reduced pressure, an oil (21 g) was obtained. This oil containing the protected prostaglandin was subjected to acidic deprotection (cat. PPTS, acetone and water) and purification (chromatography on silica gel) to provide 4.2 g (40%) misoprostol.

References

Li Y.-F. et al.; US Patent No. 5,684,177; Nov. 4, 1997; Assigned to Torcan Chemical Ltd.

"Organometallics in Synthesis: A Manual", Chapter 4, page 283-382; B. H. Lipshutz, Edited by M Schlosser, John Wiley and Sons, 1994

Lipshutz B.H., Synthesis, 325 (1987)

MITOBRONITOL

Therapeutic Function: Cancer chemotherapy

Chemical Name: 1,6-Dibromo-1,6-dideoxy-D-mannitol

Common Name: -

Structural Formula:

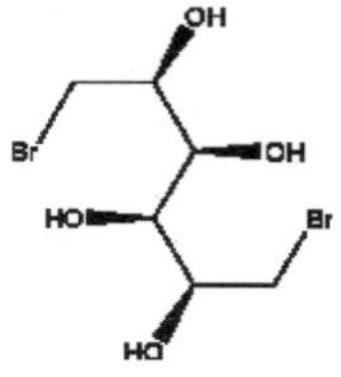

Chemical Abstracts Registry No.: 488-41-5

Trade Name	Manufacturer	Country	Year Introduced
Myelobromol	Hormonchemie	W. Germany	1967
Myelobromol	Berk	UK	1970
Myebrol	Kyorin	Japan	1978

Raw Materials

D-Mannitol
Hydrogen bromide

Manufacturing Process

750 g D-mannitol are dissolved in 4,000 ml of a 48% aqueous hydrogen bromide solution, whereupon the solution thus obtained is saturated at 0°C with gaseous hydrogen bromide until a HBr content of 69 to 70% is achieved. The reaction mixture is heated for 6 hours at 60°C in an autoclave, is then decolorized with charcoal, extracted with 1 liter chloroform twice and diluted with 7 liters of water. The pH value of the solution is adjusted by means of sodium bicarbonate to1 to 2. The crystals precipitated after cooling for a day are filtered and washed with water until free from acid. 250 g crude 1,6-dibromo-1,6-didesoxy-D-mannitol are obtained. MP 176° to 178°C. Analysis: Br % = 52 (calc.: 51.9).

250 g of the crude DBM are dissolved in 2.5 liters of hot methanol and on decolorizing and filtration 2.5 liters of dichloroethane are added. 220 g of crystalline DBM are obtained. MP 178°C. Br % = 51.9.

References

Merck Index 6076
Kleeman and Engel p. 604
I.N. p. 639
REM p. 1156
Chinoin Gyogyszeres Vegyeszeti Termekek Gyarart; British Patent 959,407; June 3,1964

MITOMYCIN

Therapeutic Function: Cancer chemotherapy

Chemical Name: Azirino(2',3':3,4)pyrrolo(1,2-a)indole-4,7-dione, 6-amino-8-(((aminocarbonyl)oxy)methyl)-1,1a,2,8,8a,8b-hexahydro-8a-methoxy-5-methyl-, (1aS,8S,8aR,8bS)-

Common Name: -

Chemical Abstracts Registry No.: 50-07-7

Structural Formula:

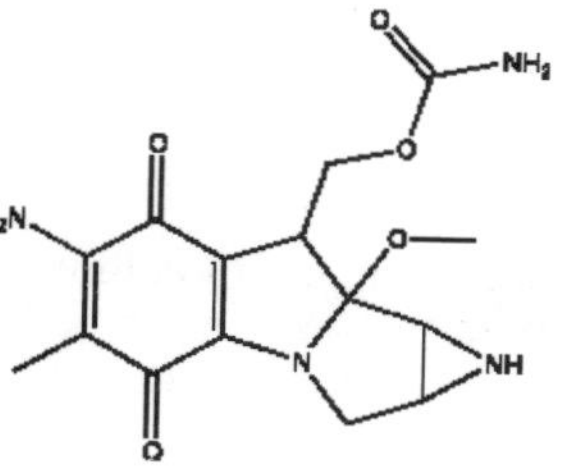

Trade Name	Manufacturer	Country	Year Introduced
Mitomycin	Medac	W. Germany	1960
Mitomycin C	Kyowa	Italy	1961
Ametycine	Choay	France	1970
Mitamycin	Bristol	US	1974
Mitomycin C	Kyowa	Japan	1980
Mitamycin	Bristol	Sweden	1983
Mitomycin C	Syntex	Switz.	1983

Raw Materials

Bacterium Streptomyces caespitosus
Nutrient broth

Manufacturing Process

The commercial production of mitomycin involves the preparation of mitomycin-containing broths by culturing a mitomycin-producing organism, e.g. Streptomyces caespitosus, in suitable media as described at length in the literature. At the end of the fermentation cycle the whole broth is usually centrifuged, filtered or otherwise treated to separate the solids (mycelia) from the supernatant which contains substantially all of the antibiotic activity.

In commercial processes there is usually a period of time intervening between the end of the fermentation cycle and the time at which the mycelia is actually removed from the broth; such a period may range from several minutes to several hours in length and may be due to a number of factors, e.g., the time necessary to conduct the actual centrifugation or filtration of large quantities of broth, or the time involved in waiting for equipment to become available for use. In the commercial preparation of mitomycin, the mitomycin-containing whole broths decrease rapidly in potency during the time following the completion of the fermentation cycle and prior to the removal of the mycelia. It has been observed that a whole broth will lose substantially all of its mitomycin activity within about 6 hours at room temperature and within about 24 hours at 10°C. It has, however, been discovered, as described in US Patent 3,042,582, that in the process for the recovery of mitomycin C from mitomycin C-containing whole broth, the step of adding about 0.1 wt % with whole broth of sodium lauryl sulfate to the whole broth at the completion of the fermentation cycle substantially eliminates such destruction of mitomycin C by mitase.

References

Merck Index 6079
Kleeman and Engel p. 604
PDR p. 724
I.N. p.640
REM p. 1156
Gourevitch, A., Chertow, B. and Lein, J.; US Patent 3,042,582; July 3,1962; assigned to Bristol-Myers Company

MITOPODOZIDE

Therapeutic Function: Antineoplastic

Chemical Name: 5,6,7,8-Tetrahydro-8-hydroxy-7-(hydroxymethyl)-5-(3,4,5-trimethoxyphenyl)naphtho[2,3d]-1,3-dioxole-6-carboxylic acid-2-ethylhydrazide

Common Name: Podophyllinic acid 2-ethylhydrazide

Structural Formula:

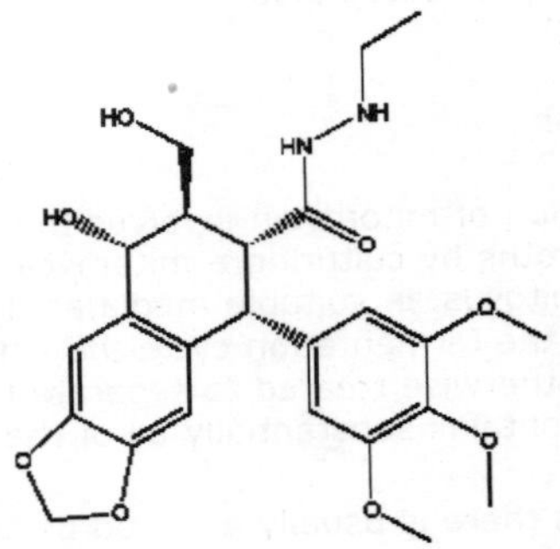

Chemical Abstracts Registry No.: 1508-45-8

Trade Name	Manufacturer	Country	Year Introduced
Proresid	Sandoz	W. Germany	1966
Proresid	Sankyo	Japan	1969

Raw Materials

Podophyllinic acid hydrazide
Acetaldehyde
Hydrogen

Manufacturing Process

500 g of podophyllinic acid hydrazide are heated together with 150 cc of acetaldehyde with 2,200 cc of methanol to 40°C. The solution obtained is filtered and then cooled. The product which crystallizes out is filtered off with suction and washed with methanol. Together with a second fraction obtained after concentration of the mother liquors there are produced 450 g of podophyllinic acid ethylidene hydrazide, having a melting point of 222°C to 224°C and a specific rotation of $[\alpha]_D$ = -285° (c. = 0.5 in ethanol).

The product is hydrogenated in 4,000 cc of ethanol at room temperature and under normal atmospheric pressure with a catalyst prepared in the usual manner from 400 g of Raney nickel alloy. The calculated amount of hydrogen is taken up in approximately 75 hours. After filtration and evaporation to a small volume, the residue is distributed between 1,000 cc of chloroform and water each. The chloroform solution is then dried over sodium sulfate and evaporated toa small volume. Precipitation of the hydrogenation product with petroleum ether yields an amorphous white powder which is filtered by suction, washed with petroleum ether and dried at 50°C in a high vacuum. 1-ethyl-2-podophyllinic acid hydrazide is obtained in a practically quantitative yield.

References

Merck Index 7414
Kleeman and Engel p. 605
I.N. p. 640
Rutschmann, J.; US Patent 3,054,802; September 18, 1962; assigned to Sandoz Ltd. (Switzerland)

MITOTANE

Therapeutic Function: Antineoplastic

Chemical Name: Ethane, 2-(o-chlorophenyl)-2-(p-chlorophenyl)-1,1-dichloro-

Common Name: Mitotane

Structural Formula:

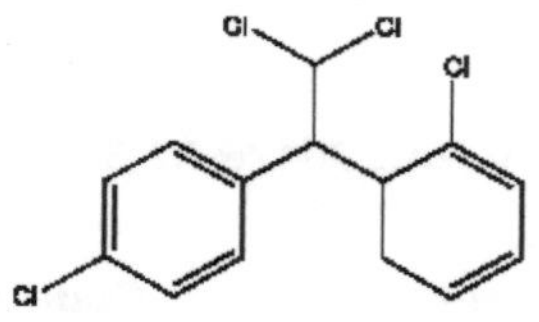

Chemical Abstracts Registry No.: 53-19-0

Trade Name	Manufacturer	Country	Year Introduced
Lysodren	Bristol Labs, Division of Bristol-Myers Squibb Canada Inc.	Canada	-

Raw Materials

Dichloroacetaldehyde
2-Chlorphenylmagnesiumbromide

Manufacturing Process

From dichloroacetaldehyde and 2-chlorphenylmagnesiumbromide was prepared 1-(2-chlorphenyl-2,2-dichloroethanol. By action of H_2SO_4 on 1-(2-chlorphenyl)-2,2-dichloroethanol in chlorobenzene was prepared 1,1-dichloro-2,2-bis(2,4'-dichlorophenyl)ethane.

References

Haller B.L. et al.; JACS 1945, 67, 1591

MITOXANTRONE DIHYDROCHLORIDE

Therapeutic Function: Antineoplastic

Chemical Name: 9,10-Anthraquinone, 5,8-bis((2-((2-hydroxyethyl)amino) ethyl)amino)-1,4-dihydroxy-, dihydrochloride

Common Name: Mitoxantrone hydrochloride; Mitozantrone hydrochloride

Structural Formula:

HO, H, N, NH, O, OH, HCl, HCl, HO, N, H, NH, O, OH

Chemical Abstracts Registry No.: 70476-82-3; 65271-80-9 (Base)

Trade Name	Manufacturer	Country	Year Introduced
Novantrone	Immunex Corporation	-	-
Oncotrone	Baxter Oncology GmbH	Germany	-

The above acid is placed in a 20 L reaction vessel with 6 L of water. 1.1 liters of soda lye (d = 1.33) and 1.848 kg of sodium bicarbonate are added. 2.1 L of dimethyl sulfate are added. After one hour, crystallisation is induced. Filtration, drying without heat and washing are carried out. Methyl benzhydrylsulphinylacetate is obtained.

1 kg of methyl benzhydrylsulphinylacetate is dissolved in 3.5 liters of anhydrous methanol in a 10-liter balloon flask. NH_3 is bubbled in at a high rate of flow for 1 hour, and then left in contact for 4 hours. Filtration, drying without heat and washing with water are then carried out. By recrystallisation from a mixture of water and methanol (4:1) and then from a mixture of water and methanol (9:1) and drying under reduced pressure, CRL 40476 is obtained in the form of a white crystalline powder; melting point 164-166°C. Total yield (calculated from the benzhydrol): 41%.

References

Lafon L.; US Patent No. 4,177,290; Dec. 4, 1989; Assigned to Laboratoire L. Lafon

MOEXIPRIL HYDROCHLORIDE

Therapeutic Function: Antihypertensive

Chemical Name: 3-Isoquinolinecarboxylic acid, 1,2,3,4-tetrahydro-6,7-dimethoxy-2-(2-((1-(ethoxycarbonyl)-3-phenylpropyl)amino)-1-oxopropyl)-, monohydrochloride, (3S-(2(R*(R*)),3R*))-

Common Name: Moexipril hydrochloride

Structural Formula:

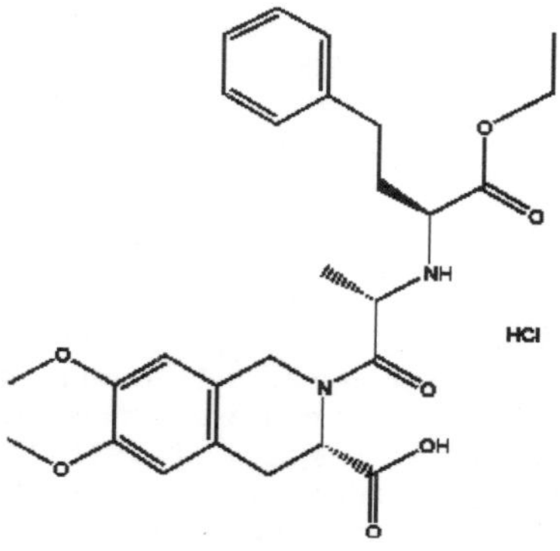

Chemical Abstracts Registry No.: 82586-52-5; 103775-10-6 (Base)

Trade Name	Manufacturer	Country	Year Introduced
Moexipril Hydrochloride	Schwarz Pharma	Germany	-
Perdix	Schwarz Pharma	Germany	-
Primox Tablets	Minipharm	-	-
Univasc	Teva Pharmaceuticals	USA	-
Univasc	Schwarz Pharma	USA	-

Raw Materials

t-Butyl alanine
Triethylamine
1-Hydroxybenzotriazole
Diclohexylcarbodiimide
Ethyl 2-bromo-4-phenylbutanoate
1,2,3,4-Tetrahydro-3-isoquinolinecarboxylic acid (S-form)

Manufacturing Process

1) A solution of 2.0 g of t-butyl alanine (S-form) and 3.78 g of ethyl 2-bromo-4-phenylbutanoate in 25 ml of DMF was treated with 1.8 ml of triethylamine and the solution was heated at 70°C for 18 hours. The solvent was removed at reduced pressure and the residue was mixed with water and extracted with ethyl ether. The organic layer was washed with water and dried over magnesium sulfate. Concentration of the solvent at reduced pressure gave the oily ethyl-α-[(1-carboxyethyl)amino]benzene-t-butanoate.

A solution of 143.7 g of this t-butyl ester in 630 ml of trifluoroacetic acid was stirred at room temperature for one hour. The solvent was removed at reduced pressure and the residue was dissolved in ethyl ether and again evaporated. This operation was repeated. Then the ether solution was treated dropwise with a solution of hydrogen chloride gas in ethyl ether until precipitation ceased. The solid, collected by filtration, was a mixture of diastereoisomers of ethyl-α-[(1-carboxyethyl)amino]benzenebutanoate hydrochloride, melting point 153-165°C; $[\alpha]_D^{23}$ = +3.6° (1% MeOH).

The free amino acid (S,S-form) was prepared by treatment of an aqueous solution of the hydrochloride with saturated sodium acetate. The product was filtered, washed efficiently with cold water and recrystallized from ethyl acetate; melting point 149-151°C; $[\alpha]_D^{23}$ = +29.7°.

2) A stirred solution of 0.0158 mole of ethyl-α-[(1-carboxyethyl)amino] benzenebutanoate hydrochloride in 200 ml of methylene chloride was treated successively with 1.60 g (0.0158 mole) of triethylamine, 0.0158 mole of 1-hydroxybenzotriazole, 0.0158 mole of 1,2,3,4-tetrahydro-6,7-dimethoxy-3-isoquinolinecarboxylic acid and then with 0.0158 mole of dicyclohexylcarbodiimide in 10 ml of methylene dichloride. Dicyclohexylurea gradually separated. The mixture was allowed to stand at room temperature overnight. Hexane (300 ml) was added and the urea was filtered. The filtrate was washed with 250 ml of saturated sodium bicarbonate, dried over sodium sulfate and concentrated to remove solvent. The viscous residue was triturated with 50 ml of ether and filtered to remove insolubles. The filtrate was concentrated to give 2-[2-[[1-(ethoxycarbonyl)-3-phenylpropyl]amino]-1-oxopropyl]-1,2,3,4-tetrahydro-6,7-dimethoxy-3-isoquinolinecarboxylic acid.

After addition of hydrochloric acid was obtained 2-[2-[[1-(ethoxycarbonyl)-3-phenylpropyl]amino]-1-oxopropyl]-1,2,3,4-tetrahydro-6,7-dimethoxy-3-isoquinolinecarboxylic acid, hydrochloride.

References

Hoefle M.L., Klutchko S.; US Patent No. 4,344,949; August 17, 1982; Assigned: Warner-Lambert Company

O'Reilly N. J., Lin H. C.; US Patent No. 4,912,221; March 27, 1990; Assigned: Occidental Chemical Corporation

Wang Z.-X., Horne S.E.; US Patent No. 6,642,384; Nov. 4, 2003; Assigned: Brantford Chemicals Inc.

MOFEBUTAZONE

Therapeutic Function: Antirheumatic, Analgesic, Antiinflammatory

Chemical Name: 3,5-Pyrazolidinedione, 4-butyl-1-phenyl-

Common Name: Mofebutazone; Monophenylbutazone; Mophebutazonum

Structural Formula:

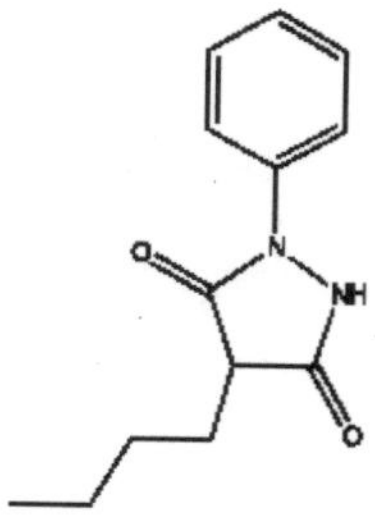

Chemical Abstracts Registry No.: 2210-63-1

Trade Name	Manufacturer	Country	Year Introduced
Butaphen	Wiedenmann	-	-
Ecasil	Laquifa	-	-
Metrogyl	Benzon	-	-
Monofen	Star	-	-

Raw Materials

Phenyl hydrazine
n-Butylmalonic acid

Manufacturing Process

A mixture comprising 108 g of phenyl hydrazine and 216 g of the diethyl ester of n-butylmalonic acid is heated on an oil bath at 170°-180°C for 12 hours. The residue is taken up with water in which an alkaline compound has been dissolved and acetic acid is added to precipitate 4-n-butyl-2-phenyl-pyrazolidine-3,5-dione. The product is a white crystalline solid having a MP: 103°C. It is soluble in acetone and benzene, soluble in hot condition in methanol and ethanol and insoluble in water.

References

Commissionara Farmaceutica Milaneze, an Italian Company, Milan Italy; G.B. Patent No. 839,057; Nov. 28, 1956

MOLINDONE

Therapeutic Function: Antipsychotic

Chemical Name: 3-Ethyl-1,5,6,7-tetrahydro-2-methyl-5-(4-morpholinylmethyl)-4H-indol-4-one

Common Name: -

Structural Formula:

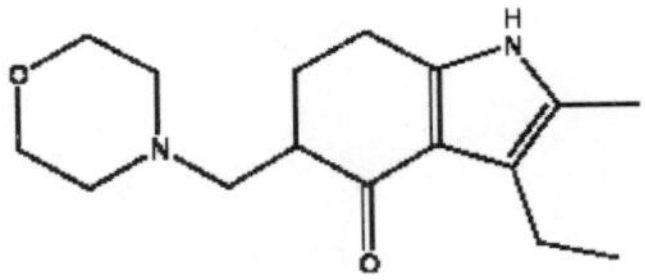

Chemical Abstracts Registry No.: 7416-34-4; 15622-65-8 (Hydrochloride salt)

Trade Name	Manufacturer	Country	Year Introduced
Moban	Endo	US	1974
Lidone	Abbott	US	1977

Raw Materials

Diethyl ketone
Paraformaldehyde
Morpholine hydrochloride
Cyclohexan-1,3-dione
Methyl nitrite

Manufacturing Process

Diethyl ketone may be reacted with methyl nitrite and that product in turn reacted with cyclohexan-1,3-dione to give 3-ethyl-4,5,6,7-tetrahydro-2-methyl-4-oxoindole.

3-ethyl-4,5,6,7-tetrahydro-2-methyl-4-oxoindole 14.1 g (0.08 mol), 14.8 g morpholine hydrochloride (0.12 mol), and 3.6 g paraformaldehyde (0.12 mol) were refluxed in 200 ml ethanol for 40 hours. The solution was evaporated to dryness in vacuo on a steam bath and the residue digested with a mixture of 150 ml water and 10 ml 2N HCl. An insoluble residue of unreacted starting material was filtered off. To the acid solution, ammonia water was added dropwise with stirring and the amine crystallized out. It was purified by dissolving in 1N HCl and addition of ammonia, then by 2 crystallizations from benzene followed by 2 crystallizations from isopropanol, to yield 3-ethyl-4,5,6,7-tetrahydro-2-methyl-4-oxoindole, melting point 180°C to 181°C.

References

Merck Index 6086
Kleeman and Engel p. 606
PDR p. 856
OCDS Vol. 2 p.455 (1980)
DOT 5 (1) 34 (1969); 9 (6) 233 (1973) and 11 (2) 60 (1975)
I.N. p. 642
REM p. 1092
Pachter, I.J. and Schoen, K.; US Patent 3,491,093; January 20, 1970; assigned to Endo Laboratories, Inc.

MOLSIDOMINE

Therapeutic Function: Coronary vasodilator

Chemical Name: Sydnone imine, N-carboxy-3-morpholino-, ethyl ester

Common Name: Molsidomine; Morsydomine

Structural Formula:

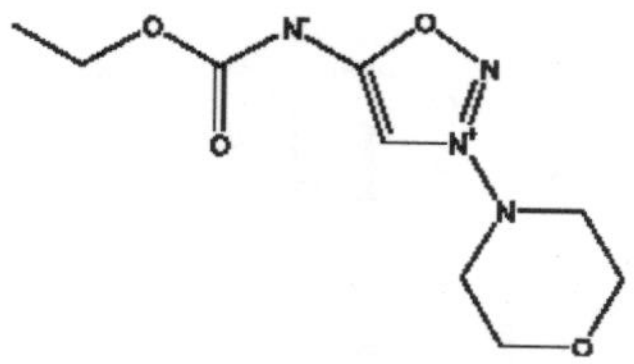

Chemical Abstracts Registry No.: 25717-80-0

Trade Name	Manufacturer	Country	Year Introduced
Corvasal	Hoechst	-	-

Raw Materials

3-Morpholinosydnonimine hydrochloride
Ethyl chloroformate
Pyridine

Manufacturing Process

1.6 parts by volume of ethyl chloroformate is stirred dropwise into a suspension of 1.0 part by weight of 3-morpholinosydnonimine hydrochloride in 5 parts by volume of pyridine, and the mixture is agitated for a while to allow reaction to take place. Pyridine is removed from the reaction mixture by evaporation, and the residue is dissolved in a small amount of water and extracted with chloroform several times. The extract is dehydrated by adding anhydrous magnesium sulfate and subjected to filtration. Chloroform is removed from the filtrate by distillation, crude crystals being obtained. Recrystallization of the crude crystals from toluene gives 0.6 part by weight of 3-morpholino-N-carboethoxysydnonimine having a melting point of 140°-141°C. Yield 51%.

References

Masuda K. et al.; US Patent No. 3,812,128; May 21, 1974; Assigned to Takeda Chemical Industries, Ltd., Osaka, Japan

MOMETASONE FUROATE

Therapeutic Function: Glucocorticoid

Chemical Name: Pregna-1,4-diene-3,20-dione, 9,21-dichloro-17-((2-furanylcarbonyl)oxy)-11-hydroxy-16-methyl-, (11β,16α)-

Common Name: Mometasone furoate

Structural Formula:

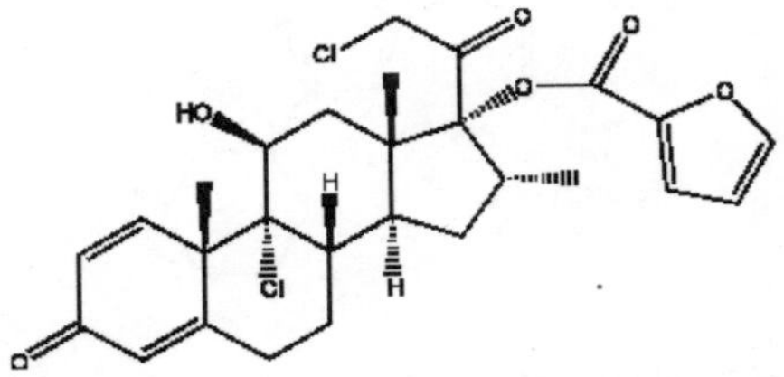

Chemical Abstracts Registry No.: 83919-23-7

Trade Name	Manufacturer	Country	Year Introduced
Asmanex	Novartis Communications	Switz.	-
Asmanex Twisthaler	Schering-Plough	Belgium	-
Cutizone-T	Crosland Research Laboratories	India	-
Elocon	Fulford GALT (India) Ltd.	India	-
Elocon	Schering-Plough	Belgium	-
Nasonex	Schering-Plough	Belgium	-

Raw Materials

9β,11β-Epoxy-17α,21-dihydroxy-16α-methyl-1,4-pregnadiene-3,20-dione
Mesyl chloride
4-Dimethylaminopyridine
2-Furoyl chloride
1,3-Dichloro-5,5-dimethylhydantoinyl

Manufacturing Process

METHOD I (Patent U.S. 4,472,393)

A. 21-Chloro-9β,11β-epoxy-17α-hydroxy-16α-methyl-1,4-pregnadiene-3,20-dione

Prepare a solution of 5.0 g. of 9β,11β-epoxy-17α,21-dihydroxy-16α-methyl-1,4-pregnadiene-3,20-dione in 20 ml of dry pyridine. Cool on an ice bath; to the stirred solution under nitrogen, add dropwise 1.1 ml of mesyl chloride. Remove the ice bath and continue stirring at room temperature for 30 min. Add 2.0 g of lithium chloride and continue stirring for a further 150 min. Add to a mixture of 150 ml ethyl acetate and 100 ml distilled water. Wash the organic phase with dilute 3% aqueous hydrochloric acid, then saturated aqueous sodium chloride solution and finally saturated sodium bicarbonate solution. Dry the organic phase over magnesium sulfate, filter and remove the solvent to give 4.62 g of 21-chloro-9β,11β-epoxy-17α-hydroxy-16α-methyl-1,4-pregnadiene-3,20-dione.

B. 21-Chloro-9β,11β-epoxy-17α-hydroxy-16α-methyl-1,4-pregnadiene-3,20-dione 17-(2'-furoate)

Prepare under argon a solution of 8 g of 4-dimethylaminopyridine in 250 ml of dry methylene chloride. Cool on an ice bath and add to the stirred solution 6.0 ml of 2-furoyl chloride. Remove from the ice bath, allow the temperature to rise to room temperature and then add 11.5 g of the 21-chloro-9β,11β-epoxy-17α-hydroxy-16α-methyl-1,4-pregnadiene-3,20-dione. After 24 hours add 500 ml of ethyl acetate saturated with water. Filter off the precipitate and then evaporate off the solvent to give the crude 21-chloro-9β,11β-epoxy-17α-hydroxy-16α-methyl-1,4-pregnadiene-3,20-dione 17-(2'-furoate).

C. 9α,21-Dichloro-11β,17α-hydroxy-16α-methyl-1,4-pregnadiene-3,20-dione 17-(2'-furoate)

To the crude 21-chloro-9β,11β-epoxy-17α-hydroxy-16α-methyl-1,4-pregnadiene-3,20-dione 17-(2'-furoate) add 50 ml of glacial acetic acid, then add a solution of 3.5 g of anhydrous hydrogen chloride in 125 ml of glacial acetic acid. Stir for 15 minutes and then quench with 500 ml of distilled water. Filter off the solids, recrystallise from methanol:water, dry for 24 hours under vacuum to give 12.6 g 9α,21-dichloro-11β,17α-hydroxy-16α-methyl-1,4-pregnadiene-3,20-dione 17-(2'-furoate) (yield 83% of theory).

Prepare under nitrogen a solution of 1.80 g of 21-chloro-17α-hydroxy-16α-methyl-1,4,9(11)-pregnatriene-3,20-dione 17-(2'-furoate). Add, with stirring, a solution of 1.15 ml of 70% perchloric acid in 2.53 ml of distilled water, and immediately thereafter 604 mg of 1,3-dichloro-5,5-dimethylhydantoin. Stir the reaction mixture for twenty minutes and then raise the temperature to ambient temperature. Monitor the consumption of starting material by thin layer chromatography of aliquots using chloroform:1,3-dichloro-5,5-dimethylhydantoinyl acetate (9:1) and hexane:ethyl acetate (1:1). When the starting material is consumed, pour the reaction mixture into 500 ml of distilled water containing the 1,3-dichloro-5,5-dimethylhydantoin and 7 g of sodium bisulphite. Add sodium chloride until the solution is saturated. Filter the precipitated solid, wash and dry at 50°C under vacuum. Purify the resulting crude product by preparative chromatography on 1000 micron silica gel plates using chloroform: ethyl acetate (19:1). Elute the desired band with ethyl acetate, filter the eluate and evaporate at room temperature to give crude product (1.3 g). Recrystallize the product by dissolving in refluxing methylene chloride, filtering and then replacing the methylene chloride at reflux with methanol and then the methanol with distilled water. Cool the suspension to room temperature, filter and dry under vacuum at 50°C to give the pure pregna-1,4-diene-3,20-dione, 9,21-dichloro-17-((2-furanylcarbonyl)oxy)-11-hydroxy-16-methyl-, (11β-,16α)-.

METHOD II

The present invention (Patent U.S. 6,177,560) refers to a new process for the preparation of mometasone furoate carried out by esterifiication of the 17 hydroxy group of mometasone without prior protection of the 11 hydroxy group. Mometasone (30 g) was suspended in methylene chloride (300 ml) and the resulting suspension was cooled to 0-5°C. At this temperature triethylamine (57 ml) was added. 2-Furoyl chloride (24 ml) was then added slowly. The mixture was then stirred at 8-12°C until the level of mometasone present was lower than 0.2% by HPLC. The reaction solution was then cooled to between -5-5°C and water (120 ml) was added with stirring. After stirring for 1 hour at 10-15°C the mixture was cooled to between 0-5°C and concentrated hydrochloric acid was added to adjust the pH of the aqueous layer between 1 and 2.

The phases were separated and the aqueous layer was extracted with methylene chloride (60 ml). To the combined organic layers concentrated hydrochloric acid (90 ml) and acetic acid (30 ml) was added at a temperature 15-25°C. Then the two phase reaction mixture was stirred until less than 0.1% of the side products remained as monitored by HPLC. The reaction mixture was cooled to 0-5°C and water (120 ml) was added. The lower organic layer was separated, water (120 ml) and 8 N aqueous sodium hydroxide solution (about 30 ml) were added to adjust the pH to between 5 and 6. After stirring for 2 hours the organic layer was separated and washed with water (120 ml). The organic solution [containing the mometasone 17-(2-furoate)] was concentrated by distillation to a volume of 120 ml. Further methanol (120 ml) was added and the mixture was concentrated to 120 ml. This procedure was repeated twice more. The reaction mixture was slowly cooled to 0-5°C and stirred for 2 hours. The crude mometasone 17-(2-furoate) was then filtered off and washed with cold methanol.

Purification of mometasone 17-(2-furoate)

The wet cake was dissolved in acetone (395 ml) and charcoal (3 g) was added. After 24 hours, the charcoal was filtered off and washed with acetone (90 ml). Charcoal (3 g) was added to the solution and the solution stirred for at least 24 hours at between 15-25°C. The charcoal was then filtered off and washed with acetone (75 ml). The solution was concentrated by distillation to a volume of 120 ml. During this concentration the mometasone 17-(2-furoate) started to crystallise. Methanol (120 ml) was added and the solution was again concentrated to 120 ml. This procedure was repeated twice. The suspension was cooled slowly to 0-5°C and stirred for about 2 hours at this temperature. The pure mometasone 17-(2-furoate) was then filtered off and washed with cold methanol. The product was dried at 60-70°C. A yield of 29.92 g was obtained.

References

Shapiro E.L.; US Patent No. 4,472,393; Sep. 18, 1984; Assigned: Schering Corporation (Kenilworth, NJ)

Heggie W., Bandarra J.; US Patent No. 6,177,560; Jan. 23, 2001; Assigned to Hovione Inter Ltd.

MONTELUKAST SODIUM

Therapeutic Function: Anti-asthmatic

Chemical Name: Cyclopropaneacetic acid, 1-((((1R)-1-(3-((1E)-2-(7-chloro-2-quinolinyl)ethenyl)phenyl)-3-(2-(1-hydroxy-1-methylethyl)phenyl) propyl)thio)methyl)-, monosodium salt

Common Name: Montelukast sodium

Structural Formula:

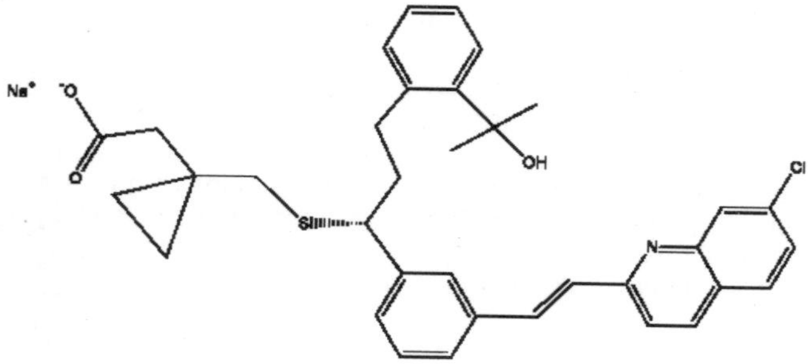

Na+ -O

O

OH

Cl

N

S

Chemical Abstracts Registry No.: 151767-02-1; 158966-92-8 (Base)

Trade Name	Manufacturer	Country	Year Introduced
Singulair	Merck Pharmaceutical Corporation	Canada	-

Raw Materials

Crotonaldehyde
4-Chloroaniline
Chloranil
Palladium (II) acetate
Cerium (III) chloride
Thionyl chloride
Sodium cyanide
Triethylamine
Potassium t-butoxide
Methyl 2-(2-iodophenyl)propanoate
Tetrabutylammonium chloride
Diethyl 1,1-cyclopropanedicarboxylate
Vinyl magnesium bromide
Methyl magnesium chloride
Borane-tetrahydrofuran complex
Diisopropylethylamine
Methanesulfonyl chloride

Manufacturing Process

Crotonaldehyde (3.23 mol) in 100 mL of 2-butanol was added dropwise to a refluxing solution of 4-chloroaniline (3.23 mol), p-chloranil (3.23 mol) and HCl conc. (808 mL) in 5.4 L of 2-butanol. After 2 hours of heating 2.7 L of solvent was removed under vacuum at 60°C. Then 2 L of toluene was added to the reaction mixture followed by removal of 2.5-3 L of solvent until a very pasty solid formed. THF (2 L) was added and the mixture heated 30 min after which it was cooled to 0°C. The solid was collected and washed with THF until pure by tlc. The solid was then dissolved in aq. K_2CO_3/EtOAc and the organic phase separated. The aqueous phase was extracted with EtOAc and the organic phases combined, dried over $MgSO_4$ and the solvent removed. The product was crystallized in the minimum amount of EtOAc to give 328.08 g (57%) of 4-chloro-2-methylquinolin.

4-Chloro-2-methylquinalin was converted into 3-(2-(7-chloro)-2-quinolinyl)ethenyl)benzaldehyde. Reaction was carried out according to a method described in U.S. Pat. No. 4,851,409

To a degassed suspension of 3-(2-(7-chloro-2-quinolinyl)ethenyl)benzaldehyde (0.34 mol) in toluene (700 mL) at 0°C was added 1.0 M vinylmagnesium bromide in toluene/THF (370 mL). After stirring for 1 hour at 0°C, the reaction was quenched by the addition of saturated NH_4Cl solution (150 ml), followed by H_2O (500 mL) and HOAc (50 mL). The product was extracted with EtOAc and the two-phase system was filtered through celite to remove an insoluble precipitate. The aqueous phase was then re-extracted with EtOAc (100 mL) and the combined organic layer was washed with H_2O, followed by brine. The solution was dried ($MgSO_4$), and evaporated to give a dark yellow residue which was purified by flash chromatography (EtOAc:hexane 1:5, then 1:3). The product was filtered from the column fractions to give a solid of 1-(3-(2-(7-chloro-2-quinolinyl)ethenyl)phenyl)-2-propen-1-ol (melting point = 110-112°C). The filtrate was concentrated and the resulting residue was recrystallized from EtOAc/hexane 1:4 to give a second crop of 15.1 g.

A degassed suspension of 1-(3-(2-(7-chloro-2-quinolinyl)ethenyl)phenyl)-2-

propen-1-ol (46.6 mmol), n-Bu_4NCl (93 mmol), $LiOAcH_2O$ (115 mmol), LiCl (93 mmol), Pd(OAc)2 (1.4 mmol) and methyl 2-(2-iodophenyl)propanoate in DMF (90 mL) was stirred for 2 hours at 100°C. The dark red solution was then cooled to 0°C and poured into saturated $NaHCO_3$ solution (500 mL). The product was extracted with EtOAc and the organic layer was washed with H_2O followed by brine. The solvent was removed under vacuum and the residue was purified by flash chromatography (EtOAc:hexane 1:10, 1:5 and 3:10) to give a pale yellow foam of ethyl 2-(3(S)-(3-(2-(7-chloro-2-quinolinyl)ethenyl) phenyl)-3-hydroxy-propyl)benzoate (18.9 g).

A mixture of anhydrous $CeCl_3$ (164 mmol) in THF (500 mL) was refluxed overnight using a Dean Stark trap filled with activated molecular sieves. Methyl magnesium chloride (3.0 Molar solution in THF, 790 mmol) was added dropwise over 30 min to the $CeCl_3$ slurry at 0°C. After stirring 2 hours, the mixture was cooled to -5°C and a toluene (600 mL) solution of the ethyl 2-(3(S)-(3-(2-(7-chloro-2-quinolinyl)ethenyl)phenyl)-3-hydroxy-propyl)benzoate (152 mmol) was added dropwise over 1 hour. The reaction mixture was stirred another hour before the addition of 2 M HOAc (600 mL) and toluene (600 mL). The organic layer was washed with saturated aq. $NaHCO_3$ and with brine. Concentration in vacuo and purification of the residue by flash chromatography (30% EtOAc in toluene) gave 63.48 g (91%) of the 2-(2-(3(S)-(3-(2-(7-chloro-2-quinolinyl)ethenyl)phenyl)-3-hydroxypropyl)phenyl)-2-propanol.

To a solution of BH_3THF complex (1 M in THF, 262 mL) was added diethyl 1,1-cyclopropanedicarboxylate (134 mmol) at 25°C under N_2. The solution was heated at reflux for 6 hours, cooled to r.t., and MeOH (300 mL) was cautiously added. The solution was stirred for 1 hour and then concentrated to an oil. The crude 2-(2-(3(S)-(3-(2-(7-chloro-2-quinolinyl)ethenyl)phenyl)-3-hydroxypropyl)phenyl)-2-propanol was dissolved in CH_2Cl_2 (234 mL) and $SOCl_2$ (15.9 g, 134 mmol) was added dropwise over a period of 15 min at 25°C. After stirring for another 15 min, the mixture was washed with aqueous $NaHCO_3$. The organic extract was dried over Na_2SO_4, filtered and concentrated to give quantitatively the 1,1-cyclopropanedimethanol cyclic sulfite.

To a solution of the 1,1-cyclopropanedimethanol cyclic sulfite (99 mmol) in DMF (83 mL) was added NaCN (199 mmol). The mixture was heated to 90°C for 20 hours. Upon cooling, EtOAc (400 mL) was added and the solution was washed with saturated $NaHCO_3$ solution (55 mL), H_2O (4 times 55 mL), saturated NaCl solution and dried over Na_2SO_4. The solution was concentrated to give 7.1 g (65%) of 1-(hydroxymethyl)cyclopropaneacetonitrile.

To a solution of 1-(hydroxymethyl)cyclopropaneacetonitrile (42 g, 378 mmol) in dry CH_2Cl_2 (450 mL) at -30°C was added Et_3N (741 mmol) followed by CH_3SO_2Cl (562 mmol) dropwise. The mixture was warmed to 25°C, washed with $NaHCO_3$, dried over Na_2SO_4 and concentrated in vacuo to give the corresponding mesylate. The mesylate was then dissolved in DMF (450 mL) and cooled to 0°C. Potassium thioacetate (55.4 g, 485 mmol) was added, and the mixture was stirred at 25°C for 18 hours. EtOAc (1.5 L) was added, the solution was washed with $NaHCO_3$, dried over Na_2SO_4 and concentrated in vacuo to give 45 g (70%) of 1-(acetythiomethyl)cyclopropaneacetonitrile.

To a solution of the 1-(acetythiomethyl)cyclopropaneacetonitrile (266 mmol) in MeOH (1.36 L) was added H_2O (84 mL) and conc. H_2SO_4(168 mL). The mixture was heated to reflux for 20 hours, cooled to 25°C, H_2O (1 L) was added and the product was extracted with CH_2Cl_2. The organic extract was washed with H_2O and dried over Na_2SO_4. Concentration of the organic solution gave 36 g (93%) of the methyl 1-(thiomethyl)cyclopropaneacetate.

To a solution of 2-(2-(3(S)-(3-(2-(7-chloro-2-quinolinyl)ethenyl)phenyl)-3-hydroxypropyl)phenyl)-2-propanol in THF was dissolved in THF (1 mL) and DMF (1 mL) at -40°C was added diisopropylethylamine (2.2 mmol) and then methanesulfonyl chloride (2.2 mmol). The mixture was stirred 2 hours with slow warming to -30°C. The methyl 1-(thiomethyl)cyclopropaneacetate (2.3 mmol) was added to the cloudy reaction mixture followed by dropwise addition of potassium tert-butoxide/THF solution (4.4 mmol). The reaction mixture was stirred at -30°C for 3.5 hours before quenching it with 25% aq NH4OAc. Extraction with EtOAc, washing the organic layer with brine and evaporation of the solvents left a residue that was purified by flash chromatography (5%-10% EtOAc in toluene) giving 658 mg (53%) of methyl 1-((((R)-(3-(2-(7-chloro-2-quinolinyl)ethenyl)phenyl)-3-(2-(2-hydroxy-2-propyl)phenyl)propyl)thio)methyl)cyclopropaneacetate.

Following the hydrolysis the methyl 1-((((R)-(3-(2-(7-chloro-2-quinolinyl) ethenyl)phenyl)-3-(2-(2-hydroxy-2-propyl)phenyl)propyl)thio)methyl) cyclopropaneacetate with NaOH was obtained the free acid: 4-((1(R)-(3-(2-(7-chloro-2-quinolinyl)ethenyl)phenyl)-3-(2-(2-hydroxy-2-propyl)-phenyl)propyl) thio)methyl)cyclopropaneacetic acid or sodium 1-(((1(R)-(3-(2-(7-chloro-2-quinolinyl)ethenyl)phenyl)-3-(2-(2-hydroxy-2-propyl)phenyl)propyl)thio) methyl) cyclopropaneacetate.

References

Zamboni R. et al.; US Patent No. 5,270,324; Dec. 14, 1993; Assigned to Merck Frosst Canada, Inc.

MOPERONE HYDROCHLORIDE

Therapeutic Function: Neuroleptic

Chemical Name: 4'-Fluoro-4-(4-hydroxy-4-p-tolylpiperidino)butyrophenone hydrochloride

Common Name: Moperone hydrochloride; Methylperidol

Chemical Abstracts Registry No.: 3871-82-7

Trade Name	Manufacturer	Country	Year Introduced
Luvatren	Cilag	-	-
Moperone hydrochloride	Janssen Pharmaceutica	-	-

Structural Formula:

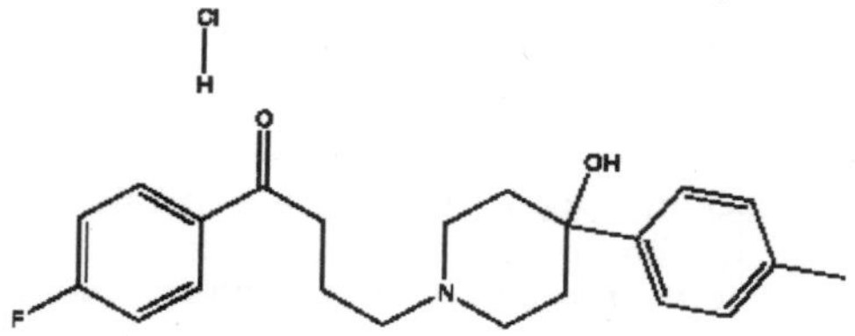

Raw Materials

Methyl bromide
Magnesium

Manufacturing Process

A solution of 95 parts of methyl bromide in 356 parts of ether was added portionwise to a refluxing suspension of 24 parts of magnesium in 214 parts of ether. The mixture was refluxed for 2 hours, and 92 parts of 4-methylacetophenone were added in the course of 90 minutes. The refluxing was continued for 3 hours, and the mixture was stirred for 24 hours at room temperature. The Grignard complex was destroyed by the addition of ammonium chloride and 10% hydrochloric acid. The mixture was extracted with ether and the ether extracts were washed with 10% sulfuric acid and then with water. Then extracts were dried over anhydrous calcium chloride, filtered, and concentrated in vacuum to remove the solvent. About 0.5 part of hydroquinone was added to the residue, which was then heated to a temperature of 100-110°C at 50 mm. The distillate was extracted with ether and the ether extracts were dried over anhydrous calcium chloride and filtered. A small quantity of hydroquinone was added to the ether. The solution was fractionated by distillation to yield 4-methyl-α-methylstyrene boiling at about 72-74°C at 80 mm.

A mixture of 856 parts of ammonium chloride and 3000 parts of 36% formaldehyde was stirred and heated to about 60°C. With cooling to maintain this temperature, 944 parts of 4-methyl-α-methylstyrene were added slowly. After the addition was completed, the mixture was stirred at room temperature until the temperature of the reaction mixture dropped to about 40°C. After 2000 parts of methanol were added, the stirring was continued for 20 hours. The methanol was removed in vacuum and the residue was diluted with 3000 parts of concentrated hydrochloric acid. For 4 hours, the mixture was heated with stirring at a temperature of 100°C. The mixture was cooled, diluted with 2000 parts of water, and made alkaline with 15 N sodium hydroxide solution. The reaction mixture was extracted with benzene, and the benzene extracts were dried over anhydrous potassium carbonate and filtered. The benzene was removed from the filtrate. The remaining residue was distilled in vacuum to yield 4-(p-tolyl)-1,2,3,6-tetrahydropyridine. This base was dissolved in benzene. Dry, gaseous hydrogen chloride was passed through the solution, whereupon there precipitated the hydrochloride, which was collected on a filter. The 4-(p-tolyl)-1,2,3,6-tetrahydropyridine hydrochloride boiling at about 162-170°C/10 mm Hg.

While the temperature was being maintained at about 10-20°C, anhydrous hydrogen bromide gas was passed for 7 hours through a solution of 160 parts of 4-(p-tolyl)-1,2,3,6-tetrahydropyridine in 500 parts of acetic acid. The mixture was stirred during the addition of the hydrogen bromide gas. The mixture was then allowed to stand at room temperature of 16 hours. The acetic acid and the excess hydrogen bromide were removed in vacuum at a bath temperature of less than 40°C. The residue was treated with ether. This solution was cooled, and the product was collected on a filter to give the 4-(p-tolyl)-4-bromopiperidine hydrobromide. A solution of 160 parts of above prepared hydrobromide in 3000 parts of water was treated with 100 parts of 20% sodium hydroxide solution. The resulting precipitate was recovered by filtration and washed with water. The precipitate was then dissolved in toluene, and the solution was dried over anhydrous potassium carbonate and filtered. The filtrate was cooled to 0°C. The crystalline product thus obtained was collected on a filter to yield 4-(p-tolyl)-piperidin-4-ol; MP: 136-137°C.

To a suspension of 341 parts of aluminium chloride in 1740 parts of carbon disulphide were added 96 parts of fluorobenzene with stirring and cooling. While the temperature was maintained at about 10°C, 141 parts of γ-chlorobutyryl chloride were added. After the addition was completed, the cooling bath was removed and the stirring was continued for 2 hours. The reaction mixture was poured into ice water. The organic layer was separated, washed with water, dried over anhydrous sodium sulfate, and filtered. The filtrate was concentrated under reduced pressure and the residue was distilled to yield γ-chloro-p-fluorobutyrophenone. BP: at about 136-142°C/6 mm Hg. 3.7 parts by weight of it, 14.16 parts of 4-(p-tolyl)-piperidin-4-ol, 0.1 parts KI and 150 parts by volume of toluene was treated in the pressure vessel at 140-150°C for 72 hours. On cooling to room temperature the reaction mixture was filtered. The solid residue was treated with mixture of water and ether and ethereal layer added to filtrate from original with water and pressed as dry as possible on the filter. It was then dissolved in 1500 parts by volume of boiling toluene to which anhydrous potassium carbonate was added to remove the remaining water. The mixture was filtered and the filtrate cooled to 0°C. The was p-fluoro-4-(4-hydroxy-4-p-tolyl-piperidino)butyrophenone. MP of chlorohydrate: 216-218°C.

References

PAUL ADRIAAN JAN JANSSEN; GB Patent No. 881,893; Nov. 8, 1961; N.V. Research Laboratorim Dr.C. Janssen, a Belgium Limited Liability Company, Belgium

MOPIDAMOL

Therapeutic Function: Platelet aggregation inhibitor

Chemical Name: 2,6-Bis(diethanolamino)-8-(N-piperidino)pyrimido[5,4-d] pyrimidine

Common Name: -

Structural Formula:

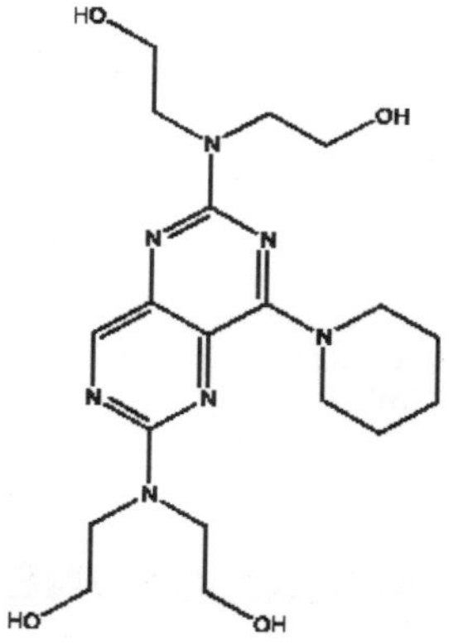

Chemical Abstracts Registry No.: 13665-88-8

Trade Name	Manufacturer	Country	Year Introduced
Rapenton	Thomae	W. Germany	1980

Raw Materials

Dipyridamole
Iodine
Zinc
Formic acid

Manufacturing Process

3.9 g (0.06 mol) of zinc powder were introduced into a solution of 5.0 g (0.01 mol) of 2,6-bis-(diethanolamino)-4,8-dipiperidino-pyrimido-[5,4-d]-pyrimidine (dipyridamole; see entry under that name for its synthesis) in 120 cc of aqueous 10% formic acid. The resulting mixture was heated on a water bath, while occasionally stirring, until the intense yellow color of the starting compound disappeared, which occurred after about 30 to 40 minutes. Thereafter, the unconsumed zinc powder was separated by vacuum filtration, the virtually colorless filtrate was essentially an aqueous solution of 2,6-bis-(diethanolamino)-8-piperidino-1,2,3,4-tetrahydropyrimido-[5,4-d]pyrimidine.

The filtrate was adjusted to a pH of 9 by adding concentrated ammonia, and then a 1 N aqueous iodine-potassium iodide solution was added dropwise, whereby the tetrahydropyrimido[5,4-d]pyrimidine obtained by hydrogenation with zinc in formic acid was converted by oxidation into 2,6-bis-(diethanolamino)-8-piperidino-pyrimido-[5,4-d]-pyrimidine. The completion of the oxidation was checked by means of a starch solution. The major amount of the oxidation product already separated out as a deep yellow crystalline precipitate during the addition of the iodine solution. After the oxidation reaction was complete, the reaction mixture was allowed to stand for a short period of time, and then the precipitate was separated by vacuum filtration, washed with water and dried. It had a melting point of 157°C to 158°C. The yield was 8.0 g, which corresponds to 95% theory.

References

Merck Index 6115
DFU 5 (11) 550 (1980)
Kleeman and Engel p. 608
DOT 17 (3) 89 (1981)
I.N. p. 644
Roch, J. and Scheffler, H.; US Patent 3,322,755; May 30,1967;assigned to Boehringer Ingelheim GmbH

MORCLOFONE

Therapeutic Function: Antitussive

Chemical Name: (4-Chlorophenyl)[3,6-dimethoxy-4-[2-(4-morpholinyl)-ethoxy]phenyl]methanone

Common Name: Dimeclophenone

Structural Formula:

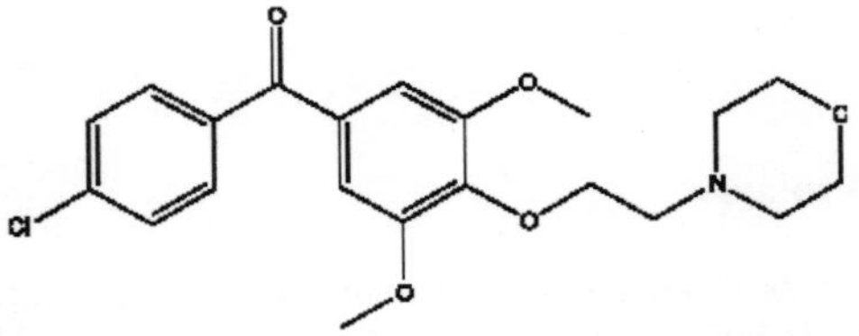

Chemical Abstracts Registry No.: 31848-01-8; 31848-02-9 (Hydrochloride salt)

Trade Name	Manufacturer	Country	Year Introduced
Plausitin	Carlo Erba	Italy	1975
Nitux	Inpharzam	Switz.	1981
Medicil	Medici	Italy	-
NovotussiI	Inpharzam	Belgium	-

Raw Materials

3,5-Dimethoxy-4'-chloro-4-hydroxybenzophenone
Sodium methoxide
β-Morpholinoethyl chloride

Manufacturing Process

Sodium methoxide (1.2 g) in dimethylformamide (150 ml) was stirred with 3,5-dimethoxy-4'-chloro-4-hydroxybenzophenone (6 g) in dimethylformamide (50 ml), for 2 hours at 120°C. The reaction mixture was then treated with β-morpholinoethyl chloride (3.4 g) and heated for 1 hour at 140C, then evaporated to dryness, and treated with water to give a solid material. The mixture was filtered, washed and crystallized from cyclohexane to give 3,5-dimethoxy-4'-chloro-4-(β-morpholinoethoxy)-benzophenone (6.5 g), MP 91°C to 92°C. The product was then reacted at about 0°C with gaseous hydrogen chloride in ether to give, after crystallization from isopropanol, the corresponding hydrochloride which had a MP of 187.9°C.

References

Merck Index 6120
Kleeman and Engel p. 609
DOT 12 (7) 269 (1976)
I.N. p. 645
Lauria, F., Vecchietti, V. and Logemann, W.; US Patent 3,708,482; January 2, 1973; assigned to Carlo Erba SpA (Italy)

MORICIZINE HYDROCHLORIDE

Therapeutic Function: Antiarrhythmic

Chemical Name: Carbamic acid, (10-(3-(4-morpholinyl)-1-oxopropyl)-10H-phenothiazin-2-yl)-, ethyl ester, hydrochloride

Common Name: Moracizine hydrochloride; Moricizine hydrochloride

Structural Formula:

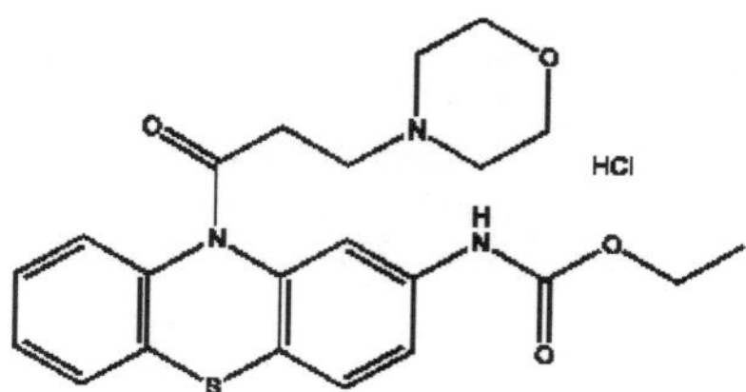

Chemical Abstracts Registry No.: 29560-58-5 ; 31883-05-3 (Base)

Trade Name	Manufacturer	Country	Year Introduced
Ethmozine	Bristol-Myers Squibb	-	-

Raw Materials

Ethyl phenthiazine-2-carbamate
3-Chloropropionyl chloride
Morpholine hydrochloride

Manufacturing Process

To a solution of 10 g (0.035 mole) of ethyl phenthiazine-2-carbamate in 30 ml of anhydrous toluene is added dropwise 5.3 g (0.042 mole) of 3-chloropropionyl chloride, and the mixture is refluxed at 110-120°C for 4 hours, followed by clarifying the mixture with activated carbon and cooling it to room temperature. A precipitate of ethyl 10-(3-chloropropionyl)-phenthiazine-2-carbamate is removed by filtration. The yield is 10.2 g (77.5% of the theoretical amount), M.P. 169-170°C.

10.2 g of ethyl 10-(3-chloropropionyl)-phenthiazine-2-carbamate ester is dissolved in 50 ml of toluene, 4.72 g of morpholine is added thereto, and the mixture is refluxed at 110-120°C for a period of 3 hours. A precipitate of morpholine hydrochloride is removed by filtration, and the filtrate is washed with water in order to remove excess morpholine, followed by acidulating with dilute hydrochloric acid to adjust the pH of the filtrate is adjusted at 3. The acidic aqueous layer is separated, clarified by treatment with activated carbon and made alkaline until the pH equals 8-9. This procedure yields the free base of ethyl 10-(β-morpholylpropionyl)-phenthiazine-2-carbamate, M.P. 156-157°C.

The free base thus obtained is extracted with toluene, the extract is dried over magnesium sulphate and to the anhydrous toluene solution is added an anhydrous ethereal solution of hydrogen chloride until the precipitation of the target compound is complete. This procedure yields 9.53 g (76.2% of the theoretical amount) of ethyl 10-(β-morpholylpropionyl)-phenthiazine-2-carbamate hydrochloride. After recrystallization from dichloroethane, the target compound melts at 189°C. (decomp.).

References

Merck Index, Monograph number: 6351, Twelfth edition, 1996, Editor: S. Budavari; Merck and Co., Inc.
Page G.O.; US Patent No. 5,202,435; April 13, 1993; Assigned to Du Pont Merck Pharmaceutical Company (Wilmington, DE)
Gritsenko A. et al.; US Patent No. 3,864,487; Feb. 4, 1975

MORINAMIDE

Therapeutic Function: Antitubercular

Chemical Name: Pyrazinecarboxamide, N-(morpholinomethyl)-

Common Name: Morfazinamida; Morinamide; Morinammide; Morphazinamide

Structural Formula:

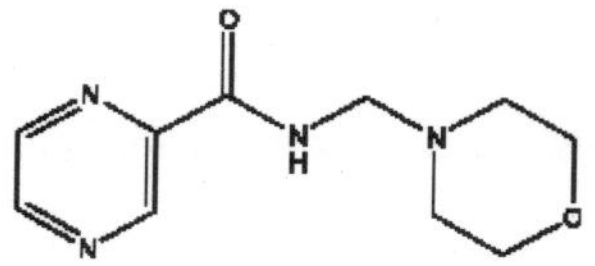

Chemical Abstracts Registry No.: 952-54-5

Trade Name	**Manufacturer**	**Country**	**Year Introduced**
Morinamide	Bracco Industria Chimica S.p.A.	-	-

Raw Materials

2-Pyrazinecarboxamide
Diethylamine
Formaldehyde
Morpholine

Manufacturing Process

98.5 parts by weight 2-pyrazinecarboxamide was mixed with 260 parts by volume of diethylamine. 91 parts by weight of 37% formaldehyde was added to above mixture by stirring for 30 minutes. The reaction mixture was spontaneously heated to 50°C. Then it heated to reflux for 5 hours on water bath. After that it was distilled to dryness at temperature between 40°-50°C. The residue was dissolved with about 200 parts by volume of ligroin (B.P. 60°C) by heating. The solution was filtered hot for removing the not reacted 2-pyrazinecarboxamide. Then it was cooled to -10°C and desired N-(diethylaminomethyl)-pyrazinecarboxylic acid amide discharged. It was filtered off and recrystallized from light petrol ester. Yield about 90%, MP: 47°-50°C.

500 parts by weight of N-(diethylaminomethyl)pyrazinecarboxylic acid amide and 2500 parts by volume of was mixed and heated by stirring to temperature 140°-150°C. At 60°-100°C a distillation begun and ended at 127°C (a boiling point of morpholine). The distillate consisted from diethylamine and morpholine. After 30-60 minutes the mixture was cooled to 50°C and distilled in vacuum to dryness. The residue was recrystallized from 400 parts by volume of benzene to give N-(morpholinomethyl)pyrazinecarboxamide, yield 92%, MP: 114°-117°C.

References

Felder E., Tiepolo U.; D.B. Patent No. 1,129,492; June 23, 1960; Bracco Industria Chimica S.p.A., Mailand (Italien)

MOROXYDINE HYDROCHLORIDE

Therapeutic Function: Antiviral

Chemical Name: 4-Morpholinecarboximidoylguanidine, hydrochloride

Common Name: Abitilguanide; Moroxydine hydrochloride; Morpholinobiguanide hydrochloride

Structural Formula:

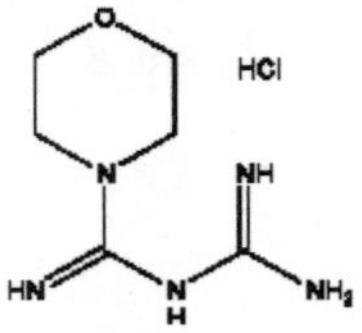

Chemical Abstracts Registry No.: 3160-91-6; 3731-59-7 (Base)

Trade Name	Manufacturer	Country	Year Introduced
Grippe	Nissin	-	-
Tamaxin	Sawai	-	-
Virobis	S.C.S.	-	-
Virustat	Delagrange	-	-

Raw Materials

Morpholine
Dicyandiamide

Manufacturing Process

43.5 g morpholine, 41.7 ml concentrated hydrochloric acid, 40 ml of water, and 42 g dicyandiamide are refluxed for 48 hours, whereupon the reaction mixture is cooled to +5°C and filtered. The filtrate is evaporated to dryness and extracted and extracted with boiling ethanol. Yield: 50 g. The formed 4-morpholinecarboximidoylguanidine hydrochloride is purified by recrystallization from methanol. The salt may be converted into the base by adding equivalent of any basic compound (triethylamine, sodium bicarbonate and so on).

In practice it is usually used as hydrochloride.

References

Aktiebolaget Kabi, a Swedish Body corporate, of Stockholm 30, Sweden; G.B. Patent No. 776,176; Sept. 15, 1953

MORPHINE SULFATE

Therapeutic Function: Narcotic analgesic, Sedative

Chemical Name: Morphinan-3,6-α-diol, 7,8-didehydro-4, 5-α-epoxy-17-methyl-, sulfate (2:1) (salt)

Common Name: Morphine sulfate

Structural Formula:

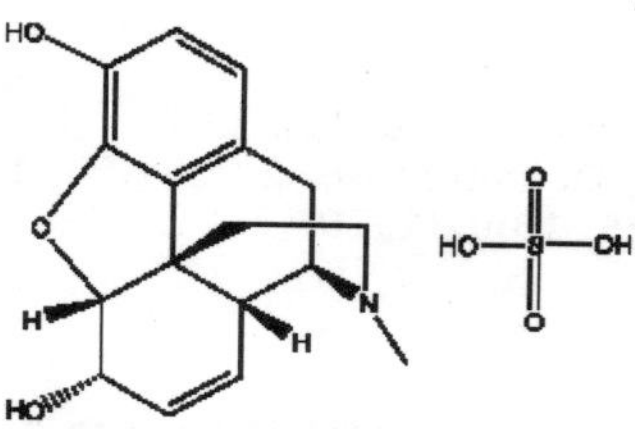

Chemical Abstracts Registry No.: 64-31-3; 57-27-2 (Base)

Trade Name	Manufacturer	Country	Year Introduced
Avinza	Aetna Inc.	-	-
Continus	Napp	-	-
Dolcontin	Kabipharmacia	-	-
Dolcontin	Norsk Ph.	-	-
Duramor	Biological E. Limited	India	-
Doloral	Atlas	-	-
Duralgin	Ethypharm	-	-
Kadian	Alpharma USHP	-	-
Kapanol	GlaxoSmithKline	Australia	-
M-Eslon	Grunenthal	Germany	-
Morcontin	Modi-Mundi Pharma Limited	India	-
Morphine Sulfate SR	Pharmascience	USA	-
MST Continus	Mundipharma	Austria	-
Oblioser	Serono	-	-
Oramorph SR	Roxane Laboratories	-	-
RMS	Upsher-Smith	-	-
Skenan	UPSA	France	-
Substitol	Mundipharma	-	-

Raw Materials

Plant vegetable
Methanol or the aqueous solution of potassium pyrosulfate

Manufacturing Process

Morphin was extracted from the plant vegetable (the poppy) by the mixture of water and methanol or the aqueous solution of potassium pyrosulfate. The precipitation of the morphin was carried out by addition to the extract the aqueous solution of sodium carbonate.

Morphin can be obtained from the extract by using the cation exchanger.

Free base of morphin was transformated to the sulfate salt.

References

Pharmazeutische Wirkstoffe, 610-611, p.610
Heropolitanski R. et al.; DE Patent No. 2,905,468, 13.02.1979
DE Patent No. 2,726,925, 13.02.1979

MOTRETINIDE

Therapeutic Function: Antipsoriatic

Chemical Name: N-Ethyl-9-(4-methoxy-2,3,6-trimethylphenyl)-3,7-dimethyl-2,4,6,8-nonatetraenamide

Common Name: -

Structural Formula:

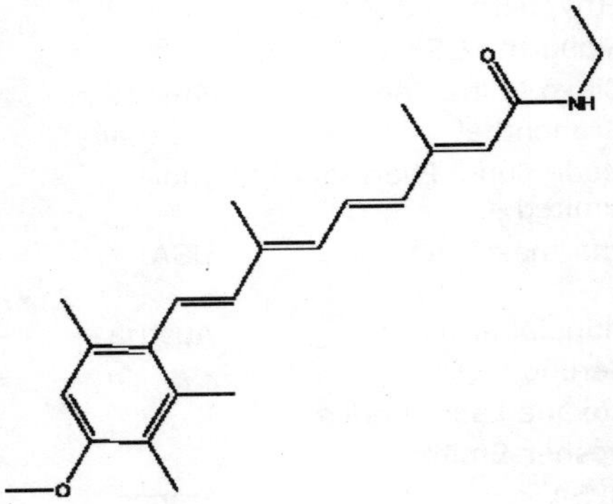

Chemical Abstracts Registry No.: 56281-36-8

Trade Name	Manufacturer	Country	Year Introduced
Tasmaderm	Roche	Switz.	1981

Raw Materials

Sodium hydride	3-Formylcrotonic acid butyl ester
Sodium hydroxide	Phosphorus trichloride
Ethylamine	5-(4-Methoxy-2,3,6-trimethylphenyl)-3-methylpenta-2,4-diene-1-triphenylphosphonium bromide

Manufacturing Process

228 g of 5-(4-methoxy-2,3,6-trimethyl-phenyl)-3-methyl-penta-2,4-diene-1-triphenylphosphonium bromide are introduced under nitrogen gassing into 910 ml of dimethylformamide and treated with cooling at 5°C to 10°C within 20 minutes with 17.5 g of a suspension of sodium hydride (about 50% by weight) in mineral oil. The mixture is stirred for 1 hour at about 10°C, then treated at 5°C to 8°C dropwise with 61.8 g of 3-formylcrotonic acid butyl ester, heated for 2 hours at 65°C, subsequently introduced into 8 liters of ice-water and, after the addition of 300 g of sodium chloride, thoroughly extracted with a total of 18 liters of hexane. The extract is washed 5 times with 1 liter of methanol/water (6:4 parts by volume) each time and 2 times with 1.5 liters of water each time, dried over sodium sulfate and evaporated under reduced pressure to leave 9-(4-methoxy-2,3,6-trimethyl-phenyl)-3,7-dimethyl-nona-2,4,6,8-tetraen-1-oic acid butyl ester, MP 80°C to 81°C as the residue.

125.8 g of 9-(4-methoxy-2,3,6-trimethyl-phenyl)-3,7-dimethyl-nona-2,4,6,8-tetraen-1-oic acid butyl ester are introduced into 2,000 ml of absolute ethanol and treated with a solution of 125.8 g of potassium hydroxide in 195 ml of water. The mixture is heated to boiling under nitrogen gassing for 30 minutes, then cooled, introduced into 10 liters of ice-water and, after the addition of about 240 ml of concentrated hydrochloric acid (pH 2-4), thoroughly extracted with a total of 9 liters of methylene chloride. The extract is washed with about 6 liters of water to neutrality, dried over calcium chloride and evaporated under reduced pressure. The residue is taken up in 700 ml of hexane. The precipitated 9-(4-methoxy-2,3,6-trimethyl-phenyl)-3,7-dimethyl-nona-2,4,6,8-tetraen-1-oic acid melts at 228°C to 230°C.

28.6 g of 9-(4-methoxy-2,3,6-trimethyl-phenyl)-3,7-dimethyl-nona-2,4,6,8-tetraen-1-oic acid are introduced into 300 ml of benzene and treated under nitrogen gassing with 12 g of phosphorus trichloride. The benzene is subsequently distilled off under reduced pressure. The remaining 9-(4-methoxy-2,3,6-trimethyl-phenyl)-3,7-dimethyl-nona-2,4,6,8-tetraen-1-oic acid chloride is dissolved in 1,200 ml of diethyl ether. The solution is added dropwise at -33°C into 500 ml of ethylamine and stirred for 3 hours. The reaction mixture is then diluted with 500 ml of diethyl ether and stirred without cooling for a further 12 hours, the ammonia evaporating. The residue is dissolved in 10 liters of methylene chloride. The solution is washed 2 times with 3 liters of water, dried over sodium sulfate and evaporated under reduced pressure. The remaining N-ethyl-9-(4-methoxy-2,3,6-trimethylphenyl)-3,7-dimethylnona-2,4,6,8-tetraen-1-oic acid amide melts, after recrystallization from ethanol, at 179°C to 180°C.

References

Merck Index 6142
DFU 3 (2) 126 (1978)
OCDS Vol. 3 p. 12 (1984)
DOT 18 (12) 653 (1982)
I.N. p. 647
Bollag, W., Ruegg, R.and Ryser, G.; US Patents 4,105,681; August 8,1978; and 4,215,215; July 29,1980; both assigned to Hoffmann-LaRoche, Inc.

MOXALACTAM DISODIUM

Therapeutic Function: Antiinfective

Chemical Name: 7-[[Carboxy(4-hydroxyphenyl)acetyl]amino]-7-methoxy-3-[[(1-methyl-1H-tetrazole-5-yl)thio]-methyl]-8-oxo-5-oxa-1-azabicyclo[4.2.0]oct-2-ene-2-carboxylic acid disodium salt

Common Name: Lamoxactam; Latamoxef

Structural Formula:

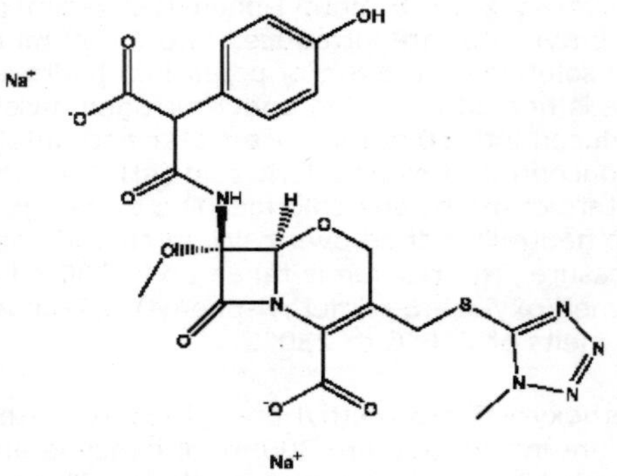

Chemical Abstracts Registry No.: 64952-97-2 (Base)

Trade Name	Manufacturer	Country	Year Introduced
Moxam	Lilly	US	1981
Moxalactam	Lilly	W. Germany	1981
Festamoxin	Shionogi	W. Germany	1981
Moxalactam	Lilly	France	1981
Moxalactam	Lilly	UK	1982
Shiomalin	Shionogi	Japan	1982

Raw Materials

p-(p-Methoxybenzyloxy)-phenylmalonic acid
Diphenylmethyl 7β-amino-7α-methoxy-3-(1-methyltetrazol-5-yl)-thiomethyl-1-oxa-dethia-3-cephem-4-carboxylate
Aluminum chloride
Sodium-2-ethylhexanoate

Manufacturing Process

To a stirred suspension of p-(p-methoxybenzyloxy)-phenylmalonic acid (125 mg) in methylene chloride (3 ml) are added triethylamine (55 l) and oxalyl chloride (26 l) at -15°C, and the suspension is stirred for 40 minutes at 0°C. The mixture is added to a solution of diphenylmethyl 7β-amino-7α-methoxy-3-(1-methyltetrazol-5-yl)thiomethyl-1-oxadethia-3-cephem-4-carboxylate (100 mg) in methylene chloride (3 ml) and pyridine (63 l), and the mixture is stirred for 30 minutes at 0°C. The reaction mixture is diluted with ethyl acetate, washed with aqueous 2N-hydrochloric acid and water, dried over sodium sulfate, and concentrated to give crude product (212 mg), which is chromatographed on silica gel (20 g) and eluted with a mixture of ethyl acetate and acetic acid (99:1) to give diphenylmethyl-7β-[α-p-(p-methoxybenzyloxy)phenyl-α-carboxyacetamido]-7α-methoxy-3-(1-methyltetrazol-5yl)thiomethyl-1-oxadethia-3-cephem-4-carboxylate as foam (71 mg). Yield: 45%.

To a solution of diphenylmethyl-7β-[α-p-(p-methoxybenzyl)-oxy-phenyl-α-p-methoxybenzyl-oxycarbonil-acetamido]-7α-methoxy-3-(1-methyltetrazol-5-yl)thiomethyl-1-oxadethia-3-cephem-4-carboxylate (1.20 g) in methylene chloride (24 ml) are added anisole (2.4 ml) and a solution of aluminum chloride (2.58 g) in nitromethane (12 ml) at 0°C under nitrogen. After stirring for 15 minutes at 0°C, the mixture is poured into cold 5% sodium hydrogen carbonate aqueous solution (100 ml) and filtered to remove the formed precipitate. The filtate is washed twice with methylene chloride (2 x 100 ml), acidified with 2N-hydrochloric acid to pH 2.60, and poured in a column of high porous polymer HP-20 (60 ml) sold by Mitsubishi Chemical Industries Ltd. The column is washed with water (300 ml) and eluted with methanol. The eluate is concentrated under reduced pressure at room temperature. The residue is dissolved in methanol, treated with active carbon, and concentrated under reduced pressure to give 7β(α-p-hydroxyphenyl-α-carboxyacetamido)-7β-methoxy-3-(1-methyl-tetrazol-5-yl)thiomethyl1-oxadethia-3-cephem-4-carboxylic acid as powder (595 mg) decomposing at 125°C to 132°C. Yield: 88.5%.

To a solution of 7β(α-p-hydroxyphenyl-α-carboxyacetamido)-7α-methoxy-3-(1-methyl-tetrazol-5-yl)thiomethyl1-oxadethia-3-cephem-4-carboxylic acid (359 mg) in methanol (7 ml) is added a solution of sodium 2-ethylhexanoate in methanol (2 mols/liter; 1.73 ml) at room temperature. After stirring for 10 minutes, the reaction mixture is diluted with ethyl acetate, stirred for 5 minutes, and filtered to collect separated solid, which is washed with ethyl acetate, and dried to give disodium salt of 7β(α-p-hydroxyphenyl-α-carboxyacetamido)-7α-methoxy-3-(1-methyl-tetrazol-5-yl)thiomethyl1-oxadethia-3-cephem-4-carboxylic acid (342 mg). Yield: 888%. Colorless powder. MP decomposition from 170°C.

References

Merck Index 6143
DFU 5 (9) 467 (1980)
PDR p. 1064
OCDS Vol. 3 p. 218 (1984)
DOT 18 (3) 132 (1982)
I.N. p. 550
Narisada, M. and Nagata, W.; US Patent 4,138,486; February 6,1979; assigned to Shionogi and Co., Ltd. (Japan)

MOXAVERINE HYDROCHLORIDE

Therapeutic Function: Spasmolytic

Chemical Name: 1-Benzyl-3-ethyl-6,7-dimethoxyisoquinoline hydrochloride

Common Name: Moxaverine hydrochloride; Meteverine hydrochloride

Structural Formula:

N

H—Cl

Chemical Abstracts Registry No.: 1163-37-7

Trade Name	Manufacturer	Country	Year Introduced
Paverin	Bracco	-	-

Raw Materials

1-Nitropropane
Sodium
Zinc
Phosphorous oxychloride
3,4-Dimethoxybenzaldehyde
Formic acid
Phenylacetic acid chloride

Manufacturing Process

166 g (1 mol) of 3,4-dimethoxybenzaldehyde were stirred for several hours (8 to 10 hours) with 180 g (2.02 mols.) of 1-nitropropane in 300 ml of methanol, in which 12 g of metallic sodium had previously been dissolved, the stirring taking place while heating to 45-50°C. After usual working up of the reaction

mixture, there were obtained 155 g of a white, crystalline product, which constituted the 1-(3,4-dimethoxyphenyl)-2-nitro-1-butanol and melted after recrystallization from isopropanol at 93-94°C (uncorrected). The composition was confirmed by elementary analysis and an infra-red spectrogram.

204 g (0.8 mol) of the above nitro alcohol were reduced at 30-35°C in 1250 g of 44% formic acid with 320 g of powdered zinc (about 4.9 at.). After working up, the base 1-(3,4-dimethoxyphenyl)-2-amino-1-butanol was obtained as a white crystalline product. After recrystallization from ethyl acetate, it melted at 91-93°C and the yield was 168 g, i.e. 93.3% of the theoretical.

90 g (0.4 mol) of the above amino alcohol are reacted at 45-50°C in 400 ml of chloroform in the presence of 95 g (1.2 mols) of pyridine by means of 139 g (0.9 mol) of phenylacetic acid chloride. After working up the reaction mixture, a yellowish-crystalline product was isolated (183.5 g, theoretical: 184.6 g), which melted at 123-125°C. After recrystallization from ethyl acetate, it yielded minute, white crystals, which melted at 129 -131°C (uncorrected). The composition of the product was confirmed by elementary analysis.

217 g of the above 1-(3,4-dimethoxyphenyl)-2-(phenylacetamido)-butanol-1-phenyl acetate (0.47 mol) were stirred in 80 1300 ml of xylene with 145 g of phosphorous oxychloride at 100-105°C. After some hours, when the evolution of hydrochloric acid gas had ceased, the reaction mixture was poured on to ice and stirred while cold until the crystallization was completed. After filtering, 144 g (89%) of the formed 1-benzyl-6,7-dimethoxy-3-ethyl isoquinoline hydrochloride were obtained in the form of yellowish crystals, which melted at 198-202°C with decomposition. From the separated aqueous mother liquors, the remainder of the formed isoquinoline base was obtained after treatment with ammonia and extraction with ether, the said base being isolated by way of the sparingly soluble and readily crystallisable acid sulfate. The salt represented a yellowish crystal powder, which melted at 239-243°C and weighed 21 g (11%). Thus, the yield of crude isoquinoline salt was almost the theoretical yield. The crude hydrochloride acid salt yielded white, lustrous prisms after recrystallisation from 96% ethanol, the said prisms melting at 208-210°C with decomposition. 1-Benzyl-6,7 -dimethoxy-3-ethyl isoquinoline may be prepared as a base from its salt by adding of equivalent of triethyl amine or any other base.

References

Orgamol S. A., a Swiss Body Corporate of Postfach, Switzerland; G.B. Patent No. 1,030,022; June 16, 1966

MOXESTROL

Therapeutic Function: Estrogen

Chemical Name: 11β-Methoxy-19-nor-17α-pregna-1,3,5(10)-trien-20-yne-3,17-diol

Common Name: 11β-Methoxy-17a-ethynylestradiol

Structural Formula:

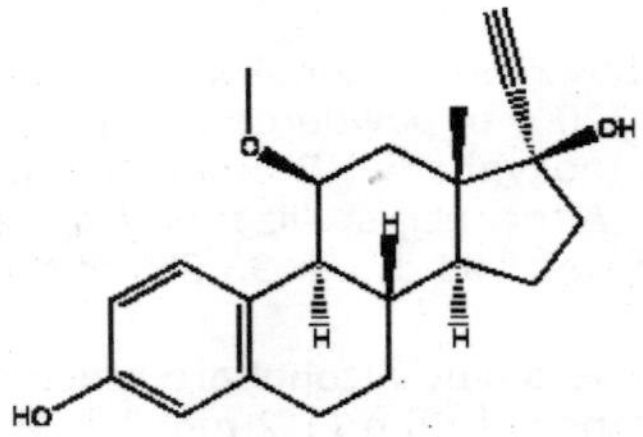

Chemical Abstracts Registry No.: 34816-55-2

Trade Name	Manufacturer	Country	Year Introduced
Surestryl	Roussel	France	1974

Raw Materials

Methanol	$\Delta^{4,9}$-Estradiene-11β-ol-3,17-dione
Acetylene	Palladium hydroxide
Potassium	

Manufacturing Process

(A) Preparation of 11β-Methoxy-$\Delta^{4,9}$-Estradiene-3,17-Dione: 0.5 g of $\Delta^{4,9}$-estradiene-11β-ol-3,17-dione were dissolved at room temperature in 25 cc of methylene chloride containing 2% of methanol and after 5 mg of p-toluenesulfonic acid were added, the reaction mixture was agitated for several minutes. Then the reaction mixture was poured into ice water, washed with water until the wash waters were neutral, and distilled to dryness under vacuum. The resulting residue was crystallized from ethyl ether to obtain 0.46 g of 11β-methoxy-$\Delta^{4,9}$-estradiene-3,17-dione having a MP of 140°C.

(B) Preparation of 11β-Methoxy-$\Delta^{1,3,5(10)}$-Estradiene-3-ol-17-one: 12.3 g of 11β-methoxy-$\Delta^{4,9}$-estradiene-3,17-dionewere dissolved in 1,230 cc of methanol and then, under an atmosphere of nitrogen, 7.38 g of palladium hydroxide were added and the mixture was held at reflux for one hour under agitation and a nitrogen atmosphere. Then the reaction mixture was cooled to 30°C, filtered, vacuum filtered and washed with methanol. The methanolic solutions were concentrated to about 50 cc, allowed to stand overnight at room temperature and filtered. The precipitate formed was triturated in methanol and dried at 80°C to obtain 10.74 g (yield = 87.5%) of 11β-methoxy-$\Delta^{1,3,5(10)}$-estradiene-3-ol-17-one having a MP of 264°C.

(C)Preparation of 11β-Methoxy-17α-Ethynyl-$\Delta^{1,3,5(10)}$-Estradiene-3,17β-Diol: Under agitation and an atmosphere of nitrogen, 12 g of potassium were heated at 80°C in 180 cc of tertiary-amyl alcohol. The mixture was agitated for 30 minutes, cooled to 20°C and after 60 cc of dioxane were added thereto, a stream of acetylene was allowed to bubble through the mixture for one hour and fifteen minutes. Then a solution of 3 g of 11β-methoxy-$\Delta^{1,3,5(10)}$-estradiene-3-ol-17-one in 50 cc of dioxane was added and the mixture was agitated for 4 hours while continuing the passage of acetylene at room temperature. Thereafter, 50 cc of a 50% aqueous acetic acid solution was added and the mixture was poured into water and extracted with ether. The organic phases were washed first with an aqueous solution containing 10% of neutral sodium carbonate, then with water until the wash waters were neutral, dried over sodium sulfate and concentrated under vacuum until crystallization started. The reaction mixture was iced for one hour, vacuum filtered and the precipitate dried under vacuum to obtain 3.8 g of the raw 17α-ethynyl derivative, which was purified by dissolution in ethyl acetate at reflux and by icing to obtain 2.33 g (yield = 77%) of 11β-methoxy-17α-ethynyl-$\Delta^{1,3,5(10)}$-estradiene-3,17β-diol, having a MP of 280°C.

References

Merck Index 6145
Kleeman and Engel p. 611
DOT 11 (4) 149 (1975)
I.N. p. 647
Bertin, D. and Pierdet, A.; US Patent 3,579,545; May 18, 1971; assigned to Roussel- UCLAF, France

MOXIFLOXACIN HYDROCHLORIDE

Therapeutic Function: Antibacterial

Chemical Name: 3-Quinolinecarboxylic acid, 1-cyclopropyl-6-fluoro-1,4-dihydro-8-methoxy-7-((4aS,7aS)-octahydro-6H-pyrrolo[3,4-b]pyridin-6-yl)-4-oxo-, monohydrochloride

Common Name: Moxifloxacin hydrochloride; Proflox

Structural Formula:

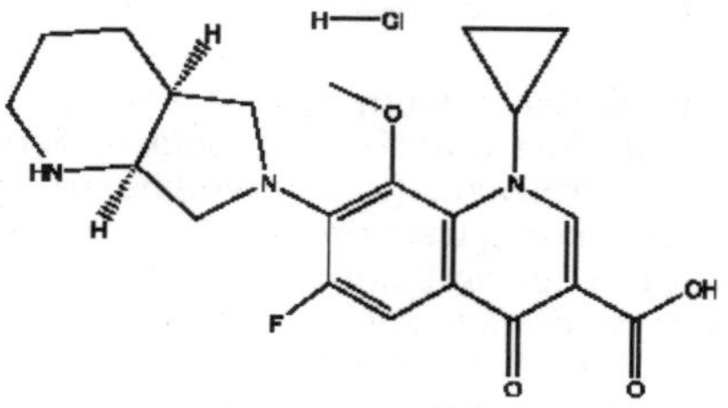

Chemical Abstracts Registry No.: 151096-09-2 (Base); 186826-86-8

Trade Name	Manufacturer	Country	Year Introduced
Avelox	Bayer	-	-
Actira	Bayer	-	-
Octegra	Lilly	-	-
Octegra	Bayer	-	-
Proflox	Div. Sigma	-	-

Raw Materials

Hydrogen
Potassium hydroxide
Copper cyanide
Hydrochloric acid
Sulfuric acid
Thionyl chloride
Diethyl malonate
Acetic anhydride
Sodium fluoride
Ethyl orthoformate
Acetic acid
Pyridine-2,3-dicarboxylic acid N-benzylimide
N-Methyl-2-pyrrolidone
Ferric chloride hexahydrate
Glycol monomethyl ether
Lithium aluminum hydride
Ruthenium on charcoal
Magnesium ethoxide
4-Toluenesulfonic acid
Sodium bicarbonate
1,2,3,4-Tetrafluorobenzene
Palladium on charcoal
Cyclopropylamine

Manufacturing Process

Synthesis of intermidate octahydropyrrolo[3,4-b]pyridine (2,8-diazabicyclo[4.3.0]nonane):

47.6 g (0.2 mol) of pyridine-2,3-dicarboxylic acid N-benzylimide (British Patent No. 1,086,637; Chem. Abstr. 68, 95695w) are hydrogenated in 400 ml of glycol monomethyl ether over 15 g of ruthenium-on-active charcoal (5% strength) at 90°C under 100 bar until the calculated amount of hydrogen has been taken up. The catalyst is then filtered off and the filtrate is concentrated on a rotary evaporator 44 g of an oily crude product are obtained.

The corresponding hydrogenation with palladium-on-active charcoal (5% strength) gives a quantitative yield of a pure 6-benzyl-5,7-dioxo-octahydropyrrolo[3,4-b]pyridine of melting point 67°-69°C.

1.52 g (40 mmol) of lithium aluminium hydride are initially introduced into 30 ml of anhydrous tetrahydrofuran, and 44 g (about 0.18 mol) of crude or pure 6-benzyl-5,7-dioxo-octahydropyrrolo[3,4-b]pyridine are added dropwise as a solution in 15 ml of anhydrous tetrahydrofuran. The mixture is then subsequently stirred at the boiling point for 10 h. 1.5 ml of water, 1.5 ml of 15% strength potassium hydroxide solution and 4.5 ml of water are added dropwise in succession to the batch and the precipitate is then filtered off with suction and washed with tetrahydrofuran. The filtrate is concentrated on a rotary evaporator and the residue is distilled. 24.4 g of a colorless oil of 6-benzyl-octahydropyrrolo[3,4-b]pyridine having a boiling point of 93°-95°C/0.06 mbar are obtained on distillation.

69 g (0.32 mol) of 6-benzyl-octahydropyrrolo[3,4-b]pyridine are hydrogenated in 450 ml of methanol over 7 g of palladium-on-active charcoal (5% strength)

at 90°C/90 bar in the course of 3 h. The catalyst is then filtered off, the filtrate is concentrated and the residue is distilled. 33.8 g (84% of theory) of a colorless solid of octahydropyrrolo[3,4-b]pyridine (2,8-diazabicyclo[4.3.0] nonane) having a melting point of 65°-67° C and a boiling point of 78°C/9 mbar are obtained.

Synthesis of intermidate 1-cyclopropyl-6,7-difluoro-1,4-dihydro-8-methoxy-4-oxo-3-quinolinecarboxylic acid:

According to the method by Bardon et al. (Tetrahedron, 22, 2541 (1966)), 1,2,3,4-tetrafluorobenzene (50 g) was brominated and methoxylated to give 1-bromo-3-methoxy-2,4,5-trifluorobenzene (22.2 g) as colorless oil.

A mixture of the oily product (22 g), cuprous cyanide (10 g) and N-methyl-2-pyrrolidone (37 ml) in sealed tube was heated for 4.5 h at 140° to 150°C. After cooling, a solution of ferric chloride hexahydrate (44 g) and concentrated hydrochloric acid (11 ml) in water (60 ml) was added to the reaction mixture and then stirred at 50° to 60°C for 20 min. The reaction mixture was extracted with ether and the organic layer was washed with dilute aqueous hydrochloric acid, with water and with saturated saline solution successively, and dried over anhydrous sodium sulfate and then concentrated. The residue was purified by distillation under reduced pressure to give 3-methoxy-2,4,5-trifluorobenzonitrile (14.25 g) as colorless oil, boiling point 94°C/8 mm Hg.

To oily product thus obtained (14.2 g) were added concentrated sulfuric acid (8.5 ml) and water (40 ml) and the mixture was stirred for 1 h at 110°C. After cooling, the reaction mixture was poured into ice water (50 ml) and the resulting precipitate was collected by filtration, washed with water, and recrystallized from a solution of dichloromethane-n-hexane to give 3-methoxy-2,4,5-trifluorobenzamide (11.59 g) as white needle, melting point 130°-133°C.

Then, to these crystals were added 18 N sulfuric acid (150 ml) and the mixture was heated for 3.5 h at 100°C. After cooling, water (400 ml) was added to the mixture and the resulting crystals were recrystallized from n-hexane to give the 3-methoxy-2,4,5-trifluorobenzoic acid (9.61 g) as colorless needle, melting point 98°-101°C.

To 3-methoxy-2,4,5-trifluorobenzoic acid (9.4 g) was added thionyl chloride (50 ml), the mixture was refluxed for 3 h and then concentrated. The residue was purified by distillation under reduced pressure to give 3-methoxy-2,4,5-trifluorobenzoyl chloride (8.86 g) as yellow oil, boiling point 108°-112°C/20 mm Hg.

To magnesium ethoxide (5.9 g) was added diethyl malonate (7 g) in anhydrous toluene (35 ml) dropwise and the mixture was warmed for 2 h at 50° to 60°C and then cooled to -10°C. To the mixture was added a solution of the acid chloride (8.86 g) in anhydrous toluene (10 ml) dropwise over 15 min. After stirring for 1 h at -5° to 0°C, ice water (30 ml) containing concentrated sulfuric acid (8 ml) was added to the mixture and the organic layer was separated. The organic layer was washed with saturated saline solution, dried over anhydrous sodium sulfate and then concentrated to give diethyl 3-methoxy-2,4,5-trifluorobenzoylmalonate (13.64 g) as brown oil.

To oily product, the diethyl 3-methoxy-2,4,5-trifluorobenzoylmalonate (13.55 g) were added water (20 ml) and p-toluenesulfonic acid (14 mg), and the mixture was refluxed for 9 h. After cooling, the reaction mixture was extracted with dichloromethane and the organic layer was washed with 7% aqueous sodium bicarbonate solution and with saturated saline solution successively, dried over anhydrous sodium sulfate and then concentrated to give ethyl 3-methoxy-2,4,5-trifluorobenzoylacetate (10.29 g).

To the ethyl 3-methoxy-2,4,5-trifluorobenzoylacetate (9.79 g) were added acetic anhydride (9.6 g) and ethyl orthoformate added cyclopropylamine (8.4 g), and the mixture was refluxed for 3 h. After supplemented further acetic anhydride (3.2 g) and ethyl orthoformate (8.8 g), the mixture was refluxed for 8 h, and then concentrated to give ethyl 2-(3-methoxy-2,4,5-trifluorobenzoyl)-3-ethoxyacrylate (9.73 g) as brown oil.

To a solution of the ethyl 2-(3-methoxy-2,4,5-trifluorobenzoyl)-3-ethoxyacrylate (9.73 g) in ethanol (20 ml) cyclopropylamine (2.0 g) was added dropwise under cooling. After stirring for 2 h at room temperature, the reaction mixture was concentrated and the residue was purified by silica gel column chromatography eluting with n-hexane-ethyl acetate (5:1) to give ethyl 2-(3-methoxy-2,4,5-trifluorobenzoyl)-3-cyclopropylaminoacrylate (7.52 g) as yellowish white crystals, melting point 56°-58°C.

The mixture of the ethyl 2-(3-methoxy-2,4,5-trifluorobenzoyl)-3-cyclopropylaminoacrylate (6.68 g), sodium fluoride (1.31 g) and anhydrous dimethylformamide (26 ml) was refluxed for 5 h. After cooling, the reaction mixture was poured into ice water (100 ml) and the resulting precipitate was collected by filtration, washed with water and recrystallized from ethyl acetate to give ethyl 1-cyclopropyl-6,7-difluoro-1,4-dihydro-8-methoxy-4-oxo-3-quinolinecarboxylate (4.53 g) as colorless needle, melting point 178°-180°C.

To the ethyl 1-cyclopropyl-6,7-difluoro-1,4-dihydro-8-methoxy-4-oxo-3-quinolinecarboxylate (4.5 g) was added a mixed solution of acetic acid (30 ml), concentrated sulfuric acid (4 ml) and water (22 ml), and the mixture was refluxed for 1 h. After cooling, ice water (100 ml) was added and the resulting precipitate was collected by filtration, washed with water and then dried to give 1-cyclopropyl-6,7-difluoro-1,4-dihydro-8-methoxy-4-oxo-3-quinolinecarboxylic acid (4 g) as colorless powder, melting point 185°-186°C.

Syntesis of 1-cyclopropyl-7-(2,8-diazabicyclo[4.3.0]non-8-yl)-6-fluoro-1,4-dihydro-8-methoxy-4-oxo-3-quinolinecarboxylic acid hydrochloride (moxifloxacin hydrochloride):

A mixture of 1-cyclopropyl-6,7-difluoro-1,4-dihydro-8-methoxy-4-oxo-3-quinolinecarboxylic acid, 2,8-diazabicyclo[4.3.0]nonane, DBU (1,8-diazobicyclo[5.4.-0]undec-7-ene and anhydrous acetonitrile was refluxed. After cooling, the resulting precipitate was collected by filtration and recrystallized from methanol to give the 1-cyclopropyl-7-(2,8-diazabicyclo[4.3.0]non-8-yl)-6-fluoro-1,4-dihydro-8-methoxy-4-oxo-3-quinolinecarboxylic acid.

The 1-cyclopropyl-7-(2,8-diazabicyclo[4.3.0]non-8-yl)-6-fluoro-1,4-dihydro-8-methoxy-4-oxo-3-quinolinecarboxylic acid are dissolved in half-concentrated hydrochloric acid by heating, the solution is concentrated and the residue is stirred with ethanol. The undissolved precipitate of 1-cyclopropyl-7-(2,8-diazabicyclo[4.3.0]non-8-yl)-6-fluoro-1,4-dihydro-8-methoxy-4-oxo-3-quinolinecarboxylic acid hydrochloride is filtered off with suction, washed with ethanol and dried at 100°C in vacuo.

References

Masuzawa K. et al.; US Patent No. 5,043,450; August 27, 1991; Assigned: Kyorin Pharmaceutical Co., Ltd., Tokuo, Japan
Petersen U. et al.; US Patent No. 4,990,517; February 5, 1991; Assigned: Bayer Aktiengesellschaft, Leverkusen, Fed. Rep. of Germany

MOXISYLYTE

Therapeutic Function: Vasodilator

Chemical Name: 4-[2-(Dimethylamino)ethoxy]-2-methyl-5-(1-methylethyl)-phenol acetate (ester)

Common Name: Thymoxamine

Structural Formula:

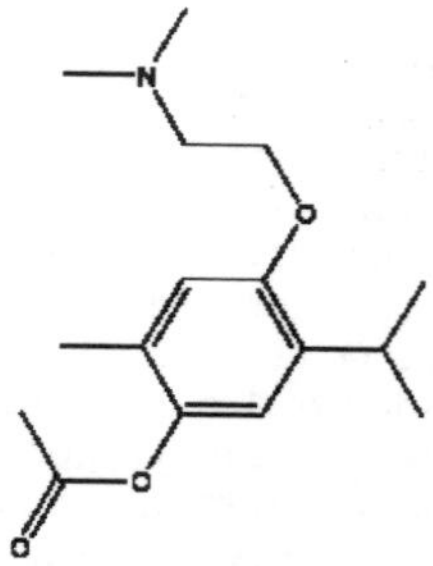

Chemical Abstracts Registry No.: 54-32-0

Trade Name	Manufacturer	Country	Year Introduced
Carlytene	Dedieu	France	1962
Vasoklin	Godecke	W. Germany	1973
Opilon	Parke Davis	Italy	1975
Apifor	Substancia	Spain	-
Arlitene	Chinoin	Italy	-
Sympal	VEB Berlin Chemie	E. Germany	-
Valyten	Landerlan	Spain	-

Raw Materials

Thymol
Sodium
Sulfuric acid
Acetic anhydride
Hydrogen chloride
Hydrogen sulfide
Dimethylaminoethyl chloride
Sodium nitrite
Ethanol

Manufacturing Process

A hydrochloric acid solution of 100 g of thymol in alcohol is reacted with 72 g of sodium nitrite, the nitrosothymol (Organic Syntheses 6, New York, 1926, p. 92) thus obtained is introduced into ammonia, and is reduced by the introduction of hydrogen sulfide to 4-aminothymol (Organic Syntheses Coll. Vol. 1, New York, 1932, p. 458). 133.3 g of this 4-aminothymol are mixed with 67 g of sodium acetate, 107 g of glacial acetic acid and 80 g of acetic acid anhydride to form 4-acetaminothymol (Plancher, Gazzetta Chimica Italiana 25, II, p. 388). 156 parts by weight of this last formed substance dissolved in 600 cc of alcohol are added to a solution of 17.6 parts by weight of sodium in 600cc of alcohol, the mixture being boiled under reflux for some time with 82 g of dimethylaminoethyl chloride. The reaction product is treated with water, and neutralized with hydrochloric acid using acid Congo reagent indicator, and the alcohol is distilled off in vacuo. The base liberated by alkali is dissolved in ether. By evaporating the ether solution the dimethylaminoethyl ether of the 4-acetaminothymol is obtained as a brownish-yellow oil. After some time this oil solidifies in a crystalline state.

100 g of this base are dissolved in a mixture of 300 cc of concentrated hydrochloric acid (density 1.19) and 400 cc of water, and the solution is boiled for one hour under a reflux condenser. Thereupon it is made alkaline, extracted with ether, and the ether is distilled off. 23.6 g of the 4-aminothymoxyethyldimethylamine thus obtained are diazotized in the presence of sulfuric acid at a temperature not exceeding 0°C using a solution of 7.2 g of sodium nitrite in 70 cc of water, and the diazo compound is heated to boiling point after the addition of 1 g of copper sulfate, until no further gas is evolved. It is then made alkaline, and carbon dioxide is introduced. The base is precipitated first in an oily state, and soon becomes crystalline. The 4-oxythymoxyethyldimethylamine forms a neutral hydrochloride which is readily soluble in water, and has a melting point of 174°C to 175.5°C.

36.8 g of 4-oxythymoxyethyldimethylamine are boiled for one hour on a water bath with 160 cc of acetic anhydride and 17.5 cc of pyridine. After this period, the solution is diluted with water, made alkaline, and the base is extracted with ether and the ether distilled off. With acids, the base obtained forms crystalline salts which are readily soluble in water. The hydrochloride melts between 208°C and 210°C.

References

Merck Index 6146
Kleeman and Engel p. 612
OCDS Vol. 1 p. 116 (1977)
I.N. p. 647
Veritas Drug Co., Ltd; British Patent 745,070; February 22,1956

MUPIROCIN

Therapeutic Function: Antibiotic

Chemical Name: Nonanoic acid, 9-((3-methyl-1-oxo-4-(tetrahydro-3,4-dihydroxy-5-((3-(2-hydroxy-1-methylpropyl)oxiranyl)methyl)-2H-pyran-2-yl)-2-butenyl)oxy)-, (2S-(2-α(E),3-β,4-β,5-α(2R*,3R*(1R*,2R*))))-

Common Name: Acidum pseudomonicum; Mupirocin; Pseudomonic acid A

Structural Formula:

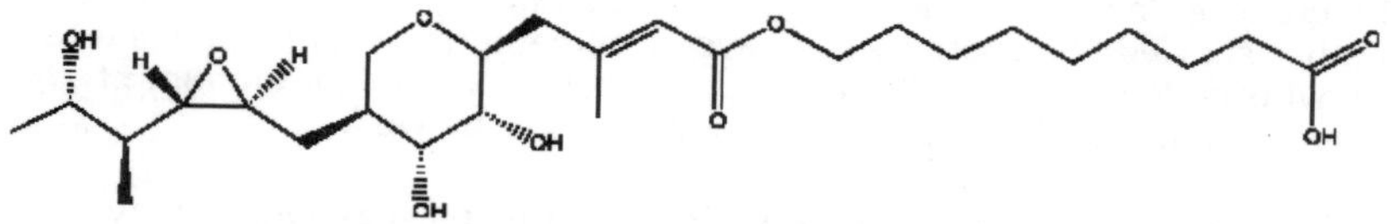

Chemical Abstracts Registry No.: 12650-69-0

Trade Name	Manufacturer	Country	Year Introduced
Bactroban	GlaxoSmithKline	UK	-
Bactroban	SmithKline Beecham	-	-
Bactroban Nasal	Beecham	-	-
Bactoderm	Pharmco Puerto Rico, Inc.	-	-
Bactoderm	Beecham	-	-
Eismycin	GlaxoSmithKline	Germany	-

Raw Materials

Pseudomonas fluorescens, strain NCIB 10586

Manufacturing Process

Production and recovery of Antibacterially active pseudomonic acid and Pseudomonic acid A

Pseudomonas fluorescens, strain NCIB 10586 was grown in submerged culture at 30°C in a medium containing 1% corn steep liquor and 0.5% glucose in a basic salts solution. The maximum yield of the antibiotic occurred after 24 hours and all of the detectable activity was in the culture fluid. After the addition of barium chloride (0.5%) the cells and precipitated non-active contaminant material were removed by centrifugation. The activity was progressively concentrated by partitioning into isobutylmethyl ketone (IBMK) (0.2 vol) at pH 4.5 water (0.8 vol) at pH 8.5, and then IBMK (0.25 vol) at pH 4.5 followed by evaporation to a small volume under reduced pressure. After a further partition into water at pH 8.5 and then adjustment to pH 7-8 the aqueous solution was freeze dried to give the sodium salt which could be stored at 0°C for several months, without loss of activity. The antibiotic extract was stable within the range pH 4-9 at 37°C for 24 hours. Outside these limits rapid loss of activity occurred. The sodium salt showed a broad antibacterial spectrum against Gram positive and Gram negative bacteria, showed low toxicity and was bacteriostatic against S. aureus (N.C.T.C. 6571) and E. coli (M.R.E. 600).

Further purification of the crude acid was effected by chromatography on Amberlite XAD-2 polystyrene resin with a linear gradient produced by adding 0.1 N methanolic ammonia, to 0.01 N aqueous ammonia. A series of low molecular weight acids was eluted first, followed by a fraction (30-60% elution) that possessed the major part of the antibacterial (biological) activity.

Purification of Pseudomonic acid and Pseudomonic Acid A

The produced biologically active material upon methylation with diazomethane in ether showed two spots by thin layer chromatography corresponding to methyl pseudomonate as the major component and a minor amount of component methyl pseudomonate-A (ratio ca 9:1 by wt.).

Methyl pseudomonate was separated from methyl pseudomonate-A by preparative layer silica gel (GF245) chromatography on development with chloroform/isopropanol (9:1). 50% of methyl pseudomonate was recovered from the impure residue by crystallization from benzene/petroleum ether to give colorless needles of m.p. 76.5-78°C.

Acetylation of the methyl ester with pyridine/acetic anhydride affords a triacetate. Reduction of the methyl ester with $LiAlH_4$ in THF afforded 1,9-dihydroxynonanoate, m.p. 46°C.

References

Barrow K. D., Mellows G.; US Patent No. 4,289,703; Sep. 15, 1981; Assigned: Beecham Group Limited

O'Hanlon P. J. et al.; US Patent No. 4,222,942; Sep. 16, 1980; Assigned: Beecham Group Limited

MUZOLIMINE

Therapeutic Function: Diuretic

Chemical Name: 3-Amino-1-(α-methyl-3,4-dichlorobenzyl)pyrazol-5-one

Common Name: -

Structural Formula:

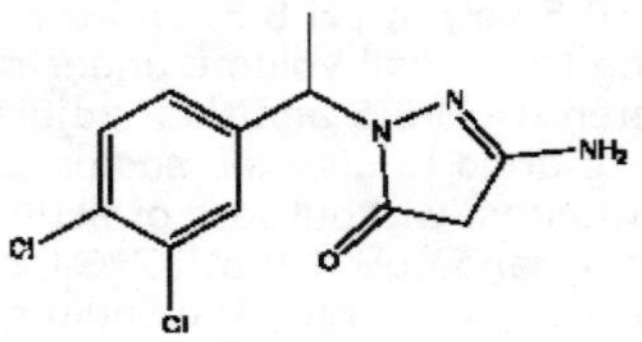

Chemical Abstracts Registry No.: 55294-15-0

Trade Name	Manufacturer	Country	Year Introduced
Edrul	Bayer	Italy	1982

Raw Materials

α-Methyl-3,4-dichlorobenzylhydrazine
β-Amino-β-ethoxyacrylic acid ethyl ester

Manufacturing Process

41 g of α-methyl-3,4-dichlorobenzylhydrazine, dissolved in absolute ethanol, were added dropwise to a solution of 31.8 g of β-amino-β-ethoxyacrylic acid ethyl ester and 1.5 g of p-toluenesulfonic acid in 150 ml of ethanol at room temperature under nitrogen gas. After stirring for 2 hours and standing overnight, the reaction solution was concentrated as far as possible on a rotary evaporator. The residue which remained was dissolved in 2 N sodium hydroxide solution. Any unconverted starting products or by-products were extracted with ether. The aqueous phase was then brought to pH 5 with acetic acid. The oil thereby produced was taken up in methylene chloride and the organic phase was dried over Na_2SO_4. After evaporating off the solvent, the reaction product crystallized out. It was recrystallized from methanol; melting point 127°C to 129°C; yield 21 g (38.5% of theory).

References

Merck Index 6165
DFU 2 (6) 387 (1977)
OCDS Vol. 3 p. 137 (1984)
DOT 18 (10) 555 (1982) and 19 (5) 267 (1983)
I.N. p. 649
Moller, E., Meng, K., Wehinger, E. and Horstmann, H.; British Patent 1,429,141; March 24, 1976; assigned to Bayer AG
Moller, E., Meng, K., Wehinger, E. and Horstmann, H.; US Patent 4,018,890; April 19, 1977; assigned to Bayer AG

MYCOPHENOLATE MOFETIL HYDROCHLORIDE

Therapeutic Function: Antiarthritic, Immunosuppressive

Chemical Name: 4-Hexenoic acid, 6-(1,3-dihydro-4-hydroxy-6-methoxy-7-methyl-3-oxo-5-isobenzofuranyl)-4-methyl-, 2-(4-morpholinyl)ethyl ester, (4E)-, hydrochloride

Common Name: Mycophenolate mofetil hydrochloride

Structural Formula:

Chemical Abstracts Registry No.: 116680-01-4; 128794-94-5 (Base)

Trade Name	Manufacturer	Country	Year Introduced
CellCept	Pharmacia	Italy	-
CellCept	Roche	Italy	-
CellCept	Syntex	-	-

Raw Materials

Sodium diethylmalonate
Diazomethane
t-Butylhypochlorite
Phosphorus
Thionyl chloride
3-Methylpent-3-en-2-on
Ammonium hydroxide
Hydroiodic acid
Silver oxide
Morpholinoethanol

Manufacturing Process

The synthesis of Mycophenolic acid (Canonica L. Et al., Tetrahedron Letters, 1971, N 28, p.2691-2692)

By condensation of sodium diethylmalonate and 3-methylpent-3-en-2-on in ethanol was obtained 2,3-dimethyl-4,6-dioxocyclohexanecarboxilic acid ethyl ester, which was aromatised to 4,6-dihydroxy-2,3-dimethylbenzoic acid ethyl ester (melting point 115-116°C). By treatment with diazomethane or with CH_3I and K_2CO_3 this compound was transformed into 2,4-dimethoxy-5,6-dimethylbenzoic acid ethyl ester (melting point 62-63°C). The hydrolysis of the ester group furnished the 2,4-dimethoxy-5,6-dimethylbenzoic acid (melting point 208-210°C), which was converted into the amide: carbamic acid 3-methoxy-4,5,6-trimethylphenyl ester (melting point 225-229°C). Treatment of the amide with t-butylhypochlorite in methylene dichloride yielded the corresponding N-chloroamide which was photolysed to the intermediate iminolactone and was immediately hydrolized to 5,7-dimethoxy-4-methyl-3H-isobenzofuran-1-one.

This compound with hydriodic acid in acetic acid in the presence of red phosphorous at reflux yielded 5,7-dihydroxy-4-methyl-3H-isobenzofuran-1-one. Condensation of 6-bromo-4-methylhex-4-enoic acid methyl ester and 5,7-dihydroxy-4-methyl-3H-isobenzofuran-1-one with silver oxide in dioxane at room temperature yielded 6-(4,6-dihydroxy-7-methyl-3-oxo-1,3-dihydro-isobenzofuran-5-yl)-4-methylhex-4-enoic acid methyl ester (36% yield). At last, monomethylation with diazomethane yield 6-(4-hydroxy-6-methoxy-7-methyl-3-oxo-1,3-dihydro-isobenzofuran-5-yl)-4-methylhex-4-enoic acid methyl ester, which was hydrolysed with aqueous sodium hydroxide to 6-(4-hydroxy-6-methoxy-7-methyl-3-oxo-1,3-dihydro-isobenzofuran-5-yl)-4-methylhex-4-enoic acid (Mycophenolic acid).

Mycophenolic acid may be obtained by the fermentation broth of Pennicillium brevicompactum. The synthesis of Mycophenolate mofetil (Patent U.S. 4,753,935). The mixture of Mycophenolic acid (32.0 g), thionyl chloride (25.0 ml) and DMF (0.3 ml) in dichloromethane (250 ml) was stirred at room temperature for 3 hours, after which the volatile components were removed under vacuum to afford mycophenolic acid chloride as an oil. The mycophenolic acid chloride oil was dissolved in dichloromethane (50.0 ml) and added to the chilled solution of morpholinoethanol (30.5 ml) in dichloromethane (250 ml). After stirring for 90 min at 4°C, the reaction mixture was washed with water and then with aqueous sodium bicarbonate. The organic solution was dried with sodium sulfate and evaporated to yield Mycophenolate mofetil: morpholinoethyl E-6-(1,3-dihydro-4-hydroxy-6-methoxy-7-methyl-3-oxo-5-isobenzofuranyl)-4-methyl-4-hexenoate (melting point 93-94°C).

The product (38.0 g) was dissolved in isopropanol (200 ml) and the solution was added to a solution of hydrogen chloride (10.0 g) in isopropanol (150 ml). The hydrochloride of Mycophenolate mofetil was collected by filtration and dried under vacuum (melting point 154-155°C).

References

Nelson P. et al.; US Patent No. 4,753,935; Jun. 28, 1988; Assigned to Syntex (USA) Inc.
Pharmazeutische Wirkstoffe, S. 613
Kida T. et al.; US Patent No. 4,452,891; June 5, 1984; Assigned: Ajinomoto Company Incorporated (Tokyo, JP)
Queener S.W. et al.; US Patent No. 4,115,197; Sep. 19, 1978; Assigned: Eli Lilly and Company (Indianapolis, IN)
Canonica L. et al.; Tetrahedron Letters, 1971, N 28, p.2691-2692

MYRTECAINE

Therapeutic Function: Local anesthetic, Spasmolytic

Chemical Name: Ethanamine, 2-(2-(6,6-dimethylbicyclo(3.1.1)hept-2-en-2-yl)ethoxy)-N,N-diethyl-

Common Name: Myrtecaine; Nopoxamine

Chemical Abstracts Registry No.: 7712-50-7

Trade Name	Manufacturer	Country	Year Introduced
Myrtecaine	Chemical Land21	-	-
Nopoxamine	Desynth	-	-
Nopoxamine	Bio Sidus	-	-

Structural Formula:

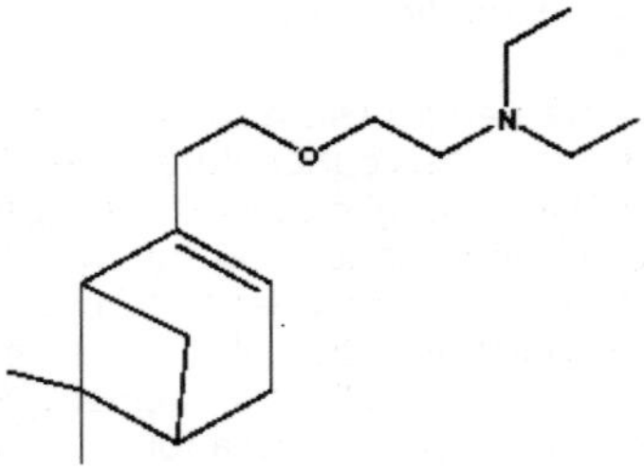

Raw Materials

Homomyrtenol
Sodium amide
Diethylaminochloroethane

Manufacturing Process

60 g (1.5 mol) of powdered sodium amide are put in suspension in 800 ml of toluene. The mixture is heated to 60°C and 166 g of homomyrtenol (1 mol) are added little by little. The reaction is continued for several hours until the homomyrtenol is entirely converted into sodium derivative. It is allowed to stand and the excess amide is filtered. The reaction is followed by titration on a sample of the decanted liquid after having removed the ammonia.

Added to the solution of the sodium derivative of the terpenic alcohol (1 mol) is a toluenic solution of 138 g (1.02 mol) diethylaminochloroethane in toluene. This mixture is refluxed in a nitrogen atmosphere for 12 hours. A precipitate of sodium chloride is formed which is dissolved in water. Two modes of extraction of the base myrtecaine are possible:

A) The toluenic solution is extracted with two times 200 ml of concentrated hydrochloric acid diluted to 20%. In this way there is obtained an aqueous solution of the hydrochloride, when the required amino base is salted out by addition of potassium carbonate. The amino ether-oxide is finally rectified under a vacuum. The fraction boiling between 135° and 140°C under 2 to 3 mm is collected; n_D^{20} = 1.477.

B) The toluenic solution is dried on potassium carbonate and then rectified. There is collected the toluene, then between 120° and 130°C under 2 mm the homomyrtenol which has not reacted, and then between 130° and 145°C a fraction which is again fractionated. The pure product is collected at 135°-140°C/2-3mm Hg.

References

Gaudin O.P.; G.B. Patent No. 861,900; Aug. 6, 1959

N

NABILONE

Therapeutic Function: Antianxiety

Chemical Name: 1-Hydroxy-3-(1',1'-dimethylheptyl)-6,6-dimethyl-6,6a,7,8,10,10a-hexahydro-9H-dibenzo[b,d]pyran-9-one

Common Name: -

Structural Formula:

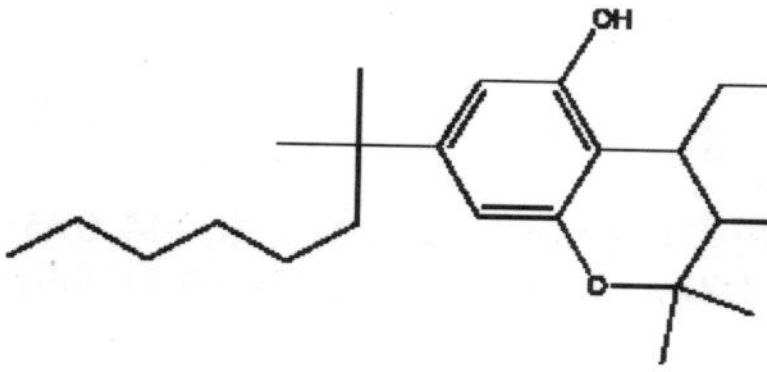

Chemical Abstracts Registry No.: 51022-71-0

Trade Name	Manufacturer	Country	Year Introduced
Cesamet	Lilly	Canada	1982
Cesametic	Lilly	W. Germany	1983
Cesamet	Lilly	UK	1983

Raw Materials

dl-3-(1',1'-Dimethylheptyl)-6,6a,7,8-tetrahydro-1-hydroxy-6,6-dimethyl-9H-dibenzo[b,d]pyran-9-one
Lithium
Ammonia

Manufacturing Process

A solution of 1.5 g of dl-3-(1',1'-dimethylheptyl)-6,6a,7,8-tetrahydro-1-hydroxy-6,6-dimethyl-9H-dibenzo[b,d]pyran-9-one in 50 ml of anhydrous tetrahydrofuran (THF) was added dropwise to a solution of lithium metal in liquid ammonia at -80°C. Excess lithium metal was added in chunks to the solution as the blue color, indicating free dissolved lithium, disappeared. After the addition was complete, ammonium chloride was added to react with any excess lithium metal still present.

The mixture was then allowed to warm to room temperature in a nitrogen atmosphere during which process the ammonia evaporated. The reaction mixture was then acidified with 1 N aqueous hydrochloric acid, and the organic constituents extracted with ethyl acetate. The ethyl acetate extracts were combined, washed with water and dried. Evaporation of the ethyl acetate under reduced pressure yielded 1.4 g of crude dl-trans-3-(1',1'-dimethylheptyl)-6,6aβ,7,8,10,10aβ-hexahydro-1-hydroxy-6,6-dimethyl-9H-dibenzo[b,d]pyran-9-one. The crude product was chromatographed over 50 g of silica gel from benzene solution and the desired product was eluted in 20 ml fractions with a benzene eluant containing 2% ethyl acetate. Fractions 200 to 240 contained 808 mg of a white crystalline solid comprising purified dl-trans-3-(1',1'-dimethylheptyl)-6,6aβ,7,8,10,10aβ-hexahydro-1-hydroxy-6,6-dimethyl-9H-dibenzo[b,d]pyran-9-one. The purified compound melted at 159°C to 160°C after recrystallization from an ethyl acetate-hexane solvent mixture.

References

Merck Index 6193
DFU 3 (3) 207 (1978)
OCDS Vol. 3, p 189 (1984)
DOT 19 (7) 415 & (8) 436 (1983)
I.N. p. 652
Archer, R.A.; US Patents 3,928,598; December 23, 1975; 3,944,673; March 16,1976; and 3,953,603; April 27, 1976; all assigned to Eli Lilly & Co.

NABUMETONE

Therapeutic Function: Antiinflammatory

Chemical Name: 2-Butanone, 4-(6-methoxy-2-naphthalenyl)-

Common Name: Nabumetone

Structural Formula:

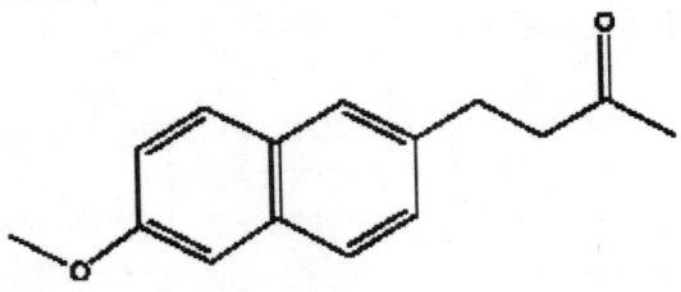

Chemical Abstracts Registry No.: 42924-53-8

Trade Name	Manufacturer	Country	Year Introduced
Artaxan	Malesci SpA Ist. Farmacobiol	Italy	-
Arthaxan	Pharm Chemical Shanghai Lansheng Corporation	China	-
Balmox	Alloga AG	Switz.	-
Dolsinal	Ferrer Internacional	Spain	-
Listran	Uriach	Spain	-
Mebutan	Bencard	-	-
Mebutan	Glaxo SmithKline Beecham	-	-
Nabuflam	Micro Labs	India	-
Nabumetone	Copley Pharmaceutical	USA	-
Nabumetone	Teva	USA	-
Nabumetone	IVAX Pharmaceuticals, Inc.	USA	-
Nabumetone	Eon Labs, Inc.	USA	-
Nabumetone	Genpharm Inc.	Canada	-
Relafen	SmithKline Beecham	-	-
Relafen	Eon Labs, Inc.	USA	-
Relifex	GlaxoSmithKline	-	-
Rodanol S	Lek	Slovenia	-

Raw Materials

Sodium methoxide
Sodium bisulfite
2-Acetyl-5-bromo-6-methoxynaphthalene
Palladium on carbon
Sodium acetate

Manufacturing Process

4-(5-Bromo-6-methoxy-2-naphthyl)-4-hydroxybut-3-en-2-one

50 grams (0.179 moles) of 2-acetyl-5-bromo-6-methoxynaphthalene and 200 ml of n-butyl acetate are placed in a flask equipped with refrigerator and stirrer and, under stirring and at the temperature of 15°C, 14.5 g (0.268 moles) of sodium methoxide are added. The temperature of the reaction mixture goes up to 25°C and is kept at this value for 30 minutes, then the mixture is warmed at 65°C for one hour, is added with 100 ml of water and is brought to pH 4 by adding a concentrated aqueous solution of hydrochloric acid. The reaction mixture is then cooled to 0°-5°C and kept at this temperature for one hour. The solid is filtered, abundantly washed with water on the filter, then washed with butyl acetate and dried in oven under vacuum obtaining 53 g of product with a yield equal to 92%.

Example 1. 4-(6-Methoxy-2-naphthyl)butan-2-one

48 grams (0.150 moles) of 4-(5-bromo-6-methoxy-2-naphthyl)-4-hydroxybut-3-en-2-one, 6.1 g of sodium acetate hydrate containing 32.4% of water, equivalent to 0.050 moles of sodium acetate, 4 g of a 50% suspension in water of 10% palladium on carbon, equivalent to 0.0019 moles of palladium, and 500 ml of methanol are put in a hydrogenator. The hydrogenator is washed with nitrogen in order to eliminate the oxygen and then hydrogen is introduced at the pressure of 2 atmospheres. The temperature of reaction is

kept at 40°C for a period of time of 6 hours, then the hydrogen is let off, the hydrogenator is washed with nitrogen and the reaction mixture is filtered to eliminate the catalyst. The solution is brought to pH 6 with a 5% aqueous solution of sodium hydroxide and concentrated under vacuum. The oily residue is dissolved into 130 ml of isopropanol and 30 ml of N,N-dimethylformamide and the solution is added with 45 ml of water and 17.6 g of sodium bisulfite obtaining a suspension that is stirred for one hour at 60°C, then is cooled to 5°C and is filtered. The obtained solid is washed with 75 ml of methanol, suspended in 200 ml of a 5% aqueous solution of sodium hydroxide and kept under stirring at room temperature for three hours. The suspension is then filtered, the solid is washed with water until neutrality and dried in oven under vacuum obtaining 18 g of product with a yield equal to 52.8%.

Example 2. 4-(6-Methoxy-2-naphthy)butan-2-one

The reaction described above is repeated with the sole changes of doubling the amount of sodium acetate hydrate containing 32.4% of water, 12.22 g equivalent to 0.100 moles of sodium acetate, and of lowering the hydrogenation time to five hours. In this way 22.5 g of product are obtained with a yield equal to 66%.

Example 3. 4-(6-Methoxy-2-naphthyl)butan-2-one

The reaction described in example 3 is repeated with the sole changes of nearly triplicating the amount of sodium acetate hydrate containing 32.4% of water, 17.60 g equivalent to 0.145 moles of sodium acetate, and of lowering the hydrogenation time to five hours. The oil obtained by evaporating the solvent at the end of the reaction is treated with 300 ml of toluene and 100 ml of water and after 15 minutes of stirring the two layers are separated. The aqueous phase is discarded while the organic phase is evaporated under vacuum at 70°C obtaining an oil that is dissolved into 100 ml of methanol. The solution is kept at 0°C for two hours and the precipitated solid is filtered, washed with 15 ml of methanol cooled to 0°C and dried in oven under vacuum. In this way 21.7 g of product are obtained. The methanolic filtrates from crystallization and washing are concentrated under vacuum to half volume so obtaining, after cooling to 0°C, the crystallization of other 4 g of product with an overall yield equal to 75.3%.

References

Cannata V. et al.; US Reissued Patent No. RE37,813; Aug. 6, 2002; Assigned to Honeywell International Inc., Morristown, NJ (US)

NADOLOL

Therapeutic Function: Antiarrhythmic

Chemical Name: 2,3-cis-1,2,3,4-Tetrahydro-5-[2-hydroxy-3-(tert-butylamino)propoxy]-2,3-naphthalenediol

Common Name: -

Structural Formula:

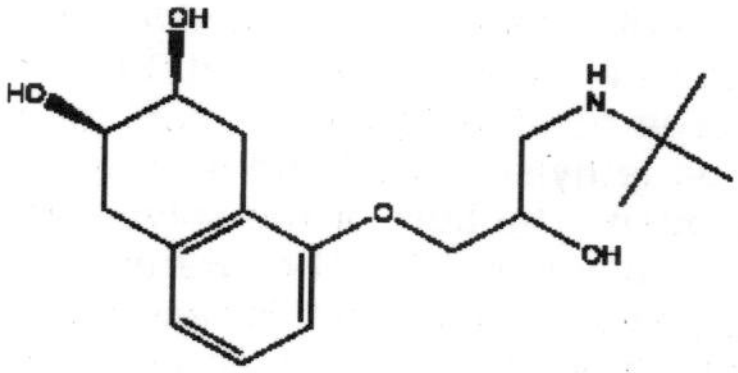

Chemical Abstracts Registry No.: 42200-33-9

Trade Name	Manufacturer	Country	Year Introduced
Solgol	Heyden	W. Germany	1978
Corgard	Squibb	Switz.	1978
Corgard	Squibb	UK	1979
Corgard	Squibb	US	1979
Corgard	Squibb	Italy	1980
Corgard	Squibb	France	1982
Betadol	Fako	Turkey	-
Corzide	Squibb	US	-

Raw Materials

5,8-Dihydro-1-naphthol
Sodium hydroxide
Acetic anhydride
Sodium methoxide
Silver acetate
Epichlorohydrin
Iodine
t-Butylamine

Manufacturing Process

(a) cis-5,6,7,8-Tetrahydro-1,6,7-naphthalenetriol: A solution of 29.2 g (0.2 mol) of 5,8-dihydro-1-naphthol and 40 ml of acetic anhydride in 100 ml of pyridine is prepared. After 16 hours the solvent is removed in vacuo and the residue dissolved in ether and washed with 200 ml of 5% hydrochloric acid, water, 200 ml of 10% sodium hydroxide, saturated salt solution and dried. Solvent removal gives 34.2 g (90.5%) of crude acetate which is dissolved in 900 ml of acetic acid and 36 ml of water. 53.3 g (0.32 mol) of silver acetate is added followed by 40.6 g (0.16 g-atom) of iodine. The slurry is heated with good stirring at 85°10°C for 3 hours under nitrogen, cooled and filtered. The filtrate is evaporated in vacuo and the residue dissolved in 250 ml of methanol and cooled to 0°C.

A solution of 40 g of sodium hydroxide in 200 ml of water is added under nitrogen and the mixture stirred overnight. The bulk of the methanol is removed in vacuo whereupon a solid forms. The solid is separated by filtration, dissolved in 150 ml of water and acidified with 20 ml of concentrated hydrochloric acid. Cooling gives a solid which is filtered and dried to give 16.5 g cis-5,6,7,8-tetrahydro-1,6,7-naphthalenetriol, melting point

184.5°C to 187°C. Three recrystallizations from absolute ethanol give the analytical sample, melting point 188°C to 188.5°C.

(b) 2,3-cis-1,2,3,4-Tetrahydro-5-[2,3-(epoxy)-propoxy]-2,3-naphthalenediol: A solution of 1.20 g (0.03 mol) of sodium methoxide and 5.4 g (0.03 mol) of cis-5,6,7,8-tetrahydro-1,6,7-naphthalenetriol in 200 ml of methanol is prepared under nitrogen. The residue obtained upon solvent removal is stirred overnight with 200 ml of dimethylsulfoxide and 4.65 g (0.05 mol) of epichlorohydrin under nitrogen. The bulk of the solvent is removed at 50°C at 0.1 mm and the residue dissolved in 100 ml of water. Extraction with chloroform (10 x 200 ml) gives a solid which is recrystallized from 150 ml of hexane-ethyl acetate to give epoxy diol of the above title.

(c)2,3-cis-1,2,3,4-Tetrahydro-5-[2-hydroxy-3-(tert-butylamino)propoxy]-2,3-naphthalenediol: A mixture of 2,3-cis-1,2,3,4-tetrahydro-5-[2,3-(epoxy)propoxy]-2,3-naphthalenediol (melting point 104°C to 107°C, one spot on TLC-alumina, 5% methanol in chloroform, iodine visualization) and 22 ml of tert-butylamine is heated at 85°C to 95°C for 15 hours in a Parr bomb and the excess amine removed in vacuo. The solid obtained by trituration of the residue with ether is filtered and recrystallized from benzene to give 3.4 g, melting point 124°C to 136°C.

References

Merck Index 6195
DFU 1 (9) 434 (1976)
Kleeman & Engel p. 614
PDR pp. 1739, 1741
OCDS Vol. 2 p. 110 (1980)
DOT 15 (9) 411 (1979)
I.N. p. 652
REM p. 905
Hauck, F.P., Cimarusti, C.M. and Narayanan, V.L.; US Patent 3,935,267; January 27, 1976; assigned to E.R. Squibb & Sons, Inc.

NAFARELIN ACETATE

Therapeutic Function: Gonadotropic

Chemical Name: 6-(3-(2-Naphthalenyl)-D-alanine)luteinizing hormone-releasing factor (pig), monoacetate (salt)

Common Name: Nafarelin acetate

Chemical Abstracts Registry No.: 76932-60-0; 76932-56-4 (Base)

Structural Formula:

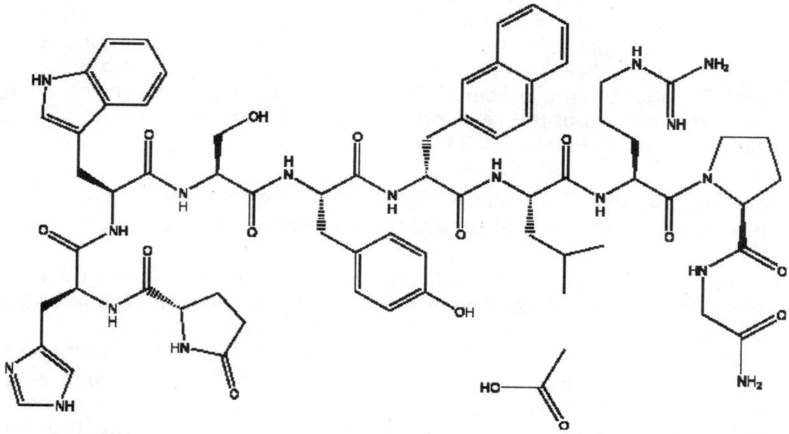

Trade Name	Manufacturer	Country	Year Introduced
Nafarelin Acetate	Bachem AG	-	-
Nasanyl	Yamanouchi Pharmaceutical Co., Ltd.	Japan	-
Synarel	Searle	France	-
Synarel	Pfizer Canada Inc.	Canada	-
Synarel Nasal Spray	Syntex	USA	-
Synrelina	Pharmacia AG	-	-
Synrelina	Pfizer AG	Switz.	-

Raw Materials

Boc-Gly-OH
Boc-Pro-OH
Boc-Leu-OH
Boc-Trp-OH
Pyroglutamic acid
Boc-His(Tosyl)-OH
N,N'-Dicyclohexylcarbodiimide
N-Boc, O-2-bromobenzoyloxycarbonyl-L-tyrosine
Benzhydrylamino-polystyrene-divinylbenzene resin
Cobalt(III) fluoride
Boc-3-(2-naphthyl)-D-alanine
Boc-Arg(Tosyl)-OH
1-Hydroxybenzotriazole
Boc-Ser(Benzyl)-OH
Hydrofluoric acid

Manufacturing Process

In the reaction vessel of a Beckman 990 Peptide Synthesizer was placed 0.8 g (0.8 mmol) of benzhydrylamino-polystyrene-divinylbenzene resin (Lab Systems, Inc.) as described by Rivaille, supra. Amino acids were added sequentially to this resin by means of the usual methods of Boc-strategy of peptide synthesis on above copolymer.

The resin was coupled sequentially with a 2.5 molar excess of each protected amino acid and DCC. Thus, the resin was treated during successive coupling cycles with 0.433 g Boc-Gly-OH, 0.432 g Boc-Pro-OH, 0.857 g Boc-Arg(Tosyl)-OH, 0.462 g Boc-Leu-OH, 0,504 g Boc-3-(2-naphthyl)-D-alanine and 0.272 g 1-hydroxybenzotriazole, 0.724 g N-Boc, O-2-bromobenzoyloxycarbonyl-L-tyrosine, 0.59 g Boc-Ser(Benzyl)-OH, 0.608 g Boc-Trp-OH, 0.654 g Boc-His(Tosyl)-OH and 0.524 g pyroglutamic acid. A coupling cycle for one amino acid and completeness of the reaction is checked by the ninhydrin method of E. Kaiser, et al., Anal. Biochem., 34, 595 (1970).

The resin was removed from the reaction vessel, washed with CH_2Cl_2, and dried in vacuo to yield 2.0 g of protected polypeptide resin.

The polypeptide product was simultaneously removed from the resin and completely deprotected by treatment with anhydrous liquid HF. A mixture of 2.0 g of protected polypeptide resin and 2 mL of anisole (scavenger) in a Kel-F reaction vessel was treated with 20 mL of redistilled (from CoF_3) anhydrous liquid HF at 0°C for 30 minutes. The HF was evaporated under vacuum and the residue of (pyro)-Glu-His-Trp-Ser-Tyr-3-(2-naphthyl)-D-alanyl-Leu-Arg-Pro-Gly-NH_2,as its HF salt, was washed with ether. The residue was then extracted with glacial acetic acid. The acetic acid extract was lyophilized to yield 0.8 g of crude material. The crude polypeptide was loaded on a 4x40 cm. Amberlite XAD-4 column (polystyrene-4% divinylbenzene copolymer) and eluted with a concave gradient from water (0.5 L) to ethanol (1 L). The tubes containing fractions from effluent volume 690 mL to 1,470 mL were pooled and stripped to dryness to yield 490 mg of partially purified polypeptide.

A 150 mg sample of the partially purified product was subjected to partition chromatography on a 3 times 50 cm. column of Sephadex G-25 using the solvent system 1-butanol/toluene/acetic acid/water containing 1.5% pyridine in the ratios 10:15:12:18. The pure fractions were pooled on the basis of thin layer chromatography (silica gel; BuOH/H_2O/HOAc/EtOAc; 1:1:1:1) and HPLC (5 micron, reverse phase, octadecylsilyl packing; 40% 0.03 M NH_4OAc/60% acetonitrile). The desired product came off the column in fractions from effluent volume 1,000 mL to 1,400 mL (Rf 0.1). The pure fractions were pooled, stripped to dryness, taken up in H_2O, and lyophilized to yield 57 mg of pure pyro-glutamyl-histidyl-tryptophylseryl-tyrosyl-3-(2-naphthyl)-D-alanyl-leucyl-arginylprolyl-glycinamide, as its acetic acid addition salt, $[\alpha]_D^{25}$-27.4° (c 0.9, HOAc), m.p. 185°-193°C (dec.).

References

Nestor et al.; US Patent No. 4,234,571; Nov. 18, 1980; Assinged to Syntex (U. S. A.) Inc., Polo Alto, Calif.

NAFCILLIN SODIUM

Therapeutic Function: Antibacterial

Chemical Name: 6-(2-Ethoxy-1-naphthamido)-3,3-dimethyl-7-oxo-4-thia-1-azabicyclo[3.2.0]heptane-2-carboxylic acid sodium salt

Common Name: 6-(2-Ethoxy-1-naphthamido)penicillin sodium salt

Structural Formula:

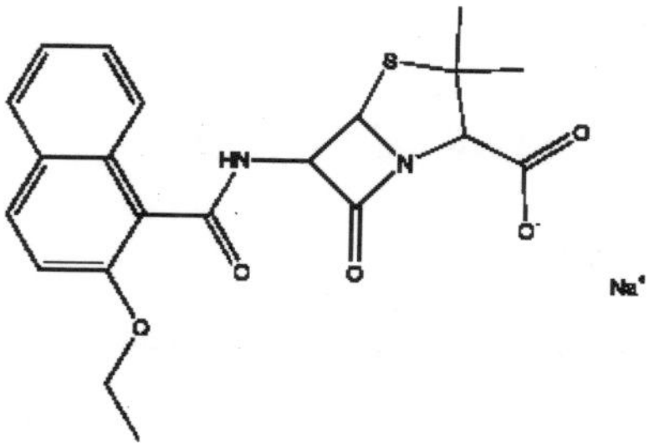

Chemical Abstracts Registry No.: 985-16-0; 147-52-4 (Base)

Trade Name	Manufacturer	Country	Year Introduced
Unipen	Wyeth	US	1964
Nafcil	Bristol	US	1976
Nallpen	Beecham	US	1983
Naftopen	Gist Brocades	-	-

Raw Materials

6-Aminopenicillanic acid
2-Ethoxy-1-naphthoyl chloride
Sodium bicarbonate

Manufacturing Process

A stirred suspension of 12.6 grams 6-aminopenicillanic acid in 130 ml dry alcohol-free chloroform was treated with 16 ml triethylamine and then with 13.8 grams of a solution of 2-ethoxy-1-naphthoyl chloride in 95 ml chloroform. After being washed successively with 58 ml each of 1 N and then 0.1 N hydrochloric acid the chloroform solution was extracted with N aqueous sodium bicarbonate (58 ml + 6 ml). The combined bicarbonate extracts were washed with 20 ml ether and then evaporated at low temperature and pressure to give the crude sodium salt of 2-ethoxy-1-naphthylpenicillin [also called sodium 6-(2-ethoxy-1-naphthamido)penicillinate] as a yellow powder (20.3 grams). This was dissolved in 20 ml water at 30°C and diluted with 180 ml n-butanol, also at 30°C, with stirring. Slow cooling to 0°C gave colorless needles of the product.

References

Merck Index 6199
Kleeman & Engel p. 615
PDR pp. 700, 1991
OCDS Vol. 1 p. 412 (1977)

I.N. p. 653
REM p. 1196
Doyle, F.P. and Nayler, J.H.C.; US Patent 3,157,639; November 17, 1964; assigned to Beecham Group Limited, England

NAFIVERINE

Therapeutic Function: Spasmolytic

Chemical Name: α-Methyl-1-naphthaleneacetic acid 1,4-piperazinediyldi-2,1-ethanediyl ester

Common Name: -

Structural Formula:

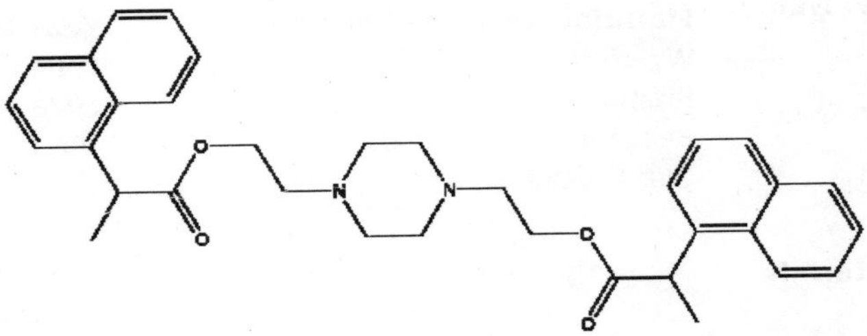

Chemical Abstracts Registry No.: 5061-22-3

Trade Name	Manufacturer	Country	Year Introduced
Naftidan	De Angeli	Italy	1969

Raw Materials

α-Methyl-1-naphthylacetic acid
Thionyl chloride
N,N'-Di-(β-hydroxyethyl)piperazine

Manufacturing Process

15 grams of α-methyl-1-naphthylacetic acid were refluxed with 50 ml of thionyl chloride during 3 hours. The excess thionyl chloride was removed under reduced pressure and the product was also isolated by distillation under reduced pressure. Yield: 15.6 grams (96%). The α-methyl-1-naphthyl acetyl chloride boils at 120° to 124°C. 1.76 grams of N,N'-di-(β-hydroxyethyl)-piperazine, 1.9 grams of sodium bicarbonate and 4.45 grams of α-methyl-1-naphthyl acetyl chloride in 30 ml of anhydrous acetonitrile were refluxed with stirring during 5 hours. After cooling the mixture was filtered and the

acetonitrile evaporated off under reduced pressure. 5.2 grams of crude ester were obtained. The hydrochloride, melting at 220° to 221°C, may be prepared by dissolving the ester in absolute ethanol and treating the solution with anhydrous gaseous hydrogen chloride.

References

Merck Index 6200
I.N. p. 653
Pala, G.; British Patent 1,016,968; Jan. 12, 1966; assigned to Instituto de Angeli, SpA, Italy

NAFRONYL OXALATE

Therapeutic Function: Vasodilator

Chemical Name: Tetrahydro-α-(1-naphthalenylmethyl)-2-furanpropanoic acid 2-(diethylamino)ethyl ester acid oxalate

Common Name: Naftidofuryl

Structural Formula:

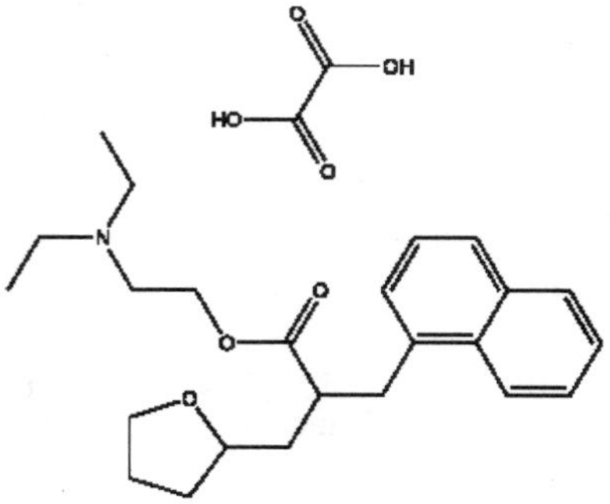

Chemical Abstracts Registry No.: 3200-06-4; 31329-57-4 (Base)

Trade Name	Manufacturer	Country	Year Introduced
Dusodril	Roland	W. Germany	1968
Praxilene	Oberval	France	1968
Prazilene	Lipha	UK	1972
Praxilene	Formenti	Italy	1973
Praxilene	Biochimica	Switz.	1980
Citoxid	Disprovent	Argentina	-

Raw Materials

Oxalic acid
β-Chloroethyl-N-diethylamine
β-(1-Naphthyl)-β'-tetrahydrofurfurylisobutyric acid

Manufacturing Process

30 grams (0.106 mol) of β-(1-naphthyl)-β'-tetrahydrofuryl isobutyric acid are heated under reflux for 8 1/2 hours in 230 cc of isopropanol with 14 grams (0.103 mol) of β-chloroethyl-N-diethylamine. After evaporation of the isopropanol in vacuo, the syrupy residue is treated with a solution of K_2CO_3. Extraction with ether is carried out after drying over Na_2SO_4.

Distillation of the extract yields 28.5 grams of a very viscous yellow liquid with a $BP_{0.95-1.09 millibar}$ = 198° to 202°C. The yield is 70.5% (theoretical quantity = 40.5 grams).1.3 grams (0.0103 mol) of dihydrated oxalic acid are dissolved while being made tepid in 8 cc of acetone. The cooled solution has added thereto 4 grams (0.0104 mol) of N-diethylaminoethyl-β-(1-naphthyl)-β'-tetrahydrofuryl isobutyrate, obtained according to the process described above and dissolved in 10 cc of acetone. The solution is brought to boiling point for 15 minutes. After cooling to ambient temperature, it is placed in a refrigerator. Crystallization occurs after 2 hours, the crystals which have formed are separated by centrifuging, and after washing in hexane and drying in vacuo 3.5 grams of white crystals are obtained. After being recrystallized three times, in alcohol and then in a mixture of alcohol and ethyl acetate, the product is analytically pure and has a MP = 110° to 111°C (heating stage).

References

Merck Index 6201
Kleeman & Engel p. 615
OCDS Vol .2 p. 213 (1980)
DOT 5 (1) 19 (1969)
I.N. p. 654
Szarvasi, E. and Bayssat, M.; US Patent 3,334,096; August 1, 1967; assigned to Lipha, Lyonnaise Industrielle Pharmaceutique, France

NAFTIFINE

Therapeutic Function: Antifungal

Chemical Name: 1-Naphthalenemethanamine, N-methyl-N-(3-phenyl-2-propenyl)-, (E)-

Common Name: Naftifine; Naftifugin

Chemical Abstracts Registry No.: 65472-88-0

Structural Formula:

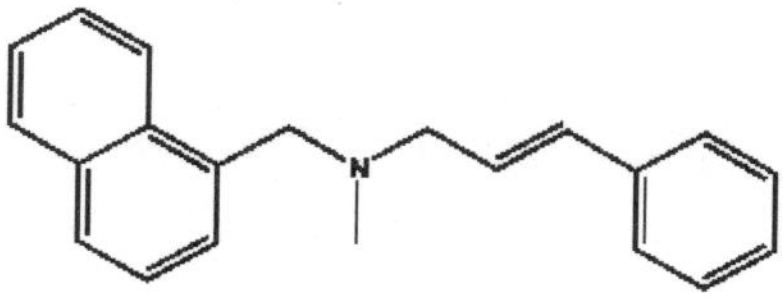

Trade Name	**Manufacturer**	**Country**	**Year Introduced**
Naftifine	Sandoz (Novartis)	-	-

Raw Materials

Methyl-(1-naphthylmethyl)amine hydrochloride
Sodium carbonate
Dimethylformamide
Cinnamyl chloride

Manufacturing Process

To a mixture of 1.42 g of methyl-(1-naphthylmethyl)amine hydrochloride, 2.89 g of sodium carbonate and 10 ml of dimethylformamide is added, at room temperature, 1.25 g of cinnamyl chloride, dropwise. After 18 hours stirring, at room temperature, the mixture is filtered and the filtrate is evaporated in vacuo. The residue is dissolved in toluene and, after drying over sodium sulphate, evaporated to obtain the trans-N-(cinnamylmethyl)-N-methyl-(1-naphthylmethyl)amine compound, boiling point 162-167°C/0.015 Torr.

The free base may be converted, with isopropanolic hydrogen chloride solution, into the hydrochloride form, melting point 177°C (from propanol).

References

Berney Daniel; US Patent No. 4,282,251; August 4, 1981; Assigned to Sandoz Ltd. (Basel, CH)

NALBUPHINE

Therapeutic Function: Analgesic

Chemical Name: N-Cyclobutylmethyl-14-hydroxydihydronormorphinone

Common Name: -

Structural Formula:

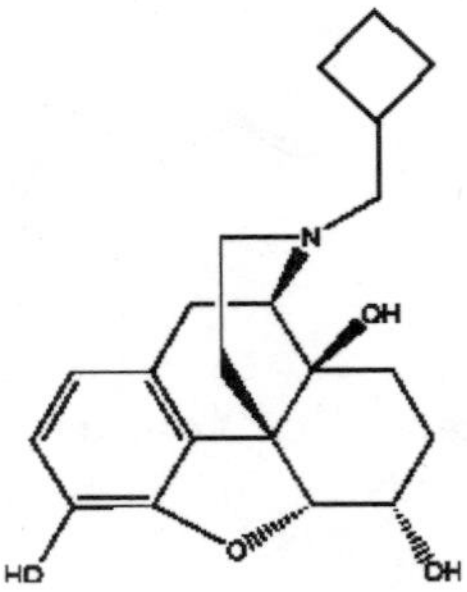

Chemical Abstracts Registry No.: 20594-83-6; 23277-43-2 (Hydrochloride salt)

Trade Name	Manufacturer	Country	Year Introduced
Nubain	Du Pont	US	1979
Nubain	Du Pont	UK	1983

Raw Materials

14-Hydroxydihydronormorphinone
Cyclobutane carboxylic acid chloride
Lithium aluminum hydride

Manufacturing Process

To a slurry of 110.5 g of 14-hydroxydihydronormorphinone in 2.5 liters of methylene chloride and 280 ml of triethylamine was added a solution of 106 g of cyclobutanecarboxylic acid chloride in 500 ml of methylene chloride. The temperature of the reaction mixture was maintained at 20°C to 25°C during the addition. After 5 minutes the reaction mixture was brought to reflux and heated for 5 hours.

It was then cooled, washed with water, dried over sodium sulfate and evaporated to dryness. The residue was crystallized from benzene and pentane to give 138.5 g of the dicyclobutanecarbonyl derivative, melting point about 112°C (dec.).

The dicyclobutanecarbonyl derivative (136.7 g) was dissolved in 200 ml of tetrahydrofuran and added dropwise to a suspension of 34.2 g of lithium aluminum hydride in 1 liters of tetrahydrofuran. The temperature of the mixture rose to reflux during the addition. Reflux was maintained for 2 hours after the addition was completed. After cooling, 110 ml of ethyl acetate was added dropwise, followed by 30 ml of water, followed by a solution of 53 g of ammonium chloride in 125 ml of water. The resulting mixture was filtered and the inorganic precipitate was washed with methanol. Evaporation of the combined filtrates gave 66 g of N-cyclobutylmethyl-14-

hydroxydihydronormorphinone, melting point 229°C to 231°C.

References

Merck Index 6203
DFU 2 (9) 613 (1977)
Kleeman & Engel p. 616
PDR p. 858
OCDS Vol. 2 p. 319 (1980)
DOT 16 (2) 51 (1980)
I.N. p. 654
REM p. 1109
Blumberg, H., Pachter, I.J. and Matossian, Z.; US Patent 3,332,950; July 25, 1967; assigned to Endo Laboratories, Inc.

NALIDIXIC ACID

Therapeutic Function: Antibacterial

Chemical Name: 1-Ethyl-1,4-dihydro-7-methyl-4-oxo-1,8-naphthyridine-3-carboxylic acid

Common Name:-

Structural Formula:

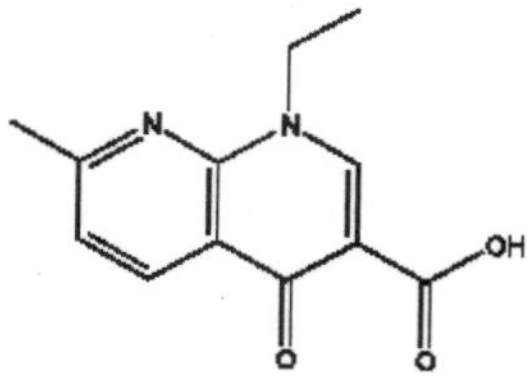

N N OH O O

Chemical Abstracts Registry No.: 389-08-2

Trade Name	Manufacturer	Country	Year Introduced
Neggram	Winthrop	US	1964
Nalidixique	Winthrop	France	1974
Jicsron	Towa Yakuhin	Japan	1981
Baktogram	Farmakos	Yugoslavia	-
Betaxina	Amelix	Italy	-
Chemiurin	Cifa	Italy	-
Cybis	Breon	US	-
Dixiben	Benvegna	Italy	-
Dixurol	I.T.I.	Italy	-
Enexina	S.I.T.	Italy	-
Entolon	Sawai	Japan	-

Trade Name	Manufacturer	Country	Year Introduced
Eucistin	San Carlo	Italy	-
Faril	Saita	Italy	-
Innoxalon	Sanko	Japan	-
Kusnarin	Kodama	Japan	-
Nali	Iltas	Turkey	-
Nalcidin	Schoum	Italy	-
Nalidicron	San-A	Japan	-
Nalidixico	Level	Spain	-
Nalidixin	Spofa	Czechoslovakia	-
Nalidixol	Hermes	Spain	-
Naligen	Sam	Italy	-
Naligram	Isis	Yugoslavia	-
Nalissina	Armour	Italy	-
Nalitucsan	Hishiyama	Japan	-
Nalix	Sigurta	Italy	-
Nalixan	Neofarma	Finland	-
Nalurin	Von Boch	Italy	-
Narigix	Taiyo	Japan	-
Naxuril	Esterfarm	Italy	-
Negabatt	Dessy	Italy	-
Nicelate	Toyo Jozo	Japan	-
Nogermin	Madaus	Spain	-
Notricel	Hortel	Spain	-
Pielos	S.T.I.P.	Italy	-
Poleon	Sumitomo	Japan	-
Renogram	Belupo Ltd.	Yugoslavia	-
Restelon	Maruishi	Japan	-
Sicmylon	Niichiko	Japan	-
Specifin	Bergamon	Italy	-
Unaserus	Isei	Japan	-
Uralgin	Ceccarelli	Italy	-
Uretrene	Mitim	Italy	-
Uriben	R.P. Drugs	UK	-
Uriclar	Crosara	Italy	-
Uri-Flor	A.G.I.P.S.	Italy	-
Urigram	Trima	Israel	-
Urisco	I.C.I.	Italy	-
Uristeril	Ripari-Gero	Italy	-
Urodixin	Italchimici	Italy	-
Urogram	Firma	Italy	-
Urolex	Sirt-B.B.P.	Italy	-
Urolgin N	Takata	Japan	-
Uromina	Ausonia	Italy	-
Uroneg	Ibirn	Italy	-
Valuren	Intersint	Italy	-
Wintomylon	Daiichi	Japan	-
Wintron	Tobishi	Japan	-

Raw Materials

2-Amino-6-methylpyridine
Ethoxymethylene malonic acid diethyl ester
Sodium hydroxide
Ethyl iodide

Manufacturing Process

A warm solution containing 41 grams of 4-hydroxy-7-methyl-1,8-naphthyridine-3-carboxylic acid and 39 grams of potassium hydroxide in 1 liter of ethanol and 200 cc of water was treated with 50 cc of ethyl iodide and the resulting mixture was refluxed gently overnight, acidified with hydrochloric acid and cooled. The resulting precipitate was collected and recrystallized twice from acetonitrile to yield 26 grams (56% yield) of 1-ethyl-7-methyl-4-oxo-1,8-naphthyridine-3-carboxylic acid, MP 229° to 230°C.

The starting material is prepared by reacting 2-amino-6-methylpyridine with ethoxymethylene-malonic acid diethyl ester and then reacting that product with sodium hydroxide.

References

Merck Index 6205
Kleeman & Engel p. 616
PDR p. 1922
OCDS Vol. 1 p. 429 (1977) & 2,370,469 (1980)
DOT 1 (1) 16 (1965)
I.N. p. 33
REM p. 1216
Lesher, G.Y. and Gruett, M.D.; US Patent 3,149,104; September 15, 1964; assigned to Sterling Drug Inc.

NALMEFENE

Therapeutic Function: Antagonist to narcotics

Chemical Name: Morphinan-3,14-diol, 17-(cyclopropylmethyl)-4,5-epoxy-6-methylene-, (5alpha)-

Common Name: Nalmefene; Nalmetrene

Chemical Abstracts Registry No.: 55096-26-9

Raw Materials

Potassium t-butoxide
Methyltriphenylphosphonium bromide
Naltrexone

Structural Formula:

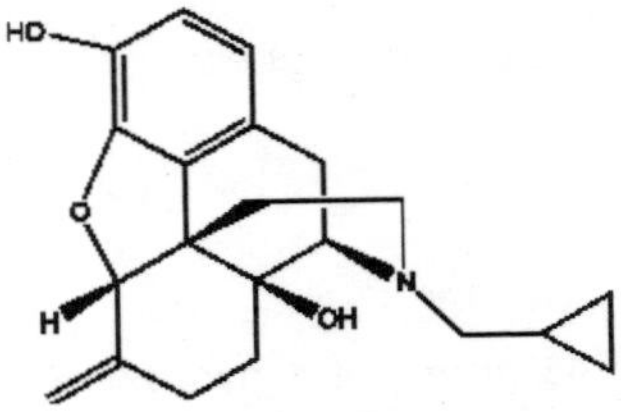

Trade Name	Manufacturer	Country	Year Introduced
Nalmefene	Mallinckrodt Inc.	-	-
Nalmefene	Somaxon Pharmaceuticals	-	-
Arthene	Ohmeda	-	-
Cervene	Baker Norton	-	-
Incystene	SSPharma	-	-
Incystene	Ivax Corporation	-	-
Revex	Baxter Healthcare Corp Anesthesia And Critical Care	-	-
Revex	Ohmeda	-	-

Manufacturing Process

A dry, 2-liter, 3-neck, round bottom flask fitted with two stoppers and a magnetic stirring bar was charged with potassium t-butoxide (61.1 g, 0.545 mol) and methyltriphenylphosphonium bromide (194.4 g, 0.544 mol). Freshly distilled tetrahydrofuran (450 ml) was introduced at 20°C. The resultant thick, bright yellow dispersion was stirred at 20°C for 0.5 h and further dry tetrahydrofuran (100 ml) was added. A solution of dry naltrexone (30 g, 0.088 mol) in dry tetrahydrofuran (200 ml) was then added dropwise over 40 min. Then the reaction mixture was stirred for a further 1.25 h, then cooled to 10°C, and quenched with 20% aqueous ammonium chloride solution (75 ml) followed by water (100 ml). The organic layer was separated and the aqueous layer extracted with four 100 ml portions of chloroform. Solvent was evaporated from the tetrahydrofuran layer and the combined chloroform extracts, the residues combined and brought to pH 2 by addition of 2 N hydrochloric acid. The resultant precipitate was filtered, washed with chloroform and suspended in a mixture of chloroform (500 ml) and water (250 ml). Ammonium hydroxide was added to attain a pH of 8 and the aqueous layer separated. The organic layer was dried over anhydrous sodium sulfate, filtered, and the solvent removed in vacuo. The resultant solid was dissolved in ethyl acetate (1400 ml), the solution filtered through a silica pad and the solvent evaporated. The product was recrystallized from chloroform and washed with hexane to yield pure 6-desoxy-6-methylenenaltrexone (also called nalmefene) as a white solid. Yield: 27.0 g, 88%.

References

Merck Index, Monograph number: 6447, Twelfth edition, 1996, Editor: S. Budavari

Merck and Co., Inc.; Meltzer P.C., Coe J. W.; US Patent No. 4,535,157; August 13, 1985; Assigned: Key Pharmaceuticals, Inc. (Miami, FL)

NALORPHINE

Therapeutic Function: Narcotic antagonist

Chemical Name: 7,8-Didehydro-4,5-epoxy-17-(2-propenyl)morphinan-3,6-diol

Common Name: N-Allylnormorphine

Structural Formula:

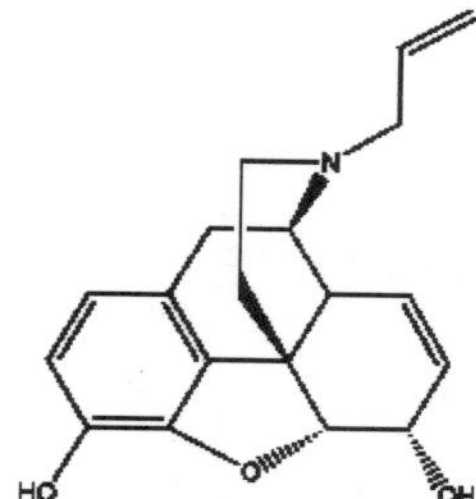

Chemical Abstracts Registry No.: 62-67-9; 57-29-4 (Hydrochloride salt)

Trade Name	Manufacturer	Country	Year Introduced
Nalline	MSD	US	1952
Lethidrone	Wellcome	W. Germany	-
Nalorphine	Clin-Comar-Byla	France	-
Norfin	Lusofarmaco	Italy	-

Raw Materials

Normorphine
Allyl bromide
Sodium bicarbonate

Manufacturing Process

6 grams of normorphine, 2.7 grams of allyl bromide, 2.65 grams of sodium bicarbonate, and 75 cc of methanol were mixed together, and the resulting mixture was heated under reflux with stirring for a period of about 5 1/2 hours. The reaction mixture was evaporated to dryness in vacuo, the residual material was extracted with 60 cc of boiling chloroform, 0.5 gram of activated charcoal was added, and the resulting mixture was filtered through a layer of diatomaceous silica. The filter cake was washed with four 10 cc portions of

boiling chloroform, and the chloroform filtrate and washings were combined and evaporated to dryness in vacuo. The residual material was triturated with 25 cc of anhydrous ether until crystalline, the ethereal mixture was cooled, maintained at a temperature of 3°C overnight, filtered, and the crystalline mixture was washed with three 10 cc portions of ice-cold ether. The resulting crystalline product was dried to give 6.0 grams of N-allylnormorphine, yield approximately 87% of theory, according to US Patent 2,891,954.

References

Merck Index 6206
Kleeman & Engel p. 617
OCDS Vol. 1 p. 288 (1977) & 2,318 (1980)
I.N. p. 655
REM p. 1106
Weijlard, J. and Erickson, A.E.; US Patent 2,364,833; December 12, 1944; assigned to Merck & Co., Inc.
Weijlard, J.; US Patent 2,891,954; June 23, 1959; assigned to Merck and Co., Inc.

NALOXONE

Therapeutic Function: Narcotic antagonist

Chemical Name: 17-Allyl-4,5α-epoxy-3,14-dihydroxy-morphinan-6-one

Common Name: N-Allylnoroxymorphone; N-Allyl-1,4-hydroxydihydronormorphinone

Structural Formula:

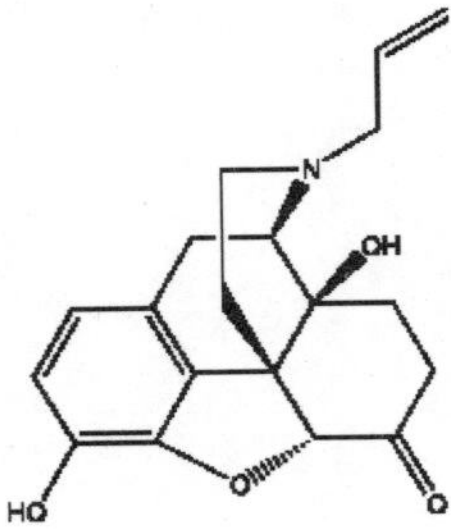

Chemical Abstracts Registry No.: 465-65-6; 357-08-4 (Hydrochloride salt)

Trade Name	Manufacturer	Country	Year Introduced
Narcan	Du Pont	US	1971
Narcan	Du Pont	UK	1975
Narcanti	Winthrop	W. Germany	1978

Trade Name	Manufacturer	Country	Year Introduced
Narcan	Winthrop	France	1980
Narcan	Crinos	Italy	1980
Nalone	Endo	US	-
Talwin	Winthrop-Breon	US	-

Raw Materials

Oxymorphone
Allyl bromide
Hydrogen chloride
Cyanogen bromide
Acetic anhydride

Manufacturing Process

10 grams of 14-hydroxydihydromorphinone (oxymorphone) was converted into its diacetate by warming it on the steam bath with 80 cc of acetic anhydride for about 2 hours. The acetic anhydride was removed on the water bath under a vacuum of about 30 mm absolute pressure. The melting point of the residue was 220°C. The residue was taken up in 100 cc of chloroform. An equal amount by weight of cyanogen bromide was added and the mixture was refluxed at about 60°C for about 5 hours. After refluxing, the mixture was washed with 100 cc of a 5% aqueous hydrochloric acid solution, dried over sodium sulfate and the chloroform removed by evaporation under a vacuum of about 30 mm. The residue had a melting point of 240°C.

The residue was then heated at about 90°C for 16 hours on a steam bath with 300 cc of 20% aqueous hydrochloric acid solution, and treated with a small amount, e.g., 1 gram of charcoal. The hydrochloric acid was then removed under a vacuum of 15 mm, the residue dissolved in 30 cc of water and precipitated by the addition of 2.4 cc of concentrated aqueous ammonia. The precipitate was filtered off and dried. It consists of 14-hydroxydihydronormorphinone. It is soluble in ethanol.

The 14-hydroxydihydronormorphinone was suspended in 200 cc of pure ethyl alcohol, half its weight of sodium bicarbonate and half its weight of allyl bromide added and the resulting mixture was refluxed at about 75°C for 48 hours. The solution was cooled, e.g., to 10°C and filtered and the alcohol removed under a vacuum of 30 mm. The residue was dissolved in chloroform and filtered. The chloroform was removed under a vacuum of 30 mm and the residue was crystallized from ethylacetate. The crystallized product, N-allyl-1,4-hydroxydihydronormorphinone, has a melting point of 184°C, is soluble in chloroform and insoluble in petroleum ether. The yield amounts to 20% based on the weight of the reacted 14-hydroxydihydromorphinone.

References

Merck Index 6208
Kleeman & Engel p. 618
PDR pp. 858, 1932
OCDS Vol. 1 p. 289 (1977) & 2,318,323 (1980)
DOT 8 (8) 295 (1972); I.N. p. 655
REM p. 1106
Lewenstein, M.J. and Fishman, J.; US Patent 3,254,088; May 31,1966

NALTREXONE

Therapeutic Function: Narcotic analgesic

Chemical Name: Morphinan-6-one, 17-(cyclopropylmethyl)-4,5-α-oxy-3,14-dihydoxy-

Common Name: Naltrexone

Structural Formula:

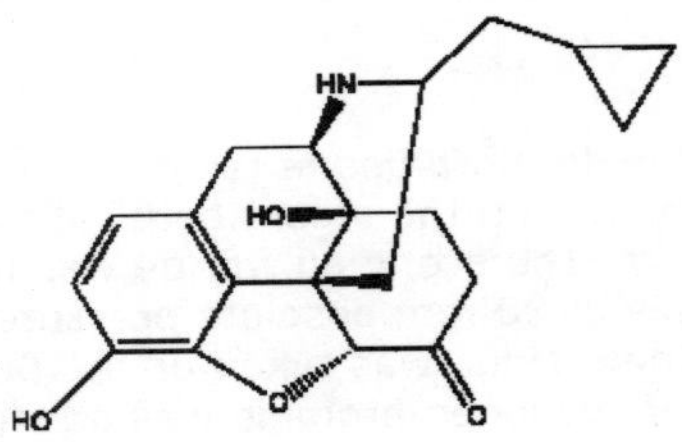

Chemical Abstracts Registry No.: 16590-41-3

Trade Name	Manufacturer	Country	Year Introduced
Antaxone	Zambon Group	Italy	-
Nalorex	DuPont Pharmaceuticals	-	-
Naltima	Intas Pharm.	-	-
Naltrexone	DuPont Merck	-	-
Nodict	Synergy (Sun)	India	-
ReVia	DuPont Pharmaceuticals	-	-

Raw Materials

Codeine
Acetic anhydride
Oxalyl chloride
Formic acid
Palladium on carbon
4-Dimethylaminopyridine
1-Chloroethyl chloroformate
(Chloromethyl)cyclopropane
3-Chloroperbenzoic acid
Boron tribromide

Manufacturing Process

Codeine is a component of gum opium and can also be produced by methylation of morphine using known prior art techniques.

A solution of codeine (30 g, 100.2 mmol), acetic anhydride (18.4 g, 180.2 mmol), triethylamine (18.25 g, 180.2 mmol) and 4-dimethylaminopyridine (0.5 g) in dry ethyl acetate (620 ml) was stirred at rt. under nitrogen for 12 hr, added saturated aqueous sodium bicarbonate solution until no acetic anhydride detected. The organic portion was separated, washed with water (3 times 120 ml), dried over anhydrous sodium sulfate, and evaporated in vacuo to dryness to give 6-acetylcodeine as white solids (34.0 g, 99% yield).

Preparation of 6-acetylnorcodeine hydrochloride.

A solution of 6-acetylcodeine (10.0 g, 29.3 mmol), 1-chloroethyl chloroformate (5.51 g, 37.8 mmol), and proton sponge (1.0 g) in methylene chloride (80 ml) was heated at reflux for 80 min. The reaction mixture was evaporated in vacuo to dryness. The residue was chromatographed on silica gel with ethyl acetate to give 6-acetyl-17-(1-chloroethoxycarbonyl)norcodeine as an oil (12.13 g), which was dissolved in methanol with a few drops of conc. HCl. The solution was heated at reflux for 1 hr and evaporated in vacuo to almost dryness. The residue was added hexane and filtered to give 6-acetylnorcodeine hydrochloride (10.7 g, 100% yield).

Preparation of norcodeine hydrochloride.

A solution of 6-acetylcodeine (10.0 g, 29.3 mmol), 1-chloroethyl chloroformate (5.56 g, 38.1 mmol), and proton sponge (1.0 g) in methylene chloride (50 ml) was heated at reflux for 50 min. The reaction mixture was evaporated in vacuo to about 30 ml. Methanol (25 ml) and concentrated HCl (2 ml) were added. The solution was heated at reflux for 40 min. and evaporated in vacuo to almost dryness. The residue was added hexane and filtered to give norcodeine hydrochloride (8.8 g, 93% yield).

Preparation of 17-cyclopropylmethylnorcodeine.

A mixture of norcodeine hydrochloride (11.48 g, 27.8 mmol), (chloromethyl)cyclopropane (5.14 g, 55.6 mmol), sodium carbonate (14.73 g, 139.0 mmol), and potassium iodide (4.61 g, 27.8 mmol) in ethanol (250 ml) was heated at reflux for 20 hr, cooled, and evaporated in vacuo to dryness. The residue was basified with NH_4OH, and extracted with methylene chloride. The extract was washed with water and evaporated in vacuo to dryness. The residue (11.7 g) was chromatographed on silica gel with a eluting solvent system of methanol/ethyl acetate (10/90) to give 17-cyclopropylmethylnorcodeine (10.68 g, 91% yield).

Preparation of 17-cyclopropylmethylnorcodeinone.

To a solution of DMSO (14.50 g, 185.6 mmol) in methylene chloride (80 ml) at -78°C, was added a solution of oxalyl chloride (11.78 g, 92.8 mmol) in methylene chloride (20 ml) in 20 min. After stirring at -78°C for 20 min., a solution of 17-cyclopropylmethylnorcodeine (9.0 g, 26.5 mmol) in methylene chloride (40 ml) was added dropwise in 50 min. The reaction mixture was stirred at -74° to -76°C for 3 hr, added triethylamine (9.39 g, 92.8 mmol), allowed to warm up to rt., added methylene chloride (200 ml), washed with water (10 times 50 ml), and evaporated in vacuo to dryness. The residue was mixed with hexane and filtered to give 17-cyclopropylmethylnorcodeinone (8.85 g, 99% yield).

Preparation of 17-cyclopropylmethylnorcodeinone dienol acetate.

A mixture of 17-cyclopropylmethylnorcodeinone (3.55 g, 10.5 mmol), acetic anhydride (20 ml, 210.4 mmol), sodium acetate (1.3 g, 15.8 mmol), and toluene (6 ml) was heated at 71°-73°C for 14 hr. The reaction mixture was cooled, added methylene chloride (250 ml), water (50 ml), and sodium bicarbonate (73.5 g), stirred for 4 hr, and filtered. The organic portion of the filtrate was separated, washed with water (30 ml), dried over anhydrous sodium sulfate, and evaporated in vacuo to dryness. The residue (3.94 g) was chromatographed on silica gel with 100% ethyl acetate to give 17-cyclopropylmethylnorcodeinone dienol acetate (2.87 g, 72% yield).

Preparation of 17-cyclopropylmethyl-14-hydroxynorcodeinone. A solution of 17-cyclopropylmethylnorcodeinone (0.20 g, 0.59 mmol), formic acid (90%, 0.304 g), water (0.504 g), EtOAc (0.27 g), and hydrogen peroxide (30%, 0.17 g) was heated at 42°-43°C for 15 hr, added water (20 ml), basified with Na_2CO_3 (1.02g), and extracted with EtOAc (80 ml and 2 times 20 ml). The combined extract was washed with water, dried over anhydrous sodium sulfate, and evaporated in vacuo to dryness to give 17-cyclopropylmethyl-14-hydroxynorcodeinone (0.10 g, 56% yield). The Rf value in TLC and the IR spectrum of the product were comparable to those obtained from an authentic sample.

Preparation of 17-cyclopropylmethyl-14-hydroxynorcodeinone.

A solution of 17-cyclopropylmethylnorcodeinone dienol acetate (1.00 g, 2.63 mmol), formic acid (8 ml, 90%), and hydrogen peroxide (0.37 g, 30%, 3.26 mmol) was heated at 44°-45°C for 6 hr, added water (20 ml) and ethyl acetate (80 ml), basified with sodium bicarbonate. The organic portion was separated, washed with water (15 ml), dried over anhydrous sodium sulfate and evaporated in vacuo to dryness, the residue (0.9 g) was chromatographed on silica gel with methanol/methylene chloride (2.5/97.5) to give 17-cyclopropylmethyl-14-hydroxynorcodeinone (0.72 g, 78% yield).

Preparation of 17-cyclopropylmethyl-14-hydroxynorcodeinone.

A solution of 17-cyclopropylmethylnorcodeinone dienol acetate (0.5 g, 1.31 mmol), 3-chloroperbenzoic acid (0.36 g, 2.10 mmol) and oxalic acid (0.27 g, 2.90 mmol) in acetic acid (7 ml) was stirred at rt. overnight, added cold water (35 ml), basified with sodium carbonate, and extracted with methylene chloride (100 ml). The extract was washed with water (2 times 30 ml), dried over anhydrous sodium sulfate, and evaporated in vacuo to dryness. The residue (0.41 g) was chromatographed on silica gel to give 17-cyclopropylmethyl-14-hydroxynorcodeinone (0.34 g, 74% yield). The Rf value in TLC and the IR spectrum of the product were comparable to those obtained from an authentic sample.

Preparation of 3-methylnaltrexone.

A mixture of 17-cyclopropylmethyl-14-hydroxynorcodeinone (0.30 g, 0.85 mmol) and Pd/C (5%, 0.45 g) in ethanol (35 ml) was hydrogenated in a Parr hydrogenator at rt. under 28 psi of hydrogen gas. The mixture was filtered. The filtrate was evaporated in vacuo to dryness to give 3-methylnaltrexone (0.30 g, 99% yield).

Preparation of naltrexone from 3-methylnaltrexone.

A solution of 3-methylnaltrexone (0.48 g, 1.35 mmol) in methylene chloride (30 ml) was cooled with an ice-water bath, and then added a solution of boron tribromide (5.4 ml, 1 M solution in methylene chloride, 5.4 mmol). The reaction mixture was stirred at rt. for 15 hr, basified with NH_4OH, and extracted with methylene chloride (60 ml). The extract was washed with water (2 times 15 ml), dried over anhydrous sodium sulfate, and evaporated in vacuo to dryness to give naltrexone (0.45 g, 98% yield).

References

Huang B.-S. et al.; US Patent No. 6,013,796; Assigned to Penick Corporation, Newark, N.J.

NANDROLONE DECANOATE

Therapeutic Function: Anabolic

Chemical Name: 17β-[(1-Oxodecyl)oxy]estr-4-en-3-one

Common Name: 19-Nortestosterone decanoate; Norandrostenolone decanoate

Structural Formula:

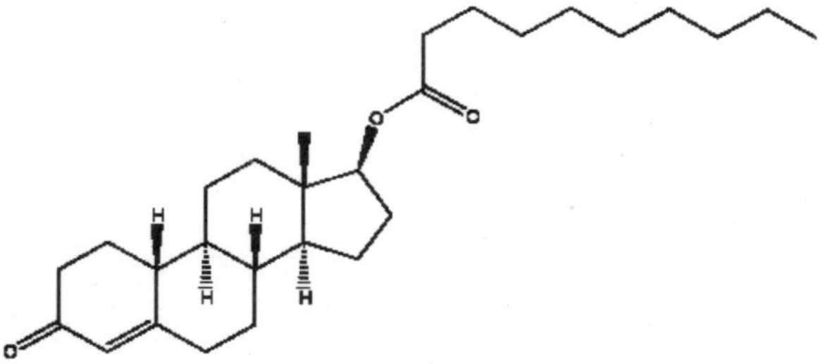

Chemical Abstracts Registry No.: 360-70-3; 434-22

Trade Name	**Manufacturer**	**Country**	**Year Introduced**
Deca-Durabolin	Organon	US	1962
Deca-Hybolin	Hyrex	US	1979
Deca-Noralone	Taro	Israel	-
Fortabolin	Deva	Turkey	-
Iebolan	I.E. Kimya Evi	Turkey	-
Kabolin	Legere	US	-
Methybol	Mepha	Switz.	-

Trade Name	Manufacturer	Country	Year Introduced
Nordecon	Ibsa	Switz.	-
Sterobolin	Neofarma	Finland	-
Turinabol-Depot	Jenapharm	E. Germany	-

Raw Materials

19-Nortestosterone
Decanoic acid chloride

Manufacturing Process

1 gram of 19-nortestosterone is dissolved in 3 ml of dry pyridine, after which the resulting solution is cooled to -20°C. A solution of 1.0 gram of decanoic acid chloride in 3 ml of dry benzene is added to the cooled solution. The mixture is maintained at -15°C for 16 hours and then poured into ice water. The aqueous liquid is extracted with benzene, the benzene solution is washed with respectively 1 N sodium hydroxide solution, 2 N hydrochloric acid and with water until neutral reaction.

Then the solution is dried on sodium sulfate, filtered, and evaporated to dryness. The residue, 1.63 grams is dissolved in hexane, this solution is filtered over 30 grams of neutral aluminum oxide, and evaporated to dryness. On paper chromatographic investigation it turned out that the obtained 19-nortestosterone 17-decanoate which at room temperature is an oil consists of a single compound, according to US Patent 2,998,423.

References

Merck Index 6212
Kleeman & Engel p. 620
PDR pp. 1033, 1286
OCDS Vol. 1 p. 171 (1977)
I.N. p. 655
REM p. 999
Donia, R.A. and Ott, A.C.; US Patent 2,798,879; July 9, 1957; assigned to The Upjohn Company
De Wit, E.D. and Overbeek, G.A.; US Patent 2,998,423; August 29, 1961; assigned to Organon Inc.

NANDROLONE PHENPROPIONATE

Therapeutic Function: Anabolic

Chemical Name: 17β-Hydroxyestr-4-en-3-one 3-phenylpropionate

Common Name: 19-Nortestosterone β-phenylpropionate

Structural Formula:

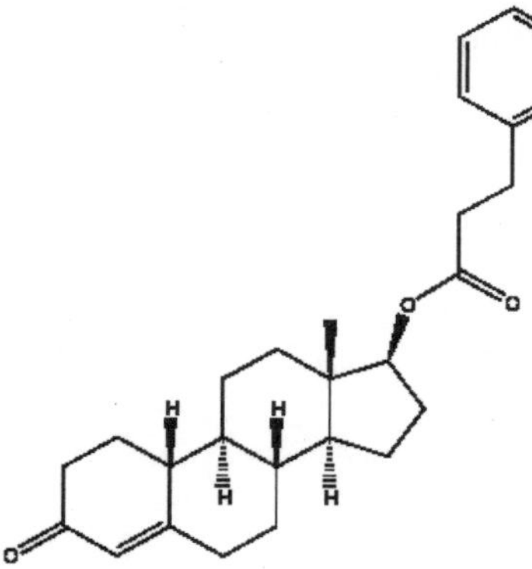

Chemical Abstracts Registry No.: 62-90-8; 434-22-0 (Base)

Trade Name	Manufacturer	Country	Year Introduced
Durabolin	Organon	US	1959
Nandrolin	Tutag	US	1979
Activin	Aristegui	Spain	-
Anticatabolin	Falorni	Italy	-
Hepa-Obaton	Noury Pharma	W. Germany	-
Hybolin Improved	Hyrex	US	-
Norabol	Pharmacia	Sweden	-
Noralone	Taro	Israel	-
Norandrol	Panther-Osfa	Italy	-
Norandros	Castillon	Spain	-
Norbalin	Bieffe	Italy	-
Noromon	Ibsa	Switz.	-
Norstenol	Ravizza	Italy	-
Sintabolin	A.F.I.	Italy	-
Strabolene	Isola-Ibi	Italy	-
Superanbolon	Spofa	Czechoslovakia	-
Superbolin	Labif	Italy	-
Turinabol	Jenapharm	E. Germany	-

Raw Materials

19-Nortestosterone
β-Phenylpropionyl chloride

Manufacturing Process

An ice-cold solution of 1.5 grams of 19-nortestosterone and 1.5 ml of dry pyridine in 10 ml of dry benzene is prepared and a solution of 1.5 ml of β-phenylpropionyl chloride in 5 ml of dry benzene is added dropwise over a period of about 2 minutes with stirring. The resulting mixture is allowed to stand overnight under an atmosphere of nitrogen and then washed successively with cold 5% aqueous hydrochloric acid solution, cold 2.5% aqueous sodium hydroxide solution, and water. After drying over anhydrous sodium sulfate, the solvent is evaporated to give an almost colorless oil. Recrystallization from methanol gives white crystals of 19-nortestosterone 17-β-phenylpropionate, MP 91° to 92.5°C.

References

Merck Index 6214
Kleeman & Engel p. 621
PDR p. 1286
OCDS Vol. 1 p. 171 (1977)
I.N. p. 656
REM p. 999
Donia, R.A. and Ott, A.C.; US Patent 2,868,809; January 13, 1959; assigned to The Upjohn Company

NAPHAZOLINE

Therapeutic Function: Nasal decongestant

Chemical Name: 4,5-Dihydro-2-(1-naphthalenylmethyl)-1H-imidazole

Common Name: 2-(1-Naphthylmethyl)imidazoline

Structural Formula:

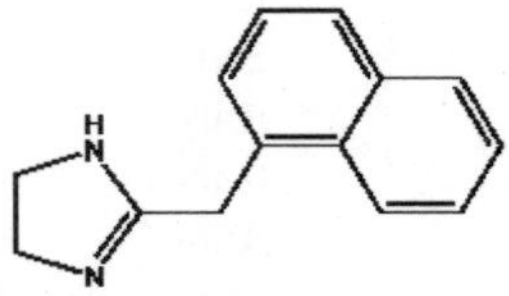

Chemical Abstracts Registry No.: 835-31-4

Trade Name	Manufacturer	Country	Year Introduced
Privine	Ciba	US	1942
Albalon	Allergan	US	1970
Naphcon Forte	Alcon	US	1975
Clera	Person Covey	US	1978
Vasoclear	Smith, Miller and Patch	US	1979
Opcon	Muro	US	1981
Nafazair	Pharmafair	US	1983

Trade Name	Manufacturer	Country	Year Introduced
Actinophtyl	Gregoire	France	-
Bactio-Rhin	Byk Liprandi	Argentina	-
Biogan	Recip	Sweden	-
Coldan	Sigmapharm	Austria	-
Degest-2	Barnes Hind	US	-
Gotinal	Promeco	Argentina	-
Imidazyl	Tubi Lux Pharma	Italy	-
Imidin	Ysat Wernigerode	E. Germany	-
Imizol	Farmigea	Italy	-
Murine	Abbott	UK	-
Naftazolina	Bruschettini	Italy	-
Naline	Ibsa	Switz.	-
Nasal Yer	Yer	Spain	-
Nomaze	Fisons	UK	-
Ocunasal	Sam-On	Israel	-
Pivanol	Tek	Turkey	-
Privin	Ciba	W. Germany	-
Proculin	Ankerwerk	E. Germany	-
Ran	Corvi	Italy	-
Rhinex S	Ysat Wernigerode	E. Germany	-
Rhinon	Petrasch	Austria	-
Rimidol	Leo	Sweden	-
Rinofug	Chimimport Export	Rumania	-
Vasoconstrictor	Pensa	Spain	-
Vistalbalon	Pharm-Allergan	W. Germany	-

Raw Materials

Naphthyl-(1)-acetonitrile
Ethanol
Methanol
Ethylene diamine

Manufacturing Process

2.7 parts of naphthyl-(1)-acetiminoethylether hydrochloride of the formula

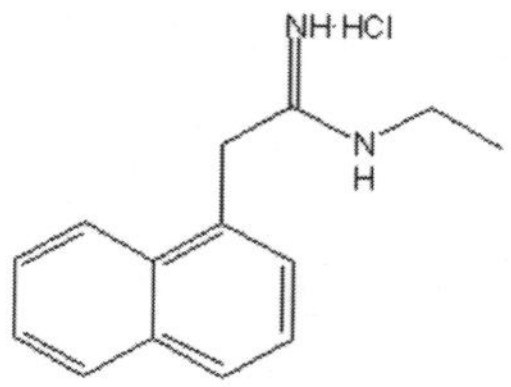

NH·HCl
N
H

(produced from naphthyl-(1)-acetonitrile and methanol) are dissolved in 12 parts of absolute alcohol. 1 part of ethylenediamine is then added and the whole is heated to gentle boiling while passing nitrogen through it and simultaneously stirring until ammonia escapes no longer. The alcohol is then distilled and the residue mixed with 40 parts of benzene and 1.8 parts of caustic potash. Stirring is continued for some time whereby the imidazoline base is dissolved in benzene. The benzene residue is recrystallized several times from toluene.

References

Merck Index 6218
Kleeman & Engel p. 622
PDR pp. 728, 809, 1549
OCDS Vol. 1 p. 241 (1977)
I.N. p. 657
REM p. 888
Sonn, A.; US Patent 2,161,938; June 13, 1939; assigned to the Society of Chemical Industry in Basle, Switzerland

NAPHAZOLINE HYDROCHLORIDE

Therapeutic Function: Vasoconstrictor, Nasal decongestant

Chemical Name: 2-Imidazoline, 2-(1-naphthylmethyl)-, monohydrochloride

Common Name: Nafazolin(a) hydrochloride; Naftazolina hydrochloride; Naphazoline hydrochloride; Naphtazoline hydrochloride; Naphthizin(um) hydrochloride

Structural Formula:

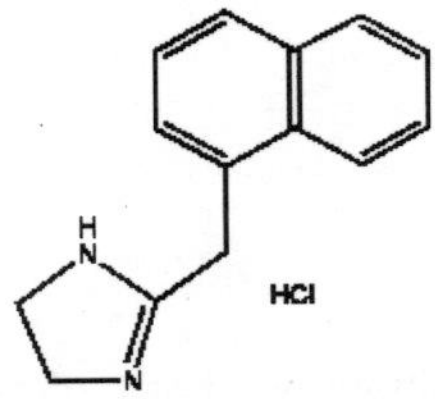

Chemical Abstracts Registry No.: 550-99-2

Trade Name	Manufacturer	Country	Year Introduced
Albalon	Allergan	Australia	-
Clear Eyes	Farnam Companies, Inc.	USA	-
Naphazoline Hydrochloride	Allergan	-	-
Niazol	Dahuachem	China	-
Rhinantin	Dahuachem	China	-

Trade Name	Manufacturer	Country	Year Introduced
Rinofug	Sc Meduman Sa	Rumania	-
Sanorin	Galena	Czech Republic	-
Vasoclear	CIBA Vision	-	-
Vasocon	Novartis	USA	-

Raw Materials

Naphtyl-(1)-acetiminoethylether
Ethylenediamine

Manufacturing Process

2.7 parts of naphtyl-(1)-acetiminoethylether (produced from naphthyl-1-acetonitrile) were dissolved in 12 parts of absolute alcohol. 1 part of ethylenediamine is then added and the mixture was heated to gentle boiling while passing it through nitrogen and simultaneously stirring until ammonia escaped no longer. The alcohol was is then distilled and the residue mixed with 40 parts of benzene and 1.8 parts of caustic potash. Stirring was continued for some time whereby the imidazoline base was dissolved in benzene. The benzene residue, is recrystallized several times from toluene. The 2-[naphthyl-1-methyl]-imidazoline represented the coloriless crystalls of melting point 252°-253°C. Its hydrochloride is easily soluble in water.

References

Sonn A., US Patent No. 2,161,938; June 13, 1939

NAPROXEN

Therapeutic Function: Antiinflammatory

Chemical Name: (+)-6-Methoxy-α-methyl-2-naphthaleneacetic acid

Common Name: d-2-(6-Methoxy-2-naphthyl)propionic acid

Structural Formula:

O
OH

Chemical Abstracts Registry No.: 22204-53-1

Trade Name	Manufacturer	Country	Year Introduced
Naprosyn	Syntex	UK	1973
Naprosyne	Cassenne	France	1975

Trade Name	Manufacturer	Country	Year Introduced
Proxen	Gruenenthal	W. Germany	1975
Naprosyn	Recordati	Italy	1975
Naprosyn	Syntex	Switz.	1975
Naprosyn	Syntex	US	1976
Naixan	Tanabe	Japan	1978
Congex	Nemi	Argentina	-
Floginax	Farmochimica	Italy	-
Gibixen	Gibipharma	Italy	-
Laser	Tosi-Novara	Italy	-
Madaprox	Madariaga	Spain	-
Naprium	Radiumpharma	Italy	-
Naprius	Magis	Italy	-
Naprux	Andromaco	Argentina	-
Naxyn	Teva	Israel	-
Novonaprox	Novopharm	Canada	-
Numide	Hosbon	Spain	-
Prexan	Lafare	Italy	-
Veradol	Schering	W. Germany	-
Xenar	Alfar Farma Clutici	Italy	-

Raw Materials

Sodium hydroxide
Magnesium
Cadmium chloride
2-Bromo-6-methoxynaphthalene
Ethyl-2-bromopropionate

Manufacturing Process

According to US Patent 3,658,858, a solution of 24 grams of 2-bromo-6-methoxynaphthalene in 300 ml of tetrahydrofuran is slowly added to 2.5 grams of magnesium turnings and 100 ml of tetrahydrofuran at reflux temperature. After the addition is complete, 20 grams of cadium chloride is added, and the resultant mixture is refluxed for 10 minutes to yield a solution of di-(6-methoxy-2-naphthyl)cadmium (which can be separated by conventional chromatography, although separation is unnecessary).

A solution of 18 grams of ethyl 2-bromopropionate in 20 ml of tetrahydrofuran is then added to the cooled reaction mixture. After 24 hours at 20°C, the product is hydrolyzed by adding 200 ml of 5 weight percent methanolic sodium hydroxide followed by heating to reflux for 1 hour. The reaction mixture is then diluted with excess 1 N sulfuric acid and extracted with ether. The ether phase is separated, evaporated to dryness and the residue is recrystallized from acetone-hexane to yield 2-(6-methoxy-2-naphthyl)propionic acid.

References

Merck Index 6269
Kleeman & Engel p. 623
PDR p. 1801

OCDS Vol. 1 p. 86 (1977)
DOT 9 (9) 384 (1973) & 10 (3) 95 (1974)
I.N. p. 658
REM p. 1119
Alvarez, F.S.; US Patent 3,637,767; January 25, 1972; assigned to Syntex Corp., Panama Harrison, I.T.; US Patent 3,658,858; April 25, 1972; assigned to Syntex Corp., Panama Alvarez, F.S.; US Patent 3,663,584; May 16, 1972; assigned to Syntex Corp., Panama

NARATRIPTAN

Therapeutic Function: Serotonin antagonist, Migraine therapy

Chemical Name: 1H-Indole-5-ethanesulfonamide, N-methyl-3-(1-methyl-4-piperidinyl)-

Common Name: Naratriptan

Structural Formula:

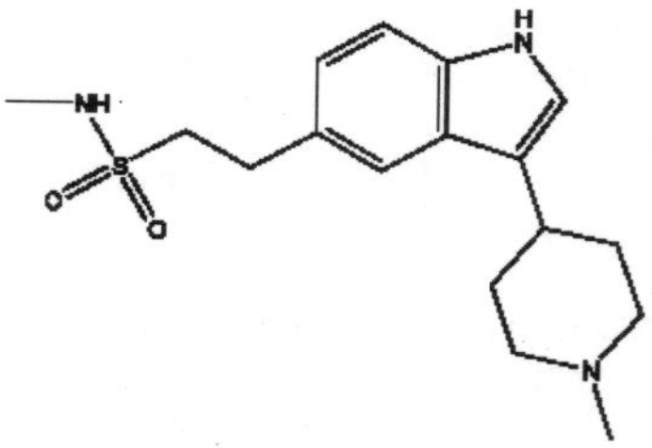

Chemical Abstracts Registry No.: 121679-13-8

Trade Name	Manufacturer	Country	Year Introduced
Amerge	Glaxo Wellcome	UK	-
GR 85548 X	GlaxoSmithKline	USA	-
Naratriptan	GlaxoSmithKline	USA	-

Raw Materials

N-Methyl-4-piperidone
Oxalic acid
Palladium on carbon
4-Hydrazino-N-methyl-benzenethanesulphonamide
N-Methyl-1H-indole-5-ethanesulphonamide
1-Methyl-4-piperidineacetaldehyde

Manufacturing Process

N-Methyl-3-(1,2,3,6-tetrahydro-1-methyl-4-pyridinyl)-1H-indole-5-ethanesulphonamide oxalate

A solution of N-methyl-1H-indole-5-ethanesulphonamide (1.0 g) in methanol (50 ml) containing potassium hydroxide (5.6 g) and N-methyl-4-piperidone (1.0 ml) was heated at reflux for 24 h, cooled, and the resulting solid filtered off (1.0 g). A sample of the solid (0.2 g) was dissolved in a hot methanolic solution of oxalic acid (0.06 g), the solution cooled, and the salt precipitated by adding ethyl acetate (20 ml) and dry ether (50 ml). The salt was filtered off, and dried in vacuo to give the title compound as a solid (0.12 g), m.p. 87°-90°C (shrinks).

Analysis Found: C,52.2; H,5.6; N,9.5. $C_{17}H_{23}N_3O_2S \cdot C_2H_2O_4 \cdot 0.6H_2O$ requires C,52.5; H,6.0; N,9.7%.

N-Methyl-3-(1-methyl-4 -piperidinyl)-1H-indole-5-ethansulphonamide

N-Methyl-3-(1,2,3,6-tetrahydro-1-methyl-4-pyridinyl)-1H-indole-5-ethanesulphonamide oxalate (as the free base) (0.36 g, 0.001 mol) in absolute alcohol (70 ml) and anhydrous dimethylformamide (5 ml) was hydrogenated, in the presence of 5% palladium on activated carbon (0.36 g) at ambient temperature and atmospheric pressure. After 20 h, hydrogen absorption (25 cm^3, theoretical = 24 cm^3) ceased. The catalyst was filtered off and the solvent removed in vacuo to given an opaque gum which solidified as a soft white solid (0.3 g). Purification by flash chromatography (Sorbsil C60 silica gel, CH_2Cl_2/EtOH/0.88 ammonia; 50:80:1) gave a colorless oil (0.21 g) that was triturated with ether to give the title compound (0.17 g) m.p. 156°-158°C. TLC SiO_2(CH_2Cl_2/EtOH/0.88 ammonia; 50:8:1) Rf 0.4.

N-Methyl-3-(1-methyl-4-piperidinyl)-1H-indole-5-ethanesulphonamide may be prepared the another way.

A solution of 4-hydrazino-N-methyl-benzenethanesulphonamide (0.5 g) and 1-methyl-4-piperidineacetaldehyde (0.35 g) in a mixture of water (10 ml) of 2 N hydrochloric acid (1.0 ml, 2.00 mmol) was stirred for 2 days at room temperature. A further quantity of the aldehyde (0.35 g) was added and stirring continued for a further 30 min. The solution was then basified with 8% sodium bicarbonate to pH 8 and extracted with chloroform (3 times 50 ml). The combined organic extracts were dried (Na_2SO_4) and evaporated in vacuo to give the crude hydrazone as an oil (1.0 g). A solution of the hydrazone (1.0 g) in chloroform (20 ml) containing polyphosphate ester (10 g) was heated at reflux for 8 min. The solution was poured onto ice (200 g), stirred for 2 h treated with 2 M sodium carbonate (20 ml) and extracted with chloroform (3 times 50 ml). The combined organic extracts were dried (Na_2SO_4), evaporated in vacuo and the residue purified by flash chromatography (silica 9385, 100 g) eluting with CH_2Cl_2/EtOH/NH_3(75:8:1) to give impure material as a yellow oil. Further flash chromatography (silica 9385, 100 g) eluting with CH_2Cl_2/EtOH/NH_3 (100:8:1) gave the product as an oil (0.05 g). This was crystallised from ethyl acetate to give the title compound solid m.p. 156°-157°C. TLC SiO_2(CH_2Cl_2/EtOH/NH_3(50:8:1)) Rf 0.6.

References

Oxford A.W. et al.; US Patent No. 4,997,841; Mar. 5, 1991; Assigned to Glaxo Group Limited, England

EPA No. 0,303,507 A2, 12.08.88

NATAMYCIN

Therapeutic Function: Antibacterial (ophthalmic)

Chemical Name: Natamycin

Common Name: Pimaricin

Structural Formula:

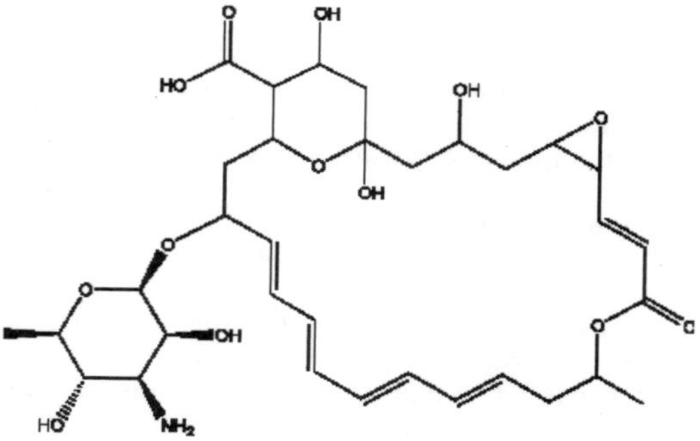

Chemical Abstracts Registry No.: 7681-93-8

Trade Name	Manufacturer	Country	Year Introduced
Pimafucine	Beytout	France	1964
Pimafucin	Brocades	UK	1965
Pimafucort	Brocades	Italy	1966
Pimafucin	Basotherm	W. Germany	1967
Natacyn	Alcon	US	1979
Myprozine	Lederle	US	-

Raw Materials

Bacterium Streptomyces gilvosporeus
Starch
Corn steep liquor

Manufacturing Process

The Fermentation Process: The process by which this antifungal substance is produced is an aerobic fermentation of an aqueous nutrient medium inoculated with a pimaricin-producing strain of Streptomyces gilvosporeus. The nutrient medium contains an assimilable source of carbon such as starch, molasses, or glycerol, an assimilable source of nitrogen such as corn steep liquor and inorganic cations such as potassium, sodium or calcium, and anions such as sulfate, phosphate or chloride. Trace elements such as boron, molybdenum or copper are supplied as needed in the form of impurities by the other constituents of the medium.

In more detail the nutrient medium used may contain sources of carbon such as starch, hydrolyzed starch, sugars such as lactose, maltose, dextrose, sucrose, or sugar sources such as molasses; alcohols, such as glycerol and mannitol; organic acids, such as citric acid and acetic acid; and various natural products which may contain other nutrient materials in addition to carbonaceous substances.

Nitrogen sources include proteins, such as casein, zein, lactalbumin; protein hydrolyzates such proteoses, peptones, peptides, and commercially available materials, such as N-Z Amine which is understood to be a casein hydrolyzate; also corn steep liquor, soybean meal, gluten, cottonseed meal, fish meal, meat extracts, stick liquor, liver cake, yeast extracts and distillers' solubles; amino acids, urea, ammonium and nitrate salts. Such inorganic elements as sodium, potassium, calcium and magnesium; and chlorides, sulfates, phosphates and combinations of these anions and cations in the form of mineral salts may be advantageously used in the fermentation.

The so-called trace elements, such as boron, cobalt, iron, copper, zinc, manganese, chromium, molybdenum and still others may also be used to advantage. Generally, these trace elements occur in sufficient quantities in the carbonaceous and nitrogenous constituents of the medium, particularly if derived from natural sources, or in the tap water, and the addition of further quantities of these trace elements may consequently be unnecessary.

The fermentation liquor is aerated in the customary manner by forcing sterile air through the fermenting mixture usually at the rate of about 1 volume of air per volume of fermentation medium per minute. To minimize contamination with foreign microorganisms, the fermentation vessels should be closed and a pressure of 2 to 15 pounds above atmospheric pressure maintained in the vessel. In addition to the agitation provided by aeration, mechanical agitation is generally desirable. Antifoaming agents, such as 1% octadecanol in lard oil, may be added from time to time as required to prevent excessive foaming. Fermentation is conducted at a temperature preferably on the order of 26°C to 30°C but may be as low as 17°C or as high as 42°C.

The time required for maximum production of the antifungal substance will vary considerably depending upon other conditions of the fermentation. Generally, about 48 hours is required before appreciable quantities of the antifungal substance are detected in the medium. The production of the antifungal substance increases with time, and the fermentation may run as long as 120 hours. The hydrogen ion conditions normally vary from about pH 6 to pH 8.0, although deviations from these values are permissible, according to British Patent 846,933. The reader is referred to the patents cited for detals of pimaricin purification.

References

Merck Index 6278
Kleeman & Engel p. 624
DOT 14 (6) 255 (1978)
I.N. p. 659
REM p. 1230
Koninkijke Nederlandsche Gist- & Spiritusfabriek N.V., Netherlands; British Patent 844,289; August 10, 1960
American Cyanamid Company; British Patent 846,933; September 7, 1960

NEDOCROMIL DISODIUM

Therapeutic Function: Antiallergic, Anti-asthmatic

Chemical Name: 4H-Pyrano(3,2-g)quinoline-2,8-dicarboxylic acid, 9-ethyl-6,9-dihydro-4,6-dioxo-10-propyl-, disodium salt

Common Name: Nedocromil sodium

Structural Formula:

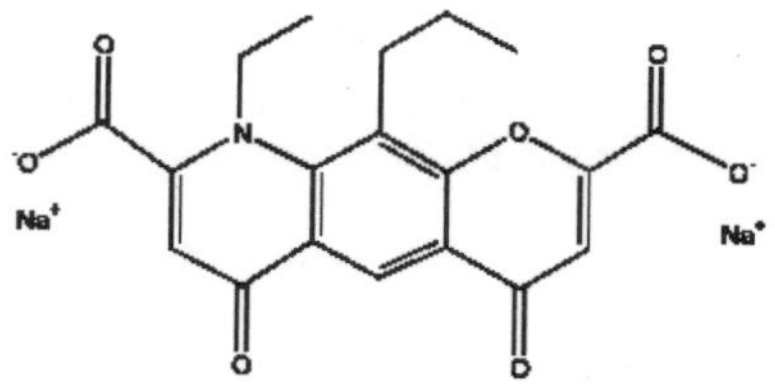

Chemical Abstracts Registry No.: 69049-74-7; 69049-73-6 (Base)

Trade Name	Manufacturer	Country	Year Introduced
Alocril	Allergan	USA	-
Cetimil	Lab. Lesvi	-	-
Irtan	Rhone-Poulenc Rorer	France	-
Halamid	Viatris	Germany	-
Halamid	ASTA Medica AWD	-	-
Kovilen	Mediolanum Farmaceutici S.p.A.	Italy	-
Nedocromil Disodium	Rhone-Poulenc Rorer	France	-
Tilade	Rhone-Poulenc Rorer	France	-
Tilade Mint.	Fisons	UK	-
Tilavist	Fisons	UK	-
Tilavist	Aventis Pharma AB	Sweden	-

Raw Materials

Allyl bromide
4-Acetamido-2-hydroxyacetophenone
Sodium ethoxide
Dimethyl acetylene dicarboxylate
Polyphosphoric acid

Manufacturing Process

4,6-Dioxo-10-propyl-4H,6H-pyrano[3,2-]quinoline-2,8-dicarbxylic acid disodium salt was prepared in 8 steps

1. 4-Acetamido-2-allylacetophenone

4-Acetamido-2-hydroxyacetophenone (19.3 g), allyl bromide (12.1 ml) and hydrous potassium carbonate (21.5 g) were stirred in dry dimethylformamide (250 ml) at room temperature for 24 hours. The reaction mixture was poured into water and the product was extracted with ethyl acetate. The organic solution was then washed well with water dried over magnesium sulphate and evaporated to dryness. The sub-title product was obtained as buff coloured solid (20.5 g). The structure of the product was confirmed by NMR and mass spectroscopy.

2. 4-Acetamido-3-allyl-2-hydroxyacetophenone

The above allyl ether (18.4 g) was heated at 200-210°C for 4 hours. 17.1 g of the thermally rearranged sub- title product was obtained as a brown solid. Again the structure was confirmed by NMR and mass spectroscopy.

3. 4-Acetamido-2-hydroxy-3-propyl acetophenone

The product of step 2 (17 g) was dissolved in glacial acetic acid and hydrogenated in the presence of Adams catalyst until hydrogen uptake had ceased. The catalyst was filtered off through a keiselguhr filter and the filtrate was evaporated to leave 13.0 g of almost colorless solid. The mass and NMR spectra confirmed the structure of product.

4. Ethyl-7-acetamido-4-oxo-8-propyl-4H-l-benzopyran-2-carboxylate

A mixture of diethyl oxalate (19.3 g; 17.9 ml) and the above product of step 3 (12.4 g) in dry ethanol (100 ml) was added to a stirred solution of sodium ethoxide in ethanol (prepared by dissolving sodium (6.1 g) in dry ethanol (200 ml)). The reaction mixture was refluxed for 3 hours and then poured into dilute hydrochloric acid and chloroform. The chloroform layer was separated, washed with water and dried. The solvent was evaporated to leave a brown solid which was dissolved in ethanol (300 ml) containing concentrated hydrochloric acid (3 ml) and the whole was refluxed for 1 hour. The reaction mixture was poured into water and the product was extracted into ethyl acetate which was washed with water and dried. The solvent was evaporated to leave 10 g of a sticky solid which had mass and NMR spectra consistent with the expected product.

5. Ethyl 7-amino-4-oxo-8-propyl-4H-1-benzopyran-2-carboxylate

A solution of the amide of step 4 (10 g) in ethanol (300 ml), containing concentrated hydrochloric acid (5 ml), was refluxed for 8 hours. The reaction mixture was diluted with water and extracted into ethyl acetate. The extract was washed with water, dried and the solvent was evaporated to leave a dark brown semi-solid. This was chromatographed on a silica gel column, using ether as eluant to give 4.8 g of the required product whose structure was confirmed by mass and NMR spectral evidence; mp 84-87°C.

6.8-Ethoxycarbonyl-2-methoxycarbonyl-4,6-dioxo-10-propyl-4H,6H-pyrano[3.2-g]quinoline

The amino benzopyran of step 5 (2.0 g) and dimethyl acetylene dicarboxylate (1.24 g; 1.01 ml) were refluxed in ethanol (30 ml) for 26 hours. The reaction mixture was cooled to 0°C and the insoluble yellow-brown solid was collected by filtration and washed with a little ethanol and dried to give 2.0 g of a product which was a mixture of maleic and fumaric esters obtained by Michael addition of the amine to the acetylene. This mixture of esters (2.0 g) was treated with polyphosphoric acid (30 ml) and heated on the steam bath with stirring for 20 minutes. The reaction mixture was then poured onto ice and stirred with ethyl acetate. The organic layer was separated, washed with water and dried. The solvent was evaporated to leave 1.6 g of a yellow orange solid. Recrystallisation of this solid from ethyl acetate gave the required product as fluffy orange needles, mp 187°-188°C.

7. 4,6-Dioxo-10-propyl-4H,6H-pyrano[3.2-g]quinoline-2,8-dicarboxylic acid

The above bis ester (2.5 g) was refluxed with sodium bicarbonate (1.64 g) in ethanol (100 ml) and water (50 ml) for 1.5 hours. The whole was poured into water and acidified to precipitate a gelatinous solid. This was collected by filtration, refluxed with ethanol and the product was separated by centrifugation (1.4 g), mp 303°-304°C dec. The structure of the product was confirmed by mass and NMR evidence.

8. Disodium 4,6-dioxo-10-propyl-4H,6H-pyrano[3,2-g]quinoline-2,8-dicarboxylate

The bis acid from step 6 (1.35 g) and sodium bicarbonate (0.661 g) in water (150 ml) were warmed and stirred until a clear solution was obtained. This solution was filtered and the filtrate was freeze dried to give 1.43 g of the required disodium salt.

References

Cairns H., Cox D.; G.B. Patent 2,022,078, 1979-12-12

NEFAZODONE HYDROCHLORIDE

Therapeutic Function: Antidepressant

Chemical Name: 3H-1,2,4-Triazol-3-one, 2,4-dihydro-2-(3-(4-(3-chlorophenyl)-1-piperazinyl)propyl)-5-ethyl-4-(2-phenoxyethyl)-, monohydrochloride

Common Name: Nefazodone hydrochloride

Structural Formula:

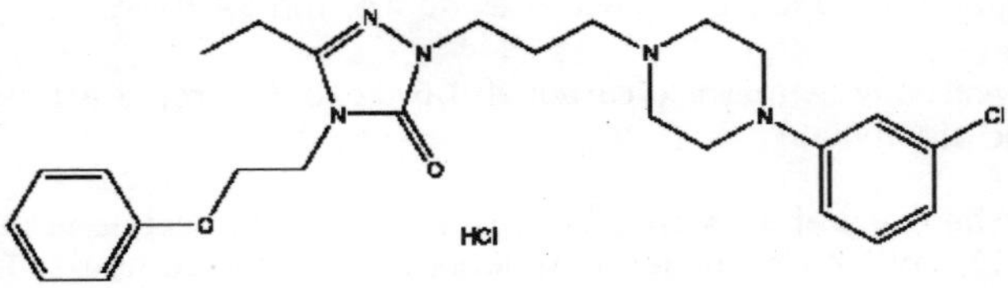

Chemical Abstracts Registry No.: 82752-99-6; 83366-66-9 (Base)

Trade Name	Manufacturer	Country	Year Introduced
Dutonin	Bristol-Myers Squibb	USA	-
Menfazona	Laboratorios Menarini, S. A.	Italy	-
Nefadar	Squibb-von Heyden	Germany	-
Nefazodone Hydrochloride	Bristol-Myers Squibb	USA	-
Reseril	Mead Johnson	USA	-
Rulivan	Europharma	-	-
Serzone	Bristol-Myers Squibb	Australia	-

Raw Materials

Hydrazine	1-(3-Chlorophenyl)-piperazine
Thionyl chloride	1-Bromo-3-chloropropane
Sodium azide	Phenoxypropionic acid
Sodium nitrite	Triethyl orthopropionate

Manufacturing Process

1. 1-(3-Chloropropyl)-4-(3-chlorophenyl)piperazine hydrochloride

A 25% NaOH solution (320 ml, 2.0 mol) is added dropwise to a stirred solution of 1-(3-chlorophenyl)-piperazine hydrochloride (196.5 g, 1.0 mol) and 1-bromo-3-chloropropane (99.0 ml, 1.0 mol) in acetone (200 ml) while maintaining temperature of 0°-10°C. After the addition is completed, the mixture is allowed to warm to room temperature and is stirred for 18 hours. The upper organic phase is then separated and concentrated under reduced pressure. The residual oil is taken up in 250 ml acetone and filtered. The filtrate is concentrated under reduced pressure and the oily residue is dissolved in 1 L of 15% boiling HCl solution. A viscous oil is separated from the cooled mixture and poured into 1 L of ice-H_2O with vigorous stirring, forming white precipitates. Recrystallization of the solid from boiling water gave 171.8 g (55.6% yield) of 1-(3-chloropropyl)-4-(3-chlorophenyl) piperazine hydrochloride; m.p 199.5°-200.5°C.

2. 1-(3-Chlorophenyl)-4-(hydrazinopropyl)piperazine

1-(3-Chloropropyl)-4-(3-chlorophenyl)piperazine hydrochloride (20.0 g, 0.065 mol) is suspended in isopropanol (65 ml) and anhydrous hydrazine (31.7 g, 0.988 mol) is added. The reaction mixture is heated at 70°-80°C for 2.5 hours and cooled to room temperature. The upper layer is separated and concentrated under reduced pressure. The residue is dissolved in isopropanol (50 ml) and the upper layer is separated, dried (Na_2SO_4), and concentrated to yield 16.5 g (94.5% yield) of 1-(3-chlorophenyl)-4-(3-hydrazinopropyl) piperazine (85% pure) as a viscous oil. The product is used directly without further purification or stored at room temperature in toluene or isopropanol solution by adding 1% MgO.

The hydrazine is dissolved in isopropanol and 1% magnesium oxide is added. The mixture is stirred for 30 min and filtered. The filtrate is cooled with ice-bath and one equivalent of anhydrous HCl in isopropanol is added under vigorous stirring. The precipitates are collected by filtration and dried at 60°C under reduced pressure to afford white powder; mp 147°-150°C.

3. Phenoxypropionic acid (249.0 g, 1.50 mol) is dissolved in four equivalents of thionyl chloride (438.0 ml, 6.0 mol) and heated to reflux until the HCl evolution has ceased. The solution is then cooled to room temperature and concentrated under reduced pressure to give 281.0 g (100% yield) of phenoxypropionyl chloride as a brown oil which solidifies on cooling.

4. Phenoxypropionyl chloride (9.23 g, 0.05 mol) is dissolved in 100 ml acetone and cooled with an ice bath as sodium azide (3.6 g, 0.055 mol) in 10 ml water is added dropwise. After addition is completed, the reaction mixture is warmed to room temperature and stirred for 30 minutes. The solution is decanted and concentrated. The residue is dissolved in 100 ml ether and washed with saturated sodium bicarbonate and brine. The organic phase is separated, dried ($MgSO_4$) and concentrated to give 6.52 g (68.0% yield) of phenoxypropionyl azide as a yellow oil which solidifies on cooling

5. Ethyl phenoxypropionate

Phenoxypropionic acid (6.64 g, 0.04 mol) is mixed with excess ethanol (10 ml) and concentrated sulfuric acid (0.5 ml) is added. The reaction mixture is refluxed for 3 hours, cooled to room temperature and concentrated. The residue is washed with 1 N NaOH and brine, dried (Na_2SO_4), and concentrated to yield 7.32 g (94.3% yield) of the ester which can be used directly for the subsequent reaction without further purification.

6. Phenoxypropionyl hydrazide

Ethyl phenoxypropionate (161 g, 0.83 mol) is cooled with an ice bath and anhydrous hydrazine (32 ml, 1 mol) is added dropwise. After the addition is completed, the solution is warmed to room temperature and stirred for 4 hours. The solution is then cooled with an ice bath under vigorous stirring.

After the white precipitate formed the mixture is kept in refrigerator for 14 hours. The solid is collected by filtration, washed with cold 10% ethanol/hexane and dried in reduced pressure at 50°C for 12 hours to give 134.7 g (90%) of phenoxypropionyl hydrazide as white powder. Mp 66°-70°C.

The hydrochloride salt of phenoxypropionyl hydrazide is prepared by dissolving the hydrazide in dichloromethane, cooling with an ice bath and bubbling through anhydrous HCl gas until pH 3. The solid is collected by filtration, washed with cold dichloromethane and air-dried to give the hydrochloride salt as fine white powder, mp 172°-174°C.

7. Phenoxyethyl isocyanate

Method A: Phenoxypropionyl azide (15.2 g, 0.08 mol) is dissolved in 50 ml toluene and heated with an external oil bath. At 75°-80°C (internal temperature) vigorous N2evolution is observed and the reaction is very exothermic. The solution is refluxed for further 30 min after the gas evolution has finished. The solution is concentrated and the residue is distilled in vacuo to give 7.8 g (60% yield) of phenoxyethyl isocyanate as a colorless oil (94°-96°C, 1 mm Hg).

Method B: Phenoxypropionyl hydrazide (125.9 g, 0.7 mol) is suspended in 650 ml ice-water and concentrated hydrochloric acid (123 ml, 1.47 mol) was added. The mixture is stirred for 20 min and toluene (350 ml) is added. A solution of sodium nitrite (53.1 g, 0.77 mol) in 200 ml water is added over a period 15 min. The internal temperature is kept below 15°C and if necessary, ice is directly added to the reaction mixture. After the addition is completed the mixture is stirred for a further 1 hour and filtered through Celite. The solid is washed with 30 ml toluene and the filtrate is separated. The aqueous layer is extracted with 200 ml toluene and the combined toluene solutions are dried over $MgSO_4$. The dried toluene solution is filtered and added dropwise to a preheated flask at 95°-100°C. Nitrogen evolution occurs as the solution is dropped in. After the addition is complete, the reaction mixture is heated to gentle reflux until nitrogen evolution has ceased. The reaction mixture is cooled to room temperature and can be used directly in subsequent reactions. 1 ml of the reaction mixture is withdrawn and evaporated to dryness, and the weight of the residue is measured. This provides an estimate of the concentration of isocyanate per ml of reaction mixture.

8.2-3-(4-[3-Chlorophenyl]-1-piperazinyl)propyl]-4-(2-phenoxyethyl)-semicarbazide)

A solution of phenoxyethyl isocyanate (89 g, 0.55 mol) in toluene (450 ml) is generated in situ (see step 7) and cooled to -20°C. To the solution is added a solution of 1-(3-chlorophenyl)-4-(3-hydrazinopropyl)piperazine (131.2 g, 0.49 mol) in 100 ml toluene at the speed that the internal temperature is below -10°C. After the addition is completed the mixture is stirred for 30 min at -20°C and for 1.5 hours at 0°C and quenched with 150 ml 1 N NaOH solution. The mixture is stirred at 0°C for 10 min and filtered through celite. The filtrate is saturated with NaCl and separated. The aqueous layer is extracted with 100 ml toluene and the combined toluene solution was dried over Na_2SO_4, filtered and concentrated to give a viscous oil. A small amount sample was purified by column chromatography (5% $MeOH/CH_2Cl_2$) to give a colorless oil. The crude product is dissolved in isopropanol, cooled with ice bath, and two equivalents of HCl/isopropanol are added. The precipitates are collected by filtration and further purified by recrystallization from ethanol to give 170.6 g (69%) of hydrochloride salt as white crystal. Mp 172°-176°C.

9.2-[3-4-(3-Chlorophenyl)-1-piperazinyl]-propyl]-5-ethyl-4-(2-phenoxyethyl)-2H-1,2,4-triazol-3(4H)-one monohydrochloride (Nefazodone monohydrochloride)

2-[3-(4-[3-Chlorophenyl-1-piperazinyl)propyl]-4-(2-phenoxyethyl)-semicarbazidedihydrochloride (23.3 g, 46 mmol) is suspended in 50 ml toluene and refluxing with Dean-Stark apparatus to remove water. The mixture is then cooled to room temperature and triethyl orthopropionate (50 ml, about 5 eq) is added. The suspension is refluxed again with Dean-Stark apparatus. As toluene is distilled the suspension becomes a clear solution which is refluxed for 48 hours. Distillation under reduced pressure removes unreacted trietyl orthopropionate and the resulting residue is dissolved in 50 ml isopropanol, treated with HCl to pH 4, stirred at 0°C for 1 hour and standed in refrigerator for 12 hours. The solid is collected with filtration and recrystallized from ethanol to give 10.5 g (45%) of nefazodone monohydrochloride as white powder (95% pure by HPLC). Further purification is achieved by fractional recrystallization to give the product with 99.5% purity. mp 183°-185°C.

References

Lei B. et al.; US Patent No. 5,900,485; May 4, 1999; Assigned to Apotex, Inc., Weston, Canada

Murthy K.S.K. et al.; US Patent No. 6,596,866 B2; Jul. 22, 2003; Assigned to Brantford Chemicals Inc., Banntford

NEFOPAM HYDROCHLORIDE

Therapeutic Function: Muscle relaxant, Antidepressant

Chemical Name: 3,4,5,6-Tetrahydro-5-methyl-1-phenyl-1H-2,5-benzoxazocine hydrochloride

Common Name: -

Structural Formula:

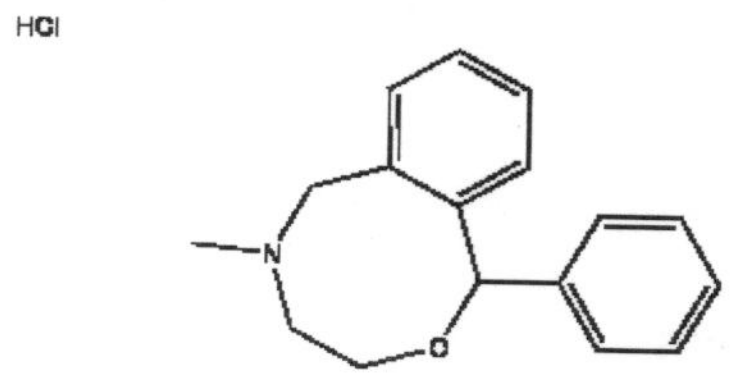

HCl

Chemical Abstracts Registry No.: 13669-70-0 (Base)

Trade Name	Manufacturer	Country	Year Introduced
Ajan	Kettelhack Riker	W. Germany	1976
Acupan	Carnegie	UK	1978
Acupan	Riker	France	1981
Lenipan	Chiesi	Italy	1981
Oxadol	I.S.I.	Italy	1982
Acupan	Boehringer Mannheim	Italy	1983

Raw Materials

2-Benzoylbenzoic acid
2-Methylaminoethanol
4-Toluenesulfonic acid
Thionyl chloride
Lithium aluminum hydride
Hydrogen chloride

Manufacturing Process

The starting material is prepared by reacting 2-benzoylbenzoic acid with thionyl chloride and then with 2-methylaminoethanol. 20.0 grams (0.07 mol) of N-(2-hydroxyethyl)-N-methyl-o-benzoylbenzamide is suspended in 100 ml tetrahydrofuran and then slowly added in small portions to a solution of 5.5 grams (0.14 mol) of lithium aluminum hydride in 150 ml tetrahydrofuran with cooling and stirring. The mixture is then refluxed for 18 hours, cooled and then to it is successively added 5.5 ml water, 5.5 ml of 3.75 N sodium hydroxide and 16 ml water. After removal of precipitated salts by filtration, the solution remaining is concentrated under reduced pressure and the residue dried to yield 19.5 grams of crude product. Yield after conversion to the hydrochloride salt and recrystallization is 17.0 grams (89%), MP 128° to 133°C.

5-methyl-1-phenyl-1,3,4,6-tetrahydro-5H-benz[f] -2,5-oxazocine is prepared as follows. 3.0 grams (0.011 mol) of 2-([N-(2-hydroxyethyl)-N-methyl]amino)methylbenzhydrol, prepared as described above, 3.0 grams p-toluenesulfonic acid and 15 ml benzene are heated together with stirring until all the benzene is distilled off. The residual oil is heated to 105°C and held at this temperature for 1 hour, then cooled and dissolved in 30 ml water. This aqueous solution is then basified to pH 10.0 with 12 N sodium hydroxide, extracted with ether, and the extracts washed with water, dried over anhydrous sodium sulfate and the solvent removed under reduced pressure. The 2.26 grams (81%) oil remaining is converted to the hydrochloride salt, MP 238° to 242°C.

References

Merck Index 6287
Kleeman & Engel p. 626
OCDS Vol. 2 p. 447 (1980)
DOT 12 (7) 275 (1976)

I.N. p. 661
Baltes, B.J.; US Patent 3,487,153; December 30, 1969; assigned to Rexall Drug and Chemical Company

NEOMYCIN

Therapeutic Function: Antibacterial

Chemical Name: O-2,6-Diamino-2,6-dideoxy-α-D-glucopyranosyl-(1->3)-O-β-D-ribofuranosyl(1->5)-O-[2,6-diamino-2,6-dideoxy-α-D-glucopyranosyl-(1->4)]-2-deoxy-D-streptamine

Common Name: Framycetin

Structural Formula:

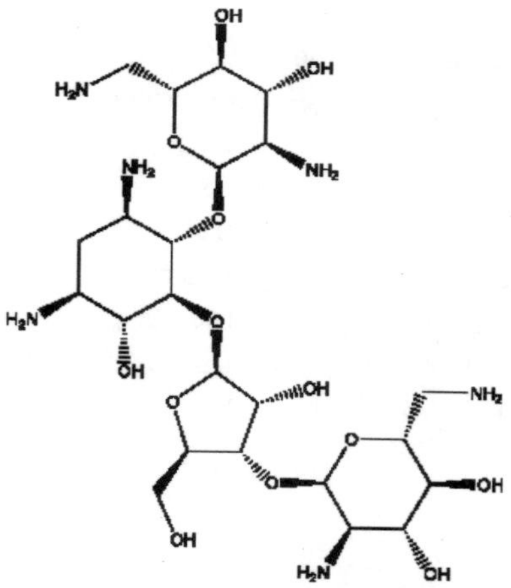

Chemical Abstracts Registry No.: 1404-04-2; 4146-30-9 (Sulfate salt)

Trade Name	Manufacturer	Country	Year Introduced
Myciguent	Upjohn	US	1951
Otobiotic	Schering	US	1954
Mycifradin	Upjohn	US	1957
Neobiotic	Pfizer	US	1958
Apokalin	A.L.	Norway	-
Biofradin	Uriach	Spain	-
Bykomycin	Byk Gulden	W. Germany	-
Cortisporin	Burroughs-Wellcome	US	-
Dexmy	Takeda	Japan	-

Trade Name	Manufacturer	Country	Year Introduced
Endomixin	Lusofarmaco	Italy	-
Fradio	Nippon Kayaku, Co.	Japan	-
Fradyl	Christiaens	Belgium	-
Ivax	Boots	UK	-
Larmicin	Larma	Spain	-
Myacyne	Werner Schnur	W. Germany	-
Mytrex	Savage	US	-
Neobretin	Norbrook	UK	-
Neodecadron	MSD	US	-
Neointestin	Hosbon	Spain	-
Neolate	Therafarm	UK	-
Neomicina Roger	Roger	Spain	-
Neomin	Glaxo	UK	-
Neo-Polycin	Merrell Dow	US	-
Neopt	Sigma	Australia	-
Neosporin	Burroughs-Wellcome	US	-
Neosulf	Protea	Australia	-
Neo-Synalar	Syntex	US	-
Octicair	Pharmafair	US	-
Otocort	Lemmon	US	-
Siquent	Sigma	Australia	-
Tampovagan	Norgine	UK	-
Topisporin	Pharmafair	US	-
Tri-Thalmic	Schein	US	-

Raw Materials

Bacterium Streptomyces fradiae
Nutrient medium

Manufacturing Process

Neomycin has been produced by growing the organism, Strepromyces No. 3535, in a suitable nutrient medium under appropriate stationary or submerged aerobic (viz shaken) conditions, and then isolating and purifying the substance, e.g., by procedure of the sort described in the figure including various steps of adsorption, recovery by elution, separation from impurities, and precipitation.

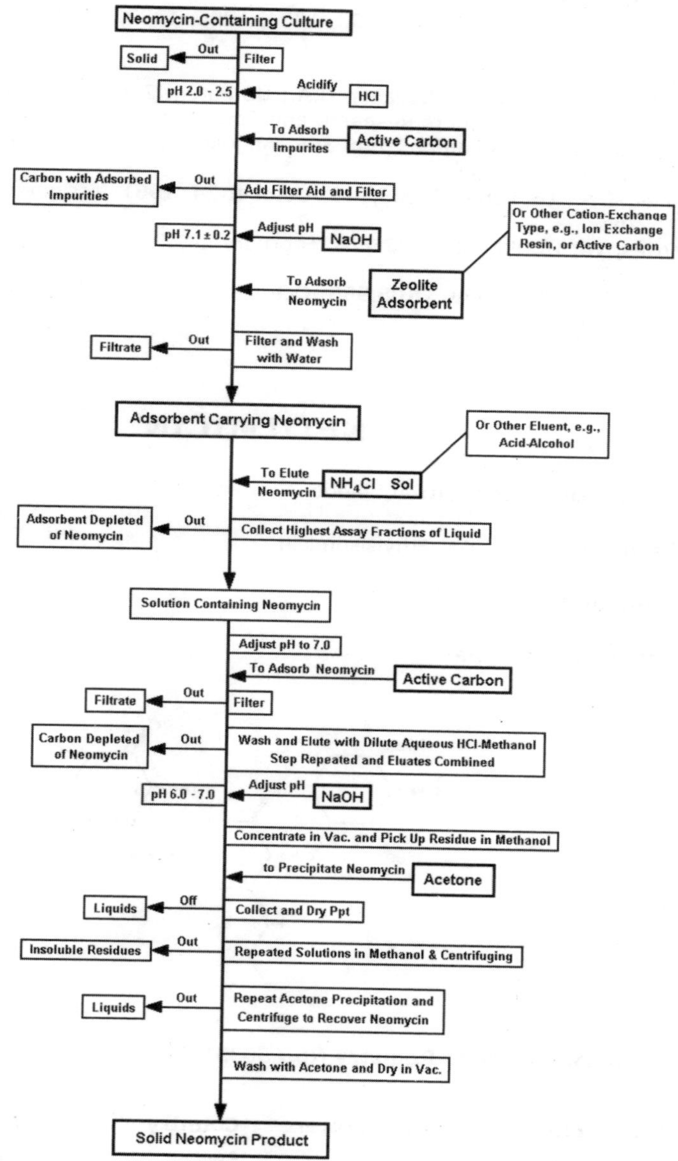

Neomycin is usually used as the sulfate.

References

Merck Index 6300

Kleeman & Engel 626
PDR pp.673, 738, 756, 888, 993, 1034, 1206, 1232, 1429, 1569, 1604, 1800
I.N. p. 663
REM p. 1181
Waksman, S.A. and Lechevalier, H.A.; US Patent 2,799,620; July 16, 1957; assigned to Rutgers Research and Educational Foundation
Jackson, W.G.; US Patent 2,848,365; August 19, 1958; assigned to The Upjohn Company
Miller, T.W.; US Patent 3,005,815; October 24, 1961; assigned to Merck & Co., Inc.
Moses, W.; US Patent 3,022,228; February 20, 1962; assigned to S.B. Penick & Company
Haak, W.J.; US Patent 3,108,996; October 29, 1963; assigned to The Upjohn Company

NETILMICIN

Therapeutic Function: Antibiotic

Chemical Name: 1-N-Ethylsisomicin

Common Name: -

Structural Formula:

Chemical Abstracts Registry No.: 56391-56-1

Trade Name	Manufacturer	Country	Year Introduced
Netromycine	Schering	Switz.	1980
Certomycin	Byk-Essex	W. Germany	1980
Netillin	Kirby-Warrick	UK	1981
Netromicine	Unicet	France	1981
Nettacin	Essex	Italy	1982
Netromycin	Schering	US	1983

Raw Materials

Sisomicin
Acetaldehyde
Sulfuric acid
Sodium cyanoborohydride

Manufacturing Process

To a solution of 5 g of sisomicin in 250 ml of water add 1 N sulfuric acid until the pH of the solution is adjusted to about 5. To the solution of sisomicin sulfuric acid addition salt thereby formed, add 2 ml of acetaldehyde, stir for 10 minutes, then add 0.85 g of sodium cyanoborohydride. Continue stirring at room temperature for 15 minutes, then concentrate solution in vacuo to a volume of about 100 ml, treat the solution with a basic ion exchange resin [e.g., Amberlite IRA 401S (OH^-)], then lyophilize to a residue comprising 1-N-ethylsisomicin.

Purify by chromatographing on 200 g of silica gel, eluting with lower phase of a chloroformmethanol-7% aqueous ammonium hydroxide (2:1:1) system. Combine the eluates as determined by thin layer chromatography and concentrate the combined eluates of the major component in vacuo to a residue comprising 1-N-ethylsisomicin (yield 1.25 g). Further purify by again chromatographing on 100 g of silica gel eluting with a chloroform-methanol-3.5% ammonium hydroxide (1:2:1) system. Pass the combined, like eluates (as determined by thin layer chromatography) through a column of basic ion exchange resin and lyophilize the eluate to obtain 1-N-ethylsisomicin (yield 0.54 g).

There is also a fermentation route to netilmicin as noted by Kleeman & Engel.

References

Merck Index 6322
DFU 3 (7) 527 (1978)
Kleeman & Engel p. 627
PDR p. 1635
DOT 17 (8) 324 (1981)
I.N. p. 666
REM p. 1183
Wright, J.J., Daniels, P.J.L., Mallams, A.K. and Nagabhushan, T.L.; US Patent 4,002,742; January 11, 1977; assigned to Schering Corp.

NEVIRAPINE

Therapeutic Function: Antiviral

Chemical Name: 5H-Dipyrido(3,2-b:2',3'-e)(1,4)diazepin-6-one, 5,11-dihydro-11-cyclopropyl-4-methyl-

Common Name: Nevirapine

Structural Formula:

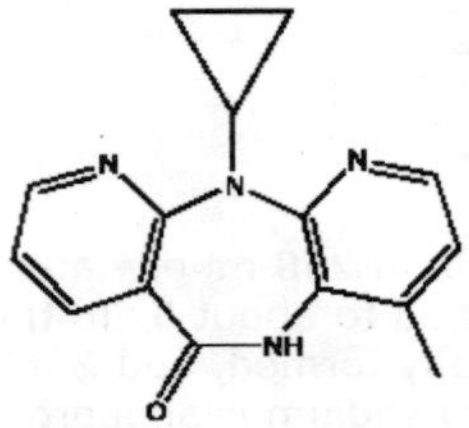

Chemical Abstracts Registry No.: 129618-40-2

Trade Name	Manufacturer	Country	Year Introduced
Neve	Le Sante	India	-
Nevimune	Cipla Limited	India	-
Nevirapine	Boehringer Ingelheim Pharma	USA	-
NVP	Roxane Laboratories	USA	-
Viramune	Boehringer Ingelheim Pharma	Germany	-
Viramune	Cipla Limited	India	-
Viramune	Roxane Laboratories	USA	-
Viramine	Boehringer Ingelheim Pharma	USA	-

Raw Materials

Calcium oxide
Sodium hydride
2-Chloro-N-(2-chloro-4-methyl-3-pyridyl)-3-pyridine carboxamide
Cyclopropylamine
Diethylene glycoldimethyl ether

Manufacturing Process

There are 3 ways for preparing of nevirapine.

117.5 kg of 2-chloro-N-(2-chloro-4-methyl-3-pyridyl)-3-pyridine carboxamide, 23.3 kg of calcium oxide and 59.4 kg of cyclopropylamine (molar ratio: 1:1:2.5) are heated to between 135° and 145°C in 235 L of diglyme (diethylene glycoldimethylether) in a 500 L VA autoclave over a period of 6 to 8 hours. The reaction mixture is then cooled to a temperature of 20°-30°C and filtered. The filter cake is washed with 58.8 L of diglyme. The filtrates are combined and initially 200 L of solvent is distilled off. The residue is then diluted with a further 117.5 L of diglyme. The resultant diluted solution is added over a period of 20 to 40 minutes to a suspension of 45.0 kg of 60% sodium hydride in 352.5 L of diglyme, heated to 130°C. The storage vessel and conduits are rinsed with a further 55.8 L of diglyme, and the mixture is stirred at a temperature of between 130° and 140°C for a further 30 to 60 minutes. The majority of the diglyme is then distilled off. Finally, the remaining residue is carefully mixed with 470 L of water. After cooling to a temperature of about 25°C, 235.0 L of cyclohexane and 57.11 of glacial acetic acid are added to the reaction mixture. The mixture is then stirred for about 1 hour at temperature of 10° to 25°C. The resultant suspension is centrifuged and the centrifuged material is then washed with 235.0 L of methyl-tert-butylether and subsequently with 353.5 L of water and finally with 235 L of ethanol. In this way, after drying, 92.5 kg (83.5% of theory) of 11-cyclopropyl-5,11-dihydro-4-methyl-6H-dipyrido[3,2-b:2',3'-e][1,4]diazepin-6-one (nevirapine) is isolated.

117.5 kg of 2-chloro-N-(2-chloro-4-methyl-3-pyridyl)-3-pyridine carboxamide, 46.7 kg of calcium oxide and 47.5 kg of cyclopropylamine (molar ratio: 1:2:2) are heated to 135° to 145°C in 235 L of diglyme (diethylene glycol dimethylether) in a 500 L VA autoclave over a period of 6 to 8 hours. The reaction mixture is then cooled to a temperature of 20° to 30°C and filtered. The filter cake is washed with 58.8 L of diglyme. The filtrates are combined and about 188 L of solvent is distilled off. The residue is then diluted with a further 117.5 L of diglyme. Over a period of 20 to 40 minutes, the resultant diluted solution is added to a suspension of 45.0 kg of 60% sodium hydride in 352.5 L of diglyme, heated to 130°C. The storage vessel and conduits are rinsed with a further 55.8 L of diglyme and the mixture is stirred at a temperature of 130° to 140°C for a further 30 to 60 minutes. The majority of the diglyme is then distilled off. Finally, the remaining residue is carefully mixed with 470.0 L of water. The reaction mixture is cooled to a temperature of about 25°C and 235.0 L of cyclohexane and 57.1 L of glacial acetic acid are added. The mixture is then stirred for about 1 hour at a temperature of 10o to 25°C. The resultant suspension is centrifuged and the centrifuged material is washed with 235.0 L of methyl tert-butylether, followed by 353.5 L of water and finally with 235 L of ethanol. In this way, after drying, 90.6 kg (81.7% of theory) of 11-cyclopropyl-5,11-dihydro-4-methyl-6H-dipyrido-[3,2-b:2',3'-e][1,4]diazepin-6-one (nevirapine) is isolated.

287.2 kg of 2-chloro-N-(2-chloro-4-methyl-3-pyridyl)-3-pyridine carboxamide, 57.0 kg of calcium oxide and 87.1 kg of cyclopropylamine (molar ratio: 1:1:1.5) are heated in 574 L of diglyme (diethylene glycol-dimethylether) to 135°-145°C for about 30 minutes in a 1200 L VA stirring apparatus. This produces a pressure of 1.2-1.5 bar and about 50% of the starting material is reacted. To this mixture, over about 30 minutes at 135°-145°C, a further 58.1 kg of cyclopropylamine is added producing a pressure of 3.0-3.5 bar, and another 25% of the starting material is reacted. The mixture is then kept at 135°-145°C for a period of 5 to 6 hours. The reaction mixture is then cooled to a temperature of 20° to 30°C and filtered. The filter cake is washed with 144 L of diglyme. The filtrates are combined and 400 L of solvent is distilled off. The residue is then diluted with a further 287 L of diglyme. Over 20-40 minutes, the resultant diluted solution is added to a suspension of 110 kg of 60% sodium hydride in 862 L of diglyme, heated to 130°C. The storage vessel and conduits are rinsed with a further 144 L of diglyme and the mixture is stirred at a temperature of 130° to 140°C for another 30 to 60 minutes. The majority of the diglyme is then distilled off. Finally, the remaining residue is carefully mixed with 1150 L of water. After the reaction mixture has been cooled to a temperature of about 25°C, 575 L of cyclohexane and 147 L of glacial acetic acid are added. The mixture is then stirred for about 1 hour at a temperature of 10°-25°C. The resultant suspension is centrifuged and the centrifuged material is then washed with 575 L of methyl-tert-butylether, followed by 862 L of water and finally with 575 L of ethanol. In this way, after drying, 225 kg (83.0% of theory) of 11-cyclopropyl-5,11-dihydro-4-methyl-6H-dipyrido[3,2-b :2',3'-e][1,4]diazepin-6-one (nevirapine) is obtained.

References

Schneider H. et al.; US Patent No. 5,559,760; Oct. 29, 1996; Assigned to Boehringer Ingelheim KG, Ingelheim am Rhein, Germany

NIACINAMIDE

Therapeutic Function: Enzyme cofactor vitamin

Chemical Name: 3-Pyridinecarboxylic acid amide

Common Name: Aminicotin; Niacinamide; Nicosylamide; Nicotilamid(e); Nicotinamide; Nicotinsaureamid; Nikotinsaureamid; Vitamin B_3; Vitamin PP; Vitaminum pellagrapraeventivum; Witamina PP

Structural Formula:

Chemical Abstracts Registry No.: 98-92-0

Trade Name	Manufacturer	Country	Year Introduced
Niacinamide	Twinlab	-	-
Niazcol	Locatelli	-	-
Nicotinamide	Endur-Amide TM, Innovite Inc.	USA	-
Vitamin B_3	Twinlab	-	-
Vitamin B_3	Biocare	-	-

Raw Materials

Gaseous ammonia
Nicotinic aci
Anhydrous ethyl acetate

Manufacturing Process

Gaseous ammonia was passed into nicotinic acid at a temperature between 200-235°C until the conversion to nicotinamide was 85%. The reaction mixture was colored light brown. The reaction mass was cooled and grounds to a fine powder. Fifty grams of this crude nicotinamide were boiled with 500 ml of anhydrous ethyl acetate until a dark solution was. obtained. A little solid remained in suspension. Gaseous ammonia was passed in below the surface of the ethyl acetate at a temperature between 60-70°C. After a short time ammonium nicotinate started to precipitate out of solution as a brown solid. Sufficient gaseous ammonia, was passed into the ethyl acetate solution to insure complete precipitation of the nicotinic acids as ammonium nicotinate. The solution was filtered at about 60-70°C. The filter cake consisted of ammonium nicotinate, which, upon drying, weighed 12.4 grams. The filtrate was stirred arid boiled for 20 minutes with one-half gram of activated carbon and two grams of activated adsorbent clay. The mixture was filtered hot. The filtrate was boiled twenty minutes with one-half gram of activated carbon and two grams of activated adsorbent clay and then filtered hot. The carbon and clay treatment was repeated once more. The final filtrate was cooled slowly with stirring to room temperature to precipitate white crystalline nieocinamide which, upon drying, weighed 26.7 grams and had a melting point of 129.5°C, and was over 99 percent pure. The mother liquor from the above filtration was boiled down to one-third of its volume and cooled to room temperature. A second crop of nicotinamide of three grams was obtained.

References

Truchan E. et al.; US Patent No. 2,993,051; July 18, 1961; Assigned to Cowles Chemical Company, Cleveland, Ohio, a corporation of Ohio

NIALAMIDE

Therapeutic Function: Antidepressant

Chemical Name: 4-Pyridinecarboxylic acid 2-[3-oxo-3-[(phenylmethyl)-amino]propyl]hydrazide

Common Name: -

Structural Formula:

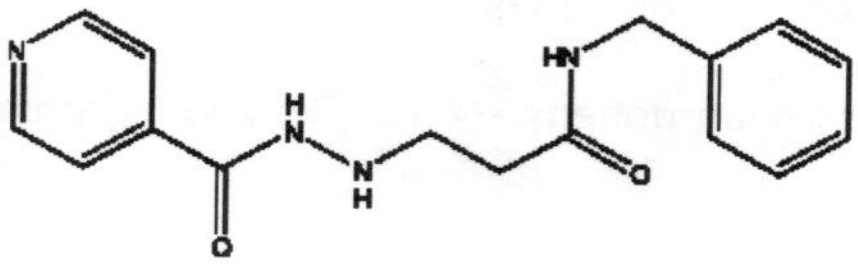

Chemical Abstracts Registry No.: 51-12-7

Trade Name	Manufacturer	Country	Year Introduced
Niamid	Pfizer	US	1959
Niamide	Pfizer	France	1960
Niamid	Taito Pfizer	Japan	-
Nuredal	EGYT	Hungary	-
Surgex	Firma	Italy	-

Raw Materials

Isoniazid
Methyl acrylate
Benzylamine

Manufacturing Process

Methyl acrylate, 28.0 g (0.4 mol) was added dropwise during one hour to a solution containing 54.8 g (0.4 mol) of isonicotinic acid hydrazide (isoniazid) and 10 ml of glacial acetic acid in 400 ml of tertiary butyl alcohol. The resulting solution then was heated for 18 hours on a steam bath. Concentration of the reaction mixture to 100 ml yielded 13.0 g of unreacted isonicotinic acid hydrazide. The filtrate was concentrated to a thick syrup which was triturated with anhydrous ether and recrystallized from isopropyl alcohol; MP 87°C to 88.5°C. Elemental analysis of the product gave 1-isonicotinyl-2-(β-carbomethoxyethyl)hydrazine.

A slurry of 7.5 g (0.034 mol) of 1-isonicotinyl-2-(carbomethoxyethyl)-hydrazine and 5 ml of benzylamine is heated with stirring at 130°C for three hours. The cooled mass is then recrystallized from ethyl acetate to yield white needles melting at 151.1°C to 152.1°C.

References

Merck Index 6330
Kleeman & Engel p. 628
OCDS Vol. 1 p. 254 (1977)
I.N. p. 667
Bloom, B.M. and Carnahan, R.E.; US Patent 2,894,972; July 14, 1959; assigned to Chas. Pfizer and Co., Inc.

NIAPRAZINE

Therapeutic Function: Antihistaminic

Chemical Name: 1-(4-Fluorophenyl)-4-[3-(3-pyridoyl)amino]butyl-piperazine

Common Name: -

Structural Formula:

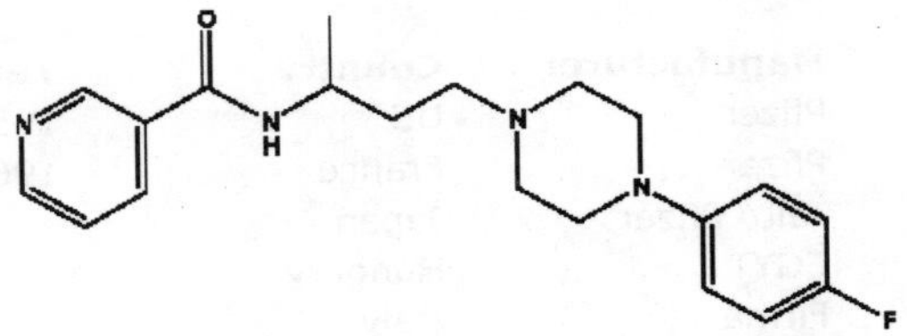

Chemical Abstracts Registry No.: 27367-90-4

Trade Name	Manufacturer	Country	Year Introduced
Nopron	Carrion	France	1976
Norpron	Riom	Italy	-

Raw Materials

Trioxymethylene
1-(4-Fluorophenyl)piperazine dihydrochloride
Acetone
Hydroxylamine hydrochloride
Nicotinic acid chloride
Lithium aluminum hydride

Manufacturing Process

1st Stage: 10 ml of concentrated (10 N) hydrochloric acid and 240 ml of acetone were added to a solution of 217.5 g (1 mol) of 1-(4-fluorophenyl)piperazine dihydrochloride in 400 ml of 96% ethanol. 50 g of powdered trioxymethylene were then added and the mixture was then slowly heated to reflux, which was maintained for 1 hour. A further 60 g of trioxymethylene were then added and heating to reflux was continued for a further 6 hours.

The mixture was then cooled, the precipitate formed was filtered off, washed with acetone and recrystallized from 96% ethanol.

The base was liberated from its salt by taking up the product in an aqueous solution of sodium bicarbonate. The precipitate of the base thus obtained was recrystallized from petroleum ether to give 160 g of the desired product; melting point 46°C; yield 64%.

2nd Stage: 45.5 g (0.65 mol) of hydroxylamine hydrochloride were added to a solution of 128 g (0.5 mol) of the amino-ketone obtained in the preceding stage in 100 ml of ethanol and 40 ml of water. The mixture was allowed to react for 15 minutes at room temperature and was then heated to reflux for ½ hour. A part of the solvent was then distilled off and the product was then allowed to crystallize on cooling. After recrystallization from 96% ethanol, 117 g of the desired product were obtained; melting point 170°C; yield 77%.

3rd Stage: 93 g (0.35 mol) of the oxime obtained in the preceding stage, in the form of the base, were added in portions to a suspension of 17 g (0.45 mol) of lithium aluminum hydride in 400 ml of anhydrous ether. The mixture was then heated to reflux for 15 hours.

10 ml of ethyl acetate and then 50 ml of dilute caustic soda were added slowly with the usual precautions to the mixture. The organic phase was separated, dried over anhydrous Na_2SO_4, the solvent was distilled off and the residue obtained was distilled under reduced pressure to give 51 g of a thick oil; boiling point (2 mm Hg), 142°C to 143°C; yield 58%.

4th Stage: 10 ml of triethylamine were added in a solution of 25.2 g (0.1 mol) of the amine obtained in the preceding stage in 100 ml of anhydrous chloroform and the mixture was cooled to 2°C to 3°C. While maintaining this temperature, 17 g (0.12 mol) of nicotinic acid chloride were added with vigorous agitation.

After evaporation of the solvent, the residue was washed with water, the product taking the form of a mass. After recrystallization from ethyl acetate, a constant melting point of 131°C was obtained.

References

Merck Index 6331
Kleeman & Engel p. 628
DOT 13 (1) 29 (1977)
I.N. p. 667
Mauvernay, R.Y., Busch, N., Simond, J. and Moleyre, J.; US Patent 3,712,893; January 23, 1973; assigned to SA Centre Europeen De Recherches Mauvernay, CERM

NICARDIPINE

Therapeutic Function: Vasodilator

Chemical Name: 2,6-Dimethyl-4-(3-nitrophenyl)-3-methoxycarbonyl-1,4-dihydropyridine-5-carboxylic acid-2(N-benzyl-N-methylamino)ethyl ester hydrochloride

Common Name: -

Structural Formula:

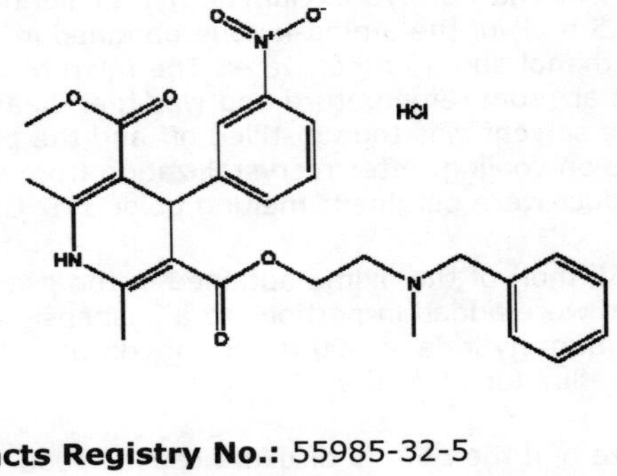

HCl

Chemical Abstracts Registry No.: 55985-32-5

Trade Name	Manufacturer	Country	Year Introduced
Nicodel	Mitsui	Japan	1981
Perdipin	Yamanouchi	Japan	1981

Raw Materials

Acetoacetic acid N-benzyl-N-methylaminoethyl ester
β-Aminocrotonic acid methyl ester
m-Nitrobenzaldehyde

Manufacturing Process

A mixture of 4.98 g of acetoacetic acid N-benzyl-N-methylaminoethyl ester, 2.3 g of β-aminocrotonic acid methyl ester, and 3 g of m-nitrobenzaldehyde was stirred for 6 hours at 100°C in an oil bath. The reaction mixture was subjected to a silica gel column chromatography (diameter 4 cm and height 25 cm) and then eluted with a 20:1 mixture of chloroform and acetone. The effluent containing the subject product was concentrated and checked by thin layer chromatography. The powdery product thus obtained was dissolved in acetone and after adjusting the solution with an ethanol solution saturated with hydrogen chloride to pH 1-2, the solution was concentrated to provide 2 g of 2,6-dimethyl-4-(3'-nitrophenyl)1,4-dihydropyridine-3,5-dicarboxylic acid 3-methylester-5-β-(N-benzyl-N-methylamino)ethyl ester hydrochloride. The product thus obtained was then crystallized from an acetone mixture, melting point 136°C to 140°C (decomposed).

References

Merck Index 6334
DFU 2 (6) 409 (1977) (as Yc-93) & 4 (12) 911 (1979)
OCDS Vol. 3 p. 150 (1984)
DOT 18 (7) 325 (1982)
I.N. p. 668
Murakami, M., Takahashi, K., Iwanami, M., Fujimoto, M., Shibanuma, T., Kawai, R. and Takenaka, T.; US Patent 3,985,758; October 12, 1976; assigned to Yamanouchi Pharmaceutical Co., Ltd.

NICERGOLINE

Therapeutic Function: Vasodilator

Chemical Name: 10-Methoxy-1,6-dimethylergoline-8β-methanol 5-bromonicotinate (ester)

Common Name: Nicotergoline; 1-Methyllumilysergol-8-(5-bromonicotinate)-10-methyl ether

Chemical Abstracts Registry No.: 27848-84-6

Structural Formula:

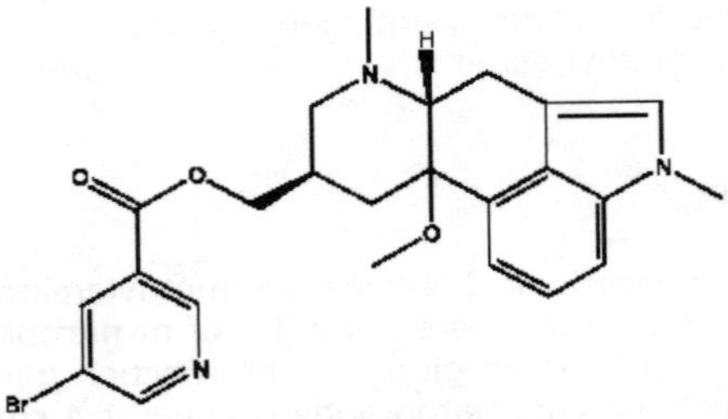

Trade Name	Manufacturer	Country	Year Introduced
Sermion	Farmitalia	Italy	1974
Sermion	Specia	France	1975
Nicergolyn	Farnex	Italy	-
Nicotergoline	Carlo Erba	Italy	-
Varson	Almirall	Spain	-
Vasospan	Exa	Argentina	-

Raw Materials

1-Methyl-lumilysergic acid
5-Bromonicotinyl chloride
Hydrogen chloride
Lithium aluminum hydride
Methanol

Manufacturing Process

Preparation of 1-Methyl Lumilysergic Acid 8-Methyl Ester-10-Methyl Ether: Into a suspension of 10 grams of 1-methyl-lumilysergic acid in 600 cc of absolute methanol a stream of anhydrous hydrogen chloride is bubbled for 1.5 hours with strong cooling. The stream of hydrogen chloride is stopped and the mixture is allowed to stand for 30 minutes at 0°C, and is evaporated in vacuo to dryness. The residue is taken up with ice-cooled water made alkaline with concentrated ammonia and extracted with chloroform. The combined chloroform extracts are washed first with a 5% aqueous solution of sodium bicarbonate, then with water, and are thereafter dried over anhydrous sodium sulfate and finally evaporated in vacuo to dryness.

Preparation of 1-Methyl Lumilysergol-10-Methyl Ether: To a boiling suspension of 2 grams of lithium aluminum hydride in 50 cc of anhydrous tetrahydrofuran, a solution of 1 gram of 1-methyl lumilysergic acid-8-methyl ester-10-methyl ether in 20 cc of anhydrous tetrahydrofuran is added dropwise and the resulting solution is refluxed for a further 2 hours. After cooling the resulting solution, aqueous tetrahydrofuran is added to destroy the excess reducing agent and the solution is filtered. Tetrahydrofuran is distilled off and the residue is recrystallized from acetone petroleum ether.

Preparation of Nicergoline: To a solution of 1-methyl lumilysergol-10-methyl ether in pyridine, 5-bromonicotinyl chloride is used as an acylating agent at room temperature. The mixture is stirred for 1 hour. Water and methanol are added and the resulting mixture is stirred for 1 hour, extracted with chloroform, and washed in sequence with 1% aqueous caustic soda, 5% aqueous sodium bicarbonate solution, and water. The resulting solution is dried over anhydrous sodium sulfate and the solvent is distilled off. By recrystallization of the residue from acetone petroleum ether, nicergoline is obtained, melting at 136° to 138°C.

References

Merck Index 6335
Kleeman & Engel p. 629
OCDS Vol. 2 p. 478 (1980)
DOT 10 (12) 342 (1974)
I.N. p. 668
Bernardi, L., Bosisio, G. and Goffredo, O.; US Patent 3,228,943; January 11, 1966; assigned to Societa Farmaceutici Italia, Italy

NICERITROL

Therapeutic Function: Antihyperlipidemic

Chemical Name: 3-Pyridinecarboxylic acid 2,2-bis[[(3-pyridinylcarbonyl) oxy]methyl]-1,3-propanediyl ester

Common Name: Pentaerythritol tetranicotinate

Structural Formula:

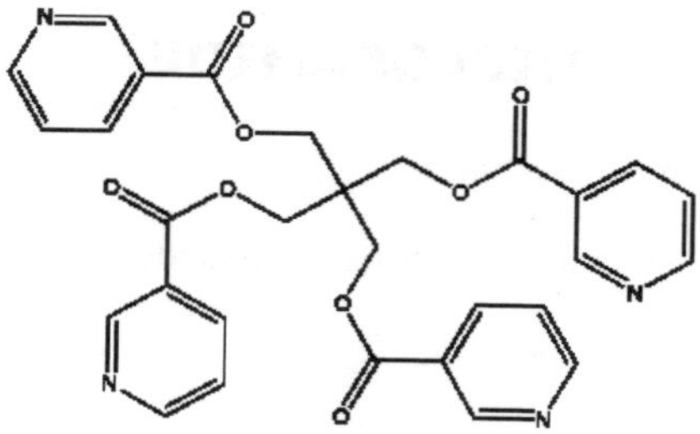

Chemical Abstracts Registry No.: 5868-05-3

Trade Name	Manufacturer	Country	Year Introduced
Cardiolipol	Gremy-Longuet	France	1972
Perycit	Sanwa	Japan	1979
Perycit	Tosi	Italy	1980
Perycit	Astra	Sweden	-

Raw Materials

Nicotinic acid chloride
Pentaerythritol
Pyridine

Manufacturing Process

160 grams of nicotinic acid chloride is charged into and made to react with 35 grams of pentaerythritol dissolved in 600 grams of dried, stabilized chloroform and 100 grams of carefully dried pyridine. Pyridine hydrochloride, pyridine and the excess of nicotinic acid chloride are removed through repeated extraction with water at a pH of approximately 3. Pentaerythritol nicotinate remains in the chloroform phase and is extracted by forming the hydrochloric acid salt of the ester using 1,000 ml of aqueous HCl at a pH of 1. The strongly acid extract is thereafter extracted several times with toluene. The acid extract is allowed to stand at room temperature for several hours in the presence of active carbon and the substance known as Versenate, i.e., the disodium salt of ethylene diamine tetraacetic acid; it is then filtered and pentaerythritol nicotinate is precipitated as a white, amorphous substance using 25% w/v aqueous ammonia, while stirring. Recrystallization of the product from ethyl alcohol gives flaky crystals, according to British Patent 1,022,880.

References

Merck Index 6336
Kleeman & Engel p. 630
I.N. p. 668
AB Bofors, Sweden; British Patent 1,022,880; March 16, 1966
AB Bofors, Sweden; British Patent 1,053,689; January 4, 1967

NICLOSAMIDE

Therapeutic Function: Anthelmintic

Chemical Name: 2',5-Dichloro-4'-nitrosalicylanilide

Common Name: -

Structural Formula:

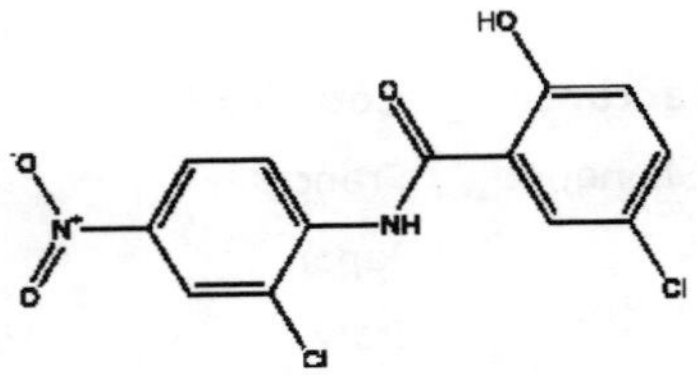

Chemical Abstracts Registry No.: 50-65-7

Trade Name	Manufacturer	Country	Year Introduced
Yomesan	Bayer	W. Germany	1960
Yomesan	Bayer	UK	1961
Yomesan	Bayer	Italy	1962
Tredemine	Roger Bellon	France	1964
Niclocide	Miles	US	1982
Anti-Tenia	Uranium	Turkey	-
Atenase	I.C.N.-Usafarma	Brazil	-
Radeverm	Arzneimittelwerk Dresden	E. Germany	-
Teniarene	A.M.S.A.	Italy	-
Tenisid	Liba	Turkey	-

Raw Materials

5-Chlorosalicylic acid
2-Chloro-4-nitroaniline
Phosphorus trichloride

Manufacturing Process

17.2 g of 5-chlorosalicylic acid and 20.8 g of 2-chloro-4-nitroaniline are dissolved in 250 ml of xylene. While boiling, there are introduced slowly 5 g of PCl_3.Heating is continued for 3 further hours. The mixture is then allowed to cool down and the crystals which separate are filtered off with suction. The crude product may be recrystallized from ethanol, melting at 233°C.

References

Merck Index 6356
Kleeman & Engel p. 630
PDR p. 1260
OCDS Vol. 2 p. 94 (1980)
I.N. p. 669
REM p. 1236
Schraufstatter, E. and Gonnert, R.; US Patent 3,147,300; September 1, 1964; assigned to Farbenfabriken Bayer A.G.

NICOMOL

Therapeutic Function: Anticholesteremic

Chemical Name: Cyclohexanol-2,2,6,6-tetrakis(hydroxymethyl) tetranicotinate

Common Name: -

Structural Formula:

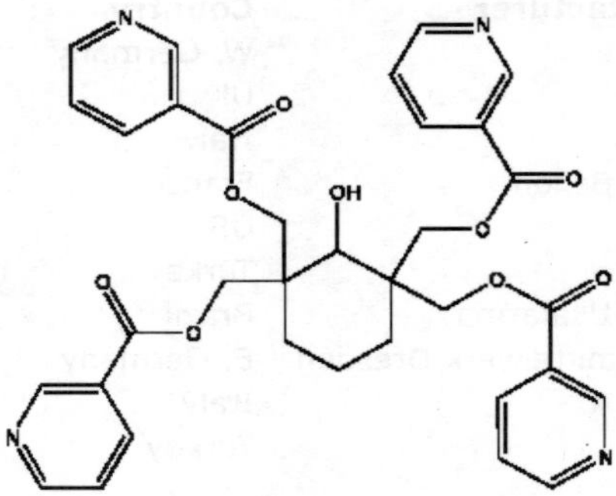

Chemical Abstracts Registry No.: 27959-26-8

Trade Name	Manufacturer	Country	Year Introduced
Cholexamine	Kyorin	Japan	1971
Acenol	Kissei Pharmaceutical Co., Ltd.	Japan	1981
Nicolanta	Sawai	Japan	-

Raw Materials

2,2,6,6-Tetramethylolcyclohexanol
Nicotinic acid chloride

Manufacturing Process

To a mixture of 60 cc of benzene, 40 cc of pyridine and 17 g of hydrochloric acid salt of nicotinic acid chloride, was added 4.5 g of 2,2,6,6-tetramethylolcyclohexanol, and the whole mixture was refuxed at 75°C to 80°C for 2.5 hours. After the mixture was cooled water was added. Precipitate formed was separated by filtration, washed thoroughly with water and dried. Recrystallization from dilute acetic acid gave 14 g of the final compound, melting point 177°C to 180°C.

References

Merck Index 6360
DOT 7 (5) 173 (1971)
I.N. p. 670
Irikura, T., Sato, S., Abe, Y. and Kasuga, K.; US Patent 3,299,077; January 17, 1967; assigned to Kyorin Seiyaku KK

NICOTINE

Therapeutic Function: Ganglion depressant, Smoking deterrent

Chemical Name: Pyridine, 3-(1-methyl-2-pyrrolidinyl)-, (S)-

Common Name: Nicotine; Nikotin

Structural Formula:

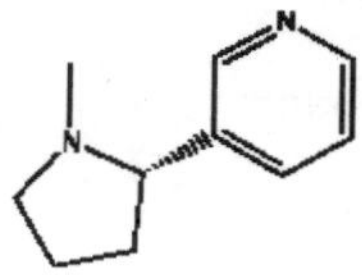

Chemical Abstracts Registry No.: 54-11-5

Trade Name	Manufacturer	Country	Year Introduced
Nicotinell TTS	Novartis	-	-
Habitrol	Ciba-Geigy	-	-
Habitrol	Novartis	-	-
Nicabate	HMR	-	-
Nicoderm CQ	SKB	-	-
Paro	Esro	-	-
Toban	Pharmacia	-	-
Nicorette	GlaxoSmithKline	-	-

Raw Materials

Extract from Nicotiana tabacum
Calcium hydroxide
Calcium sulfate

Manufacturing Process

The water extract from Nicotiana tabacum was prepared by distillation of nicotine contained liquor from tobacco leaves, as described in D.R. Patent No. 319,846; September 12, 1913.

5 kg this water extract or the same quantity of tobacco powder in water was mixed with 1.5 kg of grinded calcium hydroxide and 1.5 kg calcium sulfate. The mixture stood for 24 hours. The obtained mixture looked like a dry powder. It was extracted with ether. The ether was distilled and the residue contented 98% of clear nicotine - liquid with odor of pyridine; BP: 246C/735 mm; d_4^{20} =1.0097; $[\alpha]_d^{20}$=- 166.5.

References

Hovler H.F.; DR Patent No. 320,897; Dec. 25, 1913
Dictionary of Organic Compounds edited by I. Hielbron and H.M. Bunbury, v.3, p. 60, 1946; London

NICOTINYL ALCOHOL

Therapeutic Function: Vasodilator

Chemical Name: 3-Pyridinemethanol

Common Name: 3-Pyridylcarbinol

Structural Formula:

Chemical Abstracts Registry No.: 100-55-0

Trade Name	Manufacturer	Country	Year Introduced
Roniacol	Roche	US	1949
Danaden	Cascan	W. Germany	-
Peritard	Ikapharm	Israel	-
Ronicol	Roche	UK	-
Thilocombin	Thilo	W. Germany	-

Raw Materials

3-Cyanopyridine
Ethanol
Hydrogen
Nitrosyl chloride

Manufacturing Process

The catalyst is prepared by suspending 5 kg of catalyst grade charcoal in 200 liters of water, in a pressure vessel, and adding thereto 25 liters of 4% (as Pd metal) aqueous palladous chloride. Air is displaced from the vessel and then hydrogen is passed into the aqueous mixture at a pressure of 3 to 5 psi, while stirring, until no further absorption is noted and the chloride is completely reduced to metal.

To the aqueous suspension of the palladized charcoal catalyst thus obtained are added 20.8 kg of 3-cyano-pyridine (96% purity); and then are added 70 liters of a hydrochloric acid solution prepared by diluting 30 liters of 36% HCl with 40 liters of water. This represents approximately 1.75 mols of HCl for each mol of 3-cyano-pyridine. The suspension is maintained at 10° to 15°C and stirred continuously while introducing a current of hydrogen at a pressure of 3 to 5 psi. When absorption of hydrogen ceases and the 3-cyanopyridine is completely reduced, the reaction mixture is filtered to remove the catalyst. The filter cake is washed with 40 liters of water in two equal portions, and the wash water is added to the filtrate.

The combined liquors, which comprise an aqueous hydrochloric acid solution of 3-aminomethyl-pyridine hydrochloride, are then heated to a temperature of 60° to 65°C, and ethyl nitrite gas is passed into the heated solution. The ethyl nitrite is generated by placing 20 liters of 90% ethyl alcohol in a suitable vessel, diluting with 200 liters of water, and, while stirring, adding to the dilute alcohol 18.3 kg of nitrosyl chloride at the rate of 2.25 kg per hour. (The process using methyl nitrite is carried out by substituting a stoichiometrically equivalent quantity of methyl alcohol for the ethyl alcohol.)

When all the ethyl nitrite has been added, the reaction mixture is refluxed for approximately one hour, then concentrated to dryness under reduced pressure (25 to 30 mm Hg) and at a maximum temperature of 70°C. The crystalline residue is dissolved in 35 liters of water and adjusted to a pH of 8 to 9 by addition (with cooling and stirring) of 11 to 12 kg of caustic soda. The sodium chloride formed is filtered off, and the filter cake is washed with 20 liters of normal butyl alcohol. This wash liquid is used for the first extraction of the product from the aqueous filtrate. The filtrate is then further extracted with four successive 20-liter portions of n-butyl alcohol.

All the extracts are combined and concentrated in vacuo (100°C/20 mm) to remove the n-butyl alcohol. The residue is submitted to fractionation under reduced pressure. The forerun (up to 112°C/2 to 3 mm) consists of a small amount of n-butyl alcohol and some 3-pyridylcarbinol. The main fraction, boiling at 112° to 114°C/2 to 3 mm, consists of 3-pyridylcarbinol.

References

Merck Index 6369
Kleeman & Engel p. 633
I.N. p. 672
REM p. 852
Ruzicka, L. and Prelog, V.; US Patent 2,509,171; May 23, 1950; assigned to Ciba Limited, Switzerland
Cohen, A.; US Patent 2,520,037; August 22, 1950; assigned to Hoffmann-La Roche Inc.
Schlapfer, R.; US Patent 2,547,048; April 3, 1951; assigned to Hoffmann-La Roche Inc.
Chase, G.O.; US Patent 2,615,896; October 28, 1952; assigned to Hoffmann-La Roche Inc.

NIFEDIPINE

Therapeutic Function: Coronary vasodilator

Chemical Name: 1,4-Dihydro-2,6-dimethyl-4-(2'-nitrophenyl)-3,5-pyridinedicarboxylic acid dimethyl ester

Common Name: -

Structural Formula:

Chemical Abstracts Registry No.: 21829-25-4

Trade Name	Manufacturer	Country	Year Introduced
Adalat	Bayer	W. Germany	1975
Adalat	Bayer	Italy	1976
Adalat	Bayer	Japan	1976
Adalat	Bayer	UK	1977
Adalate	Bayer	France	1979
Procardia	Pfizer	US	1982
Alfadat	Alfa	Italy	-
Anifed	Zoja	Italy	-
Atanal	Sawai	Japan	-
Citilat	C.T.	Italy	-
Coral	Tosi	Italy	-
Corinfar	Arzneimittelwerk Dresden	E. Germany	-
Nifedicor	Schiapparelli	Italy	-
Nifedin	Gentili	Italy	-
Nifelat	Sidus	Argentina	-
Oxcord	Biosintetica	Brazil	-

Raw Materials

2-Nitrobenzaldehyde
Acetoacetic acid methyl ester
Ammonia

Manufacturing Process

45 grams 2-nitrobenzaldehyde, 80 cc acetoacetic acid methyl ester, 75 cc methanol and 32 cc ammonia are heated under reflux for several hours, filtered off, cooled and, after suction-filtration, 75 grams of yellow crystals of MP 172° to 174°C are obtained, according to US Patent 3,485,847.

References

Merck Index 6374
DFU 6 (7) 427 (1981)
Kleemen & Engel p. 633
PDR p. 1423
OCDS Vol. 2 p. 283 (1980)
DOT 8 (11) 438 (1972); 11 (4) 154 (1975) & 19 (3) 171 (1983)
I.N. p. 673
REM p. 862
Bossert, F. and Vater, W.; US Patent 3,485,847; December 23, 1969; assigned to Farbenfabriken Bayer AG, Germany
Bossert, F. and Vater, W.; US Patent 3,488,359; January 6, 1970; assigned to Farbenfabriken Bayer AG, Germany
Bossert, F. and Vater, W.; US Patent 3,511,837; May 12, 1970; assigned to Farbenfabriken Bayer AG, Germany

NIFLUMIC ACID

Therapeutic Function: Antiinflammatory

Chemical Name: 2-[[3-(Trifluoromethyl)phenyl]amino]-3-pyridinecarboxylic acid

Common Name: 2-[3-(Trifluoromethyl)anilino]nicotinic acid

Structural Formula:

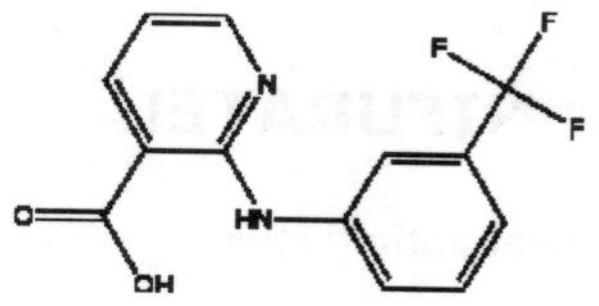

Chemical Abstracts Registry No.: 4394-00-7

Trade Name	Manufacturer	Country	Year Introduced
Nifluril	U.P.S.A.	France	1968
Actol	Von Heyden	W. Germany	1971
Flaminon	Squibb	Italy	1979

Trade Name	Manufacturer	Country	Year Introduced
Forenol	Roemmers	Argentina	-
Landruma	Landerlan	Spain	-
Nifluran	Eczacibasi	Turkey	-
Niflux	Labofarma	Brazil	-

Raw Materials

Nicotinic acid
m-Trifluoromethylaniline
Potassium iodide

Manufacturing Process

Niflumic acid is prepared as follows: Nicotinic acid, m-trifluoromethylaniline, and potassium iodide are intimately mixed and heated on an oil bath at 140°C. The mixture melts to give a dark red liquid. The temperature of the oil bath is allowed to fall to 100°C and is maintained at this temperature for an hour and a half. The mixture puffs up and forms a yellow crystalline mass. After cooling to ordinary temperature, this mass is ground up in a mortar and extracted several times with small volumes of ether to remove excess m-trifluoromethylaniline. The residue is then washed twice with 10 ml of distilled water to remove m-trifluoromethylaniline hydrochloride and potassium iodide, and finally twice with 10 ml of 95% alcohol to remove colored resinous contaminants. After drying at 100°C, 2-(m-trifluoromethylanilino)nicotinic acid is obtained as pale yellow needles (from 70% ethanol) melting at 204°C (Kofler block).

References

Merck Index 6377
Kleeman & Engel p. 634
OCDS Vol. 1 p. 256 (1977)
DOT 4 (2) 82 (1968)
I.N. p. 34
Hoffmann, C. and Faure, A.; US Patent 3,415,834; December 10, 1968; assigned to Societe anonyme dite: Laboratoires UPSA, France

NIFURATEL

Therapeutic Function: Vaginal antiinfective

Chemical Name: 5-[(Methylthio)methyl]-3-[[(5-nitro-2-furanyl)methylene]amino]-2-oxazolidinone

Common Name: Methylmercadone

Chemical Abstracts Registry No.: 4936-47-4

Structural Formula:

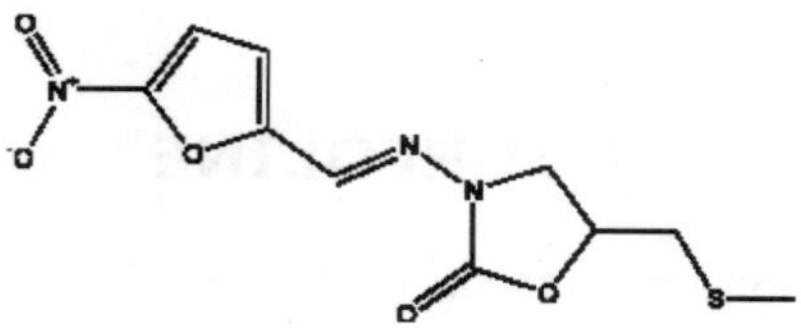

Trade Name	Manufacturer	Country	Year Introduced
Macmiror	Poli	Italy	1965
Inimur	Woelm	W. Germany	1969
Omnes	Fumouze	France	1971
Magmilor	Calmic	UK	-
Polmiror	Poli	Italy	-
Tydantil	Poli	Italy	-

Raw Materials

Methyl mercaptan
5-Nitro-2-furaldehyde
Diethyl carbonate
Hydrazine hydrate
Epichlorohydrin

Manufacturing Process

In an initial step of reactions, methyl mercaptan is reacted with epichlorohydrin to give 1chloro-3-methylthio-2-propanol. That is reacted with hydrazine hydrate to give 3-methylmercapto-2-hydroxypropyl hydrazine.

11.8 grams of diethyl carbonate (0.1 mols) and a solution of sodium methoxide prepared from 0.12 gram of sodium in 4 cc of anhydrous methanol, were added to 13.2 grams of 3-methylmercapto-2-hydroxypropyl hydrazine. After the reaction vessel had been fitted with a Liebig condenser, the reaction mixture was heated by means of an oil bath which was gradually heated up to 110°C, to remove first methyl alcohol and then ethyl alcohol formed during the reaction. After about two-thirds of the theoretical amount of ethyl alcohol had been distilled off, the heating was discontinued and the reaction mixture was diluted with 50 cc of ethyl alcohol and poured into a 5-nitro-2-furfuraldehyde solution prepared by boiling for 30 minutes 0.1 mol of nitrofurfuraldehyde diacetate in 100 ml of ethyl alcohol and 50 ml of 1:10 sulfuric acid.

A yellow crystalline precipitate was immediately formed, which, after crystallization from acetic acid, melted at 182°C and consisted of N-(5-nitro-2-furfurylidene)-3-amino-5-methyl-mercaptomethyl-2-oxazolidinone.

References

Merck Index 6380
Kleeman & Engel p. 635

I.N. p. 674
Polichimica Sap, SpA, Italy; British Patent 969,126; September 9, 1964

NIFURFOLINE

Therapeutic Function: Antibacterial

Chemical Name: 3-(4-Morpholinylmethyl)-1-[[(5-nitro-2-furanyl)-methylene] amino]-2,4-imidazolidinedione

Common Name:-

Structural Formula:

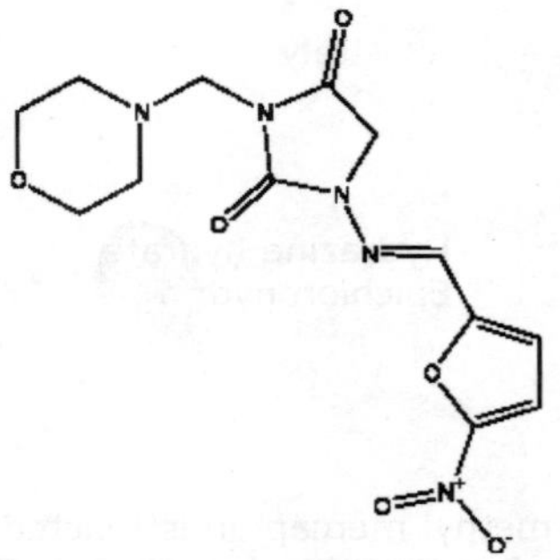

Chemical Abstracts Registry No.: 3363-58-4

Trade Name	Manufacturer	Country	Year Introduced
Furobactil	Carrion	France	1974
Urbac	Merck Clevenot	France	-

Raw Materials

Nitrofurantoin
Formaldehyde
Morpholine

Manufacturing Process

20 g of nitrofurantoin are placed in 100 cc of dimethylformamide and the solution is heated to 75°C to 80°C. This temperature is maintained and 100 cc of 40% formaldehyde are added, followed by 10 g of freshly distilled morpholine. The heating is continued for one hour, the mixture cooled and filtered and the precipitate obtained is washed with 95% alcohol. 20 g of the desired product are obtained as yellow crystals which melt at 206°C.

References

Merck Index 6381
I.N. p. 674
Laboratorios del Dr. Esteve S.A.; British Patent 1,245,095; September 2, 1971

NIFUROXAZIDE

Therapeutic Function: Antiseptic

Chemical Name: 4-Hydroxybenzoic acid [(5-nitro-2-furanyl)methylene]-hydrazide

Common Name: -

Structural Formula:

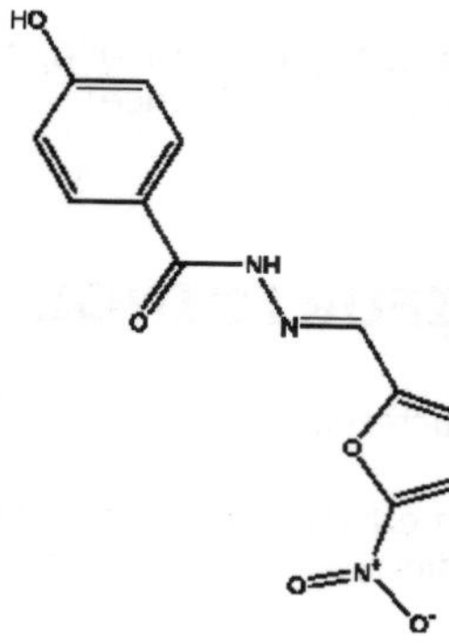

Chemical Abstracts Registry No.: 965-52-6

Trade Name	Manufacturer	Country	Year Introduced
Ercefuryl	Carriere	France	1964
Pentofuryl	Karlspharma	W. Germany	1978
Antinal	Roques	France	-
Dicoferin	Andrade	Portugal	-
Enterokod	Genekod	France	-
Mucifural	Robert and Carriere	France	-

Raw Materials

4-Hydroxybenzhydrazide
5-Nitrofurfural

Manufacturing Process

13 g (0.1 mol) of 4-hydroxybenzhydrazide were dissolved in a boiling mixture of 100 ml of water and an equal volume of dimethylformamide. 15.5 g (0.11 mol) of 5-nitrofurfural dissolved in 31 ml of dimethylformamide were added to this hot solution, and the mixture was stirred and brought to the boiling point.

The mixture was then allowed to stand for fifteen hours. The precipitate was separated, washed twice with 100 ml of water, and recrystallized by dissolving it in 250 ml of hot pyridine and pouring this solution into 250 ml of water.

The 5-nitrofurfurylidene hydrazide of 4-hydroxybenzoic acid obtained was washed with water and methanol and was dried at a moderate temperature. It weighed 23 g (83.7% yield), and melted at 298°C. The percentage nitrogen determined by the micro-Dumas method was 15.41% (theory 15.27%).

References

Merck Index 6383
Kleeman & Engel p. 636
I.N. p. 675
Carron, M.C.E.; US Patent 3,290,213; December 6, 1966; assigned to S.A. des Laboratoires Robert et Carriere (France)

NIFURTOINOL

Therapeutic Function: Antibacterial

Chemical Name: 3-(Hydroxymethyl)-1-[[(5-nitro-2-furanyl)methylene]-amino]-2,4-imidazolidinedione

Common Name: -

Structural Formula:

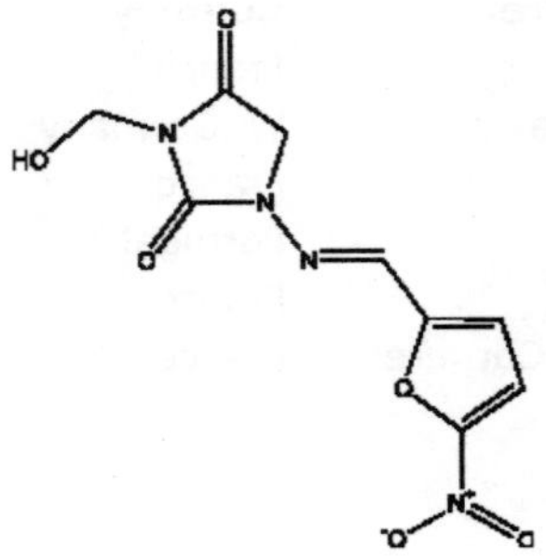

Chemical Abstracts Registry No.: 1088-92-2

Trade Name	Manufacturer	Country	Year Introduced
Urfadyne	Zambon	W. Germany	1969
Urfadyn	Arsac	France	1976
Urfadyne	Inpharzam	Switz.	1981
Levantin	Lek	Yugoslavia	-
Urfurine	Zambon	Spain	-

Raw Materials

Nitrofurantoin
Formaldehyde

Manufacturing Process

Three liters of 5% formaldehyde solution (2,625 cc water and 375 cc 40% formalin) containing 50 g of nitrofurantoin is refluxed for about 5 minutes, then filtered hot and cooled. The crystallized product is filtered and washed with 1% formaldehyde solution. It is air dried and then further dried at 65°C. There is obtained 33 g of 3-hydroxymethyl-1-(5-nitrofurfurylideneamino) hydantoin.

References

Merck Index 6388
I.N. p. 676
Michels, J.G.; US Patent 3,446,802; May 27, 1969; assigned to The Norwich Pharmacal Co.

NIFURZIDE

Therapeutic Function: Antibacterial, Antidiarrheal

Chemical Name: N1-[5'-Nitro-2'-thenoyl]-N2-[5''-nitro-2''-furylacrylidene] hydrazine

Common Name: -

Chemical Abstracts Registry No.: 39978-42-2

Trade Name	Manufacturer	Country	Year Introduced
Ricridene	Anphar	Switz.	1981
Ricridene	Lipha	France	-

Raw Materials

Hydrazine
Ethanol
5-Nitrothiophene carboxylic acid
5-Nitro-2-furylacrolein

Structural Formula:

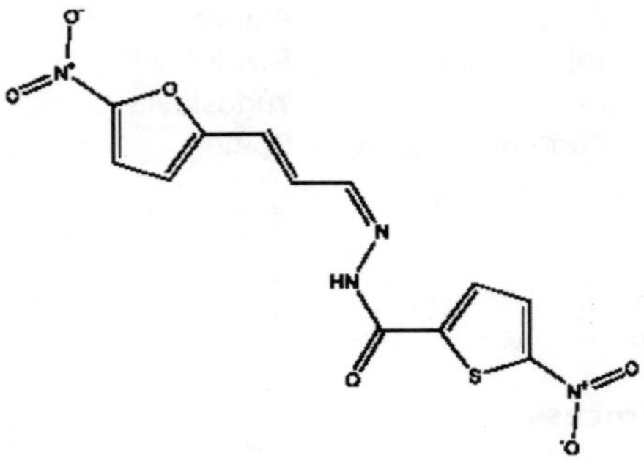

Manufacturing Process

(a) Ethyl-5-nitro-2-thiophene carboxylate:

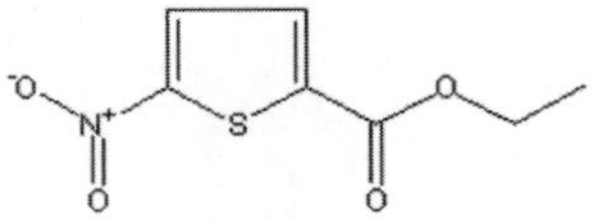

17.4 g (mol/10 = 17.31 g) of 5-nitrothiophene carboxylic acid are dissolved in 85 ml of absolute ethanol. A stream of gaseous hydrochloric acid is caused to enter the boiling solution to the point of saturation, and for 5 hours. Evaporation to dryness takes place and then the solid residue is washed with a sodium bicarbonate solution. It is suction-filtered and washed with water. After drying, there are obtained 17.7 g of a yellow product with a melting point of 63°C to 65°C and the yield is 88% (theoretical yield = 88%).

The N'-(5'-nitro-2'-thenoyl)hydrazide is prepared by reacting hydrazine with ethyl 5-nitro-2-thiophene carboxylate.

(b) 6.3 g (mol/30 = 6.5 g) of N1-[5'-nitro-2'-thenoyl]hydrazide are dissolved in 100 ml of dry tetrahydrofuran. 5.6 g (mol/30 = 5.55 g) of 5-nitro-2-furyl acrolein in 56 ml of tetrahydrofuran are added. Heating under reflux takes place for 1 hour and, 25 minutes after starting the heating, the crystallization commences; the crystals are suction-filtered, washed with ether and dried. There are obtained 7.9 g (yield 70%-theoretical yield = 11.2 g) of a yellow solid of melting point 235°C to 236°C.

Recrystallization (tepid dimethylformamide + ether) leaves the melting point unchanged.

References

Merck Index 6389
DFU 6 (6) 358 (1981)
Kleeman & Engel p. 637

DOT 17 (7) 288 (1981)
Szarvasi, E. and Fontaine, L.; US Patents 3,847,911; November 12, 1974; and 3,914,379; October 21, 1975; both assigned to Lipha, Lyonnaise Industrielle Pharmaceutique

NILUTAMIDE

Therapeutic Function: Antiandrogen

Chemical Name: 2,4-Imidazolidinedione, 5,5-dimethyl-3-(4-nitro-3-(trifluoromethyl)phenyl)-

Common Name: Nilutamide

Structural Formula:

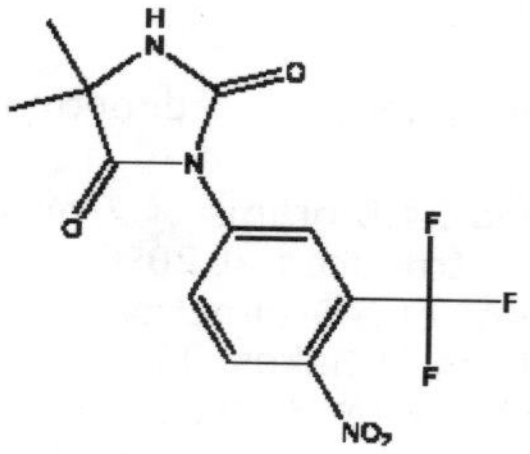

Chemical Abstracts Registry No.: 63612-50-0

Trade Name	Manufacturer	Country	Year Introduced
Anandron	Laboratoires Cassene	France	-
Anandron	Hoechst Marion Roussel	Germany	-
Nilandron	Aventis Pharmaceuticals	France	-
Nilutamide	Triquim S.A.	Argentina	-

Raw Materials

Phenyl oxide	Copper oxide
5,5-Diphenylhydantoin	Dimethyl sulfoxide
2-Nitro-5-chloro-trifluoromethylbenzene	

Manufacturing Process

There are at least five methods to prepare desired compound.

1. 1-(3'-Trifluoromethyl-4'-nitropheyl)-4,4-dimethyl-imidazoline-2,5-dione

The following were introduced into 383.52 ml of phenyl oxide: 225.60 grams of 2-nitro-5-chloro-trifluoromethylbenzene, described in the German Patent

No. DRP 637,318, 128.10 grams of 5,5-dimethylhydantoin described in Beil., Vol. 24, 289 and 198.53 grams of cuprous oxide. The mixture was heated to 200°C for 24 hours, then cooled to 20°C and filtered. The residue was rinsed with phenyl oxide, then extracted with ethyl acetate. The ethyl acetate phase was concentrated to dryness under reduced pressure at 60°C and the residue was taken up in ammoniacal dichloroethane. The crystals obtained were dried at 60°C to obtain 66.55 grams of crude product which, after purification from aqueous ethanol yielded 62.55 grams of purified desired product.

2. 1-(3'-Trifluoromethyl-4'nitrophenyl)-4,4-dimethyl-imidazoline-2,5-dione

The following were introduced into 282 ml of triglyme: 112.8 grams of 2-nitro-5-chloro-trifluoromethylbenzene, 64.1 grams of 5,5-dimethyl-hydantoin and 33.5 grams of cuprous oxide. The mixture was heated to about 215°C ± 5°C for 4 hours, then cooled to 20°C and filtered. The triglyme solution was recovered and a 22 Be ammonia solution (1 volume), toluene (1 volume) and demineralized water (4 volume) were added to the solution of triglyme (1 volume). The solution was stirred at 20°C for 15 minutes, then cooled to about -10°C and stirred again at -10°C. After washing and drying, 47.6 grams of the desired product were obtained.

3. 1-(3'-Trifluoromethyl-4'-nitrophenyl)-4,4-dimethyl-imidazoline-2,5-dione

30 ml of dimethylsulfoxide and 24.8 grams of 2-nitro-5-chloro trifluoromethylbenzene were introduced at 20°C with stirring into 100 ml of dimethylsulfoxide, 12.80 grams of 5,5-dimethyl-hydantoin and 6.28 grams of potassium hydroxide in the form of flakes. The mixture was heated to 110°C for a period of time variable between 3 and 18 hours. The product was characterized and determined by thin layer chromatography.

4. 1-(3'-Trifluoromethyl-4'-nitrophenyl)-4,4-dimethyl-imidazoline-2,5-dione

71.5 grams of copper in powder form were added to 96.10 grams of 5,5-dimethyl-hydantoin and 170.86 grams of 2-nitro-5-chloro trifluoromethylbenzene. The mixture was heated to 200°C for about 21 hours, the pressure being maintained at 450 millibars, then, was cooled to 20°C and taken up in 480 ml of ethanol. The product was characterized and determined by thin layer chromatography of the ethanol solution.

5. 1-(3'-Trifluoromethyl-4'-nitrophenyl)4,4-dimethyl-imidazoline-2,5-dione

The following were introduced into 288 ml of phenyl oxide: 96.10 grams of 5,5-dimethyl-hydantoin, 170.86 grams of 2-nitro-5-chloro trifluoromethylbenzene and 89.40 grams of cupric oxide. The mixture was heated to 190°C for about 23 hours, then cooled to 20°C and filtered. The residue was characterized in the phenyl oxide filtrate by thin layer chromatography. The analytical results obtained for these 5 examples were identical to those obtained and indicated in French Patent No. 2,329,276.

References

Seuron P. et.al.; US Patent No. 5,166,358; Nov. 24, 1992; Assigned to Rousel Uclaf, Paris, France

NILVADIPINE

Therapeutic Function: Calcium entry blocker, Antihypertensive

Chemical Name: 2-Cyano-1,4-dihydro-6-methyl-4-(3-nitrophenyl)-3,5-pyridinedicarboxylic acid 3-methyl- 5-(1-methyl ethyl) ester

Common Name: Nivadipine, Niprodipine

Structural Formula:

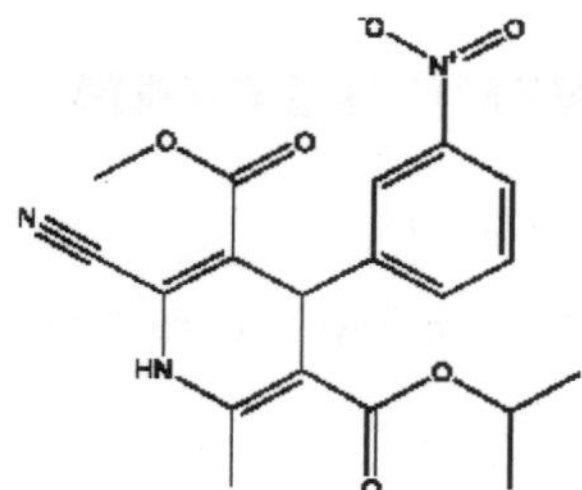

Chemical Abstracts Registry No.: 75530-68-6

Trade Name	Manufacturer	Country	Year Introduced
Nilvadipine	Fujisawa	-	-
Tensan	Klinge	-	-

Raw Materials

Isopropyl ester of 6-formyl-5-methoxycarbonyl-2-methyl-4-(3-nitrophenyl)-1,4-dihydropyridine-3-carboxylic acid
Hydroxylamine hydrochloride
Sodium acetate
Acetic anhydride

Manufacturing Process

To a solution of isopropyl ester of 6-formyl-5-methoxycarbonyl-2-methyl-4-(3-nitrophenyl)-1,4-dihydropyridine- 3-carboxylic acid (4.5 g) in acetic acid (35 ml) were added hydroxylamine hydrochloride (0.97 g) and sodium acetate (1.43 g), and the mixture was stirred at ambient temperature for 2.5 hours. After acetic anhydride (4.14 g) was added to this reaction mixture, the mixture was stirred at ambient temperature for 1.5 hours and at 95-100°C for additional 4 hours. The acetic acid and the excess of acetic anhydride were removed in vacuum, followed by adding water to the residue and it was neutralized with a saturated aqueous solution of sodium bicarbonate. This aqueous suspension was extracted twice with ethyl acetate, and the combined extract was washed with water, dried over anhydrous magnesium sulfate and evaporated to dryness under reduced pressure to give a reddish-brown oil (4.88 g), which was chromatographed over silica gel (150 g) with a mixture of benzene and ethyl acetate (10:1 by volume) as an eluent to give a crude crystals (2.99 g). These were recrystallized from ethanol to give yellow prisms (1.89 g) of isopropyl ester of 6-cyano-5-methoxycarbonyl-2-methyl-4-(3-nitrophenyl)-1,4-dihydropyridine-3-carboxylic acid melting point 148-150°C (yellow prisms from ethanol); $[\alpha]_D^{20}$ = 222.4° (c = 1 in methanol).

References

Sato Yoshinari; US Patent No. 4,338,322; July 6, 1982; Assigned to Fujisawa Pharmaceutical Co., Ltd. (Osaka, JP)

NIMETAZEPAM

Therapeutic Function: Tranquilizer

Chemical Name: 1,3-Dihydro-1-methyl-7-nitro-5-phenyl-2H-1,4-benzodiazepin-2-one

Common Name:-

Structural Formula:

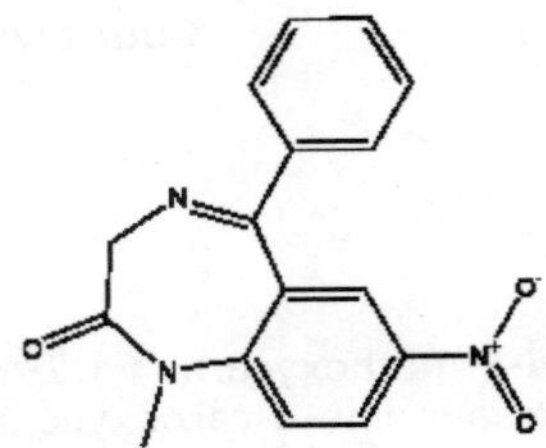

Chemical Abstracts Registry No.: 2011-67-8

Trade Name	Manufacturer	Country	Year Introduced
Erimin	Sumitomo	Japan	1977

Raw Materials

Hydrogen chloride	1-Methyl-5-nitro-3-phenylindole-2-carbonitrile
Chromic anhydride	Boron trifluoride etherate

Manufacturing Process

To a suspension of 73.9 g of 1-methyl-5-nitro-3-phenylindole-2-carbonitrile in 1.5 liters of dry tetrahydrofuran is added dropwise a solution of 126 g of boron trifluoride etherate in 220 ml of dry tetrahydrofuran with stirring for 2 hours. After addition, stirring is continued for an additional 3 hours. To the reaction mixture is added dropwise 370 ml of water and then 370 ml of concentrated hydrochloric acid with stirring under ice-cooling.

The resulting precipitate is collected by filtration, washed with water followed by ethanol, and dried to give 56.3 g of crude 2-aminomethyl-1-methyl-5-nitro-3-phenylindole hydrochloride, melting point 263°C to 267°C.

To a suspension of 6.5 g of 2-aminomethyl-1-methyl-5-nitro-3-phenylindole in 65 ml of glacial acetic acid is added dropwise a solution of 6.5 g of chromic anhydride in 6.5 ml of water at 20°C with stirring. The mixture is stirred at room temperature overnight and thereto is added 195 ml of water. To the mixture is added dropwise 100 ml of 28% ammonia water with stirring under cooling. The resultant precipitate is collected by filtration, washed with water and dried to give 5.9 g of a crude product having melting point 135°C to 140°C. Fractional recrystallization from ethanol gives 3.8 g of 1-methyl-7-nitro-5-phenyl-1,3-dihydro-2H-1,4-benzodiazepine-2-one as yellow plates, melting point 153°C to 156°C. Further recrystallization from the same solvent gives pale yellow plates having melting point 156°C to 156.5°C.

References

Merck Index 6395
Kleeman & Engel p. 637
DOT 8 (9) 350 (1972); 11 (5) 195 (1975) & 13 (1) 31 (1977)
I.N. p. 676
Yamamoto, H., Inaba, S., Okamoto, T., Hironashi, T., Ishizumi, K., Yamamoto, M., Maruyama, I., Mori, K. and Kobayashi, T.; US Patents 3,770,767; November 6, 1973; and 3,652,551; March 28, 1972; both assigned to Sumitomo Chemical Co.

NIMODIPINE

Therapeutic Function: Vasodilator

Chemical Name: 3,5-Pyridinedicarboxylic acid, 1,4-dihydro-2,6-dimethyl-4-(3-nitrophenyl)-, 2-methoxyethyl 1-methylethyl ester

Common Name: Nimodipine

Structural Formula:

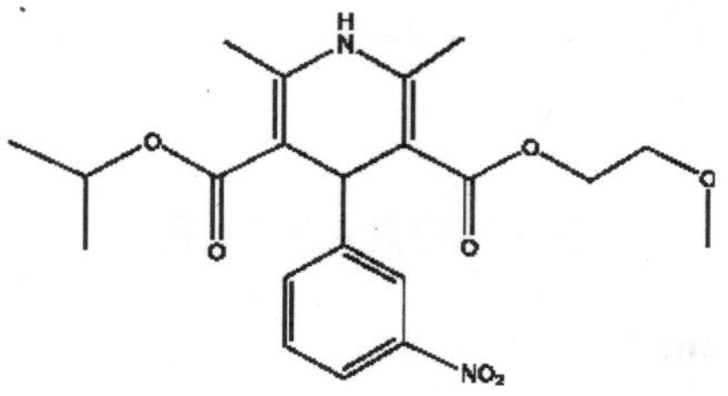

H
N
O
O
O
O
O
NO_2

Chemical Abstracts Registry No.: 66085-59-4

Trade Name	Manufacturer	Country	Year Introduced
Brainox	Euro-Labor.	-	-
Curban	Rafarm	Greece	-
Grifonmod	Laboratorio Chile S.A.	Chile	-
Modina	Pentafarma	Chile	-
Modus	Berenguer	-	-
Myodipine	Help	Greece	-
Nemotan	Medochemie Ltd.	Cyprus	-
Nimotop	Bayer AG	Germany	-
Nimotop	Miles	-	-
Nimodipine	Bayer AG	Germany	-
Norton	Farmasa	Brazil	-
Regental	Tecnofarma S.A.	Chile	-
Sobrepina	Farmoz	Portugal	-
Trinalion	Tecnimede	Portugal	-
Tropocer	Laboratorios Leti	Venezuela	-
Vasotop	Cipla Limited	India	-
Ziremex	Demo	-	-
Nimocer	Synapse (A Div. of Microlabs)	India	-

Raw Materials

3'-Nitro-benzylideneacetoacetic acid isopropylester
Acetoacetic acid β-metoxyethyl ester
Ammonium hydroxide

Manufacturing Process

After 8 hours boiling of solution of 3.8 g of 3'-nitro-benzylideneacetoacetic acid isopropylester, 8 grams of acetoacetic acid β-metoxyethyl ester and 6 ml conc ammonia in 80 ml ethanol under reflux, 2,6-dimethyl-4-(3'-nitrophenyl)-1,4-dihydropyridine 3-β-methoxyethyl ester 5-isopropyl ester of melting point 125°C (petroleum ether/ acetic ester) was obtained. Yield 49% of theory.

References

Meyer H. et al.; US Patent No. 3,799,934; March 26, 1974; Assigned to Farbenfabriken Bayer Aktiengesellschaft, Leverkusen, Germany
Meyer H. et al.; US Patent No. 4,406,906; Sep. 27, 1983; Assigned to Bayer Aktiengesellschaft, Leverkusen, Fed. Rep. of Germany

NIMORAZOLE

Therapeutic Function: Trichomonacidal

Chemical Name: N-β-Ethylmorpholino-(5)-nitroimidazole

Common Name: Nitrimidazine

Structural Formula:

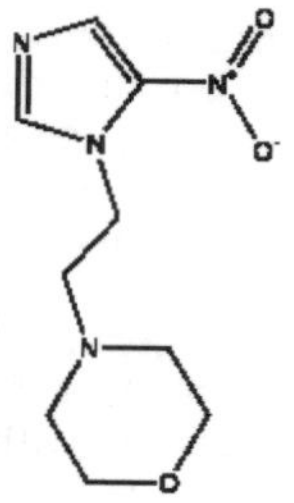

Chemical Abstracts Registry No.: 6506-37-2

Trade Name	Manufacturer	Country	Year Introduced
Naxogin	Carlo Erba	UK	1970
Naxogin	Carlo Erba	Italy	1972
Esclama	Farmitalia	W. Germany	1973
Aceterol Forte	Bristol-Myers	W. Germany	1973
Naxofem	Ikapharm	Israel	-
Nulogyl	Bristol	UK	-
Sirledi	Causyth	Italy	-

Raw Materials

Ethylene oxide
β-Chloroethyl morpholine
Morpholine
p-Toluenesulfonyl chloride
4(5)-Nitroimidazole sodium salt

Manufacturing Process

6 g 4(5)-nitroimidazole sodium salt and 9 g β-chloroethylmorpholine are allowed to react in 200 ml dry toluene. The mixture is refluxed for 50 hours, then cooled and filtered from the solid residue. The solvent is evaporated under reduced pressure. The half-solid product thus obtained solidifies by addition of petroleum ether and ethyl ether.

Crystallization from water results in N-β-ethylmorpholino-(5)-nitroimidazole (melting point 110°C to 111°C); from mother liquors N-β-ethylmorpholino-(4)-nitroimidazole (melting point 104°C to 106°C) is obtained.

The following procedure is given in US Patent 3,458,528: 78 grams (0.675 mol) of 5-nitroimidazole is dissolved in 1,500 ml of acetic acid upon the addition of 72 ml (0.57 mol) of boron trifluoride etherate. 175 ml (3.5 mols) of ethylene oxide in 175 ml of hexane, in a dropping funnel topped with a cold finger, is added slowly over 1 hour to the above solution maintained at 32° to 35°C with a water cooling bath. The mixture is concentrated under high vacuum to 100 to 150 ml volume. The residue is diluted with 500 ml of water, neutralized to pH 7 with aqueous sodium hydroxide, and extracted with 1.5 liters of ethyl acetate. The extract is dried and evaporated to yield 1-(2'-hydroxyethyl)-5-nitroimidazole.

20 grams (0.127 mols) of 1-(2'-hydroxyethyl)-5-nitroimidazole in 50 ml of dry pyridine is reacted with 75 grams of p-toluenesulfonyl chloride at 15°C for 4 hours. The reaction mixture is poured into ice and water and the crystalline precipitate is separated by filtration, washed with water and air dried to yield 1-(2'-p-toluenesulfonyloxyethyl)-5-nitroimidazole; MP 126° to 127°C.

16 grams, (0.057 mol) of 1-(2'-p-toluenesulfonyloxyethyl)-5-nitroimidazole and 9.3 ml of morpholine are heated at 95°C for 4 hours. The reaction mixture is taken up in water and extracted with ether. Evaporation of the ether yields 1-(2'-N-morpholinylethyl)-5-nitroimidazole; MP 109° to 110°C.

References

Merck Index 6398
Kleeman & Engel p. 638
OCDS Vol. 2 p. 244 (1980)
DOT 6 (5) 185 (1970) & 7 (5) 193 (1971)
I.N. p.677
Giraldi, P.N. and Mariotti, V.; US Patent 3,399,193; August 27, 1968; assigned to Carlo Erba SpA, Italy
Gal, G.; US Patent 3,458,528; July 29, 1969; assigned to Merck & Co., Inc.
Carlson, J.A., Hoff, D.R. and Rooney, C.S.; US Patent 3,646,027; February 29, 1972; assigned to Merck & Co., Inc.

NIMUSTINE

Therapeutic Function: Antitumor, Antileukemic

Chemical Name: 1-(2-Chloroethyl)-1-nitroso-3-[(2-methyl-4-aminopyrimidin-5-yl)-methyl]urea

Common Name: ACNU

Structural Formula:

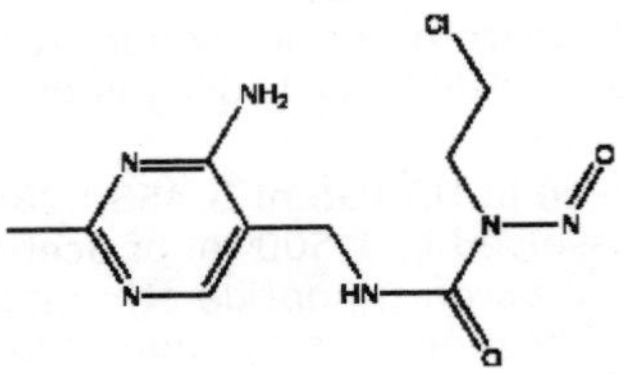

Cl
NH2
N
O
N
N
N
HN
O

Chemical Abstracts Registry No.: 42471-28-3

Trade Name	Manufacturer	Country	Year Introduced
Nidran	Sankyo	Japan	1979

Raw Materials

1-(2-Chloroethyl)-3-[(2-methyl-4-aminopyridin-5-yl)methyl]urea
Sodium nitrite
Hydrogen chloride

Manufacturing Process

0.4 g of sodium nitrite was added with stirring, at 0°C to 5°C, to a solution of 450 mg of 1-(2-chloroethyl)-3-[(2-methyl-4-aminopyridin-5-yl)methyl]urea in 8 ml of 5% hydrochloric acid, and the reaction mixture was then stirred at 0°C to 10°C for an additional 1.5 hours.

After completion of the reaction, the reaction mixture was made alkaline by the addition of sodium carbonate, whereupon crystals separated out in situ. The crystals were recovered by filtration, washed with water and then recrystallized from 6 ml of ethanol, to give 0.1 g of the pale yellow pure desired product having a decomposition point of 125°C.

References

Merck Index 6399
DFU 3 (1) 52 (1978)
Kleeman & Engel p. 639
DOT 16 (12) 426 (1980)
I.N. p. 677
Sankyo Co., Ltd.; British Patent 1,374,344; November 20, 1974
Nakao, H., Arakawa, M. and Fukushima, M.; US Patent 4,003,901; January 18, 1977; assigned to Sankyo Co., Ltd.

NISOLDIPINE

Therapeutic Function: Coronary vasodilator

Chemical Name: 3,5-Pyridinedicarboxylic acid, 1,4-dihydro-2,6-dimethyl-4-(2-nitrophenyl)-, methyl 2-methylpropyl ester

Common Name: Nisoldipine

Chemical Abstracts Registry No.: 63675-72-9

Structural Formula:

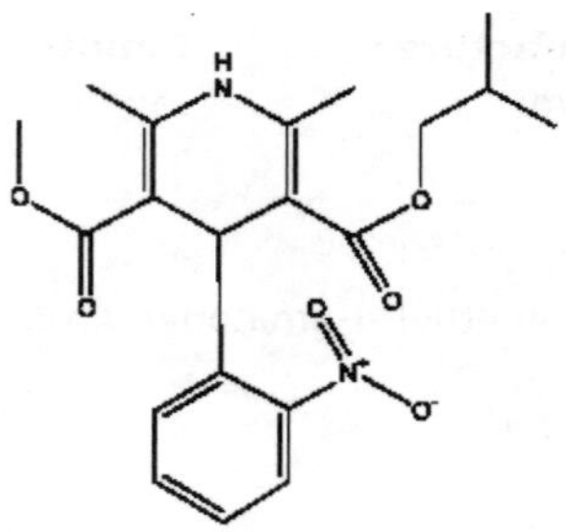

Trade Name	Manufacturer	Country	Year Introduced
Baymycard	Bayer Vital	Germany	-
Corasol	Sanitas	Chile	-
Nisoldipine	Pharmax	-	-
Norvasc	Miles	-	-
Sular	AstraZeneca	-	-
Syscor	AstraZeneca	-	-
Zadipina	SmithKline Beecham	-	-

Raw Materials

2'-Nitrobenzylideneacetoacetic acid methyl ester
β-Aminocrotonic acid isopropyl ester

Manufacturing Process

Boiling a solution of 12.7 g of 2'-nitrobenzylideneacetoacetic acid methyl ester and 7.1 g of β-amino-crotonic acid isopropyl ester in 50 ml of methanol for 10 hours yielded 2,6-dimethyl-4-(2'-nitrophenyl)-1,4-dihydropyridine-3,5-dicarboxylic acid 3-methyl ester-5-isopropyl ester of melting point 174°C (from ethanol). Yield 48% of theory.

References

Meyer H. et al.; US Patent No. 3,932,645; Jan. 13, 1976; Assigned to Farbenfabriken Bayer A G, Germany
Kutsuma T. et al.; US Patent No. 4,672,068; Jun. 9, 1987; Assigned to Fujirebio Kabushiki Kaisha, Tokyo, Japan

NITAZOXANIDE

Therapeutic Function: Anthelmintic

Chemical Name: 2-[(5-Nitro-2-thiazolyl)carbamoyl]phenyl acetate

Common Name: Nitazoxanide

Structural Formula:

Chemical Abstracts Registry No.: 55981-09-4

Trade Name	Manufacturer	Country	Year Introduced
Alinia	Romark Laboratories, L.C.	-	-
Cryptaz	Unimed Pharmaceuticals, Inc.	USA	-
Nitazoxanide	Romark Laboratories	USA	-
NTZ	Romark Laboratories	USA	-
NTZ	Unimed Pharmaceuticals, Inc.	USA	-

Raw Materials

p-Metoxy-benzoyl chloride
2-Amino-5-nitrothiazole
Triethylamine

Manufacturing Process

To a solution containing one mole p-metoxy-benzoyl chloride and one mole of carefully purified 2-amino-5-nitro-triazole in 200 ml of anhydrous tetrahydrofuran, one mole of triethylamine has been slowly added (about 10 minutes) while stirring. The reaction mixture, which became slightly warm, was stirred during 45 minutes and then poured under agitation, into 2 liters of distilled water. The stirring was continued until the precipitation of salicylamide, N-(5-nitro-2-thiazolyl)-, acetate (ester) was complete. The obtained precipitate was dried, washed with water, dried again and recrystallized from methanol. The yield about 60%; melting point 202°C.

References

Rossignol J.-F. et al.; US Patent No. 3,950,351; Apr. 13, 1976; Assigned to S.P.R.L. Phavic, Mouscron, Belgium
Rossignol J.-F.; US Patent No. 5,578,621; Nov. 26, 1996

NITRAZEPAM

Therapeutic Function: Anticonvulsant, Hypnotic

Chemical Name: 1,3-Dihydro-7-nitro-5-phenyl-2H-1,4-benzodiazepin-2-one

Common Name: -

Structural Formula:

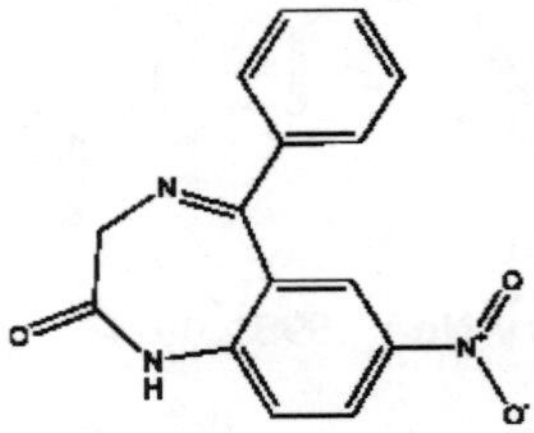

Chemical Abstracts Registry No.: 146-22-5

Trade Name	Manufacturer	Country	Year Introduced
Mogadan	Roche	W. Germany	1965
Mogadon	Roche	France	1965
Mogadon	Roche	UK	1965
Mogadon	Roche	Italy	1967
Apodorm	A.L.	Norway	-
Arem	Lennon	S. Africa	-
Atempol	Norgine	UK	-
Benzalin	Shionogi	Japan	-
Cerson	Belupo Ltd.	Yugoslavia	-
Dormicum	Glebe	Australia	-
Dormo-Puren	Klinge	W. Germany	-
Dumolid	Dumex	Denmark	-
Eatan-N	Desitin	W. Germany	-
Hipsal	Salvat	Spain	-
Hypnotin	Protea	S. Africa	-
Imadorm	Scheurich	W. Germany	-
Imeson	Desitin	W. Germany	-
Insomin	Orion	Finland	-
Ipersed	Sidus	Italy	-
Ipnozem	Biofarma	Turkey	-
Lagazepam	Lagap	Switz.	-
Lyladorm	M.P.S. Labs	S. Africa	-
Mitidin	Savoma	Italy	-
Nelbon	Sankyo	Japan	-
Nelmat	Sawai	Japan	-
Neuchlonic	Taiyo	Japan	-
Nitrados	Berk	UK	-
Nitrempax	Lafi	Brazil	-
Noctem	Alfa Farm.	Italy	-
Noctene	Rio Ethicals	S. Africa	-
Numbon	Ikapharm	Israel	-

Trade Name	Manufacturer	Country	Year Introduced
Ormodon	Ormed	S. Africa	-
Pacisyn	Medica	Finland	-
Paxisyn	Syntetic	Denmark	-
Pelson	Infale	Spain	-
Persopir	Ion	Italy	-
Prosonno	Von Boch	Italy	-
Quill	Ellea	Italy	-
Relact	Lemonier	Argentina	-
Remnos	D.D.S.A.	UK	-
Rindepres	Disprovent	Argentina	-
Somitran	Farmos	Finland	-
Somnased	Duncan Flockhart	UK	-
Somnite	Norgine	UK	-
Sonnolin	Dima	Italy	-
Surem	Galen	UK	-
Tri	Vita	Italy	-
Unisomnia	Unigreg	UK	-

Raw Materials

2-Aminobenzophenone
Glycine ethyl ester hydrochloride
Nitric acid

Manufacturing Process

A mixture of 16.8 g of 2 -aminobenzophenone, 11.9 g of glycine ethyl ester hydrochloride and 200 cc of pyridine was heated to reflux. After one hour, 20 cc of pyridine was distilled off. The solution was refluxed for 15 hours, then 11.9 g of glycine ethyl ester hydrochloride was added and the refluxing was continued for an additional 4 hours. The reaction mixture was continued for an additional 4 hours. The reaction mixture was concentrated in vacuo, then diluted with ether and water. The reaction product, 5-phenyl-3H-1,4-benzodiazepin-2(1H)-one, crystallized out, was filtered off, and then recrystallized from acetone in the form of colorless rhombic prisms, MP 182°C to 183°C.

48 g (0.2 mol) of 5-phenyl-3H-1 ,4-benzodiazepin-2(1 H)-one was dissolved in 250 cc of concentrated sulfuric acid by stirring at 15°C for ½ hour. The solution was then cooled to 0°C and a mixture of 9.1 cc of fuming nitric acid (90%, sp. gr. = 1.50) and 11.8 cc of concentrated sulfuric acid was added dropwise with stirring, keeping the temperature of the reaction mixture between -5°C and 0°C. After completion of the addition of the nitric acid-sulfuric acid mixture, stirring was continued for 1 hour and the reaction mixture was stored in the refrigerator overnight.

The mixture was then added dropwise to 2 kg of crushed ice with stirring and cooling, keeping the temperature at 0°C. After 1 hour of stirring in the cold, 640 cc of concentrated ammonium hydroxide was added dropwise at 0°C to pH 8. Stirring was continued for ½ hour and the crude product was filtered off, washed with a small amount of ice water and sucked dry overnight. The crude product was suspended in a mixture of 100 cc of methylene chloride and 1,700 cc of alcohol. 50 g of decolorizing charcoal was added and the mixture was refluxed with stirring for 2 hours. After standing overnight at room temperature 15 g of diatomaceous earth filter aid was added and the refluxing was resumed for 1½ hours. The mixture was filtered while hot. The clear, light yellow filtrate was concentrated in vacuo on the steam bath with stirring to about 600 cc. The concentrate was stirred and cooled in ice for about 2 hours; the precipitated crystalline product was filtered off, washed with some petroleum ether and sucked dry. The product, 7-nitro-5-phenyl-3H-1,4-benzodiazepin-2(1H)-one, was recrystallized from a mixture of 1,000 cc of alcohol and 50 cc of methylene chloride to obtain white prisms melting at 224°C to 225°C.

References

Merck Index 6418
Kleeman & Engel p. 640
OCDS Vol. 1 p. 366 (1977)
DOT 1 (4) 132 (1965) & 9 (6) 237 (1973)
I.N. p. 678
REM p. 1064
Kariss, J. and Newmark, H.L.; US Patent 3,116,203; December 31, 1963; assigned to Hoffmann-LaRoche, Inc.

NITROFURANTOIN

Therapeutic Function: Antibacterial (urinary)

Chemical Name: 1-[[(5-Nitro-2-furanyl)methylene]amino]-2,4-imidazolidinedione

Common Name: N-(5-Nitro-2-furfurylidene)-1-aminohydantoin

Structural Formula:

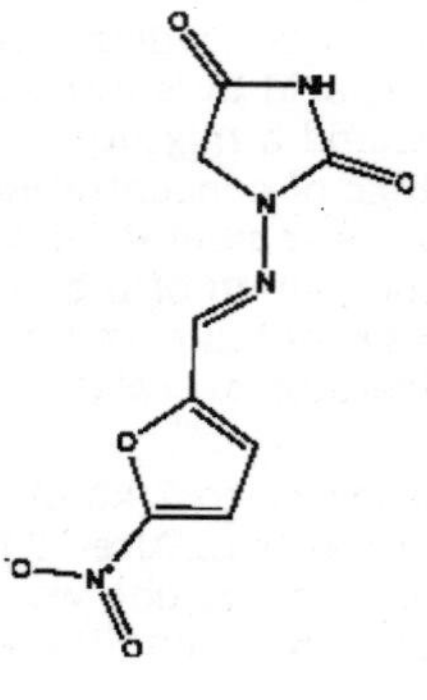

Chemical Abstracts Registry No.: 67-20-9

Trade Name	Manufacturer	Country	Year Introduced
Furadantin	Norwich Eaton	US	1953
Furadoine	Oberval	France	1954
Trantoin	McKesson	US	1969
Cyantin	Lederle	US	1970
Furachel	Rachelle	US	1970
N-Toin	Upjohn	US	1971
Parfuran	Warner Lambert	US	1974
Alfuran	Alkaloid	Yugoslavia	-
Berkfurin	Berk	UK	-
Ceduran	Cedona	Netherlands	-
Chemiofuran	Italfarmaco	Italy	-
Chemiofurin	Torlan	Spain	-
Cistofuran	Crosara	Italy	-
Cystit	Heyden	W. Germany	-
Dantafur	Norwich Eaton	US	-
Fua Med	Med	W. Germany	-
Furadoine	Oberval	France	-
Furalan	Lannett	US	-
Furaloid	Edwards	US	-
Furanex	Elliott-Marion	Canada	-
Furanite	Saunders	Canada	-
Furantoin	Spofa	Czechoslovakia	-
Furatin	Hemofarm	Yugoslavia	-
Furedan	Scharper	Italy	-
Furil	Off	Italy	-
Furobactina	Esteve	Spain	-
Furophen	Pharbil	Netherlands	-
Gerofuran	Gerot	Austria	-
Ituran	Promonta	W. Germany	-
Macrodantin	Eaton	US	-
Microdoine	Gomenol	France	-
Micturol	Liade	Spain	-
Nephronex	Cortunon	Canada	-
Nierofu	Hoyer	W. Germany	-
Nifuran	Paul Maney	Canada	-
Nifurantin	Apogepha	E. Germany	-
Nitrofur C	Leiras	Finland	-
Novofuran	Novopharm	Canada	-
Phenurin	Merckle	W. Germany	-
Profura	Rachelle	US	-
Trantoin	McKesson	US	-
Trocurine	Labatec	Switz.	-
Urantoin	D.D.S.A.	UK	-

Trade Name	Manufacturer	Country	Year Introduced
Uretoin	Tokyo Tanabe	Japan	-
Urodil	Pharma-Selz	W. Germany	-
Urodin	Streuli	Switz.	-
Urofuran	Farmos	Finland	-
Urolisa	Lisapharma	Italy	-
Urolong	Thiemann	W. Germany	-
Uro-Tablinen	Sanorania	W. Germany	-
Uvamin	Mepha	Switz.	-

Raw Materials

n-Heptaldehyde
1-Aminohydantoin
5-Nitro-2-furaldoxime

Manufacturing Process

To a solution of 18.9 grams (0.166 mol) n-heptaldehyde in 25 ml of isopropanol is added, with stirring, a solution of 19.1 grams (0.166 mol) of 1-aminohydantoin in 110 ml water acidified with concentrated HCl. The heavy white precipitate formed is filtered and washed, until acid free, with small amounts of water and ether. The yield of N-(n-heptylidene)-1-aminohydantoin is 14 grams of MP 150°C (with decomposition). This may be recrystallized from dimethylformamide.

A mixture of 2.5 grams (0.016 mol) of 5-nitro-2-furaldoxime, 3.9 grams (0.018 mol) of N-(n-heptylidene)-1-aminohydantoin and 5 cc of sulfuric acid (density 1.84) is placed in a 250 cc beaker. It is heated with stirring at steam bath temperature for about 1.5 hours. Upon cooling, a solid precipitates which is collected by filtration, washed with water, isopropanol and ether in turn and dried at 110°C for 4 hours. There is obtained N-(5-nitro-2-furfurylidene)-1-aminohydantoin in 96 to 98% yield, according to US Patent 2,927,110.

References

Merck Index 6445
Kleeman & Engel p. 641
PDR pp. 1278, 1606
OCDS Vol. 1 p. 230 (1977)
I.N. p. 680
REM p. 1215
Hayes, K.J.; US Patent 2,610,181; September 9, 1952; assigned to Eaton Laboratories, Inc.
Michels, J.G.; US Patent 2,898,335; August 4, 1959; assigned to The Norwich Pharmacal Company
Gever, G. and O'Keefe, C.; US Patent 2,927,110; March 1, 1960; assigned to The Norwich Pharmacal Company

NITROFURAZONE

Therapeutic Function: Topical antiinfective

Chemical Name: 2-[(5-Nitro-2-furanyl)methylene]hydrazinecarboxamide

Common Name: Nitrofural

Structural Formula:

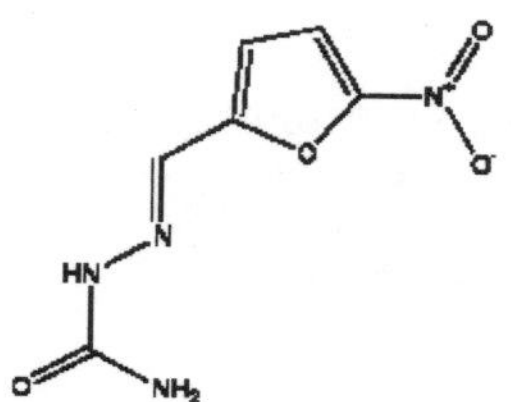

Chemical Abstracts Registry No.: 59-87-0

Trade Name	Manufacturer	Country	Year Introduced
Furacin	Norwich Eaton	US	1946
Actin-N	Chesebrough-Pond	US	1981
Amifur	Norwich Eaton	US	-
Escofuron	Streuli	Switz.	-
Furesol	A.F.I.	Norway	-
Germex	Lennon	S. Africa	-
Monofuracin	Dainippon	Japan	-
Muldacin	Mulda	Turkey	-
Nifucin	Jenapharm	E. Germany	-
Nifuzon	Pharmacia	Sweden	-
Nitrozone	Century	US	-
Yatrocin	Italfarmaco	Italy	-

Raw Materials

Semicarbazide hydrochloride
2-Formyl-5-nitrofuran

Manufacturing Process

A mixture of 43 grams of semicarbazide hydrochloride and 31 grams of sodium acetate is dissolved in 150 cc of water. The pH of this solution is approximately 5. Ethyl alcohol (95% by volume) in the amount of 250 cc is added and the mixture is stirred mechanically. A solution of 53.5 grams of carefully purified 2-formyl-5-nitrofuran in 250 cc of the said alcohol is added dropwise to the semicarbazide solution at room temperature. After completing the addition of the aldehyde solution, the mixture is stirred for another hour. The precipitate is removed from the reaction mixture by filtration. It is washed well with ethyl alcohol and dried to constant weight at 70°C in an oven. The product weighs 73 grams, corresponding to a yield of 97%. It is obtained in the form of pale yellow needles, which are not subjected to further purification, according to US Patent 2,416,234.

References

Merck Index 6446
Kleeman & Engel p. 641
PDR p. 1278
OCDS Vol. 1 p. 229 (1977)
I.N. p. 680
REM p. 1163
Stillman, W.B. and Scott, A.B.; US Patent 2,416,234; February 18, 1947; assigned to Eaton Laboratories, Inc.
Gever, G. and O'Keefe, C.; US Patent 2,927,110; March 1, 1960; assigned to The Norwich Pharmacal Company

NIZATIDINE

Therapeutic Function: Antiulcer

Chemical Name: 1,1-Ethenediamine, N-(2-(((2-((dimethylamino)methyl)-4-thiazolyl)methyl)thio)ethyl)-N'-methyl-2-nitro-

Common Name: Nizatidine

Structural Formula:

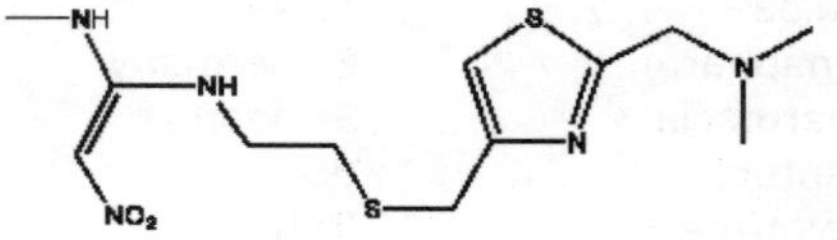

Chemical Abstracts Registry No.: 76963-41-2

Trade Name	Manufacturer	Country	Year Introduced
Acinon	Zeria	-	-
Antizid	Eli Lilly	-	-
Axid	Eli Lilly	-	-
Axid	Norgine Pharma	France	-
Calmaxid	Norgine AG	-	-
Flectar	Biomedica	Greece	-
Nizatidine	Eli Lilly	USA	-
Nizatidine	Pharmascience	Canada	-
Nizax	Eli Lilly	-	-
Nizaxid	Norgine Pharma	France	-
Panaxid	Norgine Pharma	France	-

Trade Name	Manufacturer	Country	Year Introduced
Peptodin	Kleva Ltd.	Greece	-
Tazac	Eli Lilly Australia Pty Limited	Australia	-
Ulxit	Tyrol Pharma GmbH	Germany	-
Zanizal	Bruno Farmaceutici S.p.A.	Italy	-

Raw Materials

2-Nitromethylenethiazolidine
Methylamine
4-Chloromethyl-2-dimethylaminomethylthiazole dihydrochloride

Manufacturing Process

Nizatidine may be prepared by 2 ways.

1. A mixture of (25.7 g) 2-nitromethylenethiazolidine and acetonitrile (50 ml) was stirred and heated at 40°C under nitrogen. Methylamine gas (16.0 g) was passed into the stirred mixture over 45 minutes to give a solution. A slurry of 4-chloromethyl-2-dimethylaminomethylthiazole hydrochloride (40.0 g) (prepared as described in EP 49,618) in acetonitrile (50 ml) was added to the solution over a period of 4.5 hours whilst methylamine gas was bubbled through the reaction mixture such that methylamine (38.3 g) was added over the period (total methylamine added was 54.3 g). The temperature of the reaction mixture varied between 24° and 35°C during the addition. After the addition, the mixture was diluted with acetonitrile (50 ml) and stirred at ambient temperature for 17 hours. A solid was removed by filtration and the filtrate was split into 2 equal portions.

Portion 1: The solution was evaporated to give a black oil which was partitioned between water (200 ml) and chloroform (200 ml). The separated chloroform phase was washed with saturated brine, then dried over magnesium sulphate, filtered and evaporated to give a reddish oil which was dissolved in acetone (200 ml), boiled under reflux, cooled to 40°C and then seeded with nizatidine. The mixture was left to stand at 0°-5°C for 64 hours. The mixture was filtered to give nizatidine (10.4 g, 37%) m.p. 118-122°C. The structure was confirmed by1H NMR. The product was 95.4% pure by HPLC.

Portion 2: The mixture was evaporated to give an oil which was taken up in chloroform (200 ml) then washed with water (100 ml). The chloroform solution was washed with brine (100 ml), dried over magnesium sulphate, and then concentrated under reduced pressure at 45°C to give a brown oil. The oil was dissolved in acetone (200 ml) and activated charcoal (0.5 g) was added to the solution. The mixture was boiled under reflux for 10 minutes, then cooled to 45°C and filtered at this temperature to remove the charcoal. The filtrate was cooled to 20°C, seeded with nizatidine (0.05 g), then cooled 0°-5°C for 45 minutes during which time crystallisation occurred. The mixture was filtered to give nizatidine (9.4 g, 32.2%).

2. A mixture of 2-nitromethylenethiazolidine (12.6 g) and water (30.0 ml) was stirred and heated at 40°C under argon. Methylamine (20.0 g of a 40% w/w aqueous solution) was added slowly over 30 minutes to the reaction mixture

at 40°C. The mixture was cooled at ambient temperature and further methylamine (23.6 g of 40% w/w aqueous solution) was added over 2.5 hours and a solution of 4-chloromethyl-2-dimethylaminomethylthiazole dihydrochloride (25.0 g) in water (30 ml) was added over 5.5 hours with the addition of the thiazole starting simultaneously with the addition of the methylamine. The reaction mixture was left to stir for a further 15 minutes and then was concentrated under reduced pressure. The solid obtained was dissolved in a mixture of methyl ethyl ketone (200 ml), aqueous potassium carbonate solution (43 ml, 10% w/w). The mixture was warmed slightly to obtain a solution. The mixture was separated and the aqueous layer was washed with methyl ethyl ketone (2 times 130 ml and then 1 times 100 ml). The combined organic layers were evaporated under reduced pressure to yield crude nizatidine (approximately 25.2 g), which was shown to be 89.4% pure by HPLC. The crude solid was dissolved in dichloromethane (300 ml). The solution was washed with water (3 times 75 ml). The combined aqueous layer and the washings were back extracted with dichloromethane and the combined organic layers were dried and concentrated under reduced pressure to give nizatidine (21.1 g, 74.3% yield). The solid was dissolved in ethanol (45 ml) by warming on a steam bath. The solution was removed from the steam bath treated with activated charcoal (2.3 g) and the mixture was boiled for a further 8 minutes. The mixture was hot filtered. The filtrate was cooled and filtered to give nizatidine (13.8 g, 48% yield) which was shown to be 99.8% pure by HPLC.

References

Ph. Cornwall; U. S. Patent No. 6,069,256; May 30 2000; assigned Knoll aktiengesellshaft (Lundeligshaffen, D.E.)

NOMIFENSINE MALEATE

Therapeutic Function: Psychostimulant

Chemical Name: 8-Amino-1,2,3,4-tetrahydro-2-methyl-4-phenyl-isoquinoline maleate

Common Name: -

Structural Formula:

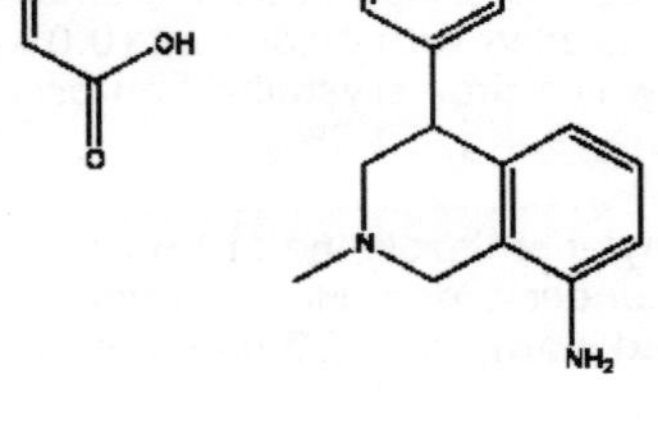

Chemical Abstracts Registry No.: 32795-47-4; 24526-64-5 (Base)

Trade Name	Manufacturer	Country	Year Introduced
Alival	Hoechst	W. Germany	1976
Merital	Hoechst	UK	1977
Alival	Hoechst	France	1977
Psicronizer	Albert Pharma	Italy	1977
Merital	Hoechst	Canada	1982
Neurolene	Magis	Italy	-
Nomival	Leiras	Finland	-

Raw Materials

Sulfuric acid
Hydrogen
Maleic acid
α-Bromoacetophenone
(2-Nitrobenzyl)methylamine
Sodium borohydride

Manufacturing Process

A solution of N-(2-aminobenzyl)-1-phenyl-2-methylaminoethanol-1 was prepared by the reaction of α-bromo-acetophenone and (2-nitrobenzyl)methylamine, followed by hydrogenation of the nitro group by means of nickel on diatomaceous earth at room temperature and reduction of the CO group by means of sodium borohydride. The intermediate thus produced was dissolved in 100 ml of methylene chloride and introduced dropwise into 125 ml of sulfuric acid at 10° to 15°C. After a short standing, the reaction mixture was poured onto ice and rendered alkaline by means of a sodium hydroxide solution. By extraction with ether, there was obtained 1,2,3,4-tetrahydro-2-methyl-4-phenyl-8-amino-isoquinoline. The base is reacted with maleic acid to give the maleate; melting point of the maleate 199° to 201°C (from ethanol).

References

Merck Index 6515
DFU 1 (2) 72 (1976)
Kleeman and Engel p. 642
PDR p. 941
DOT 13 (2) 77 (1977)
I.N. p. 685
Farbwerke Hoechst AG, Germany; British Patent 1,164,192; September 17, 1969
Ehrhart, G., Schmitt, K., Hoffmann, I. and Ott, H.; US Patent 3,577,424; May 4, 1971; assigned to Farbwerke Hoechst AG.

NONOXYNOL

Therapeutic Function: Spermatocide (vaginal)

Chemical Name: α-(Nonylphenyl)-ω-hydroxypoly(oxy-1,2-ethanediyl)

Common Name: -

Structural Formula:

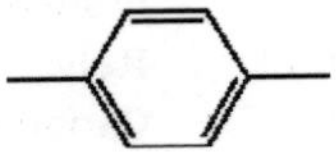

Chemical Abstracts Registry No.: 26027-38-3

Trade Name	Manufacturer	Country	Year Introduced
Ortho-Delfen	Cilag	France	1971
Semicid	Whitehall	US	1978
Intercept	Ortho	US	1980
Gynol	Ortho	US	1982
Shur-Seal	Milex	US	1983
C-Film	Hommel	Switz.	-
Emko	Emko-Schering	US	-
Encare Oval	Patentex	W. Germany	-
Glovan	Teva	Israel	-
Igepal	G.A.F.	US	-
Ortho-Creme	Cilag	US	-

Raw Materials

Isononylphenol
Sodium hydroxide
Ethylene oxide

Manufacturing Process

220 parts of isononylphenol prepared by condensation of phenol with an olefin mixture obtained by polymerization of propylene and containing essentially isononylenes are caused to react with 0.5 part of caustic alkali powder. The whole is heated to about 130°C to 135°C and the water formed is removed under reduced pressure, while stirring. Thereupon, ethylene oxide is introduced into the melt, while well stirring, during which operation care must be taken, that the temperature of the reaction mass is maintained between 180°C and 200°C. When about 300 parts of ethylene oxide are taken up, the reaction is interrupted. A water-soluble oil is obtained.

References

Merck Index 6518
PDR pp. 1661, 1900
I.N. p. 686
REM p. 1163

Steindorff, A., Balle, G., Horst, K. and Michel, R.; US Patent 2,413,477; September 3, 1940; assigned to General Aniline & Film Corp.

NORDAZEPAM

Therapeutic Function: Tranquilizer

Chemical Name: 7-Chloro-1,3-dihydro-5-phenyl-1(2H)-1,4-benzodiazepin-2-one

Common Name: Nordiazepam; Desmethyldiazepam

Structural Formula:

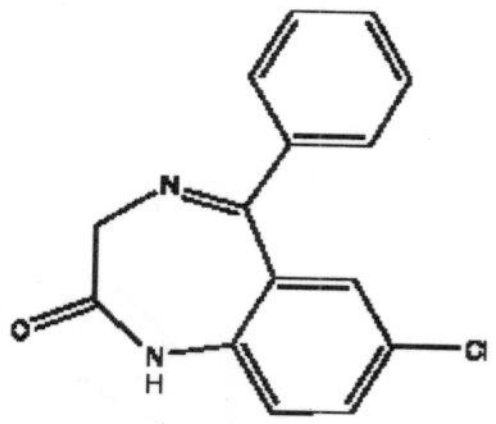

Chemical Abstracts Registry No.: 1088-11-5

Trade Name	Manufacturer	Country	Year Introduced
Madar	Ravizza	Italy	1973
Vegesan	Mack	Switz.	1981

Raw Materials

(2-Benzoyl-4-chlorophenyl-carbamoylmethyl)carbamic acid benzyl ester
Hydrogen bromide
Acetic acid

Manufacturing Process

A solution of 3.1 g of (2-benzoyl-4-chlorophenyl-carbamoylmethyl)carbamic acid benzyl ester in 30 cc of 20% hydrobromic acid in glacial acetic acid was stirred for 45 minutes at room temperature. On addition of 175 cc of anhydrous ether, a gummy solid precipitated. After several minutes the ether solution was decanted. The resultant 5-chloro-2-glycylaminobenzophenone was not isolated, but about 155 cc of ether was added to the residue and after chilling in an ice bath, 10% sodium hydroxide was added until the mixture was alkaline. The ether layer was then separated, washed twice with water and dried over sodium sulfate. After filtration, the ether solution was concentrated to dryness in vacuo. The residue was crystallized from benzene to yield 7-chloro-5-phenyl-3H-1,4-benzodiazepin-2(1H)-one.

References

Merck Index 6531
DOT 9 (6) 239 (1973)
I.N. p. 688
Stempel, A.; US Patent 3,202,699; August 24, 1965; assigned to Hoffmann-LaRoche Inc.

NORETHANDROLONE

Therapeutic Function: Androgen

Chemical Name: 17-Hydroxy-19-norpregn-4-ene-3-one

Common Name: -

Structural Formula:

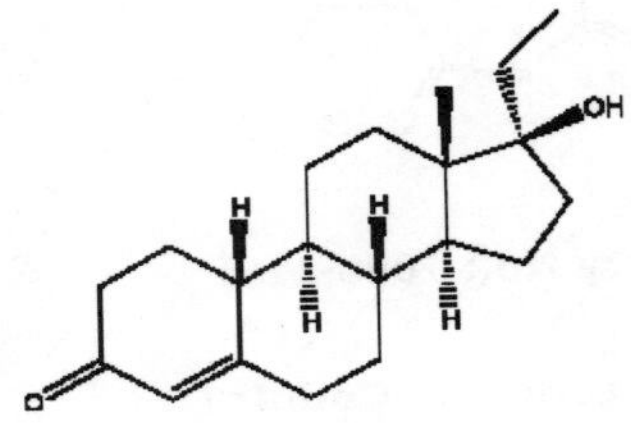

Chemical Abstracts Registry No.: 52-78-8

Trade Name	Manufacturer	Country	Year Introduced
Nilevar	Searle	US	1956
Nilevar	Searle	France	1960

Raw Materials

Norethindrone
Hydrogen

Manufacturing Process

Through a mixture of 11 parts of charcoal containing 5% palladium and 2,000 parts of dioxane a stream of hydrogen is passed for 60 minutes. Then 86 parts of 17-ethynyl-19-nortestosterone (Norethindrone) in 1,500 parts of dioxane are added and the mixture is hydrogenated until 2 mols of hydrogen are absorbed. The catalyst is then removed by filtration and the solvent is evaporated under vacuum. The crystalline residue is dissolved in 2,700 parts of benzene and thus applied to a chromatography column containing 5,000 parts of silica gel. The column is washed with 2,700 parts of benzene, 4,500 parts of a 10% solution of ethyl acetate in benzene and 27,000 parts of a 20% solution of ethyl acetate in benzene and is then eluted with 30,000 parts of a 30% solution of ethyl acetate in benzene. The resulting eluate is concentrated under vacuum and the residue is recrystallized from methanol and dried to constant weight at 75°C. The 17-ethyl-19-nortestosterone thus obtained melts at about 140°C to 141°C.

References

Merck Index 6537
Kleeman & Engel p. 644
OCDS Vol. 1 p. 170 (1977)
I.N. p. 688
Colton, F.B.; US Patent 2,721,871; October 25, 1955; assigned to G.D. Searle & Co.

NORETHINDRONE

Therapeutic Function: Progestin

Chemical Name: 17-Hydroxy-19-nor-17α-pregn-4-en-20-yn-3-one

Common Name: Norethisteron

Structural Formula:

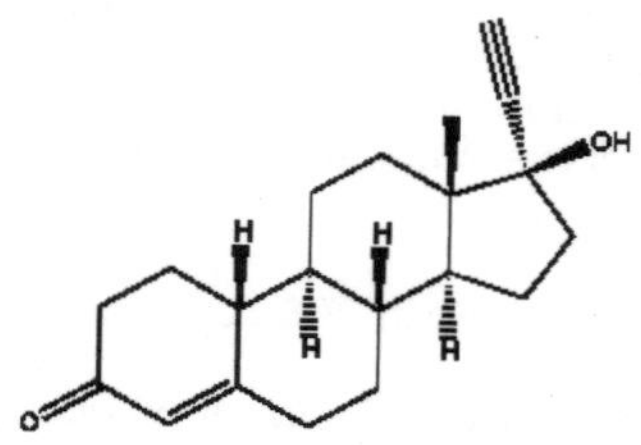

Chemical Abstracts Registry No.: 68-22-4

Trade Name	Manufacturer	Country	Year Introduced
Norlutin	Parke Davis	US	1957
Ortho-Novum	Ortho	US	1963
Norinyl	Syntex	US	1964
Nor-QD	Syntex	US	1973
Brevicon	Syntex	US	-
Conceplan	Gruenenthal	W. Germany	-
Gesta-Plan	D.A.K.	Denmark	-

Trade Name	Manufacturer	Country	Year Introduced
Micronor	Ethnor	Australia	-
Micronor	Ortho	US	-
Micronovum	Cilag	W. Germany	-
Modicon	Ortho	US	-
Monogest	Spofa	Czechoslovakia	-
Norfor	Gremy-Longuet	France	-
Norgestin	Janus	Italy	-
Noriday	Syntex	US	-
Norlestrin	Parke Davis	US	-
Ovcon	Mead Johnson	US	-
Primolut N	Schering	UK	-
Tri-Norinyl	Syntex	US	-
Utovlan	Syntex	UK	-

Raw Materials

3-Methoxyestrone
Ethyl orthoformate
Lithium
Potassium
Ammonia
Acetylene
Chromic acid

Manufacturing Process

7.5 grams of 3-methoxyestrone were dissolved in 750 cc of anhydrous dioxane in a three-neck flask, placed in a box and insulated with cotton wool. 2 liters of anhydrous liquid ammonia and 15 grams of lithium metal in the form of wire were added to the mechanically stirred solution. After stirring for one hour, 150 cc of absolute ethanol were added at such speed that no bumping occurred; when the blue color had disappeared, 500 cc of water were added in the same way. The ammonia was evaporated on the steam bath and the product collected with 2 liters of water. It was extracted with ether and then with ethyl acetate and the combined extract was washed to neutral and evaporated to dryness under vacuum, leaving 7.4 grams of a slightly yellow oil.

The oil thus obtained was dissolved in 400 cc of methanol and refluxed during one hour with 150 cc of 4N hydrochloric acid. The mixture was poured into a sodium chloride solution and extracted with ethyl acetate, washed to neutral, dried and evaporated to dryness. The product was a yellow oil which showed an ultraviolet absorption maximum characteristic of a Δ^4-3-ketone.

A solution of 2.7 grams of chromic acid in 20 cc of water and 50 cc of acetic acid was added to the stirred solution of the above oil in 100 cc of acetic acid, maintaining the temperature below 20°C. After 90 minutes standing, 50 cc of methanol were added and the mixture concentrated under vacuum (20 mm). The residue was extracted with ether, washed to neutral and evaporated to dryness. The residual semicrystalline product (7 grams) was chromatographed over alumina and the fractions eluted with ether yielded 3.2 grams of Δ^4-19-norandrosten-3,17-dione having a MP of 163° to 167°C.

A solution of 2 grams of Δ4-19-norandrosten-3,17-dione and 0.4 gram of pyridine hydrochloride in 50 cc of benzene free of thiophene was made free of moisture by distilling a small portion; 4 cc of absolute alcohol and 4 cc of ethyl orthoformate were added and the mixture was refluxed during 3 hours. 5 cc of the mixture were then distilled and after adding an additional 4 cc of ethyl orthoformate the refluxing was continued for 2 hours longer. The mixture was evaporated to dryness under vacuum and the residue was taken up in ether, washed, dried and evaporated to dryness. The residue was crystallized from hexane-acetone and then from ether to give $\Delta^{3,5}$-19-nor-3-ethoxy-androstadien-17-onewith a MP of 140° to 142°C.

One gram of potassium metal was dissolved in 25 cc of tertiary amyl alcohol by heating under an atmosphere of nitrogen. One gram of $\Delta^{3,5}$-19-nor-3-ethoxyandrostadien-17-onein 25 cc of anhydrous toluene was added and nitrogen was passed during 15 minutes. Then acetylene (especially dried and purified) was passed during 14 hours through the mechanically stirred solution, at room temperature.

The mixture was poured in water, acidified to pH 1 with dilute hydrochloric acid, heated on the steam bath for 30 minutes and then subjected to steam distillation to remove the organic solvents. The residue was filtered, dried and recystallized several times from ethyl acetate. The Δ^4-19-nor-17α-ethinylandrosten-17β-ol-3-onethus obtained had a MP of 198° to 200°C (in sulfuric acid bath), 200° to 204°C (Kofler).

References

Merck Index 6538
Kleeman & Engel p. 644
PDR pp. 1104, 1297, 1358, 1372, 1793
OCDS Vol. 1 p. 164 (1977) & 2, 145 (1980)
DOT 4 (1) 19 (1968) & 9 (4) 144 (1973)
I.N. p. 688
REM p. 992
Djerassi, C., Miramontes, L. and Rosenkranz, G.; US Patent 2,744,122; May 1, 1956; assigned to Syntex SA, Mexico
de Ruggieri, P.; US Patent 2,849,462; August 26, 1958

NORETHINDRONE ACETATE

Therapeutic Function: Chemical Name: 19-Nor-17α-pregn-4-en-20-yn-3-one, 17-hydroxy-, acetate

Common Name: -

Chemical Abstracts Registry No.: 51-98-9

Structural Formula:

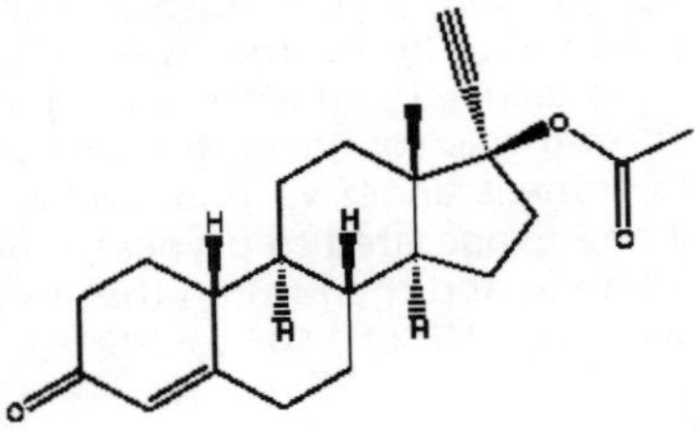

Trade Name	**Manufacturer**	**Country**	**Year Introduced**
Norlestrin	Parke Davis	US	1964
Milligynon	Schering	France	1978
Aygestrin	Ayerst	US	1982
Brevicon	Syntex	US	-
Norlutin-A	Parke Davis	UK	-
Primolut-Nor	Schering	W. Germany	-

Raw Materials

Norethindrone
Acetic anhydride
Hydrogen chloride

Manufacturing Process

2.98 grams of 17-ethinyl-19-nor-testosterone (norethindrone) are suspended in 30 cc of acetic anhydride and a solution of 1.9 grams of p-toluenesulfonic acid in 19 cc of acetic anhydride is gradually added while cooling and stirring. Complete dissolution takes place after about one hour. After additional 30 to 60 minutes, a thick, pasty mass separates. The reaction is permitted to continue for a total period of 5 hours, whereupon water is added to the reaction mixture and the 3-enol-17-diacetate which separates after stirring for 1 to 2 hours is filtered off, washed until neutral and dried in vacuo over calcium chloride at room temperature.

In order to prepare the monoacetate, the crude diacetate is suspended in 150 cc of methanol and, after adding 1.5 cc, concentrated hydrochloric acid, heated to boiling for 15 minutes in a nitrogen atmosphere. The crude monoacetate which separates upon the addition of water after cooling is filtered off, washed and dried in vacuo over calcium chloride at room temperature. The pure 17-acetete, obtained after repeated recrystallizations from methylene chloride/hexane has a MP of 161° to 162°C.

References

Merck Index 6538
Kleeman & Engel p. 645
PDR pp. 615, 1378
OCDS Vol. 1 p. 165 (1977)

I.N. p. 689
REM p. 992
Engelfried, O., Kaspar, E., Schenck, M. and Popper, A.; US Patent 2,964,537; Dec. 13, 1960; assigned to Schering AG, Germany

NORETHISTERONE ENANTHATE

Therapeutic Function: Progestin

Chemical Name: 19-Norpregn-4-en-20-yn-3-one, 17-((1-oxoheptyl)oxy)-, (17α)-

Common Name: Norethindrone enanthate; Norethisterone enanthate; Norethisterone heptanoate

Structural Formula:

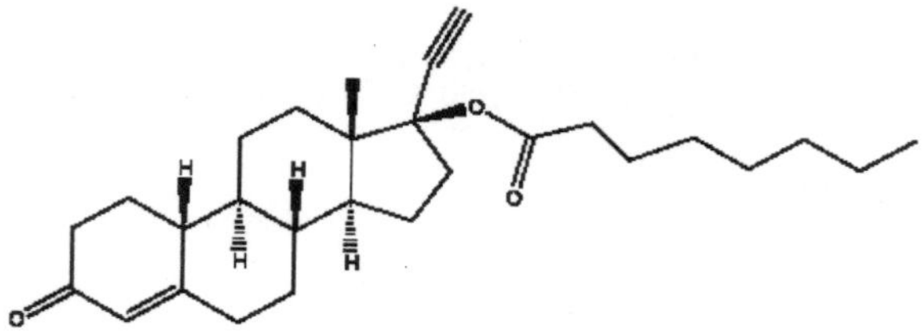

Chemical Abstracts Registry No.: 3836-23-5

Trade Name	Manufacturer	Country	Year Introduced
Norigest	Schering	-	-

Raw Materials

17-Ethinyl-19-nor-testosteron
Enanthic acid

Manufacturing Process

1 g 17-ethinyl-19-nor-testosteron was refluxed with 5 ml of enanthic acid anhydride on an oil bath at temperature 180°C 17 hours. Then the reaction mixture was distilled with water steam to the full disappearence of smell of enantic acid. After that it was washed with 2 N sodium hydroxide and finally with water to neutral, dried over sodium sulfate and evaporated to dryness. The oily residue was rubbed with some drops of methanol and stood at -8°C 24 hours to give the crystals of 3-endol diester; MP: 82°-84°C, 2 g of it was dissolved in 120 ml of methanol and heated with 1.2 ml concentrated hydrochloric acid. The partly saponified product was distilled with water steam

to full disappearence of smell of enantic acid. The residue was mixed with ether, washed with 2 N sulfuric acid, 2 N sodium hydroxide, finally with water to neutral dried over sodium sulfate and evaporated to dryness. The residue was stirred with pentane and cooled for crystallization. Pure enanthate was crystallized after repeated solution in pentane. 17-α-Ethinyl-19-nortestosterone enanthate had MP: 68°-71°C.

References

Engelfried O. et al.; D.B. Patent No. 1,017,166; June 16, 1956; Schering Aktiengesellschaft, Berlin

NORETHYNODREL

Therapeutic Function: Progestin

Chemical Name: 17-Hydroxy-19-nor-17α-pregn-5(10)-en-20-yn-3-one

Common Name: 13-Methyl-17-ethynyl-17-hydroxy-1,2,3,4,6,7,8,9,11,12,13,14,16,17-tetradecahydro-15H-cyclopenta-α-phenanthren-3-one

Structural Formula:

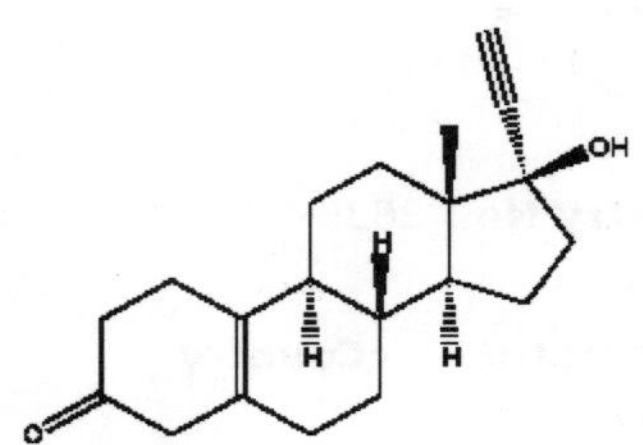

Chemical Abstracts Registry No.: 68-23-5

Trade Name	Manufacturer	Country	Year Introduced
Enovid	Searle	US	1957

Raw Materials

3-Methoxy-17-oxo-2,5-estradiene
Acetylene
Acetic acid

Manufacturing Process

Convenient starting materials are the ethers of 3-hydroxy-13-methyl-1,4,6,7,8,9,11,12,13,14,16,17-dodecahydro-15H-cyclopenta-α-phenanthren-

17-one described in US Patent 2,655,518, according to US Patent 2,691,028 where the following preparation is also described. The methyl ether is also designated as 3-methoxy-17-oxo-2,5-estradiene.

A stirred solution of 10.6 parts of 3-methoxy-13-methyl-1,4,6,7,8,9,11,12,13,14,16,17dodecahydro-15H-cyclopenta-α-phenanthren-17-one in 700 parts of anhydrous ether and 45 parts of dry toluene is cooled to 0°C and saturated with dry acetylene. While a slow stream of acetylene is passed through the reaction mixture, a solution of 20 parts of potassium t-amylate in 135 parts of anhydrous t-pentanol is added in the course of 15 minutes with stirring. Passage of acetylene and stirring are continued for an additional 4½ hours. After standing at 0°C for 16 hours, the mixture is washed with aqueous ammonium chloride solution until the aqueous phase is neutral, then with water and saturated sodium chloride solution. The organic layer is dried over anhydrous sodium sulfate, filtered and concentrated under vacuum to a residue of about 250 parts. 500 parts of petroleum ether are added and after standing at 0°C for an hour, the mixture is filtered. The collected precipitate is recrystallized from ether. The resulting 3-methoxy-13-methyl-17-ethynyl-1,4,6,7,8,9,11,12,13,14,16,17-dodecahydro-15H-cyclopenta-α-phenanthren-17-ol melts at about 181° to 182°C.

To a refluxing solution of 10 parts of 3-methoxy-17-ethynyl-17-hydroxy-13-methyl-1,4,6,7,8,9,11,12,13,14,16,17-dodecahydro-15H-cyclopenta-α-phenanthrene in 500 parts of methanol, 20 parts of glacial acetic acid are added. Refluxing is continued for 7 minutes, water is added to the point of turbidity and the reaction mixture is permitted to come to room temperature. The precipitate is collected on a filter and recrystallized from aqueous methanol. The 13-methyl-17-ethynyl-17-hydroxy-1,2,3,4,6,7,8,9,11,12,13,14,16,17-tetradecahydro-15H-cyclopenta-α-phenanthren-3-one thus obtained melts at about 169° to 170°C.

References

Merck Index 6539
Kleeman & Engel p. 647
PDR p. 1680
OCDS Vol. 1 p. 186 (1977)

DOT 4 (1) 22 (1968)
I.N. p. 689
REM p. 993
Colton, F.B.; US Patent 2,691,028; October 5, 1954; assigned to G.D. Searle & Co.
Colton, F.B.; US Patent 2,725,389; November 29, 1955; assigned to G.D. Searle & Co.

NORFENEFRINE

Therapeutic Function: Adrenergic

Chemical Name: α-(Aminomethyl)-3-hydroxybenzenemethanol

Common Name: Norphenylephrine

Structural Formula:

Chemical Abstracts Registry No.: 536-21-0; 4779-94-6 (Hydrochloride salt)

Trade Name	Manufacturer	Country	Year Introduced
Zordel	Grelan	Japan	1970
Coritat	Green Cross	Japan	-
Esbufon	Schaper and Brummer	W. Germany	-
Euro-Cir	Virgiliano	Italy	-
Molycor R	Mepha	Switz.	-
Nevadral	Pharmacia	Sweden	-
Normetolo	Selvi	Italy	-
Novadral	Goedecke	W. Germany	-
Stagural	Stada	W. Germany	-
Sympatosan	Kwizda	Austria	-
Tonolift	Teisan	Japan	-

Raw Materials

Sodium iodide
Hydrogen
Hexamethylenetetramine
m-Acetoxyacetophenone
Bromine

Manufacturing Process

100 parts of the hydrochloride of meta-hydroxy-ω-aminoacetophenone of melting point 220°C to 222°C (obtainable by brominating meta-acetoxyacetophenone, causing the bromoketone to react with sodium iodide, adding hexamethylenetetramine to the iodide in an indifferent solvent and scission of the addition product in acid solution) are shaken in aqueous solution with hydrogen in presence of 2 parts of palladium catalyst until 2 atomic proportions of hydrogen have been absorbed. The catalyst is now filtered and the filtrate evaporated in a vacuum; and the crystalline and completely dry residue is dissolved in absolute alcohol and a precipitate is produced by adding dry ether. The hydrochloride of meta-hydroxyphenylethanolamine thus obtained forms white crystals of melting point 159°C to 160°C.

References

Merck Index 6540

Kleeman & Engel p. 647
I.N. p.689
Legerlotz, H.; US Patent 2,312,916; March 2, 1943; assigned to Ciba Pharmaceutical Products Inc.

NORFLOXACIN

Therapeutic Function: Antibacterial

Chemical Name: 1-Ethyl-6-fluoro-1,4-dihydro-4-oxo-7-(1-piperazinyl)-3-quinolinecarboxylic acid

Common Name: -

Structural Formula:

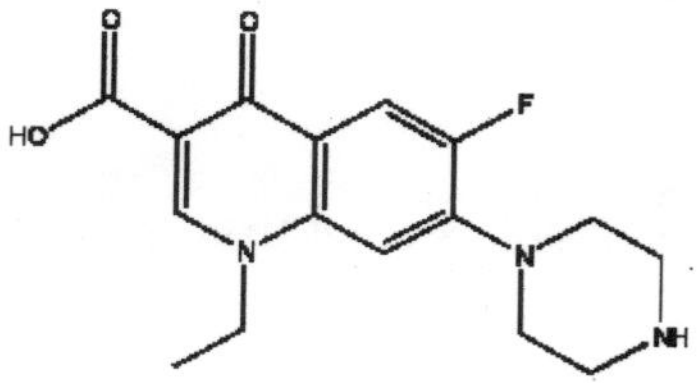

Chemical Abstracts Registry No.: 70458-96-7

Trade Name	Manufacturer	Country	Year Introduced
Noroxin	MSD	Italy	1983
Sebercim	I.S.F.	Italy	1983
Primoxin	Sharp and Dohme	W. Germany	1983
Noroxin	MSD	Switz.	1983
Fulgram	A.B.C.	Italy	-

Raw Materials

7-Chloro-1-ethyl-6-fluoro-4-oxo-1,4-dihydroquinoline-3-carboxylic acid
Piperazine

Manufacturing Process

36 g (0.134 mol) of 7-chloro-1-ethyl-6-fluoro-4-oxo-1,4-dihydroquinoline-3-carboxylic acid, 46 g of piperazine and 210 cm^3 of pyridine were heated under reflux for 6 hours, while stirring. After the starting material had dissolved, a precipitate appeared after heating for about 2 hours 30 minutes. The major part of the solvent was removed by concentration in vacuo (15 mm Hg; 100°C). In order to remove the pyridine as completely as possible, the residue was taken up in 200 cm3of water and the concentration in vacuo was repeated.

The residue, resuspended in 150 cm^3 of water, was stirred. 150 cm^3 of 2N NaOH were added thereto. The solution, which was slightly turbid, was treated with 5 g of animal charcoal and stirred for 30 minutes. After filtration, the pH was brought to 7.2 by adding acetic acid while stirring. The precipitate was filtered off, washed with water and dissolved in 250 cm^3 of a 10% aqueous acetic acid. The acid solution (pH 4.4) was filtered and then brought to pH 7.2 by gradually added 2N NaOH.

The suspension was heated to 90°C, while stirring. The crystals were separated and recrystallized from 280 cm^3 of a mixture of DMF (1 volume) and ethanol (4 volumes). After drying in vacuo over phosphorus pentoxide, 29.5 g (yield 70%) of 1-ethyl-6-fluoro-4-oxo-7-piperazinyl-1,4-dihydroquinoline-3-carboxylic acid, melting point 222°C, were obtained.

In air, this product is hygroscopic and gives a hemihydrate.

References

Merck Index 6541
DFU 7 (8) 586 (1982)
DOT 19 (6) 341 (1983)
I.N. p. 689
Pesson, M.; US Patent 4,292,317; September 29, 1981; assigned to Laboratorie Roger Bellon (France) and Dainippon Pharmaceutical (Japan)

NORGESTIMATE

Therapeutic Function: Progestin

Chemical Name: 18,19-Dinorpregn-4-en-20-yn-3-one, 17-(acetyloxy)-13-ethyl-, 3-oxime, (17α)-

Common Name: Dexnorgestrel acetime; Norgestimate

Structural Formula:

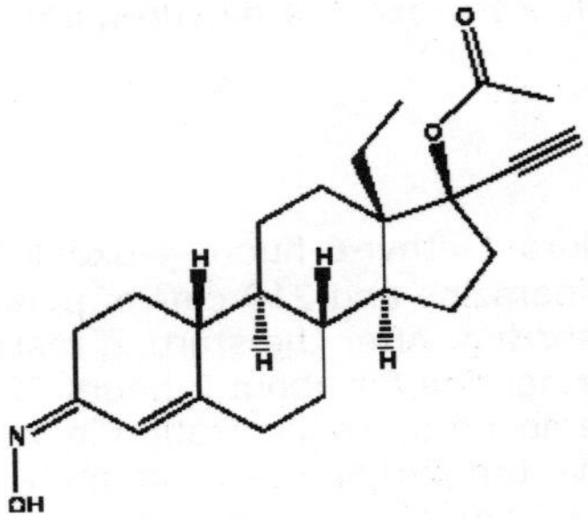

Chemical Abstracts Registry No.: 35189-28-7

Trade Name	Manufacturer	Country	Year Introduced
Ortrel	Janssen Cilag	-	-

Raw Materials

D-17β-Acetoxy-13β-ethyl-17α-ethynyl-gon-4-en-3-one
Hydroxylamine hydrochloride

Manufacturing Process

A solution of 4.5 g of D-17β-acetoxy-13β-ethyl-17α-ethynyl-gon-4-en-3-one in 15 ml of pyridine and 2.0 g of hydroxylamine hydrochloride hydroxylamine hydrochloride is heated on a steam bath for 45 min. It is then cooled and poured into a large amount of ice-water, after which the solid which is thus produced is filtered off and air dried. Recrystallization from methylene chloride-ethanol gives D-17β-acetoxy-13α-ethyl-17α-ethynyl-gon-4-en-one oxime, m.p. 214-218°C; $[\alpha]_D^{25}$ = +41°.

References

Tullar B.F., Greebbush E.; US Patent Dec. 18, 1956; Assigned to Sterling Drug Inc., Del., a corporation of Delaware

NORGESTREL

Therapeutic Function: Progestin

Chemical Name: 13-Ethyl-17-hydroxy-18,19-dinor-17α-pregn-4-en-20-yn-3-one

Common Name: 17α-Ethynyl-18-homo-19-nortestosterone

Structural Formula:

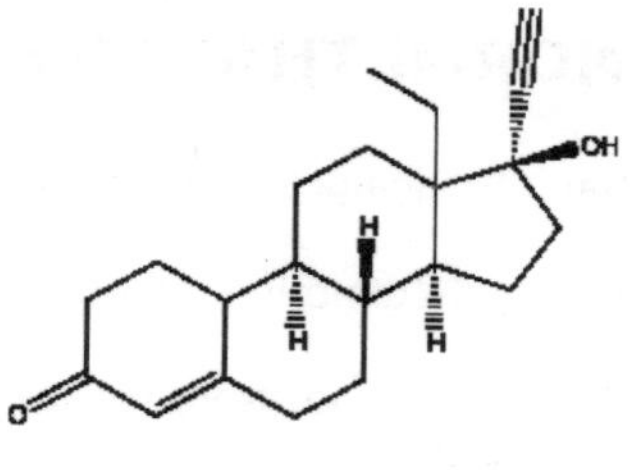

Chemical Abstracts Registry No.: 797-63-7

Trade Name	Manufacturer	Country	Year Introduced
Ovrette	Wyeth	US	1968
Eugynon	Schering	Italy	1969
Neogest	Schering	UK	1974
Microlut	Schering	W. Germany	1974
Planovar	Wyeth	Japan	1979
Duoluton	Schering	Japan	1979
Prempak	Ayerst	UK	-

Raw Materials

(+/-)-1,4-Dihydro-17α-ethynyl-18-homo-oestradiol 3-methyl ether
Hydrogen chloride

Manufacturing Process

To 0.7 gram of (+/-)-1,4-dihydro-17α-ethynyl-18-homo-oestradiol 3-methyl ether in 36 cc methanol was added 1.6 cc water and 2.4 cc concentrated hydrochloric acid. After standing at room temperature for 2 hours ether was added, and the washed and dried ethereal solution was evaporated, yielding a gum which was dissolved in 5 cc benzene and the solution absorbed on 50 grams of an activated fuller's earth. Elution with light petroleum containing increasing proportions of benzene gave a crystalline by-product: further elution with benzene containing a small proportion of ether gave a crystalline product which was recrystallized from ethyl acetate, yielding 0.11 gram of (+/-)-17α-ethynyl-18-homo-19-nortestosterone. MP 203° to 206°C.

References

Merck Index 6543
Kleeman & Engel p. 648
PDR pp. 1952, 1958, 1965
OCDS Vol. 1 p.167 (1977); 2, 151 (1980) & 3, 84 (1984)
DOT 4 (1) 24 (1968)
I.N. p. 690
REM p. 993
Hughes, G.A. and Smith, H.; British Patent 1,041,280; September 1, 1966

NORMETHADONE

Therapeutic Function: Narcotic analgesic, Antitussive

Chemical Name: 3-Hexanone, 6-(dimethylamino)-4,4-diphenyl-

Common Name: Desmethylmethadone; Noramidone; Normetadone; Normethadone; Phenyldimazone

Structural Formula:

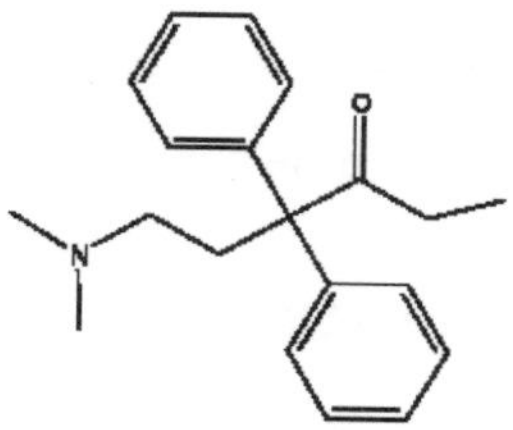

Chemical Abstracts Registry No.: 467-85-6

Trade Name	Manufacturer	Country	Year Introduced
Normethadone	Isotec, Inc.	-	-

Raw Materials

Sodium amide
Magnesium
Ethyl bromide
Diphenylacetonitrile
2-Dimethyaminoethylchloride

Manufacturing Process

65 g of thin powder sodium amide was added to a solution of 289 g diphenylacetonitrile in 300 ml benzene for 15-20 minutes at temperature 45°-50°C. Then the mixture was cooled to about 25°C, 182 g 2-dimethyaminoethylchloride was added dropwise. On ending the reaction it was heated to reflux for 15 minutes, diluted with water, the benzene layer separated and washed with diluted hydrochloric acid. The acid layer was alkalified with sodium hydroxide, extracted with ether and dried over potash. The ether was distilled off to dryness and the residue solidified to give colorless crystals of 4-dimethylamino-2,2-diphenylbutyronitrile. The C_2H_5MgBr was made from 66.5 g of magnesium, 300 ml dry ether and 33 g ethyl bromide and mixed with above prepared nitrile in 150 ml toluene. After the ending of reaction, the mixture was heated for 1.5 hour on the steam bath to give hard-grained bulk. It was mixed with 600 ml concentrate hydrochloric acid in 1500 ml water. On cooling 500 ml benzene was added and three layers arose. The middle layer crystallized shortly. The crystals were filtered off, washed with 100 ml 2 N hydrochloric acid and 3 x 100 ml acetone. 6-Dimethylamino-4,4-diphenyl-3-hexanone was prepared as a hydrochloride. MP: 231°C. The salt may be transformed into the base by adding of an equivalent of any basic compound (triethyl amine, soda and so on).

References

Bockmuhl M., Ehrhart G.; D.B. Patent No. 865,314; July 8, 1949; Farbwerke Hoechst, vormals Meister Lucius and Bruning, Frankfurt/M.-Hochst

NORTRIPTYLINE

Therapeutic Function: Antidepressant

Chemical Name: 3-(10,11-Dihydro-5H-dibenzo[a,d]cyclohepten-5-ylidene)-N-methyl-1-propanamine

Common Name: Desmethylamitriptyline; Desitriptyline

Structural Formula:

Chemical Abstracts Registry No.: 72-69-5; 894-71-3 (Hydrochloride salt)

Trade Name	Manufacturer	Country	Year Introduced
Aventyl	Lilly	UK	1963
Nortrilen	Tropon	W. Germany	1964
Aventyl	Lilly	US	1965
Psychostyl	Lilly	France	1966
Vividyl	Lilly	Italy	1967
Noritren	Dainippon	Japan	1971
Altilev	Squibb	France	1976
Pamelor	Sandoz	US	1977
Allegron	Dista	UK	-
Ateben	Sintyal	Argentina	-
Martimil	Lafarquin	Spain	-
Nortylin	Ikapharm	Israel	-
Norzepine	Bial	Portugal	-
Sensaval	Pharmacia	Sweden	-

Raw Materials

5-(3-Chloropropylidene)dibenzo[a,d]cyclohepta[1,4]diene
Methylamine

Manufacturing Process

A mixture of 114.5 g of 5-(3-chloropropylidene)dibenzo[a,d]cyclohepta[1,4] diene, 75 ml of benzene, and about 400 ml of methylamine is heated in an autoclave at 120°C for six hours. The excess methylamine is distilled from the reaction mixture under vacuum and the residue is stirred with 300 ml of water. Acidification of the mixture with hydrochloric acid causes the separation of the hydrochloride of 5-(3-methylaminopropylidene)dibenzo[a,d] cyclohepta[1,4]diene. The product is collected by filtration and is purified by recrystallization from a mixture of absolute ethanol and ethyl acetate. MP 210°C to 212°C.

References

Merck Index 6558
Kleeman & Engel p. 651
PDR p. 1588
OCDS Vol. 1 p. 151 (1977)
DOT 1 (1) 22 (1965) & 9 (6) 219 (1973)
I.N. p. 691
REM p. 1096
Peters, L.R. and Hennion, G.F.; US Patent 3,281,469; October 25, 1966; assigned to Eli Lilly & Co.

NOVOBIOCIN

Therapeutic Function: Antibiotic

Chemical Name: N-[7-[[3-O-(Aminocarbonyl)-5,5-di-C-methyl-4-O-methyl-α-L-lyxopyranosyl]oxy]-4-hydroxy-8-methyl-2-oxo-2H-1-benzopyran-3-yl]-4-hydroxy-3-(3-methyl-2-butenyl)benzamide

Common Name: Streptonivicin

Structural Formula:

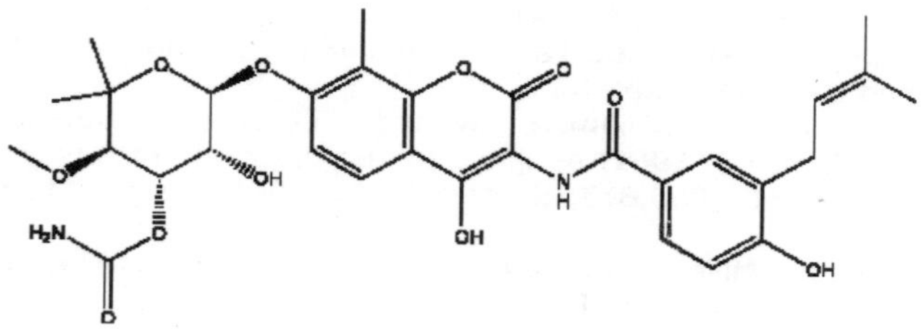

Chemical Abstracts Registry No.: 303-81-1

Trade Name	Manufacturer	Country	Year Introduced
Albamycin	Upjohn	US	1956
Cathomycin	MSD	US	1956
Cathomycine	Theraplix	France	1957
Albiocin	Upjohn	Japan	-
Inamycin	Hoechst	W. Germany	-
Robiocina	San Carlo	Italy	-
Stilbiocina	Donatello	Italy	-

Raw Materials

Bacterium Streptomyces spheroides
Soybean meal
Dextrose

Manufacturing Process

The preparation of novobiocin by fermentation is described in US Patent 3,049,534 as follows: A medium containing 2% soybean meal, 1% dextrose, 0.25% sodium chloride and 0.75% distiller's solubles was made up in tap water. About 25 ml of the prepared medium was placed in a 75 ml vial and sterilized by heating at 120°C for 20 minutes. The sterilized medium was then inoculated with a vegetative culture of Streptomyces spheroides MA-319 (NRRL 2449), and the vial loosely stoppered with cotton. The vial was then placed on a shaking machine with an amplitude of 1½ inches at 28°C for 6 days. At the end of this fermentation time, the fermented broth was assayed using the cylinder-plate method with Bacillus megatherium ATCC 9885 as the assay organism and found to have an activity of 600 units/ml or 30 mcg/ml of novobiocin. The production of larger quantities of novobiocin by submerged fermentation in suitable tanks is also described in US Patent 3,049,534.

The preparation of novobiocin by a synthetic route is described in US Patent 2,966,484, as well as in US Patent 2,925,411.

References

Merck Index 6563
Kleeman & Engel p. 652
I.N. p. 693
REM p. 1212
Stammer, C.H.; US Patent 2,925,411; February 16, 1960
Walton, E. and Spencer, C.; US Patent 2,966,484; December 27, 1960; assigned to Merck & Co., Inc.
Caron, E.L., Johnson, J.L., Hinman, J.W. and Hoeksema, H.; US Patent 2,983,723; May 9, 1961; assigned to The Upjohn Company
Wolf, F.J.; US Patent 3,000,873; September 19, 1961; assigned to Merck & Co., Inc.
Stammer, C.H. and Miller, I.M.; US Patent 3,049,475; August 14, 1962; assigned to Merck & Co., Inc.
Miller, I.M.; US Patent 3,049,476; August 14, 1962; assigned to Merck & Co., Inc.
Wallick, H.; US Patent 3,049,534; August 14, 1962; assigned to Merck & Co., Inc.
French, G.H.; US Patent 3,068,221; December 11, 1962; assigned to The Upjohn Co.

NOXIPTILIN

Therapeutic Function: Psychostimulant

Chemical Name: 10,11-Dihydro-5H-dibenzo[a,d]cyclohepten-5-one O-[2-(dimethylamino)ethyl]oxime

Common Name: Dibenzoxin

Structural Formula:

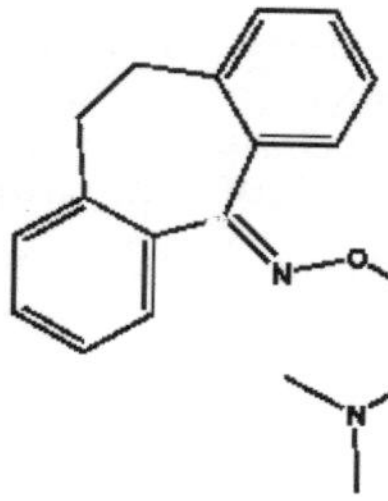

Chemical Abstracts Registry No.: 3362-45-6; 4985-15-3 (Hydrochloride salt)

Trade Name	Manufacturer	Country	Year Introduced
Agedal	Bayer	W. Germany	1969
Agedal	Bayer	Italy	1975
Nogedal	Theraplix	France	1978
Elronon	Deutsches Hydrierwerk	E. Germany	-
Sipcar	Bernabo	Argentina	-

Raw Materials

5-Keto-10,11-dihydrodibenzo[a,d]cycloheptene
Hydroxylamine hydrochloride
Sodium amide
β-(Dimethylamino)ethyl chloride

Manufacturing Process

15 grams 5-keto-10,11-dihydrodibenzo-[a,d]cycloheptene dissolved in 225 ml of pyridine was mixed with 15 grams hydroxylamine hydrochloride, and the mixture was boiled under reflux for 22 hours. The bulk of the pyridine was then distilled off under reduced pressure, the residue was poured into water, and the aqueous mixture thus formed was extracted with ether.

The ether extract was washed with water, dried and heated to distill off the ether. The solid residue was recrystallized from a mixture of benzene and light petroleum (BP 40° to 60°C). 12.8 grams of the recrystallized oxime had a MP of 167° to 169°C.

A solution of 22 grams of the above described 5-oximino-10,11-dihydrodibenzo-[a,d]cycloheptene in 120 ml benzene was treated with 7.8

grams sodamide and the mixture was stirred and heated under reflux for 2 hours. At this stage, the 14.4 grams of hydrochloride of β-(dimethylamino) ethyl chloride was added and heating under reflux was continued for 16 hours. 50 ml water was then cautiously added to decompose unreacted sodamide and the benzene layer was separated and extracted with dilute (10%) aqueous hydrochloric acid.

The aqueous acid extracts were made alkaline with concentrated aqueous potassium hydroxide solution and then extracted with ether. The ether extracts were dried, the solvent was removed and the residual oil was distilled under reduced pressure. The product was 14.5 grams of the fraction boiling at 160° to 164°C, under a pressure of 0.05 mm of mercury.

References

Merck Index 6566
Kleeman & Engel p. 653
DOT 6 (2) 56 (1970)
I.N. p. 695
Wrigley, T.I. and Leeming, P.R.; British Patent 1,045,911; October 19, 1966; assigned to Pfizer Limited, England
Schutz, S. and Hoffmeister, F.; US Patent 3,505,321; April 7, 1970; assigned to Farbenfabriken Bayer A.G.

NOXYTIOLIN

Therapeutic Function: Antifungal

Chemical Name: 1-Methyl-3-hydroxymethyl-2-thiourea

Common Name: -

Structural Formula:

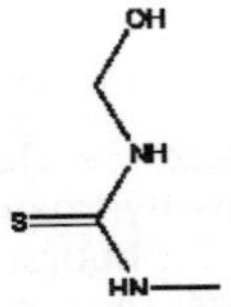

Chemical Abstracts Registry No.: 15599-39-0

Trade Name	Manufacturer	Country	Year Introduced
Noxyflex	Geistlich	UK	1964
Noxyflex	Innothera	France	1978
Gynaflex	Geistlich	Switz.	-

Raw Materials

Methyl thiourea
Formaldehyde

Manufacturing Process

400 g methyl thiourea and 2.5 g $NaHCO_3$ are dissolved in 400 ml formaldehyde solution of 35% concentration. After having been left at ordinary temperature for 2 to 3 hours, the solution is adjusted with dilute HCl to pH 7 to 7.5. After the reaction mixture had been left overnight at 15°C some of the final product crystallized and was filtered off using a Buchner funnel. The mother liquor was concentrated by evaporation in vacuo at a bath-temperature of 30°C. The crystals obtained were again collected by filtration using a Buchner funnel and were combined with the first crystalline fraction and dried in vacuo at ordinary temperature. Yield of pure substance 400 g; melting point 84°C to 86°C.

References

Merck Index 6567
Kleeman & Engel p. 653
DOT 4 (3) 106 (1968)
I.N. p. 695
Aebi, A. and Hafstetter, E.; British Patent 970,414; January 12, 1960; assigned to Ed Geistlich Sohne AG fur Chemische Industrie.

NYLIDRIN

Therapeutic Function: Vasodilator

Chemical Name: 4-Hydroxy-α-[1-[(1-methyl-3-phenylpropyl)amino]ethyl] benzenemethanol

Common Name: Buphenine

Structural Formula:

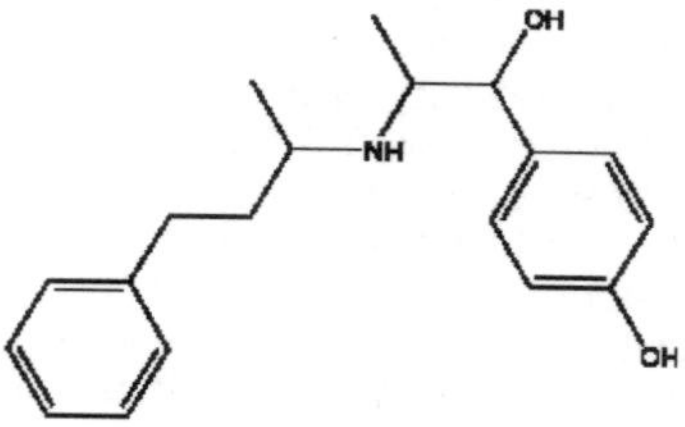

OH
NH
OH

Chemical Abstracts Registry No.: 447-41-6

Trade Name	Manufacturer	Country	Year Introduced
Arlidin	U.S.V.	US	1955
Arlibide	U.S.V.	Argentina	-
Bufedon	Cosmopharma	Netherlands	-
Buphedrin	Tatsumi	Japan	-
Dilatol	Tropon	W. Germany	-
Dilatropon	Draco	Sweden	-
Dilaver	Neopharma	Finland	-
Dilydrin	Medichemie	Switz.	-
Nyderal	Kobayashi	Japan	-
Nylin	Toho	Japan	-
Opino	Bayropharm	W. Germany	-
Penitardon	Woelm	W. Germany	-
Perdilat	Abdi Ibrahim	Turkey	-
Perdilatal	Smith and Nephew	UK	-
Pervadil	I.C.N.	Canada	-
Pharmadil	Pharmacia	Sweden	-
Rudilin	Darby	US	-
Rydrin	Kodama	Japan	-
Shatorn	Seiko	Japan	-
Tacodilydrin	Swiss Pharma	W. Germany	-
Tocodrin	Medichemie	Switz.	-
Vasiten	Crinos	Italy	-
Verina	Fujisawa	Japan	-

Raw Materials

p-Benzoxy-α-bromopropiophenone
1-Phenyl-3-aminobutane
Hydrogen

Manufacturing Process

8 grams of the hydrobromide of 1-(p-benzoxyphenyl)-2-(α-methyl-γ-phenyl-propylamino)-propanone-(1) were obtained by heating equivalent quantities of p-benzoxy-α-bromopropiophenone and 1-phenyl-3-amino-butane for an hour on the water bath in the absence of solvents. The product was purified by twice boiling with five times the quantity of acetic acid and filtration at 80°C, then shaken in contact with hydrogen with 0.8 gram of Raney nickel in 70 cc of pure methanol containing 0.96 gram (corresponding to 1 mol) of KOH. After 4 hours 2 mols of hydrogen had been taken up and the solution was filtered from the catalyst, evaporated in vacuo, and the residue triturated first with water to remove potassium bromide and then with methanol to remove potassium bromide. 3.7 grams (72% of the theoretical yield) of the compound specified, melting at 110° to 112°C, were obtained, as described in US Patent 2,661,373.

References

Merck Index 6577
Kleeman & Engel p. 123
PDR pp. 830, 993, 1606, 1809, 1999
OCDS Vol. 1 p. 69 (1977)
I.N. p. 163
REM p. 892
Schopf, C. and Kunz, K.J.; US Patent 2,661,372; December 1, 1953; assigned to Troponwerke Dinklage & Co., Germany
Kulz, F. and Schopf, C.; US Patent 2,661,373; December 1, 1953

NYSTATIN

Therapeutic Function: Antifungal

Chemical Name: Nystatin

Common Name:-

Structural Formula:

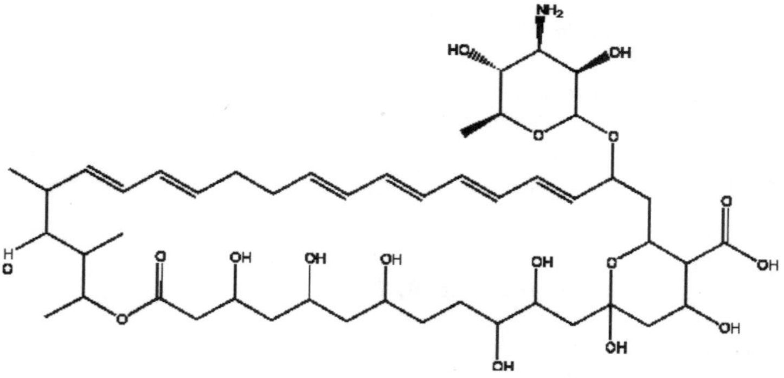

Chemical Abstracts Registry No.: 1400-61-9

Trade Name	Manufacturer	Country	Year Introduced
Mycostatin	Squibb	US	1954
Mycostatine	Squibb	France	1956
Nysta-Dome	Dome	US	1964
Nilstat	Lederle	US	1970
Nysert	Norwich Eaton	US	1979
Multilind	F.A.I.R.	UK	1979

Trade Name	Manufacturer	Country	Year Introduced
Nystex	Savage	US	1983
Biofanal	Pfleger	W. Germany	-
Candex	Dome	US	-
Candio-Hermal	Hermal	W. Germany	-
Herniocid	Mayrhofer	Austria	-
Korostatin	Holland Rantos	US	-
Mycolog	Squibb	US	-
Myco-Triacet	Lemmon	US	-
Mytrex	Savage	US	-
Nadostine	Nadeau	Canada	-
Nyaderm	K-Line	Canada	-
Nystacid	Farmos	Finland	-
Nyst-Olone	Schein	US	-
Rivostatin	Rivopharm	Switz.	-
Stereomycin	Medica	Finland	-

Raw Materials

Bacterium Streptomyces noursei
Nutrient medium

Manufacturing Process

A typical isolation and recovery procedure for nystatin is described in US Patent 2,797,183 and is shown in the following diagram:

References

Merck Index 6580
Kleeman & Engel p. 654
PDR pp.888, 1022, 1034, 1429, 1604, 1751
I.N. p. 696
REM p. 1230
Vandeputte, J. and Gold, W.; US Patent 2,786,781; March 26, 1957; assigned to Olin Mathieson Chemical Corporation
Hazen, E.L. and Brown, R.F.; US Patent 2,797,183; June 15, 1957; assigned to Research Corporation
Vandeputte, J.; US Patent 2,832,719; April 29, 1958; assigned to Olin Mathieson Chemical Corporation
Renella, J.G.; US Patent 3,517,100; June 23, 1970; assigned to American Cyanamid Co.

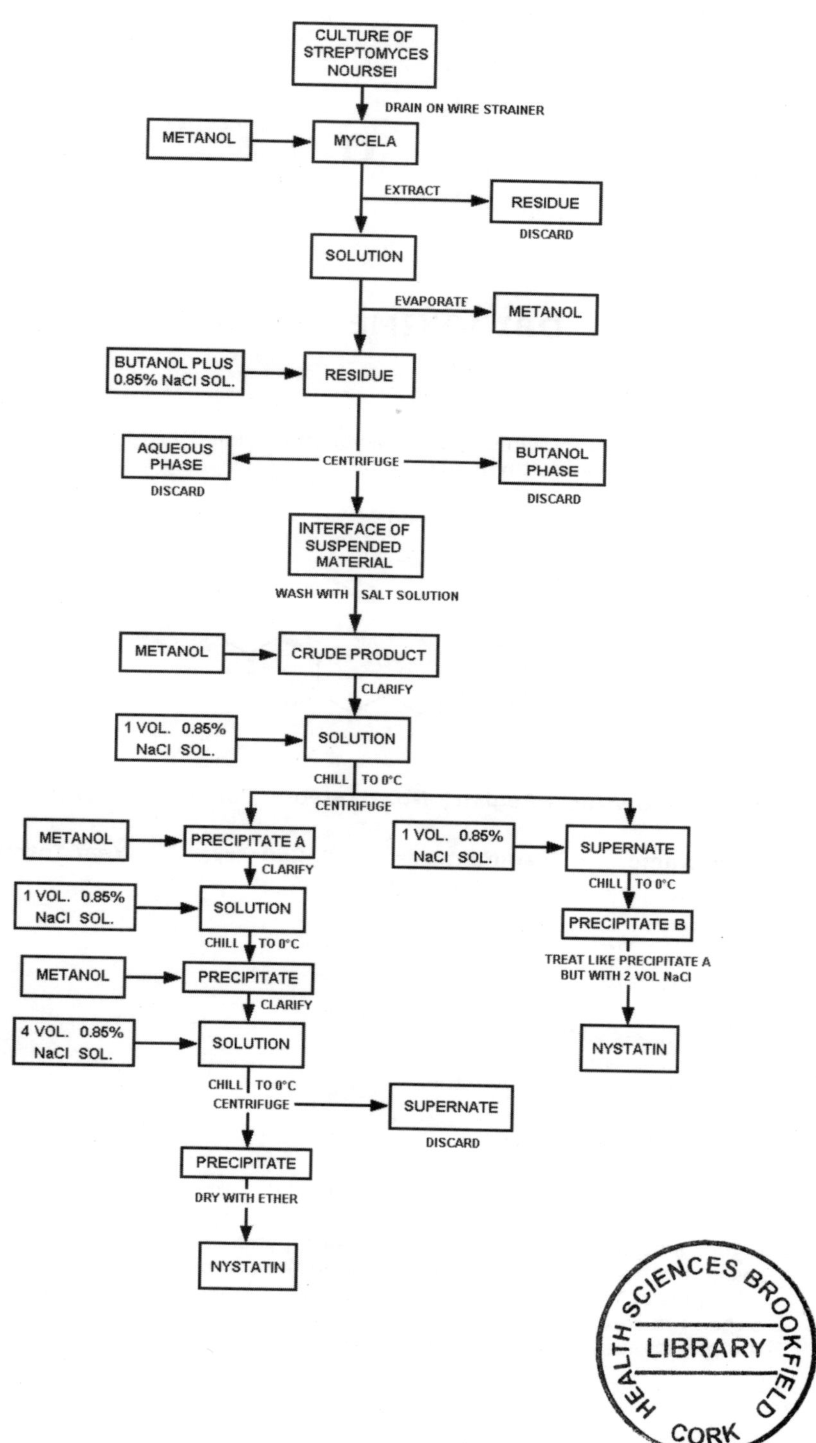
CULTURE OF STREPTOMYCES NOURSEI
DRAIN ON WIRE STRAINER
METANOL
MYCELA
EXTRACT
RESIDUE
DISCARD
SOLUTION
EVAPORATE
METANOL
BUTANOL PLUS 0.85% NaCl SOL.
RESIDUE
AQUEOUS PHASE
DISCARD
CENTRIFUGE
BUTANOL PHASE
DISCARD
INTERFACE OF SUSPENDED MATERIAL
WASH WITH SALT SOLUTION
METANOL
CRUDE PRODUCT
CLARIFY
1 VOL. 0.85% NaCl SOL.
SOLUTION
CHILL TO 0°C
CENTRIFUGE
METANOL
PRECIPITATE A
CLARIFY
1 VOL. 0.85% NaCl SOL.
SUPERNATE
CHILL TO 0°C
PRECIPITATE B
TREAT LIKE PRECIPITATE A BUT WITH 2 VOL NaCl
NYSTATIN
1 VOL. 0.85% NaCl SOL.
SOLUTION
CHILL TO 0°C
METANOL
PRECIPITATE
CLARIFY
4 VOL. 0.85% NaCl SOL.
SOLUTION
CHILL TO 0°C
CENTRIFUGE
SUPERNATE
DISCARD
PRECIPITATE
DRY WITH ETHER
NYSTATIN

O

OBIDOXIME CHLORIDE

Therapeutic Function: Antidote

Chemical Name: Pyridinium, 1,1'-(oxybis(methylene))bis(4-(hydroxyimino)methyl)-, dichloride

Common Name: Obidoxime chloride

Structural Formula:

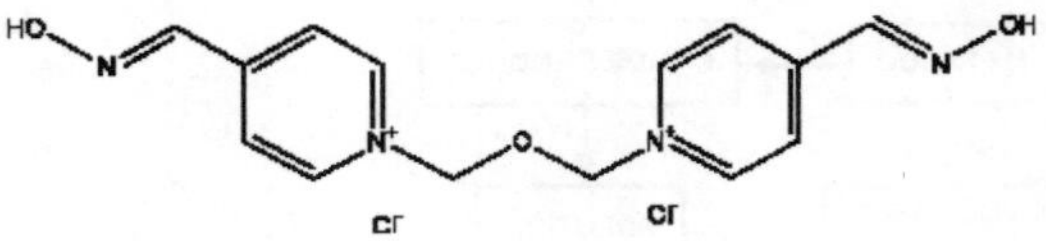

Chemical Abstracts Registry No.: 114-90-9; 7683-36-5 (Base)

Trade Name	Manufacturer	Country	Year Introduced
Toksobidin	Polfa-Starogard	-	-
Toxogonin	Merck KGaA	-	-

Raw Materials

Pyridine-4-aldoxime
bis-Chloromethyl ether
α,α-Dichloro-dimethyl-ether

Manufacturing Process

2 Methods of producing of obidoxime chloride:

1. Into a boiling agitated solution of 2.44 g pyridine-4-aldoxime in 10 ml absolute ethanol is added dropwise during the course of 25 min a solution of 1.14 g bis-chloromethyl ether in 5 ml absolute ethanol. The reaction mixture is then refluxed for 35 min, and then agitated for 5 h at room temperature. The precipitate of bis-[4-hydroxyimino-methyl-pyridinium-(1)-methyl]-ether-

dichloride is thoroughly washed with absolute acetone. The yield is 3.5 g which is 98% of the theoretical, and the melting point is 229°C. If convenient, the mother liquor can be reused to make additional product.

2. 12.2 g (0.1 mole) pyridine-4-aldoxime are dissolved with heating in 125 ml chloroform. Within 25 min, 8.5 g (0.075 mole), α,α-dichlorodimethyl ether in 20 ml chloroform are dropped while stirring into the boiling solution. The reaction mixture is heated for another 35 min. After standing for several hours, the precipitate is filtered off, washed with absolute ethanol, acetone and ether and dried at 80°C. Yield: 17.0 g, 95% of the theoretical, and the melting point is 225°C (dec.).

References

Luttringhaus A. et al.; US Patent No. 3,137,702; June 16, 1964; Assigned: E. Merck, Darmstadt, Germany

OCTOPAMINE HYDROCHLORIDE

Therapeutic Function: Hypertensive

Chemical Name: α-(Aminomethyl)-4-hydroxybenzene-methanol hydrochloride

Common Name: Norsympatol hydrochloride; Norsynephrine hydrochloride

Structural Formula:

HCl

H_2N

HO

OH

Chemical Abstracts Registry No.: 770-05-8; 104-14-3 (Base)

Trade Name	Manufacturer	Country	Year Introduced
Norfen	Morishita	Japan	1975
Depot-Norphen	Byk Gulden	W. Germany	-
Norphen	Byk Gulden	W. Germany	-

Raw Materials

Phenol
Aminoacetonitrile
Hydrogen chloride
Hydrogen

Manufacturing Process

A solution of 33 grams of anhydrous aluminum chloride in 60 grams of nitrobenzene, to which a mixture of 14 grams of phenol and 9.3 grams of hydrochloride of amino-acetonitrile was added, had dry hydrochloric acid gas introduced into it for 3 hours, while stirring and cooling to keep the temperature between 20° and 30°C. The reaction mixture was then poured, with cooling, into 70 cc of water and the deposit obtained was sucked off, washed with acetone and dissolved in 300 cc of water. The solution thus prepared was decolorized with carbon, 50 grams of 30% sodium citrate solution was added to it, and then it was made slightly alkaline with ammonia. Thereupon hydroxy-4'-phenyl-1-amino-2-ethanone crystallized out in the form of leaflets. The yield was 7.7 grams.

The hydrochloride of this base, obtained by evaporation to dryness of a solution of the base in dilute hydrochloric acid and subsequent treatment of the residue with ethyl alcohol and acetone, had a chlorine content of 18.84%, (calculated, 18.90%).

This hydrochloride, on being dissolved in water and hydrogenated with hydrogen and a nickel catalyst, gave a good yield of hydrochloride of hydroxy-4'-phenyl-1-amino-2-ethanol melting, after crystallization from a mixture of ethyl alcohol and butanone-2, at from 177° to 179°C with decomposition.

References

Merck Index 6599
Kleeman and Engel p. 655
I.N. p. 699
Asscher, M.; US Patent 2,585,988; February 19, 1952

OCTREOTIDE ACETATE

Therapeutic Function: Antiulcer, Growth hormone inhibitor

Chemical Name: L-Cysteinamide, D-phenylalanyl-L-cysteinyl-L-phenylalanyl-D-tryptophyl-L-lysyl-L-threonyl-N-(2-hydroxy-1-(hydroxymethyl)propyl)-, cyclic(2-7)-disulfide, (R-(R*,R*))-, acetate (salt)

Common Name: Octreotide acetate

Chemical Abstracts Registry No.: 79517-01-4; 83150-76-9 (Base)

Trade Name	Manufacturer	Country	Year Introduced
Sandostatin	Novartis Pharma AG	Switz.	-
Sandostatin LAR	Novartis Pharma AG	Switz.	-

Structural Formula:

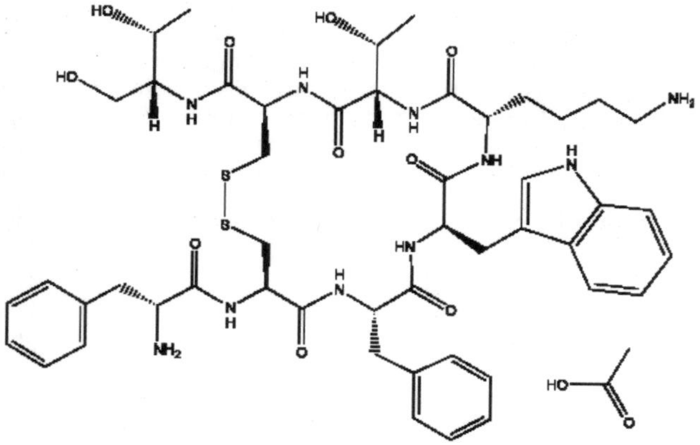

Raw Materials

Threoninol
Piperidine
Tfifluoroethanol
Iodine
Threonino
Trifluoroacetic acid

N,N-Diisopropylethylamine
Diisopropylcarbodiimide
Hydroxybenzotriazole
Boc-D-Phe; Fmoc-Cys(Trt); Fmoc-Trp; Fmoc-Lys(Boc); Fmoc-Thr(tBu)

Manufacturing Process

Synthesis of octreotide and derivative thereof can be carried out by two methods. The first method is synthesized initially by fragment condensation solution phase procedures. The synthetic process of octreotide has been described by Bauer et al. (1). The second method is the synthese by solid-phase procedures. Edward et al. (2) isolated side chain protected octreotide with a total yield of 14% by cleaving the protected peptide from the resin with threoninol. Arano et al. (3) carried out another solid phase method for octreotide. and produced it in overall 31.8% yield based on the starting Fmoc-Thr(tBu)-ol-resin. The basic difference from the other procedures already described is that the introduction of the threoninol is carried out upon the protected peptidic structure (resin-free), which, when appropriately activated, leads quantitatively and without needing to make temporary protections upon the threoninol, to the protected precursor of octreotide, which in turn, with a simple acid treatment leads to octreotide with very high yields.

At first the Fmoc-Cys-Cl-trityl-resin was prepared. The incorporation of the Fmoc-Cys(trt)-OH residue upon 2Cl-Trt resin is accomplished with an excess of 1 eq. of Fmoc-Cys(Trt) and 2.5 eq. of N,N'-diisopropylethylamine (DIEA).

2.93 g (5.0 mmol) of Fmoc-Cys(Trt) are incorporated upon 5 g of resin (f = 1.28 mmol/g of resin, 6.4 mmol). The resin and the amino acid are weighed in separate containers and left to dry in a vacuum with KOH, for a minimum of two hours. A 1/1 solution of DIEA and CH_2Cl_2 (DCM) (dry on a 4A sieve) is prepared. The already dry amino acid is dissolved with dry DCM at a concentration of 0.1 g of resin per ml, adding the minimum quantity of dry DMF to complete the dissolution. 1/3 of the 1.8 ml (12.5 mmol) DIEA solution is added to this transparent solution in 1.8 ml of DCM. This is thoroughly homogenized and added to the dry resin. It is subjected to vigorous magnetic agitation for five minutes and the rest of the DIEA is added to the reaction; the mixture is allowed to react for forty minutes more. Then, 4 ml of dry MeOH are added and allowed to react for 10 minutes, after which the resin is filtered and the washings described below are carried out.

Step	**Reagent**	**Repetitions**	**Time(min)**
1	DMF	3	1'
2	5% piperidine/(DMF/DCM)	1	10'
3	20% piperidine/DMF	1	15'
4	DMF	3	2'

The incorporation of the amino acids for obtaining of Boc-D-Phe-Cys-(Trt)-D-Trp-Lys(Boc)-Thr(tBu)-Cys(Trt)-2-Cl-trityl-resin is carried out following a synthesis program such as that described below, using an excess of 2.5 equivalents of Fmoc-amino acid, N-hydroxybenzotriazol (HOBt) and diisopropylcarbodiimide (DIPCDI). Later the Fmoc group is deprotected with 20% of piperidine/DMF for 1 min + 5 min.

Step	**Reagent**	**Repetitions**	**Time(min)**
1*	DMF	5	1'
2*	pip/DMF 20%	5	1'
3*	pip/DMF 20%	1	5'
4*	DMF	1	1'
5*	Fmoc aminoacid	-	+
6	HOBt	-	+
7	DIPCDI	-	40'
8	DMF	5	1'

[*for Thr]

Control by Ninhydrin test; if (+), return to 5; if (-) follow step 1 forward following amino acid.

The yields at the end of the synthesis are quantitative in obtaining Boc-D-Phe-Cys(Trt)-Phe-D-Trp-Lys(Boc)-Thr(tBu)-Cys(Trt)-2Cl-trityl-resin.

Preparation of Boc-D-Phe-Cys(Trt)-Phe-D-Lys(Boc)-Thr(tBu)-Cys(Trt)

250 mg (113 µmols) of Boc-D-Phe-Cys(Trt)-Phe-D-Trp-Lys(Boc)-Thr(tBu)-Cys(Trt)-2Cl-trityl-resin are treated with 6.36 ml of mixture 7/2/1 or 5.5/0.5/4 of DCM/TFE(tfifluoroethanol)/AcOH, for two hours, under magnetic agitation. The suspension is then filtered and washed 3 times with 0.2 ml of the 7/2/1 mixture of DCM/TFE/AcOH. The solution is evaporated (if it is not desired to proceed with the oxidation) until dry, at reduced pressure, and the solid obtained is washed with water. The yield is quantitative.

Obtaining of cycle $^{(2-7)}$Boc-D-Phe-Cis-Phe-D-Trp-Lys(Boc)-Thr(tBu)-Cis-OH (Oxidized 1-7 fragment).

250 mg (113 mmols) of Boc-D-Phe-Cys(Trt)-Phe-D-Trp-Lys(Boc)-Thr(tBu)-Cys(Trt)-COOH dissolved in 7 ml of the 7/2/1 mixture of DCM/TFE/AcOH, is slowly added to 290 mg (1.13 mmol) of iodine of 0.8 M concentration in the 7/2/1 mixture of DCM/TFE/AcOH. The reaction is allowed to evolve for 15 minutes. 4.3 ml of a $Na_2S_2O_7$, 1 N solution is added to eliminate the iodine excess. The aqueous phase is extracted and washed three times with 1 ml of DCM, the entirety of the organic phases is extracted with a citric acid/water solution and is evaporated at reduced pressure to dryness. The solid obtained is washed with water. The yield fluctuates between 85 and 95%.

Coupling of cycle $^{(2-7)}$Boc-D-Phe-Cis-Phe-D-Trp-Lys(Boc)-Thr(tBu)-Cis-OH (Oxidized 1-7 fragment) with Throl.

Over 250 mg (230 μmols) of $^{(2-7)}$Boc-D-Phe-Cis-Phe-D-Trp-Lys(Boc)-Thr(tBu)-Cis-OH(Oxidized 1-7 fragment), 103 mg (690 μmol) of HOBt and 72 mg (690 μmol) of threoninol are weighed out and dissolved in 10 ml of dry DMF/dry DCM (1:1); under vigorous agitation, 111 μL (690 μmols) of DIPCDI are added. The mixture is allowed to react for five hours at room temperature. It is evaporated to dryness until an oil is obtained, water is added, the mixture is well homogenized by ultrasound and lyophilized. The coupling is quantitative.

Oxidation of Boc-D-Phe-Cys(Trt)-Phe-D-Trp-Lys(Boc)-Thr(tBu)-Cys(Trt)-Throl.

Obtaining of cycle $^{(2-7)}$Boc-D-Phe-Cis-Phe-D-Trp-Lys(Boc)-Thr(tBu)-Cis-Throl(Oxidized fragment 1-8).

250 mg (147 μmols) of Boc-D-Phe-Cys(Trt)-Phe-D-Trp-Lys(Boc)-Thr(tBu)-Cys(Trt)-Throl dissolved in 8.26 ml of mixture 7/2/1 of DCM/TFE/AcOH, are slowly added to a solution of 290 mg (1.47 mmol) of iodine of 0.8 M concentration in the mixture 7/2/1 of DCM/TFE/AcOH. The reaction is allowed to evolve for 15 minutes. 4.3 ml of an $Na_2S_2O_7$ 1 N is added to eliminate the excess iodine. The aqueous phase is extracted and washed three times with 1 ml of DCM, the entirety of the organic phases is extracted with a solution of citric acid and water and is evaporated at reduced pressure to dryness. The solid obtained is washed with the help of a filter plate and water.

Removal of protecting groups. Obtention of Octreotide.

230 mmols cycle $^{(2-7)}$Boc-D-Phe-Cis-Phe-D-Trp-Lys(Boc)-Thr(tBu)-Cis-Throl (Oxidized 1-8 fragment)n are treated with 2 ml of trifluoroacetic acid (TFA) (95:5%) for five hours at ambient temperature. Later, the filtrate is dropped over 100 ml of dry and cold diethyl ether and the white precipitate obtained is once again centrifuged. The solid is resuspended in diethyl either and centrifuged again, repeating the operation five times more. The crude peptide is purified by preparative HPLC at 25% of CH_3CN/H_2O with 0.01% TFA in a 10 μml.

References

Bayer W. et al.; Eur. Patent Appl. 29,579, 1981 and Bayer W. et al.; US Patent No. 4,395,403; July 26, 1983; Assigned: Sandoz Ltd. (Basel, CH)

Edward et al.; J. Med. Chem. 1994, 37, 3749-3757

Arano et al.; Bioconjugate Chem. 1997, 8, 442-446
Obiols B. et al.; US Patent No. 6,346,601 B1; Feb. 12, 2002; Assigned to Lipotec S.A., Barcelona (ES)

OFLOXACIN

Therapeutic Function: Antibacterial

Chemical Name: 7H-Pyrido(1,2,3-de)-1,4-benzoxazine-6-carboxylic acid, 2,3-dihydro-9-fluoro-3-methyl-10-(4-methyl-1-piperazinyl)-7-oxo-, ()-

Common Name: Mefoxacin; Ofloxacin

Structural Formula:

Chemical Abstracts Registry No.: 82419-36-1; 83380-47-6

Trade Name	Manufacturer	Country	Year Introduced
Exocin	Allergan	-	-
Floxan	Janssen-Cilag	-	-
Floxil	Cilag	-	-
Floxin	Aetna Inc.	-	-
Floxin	Janssen-Ortho Inc.	-	-
Floxin	Ortho	-	-
Floxin	McNeil	-	-
Floxstat	Jannsen-Cilag	-	-
Oflo	Unique	India	-
Oflomac	Macleods Pharmaceuticals	India	-
Oflox	Allergan	-	-
Ofloxacin	JAKA-80	Macedonia	-
Ofloxacin	Nu-Pharm Inc.	Canada	-
Ofloxacin	Ranbaxy	India	-
Ofloxacin	Chemo Iberica	Spain	-
Ofloxacin	Huanguan East Asia Chemical Co.	China	-
Ofloxin 200	Leciva	Czech Republic	-
Quinoxan	Andromaco	-	-
Tarivid	Hoechst	Germany	-

Trade Name	Manufacturer	Country	Year Introduced
Tarivid	Hoechst Marion Roussel	India	-
Uroflox	Houde	-	-
Urosin	Sifar	-	-
Zanocin	Ranbaxy Guangzhow	China	-
Zanocin	Ranbaxy Ireland	Ireland	-
Zanocin	Ranbaxy	India	-

Raw Materials

3,4-Trifluoronitrobenzene
Sodium dithionite
N-Methylpiperazine
Dimethyl sulfoxide
Ethyl polyphosphate
Diethyl ethoxymethylenemalonate

Manufacturing Process

20 g of 2,3,4-trifluoronitrobenzene was dissolved in 150 ml of dimethyl sulfoxide, and to this mixture a solution of 10% potassium hydroxide was added dropwise while keeping the temperature at 18° to 20°C. Then, the mixture was stirred for 2 hours at room temperature and one liter of water was added to this reaction mixture and the mixture was shaken with chloroform. The water layer was acidified with hydrochloric acid and was extracted with chloroform. The extract was washed with water and was dried, then chloroform layer was concentrated. The residue was purified by silica gel column chromatography to provide 5.8 g of 2,3-difluoro-6-nitrophenol as yellow oil.

7.9 g of the 2,3-difluoro-6-nitrophenol, 50.1 g of 1,2-dibromoethane and 18.7 g of potassium carbonate were added to 80 ml of dimethylformamide and the mixture was stirred for 2.5 hours at from about 80° to 100°C (bath temperature). The reaction mixture was concentrated to dryness in vacuo and the residue was distributed between ethyl acetate and water. The organic solvent layer was washed with water and was dried, then the solvent was evaporated. The residue was dissolved in benzene and was purified by silica gel column chromatography to provide 7.7 g of 2-(2-bromoethoxy)-3,4-difluoronitrobenzene as light yellow oil.

1.74 g of this product was dissolved in 30 ml of methanol and a solution of 6.44 g of sodium dithionite dissolved in 15 ml of water was added thereto. The mixture was stirred for 1 hour at room temperature. Methanol was evaporated and the residue was extracted with chloroform. After the extract was washed with water and dried, the solvent was evaporated to provide 0.44 g of 2-(2-bromoethoxy)-3,4-difluoroaniline.

1.82 g of this product and 3.03 g of potassium carbonate were added to 10 ml of dimethylformamide and the mixture was stirred for 1 hour at from about 80° to 100°C (bath temperature). The reaction mixture was added to ice-cold water and was extracted with ethyl acetate. After the extract was washed with water and dried, the solvent was distilled off at room temperature to provide 1.21 g of 7,8-difluoro-2,3-dihydro-4H-[1,4]benzoxazine with m.p. 48°-54°C.

The mixture of 1.1 g of this product and 1.38 g of diethyl

ethoxymethylenemalonate was stirred for 2 hours at from about 130° to 135°C (bath temperature). The ethanol produced was evaporated and 20 g of ethyl polyphosphate was added to the residue. Then the mixture was stirred for 1.5 hours at from about 140° to 145°C (bath temperature). The reaction mixture was added to ice-cold water and was extracted with chloroform. The extract was washed fully with water. After drying, the solvent was evaporated and the residue was recrystallized from ethyl acetate. 1.3 g of ethyl 9,10-difluoro-7-oxo-2,3-dihydro-7H-pyrido[1,2,3-de][1,4]benzoxazine-6-carboxylate was obtained as colorless needles with m.p. 265°-266°C.

1.15 g of this product was added to 12 ml of mixture of concentrated hydrochloric acid and acetic acid (1:4 by volume) and the mixture was stirred for 4 hours at 100° to 110°C (bath temperature). After cooling, the precipitated crystals were collected by filtration, washed with water, methanol and chloroform to give 0.78 g of 9,10-difluoro-7-oxo-2,3-dihydro-7H-pyrido[1,2,3-de][1,4]-benzoxazine-6-carboxylic acid as colorless needles with m.p. above 300°C.

1.0 g of 9,10-difluoro-3-methyl-7-oxo-2,3-dihydro-7H-pyrido[1,2,3-de][1,4]benzoxazine-6-carboxylic acid and 2.85 g of N-methylpiperazine were added to 15 ml of dimethylsulfoxide. The mixture was stirred at a temperature of from about 100° to 110°C (bath temperature) for 12 hours and the reaction mixture was concentrated to dryness in vacuo and 40 ml of water was added to the residue. Then the product was extracted with chloroform. The extract was dried and concentrated to dryness in vacuo. The residue was recrystallized from ethanol to provide 550 mg of 9-fluoro-3-methyl-10-(4-methyl-1-piperazinyl)-7-oxo-2,3-dihydro-7H-pyrido[1,2,3-de][1,4]benzoxazine-6-carboxylic acid as colorless needles with m.p. 250°-257°C (with decomposition).

References

Hayakawa I. et al.; US Patent No. 4,382,892; May 10, 1983; Assigned to Dalichi Selyaku Co., Ltd., Tokyo, Japan

OLANZAPINE

Therapeutic Function: Antipsychotic

Chemical Name: 10H-Thieno(2,3-b)(1,5)benzodiazepine, 2-methyl-4-(4-methyl-1-piperazinyl)-

Common Name: Olanzapine

Raw Materials

Propionaldehyde
Triethylamine
2-Fluoronitrobenzene
N-Methylpiperazine
Dimethyl sulfoxide
Malononitrile
Stannous chloride

Structural Formula:

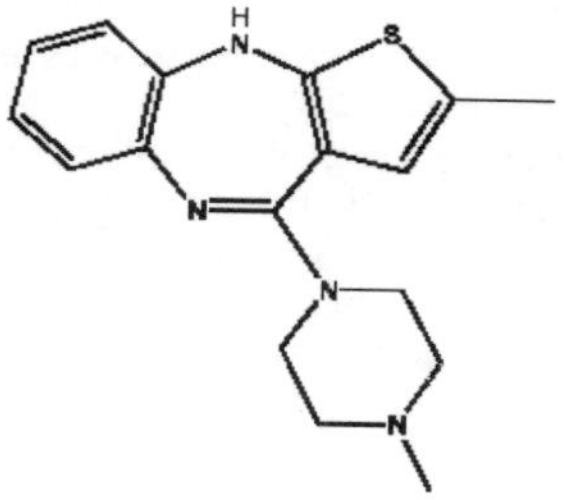

Chemical Abstracts Registry No.: 132539-06-1

Trade Name	Manufacturer	Country	Year Introduced
Joyzol	Merind Limited	India	-
Olan	Synapse (A Div. of Microlabs)	India	-
Olandus	Zydus Neurosciences	India	-
Olanex	Solus	India	-
Oleanz	Sun Pharmaceuticals Industries Ltd.	India	-
Olexa	Protech Biosystems	-	-
Onza	SPPL (Sarabhai Piramal Pharmaceuticals Ltd.)	India	-
Zyprexa	Eli Lilly	UK	-
Zyprexa Zydis	Eli Lilly	UK	-

Manufacturing Process

1. 2-Amino-5-methylthiophene-3-carbonitrile

A mixture of sulphur (217.8 g, 6.79 mol), propionaldehyde (472.5 g, 587 mL, 8.13 mol) and dimethylformamide (1350 m) was placed in a 5 liter flange-necked flask fitted with air stirrer, air condenser, thermometer and dropping funnel. Triethylamine (576 mL, 4.13 mol) was added dropwise over 30 minutes to the cooled stirred reaction mixture whilst maintaining the pot temperature between 5°-10°C with an ice-bath. After addition was complete the pot was allowed to warm up to 18°C over 50 minutes, keeping the mixture well stirred. Then a solution of malononitrile (450 g, 6.8 mol) in dimethylformamide (900 mL) was added dropwise over 70 minutes keeping the pot temperature around 20°C throughout the addition. After addition was complete the mixture was stirred at 15°-20°C for a further 45 minutes then sampled for TLC. The mixture was then poured onto ice (4 liters)/water (8 liters) with stirring and this caused the required product to precipitate. After 10 minutes the stirrer was switched off and the solid allowed to settle. The aqueous liquor was decanted away and the solid isolated by filtration. The isolated solid was well washed with water (de-ionised, 4 liters), then dried over night in vacuo at 70°-75°C to give the title compound (585 g), m.p. 100°C.

2. 2-(2-Nitroanilino)-5-methylthiophene-3-carbonitrile

To a stirred slurry of sodium hydride (14.4 g, 50% dispersion in oil, 0.3 mol) in dry tetrahydrofuran (50 mL) under nitrogen was added, dropwise, a solution of 2-fluoronitrobenzene (28.2 g, 0.2 mol) and 2-amino-5-methylthiophene3-carbonitrile (27.6 g, 0.2 mol) in dry tetrahydrofuran (250 mL). The mixture was stirred at 25°C for 24 hours, poured onto cracked ice and extracted into dichloromethane (3 times 500 mL). The combined extracts were washed with 2 N hydrochloric acid (2 times 200 mL), water (2 times 200 mL), dried over magnesium sulphate and the solvent removed under reduced pressure. The residue was crystallised from ethanol to give the title compound, (35.2 g), m.p. 99°-102°C.

3. 4-Amino-2-methyl-10H-thieno[2,3-b][1,5]benzodiazepine, hydrochloride

To a stirred slurry of 2-(2-nitroanilino)-5-methylthiophene-3-carbonitrile (3 g, 0.011 mol) in ethanol (35 mL) at 50°C was added, over 10 minutes, a solution of anhydrous stannous chloride (6.95 g, 0.037 mol) in hydrochloric acid (26 mL, 5 M). The mixture was stirred under reflux for 1 hour, concentrated under reduced pressure and allowed to crystallise over night at 5°C. The salt was filtered, washed with a small amount of water, dried (4.3 g) m.p. >250°C, and used without further purification in the next stage.

4. 2-Methyl-10-(4-methyl-1-piperazinyl)-4H-thieno[2,3-b][1,5]-benzodiazepine

Crude 4-amino-2-methyl-10H-thieno[2,3-b][1,5]benzodiazepine, hydrochloride (4.3 g) was refluxed in a mixture of N-methylpiperazine (15 mL), dimethylsulfoxide (20 mL) and toluene (20 mL) under a nitrogen atmosphere for 20 hours. The mixture was cooled to ca. 50°C, water (20 mL) added, and the product allowed to crystallise at 5°C over night. The product was filtered and crystallised from acetonitrile (30 mL) to give the title compound (1.65 g) m.p. 195°C. The structure of the compound was confirmed spectroscopically.

References

Chakrabarti J.K. et al.; US Patent No. 5,229,382; Jul. 20, 1993; Assigned to Lilly Industries, Basingstroke, England

OLEANDOMYCIN

Therapeutic Function: Antibiotic

Chemical Name: Oleandomycin

Common Name: Troleandomycin

Chemical Abstracts Registry No.: 3922-90-5

Structural Formula:

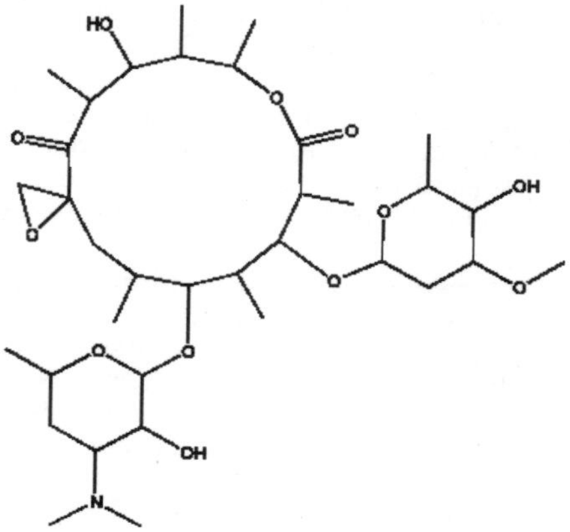

Trade Name	Manufacturer	Country	Year Introduced
Matromycin	Pfizer	US	1956
Oleandocyn	Pfizer	W. Germany	-
Oimicina	Morgan	Italy	-
Sigmamycin	Pfizer	Japan	-
Taocin-O	Sankyo	Japan	-
TAO	Roerig	US	-
Triolmicina	Ripari-Gero	Italy	-

Raw Materials

Bacterium Streptomyces antibioticus
Dextrose
Soybean meal

Manufacturing Process

A slant of S. antibioticus ATCC 11891 was cultivated on agar under controlled conditions in order to develop spores for the purpose of inoculating a nutrient medium having the following composition: 20 g Cerelose (dextrose hydrate), 15 g soybean meal, 5 g distillers' solubles, 10 g cornmeal, and tap water, in a sufficient amount for a 1,000 ml solution, adjusted to pH 7.0 to 7.2 with potassium hydroxide.

After the pH was adjusted, 5 g of calcium carbonate was added. This inoculum medium was then subjected to heat sterilization. The medium was then cooled and 2 ml of a spore suspension of an oleandomycin-producing strain of S. antibioticus was added under aseptic conditions. The cultivation of the organism was conducted in shaken flasks at 28% for a period of 48 hours.

The mixture of broth and mycelium thus formed was then transferred under aseptic conditions to a 3-liter fermentor containing 2,000 ml of a sterile fermentation medium having the following composition: 60 g Cerelose (dextrose hydrate), 18 g soybean meal, 5 g distillers' solubles, 12 g cornmeal and tap water in a sufficient amount for a 1,000 ml total volume, adjusted to pH 7.0 to 7.2 with potassium hydroxide.

After the pH had been adjusted, 5 g of calcium carbonate, 5 ml of soybean oil antifoam and 0.020 g of Acridine Orange dye were added. The mixture was then autoclaved at 20 psi (250°F) for 15 minutes in order to sterilize the contents, before transferring the broth and mycelium thereto.

After seeding the nutrient medium with the preformed inoculum previously described, the mixture was subjected to agitation and aeration under aseptic conditions for 72 hours; at 27°C to 28°C for the first 24 hours, then at 25°C to 26°C for the next 48 hours; during this period, the pH was in the range of 6.4 to 6.8. Aeration was accomplished by cultivation under submerged conditions at an air flow rate of one volume of air per volume of medium per minute. After termination of the process, the mycelium was removed by filtration and the filtered broth found to contain 450 γ of oleandomycin per ml of solution.

References

Merck Index 6703
Kleeman & Ensel P. 657
I.N. p. 701
Sobin, B.A., Routien, J.B. and Lees, T.W.; US Patent 2,757,123; July 31, 1956; assigned to Chas. Pfizer & Co., Inc.
Ratajak, E.J. and Nubel, R.C.; US Patent 2,842,481; July 8, 1958; assigned to Chas. Pfizer & Co., Inc.

OLOPATADINE

Therapeutic Function: Antiallergic

Chemical Name: 11-((Z)-3-(Dimethylamino)propylidene)-6,11-dihydrodibenz[b,e]oxepin-2-acetic acid

Common Name: Doxepadine; Olopatadine

Structural Formula:

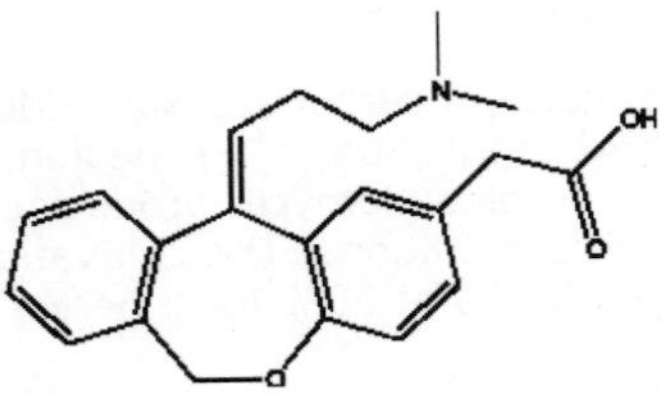

Trade Name	Manufacturer	Country	Year Introduced
Patanol	Alcon	UK	-

Chemical Abstracts Registry No.: 113806-05-6

Raw Materials

Phthalide
$POCl_3$
Butyl lithium
Sodium permanganate
Magnesium
Dibromoethane
p-Hydroxyphenyl acetic acid
Boron trifluoride-ethylether complex
Methyltriphenylphosphonium bromide
Triphenylchloromethane
3-Dimethylaminopropyl chloride
Trifluoroacetic anhydride

Manufacturing Process

402.4 g of phthalide and 200 g of sodium chloride and equal molecular quantity of p-hydroxyphenyl acetic acid are mixed with one another and stirred at 150°C for 6 hours. After completion of the reaction, the mixture is cooled until the temperature is brought back to room temperature, 4 L of aqueous 10% acetic acid solution is added thereto and the mixture is allowed to stand at room temperature overnight. After stirring the mixture at room temperature for 3 hours, deposited crystals are separated by filtration, and 6 L of water is added thereto. After stirring the mixture at room temperature for 30 minutes, the deposited crystals are separated by filtration. After the addition of 3 L of toluene to the crystals, the mixture is stirred at room temperature for one hour. The crystals are separated by filtration and dried over heating under reduced pressure to yield of 2-(4-acetoxyphenoxy)benzoic acid.

266.0 g of trifluoroacetic anhydride is added to the equal molecular quantity of 2-(4-acetoxyphenoxy)benzoic acid suspended in 5.0 L of methylene chloride and thereto. After stirring the mixture at room temperature for one hour, 19.4 g of boron trifluoride-ethylether complex is added thereto and the mixture is stirred at room temperature for two hours. The reaction solution is poured into ice water. After an organic solvent layer is separated from the mixture, the organic layer is washed with diluted aqueous sodium hydroxide solution and water, dried over anhydrous magnesium sulfate and concentrated under reduced pressure to obtain 335.3 g of methyl 11-oxodibenz[b,e]oxepin-2-carboxylate as a white crystal melting point 130°-132°C

Methyl 11-methylene-6,11-dihydrodibenz-[b,e]oxepin-2-acetate.

In 100 ml of tetrahydrofuran is suspended 25 g of methyltriphenylphosphonium bromide and 40 ml of 1.6 N n-butyl lithium hexane solution is dropwise added thereto under a nitrogen atmosphere and ice-cooling. After stirring the mixture under ice-cooling for 30 minutes, a solution obtained by dissolving equal molar quantity of 11-oxo-6,11-dihydrodibenz[b,e]oxepin-2-acetic acid in 250 ml of tetrahydrofuran is dropwise added thereto and the mixture is stirred at room temperature for two hours. The solvent is distilled away under reduced pressure and the residue is purified by column chromatography on silica gel (eluent: hexane:ethyl acetate = 3:1) to obtain the desired product as a colorless oily matter.

(11-(3-Dimethylaminopropylidene)-2-(2-triphenylmethyloxymethyl)-6,11-dihydrodibenz[b,e]oxepin.

Process A: 11-Hydroxy-2-(2-hydroxyethyl)-6,11-dihydrodibenz [b,e]oxepin

In this process, 20 g of methyl 11-oxo-6,11-dihydrodibenz[b,e]oxepin-2-acetate is dissolved in 500 ml of tetrahydrofuran. To the solution is added 6.0 g of lithium aluminum hydride and the mixture is stirred at room temperature for one hour. After decomposing an excess of the reagent by the addition of water to the solution, the mixture is filtered to remove an inorganic salts and the filtrate is concentrated to dryness under reduced pressure to obtain 17.7 g of the desired product as a white solid. Melting point: 132°-136°C.

Process B: 11-Hydroxy-2-(2-triphenylmethyloxyethyl)-6,11-dihydrodibenz[b,e]oxepin

In this process, 17.2 g of 11-hydroxy-2-(2-hydroxyethyl)-6,11-dihydrodibenz[b,e]oxepin is dissolved in 50 ml of pyridine. To the solution is added 30 g of triphenylchloromethane and the mixture is stirred at 50°C for 5 hours. After adding water and stirring the mixture for 2 hours, the solvent is distilled away under reduced pressure. The mixture is extracted with 1000 ml of ethyl acetate, washed with saturated aqueous sodium chloride solution, and dried over anhydrous sodium sulfate. The solvent is distilled away under reduced pressure and the resultant residue is purified by column chromatography on silica gel (eluent: hexane:ethyl acetate = 3:1) to obtain 21.7 g of the desired product as a colorless amorphous.

Process C: 11-Oxo-2-(2-triphenylmethyloxyethyl)-6,11-dihydrodibenz[b,e]oxepin

In this process, 10 g of 11-hydroxy-2-(2-triphenylmethyloxyethyl)-6,11-dihydrodibenz[b,e]oxepin is dissolved in a solution comprising 800 ml of acetone, 1000 ml of water, 20 ml of saturated aqueous magnesium sulfate solution and 0.2 g of disodium phosphate. To the solution is dropwise added 2.6 g of aqueous sodium permanganate solution and the mixture is stirred at room temperature for 4.5 hours. Then, 100 ml of methanol is added thereto and the mixture is heated at reflux for 3 hours. After allowing the mixture to stand for cooling, the mixture is filtered and the filtrate is extracted with 1000 ml of ethyl acetate, washed with saturated aqueous sodium chloride solution and dried over anhydrous sodium sulfate. The solvent is distilled away under reduced pressure and the resultant crude product is recrystallized from isopropanol to obtain 8.0 g of the desired product having melting point of 132°-134°C as a white crystal.

Process D: 11-(3-Dimethylaminopropyl)-11-hydroxy-2-(2-triphenylmethyloxyethyl)-6,11-dihydrodibenz[b,e]oxepin

To a solution of 3-dimethylaminopropyl magnesium chloride obtained by reacting 0.2 g of magnesium with 1.0 g of 3-dimethylaminopropyl chloride in 10 ml of tetrahydrofuran under a nitrogen atmosphere using dibromoethane as a catalyst, is dropwise added a solution obtained by dissolving 2.0 g of 11-

oxo-2-(2-triphenylmethyloxyethyl)-6,11-dihydrodibenz[b,e]oxepin in 10 ml of tetrahydrofuran under ice cooling and the mixture is stirred at room temperature for one day. Aqueous ammonium chloride solution is added thereto and the pH of the mixture is adjusted to 7.0 with aqueous 4 N hydrochloric acid solution. The solvent is distilled away under reduced pressure. The mixture is extracted with 200 ml of methylene chloride and washed with saturated aqueous sodium bicarbonate solution and saturated aqueous sodium chloride solution in order. After drying the extract over anhydrous sodium sulfate, the solvent is distilled away under reduced pressure. The resultant residue is purified by column chromatography on silica gel (eluent: hexane:ethyl acetate:triethylamine = 10:10:1) to obtain 1.2 g of the desired product as a colorless amorphous.

Process E: 11-(3-Dimethylaminopropylidene)-2-(2-triphenylmethyloxyethyl)-6,11-dihydrodibenz[b,e]oxepin

In this process, 1.2 g of 11-(3-dimethylaminopropyl)-11-hydroxy-2-(2-triphenylmethyloxyethyl)-6,11-dihydrodibenz[b,e]oxepin is dissolved in 50 ml of pyridine. To the solution is dropwise added 0.8 g of phosphorusoxychloride under a nitrogen atmosphere and ice-cooling. After stirring the mixture at room temperature for one hour, the solvent is distilled away under reduced pressure. The residue is extracted with 100 ml of methylene chloride, and washed with saturated aqueous sodium bicarbonate solution and saturated aqueous sodium chloride solution in order. After drying the mixture over anhydrous sodium sulfate, the solvent is distilled away under reduced pressure. The resultant residue is purified by column chromatography on silica gel (eluent: hexane:ethylacetate:triethylamine = 10:10:1) to obtain 0.82 g of the desired product as a colorless oily matter.

11-(3-Dimethylaminopropylidene)-2-(2-hydroxyethyl)-6,11-dihydrodibenz[b,e]oxepin

0.92 g of 11-(3-dimethylaminopropylidene)-2-(2-triphenylmethyloxyethyl)-6,11-dihydro dibenz[b,e]oxepin is dissolved in a mixed solvent of 20 ml of water and 20 ml of dioxane. To the solution is added 60 mg of p-toluene sulfonic acid and the mixture is heated at reflux for two hours. The solvent is distilled away under reduced pressure and the residue is extracted with 200 ml of ethylacetate, washed with saturated aqueous sodium bicarbonate solution and saturated aqueous sodium hydrochloride solution in oder and dried over anhydrous sodium sulfate. The solvent is distilled away under reduced pressure. The resultant residue is purified by column chromatography on silica gel (eluent: ethylacetate triethylamine = 10:1) to obtain 0.4 g of the desired product. Cis form white solid. Melting point: 100°-102°C (diethylether).

11-(3-Dimethylaminopropylidene)-6,11-dihydrodibenz [b,e]oxepin-2-acetic acid

2.2 g of 11-(3-dimethylaminopropylidene)-2-(2-hydroxyethyl)-6,11-dihydrodibenz[b,e]oxepin is dissolved in 100 ml of acetone. The Jones reagent ($Na_2Cr_2O_7$+ H_2SO_4) is added to the solution until the reaction solution shows an orange color and the mixture is stirred at room temperature for one hour. Sodium bicarbonate is added thereto and an inorganic substance is removed by filtration. The solvent of the filtrate is distilled away under reduced pressure to obtain the desired product, 11-(3-Dimethylaminopropylidene)-6,11-dihydrodibenz[b,e]oxepin-2-acetic acid. Cis form white crystal. Melting point: 118°-120°C (Isopropanol).

References

Oshima et al.; US Patent No. 5,116,863; May 26, 1992; Assigned to Kyowa Hakko Kogyo Co., Ltd., Tokyo, Japan

Hayakawa E. et al.; US Patent No. 5,641,805, Jun. 24, 1997; Assigned to Alcon Laboratories, Inc. (Fort Worth, TX); Kyowa Hakko Kogyo Co. Ltd. (Tokyo, JP)

OMEPRAZOLE

Therapeutic Function: Antiulcer

Chemical Name: 1H-Benzimidazole, 5-methoxy-2-(((4-methoxy-3,5-dimethyl-2-pyridinyl)methyl)sulfinyl)-

Common Name: Omeprazole

Structural Formula:

Chemical Abstracts Registry No.: 73590-58-6

Trade Name	Manufacturer	Country	Year Introduced
Fordex	Leti	-	-
Gastrium	Ache	-	-
Gastrotem	Temis Lostalo	-	-
Helicid 10	Leciva	Czech Republic	-
Helicid 11	Colins Laboratories Ltd.	India	-
Lenar	Chemica	-	-
Lomak	Cipla Limited	India	-
Losec	Astra	Sweden	-
Losec MUPS	AstraZeneca	Sweden	-
Lozol	Medinfar	-	-
Norpramin	Cepa	-	-
Ocid	Cadila Healthcare	India	-
Omegast	Lifesource Healthcare	India	-
Omegast	Agio Pharmaceuticals Ltd.	India	-
Omepar	Nabros Pharma	India	-

Trade Name	Manufacturer	Country	Year Introduced
Omeprazole	Balkanpharma	Bulgaria	-
Omeprazole	New Life Pharmaceuticals	India	-
Omeprazole	Chemo Iberica	Spain	-
Omeprazole	Aurobindo	India	-
Omeprazole-Richter	Lyconsa	Spain	-
Omeprol	Zdravle	Yugoslavia	-
Omez	Dr. Reddy's Laboratories Ltd.	India	-
Omizac	Torrent	India	-
OMZ	Dexa Medica	-	-
Osiren	Chemia	-	-
Pamoxan	Italmex	-	-
Pepticum	Laboratorios Andromaco	Spain	-
Prazol	Tecnoquimicos	-	-
Proseptin	Beximco	-	-
Sanamidol	Inkeysa	-	-
Ulceral	Tedec-Meiji	-	-
Ultop	Krka	Slovenia	-
Ulzol	Pliva	Horvatia	-
Zerocid	Sun Pharmaceuticals Industries Ltd.	India	-
Zolcer	Aurobindo	India	-

Raw Materials

3,5-Lutidine
Dimethyl sulfate
Thionyl chloride
Phthalic anhydride
Hydrogen peroxide
Ammonium persulfate
5-Methyl-2-mercaptobenzimidazole
Benzylammonium chloride

Manufacturing Process

3,5-Lutidine-N-Oxide

Hydrogen peroxide (45%, 200 ml) was added dropwise at 60°-70°C during 2 hours to a mixture of 3,5-Lutidine (125 g, 1.16 mole) and acetic acid (400 ml). The mixture was heated to 90°C and maintained at 90°-100°C for 2 hours after which it was cooled to 60°C. Again hydrogen peroxide (45%, 200 ml) was added dropwise at 60°-70°C during 1 hour and then the mixture was heated to 90°C and maintained at 90°-100°C for 6 hours. Thereafter, acetic acid and water was distilled off under reduced pressure and the distillation residue obtained was used as a starting product for the nitration.

3,5-Dimethyl-4-nitropyridine-N-oxide

To the distillation above obtained residue was added sulphuric acid (146 ml). Thereafter, a nitrating mixture consisting of sulphuric acid (250 ml) and nitric acid (280 ml) was added dropwise during 4 hours at 90°-100°C. The reaction mixture was heated further at 90°-100°C for 6 hours, after which it was cooled and poured over crushed ice (4 kg), Caustic lye (50%, 1150 ml) was added to the yellow solution and the precipitated crystalline compound was filtered under suction. The cake was washed with water and dried in vacuuo oven to yield the product which melted at 171°-173°C. Yield 78.5%. A sample crystallized from acetone had a melting point of 174°-174.5°C.

3,5-Dimethyl-4-nitropyridine-N-oxide-dimethyl sulfate adduct

To a suspension of 3,5-dimethyl-4-nitropyridine-N-oxide (150 g, 0.80 mole) in acetone (450 ml) was added dimethyl sulfate (90 ml, 0.95 mole). The mixture was heated to reflux until a clear solution was obtained and then allowed to cool to ambient temperature. An off-white crystalline solid separated out, which was filtered, washed with acetone and dried to yield 220 g of the adduct. Yield was 83.8% of theoretical.

3,5-Dimethyl-2-hydroxymethyl-4-nitropyridine

3,5-Dimethyl-4-nitropyridine-N-oxide-dimethyl sulfate adduct (220 g, 0.75 mole) was dissolved in methanol (1.0 ltr) and the solution heated to reflux. A solution of ammonium persulfate (140 gm) in water (200 ml) was added dropwise over 4 hours after which reflux was continued for 4 hours. Methanol was distilled off under reduced pressure and the residue was basified to pH 10 by addition of caustic lye (105 ml). The mixture was extracted with dichloromethane (2 times 400 ml). The dichloromethane layer was dried over sodium sulfate and filtered. The product was used as its solution in dichloromethane for the next reaction.

2-Chloromethyl-3,5-dimethyl-4-nitropyridine hydrochloride

To the cooled dichloromethane solution of 3,5-dimethyl-2-hydroxymethyl-4-nitropyridine was added thionyl chloride (60 ml, 0.85 mole) dropwise over a period of 2 hours and stirring was continued for a further 2 hours. Methanol (10 ml) was added to destroy excess thionyl chloride and separated product was filtered under suction and washed with dichloromethane. The cake was dried in vacuum oven to yield 55 g of a cream colored product. Melting point was 124°-126°C.

5-Methoxy-2-[(3,5-dimethyl-4-nitro-2-pyridinyl)methylthio]-1H-benzimidazole

To a suspension of 5-methyl-2-mercaptobenzimidazole (36 g, 0.2 mole), 2-chloromethyl-3,5-dimethyl-4-nitropyridine hydrochloride (47.4 g, 0.2 mole) and triethyl benzylammonium chloride (5 g) in a dichloromethane (500 ml) was added dropwise a solution of NaOH (17.6 gm, 0.44 mole) in water (30 ml). The addition was exothermic and the temperature was observed to rise to 40°C with reflux of dichloromethane - the reaction mixture was stirred for further 6 hours at ambient temperature and filtered. The cake was washed with water and dried in vacuum oven to yield 55.8 g of cream color product. Yield 81.1%; melting point 124°-128°C.

5-Methoxy-2-[(3,5-dimethyl-4-methoxy-2-pyridinyl)methylthio]-1H-benzimidazole

5-Methoxy-2-[(3,5-dimethyl-4-nitro-2-pyridinyl)methylthio]-1H-benzimidazole(50 g, 0.145 mole) was dissolved in methanol and heated to 45°C. A solution of sodium methoxide (50 g, 0.925 mole) in methanol (150 ml) was added dropwise over a period of 3 hours at 45°-60°C. Stirring was continued for another 2 hours and then methanol was distilled off under reduced pressure. To the cooled residue was added water (200 ml) followed by concentrated HCl (65 ml) until the pH of the mixture was 7.5. The reaction mixture was extracted with dichloromethane and the dichloromethane layer was washed with water (2 times 100 ml). The dichloromethane layer was dried over sodium sulfate and concentrated to yield the product as an amber color syrup. Yield was 40.1 gm, about 83.8% of theoretical. A solid sample was obtained by trituration of the syrup several times with petroleum ether. Melting point was 87°-90°C.

5-Methoxy-2-[(3,5-dimethyl-4-methoxy-2-pyridinyl)methylthio]-1H-benzimidazolehydrochloride

HCl gas was bubbled into a cooled solution of 5-methoxy-2-[(3,5-dimethyl-4-methoxy-2-pyridinyl)methylthiol]-1H-benzimidazole (50 g) in dichloromethane (250 ml) until no more precipitation was observed. The reaction mixture was warmed to 40°C and again cooled to 10°C. The solid was filtered under suction and washed with dichloromethane to yield the product (49 g) as a cream colored fine granular solid. Yield was 88.2% of theoretical. Melting point 144°-148°C.

Omeprazole from 5-methoxy-2-[(3,5-dimethyl-4-methoxy-2-pyridinyl)methylthiol]-1H-benzimidazole

To a solution of 5-methoxy-2-[(3,5-dimethyl-4-methoxy-2-pyridinyl)methylthio]-1H-benzimidazole (32.9 g, 0.1 mole) in dichloromethane (200 ml) was added phthalic anhydride (20 g, 0.135 mole) and cooled in an ice salt bath. This was followed by addition of sodium carbonate (18 g, 0.17 mole) and water (20 ml). Hydrogen peroxide (12 ml, 45%, 0.16 mm mole) was added dropwise at -5°-0°C and the reaction mixture was stirred at the same temperature. When the reaction was complete as indicated by TLC, water (200 ml) was added, cooling bath was removed and the reaction mixture was stirred for 10 mins. The organic layer was separated and washed with 5% sodium carbonate solution. The separated dichloromethane solution was charcoalised and filtered through celite. The filtrate was concentrated to 100 ml and ethyl acetate 100 ml was added thereto. The separated solid was filtered, washed with ethyl acetate and dried in vacuum oven to yield 28.20 g of omeprazole. Yield 82.4% of theoretical. Melting point was 158°-160°C (dec.).

References

Singh S. et al.; US Patent No. 6,245,913 B1; Jun. 12, 2001; Assigned to Wockhardt Europe Limited, Dublin (IE)

ONDANSETRON HYDROCHLORIDE DIHYDRATE

Therapeutic Function: Serotonin antagonist

Chemical Name: 4H-Carbazol-4-one, 1,2,3,9-tetrahydro-9-methyl-3-((2-methyl-1H-imidazol-1-yl)methyl)-, hydrochloride, hydrate (1:1:2)

Common Name: Ondansetron hydrochloride dihydrate

Structural Formula:

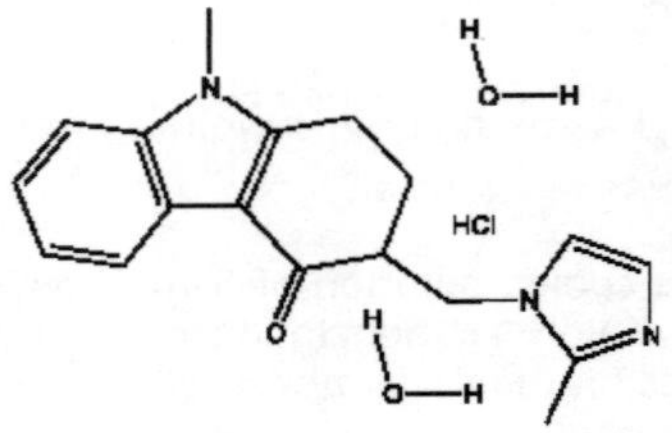

Chemical Abstracts Registry No.: 103639-04-9; 99614-02-5 (Base)

Trade Name	Manufacturer	Country	Year Introduced
Ondansetron hydrochloride	Chemo Iberica	Spain	-
Zofran	GlaxoSmithKline	UK	-

Raw Materials

Diethyl oxalate	9-Methyl-1,2,3,9-tetrahydro-4H-carbazol-4-one
Sodium	Formol solution
2-Methylimidazole	

Manufacturing Process

Preparation of 3-ethoxalyl-9-methyl-1,2,3,9-tetrahydro-4H-carbozol-4-one

3.0 g (0.13 mole) of sodium metal are portionwise added to a stirred mixture containing 19.93 g (0.1 mole) of 9-methyl-1,2,3,9-tetrahydro-4H-carbazol-4-one, 19.0 g (0.13 mole) of diethyl oxalate, 2 g of ethanol and 200 ml of dioxane. The slightly warming reaction mixture is stirred at 40° to 50°C for 4 hours, then 16 g of glacial acetic acid and finally 200 ml of water are added thereto at room temperature. After filtering off the yellow crystalline suspension, the precipitate is washed with water and dried to give the title compound in a yield of 24 g (80.2%), m.p. 118°-120°C.

Preparation of 3-hydroxymethyl-9-methyl-1,2,3,9-tetrahydro-4H-carbazol-4-one-3-glyoxylic acid lacton

After adding 0.1 g of triethylamine to a stirred suspension containing 3.00 g (0.01 mole) of the 3-ethoxalyl-9-methyl-1,2,3,9-tetrahydro-4H-carbazol-4-

one, in 20 ml of acetone, 1.13 g (0.015 mole) of formol solution are dropwise added to the mixture. The suspension becomes clear within 1 to 2 minutes and crystals begin to precipitate. After further stirring at 35° to 40°C for one hour, the reaction mixture is cooled down to room temperature, filtered off, the precipitate is washed with 50% acetone and dried to give 2.10 g (74.2%) of the title compound, m.p. 242°-244°C.

Preparation of ondansetron base (chemically 9-methyl-3-[(2-methyl-1-H-imidazol-1-yl)methyl]-1,2,3,9-tetrahidro-4-H-carbazol-4-one)

A mixture containing 2.83 g (0.01 mole) of 3-hydroxymethyl-9-methyl-2,3,9-tetrahydro-4H-carbazol-4-one-3-glyoxylic acid lactone, 15 ml of dioxane, 1.32 g of triethylamine, 1.0 g of ethanol and 1.64 g (0.02 mole) of 2-methylimidazole is boiled under reflux while stirring for 5 hours. Thereafter, the reaction mixture is diluted with 45 ml of water and cooled down. The precipitate is filtered off, washed with aqueous dioxane and dried to obtain 2.56 g (87.3%) of the title compound, m.p. 220°-223°C.

Preparation of 9-methyl-3-[(2-methyl-1-H-imidazol-1-yl)methyl]-1,2,3,9-tetrahydro-4H-carbazol-4-one hydrochloride dihydrate The process above described is followed, except that after cooling down the reaction mixture to room temperature after boiling, 20 ml of 37% aqueous hydrochloric acid are added thereto. Then, the precipitate is filtered off, washed with isopropanol and dried to obtain 2.40 g (65.6%) of the title salt, m.p. 178°-180°C. The active agent content of the product was found to be 100.3% based on potentiometric titration with sodium hydroxide solution. The theoretical water content is 9.85% (calculated for $C_{18}H_{19}N_3OHCl_2H_2O$).The water content measured is 10.03%.

References

Bod P. et al.; US Patent No. 5,478,949; Dec. 26, 1995; Assigned to Richter Gedeon Vegyeszeiti Gyar Rt., Budapest, Hungary

OPIPRAMOL

Therapeutic Function: Antidepressant, Antipsychotic

Chemical Name: 4-[3-(5H-Dibenz[b,f]azepin-5-yl)propyl]-1-piperazine-ethanol

Common Name:-

Chemical Abstracts Registry No.: 315-72-0; 909-39-7 (Dihydrochloride salt)

Raw Materials

5-(3-Toluene-p-sulfonyloxypropyl)dibenzazepine
1-(2-Hydroxyethyl)piperazine

Structural Formula:

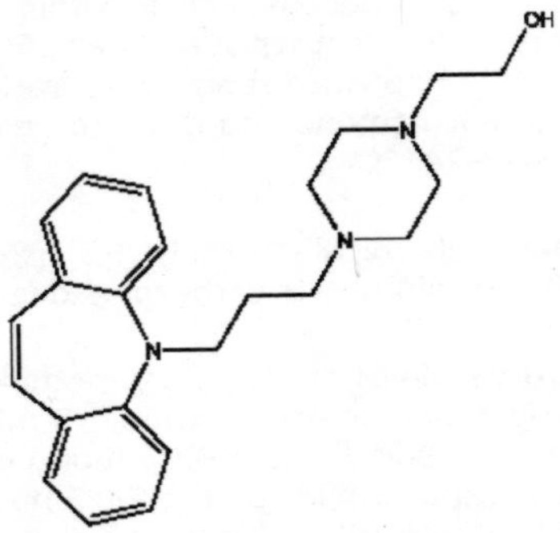

Trade Name	Manufacturer	Country	Year Introduced
Insidon	Geigy	W. Germany	1962
Insidon	Geigy	France	1962
Insidon	Geigy	Italy	1962
Deprenil	Yurtoglu	Turkey	-
Ensidon	Ciba Geigy	US	-
Oprimol	Taro	Israel	-
Pramolan	Polfa	Poland	-

Manufacturing Process

A solution of 5-(3-toluene-p-sulfonyloxypropyl)dibenzazepine (9.2g) and 1-(2-hydroxyethyl)piperazine (8.6g) in anhydrous toluene (50 cc) is heated at boiling point under reflux for 4 hours.

After cooling, distilled water (75 cc) is added. The aqueous phase is decanted. The toluene solution is washed with distilled water (25 cc) and then extracted with N hydrohloric acid (40 cc). The hydrochloric acid solution is made alkaline to phenolphthalein with sodium hydroxide (d = 1.33).The base which separates is extracted with chloroform (50 cc). The chloroform solution is dried over anhydrous sodium sulfate and then evaporated to dryness. There are obtained 5-[3-(4-β-hydroxyethylpiperazino)propyl]dibenzazepine (7.95g), the dihydrochloride of which, crystallized from ethanol, melts at about 210°C.

References

Merck Index 6727
Kleeman & Ensel P. 657
I.N. p. 703
Gaillot, P. and Gaudechon, J.; British Patent 881,398; November 1, 1961; assigned to Societe des Usines Chimiques Rhone-Poulenc

ORAZAMIDE

Therapeutic Function: Hepatoprotectant

Chemical Name: 5-Aminoimidazole-4-carboxamide orotate

Common Name: AICA orotate

Structural Formula:

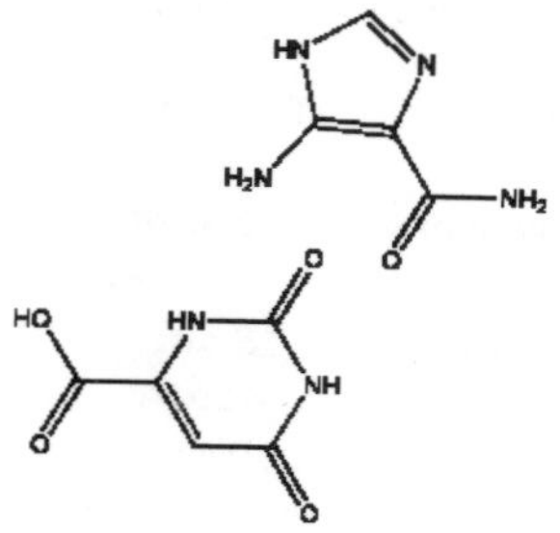

Chemical Abstracts Registry No.: 2574-78-9

Trade Name	Manufacturer	Country	Year Introduced
Aicamine	Labaz	France	1971
Aicurat	Mack	W. Germany	1962
Aicamin	Crinos	Italy	1977
Aicamin	Fujisawa	Japan	-

Raw Materials

4-Amino-5-imidazolecarboxamide
Orotic acid

Manufacturing Process

14.4 grams of 4-amino-5-imidazolecarboxamide (monohydrate) and 17.4 grams of orotic acid (monohydrate) were dissolved with heating in 600 cc of water. The solution is decolorized with Norit, cooled and then filtered off. 28.8 grams of a white crystalline salt (dihydrate) is obtained with MP 284°C (decomposition).

References

Merck Index 6739
Kleeman & Engel p. 658
I.N. p. 704
Haraoka, R. and Kamiya, T.; US Patent 3,271,398; September 6, 1966; assigned to Fujisawa Pharmaceutical Co., Ltd., Japan

ORGOTEIN

Therapeutic Function: Antiinflammatory

Chemical Name: See Structural Formula

Common Name: Ormetein

Structural Formula: Orgotein is a complex protein with a molecular weight of about 33,000. It is a divalent metal (Mg, Cu, Zn) chelated structure

Chemical Abstracts Registry No.: 9016-01-7

Trade Name	Manufacturer	Country	Year Introduced
Ontosein	Gruenenthal	W. Germany	1980
Peroxinorm	Protochemie	Switz.	1982
Peroxinorm	Gruenenthal	Japan	1982
Oxinorm	Zambeletti	Italy	-

Raw Materials

Beef blood
Ethanol
Chloroform

Manufacturing Process

Fresh beef blood was centrifuged, e.g., at about 2,600 to 5,000 x g for 10 minutes at 0°C and the plasma decanted. The red blood cells were then washed at least twice and preferably repeatedly with 2 to 3 volumes of 0.9% saline solution. The washed red blood cells were lysed by mixing with 1.1 volumes of cold deionized water containing 0.02% detergent (Saponin). After a minimum of 30 minutes at 4°C with stirring, 0.25 volume (per volume of hemolysate) of ethyl alcohol at -15°C was slowly added while stirring followed by 0.31 volume (per volume of hemolysate) of chloroform, also at -15°C. Stirring was continued for about 15 minutes at -5°C or below, at which time, the mixture was a thick paste. The hemoglobin precipitation was carried out in a cold bath which was kept at below -10°C. After the paste had stood for a further 15 minutes at 4°C, 0.2 volume of cold 0.15 M NaCl solution was added, giving an easily poured suspension. The precipitate and excess chloroform were removed by centrifuging at about 12,000 to 20,000 x g at about -10°C for 10 minutes. The supernatant liquid was removed and if desired, filtered and briefly dialyzed against cold-deionized water, prior to lyophilization.

The alcohol chloroform precipitate was dislodged, chloroform was removed, the pellet broken up and reextracted with about an equal amount of deionized water by blending the precipitate and the water in a blender and thereafter centrifuging. The reextraction solution was dialyzed and lyophilized with the main extract. If the process proceeds normally, the reextraction of the precipitated hemoglobin usually yields up to 30% of protein mixture present in the original supernatant. An additional reextraction may give an additional

5 to 15%.

The lyophilized material was redissolved in 0.025 M tris-glycine buffer containing 0.001 M Mn^{2+} at pH 7.5 (usually to a concentration of 20 mg/ml). The solution was heated at or near 65°C for about 15 minutes. This step removes the carbonic anhydrase and other heat labile proteins from the solution. After heating, the solution was rapidly cooled in an ice bath to 5°C. The solution was then centrifuged at 20,000 x g at 0°C for 10 minutes to remove the precipitate. Filtration through "Versapore" works equally well. The supernatant was thoroughly dialyzed against deionized water to remove excess metal ions and buffer and then lyophilized. The resulting solid consists largely of orgotein.

References

Merck Index 6742
DOT 9 (1) 34 (1973) 11 (3) 103 (1975) and 13 (3) 105 (1977)
I.N. p. 705
Huber, W.; US Patent 3,579,495; May 18, 1971; assigned to Diagnostic Data, Inc.
Huber, W.; US Patent 3,687927; August 29, 1972; assigned to Diagnostic Data, Inc.

ORNIDAZOLE

Therapeutic Function: Antiinfective

Chemical Name: α-(Chloromethyl)-2-methyl-5-nitro-1H-imidazole-1-ethanol

Common Name: -

Structural Formula:

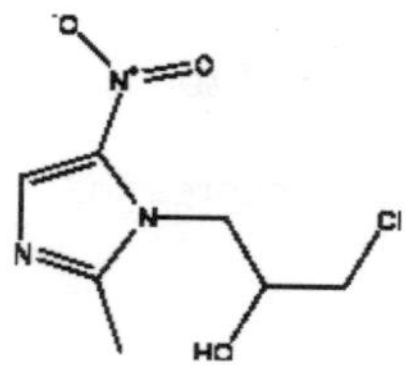

Chemical Abstracts Registry No.: 16773-42-5

Trade Name	Manufacturer	Country	Year Introduced
Tiberal	Roche	W. Germany	1977
Tiberal	Roche	Italy	1981
Tiberal	Roche	France	1981
Tiberal	Roche	Switz.	1982

Trade Name	Manufacturer	Country	Year Introduced
Tiberal	Roche	Australia	1983
Kolpicid	Roche	Sweden	1983
Madelen	Finadiet	Argentina	-
Ornidal	Selvi	Italy	-

Raw Materials

1-(2,3-Epoxypropyl)-2-methyl-5-nitroimidazole
Hydrogen chloride

Manufacturing Process

5g of 1-(2,3-epoxypropyl)-2-methyl-5-nitroimidazole was added to 30 ml of concentrated aqueous hydrochloric acid. The solution was heated to the boiling point for 20 minutes, chilled, diluted with 30 ml of water and carefully neutralized with ammonia to a pH of 7 to 8. It was then saturated with ammonium sulfate. The precipitated oil crystallized after several days. Recrystallized from toluene, there was obtained the 1-(3-chloro-2-hydroxypropyl)-2-methyl-5-nitroimidazole product melting at 77°C to 78°C.

References

Merck Index 6746
OCDS Vol. 3 p. 131 (1984)
DOT 11 (9) 369 (1975)
I.N. p. 706
REM p. 1224
Hoffer, M.; US Patent 3,435,049; March 25, 1969; assigned to Hoffmann-LaRoche, Inc.

ORNIPRESSIN

Therapeutic Function: Vasoconstrictor

Chemical Name: 8-L-Ornithinevasopressin

Common Name: -

Chemical Abstracts Registry No.: 3397-23-7

Trade Name	Manufacturer	Country	Year Introduced
POR-8	Sandoz	W. Germany	1977

Raw Materials

N-α-Carbobenzoxy-N-δ-toluenesulfonyl-L-ornithine
Glycine ethyl ester

N-Carbobenzoxy-L-proline
N-Carbobenzoxy-L-glutaminyl-L-asparaginyl-S-benzyl-L-cysteinyl-azide
N-Carbobenzoxy-S-benzyl-L-cysteinyl-L-tyrosyl-L-phenylalanine azide
Sodium
Ammonia

Structural Formula:

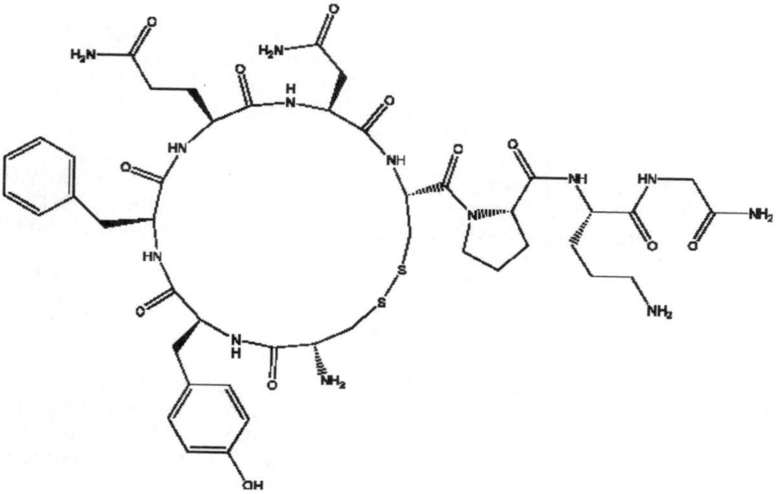

Manufacturing Process

(a)N-α-carbobenzoxy-N-δ-p-toluenesulfonyl-L-ornithyl-glycine ethyl ester: 104 g of N-α-carbobenzoxy-N-δ-p-toluenesulfonyl-L-ornithine and 27 g of glycine ethyl ester are dissolved in 450 cc of acetonitrile, the mixture is cooled at 0°C, 51 g of dicyclohexyl carbodiimide are added and the mixture is shaken at room temperature for 4 hours. Precipitated dicyclohexyl urea is filtered off and washed with acetonitrile. The whole filtrate is evaporated in a vacuum. The residue crystallizes after the addition of petroleum ether. After recrystallization from n-propanol, 93 g of N-α-carbobenzoxy-N-δ-toluenesulfonyl-L-ornithyl-glycine ethyl ester are obtained; melting point 136°C; $[\alpha]_D^{22}$ = -6.5° (96% ethanol).

(b) N-carbobenzoxy-L-prolyl-N-δ-p-toluenesulfonyl-L-ornithyl-glycinamide: 90 g of N-α-carbobenzoxy-N-δ-p-toluenesulfonyl-L-ornithyl-glycine ethyl ester are dissolved in 800 cc of anhydrous acetic acid which has been saturated with hydrogen bromide. The mixture is left to stand for one hour at 20°C, evaporated in a vacuum at a temperature below 40°C and the residue washed carefully with diethyl ether. The residue is dissolved in 500 cc of acetonitrile, 25 cc of triethylamine and 43 g of N-carbobenzoxy-L-proline are added, cooling is effected at 0°C, 355 g of dicyclohexyl carbodiimide are then added and the mixture shaken overnight at 20°C. After filtering off dicyclohexyl urea, the filtrate is evaporated in a vacuum at 30°C, the residue dissolved in ethyl acetate and this solution is washed with dilute sulfuric acid and aqueous ammonia. After drying over sodium sulfate, the ethyl acetate is removed by evaporation in a vacuum and the residue dissolved in 1 liter of absolute ethanol. The solution is cooled at 0°C, saturated with ammonia and left to stand overnight at 20°C. After evaporating in a vacuum at 30°C, the residue is recrystallized from dimethylformamide/ethyl acetate. 58 g of N-carbobenzoxy-L-prolyl-N-δ-p-toluenesulfonyl-L-ornithyl-glycinamide are obtained: melting point 122°C (with decomposition).

(c) N-carbobenzoxy-L-glutaminyl-L-asparaginyl-S-benzyl-L-cystsinyl-L-prolyl-N-δ-p-toluenesulfonyl-L-ornithyl-glycinamide: 100 g of N-carbobenzoxy-L-prolyl-N-δ-p-toluenesulfonyl-L-ornithyl-glycinamide are dissolved in 500 cc of anhydrous acetic acid which has been saturated with hydrogen bromide, the solution is left to stand for one hour at 20°C and is evaporated in a vacuum at a temperature below 40°C. The residue is carefully washed with diethyl ether and then added to a solution of 100 g of N-carbobenzoxy-L-glutaminyl-L-asparaginyl-S-benzyl-L-cysteinyl-azide and 26 cc of triethylamine in 1,000 cc of dimethylformamide. The mixture is left to stand overnight at 20°C, 3,000 cc of ethyl acetate are added thereto, the precipitate is filtered off and washing is effected with ethyl acetate. 105 g of N-carbo-benzoxy-L-glutaminyl-L-asparaginyl-S-benzyl-L-cysteinyl-L-prolyl-N-δ-p-toluenesulfonyl-L-ornithyl-glycinamide are obtained; melting point 193°C; $[\alpha]_D^{20}$ = -38.5°(dimethylformamide).

(d) N-carbobenzoxy-S-benzyl-L-cysteinyl-L-tyrosyl-L-phenyl-alanyl-L-glutaminyl-L-asparaginyl-S-benzyl-L-cysteinyl-L-prolyl-N-δ-p-toluenesulfonyl-L-ornithyl-glycinamide: 50 g N-carbobenzoxy-L-glutaminyl-L-asparaginyl-S-benzyl-L-cysteinyl-L-prolyl-N-δ-p-toluenesulfonyl-L-ornithyl-glycinamide are dissolved in 250 cc of anhydrous acetic acid which has been saturated with hydrogen bromide and the solution is left to stand for one hour at 20°C. After evaporating the solvent in a vacuum at a temperature below 40°C, the residue is carefully washed with diethyl ether and a solution of 31.5 g of N-carbobenzoxy-S-benzyl-L-cysteinyl-L-tyrosyl-L-phenylalanine-azide and 7.5 cc of triethylamine in 250 cc of dimethylformamide is added thereto. The mixture is left to stand for 2 days at 20°C. 1,000 cc of ethyl acetate are subsequently added and the precipitate is washed with ethyl acetate. After drying in a vacuum at 30°C. the product is washed with warm methanol. 45 g of N-carbobenzoxy-S-benzyl-L-cysteinyl-L-tyrosyl-L-phenylalanyl-L-glutaminyl-L-asparaginyl-S-benzyl-L-cysteinyl-L-prolyl-N-δ-p-toluenesulfonyl-L-ornithyl-glycinamide are obtained; melting point 224°C.

(e) L-cysteinyl-L-tyrosyl-L-phenylalanyl-L-glutaminyl-L-asparaginyl-L-cysteinyl-L-prolyl-L-ornirhyl-glycinamide: The necessary amount of sodium or potassium metal is added to a solution of 5 g of N-carbobenzoxy-S-benzyl-L-cysteinyl-L-tyrosyl-L-phenylalanyl-L-glutaminyl-L-asparaginyl-S-benzyl-L-cysteinyl-L-prolyl-N-δ-p-toluenesulfonyl-L-ornithyl-glycinamide in 1,200 cc of dry liquid ammonia, while stirring at the boiling temperature of the solution, to give a stable blue coloration. After the addition of 3 g of ammonium chloride, the solution is evaporated to dryness. The residue contains L-cysteinyl-L-tyrosyl-L-phenyl-alanyl-L-glutaminyl-L-asparaginyl-L-cysteinyl-L-prolyl-L-ornithyl-glycinamide.

References

Merck Index 6747

DOT 13 (11) 498 (1977)
I.N. p. 706
Boissonnas, R. and Huguenin, R.; US Patent 3,299,036; January 17, 1967; assigned to Sandoz Ltd. (Switzerland)

OROTIC ACID

Therapeutic Function: Hepatoprotectant

Chemical Name: 4-Pyrimidinecarboxylic acid, 1,2,3,6-tetrahydro-2,6-dioxo-

Common Name: Acide orotique; Acidum oroticum; Animal galactose factor; Orotic acid; Vitamin B13; Whey factor

Structural Formula:

Chemical Abstracts Registry No.: 65-86-1

Trade Name	Manufacturer	Country	Year Introduced
Lactinium	Roland	-	-

Raw Materials

Ketene	Trichloroacetyl chloride
Urea	Acetic acid
Sulfuric acid	Sodium hydroxide

Manufacturing Process

91.7 g (0.5 mol) of trichloroacetyl chloride was cooled to -35°C in a glass vessel by means of a cooling brine. In the course of 3 h, 27.0 g (0.06 mol) of pure ketene was introduced through a tube. After completion of the reaction, the vessel was immediately put under dry nitrogen to prevent penetration of moisture. So γ,γ,γ-trichloroacetoacetylchloride was produced.

The reaction mixture containing the γ,γ,γ-trichloroacetoacetyl chloride was transfered under nitrogen to a dropping funnel and in the course of 15 min was added with vigorous agitation to a suspension of 69.0 g (1.15 mole) of urea in 90.0 g of anhydrous acetic acid. Water cooling was used so that the reaction temperature would not exceed 40°C. After completion of the addition, the reaction mixture was heated as rapidly as possible to 115°C, and held at this temperature for 30 min.

Subsequently there was cooling and one more 99.0 g of glacial acetic acid and 180.0 g of water were added. The precipitated 6-trichloromethyluracil was filtered off and dried at 60°C in a vacuum drying cabinet. The yield was 91.0 g or 80%.

In a glass vessel equipped with an agitator, thermometer and pH electrode, 500 ml of water was placed and heated to 80°C. 50 g of 6-trichloromethyluracil was then added. By means of the pH electrode, the addition of sodium hydroxide was automatically controlled so that the pH value throughout the whole hydrolysis was 6.5. Into, 165 ml of 5 N NaOH was consumed. Finally, the hydrolysis solution was cooled and the precipitated sodium orotate filtered off.

The crude sodium orotate was again suspended at 80°C in water and brought into solution (pH 10.5) by addition of 30 ml of 5 N NaOH. After treatment with active charcoal, the solution was acidified with 30.0 g of 50% sulfuric acid. The solution was then cooled. The orotic acid was filtered off and carefully washed with water. After drying, 20.5 g of orotic acid, having a purity of 99.3% (titration) was obtained. This corresponds to a 60% yield.

References

Jackson B; US Patent No. 4,064,126; Dec. 20, 1977; Assigned: Lonza, Ltd., Gampel, Switzerland

ORPHENADRINE CITRATE

Therapeutic Function: Muscle relaxant

Chemical Name: N,N-Dimethyl-2-[(2-methylphenyl)phenylmethoxy] ethanamine citrate

Common Name: -

Structural Formula:

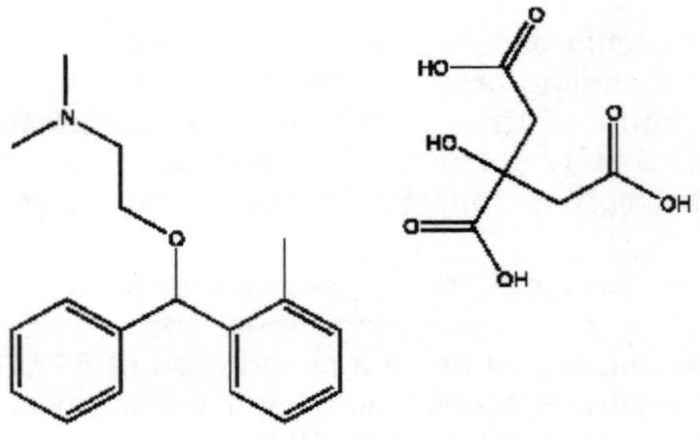

Chemical Abstracts Registry No.: 4682-36-4; 83-98-7 (Base)

Trade Name	Manufacturer	Country	Year Introduced
Norflex	Riker	US	1959
Neocyten	Central	US	1975
X-Otag	Tutag	US	1976
Banflex	O'Neal Jones	US	1980
Bio-Flex	Foy	US	-
Flexin	Taro	Israel	-
Mioflex	Formenti	Italy	-
Myotrol	Legere	US	-
Norgesic	Riker	US	-
Ro-Orphena	Robinson	US	-
Tega-Flex	Ortega	US	-

Raw Materials

o-Methylbenzhydryl bromide
β-Dimethylaminoethanol
Citric acid

Manufacturing Process

As described in US Patent 2,567,351, o-methylbenzhydryl bromide is added slowly to β-dimethylaminoethanol at refluxing temperature. After the addition has been completed the mixture is refluxed and stirred for an additional 16 hours. The mixture is cooled and the bottom layer consisting of the crude hydrobromide salt of β-dimethylaminoethanol is drawn off. The excess amino alcohol is distilled from the upper layer in vacuo and the residue is reacted with citric acid.

References

Merck Index 6752
Kleeman & Engel p. 661
PDR pp. 1033, 1452
OCDS Vol. 1 p. 42 (1977)
DOT 9 (6) 247 (1973) and 18 (2) 90 (1982)
I.N. p. 707
REM p. 932
Rieveschi, G. Jr.; US Patent 2,567,351; September 11, 195 : assigned to Parke, Davis & Company
Harms, A.F.; US Patent 2,991,225; July 4, 1961 ; assigned 1 NV Koninklijke Pharmaceutische Fabrieken, Netherlands

OSELTAMIVIR PHOSPHATE

Therapeutic Function: Antiviral

Chemical Name: 1-Cyclohexene-1-carboxylic acid, 4-(acetylamino)-5-amino-3-(1-ethylpropoxy)-, ethyl ester, (3R,4R,5S)-, phosphate (1:1)

Common Name: Oseltamivir phosphate

Structural Formula:

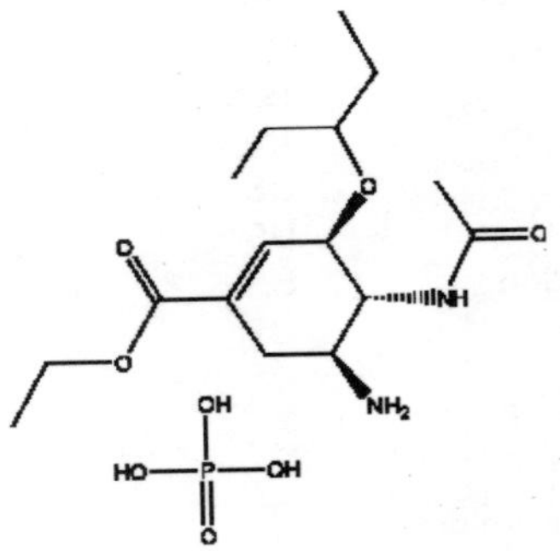

Chemical Abstracts Registry No.: 204255-11-8; 196618-13-0 (Base)

Trade Name	**Manufacturer**	**Country**	**Year Introduced**
Tamiflu	Hoffmann-La Roche Inc.	-	-

Raw Materials

Shikimic acid
Sodium azide
Dimethoxypropane
Triethylamine
Trityl chloride
Triphenylphosphine
Acetic anhydride
Ethanol
Phosphoric acid
Boron trifluoride diethyl etherate
1,8-Diazabicyclo[5.4.0]undec-7-ene
4-Toluenesulfonic acid
Methanesulfonyl chloride
N,N-Diisopropylethylamine
Chloromethyl methyl ether
Potassium hydroxide
Ammonium chloride
Dimethylaminophenol
2-(Diethylamino)ethyl(bicyclohexyl)-1-carboxylate

Manufacturing Process

To a suspension of shikimic acid (25 g, 144 mmol, Aldrich) in methanol (300 ml) was added p-toluenesulfonic acid (274 mg, 1.44 mmol, 1 mol %) and the mixture was heated to reflux for 2 h. After adding more p-toluenesulfonic acid (1 mol %) the reaction was refluxed for 26 h and was evaporated. The crude methyl ester (28.17 g) was suspended in acetone (300 ml) and was treated with dimethoxypropane (35 ml, 288 mmol) and was stirred at room temperature for 6 h and then was evaporated. The crude product was dissolved in ethyl acetate (400 ml) and was washed with saturated $NaHCO_3$ (3 times 125 ml) and saturated NaCl. The organic phase was dried ($MgSO_4$), filtered, and evaporated to afford crude 7-hydroxy-2,2-dimethyl-3a,6,7,7a-tetrahydro-benzo[1,3]dioxole-carboxylic acid methyl ester (about 2.94 g).

To a solution of 7-hydroxy-2,2-dimethyl-3a,6,7,7a-tetrahydro-benzo[1,3]dioxole-carboxylic acid methyl ester (29.4 g, 141 mmol) in CH_2Cl_2, (250 ml) at 0°C was added triethylamine (29.5 ml, 212 mmol) followed by the addition of methanesulfonyl chloride (13.6 ml, 176 mmol) over a period of 10 min. The reaction was stirred at 0°C for 1 h and ice cold water (250 ml) was added. After transfer to a separatory funnel, the organic phase was washed with water, 5% citric acid (300 ml), saturated $NaHCO_3$ (300 ml) and was dried ($MgSO_4$), filtered, and evaporated. The crude product was filtered through a short plug of silica gel on a fritted glass funnel eluting with ethyl acetate. The filtrate was evaporated to afford 7-methanesulfonyloxy-2,2-dimethyl-3a,6,7,7a-tetrahydro-benzo[1,3]dioxole-carboxylic acid methyl ester (39.5 g, 91%) as a viscous oil.

To a solution of 7-methanesulfonyloxy-2,2-dimethyl-3a,6,7,7a-tetrahydro-benzo[1,3]dioxole-carboxylic acid methyl ester (35.85 g, 117 mmol) in methanol (500 ml) was added p-toluenesulfonic acid (1.11 g, 5.85 mmol, 5 mol %) and the solution was refluxed for 1.5 h and was evaporated. The residue was redissolved in methanol (500 ml) and was refluxed an additional 4 h. The solvent was evaporated and the crude oil was triturated with diethyl ether (250 ml). After completing the crystallization overnight at 0°C, the solid was filtered and was washed with cold diethyl ether, and dried to afford 3,4-dihydroxy-5-methanesulfonyloxy-cyclohex-1-enecarboxylic acid methyl ester (24.76 g) as a white solid. Evaporation of the filtrate and crystallization of the residue from methanol/diethyl ether gave an additional 1.55 g. Obtained 26.3 g (85%) of the 3,4-dihydroxy-5-methanesulfonyloxy-cyclohex-1-ene-1-carboxylic acid methyl ester.

A suspension of 3,4-dihydroxy-5-methanesulfonyloxy-cyclohex-1-ene-1-carboxylic acid methyl ester (20.78 g, 78 mmol) in tetrahydrofuran (400 ml) at 0°C was treated with 1,8-diazabicyclo[5.4.0]undec-7-ene (11.7 ml, 78 mmol) and was stirred at room temperature for 9 h at which time the reaction was complete. The reaction was evaporated and the crude residue was dissolved in CH_2Cl_2 (200 ml) and was washed with saturated NaCl (300 ml). The aqueous phase was extracted with CH_2Cl_2 (2 times 200 ml). The combined organic extracts were dried ($MgSO_4$), filtered, and evaporated. The crude product was purified on silica gel (ethyl acetate) to afford 5-hydroxy-7-oxa-bicyclo[4.1.0]hept-3-ene-3-carboxylic acid methyl ester (12 g, 90%) as a white solid.

To a solution of 5-hydroxy-7-oxa-bicyclo[4.1.0]hept-3-ene-3-carboxylic acid methyl ester (4 g, 23.5 mmol) in CH_2Cl_2 (100 ml) was added N,N'-diisopropylethylamine (12.3 ml, 70.5 mmol) followed by chloromethyl methyl ether (3.6 ml, 47 mmol, distilled from tech. grade). The solution was refluxed for 3.5 h and the solvent was evaporated. The residue was partitioned between ethyl acetate (200 ml) and water (200 ml). The aqueous phase was extracted with ethyl acetate (100 ml). The combined organic extracts were washed with saturated NaCl (100 ml), dried ($MgSO_4$), filtered, and evaporated to afford 4.9 g of a solid residue of 5-methoxymethoxy-7-oxa-bicyclo[4.1.0]hept-3-ene-3-carboxylic acid methyl ester which was of suitable purity to use directly in the next step: melting point 62°-65°C (crude); melting point 64°-66°C (diethyl ether/hexane).

To a solution of 5-methoxymethoxy-7-oxa-bicyclo[4.1.0]hept-3-ene-3-carboxylic acid methyl ester (4.9 g, 22.9 mmol) in 8/1-MeOH/H_2O (175 ml, v/v) was added sodium azide (7.44 g, 114.5 mmol) and ammonium chloride (2.69 g, 50.4 mmol) and the mixture was refluxed for 15 h. The reaction was diluted with water (75 ml) to dissolve precipitated salts and the solution was concentrated to remove methanol. The resulting aqueous phase containing a precipitated oily residue was diluted to a volume of 200 ml with water and was extracted with ethyl acetate (3 times 100 ml). The combined organic extracts were washed with saturated NaCl (100 ml), dried ($MgSO_4$), filtered and evaporated. The crude was purified on silica gel (1/1-hexane/ethyl acetate) to afford 5-azido-4-hydroxy-3-methoxymethoxy-cyclohex-1-ene-1-carboxylic acid methyl ester (5.09 g, 86%) as a pale yellow oil. Subsequent preparations of 5-azido-4-hydroxy-3-methoxymethoxy-cyclohex-1-ene-1-carboxylic acid methyl ester provided material which was of sufficient purity to use in the next step without further purification.

To a solution of 5-azido-4-hydroxy-3-methoxymethoxy-cyclohex-1-ene-1-carboxylic acid methyl ester (6.47 g, 25.2 mmol) in CH_2Cl_2 (100 ml) at 0°C was added first triethylamine (4.4 ml, 31.5 mmol) then methanesulfonyl chloride (2.14 ml, 27.7 mmol). The reaction was stirred at 0°C for 45 min then was warmed to room temperature stirring for 15 min. The reaction was evaporated and the residue was partitioned between ethyl acetate (200 ml) and water (100 ml). The organic phase was washed with water (100 ml), saturated $NaHCO_3$ (100 ml), saturated NaCl (100 ml). The water washes were extracted with a single portion of ethyl acetate which was washed with the same $NaHCO_3$/NaCl solutions. The combined organic extracts were dried ($MgSO_4$), filtered, and evaporated. The 5-azido-4-methansulfonyloxy-3-methoxymethoxy-cyclohex-1-ene-1-carboxylic acid methyl ester was of suitable purity to be used directly in the next step.

To a solution of 5-azido-4-methansulfonyloxy-3-methoxymethoxy-cyclohex-1-ene-1-carboxylic acid methyl ester (8.56 g, 25 mmol) in THF (150 ml) at 0°C was added Ph_3P (8.2 g, 31 mmol), initially adding a third of the amount while cooling and then after removing the ice bath adding the remainder of the Ph_3P over a period of 10-15 min. After complete addition of the Ph_3P the reaction was stirred at room temperature for 3 h with the formation of a white precipitate. To this suspension was added triethyl amine (5.2 ml, 37.5 mmol) and water (10 ml) and the mixture was stirred at room temperature for 12 h. The reaction was concentrated to remove THF and the residue was partitioned between CH_2Cl_2 (200 ml) and saturated NaCl (200 ml). The aqueous phase was extracted with several portions of CH_2Cl_2 and the combined organic extracts were dried (Na_2SO_4), filtered, and evaporated to afford a crude product which was purified on silica gel (10% MeOH/EtOAc) to afford 5-methoxymethoxy-7-aza-bicyclo[4.1.0]hept-3-ene-3-carboxylic acid methyl aster (4.18 g, 78%) as an oil which typically contained trace amounts of triphenylphosphine oxide impurity.

To a solution of 5-methoxymethoxy-7-aza-bicyclo[4.1.0]hept-3-ene-3-carboxylic acid methyl aster (3.2 g, 15 mmol) in DMF (30 ml) was applied a vacuum on a rotary evaporator (40°C) for several minutes to degas the solution. To the solution was added sodium azide (4.9 g, 75 mmol) and ammonium chloride (1.6 g, 30 mmol) and the mixture was heated at 65°-

70°C for 21 h. The reaction mixture was cooled to room temperature, diluted with ethyl acetate (about 100 ml) and was filtered. The filtrate was evaporated and the residue was partitioned between diethyl ether (100 ml) and saturated NaCl (100 ml). The organic phase was washed again with saturated NaCl (100 ml), dried ($MgSO_4$), filtered, and was evaporated. Additional crude product was obtained from the aqueous washings by extraction with ethyl acetate and treated in the same manner as described above. The crude product was purified on silica gel (5% MeOH/CH_2Cl_2) to afford 4-amino-5-azido-3-methoxymethoxy-cyclohex-1-ene-1-carboxylic acid methyl ester (2.95 g) as an oil which contained a small amount of triphenylphosphine oxide impurity from the previous step.

4-Amino-5-azido-3-methoxymethoxy-cyclohex-1-ene-1-carboxylic acid methyl ester (2.59 g, 10.2 mmol) was dissolved in 5% HCl/MeOH (30 ml) and the solution was stirred for 3 h at room temperature. Additional 5% HCl/MeOH (10 ml) was added stirring 1 h and the solvent was evaporated to afford 2.52 g of the HCl salt as a tan solid after high vacuum. To a suspension of the HCl salt in CH_2Cl_2 (50 ml) at 0°C was added triethylamine (3.55 ml, 25.5 mmol) followed by the addition of solid trityl chloride (5.55 g, 12.8 mmol) in one portion. The mixture was stirred at 0°C for 1 h and then was warmed to room temperature stirring for 2 h. The reaction was cooled to 0°C, triethylamine (3.6 ml, 25.5 mmol) was added and methane sulfonyl chloride (0.97 ml, 12.5 mmol) was added, stirring the resulting mixture for 1 h at 0°C and for 22 h at room temperature. The reaction was evaporated and the residue was partitioned between diethyl ether (200 ml) and water (200 ml). The organic phase was washed with water (200 ml) and the combined aqueous phases were extracted with diethyl ether (200 ml). The combined organic extracts were washed with water (100 ml), saturated NaCl (200 ml) and were dried (Na_2SO_4), filtered, and evaporated. The crude product was purified on silica gel (1/1-hexane/CH_2Cl_2) to afford 5-azido-7-trityl-7-aza-bicyclo[4.1.0]hept-2-ene-3-carboxylic acid methyl ester (3.84 g, 86%) as a white foam.

BF_3Et_2O (43 µl, 0.35 mmol) was added to a solution of 5-azido-7-trityl-7-aza-bicyclo[4.1.0]hept-2-ene-3-carboxylic acid methyl ester (104 mg, 0.24 mmol) in 3-pentanol (2.0 ml) under argon with stirring at room temperature. The pale solution was heated at 75°C for 1.5 h and then concentrated in vacuo to give a brown residue which was dissolved in dry pyridine (2.0 ml) and treated with acetic anhydride (235 ml) and a catalytic amount of dimethylaminophenol (few crystals) at 0°C. The reaction was allowed to warm to room temperature and stirred for 1.5 h, concentrated in vacuo and partitioned between ethyl acetate and brine. The organic layer was separated and washed sequentially with dilute HCl, saturated sodium bicarbonate, brine and dried over $MgSO_4$. Concentration in vacuo followed by flash chromatography of the residue on silica gel (50% hexanes in ethyl acetate) gave 41 mg (53%) of the 4-acethylamino-5-azido-3-(1-ethyl-propoxy)-cyclohex-1-ene-1-carboxylic acid methyl ester.

To a solution of 4-acethylamino-5-azido-3-(1-ethyl-propoxy)-cyclohex-1-ene-1-carboxylic acid methyl ester (268 mg, 0.83 mmol) in THF (7.0 ml) was added aqueous KOH (1.60 ml of a 1.039 N solution) at room temperature. After stirring for 19 h at room temperature the reaction was acidified to pH 4.0 with Amberlite IR-120 (H^+) acidic resin. The resin was filtered and washed with water and ethanol. Concentration in vacuo gave the crude 4-

acethylamino-5-azido-3-(1-ethyl-propoxy)-cyclohex-1-ene-1-carboxyllic acid as a pale orange foam which was used for the next reaction without any further purification.

To a solution of 4-acethylamino-5-azido-3-(1-ethyl-propoxy)-cyclohex-1-ene-1-carboxyllic acid (crude from previous reaction, assume 0.83 mmol), ethyl alcohol (150 ml), and catalytic dimethylaminophenol in (CH_2Cl_2 (6.0 ml) was added 2-(diethylamino)ethyl(bicyclohexyl)-1-carboxylate (172 mg, 0.83 mmol) in one portion at room temperature. After several minutes a precipitate formed and after an additional 1 h of stirring the reaction was filtered and washed with CH_2Cl_2. Concentration in vacuo afforded a pale solid which was purified by flash chromatography on silica gel (50% hexanes in ethyl acetate) to give 272 mg (96%) of 4-acethylamino-5-azido-3-(1-ethyl-propoxy)-cyclohex-1-ene-1-carboxyllic acid ethyl ester as a white solid.

Triphenylphosphine (342 mg, 1.30 mmol) was added in one portion to a solution of 4-acethylamino-5-azido-3-(1-ethyl-propoxy)-cyclohex-1-ene-1-carboxyllic acid ethyl ester (272 g, 0.80 mmol) in THF (17 ml) and water (1.6 ml). The reaction was then heated at 50°C for 10 h, cooled and concentrated in vacuo to give a pale white solid. Purification of the crude solid by flash chromatography on silica gel (50% methanol in ethyl acetate) gave 242 mg (96%) of the 4-acethylamino-5-amino -3-(1-ethyl-propoxy)-cyclohex-1-ene-1-carboxyllic acid ethyl ester as a pale solid.

The racemic mixture are separated into their individual, substantially optically pure isomers through well-known techniques such as, for example, the separation of diastereomeric salts formed with optically active adjuncts, e.g. acids or bases followed by conversion back to the optically active substances. So the 4-acethylamino-5-amino-3-(1-ethyl-propoxy)-cyclohex-1-ene-1-carboxyllic acid ethyl ester,(3α,4β,5α) was obtained.

The 4-acethylamino-5-amino-3-(1-ethyl-propoxy)-cyclohex-1-ene-1-carboxyllic acid ethyl ester, (3α,4β,5α) is dissolved in H_3PO_4 to give the corresponding water soluble salt form.

References

Bischofberger N.W. et al.; US Patent No. 5,763,483; June 9, 1998; Assigned: Gilead Sciences, Inc., Foster City. Calif.

OTILONIUM BROMIDE

Therapeutic Function: Anticholinergic, Spasmolytic

Chemical Name: Ammonium, diethyl(2-hydroxyethyl)methyl-, bromide, p-(o-(octyloxy)benzamido)benzoate

Common Name: Octylonium bromide; Otilnium bromide

Chemical Abstracts Registry No.: 26095-59-0; 105360-89-2 (Base)

Structural Formula:

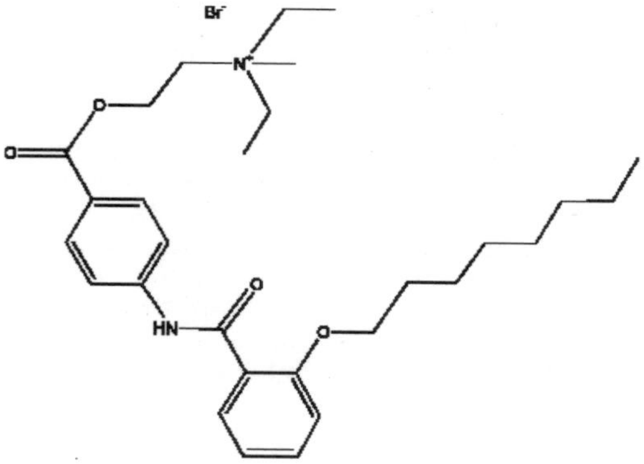

Trade Name	Manufacturer	Country	Year Introduced
Spasmomen	Minapharm Co.	-	-
Spasmomen 40	A.Menarini Pharmaceutical Industre's Group Ltd.	-	-
Doralin	Menarini Hellas A.E.	-	-
Menoctyl	Menarini	-	-
Pasminox 40	Beta	-	-
Spasen	F.I.R.M.A. S.p.A.	-	-
Spasmoctyl	Menarini	-	-

Raw Materials

Sodium hydroxide
N-Diethylaminoethanol
Methyl bromide
2-Diethylamine-ethyl-p-aminobenzoate
o-Octyloxybenzoyl chloride
2-Diethylaminoethyl chloride

Manufacturing Process

3 Methods of producing of p-[2-(n-octyloxy)benzoyl]aminobenzoate of N-diethylammoniumethanol:

1. 21.20 g (0.1 mole) of o-octyloxybenzoyl chloride and aqueous 10% NaOH are added at room temperature, with stirring and by slow dropping to 23.63 g (0.1 mole) of 2-diethylamine-ethyl-p-aminobenzoate in 100 ml of water, in such a manner as to keep the reaction mixture slightly alkaline. After concluding the slow dropping the solution is kept under stirring for 1 h and then the precipitate is collected. This precipitate, p-[2-(n-octyloxy)benzoyl] aminobenzoate of N-diethylammoniumethanol dried and recrystallized from hexane, has a melting point of 81°-82°C.

2. To 31.3 g (0.1 mole) of p-[2-(n-octyloxy)benzoyl]aminobenzoate acid in 300 ml of ethanol, are added 4.0 g (0.1 mole) of finely ground NaOH and the whole is heated to reflux for 1 h. Then 20.25 g (0.15 mol) of 2-

diethylaminoethyl chloride are slowly dropped under stirring and the heating is continued for 4 h. After cooling, the sodium chloride formed is filtered off and the solvent is separated by distillation, and the excess of the base, under a reduced pressure. The residue of p-[2-(n-tyloxy)benzoyl]aminobenzoate of N-diethylammoniumethanol, recrystallized from hexane, has a melting point of 81°-82°C.

3. 11.7 g (0.1 mole) of N-diethylaminoethanol in 200 ml of anhydrous pyridine are added by careful dropping, 34.7 g (0.1 mole) of the chloride of p-[2-(n-octyloxy)benzoyl]aminobenzoate acid and the mixture is heated in a water-bath for 3 h. The solvent is then separated by vacuum concentration, the residue is taken up with water, alkalinized and extracted with ether. The collected ether extracts, anhydridised owing to the separation of the solvent, leave a residue of p-[2-(n-tyloxy)benzoyl]aminobenzoate of N-diethylammoniumethanol which, recrystallized from hexane, has a melting point of 81°-82°C.

p-[2-(n-Octyloxy)benzoyl]aminobenzoate of N-diethylmethylammoniumethyl bromide may be prepared by reaction of the p-[2-(n-octyloxy) benzoylaminobenzoate of N-diethylammoniumethanol with methylating agents such as methylbromide.

References

Chelardoni M. et al.; US Patent No. 3,536,723; Oct. 27, 1970; Assigned: A. Menarini Societa in Accomandita Semplice, Florence, Italy, a Italian corporate body

OXACEPROL

Therapeutic Function: Antirheumatic

Chemical Name: N-Acetyl-4-hydroxy-L-proline

Common Name: Aceprolinum

Structural Formula:

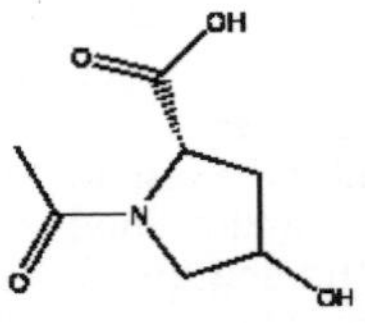

Chemical Abstracts Registry No.: 33996-33-7

Trade Name	Manufacturer	Country	Year Introduced
Jonctum	Merrell	France	1970
AHP-2000	Chephasaar	W. Germany	1975

Trade Name	Manufacturer	Country	Year Introduced
Jonctum	Merrell	Italy	1978
Tejuntivo	Valderrama	Spain	-

Raw Materials

L-Hydroxyproline
Acetic anhydride

Manufacturing Process

16.7 g (0.127 mol) of L-hydroxyproline are dissolved in 400 ml of pure boiling acetic acid. With vigorous boiling and agitation, a mixture of 13.7 ml (0.154 mol) of rectified acetic anhydride and 250 ml of pure acetic acid is added during 25 minutes. Without discontinuing the stirring, contents of the flask are cooled by simply causing fresh air to circulate externally round the flask until the temperature of the mixture is reduced to about 35°C. The acetic acid is removed by using a rotary evaporator without exceeding 35°C under a vacuum of about 15 mm Hg. After one hour, 20 ml of anhydrous toluene are added, then 10 ml of anhydrous acetone; the mixture is homogenized and concentrated again as above during 30 minutes. Then 25 ml of acetone are added again, and subsequently 20 ml of toluene, the product being concentrated again; gradually the solution is converted into an amber-colored crystallized paste. Finally, 30 ml of acetone are added to the residue, and stirring is carried out until the oily fraction surrounding the crystals is dissolved. The product is then cooled in an ice chamber, centrifuged, washed with anhydrous acetone and eventually dried, After recrystallization from acetone, crystals are obtained, melting point 132°C.

References

Merck Index 90
Kleeman & Engel p. 662
DOT 12 (1) 9 (1976)
I.N. p. 709
Coirre, P. and Coirre, B.; British Patent 1,246,141; September 15, 1971

OXACILLIN SODIUM

Therapeutic Function: Antibacterial

Chemical Name: 3,3-Dimethyl-6-(5-methyl-3-phenyl-4-isoxazolecarboxamido)-7-oxo-4-thia-1-azabicyclo[3.2.0]heptane-2-carboxylic acid, sodium salt

Common Name: 6-(5-Methyl-3-phenyl-2-isoxazoline-4-carboxamido) penicillanic acid, sodium salt; 5-Methyl-3-phenyl-4-isoxazolylpenicillin, sodium salt

Structural Formula:

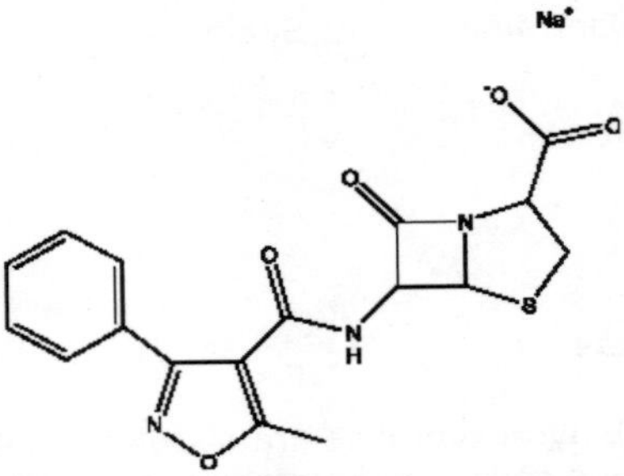

Chemical Abstracts Registry No.: 7240-38-2; 66-79-5 (Base)

Trade Name	Manufacturer	Country	Year Introduced
Resistopen	Squibb	US	1962
Prostaphlin	Bristol	US	1962
Cryptocillin	Hoechst	W. Germany	1962
Bristopen	Bristol	France	1963
Penstapho	Bristol	Italy	1966
Bactocill	Beecham	US	1972
Oxabel	Sarva	Belgium	-
Penistafil	Antibioticos	Spain	-
Stapenor	Bayer	W. Germany	-
Staphcillin V	Banyu	Japan	-

Raw Materials

Benzaldehyde
Thionyl chloride
Hydroxylamine
6-Aminopenicillanic acid
Chlorine
Sodium bicarbonate
Ethyl acetoacetate

Manufacturing Process

(A) Benzaldoxime: (Reference, Vogel, Textbook of Practical Organic Chemistry, page 883) -Materials: (Theoretical yield, 121.1 grams of free oxime), 106.1 grams (1.0 mol) of benzaldehyde (NF grade), 69.5 grams (1.0 mol) of hydroxylamine hydrochloride (practical grade), 68.0 grams (1.7 mol) of sodium hydroxide (pellet).

Procedure: The sodium hydroxide is dissolved in 200 ml water and the benzaldehyde is added. With continued stirring the hydroxylamine hydrochloride is added in portions. Some heat is developed and eventually the benzaldehyde dissolves. The solution is stirred for 15 minutes and then cooled in an ice-bath. A waxy, crystalline mass separates, and after further cooling it is collected by suction and dried in air. Yield is 86 to 149 grams. This crude material is suitable for step (B).

(B) Benzohydroximic Chloride: [Reference, G.W. Perrold et al, J. Am. Chem. Soc., 79, 462 (1957)] - Materials: 121 grams (0.77 mol) of crude benzaldoxime from step (A), 500 ml of 8.3 N hydrochloric acid, chlorine.

Procedure: The crude product from (A) is suspended in the hydrochloric acid, cooled in an ice-salt mixture, and chlorine is passed into the mixture with stirring for ½ to 1 hour. Transient blue and green colors may be noticed in the mixture during this time. The temperature will probably rise to 3° to 5°C. The solid is collected by suction filtration and dried for an hour or so on the filter before use in (C). If at all possible, it should be used on the day of preparation. Yield is 71 grams (after 1½ hours on the filter).

(C) 5-Methyl-3-Phenyl-4-Isoxazolecarboxylic Acid: [Reference, A. Quilico and R. Rusco, Gazz. Chim. Ital. 67, 589 (1937); C.A. 32, 2117[7]] - Materials: 71 grams (0.45 mol) of crude benzohydroximic chloride from (E), 78 grams (0.60 mol) of ethyl acetoacetate (practical grade), 34 grams (0.60 mol) of sodium methoxide (95% minimum), 400 ml of methanol (reagent grade).

Procedure: The sodium methoxide is cautiously added in portions to 200 ml of methanol with stirring. Some heat is evolved. To this warm solution is rapidly added the ethyl acetoacetate with continued stirring. The solution is stirred for 10 minutes and then cooled in an ice-salt-acetone mixture (-25°C). If desired a Dry Ice-acetone cooling bath may be used to shorten the addition time. The crude material from (B) is dissolved in 200 ml of methanol. At this point it is probably easier to filter this mixture by suction to remove a large amount of insoluble solid, which is probably sodium chloride. The solid may be rinsed with more methanol.

The filtrate is chilled in ice-water and added to the cooled methanolic solution of the sodium derivative of ethyl acetoacetate at a rate which keeps the temperature of the re. action mixture below 0°C. The addition time will be 15 to 20 minutes if ice-salt-acetone is used as a coolant. This reaction is extremely exothermic.

The reaction mixture is stirred overnight at room temperature and filtered to remove the sodium chloride. The filtrate is stripped in vacuo and the crude ester (literature reports MP 48°C) is dissolved in 150 ml of ethanol; 28 grams (0.70 mol of sodium hydroxide in 90 ml of water is added and the solution is refluxed for 2 hours. After removal of the ethanol in vacuo the residue is dissolved in water and extracted twice with ether. Dissolved ether is removed from the aqueous solution in vacuo and it is acidified to pH 2 with concentrated hydrochloric acid.

The crystalline crude acid is dried briefly and then recrystallized from acetonitrile to give 32 grams of white product; MP 193° to 194.5°C (literature reports 189° to 190°C). Concentration of the mother liquor gives an additional 5 grams of material having a MP of 192.5 to 194°C. The 37 grams of material represents an 18% overall yield from benzaldehyde.

(D) The acid is converted to the acid chloride by reaction with thionyl chloride.

(E) 5-Methyl-3-Phenyl-4-Isoxazolylpenicillin: A solution of 4.43 grams of 5-methyl-3-phenylisoxazole-4-carbonyl chloride in 120 ml acetone was added gradually to a stirred solution of 4.32 grams of 6-aminopenicillanic acid in 168 ml of 3% aqueous sodium bicarbonate and 50 ml acetone. When addition was complete the mixture was stirred at room temperature for 4 hours and then extracted with ether (2 x 200 ml), only the aqueous phase being retained. This aqueous solution was covered with 50 ml ether and adjusted to pH 2 by the addition of N hydrochloric acid. After separating the layers, the aqueous phase was extracted with two further 50 ml portions of ether. The combined ether solutions (which at this stage contained the free penicillin acid) were washed with water and then neutralized by shaking with 20 ml N sodium bicarbonate solution. The aqueous phase was separated, washed with ether, and evaporated at low temperature and pressure to leave the crude sodium salt of 5-methyl-3-phenyl-4-isoxazolylpenicillin as a white solid, which was finally dried in vacuo over phosphorus pentoxide and found to weigh 7.34 grams.

References

Merck Index 6777
Kleeman & Engel p. 662
PDR pp. 673, 708, 1606
OCDS Vol. 1 p. 413 (1977)
DOT 1 (3) 115 (1965)
I.N. p. 709
REM p. 1197
Doyle, F.P. and Nayler, J.H.C.; US Patent 2996,501; August 15, 1961

OXAFLOZANE HYDROCHLORIDE

Therapeutic Function: Antidepressant

Chemical Name: 2-(3-Trifluoromethyl)phenyl-4-isopropyl-tetrahydro-1,4-oxazine hydrochloride

Common Name: -

Structural Formula:

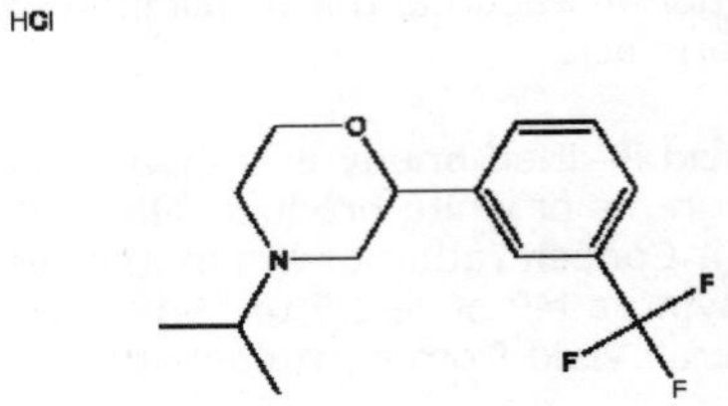

HCl

Chemical Abstracts Registry No.: 26629-86-7; 26629-87-8 (Base)

Trade Name	Manufacturer	Country	Year Introduced
Conflictan	Sarbach	France	1982
Conflictan	Riom Lab	France	-

Raw Materials

Bromine
Isopropylamine
(3-Trifluoromethyl)phenyl magnesium bromide
2-Chloroethylvinyl ether
Hydrogen chloride

Manufacturing Process

(1) 1,2-Dibromo-2-(2-chloro)ethoxyethane: 640 g of bromine (4 mols) are added dropwise, with stirring, to 426 g (4 mols) of 2-chloroethylvinyl ether dissolved in 1,040 ml of chloroform maintained at -10°C.

When addition is ended, the solvent and then the residue are distilled in vacuum to obtain 690 g of product. Yield = 65%.

(2) 2-(3-Trifluoromethyl)-2-(2-chloro)ethoxy-1-bromoethane: (3-Trifluoromethyl)phenyl magnesium bromide is prepared under the normal conditions for magnesium derivatives, from 48.6 g of magnesium turnings and 455.7 g of (3-trifluoromethyl)bromobenzeneand 1.5 liters anhydrous ether.

To the solution of the magnesium compound so obtained the following solution is added dropwise, with stirring so as to maintain a slight reflux of ether: 1,2-dibromo-2-(2-chloro)-ethoxyethane: 550 g. Anhydrous ether: 300 ml.

After the addition, reflux heating is continued for two hours, cooling is carried out and there is hydrolysis by the mixture: Ice: 500 g. Concentrated HCl: 200 ml.

The organic phase is decanted, washed in NaCl saturated water and dried on anhydrous Na_2SO_4; the ether is distilled and the residue is rectified in vacuum to obtain 361 g of the product. Yield = 54%.

According to gas phase chromatography, the product so obtained is about 95% pure and it can be used in further reactions without a second rectification.

(3) 2-(3-Trifluoromethyl)phenyl-4-isopropyl tetrahydro-1,4-oxazine hydrochloride: The following mixture is heated in an autoclave at 100°C; 2-(3-trifluoromethyl)-2-(2-chloro)-ethoxy-1-bromoethane: 33.15 g (0.1 mol); isopropylamine: 20 g (0.34 mol); toluene: 100 ml.

After filtration of the isopropylamine hydrochloride and bromohydrate, the solvent is stripped and the residue is admixed with ~ 4 N HCl and the aqueous phase is washed with ether. The aqueous phase is treated with 50% aqueous NaOH, the amine is ether-extracted and, after drying on anhydrous Na_2SO_4, the ether is distilled and the residue is rectified in vacuum to obtain 14 g of the product. Yield = 50%.

The hydrochloride is crystallized by adding ethyl acetate to the base and then adding the necessary amount of pure alcohol saturated in dry HCl. Melting point 164°C.

References

Merck Index 6780
DFU 3 (9) 667 (1978)
Kleeman & Engel p. 663
DOT 18 (10) 536 (1982)
I.N. p. 709
Mauvernay, R.Y., Busch, N., Moleyre, J. and Simond, J.; US Patent 3,637,680; January 25, 1972; assigned to Societe Anonyme: Centre Europeen De Recherches Mauvernay

OXAFLUMAZINE DISUCCINATE

Therapeutic Function: Neuroleptic, Antihistaminic, Spasmolytic

Chemical Name: N-3-(2-Trifluoromethyl-10-phenothiazinyl)-propyl-N'-2-[2-(1,3-dioxanyl)]ethyl-piperazine disuccinate

Common Name: -

Structural Formula:

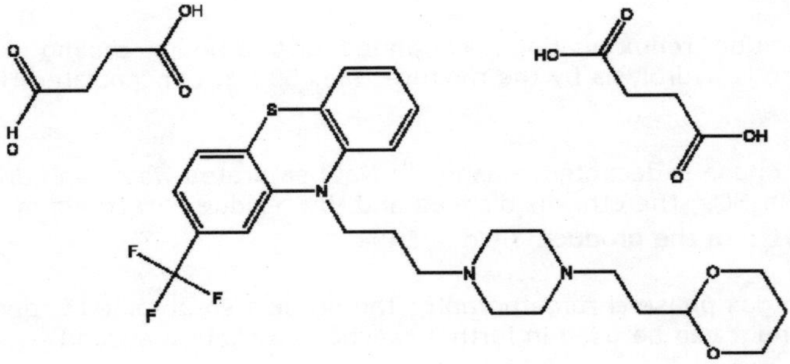

Chemical Abstracts Registry No.: 41761-40-4; 16498-21-8 (Base)

Trade Name	Manufacturer	Country	Year Introduced
Oxaflumine	Diamant	France	1970

Raw Materials

N-[2-(3,1-Dioxanyl)ethyl]piperazine
2-Trifluoromethylphenothiazine
Succinic acid
1-Bromo-3-chloropropane
Sodium

Manufacturing Process

Preparation of N-(3-chloropropyl)-N'-[2-(1,3-dioxanyl)-ethyl]-piperazine: A solution of 30 g (0.15 mol) of N-[2-(1,3-dioxanyl)-ethyl]-piperazine and 11.8 g (0.075 mol) of 1-bromo-3-chloropropane in 150 ml of dry benzene was refluxed with stirring for 5 hours. After cooling, the N-[2-(1,3-dioxanyl)-ethyl]-piperazinium bromide which had precipitated was filtered off, the filtrate was concentrated in vacuo and the residual oil was distilled. 14.1 g (68%yield) of N-(3-chloropropyl)-N'-[2-1,3-dioxanyl)-ethyl]-piperazine which occurred as a light yellow oil were obtained. Boiling point: 152°C to 155°C under 0.07 mm Hg (n_D^{23} = 1.4940). The disuccinate prepared in acetone and recrystallized from acetone melts at 104°C to 105°C on a hot stage microscope.

The sodium derivative of the 2-trifluoromethylphenothiazine was prepared from 26.7 g (0.1 mol) of 2-trifluoromethylphenothiazine and 2.3 g (0.1 g atom) of sodium in 500 ml of liquid ammonia. After the reaction was completed, the ammonia was driven off and 500 ml of dry toluene were added. A solution of 25 g (0.09 mol) of N-(3-chloropropyl)-N'-[2-(1,3-dioxanyl)-ethyl]-piperazine in 200 ml of toluene was added drop by drop to this solution which was then refluxed with stirring for 18 hours. After cooling, the precipitate which had formed was filtered and the filtrate was washed with water, dried and concentrated in vacuo. 33 g of brown oil, the N-3-(2-trifluoromethyl-10-phenothiazinyl)-propyl-N'-2-[2-(1,3-dioxanyl)]-ethyl-piperazine, were obtained.

A warm solution of 4.4 g of the base obtained in 100 ml of acetonitrile was added to a warm solution of succinic acid in 200 ml of acetonitrile. After standing for 15 hours at 0°C. the crystalline product was obtained, melting point 138°C.

References

Merck Index 6781
Kleeman & Engel p. 663
DOT 6 (3) 89 (1970)
I.N. p. 709
Societe Industrielle Pour La Fabrication Des Antibiotiques (S.I.F.A.); British Patent 1,103,311; February 14, 1968

OXAMETACINE

Therapeutic Function: Antiinflammatory

Chemical Name: 1-(4-Chlorobenzoyl)-N-hydroxy-5-methoxy-2-methyl-1H-indole-3-acetamide

Common Name: Indoxamic acid

Chemical Abstracts Registry No.: 27035-30-9

Structural Formula:

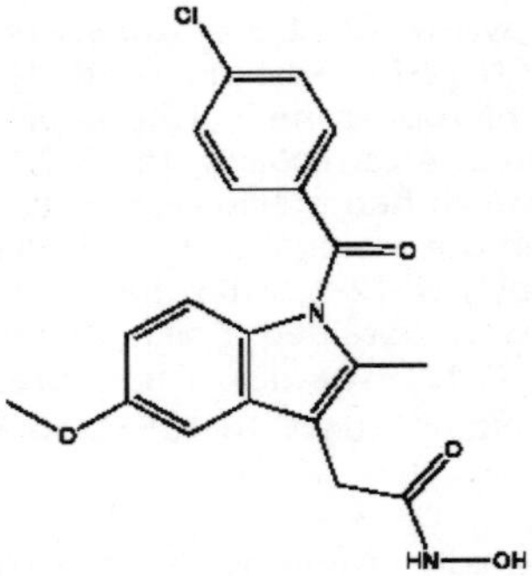

Trade Name	Manufacturer	Country	Year Introduced
Flogar	A.B.C.	Italy	1976
Flogar	U.C.B.	France	1981
Dinulcid	Pharmascience	France	1983

Raw Materials

1-p-Chlorobenzoyl-2-methyl-5-methoxy-3-indoleacetic acid
Thionyl chloride
Hydroxylamine hydrochloride

Manufacturing Process

1 g of 1-p-chlorobenzoyl-2-methyl-5-methoxy-3-indoleacetic acid [J. Am. Chem. Soc. 85, 488-489 (1963)] is treated in a nitrogen stream with 10 ml thionyl chloride in which it promptly dissolves. The solution is quickly evaporated in vacuum and the residue (which typically is of a deep brown-green color) is distempered, twice or three times, with a few ml anhydrous benzene which is removed in vacuum each time. The resulting residue is thoroughly distempered with 5 ml anhydrous ether which dissolves most of the color impurities, and separated by filtering, purified by crystallizing from plenty of anhydrous ether, yielding a crystalline mass of needles of straw-yellow color, melting point 124°C to 127°C. Yield: 0.700 g. Found: Cl % 18.62 (calculated 18.84).

The product is relatively stable towards water and aqueous alkalies in which it proves to be insoluble even after dwelling therein several hours at room temperature. It reacts, better if at elevated temperature, with lower alcohols with which it forms the corresponding esters, and with ammonia under suitable conditions for forming the amide (melting point 219°C to 221°C).

A solution of 1.330 g sodium hydroxide in 20 ml water is slowly admixed with 2.330 g hydroxylamine hydrochloride while cooling, whereupon 1 g chloride of 1-p-chlorobenzoyl-2-methyl-5-methoxy-3-indoleacetic acid is distempered in this neutral or slightly alkaline solution by vigorously stirring during a few minutes.

The acid chloride reacts with the free hydroxylamine with considerable rapidity apparently without dissolving. The reaction is completed when a sample of the suspension shows to become clear on adding aqueous alkali. The crystalline pale-yellow mass of product is separated by filtering, lavishly washed with water and dried in vacuum. The crude product yield is actually quantitative. The product is purified with excellent yields by repeatedly crystallizing from hot dioxane and washing with ether: melting point 181°C to 182°C (dec.).

References

Merck Index 6788

I.N. p. 710

De Martlis, F., Arrigoni-Martelli, E. and Tamietto, T.; US Patent 3,624,103; November 30, 1971; assigned to Instituto Biologico Chemioterapico (A.B.C.) SpA (Italy)

OXAMNIQUINE

Therapeutic Function: Antischistosomal

Chemical Name: 1,2,3,4-Tetrahydro-2-[[(1-methylethyl)amino]methyl]-7-nitro-6-quinolinemethanol

Common Name: -

Structural Formula:

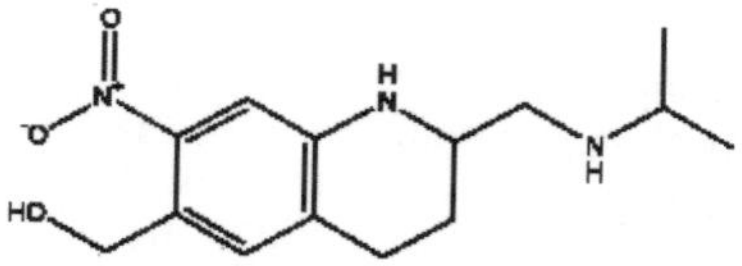

Chemical Abstracts Registry No.: 21738-42-1

Trade Name	Manufacturer	Country	Year Introduced
Vansil	Pfizer	US	1980
Vansil	Pfizer	France	1981

Raw Materials

Bacterium Aspergillus sclerotiorum Huber
Soybean meal
Glucose
2-Isopropylaminomethyl-6-methyl-7-nitro-1,2,3,4-tetrahydroquinoline

Manufacturing Process

(1) Four fermenters are set up, each one of which contained 2.0 liters of the

following medium, sterilized for 35 minutes at 15 psi, respectively:

Soybean meal	5 grams
Glucose	20 grams
NaCl	5 grams
K_2HPO_4	5 grams
Yeast extract	5 grams
Tap water to	1 liter

pH adjusted with sulfuric acid to 6.5

The fermenters are inoculated with 7.5% by volume of a 24-hour old culture of Aspergillus sclerotiorum Huber grown at 28°C in 50 ml aliquots of the above described soybean-glucose medium contained in 300 ml Erlenmeyer flasks, placed on a shaker rotating at approximately 230 rpm. The inoculated fermenters are agitated at 1,380 rpm and each aerated with 1 liter of air per minute and at a temperature of 28°C for 47 hours. A silicone antifoam is added when required. At the end of the 47 hour period, the pH of the fermentation broth rose to 6.8 to 6.9. Sulfuric acid is then added with sterile precautions to restore the pH to 6.5.

(2) 0.75 g of 2-isopropylaminomethyl-6-methyl-7-nitro-1,2,3,4-tetrahydroquinoline as hydrogen maleate, dissolved in 75 ml of sterile water, is added to each of the four fermenters and agitation and aeration are continued for a further 23 hours. The whole fermentation broths from each fermenter are pooled, the pH adjusted to 8.0 with sodium hydroxide and the 8.2 liters of fermentation broth thus obtained are extracted by agitating vigorously with 16.4 liters of methylene chloride for 10 minutes. The solvent extract is then dried over anhydrous sodium sulfate and subsequently evaporated to dryness at a temperature below 40°C (dry weight 5.567 g).

(3) The dark brown residue from (2) is extracted four times with methanol at room temperature, decanting the solution from the insoluble material. The combined methanol extracts, total volume about 200 ml, are then filtered and treated with 3 g of sodium borohydride, added in portions over a period of 30 minutes with stirring, to reduce any 6-formyl compound present to the 6-hydroxymethyl compound. The methanol solution is then allowed to stand overnight at room temperature and is thereafter diluted with 1 liter of ether. The solution is washed 4 times with 500 ml of water and the resulting pale yellow ethereal solution is dried over magnesium sulfate. The ether is next removed by vacuum distillation from a water bath at 40°C. The residue is dissolved in about 75 ml of isopropanol at 50°C, filtered to remove any insoluble particles and cooled overnight in the refrigerator. The product is collected and dried in vacuo to yield 0.5 g of 6-hydroxymethyl-2-isopropylaminomethyl-7-nitro-1,2,3,4-tetrahydroquinoline as pale yellow crystals of melting point 147°C to 149°C. A further 0.5 g of crude material is obtained from the mother liquors of the recrystallization. Total yield is therefore 1.0 g (0.0036 mol) from 3.0 g (0.0079 mol) of starting material, i.e., 45% of the theoretical amount.

References

Merck Index 6791
OCDS Vol. 2 p. 372 (1980)

DOT 17 (4) 152 (1981)
I.N. p. 710
REM p. 1236
Richards, H.C.; US Patent 3,821,228; June 28, 1974; assigned to Pfizer, Inc.

OXANDROLONE

Therapeutic Function: Androgen

Chemical Name: 17β-Hydroxy-17-methyl-2-oxa-5α-androstan-3-one

Common Name: -

Structural Formula:

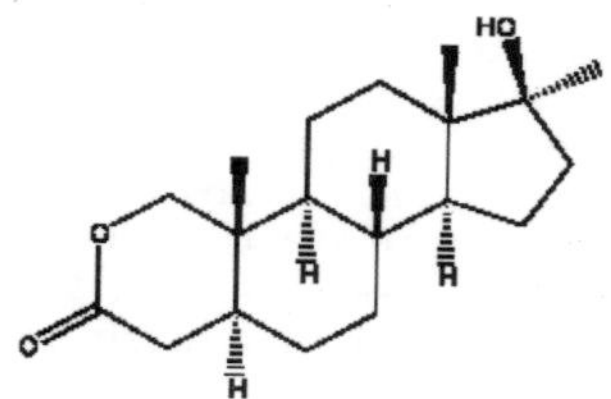

Chemical Abstracts Registry No.: 53-39-4

Trade Name	Manufacturer	Country	Year Introduced
Anavar	Searle	US	1964
Anatrophill	Searle	France	1965
Vasorome	Kowa	Japan	1969
Oxandrolone Spa	SPA	Italy	1979
Lonavar	Searle	Italy	-

Raw Materials

17β-Hydroxy-17α-methyl-5α-androst-1-en-3-one
Lead tetraacetate
Sodium borohydride

Manufacturing Process

To a solution of 6.36 parts of 17β-hydroxy-17α-methyl-5α-androst-1-en-3-one in 95 parts of acetic acid and 12 parts of water is added 40 parts of lead tetracetate and 0.6 part of osmium tetroxide. This mixture is stored at room temperature for about 24 hours, then is treated with 2 parts of lead tetracetate. Evaporation to dryness at reduced pressure affords a residue, which is extracted with benzene. The benzene extract is washed with water, and extracted with aqueous potassium bicarbonate. The aqueous extract is washed with ether, acidified with dilute sulfuric acid, then extracted with ethyl acetate-benzene. This organic extract is washed with water, dried over anhydrous sodium sulfate, and concentrated to dryness in vacuo. To a solution of the residual crude product in 20 parts of pyridine is added 10 parts of 20% aqueous sodium bisulfite and the mixture is stirred for about 20 minutes at room temperature.

This mixture is then diluted with water, washed with ethyl acetate, acidified with dilute sulfuric acid, and finally extracted with benzene. The benzene extract is washed with water, dried over anhydrous sodium sulfate, and evaporated to dryness at reduced pressure to produce crude 17β-hydroxy-17α-methyl-1-oxo-1,2-seco-A-nor-5α-androstan-2-oic acid, which after recrystallization from aqueous isopropyl alcohol melts at about 166° to 173°C (decomposition).

An aqueous slurry of 6 parts of 17β-hydroxy-17α-methyl-1-oxo-1,2-seco-A-nor-5α-androstan-2-oic acid in 200 parts of water is made alkaline to pH 10 by the addition of dilute aqueous sodium hydroxide, then is treated with 6 parts of sodium borohydride. This mixture is allowed to react at room temperature for about 3 hours. Benzene is added and the resulting mixture is acidified carefully with dilute hydrochloric acid. The benzene layer is separated, and the aqueous layer is further extracted with benzene. The combined benzene extracts are washed successively with aqueous potassium bicarbonate and water, dried over anhydrous sodium sulfate, then evaporated to dryness in vacuo. The resulting residue is triturated with ether to afford pure 17β-hydroxy-17α-methyl-2-oxa-5α-androstan-3-one, MP about 235° to 238°C, according to US Patent 3,128,283.

References

Merck Index 6794
Kleeman & Engel p. 664
PDR p. 1677
OCDS Vol. 1 p. 174 (1977)
I.N. p. 710
REM p. 999
Pappo, R.; US Patent 3,128,283; April 7, 1964; assigned to G.D. Searle and Co.
Pappo, R.; US Patent 3,155,684; November 3, 1964; assigned to G.D.Searle 81 Co.

OXAPROZIN

Therapeutic Function: Antiinflammatory

Chemical Name: 2-Oxazolepropanoic acid, 4,5-diphenyl-

Common Name: Oxaprozin

Chemical Abstracts Registry No.: 21256-18-8

Structural Formula:

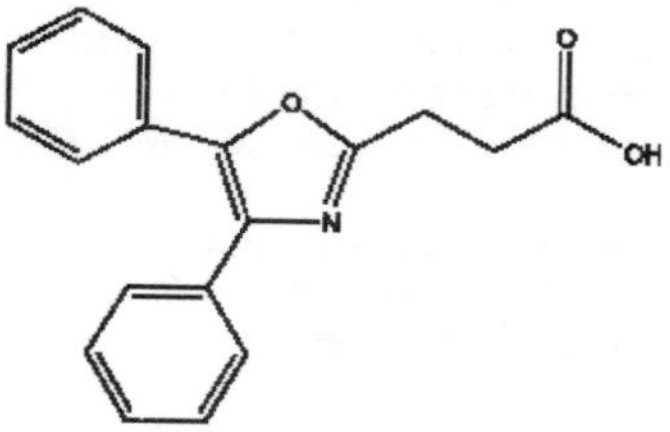

Trade Name	**Manufacturer**	**Country**	**Year Introduced**
Danoprox	TRB Chemedica	Switz.	-
Daypro	Pfizer	-	-
Daypro	Pharmacia	-	-
Dayrun	CSC Pharmaceuticals	Austria	-
Duraprox	Gerolymatos	Greece	-
Duraprox	Aventis Pasteur	France	-
Oxaprozin	Apotex Inc.	-	-
Oxaprozin	Dr. Reddy's Laboratories Ltd.	India	-

Raw Materials

Pyridine
Benzoin
Succinic anhydride
Acetic acid

Manufacturing Process

A clean dry reactor of 20 gallon (91 liters) capacity was charged with pyridine (9.25 kg), benzoin (16.5 kg) and succinic anhydride (11.7 kg.). The reactor was purged with nitrogen and a nitrogen atmosphere was maintain throughout the process. The mixture was heated without agitation until it became liquid at 85°C. Agitation was commenced and the mixture was heated at 90°-95°C for 1.5 hours. A solution of ammonium acetate (12.0 kg) in glacial acetic acid (35.0 kg) was charged to the header of the reactor and added to the reaction mixture over 15 minutes, maintaining the temperature between 90° and 95°C. The container for the solution and the header were washed with glacial acetic acid (4.0 kg) and the washing liquid was added to the reaction mixture. The reaction mixture was held at 90°-95°C for 2 hours. The reaction mixture was cooled to 50°C and transferred via a line filter to a reactor of 50 gallon (227 liters) capacity. The first reactor, lines and filter were washed with glacial acetic acid (4.0 kg.) which was combined with the reaction mixture. The reaction mixture was heated with agitation to 90°-95°C over 30 minutes and water (21.0 kg.) was added maintaining the temperature at 90°-95°C. The reaction mixture was then cooled to 20°-25°C over 55 minutes by means of water in the jacket of the reactor and then cooled to 10°-15°C by means of brine in the jacket and left overnight. The product was filtered on a ceramic filter and sucked well dry. The product on the filter was washed with a pre-filtered mixture of glacial acetic acid (25.5 kg.) and water (12.5 kg) and sucked well dry. Pre-filtered water (50.0 kg) and the filter cake were added to a reactor of 50 gallon (227 liters) capacity. The mixture was stirred at room temperature for 30 minutes and filtered on a ceramic filter and the product was sucked well dry. The product on the filter was washed twice with pre-filtered water (10 kg each time) and sucked well dry. The product was then dried in a Mitchell oven at 80°C for 16-18 hours. The yield of crude β-(4,5-diphenyloxazol-2-yl)propionic acid was 15.9 kg (69.8%). This material only just failed specification for acceptable purity because although TLC analysis showed only very faint trace impurities.

Recrystallisation of crude β-(4,5-Diphenyloxazol-2-yl)propionic acid

Methanol (62.0 kg) was added to a reactor of 50 gallon capacity (227 liters). 15.9 kg of the crude oxazole above prepared was added with agitation. The mixture was heated to reflux. All the solid dissolved. The mixture was then cooled to 50°C and transferred to a reactor of 20 gallon (91 liters) capacity. The larger reactor and transfer lines were washed through with methanol at about 40°C twice (3 kg each time). The mixture was cooled over 1 hour 50 minutes with agitation, gradually at first, to 15°-20°C by means of cooling water on the jacket of the reactor. The product was then filtered on a ceramic filter and sucked well dry. The product on the filter was washed twice with methanol (5 kg each time) and sucked well dry. The wash liquors were combined with the filtration liquors and retained. The product from the filter was dried in an air oven at 55°-60°C for 18 hours. The yield of β-(4,5-diphenyloxazol-2-yl)propionic acid was 12.1 kg. TLC investigation showed the product to be pure. Melting point 160.5°-161.5°C.

Another crop of product was obtained from the methanol liquors as follows. The liquors were added to a reactor of 20 gallon (91 liters) capacity and the solvent was distilled off for 9 hours until solid appeared. The mixture was then cooled to 15° to 20°C over 1 3/4 hours using cooling water in the jacket of the reactor. The mixture was cooled to 10°C using brine in the jacket and stirred at this temperature for 30 minutes. The product was then filtered on a ceramic filter and sucked well dry. The product on the filter was washed twice with methanol (5 kg each time) and sucked well dry. The product was dried in an air oven at 55°-60°C for 18 hours. The yield was 2.14 kg. This product may also have been acceptably pure but its purity was not investigated. It was therefore retained as crude product to be resubjected to recrystallisation with methanol. The yield for the recrystallisation was thus 12.1 kg from a consumption of 13.76 kg of crude product, that is 88%. The overall yield of pure product is 69.8% times 88%, that is, 61.4%.

References

Weston G.O.; US Patent No. 4,190,584; Feb. 26, 1980; Assigned to John Wyeth and Brother Limited, Maidenhead, England

Brown K.; US Patent No. 3,578,671; May 11, 1971; Assigned to John Wyeth and Brother Limited, Taplow, Maidenhead, Berkshire, England

OXATOMIDE

Therapeutic Function: Antiallergic

Chemical Name: 1-[3-[4-(Diphenylmethyl)-1-piperazinyl]propyl]-2-benzimidazolone

Common Name: Oxatimide

Structural Formula:

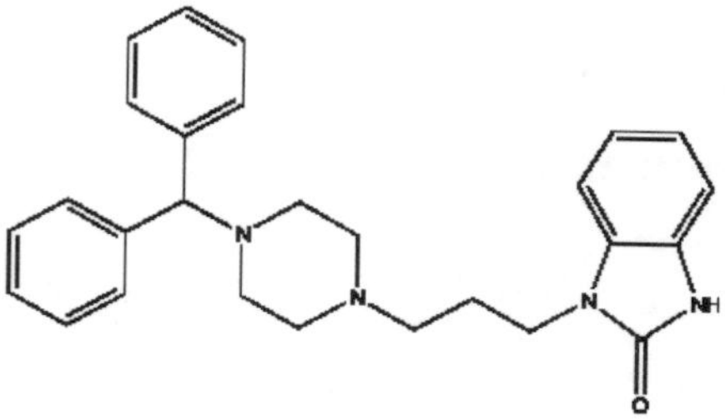

Chemical Abstracts Registry No.: 60607-34-3

Trade Name	Manufacturer	Country	Year Introduced
Tinset	Janssen	W. Germany	1981
Tinset	Janssen	UK	1982
Tinset	Janssen	Switz.	1983
Finsedyl	Microsules	Argentina	-

Raw Materials

1-(3-Chloropropyl)-2H-benzimidazol-2-one
1-(Diphenylmethyl)piperazine

Manufacturing Process

A mixture of 53 parts of 1-(3chloropropyl)-2H-benzimidazol-2one, 5 parts of 1-(diphenylmethyl)piperazine, 6.4 parts of sodium bicarbonate and 200 parts of 4-methyl-2-pentanone is stirred and refluxed overnight with water-separator. After cooling, water is added and the layers are separated. The 4-methyl-2pentanone phase is dried, filtered and evaporated. The residue is purified by column-chromatography over silica gel using a mixture of trichloromethane and 5% of methanol as eluent. The pure fractions are collected and the eluent is evaporated. The oily residue is crystallized from a mixture of 2,2'-oxybispropane and a small amount of 2-propanol. The product is filtered off and dried, yielding 1-[3-[4-(diphenylmethyl)-1-piperazinyl]propyl]-2H-benzimidazole-2-one; melting point 153.6°C.

References

Merck Index 6798

DFU 3 (6) 465 (1978)
OCDS Vol. 3 p. 173 (1984)
DOT 16 (7) 219 (1980); 18 (7) 341 and (9) 440 (1982)
I.N. p. 711
Vandenberk, J., Kennis, L.E.J., Van der Aa, M.J.M.C. and Van Heertum, A.H.M.T.; US Patent 4,200,641; April 29, 1980; assigned to Janssen Pharmaceutica N.V.

OXAZEPAM

Therapeutic Function: Tranquilizer

Chemical Name: 7-Chloro-1,3-dihydro-3-hydroxy-5-phenyl-2H-1,4-benzodiazepin-2one

Common Name: -

Structural Formula:

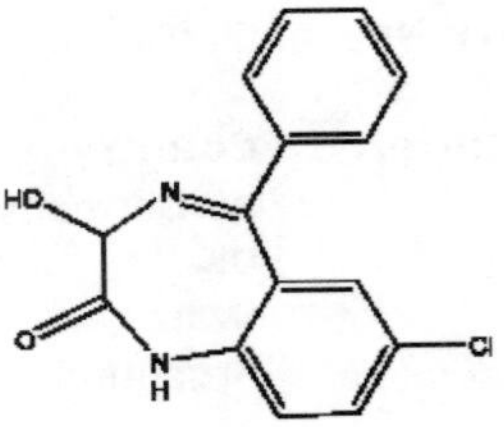

Chemical Abstracts Registry No.: 604-75-1

Trade Name	Manufacturer	Country	Year Introduced
Serax	Wyeth	US	1965
Adumbran	Thomae	W. Germany	1965
Seresta	Wyeth Byla	France	1966
Praxiten	Wyeth	UK	1966
Serpax	Wyeth	Italy	1967
Anxiolit	Gerot	Austria	-
Aplakil	Aristegui	Spain	-
Aslapax	Asla	Spain	-
Benzotran	Protea	Australia	-
Droxacepam	Jeba	Spain	-
Durazepam	Durachemie	W. Germany	-
Enidrel	Syncro	Argentina	-
Hilong	Banyu	Japan	-
Iranil	Iltas	Turkey	-
Isochin	Tosi	Italy	-
Limbial	Chiesi	Italy	-

Trade Name	Manufacturer	Country	Year Introduced
Nesontil	Promeco	Argentina	-
Noctazepam	Brenner	W. Germany	-
Oxpam	I.C.N.	Canada	-
Propax	Cipan	Portugal	-
Psicopax	Bama-Geve	Spain	-
Psiquiwas	Wassermann	Spain	-
Purata	Lennon	S. Africa	-
Quen	Ravizza	Italy	-
Quilibrex	Isnardi	Italy	-
Sedokin	Geymonat Sud	Italy	-
Serepax	Ferrosan	Denmark	-
Sigacalm	Siegfried	Switz.	-
Sobile	Lafarquin	Spain	-
Uskan	Desitin	W. Germany	-
Vaben	Rafa	Israel	-
Wakazepam	Wakamoto	Japan	-

Raw Materials

7-Chloro-1,3-dihydro-5-phenyl-2H-1,4-benzodiazepin-2-one-4-oxide
Acetic anhydride
Sodium hydroxide

Manufacturing Process

(A) Suspend 10 g of 7-chloro-1,3-dihydro-5-phenyl-2H-1,4-benzodiazepin-2-one 4-oxide in 150 ml of acetic anhydride and warm on a steam bath with stirring until all the solid has dissolved. Cool and filter off crystalline, analytically pure 3-acetoxy-7-chloro-1,3-dihydro-5-phenyl-2H-1,4-benzodiazepin-2-one, melting point 242°C to 243°C.

(B) Add to a suspension of 3.4 g of 3-acetoxy-7-chloro-1,3-dihydro-5-phenyl-2H-1,4-benzodiazepin-2-one in 80 ml of alcohol.6 ml of 4 N sodium hydroxide. Allow to stand after complete solution takes place to precipitate a solid. Redissolve the solid by the addition of 80 ml of water. Acidify the solution with acetic acid to give white crystals. Recrystallize from ethanol to obtain 7-chloro-1,3-dihydro-3-hydroxy-5-phenyl-2H-1,4-benzodiazepin-2-one, melting point 203°C to 204°C.

References

Merck Index 6799
Kleeman & Engel p. 664
PDR p. 1980
OCDS Vol. 1 p. 366 (1977) and 2, 402 (1980)
DOT 1 (3) 102 (1965) and 9 (6) 238 (1973)
I.N. p. 711
REM p. 1063
Bell, S.C.; US Patent 3,296,249; January 3, 1967; assigned to American Home Products Corp.

OXAZOLAM

Therapeutic Function: Tranquilizer

Chemical Name: 7-Chloro-5-phenyl-5'-methyltetrahydrooxazolo[5.4-b]-2,3,4,5-tetrahydro-1H-1,4-benzodiazepin-2-one

Common Name: Oxazolazepam

Structural Formula:

Chemical Abstracts Registry No.: 24143-17-7

Trade Name	Manufacturer	Country	Year Introduced
Serenal	Sankyo	Japan	1970
Quiadon	Merck	W. Germany	1980
Convertal	Roemmers	Argentina	-
Hializan	Pharma-investi	Spain	-
Tranquit	Promonta	W. Germany	-

Raw Materials

5-Chloro-2-chloroacetylaminobenzophenone
Isopropanolamine

Manufacturing Process

To a solution of 12.0 g of 5-chloro-2-chloroacetylaminobenzophenone and 3.2 g of isopropanolamine in 100 ml of ethanol was added 3.3 g of sodium acetate.

The resulting mixture was heated under reflux with stirring for 12 hours. After completion of the reaction, the solvent was distilled off and the residue was extracted with dichloromethane. The extract was washed with water, dried over anhydrous sodium sulfate and the solvent was distilled off.

The residue was recrystallized from ethanol to give 10.6 g of the desired product melting at 186°C to 188.5°C.

References

Merck Index 6801

DOT 8 (1) 18 (1972) and 9 (6) 239 (1973)
I.N. p. 712
REM p. 1064
Tachikawa, R., Takagi, H., Kamioka, T., Midayera, T., Fukunaga, M. and Kawano, Y.; US Patents 3,772,371; November 13, 1973; and 3,914,215; October 21, 1975; both assigned to Sankyo Co., Ltd.

OXELADIN

Therapeutic Function: Antitussive

Chemical Name: α,α-Diethylbenzeneacetic acid 2-[2-(diethylamino)ethoxy] ethyl ester

Common Name: -

Structural Formula:

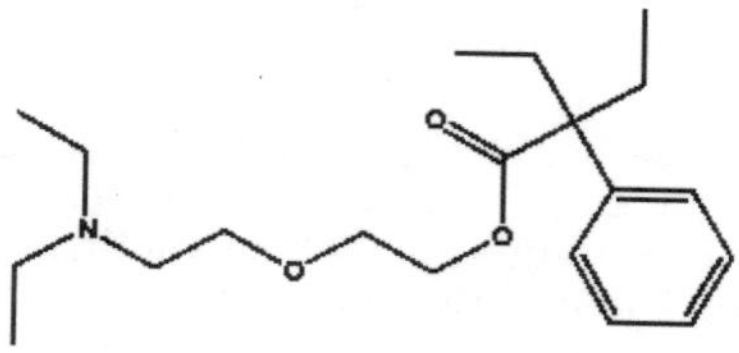

Chemical Abstracts Registry No.: 468-61-1; 16485-39-5 (Citrate)

Trade Name	Manufacturer	Country	Year Introduced
Silopentol	Schulte	W. Germany	1970
Ethochlon	Hokuriku	Japan	1970
Fustopanox	Ottia Pharm.	Japan	1970
Paxeladine	Beaufour	France	1974
Dorex	Woelm	W. Germany	-
Hihustan	Maruko	Japan	-
Hustopan	Ohta	Japan	-
Marukofon	Maruko	Japan	-
Neoasdrin	Toa	Japan	-
Neobex	Lampugnani	Italy	-
Neusedan	Nippon Zoki	Japan	-
Pectamol	Malesci	Italy	-
Pectussil	Kwizda	Austria	-
Tussilisin	Ibirn	Italy	-
Tussimol	B.D.H.	UK	-

Raw Materials

Phenylacetonitrile	Ethyl chloride
Sodium	β,β'-Dichlorodiethyl ether
Potassium hydroxide	Diethylamine

Manufacturing Process

Preparation of Diethylphenylacetonitrile: 25 grams of sodium was dissolved in 300 ml liquid ammonia containing 0.3 gram ferric chloride and 59 grams phenylacetonitrile was added slowly with stirring. After about 15 minutes a cooled solution of 80 grams of ethyl chloride in 200 ml dry ether was added and the mixture stirred for 1 hour. The ammonia was then allowed to evaporate, water added and the ether layer separated, dried, concentrated and the residual oil distilled in vacuo to yield diethylphenylacetonitrile as an oil, BP 85°C/1 mm.

Preparation of Diethylphenylacetic Acid: 46 grams of the foregoing nitrile was added to 140 ml ethylene glycol containing 36 grams potassium hydroxide and the mixture refluxed with stirring for about 20 hours. The mixture was diluted with water, extracted with light petroleum (BP 60° to 80°C) to remove traces of impurities and then acidified to yield diethylphenylacetic acid which was recrystallized from dilute ethanol (40% v/v ethanol in water).

Preparation of 2-(β-Chloroethoxy)Ethyl Diethylphenylacetate: 19.2 grams of the foregoing acid was added to a solution of 4 grams of sodium hydroxide in 40 ml ethylene glycol. 28.6 grams β,β'-dichlorodiethyl ether was added and the mixture refluxed for 1 hour. After removal of solvent under reduced pressure, 150 ml water was added to the residue and the product extracted with ether. The ethereal solution was dried, concentrated and the residue distilled in vacuo to yield the product as an oil, BP 140°C/0.7 mm.

Preparation of 2-(β-Diethylaminoethoxy)Ethyl Diethylphenylacetate: A mixture of 21 grams of 2-(β-chloroethoxy)ethyl diethylphenylacetate and 14 grams diethylamine was heated under pressure in a sealed tube at 140°C for 5 hours. After cooling, the mixture was dissolved in dilute hydrochloric acid and extracted with ether to remove traces of neutral impurities. The acid layer was then made alkaline with 10% w/v sodium hydroxide solution with cooling, and re-extracted with two portions of ether. The ether extract was dried, the ether distilled off and the residue distilled in vacuo to yield the product as an oil, BP 140°C/0.1 mm.

References

Merck Index 6803
Kleeman & Engel p. 665
OCDS Vol. 1 p. 90 (1977)
I.N. p. 712
Petrow, V., Stephenson, O. and Wild, A.M.; US Patent 2,885,404; May 5, 1959; assigned to The British Drug Houses Limited, England

OXENDOLONE

Therapeutic Function: Antiandrogen

Chemical Name: 16β-Ethyl-17β-hydroxyestr-4-ene-3-one

Common Name: -

Structural Formula:

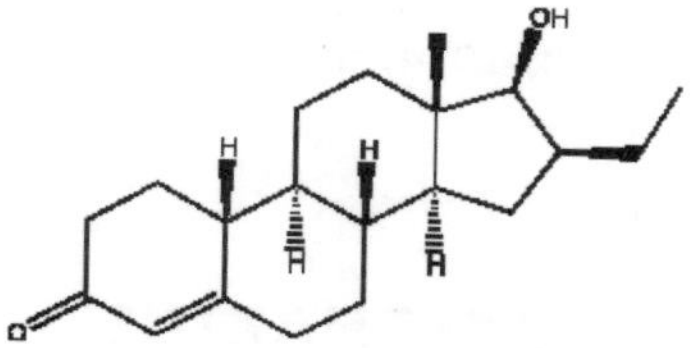

Chemical Abstracts Registry No.: 33765-68-3

Trade Name	Manufacturer	Country	Year Introduced
Prostetin	Takeda	Japan	1981

Raw Materials

Ethyl orthoformate
16β-Ethylestra-4-ene-3,17-dione
Sodium borohydride
Hydrogen chloride

Manufacturing Process

To a solution of 3.0 g of 16β-ethylestra-4-ene-3,17-dione dissolved in 150 ml of dioxane, are added 15 g of ethyl orthoformate and 0.1 g of p-toluenesulfonic acid, followed by stirring for 2 hours at room temperature. The reaction solution is poured into 300 ml of a 5% aqueous solution of sodium hydrogen carbonate and the resultant mixture is extracted with ether. The ether layer is washed with water and dried, followed by evaporation of the solvent to give crude crystals of 3-ethoxy-16β-ethylestra-3,5-diene-17-one. The crystals are recrystallized from ether to give 3.0 g of the compound melting at 114°C to 115°C.

To a solution of 3.0 g of the enol-ether compound obtained above in 50 ml of methanol, is added 1.5 g of sodium borohydride. After standing for 1.5 hours at room temperature, the reaction solution is poured into 300 ml of water. The resulting precipitates are collected by filtration and recrystallized from ether to give 2.8 g of 3-ethoxy-16β-ethylestra-3,5-dien-17β-ol melting at 131°C to 133°C.

To a solution of 2.5 g of 3-ethoxy-16β-ethylestra-3,5-diene-17β-ol dissolved in

50 ml of methanol is added 1.2 ml of concentrated hydrochloric acid, followed by stirring for 10 minutes. The reaction solution is poured into 250 ml of water. The precipitated crystals are collected by filtration and recrystallized from ether to give 2.3 g of 16β-ethyl-17β-hydroxyestra-4-en-one melting at 152°C to 153°C.

References

Merck Index 6804
DFU 5 (9) 44 (1980)
I.N. p. 712
Hiraga, K., Yoshioka, K., Goto, G., Nakayama, R. and Masuoka, M.; US Patent 3,856,829; December 24, 1974; assigned to Takeda Chemical Industries, Ltd.

OXETHAZINE

Therapeutic Function: Local anesthetic

Chemical Name: 2,2'-[(2-Hydroxyethyl)imino]bis[N-(1,1-dimethyl-2-phenylethyl)-N-methylacetamide]

Common Name: Oxetacaine

Structural Formula:

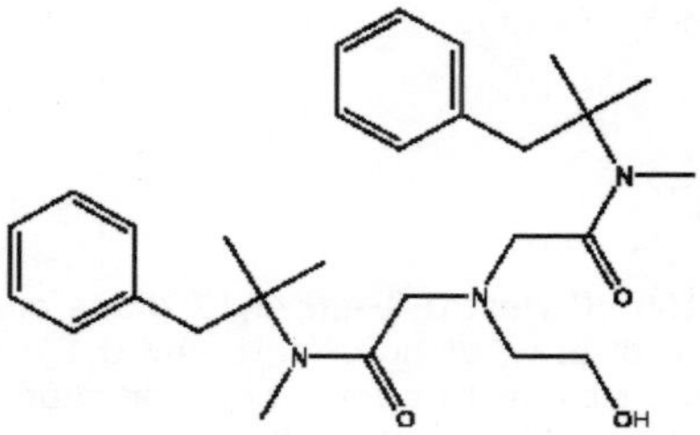

Chemical Abstracts Registry No.: 126-27-2; 13930-31-9 (Hydrochloride salt)

Trade Name	Manufacturer	Country	Year Introduced
Oxaine	Wyeth	US	1960
Emoren	Wassermann	Italy	-
Mucaine	Wyeth	UK	-
Mutesa	Wyeth Byla	France	-
Stomacain	Teisan-Pfizer	Japan	-
Strocain	Eisai	Japan	-
Tepilta	Wyeth	W. Germany	-
Topicain	Chugai	Japan	-

Raw Materials

Chloro-N-methyl-N-ω-phenyl-tert-butyl acetamide
Ethanolamine

Manufacturing Process

Chlor-N-methyl-N-ω-phenyl-tert-butyl acetamide (23.95 g) (0.1 mol) is added to n-butanol (150.0 cc) containing anhydrous potassium carbonate (50.0 g). To the stirred refluxing solution is added dropwise freshly distilled ethanolamine (3.1 g) (0.05 mol). Stirring and refluxing is maintained for twenty hours. Upon cooling the solution is filtered; the residue is washed with n-butanol. The combined filtrates are washed with aqueous sodium carbonate solution then water and finally dried over anhydrous magnesium sulfate. The solvent is distilled under vacuum leaving a dry solid residue. The residue is dissolved in dry benzene to which is added n-hexane to crystallize the product melting at 104°C to 104.5°C. Yield 71-73%. Analysis-Carbon: calc. 71.9%; found 71.93%; hydrogen: calc. 8.8%; found 8.9%; nitrogen: calc. 9.0%; found 9.0%.

To make the hydrochloride salt, the bisacetamide or, by another name, 1,11-diphenyl-2,2,3,9,10,10-hexamethyl-4,8-diketo-6-(β-hydroxyethyl)-3,6,9-triazaundecane is dissolved in n-butanol. The solution is chilled and then dry hydrogen chloride gas is passed into the solution causing an oil to separate. To the heavy oil ether is added and then stirred causing crystallization to occur. MP 146°C to 147°C. Analysis for nitrogen: calc. 83%. found 8.2%.

To make the acetate salt, the bisacetamide (4.7 g) (0.01 mol) is dissolved in ethyl acetate to which is added glacial acetic acid (0.6 g) (0.01 mol). Ether is added to precipitate the acetate as a gum which is washed with hexane, and finally added to dry ether. Allow to stand for crystallization. MP 141°C. Analysis for nitrogen: calc. 8.0%; found 8.2%.

Other salts are: sulfate, MP 56°C; acid oxalate, MP 127°C; tartrate, MP 45°C; picrate, MP 151°C to 152°C.

References

Merck Index 6806
Kleeman & Engel p. 666
OCDS Vol. 1 p.72 (1977)
I.N. p. 712
Seifter, J., Hanslick, R.S. and Freed, M.E.; US Patent 2,780,646; February 5, 1957;assigned to American Home Products Corp.

OXETORONE FUMARATE

Therapeutic Function: Serotonin antagonist, Antihistaminic

Chemical Name: 6-(3-Dimethylamino-1-propylidene)-12H-benzofuro[2,3-e] benz[b]oxepin fumarate

Common Name: -

Structural Formula:

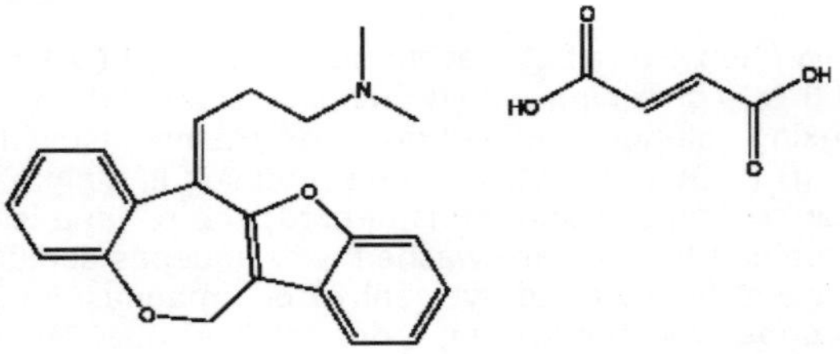

Chemical Abstracts Registry No.: 34522-46-8; 26020-55-3 (Base)

Trade Name	Manufacturer	Country	Year Introduced
Nocertone	Labaz	France	1975
Nocertone	Labaz	W. Germany	1976
Oxedix	Labaz	-	-

Raw Materials

Ethyl iodide	γ-Dimethylaminopropyl chloride
Magnesium	6-Oxo-benzo[b]benzofurano[2,3-e]oxepin
Sulfuric acid	Fumaric acid

Manufacturing Process

(A) Preparation of 6-(3-dimethylaminopropyl)-6-hydroxybenzo[b]benzofurano [2,3-e]oxepin - In a 250 ml flask equipped with a vertical condenser, a dropping-funnel, a dip thermometer and a stirrer, 1.5 g of magnesium turnings and a crystal of iodine were heated until vaporization of the iodine and then cooled, after which 20 ml of dry tetrahydrofuran were added.

The mixture was heated under reflux and a solution of 0.2 g of ethyl iodide in 5 ml of dry tetrahydrofuran was allowed to flow into the reaction medium. When the reaction started, a solution of 6.2 g of γ-dimethylaminopropyl chloride in 20 ml of dry tetrahydrofuran was added and the mixture so obtained was heated under reflux until the complete disappearance of the magnesium turnings. The reaction medium was then cooled in an ice bath, after which there was added thereto a solution in 45 ml of tetrahydrofuran of 7 g of 6-oxo-benzo[bl]-benzofurano[2,3-e]oxepin. The reaction mixture was allowed to stand for 20 hours at a temperature of 20°C, and was then poured into a saturated aqueous solution of ammonium chloride maintained at a temperature of 5°C. The mixture was extracted with ether and the organic portion was washed and dried over anhydrous sodium sulfate. After evaporation of the solvent, 9.4 g of crude product were obtained, which after recrystallization from isopropanol, provided 6.7 g of pure 6-(3-dimethylaminopropyl)-6-hydroxybenzo[b]benzofurano[2,3-e]oxepin, melting point 160°C (yield, 71%).

(8) Preparation of 6-(3-dimethylaminopropylidene)-benzo[b]benzofurano[2,3-e]oxepin and its fumarate -In an Erlenmeyer flask 6.2 g of 6-(3-dimethylaminopropyl)-6-hydroxybenzo[b]benzofurano[2,3-e]oxepin prepared as described above were dissolved in 108 ml of a 10% solution of sulfuric acid. The solution obtained was heated to boiling point for 15 minutes. After cooling, 100 ml of chloroform were added and the solution was made alkaline with a 5% solution of sodium hydroxide. The solution was then extracted with chloroform, washed with water and dried over anhydrous sodium sulfate. The solvent was evaporated and the resulting oily residue composed of 6-(3-dimethylaminopropylidene)-benzo[b]benzofurano[2,3-e]oxepin was then directly treated with a solution of fumaric acid in isopropanol to give 6.5 g of 6-(3-dimethylaminopropylidene)-benzo[b]benzofurano[2,3-e]oxepin fumarate (yield, 85%). The fumarate had a melting point of 160°C when recrystallized from isopropanol.

References

Merck Index 6807
Kleeman & Engel p. 667
OCDS Vol. 3 p. 247 (1984)
DOT 11 (1) 19 (1975)
I.N. p. 712
Binon, F. and Descamps, M.L.V.; US Patent 3,651,051; March 21, 1972; assigned to Labora. toires Labaz

OXFENDAZOLE

Therapeutic Function: Anthelmintic

Chemical Name: Carbamic acid, (5-(phenylsulfinyl)-1H-benzimidazol-2-yl)-, methyl ester

Common Name: Oxfendazole

Structural Formula:

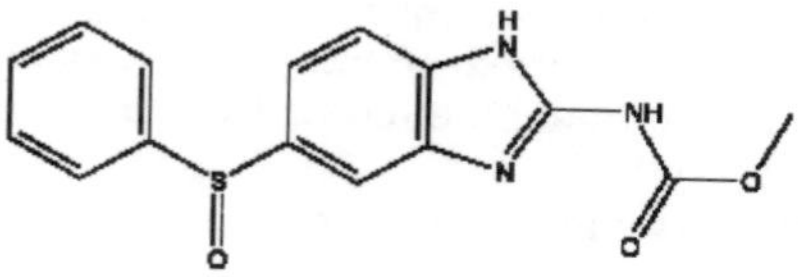

Chemical Abstracts Registry No.: 53716-50-0

Trade Name	Manufacturer	Country	Year Introduced
Autoworm	Coopers	-	-
Oxfendazole	AroKor Holdings Inc.	-	-

Trade Name	Manufacturer	Country	Year Introduced
Benzelmin	Syntex	-	-
Benzelmin	Wyeth Sante animale	-	-
Repidose	Agrovet	-	-
Interzol	Werfft	-	-
Oxfenil	Sanofi	-	-

Raw Materials

Sodium hydride	2-Amino-4-chloro-1-nitrobenzene
Thiophenol	Sodium phenyl mercaptide
Sulfuric acid	Sodium bicarbonate
Acetic anhydride	1,3-Bismethoxycarbonyl-S-methylisothiourea
Sodium acetate	Palladium on carbon
Hydrogen	Peracetic acid
Sodium bisulfite	Acetic acid

Manufacturing Process

5.0 g of 2-amino-4-chloro-1-nitrobenzene is added to a solution of sodium phenyl mercaptide, prepared under nitrogen from 2.53 g 57% sodium hydride and 6.2 ml thiophenol in 20 ml dimethylformamide, with a 10 ml dimethylformamide rinse. The mixture is stirred under nitrogen for 3 h at 20°-30°C and then diluted with water. The crude product is washed with water and hexane, then recrystallized from methanol, yielding 2-amino-4-phenylthio-1-nitrobenzene.

6.0 g of 2-amino-4-phenylthio-1-nitrobenzene is dissolved in 80 ml acetic anhydride and treated with a few drops of sulfuric acid. The mixture is left at 20°-30°C for 2 h then a little sodium acetate added and the solvent removed under vacuum. The residue is treated with water, filtered and recrystallized from methanol yielding 2-acetamido-4-phenylthio-1-nitrobenzene.

7.0 g of 2-acetamido-4-phenylthio-1-nitrobenzene is dissolved in 70 ml chloroform and treated, at -20°C to -15°C, with a solution of 5.0 g 40% peracetic acid in 10 ml methanol. The mixture is allowed to warm slowly to 20°C and stirred for 4 h. The reaction mixture is extracted with sodium bisulfite solution, then sodium bicarbonate solution, dried and evaporated. The residual gum of 2-acetamido-4-phenylsulfinyl-1-nitrobenzene is treated with 20 ml 5 N sodium hydroxide and 40 ml methanol at 20°-25°C for 1 h. Water is then added and essentially pure 2-amino-4-phenyl-sulfinyl-1-nitrobenzene filtered off. Recrystallization may be effected from benzene.

5.4 g of 2-amino-4-phenylsulfinyl-1-nitrobenzene is hydrogenated at 1 atmosphere pressure in 500 ml methanol in the presence of 5.0 g 5% palladized carbon, until the theoretical uptake of hydrogen has occurred. The catalyst is removed by filtration and the filtrate stripped under vacuum. The residue is recrystallized from methanol-benzene, yielding 1,2-diamino-4-phenylsulfinylbenzene.

A mixture of 5.5 g of 1,2-diamino-4-phenylsulfinylbenzene, 4.3 g of 1,3-bis-methoxycarbonyl-S-methylisothiourea and 1.2 ml acetic acid in 100 ml ethanol and 100 ml water is refluxed for 4 h. The mixture is cooled and essentially pure 6-phenylsulfinyl-2-carbomethoxyaminobenzimidazole filtered off and washed with methanol. Recrystallization may be effected from methanol-chloroform (melting point 253°C, dec.).

References

Beard C.C. et al.; US Patent No. 3,929,821; Dec. 30, 1975; Assigned: Syntex (U.S.A.) Inc., Palo Alto, Calif.

OXICONAZOLE NITRATE

Therapeutic Function: Antifungal

Chemical Name: 1-(2,4-Dichlorophenyl)-2-(1H-imidazol-1-yl)-O-(2,4-dichlorobenzyl)-ethanone oxime nitrate

Common Name: -

Structural Formula:

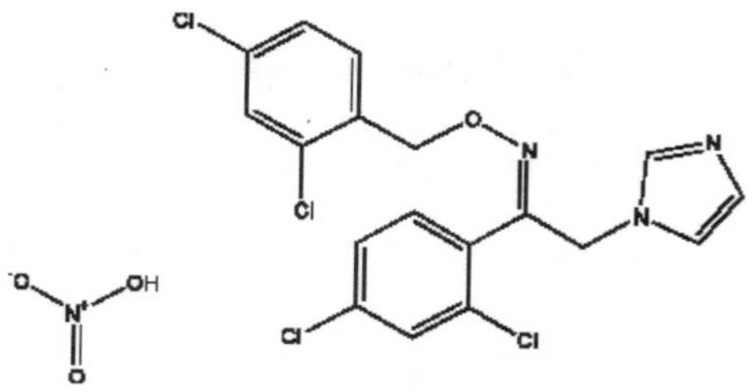

Chemical Abstracts Registry No.: 64211-46-7

Trade Name	Manufacturer	Country	Year Introduced
Myfungar	Siegfried	Switz.	1983
Oceral	Roche	Switz.	1983

Raw Materials

1-(2,4-Dichlorophenyl)-2-(1H-imidazol-1-yl)ethanone oxime
Sodium hydride
2,4-Dichlorobenzyl chloride
Nitric acid

Manufacturing Process

13.5 g of 1-(2,4-dichlorophenyl)-2-(1H-imidazol-1-yl)-ethanone oxime are dissolved in 100 ml dimethylformamide (DMF) and 1.2 g of sodium hydride are mixed in, whereupon an exothermic reaction is allowed to take place on its own with stirring. After cessation of evolution of hydrogen, a solution of 9.8 g of 2,4-dichlorobenzyl chloride in 10 cc DMF is added dropwise with continuous stirring and the stirring is carried on for 2 hours further. The reaction is then taken to completion at a bath temperature of 80°C, after which the reaction mixture is evaporated in a rotation evaporator under reduced pressure and the residue is dissolved in 100 ml ethanol. After filtering off of undissolved matter, the solution is stirred with 300 ml 2N nitric acid for the conversion of free base to the nitrate.

The liquid standing over the heavy deposits which have separated out is separated off by decanting, whereupon an isomer is obtained which after recrystallization from ethanol is obtained in a yield of 5.2 g and having a melting point of 137°C to 138°C.

References

DFU 6 (2) 99 (1981)
DOT 19 (12) 884 (1983)
I.N. p. 713
Mixich, G., Thiele, K. and Fischer, J.; US Patent 4,124,767; November 7, 1978; assigned to Siegfried AG.

OXILOFRINE

Therapeutic Function: Sympathomimetic, Analeptic

Chemical Name: Benzenemethanol, 4-hydroxy-α-(1-(methylamino)ethyl)-, (R*,S*)-

Common Name: Hydroxyephedrine; Methyloxedrine; Methylsynephrine; Oxilofrine; Oxyephedrine

Structural Formula:

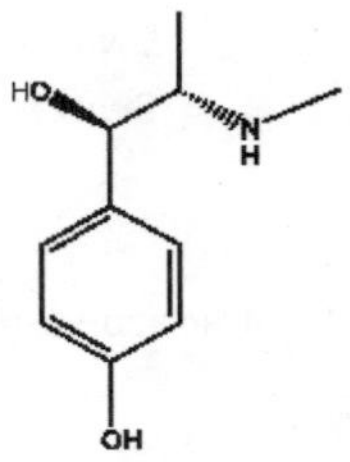

Chemical Abstracts Registry No.: 365-26-4

Trade Name	Manufacturer	Country	Year Introduced
Carnigen	Aventis Pharma Deutschland GmbH	-	-

Raw Materials

Sodium
Bromine
Caustic soda
Palladium
Ethanol absolute
p-Hydroxypropiophenone
Benzyl bromide
Methylbenzylamine
Hydrogen

Manufacturing Process

The p-benzyloxypropiophenone used for the transformation is prepared by adding to a solution of 6.9 g of sodium in 230 ml of absolute alcohol 45.0 g of p-hydroxypropiophenone and 54.0 g of benzyl bromide, and boiling for 1 h in a reflux apparatus. The excess of alcohol is then distilled and the residue is extracted with ether and water. After drying it the ethereal solution is evaporated and the residue is recrystallized from alcohol of 95% strength. The yield amounts to 60.0 g. The p-benzyloxypropiophenone melts at 100°-101°C.

55.0 g of p-benzyloxypropiophenone are dissolved in 250 ml of methylene chloride and brominated with 38.0 g of bromine. As soon as all of the bromine has been introduced drop by drop, the methylene chloride solution is washed with caustic soda solution and water. The solvent is eliminated in a vacuum and the residue is dissolved in petroleum ether. The crystalline mass which soon separates is filtered by suction and washed with petroleum ether. When recrystallized from hexahydrobenzene the p-benzyloxybromopropiophenone melts at 80°C.

40.0 g of p-benzyloxybromopropiophenone are then transformed in an alcoholic solution with 30.0 g of methylbenzylamine. After the whole has been allowed to stand for 1 day, the excess of alcohol is distilled in a vacuum and the residue is dissolved with ether. By washing with water, the benzylmethylamine hydrobrornide then formed is eliminated and the ether is removed by evaporation. The ether residue soon begins to crystallize and the p-benzyloxymethylbenzylaminopropriophenone is obtained with a good yield; when recrystallized from petroleum ether of low boiling point it melts at 58°-60°C.

15.0 g of p-benzyloxymethylbenzylaminopropriophenone are dissolved in alcohol and the solution is hydrogenated with palladium and hydrogen. As soon as the required quantity of hydrogen has been absorbed the palladium is filtered by suction and the alcohol is evaporated in a vacuum The residue is recrystallized and the p-hydroxyphenylmethylaminopropanol is obtained.

References

Bockmuhl M. et al.; US Patent No. 1,877,756; September 20, 1932

OXITEFONIUM BROMIDE

Therapeutic Function: Anticholinergic, Spasmolytic

Chemical Name: Ammonium, diethyl(2-hydroxyethyl)methyl-, bromide, α-phenyl-2-thiopheneglycolate

Common Name: Oxitefonium bromide; Oxytefonium bromide

Structural Formula:

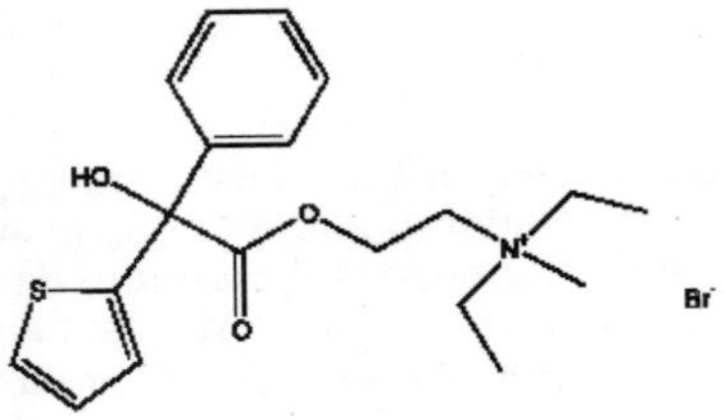

Chemical Abstracts Registry No.: 17989-37-6

Trade Name	Manufacturer	Country	Year Introduced
Oxitefonium bromide	Sintofarm Group	-	-

Raw Materials

Phenyl-(α-thienyl)hydroxyacetic acid
2-Diethylaminoethyl chloride
Methyl bromide

Manufacturing Process

Phenyl-(α-thienyl)hydroxyacetic acid (4.86 g), 2-diethylaminoethyl chloride (2.85 g) and 75 ml of isopropyl alcohol were refluxed for 15 h. After the addition of 50 ml of absolute alcohol to the cold mixture, it was treated with Norite (activated charcoal) at room temperature, filtered, and the solvents removed under reduced pressure. The residue crystallised when triturated with absolute ether. After recrystallisation from absolute alcohol, the phenyl-(α-thienyl)hydroxyacetate of 2-diethylaminoethyl hydrochloride melted at 181-182°C.

The producing of phenyl-(α-thienyl)hydroxyacetate of 2-diethylmethylaminoethyl bromide may be carried out by methylation of phenyl-(α-thienyl)hydroxyacetate of 2-diethylaminoethyl hydrochloride with methylbromide.

References

Blicke F.F., Arbor A.; US Patent No. 2,541,025; Feb. 13, 1951; Assigned: Regents of The University of Michigan, Ann Arbor, Mich., a corporation of Michigan

OXITRIPTAN

Therapeutic Function: Antidepressant, Antiepileptic

Chemical Name: 5-Hydroxytryptophan

Common Name: 5-Hydroxytryptophan

Structural Formula:

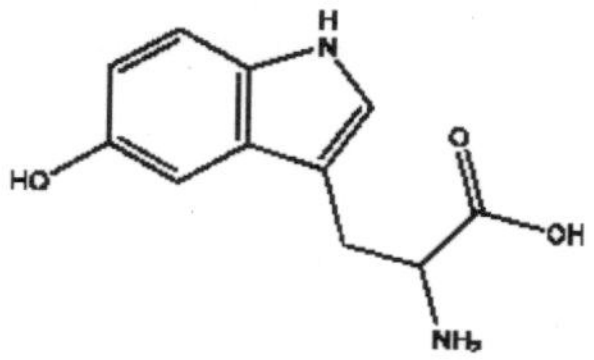

Chemical Abstracts Registry No.: 56-69-9

Trade Name	Manufacturer	Country	Year Introduced
Levotonine	Panmedica	France	1973
Pretonine	Arkodex	France	1973
Tript-Oh	Sigma Tau	Italy	1980
Levothym	Karlspharma	W. Germany	-
Quietim	Nativelle	France	-
Stimolomens	Irbi	Italy	-
Telesol	Lasa	Spain	-

Raw Materials

β-(5-Benzyloxyindolyl-3)-α-acetylamino-α-methylthiopropionic acid methanethiol ester
Hydrogen
Sulfuric acid

Manufacturing Process

β-(5-Benzyloxyindolyl-3)-α-acetylamino-α-methylthiopropionic acid methanethiol ester (449 mg) was added to 10 ml of ethanol and further 1 ml of triethylamine was added to the mixture. Then, the reaction mixture was refluxed for 17 hours, after condensation under reduced pressure and subsequent separation of the residue by column chromatography (silica gel, ethyl acetate), 353 mg of methyl β-(5-benzyloxyindolyl-3)-α-acetylamino-α-methylthiopropionate was obtained as colorless glasslike substance in the yield of 81.5%. Recrystallization of the substance from methanol water afforded 287 mg of crystals.

Raney nickel (3.5 cc) was suspended in 10 ml of ethanol and 356 mg of methyl β-(5-benzyloxyindolyl-3)-α-aminoacetyl-α-methylthiopropionate was added to the mixture together with 20 ml of ethanol. Then, the reaction mixture was stirred for 1 hour at room temperature and thereafter filtered to remove insoluble substances. The residue was washed with 100 ml of ethanol and 50 ml of acetone and both the filtrate and the wash liquid were combined and concentrated under reduced pressure. By column chromatography (silica gel and acetone), 210 mg of methyl β-(5-hydroxyindolyl-3)-α-acetylaminopropionate as colorless glasslike substance in the yield of 90%.

To 430 mg of methyl β-(5-hydroxyindolyl-3)-α-acetylaminopropionate was added 50 ml of 10% sulfuric acid and the reaction mixture was refluxed under heating for 10 hours. After condensation under reduced pressure to 15 ml volume, the reaction solution was neutralized with ammonia to pH 4, to afford the extract. The resulting extract was filtered and washed with water to afford 265 mg of 5-hydroxytryptphan in the yield of 78%.

References

Merck Index 4771
Kleeman and Engel p. 668
I.N. p. 714
Tsuchihashi, G. and Ogura, K.; U.S. Patent 4,001,276; January 4,1977; assigned to Sagami Chemical Research Center (Japan)

OXITROPIUM BROMIDE

Therapeutic Function: Anticholinergic bronchodilator

Chemical Name: (-)-N-Ethylnorscopolamine methobromide

Common Name: OTB

Structural Formula:

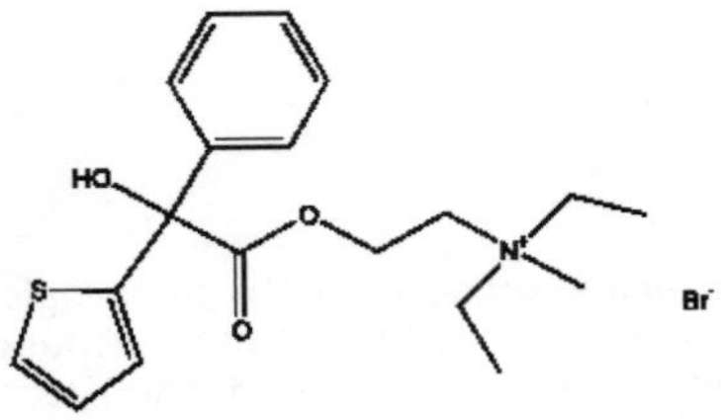

Chemical Abstracts Registry No.: 30286-75-0

Trade Name	Manufacturer	Country	Year Introduced
Ventilat	Boehringer Ingelheim	W. Germany	1983

Raw Materials

(-)-Norscopolamine
Methyl bromide
Ethyl bromide
Sodium carbonate

Manufacturing Process

14.5 g (0.05 mol) of (-)-norscopolamine and 5.4 g (0.05 mol) of ethyl bromide were dissolved in 300 cc of acetonitrile, 5.3 g (0.05 mol) of anhydrous sodium carbonate were suspended in the solution, and the suspension was heated at the boiling point for 10 hours. After a boiling time of 2.5 and 5 hours, respectively, the supply of ethyl bromide and sodium carbonate in the reaction mixture was replenished by adding each time 5.4 g (0.05 mol) of ethyl bromide and 5.3 g (0.05 mol) of anhydrous sodium carbonate. At the end of 10 hours of boiling, the inorganic sodium salts which had separated out were separated by vacuum filtration, the filter cake was washed with acetonitrile, and the acetonitrile was distilled out of the filtrate. The distillation residue was dissolved in ether, the solution was extracted with a small amount of water and then dried, and the ether was distilled off, yielding raw (-)-N-ethylnorscopolamine.

7.0 g (0.022 mol) of (-)-N-ethylnorscopolamine were dissolved in acetonitrile, 10.4 g (0.11 mol) of methyl bromide were added to the solution, and the mixture was allowed to stand at room temperature. The crystalline precipitate formed thereby was collected and recrystallized from acetonitrile.8.9 g (97.8% of theory) of white crystalline (-)-N-ethylnorscopolamine methobromide, melting point 203°C to 204°C (decomposition), were obtained.

References

Merck Index A-10
DFU4 (2) 117 (1979)
DOT 19 (7) 416 and (8) 444 (1983)
Zeile, K., Banholzer, R., Walther, G., Schulz, W. and Wick, H.; US Patent 3,472,861; Oct. 14, 1969; assigned to Boehringer Ingelheim GmbH.

OXOLAMINE CITRATE

Therapeutic Function: Antitussive, Antiinflammatory

Chemical Name: 1,2,4-Oxadiazole, 5-(2-(diethylamino)ethyl)-3-phenyl-, citrate

Common Name: Oxolamine citrate

Raw Materials

3-Chloropropionyl chloride
N-Hydroxybenzamidine
Diethyl amine

Structural Formula:

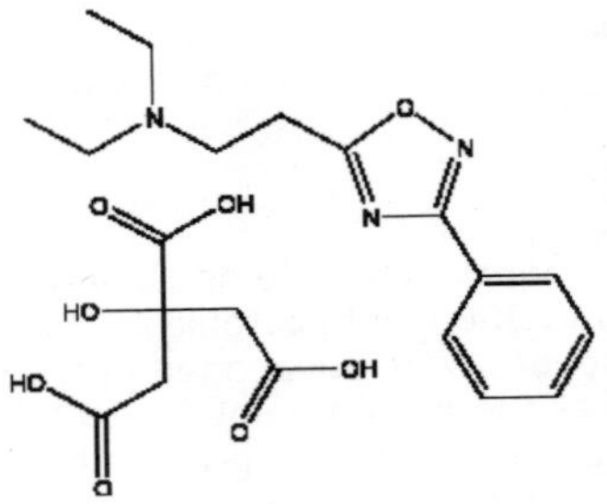

Chemical Abstracts Registry No.: 1949-20-8 ; 959-14-8 (Base)

Trade Name	Manufacturer	Country	Year Introduced
Bredon	Organon	-	-
Broncatar	Pulitzer	-	-
Perebron	Angelini Francesco	-	-
Prilon	Cassenne	-	-
Flogobron	Intersint	-	-
Oxolamine citrate	Milen	-	-
Oxolamine citrate	Yick-Vic Chemicals and Pharmaceuticals (HK) Ltd.	-	-
Oxolamine citrate	Shanghai Lansheng Corporation	-	-
Kalamin	Ilsan Ilac	-	-
Oxadron	Fustery	-	-
Regal	Andromaco	-	-
Aledron	IQFA	-	-

Manufacturing Process

18.7 g 3-chloro-propionyl chloride in 50 ml of dry ether was added dropwise with stirring to the cooled with ice solution 40 g of N-hydroxybenzamidine in 450 ml dry ester. A dense precipitate had fallen. The mixture was stirred 0.5 hour at room temperature and then was filtered off. The precipitate was thoroughly washed with water for a removal hydrochloride of starting N-hydroxybenzamidine whereas N-(amino)phenylmethylene-3-Cl-propionamide didnt solve. It was dried in vacuum over P_2O_5. Yield of clean product 94% from theoretical. MP: 98°-99°C. The solution of 9.2 g diethyl amine in 50 ml dry benzene was added dropwise to a suspension of above prepared N-(amino)phenylmethylene-3-Cl-propionamide in dry benzene with stirring and cooling.

The mixture was warmed and stirred else 2 hours after adding. Then it was cooled, washed two times with water and dried over $CaCl_2$. Then the solvent was removed. The residue was distilled in vacuum. Diethyl-[2-(3-phenyl[1,2,4]oxadiasol-5)ethyl]amine (oxolamine) had BP: 127°C/0.4 mm. Yield 10.5 g.

In practice it is usually used as citrate.

References

Angelini F.; D.B. Patent No. 1,097,998; Sept. 30, 1959

OXOLINIC ACID

Therapeutic Function: Antibacterial (urinary)

Chemical Name: 1-Ethyl-1,4-dihydro-4-oxo-1,3-dioxolo[4,5-g]quinoline-3-carboxylic acid

Common Name: -

Structural Formula:

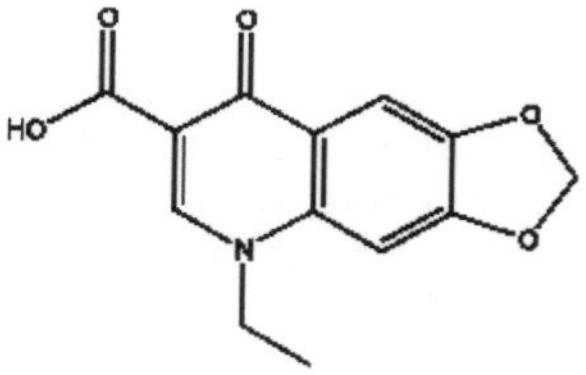

Chemical Abstracts Registry No.: 14698-29-4

Trade Name	Manufacturer	Country	Year Introduced
Prodoxol	Warner	UK	1974
Urotrate	Substantia	France	1974
Ossian	Bioindustria	Italy	1974
Utibid	Warner Lambert	US	1975
Nidantin	Sasse/Goedecke	W. Germany	1978
Decme	Poli	Italy	-
Emyrenil	Emyfar	Spain	-
Gramurin	Chinoin	Hungary	-
Oksaren	Belupo Ltd.	Yugoslavia	-
Ossion	Bioindustria	Italy	-
Oxoboi	B.O.I.	Spain	-
Oxoinex	Inexfa	Spain	-
Oxol	Casen	Spain	-
Oxolin	Prodes	Spain	-
Pietil	Argentia	Argentina	-
Tilvis	Scharper	Italy	-
Tropodil	Elea	Argentina	-
Urinox	Syncro	Argentina	-
Uro-Alvar	Alvarez-Gomez	Spain	-

Trade Name	Manufacturer	Country	Year Introduced
Uropax	Lefa	Spain	-
Uroxol	Ausonia	Italy	-

Raw Materials

3,4-Methylenedioxyaniline
Diethyl ethoxymethylenemalonate
Sodium hydroxide
Ethyl iodide

Manufacturing Process

A mixture of 27 parts by weight of 3,4-methylenedioxyaniline and 43 parts by weight of diethyl ethoxymethylenemalonate is heated at 80° to 90°C for 3 hours. The mixture is then heated at 80° to 90°C for 1 hour under about 15 mm pressure to remove the byproduct ethyl alcohol formed. The residue is recrystallized from ligroin (BP 60° to 90°C) to give diethyl[(3,4-methylenetlioxyanilino)methylene] malonate as a yellow solid melting at 100° to 102°C. The analytical sample from ligroin melts at 101° to 102°C.

A mixture of 48 parts by weight of diethyl[(3,4-methylenedioxyanilino) methylene] malonate and 500 parts by weight of diphenyl ether is refluxed for 1 hour. The mixture is allowed to cool to about 25°C with stirring and 500 parts by weight of petroleum ether are added. Filtration gives 3-carbethoxy-6,7-methylenedioxy-4-hydroxy-quinoline as a brown solid, MP 276° to 281°C. Several recrystallizations from dimethylformamide gives almost colorless analytical material, MP 285° to 286°C, (decomposes).

A mixture of 26 parts of 3-carbethoxy-6,7-methylenedioxy-4-hydroxy-quinoline,16 parts of sodium hydroxide and 50 parts of dimethylformamide is heated at 70° to 75°C for 2 hours, then 31 parts of ethyl iodide is added over 1 hour with continued heating and stirring. After an additional 3 to 4 hours of heating (at 70° to 75°C) and stirring, the mixture is diluted with 500 parts of water, refluxed for 3 to 4 hours, acidified with concentrated hydrochloric acid and filtered to yield 18 to 22 parts of 1-ethyl-1,4-dihydro-6,7-methylene-dioxy-4-oxo-3-quinoline-carboxylic acid, MP 309° to 314°C (decomposes). The analytical sample from dimethylformamide melts at 314° to 316°C (decomposes).

References

Merck Index 6814
Kleeman & Engel p. 670
OCDS Vol. 2 pp. 370, 387(1980) and 3, 185 (1984)
I.N. p. 34
Kaminsky, D. and Meltzer, R.I.; US Patent 3,287,458; November 22, 1966; assigned to Warner-Lambert Pharmaceutical Company

OXOMEMAZINE

Therapeutic Function: Antihistaminic

Chemical Name: N,N,β-Trimethyl-10-H-phenothiazine-10-propanamine 5,5-dioxide

Common Name: -

Structural Formula:

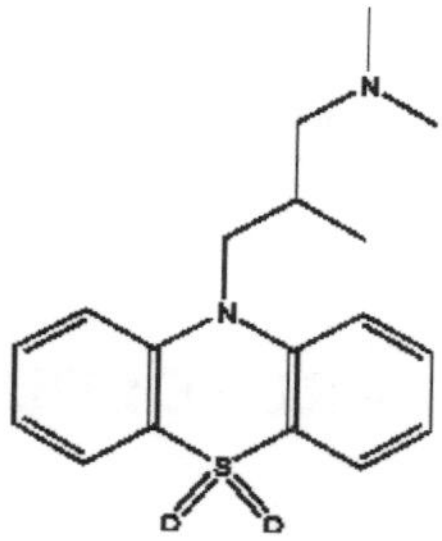

Chemical Abstracts Registry No.: 3689-50-7; 4784-40-1 (Hydrochloride salt)

Trade Name	Manufacturer	Country	Year Introduced
Doxergan	Specia	France	1964
Imakol	Rhone Poulenc	W. Germany	1965
Dysedon	Meiji	Japan	-
Rectoplexil	Specia	France	-
Toplexil	Specia	France	-

Raw Materials

Phenothiazine
Sodium amide
3-Dimethylamino-2-methylpropyl chloride
Hydrogen peroxide

Manufacturing Process

Phenothiazine is reacted with 3-dimethylamino-2-methylpropyl chloride in the presence of sodium amide to give 3-(10-phenthiazinyl)-2-methyl-1-dimethylaminopropane. 11.9 g of of this intermediate is dissolved with agitation in glacial acetic acid (120 cc). Pure sulfuric acid (d = 1.83; 0.5 cc) is added and a mixture of glacial acetic acid (10 cc) and hydrogen peroxide (8.5 cc of a solution containing 38 g of hydrogen peroxide in 100 cc) is then run in over 20 minutes. The temperature rises from 25°C to 35°C and is then kept at 60°C for 18 hours. The mixture is cooled and water (150 cc) is added and, with cooling, aqueous sodium hydroxide (d = 1.33; 220 cc). The resulting mixture is extracted with ethyl acetate (3 x 100 cc), the solvent is evaporated on a water bath and the residue is recrystallized from heptane (150 cc). 3-(9,9-dioxy-10-phenthiazinyl)-2-methyl-1-dimethylaminopropane (78 g) is obtained, MP 115°C.

The corresponding hydrochloride prepared in ethyl acetate and recrystallized from a mixture of ethanol and isopropanol melts at 250°C.

References

Merck Index 6815
Kleeman & Engel p. 670
DOT 2 (4)145 (1966)
I.N. p. 715
Jacob, R.M. and Robert, J.G.; US Patent 2,972,612; February 21, 1961; assigned to Societe des Usines Chimiques Rhone-Poulenc (France)

OXPRENOLOL

Therapeutic Function: Antiarrhythmic

Chemical Name: 1-[(1-Methylethyl)amino]-3-[2-(2-propenyloxy)phenoxy]-2-propanol

Common Name: -

Structural Formula:

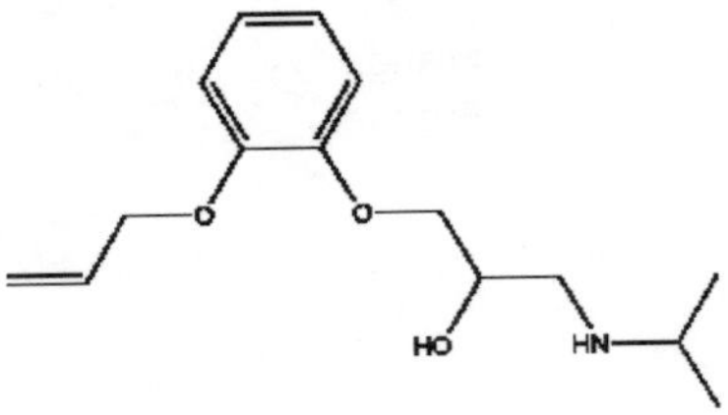

Chemical Abstracts Registry No.: 6452-71-7; 6452-73-9 (Hydrochloride salt)

Trade Name	Manufacturer	Country	Year Introduced
Trasicor	Ciba Geigy	Italy	1970
Trasicor	Ciba Geigy	W. Germany	1971
Trasicor	Ciba Geigy	UK	1972
Trasicor	Ciba Geigy	France	1975
Trasacor	Ciba Geigy	Japan	1976
Captol	Protea	Australia	-
Cordexol	Lagap	Switz.	-
Coretal	Polfa	Poland	-

Raw Materials

Epichlorohydrin
Isopropylamine
Pyrocatechol monoallyl ether

Manufacturing Process

75 grams of pyrocatechol monoallyl ether, 75 grams of epichlorohydrin, 75 grams of potassium carbonate and 400 ml of acetone are stirred and heated at the boil for 12 hours. The potassium carbonate is then filtered off. The solvent is distilled off in a water-jet vacuum. The residual oil is dissolved in ether and agitated with 2 N sodium hydroxide solution. The ether is separated, dried and distilled off. The residue is distilled in a water-jet vacuum. 3-(ortho-allyloxy-phenoxy)-1,2-epoxypropane passes over at 145° to 157°C under 11 mm Hg pressure. A solution of 15 grams of 3-(ortho-allyloxy-phenoxy)-1,2-epoxypropane and 15 grams of isopropylamine in 20 ml of ethanol is refluxed for 4 hours. The excess amine and the alcohol are then distilled off under vacuum, to leave 1-isopropylamino-2-hydroxy-3-(ortho-allyloxy-phenoxy)-propane which melts at 75° to 80°C after recrystallization from hexane.

References

Merck Index 6820
Kleeman & Engel p. 671
OCDS Vol.1 p.117 (1977) and 2, 109 (1980)
DOT 6 (1) 25 (1970)
I.N. p. 716
Ciba Limited, Switzerland; British Patent 1,077,603; August 2, 1967

OXYBUTYNIN CHLORIDE

Therapeutic Function: Spasmolytic

Chemical Name: α-Cyclohexyl-α-hydroxybenzeneacetic acid 4-(dietylamino)-2-butynyl ester hydrochloride

Common Name: -

Structural Formula:

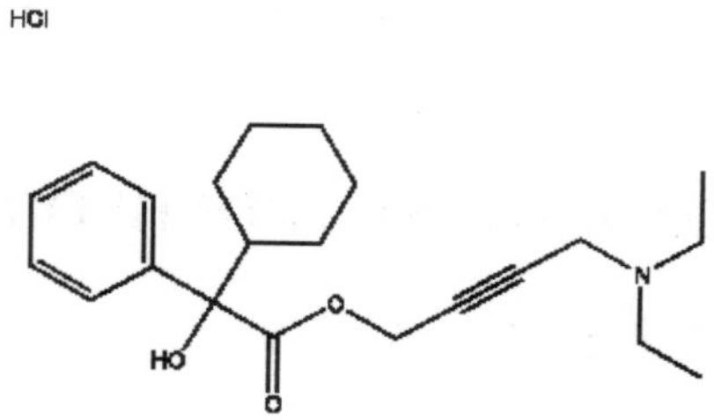

HCl

HO O O N

Chemical Abstracts Registry No.: 1508-65-2; 5633-20-5 (Base)

Trade Name	Manufacturer	Country	Year Introduced
Ditropan	Marion	US	1975
Ditropan	Scharper	Italy	-

Raw Materials

Methyl phenylcyclohexylglycolate
4-Diethylamino-2-butynyl acetate
Sodium methylate

Manufacturing Process

A mixture of 394.2 grams of methyl phenylcyclohexylglycolate and 293.1 grams of 4-diethylamino-2-butynyl acetate was dissolved with warming in 2.6 liters of n-heptane. The solution was heated with stirring to a temperature of 60° to 70°C and 8.0 grams of sodium methoxide were added. The temperature of the mixture was then raised until the solvent began to distill. Distillation was continued at a gradual rate and aliquots of the distillate were successively collected and analyzed for the presence of methyl acetate by measurement of the refractive index. The reaction was completed when methyl acetate no longer distilled, and the refractive index observed was that of pure heptane (n_D^{26} = 1.3855). About 3½ hours were required for the reaction to be completed.

The reaction mixture was then allowed to cool to room temperature, washed with water, and extracted with four 165 ml portions of 2 N hydrochloric acid. The aqueous extracts were combined and stirred at room temperature to permit crystallization of the hydrochloride salt of the desired product. Crystallization was completed by cooling the slurry in an ice bath, and the product was collected by filtration, pressed dry, and recrystallized from 750 ml of water. Yield of pure crystalline material, 323 grams.

References

Merck Index 6823
Kleeman & Engel p. 672
PDR p. 1076
OCDS Vol. 1 p. 93 (1977)
I.N. p. 716
REM p.919
Mead Johnson and Company; British Patent 940, 540; October 30, 1963

OXYCODONE HYDROCHLORIDE

Therapeutic Function: Narcotic analgesic

Chemical Name: Morphinan-6-one, 4,5α-epoxy-14-hydroxy-3-methoxy-17-methyl- hydrochloride

Common Name: Dihydrohydroxycodeinone hydrochloride; Hydrooxycodeinona; Hydroxydihydrocodeinonum hydrochloricum; Ossicodone; Oxiconum; Oxikon; Oxycodeinone hydrochloride; Oxycodone hydrochloride; Thecodin

Structural Formula:

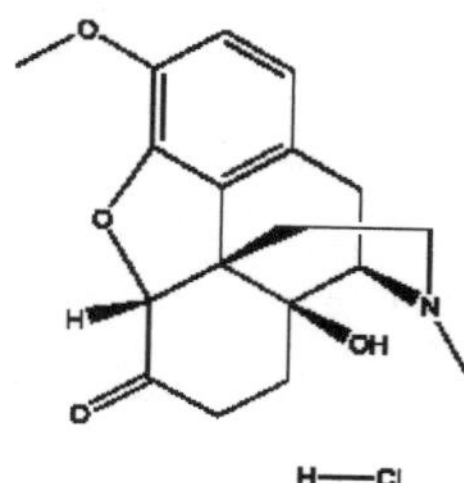

Chemical Abstracts Registry No.: 76-42-6 (Base); 124-90-3

Trade Name	Manufacturer	Country	Year Introduced
Oxyfast	Purdue Pharma, L.P.	-	-
Endone	Boots	-	-
Oxynorm	Mundipharma	-	-
Roxicodone	Roxane	-	-
Percodan	Endo Pharmaceuticals Inc.	-	-
Codeinona	Higiene	-	-

Raw Materials

Dibutylurea
$t-BuMe_2SiCl$
Peracetic acid
Acetic acid
Hydrogen
Codeine sulfate trihydrate
Ammonium hydroxide
Aluminum isopropoxide
Potassium sodium tartrate tetrahydrate
Palladium on carbon

Manufacturing Process

Preparation of codeinone from codeine:

Codeinone was prepared by oxidation of codeine sulfate trihydrate. A reaction mixture was prepared containing codeine sulfate trihydrate (10.4 g), deionized water (20 g) and isopropyl acetate (87.2 g) at ambient temperature. The reaction mixture was agitated and the resultant mixture cooled to about 20°C. Concentrated ammonium hydroxide (18.0 g) was added in several portions and the mixture was maintained at a temperature of about 20°C with stirring. Sting was continued for about 15 min, and then a small portion of the aqueous layer was withdrawn to check for pH value, which was to be advantageously maintained between 11.0 and 12.0. The aqueous layer was then separated and reextracted with isopropyl acetate (35 g). The combined organic layers (isopropyl acetate) were concentrated in vacuo to near dryness at temperature 45°C. The residual isopropyl acetate solvent was chased by adding 18 g of toluene. The concentration process was then repeated in vacuo. Codeine free base dissolved in a mixture of toluene (177 g) and cyclohexanone (47.4 g) at 45°C was then transferred to the reaction flask which was equipped with magnetic stirrer, thermocouple, Dean-Stark trap with condenser attached, addition funnel with an extender (about 4 inches height), and a nitrogen inlet adapter. The mixture was heated to boiling temperature (about 116-118°C) under a nitrogen atmosphere and 26 g (30 ml) of distillate were collected in the Dean-Stark trap. A solution of aluminum isopropoxide (3.5 g) in 35.5 g (41 ml) of toluene was then added to the addition funnel. The heating rate was adjusted and the aluminum isopropoxide/toluene solution was added into the reaction mixture at such a rate that the total volume was added over a 10-20 min period [approximately the same volume (41 ml) of distillate was collected in the Dean-Stark trap]. After completion of the addition, collection of the distillate was continued such that 57 g (66 ml) of distillate was collected in the Dean-Stark trap at a similar distillation rate. The heat source was removed and the mixture allowed to cool down to ambient temperature (under nitrogen atmosphere) over a period of about 30 min. Reaction completeness was determined by withdrawing a small sample from the batch, extracting it with a saturated sodium bicarbonate solution and ethyl acetate, concentrating the organic layer, redissolving it with the HPLC mobile phase, and analyzing the sample on HPLC.

An aqueous solution of 13 wt. % Rochelle salt was then prepared by dissolving 19.5 g of potassium sodium tartrate tetrahydrate in 130.5 g of deionized water at 20°C. The aqueous Rochelle salt solution (90 ml) was added into the reaction mixture in one portion at ambient temperature, the batch stirred for about 10 min and filtered. Both layers were saved. The organic layer was washed with 60 ml of aqueous Rochelle salt solution (both layers were saved). The organic layer was washed with a mixture of 30 ml brine solution and 30 ml 5% sodium bicarbonate solution (both layers were saved). All aqueous layers were then combined and extracted with 43 g (50 ml) of toluene. The aqueous layer was discarded. The organic layers were then combined and concentrated in vacuo at temperature 55°C to near dryness. 22 g (25 ml) of toluene was added and the resultant organic layer concentrated in vacuo twice more to remove residual cyclohexanone. Subsequently, 11.8 g (15 ml) of 2-propanol was added and the mix slurried at 0-5°C for at least eight hours under a nitrogen atmosphere. Solids were then filtered. The latter operation was repeated until no solids were left in the flask. The chilled wet cake was then rinsed with chilled (5-10°C) 2-propanol (12 g, 15 ml), and filter dried. The wet cake was then rinsed with heptane (6.8 g, 10 ml) and dried. The resulting solids were vacuum dried at 50°C to a constant weight. A yield of 5.2 to 6.45 g (65.4 to 81.2%) of white solids, with HPLC purity of about 96-99.3% was obtained. The compound was stored in a dark and cool place.

Preparation of dienolsilyl ether of codeinone:

Codeinone (6.0 g) with toluene (104 g) was added to a reaction flask equipped with a mechanical stirrer, thermocouple, Dean-Stark trap with condenser attached, and a nitrogen inlet. The batch was heated to reflux and about 27.7 g (32 ml) of distillate was collected in the Dean-Stark trap. The contents were then cooled to 20°C under a nitrogen atmosphere. A solution of dibutyl urea (DBU) (4.22 g) in toluene (3 g) was added in one portion. Subsequently, a solution of t-$BuMe_2SiCl$ (4.22 g) in toluene (5 g) was likewise added in one portion. The batch was slowly warmed to 58°C and stirred at this temperature for about 2 hours. Completion of the reaction was adjudged by withdrawing a 20 small sample from the batch, extracting it with a mixture of ethyl acetate and saturated sodium bicarbonate solution, spotting the organic layer on a TLC plate, and then eluting it with a mobile phase of 9:1 mixture of dichloromethane and methanol plus 3-4 drops of concentrated ammonium hydroxide. If the reaction was determined to be incomplete, stirring was continued at 58°C for an additional 2 hours and a TLC check performed once more. Alternatively reaction completion was accomplished by adding about 5-10% more of both DBU and t-$BuMe_2SiCl$ to the reaction mixture at the same temperature. The contents were then cooled to 20°C, and a mixture of 5% sodium bicarbonate solution (80 ml) and 60 ml of water was added in one portion. Stirring continued for about 10 min. The aqueous layer was then separated and discarded. The organic layer was washed with a mixture of 50 ml brine and 50 ml saturated ammonium chloride solution (the aqueous layers were discarded). The organic layer was concentrated to near dryness in vacuo at temperature 50°C, and the residue diluted with 33.2 g of toluene to make up a 20 wt % stock solution. Yield was approximately quantitative. The stock solution was found to be stable at ambient temperature under nitrogen atmosphere for at least 6 months.

Preparation of 14-hydroxycodeinone from denolsilyl ether of codeinone:

Peracetic acid solution (107.7 g of 9.0 wt % peracetic acid) at ambient temperature was added to a reaction flask equipped with mechanical stirrer and thermocouple, nitrogen inlet adapter and addition funnel. A 20 wt % stock solution of the dienolsilyl ether of codeinone (41.7 g) was added through the addition funnel over a period of about 5 min and the temperature of the contents maintained at 28°C. The batch was stirred at 22°C for at least 3 hours. In order to test reaction completeness, a small sample was withdrawn from the batch and quenched with saturated sodium bicarbonate solution, and extracted with ethyl acetate. The EtOAc layer was spotted onto a TLC plate and subsequently checked for the disappearance of starting dienolsilyl ether of codeinone. The TLC mobile phase was a mixture of 95:5 of dichloromethane and methanol plus 3-5 drops of concentrated ammonium hydroxide. If the reaction was adjudged incomplete, the mixture was stirred at the same temperature for an additional 2 hours then analyzed by TLC again. Alternatively completion of the reaction was pushed by the addition of 10 g of peracetic acid (9.0 wt %) and stirring for an additional 1 h (analysis was then once more performed using TLC).

Upon determination of the completion of the reaction 20.0 g of 10 wt. % of aqueous sodium hydrogen sulfate solution was added in one portion, and the resultant admixture stirred for 10 min at ambient temperature. The batch was then concentrated in vacuo at 45°C to dryness. Subsequently water (180 g), toluene (69 g), ethyl acetate (36 g) were added and vigorous stirring for about 10 min undertaken. The resulting layers were separated and the aqueous layer saved in a flask. The organic layer was washed thrice with a solution of 26 ml of 2.5% HCl. The combined aqueous layers were then filtered through a pad of wet (with water) hyflo-supercel filter aid. Subsequently, EtOAc (85 g) was added to the filtrate and concentrated ammonium hydroxide added in a quantity to adjust the pH of the aqueous layer to about 11. The mixture was stirred for 10 min at about 60°C and the layers were separated and saved. The aqueous layer was washed with EtOAc (50 g) and then discarded. The combined organic layers were concentrated in vacuo to dryness at 50°C. To the residue was added 2-propanol (13 g), and the resultant mixture stirred at 5-10°C for at least 5 hours. The solids were filtered, the flask and solids rinsed with the chilled (5°C) filtrate followed by chilled (5-10°C) 2-propanol (10 g) and heptane (8 g). The solid was then vacuum dried at 50°C to a constant weight. A yield of between 3.50-4.96 g (55%-78%) of 14-hydroxycodeinone free base with a purity of over 96A% was obtained.

Preparation of oxycodone from 14-hydroxycodeinone by catalytic hydrogenation:

14-Hydroxycodeinone (4.98 g) and acetic acid (155 g) were added to a Parr shaker equipped with hydrogen inlet and outlet connectors. The mixture was shaken for about 5 min to completely dissolve the 14-hydroxycodeinone at ambient temperature. The system was then evacuated and the Parr shaker was filled with nitrogen. In one portion, under the nitrogen atmosphere, 10% Pd/C (50% water wet, 4.0 g) was added. The system was then evacuated, and was filled with hydrogen gas to a pressure of about 38 psi. The hydrogen inlet from the supply tank was then closed and the mixture was shaken at an initial pressure of 38 psi for about 3 hours (at ambient temperature). After 3 hours of shaking, the system was evacuated and filled with nitrogen. The contents were filtered over a hyflo-supercel filtering pad (3 g, wetted with water). The Parr bottle and wet cake were then rinsed with acetic acid (2 x 21 g). The filtrate was concentrated in vacuo to dryness at 50°C. The residue was then dissolved with deionized water (50 g), and the pH adjusted to about 11.0 to 12.0 using 20% aqueous KOH solution and concentrated ammonium hydroxide (4 g). The mixture was then extracted with ethyl acetate (4 x 135 g), and the combined organic layers concentrated in vacuo to dryness. A yield of 3.51 to 4.26 g of crude oxycodone with HPLC purity of over 85A% (70.0 to 85.0% yield) was obtained.

Preparation of oxycodone from 14-hydroxycodeinone by catalytic transfer hydrogenation method:

14-Hydroxycodeinone (4.98 g) and acetic acid (137 g) were added to a reaction flask (3-neck, 250 ml) equipped with mechanical stirrer, addition funnel, thermocouple and nitrogen-inlet adapter. The system was evacuated and the flask filled with nitrogen. Subsequently, 5% Pd/C (50% water wet, 3.0 g) in one portion was added under the nitrogen atmosphere. While the mixture was stirred for about 5 min at ambient temperature, a solution of sodium hypophosphite (6.0 g) in deionized water (25 g) was prepared. The aqueous sodium hypophosphite solution was transferred into the addition funnel, and added to the reaction mixture over a period of about 30 min at about 22°C. The mixture was then warmed to about 45°C and stirred for about 1 hour.

Upon the reaction was complete, the batch was cooled to ambient temperature under the nitrogen atmosphere, and the contents filtered over a hyflo-supercel filtering pad (3.0 g, wetted with water). The flask and wet cake were rinsed with acetic acid (20 g). The filtrate was concentrated in vacuo to near dryness at temperature 50°C. The residue was dissolved with deionized water (50 g) and the pH adjusted to 11.0 to 12.0 with 20% aqueous KOH solution and concentrated ammonium hydroxide (about 4 g). The mixture was then extracted with ethyl acetate (4 x 135 g) and the combined organic layers concentrated to dryness in vacuo. Crude oxycodone with an HBLC purity of over 85% was obtained in a yield of 70.0 to 85.0% (3.51 to 4.26 g).

References

Chiu Fang-Ting, Lo Young S.; US Patent No. 6,469,170; October 22, 2002; Assigned to Boehringer Ingelheim Chemicals, Inc. (Petersburg, VA)

OXYDIBUTANOL

Therapeutic Function: Choleretic, Spasmolytic

Chemical Name: 4,4'-Oxybisbutan-2-ol

Common Name: Dihydroxydibutyl ether; Hydroxybutyloxide; Oxydibutanol

Structural Formula:

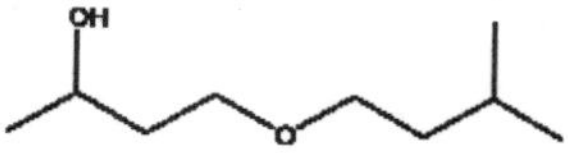

Chemical Abstracts Registry No.: 821-33-0

Trade Name	Manufacturer	Country	Year Introduced
Dyskineacutebyl	Saunier	-	-
Dis-Cinil	Lusofarmaco	-	-
Dihydroxydibutyl ether	Joulty	-	-

Raw Materials

Butyl magnesium bromide
β-Cyanoethyloxide
Nickel Raney
Hydrogen

Manufacturing Process

To a solution of 3 mols butylmagnesiumbromide in ether was added dropwise an one mol of β-cyanoethyloxide. The mixture was refluxed for 1 hour. Then after cooling to the mixture was added hydrochloric acid. An organic layer was dried under sodium sulfate and evaporated. An oil layer was distilled in vacuum. The oxo-3-butane oxide has a boiling point 123-125°C at 15 mm.

A mixture of 100 g of oxo-3-butane oxide, 500 ml ethanol and 15 g Nickel Reney was loaded in autoclave and heated at 45-50°C. In autoclave was introduced a hydrogene. After cooling the mixture was filtered and ethanol was distilled off. 1,1'-Dimethyl-3,3'-oxydipropanol was distilled at 160-161°C/18 mm.

References

Merck Index, Monograph number: 7094, Twelfth edition, 1996, Editor: S. Budavari; Merck and Co., Inc.

FR Patent No. 1,267,084; June 5, 1961; Assigned to Joulty M.A., resident of France

OXYFEDRINE

Therapeutic Function: Coronary vasodilator

Chemical Name: (R)-3-[(2-Hydroxy-1-methyl-2-phenylethyl)amino]-1-(3-methoxyphenyl)-1-propanone

Common Name: -

Structural Formula:

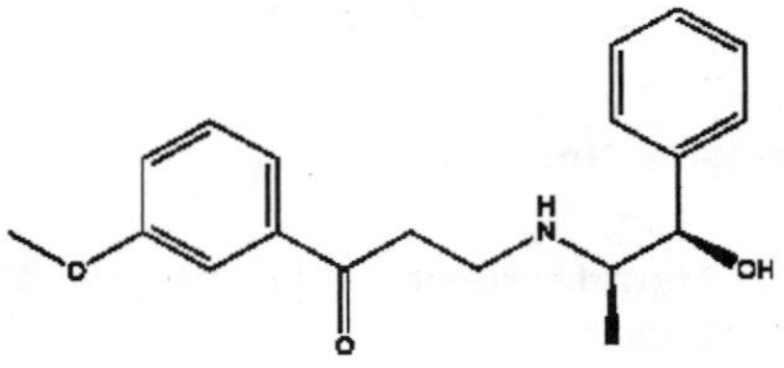

Chemical Abstracts Registry No.: 15687-41-9; 16777-42-7 (Hydrochloride salt)

Trade Name	Manufacturer	Country	Year Introduced
Ildamen	Homburg	W. Germany	1966
Ildamen	Chugai	Japan	1970
Ildamen	Homburg	Italy	1972
Ildamen	Farmades	Italy	1973
Modacor	I.S.H.	France	-
Myofedrin	Apogepha	E. Germany	-
Timoval	Homburg	W. Germany	-

Raw Materials

m-Methoxyacetophenone
L-Norephedrine
Paraformaldehyde

Manufacturing Process

45 grams of m-methoxy acetophenone, 8 grams of paraformaldehyde and 30.2 grams of 1 norephedrine were mixed with about 135 cc of isopropanol HCl solution to provide a pH of 4 and the mixture refluxed for 4 hours. The reaction mixture was cooled and the crystals filtered off on a suction filter. 3-[1-phenyl-1-hydroxypropyl-(2)-amino]-1-(m-methoxyphenyl)-propanone-(1) HCl was obtained which after recrystallization from methanol had a MP of 190° to 193°C.

References

Merck Index 6830
Kleeman & Engel p. 673
OCDS Vol. 2 p. 40 (1980)
I.N. p. 718
Thiele, K.; US Patent 3,225,095; December 21, 1965; assigned to Deutsche Gold-und Silber-Scheideanstalt, Germany

OXYMESTERONE

Therapeutic Function: Anabolic, Androgen

Chemical Name: 4,17beta-Dihydroxy-17-methylandrost-4-en-3-one

Common Name: Hydroxymethyltestosterone; Methandrostenediolone; Ossimesterone; Oximesteronum; Oxymesterone; Oxymestrone

Structural Formula:

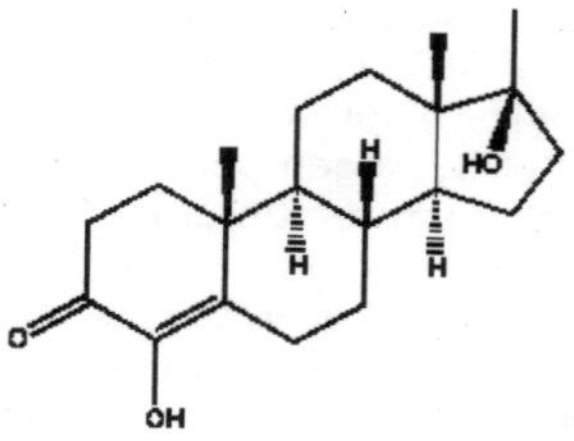

Chemical Abstracts Registry No.: 145-12-0

Trade Name	Manufacturer	Country	Year Introduced
Anamidol	Iwaki	-	-
Oranabol	Farmitalia Carlo Erba	-	-
Balnimax	Geve	-	-

Raw Materials

Sulfuric acid	4,5-Oxido-17α-methyltestosterone
Sodium chloride	Sodium bicarbonate
Potassium hydroxide	17α-Methyltestosterone
Acetic acid	Trimethylcarbinol
Hydrogen peroxide	Osmium tetroxide
Sodium bisulfite	Acetic acid

Manufacturing Process

2 Methods of producing of 4-hydroxy-17α-methyltestosterone:

1. A solution of 1.0 g of crude 4,5-oxido-17α-methyltestosterone in 50 ml of methanol is allowed to stand at room temperature overnight with 10 ml of water and 1 ml of concentrated sulfuric acid. It is then poured into water containing sodium chloride and extracted three times with ethyl acetate. The solvent is washed with water, then with 10% sodium bicarbonate solution and again with water to neutrality. The residue remaining after evaporation of the solvent is crystallized from methanol, giving 17α-methyl-androstane-4β,5α,17β-triol-3-one with a melting point of 203°-205°C.

A solution of 0.22 g of 17α-methyl-androstane-4β,5α,17β-triol-3-one in 100 ml of methanol is allowed to stand at room temperature for 22 h, under nitrogen, with 0.30 g of potassium hydroxide in 4 ml of water and 20 ml of methanol. The solution is then neutralized with acetic acid, concentrated in vacuo, diluted with water and extracted three times with ethyl acetate. The extract is washed with water and the solvent removed by distillation. The remaining residue is chromatographed over Florisil 30-60 mesh. The fractions eluted with benzene and benzene-ether (10:1) are combined and by crystallization from ether-petroleum ether give 4-hydroxy-17α-methyltestosterone (0.120 g) melting at 168°-170°C.

2. A solution of 20.0 g of 17α-testosterone in 500 ml of trimethylcarbinol is treated by addition of 56 ml of 30% hydrogen peroxide and 1.0 g of osmium tetroxide in 80 ml of trimethylcarbinol. After the mixture has stood at room temperature for 22 h, 12 ml of hydrogen peroxide are added. The reaction mixture is allowed to stand at room temperature for an additional 20 h, then concentrated in vacuo to 1/3 of its original volume, diluted with water, and the reaction product extracted with ethyl acetate. The extract is washed with water, several times with 10% sodium bisulfite solution, then with 4% sodium bicarbonate solution and finally with water to neutrality. The residue remaining after evaporation of the solvent does not show ultraviolet absorption. 1.0 g of this crude substance, by crystallization from methanol, gives l7α-methylandrostane-4,5,17β-triol-3-one (0.400 g) melting at 192°-194°C.

A solution of 20.0 g of crude 17β-methylandrostane-4,5,17β-triol-3-one in 1 L of methanol is heated under reflux in a stream of nitrogen for 20 min; then 20.0 g of potassium hydroxide in 40 ml of water and 200 ml of methanol are added. 5 min after the addition, the solution is treated by addition of 20 ml of acetic acid and concentrated in vacuo. The residue is diluted with water containing sodium chloride and extracted three times with ethyl acetate. The extract is washed with 10% sodium bicarbonate solution and then with water to neutrality. The residue remaining after evaporation of the solvent is dissolved in acetone; addition of petroleum ether gives 4-hydroxy-17α-methyl-testosterone (8.0 g) melting at 168°-170°C. The mother liquors chromatographed over Florisil 30-60 mesh yield an additional 5.0 g of the same substance melting at 168°-170°C.

References

Camerino B. et al; US Patent No. 3,060,201; Oct. 23, 1962; Assigned: Societa Farmaceutici Italia, Milan, Italy, a corporation of Italy

OXYMETAZOLINE HYDROCHLORIDE

Therapeutic Function: Nasal decongestant

Chemical Name: 3-[(4,5-Dihydro-1H-imidazol-2-yl)methyl]-6-(1,1-dimethylethyl)-2,4-dimethylphenol hydrochloride

Common Name: -

Structural Formula:

HCl

H N N OH

Chemical Abstracts Registry No.: 2315-02-8; 1491-59-4 (Base)

Trade Name	Manufacturer	Country	Year Introduced
Nasivin	Merck	W. Germany	1961
Iliadine	Merck Clevenot	France	1964
Afrin	Schering	US	1964
Nostrilla	Boehringer Ingelheim	US	1982
Alrin	Teva	Israel	-
Atomol	Allen and Hanburys	UK	-
Dristan	Whitehall	US	-
Duration	Plough	US	-
Nasivin	Bracco	Italy	-
Nasafarma	Novofarma	Spain	-
Nezeril	Draco	Sweden	-
Oxymeta	Schein	US	-
Pikorin	Medica	Finland	-
Rhinolitan	Kettelhack Riker	W. Germany	-

Trade Name	Manufacturer	Country	Year Introduced
Sinerol	Draco	Sweden	-
Utabon	Uriach	Spain	-

Raw Materials

Hydrogen chloride	2,4-Dimethyl-6-t-butylphenol
Ethylene diamine	4-Toluenesulfonic acid
Formaldehyde	Sodium cyanide
Sodium hydroxide	Hydrogen chloride

Manufacturing Process

10 grams 2,6-dimethyl-3-hydroxy-4-tertiary butylbenzylcyanide (produced by chloromethylation of 2,4-dimethyl-6-tertiary butyl-phenol with formaldehyde and HCl and conversion of the substituted benzyl chloride with NaCN; crystals, from alcohol, melting at 135° to 137°C) and 10.7 grams ethylenediamine-mono-p-toluenesulfonate are heated in an oil bath to approximately 235°C for 1½ hours, whereby ammonia is evolved. The free base is obtained from the p-toluene-sulfonic acid imidazoline salt which is difficultly soluble in water, by conversion with 50 cc of a 10% NaOH solution. Said base is recrystallized from benzene, and 7.5 grams (62% of the theoretical yield) 2-(2',6'-dimethyl-3'-hydroxy-4'-tertiary butylbenzyl)-2-imidazoline, MP 180° to 182°C, are obtained.

By dissolving the free base in an ethyl alcohol solution of hydrochloric acid and adding ether, the hydrochloride can be produced in the usual manner. Said hydrochloride melts, when recrystallized from alcoholic ether, at 300° to 303°C and is decomposed.

References

Merck Index 6834
Kleeman & Engal p. 674
PDR pp. 677, 728, 1606, 1899
OCDS Vol. 1 p. 242 (1977)
I.N. p. 719
REM p. 889
Fruhstorfer, W. and Muller-Calgan, H.; US Patent 3,147,275; September 1, 1964; assigned to E. Merck AG, Germany

OXYMORPHONE

Therapeutic Function: Narcotic analgesic

Chemical Name: 4,5α-Epoxy-3,14-dihydroxy-17-methylmorphinan-6-one

Common Name: Dihydrohydroxymorphinone

Structural Formula:

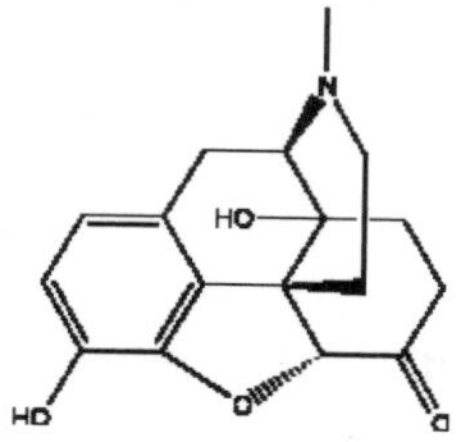

Chemical Abstracts Registry No.: 76-41-5

Trade Name	Manufacturer	Country	Year Introduced
Numorphan	Endo	US	1959

Raw Materials

Thebaine
Hydrogen peroxide
Hydrogen bromide
Hydrogen

Manufacturing Process

Thebaine is dissolved in aqueous formic acid and treated with 30% H_2O_2; neutralization with aqueous ammonia gives 14-hydroxycodeinone. It is hydrogenated to give oxycodone. 90 ml of concentrated hydrobromic acid are heated to 90°C. 9 grams of 14-hydroxydihydrocodeinone (oxycodone) are then added under stirring and the mixture is quickly heated to 116°C and kept at this temperature under reflux condenser for 20 minutes, with continued stirring. The resulting brown solution is diluted with about 90 ml of water and chilled with ice. Aqueous 10% sodium hydroxide solution is now added to alkaline reaction and the liquid is extracted 3 times with 100 cc portions of chloroform. The layers are separated and the aqueous phase is filtered and acidified by the addition of concentrated aqueous hydrochloric acid, treated with charcoal and filtered.

The filtrate is treated with concentrated aqueous ammonia until the mixture gives a pink color on phenolphthalein paper. The liquid is extracted seven times with 100 cc portions of chloroform, the extracts are combined, dried with anhydrous sodium sulfate and evaporated. The residue is dissolved in ethanol by refluxing and the ethanol evaporated nearly to dryness. 100 cc of benzene are then added, the mixture is refluxed for ½ hour and set aside for crystallization. After cooling, the desired compound is collected by filtration, 2.3 grams of a white crystalline powder are obtained; MP 245° to 247°C. This powder consisting of 14-hydroxydihydromorphinone can be purified by recrystallization from benzene, ethylacetate or ethanol. From benzene it generally forms diamond shaped platelets, while needles are obtained from ethylacetate.

On heating, the crystals are discolored from about 200°C on, and melt at 246° to 247°C to a black liquid, which decomposes with strong volume increase if the temperature is raised further by a few degrees.

References

Merck Index 6837
Kleeman & Engel p. 675
PDR p. 859
OCDS Vol. 1 p. 290 (1977) and 2, 319 (1980)
I.N. p. 719
REM p. 1105
Lewenstein, M.J. and Weiss, U.; US Patent 2,806,033; September 10, 1957

OXYPENDYL

Therapeutic Function: Antiemetic

Chemical Name: 4-[3-(10H-Pyrido[3,2-b][1,4]benzothiazin-10-yl)propyl]-1-piperazineethanol

Common Name: -

Structural Formula:

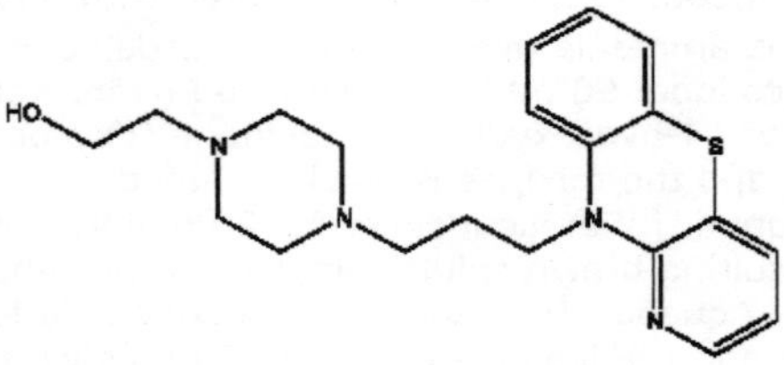

Chemical Abstracts Registry No.: 5585-93-3; 17297-82-4 (Dihydrochloride salt)

Trade Name	Manufacturer	Country	Year Introduced
Pervetral	Homburg	W. Germany	1962

Raw Materials

10-(γ-N-Piperazinopropyl)-4-azaphenthiazine
Ethylene chlorohydrin

Manufacturing Process

32 parts of 10-(γ-N-piperazinopropyl)-4-azaphenthiazine in 200 cc of butanol

with 9 parts of ethylene chlorohydrin and 14 parts of finely powdered potash are heated for 4 hours under reflux while stirring vigorously. After cooling, extraction is carried out with dilute hydrochloric acid, the substance is finally washed with water and the combined hydrochloric acid aqueous phase is washed twice with ether. The base is then liberated with concentrated sodium hydroxide solution and taken up in chloroform. The chloroform solution is dried with potash and concentrated by evaporation. 26.4 parts of (10-γ-N-B-hydroxyethylpiperazino-N^1-propyl)-4-azaphenthiazine are distilled over at 280°C to 300°C/6 mm. The dihydrochloride is obtained in isopropanol with isopropanolic hydrochloric acid. The product melts at 218°C to 220°C.

References

Merck Index 6838
Kleeman & Engel p. 676
OCDS Vol. 1 p. 430 (1977)
I.N.p. 719
Deutsche Gold-und Silber Scheideanstalt; British Patent 893,284; April 4, 1962

OXYPERTINE

Therapeutic Function: Antipsychotic, Neuroleptic

Chemical Name: 1H-Indole, 5,6-dimethoxy-2-methyl-3-(2-(4-phenyl-1-piperazinyl)ethyl)-

Common Name: Oxypertine; Oxipertinum

Structural Formula:

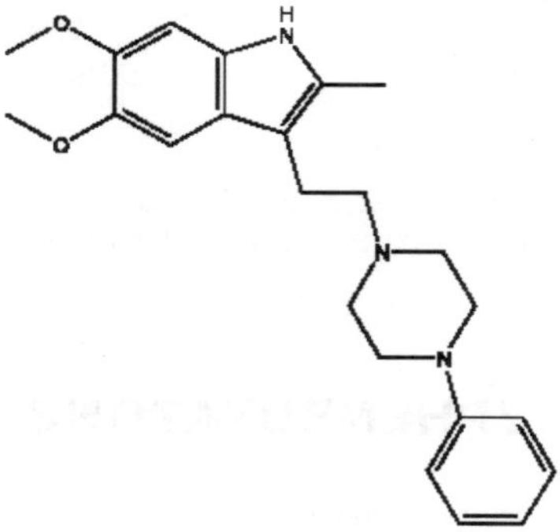

Chemical Abstracts Registry No.: 153-87-7

Trade Name	Manufacturer	Country	Year Introduced
Oxypertine capsules	Sanofi-Synthelabo	-	-

Trade Name	Manufacturer	Country	Year Introduced
Oxypertine	Bulk Drugs and Intermediates Inc.	-	-
Oxypertine	Shanghai Lansheng Corporation	-	-

Raw Materials

1-Phenylpiperazine
Acetic acid
[3-(2-Methyl-5,6-dimethoxy)indolyl] glyoxalyl chloride
Lithium aluminum hydride
Sodium hydroxide

Manufacturing Process

A cold, stirred solution of 1-phenyl-piperazine in tetrahydrofuran was treated all at once with [3-(2-methyl-5,6-dimethoxy)indolyl]glyoxalyl chloride. There was an immediate voluminous precipitate of a white crystalline solid which was removed by filtration. The filtrate was taken to dryness and the residual light brown gum was stirred and shaken with water, ethyl acetate and acetic acid. The mixture was warmed on a steam bath and the resulting solid was collected after cooling in an ice bath thus affording 1-[(3-(2-methyl-5,6-dimethoxy)indolyl)glyoxalyl]-4-phenylpiperazine as a near white solid, melting point 163°-174°C.

A solution of 1-[(3-(2-methyl-5,6-dimethoxy)indolyl)glyoxalyl]-4-phenyl piperazine in tetrahydrofuran was added over a 10 min period to a stirred suspension of lithium aluminum hydride in tetrahydrofuran. The mixture was refluxed and stirred for 6.5 h and the excess lithium aluminum hydride then destroyed by the dropwise addition of 10% sodium hydroxide solution. The mixture was filtered, the insoluble material was washed with boiling chloroform, and the filtrate dried over anhydrous sodium sulfate and concentrated to dryness. The residual light orange oil was crystallized from a benzene-hexane mixture giving 1-[(3-(2-methyl-5,6-dimethoxy)indolyl)ethyl]-4-phenyl piperazine.

References

Bethlehem S.A.; US Patent No. 3,183,313; June 8, 1965; Assigned: Sterling Drug Inc., New york, N.Y., a corporation of Delaware

OXYPHENBUTAZONE

Therapeutic Function: Antiinflammatory

Chemical Name: 4-Butyl-1-(4-hydroxyphenyl)-2-phenyl-3,5-pyrazolidinedione

Common Name: p-Hydroxyphenylbutazone

Structural Formula:

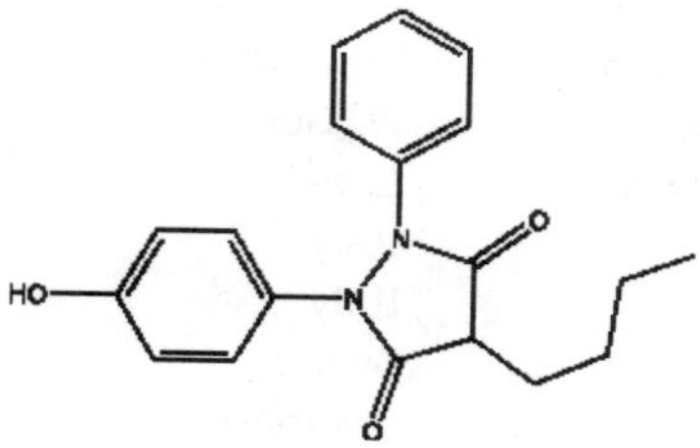

Chemical Abstracts Registry No.: 129-20-4

Trade Name	Manufacturer	Country	Year Introduced
Tanderil	Geigy	UK	1960
Tandearil	Geigy	US	1961
Tanderil	Ciba Geigy	France	1961
Tanderil	Geigy	W. Germany	1961
Tanderil	Geigy	Italy	1962
Artroflog	Magis	Italy	-
Artzone	Continental Ethicals	S. Africa	-
Butaflogin	Chemiepharma	Italy	-
Butapirone	Brocchieri	Italy	-
Buteril	Protea	S. Africa	-
Butilene	Frencia	Italy	-
Deflogin	Valeas	Italy	-
Fibutox	Phermador	S. Africa	-
Flanaril	Osfa	Italy	-
Floghene	Chibi	Italy	-
Flogistin	Scharper	Italy	-
Flogitolo	Isnardi	Italy	-
Flogodin	Firma	Italy	-
Iltazon	Iltas	Turkey	-
Imbun	Merckle	W. Germany	-
Inflamil	Leiras	Finland	-
Ipebutona	Ipecsa	Spain	-
Iridil	Farmila	Italy	-
Isobutil	Panther-Osfa	Italy	-
Miyadril	Fako	Turkey	-
Optimal	Dojin	Japan	-
Optone	Lennon	S. Africa	-
Oxalid	U.S.V.	US	-
Oxibutol	Asla	Spain	-
Oxybutazone	I.C.N.	Canada	-
Oxybuton	Streuli	Switz.	-

Trade Name	Manufacturer	Country	Year Introduced
Phlogase	Adenylchemie	W. Germany	-
Phlogistol	Helopharm	W. Germany	-
Phlogont	Azuchemie	W. Germany	-
Phloguran	Ikapharm	Israel	-
Pirabutina	Ellea	Italy	-
Piraflogin	Jamco	Italy	-
Rapostan	Mepha	Switz.	-
Rheumapax	Erco	Denmark	-
Tantal	Sawai	Japan	-
Teneral	Eczacibasi	Turkey	-
Validil	Von Boch	Italy	-
Visobutina	I.S.F.	Italy	-

Raw Materials

Sodium
Hydrogen
n-Butylmalonic acid ethyl ester
p-Benzyloxy hydrazobenzene

Manufacturing Process

43.2 parts of n-butyl malonic acid ethyl ester are added to a solution of 4.6 parts of sodium in 92 parts by volume of absolute alcohol. 39 parts of p-benzyloxy hydrazobenzene (MP 88° to 90°C) are added. About two-thirds of the alcohol is distilled off and 92 parts by volume of absolute xylene are added. Without removing the sloping condenser, the mixture is stirred for 12 hours at a bath temperature of 140° to 145°C. It is then cooled to 0° to 5°C, 100 parts of ice are added, the xylene is removed, the aqueous solution is extracted twice with chloroform and made acid to Congo red at 0° to 5°C with 6 N hydrochloric acid.

The precipitate is taken up in chloroform, the solution obtained is washed twice with water, then with saturated salt solution, dried over $Na_2SO_4$and evaporated under vacuum (bath temperature 20°C). The residue is recrystallized from alcohol and produces 1-(p-benzyloxyphenyl)-2-phenyl-4-n-butyl-3,5-dioxo-pyrazolidine (C) as tiny white needles which melt at 132° to 133°C.

16.6 parts of (C) are suspended in 166 parts by volume of ethyl acetate and, in the presence of 16.6 parts of Raney nickel, hydrogen is allowed to act at room temperature and atmospheric pressure.

After 6 hours the calculated amount of hydrogen has been taken up. The residue obtained after filtering and evaporating is taken up in benzene and extracted twice with diluted sodium carbonate solution. The alkali extract is then made acid to Congo red with 6 N hydrochloric acid and the precipitate is taken up in ethyl acetate. The solution obtained is washed twice with salt solution, dried with sodium sulfate and evaporated. The residue is recrystallized from ether/petroleum ether. 1-(p-hydroxyphenyl)-2-phenyl-4-n-butyl-3,5-dioxo-pyrazolidine melts at 124° to 125°C.

References

Merck Index 6840
Kleeman & Engel p. 677
PDR p. 1606
OCDS Vol. 1 p. 236 (1977)
I.N. p. 720
REM p. 1119
Hafliger, F.; US Patent 2,745,783; May 15, 1956; assigned to J.R. Geigy AG, Switzerland

OXYPHENCYCLIMINE

Therapeutic Function: Spasmolytic

Chemical Name: α-Cyclohexyl-α-hydroxybenzeneacetic acid (1,4,5,6-tetrahydro-1-methyl-2-pyrimidinyl)methyl ester

Common Name: -

Structural Formula:

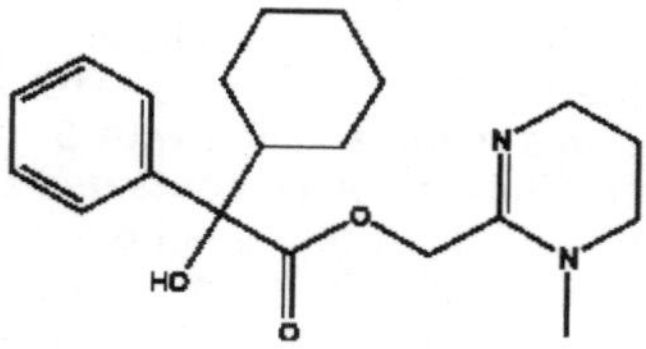

Chemical Abstracts Registry No.: 125-53-1; 125-52-0 (Hydrochloride salt)

Trade Name	Manufacturer	Country	Year Introduced
Vio-Thene	Rowell	US	1959
Daricon	Pfizer	US	1959
Setrol	Flint	US	1961
Gastrix	Rowell	US	1973
Manir	Valpan	France	1975
Caridan	B.D.H.	UK	-
Cycmin	Toyo	Japan	-
Inomaru S	Sawai	Japan	-
Norma	Sankyo	Japan	-
Oximin	A.F.I.	Norway	-
Sedomucol	Asla	Spain	-

Trade Name	Manufacturer	Country	Year Introduced
Spazamin	G.P.	Australia	-
Ulcociclina	Confas	Italy	-
Ulcomin	Remedia	Israel	-
Vagogastrin	Benvegna	Italy	-

Raw Materials

1,3-Diaminobutane
Benzoyl formic acid
Magnesium
Ethyl chlorimidoacetate
Cyclohexyl bromide

Manufacturing Process

To a stirred solution of 8.8 grams (0.1 mol) of 1,3-diaminobutane in 150 ml of ethanol maintained at 0° to 5°C, there was added 25.8 grams (0.1 mol) of ethyl chlorimidoacetate hydrochloride during a period of 20 minutes. After the mixture had been stirred at 0° to 5°C for two hours, it was acidified at this temperature by the addition of ethanolic hydrogen chloride. The mixture was warmed to room temperature and filtered to remove 4.3 grams of solid ammonium chloride. The filtrate was concentrated to approximately 40 ml, filtered and refrigerated. The solid which separated was isolated, washed with acetone and dried. There was obtained 7.4 grams (40% of the theoretical yield) of 2-chloromethyl-4-methyl-1,4,5,6-tetrahydropyrimidine hydrochloride melting at 158° to 160°C.

In a second step, cyclohexyl bromide was reacted with magnesium, then with benzoyl formic acid to give cyclohexylphenyl glycolic acid. A solution of 1.8 grams (0.01 mol) of 2-chloromethyl-1-methyl-1,4,5,6-tetrahydropyrimidine hydrochloride in 5 ml of water was made alkaline with 5 ml of 50% NaOH and extracted with ether. The ether solution, which contained the basic chloride, was dried over calcium sulfate and added to a solution of 2.3 grams (0.01 mol) of α-cyclohexylphenylglycolic acid in 75 ml of isopropanol. The solution was distilled to remove the ether, and 0.1 gram of powdered potassium iodide added to the residual isopropanol solution which was then refluxed for 6 hours. The solid which had separated was redissolved by the addition of 20 ml of ethanol and the solution charcoaled, concentrated, and cooled. The solid which separated, 1-methyl-1,4,5,6-tetrahydro-2-pyrimidylmethyl α-cyclohexylphenyl-glycolate hydrochloride, weighed 1.4 grams and melted at 228° to 229°C with decomposition after recrystallization from ethanol.

References

Merck Index 6841
Kleeman & Engel p. 677
OCDS Vol. 2 p. 75 (1980)
I.N. p. 720
REM p.917
Chas. Pfizer & Co., Inc.; British Patent 795,758; May 28, 1958

OXYPHENISATIN ACETATE

Therapeutic Function: Laxative

Chemical Name: 3,3-Bis[4-(acetyloxy)phenyl]-1,3-dihydro-2H-indol-one

Common Name: Acetphenolisatin; Endophenolphthalein; Diphesatin

Structural Formula:

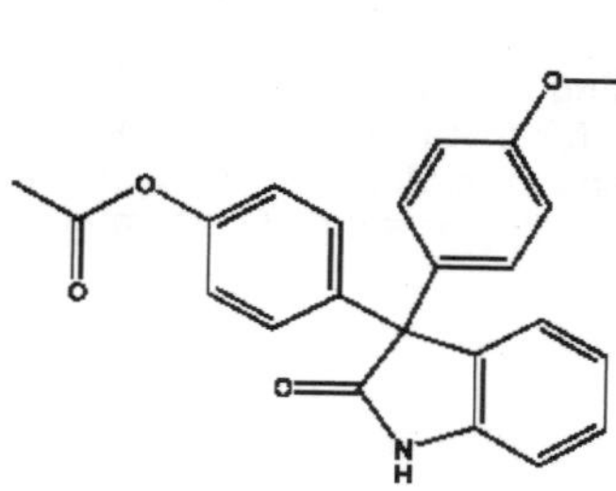

Chemical Abstracts Registry No.: 115-33-3

Trade Name	Manufacturer	Country	Year Introduced
Lavema	Winthrop	US	1959
Isalax	Vale	US	1963
Acetalax	Harvey	Australia	-
Bisco-Zitron	Biscova	W. Germany	-
Bydolax	Moore	UK	-
Darmoletten	Omegin	W. Germany	-
Eulaxin	Pliva	Yugoslavia	-
Fenisan	Chemimportexport	Rumania	-
Laxatan	Divapharma	W. Germany	-
Laxanormal	Uquifa	Spain	-
Med-Laxan	Med	W. Germany	-
Nourilax	Noury Pharma	Netherlands	-
Obstilax	Zirkulin	W. Germany	-
Promassolax	Ysat Wernigerode	E. Germany	-
Prulet	Mission	US	-
Regal	Ferrosan	Denmark	-
Sanapert	Trogalen	Austria	-
Schokolax	Dallmann	W. Germany	-
Veripaque	Winthrop	UK	-

Raw Materials

Diphenolisatin
Acetic anhydride

Manufacturing Process

235 gravimetrical parts of acetic acid anhydride (90%) are poured over 106 gravimetrical parts of diphenolisatin (Berichte der Deutschen Chemischen Gesselschsft, 18, 1885, p. 2641) and the mixture is heated on the water-bath while stirring. The solid starting material temporarily dissolves almost entirely and shortly afterwards the reaction product turns into a crystalline paste. In order to complete the reaction the heating on the water-bath is continued for a short time and then the whole is left to get cold. The reaction product may, for instance, be separated in the following manner: To the cold reaction mixture is gradually added about the same volumetrical quantity of alcohol; in this manner the excess of acetic acid anhydride is destroyed and the paste becomes thinner. Then the fluid is drawn off and the product washed with alcohol. For complete cleansing another extraction is made with warm alcohol and the product crystallized, for instance, from 10 parts of acetic acid. The product represents a light, fine crystalline powder, which is difficultly soluble or even insoluble in the usual organic solvents. Its melting point lies at 242°C.

References

Merck Index 6842
Kleeman & Engel p. 678
OCDS Vol. 2 p. 350 (1980)
I.N. p. 720
Preiswerk, E.; US Patent 1,624,675; April 12, 1927; assigned to Hoffmann-LaRoche Chemical Works

OXYPHENONIUM BROMIDE

Therapeutic Function: Anticholinergic, Spasmolytic

Chemical Name: [Ethanaminium, 2-((cyclohexylhydroxyphenylacetyl)oxy)-N,N-diethyl-N-methyl-, bromide

Common Name: Oxiphenoni bromidum; Oxyphenonium bromide

Structural Formula:

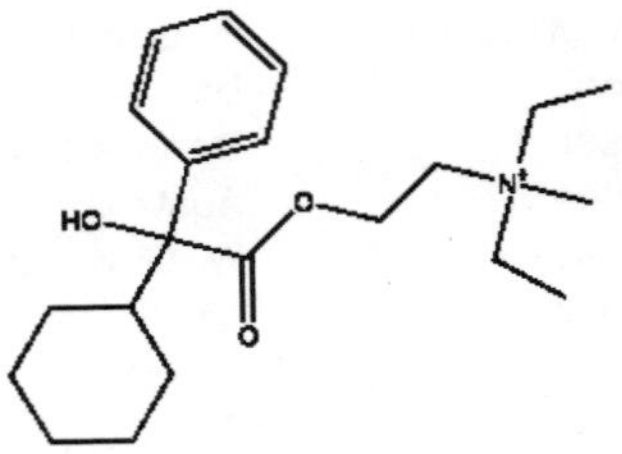

Chemical Abstracts Registry No.: 50-10-2; 14214-84-7 (Base)

Trade Name	Manufacturer	Country	Year Introduced
A-Spasm	Acme Laboratories Ltd.	-	-
Antispasmin	Pharmacia Co - Dupnitza	-	-
Oxyphenonium bromide	Yick-Vic Chemicals and Pharmaceuticals (HK) Ltd.	-	-
Oxyphenonium bromide	Pharm Products	-	-
Antrenyl	Ciba	-	-
Antrenyl	Ciba-Geigy	-	-
Antrenyl	Swiss Pharma	-	-
Calmulcer	Sons	-	-
Oxyphenon	Leciva	-	-
Spasmophen	Polfa-Pabianice	-	-
Spastrex	Propan-Generics	-	-
Antrenyl Duplex Drag (aH)	Ciba-Geigy AG Pharma Schweiz	-	-

Raw Materials

2-Diethylaminoethanol
Cycloxehylhydroxyphenylacetic acid methyl ester
Methyl bromide

Manufacturing Process

1 mol cycloxehylhydroxyphenylacetic acid methyl ester was mixed with 1 mol 2-diethylaminoethanol in presence of 1 mol sodium methylate to give cycloxehylhydroxyphenylacetic acid 2-diethylamino-ethyl ester.

5 parts by weight of it was dissolved in 50 volumes glacial acetic acid and a gaseous methyl bromide was introduced. The mixture was heated to about 50°C and crystallization of cycloxehylhydroxyphenylacetic acid diethylaminoethyl ester methyl bromide has shortly after begun. On cooling the crystals was filtered off and recrystallized from mixture of ethyl acetate ant a little ethanol to give the oxyphenonium bromide. MP: 189°-191°C.

References

Swit.R. Patent No. 259,958; Sept. 18, 1944; Assigned to Ciba

OXYPYRRONIUM BROMIDE

Therapeutic Function: Anticholinergic, Spasmolytic

Chemical Name: Pyrrolidinium, 1,1-dimethyl-2-(hydroxymethyl)-, bromide, α-phenylcyclohexaneglycolate

Common Name: Oxipyrroni bromidum; Oxypyrronium bromide

Structural Formula:

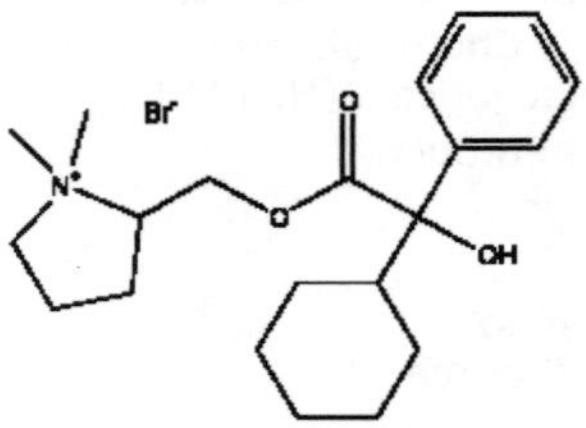

Chemical Abstracts Registry No.: 561-43-3; 116533-64-3 (Base)

Trade Name	Manufacturer	Country	Year Introduced
Oxypyrronium bromide	Shanghai Lansheng Corporation	-	-

Raw Materials

Sodium
Methanol
Methyl bromide
Sodium hydroxide
Methylphenylcyclohexylglycolate
1-Methyl-2-hydroxymethylpyrrolidine
Hydrochloric acid

Manufacturing Process

The 1-methyl-2-hydroxymethylpyrrolidine was obtained by the process of Application No 21193/56 (Serial No. 820,503).

A methanolic solution of sodium methoxide [from sodium (0.6 g) and methanol (15 ml)] was added dropwise during 3 h to a boiling solution of methyl phenylcyclohexylglycollate (33.7 g) and 1-methyl-2-hydroxymethylpyrrolidine (23.4 g) in heptane (400 ml) and the methanol that separated was removed by means of a Dean and Stark apparatus. At the end of 4 h no further separation of methanol occurred and the solvent was removed under reduced pressure. The residue was dissolved in either and the etheral solution, after washing with water (3 x 50 ml), was extracted with 5 N hydrochloric acid (3 x 100 ml). The (1-methyl-2-pyrrolidyl)methyl phenylcyclokexylglycollate hydrochloride (35.5 g 71%) crystallised out of the acid extract as colourless needles, melting point 181°-196°C. Extraction of this hydrochloride (33.0 g) with hot ethanol (150 ml) left the sparingly soluble (1-methyl-2-pyrrolidyl)methyl phenylcyclokexylglycollate hydrochloride (a-form) (7.6 g), melting point 220°-222°C.

The (1-methyl-2-pyrrolidyl)methyl phenylcyclokexylglycollate hydrochloride (a-form) (15.0 g) was dissolved in water, basified with sodium hydroxide solution and the resultant oil extracted into ether. The extracts were dried over magnesium sulfate, the ether evaporated and the residue dissolved in acetone (100 ml). Methyl bromide (7.8 g, 2 mole) was added to the acetone solution and the mixture warmed on a steam bath for 15 min. The solution was cooled and the solid filtered off, washed with a little acetone and dried to give the

(1,1-dimethyl-2-pyrrolidyl)methyl α-phenylcyclokexylglycollate bromide, melting point 185°-186°C. (86%).

References

GB Patent No. 850,260; Nov. 28, 1957; Assigned: Beecham Research Laboratories Limited, a British Company, of Brockham Park, Betchworth, Surrey

OXYQUINOL

Therapeutic Function: Antiseptic

Chemical Name: Quinolin-8-ol

Common Name: Hydroxychinolinium sulfuricum; Oxichinolini sulfas; Oxychinol; Oxyquinol; Oxyquinoline sulfate; Oxyquinolini sulfas

Structural Formula:

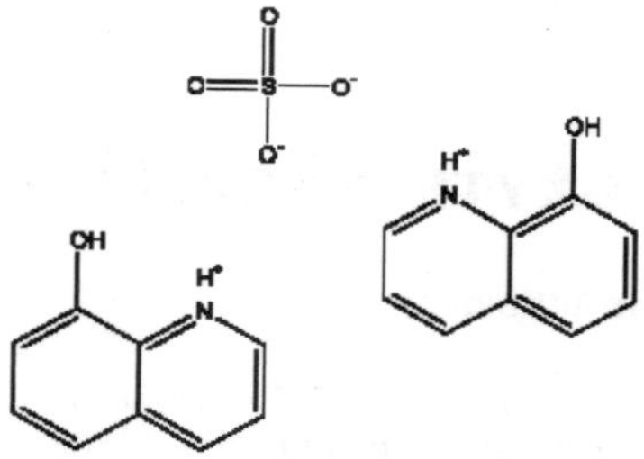

Chemical Abstracts Registry No.: 134-31-6; 148-24-3 (Base)

Trade Name	Manufacturer	Country	Year Introduced
Chinosol	Chinosolfabrik	-	-
8-Hydroxyquinoline sulfate	GFS Chemicals	-	-
Hydroxyquinoline sulfate	Shanghai Lansheng Corporation	-	-
8-Hydroxyquinoline sulfate	AroKor Holdings Inc.	-	-
8-Hydroxyquinoline sulfate	Tianjin Mid-Chem Co., Ltd.	-	-
8-Quinolinol Sulfate	Eastman Kodak Company	-	-
Oxymeria Gargle	Nile Co.	-	-

Raw Materials

o-Nitrophenol
o-Aminophenol
Sulfuric acid

Manufacturing Process

The mixture of 1.4 kg o-nitrophenol, 2.1 kg o-aminophenol, 6 kg glycerine (d = 1.26) and 5 kg sulfuric acid (d = 1.848) was heated at reflux to temperature 130°-140°C. This temperature was kept for 1.5 hours. The obtained oxyquinoline precipitated, the liquid was removed with water-steam distillation. The residue was diluted with water and alkalized with sodium hydroxide and sodium carbonate to the strong alkaline reaction. The repeated distillation with water steam gave the oil, which hardened as the long needles by cooling. MP: 75°-76°C recrystallized from diluted ethanol.

In practice it is usually used as sulfate salt.

References

Skraup Z.H.; DR Patent No. 14,976; Feb. 16, 1881; Wien
Dictionary of Organic Compounds edited by I. Heilbron and H.M. Bunbury, v.2, p.326, 1946, London

OXYTETRACYCLINE

Therapeutic Function: Antibiotic

Chemical Name: 4-(Dimethylamino)-1,4,4a,5,5a,6,11,12a-octahydro-3,5,6,10,12,12a-hexahydroxy-6-methyl-1,11-dioxo-2-naphthacenecarboxamide

Common Name: -

Structural Formula:

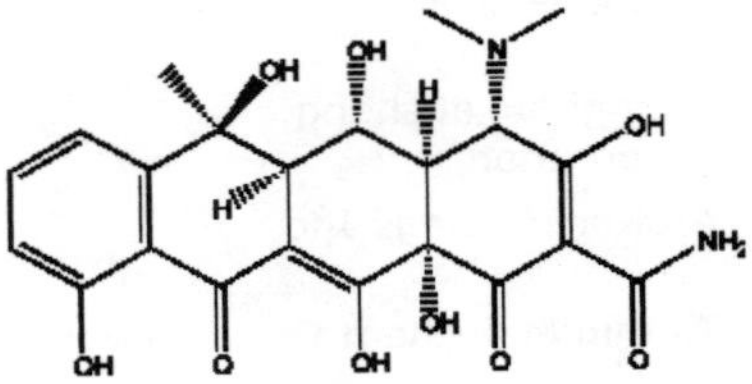

Chemical Abstracts Registry No.: 79-57-2; 2058-46-0 (Hydrochloride salt)

Trade Name	Manufacturer	Country	Year Introduced
Terramycin	Pfizer	US	1950
Gynamousse	Pfizer	France	1966
Oxy-Kesso-Tetra	McKesson	US	1970
Oxlopar	Parke Davis	US	1974
E.P. Mycin	Edwards	US	1983
Chrysocin	Pliva	Yugoslavia	-
Clinimycin	Glaxo	UK	-
Copharoxy	Cophar	Switz.	-
Crisamicin	Frumtost	Spain	-
Devacyclin	Deva	Turkey	-
Dura-Tetracyclin	Dura	W. Germany	-
Egocin	Krka	Yugoslavia	-
Elaciclina	I.F.L.	Spain	-
Galenomycin	Galen	UK	-
Geocycline	I.E. Kimya Evi	Turkey	-
Geomycin	Pliva	Yugoslavia	-
I.A.-Loxin	Inter-Alia Pharm.	UK	-
Imperacin	I.C.I.	UK	-
Macocyn	Mack	W. Germany	-
Oksisiklin	Uranium	Turkey	-
Ossitetra	Pierrel	Italy	-
Otesolut	Jenapharm	E. Germany	-
Oxacycline	Crookes	UK	-
Oxeten	Mochida	Japan	-
Oxymycin	Chelsea	UK	-
Proteroxyna	Proter	Italy	-
Stecsolin	Squibb	UK	-
Tetra-Tablinen	Sanorania	W. Germany	-
Tetrafen	Drifen	Turkey	-

Raw Materials

Soybean meal
Glucose
Bacterium Streptomyces rimosus

Manufacturing Process

Medium	Grams
Soybean meal	10
Cerelose	10
Distillers' solubles	0.5
Sodium chloride	5
Distilled water to	1,000 ml

The pH was adjusted to 7.0 with sodium hydroxide and calcium carbonate was added at the rate of 1 g/l.

500 ml portions of the above medium were added to Fernbach flasks which

were then sterilized at 121°C for 30 minutes. Upon cooling, the flasks were inoculated with a suspension of the growth of s. rimosus obtained from the surface of beef lactose agar slants, and the flasks were shaken for 4 days at 28°C on a rotary shaker having a displacement of 2" at an rpm of 200. At the end of this period the broth was found to contain 640 C.D.U/ml and 400 chloramphenicol units/ml. The mycelium was separated from the broth by filtration and the latter was adjusted to pH 9.0. The antibiotic was extracted from the broth with n-butanol, and when the ultraviolet absorption spectrum was observed on the butanol solution of the antibiotic, peaks in the absorption curve were found at 385 and 270 millimicrons.

References

Merck Index 6846
Kleeman & Engel p. 680
PDR pp. 887, 1413, 1533, 1606
OCDS Vol. 1 p. 212 (1977) and 2, 226 (1980)
I.N. p. 721
REM pp. 1206, 1260
Sobin, B.A., Finlay, A.C. and Kane, J.H.; US Patent 2,516,080; July 18, 1950; assigned to Chas. Pfizer & Co., Inc.

OXYTOCIN

Therapeutic Function: Oxytocic

Chemical Name: Oxytocin (a complex peptide)

Common Name: -

Structural Formula:

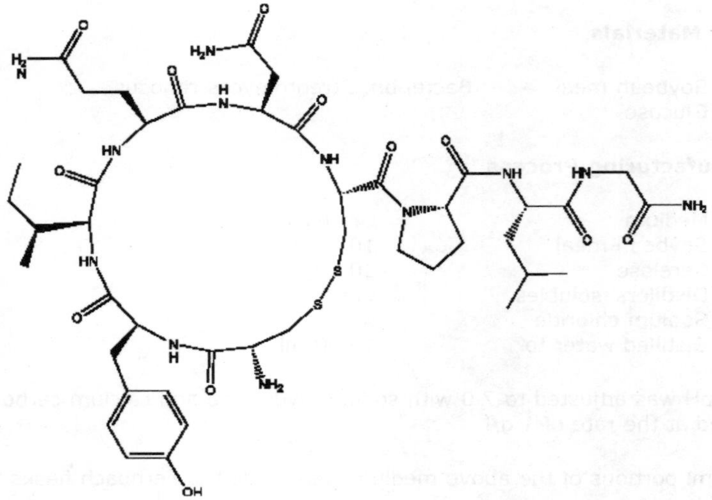

Chemical Abstracts Registry No.: 50-56-6

Trade Name	Manufacturer	Country	Year Introduced
Syntocinon	Sandoz	US	1957
Syntocinon	Sandoz	France	1958
Uteracon	Hoechst	US	1964
Atonin-O	Teikoku Zoki	Japan	-
Endopituitrina	I.S.M.	Italy	-
Orasthin	Hoechst	W. Germany	-
Oxitocin	Chinoin	Italy	-
Oxystin	Arzneimittelwerk Dresden	E. Germany	-
Oxytal	A.L.	Norway	-
Partocon	Ferring	Sweden	-
Partolact	Medica	Finland	-
Pitocin	Sankyo	Japan	-
Pituitan	Nippon Zoki	Japan	-

Raw Materials

Hydrogen
Glycine lower alkyl ester
L-Leucine lower alkyl ester
Ammonia
Hydrogen chloride
L-Tyrosine lower alkyl ester
α-Benzyl-L-aspartic acid-α-lower alkyl ester
L-Isoleucine lower alkyl ester
S,N-Ditrityl-L-cysteine diethylamine salt
N-Trityl glutamic acid-γ-lower alkyl ester
Benzyl-L-proline hydrochloride

Manufacturing Process

As described in US Patent 2,938,891, in the process for producing oxytocin, the steps comprise:

(a) Adding dicyclohexyl carbodiimide to a solution of the α-benzyl-L-aspartic acid-β-lower alkyl ester in methylene chloride, cooling the mixture to about 0°C, adding thereto the N-trityl glutamic acid-γ-lower alkyl ester, allowing the mixture to stand at room temperature to complete condensation, acidifying the reaction mixture with acetic acid, filtering off precipitated dicyclohexyl urea, and separating the resulting (N-trityl-γ-lower alkyl-L-glutamyl)-α-benzyl-L-aspartic acid-β-lower alkyl ester.

(b) Dissolving the (N-trityl-γ-lower alkyl-L-glutamyl)-α-benzyl-1-aspartic acid-β-lower alkyl ester in ethanol, adding triethylamine and palladium black to said solution, introducing hydrogen at room temperature thereinto to split off the benzyl group, and separating the (N-trityl-γ-lower alkyl-L-glutamyl)-L-aspartic acid-β-lower alkyl ester.

(c) Adding dicyclohexyl carbodiimide to a solution of the diethylamine salt of S,N-ditrityl-L-cysteine and the hydrochloride of the lower alkyl ester of L-tyrosine in methylene chloride, allowing the mixture to stand at a temperature between room temperature and about 35°C to complete condensation, acidifying the reaction mixture with acetic acid, filtering off precipitated dicyclohexyl urea, and separating the resulting lower alkyl ester of S,N-ditrityl-L-cysteinyl-L-tyrosine.

(d) Refluxing the aqueous alcoholic solution of said ester with an alcoholic alkali metal hydroxide solution to saponify the lower alkyl ester group, neutralizing the saponification mixture by the addition of hydrochloric acid, extracting the neutralized mixture with ether, and separating the resulting (S,N-ditrityl-L-cysteinyl)-L-tyrosine.

(e) Adding triethylamine to a solution of said S,N-ditrityl compound in chloroform, and precipitating the triethylamine salt of (S,N-ditrityl-L-cysteinyl)-L-tyrosine by the addition of petroleum ether.

(f) Adding dicyclohexyl carbodiimide to a solution of said triethylamine salt of (S,N-ditrityl-L-cysteinyl)-L-tyrosine and the hydrochloride of the lower alkyl ester of L-isoleucine in methylene chloride, allowing the mixture to stand at room temperature to complete condensation, acidifying the reaction mixture with acetic acid, filtering off precipitated dicylohexyl urea, and separating the resulting (S,N-ditrityl-L-cysteinyl)-L-tyrosyl-L-isoleucine lower alkyl ester.

(g) Refluxing the aqueous alcoholic solution of said ester with an alcoholic alkali metal hydroxide solution to saponify the lower alkyl ester group, neutralizing the saponification mixture by the addition of hydrochloric acid, extracting the neutralized mixture with ether, and separating the resulting (S,N-ditrityl-L-cysteinyl)-L-tyrosine-L-isoleucine.

(h) Adding dicyclohexyl carbodiimide to a solution of the diethylamine salt of S,N-ditrityl-L-cysteine and the hydrochloride of benzyl-L-proline in methylene chloride, allowing the mixture to stand at about room temperature to complete condensation, acidifying the reaction mixture with acetic acid, filtering off precipitated dicyclohexyl urea, and separating the resulting (S,N-ditrityl-L-cysteinyl)-L-prolinebenzyl ester.

(i) Refluxing said benzyl ester with an aqueous alcoholic alkali metal hydroxide solution to saponify the benzyl ester group, neutralizing the saponification mixture by the addition of hydrochloric acid, extracting the neutralized mixture with chloroform, and separating the resulting (S,N-ditrityl-L-cysteinyl)-L-proline.

(j) Adding diethylamine to a solution of said dipeptide compound in ether to yield the diethylamine salt of (S,N-ditrityl-L-cysteinyl)-L-proline.

(k) Adding dicyclohexyl carbodiimide to a solution of the diethylamine salt of (S,N-ditrityl-L-cysteinyl)-L-proline and the hydrochloride of the L-leucine lower alkyl ester in methylene chloride, allowing the mixture to stand at a temperature between about 25° and 30°C to complete condensation, acidifying the reaction mixture with acetic acid, filtering off precipitated dicyclohexyl urea, and separating the resulting (S,N-ditrityl-L-cysteinyl)-L-prolyl-L-leucine lower alkyl ester.

(l) Refluxing said lower alkyl ester with an aqueous alcoholic alkali metal hydroxide solution to saponify the lower alkyl ester group, neutralizing the saponification mixture by the addition of hydrochloric acid, extracting the neutralized mixture with ether, and separating the resulting S,N-ditrityl-L-cysteinyl-L-prolyl-L-leucine.

(m) Adding dicyclohexyl carbodiimide to a solution of the diethylamine salt of S,N-ditrityl-L-cysteinyl-L-prolyl-L-leucine and the hydrochloride of the glycine lower alkyl ester in methylene chloride, allowing the mixture to stand at a temperature between about 25° and 30°C to complete condensation, acidifying the reaction mixture with acetic acid, filtering off precipitated dicyclohexyl urea, and separating the resulting (S,N-ditrityl-L-cysteinyl)-L-prolyl-L-leucyl-glycine lower alkyl ester.

(n) Adding aqueous hydrochloric acid to a mixture of said lower alkyl ester in a solvent selected from the group consisting of acetone and acetic acid, allowing the mixture to stand at a temperature of about 35°C to complete selective detritylation of the N-trityl group, and separating the resulting (S-trityl-L-cysteinyl)-L-prolyl-L-leucyl glycine lower alkyl ester.

(o) Adding dicyclohexyl carbodiimide to a solution of the diethylamine salt of the (N-trityl-γ-lower alkyl-L-glutamyl)-L-aspartic acid-β-lower alkyl ester obtained according to step (b) and the hydrochloride of the (S-trityl-L-cysteinyl)-L-prolyl-L-leucyl glycine lower alkyl ester in methylene chloride, allowing the mixture to stand at about room temperature to complete condensation, filtering off precipitated dicyclohexyl urea, and separating the resulting (N-trityl-γ-lower alkyl-L-glutamyl)-(β-lower alkyl-L-aspartyl)-(S-trityl-L-cysteinyl)-L-prolyl-L-leucyl glycine lower alkyl ester.

(p) Adding aqueous hydrochloric acid to a mixture of said lower alkyl ester in a solvent selected from the group consisting of acetone and acetic acid, allowing the mixture to stand at room temperature to complete selective detritylation of the N-trityl group, and separating the resulting hexapeptide compound (γ-lower alkyl-L-glutamyl)-(β-lower al kyl-L-aspartyl)-(S-trityl-L-cysteinyl)-L-prolyl-L-leucyl glycine lower alkyl ester.

(q) Adding dicyclohexyl carbodiimide to a solution of the diethylamine salt of (S,N-ditrityl-L-cysteinyl)-L-tyrosyl-L-isoleucine obtained according to step (g) and the hydrochloride of (γ-lower alkyl-L-glutamyl)-(β-lower alkyl-L-aspartyl)-(S-trityl-L-cysteinyl)-L-prolyl-L-leucyl glycine lower alkyl ester in methylene chloride, allowing the mixture to stand at about room temperature to complete condensation, filtering off precipitated dicyclohexyl urea, and separating the resulting (S,N-ditrityl-L-cysteinyl)-L-tyrosyl-L-isoleucyl-(γ-lower alkyl-L-glutamyl)-(β-lower alkyl-L-aspartyl)-(S-trityl-L-cysteinyl)-L-prolyl-L-leucyl glycine lower alkyl ester.

(r) Dissolving said lower alkyl ester in a lower alkanol, saturating the resulting solution at a temperature of about -15° to -20°C with ammonia gas, allowing the mixture to stand in a sealed container at room temperature to complete replacement of the lower alkyl ester group by the amide group, and separating the resulting triamide (S,N-ditrityl-L-cysteinyl)-L-tyrosyl-L-isoleucyl-L-glutaminyl-L-asparaginyl-(S-trityl-L-cysteinyl)-L-prolyl-L-leucyl glycine amide.

(s) Dissolving said triamide in an anhydrous solvent selected from the group consisting of chloroform, a mixture of chloroform and acetic acid, and a mixture of methylene chloride and thioglycolic acid, saturating the solution with gaseous hydrochloric acid at room temperature to complete detritylation, and separating the resulting L-cysteinyl-L-tyrosyl-L-isoleucyl-L-glutaminyl-L-asparaginyl-L-cysteinyl-L-prolyl-L-leucyl glycine amide.

(t) Dissolving said nonapeptide triamide in water and agitating the solution in oxygen to cause conversion thereof into oxytocin.

References

Merck Index 6849
Kleeman & Engel p. 681
PDR pp. 1382, 1596, 1966, 1989
I.N. p. 722
REM pp. 949, 957
Velluz, L., Amiard, G., Bartos, J., Goffinet, B. and Heymes, R.; US Patent 2,938,891; May 31, 1960; assigned to Uclaf, France
Velluz, L., Amiard, G. and Heymes, R.; US Patent 3,076,797; February 5, 1963; assigned to Roussel-UCLAF SA, France

P

PACLITAXEL

Therapeutic Function: Antineoplastic

Chemical Name: 5β,20-Epoxy-1,2α,4,7,10β,13α-hexahydroxytax-11-en-9-one 4,10-diacetate 2-benzoate 13-(α-phenylhippurate)

Common Name: 7-epi-Taxol; Paclitaxel; Plaxicel

Structural Formula:

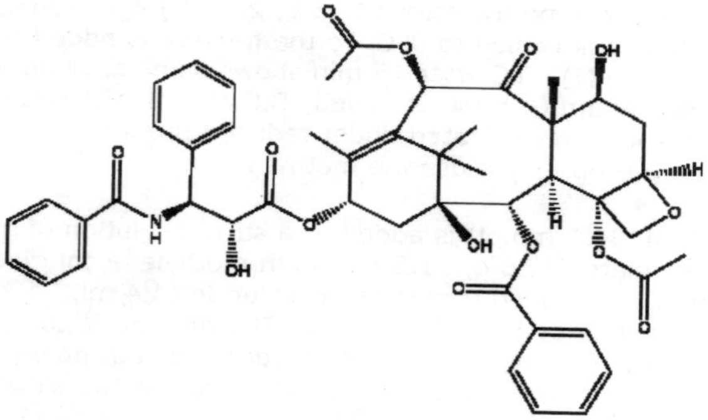

Chemical Abstracts Registry No.: 33069-62-4

Trade Name	Manufacturer	Country	Year Introduced
Altaxel	Cytomed (A div. of Alembic)	India	-
Betaxel	Biological E. Limited	India	-
Biotax	Bio Therap.	-	-
Genexol	Samyang	-	-
Intaxel	Dabur Pharmaceuticals Ltd.	India	-
Onxol	IVAX Pharmaceuticals, Inc.	USA	-
Paclitax	Cipla Limited	India	-
Paclitaxel	Dabur Pharmaceuticals Ltd.	India	-
Paclitaxel	Taihua Natural Plant Pharmaceutical Company	China	-

Trade Name	Manufacturer	Country	Year Introduced
Paclitaxel	Shanghai Jinhe Bio-TechnologyCo., Ltd.	China	-
Paclitaxel	Bristol - Myers Squibb Co.	USA	-
Paclitaxel-Ebewe	Ebewe	Austria	-
Paxen	Ivax-CR a.s.	Czech Republic	-
Taxol	Polysciences Inc.	USA	-
Taxol (A)	Bristol-Myers Squibb	Italy	-
Taxol (A)	Bristol-Myers Squibb	USA	-

Raw Materials

t-Butyl isocyanate
Triethylamine
Trimethylsilyl chloride
Hydrogen fluoride
Thiophenol
Potassium butoxide
7-SDMS Baccatin III
p-Nitrophenylsulfonyl chloride
(2R,3S)-β-Phenyl-isoserine methyl ester
Benzaldehyde dimethylacetal
4-Toluenesulfonic acid
Sodium bicarbonate
Benzoyl chloride
Triethylamine trihydrofluoride

Manufacturing Process

(2R,3S)-β-Phenyl-isoserine methyl ester (4.35 g, 22 mM) is dissolved in dry THF (100 ml) and the flask cooled to 0°C. To the mixture is added t-butyl isocyanate (2.8 ml, 25 mM). TLC after 15 min shows some starting material left so additional isocyanate (0.5 ml) is added. TLC after 1 h shows no starting material so the solvent is concentrated under reduced pressure to give the N-(t-butylaminocarbonyl)-β-phenyl isoserine methyl ester.

Triethylamine (4.8 ml, 34.4 mmol) is added to a stirred solution of methyl (2R,3S)-phenylisoserinate (7.26 g, 31.3 mmol) in methylene chloride (80 ml) at 0°C. To this slurry of is added trimethylsilyl chloride (4.4 ml, 34.7 mmol). Additional methylene chloride (45 ml) is added. The mixture is cooled to -65°C and triethylamine (9.8 ml, 70.3 mmol) is added. p-Nitrophenylsulfonyl chloride (6.93 g, 31.3 mmol) is added. The reaction rate is too slow at -65°C so the temperature is gradually raised to 0°C. Hydrogen fluoride (10% aqueous, 5 equivalents) is added. The aqueous phase is separated from the organic (methylene chloride) phase and methanol is added to the organic phase. The methylene chloride is removed under reduced pressure and the methyl (2R,3S)-3-(4-nitrobenzenesulfonamido)-3-phenyl-2-hydroxypropionate is obtained, melting point 187-189°C.

Benzaldehyde dimethylacetal (200 μl, 1.33 mmol) and a catalytic amount of p-toluenesulfonic acid (37 mg) are added to methyl (2R,3S)-3-(4-nitrobenzenesulfonamido)-3-phenyl-2-hydroxypropionate (315 mg, 0.83 mmol) in toluene 5 ml. The mixture is heated at 100°C under reduced pressure (15 mm mercury) with no condenser. After 1 h the crude reaction mixture is diluted with ethyl acetate and washed with water (2 times). After drying the organic layer over magnesium sulfate the crude material is purified by column chormatography (silica gel; eluting with ethyl acetate/cyclohexane, 35/65) to give the (2S,4S,5R)-2,4-diphenyl-3-(4-nitrobenzenesulfonamido)-5-methoxycarbonyl-1,3-oxazolidine, melting point 118°-120°C.

Water (8 ml), methanol (8 ml) and THF (8 ml) are added to (2S,4S,5R)-2,4-diphenyl-3-(4-nitrobenzenesulfonamido)-5-methoxycarbonyl-1,3-oxazolidine (1.50 g, 3.19 mmol). Potassium carbonate (1.018 g, 7.71 mmol) is then added. The resulting mixture is stirred at 20°-25°C until complete by TLC. After 5 h the reaction is complete and the reaction mixture is extracted with basic methylene chloride (2 times). The aqueous phase is then acidified with hydrochloric acid and extracted with ethyl acetate. The ethyl acetate phase is then washed with water, saline and dried over magnesiuim sulfate. Concentration of the organic phase (ethyl acetate) gives the (2S,4S,5R)-2,4-diphenyl-3-(4-nitrobenzenesulfonamido)-5-carboxy-1,3-oxazolidine, melting point 61°-65°C.

Then the (2S,4S,5R)-2,4-diphenyl-3-(4-nitrobenzenesulfonamido)-5-carboxy-1,3-oxazolidine react with the 7-SDMS Baccatin III, that is 7-(3-methylbut-2-yl)dimethylsilyl baccatin III (Baccatin III: 7,11-methano-1H-cyclodeca(3,4) benz(1,2-b)oxet-5-one,6,12b-bis(acetyloxy)-12(benzoyloxy)-1,2a,3,4,4a,6,9,10,11,12,12a,12b-dodecahydro-4,9,11-trihydroxy-4a,8,13,13-tetramethyl-, (2aR,4S,4aS,6R,9S,11S,12S,12aR,12bS), isolated from Taxus baccata).

(2S,4S,5R)-2,4-Diphenyl-3-(4-nitrobenzenesulfonamido)-5-carboxy-1,3-oxazolidine(323 mg, 0.711 mmol) is mixed with toluene (2.5 ml) at 20°-25°C. Dicyclohexylcarbodiimide (160 mg, 0.775 mmol) is then added to the reaction mixture. 7-SDMS Baccatin III (156 mg, 0.218 mmol) is added followed by 4-(dimethylamino)pyridine (35 mg, 0.286 mmol) and the reaction mixture is stirred at 20°-25°C until complete (1 h) by TLC. Sodium bicarbonate (50% aqueous, 10 ml) and more toluene (5 ml) is added to the reaction mixture and then stirred at 20°-25°C for 2 h. The reaction mixture is filtered through a medium frit to remove the urea byproduct. After filtering the phases are separated and the aqueous phase is extracted with ethyl acetate. The combined organic phases are washed with aqueous sodium bicarbonate (50%), water and saline. The organic phases are dried over magnesium sulfate, filtered and then concentrated. The concentrate is purified by column chromatograpy (silica gel; eluting with ethyl acetate/cyclohexane, 20/80) to give the 7-SDMS baccatin III 13-(2R,4S,5R)- and (2S,4S,5R)-2,4-diphenyl-3-(4-nitrobenzenesulfonamido)-1,3-oxazolidine-5-carboxylic acid ester.

THF (13.5 ml) and DMF (1.5 ml) are cooled to -35°C and degased by alternating reduced pressure and nitrogen three times. Thiophenol (0.22 ml, 2.14 mmol) is added followed by potassium butoxide/THF (1.978 M, 0.7 ml, 1.38 mmol). After 5 min, 7-SDMS baccatin III 13-(2R,4S,5R)- and (2S,4S,5R)-2,4-diphenyl-3-(4-nitrobenzenesulfonamido)-1,3-oxazolidine-5-carboxylic acid ester (877 mg, 0.762 mmol) is added. After the solids are added, the reaction mixture is slowly warmed to -10°C. The mixture is stirred at -10°C until the red color fades to yellow. After 3 h the bath is dropped allowing the mixture to warm to 20-25°C. At 20°-25°C the reaction is stirred for 1 h before assaying by TLC and HPLC. Sodium bisulfite (241 mg, 2.31 mmol) is added in water (5 ml). The mixture is stirred at 20°-25°C and after approximately 115 h the reaction is complete (by TLC) giving the free amine 7-SDMS baccatin III 13-(2R,3S)-3-amino-3-phenyl-2-hydroxypropionate.

Sodium bicarbonate (485 mg, 5.77 mmol) and water (10 ml) are added to 7-SDMS baccatin III 13-(2R,3S)-3-amino-3-phenyl-2-hydroxypropionate. The mixture is cooled to 0°C and then benzoyl chloride (150 ml, 1.3 mmol) is added. After 1 hr the reaction is complete and the reaction mixture is diluted with water and extracted with ethyl acetate. The organic phases are combined and washed with water, saline and dried over magnesium sulfate. Chromatography of the crude product (silica gel column; 20% to 100% ethyl acetate gives the 7-SDMS baccatin III 13-(2R,3S)-3-benzamido-3-phenyl-2-hydroxypropionate.

7-SDMS Baccatin III 13-(2R,3S)-3-benzamido-3-phenyl-2-hydroxypropionate (126 mg, 0.128 mmol) is dissolved in acetonitrile (2.5 ml). Triethylamine trihydrofluoride (123 mg, 0.763 mmol) is added under nitrogen and the resulting mixture is stirred at 5°C until complete by HPLC. When complete, the mixture is extracted with methyl t-butylether and washed with sodium bicarbonate solution. The aquesous washes are back extracted and combined with the organic phase. The combined organic phases are washed with water and saline, dried over magnesium sulfate, filtered and concentrated to give the Taxol (Paclitaxel), as needle from methanol with melting point 213-216°C.

References

Wuts P.G.M., Kelly R.C.; US Patent No. 6,057,452; May 2, 2000; Assigned to Pharmacia and Upjohn Company

Haugwitz R.D. et al.; US Patent No. 4,942,184; July 17, 1990; Assigned: USA as represented by the Department of Health and Human Services, Washington, D.C.

Stella V.J., Mathew A.E.; US Patent No. 4,960,790; Octouber 2, 1990; Assigned: University of Kansas, Lawrence, Kans

PAMIDRONATE SODIUM

Therapeutic Function: Bone resorption suppressant

Chemical Name: 3-Amino-1-hydroxypropane-1,1-diphosphonate, disodium salt

Common Name: Dinatrium pamidronat; Disodium pamidronate: Pamidronate sodium

Structural Formula:

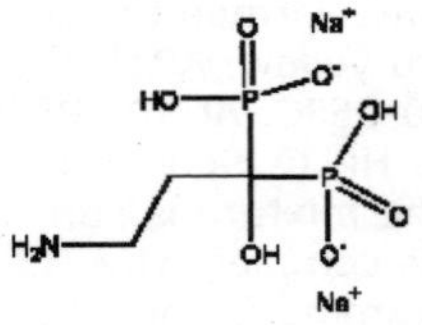

Chemical Abstracts Registry No.: 57248-88-1; 40391-99-9 (Base)

Trade Name	Manufacturer	Country	Year Introduced
Aredia	Novartis	India	-
Aredia	American Pharmaceutical Partners, Inc.	-	-
Aredia	Chiron Inc.	-	-
Aredia	Ciba-Geigy	-	-
Pamidronate Disodium	Novartis Pharmaceuticals	-	-
Pamidronate Disodium	Haorui Pharma-Chem Inc.	-	-
Pamidronate Disodium	Bedford Laboratories	-	-
Pamisol	David Bull Laboratories	-	-

Raw Materials

Mannitol
Pamidronic acid
Sodium hydroxide

Manufacturing Process

For a batch size of 5 L, 587.5 g (3.2 moles) of mannitol is dissolved in 3.5 L of water. Pamidronic acid (31.6 g, 0.133 moles) is mixed with a 1.0 L aliquot of the mannitol solution to form a slurry. The slurry is then transferred into the remainder of the mannitol solution, and stirred for at least 15 min. Aqueous 1 N sodium hydroxide (270 ml) is then added and the mixture is stirred until a clear, colorless solution results. The pH is then adjusted to 6.50.1 using either 1 M aqueous phosphoric acid or 1 N aqueous sodium hydroxide, as needed. The solution is then filtered through a 0.22 micron filter, and filled at 20°C into vials at 4.0 ml (4.172 g)/vial, under sterile conditions. The aqueous solution is frozen at -37°C and lyophilized (20 mbar, 20°-40°C) to yield 1,250 vials, each containing 30 mg of amorphous disodium pamidronate. The vials are sealed under positive nitrogen pressure. The disodium pamidronate is amorphous (noncrystalline) by X-ray diffraction and contains 0.7 wt-% water.

References

Shinal E.C.; US Patent No. 6,160,165; Dec. 12, 2000; Assigned: Aesgen, Inc., Princeton

PANCURONIUM BROMIDE

Therapeutic Function: Muscle relaxant

Chemical Name: 1,1'-[3α,17β-Bis(acetyloxy)-5α-androstane-2β,16β-diyl]bis[1-methylpiperidinium] dibromide

Common Name: -

Structural Formula:

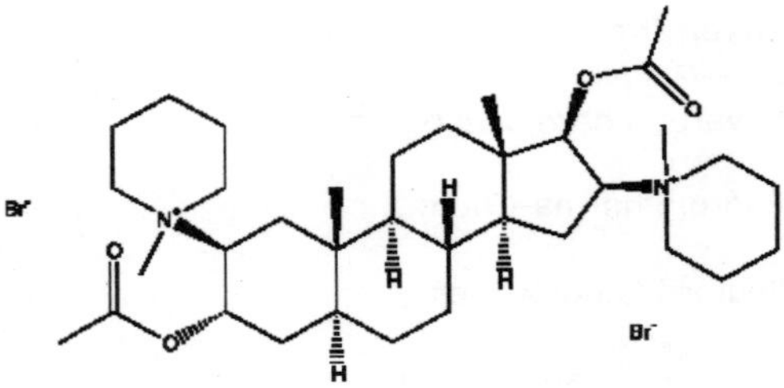

Chemical Abstracts Registry No.: 15500-66-0

Trade Name	Manufacturer	Country	Year Introduced
Pavulon	Organon-Teknika	UK	1968
Pancuronium	Organon	W. Germany	1969
Pavulon	Organon-Teknika	France	1971
Pavulon	Organon	US	1972
Myoblock	Organon-Sankyo	Japan	1973
Pavalon	Ravasini	Italy	1973

Raw Materials

Piperidine
m-Chloroperbenzoic acid
Acetic anhydride
3,17-Diacetoxy-5α-androstane-2,16-diene
Sodium borohydride
Methyl bromide

Manufacturing Process

A solution of 2α,3α,16α,17α-diepoxy-17β-acetoxy-5α-androstane (25 grams), prepared from 3,17-diacetoxy-5α-androstane-2,16-diene (Chem. Abs. 1960, 54, 8908) by treatment with m-chlor-perbenzoic acid, in piperidine (120 ml) and water (40 ml) was boiled under reflux for 5 days, the solution was concentrated and the product precipitated by the addition of water. The solid was collected, dissolved in dilute hydrochloric acid, filtered to give a clear solution and precipitated by the addition of sodium hydroxide solution. Crystallization from acetone gave 2β,16β-bis-piperidino-5α-androstan-3α-ol-17-one (18.9 grams), MP 179-185°C.

A solution of sodium borohydride (8 grams) in water (16 ml) was added to a stirred solution of 2β,16β-bis-piperidino-5α-androstan-3α-ol-17-one (17 grams) in tetrahydrofuran (70 ml) and methanol (30 ml) and the solution stirred at room temperature for 16 hours. The product was precipitated by the addition of water, filtered off, dried, and crystallized from acetone to give the diol (14.9 grams).

A solution of the piperidino-diol (9 grams) in acetic anhydride (18 ml) was heated at 90°C for 1 hour, the solution cooled, excess acetic anhydride destroyed by the careful addition of water, and the resulting solution carefully made alkaline with 2 N caustic soda solution to precipitate a solid product. The solid was dried, extracted with n-hexane and the solution filtered free of insoluble material before percolation down a column (4 x 1" diameter) of alumina. Elution with n-hexane gave a fraction (4.2 grams) which was crystallized twice from ether to give the diacetate, MP 176°-180°C.

Methyl bromide (17 grams) was added to a solution of the bis-piperidinodiacetate (4 grams) in methylene chloride (10 ml) and the resulting solution allowed to stand at room temperature for 4 days. The solution was evaporated to dryness, the residue triturated with ether, and filtered to give the bis-methobromide (5.2 grams), MP 206°C. Recrystallization from acetone-methylene chloride gave material MP 214°-217°C.

References

Merck Index 6870
Kleeman and Engel p. 681
PDR p. 1288
OCDS Vol. 2 p. 163 (1980)
DOT 5 (3) 104 (1969)
I.N.p. 726
REM p. 924
Hewett, C.L. and Savage, D.S.; US Patent 3,553,212; January 5, 1971; assigned to Organon Inc.

PANTETHINE

Therapeutic Function: Growth factor, Antihyperlipidemic

Chemical Name: Butyramide, N,N'-(dithiobis (ethyleneiminocarbonylethylene))bis(2,4-dihydroxy-3,3-dimethyl-, D-(+)-

Common Name: LBF disulfide form; Pantethine; Pantetina

Structural Formula:

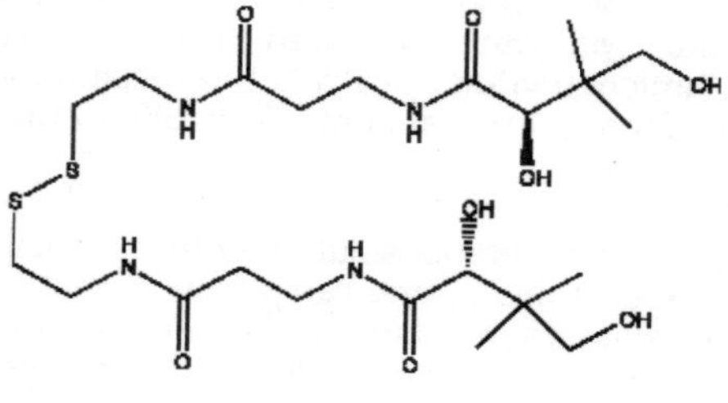

Chemical Abstracts Registry No.: 16816-67-4

Trade Name	Manufacturer	Country	Year Introduced
Atarone	Vinas	-	-
Lipodel	Shanghai Lansheng Corporation	-	-
Obliterol	Faes	-	-
Pantogen	Maruko	-	-

Raw Materials

Hydrazine hydrate
Methyl D-pantothenate
Super-filtrol
Bis(β-aminoethyl)disulfide dihydrochloride
Carbon

Manufacturing Process

To 11 g of hydrazine hydrate (85%) cooled in an ice bath are added 11.5 g of methyl d-pantothenate and the cold mixture is stirred vigorously. After the reaction takes place and the mixture is warmed to 30°C, it is allowed to stand at room temperature for two days, and then evaporated to dryness in vacuo at 50°C. The residue (14.7 g) of pantothenyl hydrazide is a clear glassy oil.

To 7.3 g of crude d-pantothenyl hydrazide dissolved in 21 ml of water and stirred in a beaker cooled on an ice bath is added sufficient. 6 N hydrochloric acid to shift the pH to 4. Then a solution of 1.7 g of sodium nitrite in 5 ml of water is added dropwise over a period of one hour, keeping the pH at 4 by additions of 6 N hydrochloric acid. After stirring for one-half hour, 2.8 g of bis(β-aminoethyl)disulfide dihydrochloride are added. The pH is then adjusted to 8.5 with 50% aqueous sodium hydroxide solution and the solution allowed to stir for one and one-half hours. It is then acidified to pH 7.5 and concentrated in vacuo to clear colorless viscous oil. The pure product can be isolated from this oil by the next method. The crude bis(N-pantothenylamidoethyl)disulfide so obtained is purified by dissolving the crude reaction product in 45 ml of anhydrous n-butanol and pouring the resulting solution through a chromatograph column containing 272 g of activated carbon. The column is washed with n-butanol and fractions are collected from time to time and the fractions containing solids assaying about 25 to 40% pure bis(N-pantothenylamidoethyl)disulfide against Lactobacillus: helveticus 80 poured onto a chromatograph column containing 136 g of an alkaline earth aluminum silicate known commercially as Super -filtrol. The column is washed thoroughly with anhydrous n-butanol and the washings and main solution discarded. N-Butanol saturated with water is poured through the column to elute the bis(N-pantothenylamidoethyl)disulfide and the resulting solution evaporated to dryness in vacuum at low temperature to obtain the desired product in pure form.

Instead of pouring the anhydrous n-butanol solution onto the alkaline earth aluminum silicate chromatograph column, one can simply repeat the treatment with a carbon chromatograph column to obtain the pure product. In some instances, the first carbon treatment produces fractions containing pure bis(N-pantothenylamidoethyl)disulfide and in those cases it is, of course, not necessary to treat the fraction with alkaline earth aluminum silicate nor again with activated carbon.

References

Snell E.E., More J.A.; US Patent No. 2,625,565; Assigned to Parke, Davis and Company, Detroit, Mich., a corporation of Michigan

PANTOPRAZOLE SODIUM

Therapeutic Function: Antiulcer

Chemical Name: 1H-Benzimidazole, 5-(difluoromethoxy)-2-(((3,4-dimethoxy-2- pyridinyl)methyl)sulfinyl)-, sodium salt

Common Name: Pantoprazole sodium

Structural Formula:

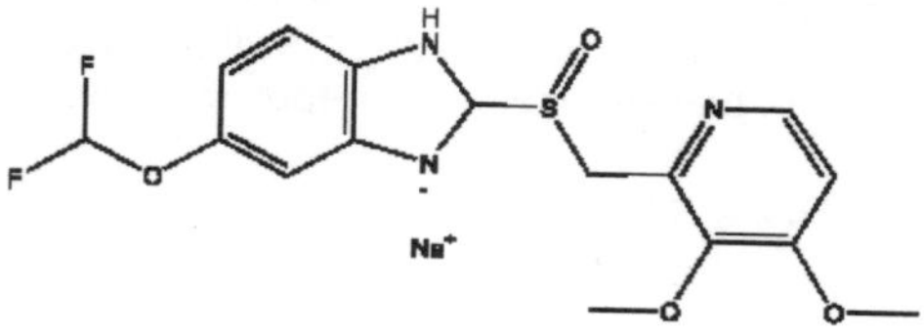

Chemical Abstracts Registry No.: 138786-67-1; 102625-70-7 (Base)

Trade Name	Manufacturer	Country	Year Introduced
Controloc	Byk Gulden	Germany	-
Pangest	Beta	-	-
Pantoloc	Solvay Pharma	-	-
Pepmark	Unimarck Pharma (India) Ltd.	India	-
Protium	Lupin Laboratories Ltd.	-	-
Protonix	Wyeth Pharmaceuticals	USA	-
Somac	Pharmacia and Upjohn	-	-

Raw Materials

Sodium hydroxide
5-Difluoromethoxy-1H-benzimidazole-2-thiol
Sodium thiosulfate
Sodium hypochlorite
2-Chloromethyl-4,5-dimethoxy-3-methylpyridinium chloride

Manufacturing Process

2-Chloromethyl-4,5-dimethoxy-3-methylpyridinium chloride (about 1.5 g) are added to a solution of 5-difluoromethoxy-1H-benzimidazole-2-thiol in 10 ml of ethanol and 10 ml of 1 N sodium hydroxide solution. The yellow reaction mixture is stirred at 20°C for 1 hour, a further 10 ml of water are added, whereupon a colorless solid precipitates out, the mixture is stirred for a further 5 hours and filtered and the residue is rinsed with 1 N sodium hydroxide solution and water and dried to constant weight. The 5-difluoromethoxy-2-[(4,5-dimethoxy-2-pyridyl)methylthio]-1H-benzimidazole is obtained as an oil.

5-Difluoromethoxy-2-[(4,5-dimethoxy-2-pyridyl)methylthio]-1H-benzimidazole (about 1 g) are dissolved in 10 ml of dioxane and 2 ml of 1 N sodium hydroxide solution. An equimolar amount of a titrated aqueous sodium hypochlorite solution, to which 1 mole per liter of sodium hydroxide solution has been added, is first added dropwise, while cooling with ice. After one hour a further equivalent and after 3 hours half the equimolar amount of sodium hypochlorite are added, to achieve complete reaction. After a reaction time of 4 hours, 5 ml of 5% strength sodium thiosulfate solution and another 25 ml of dioxane are added and the upper dioxane phase is separated off, washed once with 5 ml of sodium thiosulfate solution and concentrated on a rotary evaporator. The oily residue is dissolved in 20 ml of water and 10 ml of ethyl acetate and the solution is brought to pH 7 with about 100 ml of a buffer solution of pH 6.8. The solid which has precipitated out is filtered off with suction over a suction filter, washed with water, extracted by stirring at 0C with acetone and dried. 5-Difluoromethoxy-2-[(4,5-dimethoxy-2-pyridyl)methanesulfinyl]-1H-benzimidazole is prepared; yield about 85%.

In practice it is usually used as sodium salt.

References

Kohl B. et al, US Patent No. 4,758,579; July 19, 1988; Assigned to BYK Gulden Lomberg Chemische Fabrik GmbH (Konstanz, DE)

PANTOTHENIC ACID

Therapeutic Function: Vitamin

Chemical Name: β-Alanine, N-(2,4-dihydroxy-3,3-dimethyl-1-oxobutyl)-, (R)-

Common Name: Achromothrichiefaktor; Acidum pantothenicum; Chick antidermatitis factor; Filtrat-Faktor; Kueken-Antidermatitis-Faktor; Pantothenic acid; Pantothensaeure

Structural Formula:

OH H HO N OH O O

Chemical Abstracts Registry No.: 79-83-4

Trade Name	Manufacturer	Country	Year Introduced
Panto-250	Bio-Tech Pharmacal	-	-

Raw Materials

Isobutylaldehyde
Potassium chromate
Hydrochloric acid
β-Alanine
Formaldehyde
Sodium cyanide
α-Phenylethylamine

Manufacturing Process

Isobutylaldehyde reacted with formaldehyde in the presence potassium chromate as a result 2,2-dimethyl-3-hydroxy-propanal was obtained.

The 2,2-dimethyl-3-hydroxy-propanal was treated by sodium cyanide so 2,4-dihydroxy-3,3-dimethyl-butironitrile was prepared.

The 2,4-dihydroxy-3,3-dimethyl-butironitrile was treated hydrochloric acid and D,L-3-hydroxy-4,4-dimethyl-dihydro-furan-2-one (D,L-pantolacton) was obtained. The racemic mixture of D- and L-pantolactons was a division of D- and L- isomers by the adding of α-phenylethylamine. So D-pantolacton was isolated.

Acrylic acid contacted with $NH_3$and β-alanine was obtained.

D-Pantalacton reacted with β-alanine as a result 3-(2,4-dihydroxy-3,3-dimethyl-butyrylamino)-propanoic acid was produced.

References

Kleemann A., Engel J.; Pharmazeutische Wirkstoffe, GeorgThieme Verlag Stuttgart. New York, 1982

PAPAIN

Therapeutic Function: Enzyme, Wound adhesion inhibitor

Chemical Name: Enzyme; used to prevent wound adhesions

Common Name: -

Structural Formula: Has folded polypeptide chain of 212 residues with a molecular weight of about 23,400

Chemical Abstracts Registry No.: 9001-73-4

Trade Name	Manufacturer	Country	Year Introduced
Papain	Green Cross	Japan	1969
Panafil	Rystan	US	-
Prevenzyme	Legere	US	-

Raw Materials

Papaya fruit
Methanol

Manufacturing Process

Crude papain, obtained as the dried exudate of the fruit and leaves of Carica papaya L., Caricaceae, is usually found to have been contaminated during collection, drying, or storage by insects, rodent hair and excreta, botanical plant parts, sand, etc. and may thereby become further contaminated by harmful bacteria and enteric organisms.

Heretofore papain has been purified by dispersing the crude enzymes in water, filtering and spray-drying. In this procedure, however, the soluble contaminants are retained in the dried product. It has also been known to purify papain by dispersing it in water and adding acetone to reprecipitate the enzymes leaving many of the acetone-soluble and water-soluble impurities in the supernatant liquid. The material thus purified possesses a very disagreeable sulfidelike taste probably due to the reaction between the acetone and reactive sulfhydryl groups present in the papaya latex.

It has now been found that an enzyme mixture of high purity which contains none of the objectionable sulfidelike taste can be obtained by dispersing the crude enzymes in water, adding a quantity of a water-miscible lower-alkanol to the incipient precipitation point of the proteolytic enzymes thereby retaining the maximum proteolytic activity (i.e., the maximum amount of the proteolytic enzymes) in the solvent phase while precipitating the major portion of the lower-alkanol insoluble contaminants, removing the lower-alkanol insoluble contaminants and precipitated inert materials, for example, by filtration or centrifugation, and then adding an additional quantity of the water-miscible lower-alkanol sufficient to precipitate the proteolytic enzymes.

The following is a specific example of the conduct of the present process. 100 g of crude papain were stirred with 120 ml of 0.01 M cysteine hydrochloride for one hour during which time the papain was completely dispersed. To the dispersion was added slowly and with vigorous stirring 147 ml of methanol. The mixture, which contained 55% methanol by volume, was stirred for about thirty minutes and centrifuged and the clear supernatant liquid was removed and saved. The precipitate was washed with 50 ml of 55% aqueous methanol, and the mixture was centrifuged again. The precipitate containing the undesirable, insoluble contaminants was discarded, and the clear wash liquid was combined with the main supernatant. To the combined clear supernatant liquid was added slowly and with vigorous stirring 265 ml of methanol to give a mixture containing 75.5% methanol by volume. The enzymes were precipitated as a taffylike gum which was isolated by decantation of the supernatant liquid containing the undesirable, soluble contaminants and tray-drying. Alternatively, the precipitated enzymes can be redissolved in pure water and spray-dried.

References

Merck Index 6878
PDR pp. 1033, 1576
REM p. 1038
Losuk, A.; US Patent 3,011,952; December 5, 1961; assigned to Sterling Drug, Inc.

PAPAVERINE MONOPHOSADENINE

Therapeutic Function: Vasodilator, Platelet aggregation inhibitor

Chemical Name: Papaverine adenosine 5-monophosphate

Common Name: Papaverine adenylate

Structural Formula:

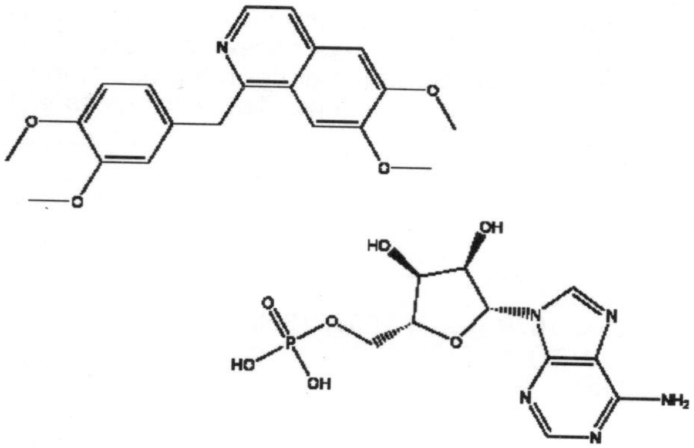

Chemical Abstracts Registry No.: 58-74-2 (Base)

Trade Name	Manufacturer	Country	Year Introduced
Lempav Ty-Med	Lemmon	US	1975
Artegodan	Artesan	W. Germany	-
Cepaverin	Eurand	Italy	-
Cerespan	U.S.V.	US	-
Dylate	Elder	US	-
Omnopon	Roche	UK	-
Pameion	Simes	Italy	-
Panergon	Mack	W. Germany	-
Papaverlumin	Pidefe	Spain	-
Papaversan	Abello	Spain	-

Trade Name	Manufacturer	Country	Year Introduced
Pavabid	Marion	US	-
Pavacron	Cenci	US	-
Pavagrant	Amfre-Grant	US	-
Pavakey	Key	US	-
Pavatym	Everett	US	-
Paver	Mulda	Turkey	-
Spastretten	Tropon	W. Germany	-
Sustaverine	I.C.N.	US	-
Udip	Marion	US	-

Raw Materials

Papaverine base
Adenosine-5'-monophosphoric acid

Manufacturing Process

To 3.65 g (0.01 mol) of monohydrated adenosine-5'-monophosphoric acid, brought into suspension in a mixture of 45 ml of water and 5 ml of ethanol, are added 339 g (0.01 mol) of papaverine base (melting point, 147°C). The mixture is gently heated until a final temperature of 40°C is reached. The solution obtained is then filtered and the filtrate is concentrated under vacuum. The remaining product quickly crystallizes. After drying to 50°C to constant weight, there are obtained 6.68 g of desired product, in the monohydrated state, as a white crystalline powder, which melts at 140°C and is very soluble in water.

References

Merck Index 6880
Kleeman and Engel p. 683
PDR pp. 830, 875, 993, 1079, 1569, 1606, 1810
OCDS Vol. 1 p. 347 (1977)
DOT 11 (8) 315 (1975)
I.N. p. 728
REM p. 852
Mauvernay, R.Y.; US Patent 3,823,234; July 9, 1974; assigned to Centre Europeen de Recherches Mauvernay C.E.R.M.

PARAFLUTIZIDE

Therapeutic Function: Diuretic

Chemical Name: 2H-1,2,4-Benzothiadiazine-7-sulfonamide, 3,4-dihydro-6-chloro-3-((4-fluorophenyl)methyl)-, 1,1-dioxide

Common Name: Paraflutizide

Structural Formula:

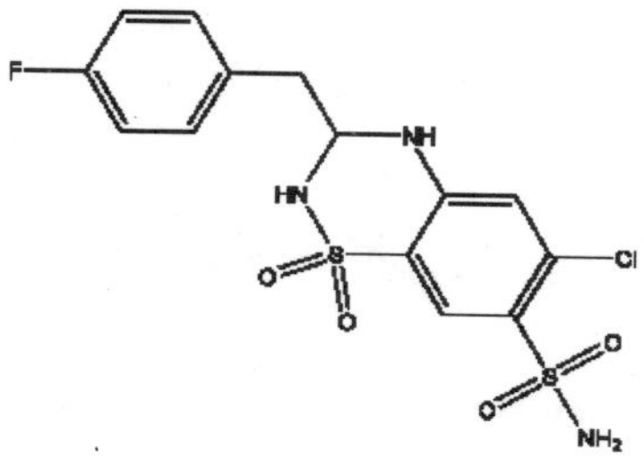

Chemical Abstracts Registry No.: 1580-83-2

Trade Name	Manufacturer	Country	Year Introduced
Paraflutizide	Shanghai Lansheng Corporation	-	-

Raw Materials

Potassium bichromate
ρ-Fluorophenylethyl alcohol
Hydrochloric acid
5-Chloro-2,4-disulphamylaniline

Manufacturing Process

A mixture of 5.0 g (0.0357 g/mol) of ρ-fluorophenylethyl alcohol, 2.0 g (0.007 g/mol) of 5-chloro-2,4-disulphamylaniline, 2.0 g (0.0068 g/mol) of potassium bichromate and 15 ml of concentrated hydrochloric acid (0.176 g/mol) and 25 ml of water is heated under reflux for 1 h. The mixture is allowed to cool, and 15 ml of ether are added to separate the excess of ρ-fluorophenylethyl alcohol. The aqueous layer is decanted and frozen for 2 h and the precipitate is separated, washed with water and dried in vacuum over phosphoric anhydride. There are collected 1.35 g (yield 47.5%) of the 1,1-dioxide of 3-ρ-fluorophenyl-methyl-7-sulphamyl-6-chloro-3,4-dihydrobenzo-1,2,4-thiadiazine, which when recrystallised from 30 ml of 50% alcohol on "Norit" active carbon takes the form of a white crystalline substance, melting point is 239°C.

References

GB Patent No. 961,641; July 31, 1962; Assigned: Les Laboboratoires Dausse, a French Body Corporate,of 58-60, Rue de la Glaciere, Paris, France

PARAMETHADIONE

Therapeutic Function: Anticonvulsant

Chemical Name: 5-Ethyl-3,5-dimethyl-2,4-oxazolidinedione

Common Name: Isoethadione

Structural Formula:

Chemical Abstracts Registry No.: 115-67-3

Trade Name	Manufacturer	Country	Year Introduced
Paradione	Abbott	US	1949

Raw Materials

Methyl ethyl ketone
Urea
Methanol
Sodium cyanide
Sodium
Dimethyl sulfate

Manufacturing Process

About 143.1 grams (one mol) of 5-methyl-5-ethyloxazolidine-2,4-dione is dissolved in 300 cc of methanol containing 23 grams of sodium. To the above mixture is added 126 grams of dimethyl sulfate in 10 cc portions while the temperature is maintained at about 50°C by external cooling. The mixture is then heated briefly to boiling, cooled, diluted with about 500 cc of water and extracted with two 250 cc portions of benzene. The benzene extract is separated, washed once with sodium bicarbonate solution and once with water. The benzene is removed by evaporation on a steam bath and the residue is fractionally distilled. The material boiling at 112° to 116°C at 25 mm pressure is taken; n_D^{25}=1.4495. Upon further fractionation, a very pure specimen boils at 101°-102°C at 11 mm.

The 5-methyl-5-ethyloxazolidine-2,4-dionemay be prepared by reacting methyl ethyl ketone with sodium cyanide and with ammonium thiocyanate followed by desulfurization. This intermediate may also be prepared by condensing α-hydroxy-α-methylbutyramide with ethyl chlorocarbonate or by condensing ethyl α-hydroxy-α-methylbutyrate with urea. Another method described (Traube and Aschar, Ber., 46, 2077-1913) consists in the condensation of ethyl α-hydroxy-α-methylbutyrate with guanidine followed by hydrolysis.

References

Merck Index 6890
Kleeman and Engel p. 685
PDR p. 545
OCDS Vol. 1 p. 232 (1977)
I.N. p. 730
REM p. 1080

Spielman, M.A.; US Patent 2,575,693; November 20, 1951; assigned to Abbott Laboratories

PARAMETHASONE ACETATE

Therapeutic Function: Glucocorticoid

Chemical Name: 6α-Fluoro-11β,17,21-trihydroxy-16α-methylpregna-1,4-diene-3,20-dione

Common Name: -

Structural Formula:

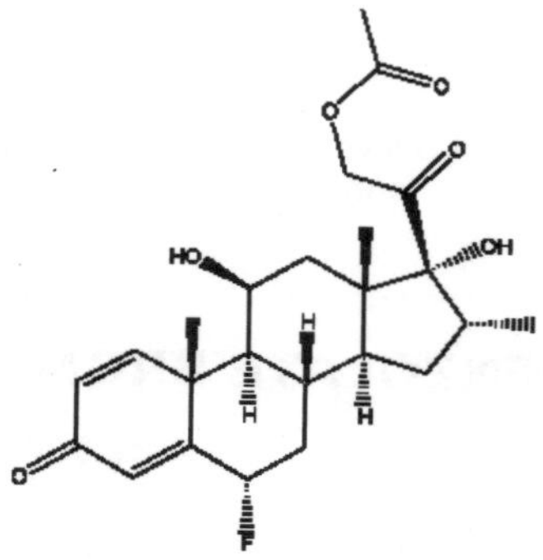

Chemical Abstracts Registry No.: 1597-82-6; 53-33-8 (Base)

Trade Name	Manufacturer	Country	Year Introduced
Haldrone	Lilly	US	1961
Dilar	Cassenne	France	1962
Paramezone	Recordati	Italy	1962
Monocortin	Gruenenthal	W. Germany	1963
Stemex	Syntex	US	1970
Cortidene	I.F.L.	Spain	-
Metilar	Syntex	UK	-
Paramesone	Tanabe	Japan	-
Sintecort	Medicamenta	Portugal	-
Triniol	I.F.L.	Spain	-

Raw Materials

Hydrogen chloride
5α,11β,17α,21-Tetrahydroxy-6β-fluoro-16α-methylallopregnane-3,20-dione-21-acetate 3-ethylene glycol ketal

Manufacturing Process

A solution of 0.144 g of the 3-ethylene glycol ketal of 5α,11β,17α,21 - tetrahydroxy-6β-fluoro-16α-methylallopregnane-3,20-dione-21 acetate in 12 ml of chloroform and 0.1 ml of absolute alcohol was cooled to -10°C in an ice-salt bath and a stream of anhydrous hydrochloric acid was gently bubbled through the solution for 2.5 hours while the temperature was maintained between -5°C and -15°C. The solution was then diluted with 25 ml of chloroform, washed with dilute sodium bicarbonate and water, dried over anhydrous sodium sulfate, and evaporated to dryness under reduced pressure at 60°C or less to give 6α-fluoro-11β,17α,21-trihydroxy-16α-methyl-4-pregnene-3,20-dione 21-acetate.

References

Merck Index 6891
Kleeman and Engel p. 686
OCDS Vol. 1 p. 200 (1977)
I.N. p. 730
REM p. 969
Lincoln, F.H., Schneider, W.P. and Spero, G.B.; US Patent 3,557,158; January 19, 1971; assigned to The Upjohn Co.

PARAPENZOLATE BROMIDE

Therapeutic Function: Antiulcer

Chemical Name: N-Methyl-4-piperidylbenzilate methobromide

Common Name: -

Structural Formula:

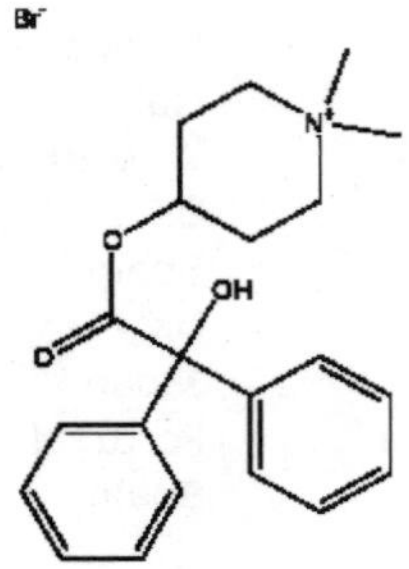

Chemical Abstracts Registry No.: 5634-41-3

Trade Name	Manufacturer	Country	Year Introduced
Spacine	Unilabo	France	1968
Vagopax	Essex	Italy	1976
Vagopax	Centrane	France	-

Raw Materials

Methyl iodide
Diphenylchloroacetyl chloride
Silver bromide
N-Methyl-4-piperidinol HCl

Manufacturing Process

N-methyl-4-piperidyl benzilate and the methiodide: An intimate mixture of 0.1 mol of N-methyl-4-piperidinol hydrochloride and 0.1 mol diphenylchloroacetyl chloride is heated at 160°C to 180°C until the evolution of hydrogen chloride ceases (usually about 4 to 5 hours). The melt is then dissolved in 500 ml of water and the resultant mixture heated on a steam bath for about ½ hour, after which time complete solution is effected. The acid solution is cooled and rendered alkaline with ammonium hydroxide solution whereupon the ester is precipitated. The ester is purified either by removal by filtration and recrystallization from benzene petroleum ether or by extracting the mixture with benzene and precipitating the ester by the addition of petroleum ether. After recrystallization there is obtained about 0.06 mol of N-methyl-4-piperidyl benzilate, melting point 162°C to 163°C.

To a solution of 0.05 mol of the above obtained ester in about 100 ml of anhydrous benzene there are added 15 ml of methyl iodide. The ensuing mixture is refluxed for several hours whereupon the quaternary salt is deposited and removed by filtration. Recrystallization from ethanol or ethanol-ether yields the quaternary salt, melting point 199°C to 200°C.

N-methyl-4piperidyl benzilate methobromide: To a suspension of 0.15 mol of freshly prepared silver bromide in 300 ml of anhydrous methanol is added a solution of 0.1 mol of quaternary iodide obtained as above. The mixture is stirred and refluxed for several hours after which time transhalogenation is complete. The mixture is cooled, the insoluble silver salt removed by filtration and the methanolic solution of the quaternaty bromide is concentrated in vacuo. The residue is recrystallized from methanol or methanol-ether yielding the quaternary bromide in quantitative amounts, melting point 237°C to 238°C.

References

OCDS Vol. 2 p. 75 (1980)
DOT6 (3) 92 (1970)
I.N. p. 731
Papa, D.; British Patent 788,126; December 23, 1957; assigned to Schering Corp.

PARGYLINE HYDROCHLORIDE

Therapeutic Function: Antihypertensive

Chemical Name: N-Methyl-N-2-propynylbenzenemethanamine hydrochloride

Common Name: N-Methyl-N-propargylbenzylamine hydrochloride

Structural Formula:

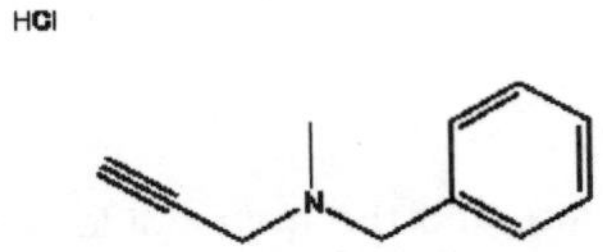

HCl

N

Chemical Abstracts Registry No.: 306-07-0; 555-57-7 (Base)

Trade Name	Manufacturer	Country	Year Introduced
Eutonyl	Abbott	US	1963

Raw Materials

N-Methylbenzylamine
Sodium carbonate
Propargyl bromide
Hydrogen chloride

Manufacturing Process

A mixture of 23.8 grams (0.2 mol) of propargyl bromide, 24.2 grams (0.2 mol) of N-methylbenzylamine and 400 ml of anhydrous ethanol in the presence of 42.4 grams (0.4 mol) of anhydrous sodium carbonate was heated at the boiling temperature and under reflux for a period of 17 hours.

The sodium carbonate was then removed by filtration and the alcohol was removed by distillation under reduced pressure. The residue was treated with 300 ml of dry ether and the resulting solution was filtered to remove sodium bromide.

The filtrate was dried and fractionally distilled under reduced pressure to obtain the desired N-methyl-N-propargylbenzylamine which boiled at 96°-97°C at 11 mm pressure.

Analysis calculated for $C_{11}H_{13}N$: C = 82.97%; H = 8.23%; N = 8.80%.
Found: C = 82.71%; H = 8.51%; N = 8.93%.

The hydrochloride salt of this amine was prepared by dissolving the amine in ether and adding ethereal hydrogen chloride to the ether solution. The solid hydrochloride salt which precipitated was recrystallized from an ethanol-ether mixture and was found to melt at 154° - 155°C.

References

Merck Index 6902
Kleeman and Engel p. 688
PDR p. 523
OCDS Vol. 1 p. 54 (1977) and 2, 27 (1980)
DOT9 (6) 217 (1973)
I.N. p. 732
REM p. 850 Martin, W.B. US Patent 3,155,584; November 3, 1964; assigned to Abbott Laboratories

PAROMOMYCIN

Therapeutic Function: Amebicidal

Chemical Name: O-2,6-Diamino-2,6-dideoxy-β-L-idopyranosyl-(1-->3)-O-β-D-ribofuranosyl-(1-->5)-O-[2-amino-2-deoxy-α-D-glucopyranosyl-(1-->4)]-2-deoxystreptamine

Common Name: Catenulin; Aminosidine; Crestomycin; Hydroxymycin; Neomycin E; Paucimycin

Structural Formula:

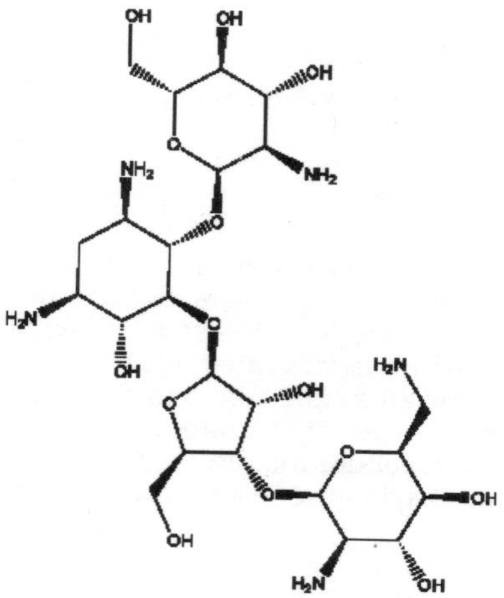

Chemical Abstracts Registry No.: 7542-37-2

Trade Name	Manufacturer	Country	Year Introduced
Humatin	Parke Davis	US	1960
Humatin	Parke Davis	W. Germany	1961
Humatin	Parke Davis	Italy	1961
Humagel	Parke Davis	France	1963

Trade Name	Manufacturer	Country	Year Introduced
Aminosidine	Kyowa	Japan	-
Aminoxidin	Farmalabor	Italy	-
Gabbromycin	Montedison	Italy	-
Gabbroral	Farmalabor	Italy	-
Paramicina	Ragionieri	Italy	-

Raw Materials

Glucose
Soybean meal
Bacterium Streptomyces rimosus forma paromomycinus

Manufacturing Process

As described in US Patent 2,916,485: 12 liters of a nutrient medium having the following composition is placed in a 30 liter fermenter equipped with stainless steel fittings including sparger, impeller, baffles and sampling lines and the medium is sterilized by heating at 121°C for two hours.

	Percent
Glucose monohydrate	0.5
Glycerol	0.5
Casein, acid hydrolyzed	0.3
Peptone	0.25
Brewer's yeast	0.1
Corn steep solids	0.25
Soybean oil meal	0.25
Acetone-butanol fermentation residue	0.25
Sodium chloride	0.5
Calcium carbonate	0.1
Water sufficient to make	100%

The medium is cooled and inoculated with 20 ml of a suspension of the spores from two Mover's sporulation agar slant cultures of Streptomyces rimosus forma paromomycinus in sterile 0.1% sodium heptadecyl sulfate solution. The inoculated culture mixture is incubated at 26°C for sixty hours during which time the mixture is stirred at 200 rpm and sterile air is passed into the medium through the sparger at the rate of 12 liters per minute. A portion of the resulting incubated culture mixture is employed for inoculation of 16 liters of a nutrient medium having the following composition:

	Percent
Glucose monohydrate	1.0
Soybean oil meal	1.0
Sodium chloride	0.5
Calcium carbonate	0.1
Ammonium chloride	0.167
Hog stomach residue, saline extracted	0.5
Water sufficient to make	100%

The pH of the latter nutrient medium is adjusted to 7.5 with 10 N sodium hydroxide solution and is placed in a 30 liter glass fermenter equipped with sparger, impeller, baffles and sampling line, The medium is sterilized by heating at 121°C for two hours, is allowed to cool and is then inoculated with 800 ml of the culture mixture obtained as described above.

The resulting culture mixture is incubated at 26°C for 94 hours during which time the mixture is stirred at 200 rpm and sterile air is passed into the medium through the sparger at the rate of 16 liters per minute. During the incubation, foaming is avoided by the addition, as needed, of crude lard and mineral oils containing mono-and diglycerides.

At the end of the incubation period the fermentation culture mixture is adjusted to pH 2 with concentrated hydrochloric acid, the solid material present is removed by filtration, and the filter cake is washed with water. The washings are combined with the main filtrate, adjusted to pH 7.0; and 15.5 liters of the filtered culture liquid is introduced into a columnar exchanger (1.5'' i.d.) packed with 380 ml of carboxylic acid resin which has been preliminarily washed in succession with two liters of an aqueous solution of 37.5 grams of sodium hydroxide and with two liters of water. The column containing paromomycin is washed with two hold-up volumes of water and is eluted with 0.5 N hydrochloric acid.

The first 19.4 liters of percolate contains little or no paromomycin and varies in pH from 6 to 7.3. When the pH of the eluate begins to fall below 6.0, two liters of the eluate are collected.

The two liter portion of the eluate, collected as indicated, is neutralized to pH 6 with 10 N sodium hydroxide solution and is filtered. The filtrate is concentrated by evaporation in vacuo to a volume of approximately one liter.

An adsorption column is prepared by pouring a slurried aqueous mixture of 65 grams of acid-washed activated charcoal (Darco G-60) and 50 grams of diatomaceous earth in a 1.5'' column and 300 ml of the concentrated filtrate is added. The column is washed with 400 ml of water and eluted successively with 325 ml of water, 425 mi of 1% aqueous acetone and 400 ml of 10% aqueous acetone. The water and acetone eluates are concentrated and lyophilized to give paromomycin hydrochloride as a powder. The product is purified by taking up the powder in methanol, adding a large excess of acetone to the solution, recovering the precipitate which forms by filtration. The product, paromomycin hydrochloride, has an optical rotation $[\alpha]_D^{25}$ = +56.5° (1% in water). By analysis it contains 35.71% carbon, 6.95% hydrogen, 8.24% nitrogen and 21.5% chlorine.

In order to obtain paromomycin in free base form, the hydrochloride is dissolved in water as a 3% solution, the solution is poured into an adsorption column containing an anion exchange resin (Amberlite IR-45 or preferably IRA-411 or IRA-400) in the hydroxyl form and the column is washed with a small amount of water.

The aqueous percolate is concentrated to dryness by lyophilization, and the solid product obtained is purified by taking up in boiling absolute ethanol, cooling and recovering the solid product paromomycin; $[\alpha]_D^{25}$ = +64° (1% in water). By analysis it contains 45.17% carbon, 7.44% hydrogen and 10.35% nitrogen.

References

Merck Index 6903
Kleeman and Engel p. 688
I.N. p. 733
REM P. 1221
Davisson, J.W. and Finlay, A.C.; US Patent 2,895,876; July 21, 1959; assigned to Chas. Pfizer and Co., Inc.
Frohardt, R.P., Haskell, T.H., Ehrlich, J. and Knudsen, M.P.; US Patent 2,916,485; Dec. 8, 1959; assigned to Parke, Davis and Company

PAROXETINE HYDROCHLORIDE

Therapeutic Function: Antidepressant

Chemical Name: Piperidine, 3-((1,3-benzodioxol-5-yloxy)methyl)-4-(4-fluorophenyl)-, hydrochloride, (3S-trans)-

Common Name: Paroxetine hydrochloride

Structural Formula:

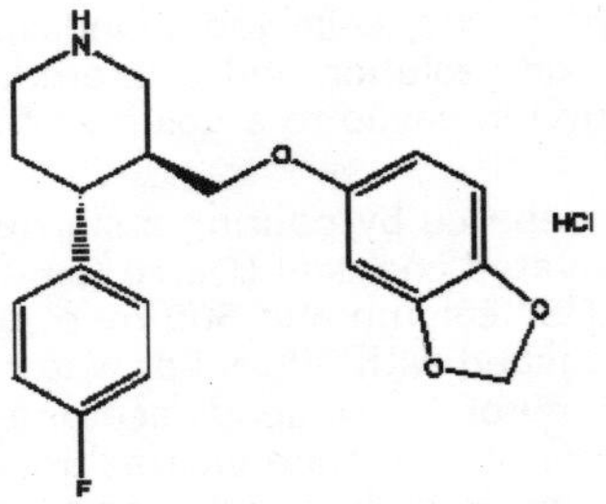

HCl

Chemical Abstracts Registry No.: 78246-49-8; 61869-08-7 (Base)

Trade Name	Manufacturer	Country	Year Introduced
Paxil	SmithKline Beecham Pharmaceuticals	France	-
Paxyl	SK Beecham	-	-

Raw Materials

Methyl-4-(4-fluorophenyl)-N-methyl-nipecotinate
Hydrochloric acid
Formaldehyde
Thionyl chloride
Sulfuric acid
4-(4-Fluorophenyl)-1-methyl-1,2,3,6-tetrahydropyridine
Sodium methoxide
Tartaric acid, dibenzoate, (-)-
Palladium on carbon
3,4-Methylenedioxyphenol

Manufacturing Process

251 g of methyl-4-(4-fluorophenyl)-N-methyl-nipecotinate, 8 g of sodium methoxide and 500 ml benzene were refluxed for 2 h. The benzene solution was washed with cold water and evaporated to give the pure α-ester which was dissolved in a mixture of 320 ml of water and 450 ml concentrated hydrochloric acid. The solution was slowly distilled to remove methanol and finally evaporated to dryness in vacuo.

400 ml thionyl chloride were added in small portions to the solid. The mixture was allowed to stand for 3 h at room temperature and was then evaporated to dryness in vacuo with tetrachloroethane giving methyl-4-(4-fluorophenyl)-N-methylnipecotic acid chloride. The acid chloride was added in small portions to a solution of 160 g (-)-menthol in 800 ml pyridine at a temperature of 0°-5°C. The mixture was allowed to stand at room temperature to the next day. Ice water and 50% sodium hydroxide were added, and the mixture was extracted with ether. The ether was dried with anhydrous magnesium sulphate, filtered and evaporated. Distillation in vacuo gave the menthol ester in a yield of 75-80%. Boiling point at 0.05 mm Hg was 165°-170°C.

Racemic 4-(4-fluorophenyl)-1-methyl-1,2,3,6-tetrahydropyridine (50 g) was dissolved in a mixture of 21.6 ml of concentrated sulfuric acid and 50 ml of water. To the solution were added 25 ml of concentrated hydrochloric acid and 22.4 ml of 37% formaldehyde solution. The mixture was refluxed for 5 h, cooled, and 125 ml of concentrated ammonia were added. The mixture was extracted with 50 ml of toluene. Drying of the toluene solution and distillation gave 38 g of 4-(4-fluorophenyl)-3-hydroxymethyl-1-methyl-1,2,3,6-tetrahydropyridine with boiling point 110°-120°C at 0.1 mm Hg.

13 g of the racemic compound and 22 g of (-)-dibenzoyltartaric acid were dissolved in 105 ml of hot methanol. On cooling, 9 g of salt of (-)-4-(4-fluorophenyl)-3-hydroxymethyl-1-methyl-1,2,3,6-tetrahydropyridine crystallized. Melting point 167°-168°C.

38 g of (-)-4-(4-fluorophenyl)-3-hydroxymethyl-1-methyl-1,2,3,6-tetrahydropyridine were dissolved in 350 ml of 99% ethanol, 5 g of 5% palladium on carbon were added, and the mixture was treated with hydrogen until 4500 ml were absorbed. The catalyst was filtered off, and the solution was evaporated to yield 37.5 g of (+)-b-4-(4-fluorophenyl)-3-hydroxymethyl-1-methylpiperidine.

To a solution of sodium in methanol (125 ml) were added 3,4-methylenedioxyphenol (29 g) and the (+)-b-4-(4-fluorophenyl)-3-hydroxymethyl-1-methylpiperidine (37,5 g). The mixture was stirred and refluxed. After removal of the solvent in vacuo, the evaporation residue was poured into a mixture of ice (150 g), water (150 ml), and ether (200 ml). The ether layer was separated, and the aqueous layer was extracted with ether. The combined ether solutions were washed with water and dried with anhydrous magnesium sulphate, and the ether was evaporated. The residue was triturated with 200 ml of 99% ethanol and 11.5 ml of concentrated hydrochloric acid, yielding 30 g of (-)-b-4-(4-fluorophenyl-3-(1,3-benzdioxolyl-(3)-oxymethyl)-1-methylpiperidine, hydrochloride were obtained. Melting point 202°C.

References

Christensen J.A., Squires R.F.; US Patent No. 4,007,196; Feb. 8, 1977; Assigned: A/S Ferrosan, Denmark

Lemmens J.M. et al.; US Patent No. 6,686,473 B2; Feb. 3, 2004; Assigned: Synthon BCT Technologies, LLC, Chapel Hill, NC (US)

PARSALMIDE

Therapeutic Function: Muscle relaxant, Antiinflammatory, Analgesic

Chemical Name: Benzamide, 5-amino-N-butyl-2-(2-propynyloxy)-

Common Name: Parsalmide; Sinovial

Structural Formula:

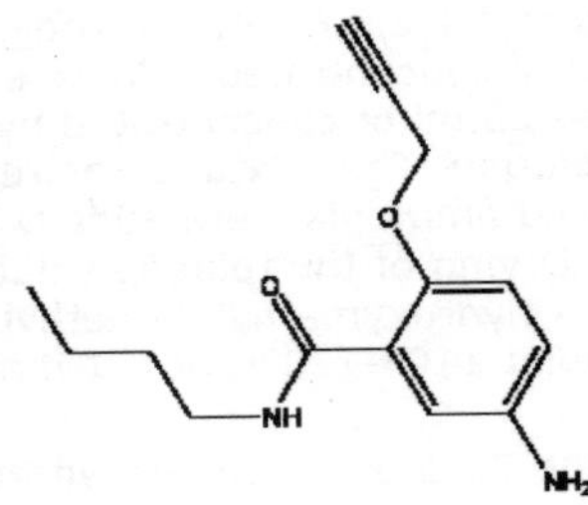

Chemical Abstracts Registry No.: 30653-83-9

Trade Name	Manufacturer	Country	Year Introduced
Parsalmide	Shanghai Lansheng Corporation	-	-
Parsal	Midy	-	-

Raw Materials

Thionyl chloride	5-Acetamido-O-salicylic acid
Butylamine	Sodium hydroxide
Sodium	Propargyl bromide
Sulfuric acid	Isopropyl alcohol

Manufacturing Process

5-Acetylamino-2-acetoxybenzoyl chloride was obtained by reaction of 5-acetylamino-2-acetoxy-benzoic acid with thionylchloride.

5-Acetylamino-N-butyl-2-hydroxybenzamide was produced in the result of treatment of 5-acetylamino-2-acetoxybenzoyl chloride with butylamine in the presence of sodium hydroxide.

5-Acetamino-N-(n-butyl)-2-propargyloxybenzamide was obtained by reaction of 5-acetylamino-N-butyl-2-hydroxybenzamide with propargylbromide in the presence of sodium, isopropyl alcohol and sulfuric acid.

28.8 g (0.1 mole) 5-acetamino-N-(n-butyl)-2-propargyloxybenzamide in 320 ml of 4 N sulfuric acid was heated, under stirring, at 90°-95°C for 2 h. The clear solution was cooled and its pH adjusted to 1 with 1 N NaOH; after filtering, further alkali was subsequently added until a pH of 10 was obtained. At this point the product was separated by filtration and recrystallized from ethanol at 60°C to give 16.6 g (a yield of 68%) of chromatographically pure 5-amino-N-(n-hutyl)-2-propargyloxybenzamide; melting point 85°-87°C.

References

Gradnik B. et al.; US Patent No. 3,739,030; June 12, 1973; Assigned: Societe d'Etudes de Recherches et d'Applications Scientifiques et Medicales E.R.A.S.M.E., Paris, France

Kleemann A., Engel J.; Pharmazeutische Wirkstoffe, GeorgThieme Verlag Stuttgart, New York, 1982

PASINIAZID

Therapeutic Function: Antitubercular

Chemical Name: Isonicotinic acid hydrazide compound with 4-amino-salicylic acid

Common Name: Pasiniazid; Umenazid

Structural Formula:

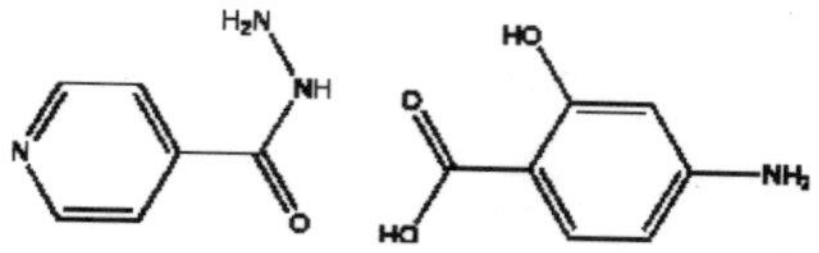

Chemical Abstracts Registry No.: 2066-89-9

Trade Name	Manufacturer	Country	Year Introduced
Pasinazid	Yick-Vic Chemicals and Pharmaceuticals (HK) Ltd.	-	-

Raw Materials

Isonicotinoyl hydrazine

4-Aminosalicylic acid

Manufacturing Process

10 parts by weight of isonicotinoyl hydrazine was dissolved in 200 by volume of methanol by stirring at 60°C. Then 11.2 parts 4-aminosalycylic acid was added at 60°C. On cooling a salt of both compounds crystallized as yellow prisms with MP: 135°-140°C (decomposed).

Isonicotinic acid hydrazide compound with 4-aminosalicylic acid may be prepared from the same components by using 750 parts of water by volume as a solvent.

10 parts 4-aminosalicylic acid was dissolved in 600 volume parts of water containing 39 parts by volume 2 N ammonium hydroxide. 9 parts by weight isonicotinoyl hydrazide was added and the mixture was heated to 30°C. The solution was acidified with 49 parts by volume of 2 N acetic acid. On cooling isonicotinic acid hydrazide compound with 4-amino-salicylic acid crystallized, which decomposed at 135°-140°C.

The yield was almost quantitative.

References

F. Hoffmann-La Roche and Co. Aktiengesellschaft, Basel (Switzerland); S.R. Patent No. 303,085; April 4, 1952

PECILOCIN

Therapeutic Function: Antibiotic

Chemical Name: 2-Pyrrolidinone, 1-(8-hydroxy-6-methyl-2,4,6-dodecatrienoyl)-, (E,E,E)-(R)-

Common Name: Pecilocin; Supral

Structural Formula:

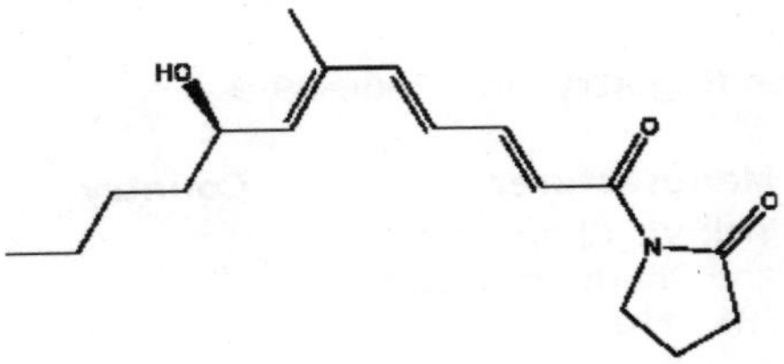

Chemical Abstracts Registry No.: 19504-77-9

Trade Name	Manufacturer	Country	Year Introduced
Variotin	Shanghai Lansheng Corporation	-	-

Raw Materials

Sucrose
Paecilomyces varioti Bainier var. antibioticus ATCC 13435
Sodium nitrate
Ferric sulfate

Manufacturing Process

Paecilomyces varioti Bainier var. antibioticus ATCC 13435 was inoculated into 10 liters of a culture medium having a pH of 6 and containing 3.0% sucrose, 0.3% sodium nitrate, 0.2% potassium dihydrogen phosphate, 0.05% magnesium sulfate, 0.05% potassium chloride and 0.001% ferrous sulfate. The cultivation was carried out in a small-aerated tank at a temperature of 25°C. After cultivation for 60 hours, the production of 30 units of pecilocin (variotin) was accomplished. The fermentation broth was then separated from the mycelium by filtration and the filtrate was extracted twice with 3 litres of ethyl acetate. The combined extracts were concentrated under reduced pressure. The concentrate was dissolved into 100 ml of methanol and, after filtering off the insoluble material which formed on refrigeration of the resulting solution, the methanolic solution was concentrated under reduced pressure. Thus 1.8 g of pecilocin having an activity of 120 u/mg were obtained.

The mycelium separated from the fermentation broth by filtration was treated with 1 liter of methanol and, after thorough grinding and stirring, was centrifugally separated. The methanol was distilled off from the methanol extract under reduced pressure and the residue extracted with ethyl acetate. The extract was concentrated under reduced pressure and the resulting concentrate was dissolved in about 100 ml of methanol. After removing the insoluble materials, which appeared on refrigeration, the methanol solution was concentrated under reduced pressure and 0.8 g of pecilocin having an activity of 90 u/mg were obtained.

100 liters of the same medium as used above were charged into the 200 liters fermentation tank. 50 g of steamed rice which had been inoculated with Paecilomyces varioti Bainier var. antibioticus ATCC 13435 and fully sporulated after cultivation for a week were seeded in the tank and cultivated with aeration and agitation at a temperature of 26°-27°C for 90 hours, said aeration being carried out by sparging of sterilized air at the rate of 90 liters per minute. At the end of 90 hours, the fermentation broth showed a pecilocin content of 16 u/ml. 86 liters of the cultured solution including the mycelium were extracted twice with 30 liters of ethyl acetate and centrifuged in a Sharples centrifugal machine. The combined extracts were concentrated under reduced pressure and about 55 g of brownish colored syrup were obtained. This syrup was dissolved in 250 ml of methanol and then refrigerated. The insoluble materials, which appeared were removed by filtration. The clarified methanol solution was then concentrated under reduced pressure and the resulting syrup was dissolved in ether and the insoluble materials were filtered off. The ether solution was concentrated under reduced pressure to a volume of about 25 ml and the concentrate mixed with ten times its volume of

petroleum ether and refrigerated. Oily material, which precipitated were separated from the solvent by decantation and washed with a small volume of petroleum ether. After drying, the treated oily materials were dissolved in 300 ml of carbon tetrachloride and then refrigerated. Brownish-red colored oily materials, which had formed were removed by decantation and the solution in carbon tetrachloride was concentrated under reduced pressure, thereby 6.6 g of a slightly yellow oily substance having an activity of 145 μ/mg were obtained.

One gram of this oily substance was subjected to a 47 tube counter-current distribution employing a 1:1 mixture of 70% methanol and carbon tetrachloride as the solvent. The results of bio-assay, ultra-violet absorption and weight measurements showed that the biologically active component was distributed mainly in tubes No 12 - 32 and that tube No 21 showed the highest concentration of active component. The samples of tubes No 15 - 26 were combined and again counter-currently distributed, 130 tubes being used. As a result of this counter-current distribution, the pecilocin was distributed in tubes No 47 - 73 of which tube No 61 showed the highest content of pecilocin. Distribution curves were plotted from bio-assay, UV-absorption and weight measurements and these curves agreed well with theoretical curve. It was thus proved that variation is a single substance. The samples of tubes No 58 - 63 were combined and concentrated under reduced pressure whereby 110 mg of colorless oily substance having a pecilocin activity of 166 u/mg were obtained.

References

Yusuke Sumiki et al.; G.B. Patent No. 866,425; April 7, 1959; Assigned to Japan Antibiotics Research Association, an incorporated body organized under laws of Japan, Tokyo, Japan and Nippon Kayaku Kabushiki Kaisha, Japan

PELARGONIC ACID

Therapeutic Function: Fungicide

Chemical Name: Nonanoic acid

Common Name: -

Structural Formula:

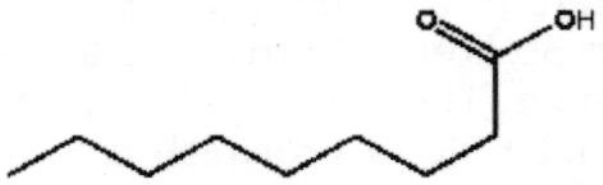

Chemical Abstracts Registry No.: 112-05-0

Trade Name	Manufacturer	Country	Year Introduced
Pellar	Crookes Barnes	US	1960

Raw Materials

Oleic acid
Oxygen

Manufacturing Process

A body of liquid, 18 inches high, comprising a 35% (by weight) solution of technical (95%) oleic acid in n-propanol, is maintained at a temperature of 86°C in a reactor. The solution also contains dissolved therein 0.042% by weight of cobalt, in the form of cobalt naphthenate. From the bottom of the reactor very fine bubbles of air are passed into and through the solution at the rate of about 0.3 cubic feet per minute, measured at standard conditions, per square foot for 72 hours. The gases leaving the reactor are first passed through an ice water reflux condenser and then vented to the atmosphere. At the end of the 72 hour period the reaction mixture is separated into its components. It is found that 60% of the oleic acid has been consumed in the reaction. For each pound of oleic acid consumed there are obtained 0.30 pound of azelaic acid (representing an efficiency of 46%, calculated on the basis that the technical oleic acid is 100% oleic acid), 0.13 pound of pelargonic acid (representing an efficiency of 23%) and 0.21 pound of 9,10-dihydroxystearic acid (representing an efficiency of 19%).

References

Merck Index 6923
MacKenzie, J.S. and Morgan, C.S. Jr.; US Patent 2,820,046; January 14, 1958; assigned to Celanese Corp. of America

PEMIROLAST POTASSIUM

Therapeutic Function: Antiallergic, Antiulcer

Chemical Name: 4H-Pyrido(1,2-a)pyrimidin-4-one-9-methyl-3-(1H-tetrazol-5-yl), potassium salt

Common Name: Artimast; Pemirolast pottassium

Structural Formula:

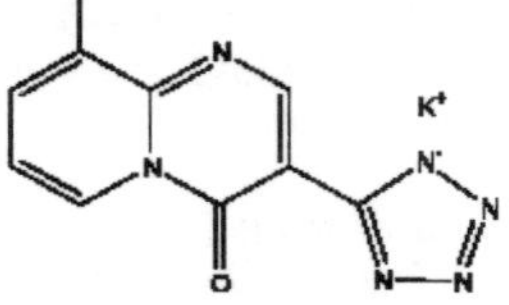

Chemical Abstracts Registry No.: 100299-08-9; 69372-19-6 (Base)

Trade Name	**Manufacturer**	**Country**	**Year Introduced**
Alamast	Santen	-	-

Raw Materials

3-Methylpyridine
N,N-Dimethylaniline
Sodium azide
Potassium hydroxide
Ferric nitrate hexahydrate
Sodium hydroxide
Ethyl ethoxymethylenecyanoacetate

Manufacturing Process

Ferrous nitrate hexahydrate (60 mg) followed by sodium (4.5 g, 0.196 g-atom) were added to liquid ammonia. To this mixture was added a solution of 3-methylpyridine (10.0 g, 0.093 mole) in N,N-dimethylaniline (21 ml) over a period of 5 min. The ammonia was allowed to evaporate and the residue heated under nitrogen by means of an oil bath maintained at 180°C for 18 h. The cooled residue was treated with ice (50 g) followed by 2 N sodium hydroxide (50 ml). The mixture was triturated for 2 h and then filtered. The collected solid was washed with boiling toluene (2 times 100 ml). The toluene layer was separated from the combined filtrate and washings, concentrated to about 50 ml and extracted with 5% aqueous acetic acid (5 times 20 ml). The combined extracts were filtered and reduced to dryness. The residue was recrystallized from methylcyclohexane to give 2-amino-3-methylpyridine acetate (4.9 g, 29%), melting point 85°-95°C. The acetate (2.5 g, 1.37 mmoles) was briefly suspended in 1 N sodium hydroxide (50 ml). The mixture was extracted with methylene chloride. The extract was washed with water, dried, and concentrated to give 2-amino-3-methylpyridine as an oil.

A solution of 2-amino-3-methylpyridine (5.0 g, 0.0462 mole) and ethyl ethoxymethylenecyanoacetate (7.82 g, 0.0462 mole) in toluene (4 ml) was heated for 15 min by means of an oil bath maintained at 100°C. The solution was cooled and the crude product (9.1 g, 85%) collected by filtration. The product was recrystallized from 2-propanol to give an analytical sample of ethyl 2-cyano-3-(3-methyl-2-pyridylamino)acrylate, melting point 144°-146°C.

Aluminum chloride (3.51 g, 0.0263 mole) was added to cold (-30°C) tetrahydrofuran (180 ml). Sodium azide (5.12 g, 0.0788 mole) was added and the mixture heated under reflux for 30 min. The mixture was cooled to 5°C. Ethyl 2-cyano-3-(3-methyl-2-pyridylamino)acrylate (5.0 g, 0.0216 mole) was added and the mixture heated under reflux for 18 h. The tetrahydrofuran was removed under reduced pressure. The residue was treated with ice water (100 ml) and acidified to pH 3 with 6 N hydrochloric acid. The mixture was filtered and the collected solid recrystallized from N,N-dimethylformamide to give the 9-methyl-3-(1H-tetrazol-5-yl)-4H-pyrido[1,2-a]pyrimidin-4-one (2.5 g, 50.7%). Melting point 310°-311°C, dec.

Potassium hydroxide was added dropwise to a stirred mixture of 9-methyl-3-(1H-tetrazol-5-yl)-4H-pyrido[1,2-a]pyrimidin-4-one in water .The mixture was diluted with water to a volume of about 300 ml and was then heated to a temperature of 50°C during 2 min. The mixture was filtered and the water removed from the filtrate by lyophilization. The residue was recrystallized from water:ethanol to give the 9-methyl-3-(1H-tetrazol-5-yl)-4H-pyrido[1,2-

a]pyrimidin-4-one potassium salt.

References

Juby P.F.; US Patent No. 4,122,274; Oct. 24, 1978; Assigned: Bristol-Myers Company, New York, N.Y.

PEMOLINE

Therapeutic Function: Psychostimulant

Chemical Name: 2-Imino-5-phenyl-4-oxazolidinone

Common Name: Phenoxazole; Phenylisohydantoin

Structural Formula:

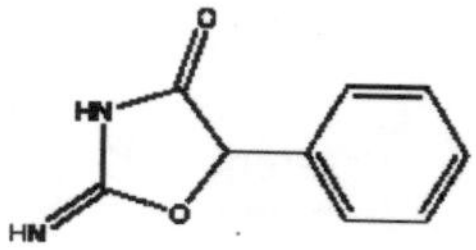

Chemical Abstracts Registry No.: 2152-34-3

Trade Name	Manufacturer	Country	Year Introduced
Deltamine	Aron	France	1960
Cylert	Abbott	UK	1975
Cylert	Abbott	US	1975
Antimeran	Nichiiko	Japan	-
Betanamin	Sanwa	Japan	-
Dynalert	Restan	S. Africa	-
Hyton	Pharmacia	Sweden	-
Kethamed	Medo	UK	-
Nitan	Teva	Israel	-
Phenoxine	P.C.B.	Belgium	-
Pioxol	Horner	Canada	-
Pondex	Chinoin	Hungary	-
Revibol	Pliva	Yugoslavia	-
Ronyl	Rona	UK	-
Sigmadyn	Spemsa	Italy	-
Sofro	Thilo	W. Germany	-
Stimul	Nadrol	W. Germany	-
Tradon	Beiersdorf	W. Germany	-
Vidil	Waldheim	Austria	-

Raw Materials

Mandelic acid ethyl ester
Guanidine

Manufacturing Process

It is preferably prepared by reacting mandelic acid ethyl ester with guanidine in boiling alcoholic solution whereby it is obtained as difficultly soluble precipitate with a yield of 90%.

This compound is a white, crystalline compound melting at 256°-257°C with decomposition. It is readily soluble in concentrated aqueous alkali hydroxide solutions and in concentrated aqueous mineral acids.

References

Merck Index 6931
Kleeman and Engel p. 690
PDR p. 509
DOT 9 (6) 212 (1973)
I.N. p. 736
REM p. 1137
Schmidt, L. and Scheffler, H.; US Patent 2,892,753; June 30, 1959; assigned to C.H. Boehringer Sohn, Germany

PENBUTOLOL

Therapeutic Function: Beta-adrenergic blocker

Chemical Name: 1-(2-Cyclopentylphenoxy)-3-[(1,1-dimethylethyl)amino]-2-propanol

Common Name: -

Structural Formula:

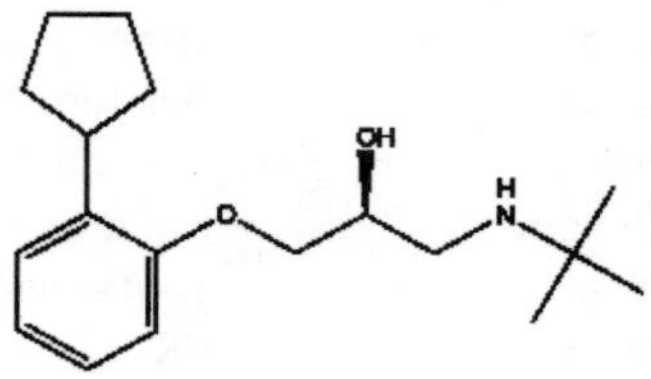

Chemical Abstracts Registry No.: 38363-40-5

Trade Name	Manufacturer	Country	Year Introduced
Betapressin	Hoechst	W. Germany	1980

Trade Name	Manufacturer	Country	Year Introduced
Betapressin	Hoechst	Switz.	1982
Betapressin	Hoechst	Italy	1983

Raw Materials

2-Cyclopentylphenol
Epichlorohydrin
t-Butylamine

Manufacturing Process

21.8 g (0.1 mol) of 1,2-epoxy-3-(2'-cyclopentylphenoxy)propane, boiling at 113°C to 115°C/0.2 mm Hg (prepared from 2-cyclopentylphenol and epichlorhydrin in the presence of alkali) were dissolved in 250 ml of ethanol; to this solution, there were added dropwise, while stirring.8.9 g (0.15 mol) of t-butylamine. The reaction mixture was stirred for 2 hours at 60°C and then the solvent and the excess t-butylamine were removed by distillation. The residue which had been purified via the aqueous hydrochloride, crystallized, after removal of the ether by evaporation, upon rubbing or inoculation and yielded, after recrystallization from n-heptane, the 1-t-butylamino-2-hydroxy-3-(2'-cyclopentylphenoxy)propane which was found to melt at 69°C to 70°C.

References

Merck Index 6935
DFU 1 (10) 494 (1976)
Kleeman and Engel p. 691
DOT 17 (12) 555 (1981) and 18 (10) 551 (1982)
I.N. p. 737
Ruschig, H., Schmitt, K., Lessenich, H. and Hartfelder, G.; US Patent 3,551,493; Dec. 29, 1970; assigned to Farbwerke Hoechst A.G. (W. Germany)

PENCICLOVIR

Therapeutic Function: Antiviral

Chemical Name: 6H-Purin-6-one, 1,9-dihydro-2-amino-9-(4-hydroxy-3-(hydroxymethyl)butyl)-

Common Name: Penciclovir

Chemical Abstracts Registry No.: 39809-25-1

Structural Formula:

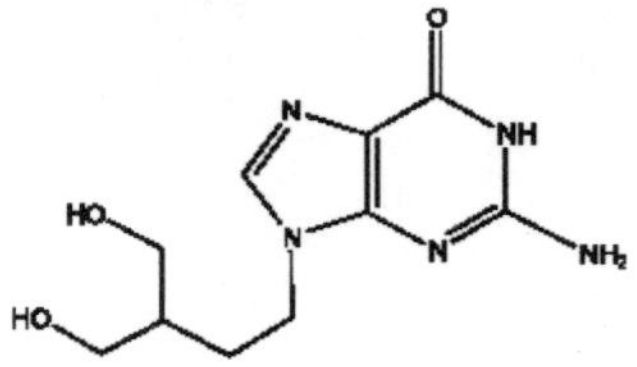

Trade Name	Manufacturer	Country	Year Introduced
Denavir	Novartis Consumer Health	-	-
Vectavir	SmithKline Beecham Consumer Healthcare	France	-
Vectavir	Beecham	UK	-

Raw Materials

Triphenylphosphine
2,2-Dimethoxypropane
Carbon tetrabromide
Hydrochloric acid
Triethyl 1,1,2-ethanetricarboxylate
4-Toluenesulfonic acid monohydrate
Lithium aluminum hydride
Sodium hydroxide

Manufacturing Process

To a suspension of lithium aluminum hydride (2.87 g, 76 mmol) in tetrahydrofuran (125 ml), a solution of triethyl 1,1,2-ethanetricarboxylate (9.2 ml, 9.85 g, 40 mmol) in tetrahydrofuran (25 ml) was added dropwise with stirring over 2 hours. The inorganic salts were filtered off and washed with ethanol (100 ml). The filtrate and washings were combined and the solvent was evaporated under reduced pressure to afford a colourless oil (4.85 g). To a suspension of this oil in acetone (100 ml) 2,2-dimethoxypropane (25 ml) and p-toluenesulphonic acid monohydrate (2.3 g, 12 mmol) were added. The mixture was stirred for 1 hour. The resulting solution was neutralised with Amberlite IR 45 (methanol washed), filtered and the solvent evaporated under reduced pressure. The residue was purified by column chromatography on silica gel, eluting with chloroform-methanol mixtures (40:1 and 25:1) to afford 5-(2-hydroxyethyl)-2,2-dimethyl-1,3-dioxan as a colourless liquid (3.01 g, 47%).

To an ice-cooled solution of 5-(2-hydroxyethyl)-2,2-dimethyl-1,3-dioxan (1.92 g, 12 mmol) and carbon tetrabromide (7.96 g, 24 mmol) in dimethylformamide (100 ml) triphenylphosphine (6.30 g, 24 mmol) was added and the solution was left at 4°C overnight. To this solution methanol (20 ml) was added and the solvent was then evaporated under reduced pressure. The residue was purified by column chromatography on silica gel, eluting with hexane-acetone (12:1) to afford 5-(2-bromoethyl)-2,2-dimethyl-1,3-dioxan as a clear colourless liquid (0.89 g, 40%).

To a solution of 5-(2-bromoethyl)-2,2-dimethyl-1,3-dioxan (0.75 g, 3.7 mmol) in dry dimethylformamide (12 ml) 2-amino-6-chloropurine (0.68 g, 4.0 mmol) and then anhydrous potassium carbonate (0.83, 6.0 mmol) were added. The solution was stirred at room temperature for 5 hours and left at 4°C overnight. The solution was filtered and the solvent removed. The residue was purified by column chromatography on silica gel, eluting with chloroform-methanol mixtures (80:1 and 60:1) to afford 2-amino-6-chloro-9-[2-(2,2-dimethyl-1,3-dioxan-5-yl)ethyl]purine as a white crystalline solid (0.74 g, 64%), melting point 125°-126°C.

2-Amino-6-chloro-9-[2-(2,2-dimethyl-1,3-dioxan-5-yl)-ethyl]purine (0.59 g, 1.9 mmol) in hydrochloric acid (1.0 M, 4 ml) was stirred at 60°C for 24 hours. The solution was diluted with water and neutralised with Amberlite IR 45. The mixture was filtered, the resin washed with water and the solvent evaporated under reduced pressure. The residue was recrystallised from water to afford 9-(4-hydroxy-3-hydroxymethylbut-1-yl)guanine (238 mg, 49%), melting point 275°-277°C.

References

Javest R.L., Harnden M.R.; US Patent No. 5,075,445; Dec. 24, 1991; Assigned: Beecham Group p.l.c., Middlesex, United Kingdom

PENFLURIDOL

Therapeutic Function: Antipsychotic

Chemical Name: 1-[4,4-Bis(4-fluorophenyl)butyl]-4-[4-chloro-3-(trifluoromethyl)phenyl]-4-piperidinol

Common Name: -

Structural Formula:

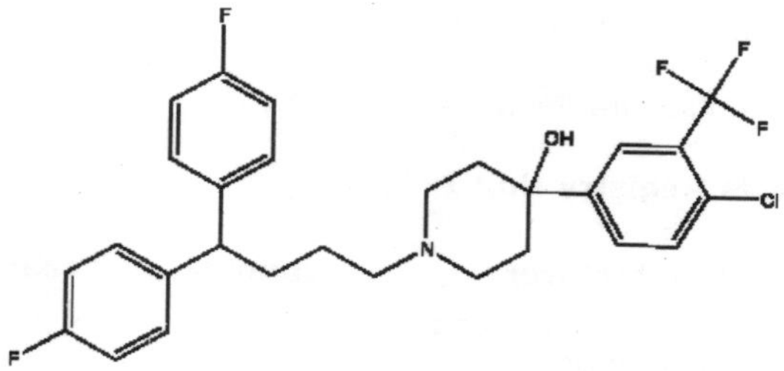

Chemical Abstracts Registry No.: 26864-56-2

Trade Name	Manufacturer	Country	Year Introduced
Semap	Janssen-Le Brun	W. Germany	1975
Semap	Janssen	France	1975
Flupidol	Zambeletti	Italy	1979
Longoran	Isis	Yugoslavia	-
Micefal	Spofa	Czechoslovakia	-
Semap	Abic	Israel	-

Raw Materials

4,4-Bis(p-fluorophenyl)butyl chloride
4-(4-Chloro-α,α,α-trifluoro-m-tolyl)-4-piperidinol

Manufacturing Process

A mixture of 24 parts of 4,4-bis(p-fluorophenyl)butyl chloride, 20.9 parts of 4(4-chloro-α,α,α-trifluoro-m-tolyl)-4-piperidinol, 13.8 parts of sodium carbonate, a few crystals of potassium iodide in 600 parts of 4-methyl-2-pentanone is stirred and refluxed for 60 hours. The reaction mixture is cooled and 150 parts of water is added. The organic layer is separated, dried, filtered and evaporated. The oily residue is crystallized from diisopropylether, yielding 4-(4chloro-α,α,α-trifluoro-m-tolyl)-1-[4,4-bis(p-fiuorophenyl)butyl]-4-piperidinol; melting point 106.5°C.

References

Merck Index 6939
Kleeman and Engel p. 691
OCDS Vol. 2 p. 334 (1980)
DOT 10 (5) 167 (1974)
I.N. p. 737
Hermans, H.K.F. and Niemegeers, C.J.E.J.; US Patent 3,575,990; April 20, 1971; assigned to Janssen Pharmaceutica N.V. (Belgium)

PENGITOXIN

Therapeutic Function: Cardiotonic

Chemical Name: Gitoxin pentaacetate

Common Name: Pengitoxin; Pentaacetylgitoxin

Chemical Abstracts Registry No.: 7242-04-8

Trade Name	Manufacturer	Country	Year Introduced
Pengitoxin	Shanghai Lansheng Corporation	-	-

Raw Materials

Gitoxin
Acetic anhydride
Potassium bicarbonate
Hydrochloric acid

Structural Formula:

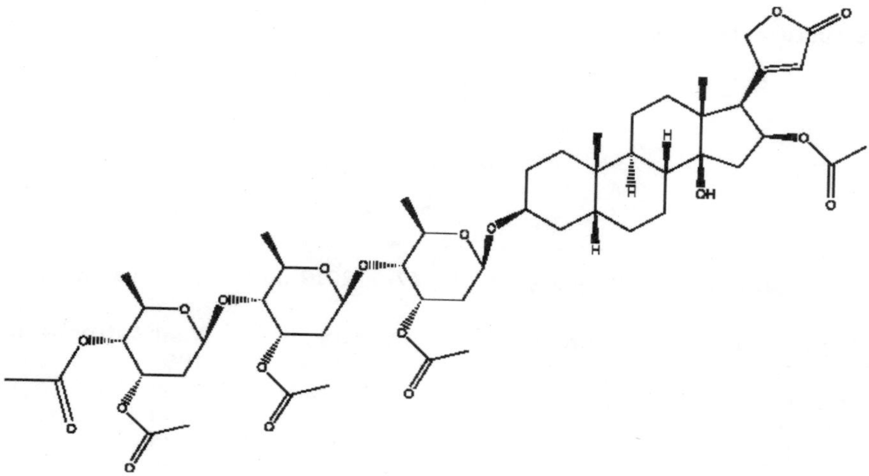

Manufacturing Process

Gitoxin is isolated from leaves of Digitalis purpurea L. genus Scrofphulariaceae.

10.0 g pure gitoxin are boiled under reflux with 1 L pure acetic anhydride. Gitoxin thereby goes into solution in the course of 1 h. The boiling is discontinued after 1 h and the acetic anhydride distilled off in a vacuum as completely as possible. After taking up the oily residue with 500 ml chloroform, the solution is successively washed with 200 ml of 2 N potassium bicarbonate solution, 0.1 N hydrochloric acid and water. After drying the chloroform solution with anhydrous sodium sulfate, the chloroform is first distilled off over an open flame and then on a water bath at 40°C in a vacuum, a substantially crystalline residue thereby being obtained. The 12.8 g of crude product obtained are recrystallized from 12 times the amount of a mixture of pyridine, methanol and water (25:10:65) to give rhombic crystals, melting point 151°-155°C.

References

GB Patent No. 1,043,029; Sept. 21, 1966; Assigned: VEB Arzneimittelwerk Dresden, of 35, Wilhelm-Pieck-Strasse, 8122 Radebeul 1, Germany

PENICILLAMINE

Therapeutic Function: Antiarthritic

Chemical Name: 3-Mercapto-D-valine

Common Name: Dimethylcysteine

Structural Formula:

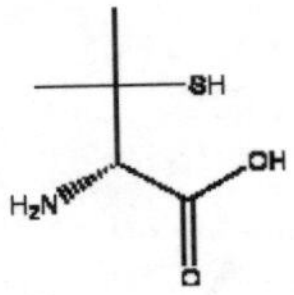

Chemical Abstracts Registry No.: 52-67-5; 2219-30-9 (Hydrochloride salt)

Trade Name	Manufacturer	Country	Year Introduced
Cuprimine	MSD	US	1963
Trolovol	Bayer	W. Germany	1963
Pendramine	B.D.H.	UK	1973
Pemine	Lilly	Italy	1975
Trolovol	Bayer	France	1979
Depen	Wallace	US	1979
Artamin	Biochemie	Austria	-
Cuprenil	Polfa	Poland	-
Cupripen	Rubio	Spain	-
Depamine	Berk	UK	-
Distamine	Dista	UK	-
Gerodyl	Gea	Denmark	-
Metalcapase	Knoll	W. Germany	-
Reumacillin	Medica	Finland	-
Rhumantin	Gea	Denmark	-
Sufortanon	Lacer	Spain	-

Raw Materials

Sodium hydroxide
Potassium benzyl penicillin
Hydrogen sulfide
Phenylhydrazine
Mercuric chloride

Manufacturing Process

(a) Preparation of mercuric chloride complex of penicillamine: To a solution of 372 g (1 mol) of potassium benzyl-penicillin in 940 ml of distilled water at room temperature is added a solution of 40 g (1 mol) of sodium hydroxide in 180 ml of distilled water over a period of one-half hour. The solution is then stirred for two hours at room temperature. While maintaining room temperature, 67 ml of concentrated hydrochloric acid is added at a slow rate. This solution is then added, over a period of time of one-half hour, to a solution of 271 g (1 mol) of $HgCl_2$ in 3.52 liters of distilled water in the presence of 50 g of Hyflo and 5 ml of octyl alcohol. After one hour of agitation, the resulting mixture is treated with 185 ml of concentrated hydrochloric acid and filtered.

(b) Removal of benzylpenilloaldehyde: To the filtrate obtained in step (a), warmed to 50°C is slowly added 108 g (1 mol) of phenyl hydrazine. The mixture is cooled to room temperature and 84 ml of concentrated hydrochloric acid are added. The mixture is agitated briefly and the precipitated benzylpenilloaldehyde phenyl hydrazone is filtered off.

(c) Preparation of isopropylidene penicillamine hydrochloride: To the filtrate obtained in step (b) is added at 20°C to 25°C a total of 85 g of hydrogen sulfide. The precipitated HgS is filtered off and the filtrate is concentrated under reduced pressure to a volume of 200 to 500 ml. Following a polish filtration, the product-rich concentrate is mixed with 1.5 liters of isobutyl acetate. The mixture is refluxed at about 40°C under reduced pressure in equipment fitted with a water separation device. When no further water separates, the batch is cooled to 30°C and filtered. The reactor is washed with 1 liter of acetone, which is used also to wash the cake. The cake is further washed with 200 ml of acetone. The acetone washes are added to the isobutyl acetate filtrate and the mixture is refluxed for 20 to 30 minutes. After a holding period of one hour at 5°C. the crystals of isopropylidene penicillamine hydrochloride are filtered and washed with 200 ml of acetone. On drying for twelve hours at 25°C this product, containing 1 mol of water, weighs about 178 g (73%).

(d) Preparation of penicillamine hydrochloride: The 178 g of isopropylidene penicillamine hydrochloride obtained in step (c) is dissolved in 350 ml of distilled water. The solution is heated at 90°C to 95°C for one to one and one-half hours, removing acetone by distillation through an efficient column. There is then added 2.6 liters of isobutyl acetate. The mixture is refluxed at a temperature of about 40°C under reduced pressure in equipment fitted with a water separation device. When no further water separates, the pressure is adjusted so that the mixture distills at a vapor temperature of 83°C to 88°C. A total of 650 ml of distillate is collected. The batch is allowed to cool to 50°C and then filtered. The crystals are washed with isobutyl acetate and then dried at 35°C for 24 hours. The virtually anhydrous penicillamine hydrochloride obtained weighs about 128 g (69% from potassium benzyl-penicillin).

References

Merck Index 6940
Kleeman and Engel p. 693
PDR pp. 1153, 1872
DOT 9 (7) 302 (1973)
I.N. p. 738
REM p. 1225
Restivo, A.R., Dondzila, F.A. and Murphy, H. Jr.; US Patent 3,281,461; October 25, 1966; assigned to E.R. Squibb and Sons, Inc.
Sota, K., Ogawa, T. and Sawada, J.; US Patent 4,150,240; April 15, 1979; assigned to Taisho Pharmaceutical Co., Ltd. (Japan)

PENICILLIN G BENZATHINE

Therapeutic Function: Antibacterial

Chemical Name: Penicillin G compound with N,N'-dibenzylethylenediamine

Common Name: Benzethacil

Structural Formula:

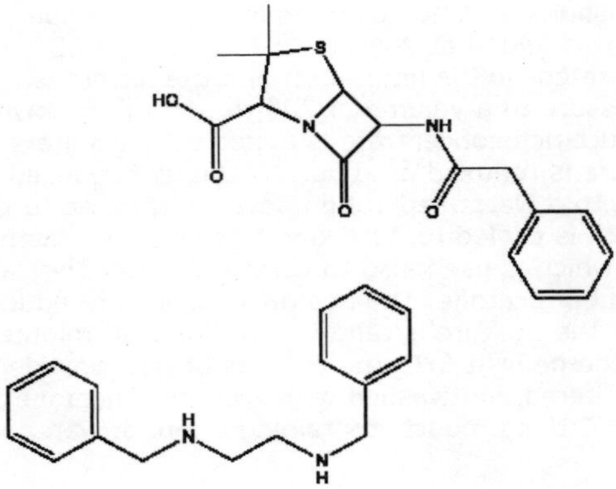

Chemical Abstracts Registry No.: 1538-09-6

Trade Name	Manufacturer	Country	Year Introduced
Bicillin	Wyeth	US	1951
Permapen	Pfizer	US	1953
Neolin	Lilly	US	1954
Extencilline	Specia	France	-
Benzetacil-Simple	Antibioticos	Spain	-
Brevicilina-Simple	Wassermann	Spain	-
Brunocillin	Mepha	Switz.	-
Cepacilina	Cepa	Spain	-
Depotpen	Dauelsberg	W. Germany	-
Diaminocillina	Farmalabor	Italy	-
Durabiotic	Teva	Israel	-
Longacillin	Besy	Brazil	-
LPG	C.S.L.	Australia	-
Megacillin	Merck-Frosst	Canada	-
Pen-Di-Ben	Bago	Argentina	-
Pendysin	Jenapharm	E. Germany	-
Penidural	Wyeth	UK	-
Peniroger Retard	Roger	Spain	-
Pipercilina	Iskia	Spain	-
Retarpen	Biochemie	Austria	-
Tardocillin	Bayer	W. Germany	-
Tardopenil	Farmabion	Spain	-

Raw Materials

Ethylenediamine
Benzaldehyde
Sodium penicillin G

Manufacturing Process

Ethylenediamine (15 g, 0.25 mol) was added dropwise to 100 ml 98-100% formic acid in a two-necked 500 ml flask, fitted with an addition tube and reflux condenser with drying tube, cooled in an ice-bath. After complete addition of the base, 53 g of benzaldehyde (0.5 mol) was added in one lot. The ice-bath was removed and the flask was heated to the refluxing temperature. The initial rate of carbon dioxide evolution was too rapid to measure. After twenty minutes, the rate was circa 100 ml per minute and decreased rapidly to 8 ml per minute in one hour. Heating at reflux was continued for 35 hours.

Following the refluxing most of the excess formic acid was removed under reduced pressure. Hydrochloric acid (200 ml 6 N) was added to the viscous amber residue and heated under reflux, After 15 minutes, bumping necessitated cooling and filtering to remove crystalline dihydrochloride, which after washing with isopropanol was dried, MP circa 300°C. The mother liquors were refluxed one hour and cooled, obtaining an additional amount of product, MP circa 300°C. The filtrate was concentrated in vacuo to 100 ml, cooled and made alkaline with 40% NaOH. The supernatant oil was extracted with ether, dried, and fractionated from a stillpot packed with glass wool and heated in a sand-bath at 320°C. The first fraction at 106°C at 0.6-0.7 mm was N-benzylethylenediamine (dipicrate, MP 222°C). The N,N'-dibenzylethylenediamine was collected at 177°C to 206°C at 0.6-1.0 mm as a colorless liquid.

To a solution of 60 g of sodium penicillin G in 800 cc of distilled water cooled to 0°C to 4°C in an ice-bath, a solution of 35 g of N,N'-dibenzylethylenediamine diacetate in 200 cc of distilled water is added dropwise with stirring. The thick slurry is filtered with suction, washed twice with 100 cc of cold water, dried by suction and spread out in a thin layer for completion of drying. The product weighed 80 g.

The air-dried powder has a broad melting point, sintering at 100°C, melting above 110°C to a cloudy liquid becoming clear at 135°C.

References

Merck Index 6948
Kleeman and Engel p. 85
PDR pp. 1406, 1941, 1989
I.N. p. 126
REM p. 1197
Szabo, J.L.and Bruce,W.F.; US Patent 2,627,491; February 3, 1953; assigned to Wyeth, Inc.

PENICILLIN G HYDRABAMINE

Therapeutic Function: Antibacterial

Chemical Name: N,N'-Bis(dehydroabietyl)ethylenediamine dipenicillin G

Common Name: -

Structural Formula:

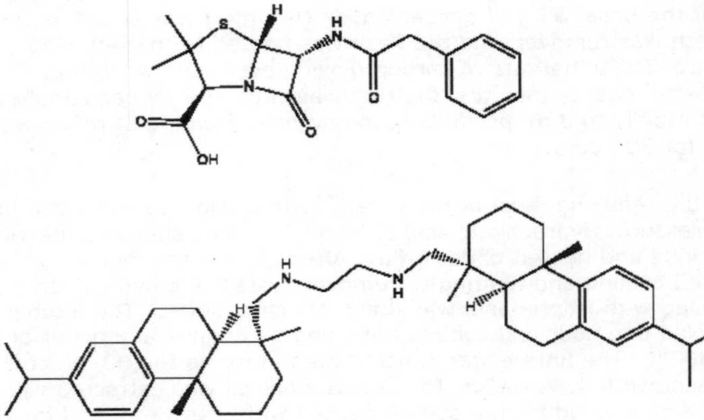

Chemical Abstracts Registry No.: 3344-16-9

Trade Name	Manufacturer	Country	Year Introduced
Compocillin	Abbott	US	1954

Raw Materials

Dehydroabietylamine
Ethylene dibromide
Penicillin G

Manufacturing Process

A mixture of 142.5 g of "Rosin Amine D" containing about 70% dehydroabietylamine and 30% dihydro and tetrahydroabietylamine, 47.0 g of ethylene dibromide, and 60.6 g of triethylamine is dissolved in 350 cc of anhydrous xylene and refluxed for about 16 hours. Thereafter the triethylamine dibromide salt formed is separated from the solution by filtering the cool reaction mixture and washing with ether. The solution is then concentrated under reduced pressure to dryness to remove the ether, xylene and excess triethylamines present. The viscous oil resin is slurried twice with 250 cc portions of methanol to remove any unreacted primary amines. The oil residue after being washed with methanol is dissolved in ethyl alcohol and 75 cc of concentrated hydrochloric acid is added dropwise to the warm alcohol solution of the base. The dihydrochloride salts of the several hydroabietyl ethylenediamines precipitates immediately from solution. The salt is then separated by filtering and is washed twice with 100 cc portions of cooled ethyl alcohol. The dihydrochloride salts of the dehydroabietyl, dihydroabietyl and tetrahydroabietyl ethylenediamine mixture have a melting point of about 292°C to 295°C. On subjecting the mixture to solubility analyses it is found that the dehydroabietyl ethylenediamine is present in substantially the same proportion as is the dehydroabietylamine in the original "Rosin Amine D."

An amyl acetate-penicillin acid solution (10 liters) having a potency of 100,000 U/ml which is sufficient to supply 565 g (2 mols) of penicillin acid is added with constant agitation to 505 g of crude N,N'-bis-(dehydroabietyl)-ethylenediamine dissolved in 500 ml of amyl acetate. A slight excess of the ethylenediamine bases is added to the mixture until precipitation is completed. The reaction is preferably carried out in a cold room having a temperature of about 5°C. The precipitation salts comprise about 70% N,N'-bis-(dehydroabietyl)-ethylenediamine-dipenicillin salt and approximately 25-30% of the N,N'-bis-(dihydroabietyl)-ethylenediamine-and N,N'-bis-(tetrahydroabietyl)-ethylenediamine-dipenicillin salts are recovered by filtration and are washed with about 1/10 solution volume of amyl acetate. The crude preparation is further washed with 1/10 solution volume of diethyl ether and dried. The melting point of the product is about 153°C when taken on a microblock.

The total yield of the crude precipitation obtained in the above manner comprising about 1 kg is then dissolved in chloroform so as to form a 15% solution of a crude penicillin salt. To the filtered chloroform solution is added ethyl acetate slowly and with agitation until the solution becomes turbid as crystallization begins. Thereafter crystallization is allowed to proceed undisturbed for about 30-60 minutes in a cold room having a temperature of about 5°C. Sufficient ethyl acetate is slowly added to provide a final concentration of about 50% ethyl acetate and the mixture is allowed to stand in the cold room for one hour to complete crystallization. The precipitate is filtered and washed with about 750 ml of ethyl acetate and thereafter washed with the same volume of ether. The crystals are dried in vacuo and a yield of about 900 g of N,N'-bis-(dehydroabietyl)-ethylenediamine-dipenicillin G is obtained. The penicillin product melts with decomposition at a temperature of 170°C to 172°C on a Kofler hot stage. Solubility analysis of the product shows the product to be 95.3% pure.

References

Merck Index 6951
I.N. p. 739
De Rose, A.F.; US Patent 2,812,326; November 5, 1957; assigned to Abbott Laboratories

PENICILLIN G PROCAINE

Therapeutic Function: Antibacterial

Chemical Name: Penicillin G compound with 2-(diethylamino)ethyl p-aminobenzoate

Common Name: -

Structural Formula:

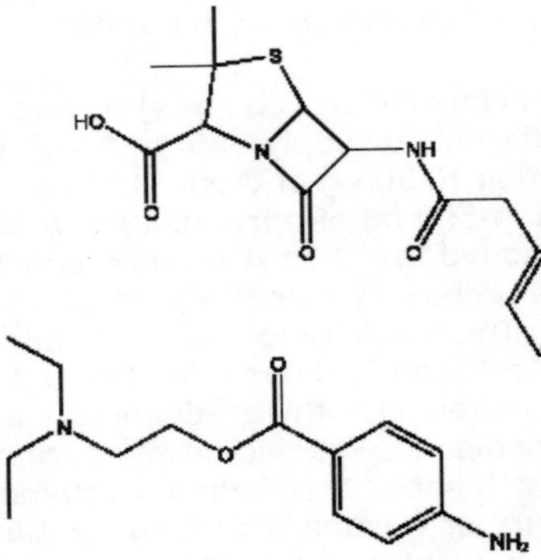

Chemical Abstracts Registry No.: 54-35-3

Trade Name	Manufacturer	Country	Year Introduced
Duracillin	Lilly	US	1948
Flo-Cillin	Bristol	US	1949
Ledercillin	Lederle	US	1949
Wycillin	Wyeth	US	1949
Diurnal Penicillin	Upjohn	US	1950
Abbocillin	Abbott	US	1951
Ampin-Penicillin	Badische Arzneim.	W. Germany	-
Aquacaine	C.S.L.	Australia	-
Aquasuspen	SK Kauelsberg	W. Germany	-
Aqucilina	Antibioticos	Spain	-
Cilicaine	Sigma	Australia	-
Distaquaine	Distillers	UK	-
Excolicin	Jenapharm	E. Germany	-
Farmaproina	Cepa	Spain	-
Francacilline	Franca	Canada	-
Hypercillin	Cutter	US	-
Hypropen	Biochemie	Austria	-
Intrasept	Streuli	Switz.	-
Klaricina	Clariana	Spain	-
Novocillin	Solac	France	-
Penifasa	Lifasa	Spain	-
Peniroger Procain	Roger	Spain	-
Premocillin	Premo	US	-
Procapen	Orion	Finland	-
Prokapen	Weifa	Norway	-
Retardillin	EGYT	Hungary	-

Trade Name	Manufacturer	Country	Year Introduced
Sanciline Procaina	Santos	Spain	-
Therapen I.M.	Therapex	Canada	-

Raw Materials

Penicillin G
Procaine

Manufacturing Process

There was added to 250 ml of a concentrated butyl acetate extract containing 74,000 units of the acid form of penicillin per ml, 50 ml of a butyl acetate solution containing 0.238 g per ml of procaine base. The solution was agitated for one hour. The precipitate which formed was very gummy and not in the form of discrete crystals. This precipitate was crystallized by scratching the side of the vessel and agitating further. After this treatment 18.25 g of crystalline procaine penicillin was obtained which assayed 1010 units per mg representing a yield of 99.6% of the activity contained in the concentrated extract.

References

Merck Index 6953
PDR pp. 1408, 1742, 1941, 1989
I.N. p. 739
REM p. 1198
Bardolph, M.P.; US Patent 2,739,962; March 27, 1956; assigned to Commercial Solvents Corp.

PENICILLIN O

Therapeutic Function: Antibacterial

Chemical Name: 3,3-Dimethyl-7-oxo-6-[[(2-propenylthio)acetyl]amino]-4-thia-1-azabicyclo[3.2.0]-heptane-4-carboxylic acid

Common Name: Allylmercaptomethylpenicillin

Structural Formula:

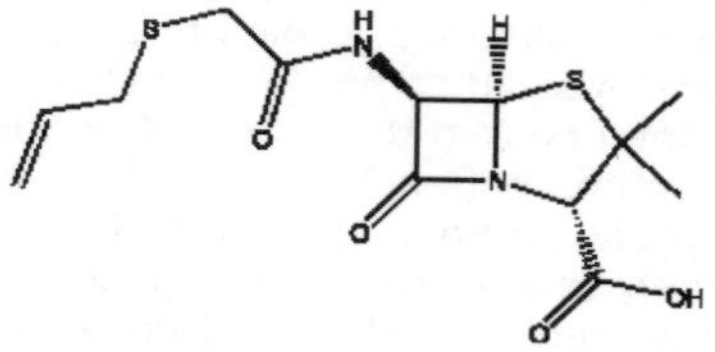

Chemical Abstracts Registry No.: 87-09-2

Trade Name	Manufacturer	Country	Year Introduced
Cero-O-Cillin	Upjohn	US	1950

Raw Materials

Lactose
Bacterium Penicillium
Corn steep liquor
N-(2-Hydroxyethyl)allylmercaptoacetamide

Manufacturing Process

A culture medium is prepared in the following proportions:

Lactose	125 g
Corn steep solids	150 g
Calcium carbonate	25 g
N-(2-Hydroxyethyl)-allylmercaptoacetamide	0.140 g
Water	5,000 cc

The culture medium is distributed in 200 cc portions in 1 liter Erlenmeyer flasks, sterilized, inoculated with a spore suspension of Penicillium mold strain Q-176, and stoppered with cotton plugs. The flasks are maintained at a temperature of about 23°C to 26°C and shaken constantly for five days. The flask contents are then filtered to remove the mold mycelium, the filtrate cooled to about 0°C, acidified to about pH 2.2 with o-phosphoric acid and shaken with an equal volume of amyl acetate. The amy acetate layer is separated and extracted with three 100 cc portions of cold water to which cold N/10 sodium bicarbonate solution is added during the course of each extraction until a pH of about 7.1 to 7.3 is attained in the aqueous phase. The aqueous extracts are combined, cooled to about 0°C. acidified to about pH 2.2 with o-phosphoric acid and extracted with three 100 cc portions of ether. The ether extracts are combined, and are passed through a chromatographic type silica adsorption column about 30 mm in diameter and 300 mm long, and containing a pH 6.2 phosphate buffer. The silica column is developed by percolation with six 100 cc portions of ether containing successively increasing amounts of methanol in the order of 0.5, 1.5%, 2, 2.5, and 3 percent.

The developed silica column is divided into about 12 equal sections and each section is eluted with three 30 cc portions of M/15 phosphate buffer of pH 7.0. The eluates are assayed bacteriologically to determine their penicillin content. Most of the antibiotic activity originates in a single bank in the silica column and results from the presence of allylmercaptomethylpenicillin. The eluates obtained from this band are combined, cooled to about 0°C, acidified to about pH 2.2and extracted with three 50 cc portions of chloroform. The combined chloroform extracts are then passed through a silica adsorption column containing a pH 6.2 phosphate buffer. This silica gel column is developed by percolation with three 100 cc portions of chloroform containing successively increasing amounts of methanol in the order of 1, 2 and 3 percent. The developed silica column is then divided into 12 equal sections and each section is eluted with three 30 cc portions of M/15 phosphate buffer of pH 7.0.

Again, most of the total antibiotic activity originates in a single band in the silica column. The eluates obtained by extraction of the silica column sections which comprise this band are combined, cooled to about 0°C, acidified to about pH 2.2 and extracted with three 100 cc portions of ether. The ether extracts are combined and extracted with about 75 cc of a cool dilute aqueous solution of sodium hydroxide to which N/10 sodium hydroxide solution is added during the course of the extraction so that a final pH of about 7.0 is obtained in the aqueous phase. From this aqueous solution the sodium salt of allylmercaptomethylpenicillin is separated, for example, by freezing and evaporation in vacuo from the frozen state.

References

Merck Index 6955
I.N. p. 58
Behrens, O.K., Jones, R.G., Soper, Q.F. and Corse, J.W.; US Patent 2,623,876; December 30, 1952; assigned to Eli Lilly and Co.

PENICILLIN V

Therapeutic Function: Antibacterial

Chemical Name: 3,3-Dimethyl-7-oxo-6-[(phenoxyacetyl)amino]-4-thia-1-azabicyclo[3.2.0]-heptane-2-carboxylic acid

Common Name: 6-Phenoxyacetamidopenicillanic acid; Phenoxymethylpenicillin

Structural Formula:

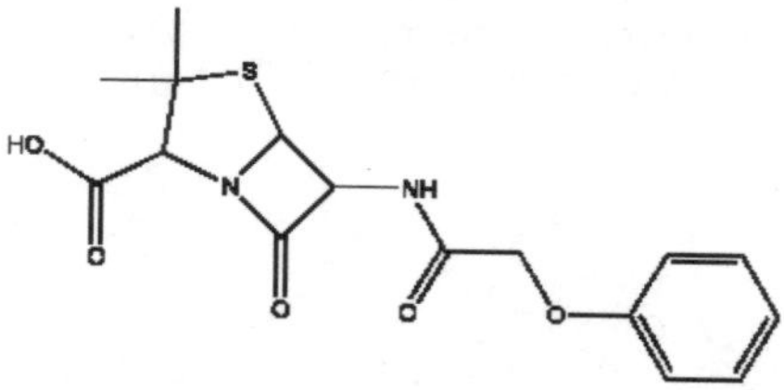

Chemical Abstracts Registry No.: 87-08-1

Trade Name	Manufacturer	Country	Year Introduced
Oracilline	Theraplix	France	1954
V-Cillin	Lilly	US	1955
Pen-Vee	Wyeth	US	1955
Calcipen	Farmabion	Spain	-
Fenocin	Dumex	Denmark	-
Fenospen	Farmalabor	Italy	-
Ibaden	Lek	Yugoslavia	-
Intalpen	Inter-Alia Pharm.	UK	-

Trade Name	Manufacturer	Country	Year Introduced
Ospen	Biochemie	Austria	-
Penorline	Allard	France	-
Rivopen V	Rivopharm	Switz.	-
V-Tablopen	Arzneimittelwerk Dresden	E. Germany	-
Weifapenin	Weifa	Norway	-

Raw Materials

Phenoxyacetyl chloride
6-Aminopenicillanic acid

Manufacturing Process

The following description is taken from US Patent 2,941,995. A solution of phenoxyacetyl chloride (360 mg) in dry acetone (5 ml) was added dropwise during 10 minutes to a stirred solution of 6-aminopenicillanic acid (450 mg, approximately 75% pure) in 3% aqueous bicarbonate (18 ml), and acetone (12 ml). When addition was complete the mixture was stirred at room temperature for 30 minutes and then extracted with ether (30 ml in 3 portions), only the aqueous phase being retained. This aqueous solution was covered with butanol (5 ml) and adjusted to pH 2 by the addition of N hydrochloric acid. After separating the layers, the aqueous phase was extracted with two 2.5 ml portions of butanol, adjusting to pH 2 each time. The combined butanol solutions (which at this stage contained the free penicillanic acid) were washed with water (3 x 2 ml) and then shaken with water (10 ml) to which sufficient 3% sodium bicarbonate solution was added to bring the aqueous phase to pH 7. The butanol solution was further extracted with two 5 ml portions of water to each of which was added enough bicarbonate solution to produce an aqueous phase of pH 7. The combined aqueous solutions were washed with ether (20 ml) and then evaporated at low temperature and pressure to leave the crude sodium salt of phenoxymethyl penicillin which, after drying in a vacuum desiccator, was obtained as a slightly hygroscopic powder (591 mg).

References

Merck Index 6957
Kleeman and Engel p. 716
PDR pp. 673, 694, 1071, 1381, 1606, 1723, 1770, 1968
I.N. p. 760
REM p. 1199
Behrens, O.K., Jones, R.G., Soper, Q.F. and Corse, J.W.; US Patent 2,562,410; July 31, 1951; assigned to Eli Lilly and Company
Sheehan, J.C.; US Patent 3,159,617; December 1, 1964; assigned to Arthur D. Little, Inc.
Doyle, F.P., Nayler, J.H.C. and Rolinson, G.N.; US Patent 2,941,995; June 21, 1960; assigned to Beecham Research Laboratories Limited, England

PENICILLIN V HYDRABAMINE

Therapeutic Function: Antibacterial

Chemical Name: N,N'-Bis(dehydroabietyl)ethylendiamine bis(phenoxymethylpenicillin)

Common Name: -

Structural Formula:

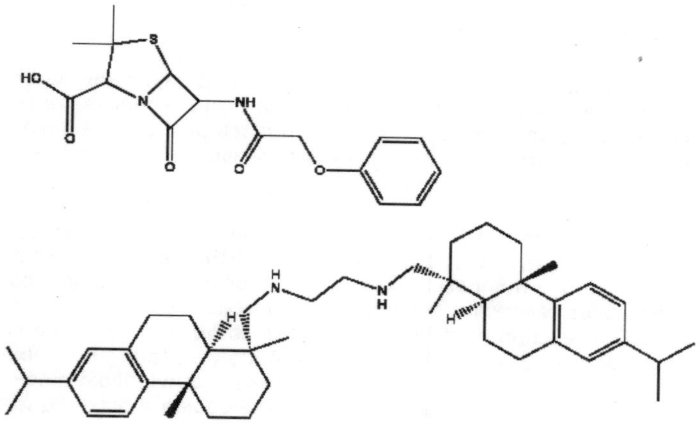

Chemical Abstracts Registry No.: 6591-72-6

Trade Name	Manufacturer	Country	Year Introduced
Compocillin-V	Abbott	US	1954
Flavopen	G.P.	Australia	-

Raw Materials

Penicillin V
Dehydroabietylethylenediamine

Manufacturing Process

The crude dihydrochlorides of dehydroabietylethylenediamine bases (985 g) are extracted with a solution of about 3 liters of chloroform and 3 liters of water which is adjusted to about pH 10 and a second extraction is performed using a solution of about 2 liters of chloroform and the mixture readjusted to about pH 10 with 6 N NaOH if necessary. The chloroform layer containing the mixed free bases is separated from the aqueous layer containing NaCl and is washed with about 1/10 its volume of water to remove any NaCl in the wet chloroform solution. The chloroform solution containing a mixture of the free bases having a volume of about 5 liters is dried with anhydrous Na_2SO_4 and then filtered to obtain a clear solution containing about 0.85 kg of the mixed free bases.

Approximately 1,000 g of phenoxymethylpenicillin acid (Penicillin V) is dissolved directly in about 5 liters of ethyl acetate to a concentration of 20% w/v. The resulting solution is filtered to remove any insoluble salts. The penicillin V acid (1,000 g) may also be obtained by extracting an aqueous solution of 1,110 g of the potassium salt of phenoxymethylpenicillin at a temperature of about 5°C, this solution being adjusted to pH 2-3 by the addition of 6 N sulfuric acid, twice with a total of 5 liters of ethyl acetate so that the final washed combined volume will have a concentration of about 20% w/v. The abovementioned ethyl acetate solution having a volume of about 5 liters is then dried with anhydrous Na_2SO_4 and filtered to obtain a clear ethyl acetate solution of phenoxymethylpenicillin acid.

In place of the hydrochlorides of the above described bases any other acid salt thereof can be used, including both inorganic and organic salts such as phosphoric, sulfuric, and acetic acids. Also, in place of the mentioned penicillin, any of the other common salts of penicillin can be used as a source of penicillin acid.

The chloroform solution of the free bases prepared in the above manner is then slowly added to the ethyl acetate solution of the penicillin V acid prepared in the above manner. A clear solution forms which rapidly becomes turbid as the bases react with the penicillin acid and crystallization commences. The reaction mixture is allowed to stand overnight in a cool room having a temperature of about 5°C after thoroughly agitating the mixture. Thereafter, the crystalline N,N'-bis-(dehydroabietyl)-ethylenediamine-dipenicillin V is filtered to separate therefrom the cooled mother liquor which contains the unprecipitated N,N'-bis-(dihydroabietyl)-ethylenediamine-dipenicillin salt and N,N'-bis-(tetrahydroabietyl)-ethylenediamine-dipenicillin salt and other impurities. The precipitate is washed thoroughly with about 4 liters of a mixture of chloroform and ethyl acetate (1:1) which is divided into three separate portions. After the final washing, the crystals are substantially colorless. The crystalline penicillin salt is thoroughly dried under vacuum at a temperature of about 50°C. The N,N'-bis-(dehydroabietyl)-ethylenediamine-dipenicillin V salt is obtained having purity as determined by solubility analysis in excess of about 90% and melts with decomposition at 163°C to 165°C on a Kofler hot stage.

References

Merck Index 6959
I.N. p. 494
De Rose, A.F.; US Patent 2,812,326; November 5, 1957; assigned to Abbott Laboratories

PENIMEPICYCLINE

Therapeutic Function: Antibiotic

Chemical Name: 4-(Dimethylamino)-1,4,4a,5,5a,6,11,12a-octahydro-

3,6,10,12,12a-pentahydroxy-N-((4-(2-hydroxyethyl)-1-piperazinyl) methyl)-6-methyl-1,11-dioxo-2-naphthacenecarboxamide salt with phenoxymethylpenicillin

Common Name: Mepenicycline; Penimepiciclina; Penimecycline

Structural Formula:

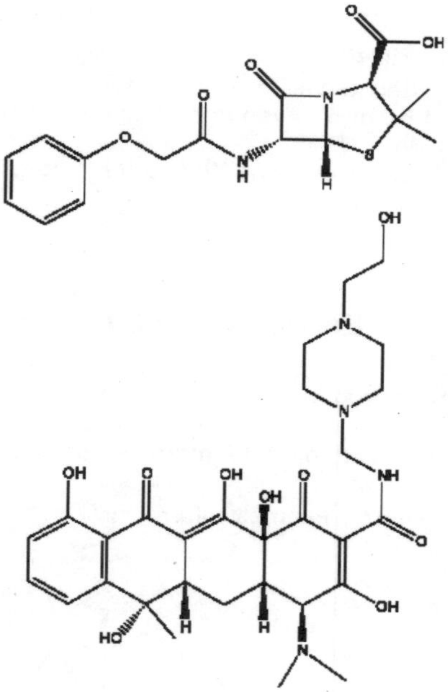

Chemical Abstracts Registry No.: 4599-60-4

Trade Name	Manufacturer	Country	Year Introduced
Penimepicyclina	Elenco Farmaci	-	-
Duamin	Luso-Farmaco	-	-
Tonsil	Gap	-	-
Ultrabiotic	Latino	-	-

Raw Materials

Tetracycline
Formaldehyde
1-(2-Hydroxyethyl)piperazine
Penicillin V

Manufacturing Process

N-(4^1-β-hydroxyethyl-1^1-piperazinylmethyl)tetracycline was obtained by reaction of tetracycline with 1-(2-hydroxyethyl)piperazine in the presence of formaldehyde.

8.6 g of N-(4^1-β-hydroxyethyl-1^1-piperazinylmethyl)tetracycline and 35.0 g phenoxymethyl-penicilline, were dissolved in 300 ml methanol, with agitation. The solution was filtered on a Buchner filter, and the filtrate was taken up with 900 ml anhydrous ether with strong agitation, again filtered under pressure, and the filter cake was washed twice with 50 ml anhydrous ether. The product was; dried in vacuum. 84.0 g were obtained of a penimepicycline as yellowish white powder.

References

Gradnik B., Pedrazzoli A.; GB Patent No. 891,004; March 7, 1962; Assigned: Societe D'Etudes de Recherches et D'Applications Scientifiques et Medicales E.R.A.S.M.E., of 67 Avenue de Wagram, Paris 17, France, a French body corporate

Kleemann A., Engel J.; Pharmazeutische Wirkstoffe, GeorgThieme Verlag Stuttgart, New York, 1982

PENMESTEROL

Therapeutic Function: Androgen

Chemical Name: 3-(Cyclopentyloxy)-17-methyl-androsta-3,5-dien-17β-ol

Common Name: Penmesterol; Penmestrol

Structural Formula:

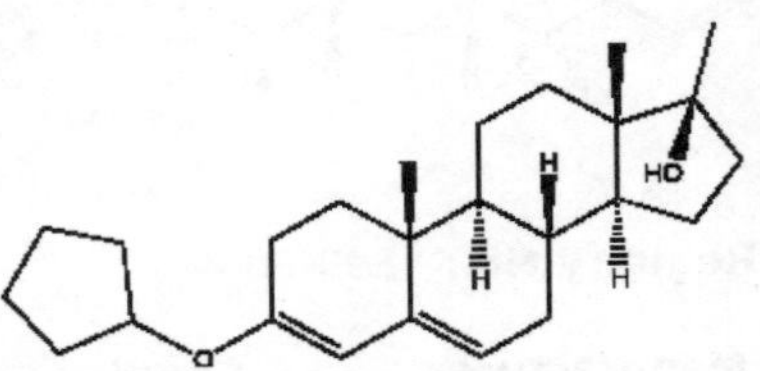

Chemical Abstracts Registry No.: 67-81-2

Trade Name	Manufacturer	Country	Year Introduced
Penmesterol	Shanghai Lansheng Corporation	-	-

Raw Materials

Androsteaedione
4-Toluenesulfonic acid
Methyl magnesium bromide
Orthoformate
Cyclopentanol

Manufacturing Process

A) 14 g of ethyl enolether of androstenedione, melting at 149°-151°C (obtained in a yield of 85% of the theoretical amount by treating androsteaedione with ethyl orthoformate), were added to a boiling solution of

37 ml of cyclopentanol and 0.450 g of p-toluenesulfonic acid in 2.5 L of benzene. The mixture was distilled over an approximately 40 minute period, so that the ethanol, which evolved during the exchange reaction, was evaporated off completely.

Then 0.5 ml of pyridine was added to the remaining solution and the mixture was concentrated under vacuum to dryness. The residue, taken up with a mixture of methanolmethylene chloride containing a few drops of pyridine, gave 13.8 g of cyclopentyl enolether of androstenedione melting at 181°-183°C. Yield about 85%.

B) The cyclopentyl enolether of androstenedione was converted to the corresponding cyclopentyl enolether of 17α-methyl testosterone as follows:

In a 3-necked flask fitted with a dropping funnel, reflux condenser, stirrer and nitrogen inlet tube, there was placed a solution of 25 g of methyl magnesium bromide in 150 ml of ether. With stirring and under an atmosphere of nitrogen, a solution of 4.1 g of androstenedione 3-cyclopentyl enolether in 80 ml of anhydrous benzene was added slowly. The reaction mixture was refluxed for 1 hour and allowed to stand overnight at room temperature. The mixture was then treated with an aqueous solution of 30% ammonium chloride, the organic layer separated off, washed with water and dried over anhydrous sodium sulfate. The solvent was evaporated and the residue taken up with dilute methanol to yield 3.2 g of a white product. Crystallization from methanol containing few drops of pyridine give the pure 17α-methyl-testosterone 3-cyclopentyl enolether; MP: 148-152°C; $[\alpha]^D$ =-150° (dioxane).

References

Ercoli A.; US Patent No. 3,019,241; Jan. 30, 1962

PENTAERYTHRITOL TETRANITRATE

Therapeutic Function: Coronary vasodilator

Chemical Name: 2,2-Bis[(nitroxy)methyl]-1,3-propanediol dinitrate

Common Name: PETN; Pentanitrolum

Structural Formula:

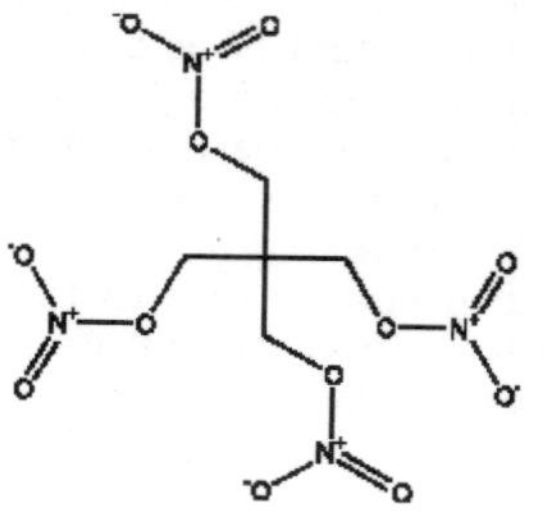

Chemical Abstracts Registry No.: 78-11-5

Trade Name	Manufacturer	Country	Year Introduced
Pentanitrine	Promedica	France	1948
Peritrate	Warner Lambert	US	1952
Pentritol	Armour	US	1955
Pentafin	Tutag	US	1956
Vasodiatol	Rowell	US	1958
Metranil	Meyer	US	1960
Pentryate	Fellows-Testagar	US	1960
Tranite D-Lay	Westerfield	US	1961
Peridex	Robins	US	1962
Antime	Century	US	1962
SK-Petin	SKF	US	1971
Perispan	U.S.V.	US	1971
Pentraspan	Glenwood	US	1980
Pentraspan	Vitarine	US	1983
Cardiacap	Consol. Chem	UK	-
Dilcoran	Godecke	W. Germany	-
Duotrate	Marion	US	-
Hasethrol	Shionogi	Japan	-
Hypothurol	Nissin	Japan	-
Lentrat	Medinova	Switz.	-
Neo-Corodil	Ethica	Canada	-
Neo-Corovas	Amfre-Grant	US	-
Nitrodex	Dexo	France	-
Nitropent	A.C.O.	Sweden	-
Pectolex	Shionogi	Japan	-
Penritol	Langley	Australia	-
Pentalong	Isis-Chemie	E. Germany	-
Peritrine	Norgine	Belgium	-
Perynitrate	Barlow Cote	Canada	-

Raw Materials

Pentaerythritol
Nitric acid

Manufacturing Process

Cooling water was turned on and 420 parts nitric acid of 94% strength was introduced into the nitrator. The amount of acid was such that the ratio of nitric acid to pentaerythritol was 4.29. The agitator was started and the agitator speed adjusted to 120 rpm. 92 parts pentaerythritol, which had been screened previously through a 14-mesh screen was used in each charge. About 45 parts pentaerythritol was added to the nitrator at such a rate that the temperature in the nitrator gradually rose to 110°F. This required about 12 minutes. Time was allowed for the temperature rise to cease before each succeeding increment of material was added.

After reaching 110°F the charge was maintained at about said temperature from 12 to 14 minutes during which time approximately 30 parts pentaerythritol was added to the nitrator. During the following 14 minutes, approximately, the remainder of the 92 parts pentaerythritol was added in like manner to the charge and the temperature gradually reduced. The pentaerythritol was introduced into the acid in finely divided and well-dispersed particles and not in large unitary quantities. The entire 92 parts of pentaerythritol tetranitrate was introduced in 35 to 40 minutes. The pentaerythritol thus obtained was separated from the spent acid by filtering or drowning in water. To recover the spent acid the charge was passed onto a nutsch and filtered. The crude product was washed with water, then with a weak water-soluble alkali solution, such as sodium carbonate for example, and subsequently with water in order to remove the acid.

After the removal of acid, the nitrate was dried by suction on the nutsch for about 15 minutes. The dried material was refined by means of acetone treatment or other suitable refining means. About 210 parts refined pentaerythritol tetranitrate per charge was obtained.

References

Merck Index 6977
DFU 4 (5) 351 (1979)
Kleeman and Engel p. 695
PDR pp. 1382, 1606
I.N. p. 741
REM p. 854
Acken, M.F. and Vyverberg, J.C. Jr.; US Patent 2,370,437; February 27,1945; assigned to E.I. du Pont de Nemours and Co.

PENTAGASTRIN

Therapeutic Function: Gastrosecretory hormone

Chemical Name: N-Carboxy-β-alanyl-L-tryptophyl-L-methionyl-L-aspartylphenyl-L-alaninamide N-tert-butylester

Common Name: -

Chemical Abstracts Registry No.: 5534-95-2

Trade Name	Manufacturer	Country	Year Introduced
Peptavlon	I.C.I.	UK	1967
Gastrodiagnost	Merck	W. Germany	1970
Pentagastrin	I.C.I.	Japan	1973
Peptavlon	Ayerst	US	1976
Peptavlon	I.C.I.	France	1981
Acignost	VEB Berlin Chemie	E. Germany	-

Structural Formula:

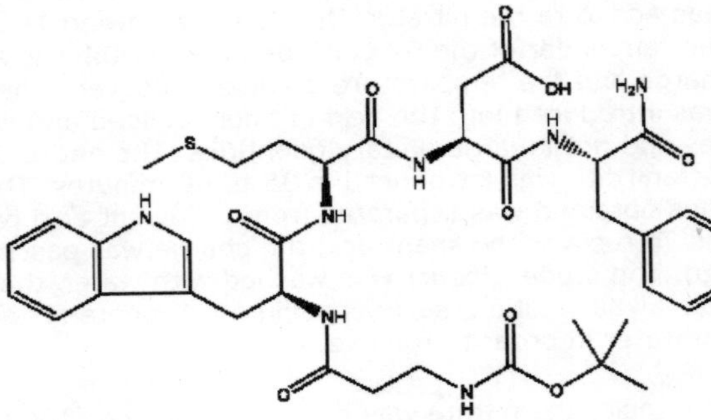

Raw Materials

L-Tryptophanyl-L-methionyl-L-aspartyl-L-phenylalanine amide trifluoroacetate
N-t-Butyloxycarbonyl-β-alanine 2,4,5-trichlorophenyl ester

Manufacturing Process

A solution of 3.55 parts of L-tryptophanyl-L-methionyl-L-aspartyl-L-phenylalanine amide trifluoroacetate in 30 parts of dimethylformamide is cooled to 0°C, and 1.01 parts of triethylamine are added. The mixture is stirred while 1.84 parts of N-tert-butyloxycarbonyl-β-alanine 2,4,5-trichlorophenyI ester are added at 0°C. The reaction mixture is kept at 0°C for 48 hours and then at 20°-23°C for 24 hours. The mixture is added to a mixture of 100 parts of ice-water, 0.37 part of concentrated hydrochloric acid (SG 1.18), 1.2 parts of acetic acid and 20 parts of ethyl acetate. The mixture is stirred for 15 minutes at 0°-10°C and is then filtered. The solid residue is washed with water and then with ethyl acetate, and is dried at 40°-50°C under reduced pressure. There is thus obtained N-tert-butyloxycarbonyl-β-alanyl-L-tryptophanyl-L-methionyl-L-aspartyl-L-phenylalanine amide, MP 213°C with de composition.

References

Merck Index 6978
PDR p. 2004
DOT 3 (4) 150 (1967)
I.N. p. 742
REM p. 1273
Hardy, P.M., Kenner, G.W., Sheppard, R.C., MacLeod, J.K. and Morley, J.S.; British Patent 1,042,487; assigned to Imperial Chemical Industries Limited, England
Hardy, P.M., Kenner, G.W., Sheppard, R.C., Morley, J.S. and MacLeod, J.K.; US Patent 3,896,103; July 22, 1975; assigned to Imperial Chemical Industries Ltd.

PENTAMIDINE ISETHIONATE

Therapeutic Function: Antiprotozoal

Chemical Name: Ethanesulfonic acid, 2-hydroxy-, compd. with 4,4'-(1,5-pentanediylbis(oxy))bis(benzenecarboximidamide) (2:1)

Common Name: Pentamidine isethionate

Structural Formula:

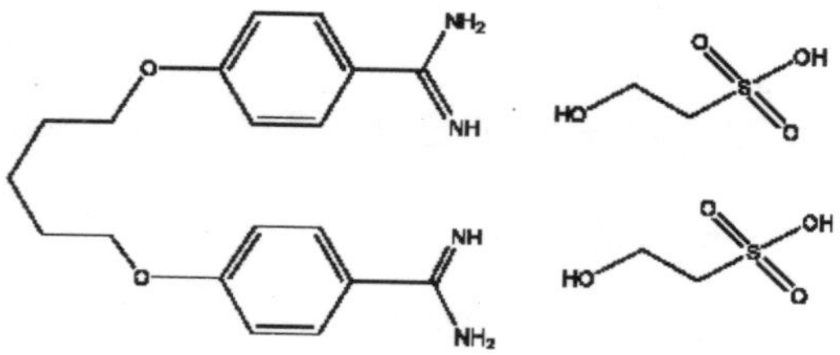

Chemical Abstracts Registry No.: 140-64-7; 100-33-4 (Base)

Trade Name	Manufacturer	Country	Year Introduced
Nebupent	Fujisawa Healthcare Inc	USA	-
Nebupent	American Pharmaceutical Partners, Inc.	USA	-
Pentacarinat	Rhone - Poulenc Rorer	USA	-
Pentamidine isethionate	Fujisawa Healthcare Inc	USA	-
Pentam 300	Fujisawa Healthcare Inc	USA	-
Pneumopent	Fisons Corporation	-	-

Raw Materials

p,p'-Dicyano-1:5-diphenoxy-pentane
Ethyl alcoholic ammonia

Manufacturing Process

2.5 g of p,p'-dicyano-1,5-diphenoxy-pentane (obtained by the interaction of p-hydroxybenzonitrile and pentamethylene-dibromide in aqueous alkaline solution, melting point 114°C) are dissolved in 15 cc of nitrobenzene and 2.5 cc of absolute ethyl alcohol added. The solution is saturated with dry hydrochloric acid gas at 0°C and allowed to stand for 48 h. It is then diluted with dry ether and the precipitated 1,5-diphenoxypentane, 4,4'-di(ethoxycarbonimidoyl) dihydrochlorid is filtered and washed with ether.

4 g of 1,5-diphenoxypentane, 4,4'-di(ethoxycarbonimidoyl) dihydrochloride are mixed with 30 cc. of 6 % ethyl alcoholic ammonia and heated in a closed vessel at 50°C for 5 h. The alcohol is removed and the residual 1,5-diphenoxypentane, 4,4'-diamidino dihydrochloride is twice recrystallised from

dilute hydrochloric acid and finally purified by dissolving in water and precipitating with acetone. Its melts at 236°C, dec.

Pentamidine isetionate salt may be produced by the reaction pentamidine base with isethionic acid.

References

Ewins A.J. et al.; US Patent No. 2,277,861; March 31, 1942; Assigned: May and Baker, Limited, London, England, a company of Great Britain and Northern Ireland

PENTAPIPERIDE METHOSULFATE

Therapeutic Function: Spasmolytic

Chemical Name: α-(1-Methylpropyl)benzeneacetic acid 1-methyl-4-piperidinyl ester methosulfate

Common Name: Pentapiperium methosulfate

Structural Formula:

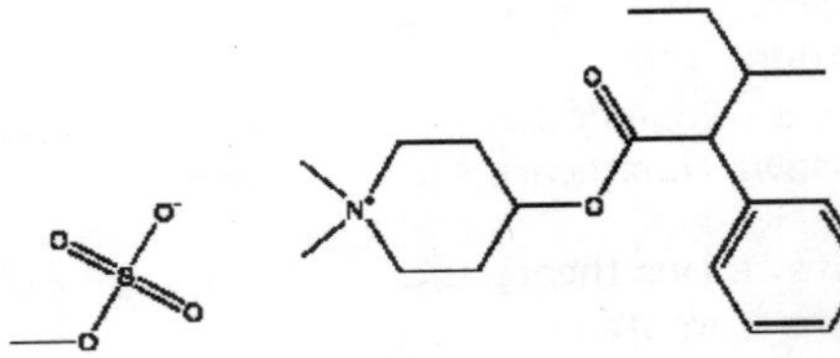

Chemical Abstracts Registry No.: 7681-80-3; 7009-64-3 (Base)

Trade Name	Manufacturer	Country	Year Introduced
Quilene	Warner Lambert	US	1969
Crylene	Auclair	France	1971
Crilin	Ayerst	Italy	1973
Perium	Rover	US	-
Togestal	Biosedra	France	-

Raw Materials

Phenylacetonitrile
Thionyl chloride
Sodium amide
1-Methyl-4-piperidinol
Sec-Butyl bromide
Dimethyl sulfate
Sodium hydroxide

Manufacturing Process

Phenylacetonitrile is alkylated with secondary butyl bromide and the resultant nitrile is hydrolyzed to 3-methyl-2-phenylvaleric acid. The acid is converted to the acid chloride with thionyl chloride and the acid chloride is in turn reacted with 1-methyl-4-piperidinol. Finally dimethyl sulfate is reacted with the ester.

References

Merck Index 6988
Kleeman and Engel p. 697
OCDS Vol. 2 p. 76 (1980)
DOT 6 (2) 61 (1970)
I.N. p. 743
Martin, H. and Habicht, E.; US Patent 2,987,517; June 6, 1961; assigned to Cilag Chemie Limited, Switzerland

PENTAZOCINE HYDROCHLORIDE

Therapeutic Function: Analgesic

Chemical Name: 2,6-Methano-3-benzazocin-8-ol, 1,2,3,4,5,6-hexahydro-6,11-dimethyl-3-(3-methyl-2-butenyl)-, hydrochloride, (2R,6R,11R)-rel-

Common Name: Pentazocine hydrochloride

Structural Formula:

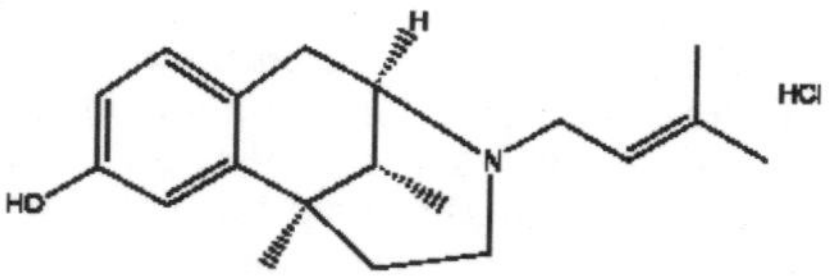

Chemical Abstracts Registry No.: 64024-15-3; 68964-90-9; 359-83-1 (Base)

Trade Name	Manufacturer	Country	Year Introduced
Pentazocine Hydrochloride	Mallinckrodt Inc.	-	-

Raw Materials

Methyliodide
Hydrogen
Hydrobromic acid
Hydrochloric acid
4-Methoxybenylmagnesium chloride
3,4-Dimethylpyridine
Palladium on charcoal

Manufacturing Process

A solution of 3,4-dimethylpyridine was added to a methyliodid. Then to the resulting solution containing 1,3,4-trimethylpyridinium iodide the 4-methoxybenzylmagnesium chloride was added. After reaction process the 1,3,4-trimethyl-2-(4-methoxy-benzyl)-pyridine was obtained.

To the solution of 1,3,4-trimethyl-2-(4-methoxy-benzyl)-pyridine was reduced by hydrogen over 10% palladium-on-charcoal, and when reduction was complete, the catalyst was removed by filtration and the filtrate taken to dryness. The residue was recrystallized to give 2-(4-methoxybenzyl)-1,3,4-trimethyl-1,2,5,6-tetrahydropyridine.

To the 2-(4-methoxybenzyl)-1,3,4-trimethyl-1,2,5,6-tetrahydropyridine the solution of hydrobromic acid was added and heated under reflux.The product obtained was recrystallized and yield N-methyl-1,2,3,4,5,6-hexahydro-6,11-dimethyl-8-hydroxy-2,6-methano-3-benzazocine (2'-hydroxy-2,5,9-trimethylbenzo-6-morphen), which then was demethylated by bromcyan (BrCN). As a result the racemic cis-1,2,3,4,5,6-hexahydro-6,11-dimethyl-8-hydroxy-2,6-methano-3-benzazocine was obtained (that is a. 2'-hydroxy-5,9-dimethyl-6,7-benzomorphen).

A mixture of 8.7 g racemic cis-1,2,3,4,5,6-hexahydro-6,11-dimethyl-8-hydroxy-2,6-methano-3-benzazocine, 6.0 g of 1-bromo-3-methyl-2-butene, 5.0 g of sodium bicarbonate, and 125 ml of N,N-dimethylformamide was stirred and refluxed for approximately 4.5 hours. The reaction mixture was then filtered, and the solid on the filter was washed with ethanol. The filtrate and the wash liquor were combined, concentrated under reduced pressure, and then extracted with chloroform. The chloroform extract was concentrated under reduced pressure to yield a syrup which weighed 15.8 g. This syrup was dissolved in 120 ml of diethyl ether and the resulting solution was filtered to remove approximately 0.5 g of a brown amorphous solid. The filtrate was extracted with a mixture of 5 ml of concentrated hydrochloric acid and 20 ml of water. To the extract there was added 5 ml of concentrated ammonium hydroxide solution and ice. A pale tan syrup separated from solution and after stirring, this syrup solidified. The resulting pale tan solid was collected and dried; it weighed 10.6 g. After two recrystallizations from a mixture of methyl alcohol and water, with charcoaling, the 1,2,3,4,5,6-hexahydro-3-(3-methyl-2-butenyl)-6,11-dimethyl-8-hydroxy-2,6-methano-3-benzazocine weighed 8.2 g and melted at 145°-147°C.

The1,2,3,4,5,6-hexahydro-3-(3-methyl-2-butenyl)-6,11-dimethyl-8-hydroxy-2,6-methano-3-benzazocinewas soluble in a mixture of 0.35 ml of 2 N hydrochloric acid and 0.15 ml of water to the extent of 10%, the pH of the 1% solution being 2.80; and when the pH of the 1% solution was gradually raised by addition of 10 N sodium hydroxide solution, a precipitate formed at pH 5.4. The 1,2,3,4,5,6-hexahydro-3-(3-methyl-2-butenyl)-6,11-dimethyl-8-hydroxy-2,6-methano-3-benzazocine hydrochloride melted at 245°-247°C, dec.

References

Albertson N. F.; US Patent No. 3,936,462; Feb. 3, 1976; Assigned: Sterling Drug Inc., New York, N.Y.

Archer S.; US Patent No. 4,105,659; Aug. 8, 1978; Assigned: Sterling Drug Inc., New York, N.Y.

PENTETRAZOL

Therapeutic Function: Analeptic, Central stimulant

Chemical Name: 5H-Tetrazolo[1,5-a]azepine, 6,7,8,9-tetrahydro-

Common Name: Corazol; Leptazol; Pentamethylentetrazol; Pentetrazol; Pentylenetetrazol

Structural Formula:

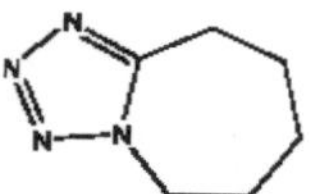

Chemical Abstracts Registry No.: 54-95-5

Trade Name	Manufacturer	Country	Year Introduced
Pentylenetetrazole	Spectrum Chemicals and Laboratory Products, Inc.	-	-
Angioton	Leo	-	-
Cardiamine	Chropi	-	-
Cardional	Oliveira	-	-
Cardiotonico	C.F.	-	-
Cardiotonicum	Blomberg	-	-
Inocor	Bengen	-	-
Analeptin	Biotica	-	-

Raw Materials

Cyclohexanone
Nitric acid
Sulfuric acid

Manufacturing Process

A solution of 9.8 g cyclohexanone and 8.6 g HNO_3 in about 250 ml benzene were slowly added dropwise to 20 ml concentrate sulfuric acid in 100 ml benzene by ice cooling and stirring. After ending of a generation of N_2 (0.1 moles) to a corresponding quantity of cyclohexane (0.1 moles). Acid layer was diluted with ice, and made neutral with strong alkaline to give a reaction product as oil. Then it was exrtacted with chloroform, all solvents were distilled and the residue was diluted with water. The desired 6,7,8,9-tetrahydro-5H-tetrazoloazepine dropped. Yield was 7.5 g after recrystallization from ester or distillation. M.P: 65°C.

References

Firma Knoll and Co. in Ludwigshafen a. Rh. und Dr. Karl Fridrich Shmidt in Heidelberg; D.R. Patent No. 427,858; July 20, 1923

PENTHIENATE BROMIDE

Therapeutic Function: Anticholinergic

Chemical Name: 2-[(Cyclopentylhydroxy-2-thienylacetyl)oxy]-N,N-diethyl-N-methylethanaminium bromide

Common Name: -

Structural Formula:

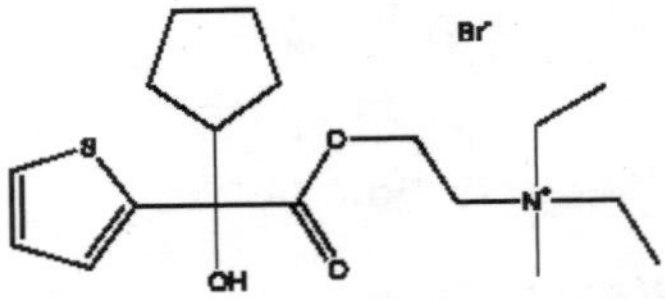

Br⁻ ... OH ... O ... S ... N⁺

Chemical Abstracts Registry No.: 60-44-6

Trade Name	Manufacturer	Country	Year Introduced
Monodral	Winthrop	US	1954
Monodral	Kanebo, Ltd.	Japan	1970

Raw Materials

2-Diethylaminoethyl chloride
Cyclopentyl-(α-thienyl)hydroxyacetic acid
Methyl bromide

Manufacturing Process

An aqueous solution of 13.8 g of 2-diethylaminoethyl chloride hydrochloride was neutralized with sodium hydroxide, and the free 2-diethylaminoethyl chloride was extracted with ether. The ether extracts were dried over anhydrous magnesium sulfate, filtered, and the filtrate was added to a solution of 13.6 g of cyclopentyl-(α-thienyl)hydroxyacetic acid in 100 ml of isopropyl alcohol. The mixture was then distilled through a 25-cm Vigreaux-type column until the temperature of the vapors reached 80°C. The residual solution was refluxed overnight and then transferred to a beaker along with 350 ml of isopropyl alcohol. The crystalline hydrochloride had meanwhile separated out, and this was filtered, washed with isopropyl alcohol, ether and then dried, giving 23 g, melting point 172°C to 173.5°C. Recrystallization from 400 ml of isopropyl alcohol gave 20.3 g of 2-diethylaminoethyl cyclopentyl-(α-thienyl)hydroxyacetate hydrochloride, melting at 174°C to 175°C; deep yellow-orange color with concentrated sulfuric acid.

The hydrochloride may then be converted to the methobromide by reaction with methyl bromide.

References

Merck Index 6996
Kleeman and Engel p. 699
I.N. p. 744
Blicke, F.F.; US Patent 2,541,634; February 13, 1951; assigned to Regents of the University of Michigan

PENTIFYLLINE

Therapeutic Function: Vasodilator, Diuretic

Chemical Name: 1H-Purine-2,6-dione, 3,7-dihydro-1-hexyl-3,7-dimethyl-

Common Name: Hexyltheobromine; Pentifylline

Structural Formula:

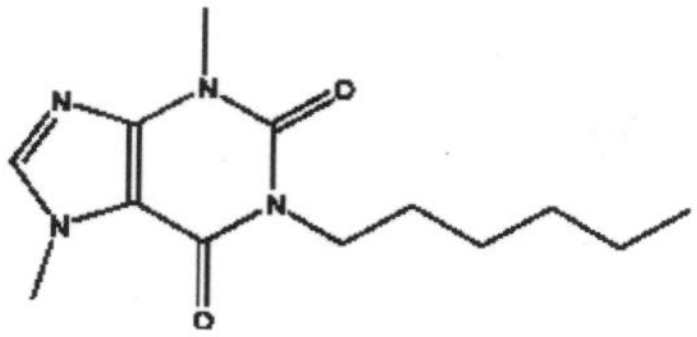

Chemical Abstracts Registry No.: 1028-33-7

Trade Name	Manufacturer	Country	Year Introduced
Pentifylline	Shanghai Lansheng Corporation	-	-
Cosaldon retard mono	Aventis	-	-
1-Hexyltheobromine	Fluorochem	-	-
1-Hexyltheobromine	Trans World Chemicals, Inc.	-	-

Raw Materials

Theobromine
Theobromine sodium
n-Hexyl chloride or n-hexyl bromide

Manufacturing Process

The mixture 25 g theobromine, 38 ml 4 N sodium hydroxide, 60 ml isopropanol, and 17 g n-hexyl chloride were heated 24 hours to 100°C in autoclave. The solvent was removed and the residual alkaline solution was extracted with chloroform, water layer was acidified. Yield of 1-hexyl-3,7-dimethylxanthine was 88%; MP: 82°-83°C. The product may be prepared from theobromine sodium. 20.2 g theobromine sodium, 20 n-hexyl bromide and 100 ml toluene were ground 10 hours at 100°C in a ball mill. After above written treatment 22.3 g (84.5%) 1-hexyltheobromine was prepared; MP: 84°C.

References

Eidebenz E., von Schuh H.G.; D.B. Patent No. 860,217; Oct. 28, 1950; Chemishe Werke Albert, Wiesbaden-Biebrich.

PENTOBARBITAL SODIUM

Therapeutic Function: Hypnotic, Sedative

Chemical Name: 5-Ethyl-5-(1-methylbutyl)-2,4,6-(1H,3H,5H)-pyrimidinetrione monosodium salt

Common Name: -

Structural Formula:

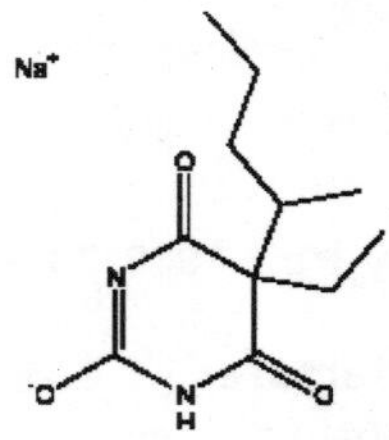

Chemical Abstracts Registry No.: 57-33-0; 76-74-4 (Base)

Trade Name	Manufacturer	Country	Year Introduced
Nembutal	Abbott	US	1941
Butylone	Hartz	Canada	-
Hypnol	Stickley	Canada	-
Mintal	Tanabe	Japan	-
Nebralin	Dorsey	US	-
Neodrom	Minden	W. Germany	-
Novopentobarb	Novopharm	Canada	-
Penbon	Adams	Australia	-

Trade Name	Manufacturer	Country	Year Introduced
Pentanca	Anca	Canada	-
Pentogen	Paul Maney	Canada	-
Pentone	Faulding	Australia	-
Prodormol	Teva	Israel	-
Repocal	Desitin	W. Germany	-
Sombutol	Farmos	Finland	-
Somnotol	M.T.C.	Canada	-
Sopental	Continental Ethicals	S. Africa	-

Raw Materials

Sodium
di-n-Butyl ethyl 1-methyl-n-butylmalonate
Butanol
Urea

Manufacturing Process

Sodium (9.6 parts) was dissolved in butanol (192 parts) and di-n-butyl ethyl 1-methyl-n-butylmalonate (62.8 parts) and urea (14.4 parts) were added to the warm solution with agitation. The mixture was then heated to reflux temperature in three quarters of an hour and maintained for 2 hours. The reaction mass was kept, water (150 parts) added, the aqueous portion separated, and the butanol layer extracted with water (3 x 50 parts). The combined aqueous extracts were then given 3 small extractions with benzene, the aqueous liquors separated, charcoaled, filtered and precipitated with concentrated hydrochloric acid (acid to congopaper). The solid was collected, washed with water, dissolved in N-sodium hydroxide and reprecipitated with carbon dioxide. On recrystallization, from aqueous alcohol, the pentobarbitone was obtained.

References

Merck Index 6998
Kleeman and Engel p. 700
PDR pp. 531, 872, 1989
OCDS Vol. 1 p. 268 (1977)
I.N. p. 745
REM p. 1067
The Geigy Co. Ltd.; British Patent 650,354; February 21, 1951

PENTOREX TARTRATE

Therapeutic Function: Anorexic

Chemical Name: Benzeneethanamine, α,α,β-trimethyl-, tartrate (1:1)

Common Name: Pentorex tartrate; Phenpentermine tartrate

Structural Formula:

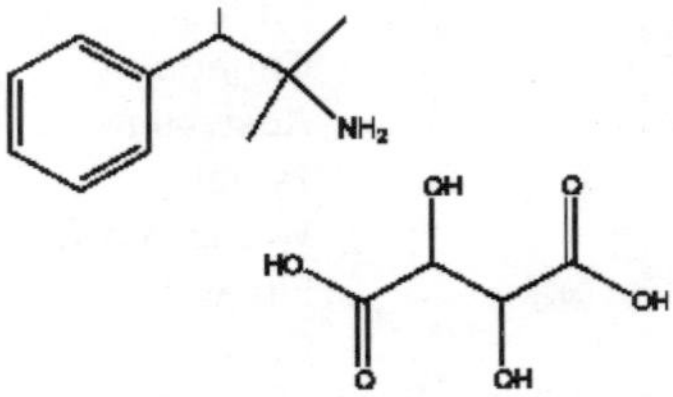

Chemical Abstracts Registry No.: 434-43-5 (Base); 22876-60-4

Trade Name	Manufacturer	Country	Year Introduced
Liprodene	Anphar-Rolland	-	-
Modatrop	Nordmark	-	-

Raw Materials

2-Phenylbutanone-3
Sodium cyanide
Sulfuric acid
Methyl magnesium bromide
Acetic acid

Manufacturing Process

To a solution of 105.6 g of 2-phenylbutanone-3 in 110 ml ether was added dropwise a solution of methylmagnesium bromide (prepared from 19.4 g magnesium and 94.7 g methyl bromide in 400 ml of ether) for 60-90 min. Then the mixture was refluxed for 1 hour. After cooling to the mixture was added diluted sulfuric acid and then extracted with ether. Organic layer was dries with sodium sulfate. After evaporation of ether the 2-phenyl-3-methylbutanol was distilled, B.P. 116-118°C/20 mm, yield 105 g, n_d^{22} 1.5152.

To 25.5 g of NaCN at 10-20°C ware added dropwise under stirring 64 ml of glacial acetic acid and ten at 20°C a mixture of 70 ml concentrated sulfuric acid and 64 ml of glacial acetic acid. To the prepared mixture at 20-25°C was added dropwise 82 g of 2-phenyl-3-methylbutanol. The mixture was stirred at 45-50°C for 10-20 min and then at 75°C for 30 min. To the reaction mixture was added 750 ml of water. The acids was neutralized with sodium carbonate. Product was extracted with ether and distilled. Boiling point of (dimethylbenzylcarbinyl)formamide 173-176°C/0 mm, yield 63 g.

52.3 g of (dimethylbenzylcarbinyl)formamide, 245 ml concentrated hydrochloric acid and 196 ml of water were refluxed for 6 hours. The unreacted compounds was extracted with ether. The residuum was stirred with sodium hydroxide and extracted with ether. By distillation was obtained 48.2 g of 2-amino-2-methyl-3-phenylbutane; B.P. 109-111°C/20 mm.

Hydrochloride of 2-amino-2-methyl-3-phenylbutane have melting point 164-166°C.

In practice it is usually used as tartrate salt.

References

Brevet Special de Medicament 931,804, April 17, 1963; Assigned to Nordmak Werke Gesellschaft mit Beschraenkter Haftung, residant en Allemagne

PENTOXIFYLLINE

Therapeutic Function: Vasodilator

Chemical Name: 3,7-Dihydro-3,7-dimethyl-1-(5-oxohexyl)-1H-purine-2,6-dione

Common Name: Oxpentifylline; Vazofirin

Structural Formula:

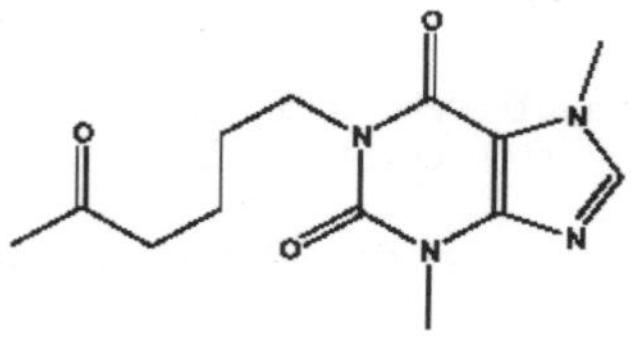

Chemical Abstracts Registry No.: 6493-05-6

Trade Name	Manufacturer	Country	Year Introduced
Trental	Albert Roussel	W. Germany	1972
Torental	Hoechst	France	1974
Trental	Hoechn	UK	1975
Trental	Albert Pharma	Italy	1976
Trental	Hoechst	Japan	1977
Agapurin	Spofa	Czechoslovakia	-
Techlon	Sawai	Japan	-

Raw Materials

1-Bromo-5-hexanone
Theobromine sodium

Manufacturing Process

A solution of 35.4 g of 1-bromohexanone-5 in 200 ml of ethanol was gradually mixed at the reflux temperature with vigorous stirring with 39.7 g of theobromine-sodium in 100 ml of water. After 3 hours reflux the unreacted theobromine was filtered off with suction, the filtrate was evaporated to dryness, the residue was dissolved in water and the solution was extracted with chloroform. The chloroform was distilled off and 1-(5'-oxohexyl)-3,7-

dimethylxanthine was obtained as residue; after recrystallization from isopropanol, it melted at 102°C to 103°C (about 25% yield, calculated on the reacted theobromine).

References

Merck Index 7002
Kleeman and Engel p. 701
PDR p. 947
OCDS Vol. 2 p. 466 (1980)
I.N. p. 746
Mohler, W., Reiser, M. and Popendiker, K.; US Patent 3,737,433; June 5, 1973; assigned to Chemische Werke Albert A.G. (W. Germany)

PENTOXYVERINE CITRATE

Therapeutic Function: Antitussive

Chemical Name: Cyclopentanecarboxylic acid, 1-phenyl-, 2-(2-(diethylamino)ethoxy)ethyl ester, citrate (1:1)

Common Name: Carbapentane citrate; Carbetopentane citrate; Pentoxiverini citras; Pentoxyverine citrate

Structural Formula:

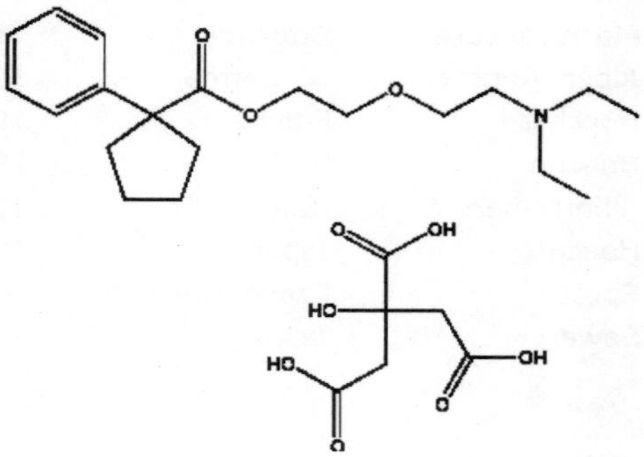

Chemical Abstracts Registry No.: 23142-01-0; 77-23-6 (Base)

Trade Name	Manufacturer	Country	Year Introduced
Asthma	Nichiiko	-	-
Carbex	Etna	-	-

Raw Materials

Sodium amide
Sulfuric acid
Phenylacetonitrile
1,4-Dibromobutane

Thionyl chloride 2-(2-Diethylaminoethoxy)ethanol

Manufacturing Process

1-Phenylcyclopentane carbonitrile was obtained by treatment of phenylacetonitrile with sodium amide and 1,4-dibrombutane.

1-Phenyl-1-cyclopentane carboxylic acid was produced in the result of reaction of 1-phenylcyclopentane carbonitrile with sulfuric acid.

1-Phenyl-1-cyclopentanecarbonyl chloride was obtained by treatment of 1-phenyl-1-cyclopentane carboxylic acid with thionyl chloride.

A mixture of 0.5 mol of 1-phenyl-1-cyclopentanecarbonyl chloride and of 0.5 mol of 2-(2-diethylaminoethoxy)ethanol (herein-after referred to as the amino alcohol) in 300 ml of toluene is heated under reflux for 20 h. The mixture is thereafter made alkaline by means of an aqueous solution of caustic soda and decanted; the toluenic layer is washed with water and concentrated in vacuo. The residue is distilled under high vacuum. After two fractional distillations, the 2-(2-diethylaminoethoxy)ethyl 1-phenylcyclopentane-carboxylate is obtained, in 85% yield. Boiling point 164°C/0.1 mm. Hg.

References

Morren H.; GB Patent No. 753,779; August 1, 1956
Kleemann A., Engel J.; Pharmazeutische Wirkstoffe, GeorgThieme Verlag Stuttgart. New York, 1982

PEPLOMYCIN SULFATE

Therapeutic Function: Antineoplastic

Chemical Name: 3-[(S)-1'-Phenylethylamino]propylaminobleomycin sulfate

Common Name: -

Chemical Abstracts Registry No.: 68247-85-8 (Base)

Trade Name	Manufacturer	Country	Year Introduced
Pepleo	Nippon Kayaku, Co.	Japan	1981

Raw Materials

Bleomycinic acid
N-[(S)-1'-Phenylethyl]-1,3-diaminopropane
Sulfuric acid

Structural Formula:

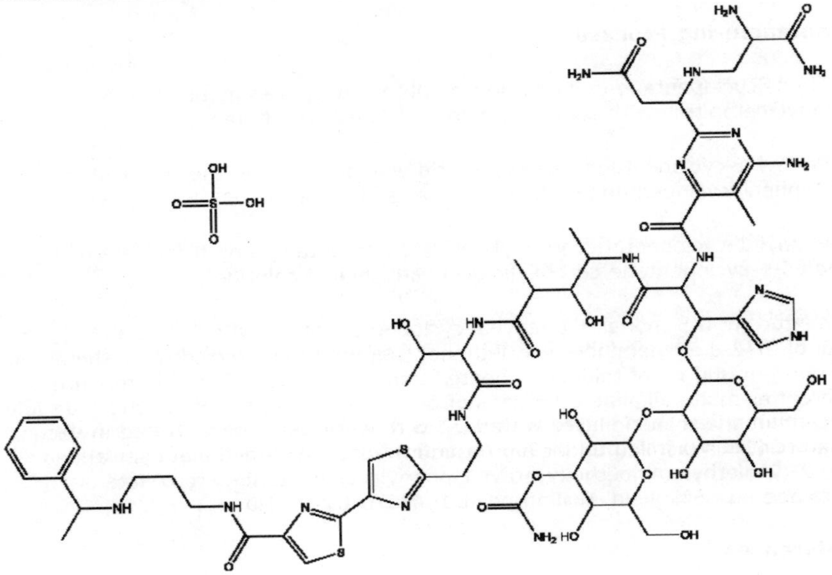

Manufacturing Process

In 400 ml of dimethylformamide was dissolved 15.0 g of bleomycinic acid (copper-containing form). To the solution kept at 0°C by cooling were added 1.1 ml of N-methylmorpholine and 10.3 g of 6-chloro-1-p-chlorobenzenesulfonyloxybenzotriazole (CCBT) as an activating compound. The mixture was stirred for 5 minutes at 0°C, then admixed with 5.3 g of N-[(S)-1'-phenylethyl]-1,3-diaminopropane and further stirred for 1 hour.

After termination of the reaction by adding 200 ml of a 25% aqueous acetic acid solution, the reaction mixture was mixed with 5 liters of cold acetone to precipitate the reaction product. The precipitate was collected by filtration, washed with acetone, and dissolved in 500 ml of distilled water. The resulting aqueous solution was immediately adjusted to pH 6.0 and poured into a column containing 2 liters of CM-Sephadex C-25 (NH_4^+type) packed in 0.05 M aqueous ammonium chloride solution to adsorb bleomycins.

Using aqueous ammonium chloride solution, elution was performed by passing through the column 20 liters of eluent in which the concentration of ammonium chloride was continually increased from 0.05 to 1.0 M. The unreacted bleomycinic acid was found in the effluent at the ammonium chloride concentration of about 0.05 M and NK631 at the ammonium chloride concentration of about 0.45 M. Both fractions, which showed UV absorption at 292 nm;, were separately collected.

The NK631-containing fraction was poured into a resin column containing 2.6 liters of Amberlite XAD-2. The column was then washed thoroughly with water and eluted with 0.01 N hydrochloric acid in methanol-water (4:1 v/v). A total of 2.5 liters of the blue fraction, which showed UV absorption at 292 mμ, was collected. After evaporating off the methanol from the eluent fraction, the concentrate was adjusted to pH 6.0 with Dowex 44 (OH^- type, an anion-exchange resin composed of a copolymer of epichlorohydrin and ammonia) and was freeze-dried to obtain 16.1 g (92% yield) of NK631 dihydrochloride (copper-containing form) in the form of blue amorphous powder.

By similar treatment, 280 mg of the unreacted bleomycinic acid (copper-containing form) were recovered.

In 200 ml of distilled water was dissolved 10.0 g of the NK631 dihydrochloride (copper-containing form). The solution was poured into a column containing 600 ml of Amberlite XAD-2 packed in distilled water. The column was washed successively with 2 liters of an aqueous solution containing 5% of EDTA-Na_2, 2.5 liters of a 5% aqueous sodium sulfate solution, and 630 ml of distilled water.

The column was then eluted with 0.0025 N sulfuric acid in methanol-water mixture (1:1 v/v). A total of 900 ml of fractions containing a substance which showed UV absorption at 290 mμ was collected. After removal of methanol by distillation, the residual liquid was adjusted to pH 6.0 with Dowex 44 (OH^- type) and freeze-dried to obtain 9.3 g (95% yield) of NK631 monosulfate (copper-free form) in the form of pale yellowish-white amorphous powder.

References

Merck Index 7011
DFU 6 (2) 101 (1981)
DOT 17 (8) 331 (7981)
Takita, T., Fujii, A., Fukuoka, T., Muraoka, Y., Yoshioka, O. and Umezawa, H.; US Patent 4,195,018; March 25, 1980; assigned to Nippon Kayaku K.K.
Umezawa, H., Maeda, K., Takita, T., Nakayama, Y., Fujii, A. and Shimada, N.; US Patent 3,846,400; November 5, 1974; assigned to Zaidan Hojin Biseibutsu Kagaku Kenkyu Kai.

PERGOLIDE MESYLATE

Therapeutic Function: Dopamine agonist

Chemical Name: Ergoline, 8-β-((methylthio)methyl)-6-propyl-, methanesulfonate (1:1)

Common Name: Pergolide mesilate

Chemical Abstracts Registry No.: 66104-23-2; 66104-22-1 (Base)

Trade Name	Manufacturer	Country	Year Introduced
Celance	Eli Lilly	-	-
Permax	Eli Lilly	-	-
Nopar	Lilly	-	-

Structural Formula:

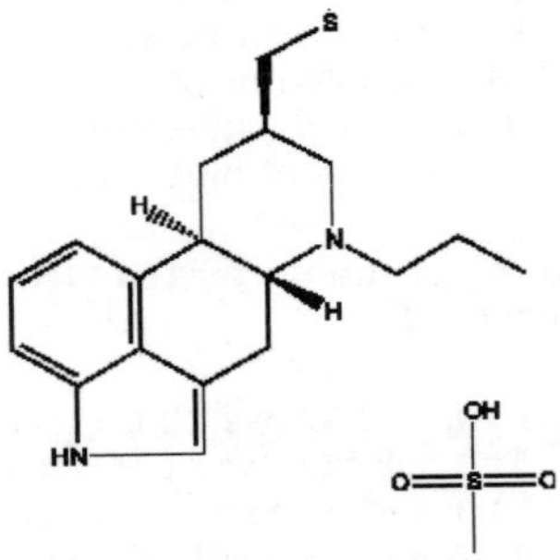

Raw Materials

9,10-Dihydrolysergol
Propionic anhydride
Dimethyl disulfide
Sodium hydroxide
Sodium iodide
Lithium aluminum hydride
Tri-n-butylphosphine

Manufacturing Process

Dimethyl disulfide (73.6 ml, 0.79 mol) and tri-n-butylphosphine (79.6 ml, 0.32 mol) were added to a solution of 9,10-dihydrolysergol in (8.1 g, 0.032 mol) in the 150 ml of anhydrous DMF and were stirred at room temperature for 6 hours under a nitrogen atmosphere. Dimethyl disulfide of the reaction mixture was removed under vacuo. A solution of the residue in ethyl acetate was extracted with 3.7% HCl (aq.). The aqueous layer was basified with ammonium hydroxide to a pH of 10 and then extracted with ethyl acetate. Removal of ethyl acetate followed by a silica gel column purification eluting with 10% $MeOH/CH_2Cl_2$ gave5.5.g of D-6-methyl-8β-(methylthiomethyl)ergoline (60%).

A solution of D-6-methyl-8β-(methylthiomethyl)ergoline (0.4 g, 0.0014 mol) and NaI (0.63 g, 0.0042 mol) in 10 ml of propionic anhydride was refluxed for 40 hours. The reaction mixture was guenched with a 10% Na_2CO_3 solution and extracted by ethyl acetate. The combined organic layers were washed with a saturated brine solution, dried with magnesium sulfate and concentrated to produce oil. The oil was purified by silica gel column, eluting with 10% $MeOH/CH_2Cl_2$ to give 0.33 g of D-1,6-dipropionyl-8β-(methylthiomethyl)ergoline.

$LiAlH_4$ (0.6 g, 0.0156 mol) was slowly added to a solution of D-1,6-dipropionyl-8β-(methylthiomethyl)ergoline in the 20 ml anhydrous THF at 0°C under nitrogene atmosphere. The mixture was stirred at 0°C for 30 min and then at room temperature for 4 hours. The reaction was cooled to 0°C and 0.6 ml of water was slowly added. The mixture was stirred at 0°C for 10 min and 1.8 ml of 15% NaOH (aq.) and 2.5 ml of water were added respectively. The mixture was stirred for 30 min at room temperature and then filtered. Excess of the solvent was removed under reduced pressure to give 150 mg of 8β-((methylthio)methyl)-6-propyl-ergoline or pergolide (yield: 68%). Ergoline,

8-((methylthio)methyl)-6-propyl-, monomethanesulfonate, (8β)- may be prepared by mixing of components in solution.

References

Wu E.S.C., Wu M.; US Patent No. 6,388,079, May 14; Assigned to Scinopharm Singapore Pte Ltd., Singapore (SG)

PERHEXILINE MALEATE

Therapeutic Function: Coronary vasodilator

Chemical Name: 2-(2,2-Dicyclohexylethyl)piperidine maleate

Common Name: -

Structural Formula:

NH

HO O HO O

Chemical Abstracts Registry No.: 6724-53-4; 6621-47-2 (Base)

Trade Name	Manufacturer	Country	Year Introduced
Pexid	Merrell-Tourade	France	1973
Pexid	Merrell	W. Germany	1974
Pexid	Merrell	Italy	1974
Pexid	Merrell	UK	1975
Corzepin	Prodes	Spain	-
Daprin	Gerardo Ramon	Argentina	-

Raw Materials

Ethyl formate
α-Picoline
Sodium hydroxide
Maleic acid
Cyclohexylmagnesium bromide
Hydrogen chloride
Hydrogen

Manufacturing Process

1,1-Dicyclohexyl-2-(2'-pyridyl)ethanol hydrochloride (5 grams) was dehydrated by heating with 25 ml of concentrated hydrochloric acid at steam bath temperature for 10 minutes. 70 ml of water were added to the reaction mixture to give the crystalline hydrochloride salt. The product, 1,1-dicyclohexyl-2-(2'-pyridyl)ethylene hydrochloride, was recrystallized from methanol-ethyl acetate to yield a white solid melting at 150°-151.5°C.

1,1-Dicyclohexyl-2-(2'-pyridyl)ethylene hydrochloride (15 grams) in 150 ml of ethanol was hydrogenated in the presence of platinum oxide at about 60 pounds per square inch of hydrogen pressure. The product, 1,1-dicyclohexyl-2-(2'-piperidyl)ethane hydrochloride, crystallized from a mixture of methanol and methyl ethyl ketone as a white solid melting at 243° to 245.5°C.

The hydrochloride salt was neutralized with 10% sodium hydroxide solution and the free base so produced was dissolved in ether. The ether solution was dried over anhydrous magnesium sulfate. Addition of an excess of maleic acid in methanol to the solution yielded the acid maleate salt which melted at 188.5°-191°C.

The starting material was obtained by reacting ethyl formate with cyclohexylmagnesium bromide to give dicyclohexylcarbinol. That is oxidized to dicyclohexylketone and then reacted with α-picoline.

References

Merck Index 7026
Kleeman and Engel p. 703
DOT 10 (8) 299 (1974)
I.N. p. 747
REM p. 854
Richardson-Merrell Inc.; British Patent 1,025,578; April 14, 1966
Horgan, S.W., Palopoli, F.P. and Schwoegler, E.J.; US Patent 4,069,222; January 17, 1978; assigned to Richardson-Merrell Inc.

PERICIAZINE

Therapeutic Function: Neuroleptic

Chemical Name: 10H-Phenothiazine-2-carbonitrile, 10-(3-(4-hydroxypiperidino)propyl)-

Common Name: Periciazine; Pericyazine

Chemical Abstracts Registry No.: 2622-26-6

Structural Formula:

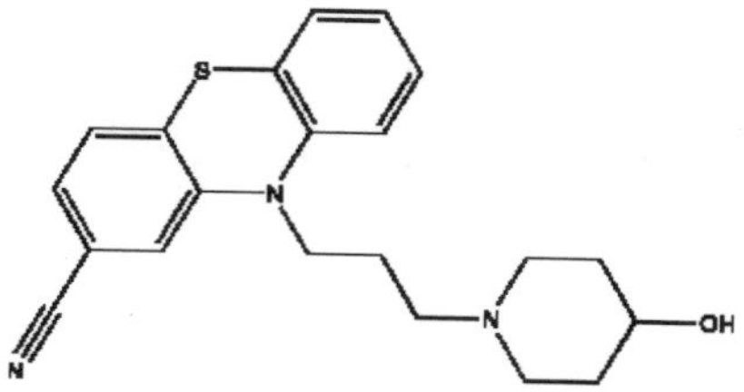

Trade Name	Manufacturer	Country	Year Introduced
Aolept	Bayer Vital	-	-
Neuleptil	Rhone-Poulenc Rorer	-	-
Neuleptil	Vitoria	-	-
Neuleptil	Gerot	-	-
Neuleptil	Alkaloid	-	-
Nemactil	Rhone-Poulenc Rorer	-	-
Neulactil	Rhone-Poulenc Rorer	-	-
Neuperil	Orion	-	-
Neuleptil	Aventis Pharma B.V.	-	-

Raw Materials

Hydrochloric acid	4-Hydroxypiperidine
Sodium hydroxide	2-Cyano-10-(3-methanesulfonyloxypropyl) phenthiazine

Manufacturing Process

2-Cyano-10-(3-methanesulfonyloxypropyl)phenthiazine and 4-hydroxypiperidine in toluene were heated under reflux with stirring. The reaction mixture was allowed to cool and water was added. The resulting toluene solution layer was decanted and washed twice with water. The toluene solution was then stirred with 5% hydrochloric acid. The hydrochloride of the desired phenthiazine base precipitated in gummy condition in the aqueous layer. This was decanted and treated with sodium hydroxide (density 1.33). It was then extracted three times with ethyl acetate. The extracts were dried over sodium sulfate, filtered and concentrated in vacuum. A resinous product was obtained. This product was dissolved in a mixture of benzene and cyclohexane and chromatographed on a column containing alumina. The chromatographed product was eluted successively with mixtures of benzene and cyclohexane and then with benzene and finally with a mixture of benzene and ethyl acetate. The eluates were evaporated to yield a crude product. This product was recrystallised from aqueous ethanol (40% water) and yielded 2-cyano-10-[3-(4-hydroxy-1-piperidyl)propyl]phenthiazine as white crystals.

References

Jacob R.M., Robert J.G.; US Patent No. 3,150,129; Sept. 22, 1964; Assigned: Rhone-Poulenc S.A., Paris, France, a corporation of France

PERIMETHAZINE

Therapeutic Function: Tranquilizer

Chemical Name: 1-[3-(2-Methoxyphenothiazin-10-yl)-2-methylpropyl]-4-piperidinol

Common Name: -

Structural Formula:

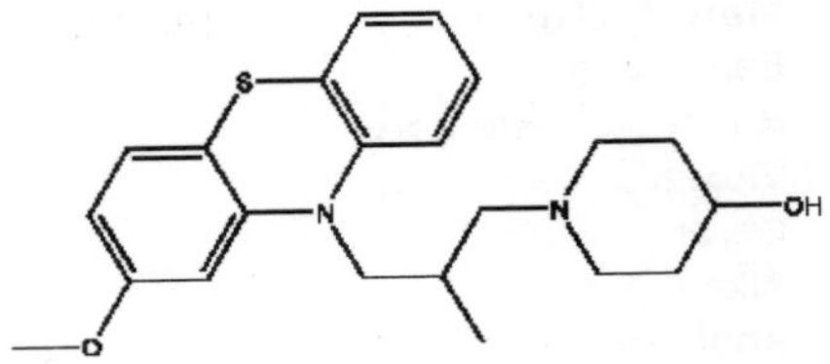

Chemical Abstracts Registry No.: 13093-88-4

Trade Name	Manufacturer	Country	Year Introduced
Leptryl	Roger Bellon	France	1970

Raw Materials

3-Methoxy-10-(3-chloro-2-methylpropyl)phenthiazine
4-Hydroxypiperidine

Manufacturing Process

A solution of 3-methoxy-10-(3-chloro-2-methylpropyl)phenthiazine (9.65 grams) and 4-hydroxypiperidine (6.1 grams) in xylene (10 cc) is heated under reflux for 5 hours. After cooling the mixture is diluted with ether (60 cc) and the basic compounds are extracted by agitation with water (30 cc) and 4 N hydrochloric acid (20 cc). The aqueous acid phase is made alkaline with 4 N sodium hydroxide solution (23 cc) and the liberated base is extracted with ether. The ethereal solution is washed with water (60 cc) and dried over sodium sulfate. Finally the solvent is distilled off on a water-bath.

The solid residue obtained is recrystallized from a mixture (15:85) of benzene and cyclohexane and there is obtained 3-rnethoxy-10-[2-methyl-3-(4-hydroxy-1-piperidyl)-propyl]-phenthiazine (5.7 grams) as a white crystalline powder, MP 137°-138°C.

References

Merck Index 7030
Kleeman and Engel p. 704
DOT 6 (4) 190 (1970)

I.N. p. 748
Jacob, R.M. and Robert, J.G.; US Patent 3,075,976; January 29, 1963; assigned to Societe des Usines Chimiques Rhone-Poulenc, France

PERINDOPRIL ERBUMINE

Therapeutic Function: Antihypertensive

Chemical Name: 1H-Indole-2-carboxylic acid, octahydro, 1-{2-[(1-ethoxycarbonyl)butyl)amino]-1-oxopropyl}-, (2S-(1(R*(R*)),2-α,3a-β,7a-β))-, compd. with 2-methyl-2-propanamine (1:1)

Common Name: Perindopril erbumine

Structural Formula:

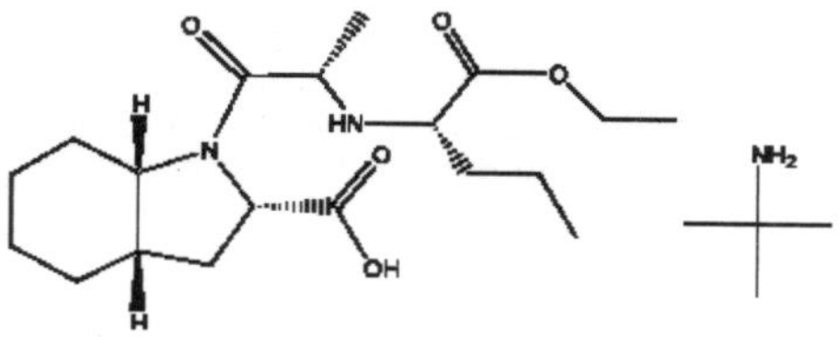

Chemical Abstracts Registry No.: 107133-36-8; 82834-16-0 (Base)

Trade Name	Manufacturer	Country	Year Introduced
Aceon	Solvay Pharmaceuticals Inc.	USA	-
Acertil	Les Laboratoires Servier	France	-
Coverex	Egis Pharmaceuticals Ltd.	Hungary	-
Coversum	Les Laboratoires Servier	France	-
Coversum	Itherapia	Germany	-
Coversyl	Servier	France	-
Coversyl	Serdia Pharmaceuticals (India) Ltd.	India	-
Coversyl	Paranova Oy	Finland	-
Perigard	Glenmark Pharmaceuticals Ltd.	India	-
Perindopril Erbumine	Les Laboratoires Servier	France	-
Prestarium	Les Laboratoires Servier	France	-
Prexanil	Servier Pharma	France	-

Raw Materials

2-Carboxyindole
Sodium hydroxide
Rhodium
Sulfuric acid
(+)-α-Methylbenzylamine
Pyruvic acid

Palladium on charcoal
4-Toluenesulfonic acid
L-Norvaline
1-Hydroxybenzotriazole
Hydrogen
Triethylamine
Tin
Thionyl chloride
Dicyclohexylcarbodiimide

Manufacturing Process

Heat 5 kg of 2-carboxyindole suspended in ethanol in the presence of sulfuric acid to boiling for 8 hours. Evaporate off take up the crystalline mass with hexane. After filtering off and drying, 5.3 kg of 2-ethoxycarbonylindole crystals are obtained. Melting point: 123°-125°C.

Suspend, in a reactor, 10 kg of 2-ethoxycarbonylindoline obtained previously in 110 liters of hydrochloric ethanol. Next, add 20 kg of granulated tin. Keep stirring for approximately 2 days at room temperature. Evaporate off the ethanol, take up the residue with water and add 110 liters of toluene. Stir for approximately 20 min. Alkalify with aqueous ammonia. Separate off the aqueous phase and extract once again with 150 liters of toluene. Combine the toluene phases and wash them with water. Separate off the toluene phases, filter. Remove the water by distilling the water-toluene azeotrope. Cool and pass through a stream of anhydrous HCl gas. Cool. Evaporate down and wash with pure toluene. Weight obtained of (R,S)-2-ethoxycarbonylindoline 10.11 kg. Yield: 84%.

2.15 kg of (R,S)-2-ethoxycarbonylindoline dissolved in ethanol are saponified with 12.5 liters of sodium hydroxide with stirring for 24 hours. After washing the alkaline solution, neutralize with concentrated hydrochloric acid. After filtering off, washing and drying, 1.57 kg of white crystals of the (R,S)-2-carboxyindoline are obtained. Yield: 86%. Melting point: 188°-189°C.

6.05 kg of (R,S)-2-carboxyindoline are added to a solution of 4.49 kg of (+)-α-methylbenzylamine in anhydrous ethanol. A white precipitated product is obtained which, after filtering off, is digested in refluxing isopropanol. After cooling, the solid is filtered off and washed with a little isopropanol. 1 kg of the obtained salt was dissolved in 5 liters of water and neutralizing with an aqueous hydrochloric acid solution. The precipitate is filtered off, washed with water and dried and (S)-2-carboxyindoline was prepared.

Place 25 kg of (S)-2-carboxyindoline, obtained previously, in 110 liters of methanol in a vessel. Keep stirred. Charge the rhodium (5% dry) catalyst into a mixer. Start up the stirring in a hydrogenator, charge the methanolic suspension of (S)-2-carboxyindoline by passing it through the mixer and rinse the assembly with water. Heat to 60°C and pressurize with hydrogen (30 bars). Filter off the catalyst on a single-plate filter. Collect the hydroalcoholic liquors in a reactor and evaporate the methanol off under vacuum. After concentrating, charge approximately 300 kg of dioxane. Heat to boiling and add water until a solution is obtained. Allow to cool. Filter off and dry. 22.3 kg of crystals of (2S,3aS,7aS)-2-carboxyoctahydroindole are obtained. Yield: 86.1%.

Place 35 kg of L-norvaline in approximately 300 kg of denatured ethanol in a reactor. Introduce approximately 60 kg of thionyl chloride, slowly and gradually. After stirring for a quarter of an hour, heat to reflux for 3 hours and then evaporate off the ethanol under vacuum. Take up the residue with 300 liters of cyclohexane and heat to boiling. Allow to cool, filter, wash with cyclohexane and dry. 52.9 kg of ethyl L-norvalinate hydrochloride are obtained, that is a 97.6% yield.

Place 45 kg of ethyl N-norvalinate hydrochloride approximately 110 liters of water in a vessel equipped with a stirrer. Alkalify, then pour 23 kg of pyruvic acid very gradually into the solution obtained previously and stir the reaction mixture for 30 min. Place an aqueous suspension of charcoal containing 5% palladium and the alkaline solution of ethyl L-norvalinate obtained previously in a hydrogenation apparatus. Hydrogenate under pressure (30 bars) at room temperature for approximately one day. Filter under vacuum and evaporate the filtrate under reduced pressure, filter off and dry. Treat the residue obtained with ethanol; remove the insoluble material, consisting of sodium chloride, by filtration and rinse it with ethanol. Combine the ethanolic solutions; evaporate off the ethanol under reduced pressure and crystallize the residue from acetonitrile 34.3 kg of N-[(S)-1-carbethoxybutyl]-(S)-alanine are obtained, that is a 63.9% yield.

In a 30-liter reactor, reflux 12.5 kg of (2S,3aS,7aS)-2-carboxyperhydroindole, 50 kg of para-toluenesulfonic acid and 14.2 kg of benzyl alcohol and 38.4 kg of toluene, removing the water formed with the aid of a continuous separator. When no more water separates out, cool, filter off the precipitate of para-toluenesulfonate of the benzyl ester of (2S,3aS,7aS)-2-carboxyoctahydroindole formed, and dry. Yield: 91.3%.

Add approximately 3.5 kg of triethylamine to a suspension of approximately 5 kg of para-toluenesulfonate of the benzyl ester of (2S,3aS,7aS)-2-carboxyoctahydroindole in approximately 60 kg of ethyl acetate, followed by approximately 6 kg of 1-hydroxybenzotriazole, approximately 7.5 kg of the N-[(S)-1-carbethoxybutyl]-(S)-alanine and approximately 7.0 kg of dicyclohexylcarbodiimide. Stir, cooling slightly for approximately 3 hours, then filter off the dicyclohexylurea formed and wash the organic phase with water. The dried organic phase is evaporated to dryness and benzyl ester of (2S,3aS,7aS)-1-{2-[1-(ethoxycarbonyl)-(S)-butylamino]-(S)-propionyl}octahydroindole-2-carboxylic acid was obtained. Yield: 92.3%.

Dissolve, in a hydrogenator, 14 kg of benzyl ester of the (2S,3aS,7aS)-1-{2-[1-(ethoxycarbonyl)-(S)-butylamino]-(S)-propionyl}octahydroindole-2-carboxylic acid in cyclohexane. Add the charcoal containing 5% palladium and approximately 50 liters of water. Hydrogenate at ordinary temperature and pressure until the theoretical volume of hydrogen has been absorbed. Filter, wash the insoluble material with cyclohexane, separate off the organic phase and wash the aqueous phase again with cyclohexane. Isolate the (2S,3aS,7aS)-1-{2-[1-(ethoxycarbonyl)-(S)-butylamino]-(S)-propionyl}octahydroindole-2-carboxylic acid from the aqueous phase by freeze-drying.

In practice it is used combined with 2-methyl-2-propanamine.

References

Vincent M. et al.; US Patent No. 4,914,214; April 3, 1990; Assigned:Adir Et Cie, Neuilly-sur-Seine, France

PERISOXAL CITRATE

Therapeutic Function: Antiinflammatory, Analgesic

Chemical Name: 3-(2-Piperidino-1-hydroxyethyl)-5-phenylisoxazole citrate

Common Name: -

Structural Formula:

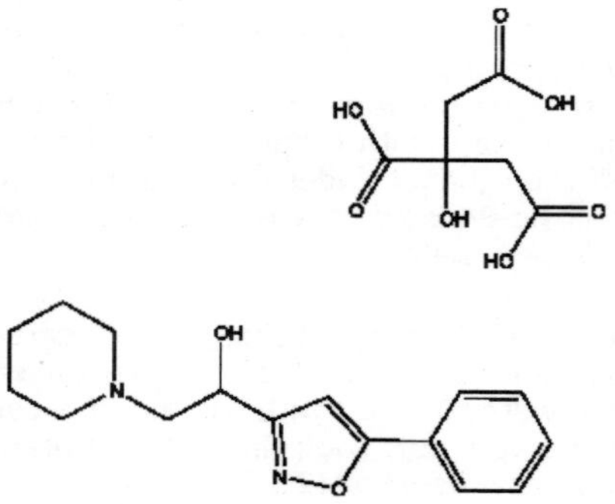

Chemical Abstracts Registry No.: 2055-44-9 (Base)

Trade Name	Manufacturer	Country	Year Introduced
Isoxal	Shionogi	Japan	1979

Raw Materials

Citric acid
3-(2-Methylthio-2-piperidinoacetyl)-5-phenylisoxazole
Sodium borohydride

Manufacturing Process

Crude crystals of 3-(2-methylthio-2-piperidinoacetyl)-5-phenilisoxazole (1.631 g) are suspended in 20 ml of methanol without being further purified and the suspension is stirred after a portionwise addition (in about 10 minutes) of 143 mg (3.78 mmol) of sodium borohydride at room temperature for about 30 minutes.

The methanol in the reaction mixture (pale yellow solution) is then removed by evaporation under reduced pressure to leave a residue which is subsequently dissolved in 30 ml of benzene. The benzene solution is shaken four times with 20 ml of 4 N hydrochloric acid each time to extract the basic substance. Each of the hydrochloric acid layers is washed once with 20 ml of benzene and combined together to be neutralized with potassium carbonate while being ice-cooled until it becomes basic (pH = 10).

The liberated crystalline substance is extracted twice with 50 ml of dichloromethane each time. After being separated, the dichloromethane layers are combined and washed once with 30 ml of water and dried over sodium sulfate. The solvent of the layer is removed by evaporation under reduced pressure to leave a crystalline residue (72.56 mg, 53% crude yield).

Recrystallization of this product from dichloromethane-ether (1:4) affords needles of 3-(2-piperidino-1-hydroxyethyl)-5-phenylisoxazole (701 mg, 51.3% as an overall yield calculated based on the starting material, melting point 104°C to 106°C. The product thus obtained may be reacted with citric acid to give the citrate.

References

Merck Index 7038
DFU 4 (4) 269 (1979)
I.N. p. 748
Hirai, S. and Kawata, K.; US Patent 3,939,167; February 17, 1976; assigned to Shionogi and Co., Ltd.

PERLAPINE

Therapeutic Function: Hypnotic

Chemical Name: 6-(4-Methyl-1-piperazinyl)-11H-dibenz[b,e]azepine

Common Name: 6-(4-Methyl-1-piperazinyl)morphanthridine

Structural Formula:

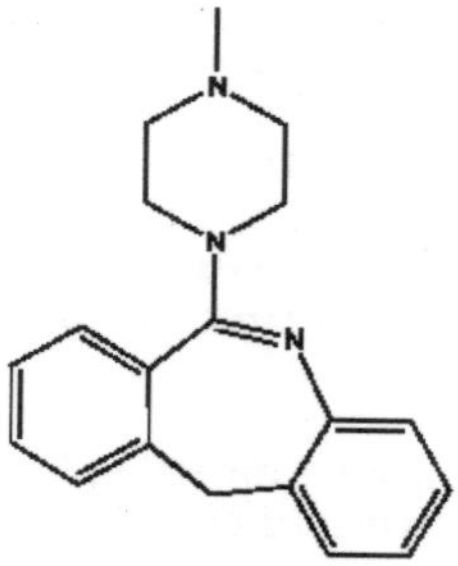

Chemical Abstracts Registry No.: 1977-11-3

Trade Name	Manufacturer	Country	Year Introduced
Hypnodin	Takeda	Japan	1974
Pipnodine	Takeda	Japan	-

Raw Materials

o-Aminodiphenylmethane
N-Methylpiperazine
Phosphorus oxychloride
Aluminum chloride
Phosgene

Manufacturing Process

The 5.6-dihydro-6-oxo-morphanthridine used as a starting material is usefully obtained in the following way. 30.2 grams of o-aminodiphenylmethane are dissolved in 65 ml of absolute toluene and, while stirring and at a temperature of between 0° and -10°C, 140 ml of 20% phosgene solution in toluene are added drop by drop. By bubbling phosgene slowly through it the milky mixture is heated within 30 minutes to reflux temperature, which is maintained during some 20 minutes. While stirring vigorously, dry nitrogen is passed into the boiling reaction mixture for 10 minutes. After evaporation of the solvent there are obtained by vacuum distillation 29.7 grams (86% of the theory of o-isocyanatodiphenylmethane of boiling point 169°C/12 mm Hg.

21.1 grams of aluminum chloride are heated in 110 ml of o-dichlorobenzene to 80°C and, while stirring, a solution of 29.7 grams of o-isocyanatodiphenylmethane in 60 ml of o-dichlorobenzene is added drop by drop, whereupon the temperature of the mixture rises to 120°C. This temperature is maintained for one hour while stirring. After cooling the reaction mixture is poured into 200 ml of 2 N hydrochloric acid, whereupon a brown precipitate is formed. After steam distillation the residue is isolated by filtration and crystallized from acetone/water. There are obtained 28.6 grams (97% of the theory) of 5,6-dihydro-6-oxo-morphanthridine of melting point 201°-203°C.

A mixture of 4.9 grams of 5,6-dihydro-6-oxo-morphanthridine, 37 ml of phosphorus oxychloride and 1.5 ml of dimethylaniline is heated for 3 hours at reflux, The viscous oil, obtained by evaporation of the reaction mixture in vacuo at 60°C, is diluted with 20 ml of absolute dioxane and, after adding 30 nil of N-methylpiperazine, heated for 4 hours at reflux. The resulting clear solution is evaporated in vacuo at 60°C to dryness. The residue is distributed between ether and ammonia water. The ethereal solution is separated, washed with water and then extracted with 1 N acetic acid. The acetic acid extract is mixed with ammonia water and then extracted with ether. The ethereal solution is washed with water, dried over sodium sulfate, filtered through alumina and evaporated.

The residue is caused to crystallize from ether/petroleum ether, and recrystallized from acetone/petroleum ether. 6.0 grams (88% of the theory) of 6-(4-methyl-1-piperazinyl)-morphanthridine of melting point 138°-138.5°C are obtained.

References

Merck Index 7040
Kleeman and Engel p. 705
OCDS Vol. 2 p. 425 (1980)
DOT 11 (2) 76 (1975)
I.N. p. 748

Schmutz, J., Hunziker, F. and Kunzle, F.M.; US Patent 3,389,139; June 18, 1968; assigned to Dr. A. Wander, SA, Switzerland

PERPHENAZINE

Therapeutic Function: Tranquilizer

Chemical Name: 4-[3-(2-Chlorophenothiazin-10-yl)propyl]-1-piperazineethanol

Common Name: Chlorpiprazine

Structural Formula:

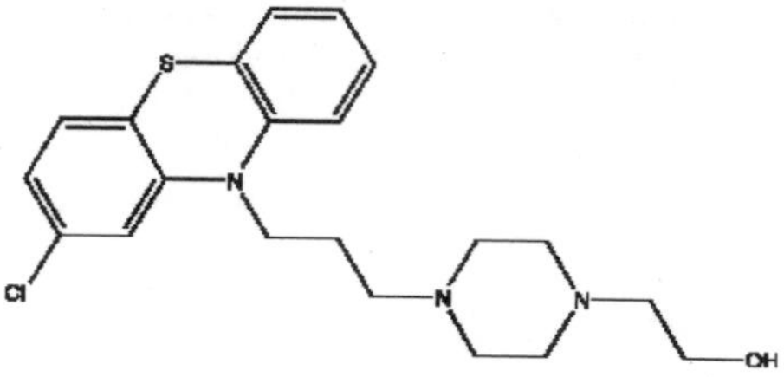

Chemical Abstracts Registry No.: 58-39-9

Trade Name	Manufacturer	Country	Year Introduced
Trilafon	Schering	US	1957
Decentan	Merck	W. Germany	-
Etrafon	Schering	US	-
Fentazin	Allen and Hanburys	UK	-
F-Mon	Nippon Shinyaku	Japan	-
Peratsin	Farmos	Finland	-
Perfenil	Scalari	Italy	-
Perphenan	Taro	Israel	-
Phenazine	I.C.N.	Canada	-
Triavil	MSD	US	-
Trilifan	Cetrane	France	-
Triomin	Yamanouchi	Japan	-

Raw Materials

2-Bromoethanol
Piperazine
1-Bromo-3-chloropropane
2-Chlorophenothiazine

Manufacturing Process

A mixture of 155 parts of 2-chloro-10-(γ-chloropropyl)phenothiazine, 76 parts of sodium iodide, 216 parts of piperazine and 2,000 parts of butanone is refluxed for 8 hours, concentrated and extracted with dilute hydrochloric acid. The extract is rendered alkaline by addition of dilute potassium carbonate and benzene or chloroform extracted. This extract is washed with water, dried over anhydrous potassium carbonate, filtered and evaporated. Vacuum distillation at 0.1 mm pressure yields 2-chloro-10-[γ-(N-piperazino)propyl]phenothiazine at about 214°-218°C.

A stirred mixture of 5 parts of 2-chloro-10-[γ-(N-piperazino)propyl]phenothiazine, 1.92 parts of 2-bromoethanol, 2.11 parts of potassium carbonate and 35 parts of toluene is refluxed for 5 hours. The mixture is treated with water and benzene and the organic layer is separated, washed with water, dried over anhydrous potassium carbonate, filtered and evaporated. The residue is distilled at about 240°-244°C and 0.15 mm pressure to yield 2-chloro-10-[γ-(N'-β-hydroxyethyl-N-piperazino)-propyl]phenothiazine according to US Patent 2,838,507.

The 2-chloro-10-(γ-chloropropyl)phenothiazine starting material is produced from 2-chlorophenothiazine and 1-bromo-3-chloropropane.

References

Merck Index 7044
Kleeman and Engel p. 705
PDR pp. 1217, 1617, 1655
OCDS Vol. 1 p. 383 (1977)
DOT 9 (6) 228 (1973)
I.N. p. 749
REM p. 1090
Cusie, J.W. and Hamilton, R.W.; US Patent 2,838,507; June 10, 1958; assigned to G.D. Searle and Co.
Sherlock, M.H. and Sperber, N.; US Patent 2,860,138; November 11, 1958; assigned to Schering Corporation

PERUVOSIDE

Therapeutic Function: Cardiotonic

Chemical Name: Card-20(22)-enolide, 3-((6-deoxy-3-O-methyl-α-L-glucopyranosyl)oxy)-14-hydroxy-19-oxo-, (3β,5β)-

Common Name: Peruvoside

Chemical Abstracts Registry No.: 1182-87-2

Trade Name	Manufacturer	Country	Year Introduced
Peruvoside	Indena	-	-
Peruvoside	ZYF Pharm Chemical	-	-

Structural Formula:

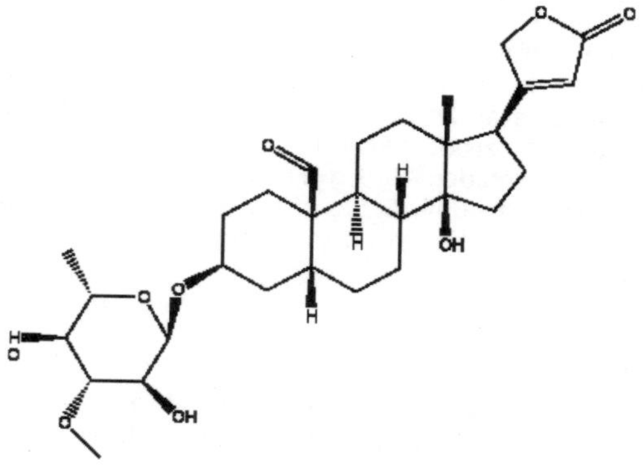

Raw Materials

Fruits or seeds of ApocynaceaThevetia peruviana
Column of silica gel

Manufacturing Process

Peruvoside is obtained employing the fruit or seeds of ApocynaceaThevetia peruviana by the fermentation and the separation the extracted glucoside mixture by chromatography.

Two kilograms of the ground fruit of ApocynaceaThevetia peruviana is mixed with 100 g of grain chaff and moistened with 900 ml of hot water (approximately 60°C). The mixture is mixed with 20 ml of toluene and maintained at 45°-55°C in a closed vessel for 5 days. The thus fermented material is extracted six times with 1200 ml portions of acetone. The combined extracts are concentrated at 30°C under reduced pressure, until there remains about 600 ml of a dark colored aqueous concentrate. The latter is shaken out with an equal volume of petroleum ether. The organic phase is discarded after recovering the petroleum ether. The thus degreased aqueous concentrate is extracted six times with 500 ml portions of dichloromethane. The dichloromethane extracts are dried over sodium sulfate, combined, and concentrated to about 100 ml. Then the residue is stirred into 250 ml of petroleum ether. After standing overnight, the thus-separated crystallized product is vacuum-filtered, washed petroleum ether and dried at about 40°C. The thus obtained crude glycoside mixture (21 g) is dissolved in a mixture of chloroform/methanol and chromatographed on a column of silica gel. There is thus isolated a small amount of oil and fat, a total of 15.5 g of cerberine, acetylperuvoside and neriifolin, 2.2 g of pure peruvoside, MP: 160°-163°C; $[\alpha]_{22}^{d}$=70° (c = 1.3 in CH_3OH), corresponding to a yield 0.11%, based on the quantity of fruit of Thevetia peruviana employed as the starting material.

One kilogram of ground seeds of Thevetia peruviana is mixed with 300 g of grain chaff and moistened with 500 ml of hot water (60°C). After addition of 10 ml of toluene the mixture is allowed to stand in a sealed vessel for 5 days at 45°-55°C. Thereafter, the moist drug material is extracted six times with

500 ml portions of methanol. The extracts are concentrated at about 30°C to about 300 ml. Without degreasing, glycosides are extracted from this aqueous concentrate with three 300 ml portions followed by three 100 ml portions of chloroform. The chloroform extracts are concentrated at about 40°C to a volume of about 200 ml. This concentrate is then mixed with 500 ml of petroleum ether. After standing overnight, the crystallized product, which separated, is vacuum-filtered, washed with petroleum ether, and dried at 40°C. Yield of rude glycoside: 46.9 g. From this product, after the usual separation by chromatography, 5.1 g of pure peruvoside is obtained, MP: 161°-164°C. Yield 0.51%, based on the weight of the non-degreased seeds of Thevetia peruviana.

References

Balsam et al.; US Patent No. 3,713,980; Jan. 30, 1973; Assigned to Merck Patent Gesellshaft mit beschrankter Haftung, Darmstadt, Germany

PHANQUINONE

Therapeutic Function: Antiamebic

Chemical Name: 4,7-Phenanthroline-5,6-quinone

Common Name: Fanquinonum; Phanchinonum; Phanquinone; Phanquone

Structural Formula:

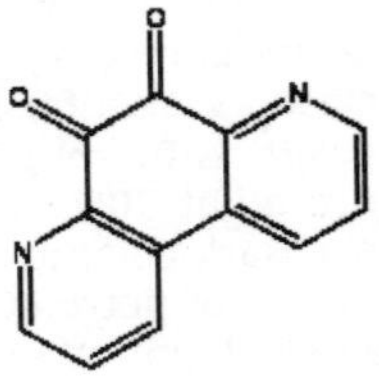

Chemical Abstracts Registry No.: 84-12-8

Trade Name	Manufacturer	Country	Year Introduced
Phanquinone	Yick-Vic Chemicals and Pharmaceuticals (HK) Ltd.	-	-
Phanquinone	ZYF Pharm Chemical	-	-

Raw Materials

Sulfuric acid
6-Methoxy-4:7-phenanthroline
Nitric acid
Caustic soda

Manufacturing Process

2 parts of 6-methoxy-4:7-phenanthroline are mixed with 10 parts by volume of concentrated sulfuric acid and while cooling with a mixture of ice and sodium chloride, with 6 parts by volume of fuming nitric acid (density = 1.51), and the whole is heated for 2 h at 120°C. The reaction solution is poured on to ice, its pH value is adjusted to 7 by means of a 10 N solution of caustic soda, after standing for 2 h the whole is filtered with suction to remove the precipitate which separates, and the latter is washed with hot water. After recrystallising the product from methyl alcohol and drying it at 100°C under 0.1 mm pressure, there are obtained 1.8 parts (i.e. 90 % of the calculated yield) of 4:7-phenanthroline-5:6-quinone in the form of pale yellow crystals melting at 295°C.

References

GB Patent No. 688,802; March 11, 1953; Assigned: CIBA Limited, a body corporate, Basle, Switzerland

PHENACAINE

Therapeutic Function: Local anesthetic

Chemical Name: Ethanimidamide, N,N'-bis(4-ethoxyphenyl)-

Common Name: Fenacaine; Phenacaine; Tanicaine

Structural Formula:

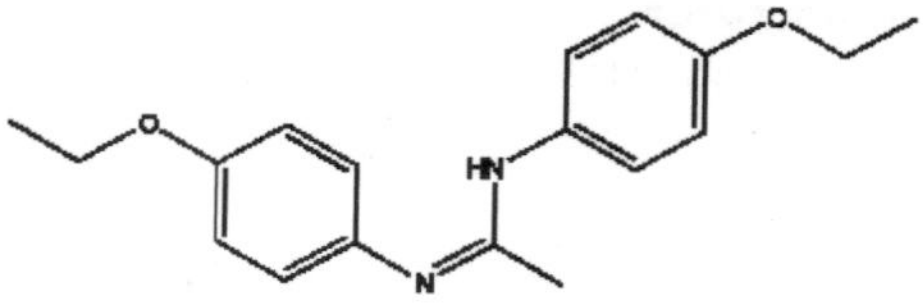

Chemical Abstracts Registry No.: 101-93-9

Trade Name	Manufacturer	Country	Year Introduced
Phenacaine	ZYF Pharm Chemical	-	-

Raw Materials

Hydrogen chloride
Phenacetin (4-ethoxyacetanilide)

Manufacturing Process

Phenacetin (4-ethoxyacetanilide) was treated with slowly current of gaseous hydrogen chloride for 15 hours at 150°C. On cooling the product was poured into 20 volumes of water and heated. Then it was cooled to room temperature, filtered off and excess of alkali was added. The precipitated crude product was at first re-crystallized from 60% ethanol, then from benzene-ligroin to give pure acetamidine as a white needles; MP: 121°C.

References

Tauber E.; D.R. Patent No. 79,868; March 16, 1894

PHENACETIN

Therapeutic Function: Analgesic

Chemical Name: Acetamide, N-(4-ethoxyphenyl)-

Common Name: Acetophenetidin

Structural Formula:

Chemical Abstracts Registry No.: 62-44-2

Trade Name	Manufacturer	Country	Year Introduced
Phenacetin	Environmental Health and Safety	-	-

Raw Materials

4-Ethoxyaniline
Acetic anhydride
Sodium hydrosulfite

Manufacturing Process

A mixture of 10 g of 4-ethoxyaniline and 8.6 g of acetic anhydride in 28 g of dry benzene was refluxed for 4 hours. To the reaction mixture was added a small amount of $Na_2S_2O_4$. After cooling the phenacetin was crystallized; yield 12.5 g (96%), M.P. 136°C.

References

Merck Index, Monograph number: 7344, Twelfth edition, 1996, Editor: S. Budavari; Merck and Co., Inc.

Friedlander, Berichte, 1893, 25, 178
Lumiere A., 1906, [3], 33, 785

PHENACTROPINIUM CHLORIDE

Therapeutic Function: Antihypertensive

Chemical Name: α-Hydroxybenzeneacetic acid 8-methyl-8-[(2-oxo-2-phenyl)-ethyl]-8-azoniabicyclo[3.2.1]-oct-3-yl ester chloride

Common Name: -

Structural Formula:

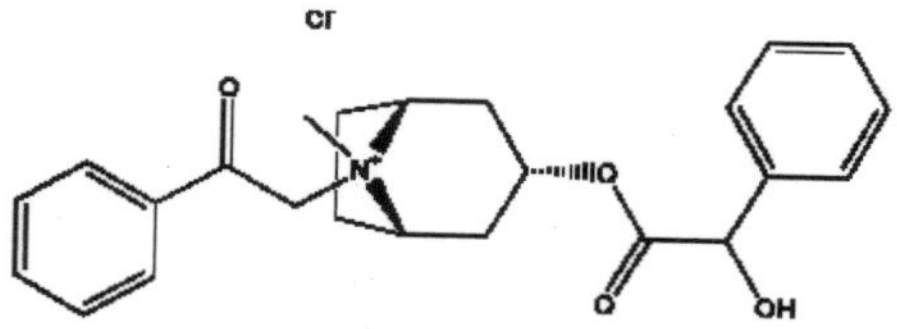

Chemical Abstracts Registry No.: 3784-89-2

Trade Name	Manufacturer	Country	Year Introduced
Trophenium	American Cyanamid (AHP)	US	1961
Trophenium	Duncan Flockhart	UK	-

Raw Materials

Homatropine
Phenacyl chloride

Manufacturing Process

330 g (1.2 M) of homatropine were dissolved in 1 liter of dry methyl ethyl ketone and gently refluxed on a water-bath during the gradual addition of a solution of 204 g (1.32 M) redistilled phenacyl chloride in 200 ml of the same solvent. After 10 to 15 minutes 1 g of previously prepared homatropine phenacyl chloride was added to avoid formation of a supersaturated solution of the quaternary compound. Reflux was continued for 9 hours, then the thick suspension was allowed to cool, filtered and washed with 200 ml methyl ethyl ketone to yield 490 g (95%) slightly creamy solid, MP 188°C to 191°C.

For purification the crude quaternary salt was dissolved in hot ethyl alcohol (2 ml/g) and warm dry acetone (8 ml/g) was stirred into the clear filtrate. On cooling, 387 g (78 % recovery) of a pure white powder, MP 195°C to 197°C, were obtained, in which the ionizable chlorine assayed at 99.7% of the theoretical value.

References

Merck Index 7067
I.N. p. 752
Johnston, R.G. and Spencer, K.E.V.; US Patent 2,828,312; March 25, 1958; assigned to T. and H. Smith, Ltd. (UK)

PHENAGLYCODOL

Therapeutic Function: Tranquilizer

Chemical Name: 2-(4-Chlorophenyl)-3-methyl-2,3-butanediol

Common Name: -

Structural Formula:

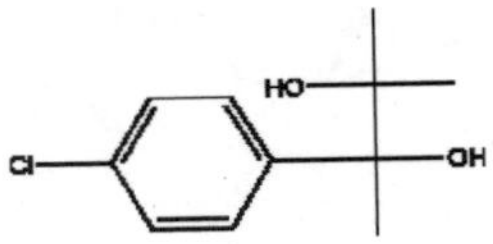

Chemical Abstracts Registry No.: 79-93-6

Trade Name	Manufacturer	Country	Year Introduced
Ultran	Lilly	US	1975
Felixyn	Radiumpharma	Italy	-

Raw Materials

Sodium cyanide
Hydrogen chloride
Ethanol
Magnesium
p-Chloroacetophenone
Sodium hydroxide
Methyl iodide

Manufacturing Process

To a mixture of 460 g of p-chloroacetophenone, 350 ml of ether and 500 ml of water are added 410 g of sodium cyanide, with vigorous stirring. The reaction mixture is cooled to about 5°C to 10°C and 700 ml of concentrated hydrochloric acid are added at such a rate that no hydrogen cyanide is formed and the temperature of the mixture does not rise above 10°C. After the addition of the acid is complete, the reaction mixture is stirred for about three hours at room temperature, and allowed to separate into an aqueous and an organic phase. The organic phase is removed from the aqueous phase, and the aqueous phase and any salt which may have separated in the course of the reaction are washed with about 300 ml of ether. The combined ether washings and organic phase are dried over anhydrous magnesium sulfate, and the ether is removed by evaporation in vacuo at room temperature. The residue is poured with stirring into 800 ml of concentrated hydrochloric acid kept at about 0°C by cooling with solid carbon dioxide. The acid mixture is saturated with gaseous hydrogen chloride at 0°C, and stirred at room temperature overnight. The resulting precipitate of p-chloroatrolactamide is removed by filtration, washed by slurrying with water and dried. After recrystallization from ethanol, p-chloroatrolactamide melts at about 105°C to 107°C.

A mixture of 200 g of p-chloroatrolactamide and 1 liter of 25% sodium hydroxide solution is refluxed with stirring for about sixteen hours. The reaction mixture is then poured over cracked ice and diluted with water to a volume of about 3 liters. The aqueous solution is washed with two 1 liter portions of ether, and acidified with concentrated hydrochloric acid, whereupon a precipitate of p-chloroatrolactic acid forms. The precipitated acid is removed by filtration, and is dissolved in 500 ml of ether, washed with two 250 ml portions of water and dried. The ether is removed by evaporation. p-chloroatrolactic acid thus prepared melts at about 117°C to 120°C.

A mixture of 185 g of p-chloroatrolactic acid, 600 ml of ethanol and 60 ml of concentrated sulfuric acid is refluxed for about twelve hours. About half the solvent is then removed by evaporation in vacuo at room temperature, the residue is poured over cracked ice, and diluted with water to a volume of about 2 liters. The ethyl p-chloroatrolactate formed in the reaction is extracted with two 1 liter portions of ether. The combined ether extracts are washed with successive 200 ml portions of water, 5% sodium carbonate solution, and water, and are dried over anhydrous magnesium sulfate. The dried ether solution is subjected to fractional distillation, and the fraction boiling at about 90°C to 100°C at a pressure of 0.1 mm of mercury, is collected. The distillate consists of ethyl p-chloroatrolactate.

To a solution of 2 mols of methylmagnesium iodide in 1.5 liters of ether are added with vigorous stirring 107 g (0.5 mol) of ethyl p-chloroatrolactate. The reaction mixture is stirred for about sixteen hours, and is then decomposed by the addition of about 320 ml of saturated aqueous ammonium chloride solution. After standing, the ether layer is decanted from the mixture and the aqueous phase and the precipitated salts are washed with several 500 ml portions of ether. The combined ether solution and washings are washed with successive 500 ml portions of 5% ammonium chloride solution and water, are dried over anhydrous magnesium sulfate, and are evaporated to dryness in vacuo. The crystalline residue consisting of 2-p-chlorophenyl-3-methyl-2,3-butanediol, is recrystallized from a mixture of benzene and petroleum ether.

2-p-chlorophenyl-3-methyl-2,3-butanediol thus prepared melts at about 66°C to 67°C.

References

Merck Index 7070
Kleeman and Engel p. 709
OCDS Vol. 1 p. 219 (1977)
I.N. p. 752
Mills, J.; US Patent 2,812,363; November 5, 1957; assigned to Eli Lilly and Co.

PHENAZOCINE

Therapeutic Function: Narcotic analgesic

Chemical Name: 2,6-Methano-3-benzazocin-8-ol, 1,2,3,4,5,6-hexahydro-6,11-dimethyl-3-(2-phenylethyl)-

Common Name: Fenatsokin; Phenazocine; Phenobenzorphan; Xenagol

Structural Formula:

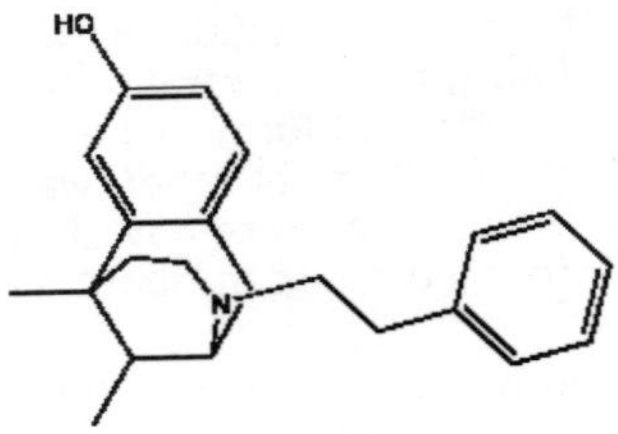

Chemical Abstracts Registry No.: 127-35-5

Trade Name	Manufacturer	Country	Year Introduced
Phenazocine	SmithKline French (GSK)	-	-

Raw Materials

3,4-Lutidine
Methyl iodide
Hydrobromic acid
Ethereal solution of p-methoxybenzylmagnesium chloride
Lithium aluminum hydride
Palladium on barium
Acetic anhydride

Manufacturing Process

25.0 g 3,4-lutidine methyl iodide in 60 ml of dry ethyl ether is stirred while 400 ml of a 0.3958 N ethereal solution of p-methoxybenzylmagnesium chloride is added at room temperature. The mixture is stirred for 30 minutes and then decomposed with a solution of 100 ml of water containing 25 g of ammonium chloride and 10 ml of concentrated ammonium hydroxide. The layers are separated. The organic layer is extracted with a solution of 75 ml of water and 17 ml of concentrated hydrochloric acid. The extracts are neutralized and taken into ether. The volatiles are evaporated to leave a light yellow oil, the dehydro base.

The oily residue is then hydrogenated at 17 p.s.i. of hydrogen with 5% palladium-on-barium sulfate in 100 ml of 2 N hydrochloric acid for six hours. The reaction mixture is filtered, made alkaline and taken through ether to give the tetrahydro base as a clear oil.

The oily tetrahydro base (about 10.0 g) in 150 ml of 48% hydrobromic acid is heated at 135°C for 24 hours, and then quenched in an ice Treating with base and taking through chloroform gives a brown residue of the isomeric mixture of 2'-hydroxy-2,5,9-trimethyl-6,7-benzmorphan.

This residue is triturated with ether, cooled and the resulting slurry filtered. The solid product is dissolved in a minimum of dry ethanol and made acid with ethereal hydrogen chloride. The cooled mixture is filtered to give the hydrochloride salt of the N-methyl-iso-benzmorphan, MP: 279-282°C, after recrystallization from ethanol. The base melts at 215°C.

The ethereal filtrate is evaporated. A residue is neutralized to give the crude normal N-methyl-benzmorphan isomer, MP: 229-230°C. The hydrochloride salt of this isomer is formed, MP: 196-198°C, as a hydrate. A mixture of 10.0 g of the N-methyl-iso-benzmorphan isolated above in 15 ml of acetic anhydride is heated on the steam bath for about an hour, then quenched in an ice slurry. The mixture is then neutraliized and taken through ether to give the O-acetate derivative, iso-2'-acetoxy-2,5,9-trimethyl-6,7-benzomorphan. The crude acetate (9.5 g) is reacted with 5.0 g of cyanogen bromide in 100 ml of chloroform at reflux for several hours. The volatiles are removed in vacuo to leave a residue, which is refluxed in 150 ml of dilute hydrochloric acid for 24 hours. The mixture is cooled, neutralized and taken through chloroform to give the desired base with two methyl groups as a viscous syrup which crystallized slowly, MP: 173-175°C from methanol. The base, 6.5 g, is reacted with 5.0 g of phenylacetyl chloride in the presence of an excess of sodium carbonate in water. The mixture is stirred for several hours, diluted with water and taken into ether to give the N-phenacetylated compound. This compound in ether (250 ml) is reacted with an excess of 1.5 M ethereal lithium aluminum hydride at reflux overnight. The reaction mixture is evaporated to dryness, after quenching carefully with water and hydrobromic acid, to give the crude 2'-hydroxy-5,9-dimethyl-2-phenethyl-6,7-benzomorphan hydrobromide salt which is optionally recrystallized from ethanol, MP: 272-273°C. The hydrobromide salt in the normal series melts at 170-173°C. The base is isolated by neutralizing of the hydrobromide salt in an ether alkali mixture, with following separating and evaporating the organic solvent.

References

Gordon M. et al.; US Patent No. 2,959,594; November 8, 1960; Assigned to Smith Kline and French Laboratories, Philadelphia, Pa., a corporation of Pennsylvania

PHENAZOPYRIDINE HYDROCHLORIDE

Therapeutic Function: Urinary analgesic, Antiseptic, Diagnostic aid

Chemical Name: 2,6-Pyridinediamine, 3-(phenylazo)-, monohydrochloride

Common Name: Phenazopyridine hydrochloride; Azopirin

Chemical Abstracts Registry No.: 94-78-0 (Base); 136-40-3

Structural Formula:

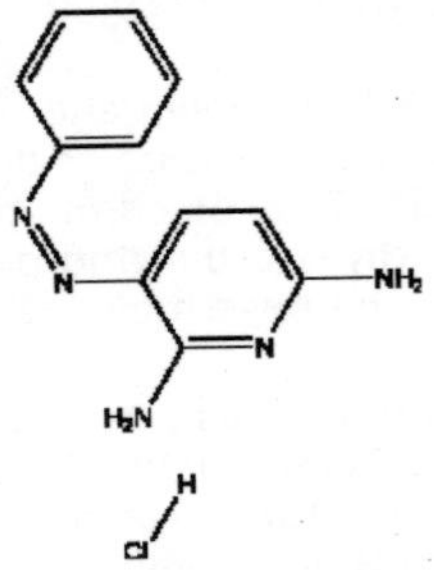

Trade Name	Manufacturer	Country	Year Introduced
Phenazopyridine Hydrochloride	AroKor Holdings Inc.	-	-
Azopirin	Barcymex	-	-
Cystamine	Mc-Clung	-	-
Phenazopyridine Hydrochloride	Xi'an Boojie Pharmaceutical and Chemical Technology Co., Ltd.	-	-
Phenazopyridine Hydrochloride	Azide Chemical Co., Ltd.	-	-
Prodium	Breckenridge	-	-
Sedural	Teva	-	-
Urisept	Kahira	-	-
Urophenyl	Nadeau	-	-
Urologin	Delta	-	-

Raw Materials

Phenyldiazene
2,6-Diaminopyridine

Manufacturing Process

Phenyldiazene chloride reacted with 2,6-diaminopyridine and in the result 2,6-diamino-3-(phenylazo)pyridine was obtained.

In practice it is usually used as monohydrochloride.

References

Kleemann A., Engel J.; Pharmazeutische Wirkstoffe, GeorgThieme Verlag Stuttgart. New York, 1982

PHENDIMETRAZINE TARTRATE

Therapeutic Function: Antiobesity

Chemical Name: 3,4-Dimethyl-2-phenylmorpholine bitartrate

Common Name: 3,4-Dimethyl-2-phenyltetrahydro-1,4-oxazine bitartrate

Structural Formula:

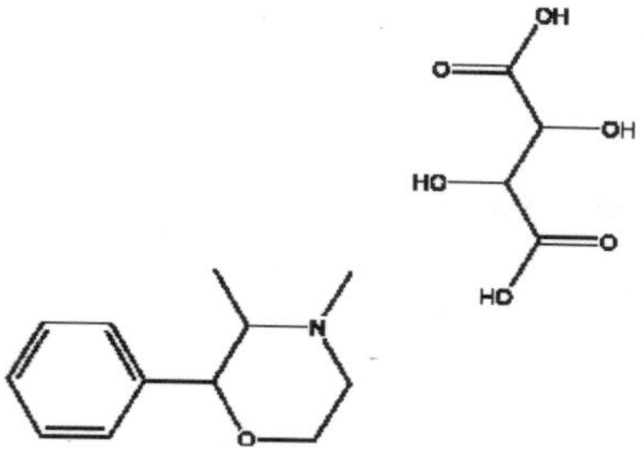

Chemical Abstracts Registry No.: 50-58-8; 634-03-7 (Base)

Trade Name	Manufacturer	Country	Year Introduced
Plegine	Ayerst	US	1961
Statobex	Lemmon	US	1972
Bacarate	Tutag	US	1972
Prelu-2	Boehringer Ingelheim	US	1980
Sprx 105	Tutag	US	1980
Obezine	Western Research	US	1981
X-Trozine	Rexar	US	1981
Hyrex-105	Hyrex	US	1983
Adipost	Ascher	US	1983
Slyn-LL	Edwards	US	1983
Trimcaps	Mayrand	US	1983
Adipo II	Sig	US	-
Adphen	Ferndale	US	-
Amphasub	Palmedico	US	-
Anoxine T	Winston Pharm.	US	-
Arcotrol	Arco	US	-
Bacarate	Reid-Provident	US	-
Bontril	Carnrick	US	-
Di-Ap-Trol	Foy	US	-
Dyrexan	Trimen	US	-
Ephemet	Canright	US	-
Fringanor	Sobio	France	-
Melfiat	Reid-Rowell	US	-
Neo-Nilorex	A.V.P.	US	-

Trade Name	Manufacturer	Country	Year Introduced
Obe-Del	Marlop	US	-
Obepar	Parmed	US	-
Obesan	SCS Pharmalab	S. Africa	-
Obex-LA	Rio Ethicals	S. Africa	-
Pan-Rexin	Pan American	US	-
Phenazine	Jenkins	US	-
Reducto	Arcum	US	-
Reton	Tri-State	US	-
Stodex	Jalco	US	-
Symetra	Westerfield	US	-
Trimstat	Laser	US	-
Wehless	Hauck	US	-
Weightrol	N. Amer. Pharm.	US	-
X-Trozine	Rexar	US	-

Raw Materials

Propiophenone
2-Methylaminomethanol
Bromine
Formic acid

Manufacturing Process

A mixture of 61 grams 1-phenyl-1-oxo-2-(N-methyl-N-ethanolamino)-propane hydrochloride and 100 cc 98-100% formic acid was refluxed at the boiling point at atmospheric pressure for 45 minutes on an oil bath. Thereafter, the oil bath temperature was increased to 180°C and as much of the excess unreacted formic acid as possible was distilled off. A vigorous evolution of carbon dioxide developed during the distillation, which ceased after approximately 45 additional minutes. The honey-yellow syrup which remained as the distillation residue was worked up by admixing it with about six volumes of water and adjusting the aqueous mixture to alkaline reaction with concentrated sodium hydroxide. An oily phase separated out which was extracted with ether. The ether extract was washed with water and dried over potassium carbonate. The solvent was distilled off and the distillation residue was fractionally distilled in vacuo. The base boils at 132°-133°C at 12 mm. The yield was 93% of theory. Reaction with tartaric acid gave the final product.

The starting material is produced by reacting propiophenone with bromine and then reacting the α-bromopropiophenone produced with 2-methylaminomethanol.

References

Merck Index 7088
Kleeman and Engel p. 711
PDR pp. 633, 679, 778, 928, 948, 992, 1448, 1450, 1807
OCDS Vol. 1 p. 260 (1977) and 2, 261 (1980)
I.N. p. 754
REM p. 892

Heel, W. and Zeile, K.; US Patent 2,997,469; August 22, 1961; assigned to C.H. Boehringer Sohn, Germany

PHENELZINE SULFATE

Therapeutic Function: Psychostimulant

Chemical Name: (2-Phenethyl)hydrazine sulfate

Common Name: -

Structural Formula:

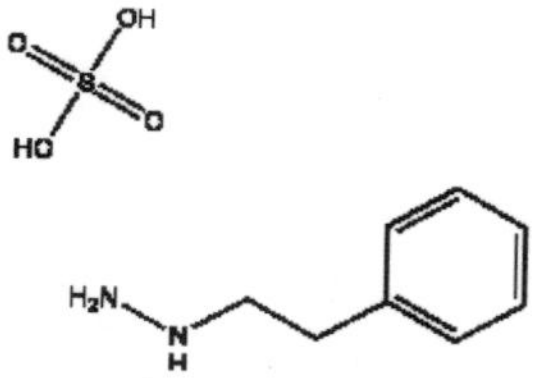

Chemical Abstracts Registry No.: 156-51-4; 51-71-8 (Base)

Trade Name	Manufacturer	Country	Year Introduced
Nardil	Parke Davis	US	1959
Nardelzine	Substantia	France	-

Raw Materials

Phenethyl bromide
Hydrazine hydrate

Manufacturing Process

To a refluxing solution containing 147.5 grams of 85% hydrazine hydrate in 500 cc of ethanol was added, during a period of 5 hours, 92.5 grams of phenethylbromide (0.50 mol) in 150 cc of ethanol. Stirring and refluxing were continued for two hours. The ethanol was removed by distillation and the residue extracted repeatedly with ether. The ether was dried with potassium carbonate and the product base collected by distillation, BP 74°C/0.1 mm, yield 52.3 grams (77%). The base is reacted with sulfuric acid in propanol to give the sulfate.

References

Merck Index 7089
Kleeman and Engel p. 711

PDR p. 1368
OCDS Vol. 1 p. 74 (1977)
I.N. p. 754
REM p. 1096
Biel, J.H.; US Patent 3,000,903; September 19, 1961; assigned to Lakeside Laboratories, Inc.

PHENETHICILLIN POTASSIUM

Therapeutic Function: Antibacterial

Chemical Name: 3,3-Dimethyl-7-oxo-6-[(1-oxo-2-phenoxypropyl)amino]-4-thia-1-azabicyclo[3.2.0]heptane-2-carboxylic acid potassium salt

Common Name: Penicillin MY

Structural Formula:

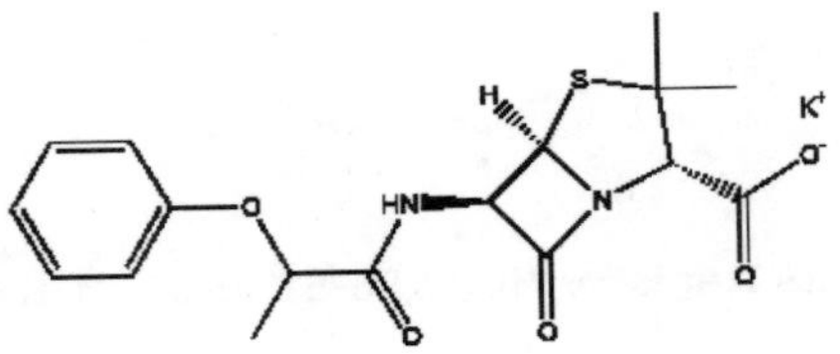

Chemical Abstracts Registry No.: 132-93-4; 147-55-7 (Base)

Trade Name	Manufacturer	Country	Year Introduced
Syncillin	Bristol	US	1959
Ro-Cillin	Rowell	US	1960
Chemiphen	Squibb	US	1960
Semopen	Massengill	US	1960
Dramcillin-S	White	US	1960
Maxipen	Roerig	US	1960
Darcil	Wyeth	US	1960
Alpen	Schering	US	1960
Altocillin	Caber	Italy	-
Bendralan	Antibioticos	Spain	-
Broxil	Beecham	UK	-
Metilpen	Boniscontro-Gazzone	Italy	-
Optipen	C.S.L.	Australia	-
Pen-200	Pfizer	W. Germany	-
Peniplus	Fumouze	France	-
Penopen	Pliva	Yugoslavia	-
Penorale	Lusofarmaco	Italy	-
Synthecilline	Bristol	France	-
Synthepen	Meiji	Japan	-

Raw Materials

α-Phenoxypropionic acid
Isobutyl chloroformate
6-Aminopenicillanic acid
Potassium 2-ethylhexanoate

Manufacturing Process

Triethylamine (1.5 ml) was added to a cold solution (10°C) of α-phenoxypropionic acid (1.66 g, 0.01 mol) in 15 ml of pure dioxane, with stirring and cooling to 5°C to 10°C while isobutyl chloroformate (1.36 g, 0.01 mol) in 5 ml of dioxane was added dropwise. Then the mixture was stirred for ten minutes at 5°C to 8°C. A solution of 6-amino-penicillanic acid (2.16 g, 0.01 mol) in 15 ml of water and 2 ml of triethylamine was then added dropwise while the temperature was maintained below 10°C. The resulting mixture was stirred in the cold for 15 minutes then at room temperature for 30 minutes, diluted with 30 ml of cold water and extracted with ether which was discarded. The cold aqueous solution was then covered with 75 ml of ether and acidified to pH 2 with 5 N H_2SO_4. After shaking, the ether layer containing the product 6-(α-phenoxypropionamido)penicillanic acid, was dried for ten minutes over anhydrous sodium sulfate and filtered. Addition of 6 ml of dry n-butanol containing 0.373 g/ml of potassium 2-ethylhexanoate precipitated the potassium salt of the product as a colorless oil which crystallized on stirring and scratching and was collected, dried in vacuo and found to weigh 2.75 g, to melt at 217°C to 219°C.

References

Merck Index 7093
Kleeman and Engel p. 712
OCDS Vol. 1 p. 410 (1977)
I.N. p. 755
Beecham Research Laboratories, Ltd.; British Patent 877,120; September 13, 1961

PHENFORMIN

Therapeutic Function: Antidiabetic

Chemical Name: N-(2-Phenylethyl)imidodicarbonimidic diamide

Common Name: Phenethyldiguanide

Chemical Abstracts Registry No.: 114-86-3

Raw Materials

β-Phenylethylamine
Dicyandiamide
Hydrogen chloride

Structural Formula:

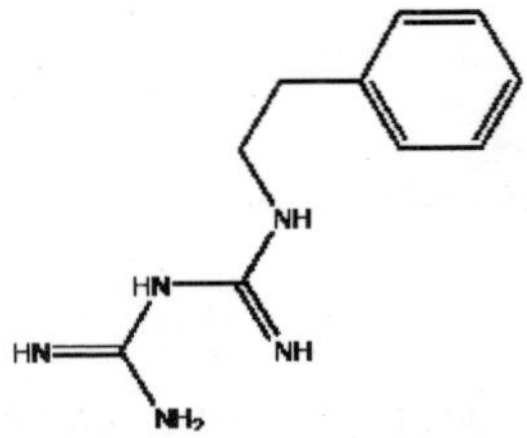

Trade Name	Manufacturer	Country	Year Introduced
DBI	Geigy	US	1959
Meltrol	U.S.V. Pharm.	US	1971
Adiabetin	Arcana	Austria	-
Antipond	Arcana	Austria	-
Cronoformin	Guidotti	Italy	-
De Be J	Isa	Brazil	-
Debeone	U.S.V.	US	-
Diabis	Funk	Spain	-
Dibein	Pharmacia	Sweden	-
Dibophen	Polfa	Poland	-
Insoral	U.S.V.	US	-
Kataglicina	Marxer	Italy	-
Prontoformin	Guidotti	Italy	-

Manufacturing Process

15.76 g of β-phenylethylamine hydrochloride and 8.4 g of dicyandiamide were ground and intimately mixed. The mixture was heated in an oil bath in a 3-neck flask fitted with a thermometer and stirrer, and the mixture began to melt at a bath temperature of 125°C and was completely fluid at 130°C. Further heating at 145°C to 150°C initiated an exothermic reaction and the temperature of the fusion mixture (156°C) exceeded the oil bath temperature (150°C) by 6°. Heating was continued for one hour at bath temperature of 148°C to 150°C. The reaction mixture was cooled, dissolved in about 100 cc of methanol and filtered. The methanol filtrate was concentrated under reduced pressure, cooled and the product (β-phenylethylbiguanide hydrochloride) filtered off and recrystallized from 95% isopropanol.

References

Merck Index 7099
OCDS Vol. 1 p. 75 (1977)
I.N. p. 755
Shapiro, S.L. and Freedman, L.; US Patent 2,961,377; November 22, 1960; assigned to US Vitamin and Pharmaceutical Corp.

PHENGLUTARIMIDE HYDROCHLORIDE

Therapeutic Function: Anticholinergic, Antiparkinsonian

Chemical Name: Glutarimide, 2-(2-(diethylamino)ethyl)-2-phenyl-, hydrochloride

Common Name: Phenglutarimide hydrochloride

Structural Formula:

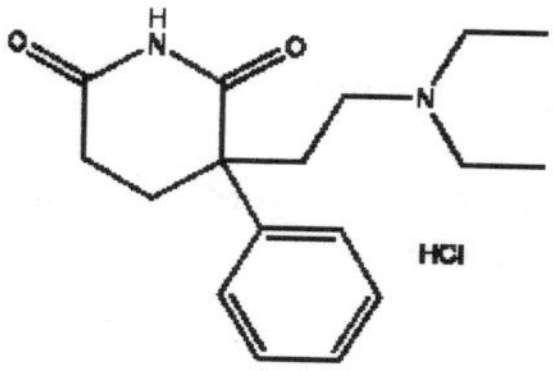

Chemical Abstracts Registry No.: 1674-96-0; 1156-05-4 (Base)

Trade Name	Manufacturer	Country	Year Introduced
Ciba 10870	Ciba Pharmaceutical Products, Inc.	-	-

Raw Materials

Acetic acid
Acetic anhydride
Sulfuric acid
Hydrochloric acid
Potassium salt of 2-phenyl-2-(β-diethylaminoethyl)-pentane-1,5-diacid mononitrile

Manufacturing Process

350 parts by weight of the potassium salt of 2-phenyl-2-(β-diethylaminoethyl)-pentane-1,5-diacid mononitrile are dissolved with heating in 700 parts by volume of glacial acetic acid, 850 parts by volume of acetic anhydride are added, and then 250 parts by volume of concentrated sulfuric acid introduced portionwise. The temperature of the reaction mixture in this operation rises to 120-130°C. When the reaction subsides, the whole is finally maintained for a further 15 min on the boiling water bath. The solvent is removed on the water bath under reduced pressure, the residue poured onto ice and caustic soda solution, and the whole extracted with chloroform. The chloroform solution washed with water, dried over potassium carbonate and the solvent evaporated. The crystalline residue, consisting of 3-phenyl-3-(β-diethylaminoethyl)-2,6-dioxopiperidine. After recrystallization from a mixture of ethyl acetate and ligroin, melts at 118-120°C.

The hydrochloride (produced by dissolving the base in ethyl acetate and adding an equivalent quantity of hydrochloric acid gas dissolved in ethyl acetate) melts, after recrystallization from a mixture of methyl alcohol and ethyl acetate, at 168-172°C.

References

Hoffmann K., Tadmann E.; US Patent No. 2,664,424; Dec. 29, 1953; Assigned to Ciba Pharmaceutical Products, Inc., Summit, N.J.

PHENINDAMINE TARTRATE

Therapeutic Function: Antihistaminic

Chemical Name: 2,3,4,9-Tetrahydro-2-methyl-9-phenyl-1H-indeno[2,1,c]pyridine tartrate

Common Name: 2-Methyl-9-phenyl-2,3,4,9-tetrahydro-1-pyridindene tartrate

Structural Formula:

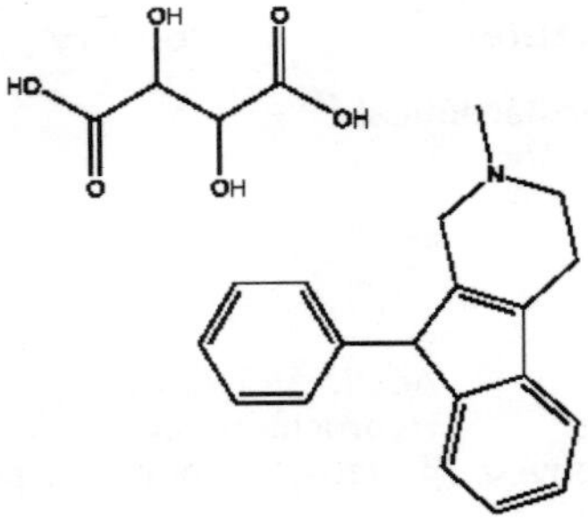

Chemical Abstracts Registry No.: 569-59-5; 82-88-2 (Base)

Trade Name	Manufacturer	Country	Year Introduced
Thephorin	Roche	US	1947
Nolahist	Carnrick	US	-
Nolamine	Carnrick	US	-
Pernovin	Chinoin	Hungary	-
PV-Tussin	Reid-Rowell	US	-

Raw Materials

Acetophenone
Formaldehyde
Hydrogen bromide
Potassium thiocyanate
Methylamine
Sodium hydroxide
Hydrogen

Manufacturing Process

A mixture of 750 grams of 1-methyl-3-benzoyl-4-hydroxy-4-phenylpiperidine and 2,500 cc of 48% hydrobromic acid is refluxed for about 20 minutes. It is then poured into 8 liters of water. An oily precipitate appears which on standing crystallizes. It is filtered and crystallized from about 3.5 liters of alcohol. 2-Methyl-9-phenyl-2,3-dihydrel-pyridindene hydrobromide, MP 201°-203°C, is obtained.

A mixture of 680 grams of 2-methyl-9-phenyl-2,3-dihydrol-pyridindene hydrobromide, 6,000 cc of water and about 100 grams of Raney-nickel catalyst is hydrogenated at room temperature and at about 1,000 lb pressure for a period of three hours. The catalyst is filtered. The clear filtrate is treated with a solution of 240 grams potassium thiocyanate in 400 cc of water. A heavy solid precipitates from which the supernatant liquid is decanted.

The residue is dissolved in 10 liters of boiling alcohol with stirring in the presence of nitro gen. The solution is cooled to room temperature under nitrogen, and then allowed to stand overnight. 2-Methyl-9-phenyl-tetrahydro-1-pyridindene thiocyanate separates in crystals of MP 188°-189°C. From the concentrated filtrate an additional amount is obtained. The corresponding free base, prepared by treating the slightly soluble thiocyanate in aqueous suspension with sodium hydroxide and extracting with ether, has a MP of 90°-91°C. It forms a tartrate of MP 160°C.

The starting material was prepared by reacting acetophenone, methylamine and formaldehyde followed by treatment of the intermediate with sodium hydroxide.

References

Merck Index 7103
Kleeman and Engel p. 713
PDR pp. 781, 1448
I.N. p. 756
Plati, J.T. and Wenner, W.; US Patent 2,470,108; May 17,1949; assigned to Hoffmann-La Roche Inc.

PHENIPRAZINE

Therapeutic Function: Antihypertensive

Chemical Name: (1-Methyl-2-phenylethyl)hydrazine

Common Name: -

Chemical Abstracts Registry No.: 55-52-7

Structural Formula:

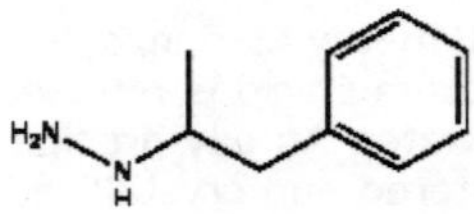

Trade Name	**Manufacturer**	**Country**	**Year Introduced**
Catron	Lakeside	US	1959
Catroniazide	Lakeside	-

Raw Materials

1-Phenyl-2-propylidenylhydrazine
Acetic acid
Hydrogen

Manufacturing Process

A solution containing 741 g (5.0 mols) of 1 -phenyl-2-propylidenylhydrazine, 300 g (5.0 mols) of glacial acetic acid and 900 cc of absolute ethanol was subjected to hydrogenation at 1,875 psi of hydrogen in the presence of 10 g of platinum oxide catalyst and at a temperature of 30°C to 50°C (variation due to exothermic reaction). The catalyst was removed by filtration and the solvent and acetic acid were distilled. The residue was taken up in water and made strongly alkaline by the addition of solid potassium hydroxide. The alkaline mixture was extracted with ether and the ether extracts dried with potassium carbonate. The product was collected by fractional distillation, BP 85°C (0.30 mm); yield 512 g (68%).

The hydrochloride salt was formed in a mixture of 1:10 isopropyl alcohol:diisopropyl ether and recrystallized from acetonitrile, yield 87%, MP 124°C to 125°C.

References

Merck Index 7105
OCDS Vol. 1 p. 74 (1977)
I.N. p. 757
Biel, J.H.; US Patent 2,978,461; April 4, 1961; assigned to Lakeside Laboratories, Inc.

PHENIRAMINE MALEATE

Therapeutic Function: Antihistaminic

Chemical Name: N,N-Dimethyl-γ-phenyl-2-pyridine-propanamine maleate

Common Name: Prophenpyridine

Structural Formula:

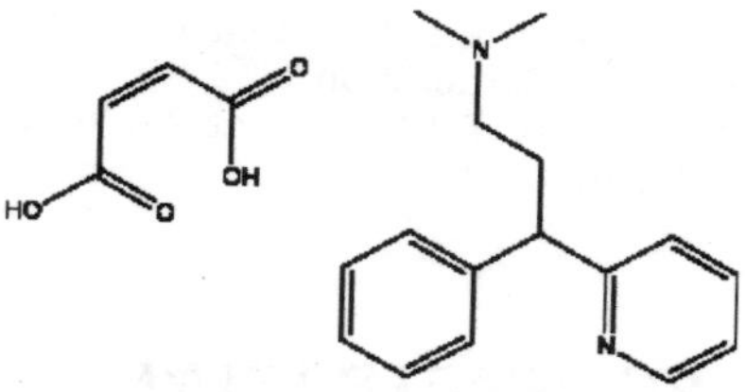

Chemical Abstracts Registry No.: 132-20-7; 86-21-5 (Base)

Trade Name	Manufacturer	Country	Year Introduced
Trimeton Maleate	Schering	US	1948
Avil	Albert Roussel	W. Germany	-
Citra Forte	Doyle	US	-
Daneral	Hoechst	UK	-
Dristan	Whitehall	US	-
Fenamine	Fawns and McAllan	Australia	-
Fiogesic	Sandoz	US	-
Inhiston	Upjohn	US	-
Poly-Histine	Bock	US	-
Ru-Tuss	Boots	US	-
S.T. Forte	Scot-Tussin	US	-
Triaminic	Dorsey	US	-
Tussirex	Scot-Tussin	US	-

Raw Materials

2-Benzylpyridine
β-Dimethylaminoethyl chloride
Potassium amide
Maleic acid

Manufacturing Process

According to US Patent 2,676,964: to 1.0 mol of potassium amide in 3 liters of liquid ammonia, is added 1.0 mol of 2-benzylpyridine. After 15 minutes, 1.1 mols of β-dimethylaminoethyl chloride are added. The ammonia is allowed to evaporate and the reaction product decomposed with water and ether extracted. The ether layer is dried over sodium sulfate and after evaporation the residue is distilled, giving the 3-phenyl-3-(2-pyridyl)-N,N-dimethylpropylamine, BP 139°-142°C/1-2 mm. The maleate is produced by reaction with maleic acid.

References

Merck Index 7106

Kleeman and Engel p. 713; PDR pp. 674, 688, 692, 849, 1583, 1662, 1899
OCDS Vol. 1 p. 77 (1977)
I.N. p. 757
REM p. 1131
Sperber, N., Papa, D. and Schwenk, E.; US Patent 2,567,245; September 11, 1951; assigned to Schering Corporation
Sperber, N., Papa, D. and Schwenk, E.; US Patent 2,676,964; April 27, 1954; assigned to Schering Corporation

PHENMETRAZINE

Therapeutic Function: Antiobesity

Chemical Name: 3-Methyl-2-phenylmorpholine

Common Name: Oxazimdrine

Structural Formula:

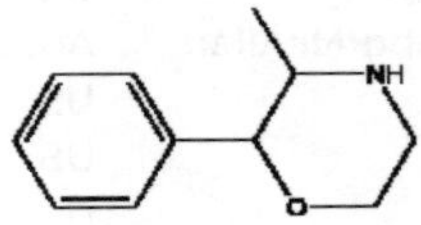

NH
O

Chemical Abstracts Registry No.: 134-49-6; 1707-14-8 (Hydrochloride salt)

Trade Name	Manufacturer	Country	Year Introduced
Preludin	Boehringer Ingelheim	US	1956
Anorex	Pfizer	US	-
Cafilon	Yamanouchi	Japan	-
Marsin	Ikapharm	Israel	-

Raw Materials

Bromopropiophenone
Hydrogen
Benzyl ethanolamine
Hydrogen chloride

Manufacturing Process

10 grams of β-phenyl-α-methyl-β,β'-dihydroxy-diethylamine hydrochloride (produced by hydrogenation in the presence of palladium and charcoal of β-phenyl-α-methyl-β-keto-β'-hydroxy-N-benzyl-diethylamine hydrochloride obtained from bromopropiophenone by reacting with benzyl-ethanolamine), are warmed with 10% hydrochloric acid for 6 hours on a water bath.

After working up in the usual manner, the hydrochloride of the 2-phenyl-3-methyl-morpholine crystallizes out from methanolic hydrochloric acid and acetone, MP = 182°C, according to US Patent 2,835,669.

References

Merck Index 7108
Kleeman and Engel p. 714
PDR p. 678
OCDS Vol. 1 p. 260 (1977)
I.N. p. 757
REM p. 892
Thoma, O.;US Patent 2,835,669; May 20, 1958; assigned to C.H. Boehringer Sohn, Germany
Siemer, H. and Hengen, O.; US Patent 3,018,222; January 23, 1962; assigned to Ravensberg GmbH, Germany

PHENOBARBITAL

Therapeutic Function: Anticonvulsant, Antiepileptic, Hypnotic, Sedative

Chemical Name: Barbituric acid, 5-ethyl-5-phenyl-

Common Name: Phenobarbital; Fenemal; Fenobarbital; Fenobarbiton; Phenylethylbarbituric acid

Structural Formula:

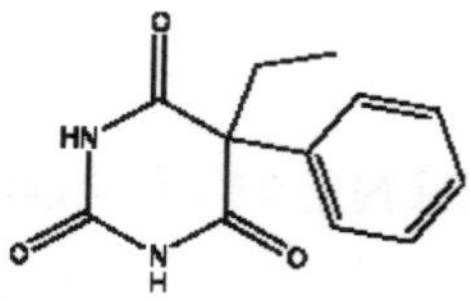

Chemical Abstracts Registry No.: 50-06-6

Trade Name	Manufacturer	Country	Year Introduced
Phenobarbital	Inter-Chemical Ltd.	-	-
Phenobarbital	Zxchem	-	-
Dormital	Acromax	-	-
Hypnogen	Fragner	-	-
Leonal	Leo	-	-
Noctinal	Faes	-	-
Sedabar	Saunders	-	-
Sednotic	Medical Arts	-	-
Sedo	Avicopharma	-	-
Sedonal	Assia	-	-

Raw Materials

Urea
Phenylethylmalonic diethyl ester
Sodium
Ethanol absolute
Sulfuric acid

Manufacturing Process

528 g phenylethyl malonic diethyl ester is dissolved in 500 ml of absolute alcohol. There is then added 140 g urea to the mixture. To this mixture is then added a solution of 57.5 g sodium in 1000 ml absolute alcohol, at such rate that one-half the solution is added during the first hour, a quarter the second hour; an eighth the third hour, and the final eighth during the 4 hours. Then the alcohol is distilled from the reaction mixture. When the alcohol has all been removed, 250 ml xylol is added to the mixture. The reaction mixture is cooled to room temperature and 3 L of water added. The xylene layer was separated and the water solution washed with another 200 ml portion of xylene There is then added to the water solution a 10% excess of a 50% by weight solution of sulfuric acid. The phenobarbital is precipitated as nearly white fluffy crystals, which are filtered off. When dried, they showed 100% phenobarbital by titration. This product may be purified by recrystallization. The unreacted ester in the xylene solution was recovered by distilling off the xylene, and then the phenylethyl malonic ester.

References

Inman M.T., Bitler W.P.; US Patent No. 2,358,072; Sept. 12, 1944; Assigned to Kay-Fries Chemicals, Inc., West Haverstraw, N.Y., a corporation of New York

PHENOPERIDINE HYDROCHLORIDE

Therapeutic Function: Analgesic

Chemical Name: 1-(3-Hydroxy-3-phenylpropyl)-4-phenyl-4-piperidinecarboxylic acid ethyl ester hydrochloride

Common Name: 3-(4-Carboethoxy-4-phenylpiperidino)-1-phenyl-1-propanol hydrochloride

Chemical Abstracts Registry No.: 3627-49-4; 562-26-5 (Base)

Trade Name	Manufacturer	Country	Year Introduced
Operidine	Janssen	US	1965
Lealgin	Leo	Sweden	-
R-1406	Le Brun	France	-

Structural Formula:

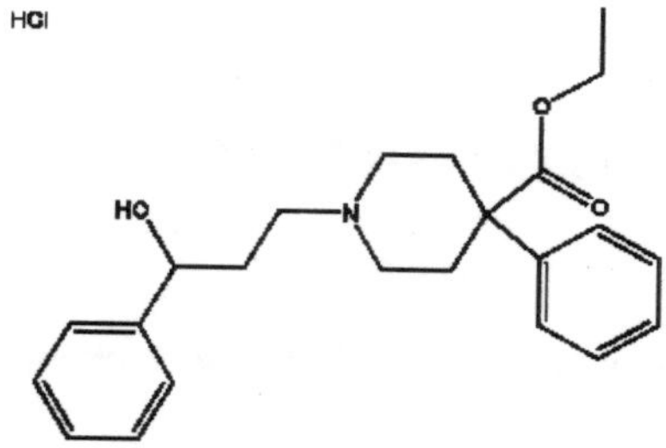

Raw Materials

Hydrogen
Phenylacetonitrile
Benzoylethylene
Bis-chloroethyl toluene sulfonyl amide

Manufacturing Process

The starting materials for the overall process are phenylacetonitrile with bis-chloroethyl toluene sulfonyl amide. These react to give a product which hydrolyzes to normeperidine (4-carboethoxy-4-phenylpiperidine). Condensation of that material with benzoylethylene gives the ketone: β-(4-carboethoxy-4-phenylpiperidino)propiophenone.

A reaction mixture was prepared containing 4 grams of β-(4-carboethoxy-4-phenylpiperidino)-propiophenone hydrochloride, 100 ml of methanol and about 0.5 gram of platinum oxide catalyst. The mixture was placed in a low pressure hydrogenation apparatus and was hydrogenated at a temperature of about 27°C and a pressure of about 3.5 atmospheres of hydrogen to convert the keto group of the β-(4-carboethoxy-4-phenylpiperidino)-propiophenone to a hydroxy group, and to form 3-(4-carboethoxy-4-phenylpiperidino)-1-phenyl-1-propanol hydrochloride. After the hydrogenation was complete, the catalyst was separated from the reaction mixture by filtration, and the filtrate was evaporated to dryness in vacuo leaving a residue containing 3-(4-carboethoxy-4-phenylpiperidino)-1-phenyl-l-propanol hydrochloride. The residue was digested with ethyl acetate thereby causing 3-(4-carboethoxy-4-phenylpiperidino)-1-phenyl-1-propanol hydrochloride to crystallize. This compound melted at about 188°-189°C after being recrystallized three times from an ethyl acetate-methanol solvent mixture, according to US Patent 2,951,080.

References

Merck Index 7125
Kleeman and Engel p. 715
OCDS Vol. 1 p. 302 (1977)
I.N. p. 759
Pohland, A.; US Patent 2,951,080; August 30, 1960; assigned to Eli Lilly and Company

Cutler, F.A., Jr. and Fisher, J.F.; US Patent 2,962,501; November 29, 1960; assigned to Merck and Co., Inc.

PHENOXYBENZAMINE HYDROCHLORIDE

Therapeutic Function: Adrenergic blocker

Chemical Name: N-(2-Chloroethyl)-N-(1-methyl-2-phenoxyethyl) benzenemethanamine hydrochloride

Common Name: -

Structural Formula:

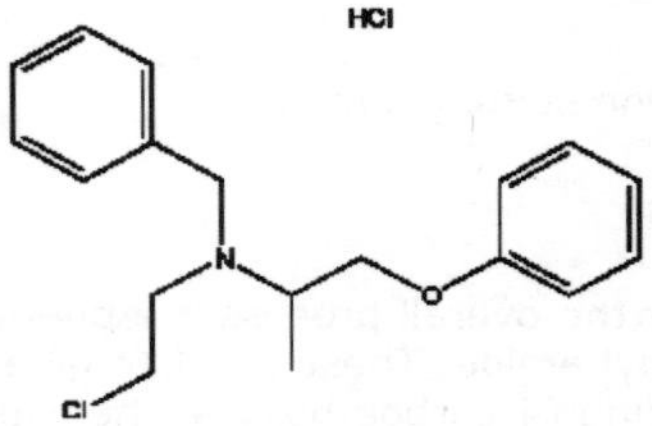

HCl

Cl N O

Chemical Abstracts Registry No.: 63-92-3; 59-96-1 (Base)

Trade Name	Manufacturer	Country	Year Introduced
Dibenzyline	SKF	US	1953
Dibenzyran	Rohm Pharma	W. Germany	-

Raw Materials

1-Phenoxy-2-propanol
Ethanolamine
Hydrogen chloride
Thionyl chloride
Benzyl chloride

Manufacturing Process

Step 1: In a 500 ml flask equipped with gas inlet tube, dropping funnel and reflux condenser is placed 139 grams of 1-phenoxy-2-propanol. A stream of dry air is bubbled through the alcohol while 55 grams of thionyl chloride is added dropwise with external cooling. The stream of dry air is continued for about six hours or until most of the hydrogen chloride has been expelled and then another 55 grams of thionyl chloride is added. The reaction mixture is allowed to stand twenty-four hours, a few drops of pyridine are added and the mixture heated 4 hours on the steam bath. The cooled reaction mixture is poured into water, the crude product is washed with dilute sodium bicarbonate solution and finally taken up in benzene. The benzene is distilled at ordinary pressure and the residue distilled in vacuo to yield 60-70% of 1-phenoxy-2-chloropropane, BP 93°-94°C/5 mm.

Step 2: To 494 grams of ethanolamine, heated to approximately 150°C in a 500 ml flask equipped with stirrer, condenser and dropping funnel, is added 465 grams of 1-phenoxy-2-chloropropane with mechanical stirring. The reaction mixture is then heated to reflux for 3 hours, cooled and poured into a liter of water. The organic layer is extracted into ether and the ether solution is extracted with dilute hydrochloric acid. The aqueous acid solution is then made alkaline with 40% sodium hydroxide solution and the organic base is extracted into ether. Removal of the ether leaves N-(phenoxyisopropyl)-ethanolamine which, after recrystallization from hexane, melts at 70.5°-72°C.

Step 3: To 43 grams of N-(phenoxyisopropyl)ethanolamine dissolved in 500 ml of alcohol in a 1,000 ml flask equipped with stirrer and condenser is added 28 grams of benzyl chloride and 18.5 grams of sodium bicarbonate. The mixture is stirred and refluxed for 10 hours and then approximately half the alcohol is removed by distillation. The remaining solution is poured into 500 ml of water and the organic material extracted with 3 100-ml portions of ether. The combined ether extracts are washed with water, dried over anhydrous potassium carbonate and filtered. After removal of the ether, the residue is distilled in vacuo to yield N-(phenoxyisopropyl)-N-benzylethanolamine, BP 163°-168°C/0.2 mm.

Step 4: A solution of 20 grams of the above amino alcohol is dissolved in 50 ml of dry chloroform and treated with dry hydrogen chloride until acid. Then a solution of 9 grams of thionyl chloride in 50 ml of dry chloroform is added and the reaction mixture is heated on a water bath at 50°-60°C for 2 hours. Most of the chloroform is removed by distillation under reduced pressure. Addition of ether to the residue causes the product to crystallize. After recrystallization from a mixture of alcohol and ether, the N-(phenoxyisopropyl)-N-benzyl-β-chloroethylamine hydrochloride melts at 137.5°-140°C.

References

Merck Index 7134
Kleeman and Engel p. 716
PDR p. 1713
OCDS Vol. 1 p. 55 (1977)
I.N. p. 760
REM p. 905
Kerwin, J.F. and Ullyot, G.E.; US Patent 2,599,000; June 3, 1952; assigned to Smith, Kline and French Laboratories

PHENPROCOUMON

Therapeutic Function: Anticoagulant

Chemical Name: 4-Hydroxy-3-(1-phenylpropyl)-2H-1-benzopyran-2-one

Common Name: 3-(1-Phenylpropyl)-4-hydroxycoumarin

Structural Formula:

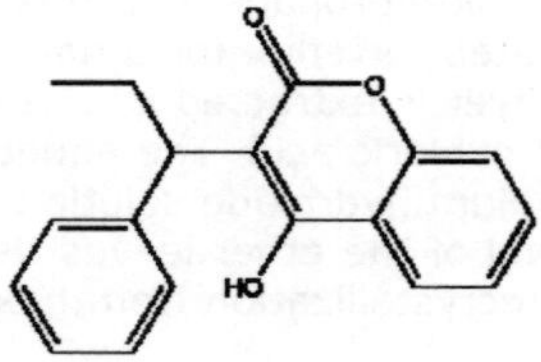

Chemical Abstracts Registry No.: 435-97-2

Trade Name	Manufacturer	Country	Year Introduced
Liquamar	Organon	US	1958
Falithrom	Fahlberg-List	E. Germany	-
Fencumar	Medica	Finland	-
Marcumar	Roche	W. Germany	-

Raw Materials

Methanol
Sodium
Sodium hydroxide
Diethyl-(1'-phenylpropyl)malonate
Acetylsalicylic acid chloride

Manufacturing Process

8.3 parts by weight of powdered sodium in 300 parts by volume of benzene, 100 parts by weight of diethyl (1'-phenylpropyl)-malonate and 72 parts by weight of acetylsalicylic acid chloride are reacted together to form diethyl 1-(o-acetoxybenzoy1)-1-(1'-phenylpropyl)malonate, which boils at 195°-198°C/0.03 mm Hg.

10.3 parts of weight of diethyl 1-(o-acetoxybenzoyl)-1-(1'-phenylpropyl)-malonate are dissolved in 60 parts by volume of absolute ether and to this solution are added portion. wise at 10°C, while stirring, 2.6 parts by weight of sodium methylate. The reaction mixture is stirred for 4 hours, whereupon it is poured into ice water. The ether solution is washed neutral with ice water. After having distilled off the ether, a thick oil consisting of 3-carbethoxy-3-(1'-phenylpropyl)-4-oxo-dihydrocoumarinis obtained. This compound crystallized in butyl oxide and has a MP of 108°-109°C.

The 3-carbethoxy-3-(1'-phenylpropyl)-4-oxo-dihydrocoumarinmay be hydrolyzed and decarboxylated as follows. The crude product is heated to 85°C for 1/2 hour with 100 parts by volume of 5% aqueous sodium hydroxide, while agitating or stirring. To remove traces of undissolved oil, the cooled solution is treated with 1 part by weight of charcoal, whereupon it is filtrated and acidified to Congo reaction with dilute sulfuric acid. The 3-(1'-phenylpropyl)-4-hydroxycoumarin formed is separated off and recrystallized in 80% ethanol, whereupon it melts at 178°-179°C according to US Patent 2,701,804.

References

Merck Index 7139
Kleeman and Engel p. 718
I.N. p. 761
REM p. 827
Hegedus, B. and Grussner, A.; US Patent 2,701,804; February 8, 1955; assigned to Hoffmann-La Roche Inc.
Schroeder, C.H. and Link, K.P.; US Patent 2,872,457; February 3, 1959; assigned to Wisconsin Alumni Research Foundation
Preis, S., West, B.D. and Link, K.P.; US Patent 3,239,529: March 8, 1966; assigned to Wisconsin Alumni Research Foundation

PHENSUXIMIDE

Therapeutic Function: Anticonvulsant

Chemical Name: 1-Methyl-3-phenyl-2,5-pyrrolidinedione

Common Name: N-Methyl-α-phenylsuccinimide

Structural Formula:

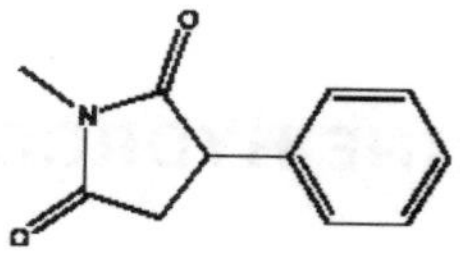

Chemical Abstracts Registry No.: 86-34-0

Trade Name	Manufacturer	Country	Year Introduced
Milontin	Parke Davis	US	1953
Lifene	Debat	France	-
Petimid	Dincel	Turkey	-
Succitimal	Katwijk	Netherlands	-

Raw Materials

Phenylsuccinic anhydride
Methyl amine
Acetyl chloride

Manufacturing Process

10 grams of phenylsuccinic anhydride is dissolved in 250 ml of absolute ether and the solution is treated with dry methylamine until a precipitate ceases to form. After standing for ½ hour the ether is decanted off and the residue is

washed with 40 ml of water by decantation. The mixture is filtered and the precipitate washed with 10 ml of water. By acidification of the filtrate, a white precipitate is obtained. After drying it weighs 8 grams and melts at 136°-140°C. The two precipitates are combined and recrystallized from aqueous alcohol to give β-N-methylphenylsuccinamic acid which melts at 158°-160°C.

9 grams of β-N-methylphenylsuccinamic acid and 200 ml of acetyl chloride are heated together on a steam bath for ½ hour. The excess acetyl chloride is removed by distillation and 50 ml of water are added to the thick residue. After allowing for hydrolysis of the excess acetyl chloride the water is decanted and the yellow residue dissolved in 75 ml of ether. The resulting solution is treated with charcoal twice and dried over anhydrous magnesium sulfate. On partial evaporation of the ether a white solid precipitates. There is obtained 4 grams of N-methyl-α-phenylsuccinimide which melts at 71°-73°C.

References

Merck Index 7140
Kleeman and Engel p. 718
PDR p. 1367
OCDS Vol. 1 p. 226 (1977)
I.N. p. 762
REM p. 1080
Miller, C.A. and Long, L.M.; US Patent 2,643,258; June 23, 1953; assigned to Parke, Davis and Company

PHENTERMINE HYDROCHLORIDE

Therapeutic Function: Antiobesity

Chemical Name: α,α-Dimethylbenzeneethanamine hydrochloride

Common Name: α-Benzylisopropylamine hydrochloride; Phenyl-tert-butylamine hydrochloride

Structural Formula:

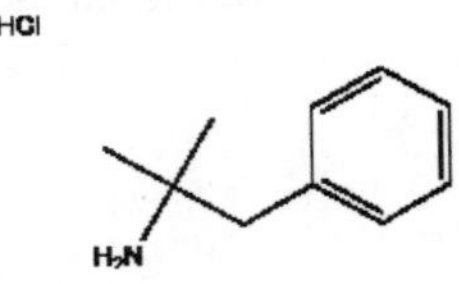

Chemical Abstracts Registry No.: 1197-21-3; 122-09-8 (Base)

Trade Name	Manufacturer	Country	Year Introduced
Wilpo	Dorsey	US	1961
Linyl	Roussel	France	1962
Fastin	Beecham	US	1973

Trade Name	Manufacturer	Country	Year Introduced
Adipex-P	Lemmon	US	1976
Ona Mast	Mast	US	1980
Obestin	Ferndale	US	1980
Oby-Trim	Rexar	US	1982
Duromine	Riker	UK	-
Ex-Adipos	Eurand	Italy	-
Ionamin	Pennwalt	UK	-
Jonakraft	Kraft Pharm	US	-
Lipopil	Roussel Maestretti	Italy	-
Minobese	Restan	S. Africa	-
Mirapront	Bracco	Italy	-
Netto-Longcaps	Heyden	W. Germany	-
Panbesy	Asperal	Belgium	-
Panshade	Pan American	US	-
Parmine	Parmed	US	-
Phentermine	Schein	US	-
Phentermyl	Diethelm	W. Germany	-
Regulin	Kwizda	Austria	-
Span R/D	Metro Med	US	-
Teramine	Legere	US	-

Raw Materials

Isobutyryl chloride
Ammonia
Hydrogen chloride
Bromine
Calcium hydroxide
Sodium
Benzyl bromide
Benzene
Potassium hydroxide

Manufacturing Process

Preparation of isobutyrophenone: In a 12 liter, 3-necked flask, 1,280 grams of aluminum chloride was covered with 2,000 cc of dry thiophene-free benzene and a solution of 919 grams of isobutyryl chloride, (BP 92°-94°C) in 1 liter of benzene was added slowly with stirring. After heating for 3 hours at reflux, the solution was cooled and poured over a mixture of 1 liter of concentrated hydrochloric acid and 5 kg of ice. The benzene layer was separated, the aqueous layer extracted with benzene, and the combined benzene solutions were washed, dried and concentrated in vacuo. The residue was distilled rapidly to give 1,051 grams of isobutyrophenone, boiling at 81°-89°C at 1 mm, yield 83.4%.

Preparation of 1,3-Diphenyl-2,2-Dimethylpropanone-1: Sodamide was prepared from 12.5 grams of sodium added in small portions to 600 cc of liquid ammonia with 1 gram of hydrous ferric chloride as catalyst. The ammonia was replaced by 200 cc of dry toluene and without delay a solution of 74 grams of isobutyrophenone and 76.5 grams of benzyl bromide in 200 cc of benzene was slowly added with stirring. The reaction mixture was heated on a boiling water bath for 48 hours. Water was then added, the organic layer separated and the product isolated by distillation. The 1,3-diphenyl-2,2-

dimethylpropanone-1 boiled from 142°-143°C at a pressure of 3 mm, n_D^{20}1.5652.

Preparation of α,α-Dimethyl-β-Phenylpropionamide: Sodamide was prepared from 7.6 grams of sodium in 350 cc of liquid ammonia with 0.9 gram of hydrous ferric chloride. The ammonia was replaced by 250 cc of toluene, the mixture was heated to 60°C and 71.4 grams of 1,3-diphenyl-2,2-dimethyl propanone-1 dissolved in 150 cc of toluene was added. The mixture was stirred and heated on a steam bath for 5 hours. A clear red color appeared in 15 minutes and disappeared after about an hour. After cooling, water was added, the organic layer was washed, dried, and concentrated to give 36.5 grams of α,α-dimethyl-β-phenyl propionamide which crystallized slowly after the addition of an equal volume of petroleum ether. The product melted at 62°C after crystallization from benzene-petroleum ether.

Preparation of Di-(β-Phenyl-α,α-Dimethylethyl)Urea: 3.5 grams of α,α-dimethyl-β-phenylpropionamide in 420 cc of water was added to a solution of 87.5 grams of potassium hydroxide and 35 grams of bromine in 350 cc of water. After 2 hours at 60°C, the product was obtained on crystallization from ethanol, melting at 184°C.

Preparation of ω-Phenyl-tert-Butylamine: 24 grams of the urea derivative obtained as indicated above, were well mixed with 96 grams of calcium hydroxide in a flask immersed in an air bath and provided with a dropping funnel the stem of which reached the bottom of the flask. The mixture was heated to 240°-260°C (inside temperature) for 7 hours during which time 86 cc of water was slowly added. The vapors were collected in a receiver cooled with ice. After extraction with ether and distillation, the product was obtained as a colorless liquid boiling from 80°-84°C at 9 mm according to US Patent 2,590,079.

The ether solution may be dried and saturated with hydrogen chloride and the precipitated hydrochloride recrystallized from a mixture of 50 parts alcohol and 100 parts of acetone.

The pure hydrochloride is thus obtained as a white crystalline substance having a MP of 195°-196°C, according to US Patent 2,408,345.

References

Merck Index 7141
Kleeman and Engel p. 719
PDR pp. 660, 1033, 1034, 1246, 1450, 1606, 1999
OCDS Vol. 1 p. 72 (1977)
I.N. p. 762
REM p. 892
Shelton, R.S. and Van Campen, M.G., Jr.; US Patent 2,408,345; September 24, 1946; assigned to The Wm. S. Merrell Company
Abell, L.L., Bruce, W.F. and Seifter, J.; US Patent 2,590,079; March 25, 1952; assigned to Wyeth Incorporated

PHENTOLAMINE HYDROCHLORIDE

Therapeutic Function: Adrenergic blocker

Chemical Name: 3-[[(4,5-Dihydro-1H-imidazol-2-yl)methyl](4-methylphenyl) amino]phenol hydrochloride

Common Name: 2-(m-Hydroxy-N-p-tolylanilinomethyl)-2-imidazoline hydrochloride

Structural Formula:

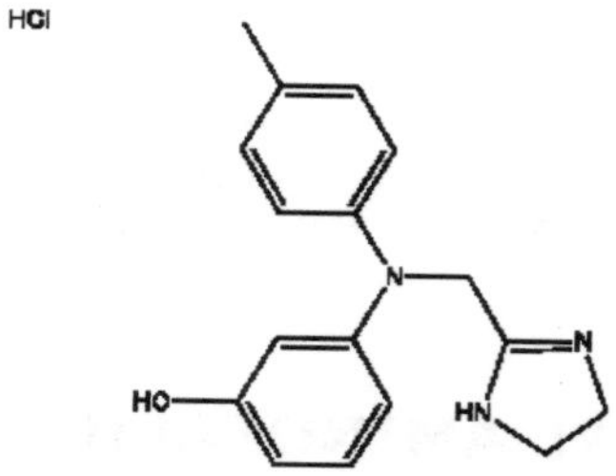

HCl

Chemical Abstracts Registry No.: 73-05-2; 50-60-2 (Base)

Trade Name	Manufacturer	Country	Year Introduced
Regitine	Ciba	US	1952
Regitine	Ciba Geigy	Japan	-
Regitine	Ciba	UK	-

Raw Materials

Hydrogen chloride
N-(p-Methylphenyl)-m'-hydroxyphenylamine
2-Chloromethylimidazoline HCl

Manufacturing Process

199.24 parts of N-(p-methylphenyl)-m'-hydroxyphenylamine and 77.52 parts of 2-chloromethylimidazoline hydrochloride are heated for sixteen hours in an oil bath having a temperature of 150°C, while stirring and introducing a current of nitrogen. The viscous contents of the flask are then cooled to about 100°C, mixed with 400 parts by volume of hot water, and stirred for a short time.

After further cooling to about 60°C, 200 parts by volume of water and 500 parts by volume of ethyl acetate at 60°C are added, and the aqueous layer is separated. The excess of starting material may be recovered from the ethyl acetate.

The aqueous portion is chilled in a cooling chamber at -10°C, whereupon the hydrochloride of 2-[N-(p-methylphenyl)-N-(m'-hydroxyphenyl)-aminomethyl]-imidazoline crystallizes. Upon being concentrated and cooled the mother liquor yields a further quantity of the hydrochloride. The combined quantities of hydrochloride are treated with a small quantity of cold water, dried with care, and washed with ethyl acetate. The product is then crystallized from a mixture of alcohol and ethyl acetate, and there is obtained a hydrochloride melting at 239°-240°C.

References

Merck Index 7143
Kleeman and Engel p. 719
PDR p. 809
OCDS Vol. 1 p. 242 (1977)
I.N. p. 762
REM p. 906
Miescher, K., Marxer, A. and Urech, E.; US Patent 2,503,059; April 4, 1950; assigned to Ciba Pharmaceutical Products, Inc.

PHENYL AMINOSALICYLATE

Therapeutic Function: Antibacterial (tuberculostatic)

Chemical Name: 4-Amino-2-hydroxybenzoic acid phenyl ester

Common Name: Fenamisal

Structural Formula:

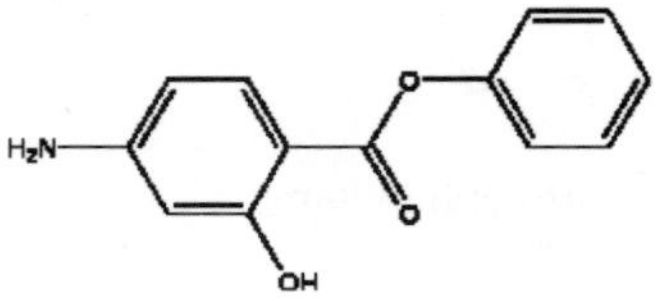

H_2N OH O O

Chemical Abstracts Registry No.: 133-11-9

Trade Name	Manufacturer	Country	Year Introduced
Pheny-Pas-Teb-Amin	Purdue Frederick	US	1959
Fenil-PAS	Farmabion	Spain	-

Raw Materials

p-Nitrosalicylic acid
Phosphorus oxychloride
Phenol
Hydrogen

Manufacturing Process

183 g of p-nitrosalicylic acid are dissolved in 564 g of phenol by heating to 140°C to 150°C on an oil bath. When all the p-nitrosalicylic acid is dissolved, 153 g of phosphorus oxychloride are run in, drop by drop, over a period of about 2 hours, while maintaining the temperature at about 150°C. The still warm mixture is run into 2 liters of water with agitation. The precipitate formed is filtered off, washed with water until phenol is removed and then dried.

There are thus obtained 250 g of 2-hydroxy-4-nitrophenylbenzoate which melts at 154°C to 155°C.

In a hydrogenation autoclave are introduced 92 g of 2-hydroxy-4-nitrophenylbenzoate preceded by 200 cc of ethyl acetate; Raney nickel, obtained from 30 g of alloy, is added with 300 cc of ethyl acetate. Hydrogenation under pressure (100 to 120 kg) at ordinary temperature is carried out during a period of about 12 hours. The nickel is filtered off and the ethyl acetate is removed by distillation on the water bath under a vacuum of 300 mm. There is thus obtained 80 g of crude damp 2-hydroxy-4-amnophenylbenzoate which after recrystallization from isopropyl alcohol melts at 153°C.

References

Merck Index 7151
OCDS Vol. 2 p. 89 (1980)
I.N. p. 415
Freire, SA.; US Patent 2,604,488; July 22, 1952; assigned to Soc. des Usines Chimiques Rhone-Poulenc (France)

PHENYLBUTAZONE

Therapeutic Function: Antiinflammatory, Antiarthritic

Chemical Name: 4-Butyl-1,2-diphenyl-3,5-pyrazolidinedione

Common Name: 3,5-Dioxo-1,2-diphenyl-4-n-butylpyrazolidine

Structural Formula:

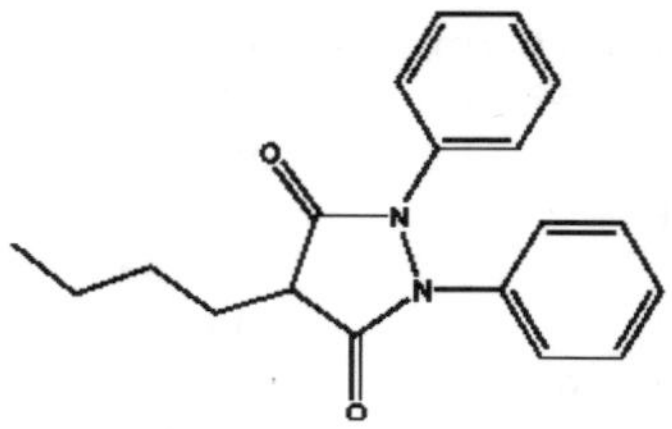

Chemical Abstracts Registry No.: 50-33-9

Trade Name	Manufacturer	Country	Year Introduced
Butazolidin	Geigy	US	1952
Butazolidin	Ciba Geigy	France	1954
Azolid	U.S.V. Pharm.	US	1971
Acrizeal	S.S. Pharm	Japan	-
Alkabutazona	Lovens	Denmark	-
Anuspiramin	Farbios	Spain	-
Artropan	Polifarma	Italy	-
Bulentin	Sanwa	Japan	-
Butacal	Langley	Australia	-
Butacote	Geigy	UK	-
Butadion	Streuli	Switz.	-
Butadiona	Miquel	Spain	-
Butadyne	Bio-Chimique	Canada	-
Butalan	Lancet	Australia	-
Butalgin	Fawns and McAllan	Australia	-
Butalgina	Esteve	Spain	-
Butaluy	Miluy	Spain	-
Butaphen	Mulda	Turkey	-
Butapirazol	Polfa	Poland	-
Butarex	Adams	Australia	-
Butartril	Chiesi	Italy	-
Butazina	Vis	Italy	-
Butazone	DDSA	UK	-
Butiwas Simple	Wassermann	Spain	-
Butoroid	Virax	Australia	-
Butrex	SCS Pharmalab	S. Africa	-
Carudol	Lab. Franc. Therap.	France	-
Chembuzone	Chemo-Drug	Canada	-
Demoplas	Adenylchemie	W. Germany	-
Digibutina	Bicsa	Spain	-
Diossidone	Eliovit	Italy	-
Ecobutazone	I.C.N.	Canada	-
Elmedal	Thiemann	W. Germany	-
Equi Bute	Fort Dodge Labs	US	-
Eributazone	Eri	Canada	-
Fenibutasan	Santos	Spain	-
Fenibutol	Atral	Portugal	-
Flexazone	Berk	UK	-
IA-But	Inter-Alia Pharm.	UK	-
Intalbut	Inter-Alia Pharm.	UK	-
Kadol	Midy	Italy	-
Merizone	Meriot	Canada	-
Neo-Zoline	Neo	Canada	-
Neuplus	Toyo	Japan	-
Novobutazone	Novopharm	Canada	-
Novophenyl	Novopharm	Canada	-

Trade Name	Manufacturer	Country	Year Introduced
Panazone	Propan-Lipworth	S. Africa	-
Phenbutazol	Smallwood	Canada	-
Phenyl Betazone	Barlow Cote	Canada	-
Phenylone	Medic	Canada	-
Pilazon	Kobayashi	Japan	-
Pirarreumol	Hermes	Spain	-
Praecirheumin	Pfleger	W. Germany	-
Rectofasa	Lifasa	Spain	-
Reumasyl	Leiras	Finland	-
Reumazin	Mohan	Japan	-
Reumuzol	Farmos	Finland	-
Reupolar	Farmos	Finland	-
Rheumaphen	Reiss	W. Germany	-
Schemergen	Azusa	Japan	-
Sedazole	Toho	Japan	-
Servizolidin	Servipharm	Switz.	-
Shigrodin	Ikapharm	Israel	-
Spondyril	Dorsch	W. Germany	-
Tetnor	Drugs, Ltd.	UK	-
Tevcodyne	Tevcon	US	-
Therazone	Western Serum	US	-
Ticinil	De Angeli	Italy	-
Todalgil	Lopez-Brea	Spain	-
Tokugen	Sawai	Japan	-
Uzone	Kempthorne Prosser	New Zealand	-
Wescozone	Saunders	Canada	-
Zolidinium	Kwizda	Austria	-

Raw Materials

Hydrazobenzene
Diethyl-n-butyl malonate
Sodium
Ethanol

Manufacturing Process

7.6 parts of sodium are dissolved in 190 parts by volume of absolute alcohol; 65 parts of diethyl-n-butyl malonate and 65 parts of hydrazobenzene are added. The alcohol is slowly distilled off and the reaction mixture heated for 12 hours at a bath temperature of 150°C and finally in vacuo, until no more alcohol comes off.

The product is dissolved in water, clarified with a little animal charcoal and 15% hydrochloric acid is slowly added until an acid reaction to Congo red paper is produced. 1,2-Diphenyl-3,5-dioxo-4-n-butyl-pyrazolidine separates as an oil, which rapidly become crystalline. It crystallizes from alcohol as colorless needles with a MP of 105°C.

References

Merck Index 7157
Kleeman and Engel p. 720
PDR pp. 830, 891, 1606, 1999
OCDS Vol. 1 p. 236 (1977) and 2, 388, 474 (1980)
I.N. p. 763
REM p. 1120
Stenzl, H.; US Patent 2,562,830; July 31, 1951; assigned to J.R. Geigy AG, Switzerland

PHENYLEPHRINE HYDROCHLORIDE

Therapeutic Function: Adrenergic

Chemical Name: (R)-3-Hydroxy-α-[(methylamino)methyl]benzenemethanol hydrochloride

Common Name: m-Methylaminoethanolphenol hydrochloride; Metaoxedrin

Structural Formula:

HCl

HO ... N H ... OH

Chemical Abstracts Registry No.: 61-76-7

Trade Name	Manufacturer	Country	Year Introduced
Neosynephrine	Badrial	France	1953
Mydfrin	Alcon	US	1979
Nostril	Boehringer Ingelheim	US	1982
Adrianol	Anasco	W. Germany	-
Atrohist	Adams	US	-
Bromphen	Schein	US	-
Codimal	Central	US	-
Comhist	Norwich Eaton	US	-
Congespirin	Bristol-Myers	US	-
Coryban	Pfipharmecs	US	-
Dallergy	Laser	US	-
Deconsal	Adams	US	-
Decontabs	Zenith	US	-
Degest	Barnes Hind	US	-
Derizene	Hollister-Stier	US	-

Trade Name	Manufacturer	Country	Year Introduced
Donatussin	Laser	US	-
Dristan	Whitehall	US	-
Dura-Vent	Dura	US	-
E.N.T.	Springbok	US	-
Entex	Norwich Eaton	US	-
Extendryl	Fleming	US	-
Fenilfar	Farmila	Italy	-
Histalet	Reid-Rowell	US	-
Histamic	Metro Med	US	-
Histaspan	U.S.V. Pharm.	US	-
Histor	Hauck	US	-
Hycomine	Du Pont	US	-
Isonefrine	Tubi Lux Pharma	Italy	-
Isophrine	Broemmel	US	-
Isotropina	Tubi Lux Pharma	W. Germany	-
Korigesic	Trimen	US	-
Matafa-Lind	Anasco	US	-
Naldecon	Bristol	Spain	-
Nasophen	Premo	US	-
Neosinefrina	Reunidos	US	-
Newphrine	Vitarine	US	-
Nostril	Boehringer Ingelheim	US	-
Pediacof	Winthrop-Breon	US	-
Phenergan	Wyeth	US	-
Protid	La Salk	US	-
PV-Tussin	Reid-Rowell	US	-
Quelidrine	Abbott	US	-
Rinisol	Farmos	Finland	-
Ru-Tuss	Boots	US	-
Singlet	Lakeside	US	-
S-T Forte	Scot-Tussin	US	-
Synasal	Texas Pharm	US	-
Tear-Efrin	Tilden Yates	US	-
Tussar	U.S.V. Pharm.	US	-
Tussirex	Scot-Tussin	US	-
Tympagesic	Adria	US	-
Visopt	Sigma	Australia	-
Zeph	Scott and Turner	Australia	-

Raw Materials

Hydrogen
m-Hydroxymethylaminoacetophenone
Hydrogen chloride

Manufacturing Process

4.5 g of the hydrochloride of m-hydroxymethylaminoacetophenone are dissolved in a small amount of water; to the solution a solution of colloidal palladium obtained from palladiumchloride is added, and the mixture is treated with hydrogen.

After diluting the reaction liquid with acetone it is filtered, and the residue obtained after the evaporation of the filtrate in vacuo, and complete drying over pentoxide of phosphorus is then dissolved in absolute alcohol, and to this is added about the same volume of dry ether, until turbidity just commences to occur. After a short time the hydrochloride of the m-hydroxyphenylethanol-methylamine will separate out as a colorless mass of crystals at a melting point of 142°C to 143°C.

References

Merck Index 7167
PDR pp. 555, 562, 570, 677, 688, 701, 727, 784, 855, 865, 880, 928, 991, 1246, 1272, 1276, 1404, 1447, 1606, 1662, 1735, 1807, 1813, 1824, 1899, 1923, 1973, 1999
OCDS Vol. 1 p. 63 (1977); 2, 265 (1980) and 3, 20 (1984)
I.N. p. 764
REM p. 889
Legerlotz, H.; US Patent 1,932,347; October 24, 1933; assigned to Frederick Stearns and Co.

PHENYLPROPANOLAMINE HYDROCHLORIDE

Therapeutic Function: Nasal decongestant, Anorexic

Chemical Name: α-(1-Aminoethyl)benzenemethanol hydrochloride

Common Name: dl-Norephedrine hydrochloride; 2-Amino-1-phenyl-1-propanol hydrochloride

Structural Formula:

HCl
NH_2
OH

Chemical Abstracts Registry No.: 154-41-6; 14838-15-4 (Base)

Trade Name	Manufacturer	Country	Year Introduced
Propadrine	MSD	US	1941
Dexatrim	Thompson	US	1980
Dietac	Menley and James	US	1980
Obestat	Lemmon	US	1980
Permatrim	Lee	US	1980

Trade Name	Manufacturer	Country	Year Introduced
Nobese	O'Neal Jones	US	1981
Dexatrim Extra	Thompson	US	1981
Propagest	Carnrick	US	1982
Acutr im	Ciba Geigy	US	1983
Help	Verex	US	1983
Appedrine	Thompson	US	-
Bromphen	Schein	US	-
Codimal	Central	US	-
Comtrex	Bristol-Myers	US	-
Congespirin	Bristol-Myers	US	-
Control	Thompson	US	-
Corvban-D	Pfipharmecs	US	-
Co-Tylenol	McNeil	US	-
Cremacoat	Vicks	US	-
Decontabs	Zenith	US	-
Dietrim	Legere	US	-
Dimetane-D.C.	Robins	US	-
Dura Vent	Dura	US	-
E.N.T.	Springbok	US	-
Entex	Norwich Eaton	US	-
Fiogesic	Sandoz	US	-
Head and Chest	Procter and Gamble	US	-
Histaminic	Metro Med	US	-
Hycomine	Du Pont	US	-
Korigesic	Trimen	US	-
Kronohist	Ferndale	US	-
Monydrin	Draco	Sweden	-
Naldecon	Bristol	US	-
Nolamine	Carnrick	US	-
Ornade	SKF	US	-
Poly-Histine	Bock	US	-
Prolamine	Thompson	US	-
Rhindecon	McGregor	US	-
Rhinolar	McGregor	US	-
Ru-Tuss	Boots	US	-
Sinubid	Parke Davis	US	-
Sinulin	Carnrick	US	-
Tinaroc	Remeda	Finland	-
Triaminic	Dorsey	US	-
Tuss-Ornade	SKF	US	-

Raw Materials

Benzaldehyde
Nitroethane
Sodium bisulfite
Hydrogen
Hydrogen chloride

Manufacturing Process

In one route as described in US Patent 2,151,517, 10.7 kg of technical benzaldehyde is vigorously agitated with a solution of 11.0 kg of sodium bisulfite in 50.0 liters of water until the formation of the addition-product is complete. Simultaneously, 8.25 kg of nitroethane is dissolved in a solution of 4.5 kg of caustic soda in 20.0 liters of water and the resultant warm solution is added with vigorous stirring to the magma of benzaldehyde sodium bisulfite. The mixture is agitated for 30 minutes and then allowed to stand overnight.

The aqueous portion of the mixture is now siphoned off from the supernatant layer of oily phenylnitropropanol and replaced with a fresh solution of 11.0 kg of sodium bisulfite in 50.0 liters of water. The mixture of phenylnitropropanol and bisulfite solution is now vigorously agitated for 15 minutes in order to remove and recover small amounts of unreacted benzaldehyde, and is then again allowed to stratify. This time, the phenylnitropropanol is siphoned off and filtered to remove a small amount of resinous material. The aqueous solution of sodium bisulfite remaining behind is reacted with benzaldehyde, as described above, thus making the process continuous.

The 1-phenyl-2-nitropropanol thus obtained is a colorless oil, specific gravity 1.14 at 20°C, odorless when pure, volatile with steam and boiling at 150° to 165°C under a pressure of 5 mm of mercury. It is soluble in alcohol, ether, acetone, chloroform, carbon tetrachloride, benzene and glacial acetic acid. The yield of 1-phenyl-2-nitropropanolobtained by this procedure is 17.1 to 17.7 kg.

It is hydrogenated and converted to the hydrochloride in subsequent steps. The hydrogen chloride has a melting point of 192°-194°C.

In an alternative route described in US Patent 3,028,429 propiophenone may be reacted with an alkyl nitrite to give isonitrosopropiophenone which is then hydrogenated and finally converted to the hydrochloride.

References

Merck Index 7189
Kleeman and Engel p. 721
PDR pp. 674, 688, 702, 727, 781, 784, 850, 854, 865, 875, 1033, 1084, 1246, 1277, 1388, 1404, 1431, 1454, 1583, 1606, 1719, 1730, 1735, 1805, 1807, 1869, 1999
I.N. p. 766
REM p. 889
Kamlet, J.; US Patent 2,151,517 March 21, 1939
Wilbert, G. and Sosis, P.; US Patent3,028,429; April 3, 1962; assigned to Nepera Chemical Co., Inc.

PHENYLTOLOXAMINE

Therapeutic Function: Antihistaminic

Chemical Name: N,N-Dimethyl-2-[2-(phenylmethyl)phenoxy]ethanamine

Common Name: Bistrimin

Structural Formula:

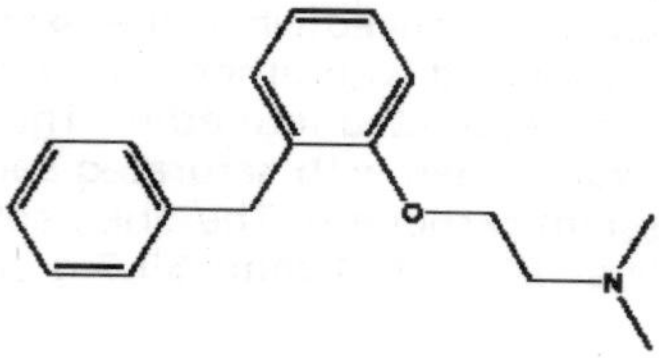

Chemical Abstracts Registry No.: 92-12-6; 6152-43-8 (Hydrochloride salt)

Trade Name	Manufacturer	Country	Year Introduced
Bristalin	Bristol	US	1952
Bristamine	Banyu	Japan	-
Codipront	Mack	W. Germany	-
Ephepect	Bolder	W. Germany	-
Floxamine	Durst	US	-
Fluidol	Metadier-Tours	France	-
Histionex	Strasenburgh	US	-
Netux	Roussel	France	-
Pholtex	Riker	UK	-
Quadrahist	Schein	US	-
Rinurel	Warner	UK	-
Tussionex	Pennwalt	US	-

Raw Materials

o-Benzylphenol
Methanol
Sodium
Dimethylaminoethyl chloride

Manufacturing Process

Sodium methylate is made by dropping 11.7 g of sodium strips into 199 ml of absolute methanol in a 1-liter three-necked flask. 93.9 g of o-benzylphenol are dissolved in 200 ml of dry toluene and added to the sodium methylate solution. The solution is distilled until the boiling point of toluene is reached. At the end of the distillation, enough toluene is added to restore the original volume of solvent.

109.5 g of dimethylaminoethyl chloride hydrochloride and 200 ml of toluene are placed in a 1-liter Erlenmeyer flask, cooled in an ice bath, and decomposed with 167.5 g of 20% sodium hydroxide solution. The toluene and water layers are separated, and the water layer is extracted again with 50 ml of toluene. The toluene layers are combined, washed with saturated salt solution, and dried over anhydrous potassium carbonate.

The dried dimethylaminoethyl chloride solution is poured into the toluene solution of the sodium salt of o-benzylphenol, heated to reflux, and refluxed 16 hours. After refluxing, enough water is added to the mixture to dissolve the precipitated solid. The layers are separated, and the toluene layer is further washed with water until the water extract is just slightly alkaline. The toluene solution is then made acid with 6N hydrochloric acid and extracted with water until no cloudiness is produced when the extract is made alkaline. The acidic aqueous extract is washed with ether, then made alkaline with 20% sodium hydroxide solution, and extracted into ether. The ether solution is washed several times with water, then with saturated salt solution, and is dried over anhydrous potassium carbonate. The dried solution is filtered and distilled. The product distills at 143.5°C/1 mm; 69.7 g of pale yellow oil are recovered.

57.1 g of the free base are dissolved in ether and precipitated with dry HCl. 66.0 g of crude hydrochloride are recovered. The hydrochloride is dissolved in 130 ml of reagent acetone by boiling, filtered hot, and allowed to cool. The crystalline material obtained on cooling is filtered, washed with a little acetone, washed with ether, and dried in vacuo. 44.8 g, MP 119.5°C to 121°C, are recovered from the first crop of crystals. Ethyl acetate may also be used as the solvent for recrystallization.

References

Merck Index 7197
Kleeman and Engel p. 721
PDR p. 1606
OCDS Vol. 1 p. 115 (1977);
I.N. p. 766
Binkley, S.B. and Cheney, L.C.; US Patent 2,703,324; March 1, 1955; assigned to Bristol Laboratories, Inc.

PHENYRAMIDOL

Therapeutic Function: Analgesic, Muscle relaxant

Chemical Name: α-[(2-Pyridinylamino)methyl]benzenemethanol

Common Name: Fenyramidol

Structural Formula:

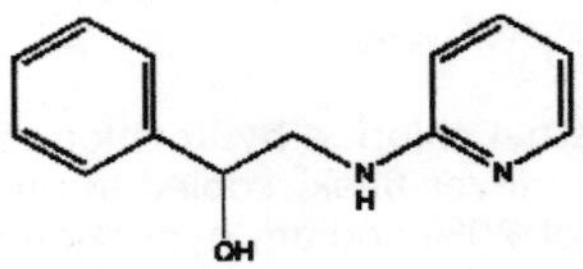

N
H
N
OH

Chemical Abstracts Registry No.: 553-69-5; 326-43-2 (Hydrochloride salt)

Trade Name	Manufacturer	Country	Year Introduced
Analexin	Mallinckrodt Inc.	US	1960
Cabral	Kali-Chemie	W. Germany	1962
Fenprin	RBS	Italy	1962
Anabloc	Irbi	Italy	-
Aramidol	A.B.C.	Italy	-
Bonapar	Minerva-Chemie	Netherlands	-
Evasprine	Millot	France	-
Firmalgil	Firma	Italy	-
Miodar	I.S.M.	Italy	-
Pheniramidol	Pulitzer	Italy	-
Vilexin	Vitrum	Sweden	-

Raw Materials

2-Aminopyridine
Lithium amide
Styrene oxide

Manufacturing Process

A mixture containing 188 g (0.20 mol) of 2-arninopyridine, 0.55 g of lithium amide and 75 cc of anhydrous toluene was refluxed for 1.5 hours. Styrene oxide (12.0 g = 0.10 mol) was then added to the reaction mixture with stirring over a period of ten minutes. The reaction mixture was stirred and refluxed for an additional 3.5 hours. A crystalline precipitate was formed during the reaction which was removed by filtration, MP 170°C to 171°C. 1.5 g. The filtrate was concentrated to dryness and a dark residue remained which was crystallized from anhydrous ether; yield 6.0 g. Upon recrystallization of the crude solid from 30 cc of isopropyl alcohol, 2.0 g of a light yellow solid was isolated; MP 170° to 171°C.

References

Merck Index 7203
Kleeman and Engel p. 399
OCDS Vol. 1 p. 165 (1977)
I.N. p. 422
Biel, J.H.; US Patent 3,040,050; June 19, 1962; assigned to Lakeside Laboratories, Inc.

PHENYTOIN

Therapeutic Function: Antiepileptic

Chemical Name: 5,5-Diphenyl-2,4-imidazolidinedione

Common Name: Diphenylhydantoin

Structural Formula:

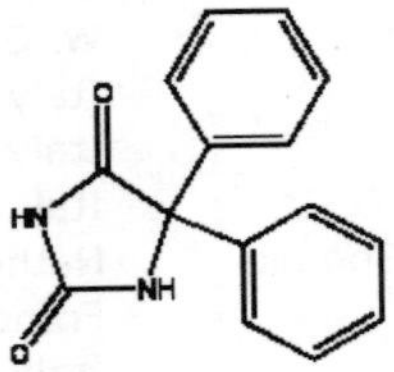

Chemical Abstracts Registry No.: 57-41-0

Trade Name	Manufacturer	Country	Year Introduced
Dilantin	Parke Davis	US	1938
Ditan	Mallard	US	1980
Aleviatin	Dainippon	Japan	-
Citrullamon	Sudmedica	W. Germany	-
Didan	Canfield	US	-
Difhydan	Leo	Sweden	-
Dihydan	Carrion	France	-
Dihydantoin	Orion	Finland	-
Dintoina	Recordati	Italy	-
Diphentyn	I.C.N.	Canada	-
Enkefal	Leiras	Turkey	-
Epanutin	Parke Davis	W. Germany	-
Epinat	Nyegaard	Norway	-
Fenantoin	A.C.O.	Sweden	-
Hydantin	Medica	Finland	-
Hydantol	Fujinaga	Japan	-
Lehydan	Leo	Sweden	-
Novophenytoin	Novopharm	Canada	-
Phenhydan	Desitin	W. Germany	-
Pyoredol	Roussel	France	-
Solantyl	Roussel	France	-
Tacosal	Helvepharm	Switz.	-
Zentropil	Nordmark	W. Germany	-

Raw Materials

Benzophenone
Potassium cyanide
Ammonium carbonate

Manufacturing Process

10 g of benzophenone (1 mol), 4 g of potassium cyanide (1.22 mols) and 16 g of ammonium carbonate (3.3 mols) are dissolved in 100 cc of 60% (by volume) ethyl alcohol and the mixture warmed under a reflux condenser without stirring at 58° to 62°C. After warming the mixture for 10 hours a

partial vacuum is applied and the temperature is raised enough to permit concentration of the reaction mixture to two-thirds of its initial volume.

A slight excess of mineral acid, such as sulfuric or hydrochloric acid is added to acidify the mixture which is then chilled and the solid which separates is filtered off. It is then treated with an aqueous solution of dilute sodium hydroxide to dissolve the hydantoin from the solid unreacted benzophenone. After filtration, the alkaline extract is then acidified to cause the separation of solid pure diphenylhydantoin which is filtered off and dried. It melts at 293° to 296°C.

A net yield of about 95% is obtained by the procedure described above. If the time of warming the reaction mixture is increased three-or four-fold, practically 100% net yields are obtained. The same high net yields are also obtained by heating for even longer periods of time. For example, by heating for 90 hours, a 100% net yield, or 67% gross yield, is obtained.

References

Merck Index 7204
Kleeman and Engel p. 722
PDR pp. 1334, 1337
DOT 9 (6) 245 (1973)
I.N. p. 767
REM p. 1081
Henze, H.R.; US Patent 2,409,754; October 22, 1946; assigned to Parke, Davis and Company

PHETHENYLATE SODIUM

Therapeutic Function: Anticonvulsant

Chemical Name: 5-Phenyl-5-(2-thienyl)-2,4-imidazolidinedione monosodium salt

Common Name: -

Structural Formula:

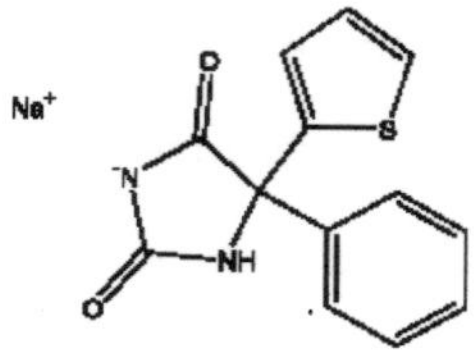

Chemical Abstracts Registry No.: 510-34-9

Trade Name	Manufacturer	Country	Year Introduced
Thiantoin	Lilly	US	1950

Raw Materials

Phenyl-(2-thienyl)ketone
Potassium cyanide
Ammonium carbonate

Manufacturing Process

The 5-phenyl-5-(2-thienyl)hydantoin is prepared by heating a mixture of 5.64 g (0.03 mol) of phenyl-(2-thienyl)ketone, 3.25 g (0.03 mol) of potassium cyanide and 10.2 g (0.09 mol) of ammonium carbonate in 75 cc of 50% ethanol for 28 hours at a temperature of about 110°C. An additional 3.25 g of potassium cyanide and 3 g of ammonium carbonate are added and the mixture heated for 24 hours at about 110°C.

The reaction mixture is removed and about half of the liquid evaporated, an oil separating during the process. The mixture is acidified with concentrated hydrochloric acid and extracted with two 100 cc portions of ether. The extracts, which contain the 5-phenyl-5-(2-thienyl)hydantoin, are combined and the combined ether extracts are shaken with two 25 cc portions of 5% potassium hydroxide solution. The alkaline solution, which dissolves the 5-phenyl-5-(2-thienyl)hydantoin to form the potassium salt thereof, is acidifed with hydrochloric acid and heated to expel ether.

By the process of purification, 4.3 g of 5-phenyl-5-(2-thienyl)hydantoin is obtained, and from the ether layer, 2.2 g of unreacted ketone. The yield of the 5-phenyl-5-(2-thienyl)hydantoin is about 56%. The melting point of the purified 5-phenyl-5-(2-thienyl)hydantoinis about 256°C to 257°C.

References

Merck Index 7206
Spurlock, J.J.; US Patent 2,366,221; January 2, 1945

PHOLEDRINE SULFATE

Therapeutic Function: Sympathomimetic, Mydriatic, Analeptic, Vasopressor

Chemical Name: Phenol, p-(2-(methylamino)propyl)- sulfate

Common Name: Foledrine; Methylparedrine; Pholedrine sulfate

Chemical Abstracts Registry No.: 370-14-9 (Base); 6114-26-7

Structural Formula:

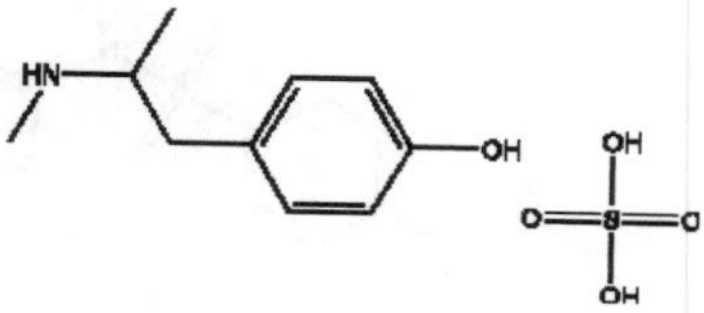

Trade Name	Manufacturer	Country	Year Introduced
Pholedrine Sulfate	ZYF Pharm Chemical	-	-
Paredrinol	Sigma-Aldrich	-	-
Pressitan	Allard	-	-
Pulsotyl	CHINOIN - BUDAPES	-	-

Raw Materials

p-Methoxybenzylmethylketone
Hydrogen
Nickel

Manufacturing Process

100 g β-(methoxyphenyl)isopropylamine prepared, for example by reduction of p-methoxybenzylmethylketone with hydrogen and nickel catalyst in presence of ammonia or by Mannix's method, 250 ml ethanol and calculated quantity of solution of formaldehyde and 70 g activated aluminum shaving were heated for 6 hours by stirring. The mixture was filtered, the solvent was removed in vacuum and the residue was dissolved in ethanol contained hydrochloric acid. The prepared hydrochloride of β-(p-methoxyphenyl)isopropylmethylamine had MP: 174°C.

In practice it is usually used as sulfate.

References

Knoll A.-G.; D.R. Patent No. 665,793; May 27, 1936; Assigned to Chemishe Fabriken in Ludvigshafen, Rhein, Germany
Knoll A.-G.; D.R. Patent No. 674,753; May 27, 1936; Assigned to Chemishe Fabriken in Ludvigshafen, Rhein, Germany

PHTHALYLSULFATHIAZOLE

Therapeutic Function: Antibacterial (intestinal)

Chemical Name: 2-[[[4-[(2-Thiazolylamino)sulfonyl]phenyl]amino] carbonyl]benzoic acid

Common Name: -

Structural Formula:

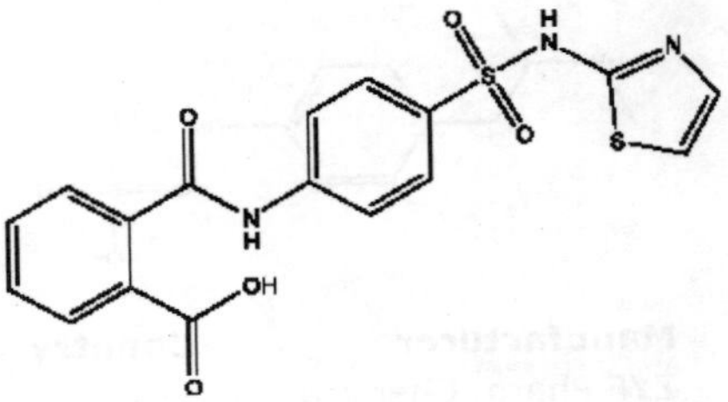

Chemical Abstracts Registry No.: 85-73-4

Trade Name	Manufacturer	Country	Year Introduced
Sulfathalidine	MSD	US	1946
Talidine	Clin Midy	France	1948
AFI-Ftalyl	A.F.I.	Norway	-
Colicitina	Panthox and Burck	Italy	-
Enterosteril	Ripari-Gero	Italy	-
Ftalysept	Ferrosan	Denmark	-
Gelotamide	Choay	France	-
Lyantil	Syntex Daltan	France	-
Novosulfina	Medosan	Italy	-
Phtalazol	Geistlich	Switz.	-
Phthalazol	Knoll	Australia	-
Sulfatalyl	Pharmacia	Sweden	-
Talisulfazol	Chemiek	E. Germany	-
Thalazole	May and Baker	UK	-

Raw Materials

Phthalic anhydride
Sulfathiazole

Manufacturing Process

5 g of phthalic anhydride was added to a boiling suspension of 10 g of sulfathiazole in 100 cc of alcohol. The mixture was then refluxed for 5 minutes after the addition was complete at which time all of the solids were in solution. The solution was then cooled and diluted with an equal volume of water, The white solid precipitate which formed was filtered and recrystallized from dilute alcohol, yielding 2-N^4-phthalylsulfanilamidothiazole,which decomposes above 260°C, according to US Patent 2,324,015.

References

Merck Index 7261
Kleeman and Engel p. 723
OCDS Vol. 1 p. 132 (1977)
I.N. p. 769

Moore, M.L.; US Patent 2,324,013; July 13, 1943; assigned to Sharp and Dohme, Incorporated
Moore, M.L.; US Patent 2,324,014; July 13, 1943; assigned to Sharp and Dohrne, Incorporated
Moore, M.L.; US Patent 2,324,015; July 13, 1943; assigned to Sharp and Dohme, Incorporated

PHYTATE SODIUM

Therapeutic Function: Hypocalcemic

Chemical Name: myo-Inositol hexakis(dihydrogen phosphate) sodium salt

Common Name: -

Structural Formula:

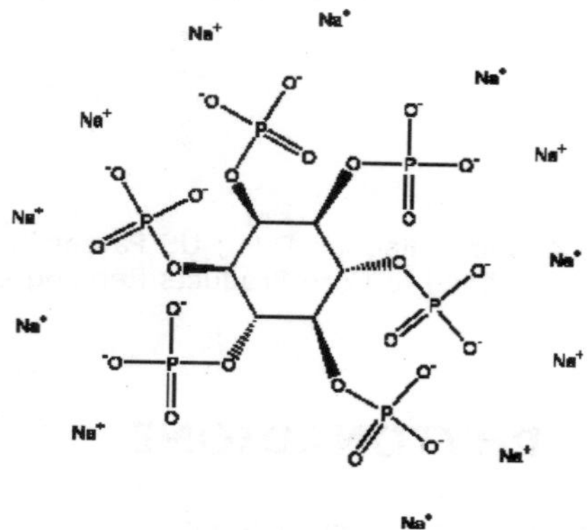

Chemical Abstracts Registry No.: 83-86-3 (Acid)

Trade Name	Manufacturer	Country	Year Introduced
Rencal	Squibb	US	1962
Iliso	Made	Spain	-

Raw Materials

Corn steep water
Lime
Cation exchange resin

Manufacturing Process

Cereal grains are particularly rich in phytates; corn steep water produced in the wet milling of corn, is one of the best sources of such material. To recover the phytate from corn steep water it is customary to neutralize the same with an alkaline material, suitably lime, causing the phytate to precipitate as a crude salt which can be removed readily by filtration. This material contains substantial amounts of magnesium, even though lime may have been employed as precipitant, and traces of other metallic ions, as well as some proteinaceous materials and other contaminants from the steep water. It may be partially purified by dissolving in acid and reprecipitating but, nevertheless, such commercial phytates do not represent pure salts. They always contain some magnesium, appreciable amounts of iron and nitrogenous materials, and traces of heavy metals, such as copper.

Heretofore, no economical method for preparing pure phytic acid was known. The classical method was to dissolve calcium phytate in an acid such as hydrochloric acid, and then add a solution of a copper salt, such as copper sulfate to precipitate copper phytate. The latter was suspended in water and treated with hydrogen sulfide, which formed insoluble copper sulfide and released phytic acid to the solution. After removing the copper sulfide by filtration, the filtrate was concentrated to yield phytic acid as a syrup.

The phytic acid in the form of a calcium phytate press cake may however be contacted with a cation exchange resin to replace the calcium with sodium to yield phytate sodium.

References

Merck Index 7269
I.N. p. 25
Baldwin, A.R., Blatter, L.K. and Gallagher, D.M.; US Patent 2,815,360; December 3, 1957; assigned to Corn Products Refining Co.

PHYTONADIONE

Therapeutic Function: Prothrombogenic vitamin

Chemical Name: 2-Methyl-3-(3,7,11,15-tetramethyl-2-hexadecenyl)-1,4-naphthalenedione

Common Name: Vitamin K; Phytomeanadion; Phylloquinone

Structural Formula:

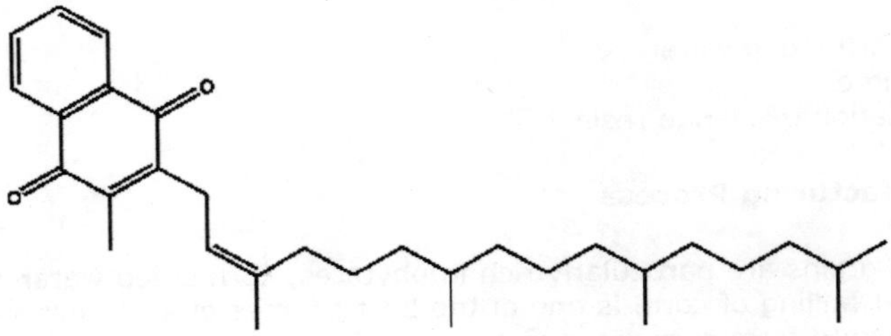

Chemical Abstracts Registry No.: 84-80-0

Trade Name	Manufacturer	Country	Year Introduced
Mephyton	MSD	US	1941
Konakion	Roche	US	1959
Aquamephyton	MSD	US	1960
Mono-Kay	Abbott	US	1961
Eleven-K	Nippon Shinyaku	Japan	-
Hymeron	Yamanouchi	Japan	-
Kanavit	Spofa	Czechoslovakia	-
Kativ-N	Takeda	Japan	-
Kayeine	Kanto	Japan	-
Kaywan	Eisai	Japan	-
K-Eine	Hokuriku	Japan	-
Keipole	Kyowa	Japan	-
Kennegin	Kowa	Japan	-
Kephton	Toyo Jozo	Japan	-
Kinadione	Chugai	Japan	-
Kisikonon	Kyorin	Japan	-
K-Top Wan	Sawai	Japan	-
Monodion	Maruko	Japan	-
Nichivita-K	Nichiiko	Japan	-
One-Kay	Mohan	Japan	-
Synthex P	Tanabe	Japan	-
Vita-K	Kobayashi	Japan	-
Vitamine K1	Delagrange	France	-

Raw Materials

2-Methyl-1,4-naphthohydroquinone
Phytol
Hydrogen

Manufacturing Process

11 parts by weight of 2-methyl-1,4-naphthohydroquinone, 30 parts by volume of water-free dioxane and 1.5 parts by volume of boron trifluoride etherate are heated to 50°C. While agitating and introducing nitrogen, 10 parts by weight of phytol dissolved in 10 parts by volume of dioxane are added in the course of 15 minutes. Thereupon, the dark colored reaction mixture is stirred for 20 additional minutes at 50°C, cooled down and 60 parts by volume of ether are added. The reaction mixture is washed first with water, then with a mixture of 3 parts of N-sodium hydroxide and 2 parts of a 2.5% solution of sodium hydrosulfite and again with water. The aqueous extracts are washed with ether. The ether solutions are collected, dried over sodium sulfate and concentrated, toward the end under reduced pressure.

The waxlike condensation product so obtained is mixed with 60 parts by volume of petroleum ether (boiling limits 30°C to 40°C) and agitated with hydrogen in the presence of a little active palladium lead catalyst (Pd-$CaCO_3$

catalyst, the activity of which is reduced by the addition of lead and quinoline). During the operation, the condensation product separates in the form of a voluminous white precipitate. The latter is separated by filtration in the absence of air while adding an inert coarse-grained adsorption agent (for example, aluminum silicate salt for filter purposes), and washed with cooled petroleum ether. Thereupon, the 2-methyl-3-phytyl-1,4-naphthohydroquinone is extracted from the filter cake by means of ether, the ethereal solution is concentrated to 100 parts by volume and the reaction product is oxidized by stirring the solution with 6.6 parts by weight of silver oxide during 30 minutes. The solution is filtered through sodium sulfate, the latter is rinsed with ether and the solvent is evaporated. There are obtained 5.7 parts by weight of 2-methyl-3-phytyl-1,4-naphthoquinone (vitamin K1) in the form of a golden yellow oil.

References

Merck Index 9834
Kleeman & Engel p. 724
PDR pp. 1140, 1488
I.N. p. 770
REM p. 1011
Isler, O. and Doebel, K.; US Patent 2,683,176; July 6, 1954; assigned to Hoffmann-La Roche, Inc.

PICOPERINE

Therapeutic Function: Antitussive

Chemical Name: N-(2-Piperidinoethyl)-N-(2-pyridylmethyl)aniline

Common Name: Picoperamidine

Structural Formula:

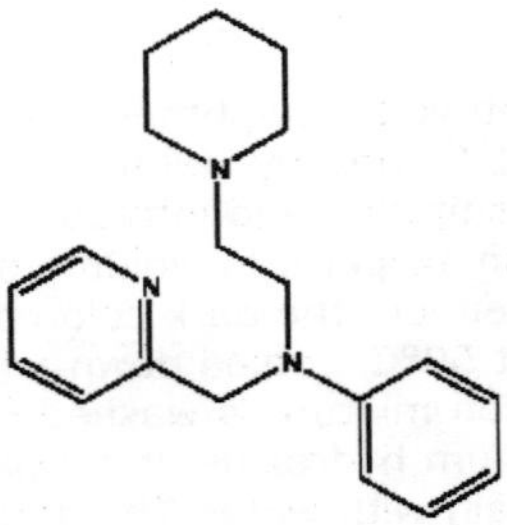

Chemical Abstracts Registry No.: 21755-66-8

Trade Name	Manufacturer	Country	Year Introduced
Coben	Takeda	Japan	1971

Raw Materials

N-(2-Pyridylmethyl)aniline
Sodium amide
2-Piperidinoethyl chloride

Manufacturing Process

To a simultaneously stirred and refluxed suspension of 5.6 parts by weight of sodamide in 60 parts by volume of anhydrous toluene, there is added dropwise a solution of 18.4 parts by weight of N-(2-pyridylmethyl)aniline in 20 parts by volume of anhydrous toluene. After the addition is complete, the mixture is refluxed for two hours under constant stirring.

To the resulting mixture there is added dropwise a solution of 14.9 parts by weight of 2-piperidinoethyl chloride in 20 parts by volume of anhydrous toluene and the whole mixture is stirred and refluxed for another two hours. After cooling, water is added carefully to decompose the unreacted sodamide, the separated toluene layer is dried over anhydrous sodium sulfate and the solvent removed under reduced pressure.

The residual oil is subjected to distillation under reduced pressure, the fraction boiling in the range of 185°C to 198°C/4 mm Hg being collected. Purification of the fraction by redistillation under reduced pressure gives 22.5 parts by weight of N-(2-piperidinoethyl)-N-(2-pyridylmethyl)-aniline which boils at 195°C to 196°C/4 mm Hg. Yield 76.3%.

References

Merck Index 7285
DOT 8 (5) 185 (1972)
I.N. p. 771
Mitano, S. and Kase, Y.; US Patent 3,471,501; October 7, 1969; assigned to Takeda Chemical Industries, Ltd.

PICOSULFATE SODIUM

Therapeutic Function: Laxative

Chemical Name: 4,4'-(2-Pyridinylmethylene)bisphenol-bis(hydrogen sulfate) (ester) disodium salt

Common Name: Picosulfol

Chemical Abstracts Registry No.: 10040-45-6

Raw Materials

2-Pyridine aldehyde
Sodium hydroxide
2-Chlorophenol
Chlorosulfonic acid

Structural Formula:

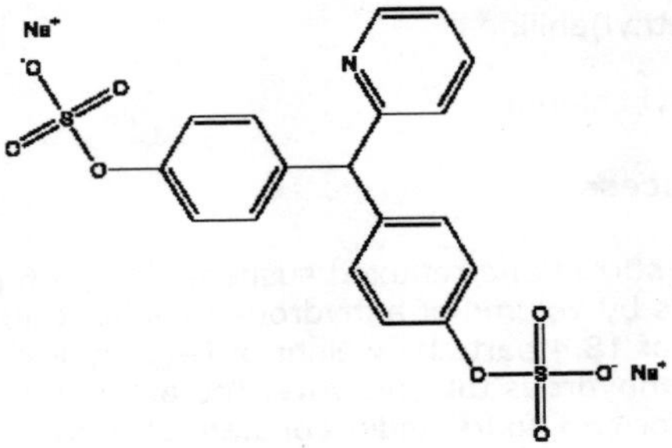

Trade Name	Manufacturer	Country	Year Introduced
Guttalax	De Angelini	Italy	1967
Laxoberal	Thomae	W. Germany	1972
Laxoberal	W.B. Pharm.	UK	1975
Laxoberon	Teijin	Japan	1980
Contumax	Casen	Spain	-
Evacuol	Almirall	Spain	-
Gocce Euchessina	Antonetto	Italy	-
Gocce Lassative Aicardi	Aicardi	Italy	-
Laxante Azoxico	Bescansa	Spain	-
Laxidogol	Dolorgiet	W. Germany	-
Picolax	Falqui	Italy	-
Skilax	Prodes	Spain	-
Trali	Sintyal	Argentina	-

Manufacturing Process

Preparation of 3,3'-Dichloro-4,4'-Dioxy-Diphenyl-(2-Pyridyl)-Methane: 75 g (0.7 mol) of 2-pyridinaldehyde are dropped during about 1 hour to a homogeneous mixture [obtained between 0° and 10°C from 107 ml of concentrated sulfuric acid and 292.9 g (2.28 mols) of 2-chlorophenol], maintaining the temperature between 0° and 5°C. The mixture is stirred for ½ hour at this temperature, which is then allowed to rise spontaneously, taking care not to exceed 30°C. After stirring for 1½ hours, the mixture is maintained overnight at room temperature, then it is dissolved, with external cooling, with a 10% sodium hydroxide solution, filtered with charcoal and neutralized with 5% hydrochloric acid. The precipitate obtained, consisting of crude product, filtered, washed with water, dried, triturated with ether and dried again, weighs 211 g.

The isomer 2,4'-dioxy-3,3'-dichloro-diphenyl-(2-pyridyl)-methane is removed by thoroughly washing with 430 ml of 95°C boiling alcohol, obtaining 167 g of isomer-free product (yield 69%). The 3,3'-dichloro-4,4'-dioxy-diphenyl-(2-pyridyl)-methane is a white solid, crystallizing from 95% alcohol; MP 212° to 215°C.

Preparation of 4,4'-Dioxy-Diphenyl-(2-Pyridyl)-Methane: 100 g of 3,3'-dichloro-4,4'-dioxydiphenyl-(2-pyridyl)-methane, obtained as above described, are dissolved in 660 ml of 10% sodium hydroxide and 49 g of Raney-nickel alloy are added to the solution with vigorous stirring, at room temperature and during 4 hours. The mixture is stirred overnight at room temperature, then it is filtered and brought to pH 5 with 10% acetic acid. The precipitate obtained, filtered, washed and dried is then dissolved in 1,500 ml of 95°C boiling alcohol to eliminate the insoluble salts. The residue obtained after the evaporation of the alcoholic solution weighs 74 g (yield 92%). The yield in respect to 2-pyridinaldehyde is 63.5%. The compound is a white solid, crystallizing from 95% alcohol; MP 248° to 250.5°C, according to US Patent 3,558,643.

Preparation of Disodium 4,4'-Disulfoxy-Diphenyl-(2-Pyridyl)-Methane: In ½ hour, 102 g chlorosulfonic acid are added to a solution of 100 g 4,4'-dihydroxydiphenyl-(2-pyridyl)methane in 750 ml of anhydrous pyridine, the temperature being maintained at between 0° and 5°C. Towards the end of the addition of acid, a precipitate is formed which is slowly redissolved during subsequent agitation.

Upon completion of the addition, the mixture is agitated for 7 hours at ambient temperature. The solution is then poured into 3 liters of water/ice obtaining a clear solution of dark yellow color which is rendered alkaline upon phenolphthalein with 30% NaOH and extracted with ethyl ether to eliminate the majority of the pyridine. The mixture is filtered with active charcoal, the pH adjusted to 8 with hydrochloric acid 1:1 and extracted with chloroform to remove the 4,4'-dihydroxydiphenyl-(2-pyridyl)-methane which has not reacted.

The aqueous solution is then concentrated to dryness at an outside temperature of 40° to 45°C and at low pressure. The residue, obtained by drying in a vacuum at 40° to 45°C is triturated in a mortar with ethyl ether and, after filtration, is extracted with 3,400 ml boiling absolute ethanol. The ethanol extract is separated from the undissolved part by filtration, cooled and the product which crystallizes by cooling is filtered and dried at 40°C in a vacuum. In that manner the disodium (4,4'-disulfoxy-diphenyl)-(2-pyridyl)methane bi-hydrate is obtained, which takes the form of a white solid, according to US Patent 3,528,986.

References

Merck Index 7286
Kleeman & Engel p. 725
DOT 8 (8) 302 (1972)
I.N. p. 771
Pala, G.; US Patent 3,528,986; September 15, 1970; assigned to Istituto de Angeli S.p.A., Italy
Pala, G.; US Patent 3,558,643; January 26, 1971; assigned to Istituto de Angeli S.p.A., Italy

PIFARNINE

Therapeutic Function: Antiulcer

Chemical Name: 1-(1,3-Benzodioxol-5-ylmethyl)-4-(3,7,11-trimethyl-2,6,10-dodecatrienyl)piperazine

Common Name: -

Structural Formula:

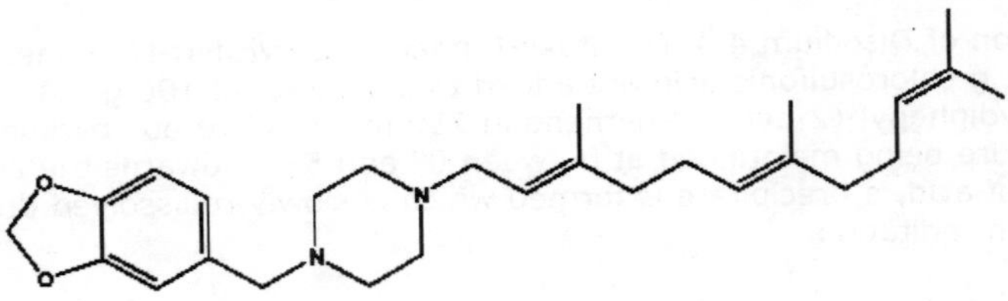

Chemical Abstracts Registry No.: 56208-01-6

Trade Name	Manufacturer	Country	Year Introduced
Pifazin	Pierrel	Italy	1983

Raw Materials

1-Bromo-3,7,11-trimethyl-2,6,10-dodecatriene
Piperonylpiperazine
Triethylamine

Manufacturing Process

A solution of 45 mmols of 1-bromo-3,7,11-trimethyl-2,6,10-dodecatriene (obtained from synthetic farnesol, commercially available and containing four isomers) in 10 ml of benzene was added dropwise at 0°C to a stirred solution of 45 mmols of piperonylpiperazine in 60 ml of benzene containing 5 g of triethylamine. The mixture was stirred for 2 hours and then the precipitated triethylammonium bromide was filtered off. The benzene solution was washed first with water and then with K_2CO_3 solution and finally dried (K_2CO_3). Removal of benzene under reduced pressure gave a crude oily residue which was dissolved in acetone and treated at 5°C to 8°C with a slight excess of 37% HCl solution. The precipitated hydrochloride was filtered, washed with acetone and with absolute ethanol. The corresponding base was purified on a silica gel column and the purity of all fractions was checked by thin layer chromatography and gas liquid chromatography. Thin layer chromatography on silica gel gave three spots in the solvent system ethylacetate-petrol ether 1:1. Gas liquid chromatography showed three peaks indicating the presence of four possible isomers. The pure product was a colorless oil.

References

Merck Index 7299
DFU 2 (12) 829 (1977)
Kleeman and Engel p. 725
I.N. p. 772
Zumin, S.T., Riva, M. and Iafolla, G.; US Patent 3,875,163; April 1, 1975; assigned to Pierrel S.p.A. (Italy)

PILOCARPINE HYDROCHLORIDE

Therapeutic Function: Cholinergic

Chemical Name: 2(3H)-Furanone, 3-ethyldihydro-4-[(1-methyl-1H-imidazol-5-yl)methyl]-, (3S-cis)-, monohydrochloride

Common Name: Pilocarpine hydrochloride; Pilokarpin hydrochloride

Structural Formula:

HCl

Chemical Abstracts Registry No.: 54-71-7; 92-13-7 (Base)

Trade Name	Manufacturer	Country	Year Introduced
Andre Carpine	Andre Laboratries Pvt. Ltd.	India	-
Pilokarpine hydrochloride	Merck KGaA	Germany	-
Pilokarpine hydrochloride	Boehringer Ingelheim Pharma KG	Germany	-
Pilokarpine hydrochloride	Sigma	-	-

Raw Materials

Sodium hydride
Palladium on carbon
Diisobutylaluminum hydride
Hydrogen chloride
Ethyl diethylphosphonoethoxyacetate
1-Methylimidazole-5-aldehyde
2-Diethylphosphonobutyric acid
4-Dimethylaminopyridine

Manufacturing Process

The 1-methylimidazole-5-aldehyde is easily accessible from sarcosine methyl ester hydrochloride and dimethylamino-2-azaprop-2-en-1-

ylidenedimethylammonium.

0.14 mol of ethyl diethylphosphonoethoxyacetate is slowly added dropwise with stirring and under inert gas to a suspension of 0.14 mol of NaH (paraffin-free) in 250 ml of abs. THF, the mixture is stirred for 1 h at 20°C and a solution of 0.093 mol of 1-methylimidazole-5-aldehyde in 100 ml of abs. THF is added dropwise. After stirring at 20°C for 10 min, the solvent is distilled off in vacuo, the residue is taken up in a little H_2O, and the solution is acidified with 1 N HCl and washed several times with ether. The aqueous phase is rendered alkaline using 2 N NaOH with cooling (0°-5°C) and extracted several times with CH_2Cl_2. After drying of the organic extracts with Na_2SO_4, the solvent is removed in vacuo and 2-ethoxy-3-[(1-methyl-1H-imidazol-5-yl)methyl]-acrilic acid ethyl ester. Yield: 99% of theory.

122 ml of 45% diisobutylaluminum hydride solution (328 mmol) are slowly added dropwise under inert gas, with stirring and ice cooling, to a solution of 137 mmol of 2-ethoxy-3-[(1-methyl-1H-imidazol-5-yl)methyl]-acrilic acid ethyl ester in 600 ml of abs. C_6H_6. Stirring of the mixture is continued for a further 30 min at 0°-5°C and 600 ml of CH_3OH, then 100 ml of H_2O, are slowly added. The hydroxide precipitate is filtered off with suction and washed several times with hot CH_3OH. After drying of the combined filtrates the solvents are distilled off in vacuo and the residue is crystallized using C_2H_5OH. 2-ethoxy-3-[(1-methyl-1H-imidazol-5-yl)methyl]-prop-2-en-1-ol was obtained. Yield: 100% of theory. The crude product is pure enough for the subsequent reaction. Recrystallization of an analytical sample from CH_3OH/acetone: melting point 129°C.

A solution of 58 mmol of 2-ethoxy-3-[(1-methyl-1H-imidazol-5-yl)methyl]-prop-2-en-1-ol in 116.6 ml of HCl (= 116.6 mmol) is stirred at 30°-35°C for 1.5 h and concentrated in vacuo at the same temperature. The residual HCl is removed by distillation with $CHCl_3$ in vacuo. After seeding, the residue crystallizes at 20°C (15 h) 1-hydroxy-3-[(1-methyl-1H-imidazol-5-yl)methyl]-propan-2-one hydrochloride. The crystallizate is filtered off with suction, washed with a little CH_3OH and dried in vacuo. Yield: 86% of theory; melting point 190°C.

About 80-90% of the equivalent amount of $NaOCH_3$ solution in CH_3OH is slowly added dropwise at 20°C with stirring and exclusion of moisture to a suspension of 21.24 mmol of 1-hydroxy-3-[(1-methyl-1H-imidazol-5-yl)methyl]-propan-2-one hydrochloride in 80 ml of CH_3OH, in the course of which the pH of 6.5 is not to be exceeded. The solvent is distilled off in vacuo at a maximum of 30°C and the residue of 1-hydroxy-3-[(1-methyl-1H-imidazol-5-yl)methyl]-propan-2-one is purified by flash chromatography (silica gel; $CHCl_3$/CH_3OH). Yield: 100% of theory; viscous, orange-colored oil.

Catalytic amounts of 4-dimethylaminopyridine and a solution of 21.3 mmol of 1-hydroxy-3-[(1-methyl-1H-imidazol-5-yl)methyl]-propan-2-one in 80 ml of CH_2Cl_2 are added to a solution of 26.44 mmol of 2-diethylphosphonobutyric acid in 40 ml of purified CH_2Cl_2. After cooling to 0°-5°C, a solution of 23.5 mmol of dicyclohexylcarbodiimide in 60 ml of CH_2Cl_2 is added dropwise and the mixture is stirred for 1 h at 0°-5°C and for 2 h at 20°C. The crystallized dicyclohexylurea is filtered off with suction and the filtrate is washed with H_2O and saturated $NaHCO_3$ solution. After drying of the organic phase the solvent is distilled off at 30°C in vacuo and the residue of 2-diethoxy-phosphoryl)-butyric acid 3-[(1-methyl-1H-imidazol-5-yl)methyl]-2-oxo-propyl ester is purified by flash chromatography (silica gel; ethyl acetate/CH_3OH). Yield: 95% of theory of a viscous, orange-colored oil.

A mixture of 5 mmol each of 80% NaH and 15-crown-5 in 50 ml of absol. toluene is stirred at 20°C under inert gas for 10 min and a solution of 5 mmol of 2-diethoxy-phosphoryl)-butiric acid 3-[(1-methyl-1H-imidazol-5-yl)methyl]-2-oxo-propyl ester in 50 ml of absol. toluene is then added dropwise. Stirring is continued for a further 15 min under inert gas and the mixture is hydrolyzed with a little water until phase separation is detectable. After separating off the organic phase, the aqueous layer is saturated with NaCl and extracted several times with $CHCl_3$. The combined organic phases are the solvent is distilled off at 40°C in vacuo and the residue 3-ethyl-4-[(1-methyl-1H-imidazol-5-yl)methyl]-5H-furan-2-one is purified twice by flash chromatography (silica gel; ethyl acetate/CH_3OH). Yield: 52% of theory; virtually colorless oil.

1.36 mmol of 3-ethyl-4-[(1-methyl-1H-imidazol-5-yl)methyl]-5H-furan-2-one in 15.5 ml of CH_3OH are hydrogenated for 5 h at 50 bar and 60°C using 210 mg of Pd/carbon (10%). After filtering off the catalyst and distilling off the solvent at 30°C in vacuo, the oily residue (about 250 mg) is treated with 10 ml of 1 N HCl and the mixture is stirred for 3 h at 20°C. The hydrochloric acid is distilled off in vacuo at 35°-40°C, the oily residue is taken up in a little CH_3OH and ether is added. The precipitate of pilocarpine hydrochloride is recrystallized from CH_3OH/ether. Yield: 73% of theory; melting point 210°C.

References

Reimann E.; US Patent No. 5,530,136; Jan. 25, 1996; Assigned: Merck Patent Gesellschaft Mit Beschrankter Haftung, Darmsstadt, Germany

Reuther G.R.; US Patent No. 5,059,531; Oct. 22, 1991; Assigned: Merck Patent Gesellschaft Mit Beschrankter Haftung, Darmsstadt, Germany

Courtois D. et al.; US Patent No. 5,569,593; Oct. 29, 1996; Assigned: Nestec S.A.,Vevey, Switzerland

PIMEFYLLINE NICOTINATE

Therapeutic Function: Coronary vasodilator

Chemical Name: 3,7-Dihydro-1,3-dimethyl-7-[2-[(3-pyridinylmethyl)amino] ethyl]-1H-purine-2,6-dione nicotinate

Common Name: 7-(β-3'-Picolylaminoethyl)theophylline nicotinate

Structural Formula:

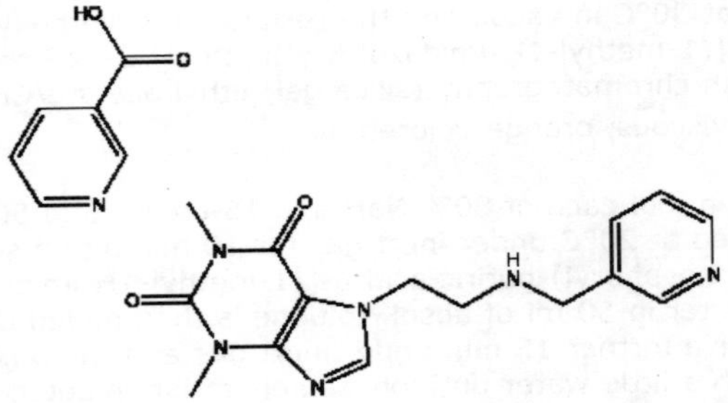

Chemical Abstracts Registry No.: 10058-07-8; 10001-43-1 (Base)

Trade Name	Manufacturer	Country	Year Introduced
Teonicon	Bracco	Italy	1975
Teonicon	Neopharmed	Japan	-

Raw Materials

7-(2-Bromethyl)theophylline
3-Picolylamine
Nicotinic acid

Manufacturing Process

77 g 7-(β-bromoethyl)-theophylline (C.A. 50, 12071f) and 57.8 g 3-picolylamine in 750 ml toluene were refluxed 16 hours with vigorous agitation. The 3-picolylamine hydrobromide formed was filtered off, and the filtrate was evaporated in a vacuum to about one-third of its original volume. About 300 to 400 ml diisopropyl ether were added, and the solution was seeded with a few pure crystals of the desired product.

7-(β-3'-picolylaminoethyl)-theophylline crystallized over a period of a few hours. It was filtered off with suction, washed with a little diisopropyl ether, and dried. The yield of crude product was 69.3 g (82%), its MP 103° to 106°C. The MP was 111° to 112°C after recrystallization from isopropyl acetate. The compound was identified by microanalysis.

39.3 g 7-(β-3'-picolylaminoethyl)-theophylline were dissolved in 300 ml boiling isopropanol, and 15.4 g nicotinic acid were added to the solution in which the acid promptly dissolved. The nicotinate formed crystallized after a short time. It was filtered with suction and dried. The yield was 52.3 g (95.5%). The MP of 159° to 160°C was not significantly changed by recrystallization from ethanol.

References

Merck Index 7306

Kleeman & Engel p. 727
Suter, H. and Zutter, H.; US Patent 3,350,400; October 31, 1967; assigned to Eprova Limited, Switzerland

PIMOZIDE

Therapeutic Function: Antipsychotic

Chemical Name: 1-[1-[4,4-Bis(4-fluorophenyl)butyl]-4-piperidinyl]-1,3-dihydro-2H-benzimidazol-2-one

Common Name: -

Structural Formula:

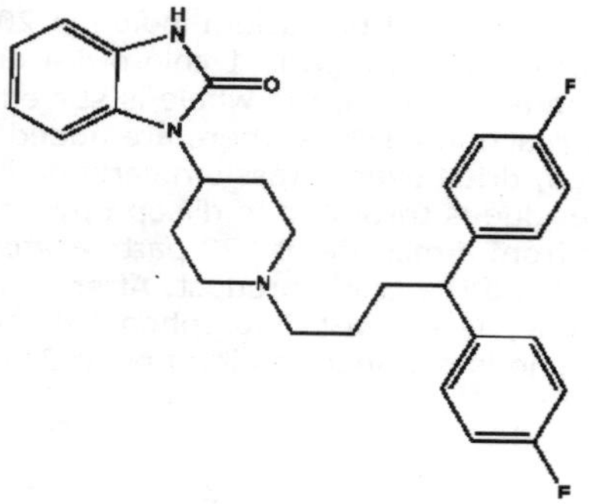

Chemical Abstracts Registry No.: 2062-78-4

Trade Name	Manufacturer	Country	Year Introduced
Orap	Janssen	W. Germany	1971
Opiran	Cassenne	France	1971
Orap	Janssen	UK	1971
Orap	Fujisawa	Japan	1974
Orap	Janssen	Italy	1977
Norofren	Dif-Dogu	Turkey	-
Oralep	Abic	Israel	-
Pimotid	Medica	Finland	-

Raw Materials

Thionyl chloride
Cyclopropyl-di-(4-fluorophenyl)-carbinol
Hydrogen
4-(2-Oxo-1-benzimidazolinyl)piperidine

Manufacturing Process

To a solution of 130 parts cyclopropyl-di-(4-fluorophenyl)-carbinol in 240 parts benzene are added dropwise 43 parts thionyl chloride. The whole is refluxed until no more gas is evolved. The reaction mixture is then evaporated. The residue is distilled in vacuo, yielding 4-chloro1,1-di-(4-fluorophenyl)-1-butene, boiling point 165°C to 167°C at 6 mm pressure; n_D^{20}: 1.5698; d_{20}^{20}:1.2151.

A solution of 61 parts 4-chloro-1,1-di-(4-fluorophenyl)-1-butene in 400 parts 2-propanol is hydrogenated at normal pressure and at room temperature in the presence of 5.5 parts palladium-on-charcoal catalyst 10% (exothermic reaction: temperature rises to about 30°C). After the calculated amount of hydrogen is taken up, hydrogenation is stopped. The catalyst is filtered off and the filtrate is evaporated. The oily residue is distilled in vacuo, yielding 1-chloro-4,4-di-(4-fluorophenyl)-butane, boiling point 166°C to 168°C at 6 mm pressure; n_D^{20}: 1.5425; d_{20}^{20}:1.2039.

To a mixture of 4.4 parts of 4-(2-oxo-1-benzimidazolinyl)-piperidine, 3.3 parts sodium carbonate, a few crystals of potassium iodide in 200 parts 4-methyl-2-pentanone are added portionwise 6.2 parts 1-chloro-4,4-di-(4-fluorophenyl)-butane. After the addition is complete, the whole is stirred and refluxed for 65 hours. After cooling the reaction mixture, there are added 70 parts water. The organic layer is separated, dried over potassium carbonate, filtered and evaporated. The solid residue is triturated in diisopropyl-ether, filtered off again and recrystallized from a mixture of 120 parts acetone and 80 parts 4-methyl-2-pentanone, yielding the crude product. After recrystallization of this crop from 80 parts acetone, 1-[4,4-di-(4fluorophenyl)-butyl]-4-(2-oxo-1-benzimidazolinyl)-piperidine is obtained, melting point 217°C to 219°C.

References

Merck Index 7310
Kleeman & Engel p. 727
PDR p. 1091
OCDS Vol. 2 p. 390 (1980)
DOT 5 (1) 36 (1969); 7 (5) 176 (1971); and 9 (6) 235 (1973)
I.N. p. 774
REM p. 1092
Janssen, P.A.J.; US Patent 3,196,157; July 20, 1965; assigned to Research Laboratorium Dr. C. Janssen N.V. (Belgium)

PINAZEPAM

Therapeutic Function: Antidepressant

Chemical Name: 7-Chloro-1,3-dihydro-5-phenyl-1-(2-propynyl)-2H-1,4-benzodiazepin-2-one

Common Name: -

Structural Formula:

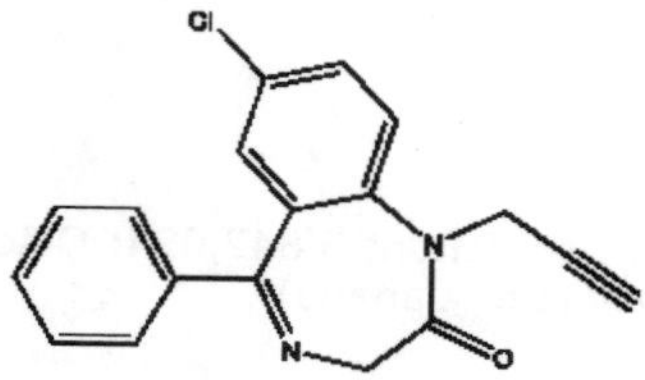

Chemical Abstracts Registry No.: 52463-83-9

Trade Name	Manufacturer	Country	Year Introduced
Domar	Zambeletti	Italy	1975
Duna	Zambeletti	Italy	-

Raw Materials

Propargyl bromide
2-Amino-5-chlorobenzophenone
Hydrazine hydrate
Phthalimidoacetyl chloride

Manufacturing Process

46.3 g (0.2 mol) of 2-amino-5-chlorobenzophenone were dissolved in 100 ml (1.28 mols) of propargyl bromide and the mixture refluxed for 4 hours. Thereafter, the whole was evaporated to dryness and the residue recrystallized from methanol to give 32.4 g (60.2%) of the desired 2-propargylamino-5-chlorobenzophenone; melting point 92°C to 93°C.

2.7 g (0.01 mol) of the 2-propargylamino-5-chlorobenzophenone obtained as above and 2.23 g (0.01 mol) of phthalimidoacetyl chloride were added to 30 ml of chloroform and the whole was refluxed overnight. Thereafter, the reaction mixture was evaporated to dryness and the residue recrystallized from methanol to give 2.66 g (58.3%) of the desired 2-(N-propargyl)phthalimidoacetamide-5-chlorobenzophenone. Melting point: 176°C.

A suspension of 22.8 g (0.05 mol) of 2-(N-propargyl)-phthalimidoacetamido-5-chlorobenzophenone in 250 ml ethanol containing 7.5 g hydrazine hydrate (0.15 mol) was heated under reflux for 2 hours, at the end of which time the reaction mixture was set aside overnight at ambient (25°C) temperature. Thereafter, the crystalline phthalyl hydrazide which had precipitated out was removed by filtration and washed with 3 x 50 ml aliquots of chloroform. The filtrate and washings were diluted with water and exhaustively extracted with chloroform. The chloroform extract was then evaporated and the residue washed with 100 ml hexane to promote crystallization. The crude 7-chloro-1-propargyl-3H-1,4-benzodiazepine-2(1H)-one was recrystallized from a methanol-water mixture to give 10.5 g (71.4%) of the pure product. Melting point: 140°C to 142°C.

References

Merck Index 7316
Kleeman & Engel p. 728
DOT 12 (4) 147 (1976)
I.N. p. 774
Podesva, C. and Vagi, K.; US Patent 3,842,094; October 15, 1974; assigned to Delmar Chemicals Ltd. (Canada)

PINDOLOL

Therapeutic Function: Beta-adrenergic blocker

Chemical Name: 2-Propanol, 1-(1H-indol-4-yloxy)-3-((1-methylethyl)amino)-

Common Name: Pindolol: Prindolol

Structural Formula:

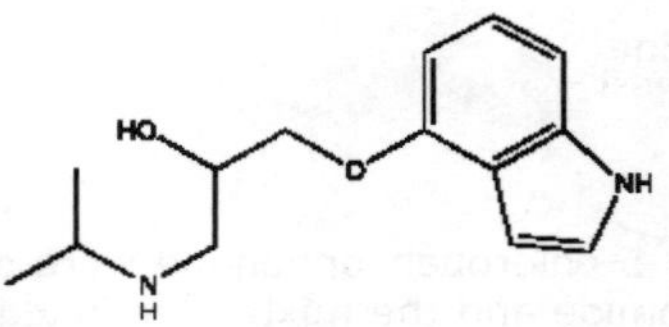

Chemical Abstracts Registry No.: 13523-86-9

Trade Name	Manufacturer	Country	Year Introduced
Betadren	Lagap	-	-
Cardilate	Teikoku	-	-
Lizenil	Nippon Kayaku	-	-

Raw Materials

Sodium amide	Ammonia
Hydrogen	Palladium on aluminum oxide
Epichlorohydrin	Sodium hydroxide
Isopropylamine	Tartaric acid

Manufacturing Process

4-Hydroxyindole is obtained by debenzylation of 4-benzyloxyindole with hydrogen in the presence of a 5% palladium catalyst on aluminium oxide.

10.0 g of 4-hydroxyindole and subsequently 7.4 ml of epichlorohydrin are added while stirring in an atmosphere of nitrogen to a solution of 2.73 g of sodium hydroxide in 65 ml of water. Stirring is effected at room temperature for a further 15 h, the reaction mixture is extracted 4 times with 50 ml of methylene chloride and the combined organic layers which have been dried over magnesium sulfate are evaporated at reduced pressure. So 3-chloro-1-(4-indolyloxy)-2-propanol is obtained.

The 3-chloro-1-(4-indolyloxy)-2-propanol is dissolved in 50 ml of toluene and 50 ml of isopropylamine and heated to the boil for 45 h. Evaporation to dryness is effected in a vacuum, the residue is shaken out thrice between ethyl acetate and a 1 N tartaric acid solution and a 5 N sodium hydroxide solution is then added to the combined tartaric acid phases until an alkaline reaction is obtained. The alkaline solution is shaken out thrice with 50 ml of methylene chloride, the extracts are dried over magnesium sulfate and the solvent evaporated in vacuum. The residue is crystallized from ethyl acetate/ether to give the 4-(2-hydroxy-3-isopropylaminopropoxy)indole.

References

Troxler F., Hofmann A.; GB Patent No. 1,138,968; Jan. 13, 1966; Assigned: Sandoz Ltd., of Lichtstrasse 35, Basle, Switzerland, a Swiss Body Corporate

PIOGLITAZONE HYDROCHLORIDE

Therapeutic Function: Antidiabetic

Chemical Name: (+-)-2,4-Thiazolidinedione, 5-((4-(2-(5-ethyl-2-pyridinyl) ethoxy)phenyl)methyl), monohydrochloride

Common Name: Pioglitazone hydrochloride

Structural Formula:

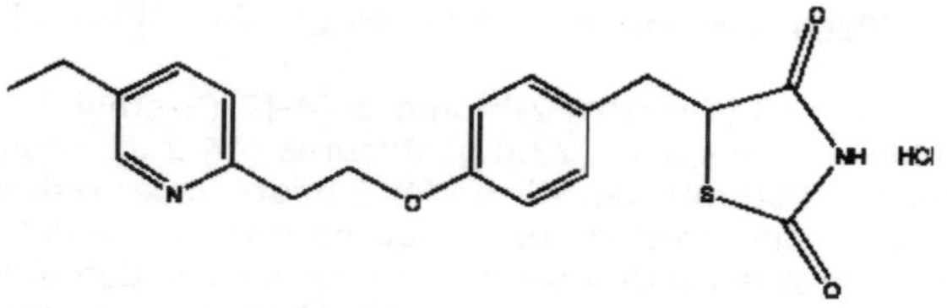

Chemical Abstracts Registry No.: 112529-15-4; 111025-46-8 (Base)

Trade Name	Manufacturer	Country	Year Introduced
Actos	Eli Lilly	USA	-
Actos	Takeda Pharmaceuticals	USA	-
Actos	Takeda Chemical Industry	Japan	-

Raw Materials

Palladium on carbon	2-(5-Ethyl-2-pyridyl)ethanol
Hydrogen bromide	Copper oxide
4-Fluoronitrobenzene	Sodium hydride
$NaNO_2$	Sodium methylate
Methyl acrylate	Sodium acetate

Manufacturing Process

To a solution of 2-(5-ethyl-2-pyridyl)ethanol (53.0 g) and 4-fluoronitrobenzene (47.0 g) in DMF (500 ml) was added portionwise under ice-cooling 60% sodium hydride in oil (16.0 g). The mixture was stirred under ice-cooling for one hour, then at room temperature for 30 min, poured into water and extracted with ether. The ether layer was washed with water and dried ($MgSO_4$). The solvent was evaporated off to give 4-[2-(5-ethyl-2-pyridyl)ethoxy]nitrobenzene as crystals (62.0 g, 62.9%). Recrystallization from ether-hexane gave colorless prisms, melting point 53°-54°C.

A solution of 4-[2-(5-ethyl-2-pyridyl)ethoxy]nitrobenzene (60.0 g) in methanol (500 ml) was hydrogenated at room temperature under one atmospheric pressure in the presence of 10% Pd-C (50% wet, 6.0 g). The catalyst was removed by filtration and the filtrate was concentrated under reduced pressure. The residual oil was dissolved in acetone (500 ml)-methanol (200 ml). To the solution was added a 47% HBr aqueous solution (152 g). The mixture was cooled, to which was added dropwise a solution of $NaNO_2$ (17.3 g) in water (30 ml) at a temperature not higher than 5°C. The whole mixture was stirred at 5°C for 20 min, then methyl acrylate (112 g) was added thereto and the temperature was raised to 38°C. Cuprous oxide (2.0 g) was added to the mixture in small portions with vigorous stirring. The reaction mixture was stirred until nitrogen gas evolution ceased, and was concentrated under reduced pressure. The concentrate was made alkaline with concentrated aqueous ammonia, and extracted with ethyl acetate. The ethyl acetate layer was washed with water and dried ($MgSO_4$) The solvent was evaporated off to leave methyl 2-bromo-3-{4-[2-(5-ethyl-2-pyridyl)ethoxy]phenyl}propionate as a crude oil (74.09 g, 85.7%).

A mixture of the crude oil of methyl 2-bromo-3-{4-[2-(5-ethyl-2-pyridyl)ethoxy]phenyl}propionate (73.0 g) thiourea (14.2 g), sodium acetate (15.3 g) and ethanol (500 ml) was stirred for 3 hours under reflux. The reaction mixture was concentrated under reduced pressure, and the concentrate was neutralized with a saturated aqueous solution of sodium hydrogencarbonate, to which were added water (200 ml) and ether (100 ml). The whole mixture was stirred for 10 min to yield 5-{4-[2-(5-ethyl-2-pyridyl)ethoxy]benzyl}-2-imino-4-thiazolidinone as crystals (0.3 g, 523.0%). Recrystallization from methanol gave colorless prisms, melting point 187°-188°C, dec.

A solution of 5-{4-[2-(5-ethyl-2-pyridyl)ethoxy]benzyl}-2-imino-4-thiazolidinone (23.5 g) in 2 N HCl (200 ml) was refluxed for 6 hours. The solvent was evaporated off under reduced pressure, and the residue was neutralized with a saturated aqueous solution of sodium hydrogencarbonate. The crystals (23.5 g, 97.5%) which precipitated were collected by filtration and recrystallized from DMF-H_2O to give 5-{4-[2-(5-ethyl-2-pyridyl)ethoxy]benzyl}-2,4-thiazolidinedione as colorless needles (20.5 g, 86.9%), melting point 183°-184°C.

In practice it is usually used as hydrochloride salt.

References

Meguro K., Fujita T.; US Patent No. 4,687,777; August 18, 1987; Assigned: Takeda Chemical Industries, LTD., Osaka, Japan

PIPAMAZINE

Therapeutic Function: Antiemetic

Chemical Name: 1-[3-(2-Chloro-10H-phenothiazin-10-yl)propyl]-4-pyridinecarboxamide

Common Name: -

Structural Formula:

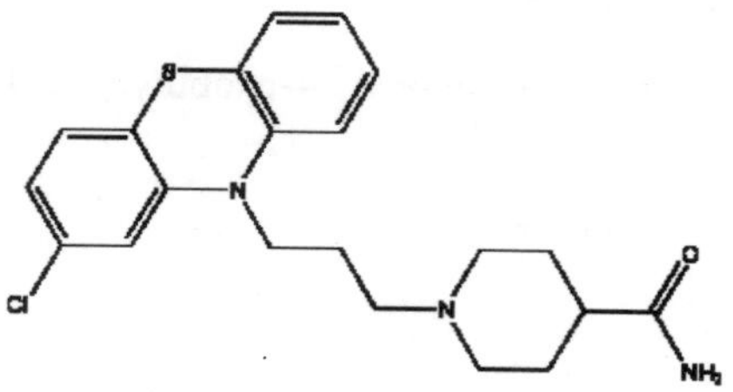

Chemical Abstracts Registry No.: 84-04-8

Trade Name	Manufacturer	Country	Year Introduced
Mornidine	Searle	US	1959
Nausidol	Gremy-Longuet	France	-

Raw Materials

4-Piperidinecarboxamide
2-Chloro-10-(γ-chloropropyl)phenothiazine

Manufacturing Process

To a stirred and refluxing suspension of 4.95 parts of 4-piperidinecarboxamide, 1 part of sodium iodide and 8.4 parts of potassium carbonate in 40 parts of butanone there are added in the course of 30 minutes 9.3 parts of 2-chloro-10-(γ-chloropropyl)phenothiazine in 40 parts of butanone. Stirring and refluxing are continued for 12 hours after which the mixture is cooled and filtered. The filtrate is concentrated under vacuum to give a residue which is recrystallized from a mixture of 2-propanol and petroleum ether. The 1-[γ-(2'-chloro-10'-phenothiazine)propyl]piperidine-4-carboxamide thus obtained melts at approximately 139°C.

This base is dissolved in a small amount of 2-propanol and treated with a 25% solution of hydrogen chloride in 2-propanol. Upon treatment of this solution with anhydrous ether a hydrochloride precipitates as a white solid melting at about 196°C to 197°C with formation of bubbles.

References

Merck Index 7326
Kleeman & Engel p. 729
OCDS Vol. 1 p. 385 (1977)
I.N. p. 775
Cusic, J.W. and Sause, H.W.; US Patent 2,957,870; October 25, 1960; assigned to G.D. Searle & Co.

PIPAMPERONE

Therapeutic Function: Antipsychotic

Chemical Name: 1'-[4-(4-Fluorophenyl)-4-oxobutyl]-[1,4'-bipiperidine]-4'-carboxamide

Common Name: Floropipamide

Structural Formula:

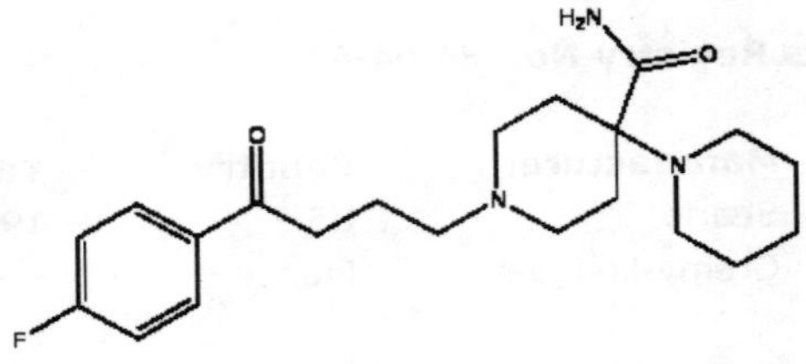

Chemical Abstracts Registry No.: 1893-33-0

Trade Name	Manufacturer	Country	Year Introduced
Dipiperon	Janssen	W. Germany	1961
Dipiperon	Janssen-Le Brun	France	1968
Piperonil	Lusofarmaco	Italy	1970
Propitan	Eisai	Japan	-

Raw Materials

Piperidine hydrochloride	1-Benzyl-4-piperidone
Potassium cyanide	γ-Chloro-4-fluorobutyrophenone
Sulfuric acid	Hydrogen

Manufacturing Process

To a stirred solution of 130.4 parts of potassium cyanide and 243.2 parts of piperidine hydrochloride in a mixture of 800 parts of water and 320 parts of ethanol is added portionwise 378 parts of 1-benzyl-4-piperidone. After about one hour a solid starts to precipitate. Stirring is continued for 24 hours. The reaction mixture is filtered and the solid is recrystallized from 1,200 parts of diisopropyl ether. On cooling to room temperature a first crop of 1-benzyl-4-cyano-4-piperidinopiperidine melting at about 104°C to 106°C is obtained. By concentrating and further cooling of the mother liquor a second crop of the above compound is obtained.

A mixture of 14.1 parts of 1-benzyl-4-cyano-4-piperidinopiperidine and 40 parts of 90% sulfuric acid is heated on a steam bath for 10 minutes. Without further heating, the mixture is stirred until a temperature of about 20°C is obtained. The mixture is then poured into 150 parts of ice-water and the resultant solution is alkalized with excess ammonium hydroxide solution. The aqueous solution is decanted from the precipitated oil. On treating this oil with 80 parts of acetone, crystallization sets in. After one hour the solid is filtered off and dried to yield 1-benzyl-4-piperidinopiperidine-4-carboxamide melting at about 137.5°C to 140°C.

A mixture of 215 parts of 1-benzyl-4-piperidinopiperidine-4-carboxamide, 1,200 parts of isopropyl alcohol, 1,000 parts of distilled water and 157 parts of hydrogen chloride is debenzylated under atmospheric pressure and at a temperature of about 40°C in the presence of 40 parts of a 10% palladium-on-charcoal catalyst. After the calculated amount of hydrogen is taken up, hydrogenation is stopped. The mixture is filtered and the filtrate is evaporated. The semisolid residue is treated with a mixture of 80 parts of acetone and 80 parts of benzene and evaporated again. The residue is triturated in 200 parts of methanol and filtered, yielding the dihydrochloride of 4-piperidinopiperidine-4-carboxamide melting at about 299°C to 300.8°C with decomposition. A sample of 20 parts of the dihydrochloride is dissolved in 30 parts of water. The aqueous solution is alkalized with 15 parts of 44% sodium hydroxide and stirred for a short time. The solid obtained is filtered off yielding crude product. To separate the free base from organic and inorganic salts, it is extracted overnight in a Soxhlet apparatus with toluene. The toluene extract is evaporated and the solid residue is filtered off, yielding 4-piperidinopiperidine-4-carboxamide melting at about 118.5°C to 119.5°C.

To a mixture of 4.1 parts of 4-piperidinopiperidine-4-carboxamide, 6.4 parts of sodium carbonate, and a few crystals of potassium iodide in 100 parts of anhydrous toluene is added dropwise a solution of 5.6 parts of γ-chloro-4-fluorobutyrophenone and 40 parts of anhydrous toluene at a temperature of 30°C to 40°C. The mixture is stirred and refluxed for 48 hours. The reaction mixture is cooled and divided between 50 parts of water and 60 parts of chloroform. The combined organic layers - toluene and chloroform - are dried over potassium carbonate, filtered, and evaporated. The oily residue solidifies on treatment with 80 parts of ether. After cooling for 30 minutes at 0°C, there is obtained 1-[γ-(4-fluorobenzoyl)propyl]-4-piperidinopiperidine-4-carboxamide melting at about 124.5°C to 126°C.

References

Merck Index 7327
Kleeman & Engel p. 729
OCDS Vol. 2 p. 388 (1980)
I.N. p. 775
Janssen, P.A.J.; US Patent 3,041,344; June 26, 1962; assigned to Research Laboratorium Dr. C. Janssen N.V. (Belgium)

PIPAZETHATE

Therapeutic Function: Antitussive

Chemical Name: 10H-Pyrido[3,2-b][1,4]benzothiadiazine-10-carboxylic acid 2-(2-piperidinoethoxy)ethyl ester

Common Name: -

Structural Formula:

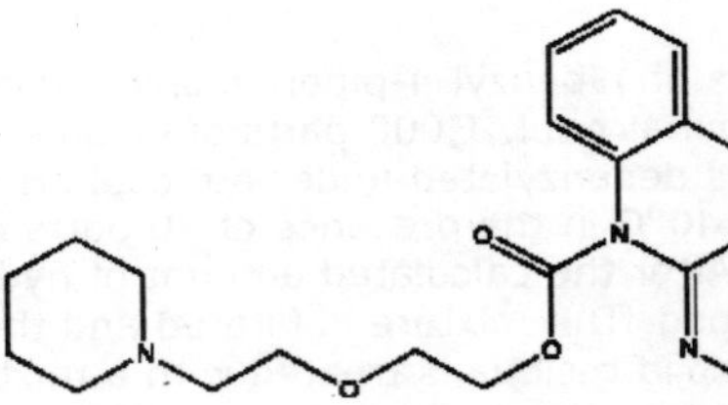

Chemical Abstracts Registry No.: 2167-85-3; 6056-11-7 (Hydrochloride salt)

Trade Name	Manufacturer	Country	Year Introduced
Theratuss	Squibb	US	1962
Dipect	Draco	Sweden	-
Lenopect	Draco	Sweden	-
Selvigon	Homburg	W. Germany	-

Raw Materials

1-Azaphenothiazine carboxylic acid chloride
Piperidinoethoxy ethanol

Manufacturing Process

8.5 parts of 1-azaphenothiazine carboxylic acid chloride and 14 parts of piperidino-ethoxyethanol were introduced into 100 parts of chlorobenzene and the mixture boiled under reflux for 5 minutes. After cooling off the precipitated hydrochloride salt of piperidino-ethoxyethanol was filtered off on a suction filter. Water was added to the filtrate and the pH thereof adjusted to 5 to 6 with dilute HCl. The aqueous phase was then removed, a caustic soda solution added thereto and then extracted with ether. The ethyl extract was washed with water, then dried with potash and the ether distilled off. 9.4 parts of the piperidino-ethoxy-ethyl ester of 1-azaphenothiazine carboxylic acid were obtained. This product was dissolved in 20 parts of isopropanol and the solution neutralized with isopropanolic HCl. The monohydrochloride which precipitated out after recrystallization from isopropanol had a melting point of 160°C to 161°C.

References

Merck Index 7328
Kleeman & Engel p. 730
OCDS Vol. 1 p. 390 (1977)
I.N. p. 775
Schuler, W.A.; US Patent 2,989,529; June 20, 1961; assigned to Degussa (Germany)

PIPEBUZONE

Therapeutic Function: Antiinflammatory

Chemical Name: 1,2-Diphenyl-3,5-dioxo-4-n-butyl-4-(N'-methylpiperazinomethyl)pyrazolidine

Common Name: -

Structural Formula:

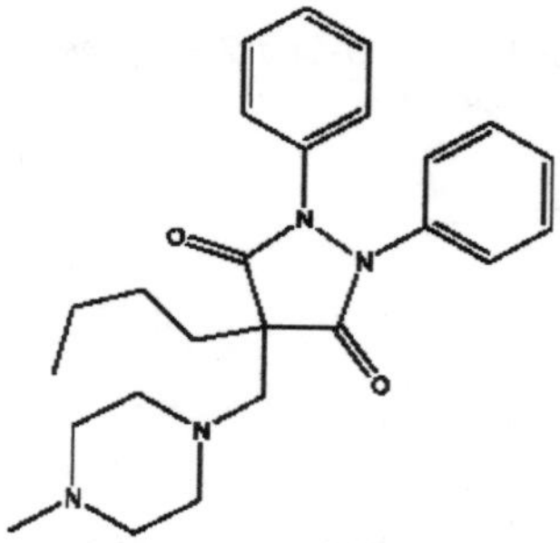

Chemical Abstracts Registry No.: 27315-91-9

Trade Name	Manufacturer	Country	Year Introduced
Elarzone	Dausse	France	1973

Raw Materials

Phenylbutazone
Formaldehyde
N-Methylpiperazine

Manufacturing Process

77 g (0.25 mol) of phenylbutazone, 30 ml of a 30% strength solution of formaldehyde and 50 ml of ethyl alcohol are introduced into a 500 ml flask, 25 g (0.25 mol) of N-methylpiperazine are slowly added to this mixture which is stirred mechanically. The mixture is then heated for one hour on a water bath, left to cool, and crystallization started by scratching.

After being left in the refrigerator overnight the mixture, which has set solid, is triturated with 50 ml of isopropyl alcohol and the solid product filtered off and dried in vacuo over phosphorus pentoxide. 63 g (60% yield) of 1,2-diphenyl-3,5-dioxo-4-n-butyl-4-(N'-methylpiperazinomethyl)pyrazolidine are obtained, melting at 129°C after recrystallization from 150 ml of isopropyl alcohol.

References

Merck Index 7329
Kleeman & Engel p. 730
DOT 9 (11) 476 (1973)
I.N. p. 775
Dausse, S.A.; British Patent 1,249,047; October 6, 1971

PIPEMIDIC ACID

Therapeutic Function: Antibacterial (urinary)

Chemical Name: 8-Ethyl-5,8-dihydro-5-oxo-2-(1-piperazinyl)pyrido[2,3-d]pyrimidine-6-carboxylic acid

Common Name: Piperamic acid

Structural Formula:

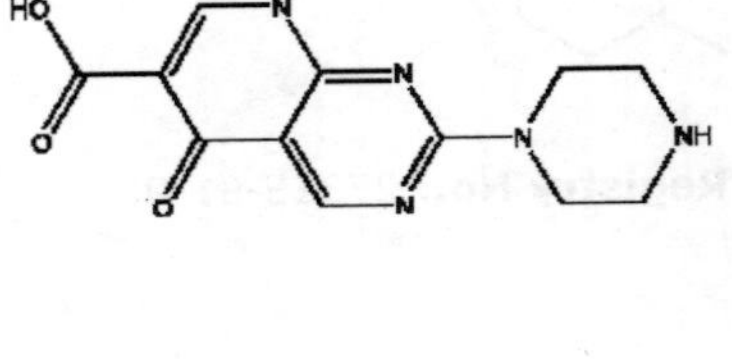

Chemical Abstracts Registry No.: 51940-44-4

Trade Name	Manufacturer	Country	Year Introduced
Pipram	Bellon	France	1975
Deblaston	Madaus	W. Germany	1975
Pipram	RBS Pharma	Italy	1978
Dolcol	Dainippon	Japan	1979
Pipram	Bellon	Italy	1979
Pipedac	Mediolanum	Italy	1980
Deblaston	Madaus	Switz.	1981
Filtrax	Biomedica Foscama	Italy	-
Gastrurol	Gibipharma	Italy	-
Memento	Volpino	Argentina	-
Nuril	Prodes	Spain	-
Pipedase	Scalari	Italy	-
Pipemid	Gentili	Italy	-
Pipurin	Brocchieri	Italy	-
Priper	Syncro	Argentina	-
Septidron	Ethimed	S. Africa	-
Tractur	Baldacci	Italy	-
Uropimid	C.T.	Italy	-
Urotractin	Zambeletti	Italy	-
Uroval	Firma	Italy	-

Raw Materials

Sodium hydroxide
Diethyl sulfate
Piperazine hydrate
6-Amino-2-methylthiopyrimidine
Ethoxymethylene malonic acid diethyl ester

Manufacturing Process

A mixture containing 1.33 g of 5,8-dihydro-8-ethyl-2-methylthio-5-oxopyridol[2,3-d]pyrimidine-6-carboxylic acid, 1.94 g of piperazine hexahydrate and 20 ml of dimethyl sulfoxide was heated at 110°C for 1 hour with stirring. The separated solid was collected by filtration, washed with ethanol, and then dried at such a temperature that did not rise above 50°C to give 1.57 g of the trihydrate of the product as nearly colorless needles, MP 253° to 255°C.

The starting material may be produced by reacting 6-amino-2-methylthiopyrimidine with ethoxymethylene malonic acid diethyl ester. The intermediate thus produced is converted by boiling in diphenyl ether to 6-ethoxycarbonyl-2-methylthio-5-oxo-5,8-dihydropyrido[2,3-d]pyrimidine. That is hydrolyzed by sodium hydroxide to cleave the ethoxy group and then ethylated with diethyl sulfate to give the starting material.

References

Merck Index 7332
Kleeman & Engel p. 731

DOT 11 (10, 408 (1975) & 12 (3) 99 (1976)
I.N. p. 36
Minami, S., Matsumoto, J.-I., Kawaguchi, K., Mishio, S., Shimizu, M., Takase, Y. and Nakamura, S.; US Patent 3,887,557; June 3, 1975; assigned to Dainippon Pharmaceutical Co. Ltd., Japan
Minami, S., Matsumoto, J.-I., Kawaguchi, K., Mishio, S., Shimizu, M., Takase, Y. and Nakamura, S.; US Patent 3,962,443; June 8, 1976; assigned to Dainippon Pharmaceutical Co. Ltd., Japan

PIPENZOLATE BROMIDE

Therapeutic Function: Spasmolytic

Chemical Name: 1-Ethyl-3-[(hydroxydiphenylacetyl)oxy]-1-methylpiperidinium bromide

Common Name: N-Ethyl-3-piperidyl benzilate methobromide

Structural Formula:

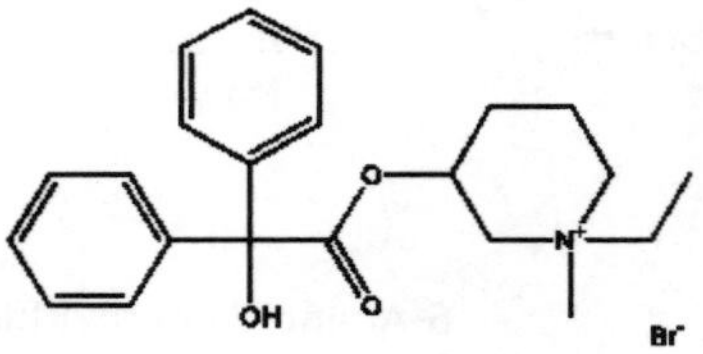

Chemical Abstracts Registry No.: 125-51-9

Trade Name	Manufacturer	Country	Year Introduced
Piptal	Merrell National	US	1955
Piptal	Roger Bellon	France	1960
Piper	Panthox and Burck	Italy	-

Raw Materials

N-Ethyl-3-chloropiperidine
Benzilic acid
Methyl bromide

Manufacturing Process

N-ethyl-3-chloropiperidine was prepared according to the method of Fuson and Zirkle described in Volume 70, J. Am. Chem. Soc., p 2760. 12.0 g (0.081 mol) of N-ethyl-3-chloropiperidine was mixed with 18.6 g (0.081 mol) of benzilic acid and 80 cc of anhydrous isopropyl alcohol as a solvent. The mixture was refluxed for 72 hours. The solution was then filtered and concentrated at 30 mm of mercury. The concentrate was dissolved in water, acidified with hydrochloric acid and extracted with ether to remove the unreacted benzilic acid.

The aqueous layer was neutralized with sodium bicarbonate and the product was extracted with ether. The ethereal solution of the product was dried with potassium carbonate, the ether was removed by distillation and the residue was distilled at 0.12 to 0.18 mm of mercury, the BP being 194° to 198°C. A yield of 16.5 g (60% of theoretical) of N-ethyl-3-piperidyl-benzilate was obtained.

34 g (0.1 mol) of the basic ester is dissolved in 75 cc of isopropyl alcohol and treated with 9.5 g (0.1 mol) of methyl bromide. The mixture is allowed to stand at room temperature until precipitation is complete. The product is removed by filtration and washed with isopropyl alcohol, yield 33 g, MP 175° to 177°C. On recrystallization from isopropyl alcohol, the MP was raised to 179° to 180°C dec.

References

Merck Index 7333
Kleeman & Engel p. 732
I.N. p 776
Biel, J.H.; US Patent 2,918,406; December 22, 1959; assigned to Lakeside Laboratories, Inc.

PIPERACILLIN SODIUM

Therapeutic Function: Antibiotic

Chemical Name: Sodium salt of 6-[D(-)-α-(4-ethyl-2,3-dioxo-1-piperazinocarbonylamino)phenylacetamido]penicillanic acid

Common Name: -

Structural Formula:

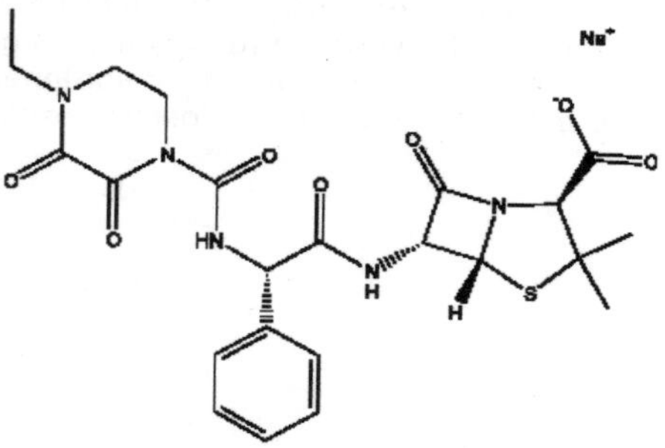

Chemical Abstracts Registry No.: 59703-84-3; 61477-96-1 (Base)

Trade Name	Manufacturer	Country	Year Introduced
Pentcillin	Toyama	Japan	1980
Pipril	Lederle	W. Germany	1980
Pipril	Lederle	Switz.	1980
Piperallin	Toyama	France	1981
Pipril	Lederle	UK	1982
Avocin	Cyanamid	Italy	1982
Pipracil	Lederle	US	1982
Pentocillin	Sankyo	Japan	-

Raw Materials

Diethyl oxalate
Trimethylsilyl chloride
Phosgene
N-Ethylethylenediamine
Sodium 2-ethylhexanoate
6-[D(-)-α-Aminophenylacetamido] penicillanic acid

Manufacturing Process

To a suspension of 0.9 g of 6-[D(-)-α-aminophenylacetamido]penicillanic acid in 30 ml of anhydrous ethyl acetate were added at 5°C to 10°C 0.55 g of triethylamine and 0.6 g of trimethylsilyl chloride. The resulting mixture was reacted at 15°C to 20°C for 3 hours to form trimethylsilylated 6-[D(-)-α-aminophenylacetamido]penicillanic acid.

To this acid was then added 1 g of 4-ethyl-2,3-dioxo-1-piperazinocarbonyl chloride (from the reaction of N-ethylethylenediamine and diethyl oxalate to give 2,3-dioxo-4-ethyl-piperazine which is then reacted with phosgene) and the resulting mixture was reacted at 15°C to 20°C for 2 hours. After the reaction, a deposited triethylamine hydrochloride was separated by filtration, and the filtrate was incorporated with 0.4 g of n-butanol to deposit crystals. The deposited crystals were collected by filtration to obtain l.25 g of white crystals of 6-[D(-)α-(4-ethyl-2,3-dioxo-1-piperazinocarbonylamino) phenylacetamido]penicillanic acid. Into a solution of these crystals in 30 ml of tetrahydrofuran was dropped a solution of 0.38 g of a sodium salt of 2-ethyl-hexanoic acid in 10 ml of tetrahydrofuran, upon which white crystals were deposited. The deposited crystals were collected by filtration, sufficiently washed with tetrahydrofuran and then dried to obtain 1.25 g of sodium salt of 6-[D(-)-α-(4-ethyl-2,3-dioxo-1-piperazinocarbonylamino)phenylacetamido] penicillanic acid, melting point 183°C to 185°C (decomposition), yield 90%.

References

Merck Index 7335
DFU 3 (11) 829 (1978)
Kleeman & Engel p. 732
PDR p. 1026
OCDS Vol. 3 p. 207 (1984)
DOT 17 (1) 29 (1981)
I.N. p. 776
REM p. 1199

Saikawa, I,, Takano, S., Yoshida, C., Takashima, O., Momonoi, K., Kuroda, S., Komatsu, M., Yasuda, T. and Kodama, Y.; US Patents 4,087,424; May 2, 1978; 4,110,327; Aug. 29, 1978; 4,112,090; September 5, 1978; all assigned to Toyama Chemical Co., Ltd.

PIPERIDOLATE

Therapeutic Function: Spasmolytic

Chemical Name: α-Phenylbenzeneacetic acid 1-ethyl-3-piperidinyl ester

Common Name: N-Ethyl-3-piperidyl diphenylacetate

Structural Formula:

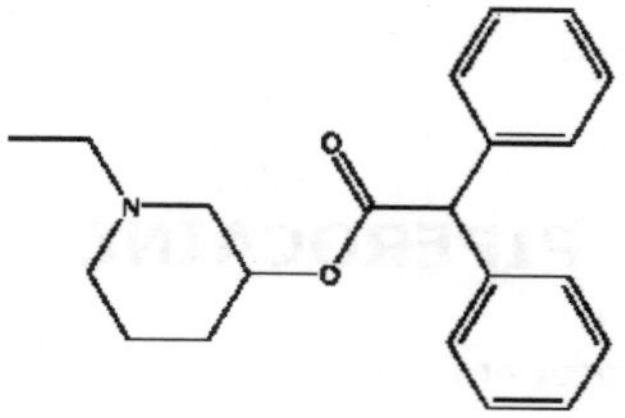

Chemical Abstracts Registry No.: 82-98-4; 129-77-1 (Hydrochloride salt)

Trade Name	Manufacturer	Country	Year Introduced
Dactil	Merrell National	US	1954
Dactil	Roger Bellon	France	1958
Cactiran	Kyorin	Japan	-
Crapinon	Sanzen	Japan	-
Dactylate	Sawai	Japan	-
Edelel	Mochida	Japan	-

Raw Materials

Furfural
Acetic acid
Hydrogen bromide
Hydrogen
Ethylamine
Diphenylacetyl chloride

Manufacturing Process

To obtain the free base, 34 g (0.256 mol) of N-ethyl-3-piperidinol and 20 g (0.22 mol) of diphenylacetyl chloride were mixed in 80 cc of isopropanol and the solution was refluxed for 2 hours. The isopropanol was evaporated in vacuo at 30 mm pressure, the residue was dissolved in 150 cc of water and the aqueous solution was extracted several times with ether. The aqueous solution was then neutralized with potassium carbonate and extracted with ether. The ethereal solution was dried over anhydrous potassium carbonate and the ether removed by distillation. The product was then distilled at its boiling point 180° to 181°C at 0.13 mm of mercury whereby 14 g of a clear yellow, viscous liquid was obtained. The nitrogen content for $C_{21}H_{25}NO_2$ was calculated as 4.33% and the nitrogen content found was 4.21%.

The starting material was produced by the reaction of furfural with ethylamine followed by hydrogenation to give N-ethyl-N-(2-tetrahydrofurfuryl)amine. Treatment of that material with hydrogen bromide in acetic acid gives N-ethyl-3-piperidinol.

References

Merck Index 7345
Kleeman & Engel p. 733
OCDS Vol. 1 p.91 (1977)
I.N. p. 778
Biel, J.H.; US Patent 2,918,407; December 22, 1959; assigned to Lakeside Laboratories, Inc.

PIPEROCAINE

Therapeutic Function: Local anesthetic

Chemical Name: 1-Piperidinepropanol, 2-methyl-, benzoate (ester)

Common Name: Piperocaine

Structural Formula:

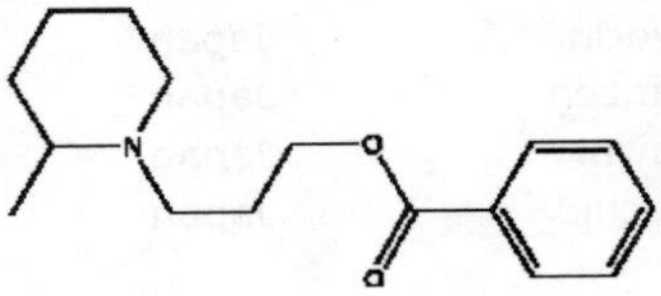

Chemical Abstracts Registry No.: 136-82-3

Trade Name	Manufacturer	Country	Year Introduced
Piperocaine	ZYF Pharm Chemical	-	-
Isocaine	Iso-Sol	-	-

Raw Materials

2-Methylpiperidine
Hydrogen chloride
Benzoyl chloride
γ-Chloropropylbenzoate
Sodium hydroxide
3-Chloropropanol

Manufacturing Process

The γ-chloropropylbenzoate was obtained by treatment of benzoyl chloride with 3-chloropropanol.

15.0 g of 2-methyl piperidine and 15.0 g of γ-chloropropylbenzoate are mixed, and heated under a reflux at a temperature of 120°-140°C for 30-40 min. The reaction mixture is then cooled, and treated with 100 ml of ether, and the precipitated secondary amine hydrochloride (i.e., some of the hydrochloride of the unchanged piperidine) filtered off. Hydrogen chloride gas is passed into the filtrate, and crude γ-(2-methylpiperidino)propylbenzoate hydrochloride thereby precipitated. The ether is decanted from the precipitate, and the latter is dissolved in 20 ml of cold water. This solution is treated with 5 ml of 40% sodium hydroxide and 5 ml of benzoyl chloride and the resulting mixture shaken vigorously until the odor of the benzoyl chloride has disappeared. In this manner any unchanged secondary amine is converted into an amide. The alkaline solution, which contains the free base, γ-(2-methylpiperidino)propylbenzoate in suspension is extracted with ether, and the ether extract is evaporated and dried.

References

McElvain S.M.; US Patent No. 1,784,903; Dec. 16, 1930
Kleemann A., Engel J.; Pharmazeutische Wirkstoffe, GeorgThieme Verlag Stuttgart, New York, 1982

PIPERYLONE

Therapeutic Function: Analgesic

Chemical Name: 4-Ethyl-1-(1-methyl-4-piperidyl)-3-phenyl-3-pyrazolin-5-one

Common Name: Piperylone

Structural Formula:

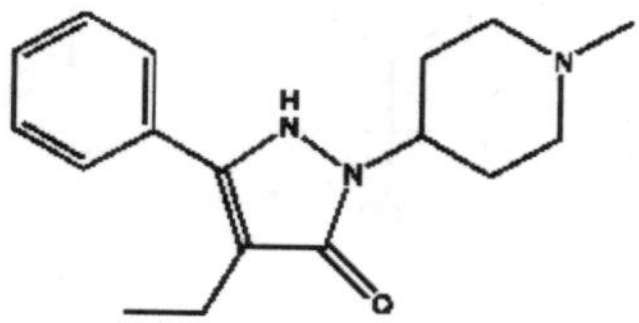

Chemical Abstracts Registry No.: 2531-04-6

Trade Name	Manufacturer	Country	Year Introduced
Palerol	Novartis	-	-

Raw Materials

α-Ethylbenzoylacetic acid ethyl ester
N-Methyl-piperidyl-4-hydrazine

Manufacturing Process

A mixture of 8.8 parts of α-ethylbenzoylacetic acid ethyl ester and 5,3 parts of N-methyl-piperidyl-4-hydrazine is allowed to stand for 30 min at 22°C, after which the mixture is heated for 5 hours to 130°C under a pressure of 12 mm. After 4 hours 1-(N-methylpiperidyl-4)-3-phenyl-4-ethylpyrazolone-5 begins crystallize out. The reaction mixture is allowed to cool, after which the crystal mass is triturated with ether and then recrystallized from methanol-ether or from acetone. Melting point 159-161°C.

References

Merck Index, Monograph number: 7632, Twelfth edition, 1996, Editor: S. Budavari Merck and Co., Inc.
Jucker E., Erbnoether A., Lindenmann J.; US Patent No. 2,903,460, Assigned to Sandoz A.G., Basel, Switzerland

PIPETHANATE ETHOBROMIDE

Therapeutic Function: Anticholinergic, Antiulcer

Chemical Name: Benzilic acid, 2-piperidinoethyl ester ethobromide

Common Name: Piperilate ethyl bromide

Structural Formula:

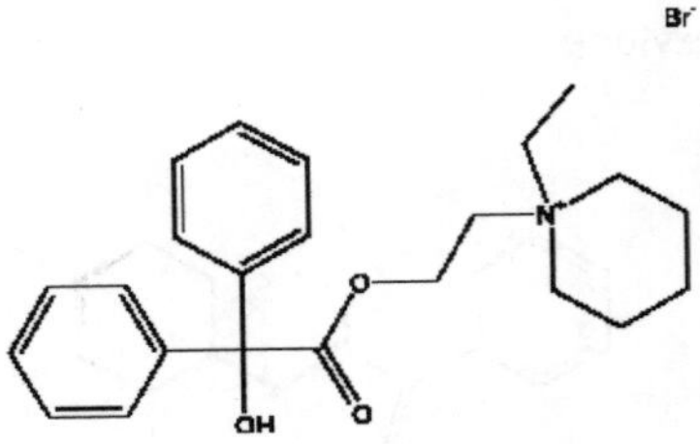

Chemical Abstracts Registry No.: 4546-39-8 (Base)

Trade Name	Manufacturer	Country	Year Introduced
Panpurol	Nippon Shinyaku	Japan	-

Raw Materials

Pipethanate hydrochloride
Sodium hydroxide
Ethyl bromide

Manufacturing Process

Pipethanate hydrochloride is dissolved in water and the solution is made alkaline by adding 10% sodium hydroxide solution. The crystals that are separated are filtered off and recrystallized from dilute ethanol. The monohydrate thereby obtained is dehydrated at 100°C under reduced pressure for 20 minutes. The products that are now in the form of a syrup due to loss of water of crystallization are further dehydrated for 2 days in a desiccator over phosphorus pentoxide whereupon the anhydrous pipethanate is obtained.

3.8 g of the anhydrous pipethanate prepared by the method described is dissolved in 15 cc of acetone, 18 g of purified ethyl bromide is added, and the mixture heated for 8 hours in a sealed tube at 100°C to 110°C. After cooling the crystals are separated and isolated by filtration. They are then washed with acetone to give 5.2 g (95.6%) of pipethanate ethylbromide with a decomposition point of 218°C to 220°C. The crystals are almost pure.

References

Merck Index 7346
DOT 7 (1) 23 (1971)
I.N. p. 779
Nippon Shinyaku Co., Ltd.; British Patent 1,148,858; April 16, 1969

PIPOBROMAN

Therapeutic Function: Antineoplastic

Chemical Name: 1,4-Bis-(3-bromo-1-oxopropyl)piperazine

Common Name: -

Structural Formula:

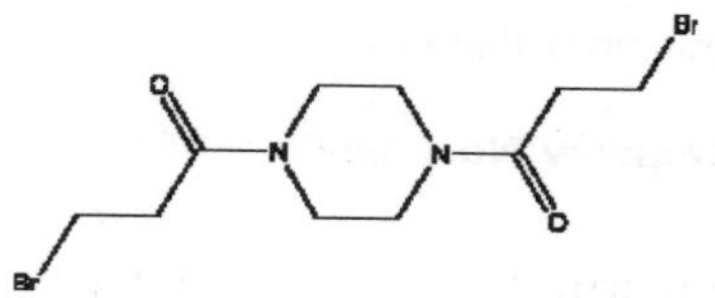

Chemical Abstracts Registry No.: 54-91-1

Trade Name	Manufacturer	Country	Year Introduced
Vercyte	Abbott	US	1966
Vercyte	Abbott	France	1970
Vercite	Abbott	Italy	1972
Amedel	Dainippon	UK	1973

Raw Materials

Piperazine
3-Bromopropionyl chloride

Manufacturing Process

To a solution of 17.2 g (0.10 mol) of 3-bromopropionyl chloride in 100 ml of anhydrous benzene was added dropwise with stirring a solution of 8.6 g (0.10 mol) of anhydrous piperazine in 20 ml of dry chloroform over a period of 30 minutes. The temperature rose spontaneously to 45°C during the addition. After the temperature ceased to rise, stirring was continued for another hour. The reaction mixture was then filtered to remove the piperazine hydrochloride by-product. The filtrate was evaporated to dryness and the residue recrystallized from ethanol to obtain the desired N,N'-bis-(3-bromopropionyl)piperazine as a white crystalline solid melting at 103°C to 104°C. The identity of the product was further established by elemental analysis.

References

Merck Index 7355
Kleeman & Engel p. 735
OCDS Vol. 2 p. 299 (1980)
I.N. p. 779
REM p. 1156
Abbott Laboratories; British Patent 921,559; March 20, 1963

PIPOTIAZINE

Therapeutic Function: Neuroleptic

Chemical Name: 10H-Phenothiazine-2-sulfonamide, 10-(3-(4-(2-hydroxyethyl)-1-piperidinyl)propyl)-N,N-dimethyl-

Common Name: Pipothiazine; Pipotiazine

Chemical Abstracts Registry No.: 39860-99-6

Trade Name	Manufacturer	Country	Year Introduced
Piportil	Rhone-Poulenc Rorer	-	-
Piportil	Aventis	-	-

Structural Formula:

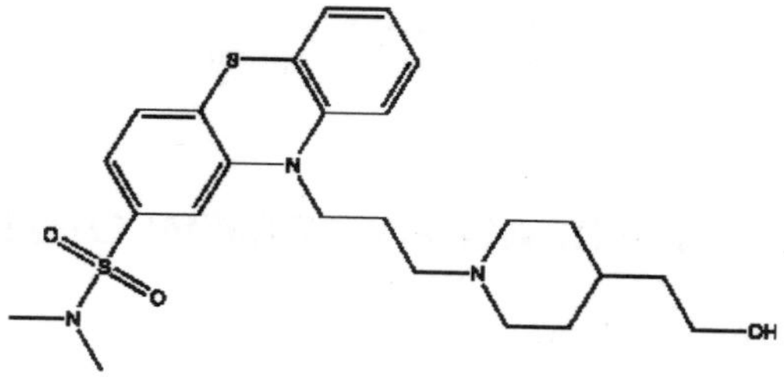

Raw Materials

Hydrochloric acid
Sodium amide
4-Hydroxyethyl piperidine
Sodium hydroxide
1-Chloro-3-tetrahydropyranyloxy propane
Methanesulfonyl chloride
Phenthiazine-2-sulfonic acid dimethylamide

Manufacturing Process

10-(3-Tetrahydropyranyloxypropyl)phenthiazine-2-sulfonic acid dimethylamide prepared by condensing 1-chloro-3-tetrahydropyranyloxy propane with phenthiazine-2-sulfonic acid dimethylamide (melting point 140°C) in xylene in the presence of sodamide.

10-(3-Hydroxypropyl)phenthiazine-2-sulfonic acid dimethylamide was prepared by the action of hydrochloric acid in ethanol on 10-(3-tetrahydropyranyloxypropyl)phenthiazine-2-sulfonic acid dimethylamide.

10-(3-Methanesulphonyloxypropyl)phenthiazine-2-sulphonic acid dimethylamide, was obtained by condensing methanesulphonyl chloride in anhydrous pyridine with 10-(3-hydroxypropyl)phenthiazine-2-sulfonic acid dimethylamide.

10-(3-Methanesulphonyloxypropyl)phenthiazine-2-sulfonic acid dimethylamide and 4-hydroxyethyl piperidine in toluene were heated under reflux with stirring. The reaction mixture was allowed to cool and water was added. The resulting toluene solution layer was decanted and washed twice with water. The toluene solution was then stirred with 5% hydrochloric acid. The hydrochloride of the desired phenthiazine base precipitated in gummy condition in the aqueous layer. This was decanted and treated with sodium hydroxide. It was then extracted three times with ethyl acetate. The extracts were dried over sodium sulfate, filtered and concentrated in vacuum. A resinous product was obtained. This product was dissolved in a mixture of benzene and cyclohexane and chromatographed on a column containing alumina. The chromatographed product was eluted successively with mixtures of benzene and cyclohexane and then with benzene and finally with a mixture of benzene and ethyl acetate. The eluates were evaporated to yield a crude product. This product was recrystallised from aqueous ethanol and yielded 10-[3-[4-(2-hydroxyethyl)piperidyl]propyl]phenthiazine-2-sulfonic acid dimethylamide.

References

Jacob R.M., Robert J.G.; US Patent No. 3,150,129; Sept. 22, 1964; Assigned: Rhone-Poulenc S.A., Paris, France a corporation of France

PIPOXOLAN HYDROCHLORIDE

Therapeutic Function: Spasmolytic

Chemical Name: 5,5-Diphenyl-2-[2-(1-piperidinyl)ethyl]-1,3-dioxolan-4-one hydrochloride

Common Name: -

Structural Formula:

HCl

Chemical Abstracts Registry No.: 18174-58-8; 23744-24-3 (Base)

Trade Name	Manufacturer	Country	Year Introduced
Rowapraxin	Rowa/Wagner	W. Germany	1969

Raw Materials

Benzilic acid
β-Chloropropionaldehyde diethylacetal
Piperidine
Hydrogen chloride

Manufacturing Process

33 g (0.14 mol) of benzilic acid and 22 g (0.13 mol) of β-chloropropionaldehyde diethyl acetal were dissolved in 100 ml of glacial acetic acid by heating. After cooling to 40°C, a slow stream of dry HCl gas was introduced while stirring for 2½ hours. After evaporating the glacial acetic acid in vacuo, the reforming oil was taken up in CH_2Cl_2 and treated with solid $KHCO_3$. After the evolution of CO_2 had ended, water was added and the organic phase was neutralized by means of $KHCO_3$ solution. After drying, the solvent was removed; the remaining oil distilled over under high vacuum at 0.001 mm and at 120° to 130°C to yield the compound 2-(β-chloroethyl)-4,4-

diphenyl-1,3-dioxolan-5-one hydrochloride.

This compound was boiled with 12 g of dry piperidine in 120 ml of absolute benzene for 12 hours under reflux, a total of 6 g of piperidine hydrochloride being separated out. This was filtered off and the benzene solution was concentrated by evaporation. The residue was taken up in a little chloroform and the solution was applied to a dry aluminum oxide column (according to Brockmann); it was thereafter extracted with chloroform. After concentrating the solution by evaporation, an oil was obtained, which was taken up in absolute diethylether. Introduction of dry HCl gas into the cooled solution gave a precipitate which was dissolved and allowed to crystallize from isopropanol/ether. MP 193° to 199°C.

References

Merck Index 7358
Kleeman & Engel p. 736
DOT 6 (3) 95 (1970)
I.N. p. 780
Rowa-Wagner Kommanditgesellschaft Arzneimittelfabrik, Germany; British Patent 1,109,959; April 18, 1968

PIPRADROL HYDROCHLORIDE

Therapeutic Function: Central stimulant

Chemical Name: 2-Piperidinemethanol, α,α-diphenyl-, hydrochloride

Common Name: Pipradrol hydrochloride

Structural Formula:

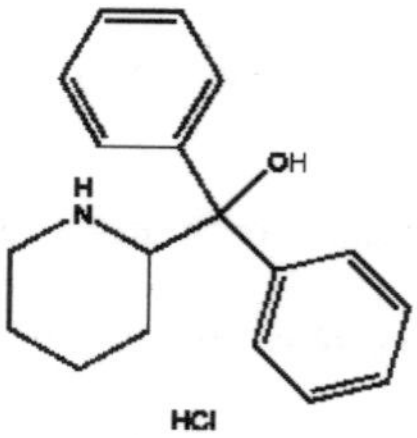

HCl

Chemical Abstracts Registry No.: 71-78-3; 467-60-7 (Base)

Trade Name	Manufacturer	Country	Year Introduced
Alertonic	Adcock Ingram Ltd.	-	-

Raw Materials

Hydrogen	α,α-Diphenyl-2-pyridinemethanol hydrochloride

Formaldehyde	Adams' platinum catalyst
Formic acid	Hydrochloric acid

Manufacturing Process

A mixture of 48 g (0.167 mole) of α,α-diphenyl-2-pyridinemethanol hydrochloride (Emraert et al., Ber. 72B, 1188 (1939); 74B, 714 (1940), 160 ml of ethanol, and 3 0.5 g of Adams' platinum catalyst was shaken under an initial hydrogen pressure of 60 pounds. The theoretical amount of hydrogen was absorbed in 5 hours. The reaction mixture was refluxed, diluted with enough water to dissolve all the white solid, and filtered hot from the catalyst. The filtrate was cooled and filtered; yield of 38 g of α,α-diphenyl-alpha-(2-piperidyl)methanol white product melting at 308-309°C with decomposition.

A mixture of 3.5 grams (0.013 mole) of the above α,α-diphenyl-alpha-(2-piperidyl)methanol, 4 g (0.05 mole) of formaldehyde (37%), and 6 grams (0.1 mole) of formic acid was refluxed for 2 days. The reaction mixture was treated with 1.3 g (0.013 mole) of conc. hydrochloric acid and vacuum distilled on the steam bath. The residue was recrystallized from butanone to give the α,α-diphenyl-alpha-(2-piperidyl)methanol hydrochloride which melted at 228-229°C (dec.).

References

Merck Index, Monograph number: 7638, Twelfth edition, 1996, Editor: S. Budavari; Merck and Co., Inc.

Werner H.W., Tilford Ch.H.; US Patent No. 2,624,739; Jan. 6, 1953; Assigned to The Wm. Merrel Company, Ohio, a corporation of Delavere

PIPRINHYDRINATE

Therapeutic Function: Antihistaminic, Antiemetic

Chemical Name: 4-Diphenylmethoxy-1-methylpiperidine-, compd. with 8-chlorotheophylline (1:1)

Common Name: Diphenylpyraline teoclate; Piprinhydrinate

Chemical Abstracts Registry No.: 606-90-6

Trade Name	Manufacturer	Country	Year Introduced
Piprinhydrinate	Kraeber and Co. GmbH	-	-
Piprinhydrinate	Transo-Pharm	-	-

Raw Materials

1-Methyl-4-piperidinol
Benzhydryl bromide
8-Chlorotheophylline

Structural Formula:

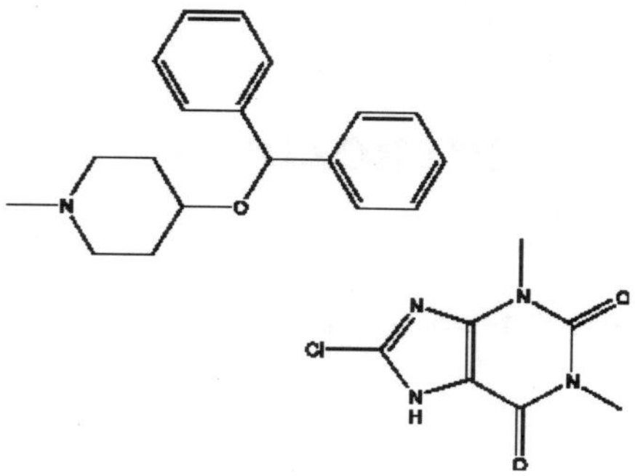

Manufacturing Process

A mixture of 46 g of 1-methyl-4-piperidinol (0.4 mol), 49.4 g of benzhydryl bromide (0.2 mol) and 100 ml of xylene was refluxed for approximately 24 hours. The reaction mixture separated into two phases with the upper phase containing the desired ether compound dissolved in xylene. The lower phase consisted of the hydro bromide salt of the excess 1-methyl-4-piperidinol. The upper phase was separated from the lower phase and the desired benzhydryl ether recovered in the crude state by distilling off the xylene under reduced pressure. The crude benzhydryl ether was a clear reddish oil. It was dissolved in 75 ml of 20% hydrochloric acid and the aqueous acid solution then washed three times with 50 ml portions each of ethyl ether. The aqueous acid solution was then decolorized with activated carbon and thereafter slowly admixed with 75 ml of 28% aqueous ammonia. The benzhydryl ether separated as an oily material and was removed from the aqueous mixture by extraction with three 50 ml portions of ethylether. On evaporation of the ethyl ether from the ethyl ether solution, the benzhydryl ether was recovered as a pale yellow oil. The benzhydryl ether was dissolved in 60 ml of isopropanol and the isopropanol solution acidified to a PH of 3 with dry hydrogen chloride-methanol solution. The acidic propanol solution was then diluted with ethyl ether until a faint turbidity was observed. In a short time, the crystalline hydrochloride salt of the benzhydryl ether separated from the propanol solution. The crystallized salt was recrystallized once from 75 ml of isopropanol with the aid of ethyl ether in order to further purify the material. A yield 24.5 g of the pure hydrochloride salt 1-methylpiperldyl-4-benzhydryl ether (diphenylpyraline) was obtained. This was 39% of the theoretical yield. The pure material had a melting point of 206°C.

107 g (0.5 mole) 8-chlorotheophylline was dissolved in the diluted solution of ammonia contained 0.5 moles NH_3. 1 equivalent of this ammonim salt was mixed with 1 equivalent hydrochloride of 1-methylpiperldyl-4-benzhydryl ether in 150 ml of water. 4-Diphenylmethoxy-1-methylpiperidine compound of 8-chlorotheophylline precipitated, filtered off, washed, dried. Yield was quantitative. MP: 151-152°C.

References

Howland L. et al.; US Patent No. 2,479,843; August 23, 1949; Assigned to Nopco Chemical Company, Harrison,N.J., a corporation of New Jersey

Schuler W.A.; D.B. Patent No. 934,890; July 8, 1949; Chemische Fabrik Promonta Gesellschaft mit beschrunkter Haftung, Hamburg

PIPROZOLIN

Therapeutic Function: Choleretic

Chemical Name: [3-Ethyl-4-oxo-5-(1-piperidinyl)-2-thiazolidinylidene]acetic acid ethyl ester

Common Name: -

Structural Formula:

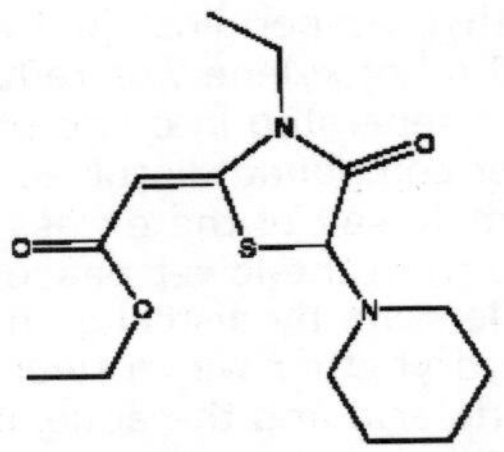

Chemical Abstracts Registry No.: 17243-64-0

Trade Name	Manufacturer	Country	Year Introduced
Probilin	Goedecke	W. Germany	1977
Probilin	Parke Davis	Italy	1979
Coleflux	Finadiet	Argentina	-
Epsyl	Exa	Argentina	-
Secrebil	Isnardi	Italy	-

Raw Materials

Ethyl thioglycolate
Piperidine
Diethyl sulfate
Sodium ethylate
Ethyl cyanoacetate

Manufacturing Process

Ethyl thioglycolate and ethyl cyanoacetate are first reacted in the presence of sodium ethylate to give 4-oxo-thiazolidin-2-ylideneacetic acid ethyl ester. That is reacted with diethyl sulfate and then with piperidine to give piprozolin.

References

Merck Index 7361
DFU 2 (10) 681 (1977)

Kleeman & Engel p. 737
OCDS Vol. 2 p. 270 (1980)
DOT 14 (1) 26 (1976)
I.N. p. 781
Satzinger, G., Herrmann, M. and Vollmer, K.O.; US Patent 3,971,794; July 27, 1976; assigned to Warner-Lambert Co.

PIRACETAM

Therapeutic Function: Psychotropic

Chemical Name: 2-Oxo-1-pyrrolidineacetamide

Common Name: -

Structural Formula:

Chemical Abstracts Registry No.: 7491-74-9

Trade Name	Manufacturer	Country	Year Introduced
Nootropyl	UCB	France	1972
Nootropil	UCB-Smit	Italy	1974
Nootrop	UCB Chemie	W. Germany	1974
Normabrain	Cassella-Riedel	W. Germany	1974
Gabacet	Carrion	France	1980
Ciclocetam	Callol	Spain	-
Ciclofalina	Almirall	Spain	-
Encefalux	Bama-Geve	Spain	-
Eumental	Wassermann	Spain	-
Genogris	Vita	Spain	-
Gericetam	Level	Spain	-
Huberdasen	Hubber	Spain	-
Ideaxan	Millot	France	-
Merapiran	Finadiet	Argentina	-
Nootron	Biosintetica	Brazil	-
Nootropicon	Sidus	Argentina	-
Norotrop	Drifen	Turkey	-
Norzetam	Albert Pharma	Spain	-
Oikamid	Pliva	Yugoslavia	-
Pirroxil	S.I.T.	Italy	-
Pyramen	Pharmachim	Bulgaria	-
Stimubral	Lusofarmaco	Portugal	-
Stimucortex	Kalifarma	Spain	-

Raw Materials

2-Pyrrolidone	Ethyl chloroacetate
Sodium hydride	Ammonia

Manufacturing Process

2-Pyrrolidone is first reacted with sodium hydride, then with ethyl chloroacetate to give ethyl 2-oxo-1-pyrrolidine acetate.

A solution of 0.3 mol of ethyl 2-oxo-1-pyrrolidine acetate in 300 ml of methanol, saturated with ammonia at 20° to 30°C, is heated at 40° to 50°C for 5 hours, while continuously introducing ammonia. The reaction mixture is evaporated to dryness and the residue recrystallized from isopropanol. 2-Oxo-1-pyrrolidineacetamide is obtained in a yield of 86%. MP 151.5° to 152.5°C.

References

Merck Index 7363
Kleeman & Engel p. 737
DOT 9 (6) 215 (1973) & (8) 327 (1973)
I.N. p. 781
Morren, H.; US Patent 3,459,738; August 5, 1969; assigned to UCB (Union Chimique-Chemische Bedrijven), Belgium

PIRBUTEROL

Therapeutic Function: Bronchodilator

Chemical Name: 2-Hydroxymethyl-3-hydroxy-(1-hydroxy-2-tert-butylaminoethyl)pyridine

Common Name: -

Structural Formula:

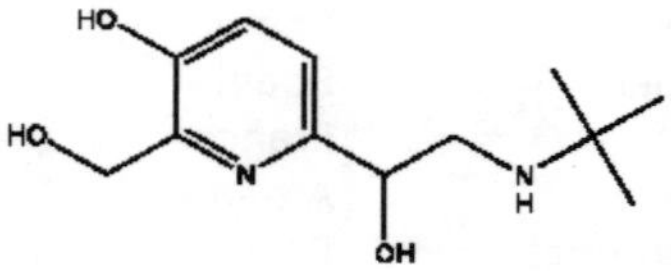

Chemical Abstracts Registry No.: 38677-81-5

Trade Name	Manufacturer	Country	Year Introduced
Exirel	Pfizer Taito	Japan	1982
Exirel	Pfizer	UK	1983
Exirel	Pfizer	Switz.	1983

Raw Materials

N-tert-Butyl-2-(5-benzyloxy-6-hydroxymethyl-2-pyridyl)-2-hydroxyacetamide
Diborane
Hydrogen

Manufacturing Process

To 78 ml of a 1 M solution of diborane in tetrahydrofuran under nitrogen and cooled to 0°C is added dropwise over a period of 40 minutes 13.5 g of N-tert-butyl-2-(5-benzyloxy-6-hydroxymethyl-2-pyridyl)-2-hydroxyacetamide in 250 ml of the same solvent. The reaction mixture is allowed to stir at room temperature for 3.5 hours, and is then heated to reflux for 30 minutes and cooled to room temperature. Hydrogen chloride (70 ml, 1.34 N) in ethanol is added dropwise, followed by the addition of 300 ml of ether. The mixture is allowed to stir for 1 hour and is then filtered, yielding 11.0 g, melting point 202°C (dec.). The hydrochloride dissolved in water is treated with a sodium hydroxide solution to pH 11 and is extracted into chloroform (2 x 250 ml). The chloroform layer is dried over sodium sulfate, concentrated to dryness in vacuo, and the residue recrystallized from isopropyl ether, 3.78 g, melting point 81°C to 83.5°C.

A solution of 1.7 g of 2-hydroxymethyl-3-benzyloxy-(1-hydroxy-2-tert-butyl-aminoethyl)pyridine in 30 ml of methanol containing 1.2 ml of water is shaken with 700 mg of 5% palladiumon-charcoal in an atmosphere of hydrogen at atmospheric pressure. In 17 minutes the theoretical amount of hydrogen has been consumed and the catalyst is filtered. Concentration of the filtrate under reduced pressure provides 1.4 g of the crude product as an oil. Ethanol (5 ml) is added to the residual oil followed by 6 ml of 1.75 N ethanolic hydrogen chloride solution and, finally, by 5 ml of isopropyl ether. The precipitated product is filtered and washed with isopropyl ether containing 20% ethanol, 1.35 g, melting point 182°C (dec.).

References

Merck Index 7364
DFU 2 (1) 60 (1977)
OCDS Vol. 2 p. 280 (1980)
DOT 19 (2) 113 (1983) & (7) 384 (1983)
I.N. p. 782
Barth, W.E.; US Patents 3,700,681; October 24, 1972; 3,763,173; October 2, 1973; 3,772,314; November 13, 1973; all assigned to Pfizer, Inc.

PIRENZEPINE HYDROCHLORIDE

Therapeutic Function: Antiulcer, Antiemetic

Chemical Name: 6H-Pyrido(2,3-b)(1,4)benzodiazepin-6-one, 5,11-dihydro-11-((4-methyl-1-piperazinyl)acetyl)-, dihydrochloride

Common Name: Pirenzepine hydrochloride

Structural Formula:

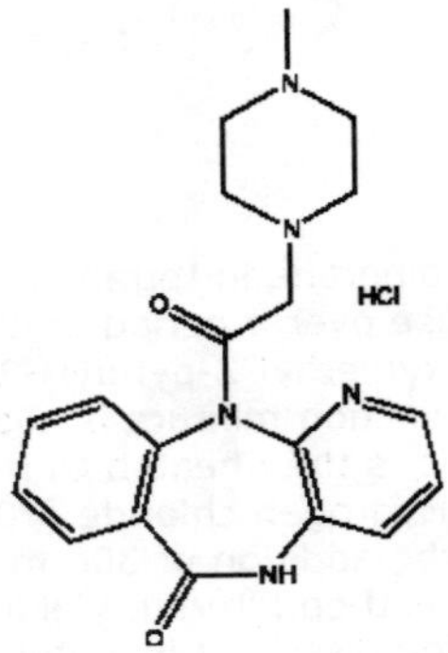

Chemical Abstracts Registry No.: 29868-97-1; 28797-61-7 (Base)

Trade Name	Manufacturer	Country	Year Introduced
Droxol	Microsules Bernabo	-	-
Gasteril	Ripari-Gero	-	-
Gastril	Torrent	-	-
Gastropin	Boehringer-Ingelheim	-	-
Gastrosed	Amsa	-	-
Ulcosan	Dompe	-	-

Raw Materials

Triethylamine
5,11-Dihydro-6H-pyrido[2,3-b][1,4]benzo-diazepin-6-one
N-Methylpiperazine
Chloroacetyl chloride

Manufacturing Process

48.4 g of 5,11-dihydro-6H-pyrido[2,3-b][1,4]benzo-diazepin-6-one were refluxed in 900 ml of absolute dioxane for 15 minutes. Thereafter, over a period of 45 minutes, 28 ml of chloroacetyl chloride and 52 ml of triethylamine were simultaneously added dropwise to the mixture. The mixture was refluxed for eight hours and then vacuum-filtered after having cooled. The filtrate was evaporated in vacuum. The crystalline residue was recrystallized from acetonitrile in the presence of activated charcoal. MP: 212°-213°C (with decomposition). Yield: 85% of theory.

A mixture of 67.5 g of 11-chloroacetyl-5,11-dihydro-6H-pyrido[2,3-b][1,4]benzodiazepin-6-one, 183 ml of N-methylpiperazine and 1.37 liters of absolute benzene was refluxed for 18 hours. Thereafter, the crystalline precipitate was vacuum filtered off, dissolved in aqueous 20% hydrochloric acid, the solution was evaporated in vacuum, the crystalline residue was dissolved in 250 ml of water while heating, the solution was admixed with 150 ml of isopropanol and active charcoal, filtered, and 2.5 liters of isopropanol were added to the filtrate. After cooling, the precipitate was vacuum filtered off, yielding 70% of theory of the 5,11-dihydro-11-[(4'-methyl-1'-piperazinyl)-acetyl]-6H-pyrido[2,3-b][1,4]benzodiazepin-6-one dihydrochloride, M.P. 257-259°C (decomp.).

The free base of pirenzepine, obtained from the dihydrochloride by making an aqueous solution thereof alkaline with dilute sodium hydroxide and extracting it with chloroform, had MP: 226°-228°C after recrystallization from methanol/ether.

References

Schmidt G. et al.; US Patent No. 3,743,734; July 3, 1973; Assigned to Boehringer Ingelheim G.m.b.H., am Rhein, Germany

PIRETANIDE

Therapeutic Function: Diuretic

Chemical Name: 3-N-Pyrrolidino-4-phenoxy-5-sulfamylbenzoic acid

Common Name: -

Structural Formula:

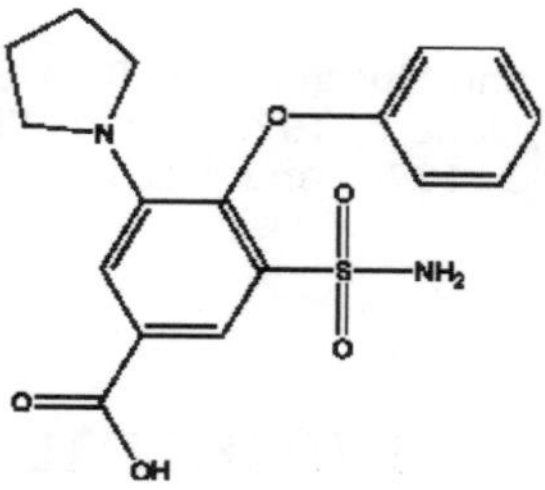

N O S NH2 O O O OH

Chemical Abstracts Registry No.: 55837-27-9

Trade Name	Manufacturer	Country	Year Introduced
Arelix	Hoechst	Italy	1980
Arelix	Cassella-Riedel	W. Germany	1982
Tauliz	Hoechst	W. Germany	-

Raw Materials

3-N-Succinimido-4-phenoxy-5-sulfamylbenzoic acid methyl ester
Sodium borohydride
Sodium hydroxide

Manufacturing Process

12.3 g (0.03 mol) of 3-N-succinimido-4-phenoxy-5-sulfamylbenzoic acid methyl ester are dissolved or suspended in 100 ml of absolute diglyme. 9 g of boron trifluoride etherate are added direct to this mixture and a solution of 2.4 g (~0.063 mol) of $NaBH_4$ in 80 ml of diglyme is then added dropwise at room temperature with stirring. As the reaction proceeds exothermically, it is necessary to cool with ice water. The reaction is normally complete after the dropwise addition and a short period of stirring thereafter.

The excess reducing agent is then decomposed by means of a little water (foaming), the solution is filtered and about 300 ml of water are added while stirring. The 3-N-pyrrolidino-4phenoxy-5-sulfamylbenzoic acid methyl ester which has crystallized out is recrystallized from methanol in the form of colorless crystals, melting point 191°C to 192°C.

61 g of 3-N-pyrrolidino-4-phenoxy-5-sulfamylbenzoic acid methyl ester are suspended in 350 ml of 1 N NaOH and the suspension is heated for one hour on the waterbath. 3-N-pyrrolidino4-phenoxy-5-sulfamylbenzoic acid is precipitated from the clear solution by means of 2 N HCl while stirring well. The almost pure crude product can be recrystallized from methanol/water in the form of light yellow platelets, melting point 225°C to 227°C, with decomposition.

References

Merck Index 7366
DFU 2 (6) 393 (1977)
OCDS Vol. 3 p. 58 (1984)
DOT 18 (6) 274 (1982) & (10) 555 (1982)
I.N. p. 782
Bormann, D., Merkel, W. and Muschaweck, R.; US Patents 4,010,273; March 1, 1977; 4,093,735; June 6, 1978; 4,111,953; September 5, 1978; 4,118,397; October 3, 1978; and 4,161,531; July 17, 1979; all assigned to Hoechst AG

PIRIBEDIL

Therapeutic Function: Vasodilator

Chemical Name: 2-[4-(1,3-Benzodioxol-5-ylmethyl)-1-piperazinyl]pyrimidine

Common Name: -

Structural Formula:

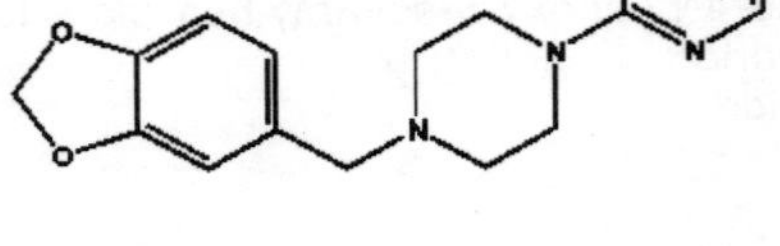

Chemical Abstracts Registry No.: 3605-01-4

Trade Name	Manufacturer	Country	Year Introduced
Trivastal	Eutherapie	France	1969
Trivastan	Servier	Italy	1975
Trivastal	Pharmacodex	W. Germany	1975
Circularina	Searle	-	-

Raw Materials

2-Chloropyrimidine
1-(3':4'-Methylenedioxybenzyl)-piperazine

Manufacturing Process

To a solution of 21 g of 1-(3':4'-methylenedioxybenzyl)-piperazine in solution in 300 cc of anhydrous xylene there were added 28 g of anhydrous potassium carbonate and then 11.3 g of 2-chloropyrimidine. The suspension was then heated for 9 hours at boiling point (130°C). After this time, the mixture was cooled and extracted several times with 10% hydrochloric acid. The acid solution obtained was washed with ether and then rendered alkaline with potassium carbonate; the oily product which was separated was extracted with chloroform and this, after drying with potassium carbonate and evaporation, gave an oily residue weighing 20 g. By dissolution in boiling ethanol and crystallization, 15 g of crystals melting at 96°C were recovered.

References

Merck Index 7368
Kleeman & Engel p. 739
DOT (As ET-495) 6 (1) 29 (1970) & 10 (9) 324, 340 (1974)
I.N. p. 783
Regnier, G., Canevari, R. and Laubie, M.; US Patent 3,299,067; January 17, 1967; assigned to Science Union Et Cie, Societe Francaise De Recherche Medicale (France)

PIRITRAMIDE

Therapeutic Function: Analgesic

Chemical Name: 1-(3,3-Diphenyl-3-cyanopropyl)-4-piperidino-4-piperidinecarboxamide

Common Name: Pirinitramide

Chemical Abstracts Registry No.: 302-41-0

Structural Formula:

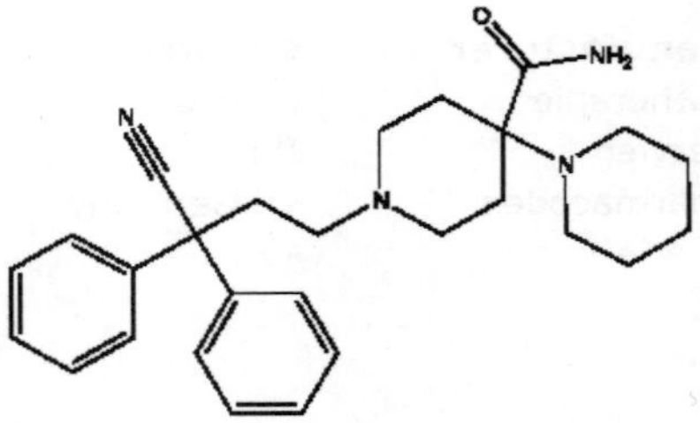

Trade Name	Manufacturer	Country	Year Introduced
Dipidolor	Janssen	W. Germany	1969
Dipidolor	Janssen	UK	1972
Piridolan	Leo	Sweden	-

Raw Materials

3,3-Diphenyl-3-cyanopropyl bromide
4-Piperidino-4-piperidinecarboxamide

Manufacturing Process

A mixture of 84 parts of 3,3-diphenyl-3-cyanopropyl bromide, 41 parts of 4-piperidino-4-piperidinecarboxamide, 64 parts of sodium carbonate, a small amount of potassium iodide and 1,200 parts of anhydrous toluene was stirred, and heated under reflux for 48 hours. At the end of this time the reaction mixture was allowed to cool to room temperature, and 500 parts of water were added. The resultant precipitate was removed by filtration, and triturated with diisopropyl ether. The crystalline material thus obtained was removed by filtration, and recrystallized from 320 parts of acetone, to give 1-(3,3-diphenyl-3-cyanopropyl)-4-piperidino-4-piperidinecarboxamide, melting at about 149°C to 150°C.

References

Merck Index 7373
Kleeman & Engel p. 739
OCDS Vol. 1 p. 308 (1977)
DOT 5 (3) 107 (1969)
I.N. p. 783
N.V. Research Laboratorium Dr. C. Janssen; British Patent 915,835; January 16, 1963
Janssen, P.A.J.; US Patent 3,080,360; March 5, 1963; assigned to Research Laboratorium Dr. C. Janssen N.V.

PIROHEPTINE

Therapeutic Function: Antiparkinsonian

Chemical Name: 3-(10,11-Dihydro-5H-dibenzo[a,d]cyclohepten-5-ylidene)-1-ethyl-2-methylpyrrolidine

Common Name: -

Structural Formula:

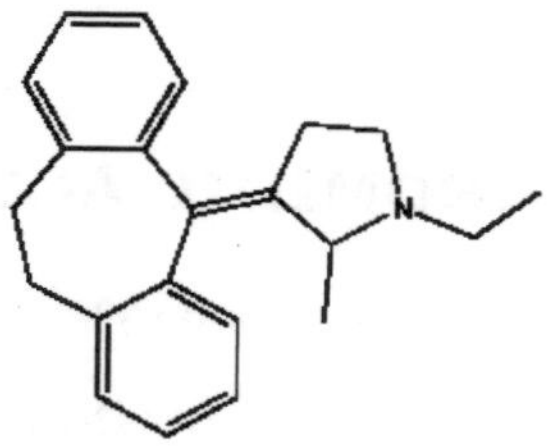

Chemical Abstracts Registry No.: 16378-21-5

Trade Name	Manufacturer	Country	Year Introduced
Trimol	Fujisawa	Japan	1974

Raw Materials

2-Methyl-3-(10,11-dihydro-5H-dibenzo[a,d]cycloheptene-5-ylidene)-1-pyrroline
Ethyl iodide
Sodium borohydride

Manufacturing Process

(1) To 3.8 g of 2-methyl-3-(10,11-dihydro-5H-dibenzo[a,d]cycloheptene-5-ylidene)-1-pyrroline, there were added 8 g of ethyl iodide. This mixture was placed into a closed vessel and heated at 80°C in a water-bath for one hour. After completing the reaction, the reaction mixture was cooled and the unreacted ethyl iodide was distilled off to yield 5.5 g of 1-ethyl-2methyl-3-(10,11-dihydro-5H-dibenzo[a,d]cycloheptene-5-ylidene)-1-pyrrolinium iodide in the form of yellow crystals. These crystals were recrystallized from a mixture of acetone and ether to yield yellow needles of the melting point 223°C.

(2)1-Ethyl-2-methyl-3-(10,11)-dihydro-5H-dibenzo[a,d]cycloheptene-5-ylidene)-1-pyrroliniumiodide (4.7 g) was dissolved in 7 cc of methanol. To this solution there were added 1.4 g of sodium boron hydride within about 80 minutes with stirring and stirring of the solution was continued for two hours to complete the reaction. The reaction mixture was acidified with 10% aqueous hydrochloric acid solution and then the methanol was distilled off. The residual solution was alkalized with 20% aqueous sodium hydroxide solution and extracted with ether. The ether layer was dried over magnesium sulfate and the ether was distilled off. The resulting residue was further distilled under reduced pressure to yield 2.0 g of 1-ethyl-2-methyl-3-(10,11)-dihydro-5H-dibenzo[a,d]cycloheptene-5-ylidene)pyrrolidine (boiling point 167°C/4 mm Hg.).

References

Merck Index 7375
DOT 9 (6) 247 (1973) & 10 (9) 325 (1974)
I.N. p. 784
Deguchi, Y., Nojima, H. and Kato, N.; US Patent 3,454,495; July 8, 1969; assigned to Fujisawa Pharmaceutical Co., Ltd. (Japan)

PIROMIDIC ACID

Therapeutic Function: Antibacterial (urinary)

Chemical Name: 8-Ethyl-5,8-dihydro-5-oxo-2-(1-pyrrolidinyl)pyrido[2,3-d]pyrimidine-6-carboxylic acid

Common Name: -

Structural Formula:

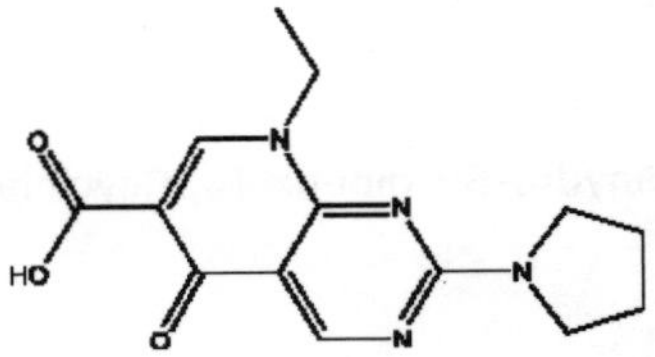

Chemical Abstracts Registry No.: 19562-30-2

Trade Name	Manufacturer	Country	Year Introduced
Panacid	Dainippon	Japan	1972
Pirodal	I.S.F.	Italy	1977
Bactramyl	Carrion	France	1978
Septural	Gruenenthal	W. Germany	1978
Adelir	Teikoku	Japan	-
Coltix	Gerardo Ramon	Argentina	-
Panerco	Erco	Denmark	-
Purim	Mayoly-Spindler	France	-
Reelon	Sanken	Japan	-
Uriclor	Almirall	Spain	-
Urisept	Srbolek	Yugoslavia	-
Zaomeal	Isei	Japan	-

Raw Materials

Pyrrolidine
Sodium hydroxide
Diethyl sulfate
6-Amino-2-methylthiopyrimidine
Ethoxymethylene malonic acid diethyl ester

Manufacturing Process

150 mg of 6-carboxy-5,8-dihydro-8-ethyl-2-methylthio-5-oxopyrido[2,3-d]pyrimidine was added to 30 ml of absolute ethanol containing 1.1 g of dissolved pyrrolidine, and the mixture was reacted for 5 hours at 95°C in a sealed tube. The solvent was removed by distillation, and the residue was recrystallized from methanol-chloroform. There were obtained 111 mg of 6-carboxy-5,8-dihydro-8-ethyl-5-oxo-2-pyrrolidino-pyrido[2,3-d]pyrimidine having a MP of 314° to 316°C.

The starting material is produced by reacting 6-amino-2-methylthiopyrimidine with ethoxymethylene malonic acid diethyl ester. That intermediate is thermally treated in diphenyl ether to give 6-ethoxycarbonyl-2-methylthio-5-oxo-5,8-dihydro-pyrido[2,3-d]pyrimidine. The ethoxy group is hydrolyzed off with sodium hydroxide and one nitrogen is ethylated with diethyl sulfate to give the starting material. These are the same initial steps as used in the pipemidic acid syntheses earlier in this volume.

References

Merck Index 7377
Kleeman and Engel p. 739
OCDS Vol. 2 p. 470 (1980)
DOT 7 (5) 188 (1971)
I.N. p. 36
Dainippon Pharmaceutical Co. Ltd., Japan; British Patent 1,129,358; October 2, 1968
Minami, S., Shono, T., Shmmizu, M. and Takase, Y.; US Patent 3,673,184; June 27, 1972; assigned to Dainippon Pharmaceutical Co. Ltd.
Pesson, M.E. and Geiger, S.W.; US Patent 4,125,720; November 14, 1978; assigned to Laboratoire Roger Bellon

PIROXICAM

Therapeutic Function: Antiinflammatory, Analgesic

Chemical Name: 2H-1,2-Benzothiazine-3-carboxamide, 4-hydroxy-2-methyl-N-2-pyridinyl-, 1,1-dioxide

Common Name: Piroksikam; Pyroxycam

Structural Formula:

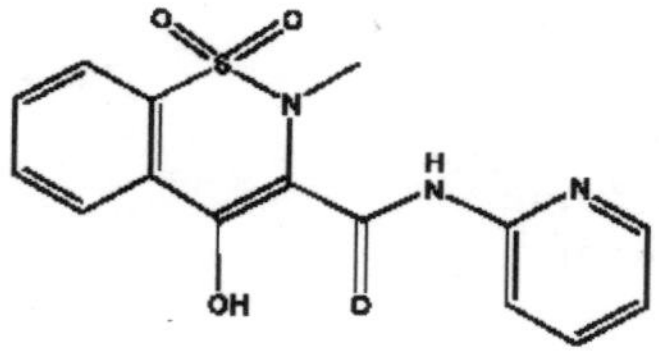

Chemical Abstracts Registry No.: 36322-90-4

Trade Name	Manufacturer	Country	Year Introduced
Amida	Euphoric Pharmaceuticals Pvt. Ltd.	-	-
Anartrir	Qif	-	-
Apo-Piroxicam	Apotex Inc.	Canada	-
Artril	Pharmaetica	-	-
Dacam	Star	-	-
Desinflam	Sintyal	-	-
Erazon	Krka	Slovenia	-
Feldene gel	Pfizer	USA	-
Feldoral Sedico	Sedico	Egypt	-
Flogostop	Szabo	-	-
Foldox	Recalcine	-	-
Hotemin	Egis	Hungary	-
Oxa	Beta	-	-
Piroflam	Intas	-	-
Piroflam	Lichtenstein	-	-
Piroflam	Opus	-	-
Pirox	Cipla Limited	India	-
Piroxan	Diba	-	-
Piroxicam	Glaxo Wellcome Poznan S.A.	Poland	-
Piroxicam	Norton Healthcare Ltd.	UK	-
Piroxicam	IPCA laboratories Ltd.	India	-
Piroxicam	Pharmachim Holding EAD, Sopharma AD	Bulgaria	-
Piroxicam	Chemo Iberica	Spain	-
Piroxicam	Nantong General Pharmaceutical Factory	China	-
Piroxicam	Pharmaline	Livan	-
Piroxicam	Jelfa S.A.	Poland	-
Piroxicam	LaborMed Pharma	Rumania	-
Piroxicam	Zdravle	Yugoslavia	-
Piroxicam	Darou Paksh Pharmaceutical Company	Iran	-
Piroxicam	Leciva	Czech Republic	-
Piroxicam Jenapharm	Jenapharm	Germany	-
Piroxicam-Ratiopharm	Ratiopharm	Germany	-
Piroxicam Stada	Stada Arzneimittel AG	Germany	-
Piroxicam-Teva	Teva	Israel	-
Proxigel	Procaps	-	-
Reumador	Slovakofarma	-	-
Reumaplus	Medichrom	-	-
Roxicam	Zdravle	Yugoslavia	-
Roxikam	Rolab	-	-

Trade Name	Manufacturer	Country	Year Introduced
Roxikam	Zdravje	-	-
Sinalgico	Finadiet	-	-
Tetram	Nycomed	-	-

Raw Materials

Methyl iodide
Methyl 3-oxo-1,2-benzoisothyazolin-2-acetate 1,1-dioxide
Sodium hydroxide
Sodium methoxide
Hydrochloric acid

Manufacturing Process

189.6 g (3.51 mol) of sodium methoxide in 1.4 L of dry dimethylsulfoxide was stirred at room temperature (~ 25°C), while under a dry nitrogen atmosphere. To the stirred slurry, there were then added in one complete portion 300 g (1.17 moles) of methyl 3-oxo-1,2-benzoisothyazolin-2-acetate 1,1-dioxide (Chemische Berichte, vol. 30, p. 1267 (1897)) and flask containing the system was then immediately immersed in an ice-methanol bath. The resulting deep red solution was cooled to 30°C and the ice bath removed. The solution was then stirred under dry nitrogen at 30°C for 4 min, cooled quickly to 18°C and then immediately poured into 4.8 L of 3 N hydrochloric acid solution admixed with ice. The resulting slurry was stirred for 15 min, filtered, then washed with water to give 250 g of crude product. Recrystallization from a chloroform-ethanol mixture (1:1) in the presence of charcoal, then afforded a 61% yield of methyl 3,4-dihydro-4-oxo-2H-1,2-benzothiazine-3-carboxylate 1,1-dioxide, melting point 173-174°C after two recrystallizations from isopropanol.

A 22 L round-bottomed flask charged with 800 g (3.13 moles) of methyl 3,4-dihydro-4-oxo-2H-1,2-benzothiazine-3-carboxylate 1,1-dioxide, 3.2 l of water, 9.6 l of 95% ethanol, 673 ml of methyl iodide (1.53 kg, 10.87 moles) and 3.14 L of 1 N aqueous sodium hydroxide. The reaction mixture was then stirred for 30 min at room temperature, under nitrogen atmosphere and then solution was stored for 23 h. The slurry was then chilled at 0°C and filtered. After washing the filter cake twice with water, ethanol and then diethyl ether there were obtained 537 g of methyl 3,4-dihydro-2-methyl-4-oxo-2H-1,2-benzothiazine-3-carboxylate 1,1-dioxide, melting point 165°-168°C after recrystallization from 1.25 L of acetonitrile.

In 3 L round-bottomed flask there were placed methyl 3,4-dihydro-2-methyl-4-oxo-2H-1,2-benzothiazine-3-carboxylate 1,1-dioxide, 2-aminopyridin and dry xylene. Nitrogen gas was then bubbled into the suspension for 5 min, then the reaction mixture was heated to begin a period of slow distillation, with complete solution effected during the first 10 min of heating. After 5.5 h, the period of slow distillation was discontinued and reaction mixture was allowed to heat at reflux for approximately 16 h. After that the reaction mixture was cooled to room temperature and filtered. The solid material was crystallized from chloroform with methanol and againe from methanol and then there were obtained piroxicam, melting point 197°-200°C, dec.

References

Lombardino J.G.; US Patent No. 3,591,584; July 6, 1971; Assigned: Pfizer Inc., New York, N.Y.

PIROZADIL

Therapeutic Function: Antihyperlipidemic, Platelet aggregation inhibitor

Chemical Name: 2,6-Pyridinemethanol-bis(3,4,5-trimethoxybenzoate)

Common Name: -

Structural Formula:

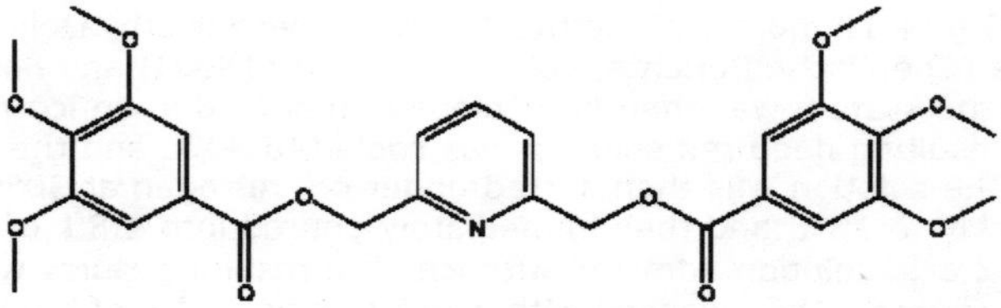

Chemical Abstracts Registry No.: 54110-25-7

Trade Name	Manufacturer	Country	Year Introduced
Pemix	Prodes	Spain	1982

Raw Materials

3,4,5-Trimethoxybenzoic acid
Thionyl chloride
Pyridine-2,6-dimethanol

Manufacturing Process

15 kg (70.7 mols) of 3,4,5-trimethoxybenzoic acid and 65 liters of benzene were introduced into a reactor, to which mixture was added 27.4 liters of thionyl chloride. The mass was heated to 56°C to 70°C during a period of 5 hours. The excess of benzene and thionyl chloride was distilled under vacuum. The residue was kept under vacuum at 120°C to 123°C for 1 hour, to obtain a hard crystalline solid.

A solution comprising 3.24 kg (23.3 mols) of pyridine-2,6-dimethanol in 35 liters of pure pyridine was added to the residue and the mass was heated to 80°C for 2½ hours. The reaction mass became brown in color. The chlorhydrate of pyridine so formed was cooled and crystallized. The resulting reaction mass was then poured into water. The precipitate obtained was filtered, repeatedly rinsed with water, and dissolved in 400 liters of methanol. The resulting solution was filtered with activated charcoal. From this filtration 50 liters of methanol were distilled at normal pressure and then crystallized. 8.35 kg (15.8 mols) of pyridine-2,6dimethanol trimethoxybenzoate were obtained, which represented a yield of 68%.

The product was a white crystalline solid which melted at 119°C to 126°C. Recrystallization in methanolone gave a product which melted at 126°C to 127°C.

References

Merck Index 7379
DFU 6 (5) 290 (1981)
DOT 18, Suppl. 1
Instituto International Terapeutico; British Patent 1,401,608; July 30, 1975

PIRPROFEN

Therapeutic Function: Antiinflammatory

Chemical Name: α-(3-Chloro-4-pyrrolinophenyl)-propionic acid

Common Name: -

Structural Formula:

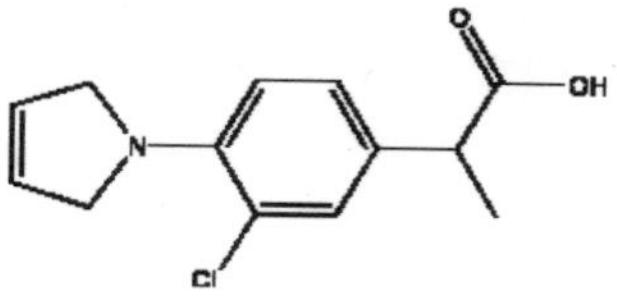

Chemical Abstracts Registry No.: 31793-07-4

Trade Name	Manufacturer	Country	Year Introduced
Rengasil	Ciba Geigy	France	1981
Rengasil	Ciba Geigy	Switz.	1981

Raw Materials

Ethyl α-(3-chloro-4-aminophenyl)-propionate hydrochloride
1,4-Dibromo-2-butene

Manufacturing Process

To the mixture of 85.5 g ethyl α-(3-chloro-4-aminophenyl)-propionate hydrochloride, 142 g sodium carbonate and 600 ml dimethyl formamide, 107 g 1,4-dibromo-2-butene are added dropwise while stirring and the whole is refluxed for 5 hours and allowed to stand overnight at room temperature. The mixture is filtered, the filtrate evaporated in vacuo, the residue is triturated with hexane, the mixture filtered, the residue washed with petroleum ether and the filtrate evaporated. The residue is combined with 280 ml 25% aqueous sodium hydroxide and the mixture refluxed for 8 hours. After cooling, it is diluted with water, washed with diethyl ether, the pH adjusted to 5 to 5.2 with hydrochloric acid and extracted with diethyl ether. The extract is dried, filtered, evaporated and the residue crystallized from benzene-hexane, to yield the α-(3-chloro-4-pyrrolinophenyl)-propionic acid melting at 94°C to 96°C.

References

Merck Index 7380
DFU 1 (1) 23 (1976)
OCDS Vol. 2 p. 69 (1980)
DOT 11 (3) 103 (1975)
I.N. p. 784
Carney, R.W.J. and De Stevens, G.; US Patent 3,641,040; February 8, 1972; assigned to Ciba Geigy Corp.

PIVAMPICILLIN

Therapeutic Function: Antibacterial

Chemical Name: 6-[(Aminophenylacetyl)amino]-3,3-dimethyl-7-oxo-4-thia-1-azabicyclo[3.2.0]heptane-2-carboxylic acid (2,2-dimethyl-1-oxopropoxy)methyl ester

Common Name: -

Structural Formula:

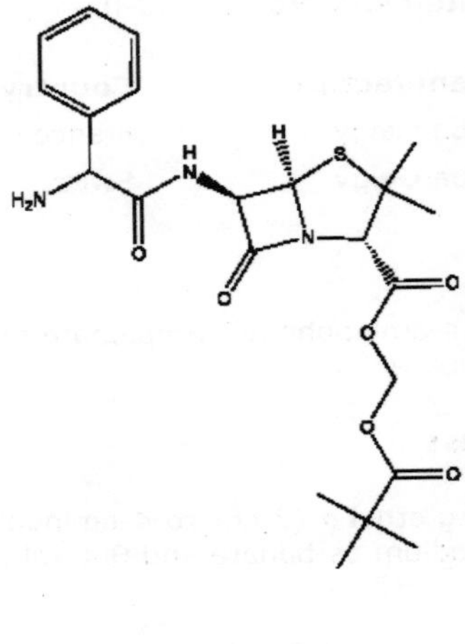

Chemical Abstracts Registry No.: 33817-20-8; 26309-95-5 (Hydrochloride salt)

Trade Name	Manufacturer	Country	Year Introduced
Maxifen	Sharp and Dohme	W. Germany	1972
Berocillin	Boehringer Ingelheim	W. Germany	1972
Pondocillina	Sigma Tau	Italy	1972
Pivatil	MSD	Italy	1972
Pivatil	Chibret	France	1973
Pondocillin	Burgess	UK	1980
Acerum	Jeba	Spain	-
Bensamin	Turro	Spain	-
Brotacilina	Escaned	Spain	-
Co-Pivam	Sanchez-Covisa	Spain	-
Crisbiotic	Crisol	Spain	-
Dancilin	Hemofarm	Yugoslavia	-
Devonian	Perga	Spain	-
Diancina	Septa	Spain	-
Inacilin	Inibsa	Spain	-
Isvitrol	Therapia	Spain	-
Kesmicina	Kessler	Spain	-
Lancabiotic	Lanzas	Spain	-
Novopivam	Osiris	Argentina	-
Oxidina	Sanitas	Argentina	-
Penimenal	Alalan	Spain	-
Pibena	Jebena	Spain	-
Piva	Efesal	Spain	-
Pivabiot	Galepharma Iberica	Spain	-
Pivadilon	De La Cruz	Spain	-
Pivambol	B.O.I.	Spain	-
Pivamkey	Pereira	Spain	-
Pivapen	Juste	Spain	-
Pivastol	Graino	Spain	-
Piviotic	Miquel	Spain	-
Sanguicillin	Zdravlje	Yugoslavia	-
Tam-Cilin	Quimia	Spain	-
Tryco	Durban	Spain	-
Vampi-Framan	Oftalmiso	Spain	-

Raw Materials

Hydrogen
Potassium D(-)-α-azidobenzylpenicillinate
Chloromethyl pivalate

Manufacturing Process

(A) Pivaloyloxymethyl D(-)-α-azidobenzylpenicillinate: To a suspension of potassium D(-)α-azidobenzylpenicillinate (4.14 g) and potassium dicarbonate

(1.5 g) in acetone (100 ml) and 10% aqueous sodium iodide (2 ml), chloromethyl pivalate (2.7 ml) was added and the mixture refluxed for 2 hours. After cooling, the suspension was filtered and the filtrate evaporated to dryness in vacuo. The remaining residue was washed repeatedly by decantation with petroleum ether to remove unreacted chloromethyl pivalate. The oily residue was taken up in ethyl acetate (100 ml), and the resulting solution washed with aqueous sodium bicarbonate and water, dried and evaporated in vacuo to yield the desired compound as a yellowish gum, which crystallized from ether, melting point 114°C to 115°C.

(B) Pivaloyloxymethyl D(-)-α-aminobenzylpenicillinate, hydrochloride: To a solution of pivaloyloxymethyl D(-)-α-azidobenzylpenicillinate (prepared as described above) in ethyl acetate (75 ml) a 0.2 M phosphate buffer (pH 2.2) (75 ml) and 10% palladium on carbon catalyst (4 g) were added, and the mixture was shaken in a hydrogen atmosphere for 2 hours at room temperature. The catalyst was filtered off, washed with ethyl acetate (25 ml) and phosphate buffer (25 ml), and the phases of the filtrate were separated. The aqueous phase was washed with ether, neutralized (pH 6.5 to 7.0) with aqueous sodium bicarbonate, and extracted with ethyl acetate (2 x 75 ml). To the combined extracts, water (75 ml) was added, and the pH adjusted to 2.5 with 1 N hydrochloric acid. The aqueous layer was separated, the organic phase extracted with water (25 ml), and the combined extracts were washed with ether, and freeze-dried. The desired compound was obtained as a colorless, amorphous powder.

The purity of the compound was determined iodometrically to be 91%. A crystalline hydrochloride was obtained from isopropanol with a melting point of 155°C to 156°C (dec.).

References

Merck Index 7387
Kleeman & Engel p. 741
OCDS Vol. 1 p. 414 (1977)
DOT 8 (4) 148 (1972) & 19 (6) 331 (1983)
I.N. p. 785
REM p. 1201
Frederiksen, E.K. and Godtfredsen, W.O.; US Patent 3,660,575; May 2, 1972; assigned to Lovens Kemiske Fabrik Produktionsaktieselskab (Denmark)
Binderup, E.T., Petersen, H.J, and Liisberg, S.; US Patent 3,956,279; May 11, 1976; assigned to Leo Pharmaceutical Products Ltd. (Denmark)

PIVMECILLINAM

Therapeutic Function: Antibacterial

Chemical Name: 6-[[(Hexahydro-1H-azepin-1-yl)methylene]amino]-3,3-dimethyl-7-oxo-4-thia-1-azabicyclo[3.2.0]heptane-2-carboxylic acid (2,2-dimethyl-1-oxopropoxy)methyl ester

Common Name: Amdinocillin pivoxil

Structural Formula:

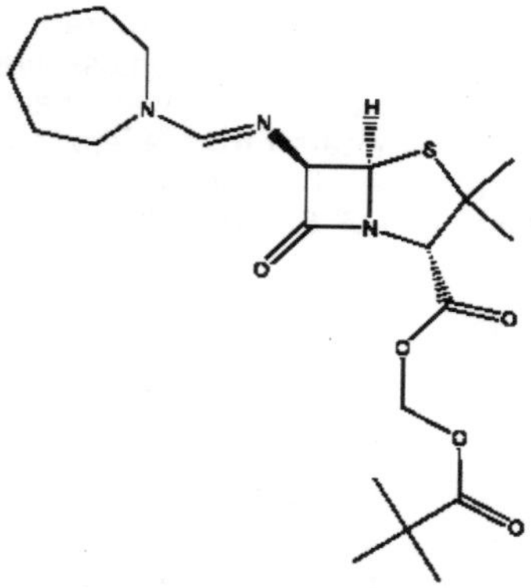

Chemical Abstracts Registry No.: 32886-97-8

Trade Name	Manufacturer	Country	Year Introduced
Selexid	Leo	UK	1977
Melysin	Takeda	Japan	1979
Selexid	Leo	Switz.	1980
Negaxid	Sigma Tau	Italy	1980

Raw Materials

N-Formylhexamethyleneimine
Oxalyl chloride
Pivaloyloxymethyl 6-aminopenicillinate tosylate
Sodium bicarbonate

Manufacturing Process

The starting material N-formylhexamethyleneimine was prepared from hexamethyleneimine and chloral.

12.7 g of N-formylhexamethyleneimine were dissolved in 250 ml of dry ether. While stirring and cooling, 8.5 ml of oxalyl chloride in 50 ml of dry ether were added dropwise, whereafter the mixture was stirred overnight at room temperature. The precipitated amide chloride was filtered off and washed with dry ether, and was placed in an exsiccator.

27.5 g of pivaloyloxymethyl 6-aminopenicillinate tosylate was suspended in 1,500 ml of ethyl acetate with continuous stirring and cooling in an ice bath and 950 ml of ice-cold aqueous sodium bicarbonate (2%) were added. The ethyl acetate layer was separated and was shaken with 750 ml of ice-water containing 25 ml of aqueous sodium bicarbonate (2%), whereafter it was dried over magnesium sulfate at 0°C. After filtration, the solution was evaporated to dryness in vacuo. The residue was dissolved in a solution of 15.5 ml of dry triethylamine in 75 ml of dry alcohol-free chloroform. To this solution, 10 g of the above prepared amide chloride dissolved in 75 ml of dry alcohol-free chloroform were added dropwise at a temperature of about -20°C. After standing for half an hour at -20°C, the temperature was raised to 0°C within 15 minutes and the solution was evaporated to dryness in vacuo. The residue was stirred with 750 ml of ether. Undissolved triethylamine hydrochloride was filtered off, and the filtrate was again evaporated to dryness in vacuo. The residue was reprecipitated from acetone (200 ml) - water (150 ml). After recrystallization from cyclohexane an analytically pure product was obtained with a melting point of 118.5°C to 119.5°C.

References

Merck Index 391
Kleeman & Engel p. 741
DOT 19 (6) 331 (1983)
I.N. p. 786
REM p. 1201
Lund, F.J.; US Patent 3,957,764; May 18, 1976; assigned to Lovens Kemiske Fabrik Produktionsartieselskab (Denmark)

PIXIFENIDE

Therapeutic Function: Antiinflammatory

Chemical Name: 1-[[4-[1-(Hydroxyimino)ethyl]phenoxy]acetyl]piperidine

Common Name: N-(p-1-Nitrosoethyl)phenoxyacetylpiperidine; Pifoxime

Structural Formula:

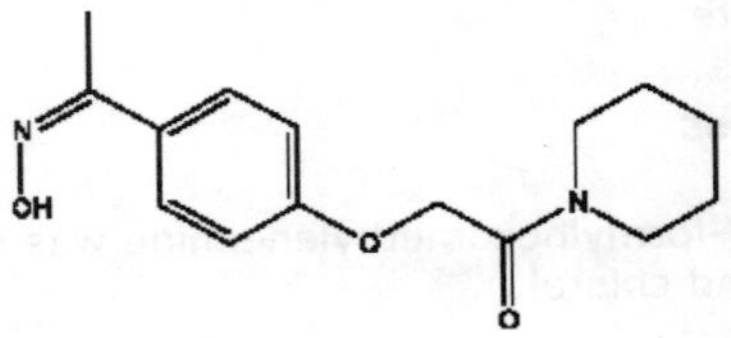

Chemical Abstracts Registry No.: 31224-92-7

Trade Name	Manufacturer	Country	Year Introduced
Flamanil	Salvoxyl-Wander	France	1975

Raw Materials

p-Hydroxyacetophenone
Methanol
Hydroxylamine
Chloroacetic acid
Piperidine

Manufacturing Process

(A) Preparation of p-Acetylphenoxyacetic Acid: p-Hydroxy-acetophenone is treated with chloroacetic acid in aqueous solution in the presence of sodium hydroxide. The desired acid is then isolated from its sodium salt in a total yield of 80 to 82%, excess of p-hydroxy-acetophenone having been extracted with methylene chloride.

(B) Preparation of Methyl p-Acetylphenoxy-Acetate: A mixture of 80 g of the acid obtained in (A) and 200 ml of methyl alcohol in 600 ml of dichloromethane is refluxed in the presence of sulfuric acid. The desired ester is isolated in accordance with a method known per se, and recrystallized. When the refluxing period is 12 hours, the ester is obtained with a yield of 70%. When the refluxing period is 18 hours, the yield for this ester is 85%.

(C) Preparation of N-(p-Acetylphenoxy-Acetyl)-Piperidine: The ester from (B) is refluxed for 8 hours with 2.5 mols of thoroughly dried piperidine. Then 1 volume of water is added and the product is left to crystallize in the cold. The desired amide is obtained in an 80% yield.

(D) Preparation of N-(p-[1-Isonitrosoethyl]-Phenoxy-Acetyl)-Piperidine: The amide from (C) is refluxed for 5 hours with technical (98%) hydroxylamine and alcohol denatured with methanol. The desired product is obtained in a 75% yield.

In semiindustrial synthesis, to achieve better yields, it is possible to omit (A), by directly preparing the ester (B) by reaction of p-hydroxy acetophenone on ethyl 2-bromoacetate in the presence of potassium carbonate in butanone. The yield of ester is 90%, and elimination of excess of p-hydroxyacetophenone is effected by washing with sodium hydroxide.

References

Merck Index 7300
Kleeman & Engel p. 725
DOT 12 (2) 50 (1976)
Mieville, A.; US Patent 3,907,792; September 23, 1975

PIZOTYLINE HYDROCHLORIDE

Therapeutic Function: Migraine therapy

Chemical Name: 4-(9,10-Dihydro-4H-benzo[4,5]cyclohepta[1,2-b]thien-4-ylidene-1-methylpiperidine hydrochloride

Common Name: Pizotifen

Chemical Abstracts Registry No.: 15574-96-6 (Base)

Structural Formula:

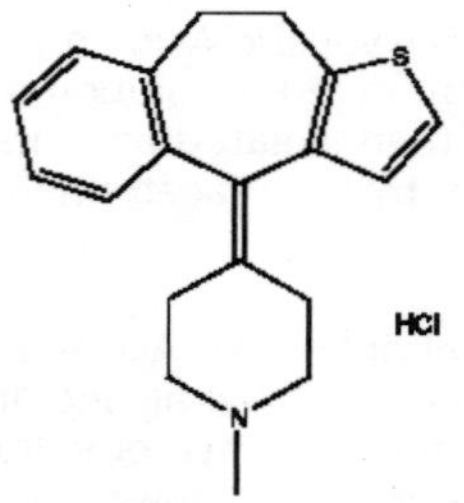

Trade Name	Manufacturer	Country	Year Introduced
Sandomigran	Sandoz	Italy	1972
Sandomigran	Sandoz	W. Germany	1974
Sanomigran	Wander	UK	1975
Mosegor	Wander	W. Germany	1976
Sanmigran	Salvoxyl-Wander	France	1976
Polomigran	Polfa	Poland	-

Raw Materials

Phosphorus
Hydrogen chloride
Phthalic anhydride
Magnesium
Thienyl-(2)-acetic acid
1-Methyl-4-chloropiperidine
Phosphorus pentoxide

Manufacturing Process

(A) Preparation of Thenylidene-(2)-Phthalide: 24.2 g of thienyl-(2)-acetic acid, 52.0 g of phthalic acid anhydride, 4.0 g of anhydrous sodium acetate and 125 ml of 1-methylpyrrolidone-(2) are heated while stirring in an open flask for 3 hours to 205° to 208°C, while nitrogen is passed through. It is then cooled and the viscous reaction mixture poured into 1 liter of water. The precipitated substance is filtered off, washed with water and then dissolved in 200 ml of chloroform. After filtering off some undissolved substance, shaking is effected twice with 100 ml of 2 N sodium carbonate solution and then with water, drying is then carried out over sodium sulfate and the volume is reduced by evaporation. The crude phthalide is repeatedly recrystallized from ethanol, while treating with animal charcoal. It melts at 114° to 115°C.

(B) Preparation of o-[2-Thienyl-(2')-Ethyl]Benzoic Acid: 24.0 g of thenylidene-(2)-phthalide, 8.8 g of red pulverized phosphorus, 240 ml of hydrochloric acid (d = 1.7) and 240 ml of glacial acetic acid are heated to boiling under nitrogen and while stirring vigorously. 70 ml toluene are then added and 6.0 g of red phosphorus added in small portions over a period of 1 hour. It is then poured into 3 liters of ice water, stirred with 300 ml of chloroform and the phosphorus removed by filtration.

The chloroform phase is then removed, the aqueous phase extracted twice more with 200 ml of chloroform and the united extracts shaken out 4 times, each time with 200 ml of 2 N sodium hydroxide solution. The alkaline solution is then rendered acid to Congo red reagent, using hydrochloric acid and extracted 3 times with chloroform. After drying over sodium sulfate and evaporating the solvent, the residue is chromatographed on aluminum oxide (Activity Stage V). The substance eluted with benzene and benzene/chloroform (1:1) is recrystallized from chloroform/hexane (1:1); MP 107° to 109°C.

(C) Preparation of 9,10-Dihydro-4H-Benzo[4,5]Cyclohepta[1,2-b]Thiophen-(4)-One: 200 ml of 85% phosphoric acid and 112 g of phosphorus pentoxide are heated to 135°C. 7.0 g of o-[2-thienyl-(2')-ethyl]benzoic acid are then introduced while stirring thoroughly over a period of 30 min. Stirring is then continued for another hour at 135°C and the reaction mixture is then stirred into 1 liter of ice water. Extraction is then effected 3 times, using 250 ml ether portions, the ethereal extract is washed with 2 N sodium carbonate solution, dried over sodium sulfate and reduced in volume by evaporation. The residue is boiled up with 55 ml of ethanol, the solution freed of resin by decanting and then stirred at room temperature for 6 hours with animal charcoal. It is then filtered off, reduced in volume in a vacuum and the residue distilled. BP 120° to 124°C/0.005 mm, $n_D^{24.5}$ = 1.6559.

(D) Preparation of 4-[1'-Methyl-Piperidyl-(4')]-9,10-Dihydro-4H-Benzo[4,5]Cyclohepta[1,2b]Thiophen-(4)-ol: 0.94 g of magnesium filings which have been activated with iodine are covered with a layer of absolute tetrahydrofuran and etched with a few drops of ethylene bromide. A solution of 5.0 g of 1-methyl-4-chloropiperidine in 5 ml of tetrahydrofuran is then added dropwise and boiling then effected for a further hour under reflux. After cooling to room temperature, the solution of 4.5 g of 9,10-dihydro-4H-benzo[4,5]cyclohepta[1,2-b]thiophen-(4)-one in 5 ml of tetrahydrofuran is added dropwise.

Stirring is carried out first for 3 hours at room temperature and then for 2 hours at boiling temperature, it is then cooled and poured into 300 ml of ice-cold 20% ammonium chloride solution. It is then shaken out with methylene chloride, the methylene chloride solution washed with water and shaken 3 times with 30 ml portions of aqueous 2 N tartaric acid solution. The tartaric acid extract is rendered alkaline while cooling thoroughly and then extracted twice with methylene chloride. After washing with water, drying over potassium carbonate and reducing in volume by evaporation, the residue is recrystallized from ethanol. MP 197° to 199°C.

(E) Preparation of 4-[1'-Methyl-Piperidylidene-(4')]-9,10-Dihydro-4H-Benzo[4,5]Cyclohepta[1,2-b]Thiophene Hydrochloride: 2 g of 4-[1'-methyl-piperidyl-(4')]-9,10-dihydro4H-benzo[4,5]cyclohepta[1,2-b]thiophen-(4)-ol, 60 ml of glacial acetic acid and 20 ml of concentrated hydrochloric acid are boiled for 30 minutes under reflux. After evaporating in a vacuum, the residue is triturated with 3 ml of acetone, the precipitated hydrochloride is then filtered off and it is recrystallized from isopropanol/ether. MP 261° to 263°C (decomposition).

References

Merck Index 7389
Kleeman & Engel p. 742

DOT 9 (6) 221 (1973)
I.N. p. 786
Jucker, E., Ebnother, A., Stoll, A., Bastian, J.-M. and Rissi, E.; US Patent 3,272,826; September 13, 1966; assigned to Sandoz Ltd., Switzerland

POLOXALKOL

Therapeutic Function: Pharmaceutic aid (surfactant)

Chemical Name: Poly(oxyethylene)-poly(oxypropylene)-poly(oxyethylene)

Common Name: Poloxalene

Structural Formula: $HO(CH_2CH_2O)_a[CH(CH_3)CH_2O]_b(CH_2CH_2O)_cH$
Average values for a, b, c are a= 12, b = 34, c = 12

Chemical Abstracts Registry No.: 9003-11-6

Trade Name	Manufacturer	Country	Year Introduced
Polykol	Upjohn	US	1958
Therabloat	Norden	US	-

Raw Materials

Propylene glycol
Ethylene oxide

Manufacturing Process

(A) In a 1-liter 3-necked round bottom flask equipped with a mechanical stirrer, reflux condenser, thermometer and propylene oxide feed inlet, there were placed 57 g (0.75 mol) of propylene glycol and 7.5 g of anhydrous sodium hydroxide. The flask was purged with nitrogen to remove air and heated to 120°C with stirring and until the sodium hydroxide was dissolved. Then sufficient propylene oxide was introduced into the mixture as fast as it would react until the product possessed a calculated molecular weight of 2,380. The product was cooled under nitrogen, the NaOH catalyst neutralized with sulfuric acid and the product filtered. The final product was a water-insoluble polyoxypropylene glycol having an average molecular weight of 1,620 as determined by hydroxyl number or acetylation analytical test procedures.

(B) The foregoing polyoxypropylene glycol having an average 1,620 molecular weight was placed in the same apparatus as described in procedure (A), in the amount of 500 g (0.308 mol), to which there was added 5 g of anhydrous sodium hydroxide. 105 g of ethylene oxide was added at an average temperature of 120°C, using the same technique as employed in (A). The amount of added ethylene oxide corresponded to 17.4% of the total weight of the polyoxypropylene glycol base plus the weight of added ethylene oxide.

References

Merck Index 7431
I.N. p. 789
REM p. 1320
Lundsted, L.G.; US Patent 2,674,619; April 6, 1954; assigned to Wyandotte Chemicals Corporation

POLYESTRADIOL PHOSPHATE

Therapeutic Function: Estrogen

Chemical Name: Estra-1,3,5(10)-triene-3,17 diol (17-beta)-, polymer with phosporic acid

Common Name: Polymeric ester of phosphoric acid and estradiol

Structural Formula: Estradiol phosphate polymer

Chemical Abstracts Registry No.: 28014-46-2

Trade Name	Manufacturer	Country	Year Introduced
Estradurin	Ayerst	US	1957
Estradurin	Abello	Spain	-
Estradurin	Leo	Sweden	-

Raw Materials

Estradiol
Phosphorus oxychloride

Manufacturing Process

3 g of estradiol was dissolved in 75 ml of anhydrous pyridine. The solution was cooled to -10°C, whereupon a solution of 1.1 ml of phosphorus oxychloride in 10 ml of anhydrous pyridine was added with agitation. After the addition, which required 7 minutes, the reaction mixture was kept at -10°C for a further period of 3 hours, and then it was left standing at room temperature for 15 hours. A clear solution thus resulted, to which finely crushed ice was then added. The resulting solution was evaporated in vacuum to dryness. After drying in a vacuum desiccator, 3.8 g of a white powder was obtained. This powder was suspended in 2 ml of pyridine, and 25 ml of 0.5 N sodium hydroxide was added, whereupon a solution was obtained which was then diluted with water to 100 ml.

The solution was then dialyzed through a cellophane membrane against 4 liters of water for 10 hours, with stirring. The dialysis was repeated 2 additional times, with fresh amounts of water. To the dialyzed solution there was added 2 ml of 1 N hydrochloric acid, whereupon polyestradiol phosphate was precipitated as a white bulky precipitate. This was centrifuged off and washed repeatedly with 0.1 N hydrochloric acid. Thereafter it was dried in a vacuum desiccator. The yield was 3 g of polyestradiol phosphate. The analysis shows 0.65% of water, 1.35% of pyridine and 9.3% of phosphorus (calculated on a dry sample).

References

Merck Index 7439
PDR p. 618
I.N. p. 790
REM p. 987
Diczfalusy, E.R., Ferno, O.B., Fex, H.J., Hogberg, K.B. and Linderot, T.O.E.; US Patent 2,928,849; March 15, 1960; assigned to Leo AB, Sweden

POLYETHYLENE GLYCOL 3350

Therapeutic Function: Laxative

Chemical Name: Synthetic polyglycol having an average molecular weight of 3350

Common Name: -

Structural Formula:

Chemical Abstracts Registry No.: 25322-68-3

Trade Name	Manufacturer	Country	Year Introduced
MiraLax	Braintree Laboratories	-	-
Polyethylene Glycol 3350	Schwarz Pharma	-	-

Raw Materials

Ethylene oxide
Polyethylene glycol 400
Potassium hydroxide

Manufacturing Process

Polyethylene glycol 3350 was obtained by polymerization of ethylene oxide in an autoclave at 80-100°C using as a catalyst dipotassium alcogolate of polyethylene glycol 400.

Dipotassium alcogolate of polyethylene glycol 400 was synthesized by a heating of the dry mixture of polyethylene glycol 400 and potassium hydroxide. The molecular weight of polymer was regulated by the ratio of monomer:catalyst.

References

Bailey F.E. and Koleske J.// Poly(ethylene oxide). N.Y., Acad. Press, 1976

POLYMYXIN

Therapeutic Function: Antibacterial

Chemical Name: See structure

Common Name: -

Structural Formula: Complex antibiotic

Chemical Abstracts Registry No.: 1406-11-7

Trade Name	Manufacturer	Country	Year Introduced
Aerosporin	Burroughs-Wellcome	US	1951
Cortisporin	Burroughs-Wellcome	US	-
Mastimyxin	Chassot	Switz.	-
Neo-Polycin	Merrell Dow	US	-
Neosporin	Burroughs-Wellcome	US	-
Octicair	Pharmafair	US	-
Ophthocort	Parke Davis	US	-
Otobiotic	Schering	US	-
Otocort	Lemmon	US	-
Polyfax	Pitman-Moore	US	-
Polysporin	Burroughs-Wellcome	US	-
Pyocidin	Berlex	US	-
Topisporin	Pharmafair	US	-
Tri-Thalmic	Schein	US	-

Raw Materials

Bacterium Bacillus polymyxa
Nutrient medium
Corn meal

Manufacturing Process

As described in US Patent 2,595,605, in a pilot plant tank 225 liters of a medium containing the following ingredients was prepared: 2% ammonium sulfate, 0.2% potassium dihydrogen phosphate, 0.05% magnesium sulfate heptahydrate, 0.005% sodium chloride, 0.001% ferrous sulfate heptahydrate, 0.5% yeast extract, 1% dextrose, 1% calcium carbonate and 3% corn meal. The fermentation medium was adjusted to pH 7.3 to 7.4. It was then sterilized for 30 minutes at 110°C. After sterilization the pH was about 7. To the medium was added 225 ml of mineral oil.

The fermentation medium was inoculated with Bacillus polymyxa prepared as follows: A culture of Bacillus polymyxa in a tube with Trypticase soybean broth was incubated overnight at 25°C. 5 ml of this culture was transferred to 100 ml of the tank medium in a 500 ml Erlenmeyer flask which was incubated for 48 hours at room temperature. This 100 ml culture served as inoculum for one tank. During the course of fermentation the medium was aerated at the rate of 0.3 volume of air per volume of mash per minute. The temperature was maintained at about 27°C. Samples of mash were taken every 8 hours in order to determine pH and the presence of contaminants and spores. After 88 hours of fermentation the pH was about 6.3 and an assay using Escherichia coli showed the presence of 1,200 units of polymyxin per cubic centimeter. The polymyxin was extracted and purified by removing the mycelia, adsorbing the active principle on charcoal and eluting with acidic methanol.

Polymyxin is usually used as the sulfate.

References

Merck Index 7445
Kleeman & Engel p. 743
PDR pp.671, 732, 738, 757, 888, 1034, 1232, 1380, 1415, 1429, 1606, 1645
DOT 8 (1) 21 (1972)
I.N. p. 790
REM p. 1202
Ainsworth, G.C. and Pope, C.G.; US Patent 2,565,057; August 21, 1951; assigned to Burroughs Wellcome & Co. (U.S.A.) Incorporated
Petty, M.A.; US Patent 2,595,605; May 6, 1952; assigned to American Cyanamid Company
Benedict, R.G. and Stodola, F.H.; US Patent 2,771,397; November 20, 1956; assigned to the US Secretary of Agriculture

POLYTHIAZIDE

Therapeutic Function: Diuretic

Chemical Name: 6-Chloro-3,4-dihydro-2-methyl-3-[[(2,2,2-trifluoroethyl) thio]methyl]-2H-1,2,4-benzothiadiazine-7-sulfonamide-1,1-dioxide

Common Name: -

Chemical Abstracts Registry No.: 346-18-9

Structural Formula:

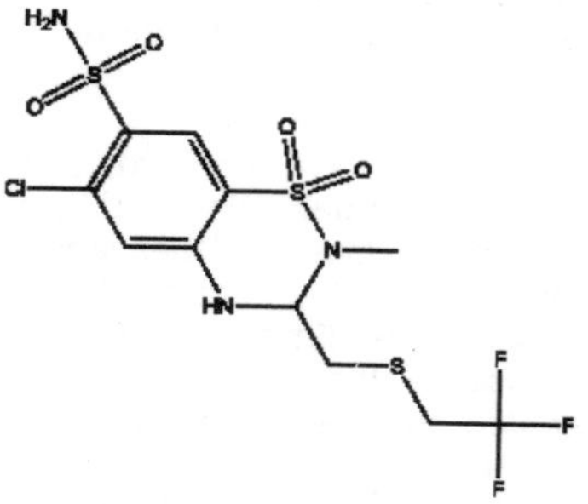

Trade Name	**Manufacturer**	**Country**	**Year Introduced**
Renese	Pfizer	US	1961
Drenusil	Pfizer	W. Germany	1962
Renese	Pfizer	Italy	1962
Renese	Pfizer	France	1965
Envarese	Pfizer	France	-
Minizide	Pfizer	US	-
Nephril	Pfizer	UK	-
Polyregulon	Yamanouchi	Japan	-
Toleran	Medica	Finland	-

Raw Materials

4-Amino-2-chloro-5-(methylsulfamyl)benzenesulfonamide
Sodium
Mercaptoacetaldehyde dimethylacetal
Trifluoroethyl iodide

Manufacturing Process

(A) Preparation of trifluoroethylthioacetaldehyde dimethylacetal: To 4.6 g (0.2 mol) of metallic sodium dissolved in 75 ml of absolute methanol is rapidly added 24.4 g (0.2 mol) of mercaptoacetaldehyde dimethylacetal followed by dropwise addition of 42.0 g (0.2 mol) of trifluoroethyl iodide.

The resulting reddish mixture is refluxed on a steam bath for one hour. One half of the alcohol is removed by concentration and the remainder diluted with several volumes of water and extracted with ether. The combined ether extracts are dried over sodium sulfate, the ether then removed at reduced pressure and the residue distilled to about 30 g (BP 82°C/25 mm).

(B) Preparation of 4-Amino-2-Chloro-5-(Methylsulfamyl)Benzenesulfonamide: The 5-substituted-2,4-disulfamyl anilines may be prepared by procedures described in the literature, for example, the general procedures in Monatsch. Chem. vol. 48, p 87 (1927), which involves the treatment of a m-substituted aniline with from 10 to 20 parts by weight of chlorosulfonic acid followed by the gradual addition of from about 90 to 170 parts by weight of sodium chloride. The resultant mixture is heated at approximately 150°C for about 2 hours after which the reaction mixture is poured into water and the resultant 5substituted aniline-2,4-disulfonyl chloride is filtered and is then treated with concentrated ammonium hydroxide or suitable amine by standard procedures to obtain the corresponding disulfonamide.

(C) Preparation of 2-Methyl-3-(2,2,2-Trifluoroethyl)Thiomethyl-6-Chloro-7-Sulfamyl-3,4-Dihydro-1,2,4-Benzothiadiazine-1,1-Dioxide: To 4.6 g (0.015 mol) of 4-amino-2-chloro5-(methylsulfamyl)benzenesulfonamide in 30 ml of the dimethyl ether of ethylene glycol is added 4.08 g (0.02 mol) of 2,2,2-trifluoroethylmercaptoacetaldehyde dimethylacetal followed by 1 ml of ethyl acetate saturated with hydrogen chloride gas. The resulting solution is refluxed for 1.5 hours, cooled and then slowly added to cold water dropwise with stirring. The crude product is filtered, dried and recrystallized from isopropanol (3.2 g), MP 202° to 202.5°C. A second recrystallization from isopropanol raised the MP to 202° to 203°C.

References

Merck Index 7457
Kleeman & Engel p. 743
PDR pp. 1409, 1421
OCDS Vol. 1 p. 360 (1977)
I.N. p. 791
REM p. 940
McManus, J.M.; US Patent 3,009,911; November 21, 1961; assigned to Chas. Pfizer & Co., Inc.

PORFIMER SODIUM

Therapeutic Function: Antineoplastic, Photosensitizer

Chemical Name: Photofrin porfimer sodium (porfimer sodium disigned also as polyporphin oligomer containing ester and ether linkage)

Common Name: Porfimer Sodium; Photofrin II

Trade Name	Manufacturer	Country	Year Introduced
PhotoBarr	Axcan Pharma Inc.	Canada	-
Photofrin II	Lederle Parenterals, Inc.	USA	-
Photofrin II	QLT Phototherapeutics Inc.	-	-
Photofrin II	Wyeth-Ayerst Lederle Parenterals, Inc.	-	-
Photofrin II	Axcan Scandipharm Inc.	USA	-

Raw Materials

Acetic acid
Sodium hydroxide
Sodium acetate
Hematoporphyrin hydrochloride
Sulfuric acid

Structural Formula:

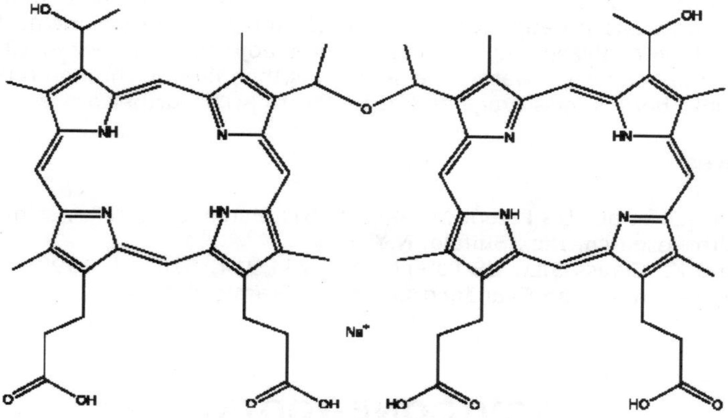

Chemical Abstracts Registry No.: 87806-31-3

Manufacturing Process

285 ml of acetic acid was added to a 1000 ml Erlenmeyer flask containing Teflon-coated magnetic stirring bar. Stirring the acetic acid, slowly 15 ml of concentrated sulfuric acid was added; weighing out 15.0 g of hematoporphyrin hydrochloride (preferably obtained from Roussel Corporation, Paris, France); adding to the acid solution; the reaction acetic mixture was stirred for 1 h. To the reaction acetic mixture 3 L of 5% sodium acetate was added.

The 5% sodium acetate solution now contains a dark red precipitate which is preferably allowed to stand for 1 h with occasional stirring; the dark red precipitate is then again filtered, preferably using the above-identified filter mechanism; the filter cake from the filtering process is then washed with glass-distilled water until the filtrate is at pH of 5.5-6.0 (1500-2500 ml of wash water may be required); and the filter cake is then preferably allowed to dry in air at room temperature. The air-dried precipitate is ground, using for instance, a mortar and pestle until a fine powder is obtained. The powder may then be transferred to a 250 ml round bottom flask. The flask is then attached to a rotating evaporator and rotation under vacuum is maintained at room temperature for preferably 24 h and hemoporphyrin acetate was obtained.

Acetylated hematoporphyrin (1 part by weight) is dissolved in 0.1 N sodium hydroxide (50 parts by volume) and stirred for 1 h at room temperature. After the stir period the solution is adjusted to pH 9.4 to 9.6 with 1 N hydrochloric acid. It is filtered through a 5 μm filter and then concentrated to (12.5 parts) of its original volume in an ultrafilter with 10,000 molecular weight cut off membranes. The solution is then purified via diafiltration maintaining constant volume with 120 volumes of water and keeping the pH at 9.4 to 9.6 with 0.1 N sodium hydroxide. This is also done at room temperature. After the purification, the solution is removed from the ultrafilter, diluted to 3/8 (18.8 parts) of its original volume and pH adjusted to 7.5 to 7.7 with 1 N hydrochloric acid. The solution is then stored at 4°C for 14 to 21 days. After storage, the solution is pH adjusted to 9.4 to 9.6 with 0.1 N sodium hydroxide and concentrated to 1/4 (12.5 parts) of its original volume. The solution is then repurified as above. The solution is diluted to 3/8 (18.8 parts) of its original volume and pH adjusted to 7.5 to 7.7 with 1 N hydrochloric acid. The solution is then analyzed and, if necessary, it is adjusted to between 13 to 18 mg/ml by the addition of water. The solution is then filtered through a 0.22 μm filter into bottles for storage at 1°C to 4°C to await further processing.

References

Dougherty T.J. et al.; US Patent No. 4,649,151; March 10, 1987; Assigned: Health Research, Inc., Buffalo, N.Y.

Zawadzki R.K., Clauss S.L.; US Patent No. 5,244,914; Sep. 14, 1993; Assigned: American Cyanamid Company, Stamford, Conn.

POVIDONE-IODINE

Therapeutic Function: Topical antiinfective

Chemical Name: 1-Ethenyl-2-pyrrolidinone homopolymer compound with iodine

Common Name: PVP-I

Structural Formula:

$$\left[\text{-CH(N-pyrrolidin-2-one)-CH}_2\text{-} \right]_n \cdot xI_2$$

Chemical Abstracts Registry No.: 25655-41-8

Trade Name	Manufacturer	Country	Year Introduced
Betadine	Purdue Frederick	US	1957
Betadine	Sarget	France	1970
Efodine	Fougera	US	1978
Vagidine	Beecham	US	1981
Clinidine	Clinipad	US	1982
Mallisol	Mallard	US	1983
ACU-Dyne	Acme Laboratories Ltd.	US	-
Batticon	Trommsdorff	W. Germany	-
Betadine Ginecologico	Chinoin	Italy	-
Betaisodona	Mundipharma	Austria	-

Trade Name	Manufacturer	Country	Year Introduced
Braunol	Braun	W. Germany	-
Chem-O-Dine	Remedia	S. Africa	-
Difexon	Bago	Argentina	-
Disadine	Stuart	UK	-
Isodine	Purdue Frederick	US	-
Jodobac	Bode	W. Germany	-
Jodocur	Farm. Milanese	Italy	-
Neojodin	Iwaki	Japan	-
Nutradine	Restan	S. Africa	-
Pevidine	Berk	UK	-
Polydine	Fischer	Israel	-
Povadyne	Chaston	US	-
Proviodine	Rougier	Canada	-
Summer's Eve	Fleet	US	-
Topionic	Rius	Spain	-

Raw Materials

Polyvinylpyrrolidone
Iodine

Manufacturing Process

12 g of dry polyvinylpyrrolidone having a K value of 90 (water content about 2 to 3%) was added to 6 g of solid iodine crystals in a glass bottle containing a few pebbles and beads. This was rolled for 3 days on a roller mill with occasional manual stirring to loosen the material caked on the sides of the bottle. Analysis showed that the thus-obtained product contained 35.4% total iodine and 31.91% available iodine. The material was heat-treated at 95°C for 64 hours in a closed glass bottle with occasional stirring. On completion of this treatment, analysis showed that the material contained 35.3% total iodine, 25.7% available iodine, according to US Patent 2,706,701.

References

Merck Index 7595
PDR pp. 880, 888, 1432
DOT 7 (4) 149 (1971)
I.N. p. 793
REM p. 1164
Beller, H. and Hosmer, W.A.; US Patent 2,706,701; April 19, 1955; assigned to General Aniline & Film Corporation
Hosmer, W.A.; US Patent 2,826,532; March 11, 1958; assigned to General Aniline & Film Corporation
Siggia, S.; US Patent 2,900,305; August 18, 1959; assigned to General Aniline & Film Corporation

PRACTOLOL

Therapeutic Function: Antiarrhythmic

Chemical Name: N-[4-[2-Hydroxy-3-[(1-methylethyl)amino]propoxy]phenyl] acetamide

Common Name: 1-(4-Acetamidophenoxy)-3-isopropylamino-2-propanol

Structural Formula:

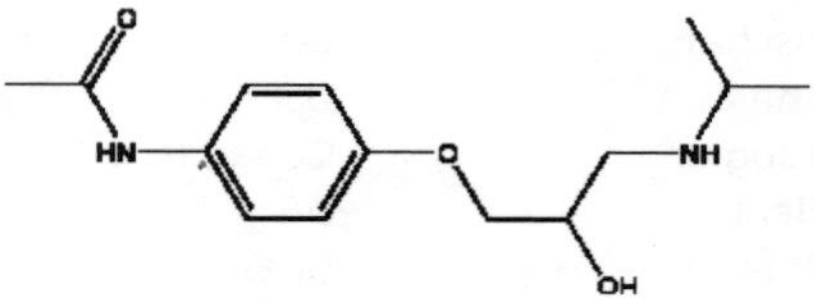

Chemical Abstracts Registry No.: 6673-35-4

Trade Name	Manufacturer	Country	Year Introduced
Eraldin	I.C.I.	UK	1970
Eraldin	I.C. Pharma	Italy	1972
Dalzic	Rhein Pharma	W. Germany	1973
Eraldine	I.C.I. Pharma	France	1973
Cardiol	Orion	Finland	-
Pralon	Farmos	Finland	-

Raw Materials

4-Acetamidophenol
Epichlorohydrin
Isopropylamine

Manufacturing Process

The 1-(4-acetamidophenoxy)-2,3-epoxypropane used as starting material may be obtained as follows. To a solution of 4.5 parts of 4-acetamidophenol and 1.5 parts of sodium hydroxide in 50 parts of water at 15°C, there is added 3.5 parts of epichlorohydrin. The mixture is stirred for 16 hours at ambient temperature, filtered and the solid residue is washed with water. There is thus obtained 1-(4-acetamidophenoxy)-2,3-epoxypropane, MP 110°C.

A mixture of 2 parts of 1-(4-acetamidophenoxy)-2,3-epoxypropane and 10 parts of isopropylamine is stirred at ambient temperature for 16 hours. The resulting solution is evaporated to dryness under reduced pressure and the residue is crystallized from butyl acetate. There is thus obtained 1-(4-acetamidophenoxy)-3-isopropylamino-2-propanol, MP 134° to 136°C.

References

Merck Index 7597
OCDS Vol. 2 pp. 106, 108 (1980)
DOT 6 (5) 188 (1970)
I.N. p. 794
Howe, R. and Smith, L.H.; US Patent 3,408,387; October 29, 1968; assigned to Imperial Chemical Industries Limited, England

PRAJMALINE BITARTRATE

Therapeutic Function: Antiarrhythmic

Chemical Name: 17R,21α-Dihydroxy-4-propylajmalanium bitartrate

Common Name: -

Structural Formula:

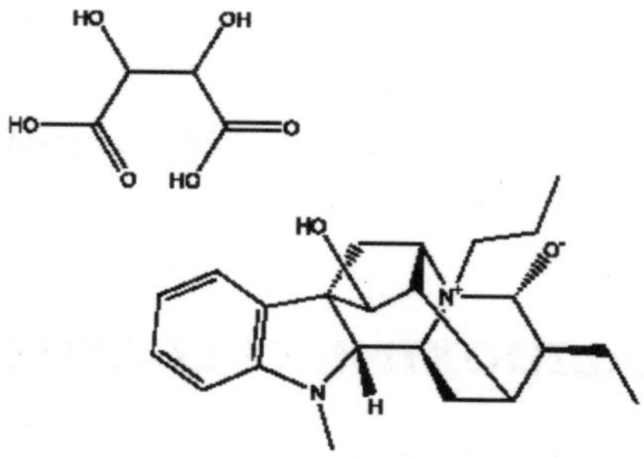

Chemical Abstracts Registry No.: 2589-47-1; 35080-11-6 (Base)

Trade Name	Manufacturer	Country	Year Introduced
Neo-Gilurtymal	Giulini	W. Germany	1973
Neo-Aritmina	Byk Gulden	Italy	1979

Raw Materials

Ajmaline
Sodium bicarbonate
Allyl bromide
Tartaric acid

Manufacturing Process

1 g of ajmaline was dissolved in 4 cc of chloroform, and 1 cc of allyl bromide

was added to the resulting solution. The reaction mixture thus obtained was allowed to stand for 24 hours at room temperature. Thereafter, the clear reaction solution was briefly cooled to a temperature below 0°C, whereby crystallization set in. The crystals were filtered off and were then recrystallized from a mixture of absolute methanol and absolute ether. The purified colorless crystalline product was identified to be N-(b)-allyl-ajmalinium-bromide having a melting point of 252°C to 254°C.

75 g of N-(b)-n-propyl-ajmalinium-bromide were suspended in 3 liters of an aqueous saturated solution of sodium bicarbonate, and the suspension was admixed with 3 liters of chloroform. The resulting mixture was vigorously stirred for six to eight hours. Thereafter, the chloroform phase was separated and evaporated to dryness. 68 g of a yellow syrup remained as a residue. The aldehyde base was dissolved in about 150 cc of acetone and, while stirring and cooling on an ice bath, the solution was slowly admixed with a solution of 25 g of tartaric acid in 2 liters of acetone. The fine white precipitate formed thereby was separated by vacuum filtration, washed with ether and dried. The raw product, weighing 80 g, was recrystallized once from a mixture of ethanol and ether, yielding 50 g of N-(b)-n-propyl-ajmalinium hydrogen tartrate having a melting point of 149°C to 152°C (decomposition).

References

Merck Index 7598
Kleeman & Engel p. 744
I.N. p. 794
Keck, J.; US Patent 3,414,577; December 3, 1968; assigned to Boehringer Ingelheim G.m.b.H. (Germany)

PRALIDOXIME CHLORIDE

Therapeutic Function: Antidote (nerve gas)

Chemical Name: 2-[(Hydroxyimino)methyl]-1-methylpyridinium chloride

Common Name: 2-PAM chloride

Structural Formula:

Cl^- N^+ N—OH

Chemical Abstracts Registry No.: 51-15-0; 495-94-3 (Base)

Trade Name	Manufacturer	Country	Year Introduced
Contrathion	Specia	France	1961
Protopam	Ayerst	US	1964
Combo Pen	Rodana Res. Corp.	US	-

Raw Materials

2-Pyridinealdoxime
Nitrosyl chloride
Methyl chloride
α-Picoline
Dimethyl sulfate
Sodium hydroxide

Manufacturing Process

As described in US Patent 3,123,613, the preparation of the intermediate product, 2-pyridinealdoxime methomethylsulfate, is as follows. 1 kg of 2-pyridinealdoxime is dissolved in 6 liters of acetone and filtered until clear. 2 kg (2 equivalents) of freshly distilled dimethyl sulfate are added and the solution mixed. In about 30 minutes crystals start to appear, after which a cooling bath is used to keep the temperature at about 30° to 35°C until the reaction is nearly complete (about 2 hours).

The mixture is allowed to stand at room temperature overnight, the crystals filtered off and washed on a filter with acetone. The product is obtained as colorless needles, which melt at 111° to 112.5°C. The methylsulfate is not stable indefinitely. For preparation of pure chloride salt it is desirable to use methylsulfate which gives no titratable acidity with sodium hydroxide using bromophenol blue as indicator.

10 g of 2-pyridinealdoxime methomethylsulfate are then dissolved in 6 cc of concentrated hydrochloric acid, and 60 cc of isopropanol is added with stirring. Crystals appear almost instantly. After 2 hours standing at room temperature, the crystals are separated by filtration and washed with acetone. The product had a melting point of 227° to 228°C and the yield was 85%.

An alternative route is described in US Patent 3,155,674.

(A) Preparation of 1-Methyl-2-Picolinium Chloride: 98 ml of α-picoline is dissolved in 200 ml of methanol, cooled and 85 ml (at -68°C) of methyl chloride is added. The solution is charged to an autoclave, sealed and the nitrogen pressure of 300 psig is established. The mixture is heated at 120° to 130°C for 2 hours, cooled and opened. The resulting solution is then evaporated to dryness in vacuo, yielding a residue of 110 g. This residue is then dissolved in 50 ml of water and extracted with two 50 ml portions of ether. The aqueous phase is then diluted to 150 ml with water and an assay for ionic chloride is performed which indicates the presence of chloride ion equivalent to 721 mg/ml of l-methyl-2picolinium chloride.

(B) Preparation of 2-(Hydroxyiminomethyl)-1-Methyl Pyridinium Chloride: An aqueous solution of 15 ml of 1-methyl-2-picolinium chloride having a concentration of 477 mg/ml is covered with 50 ml of benzene in an atmosphere of nitrogen and cooled to below 10°C. An aqueous solution of sodium hydroxide is added dropwise and the mixture is stirred for 5 minutes and allowed to stratify. The aqueous phase is then drawn off and the benzene solution is added slowly to a solution of 3 ml of nitrosyl chloride in 175 ml of benzene containing 0.5 ml of dimethyl formamide at about 10°C in an atmosphere of nitrogen with good agitation. The mixture is then stirred for 1.5 hours and then extracted with four 5 ml of portions of water. The aqueous extracts are then concentrated in vacuo, 30 ml of isopropanol is added and the concentration is repeated. 20 ml of isopropanol is then added to the concentrated mixture, and the mixture is cooled to room temperature and filtered, yielding 3.04 g of crude 2-(hydroxyiminomethyl)-1-methylpyridinium chloride, melting at 202° to 214°C with decomposition. The filtrate is then further concentrated to a 7 g residue which is crystallized from absolute alcohol and yields 0.9 g of 2-(hydroxyiminomethyl)-1-methyl pyridinium chloride melting at 221° to 225°C with decomposition.

References

Merck Index 7599
Kleeman & Engel p. 744
PDR p. 648
I.N. p. 794
REM p. 901
Bloch, L.P.; US Patent 3,123,613; March 3, 1964; assigned to Campbell Pharmaceuticals, Inc.
Ellin, R.I., Easterday, D.E. and Kondritzer, A.A.; US Patent 3,140,289; July 7, 1964; assigned to the US Secretary of the Army
McDowell, W.B.; US Patent 3,155,674; November 3, 1964; assigned to Olin Mathieson Chemical Corporation

PRAMIPEXOLE DIHYDROCHLORIDE

Therapeutic Function: Antiparkinsonian, Antipsychotic

Chemical Name: 2,6-Benzothiazolediamine-4,5,6,7-tetrahydro-N6-propyl, dihydrochloride, (6S)-

Common Name: Pramipexole dihydrochloride

Structural Formula:

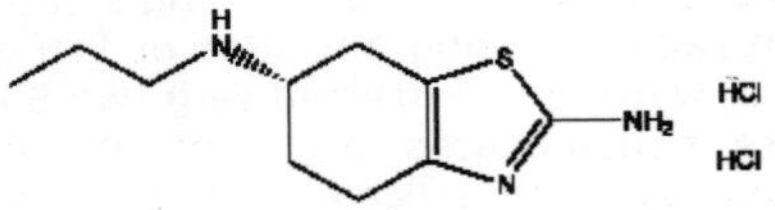

Chemical Abstracts Registry No.: 104632-25-9; 104632-26-0 (Base)

Trade Name	Manufacturer	Country	Year Introduced
Mirapex	Pharmacia and Upjohn	USA	-

Raw Materials

Phthalic anhydride
Ethyldiisopropyl amine
Acetic acid
Thiourea
n-Propanal
Hydrobromic acid
4-Aminocyclohexanol hydrochloride
Potassium dichromate
Bromine
Hydrazine hydrate
Hydrochloric acid
Sodium borohydride

Manufacturing Process

75.5 g (0.5 mol) of 4-aminocyclohexanol hydrochloride and 74.0 g (0.5 mol) of phthalic acid anhydride are mixed with 65 g (0.5 mol) of ethyldiisopropyl amine and 1000 ml of toluene and boiled for 36 hours with a water separator. Then water is added, the toluene phase is separated off and the aqueous phase is extracted several times with chloroform. The organic phases are combined, dried and concentrated. The concentrated residue is recrystallised from isopropanol and 4-(phthalimido)-cyclohexanol was obtained. Yield: 95 g (77.8%). Melting point 175°-176°C.

95 g (0.388 mol) of 4-(phthalimido)-cyclohexanol are dissolved in 600 ml of chloroform and, after the addition of 450 ml of water and 120 ml of sulfuric acid, 90 g (0.3 mol) of potassium dichromate are added in batches. The internal temperature of the mixture is maintained at between 25° and 30°C by slight cooling. The mixture is stirred for a further 3 hours, then the chloroform phase is separated off and the mixture extracted twice more with chloroform. After drying and concentration of the extracts 82 g (86.9%) of 4-(phthalimido)-cyclohexanone was obtained.

48.6 g (0.2 mol) of 4-(phthalimido)cyclohexanone are dissolved in glacial acetic acid, mixed with 36% of hydrobromic acid in glacial acetic acid and then 32 g (0.2 mol) of bromine in glacial acetic acid is added dropwise with cooling. The mixture is then concentrated by evaporation in vacuo and the residue is triturated several times with diethylether. The ether extracts are discarded and the residue is dissolved in of ethanol. After thiourea have been added the mixture is refluxed for 5 hours. It is then concentrated by evaporation, made alkaline with sodium hydroxide solution and extracted with chloroform. After drying and concentration of the extracts, the residue is purified by column chromatography on silica gel (eluant: chloroform/methanol = 1/1). The 2-amino-6-phthalimido-4,5,6,7-tetrahydro-benzthiazol was obtained. Melting point 244-246°C, dec. Yield: 30 g (50%).

9.5 g (31.7 mmol) of 2-amino-6-phthalimido-4,5,6,7-tetrahydro-benzthiazole are suspended in 100 ml of ethanol and, after the addition of 1.8 g (36 mmol) of hydrazine hydrate, refluxed for 2 hours. The mixture is then concentrated and purified by column chromatography on silica gel using methanol as eluant. The 2,6-diamino- 4,5,6,7-tetrahydro-benzthiazole was obtained.

To a solution of 2,6-diamino- 4,5,6,7-tetrahydro-benzthiazole in dimethylformamide are added n-propanal and the mixture is heated to 50°C for 1 hour. After cooling, the reaction solution is mixed with sodium borohydride and heated to 50°C for 30 min. The solvent is largely eliminated in vacuo. Whilst cooling with ice, the residue is mixed with water and 2 N hydrochloric acid until a pH of 1 is obtained. The aqueous solution is extracted with ethylacetate and the organic phase discarded. The aqueous phase is mixed with potassium carbonate until an alkaline reaction is obtained and then extracted with ethyl acetate. The organic phase is dried and concentrated. The 2-amino-6-n-propylamino-4,5,6,7-tetrahydro-benzthiazole dihydrochloride crystallizes out when ethereal hydrochloric acid is added. Yield: 42%. Melting point: 286°-288°C.

References

Griss G. et al.; US Patent No. 4,886,812; Dec. 12, 1989; Assigned: Dr. Karl Thomae GmbH, Biberach an der Riss, Fed. Rep. of Germany

PRAMIVERIN

Therapeutic Function: Spasmolytic

Chemical Name: N-(1-Methylethyl)-4,4-diphenylcyclohexanamine

Common Name: Primaverine; Propaminodiphen

Structural Formula:

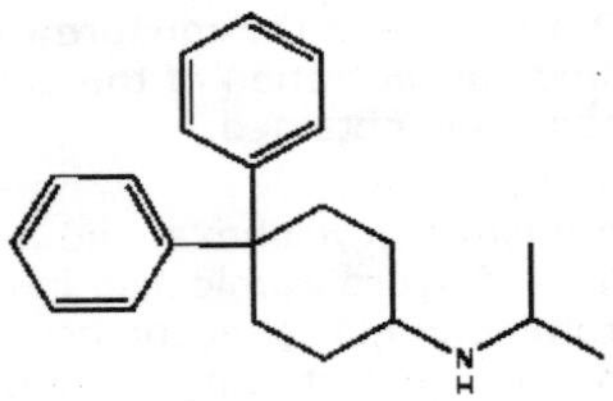

Chemical Abstracts Registry No.: 14334-40-8; 14334-41-9 (Hydrochloride salt)

Trade Name	Manufacturer	Country	Year Introduced
Sistalgin	Bracco	Italy	1974
Sistalgin	Cascan	W. Germany	1976

Raw Materials

Isopropylamine
4,4-Diphenyl-cyclohexen-(2)-one
Hydrogen

Manufacturing Process

20 g 4,4-diphenyl-cyclohexen-(2)-one, 10 g isopropylamine, and 50 ml tetrahydrofuran are agitated for 10 hours in a bomb tube at 200°C. Subsequently, the reaction mixture is cooled, and the tetrahydrofuran and the

excess isopropylamine are distilled off. The remaining Schiff base is dissolved in methanol and after the addition of 2 g platinum oxide, the base is hydrogenated at normal pressure and room temperature until a quantity of hydrogen corresponding to 2 mols has been absorbed.

The mixture is filtered off from the catalyst, made acidic with dilute hydrochloric acid, and the methanol is removed under vacuum. The remaining aqueous solution is made alkaline with solution of sodium hydroxide and extracted with ether. After drying and concentrating the ether extract, there is obtained 17 g 1-isopropylamino-4,4-diphenyl-cyclohexane, boiling point 164°C to 165°C/0.05 mm. The hydrochloride melts at 230°C.

References

Merck Index 7602
Kleeman & Engel p. 745
DOT 11 (8) 320 (1975)
I.N. p. 795
Unger, R., Sommer, S., Schorscher, E. and Encakel, H.J.; US Patent 3,376,312; April 2, 1968; assigned to E. Merck A.G. (Germany)

PRAMOXINE HYDROCHLORIDE

Therapeutic Function: Local anesthetic

Chemical Name: 4-[3-(4-Butoxyphenoxy)propyl]morpholine hydrochloride

Common Name: Pramocaine hydrochloride; Proxazocain hydrochloride

Structural Formula:

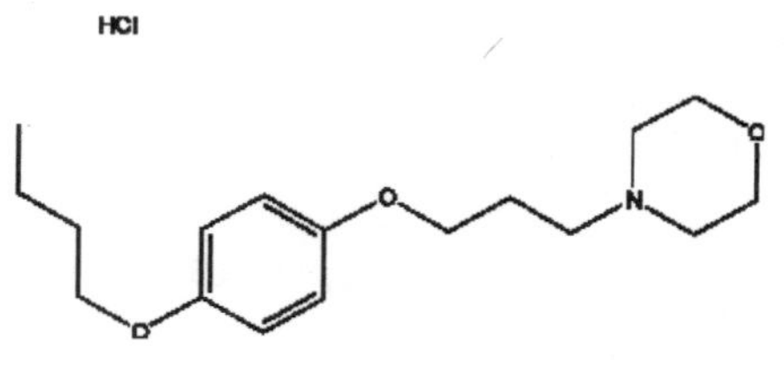

HCl

Chemical Abstracts Registry No.: 637-58-1; 140-65-8 (Base)

Trade Name	Manufacturer	Country	Year Introduced
Tronothane	Abbott	US	1954
Tronothane	Abbott	France	1956
Proctofoam	Reed Carnrick	US	1975
Prax	Ferndale	US	1980
Analpram	Ferndale	US	-
Anusol	Parke Davis	US	-

Trade Name	Manufacturer	Country	Year Introduced
F.E.P.	Boots	US	-
Fleet Relief	Fleet	US	-
Otic-HC	Hauck	US	-
Pramosone	Ferndale	US	-
Tronolane	Ross	US	-
Zone-A	U.A.D. Labs	US	-

Raw Materials

Hydroquinone monobutyl ether
Potassium hydroxide
γ-Morpholinopropyl chloride
Hydrogen chloride

Manufacturing Process

About 5.6 g of potassium hydroxide is dissolved in about 150 cc of refluxing ethanol, and then about 16.6 g of hydroquinone monobutyl ether is added to the alcoholic solution. When the hydroquinone is dissolved, about 16.3 g of γ-morpholinopropyl chloride (dissolved in a small amount of ethanol) is added to the refluxing solution. The solution is refluxed for about 24 hours and then cooled. The product is recovered by filtering the reaction mixture and then removing the solvent by vacuum distillation. The oily residue is acidified and shaken with ether. The acidic phase is made strongly alkaline with 40% sodium hydroxide, and the oil which separates is extracted into ether. The ethereal phase is dried, and the solvent removed by vacuum distillation. The product distills at 183° to 184°C at a pressure of 2.8 mm. The hydrochloride salt of the foregoing base is prepared by dissolving the base in ether and acidifying with hydrochloric acid and is found to have a MP of 181° to 183°C.

References

Merck Index 7603
Kleeman & Engel p. 745
PDR pp. 684, 875, 880, 928, 1316, 1565, 1808
OCDS Vol. 1 p. 18 (1977)
I.N. p. 795
REM p. 1057
Wright, H.B. and Moore, M.B.; US Patent 2,870,151; January 20, 1959; assigned to Abbott Laboratories

PRANOPROFEN

Therapeutic Function: Analgesic, Antiinflammatory

Chemical Name: 2-(5H-[1]Benzopyrano[2,3-b]-pyridin-7-yl)propionic acid

Common Name: -

Structural Formula:

Chemical Abstracts Registry No.: 52549-17-4

Trade Name	Manufacturer	Country	Year Introduced
Niflan	Yoshitomi	Japan	1981

Raw Materials

Ethyl 2-cyano-2-(5H-[1]benzopyrano[2,3-b]-pyridin-7-yl)propionate
Hydrogen chloride

Manufacturing Process

A mixture of 100 g of ethyl 2-cyano-2-(5H-[1]benzopyrano[2,3-b]-pyridin-7-yl)propionate, 500 ml of glacial acetic acid and 200 g of concentrated hydrochloric acid is refluxed for 48 hours. The reaction mixture is concentrated, and the residue is dissolved in hot water. The solution is adjusted to pH 2 to 3 by addition of 10% sodium hydroxide. The resulting crystalline precipitate is washed thoroughly with water, and recrystallized from aqueous dioxane to give 74 g of 2-(5H-[1]benzopyrano[2,3-b]-pyridin-7-yl)propionic acid as white crystals melting at 183°C to 183.5°C.

References

Merck Index 7604
DFU 2 (3) 217 (1977) (As Y-8004) & 2 (12) 829 (1977)
Nakanishi, M., Oe, T. and Tsuruda, M.; US Patent 3,931,205; January 6, 1976; assigned to Yoshitomi Pharmaceutical Industries, Ltd.

PRASTERONE

Therapeutic Function: Glucocorticoid

Chemical Name: Androst-5-en-17-one, 3-hydroxy-, (3β)-

Common Name: Dehydroandrosterone; Dehydroepiandrosterone; Dehydroisoandrosterone; Prasterone

Chemical Abstracts Registry No.: 53-43-0

Structural Formula:

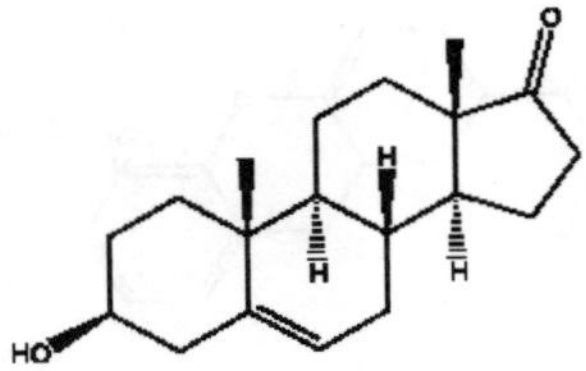

Trade Name	Manufacturer	Country	Year Introduced
Aslera	Genelabs Technologies, Inc.	-	-
Prasterone	Proquina SA	-	-
Deandros	Schering AG	-	-

Raw Materials

16-Dehydropregnenolon-3β-acetate
Hydroxylamine hydrochloride
Phosphorus pentachloride

Manufacturing Process

To a solution of 1 gram of 16-dehydropregnenolon-3β-acetate in 10 ml pyridine is added 0.22 gram of hydroxylamine hydrochloride, and the mixture is allowed to stand at room temperature for four days. One gram of 16-dehydropregnenolon-3β-acetate oxime is dissolved in 30 ml of hot dioxane, and then the solution is cooled in an ice bath until about one-half of the dioxane has solidified. Then 1 gram of phosphorus pentachloride is added and the mixture is shaken until all the dioxane has melted. The mixture is maintained at 35°C, for seventy-five minutes, then an excess of ice is added and the solution is again allowed to stand at 35°C. After about thirty minutes, a solution of 5 ml of concentrated hydrochloric acid in 10 ml of water is added, and the mixture is diluted with water, extracted with ether and the ethereal extract washed with dilute sodium hydroxide solution. The ether is removed on a steam bath and the residue is worked up to yield dehydro-isoandrosterone.

References

Tendick F.H., Lawson E.J.; US Patent No. 2,335,616; Nov. 30, 1943; Assigned to Parke Davis and Company, Detroit, Mich., a corporation of Michigan

PRAVASTATIN SODIUM

Therapeutic Function: Antihyperlipidemic

Chemical Name: (1S-(1-α(β-S*,δ-S*),2α,6α,8β(R*),8a-α))-1-Naphthaleneheptanoic acid 1,2,6,7,8,8a-hexahydro-2-methyl-8-(2-methyl-1-oxobutoxy)-β,δ,6-trihydroxy-, monosodium salt

Common Name: Eptastation sodium; Pravastatin sodium

Structural Formula:

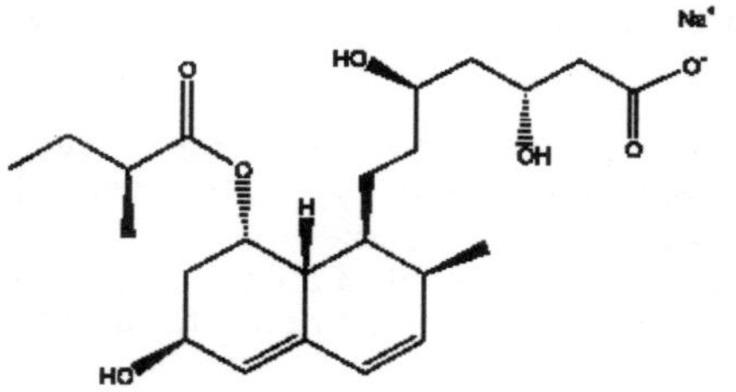

Chemical Abstracts Registry No.: 81131-70-6; 81093-37-0 (Base)

Trade Name	Manufacturer	Country	Year Introduced
Apo-Pravastatin	Apotex Inc.	Canada	-
Apo-Pravastatin	Cobalt Pharmaceuticals Inc.	-	-
Lipostat	Bristol-Myers Squibb	France	-
Prastatin	Emcure Pharmaceuticals Ltd.	India	-
Prava	B.-M. Sq./South Africa	S. Africa	-
Pravachol	Bristol-Meyers Squibb	France	-
Pravachol	Silanes	-	-
Pravachol	Generic	-	-
Pravator	Solus	-	-

Raw Materials

Nocardia autotrophica subsp. amethystina FERM P-6183
2-Methyl-8-(2-methyl-1-oxobutoxy)-β,δ-dihydroxy(1S-(1-α(β-S*,δ-S*),2-α,6-α,8-β(R*),8a-α))-1-1,2,6,7,8,8a-hexahydronaphthaleneheptanoate sodium
Sodium hydrogen carbonate
Mortierella maculata nov. spec. E-97 [NCAIM(P)F 001266]

Manufacturing Process

Pravastatin was isolated as products of enzymatic hydroxylation by some kinds of microorganisms of [1S-[1-α(R*),7β,8β(2S*,4S*)8αβ]]-2-methylbutanoic acid 1,2,3,7,8,8a-hexahydro-7-methyl-8-[2-(tetrahydro-4-hydroxy-6-oxo-2))-pyran-2-yl)ethyl]-1-naphthalenic lactone (campactin) or their carboxylic acid or their salts (products of animal metabolism of microorganisms from the genera Nocardia, Streptomyces et cetera).

Pravastatin may be prepared by using the microorganisms of genera Nocardia (method 1) and Mortierella (method 2).

Method 1

Cultivation of Nocardia autotrophica subsp. amethystine

Cells of Nocardia autotrophica subsp. amethystina FERM P-6183 was inoculated from a slant culture by means of a platinum loop into each of twenty 500 ml Erlenmeyer flasks, each containing 100 ml of a culture medium having the following composition: glucose - 1.0%, peptone - 0.2%, meat extract - 0.1%, yeast extract - 0.1%, corn steep liquor 0.3%, tap water balance..

Shaking was then carried out at 26°C and 220 r.p.m. for 2 days, at which time sodium 2-methyl-8-(2-methyl-1-oxobutoxy)-β,δ-dihydroxy(1S-(1-α(β-S*,δ-S*),2-α,6-α,8-β(R*),8a-α))-1-1,2,6,7,8,8a-hexahydronaphthaleneheptanoate was added to a final concentration of 0.05% w/v. Incubation was continued at 26°C and 220 r.p.m. for a further 5 days.

Preparation of pravastatin

After completion of the cultivation, the reaction mixture was filtered and the pH of the filtrate was adjusted to a value of 3 by the addition of trifluoroacetic acid. The acidified filtrate was then extracted three times, each with 1 liter of ethyl acetate, to give extracts containing a mixture (6-α and 6-β) of (1S-(1-α,β- S*,δ-S*),2-α,8-βR*),8a-α))-1-naphthaleneheptanoic acid 1,2,6,7,8,8a-hexahydro-2-methyl-8-(2-methyl-1-oxobutoxy)-β,δ,6-trihydroxy.

This extract was then immediately transferred into a 5% w/v aqueous solution of sodium hydrogen carbonate, and the pH of the mixture was adjusted to a value of 7.0 by the addition of 2 N hydrochloric acid. The mixture was then adsorbed on a Diaion HP-20 column. The column was washed with water and then eluted with 50% v/v aqueous acetone to give a fraction containing (1S-(1-α,β- S*,δ-S*),2-α,6-α,8-βR*),8a-α))-1-naphthaleneheptanoic acid 1,2,6,7,8,8a-hexahydro-2-methyl-8-(2-methyl-1-oxobutoxy)-β,δ,6-trihydroxy-, monosodium salt (pravastatin). This was freeze-dried, to give 200 mg of pravastatine.

Method 2

Cultivation of Mortierella maculata nov. spec. E-97 [NCAIM(P)F 001266]

A spore suspension was prepared with 5 ml of a 0.9% sodium chloride solution obtained from a 7-10 day old, malt extract-yeast extract agar slant culture of Mortierella maculata nov. spec. E-97 [NCAIM(P)F 001266] strain able to 6-β-hydroxylate compactin and the suspension was used to inoculate 100 ml inoculum medium PI (glucose-50 g, soybean meal-20 g, in 1000 ml tap water) sterilized in a 500 ml Erlenmeyer flask.

5 liters working volume a bioconversion culture medium is prepared (glucose-20 g, glycerine-20 g, soybean meal-20 g, peptone-5 g, potassium dihydrogen phosphate-0.5 g, polypropylene glycol 2000-1 g, in 1000 ml tap water); the components of the culture medium are added corresponding to 5 liters. Then it was sterilized for 45 min at 121°C and seeded with 500 ml of the inoculum culture.

Before sterilization the pH of the medium was adjusted to 7.0 value.

The fermentation was carried out at 28°C, with a stirring rate of 400 rpm and with an aeration rate from bottom direction 60 liters/hour for 4 days. At the 2nd day after the transfer the culture started to foam heavily, which can be decreased by the addition of further polypropylene glycol 2000. The pH reached 6.3-7.5 by the 4th day. The feeding of the sodium 2-methyl-8-(2-methyl-1-oxobutoxy)-β,δ-dihydroxy(1S-(1-α(β- S*,δ-S*),2-α,6-α,8-β(R*),8a-α))-1-1,2,6,7,8,8a-hexahydronaphthaleneheptanoate substrate is allowed to be started if the pH of the broth is above 6.3.

Preparation of pravastatin

At the 4th day of the fermentation 2.5 g compactin substrate is added in sterile filtered aqueous solution. Calculated for the volume of the broth 0.5-1.0% glucose was added into the culture depending on the pH in the form of 50% solution sterilized at 121°C for 25 min in parallel with the substrate feeding. After 24 hours the compactin substrate is consumed from the culture (is detected by HPLC) and was converted to pravastatin. By lyophilization of the aqueous residue 1.3 g pravastatin was obtained. The chromatographically pure product was crystallized from a mixture of ethanol and ethyl acetate. Melting point: 170-173°C (decomp.).

References

Terahara A., Tanaka M.; US Patent No. 4,537,859; Aug. 27, 1985; Assigned: Sankyo Company, Limited, Tokyo, Japan

Jekkel A., et al.; US Patent No. 6,682,913 B1; Jan. 27, 2004; Assigned: Institute for Drug Research Ltd.

PRAZEPAM

Therapeutic Function: Tranquilizer

Chemical Name: 7-Chloro-1-(cyclopropylmethyl)-1,3-dihydro-5-phenyl-2H-1,4-benzodiazepin-2-one

Common Name: -

Structural Formula:

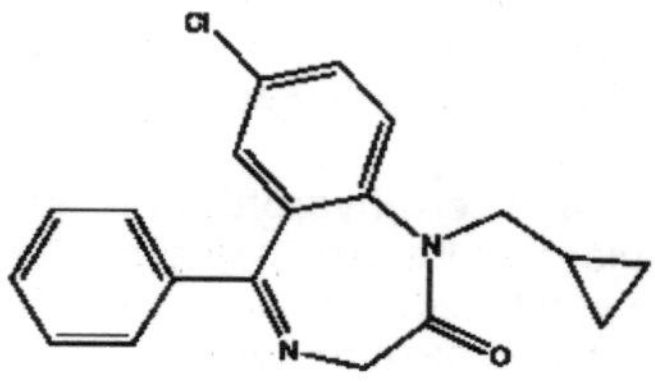

Chemical Abstracts Registry No.: 2955-38-6

Trade Name	Manufacturer	Country	Year Introduced
Demetrin	Goedecke	W. Germany	1973
Centrax	Parke Davis	US	1977
Demetrin	Cosmopharma	Switz.	1978
Lysanxia	Substantia	France	1979
Prazene	Parke Davis	Italy	1980
Trepidan	Sigma Tau	Italy	1980
Centrax	Warner William	UK	1981
Demetrin	Parke Davis	France	1982
Reapam	Goedecke	W. Germany	-
Verstran	Warner-Chilcott	US	-

Raw Materials

Phthalimidoacetyl chloride	Lithium aluminum hydride
Manganese dioxide	2-Amino-5-chlorobenzophenone
Hydrazine hydrate	Cyclopropane carboxylic acid chloride

Manufacturing Process

Preparation of 2-Cyclopropylcarbonylamido-5-Chlorobenzophenone: To 400.5 g (1.73 mols) of 2-amino-5-chlorobenzophenone dissolved in 220 g (2.18 mols) of triethylamine and 3.5 liters of tetrahydrofuran is added cautiously 181 g (1.73 mols) of cyclopropanecarboxylic acid chloride. The reaction is refluxed 2½ hours and allowed to cool to room temperature. The solvent is then removed under vacuum to obtain 2-cyclopropylcarbonylamido-5-chlorobenzophenone as a residue which is dissolved in 1 liter of methylene chloride, washed twice with 5% hydrochloric acid, and then twice with 10% potassium hydroxide. The methylene chloride solution is then dried over anhydrous magnesium sulfate, filtered and the solvent removed under vacuum. The residue is recrystallized from 1,500 ml of methanol, charcoal-treating the hot solution to give 356 g of 2-cyclopropylcarbonylamido-5-chlorobenzophenone, MP 105° to 105.5°C (69% yield).

Preparation of 2-Cyclopropylmethylamino-5-Chlorobenzhydrol: To a slurry of 94.8 g (2.47 mols) of lithium aluminum hydride in 1.2 liters of tetrahydrofuran is added with stirring a solution of 356 g (1.18 mols) of 2-cyclopropylcarbonylamido-5-chlorobenzophenone in 1.8 liters of tetrahydrofuran. The addition takes 80 minutes while maintaining gentle refluxing, and the reaction mixture is then refluxed overnight and allowed to cool to room temperature over a period of 3 days. The complex formed in the reaction mixture is then hydrolyzed with water.

During the hydrolysis, 500 ml of tetrahydrofuran is added to facilitate stirring. At a point where the flocculant white precipitate settles quickly when stirring is interrupted, the mixture is filtered, the filter cake washed with solvent, the combined filtrates dried over magnesium sulfate, filtered and the solvent removed under vacuum to obtain 2-cyclopropylmethylamino-5-chlorobenzhydrol as a residue. The residue is recrystallized from 1,300 ml of Skelly B, giving 315 g of 2-cyclopropylmethylamino-5-chlorobenzhydrol, MP 85° to 85.5°C (93% yield).

Preparation of 2-Cyclopropylmethylamino-5-Chlorobenzophenone: To a solution of 315 g (1.09 mols) of 2-cyclopropylmethylamino-5-chlorobenzhydrol in 4 liters of benzene is added 453.6 g (5.22 mols) of manganese dioxide, freshly prepared according to the method of Attenburrow et al, J.C.S. 1952, 1104. The mixture is then refluxed for 1¼ hours, filtered, and the filtrate evaporated under vacuum. The reddish residue is recrystallized from 510 ml of 90% acetone-10% water, giving 181 g of pure 2-cyclopropylmethylamino-5-chlorobenzophenone, MP 79° to 80°C (58% yield). Upon concentration of the mother liquor a second crop of 2-cyclopropylmethylamino-5-chlorobenzophenone weighing 34.1 g and melting at 76.5°-78°C are obtained.

Preparation of 2-(N-Phthalimidoacetyl-N-Cyclopropylmethyl) -Amino-5-Chlorobenzophenone: To a solution of 36.0 g (0.126 mol) of 2-cyclopropylmethylamino-5-chlorobenzophenone in 500 ml of tetrahydrofuran is added 50.7 g (0.252 mol) of phthalimidoacetyl chloride. The resulting solution is refluxed for 16 to 24 hours, the solvent removed under vacuum, the residual oil crystallized from 200 ml of ethanol and recrystallized from 500 ml of 80% ethanol-20% tetrahydrofuran giving 44.7 g of 2-(N-phthalimidoacetyl-N-cyclopropylmethyl)-amino-5-chlorobenzophenone, MP 163° to 164°C (75% yield).

Preparation of 1-Cyclopropylmethyl-5-Phenyl-7-Chloro-1H-1,4-Benzodiazepine-2(3H)-one: To a solution of 39.5 g (0.0845 mol) of 2-(N-phthalimidoacetyl-N-cyclopropylmethyl)amino5-chlorobenzophenone in a mixture of 423 ml of chloroform and 423 ml of ethanol is added 9.52 g (0.1903 mol) of hydrazine hydrate and 9.52 ml of water. This solution is allowed to stand at room temperature. In 3 hours a precipitate begins to form in the solution. After standing 16 to 24 hours a voluminous pulpy white precipitate forms. The solvents are removed under vacuum while keeping the temperature under 40°C and the residue is partitioned between dilute ammonia water and ether.

The aqueous layer is separated and washed with ether, the ether extracted with 5% hydrochloric acid, the acidic solution is made basic with 10% sodium hydroxide and again extracted with ether. Since some spontaneous crystallization occurs in the ether, the solvent is removed without drying under vacuum and the residue is recrystallized from 35 ml of ethanol giving 18.0 g of 1-cyclopropylmethyl-5-phenyl-7-chloro-1H-1,4-benzodiazepine-2(3H)-one, MP 145° to 146°C (65% yield), according to US Patent 3,192,199.

References

Merck Index 7608
Kleeman & Engel p. 747
PDR p. 1320
OCDS Vol. 2 p. 405 (1980)
DOT 2 (3) 119 (1966); 9 (6) 237 (1973); & 10 (5) 179 (1974)
I.N. p. 796
REM p. 1063
McMillan, F.H. and Pattison, I.; US Patent 3,192,199; June 29, 1965
Wuest, H.M.; US Patent 3,192,200; June 29, 1965

PRAZIQUANTEL

Therapeutic Function: Anthelmintic

Chemical Name: 2-(Cyclohexylcarbonyl)-1,2,3,6,7,11b-hexahydro-4H-pyrazino[2,1-a]isoquinolin-4-one

Common Name: -

Structural Formula:

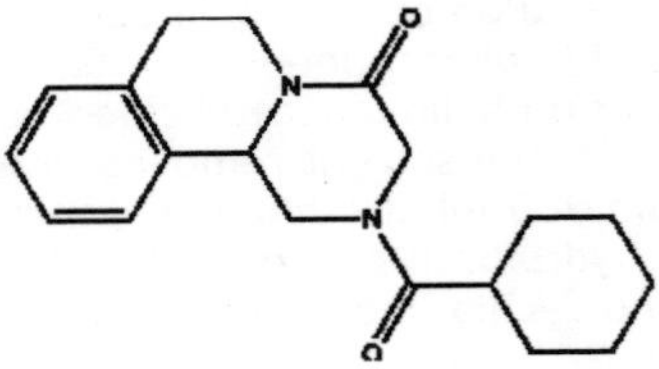

Chemical Abstracts Registry No.: 55268-74-1

Trade Name	Manufacturer	Country	Year Introduced
Cesol	Merck	W. Germany	1980
Biltricide	Bayer	W. Germany	1980
Cenaride	Merck Clevenot	France	1981
Biltricide	Bayer	France	1983
Biltricide	Miles	US	1983
Droncit	Bayvet	US	-

Raw Materials

2-Cyclohexylcarbonyl-4-oxo-2,3,6,7-tetrahydro-4H-pyrazino[2,1-a] isoquinoline
Hydrogen

Manufacturing Process

15 g of a nickel-aluminum alloy (1:1) is introduced in incremental portions and under agitation into 200 ml of 20% sodium hydroxide solution within 5 minutes; the mixture is maintained at 80°C for 45 minutes, then allowed to settle, decanted off, washed with water, and 1,000 ml of 1% (-)-tartaric acid solution is added thereto, adjusted to pH 5 with 1 N sodium hydroxide solution. The mixture is heated under agitation for 90 minutes to 80°C, decanted, and washed with water and methanol. The thus-obtained (-)-tartaric acid-Raney nickel catalyst is added to a solution of 2-cyclohexylcarbonyl-4-oxo-2,3,6,7-tetrahydro-4H-pyrazino[2,1-a]isoquinoline. The reaction mixture is hydrogenated under normal pressure and at room temperature. After the catalyst has been filtered off and the solvent evaporated, 2-cyclohexylcarbonyl-4-oxo-1,2,3,6,7,11b-hexahydro-4H-pyrazino[2,1-a]isoquinoline, melting point 136°C to 138°C, is produced.

References

Merck Index 7609
Kleeman & Engel p. 748
PDR p. 1249
DOT 13 (3) 121 (1977) & 17 (10) 429 (1981)
I.N. p. 796
REM p. 1237
Seubert, J., Thomas, H. and Andrews, P.; US Patent 4,001,411; January 4, 1977; assigned to Merck Patent G.m.b.H. (Germany)

PRAZOSIN

Therapeutic Function: Antihypertensive

Chemical Name: 1-(4-Amino-6,7-dimethoxy-2-quinazolinyl)-4-(2-furanylcarbonyl)piperazine

Common Name: Furazosin

Structural Formula:

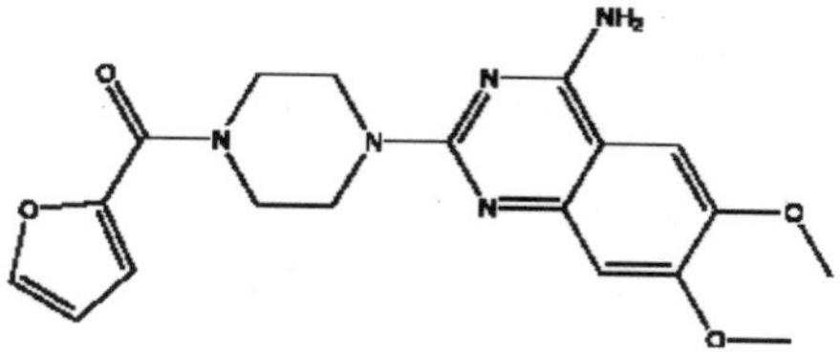

Chemical Abstracts Registry No.: 19216-56-9; 19237-84-4 (Hydrochloride salt)

Trade Name	Manufacturer	Country	Year Introduced
Hypovase	Pfizer	UK	1974
Minipress	Pfizer	US	1976
Minipress	Pfizer	W. Germany	1977
Minipress	Pfizer	Italy	1978
Minipress	Pfizer	France	1979
Minipress	Pfizer Taito	Japan	1981
Adversuten	Arzneimittelwerk Dresden	E. Germany	-
Orbisan	Mack	W. Germany	-
Pratsiol	Orion	Finland	-
Prazac	Erco	Denmark	-
Sinetens	Carlo Erba	UK	-
Vasoflex	Alkaloid	Yugoslavia	-

Raw Materials

Piperazine
2,4-Dichloro-6,7-dimethoxyquinazoline
Ammonia
2-Furoyl chloride

Manufacturing Process

Preparation of 2-Chloro-4-Amino-6,7-Dimethoxyquinazoline: To 800 ml of a solution of anhydrous ammonia in tetrahydrofuran at room temperature is added 30 g of 2,4-dichloro-6,7-dimethoxyquinazoline [F.H.S. Curd et al., J. Chem. Soc., p 1759 (1948)]. The mixture is stirred for 44 hours. The precipitate (29 g, MP 267° to 268°C) is filtered and recrystallized from methanol to yield 19 g of 2-chloro-4-amino-6,7-dimethoxyquinazoline, MP 302°C (dec.).

Preparation of 2-(1-Piperazinyl)-4-Amino-6,7-Dimethoxyquinazoline: To 5 g of 2-chloro-4-amino-6,7-dimethoxyquinazoline, is added 20 g of a 25% solution of piperazine in ethanol. The mixture is heated at 160°C for 16 hours in a pressure bottle. The solvent is then evaporated and the residue is recrystallized from methanol/water.

Preparation of 2[4-(2-Furoyl)-Piperazinyl]-4-Amino-6,7-Dimethoxyquinazoline: To 0.10 mol 2-(1-piperazinyl)-4-amino-6,7-dimethoxyquinazoline in 300 ml methanol is added with vigorous stirring, 0.10 mol 2-furoyl chloride. After addition is complete, the mixture is stirred for 3 hours at room temperature. The solids are filtered to give the desired product, MP 278° to 280°C.

References

Merck Index 7610
Kleeman & Engel p. 748
PDR pp. 1420, 1421
OCDS Vol. 2 p. 382 (1980) & 3, 194 (1984)
DOT 11 (2) 67, 80 (1975)
I.N. p. 796
REM p. 844
Hess, H.-J.E.; US Patent 3,511,836; May 12, 1970; assigned to Chas. Pfizer & Co., Inc.

PREDNIMUSTINE

Therapeutic Function: Cancer chemotherapy

Chemical Name: Prednisolone 21-[4'-[p-bis(2-chloroethyl)amino]phenyl] butyrate

Common Name: Prednisolone chlorambucil ester

Structural Formula:

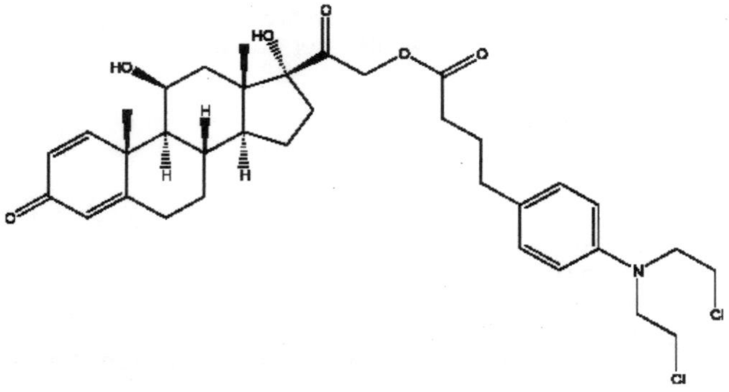

Chemical Abstracts Registry No.: 29069-24-7

Trade Name	Manufacturer	Country	Year Introduced
Stereocyt	Bellon	France	1978
Sterecyt	Leo	Switz.	1981
Mostarina	Abello	Spain	-

Raw Materials

Prednisolone
p-[N-Bis(β-chloroethyl)amino]phenyl butyric acid
Thionyl chloride

Manufacturing Process

p-[N-bis(β-chloroethyl)amino] phenyl butyric acid was dissolved in a mixture of 150 ml dry benzene and 8.04 ml dry pyridine. The solution was cooled in an ice bath, and a solution of thionyl chloride in 30 ml dry benzene was slowly added with stirring under anhydrous conditions.

The reaction mixture was then kept at room temperature for 1 hour and thereafter poured into a mixture of 5.0 N HCl and crushed ice. The benzene solution was immediately washed with water, with cold 1.0 N $NaHCO_3$ and finally with cold water. After drying over anhydrous sodium sulfate, the benzene was removed in vacuo. The residue is the p-[N-bis(β-chloroethyl)amino]phenyl butyric anhydride which could be used without any further purification.

To a solution of 42.0 g of p-[N-bis(β-chloroethyl)amino]phenyl butyric anhydride in 500 ml dry pyridine was added 24.4 g of prednisolone. The reaction mixture was kept at room temperature for 24 hours under anhydrous condition. It was then poured into a mixture of concentrated HCl and crushed ice and extracted with ether-ethyl acetate (1:1).

The organic phase was washed several times with cold 1.0 N K_2CO_3 and finally water. After drying over $CaCl_2$ the solvent was removed in vacuo.

The residue is prednisolone 21-[4'-[p-bis(β-chloroethyl)amino]phenyl]butyrate which after crystallization from methanol/water had a melting point of 163°C to 164°C.

References

Merck Index 7612
DFU 1 (3) 137 (1976)
Kleeman & Engel p. 749
OCDS Vol. 3 p. 93 (1984)
DOT 16 (3) 84 (1980)
I.N. p. 797
Fox, H.J., Hogberg, K.B. and Konyves, I.; US Patent 3,732,260; May 8, 1973; assigned to A.B. Leo

PREDNISOLAMATE

Therapeutic Function: Glucocorticoid

Chemical Name: 11β,17,21-Trihydroxypregna-1,4-diene-3,20-dione 21-N,N-diethylglycine ester

Common Name: Prednisolamate; Prednisolone diethylaminoacetate

Structural Formula:

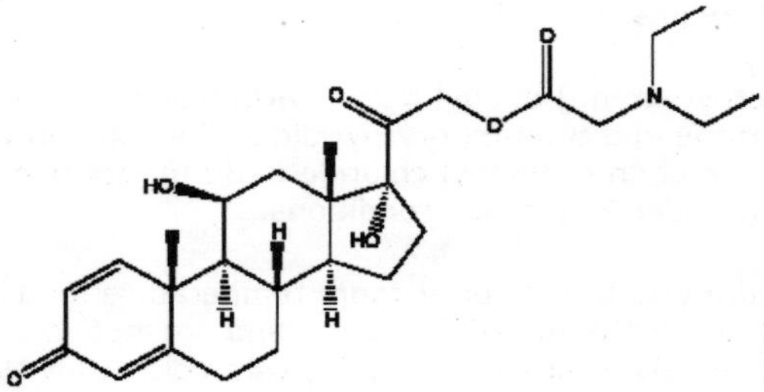

Chemical Abstracts Registry No.: 5626-34-6

Trade Name	Manufacturer	Country	Year Introduced
Prednisolamate	Pfizer and Co.	-	-

Raw Materials

Prednisolone
Chloroacetyl chloride
Diethylamine

Manufacturing Process

To a solution of 30 g prednisolone and 10.2 ml pyridine in 99 ml dimethylformamide, cooled to 0°C in an ice bath and protected from atmospheric moisture, was added dropwise with stirring 9.6 ml chloroacetyl chloride. Stirring was continued for 1 hour at 0°C, then stopped and the reaction allowed to come to room temperature and stand overnight. The following morning the reaction mixture was poured with vigorous stirring into a vessel containing 360 ml 1 N sulfuric acid, and the product washed with fresh portions of water to neutral wash was obtained. The product was thoroughly dried in a vacuum desiccator and recrystallized from isopropyl alcohol, using 200 ml solvent. The yield was better than 80%, and the compound had MP: 240.6°-242.8°C and $[\alpha]^D$ = +114.6° (Dioxane).

Two grams of this prednisolone chloracetate and 40 ml of colorless, freshly distilled diethylamine were refluxed, with stirring, under nitrogen for one hour. The excess diethylamine was removed in vacuum at room temperature. The residue was taken up in 100 ml $CHCl_3$, and a small amount of water; the $CHCl_3$, was washed with one 50 ml portion of 5% aqueous sodium bicarbonate and two 50 ml portions of water, and dried over sodium sulfate. The chloroform was then concentrated to dryness in vacuum. The residue was recrystallized from acetone-hexane yielded prednisolone 21-diethylaminoacetate 1.67 g; MP: 175.0°-197.2°C. This product (1 g) was suspended in a mixture of 15 ml acetone and 1.5 ml chloroform. The suspension was cooled to 0°C in an ice bath, and with vigorous stirring the ethereal HCl solution was slowly added until the resulting mixture gave an acid reaction to the congo red paper. The product was removed by filtration and recrystallized from ethanol. The yield was approximately 80%; MP: 239.4°-239.8°C; $[\alpha]^D$ = +120.7° (water). The product was the hydrochloride salt of the prednisolone-N,N-diethylaminoacetate, very active therapeutically and excellently water-soluble.

References

Ch.Pfizer and Co., Inc., a corporation of the State Delaware, USA; G.B. Patent No. 862,370; Aug. 31, 1956

PREDNISOLONE

Therapeutic Function: Glucocorticoid

Chemical Name: 11β,17,21-Trihydroxypregna-1,4-diene-3,20-dione

Common Name: Metacortandralone; Δ^1-Hydrocortisone

Chemical Abstracts Registry No.: 50-24-8

Raw Materials

Hydrocortisone	Bacterium Corynebacterium simplex

Structural Formula:

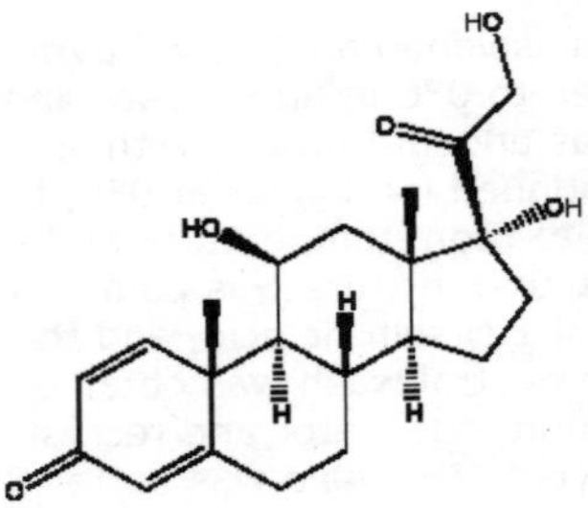

Trade Name	Manufacturer	Country	Year Introduced
Sterane	Pfizer	US	1955
Meticortelone	Schering	US	1955
Delta-Cortef	Upjohn	US	1955
Hydeltra	MSD	US	1955
Paracortol	Parke Davis	US	1957
Sterolone	Rowell	US	1957
Prednis	U.S.V. Pharm.	US	1957
Ulacort	Fellows-Testagar	US	1960
Cosilone	Person Covey	US	1963
Adnisolone	Adams	Australia	-
Aprednislon	Arcana	Austria	-
Caberdelta	Caber	Italy	-
Cordrol	Vita Elixir	US	-
Cortalone	Halsey	US	-
Cortisolone	S.I.T.	Italy	-
Cotolone	Truxton	US	-
Dacortin	Igoda	Spain	-
Decaprednil	Dorsch	W. Germany	-
Decortasmyl	Larec	Ecuador	-
Delta-Hycortol	Medica	Finland	-
Delta-Larma	Larma	Spain	-
Deltalone	D.D.S.A.	UK	-
Deltasolone	Knoll	Australia	-
Deltidrosol	Poli	Italy	-
Deltisolon	Ferring	Sweden	-
Domucortone	Medici Domus	Italy	-
Encortolone	Polfa	Poland	-
Fernisolon	Ferndale	US	-
Ibisterolon	I.B.I.	Italy	-
Keteocort -H	Desitin	W. Germany	-
Neodelta	Amelix	Italy	-
Normosona	Normon	Spain	-
Novoprednisolone	Novopharm	Canada	-
Panafcortelone	Glebe	Australia	-
Predartrina	Farmochimica	Italy	-
Prednicen	Central	US	-

Trade Name	Manufacturer	Country	Year Introduced
Predni-Coelin	Pfleger	W. Germany	-
Prednicort	Cortec	Denmark	-
Predni-Helvacort	Helvepharm	Switz.	-
Predni-H-Tablinen	Sanorania	W. Germany	-
Predniretard	Boots-Dacour	France	-
Prelone	Langley	Australia	-
Ropredlone	Robinson	US	-
Scherisolon	Schering	W. Germany	-
Serilone	Serpero	Italy	-
Stermin	Schlicksup	US	-
Vitacort	Vitarine	US	-

Manufacturing Process

The following procedure is described in US Patent 2,837,464: from a solution of 3 grams of yeast extract (Difco) in 3.0 liters of tap water containing 13.2 grams of potassium dihydrogen phosphate and 26.4 grams disodium hydrogen phosphate (pH of the solution, 6.9) 27 portions of 100 ml each are withdrawn, placed in 300 ml Erlenmeyer flasks and sterilized by autoclaving for 15 minutes at 15 pounds steam pressure (120°C). After autoclaving and cooling of the broth, one ml of suspension of Corynebacterium simplex (ATCC 6946) is placed in each flask. The flasks are then shaken on a shake table at 220 rpm and 28°C for 24 hours.

Into each of 27 Erlenmeyer flasks are placed 150 mg of Kendall's Compound F (hydrocortisone). The flasks and contents are then sterilized for 15 minutes at 15 pounds steam pressure (120°C). To each flask are then added 5.0 ml of ethanol. The 24-hour bacterial culture is then transferred aseptically and the resulting suspensions are shaken on a shake table at 220 rpm and 28°C for 48 hours. The pH at the end of the shake period is 7.0.

The contents of all the flasks are combined and extracted with a total of 9.0 liters of chloroform in 3 equal portions. The combined extracts are then concentrated to a residue which weighs 3.75 grams. The MP of the residue is 227°-232°C. From 2.75 grams of this crude material on sludging with 50 ml of acetone and cooling, there is recovered on filtration 1.35 grams of $\Delta^{1,4}$-pregnadiene-11β,17α,21-triol-3,20-dione,MP 237°-239°C (dec.). Additional product can be recovered from the mother liquor. Recrystallization from acetone raised the MP to 239°-241°C (dec.).

References

Merck Index 7613
Kleeman & Engel p. 750
PDR pp. 830, 1569, 1606
OCDS Vol. 1 p. 192 (1977) & 2, 178 (1980)
I.N. p. 797
REM p. 969
Nobile, A.; US Patent 2,837,464; June 3, 1958; assigned to Schering Corporation

Oliveto, E.P. and Gould, D.H.; US Patent 2,897,216; July 28, 1959; assigned to Schering Corporation

PREDNISOLONE ACETATE

Therapeutic Function: Glucocorticoid

Chemical Name: 11β,17,21-Trihydroxypregna-1,4-diene-3,20-dione 21-acetate

Common Name: -

Structural Formula:

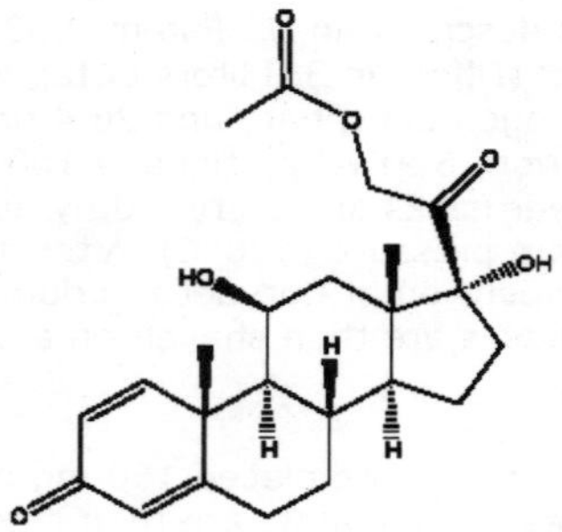

Chemical Abstracts Registry No.: 52-21-1

Trade Name	Manufacturer	Country	Year Introduced
Sterane	Phipharmex	US	1955
Nisolone	Ascher	US	1962
Savacort	Savage	US	1969
Econapred	Alcon	US	1973
Pred Mild	Allergan	US	1974
Pred Cor 100	Hauck	US	1977
Alto-Pred	Alto	US	-
Cortipred	Italsuisse	Italy	-
Deltacortilen	S.I.F.I.	Italy	-
Dermo-Nydol	Brichard	France	-
Durapred	Federal	US	-
Hexacorton	Spirig	Switz.	-
Ibisterolon-Pommada	I.B.I.	Italy	-
Inflanefran	Allergan	W. Germany	-
Key-Pred	Hyrex	US	-
Metimyd	Schering	US	-
Meticortelone	Essex	Italy	-

Trade Name	Manufacturer	Country	Year Introduced
Predate	Legere	US	-
Predicort	Dunhall	US	-
Prednifor	Vifor	Switz.	-
Prenema	Nortech	US	-
Pricortin	Premedics	US	-
Sigpred	Sig	US	-
Ulacort	Fellows-Testagar	US	-
Ultracortenol	Dispersa	Switz.	-

Raw Materials

Prednisolone
Acetic anhydride

Manufacturing Process

To a solution of 0.85 gram of 1,4-pregnadiene-11β,17α,21-triol-3,20-dione (prednisolone) in 5 ml of pyridine are added 3 ml of acetic anhydride. The reaction mixture is allowed to stand at room temperature overnight and is then diluted with ice water. The resulting precipitate is filtered from the mixture and recrystallized from acetone-hexane. There is recovered 0.45 gram of 1,4-pregnadiene-11β,17α,21-triol-3,20-dione 21-acetate, MP 235°239°C. On recrystallization, the MP rose to 237°-239°C.

References

Merck Index 7613
Kleeman & Engel p. 750
PDR pp. 1033, 1633
OCDS Vol. 1 p. 192 (1977)
I.N. p. 798
REM p. 969
Nobile, A.; US Patent 3,134,718; May 26, 1964; assigned to Schering Corporation

PREDNISOLONE PHOSPHATE SODIUM

Therapeutic Function: Glucocorticoid

Chemical Name: 11β,17,21-Trihydroxypregna-1,4-diene-3,20-dione 21-(dihydrogen phosphate) disodium salt

Common Name: -

Chemical Abstracts Registry No.: 125-02-0

Structural Formula:

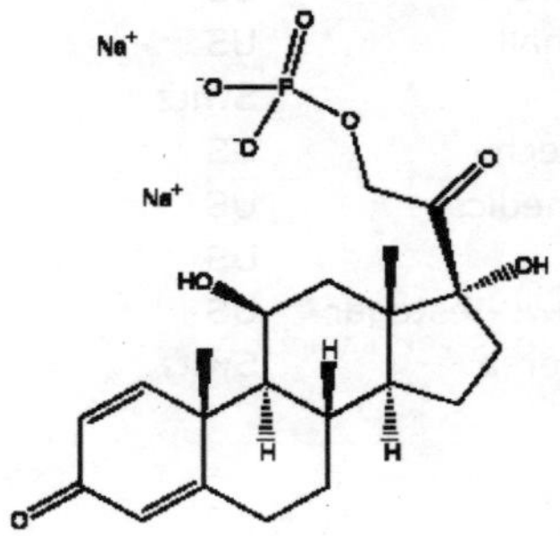

Trade Name	Manufacturer	Country	Year Introduced
Hydeltrasol	MSD	US	1957
Inflamase	Cooper Vision	US	1969
Optival	White	US	1969
PSP-IV	Tutag	US	1972
Alto-Pred	Alto	US	-
Caberdelta	Caber	Italy	-
Codelsol	MSD	UK	-
Hydrosol	Rocky Mtn.	US	-
Key-Pred S.P.	Hyrex	US	-
Metreton	Schering	US	-
Nor-Preds	North Amer. Pharm.	US	-
Parisolon	Riker	US	-
Predate S	Legere	US	-
Prednesol	Glaxo	US	-
Savacort	Savage	US	-
Sodasone	Fellows-Testagar	US	-
Solucort	Chibret	France	-
Solu-Pred	Myers-Carter	US	-

Raw Materials

Prednisolone
Sodium hydroxide
Phosphoric acid
Sodium iodide
Methanesulfonyl chloride

Manufacturing Process

Preparation of Prednisolone 21-Methanesulfonate: Seventy liters of dry pyridine and 7.5 kg of prednisolone are charged to a 30-gallon jacketed glass-lined still. The mixture is agitated until complete solution is obtained. About 40 liters of pyridine are distilled at high vacuum while maintaining the batch temperature below 40°C. The solution is cooled to 0°C, and 2.2 liters of methanesulfonyl chloride are charged. The batch temperature is maintained between 0°C and +3°C during charging of the methanesulfonyl chloride. An atmosphere of flowing nitrogen is maintained in the still, and the mixture is agitated during the last stages of the addition. The mixture is then aged for one hour, and 15 gallons of ice water are added cautiously to the still while maintaining the temperature between 0° and 5°C.

The still contents are then transferred to a jacketed kettle equipped with an agitator, and 62 kg of cracked ice in 15 gallons of deionized water are added. The batch is aged one hour and a solution of 2 liters of concentrated (37%) hydrochloric acid in 4 gallons of deionized water is added. The batch is centrifuged and the centrifuge cake washed free of pyridine with deionized water. The centrifuge cake is then vacuum-dried at 50°C to a moisture content of about 1%, which requires about 3 days of drying. Yield about 7.77 kg (92%), according to US Patent 2,932,657.

Preparation of Prednisolone 21-Iodide: To a 30-gallon jacketed glass-lined still 64.5 lb (31.0 liters) of dimethylformamide are charged by vacuum. The still contents are agitated as 7.74 kg of dry (less than 1% moisture) prednisolone 21-methanesulfonate are charged. Then 4.02 kg of sodium iodide are charged. The still contents are heated to 57° to 60°C by means of a steam jacket and held at this temperature for 30 minutes. The batch is cooled to 35°C and 12 gallons of deionized water are added at the rate of about 1 gallon per minute. In the event the solution becomes cloudy, addition of water is interrupted and the mixture agitated for five minutes before resumption of water addition. After all of the water is added, the batch is transferred to a 50 gallon kettle equipped with agitator and an additional 16.7 gallons of deionized water are added. The batch is cooled to 0° to 5°C and aged for one hour. The batch is filtered and the filter cake washed and vacuum dried at 30° to 35°C to a moisture content of less than 1%. Yield about 7.95 kg (96%), according to US Patent 2,932,657.

Preparation of Prednisolone 21-Disodium Phosphate: Acetonitrile (50.0 ml) containing phosphoric acid (90%; 1.0 ml) was treated with triethylamine (3.0 ml) and the solution added to 11β,17α-dihydroxy-21-iodopregna-1,4-diene-3,20-dione (1.0 gram; powdered). The mixture was refluxed for 2.75 hours and the solvent was then evaporated under reduced pressure to give a yellow oil. The oil was taken up in methanol (25 ml) and titrated to pH 10.9 with sodium hydroxide in methanol (N) using a pH meter. The precipitate was filtered off and the filtrate evaporated to a gum under reduced pressure. The gum was taken up in methanol (5 ml), filtered through filter paper and acetone (100 ml) was added to the filtrate. The precipitate was filtered off, washed with acetone and dried at 100°C/1 mm for 0.75 hour giving a pale yellow solid, prednisolone disodium phosphate (0.74 gram), which was completely soluble in water, according to US Patent 2,936,313.

References

Merck Index 7615
Kleeman & Engel p. 752
PDR pp. 1033, 1633
I.N. p. 798
REM p. 970
Sarett, L.H.; US Patent 2,789,117; April 16, 1957; assigned to Merck & Co., Inc.
Christensen, B.G., Hirschmann, R.F. and Putter, I.; US Patent 2,932,657; April 12, 1960; assigned to Merck & Co., Inc.

Elks, J. and Phillipps, G.H.; US Patent 2,936,313; May 10, 1960; assigned to Glaxo Laboratories Limited, England

PREDNISOLONE STEAROYLGLYCOLATE

Therapeutic Function: Glucocorticoid

Chemical Name: 11β,17-Dihydroxy-21-[[[(1-oxoctadecyl)oxy]acetyl]oxy] pregna-1,4-diene-3,20-dione

Common Name: Prednisolone steaglate

Structural Formula:

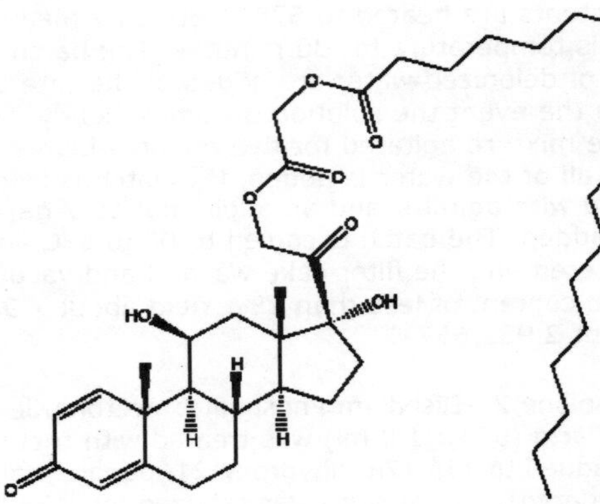

Chemical Abstracts Registry No.: 5060-55-9

Trade Name	Manufacturer	Country	Year Introduced
Deturgylone	Dausse	France	1970
Erbacort	Erba	Italy	-
Estilsona	Erba	Italy	-
Glistelone	Erba	Italy	-
Glitisone	Vis	Italy	-
Prenisol	Cifa	Italy	-
Rollsone	Bellon	France	-
Sintisone	Erba	Italy	-
Verisone	Tiber	Italy	-

Raw Materials

Prednisolone	Prednisolone-21-chloroacetate
Potassium stearate	Stearoyl-glycolyl chloride

Manufacturing Process

This material can be prepared, e.g., by reaction of prednisolone-21-chloroacetate in solvent with the sodium or potassium salt of the corresponding aliphatic or aromatic acid, or by reaction of prednisolone with the chloride of the corresponding acyl-glycolic acid, in the presence of a hydrochloric acid acceptor.

Alternative (A): 3 grams (0.0068 mol) prednisolone chloroacetate dissolved in 200 ml tetrahydrofuran and 10 ml H_2O are added with 2.7 grams (0.0084 mol) K stearate and 0.06 g NaI and heated to boiling, under stirring, for 36 hours, then evaporated in vacuum to dryness.

The residue is washed with H_2O to disappearance of the Cl-ion from the filtrate. Crystallization from diluted alcohol results in prednisolone-21-stearoyl-glycolate (MP 104°-105°C).

Alternative (B): 3.6 grams (0.01 mol) prednisolone and 4.32 grams (0.012 mol) stearoylglycolyl-chloride, separately dissolved in dry dioxane, are added with 0.89 ml (0.011 mol) dry pyridine. The mixture is kept at 60°C for 20 hours, then poured into water-ice and filtered. Crystallization from diluted ethanol results in prednisolone-21-stearoyl-glycolate (MP 104°-105°C).

References

Merck Index 7618
Kleeman & Engel p. 753
DOT 3 (1) 18 (1967)
I.N. p. 799
Giraldi, P.N. and Nannini, G.; US Patent 3,171,846; March 2, 1965; assigned to Carlo Erba SpA, Italy

PREDNISOLONE TEBUTATE

Therapeutic Function: Glucocorticoid

Chemical Name: 21-(3,3-Dimethyl-1-oxobutoxy)-11β,17-dihydroxypregna-1,4-diene-3,20-dione

Common Name: Prednisolone-21-tert-butyl acetate

Chemical Abstracts Registry No.: 7681-14-3

Raw Materials

tert-Butyl acetyl chloride
Prednisolone

Structural Formula:

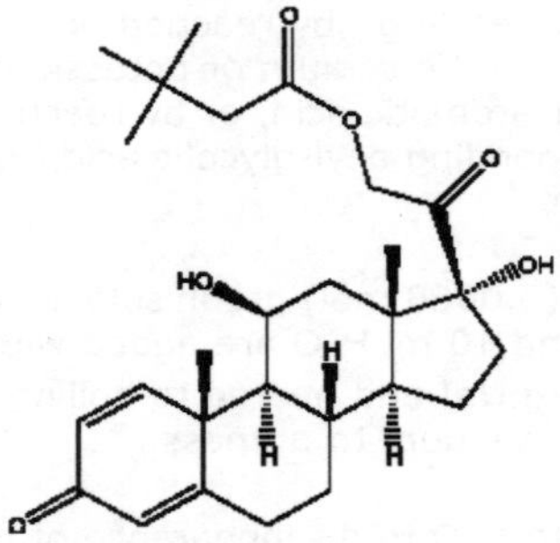

Trade Name	Manufacturer	Country	Year Introduced
Hydeltra TBA	MSD	US	1956
Codelcortone TBA	MSD	US	-
Predate TBA	Legere	US	-
Prednisol TBA	Pasadena	US	-
Rodelta TBA	Rocky Mtn.	US	-

Manufacturing Process

A solution of about 10 parts of tertiary-butyl acetyl chloride in 45 parts of dry chloroform is added portionwise to a cold solution of 25 parts of $\Delta^{1,4}$-3,20-diketo-11β,17α,21-trihydroxy-pregnadiene(prednisolone) in 125 parts of anhydrous pyridine. The resulting solution is allowed to stand for about 15 hours at 0° to 5°C, and the reaction solution is poured into 750 parts of water. The resulting aqueous mixture is extracted four times with 250 parts of chloroform each extraction. The combined chloroform layers are washed with water, dilute aqueous hydrochloric acid solution, water, 5% aqueous sodium bicarbonate solution, and finally with water. The chloroform extract is dried over magnesium sulfate, and the chloroform is evaporated in vacuo to give a residual oil. This oil is triturated with alcohol until it crystallizes, and is then recrystallized from ethanol to give substantially pure $\Delta^{1,4}$3,20-diketo-11β,17α,21-trihydroxy-pregnadiene21-tertiary-butyl acetate.

References

Merck Index 7619
Kleeman & Engel p. 754
PDR pp. 1033, 1183
I.N. p. 798
REM p. 970
Sarett, L.H.; US Patent 2,736,734; February 28, 1956; assigned to Merck & Co., Inc.

PREDNISONE

Therapeutic Function: Glucocorticoid

Chemical Name: 17α,21-Dihydroxy-pregna-1,4-diene-3,11,20-trione

Common Name: Deltacortisone

Structural Formula:

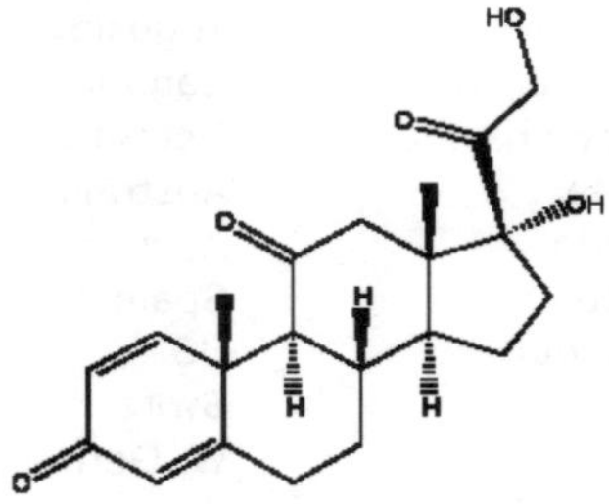

Chemical Abstracts Registry No.: 53-03-2

Trade Name	Manufacturer	Country	Year Introduced
Meticorten	Schering	US	1955
Deltasone	Upjohn	US	1955
Deltra	MSD	US	1955
Paracort	Parke Davis	US	1960
Lisacort	Fellows-Testagar	US	1960
Servisone	Lederle	US	1970
Orasone	Rowell	US	1972
Wojtab	Philips Roxane	US	1981
Adasone	Adams	Australia	-
Alto-Pred	Alto	US	-
Colisone	Merck-Frosst	Canada	-
Cortan	Halsey	US	-
Cortancyl	Roussel	France	-
Cortialper	Santos	Spain	-
Dacortin	Igoda	Spain	-
Decortin	Merck	W. Germany	-
Decortisyl	Roussel	UK	-
Decorton	Salfa	Italy	-
Deidrocortisone	Stip	Italy	-
Deltacortene	Lepetit	Italy	-
Delta Dome	Dome	US	-
Delta Prenovis	Vister	Italy	-
Deltison	Ferring	Sweden	-
Erftopred	Erfto	W. Germany	-
Fernisone	Ferndale	US	-

Trade Name	Manufacturer	Country	Year Introduced
Hostacortin	Hoechst	W. Germany	-
Inocortyl	Liposeptine	France	-
Keteocort	Desitin	W. Germany	-
Keysone	Key	US	-
Liquid Pred	Muro	US	-
Marnisonal	Juan Martin	Spain	-
Marvidiene	Panther-Osfa	Italy	-
Me-Korti	Farmos	Finland	-
Nisone	Llorente	Spain	-
Nizon	Bosnalijek	Yugoslavia	-
Novoprednisone	Novopharm	Canada	-
Nurison	Noury Pharma	Netherlands	-
Panafcort	Protea	Australia	-
Parmenison	Kwizda	Austria	-
Predniartrit	Maipe	Spain	-
Prednicen-M	Seymour	US	-
Prednifor	Vifor	Switz.	-
Prednilonga	Dorsch	W. Germany	-
Predni-Tablinen	Sanorania	W. Germany	-
Predni-Wolner	Wolner	Spain	-
Prednovister	Substancia	Spain	-
Predsol	Morgan	Italy	-
Predsone	Century	US	-
Presone	Langley	Australia	-
Pronison	Galenika	Yugoslavia	-
Propred	Medac	Australia	-
Rectodelt	Trommsdorff	W. Germany	-
Ropred	Robinson	US	-
Sarogesic	Saron	US	-
Sone	Fawns and McAllan	Australia	-
Sterapred	Mayrand	US	-
Supopred	Europa	Spain	-
Urtilone	Recherche Therap.	France	-
Wescopred	Saunders	Canada	-
Winpred	I.C.N.	Canada	-

Raw Materials

Bacterium Corynebacterium simplex
Cortisone

Manufacturing Process

From a solution of 30 grams of yeast extract (Difco) in 3.0 liters of tap water containing 13.2 grams of potassium dihydrogen phosphate and 26.4 grams of disodium hydrogen phosphate (pH of the solution 6.9) 27 portions of 100 ml each are withdrawn, placed in 300 ml Erlenmeyer flasks and sterilized by autoclaving for 15 minutes at 15 pounds steam pressure (120°C). After autoclaving and cooling of the broth one ml of a suspension of

Corynebacterium simplex (ATCC 6946) is placed in each flask. The flasks are then shaken on a shake table at 220 rpm and 28°C for 24 hours.

Into each of 27 Erlenmeyer flasks are placed 150 mg of Kendall's Compound E (cortisone). The flasks and contents are then sterilized for 15 minutes at 15 pounds steam pressure (120°C). To each flask are then added 5.0 ml of ethanol. The 24-hour bacterial culture is then transferred aseptically and the resulting suspensions are shaken on a shake table at 220 rpm and 28°C for 48 hours. The final pH is 7.2.

The contents of all the flasks are combined and extracted with a total of 9.0 liters of chloroform in three equal portions. The combined extracts are then concentrated to a residue which is crystallized from acetone-hexane. There results 1.1 grams of $\Delta^{1,4}$-pregnadiene-17α,21-diol-3,11,20-trione,MP 210°-215°C (dec.). Several additional recrystallizations raised the MP to 230°-232°C (dec.).

References

Merck Index 7621
Kleeman & Engel p. 755
PDR pp.830, 993, 1268, 1573, 1606, 1723, 1837
OCDS Vol. 1 p. 192 (1977)
I.N. p. 799
REM p. 970
Djerassi, C., Rosenkranz, G. and Berlin, J.; US Patent 2,579,479; December 25, 1951; assigned to Syntex SA, Mexico
Nobile, A.; US Patent 2,837,464; June 3, 1958; assigned to Schering Corporation
Oliveto, E.P. and Gould, D.H.; US Patent 2,897,216; July 28, 1959; assigned to Schering Corporation

PRENALTEROL

Therapeutic Function: Adrenergic

Chemical Name: 4-[2-Hydroxy-3-[(1-methylethyl)amino]propoxy]phenol

Common Name: -

Structural Formula:

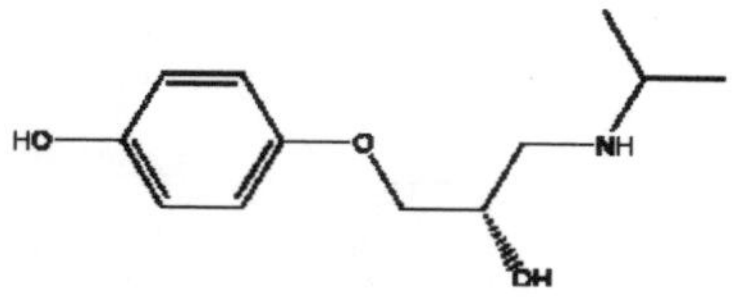

HO … O … NH … OH

Chemical Abstracts Registry No.: 57526-81-5

Trade Name	Manufacturer	Country	Year Introduced
Coleb	Astra	W. Germany	1981
Hyprenan	Astra	UK	1981
Varbian	Ciba	UK	1981

Raw Materials

4-Hydroxyphenoxypropylene oxide
Isopropylamine

Manufacturing Process

A solution of 100 g (1.7 mols) of isopropylamine in 60 cc of water was stirred into a solution of 4-hydroxyphenoxypropylene oxide. After the exothermic reaction has subsided, the reaction mixture was heated for two hours at 60°C. Thereafter, the aqueous ethanol was distilled off, and the solid residue was dissolved in aqueous hydrochloric acid comprising more than the theoretical stoichiometric molar equivalent of hydrochloric acid. The aqueous acid solution was extracted with ether and was then made alkaline with sodium hydroxide, whereby a solid crystalline precipitate was formed which was filtered off and dried over phosphorus pentoxide. The product was 1,1-(4'-hydroxyphenoxy)-2-hydroxy-3-isopropylamino-propane. Its hydrochloride had a melting point of 166°C to 169°C.

References

Merck Index 7639
DFU 4 (1) 46 (1979)
OCDS Vol. 3 p. 30 (1984)
DOT 17 (5) 199 (1981) & 18 (4) 190 (1982)
I.N. p. 801
Koppe, H., Engelhardt, A., Ludwig, G. and Zeile, K.; US Patent 3,637,852; January 25, 1972; assigned to Boehringer Ingelheim G.m.b.H. (Germany)

PRENYLAMINE

Therapeutic Function: Coronary vasodilator

Chemical Name: N-(1-Methyl-2-phenylethyl)-γ-phenylbenzenepropanamine

Common Name: -

Structural Formula:

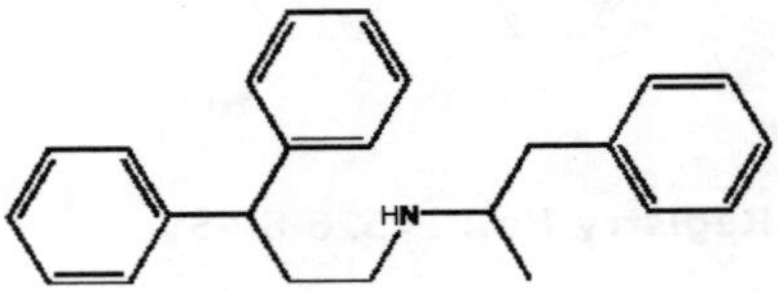

HN

Chemical Abstracts Registry No.: 390-64-7

Trade Name	Manufacturer	Country	Year Introduced
Synadrin	Hoechst	UK	1961
Segontin	Hoechst	Italy	1962
Segontin	Hoechst	W. Germany	1964
Segontine	Hoechst	France	-
Agozol	Tableta	Rumania	-
Angiovigor	Violani-Farmavigor	Italy	-
Angorsan	Isola-Ibi	Italy	-
Cardional	Unipharm	Israel	-
Corditin-Same	Savoma	Italy	-
Coredamin	Meiji	Japan	-
Crepasin	Hoei	Japan	-
Daxauten	Woelm Pharma	W. Germany	-
Epocol	Teisan-Nagase	Japan	-
Eucardion	Vita	Italy	-
Falicor	Fahlberg-List	E. Germany	-
Herzcon	Sana	Japan	-
Incoran	I.T.A.	Italy	-
Irrorin	Alfa Farm.	Italy	-
Lactamine	Daisan	Japan	-
Newsantin	Sawai	Japan	-
NP 30	Sanken	Japan	-
Nyuple	Ohta	Japan	-
Onlemin	Ono	Japan	-
Plactamin	Morishita	Japan	-
Prectolact	Showa Yakuhin	Japan	-
Rausetin	Tanabe	Japan	-
Reocorin	Farmochimica	Italy	-
Roinin	Mohan	Japan	-
Seccidin	Nippon Kayaku, Co.	Japan	-
Wasangor	Wassermann	Italy	-

Raw Materials

Phenyl acetone
1,1-Diphenyl-propylamine-(3)
Palladium
Hydrogen

Manufacturing Process

10.6 g of 1,1-diphenylpropylamine-(3) are hydrogenated by means of palladium with 6.7 g of phenyl acetone in 200 cc of methanol at 50°C. The calculated amount of hydrogen is taken up. The separated oily base is dissolved by heating with alcohol. After filtration water is added until turbidity sets in. 24.5 g of 2-(1',1'-diphenylpropyl-3'-amino)-3-phenyl-propane are obtained with a boiling point at 195°C to 198°C under a pressure of 0.5 mm of mercury, which after prolonged standing crystallizes out. Melting point about 38°C to 40°C. Hydrochloride (prepared in usual manner): melting point 188°C to 190°C.

References

Merck Index 7641
Kleeman & Engel p. 759
OCDS Vol. 1 p. 76 (1977)
I.N. p. 801
Ehrhart, G., Ott, H. and Lindner, E.; US Patent 3,152,173; October 6, 1964; assigned to Farbwerke Hoechst A.G. (Germany)

PRIDINOL HYDROCHLORIDE

Therapeutic Function: Antiparkinsonian, Anticholinergic

Chemical Name: 1,1-Diphenyl-3-piperidino-1-propanol hydrochloride

Common Name: Pridinol; Ridinol

Structural Formula:

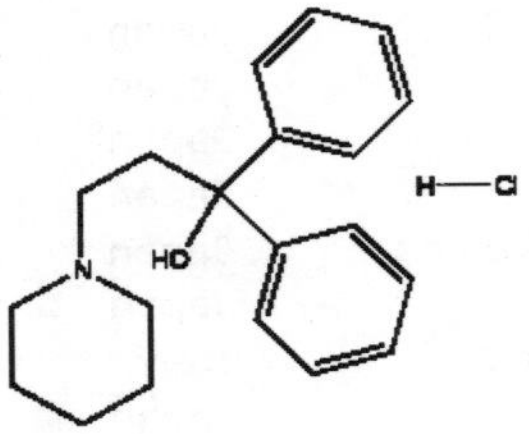

Chemical Abstracts Registry No.: 968-58-1

Trade Name	Manufacturer	Country	Year Introduced
Hikiceton	Tatsumi	-	-
Konlax	Nippon Shinyaku	-	-
Loxeen	Hommel	-	-
Myoson	Strathmann	-	-

Raw Materials

Bromobenzene
1-Piperidinopropionic acid n-butyl ester
Magnesium

Manufacturing Process

24 parts by weight of 1-piperidinopropionic acid n-butyl ester (BP: 137-

138°C) was added dropwise to solution of phenyl magnesium bromide from 157 parts by weight of bromobenzene and 24 parts of magnesium.

There was a spontaneous heating and the ether boiled. The mixture refluxed for 2 hours after the completion adding. Then it was poured into mixture of 200 parts of 37% hydrochloric acid and 800 parts (by weight) of ice with stirring. Hydrochloride of α,α-diphenyl-1-piperidinepropanol precipitated. It was filtered off, washed with ether, diluted hydrochloric acid and dried over sodium hydroxide in vacuum dessicator. The colorless crystals were light dissolved in hot water and had MP: 216°C. The base is the colorless powder with MP: 119-120°C.

References

Eisleb O.; D.B. Patent No. 875,660; May 4 1953; Fabwerke Hoechst, vormals Meister Lucius and Bruning, Frankfurt/M/-Hoechst

PRILOCAINE HYDROCHLORIDE

Therapeutic Function: Local anesthetic

Chemical Name: N-(2-Methylphenyl)-2-(propylamino)-propanamide hydrochloride

Common Name: Propitocaine hydrochloride

Structural Formula:

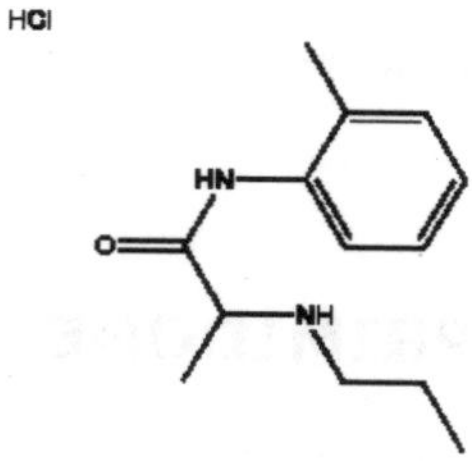

HCl

Chemical Abstracts Registry No.: 1786-81-8; 721-50-6 (Base)

Trade Name	Manufacturer	Country	Year Introduced
Xylonest	Astra	W. Germany	1963
Citanest	Astra	UK	1974
Citanest	Astra	US	1966
Citanest	Pierrel	Italy	1968
Citanest	Bellon	France	1973

Raw Materials

o-Toluidine
α-Bromopropionyl bromide
n-Propylamine

Manufacturing Process

One mol of ortho-toluidine is dissolved in 800 ml of glacial acetic acid. The mixture is cooled to 10°C whereupon 1.1 mols of α-bromopropionylbromide is added. The mixture is vigorously stirred for about a minute and a solution of sodium acetate (330 grams of $CH_3COONa \cdot 3H_2O$ in 1,380 ml of water) or another buffering or alkalizing substance or solution is added in one portion. The reaction mixture is then shaken for half an hour. The precipitate formed is filtered off, washed with water and dried. The product is sufficiently pure for further processing. Yield: 70-80% of theory. MP 133°-134°C.

One mol of α-bromopropio-ortho-toluidine is mixed with a solution of 3 mols of n-propylamine in 500 ml of water-free benzene and the reaction mixture is heated in an autoclave to 80°C for 8 hours. After cooling the reaction mixture is treated as described above. The base is obtained as a colorless oil. BP 159°-162°C/0.1 mm. Yield 55%. The base is then converted to the hydrochloride by reaction with HCl.

References

Merck Index 7646
DFU 8 (12) 1021 (1983)
Kleeman & Engel p. 760
OCDS Vol. 1 p. 17 (1977)
I.N. p. 802
REM p. 1053
Aktiebolaget Astra: Apotekarnes Kemiska Fabriker, Sweden; British Patent 839,943; June 29, 1960

PRIMIDONE

Therapeutic Function: Anticonvulsant

Chemical Name: 5-Ethyldihydro-5-phenyl-4,6(1H,5H)-pyrimidinedione

Common Name: 2-Desoxyphenobarbital; Primaclone

Chemical Abstracts Registry No.: 125-33-7

Raw Materials

α,α-Phenylethylmalonic acid diamide
Formamide

Structural Formula:

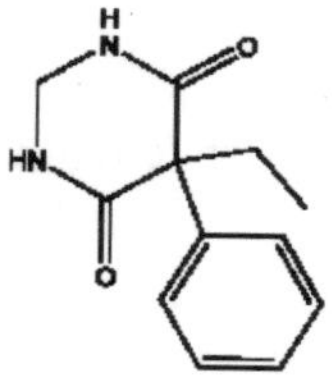

Trade Name	**Manufacturer**	**Country**	**Year Introduced**
Mysoline	I.C.I.	France	1953
Mysoline	Ayerst	US	1954
Cyral	Gerot	Austria	-
Liskantin	Desitin	W. Germany	-
Majsolin	Pliva	Yugoslavia	-
Midone	Protea	Australia	-
Mylepsinum	ICI Pharma	W. Germany	-
Mysedon	Medica	Finland	-
Primidone	Schein	US	-
Primoline	Darby	US	-
Primron	Fujinaga	Japan	-
Prysoline	Abic	Israel	-
Resimatil	Labaz	W. Germany	-
Sertan	Chinoin	Hungary	-

Manufacturing Process

50 parts of α,α-phenylethylmalondiamide and 150 parts of formamide are boiled together under reflux for 2 hours. The mixture is then cooled to 0°C and filtered. The solid residue is washed with 50 parts of ethanol and then crystallized from 660 parts of an 80% ethanol water mixture. There is obtained 5-phenyl-5-ethylhexahydropyrimidine-4,6-dione, MP 281°C-282°C.

References

Merck Index 7649
Kleeman & Engel p. 761
PDR pp. 631, 830, 1606
OCDS Vol. 1 p. 276 (1977)
I.N. p. 803
REM p. 1081
Boon, W.R., Carrington, H.C. and Vasey, C.H.; US Patent 2,578,847; December 18, 1951; assigned to Imperial Chemical Industries Limited, England

PROBENECID

Therapeutic Function: Antiarthritic

Chemical Name: 4-[(Dipropylamino)sulfonyl]benzoic acid

Common Name: -

Structural Formula:

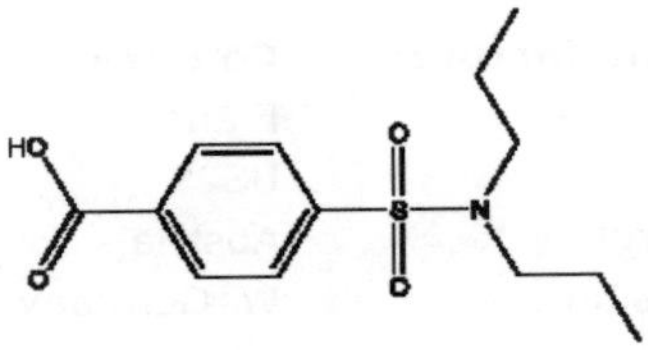

Chemical Abstracts Registry No.: 57-66-9

Trade Name	Manufacturer	Country	Year Introduced
Benemid	MSD	US	1952
Benemide	Theraplix	France	1954
Benecid	Kaken	Japan	-
Benuryl	I.C.N.	Canada	-
Colbenemid	MSD	UK	-
Panuric	Propan-Lipworth	S. Africa	-
Perdurine	Pharma-Union	Belgium	-
Probemid	Lefa	Spain	-
Probenicid	Lederle	US	-
Probenemid	Merck-Banyu	Japan	-
Procid	Protea	Australia	-
Solpurin	Salfa	Italy	-
Urecid	Frosst	Australia	-
Uroben	Mitim	Italy	-

Raw Materials

p-Carboxybenzene sulfonyl chloride
Di-n-propylamine

Manufacturing Process

24.0 grams (0.11 mol) of p-carboxybenzenesulfonyl chloride was added in small portions to a suspension of 20.0 grams (0.146 mol) of di-n-propylamine in 100 milliliters of 10% sodium hydroxide with vigorous stirring at a temperature of 15°-25°C. Stirring was continued for 15 minutes after the final addition. The clear solution was treated with decolorizing carbon and filtered.

The product was precipitated by the addition of an excess of hydrochloric acid. The crude product was purified by reprecipitation from bicarbonate solution and recrystallization from dilute alcohol. The yield was 20.0 grams (64%) melting at 194°-196°C.

References

Merck Index 7656
Kleeman & Engel p. 761
PDR pp. 705, 830, 993, 1142, 1150, 1606, 1999
OCDS Vol. 1 p. 135 (1977)
I.N. p. 804
REM p. 944
Miller, C.S.; US Patent 2,608,507; August 26, 1952; assigned to Sharp & Dohme, Inc.

PROBUCOL

Therapeutic Function: Antihyperlipidemic

Chemical Name: Bis(3,5-di-tert-butyl-4-hydroxyphenyl)acetone mercaptole

Common Name: -

Structural Formula:

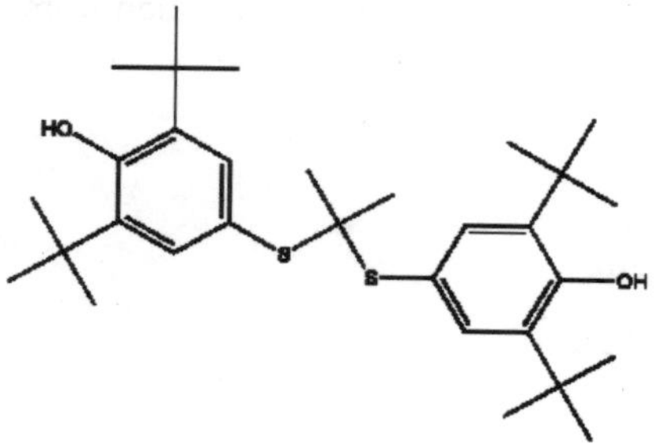

Chemical Abstracts Registry No.: 23288-49-5

Trade Name	Manufacturer	Country	Year Introduced
Lorelco	Merrell Dow	US	1977
Lurselle	Lepetit	France	1980
Lurselle	Lepetit	UK	1980
Lurselle	Dow-Lepetit	Switz.	1980
Lurselle	Merrell	W. Germany	1980
Lurselle	Lepetit	Italy	1982
Biphenabid	Merrell Dow	-	-
Lesterol	Lepetit	-	-

Raw Materials

Acetone
2,6-Di-tert-butyl-4-mercaptophenol

Manufacturing Process

Bis(3,5-di-tert-butyl-4-hydroxyphenyl) acetone mercaptole, melting at 125°C to 126°C is prepared by employing 2,6-di-tert-butyl-4-mercaptophenol and acetone as starting materials. In one representative procedure, the 2,6-di-tert-butyl-4-mercaptophenol (47.5 g, 0.2 mol) is dissolved in methanol (50 ml) heated at a temperature of 50°C. A catalytic amount of concentrated hydrochloric acid (1 ml) is added, followed by acetone (5.8 g, 0.1 mol). The temperature of the mixture rises to about 60°C, and is maintained at about 60°C to 65°C for 1.5 hours. The mixture is cooled, diluted with water and about 10 ml of aqueous sodium bicarbonate and extracted with ether. The ether extract is evaporated, and the product is obtained as a residue, which is recrystallized from ethanol and then from isopropanol to obtain the bis(3,5-di-tert-butyl-4-hydroxyphenyl) acetone mercaptole as a crystalline solid melting at about 125°C to 126°C.

In another representative procedure about 2.3 mols of 2,6-di-tert-butyl-4-mercaptophenol is dissolved in about 1,700 ml of methanol under a nitrogen atmosphere; about 100 ml of concentrated hydrochloric acid and 180 ml of acetone are added, and the mixture is stirred and maintained at a temperature of about 35°C to 50°C, for 1.5 hours. The mixture is then cooled to room temperature and filtered, and the bis(3,5-di-tert-butyl-4-hydroxyphenyl) acetone mercaptole product is collected as a colorless crystalline solid filter cake. The product is washed with water and aqueous sodium bicarbonate and purified by recrystallization from ethanol.

References

Merck Index 7657
DFU 2 (2) 128 (1977)
Kleeman & Engel p. 762
PDR p. 1229
OCDS Vol. 2 p. 126 (1980)
DOT 14 (1) 33 (1978)
I.N. p. 804
REM p. 864
Barnhart, J.W. and Shea, P.J.; US Patent 3,862,332; January 21, 1975; assigned to The Dow Chemical Co.

PROCAINAMIDE HYDROCHLORIDE

Therapeutic Function: Antiarrhythmic

Chemical Name: Benzamide, p-amino-N-(2-(diethylamino)ethyl)-, monohydrochloride

Common Name: Amidoprocaine; Novocainamidum; Procainamide hydrochloride; Prokainamid

Structural Formula:

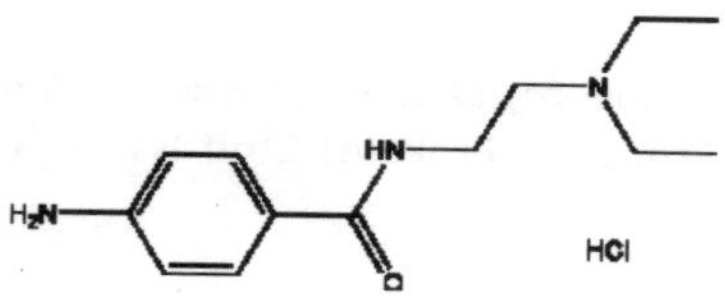

Chemical Abstracts Registry No.: 614-39-1; 51-06-9 (Base)

Trade Name	Manufacturer	Country	Year Introduced
PAD	Astra	-	-
Procamide	Zambon	-	-
Procan	Parke-Davis	-	-
Procanbid	Warner Lambert	USA	-
Promine	Major	-	-
Pronestyl	Sarabhai Chemicals	India	-
Roxyl	Star	-	-

Raw Materials

Sodium ethoxide
Diethyl aminoethane
Hydrogen
4-Nitrobenzoyl chloride
Nickel

Manufacturing Process

To the solution of 4-nitro-benzoylchloride the diethyl aminoethane and sodium ethoxide were added and mixed. As a result of reaction a N-(2-diethylamino-ethyl)-4-nitro-benzamide was obtained.

The N-(2-diethylamino-ethyl)-4-nitro-benzamide was reduced by hydrogen Ni as catalyst to give N-(2-diethylamino-ethyl)-4-amino-benzamide (procainamide).

In practice it is usually used as hydrochloride salt.

References

Baltzy R. et al.; J. Am. Chem. Soc. 64, 2231 (1942); Yamazaki M.Y. et al.; J. Pharm. Soc. Japan 73, 294 (1953)

PROCAINE

Therapeutic Function: Local anesthetic, Analgesic, Antiviral

Chemical Name: Benzoic acid, 4-amino-, 2-(diethylamino)ethyl ester

Common Name: Cocainum novum; Factor H_3; Novocainum; PABA-diethylaminoethanol; Procaine; Prokain; Stoff H_3; Vitamin H_3

Structural Formula:

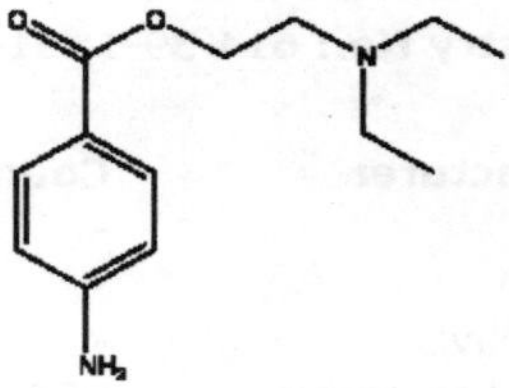

Chemical Abstracts Registry No.: 59-46-1

Trade Name	Manufacturer	Country	Year Introduced
Endocaina	Lafage	-	-
Isocain	Bernburg	-	-
Novadren	Sopharma	-	-
Polocaine	Polfa	-	-
Unicaine	Kay	-	-

Raw Materials

Ethylene chlorohydrin
p-Nitrobenzoyl chloride
Diethyl amine
Tin

Manufacturing Process

The equal quantity of ethylene chlorohydrin and p-nitrobenzoyl chloride was heated on an oil bath at temperature 120°-125°C till the ending of the isolation of hydrogen chloride. The mixture was poured into water to give the oil, which has soon hardened. It was recrystallized from diluted ethanol as white needles of p-nitrobenzoylchloro ethanol; MP: 56°C.

2 g of above prepared product and 2 g of diethyl amine were heated at 100°-120°C for 10 hours in the soldered tube to give 4-nitobenzoic acid diethylaminoethyl ether as an oily viscous mass. It was dissolved in hydrochloric acid and reduced with tin to procaine, which has precipitated as oil after adding sodium carbonate. 4-Aminobenzoic acid diethylaminoethyl

ester crystallized from diluted ethanol with two molecules of water as white needles; MP: 51°C. It crystallized without water from napthaline; MP: 58°-60°C, and formed monohydrochloride with MP: 156°C.

References

Farbwerke vorm. Meister Lucius and Bruning in Hochst a. M.; D.R. Patent No. 179,627; November 27, 1904

PROCARBAZINE HYDROCHLORIDE

Therapeutic Function: Cancer chemotherapy

Chemical Name: N-(1-Methylethyl)-4-[(2-methylhydrazino)methyl] benzamide hydrochloride

Common Name: Ibenmethyzin

Structural Formula:

HCl

Chemical Abstracts Registry No.: 366-70-1; 671-16-9 (Base)

Trade Name	Manufacturer	Country	Year Introduced
Natulan	Roche	France	1965
Natulan	Roche	W. Germany	1966
Natulan	Roche	UK	1966
Natulan	Roche	Italy	1967
Matulane	Roche	US	1969
Natulan	Nippon Roche	Japan	1973

Raw Materials

Methanol
Sodium hydroxide
Hydrogen bromide
Bromine
Isopropylamine
1-Methyl-1,2-dicarbobenzoxyhydrazine
4-Methylbenzoic acid
Thionyl chloride
Sodium hydride
Hydrogen chloride

Manufacturing Process

544 grams of 4-methylbenzoic acid was boiled with 550 ml of thionyl chloride until a clear solution was obtained. After the excess thionyl chloride was distilled off, the residue was fractionated, yielding 605 g of 4-methylbenzoyl chloride; BP 91°C/9 mm Hg, n_D^{24} = 1.5532. This was dissolved in 550 ml of absolute benzene and the so-formed solution added to a mixture of 248 ml of absolute methanol and 550 ml of absolute benzene. After the exothermic reaction had terminated, the reaction mixture was boiled for a further 20 hours, then concentrated in vacuo and the product, 4-methylbenzoic acid methyl ester, isolated by conventional means. It could be purified by distillation, and the purified product boiled at 91°C/9 mm Hg, MP 32°C.

574 grams of this ester were dissolved in 1200 ml of carbon tetrachloride and, while boiling and exposing to a UV lamp, treated dropwise with a solution of 109 ml of bromine in 400 ml of carbon tetrachloride. After all of the bromine had been dropped in, the mixture was heated for a further hour, concentrated in vacuo and the residue crystallized from low boiling petroleum ether, yielding as colorless fine crystals, 4-(bromo-methyl)-benzoic acid methyl ester, which melted at 52°C. For the reaction of this ester with 1-methyl-1,2-dicarbobenzoxy-hydrazine, the following procedure was followed.

309 grams of a 27% suspension of sodium hydride in an inert solvent was treated with 300 ml of dimethylformamide, and a solution of 1095 grams of 1-methyl-1,2-dicarbobenzoxy-hydrazine in dimethylformamide was added thereto. When all the material had been added and the hydrogen evolution had nearly come to a standstill, the mixture was heated for an hour at about 80°C in order to carry the formation of the sodium salt to completion. A mixture of 759 grams of 4-(bromo-methyl)-benzoic acid methyl ester in 700 ml of dimethylformamide was then dropped in, and finally the reaction mixture was heated for an hour at 80°C. After cooling, the reaction mixture was poured into 10 liters of ice water and the condensation products taken up in ether. The thereby obtained crude methyl ester (n_D^{24} = 1.1558) was used without further purification for the next step. It was dissolved in about 2,200 ml of dioxane, treated with a solution of 133 grams of sodium hydroxide in 870 ml of water, and the resulting mixture stirred for about 24 hours at room temperature. It was then poured into 10 liters of ice water and neutral materials were extracted with ether.

The aqueous phase was rendered acid with concentrated hydrochloric acid (weak Congo red) and the separated acid taken up in ether. The isolated crude acid was recrystallized from dibutyl ether, yielding colorless crystals of 4-[(2-methyl-1,2-dicarbobenzoxy-hydrazino)-methyl]-benzoic acid, which melted at 112°C. The so-obtained product was sufficiently pure for further reaction.

15 grams of 4-[(2-methyl-1,2-dicarbobenzoxy-hydrazino)-methyl]-benzoic acid were boiled with an excess of thionyl chloride for 1 hour under reflux. The unconverted thionyl chloride was distilled off in vacuo, the residue twice dissolved each time in 75 ml of absolute benzene and then concentrated in vacuo. The so-obtained 4-[(2-methyl-1,2-dicarbobenzoxyhydrazino)-methyl]-benzoyl chloride, a viscous light yellow oil, was dissolved in 50 ml of absolute benzene and with stirring mixed with a solution of 4.45 grams of isopropylamine in 100 ml of absolute benzene. By cooling, the temperature of

the reaction mixture was kept below 30°C. After the mixing had been completed, the reaction mixture was maintained first at room temperature for 3 hours and then for ½ hour at 40°C. It was then cooled down and poured into about 100 ml of ice water. After the addition of a mixture of methylene chloride and ether (40 ml + 200 ml), the organic phase was separated and then washed with water, dilute hydrochloric acid, water, dilute sodium hydroxide and again with water.

The solvents were then evaporated, yielding 4-[(2-methyl-1,2-dicarbobenzoxyhydrazino)-methyl]-benzoic acid isopropylamide as a yellow oil, which crystallized upon triturating with ether; MP 90°-92°C. This product was then covered with 70 ml of a 33% solution of hydrogen bromide in glacial acetic acid, and then permitted to stand for 2 hours with occasional swirling, whereupon a thick slurry of crystals was formed. The precipitate was filtered off, washed with 20 ml of glacial acetic acid and finally with ether, yielding crystals of 4-[(2-methyl-hydrazino)-methyl]-benzoic acid isopropylamide hydrobromide, which after recrystallization from methanol/ether melted at 216°-217°C (dec.).

87.5 grams of 4-[(2-methyl-hydrazino)-methyl]-benzoic acid isopropylamide hydrobromide (obtained as described above) were dissolved in 550 ml of water. To this solution, there were added 1,000 ml of methylene chloride and, while cooling with ice and stirring under nitrogen atmosphere, 1,200 grams of potassium carbonate portionwise. The methylene chloride layer was separated and the aqueous slurry extracted three times with 500 ml of methylene chloride in a nitrogen atmosphere. The united methylene chloride extracts were concentrated in vacuo. The residue was dissolved under nitrogen in 100 ml of methanol and treated, while cooling with ice, with 40 ml of a 45% methanolic hydrochloric acid solution, which induces immediate crystallization. The crystals were filtered off and recrystallized from methanol, yielding 4-[(2-methyl-hydrazino)-methyl]-benzoic acid isopropylamide hydrochloride melting at 223°-226°C.

References

Merck Index 7662
Kleeman & Engel p. 763
PDR p. 1491
OCDS Vol. 2 p. 27 (1980)
I.N. p.805
REM p. 1153
Bollag, W., Gutmann, H., Hegedus, B., Kaiser, A., Langemann, A., Muller, M. and Zeller, P.; US Patent 3,520,926; July 21, 1970; assigned to Hoffmann-La Roche Inc.

PROCATEROL

Therapeutic Function: Bronchodilator

Chemical Name: 8-Hydroxy-5-[1-hydroxy-2-[(1-methylethyl)amino]butyl]-2(1H)-quinolinone

Common Name: -

Structural Formula:

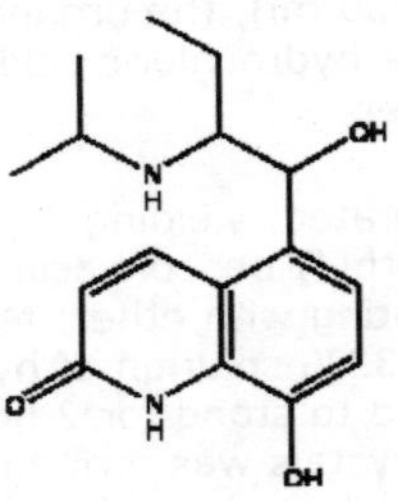

Chemical Abstracts Registry No.: 72332-33-3

Trade Name	Manufacturer	Country	Year Introduced
Meptin	Otsuka	Japan	1981

Raw Materials

α-Bromobutyric acid bromide
Isopropylamine
8-Hydroxycarbostyril
Lithium aluminum hydride

Manufacturing Process

50 g of α-bromobutyric acid bromide, 50 g of anhydrous aluminum chloride and 400 ml of carbon disulfide were added to 20 g of 8-hydroxycarbostyril. The resulting mixture was heated at a temperature of 50°C for 13 hours and the carbon disulfide layer was removed by decantation. Crushed ice was added to the residue, and the precipitated crystals were filtered, washed with water and recrystallized from methanol to obtain 27 g of 5-(α-bromobutyryl)-8-hydroxycarbostyril having a melting point of 218°C to 219°C (with coloring and decomposition). To 5 g of the thus obtained 5-(α-bromobutyryl)-8-hydroxycarbostyril was added 100 ml of isopropylamine, and the mixture was heated at a temperature of 50°C for 4 hours followed by concentration to dryness. Crystals which formed upon addition of water were filtered, washed with water and then recrystallized from methanol to obtain 4.6 g of a methanol solvate of 5-(α-isopropylaminobutyryl)-8-hydroxycarbostyril having a melting point of 136°C to 137°C (with foaming and decomposition).

20 g of tetrahydrofuran was added to 1 g of 5-(α-isopropylaminobutyryl)-8-hydroxycarbostyril hydrochloride, and the resulting mixture was added dropwise to a suspension of 0.12 g of lithium aluminum hydride in 10 ml of tetrahydrofuran while stirring at room temperature. After completion of the addition, a small amount of water was added to the reaction mixture to decompose any excess of lithium aluminum hydride. The reaction mixture was then poured into 50 ml of ice-water and the aqueous layer of the resulting solution was separated and concentrated to dryness. The precipitated crystals

were filtered, washed with acetone and dissolved in water. The solution was adjusted to pH of 8 with aqueous sodium hydroxide to precipitate crystals which were then filtered and recrystallized from ethanol to obtain 0.8 g of 5-(1-hydroxy-2-isopropylamino)butyl-8-hydroxycarbostyril monohydrate having a melting point of 141°C to 142°C (with cooling and decomposition).

References

Merck Index 7663
DFU 3 (2) 135 (1978)
OCDS Vol. 3 p. 184 (1984)
DOT 17 (6) 256 (1981)
Nakagawa, K., Yoshizaki, S., Tanimura, K. and Tamada, S.; US Patent 4,026,897; May 3, 1977; assigned to Otsuka Pharmaceutical Co. (Japan)

PROCHLORPERAZINE

Therapeutic Function: Antiemetic, Antipsychotic

Chemical Name: 2-Chloro-10-[3-(4-methyl-1-piperazinyl)propyl]-10H-phenothiazine

Common Name: Chlormeprazine

Structural Formula:

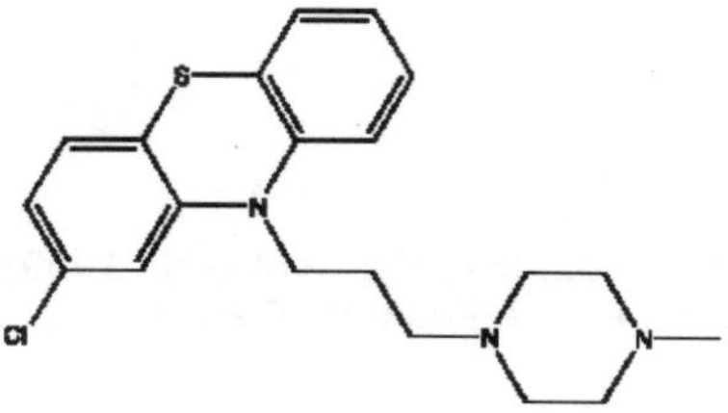

Chemical Abstracts Registry No.: 58-38-8; 84-02-6 (Maleate)

Trade Name	Manufacturer	Country	Year Introduced
Compazine	SKF	US	1956
Tementil	Specia	France	1957
Anti-Naus	Protea	Australia	-
Combid	SKF	US	-
Klometil	Farmos	Finland	-
Mitil	Lennon	S. Africa	-
Nibromin-A	Maruko	Japan	-
Normalmin	Sawai	Japan	-
Novamin	Shionogi	Japan	-

Trade Name	Manufacturer	Country	Year Introduced
Pasotomin	Yoshitomi	Japan	-
Stemetil	May and Baker	UK	-
Vertigon	SKF	UK	-

Raw Materials

3-Chloro-10-[3-(di-N-2-chloroethyl)aminopropyl]phenthiazine hydrochloride
1-Methylpiperazine

Manufacturing Process

3-Chloro-10-[3-(di-N-2-chloroethyl)aminopropyl]phenthiazine hydrochloride (1.8 g) is heated in a sealed tube for 4 hours at 140°C with a 290 g/l aqueous solution (9 cc) of monomethylpiperazine. The contents of the tube are treated with chloroform (40 cc). The aqueous layer is decanted and the chloroform layer is shaken with N hydrochloric acid (15 cc followed by 2 cc). The aqueous solution is treated with sodium hydroxide (d = 1.33, 10 cc) and chloroform (20 cc). After evaporation of the solvent, the base (1.5 g) is obtained. A solution of maleic acid (1 g) in ethanol (5 cc) is added and after recrystallization from water, 3-chloro10-[3-(4'-methyl-1'-piperazinyl)propyl]phenothiazine dimaleate is obtained, melting point 228°C (inst .).

References

Merck Index 7665
Kleeman & Engel p. 764
PDR pp. 1606, 1706
OCDS Vol. 1 p. 381 (1977)
DOT 9 (6) 228 (1973)
I.N. p. 806
REM p. 809
Horclois, R.J.; US Patent 2,902,484; September 1, 1959; assigned to Societe des Usines Chimiques Rhone-Poulenc, France

PROCYCLIDINE HYDROCHLORIDE

Therapeutic Function: Antiparkinsonian

Chemical Name: α-Cyclohexyl-α-phenyl-1-pyrrolidinepropanol hydrochloride

Common Name: -

Chemical Abstracts Registry No.: 1508-76-5; 77-37-2 (Base)

Structural Formula:

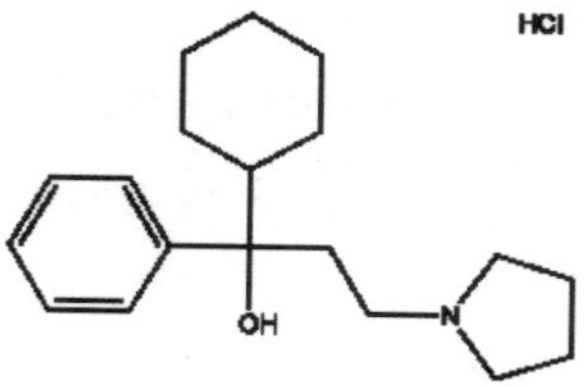

Trade Name	**Manufacturer**	**Country**	**Year Introduced**
Kemadrin	Burroughs-Wellcome	US	1956
Kemadrine	Wellcome	France	1965
Arpicolin	R.P. Drugs	UK	-
Kemadren	Gayoso Wellcome	Spain	-
Osnervan	Wellcome	W. Germany	-
Procyclid	I.C.N.	Canada	-

Raw Materials

Acetophenone
Magnesium
Paraformaldehyde
Hydrogen
Pyrrolidine
Hydrogen chloride
Bromobenzene

Manufacturing Process

1,1-Diphenyl-3-pyrrolidinopropan-1-ol (30 grams) was dissolved in glacial acetic acid (120 ml), Adams' platinum catalyst (6 grams) added, and the mixture shaken in an atmosphere of hydrogen until the equivalent of 3.4 molecules had been taken up per molecule of compound. Water was added, the catalyst removed by filtration, excess of ammonia added, and the liberated base extracted with ether. The ethereal extract was dried and evaporated and the residue recrystallized from light petroleum (BP 40°-60°C). The 1-cyclohexyl-1-phenyl3-pyrrolidinopropan-1-ol (19.3 grams) so obtained had a melting point of 85.5°-86.5°C. The hydrochloride recrystallized from a mixture of ethanol and ethyl acetate, melted with decomposition at 226°-227°C according to US Patent 2,891,890.

The starting material is prepared by the reaction of acetophenone, paraformaldehyde and pyrrolidine to give ω-pyrrolidinopropiophenone. That is in turn reacted with phenyl magnesium bromide to give 1,1-diphenyl-3-pyrrolidinpropan-1-ol.

References

Merck Index 7667
Kleeman & Engel p. 765
PDR p. 745
OCDS Vol. 1 p. 47 (1977)
DOT 18 (2) 88 (1982)

I.N. p. 806
REM p. 932
Bottorff, E.M.; US Patent 2,826,590; March 11, 1958; assigned to Eli Lilly and Company
Harfenist, M. and Magnien, E.G.; US Patent 2,842,555; July 8, 1958; assigned to Burroughs Wellcome & Co. (U.S.A.) Inc.
Adamson, D.W.; US Patent 2,891,890; June 23, 1959; assigned to Burroughs Wellcome & Co. (U.S.A.) Inc.

PROGLUMETACIN MALEATE

Therapeutic Function: Antiinflammatory

Chemical Name: N'-2-[1-(p-Chlorobenzoyl)-5-methoxy-2-methyl-3-indoleacetoxy]-ethyl-N-3-(N-benzoyl-N',N'-di-n-propyl-DL-isoglutaminoyl)-oxypropyl piperazine dimaleate

Common Name: Protacine

Structural Formula:

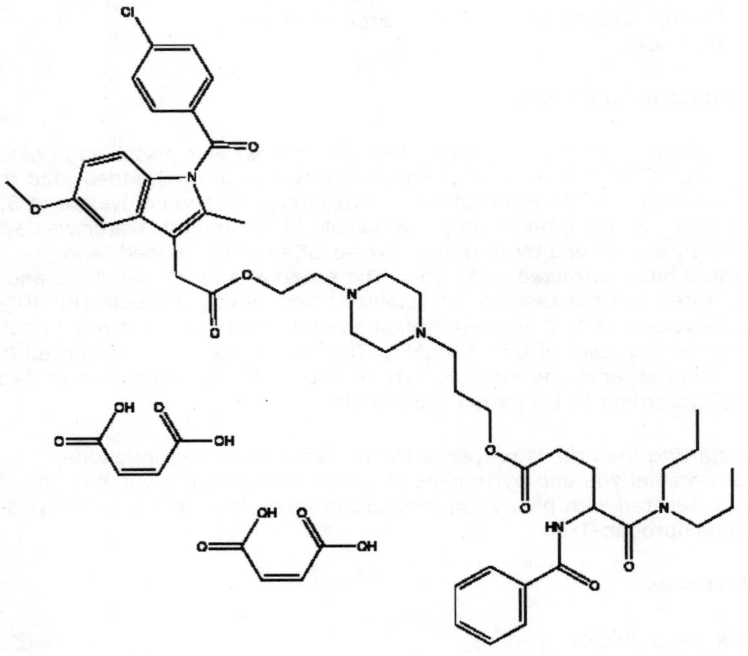

Chemical Abstracts Registry No.: 57132-53-3 (Base)

Trade Name	Manufacturer	Country	Year Introduced
Afloxan	Rotta	Italy	1981
Proxil	Rorer	Italy	1981

Raw Materials

N'-(2-Hydroxyethyl)-N-3-(N-benzoyl-N',N'-di-n-propyl-DL-isoglutaminoyl)-oxypropylpiperazine
1-(p-Chlorobenzoyl)-5-methoxy-2-methyl-3-indoleacetic acid
N,N'-Dicyclohexylcarbodiimide
Maleic acid

Manufacturing Process

To a titrated solution of 400 cc of ethyl acetate containing 0.1 mol of N'-(2-hydroxyethyl)-N-3-(N-benzoyl-N',N'-di-n-propyl-DL-isoglutaminoyl)-oxypropyl piperazine [obtained by dissolving 71.9 g (0.105 mol) of the corresponding di-oxalate in 500 cc of water, bringing this solution to a pH of between 9 and 10 with sodium bicarbonate and finally extracting the oily emulsion thus formed twice in succession with a total of 400 cc of ethyl acetate], there are added successively 35.8 g (0.1 mol) of 1-(p-chlorobenzoyl)-5-methoxy-2-methyl-3-indoleacetic acid and 20.6 g (0.1 mol) of N,N'-dicyclohexylcarbodiimide. This is left at room temperature for 24 hours, and after having filtered the N,N'-dicyclohexyl urea precipitate the organic phase is then washed with dilute HCl, a solution of sodium bicarbonate and a saturated solution of sodium chloride.

The ethyl acetate is dried with anhydrous sodium sulfate, filtered and dried off. The oily residue is dissolved in 600 cc of methanol; the di-oxalate is precipitated by the addition of a solution of oxalic acid in methanol. Yield 85%, melting point 190°C to 192°C (crystallized by methanol). Microcrystalline substance, creamy white color.

By the same method one can obtain the dimaleate. Yield, 83%; melting point, 146°C to 148°C (crystallized by ethanol). Microcrystalline pale cream colored substance.

References

Merck Index 7679
DFU 5 (3) 142 (1980)
DOT 17 (4) 157 (1981)
Makovec, F., Senin, P. and Rovati, L.; US Patent 3,985,878; October 12, 1976; assigned to Rotta Research Laboratorium S.p.A.

PROGLUMIDE

Therapeutic Function: Gastric antisecretory

Chemical Name: Glutaramic acid, 4-benzamido-N,N-dipropyl-, DL-

Common Name: Proglumide; Xilamida; Xylamide

Structural Formula:

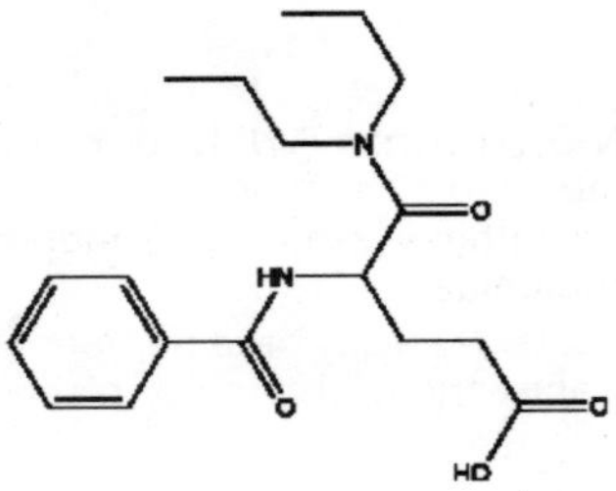

Chemical Abstracts Registry No.: 6620-60-6

Trade Name	Manufacturer	Country	Year Introduced
Milid	Popular	-	-
Milid	Rotta Research Laboratorium spa	-	-
Promid	Kaken	-	-
Promid	Opfermann	-	-
Snol	Inexfa	-	-

Raw Materials

Glutamic acid
Benzoyl chloride
Dipropylamine
Acetic anhydride

Manufacturing Process

588 g L-(+)-glutamic acid [commercial grade], are gradually added, in small portions, while stirring, to 2400 ml 2 N NaOH, in such a manner that the internal temperature does not exceed 5°C by external cooling (ice or brine). When all the glutamic acid has been added and is dissolved there is added to the reaction mixture with continued stirring and in such a manner that the internal temperature does not exceed 15°C, 471 ml benzoyl chloride and 1600 ml 3 N NaOH from two separatory funnels, the addition being made in the following manner: add at once 94.2 ml benzoyl chloride, then, dropwise 160 ml 3 N NaOH (from the other funnel), the speed of addition is regulated so that the pH of the mixture does not exceed 8 (universal indicator paper) and the temperature does not exceed 15°C. When the 160 ml of the 3 N NaOH solution have all been added, add 47.1 ml benzoyl chloride, then slowly add 160 ml 3 N NaOH, again add 47.1 ml of benzoyl chloride, followed by the dropwise addition of the same volume of 3 N NaOH solution. This procedure of alternate addition is continued until the benzoyl chloride and 3 N NaOH solution have all been added. At this point one adds an additional 1125 ml of the 3 N NaOH solution at a speed, which keeps the temperature below 15°C

and the pH under 8, using universal indicator paper for testing the pH. When all has been added, including the last addition of NaOH, stirring is continued for an additional 30 minutes. The reaction mixture is then acidified, dropwise, with concentrated HCI until congo red paper turns blue. The acid solution is stirred for 5 additional minutes, then transferred to a suitable container and stored for 10-18 hours at +5°C. The solids are filtered repulped in a mortar with 600 ml ice water and filtered again. The solids are washed on the filter with 400 ml ice water and pressed dry. The material is then spread out in a thin layer and dried in the air to obtain N-benzoylaminoglutamic acid, MP: 136°-140°C.

1500 g of N-benzoylaminoglutamic acid, obtained above, are added under stirring to 6 liters of acetic anhydride, previously placed in a flask, equipped with a reflux condenser and a stirrer. The stirred mixture is maintained at room temperature for 8 hours without cooling bath and let stand overnight at room temperature. The reaction mixture is filtered, pressed dry, then dried in an air current for one hour at 60°-70°C, and one hour at 100°C to obtain N-benzoylaminoglutamic acid anhydride. Yield: 8.50 g (61%).

To an aqueous solution of dipropylamine (334 ml of amine in sufficient water to yield 1400 ml aqueous solution), are added over a period of 60-75 minutes, under efficient stirring and with cooling to -3°C, 312 g of N-benzoylaminoglutamic acid anhydride in such a manner that the temperature remains between -2°C and -4°C. When the addition is completed, stirring is continued for 10-15 minutes at -3°C and 650 ml glacial acetic acid are added. The temperature is allowed to rise to 6°C. The stirring is continued for 60-80 minutes. The reaction mixture is seeded by adding 2-3 g of previously prepared 4-benzamido-N,N-dipropylglutaramic acid which initiates precipitation of the desired product. The product is purified by dissolving the crude material in 20 times by weight of water and adding a stoichiometric amount of $NaHCO_3$, or a slight excess at 60°-70°C. The mixture is acidified with 20% acetic acid with vigorous stirring at room temperature to obtain a pH of 5.5. The stirring is continued for an additional 10-15 minutes, the product 4-benzamido-N,N-dipropylglutaramic acid is filtered, washed with stirring with 700 ml of water for 15 minutes, filtered again and dried in air current at 25°C to constant weight. Yield 140 g; MP: 142°-145°C.

References

Rotta Research Laboratorium S.p.A., an Italian Joint Stock Company, of San Fruttuoso di Monza, Milan, Italy; G.B. Patent No. 1,108,819; July 31, 1964

PROLINTANE HYDROCHLORIDE

Therapeutic Function: Analeptic, Stimulant, Antidepressant

Chemical Name: Pyrrolidine, 1-(1-(phenylmethyl)butyl)-, hydrochloride

Common Name: Prolintane hydrochloride; Promotil

Structural Formula:

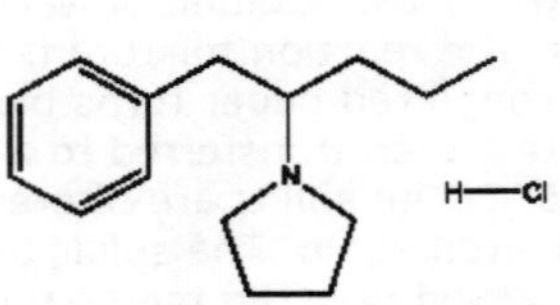

Chemical Abstracts Registry No.: 1211-28-5

Trade Name	Manufacturer	Country	Year Introduced
Catorid	Boehringer Ingelheim	-	-
Catovit	Boehringer Ingelheim	-	-
Katovit	Thomae	-	-
Promotil	Boehringer Ingelheim	-	-

Raw Materials

Benzyl chloride
Magnesium
α-Pyrrolidinovalero nitrile

Manufacturing Process

390 g benzyl chloride was added dropwise to 72 g of magnesium powder in the mixture of 1:1 of benzene and tetrahydrofuran by stirring and at temperature not above 40°C. The Grignard reagent was diluted with 750 ml of benzene-tetrahydrofuran (1:1). Then a solution of α-pyrrolidinovalero nitrile in 400 ml of benzene-tetrahydrofuran was added dropwise by stirring at temperature not above 40°C. After that the mixture was stirred 3 hours at 40°C and some hours at room temperature. At last the main part of solvents was distilled off in vacuum and ice with hydrochloric acid was added to the residue to an acidic pH. 1 L of benzene was added. The acid layer was separated. Benzene layer was shook with diluted hydrochloric acid. The combined acidic water layers were alkalized with ammonia to alkaline reaction. The oil dropped out was dissolved in benzene.

The solvent was removed and the residue was distilled in vacuum to give 275 g 1-phenyl-2-pyrrolidinylpentane (prolintane); BP: 90-92°C at 0.36 mm.

In practice it is usually used as hydrochloride salt.

References

Kottler A., Seeger E.; D.B. Patent No. 1,088,962; April 26, 1956

PROLONIUM IODIDE

Therapeutic Function: Iodine source

Chemical Name: 1,3-Propanediaminium, 2-hydroxy-N,N,N,N',N',N'-hexamethyl-, diiodide

Common Name: Diiodomedrine; Hydroxytrimethonium iodide; Ksameprol; Prolonii iododum; Prolonium iodide

Structural Formula:

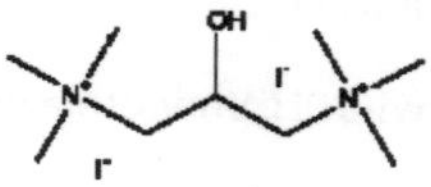

Chemical Abstracts Registry No.: 123-47-7

Trade Name	Manufacturer	Country	Year Introduced
Prolonium iodide	Yick-Vic Chemicals and Pharmaceuticals (HK) Ltd.	-	-
Prolonium iodide	AJAY NORTH AMERICA, L.L.C.	-	-
Prolonium iodide	Alfa Chem	-	-
Hexajodin	Galenika	-	-
Prolonium iodide	ZYF Pharm Chemical	-	-
Micoiodina	VETEM TECNICA SpA	-	-

Raw Materials

Dimethylamine
Epichlorohydrin
Methyliodide

Manufacturing Process

Tetramethyldiaminoisopropanol can be produced by reaction of dimethylamine with epichlorhydrin.

146 parts of symmetrical tetramethyldiaminoisopropanol are mixed with 600 parts of benzene and 284 parts of methyliodide are slowly added whilst cooling and stirring. An oil which crystallizes after some time separates. It is separated from the benzene and crystallized from hot alcohol. The symmetrical hexamethyldiaminoisopropylalcoholiodide forms white crystals, melting point 270°-275°C (dec.).

References

Callsen J.; US Patent No. 1,526,627; Feb. 17, 1925; Assigned: Farbenfabriken Vorm. Friedr. Bayer and Co., of Leverkusen, near Cologne on the Rhine, Germany

Kleemann A., Engel J.; Pharmazeutische Wirkstoffe, GeorgThieme Verlag Stuttgart, New York, 1982

PROMAZINE HYDROCHLORIDE

Therapeutic Function: Tranquilizer

Chemical Name: N,N-Dimethyl-10H-phenothiazine-10-propanamine hydrochloride

Common Name: -

Structural Formula:

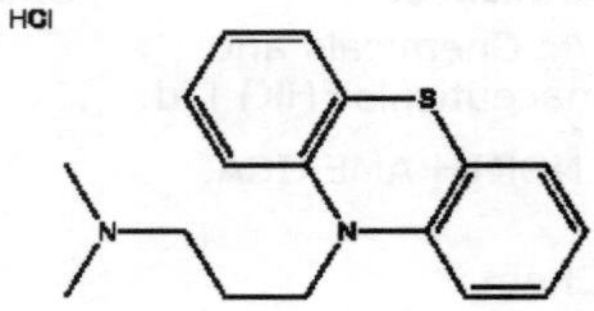

HCl

Chemical Abstracts Registry No.: 53-60-1; 58-40-2 (Base)

Trade Name	Manufacturer	Country	Year Introduced
Sparine	Wyeth	US	1956
Atarzine	Saunders	Canada	-
Calmotal	S.I.T.	Italy	-
Eliranol	Wyeth	Italy	-
Frenil	Polfa	Poland	-
Neuroplegil	Gentili	Italy	-
Promanyl	Paul Maney	Canada	-
Promazettes	Barlow Cote	Canada	-
Promezerine	Barlow Cote	Canada	-
Protactyl	Wyeth	W. Germany	-
Savamine	Banyu	Japan	-
Sediston	Serono	Italy	-
Starazine	Star	Finland	-
Talofen	Pierrel	Italy	-
Tranquazine	Anthony	US	-

Raw Materials

Phenothiazine

3-Dimethylamino-1-chloropropane
Sodium amide
Hydrogen chloride

Manufacturing Process

30 grams of phenothiazine, 120 grams of xylene and 7 grams of sodamide (80%) are mixed and heated under reflux. 23 grams of 3-dimethylamino-1-chloropropane, diluted with its own weight of xylene, is then added little by little during one hour, while maintaining the temperature of the reaction mixture; heating under reflux is then continued for a further hour. After cooling, the mixture is taken up in 400 cc of water and rendered slightly acid with hydrochloric acid. The xylene is decanted, the aqueous layer is rendered strongly alkaline with caustic soda and the base which separates is extracted with ether. On rectification of the ether extract, there is obtained N-(3'-dimethyl-amino-propyl)-phenothiazine which boils at 208°-210°C under 3 mm. The hydrochloride of this base melts at 181°C (Maquenne block).

References

Merck Index 7688
Kleeman & Engel p. 768
PDR p. 1989
OCDS Vol. 1 p. 377 (1977)
I.N. p. 810
REM p. 1090
Charpentier, P.; US Patent 2,519,886; August 22, 1950; assigned to Societe des Usines Chimiques Rhone-Poulenc, France

PROMEGESTONE

Therapeutic Function: Progestin

Chemical Name: 17α,21-Dimethyl-19-nor-$\Delta^{4,9}$-pregnadiene-3,20-dione

Common Name: -

Structural Formula:

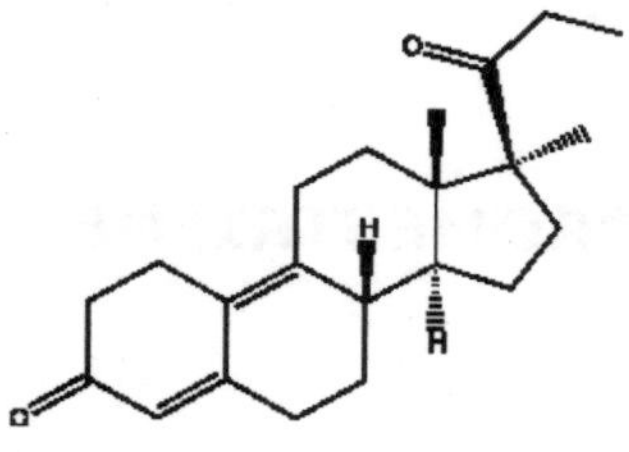

Chemical Abstracts Registry No.: 34184-77-5

Trade Name	Manufacturer	Country	Year Introduced
Surgestone	Cassenne	France	1983

Raw Materials

17α-Methyl-19-nor-Δ($^{5(10)}$)-pregnene-3,20-dione
Bromine
Pyridine

Manufacturing Process

16.3 cc of a solution of 29% of bromine in methanol were added with agitation under a nitrogen atmosphere to a solution of 8.50 g of 17α-methyl-19-nor-$\Delta^{5(10)}$-pregnene-3,20-dionein 85 cc of pyridine cooled to 0°C and the mixture was stirred for 30 minutes at 0°C. The temperature was allowed to return to room temperature and the mixture was stirred for 16 hours.

The mixture was added to 850 cc of water-ice mixture and 82 cc of hydrochloric acid were added thereto. The mixture was extracted with methylene chloride and the combined extracts were washed with water until the wash waters were neutral, were dried over magnesium sulfate and distilled to dryness to obtain 8.480 g of crude product which is purified by crystallion from isopropyl ether to obtain 5.810 g of 17α-methyl-19-nor-$\Delta^{4,9}$-pregnadiene-3,20-dione melting at 106°C.

The mother liquors from the purification of the product were combined and evaporated to dryness. The residue was fractionated by chromatography over silica gel (Kieselgel) and elution with a 7:3 mixture of benzene-ethyl acetate. The first fractions were discarded and the ensuing fraction was evaporated to obtain colorless crystals. The product was purified by mixing with five volumes of boiling isopropyl ether and the crystals formed after cooling were recovered by vacuum filtration, were washed twice with two volumes of isopropyl ether and dried in a ventilated atmosphere to obtain 17α,21-dimethyl-19-nor-$\Delta^{4,9}$-pregnadiene-3,20-dione melting at 152°C.

References

DFU 3 (6) 469 (1978)
DOT 19 (7) 416 (1983)
I.N. p. 810
Warnant, J. and Farcilli, A.; US Patents 3,679,714; July 25, 1972; and 3,761,591; Sept. 25, 1973; both assigned to Roussel UCLAF

PROMESTRIENE

Therapeutic Function: Glucocorticoid

Chemical Name: 17β-Methoxy-3-propoxyestra-1,3,5(10)-triene

Common Name: Promestriene

Structural Formula:

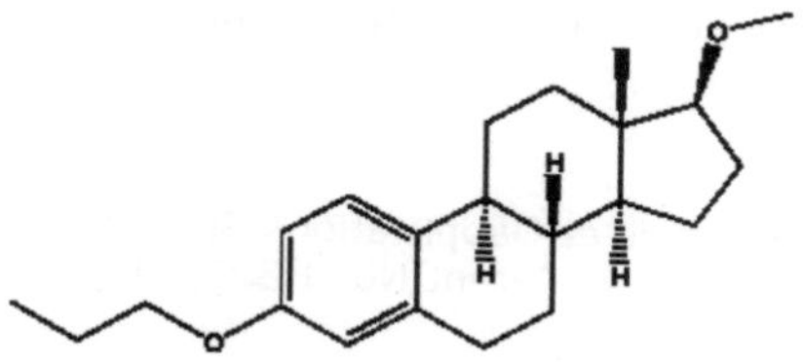

Chemical Abstracts Registry No.: 39219-28-8

Trade Name	Manufacturer	Country	Year Introduced
Colpotrophine	Chengdu Yuyang Hige-tech Developing Co.,Ltd.	-	-
Colpotrophine 1% creme	Theramex	-	-

Raw Materials

Estradiol
n-Propyl bromide
Dimethyl sulfate
Sodium
Dimethyl sulfoxide

Manufacturing Process

An ethanolic solution of sodium ethoxide is prepared by reacting 3 g of sodium with 300 ml of absolute ethanol. 30 g of estradiol are dissolved in the resultant solution and there are then added thereto, with stirring, 30 ml of n-propyl bromide. Reaction is continued for 3 hours with stirring at 60°C and then the reaction mixture is concentrated under vacuum at 30°C to about 50 ml. The residue is taken up in 500 ml of benzene and then washed twice with 250 ml of a 0.25 N solution of sodium hydroxide and then with distilled water to neutrality. The solution is then dried over sodium sulphate and concentrated to give 32 g of crude product (yield 93%), which on recrystallization from 100 ml of methanol gives 31 g (yield 89%) of pure product; MP: 100°-101°C.

40 g of the 3-propyl-ether obtained above is dissolved in 400 ml of anhydrous dimethyl sulphoxide. Several crystals of triphenylmethane are added to the solution (as a coloured indicator) followed by freshly prepared until a permanent red colour is obtained. There is then added about 50% excess dimethylsulfinyl sodium. The reaction mixture is allowed to stand for about 15 minutes at ambient temperature and is then cooled on an ice bath. 40 ml of redistilled dimethyl sulphate are then slowly added to the mixture, which is then stirred for 15 minutes at room temperature. The excess methyl sulphate is then destroyed by the addition of about 2 L of 2 N sodium hydroxide and the mixture stirred for about 2 hours. The pH of the reaction mixture must be alkaline at the end of the operation. The reaction mixture is extracted with benzene and the benzene fractions are washed with distilled water to neutrality. The benzene extracts are then dried over sodium sulphate and concentrated under vacuum at 33°C to give 42 g of crude product which on

repeated recrystallisation from ethanol gives a pure white product promestriene in a yield of 78%; MP: 66°-67°C. The purity of the end product is controlled by gas phase and thin layer chromatography.

References

Societe Generale de Recherches et Dapplications Sogeras, a French body Corporate, Paris, France G.B. Patent No. 1,337,198; March 17, 1972

PROMETHAZINE HYDROCHLORIDE

Therapeutic Function: Antihistaminic

Chemical Name: N,N,α-Ttrimethyl-10H-phenothiazine-10-ethanamine hydrochloride

Common Name: Proazamine hydrochloride

Structural Formula:

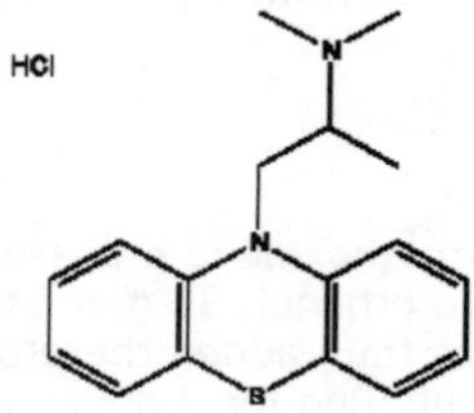

HCl

Chemical Abstracts Registry No.: 58-33-3; 60-87-7 (Base)

Trade Name	Manufacturer	Country	Year Introduced
Phenergan	Wyeth	US	1951
Ganphen	Tutag	US	1971
Remsed	Endo	US	1973
Lemprometh	Lemmon	US	1974
Bromethacon	Alcon	US	1981
Baymethazine	Bay	US	1982
Atosil	Bayer	W. Germany	-
Avomine	May and Baker	UK	-
Diphergan	Polfa	Poland	-
Dorme	A.V.P.	US	-
Fargan	Farmitalia	Italy	-
Fellozine	Fellows-Testagar	US	-
Fenazil	Sella	Italy	-
Fenergan	Rhodia Iberica	Spain	-

Trade Name	Manufacturer	Country	Year Introduced
Hiberna	Yoshitomi	Japan	-
Lenazine	Lennon	S. Africa	-
Lergigan	Recip	Sweden	-
Mopergan	Wyeth	US	-
Pelpica	P.C.B.	Belgium	-
Perduretas	Medea	Spain	-
Phencen	Central	US	-
Pipolphen	Nakataki	Japan	-
Progan	Adams	Australia	-
Promet	Legere	US	-
Promethapar	Parmed	US	-
Promethazine	Lederle	US	-
Promine	Laser	US	-
Prorex	Hyrex	US	-
Prothazine	Knoll	Australia	-
Prothia	Kanto	Japan	-
Prothiazine	Novis	Israel	-
Provigan	Reid-Provident	US	-
Pyrethia	Shionogi	Japan	-
Quadnite	Reid-Provident	US	-
Rivozine	Rivopharm	Switz.	-
Sayamol	Cinfa	Spain	-
V-Gan	Hauck	US	-
Zipan	Savage	US	-

Raw Materials

Phenothiazine
1-Dimethylamino-2-propyl chloride
Sodium amide
Hydrogen chloride

Manufacturing Process

30 grams of phenothiazine, 120 grams of xylene, and 7 grams of sodamide (85%) are mixed and heated under reflux. A solution of 23 grams of the base obtained by the action of sodium hydroxide on the hydrochloride of 1-dimethylamino-2-chloropropane, in 25 grams of xylene, is then added little by little during one hour, while maintaining the temperature of the reaction mixture; heating under reflux is then continued for a further hour. After cooling, the mixture is taken up in 400 cc of water and rendered slightly acid with hydrochloric acid. The xylene is decanted, the aqueous layer is rendered strongly alkaline with caustic soda and the base which separates is extracted with ether. The ethereal extract is rectified, the fraction which boils at 190°-192°C under 3 mm being recovered. This is diluted with acetone or ethyl acetate and dry hydrochloric acid is added. The hydrochloride of N-(2'-dimethylamino-2'-methyl-ethyl)-phenothiazine separates, according to US Patent 2,530,451.

References

Merck Index 7691
Kleeman and Engel p. 769
PDR pp.861, 993, 1033, 1959, 1968, 1989
OCDS Vol. 1 pp. 373, 377 (1977)
I.N. p. 811
REM p. 1129
Charpentier, P.; US Patent 2,530,451; November 21, 1950; assigned to Societe des Usines Chimiques Rhone-Poulenc, France
Berg, S.S. and Ashley, J.N.; US Patent 2,607,773; August 19, 1952; assigned to Societe des Usines Chimiques Rhone-Poulenc, France

PROPAFENONE HYDROCHLORIDE

Therapeutic Function: Antiarrhythmic

Chemical Name: 2'-(2-Hydroxy-3-propylaminopropoxy)-3-phenylpropiophenone hydrochloride

Common Name: Fenoprain

Structural Formula:

HO O NH HCl

Chemical Abstracts Registry No.: 34183-22-7; 54063-53-5 (Base)

Trade Name	Manufacturer	Country	Year Introduced
Rytmonorm	Knoll	W. Germany	1978
Rytmonorm	Knoll	Italy	1983
Rytmonorm	Knoll	Switz.	1983
Baxarytmon	Helopharm	W. Germany	-
Normorytmin	Knoll	W. Germany	-

Raw Materials

n-Propylamine
2'-Hydroxy-3-phenylpropiophenone
Epichlorohydrin
Hydrogen chloride

Manufacturing Process

2'-(2,3-epoxypropoxy)-3-phenylpropiophenone - 24.8 g of the sodium salt of 2'-hydroxy-3-phenylpropiophenone were mixed with 40 cm^3 of 1-chloro-2,3-epoxypropane (epichlorohydrin) and the mixture heated on a boiling water bath while stirring, using a reflux condenser. The initially pasty-to-solid mixture liquefied after about 2 hours, sodium chloride separating out. Thereafter it was heated for a further 2 hours while stirring, using a reflux condenser. The mixture was then allowed to cool and subsequently freed, by filtration, from the sodium chloride formed. The filtrate was concentrated in vacuo, and the excess 1-chloro-2,3-epoxypropane thus separated from the desired 2'-(2,3-epoxypropoxy)-3-phenylpropiophenone. The latter remained as a yellowish oil which solidified in the cold, but did not crystallize. Purification of the intermediate product, by distillation in vacuo, was not necessary, particularly as the substance only boiled at a temperature of 280°C/12 mm Hg and at the same time decomposed.

2'-(2-hydroxy-3-propylaminopropoxy)-3-phenylpropiophenone hydrochloride - The above product was treated with 20 cm^3 of n-propylamine and the mixture warmed on a water bath for approximately 4 hours, while stirring, using a reflux condenser. Thereafter, the excess n-propylamine was distilled off. On cooling, the residue solidified to give a viscous yellow mass. 20 cm^3 of 1 M aqueous hydrochloric acid were added to it, and the whole was boiled for 1 hour under reflux, while stirring. The mixture was then poured into a suitable vessel and allowed to crystallize at room temperature. The crude product was drained thoroughly by suction and subsequently crystallized from a mixture of acetone/methanol (80:20, v/v).

Approximately 25 g (66.2% of theory) of a white crystalline substance were obtained. The melting point of the hydrochloride was 173°C to 174°C.

References

Merck Index 7698
DFU 2 (5) 325 (1977)
Kleeman and Engel p. 770
I.N. p. 812
Sachse, R.; British Patent 1,307,455; February 21, 1973; assigned to Helopharm W. Petrick & Co. K.G.

PROPALLYLONAL

Therapeutic Function: Hypnotic

Chemical Name: 2,4,6(1H,3H,5H)-Pyrimidinetrione, 5-(2-bromo-2-propenyl)-5-(1-methylethyl)-

Common Name: Acidum isopropyl-bromallyl-barbituricum; Bromoaprobarbital; Ibomalum; Propallilonalum; Propallylonal; Propyallylonal

Structural Formula:

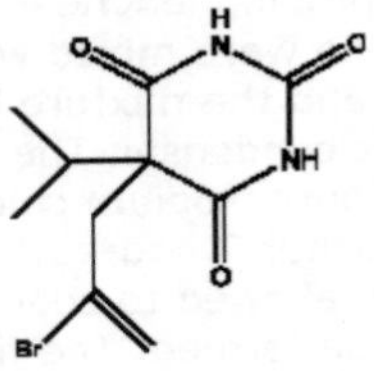

Chemical Abstracts Registry No.: 545-93-7

Trade Name	Manufacturer	Country	Year Introduced
Noctal	Cassella-Riedel	-	-
Noctal	UCB	-	-

Raw Materials

Isopropylbarbituric acid
Sodium
1,2-Dibrom-2.3-propylene

Manufacturing Process

1). 170 parts of isopropylbarbituric acid are gradually added at room temperature to a sodium ethylate solution prepared from 23 parts of sodium and warming at 80°-85°C with brisk agitation, and finally 240 parts of 1,2-dibrom-2.3-propylene are permitted to flow in slowly. After heating for several hours at 90°-100°C the reaction is completed. After blowing off the alcohol the 5-(2-bromoallyl)-5-isopropylbarbituric acid is recovered in almost quantitative yield in the form of colorless crystals. After re-crystallization from dilute acetic acid the acid shows a melting point of 181°C.

2). 250 parts of dibrompropylene are added to a clear solution of 170 parts of isopropylbarbituric acid in dilute caustic soda solution containing 40 parts of sodium hydroxide, in the cold, and the mixture is briskly shaken. After a short time colorless crystals begin to separate out and the separation continues steadily with further shaking but only gradually. The precipitated 5-(2-bromoallyl)-5-isopropylbarbituric acid is filtered by suction and re-crystallized from water or dilute acetic acid. MP: 181°C.

Any untransformed isopropylbarbituric acid is recovered from the filtrate after separating out the unchanged dibrompropylene, by precipitation with concentrated hydrochloric acid.

References

Boedecker F.; US Patent No. 1,622,129; March 22, 1927; Assigned to the firm I.D. Riedel A.G., of Berlin-Britz, Germany

PROPANIDID

Therapeutic Function: Anesthetic

Chemical Name: 4-[2-(Diethylamino)-2-oxoethoxy]-3-methoxybenzene-acetic acid propyl ester

Common Name: -

Structural Formula:

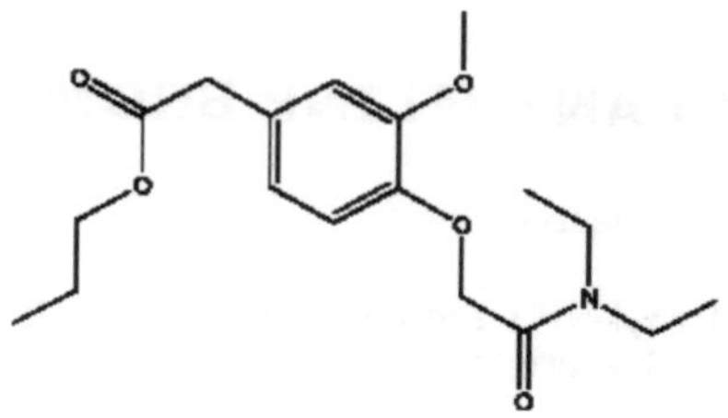

Chemical Abstracts Registry No.: 1421-14-3

Trade Name	Manufacturer	Country	Year Introduced
Epontol	Bayer	W. Germany	1965
Epontol	Bayer	Italy	1967
Epontol	Theraplix	France	1967
Epontol	Bayer	Japan	1970
Fabontal	Bayer	-	-
Sombrevin	Gedeon Richter	Hungary	-

Raw Materials

Sodium
Homovanillic acid n-propyl ester
Chloracetic acid-N,N-diethylamide

Manufacturing Process

To a solution of 4 g of sodium in 200 ml of n-propanol is added 39 g of homovanillic acid-n-propyl ester (boiling point 160°C to 162°C/4 mm Hg) and the mixture is concentrated by evaporation under vacuum. After dissolving the residue in 200 ml of dimethylformamide and the addition of 0.5 g of sodium iodide,26.2 g of chloracetic acid-N,N-diethylamide are added dropwise with stirring at an internal temperature of 130°C, and the mixture is further heated at 130°C for three hours. From the cooled reaction mixture the precipitated salts are removed by filtering off with suction. After driving off the dimethylformamide under vacuum, the product is fractionated under vacuum, and 44.3 g of 3-methoxy-4-N,N-diethylcarbamido-methoxyphenylacetic acid-n-propylester are obtained as a yellowish oil of boiling point 210°C to 212°C/0.7 mm Hg.

References

Merck Index 7705
OCDS Vol. 2 p. 79 (1980)
DOT 2 (3) 110 (1966)
I.N. p.813
REM p. 1047
Hiltman, R., Wollweber, H., Hoffmeister, F. and Wirth, W.; US Patent 3,086,978; April 23, 1963; assigned to Farbenfabriken Bayer A.G. (Germany)

PROPANTHELINE BROMIDE

Therapeutic Function: Spasmolytic

Chemical Name: N-Methyl-N-(1-methylethyl)-N-[2-[(9H-xanthen-9-ylcarbonyl)oxy]ethyl]-2-propanaminium bromide

Common Name: Diisopropylaminoethyl xanthene-9-carboxylate methobromide

Structural Formula:

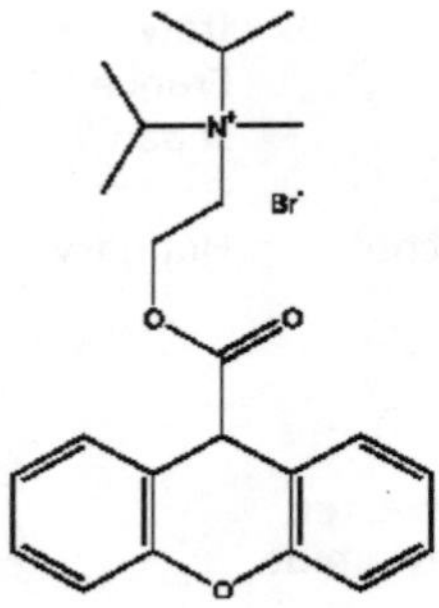

Chemical Abstracts Registry No.: 50-34-0

Trade Name	Manufacturer	Country	Year Introduced
Pro-Banthine	Searle	US	1953
Probanthine	Searle	France	1981
Apopant	A.L.	Norway	-
Banlin	Paul Maney	Canada	-
Corigast	Searle	W. Germany	-
Ercoril	Erco	Denmark	-
Giquel	Danal	US	-
Ketaman	Desitin	W. Germany	-
Neo-Banex	Neo	Canada	-

Trade Name	Manufacturer	Country	Year Introduced
Neo-Dexabine	Noury Pharma	Netherlands	-
Neo-Gastrosedan	Star	Finland	-
Neo-Metantyl	Zambon	Italy	-
Pantheline	Protea	Australia	-
Panthene	Vangard	US	-
Pervagal	Zambeletti	Italy	-
Probital	Searle	US	-
Prodixamon	A.L.	Norway	-
Propanthel	I.C.N.	Canada	-
Suprantil	Prodotti Erma	Italy	-
Tensilan	Desitin	W. Germany	-

Raw Materials

Xanthene-9-carboxylic acid
β-Diisopropylaminoethyl chloride
Methyl bromide

Manufacturing Process

365 parts of β-diisopropylaminoethyl chloride and 565 parts of xanthene-9-carboxylic acid dissolved in 800 parts of isopropanol is heated to reflux for 5 hours. The solution is then cooled, diluted with dry ether and the crystalline precipitate of β-diisopropylaminoethyl xanthene-9-carboxylate hydrochloride is collected on a filter and dried. This salt melts at 111°-112°C. 38 parts of the foregoing salt are dissolved in the minimum of water and treated with an aqueous solution of potassium carbonate. The suspension of β-diisopropylaminoethylxanthene-9-carboxylate thus formed is extracted with ether and the ether extract is dried and evaporated. There is thus obtained 33 parts of the free base which are treated with 10 parts of methyl bromide in 100 parts of chloroform for 22 hours at 70°-80°C. The reaction mixture is chilled, diluted with anhydrous ether and the quaternary salt thus precipitated is collected on a filter and washed with dry ether and then with butanone. β-Diisopropylaminoethyl xanthene-9-carboxylate methobromide thus obtained melts at 152°-153°C. After recrystallization from isopropanol it melts at 157°-155°C.

References

Merck Index 7708
Kleeman & Engel p. 771
PDR pp. 830, 1569, 1606, 1694, 1723
OCDS Vol. 1 p. 394 (1977)
I.N.p.813
REM p. 919
Cusic, J.W. and Robinson, R.A.; US Patent 2,659,732; November 17, 1953; assigned to G.D. Searle & Co.

PROPARACAINE HYDROCHLORIDE

Therapeutic Function: Local anesthetic

Chemical Name: Benzoic acid, 3-amino-4-propoxy-, 2-(diethylamino)ethyl ester, monohydrochloride

Common Name: Proparacaine hydrochloride; Proparakain hydrochloride; Proximetacainum hydrochloride; Proxymetacaine hydrochloride

Structural Formula:

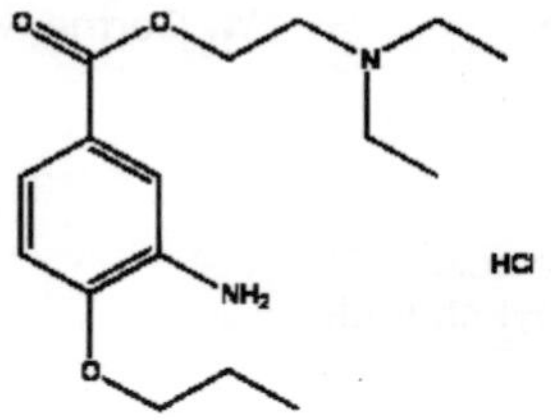

Chemical Abstracts Registry No.: 5875-06-9

Trade Name	Manufacturer	Country	Year Introduced
Alcaine	Alcon	-	-
Ophthaine	AmeriSource	USA	-
Ophthetic	Allergan	-	-

Raw Materials

Thionyl chloride	3-Nitro-4-propyl-oxy-benzoic acid
γ-Diethylaminoethanol	Hydrochloric acid

Manufacturing Process

3-Nitro-4-propyl-oxy-benzoic acid is first converted to the acid chloride by refluxing with thionyl chloride, and after removal of the excess thionyl chloride the acid chloride is reacted with γ-diethylaminoethanol in benzene solution. The γ-diethylaminoethanol-3-nitro-4-propylbenzoate hydrochloride obtained is then reduced in the presence Fe and HCl, to give the γ-diethylaminoethanol-3-amino-4-propylbenzoate.

The hydrochloride of the free base (prepared by reacting the free base with hydrochloric acid) may be recrystallized from acetone.

References

Vliet E. B. et al.; US Patent No. 2,288,334; June 30, 1942; Assigned: Abbott Laboratories, North Chicago, Ill., a corporation of Illinois

PROPICILLIN POTASSIUM

Therapeutic Function: Antibiotic

Chemical Name: 4-Thia-1-azabicyclo(3.2.0)heptane-2-carboxylic acid, 3,3-dimethyl-7-oxo-6- (2-phenoxybutyramido)-, monopotassium salt

Common Name: Phenoxypropylpenicillin potassium; Propicillin potassium

Structural Formula:

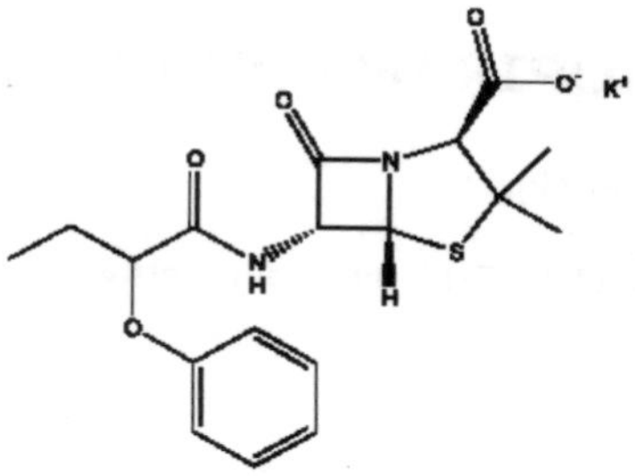

Chemical Abstracts Registry No.: 1245-44-9; 551-27-9 (Base)

Trade Name	Manufacturer	Country	Year Introduced
Oracillin	Takeda	-	-

Raw Materials

Triethylamine
2-Phenoxybuturic acid
Isobutyl chloroformate
6-Aminopenicillanic acid

Manufacturing Process

Triethylamine (1.5 ml) was added to a cold solution (10°C) of 2-phenoxybuturic acid (2.16 g, 0.01 mole) in 15 ml of pure dioxane, with stirring and cooling to 5°-10°C, then isobutyl chloroformate (1.36 g, 0.01.mole) in 5 ml of dioxane was added dropwise. Then the mixture was stirred for ten minutes at 5°-8°C. A solution of 6-aminopenicillanic acid (2.16 g, 0.01 mole) in 15 ml of water and 2 ml of triethylamine was then added dropwise while the temperature was maintained below 10°C. The resulting mixture was stirred in the cold for 15 minutes then at room temperature for 30 minutes, diluted with 30 ml of cold water and extracted with ether, which was discarded. The cold aqueous solution was then covered with 75 ml of ether and acidified to pH 2 with 5 N sulfuric acid. After shaking, the ether layer containing the product 6-(2-phenoxybutyramido)penicillanic acid, was separeted, dried over anhydrous sodium sulphate, solvent was removed to give 1.03 g of product propicillin; MP: 175°-179°C (became dark at 170°C), contained the β-lactam structure as shown by infrared analysis and inhibited Staph. Aureus Smith at concentration 0.05 mcg/ml.

In practice it is usually used as potassium salt.

References

Beecham Research Laboratories Ltd., Brendford, Middlesex, England; G.B. Patent No. 3,316,248; April 25, 1967

PROPIRAM FUMARATE

Therapeutic Function: Analgesic

Chemical Name: N-[1-Methyl-2-(1-piperidinyl)ethyl]-N-2-pyridinylpropanamide fumarate

Common Name: -

Structural Formula:

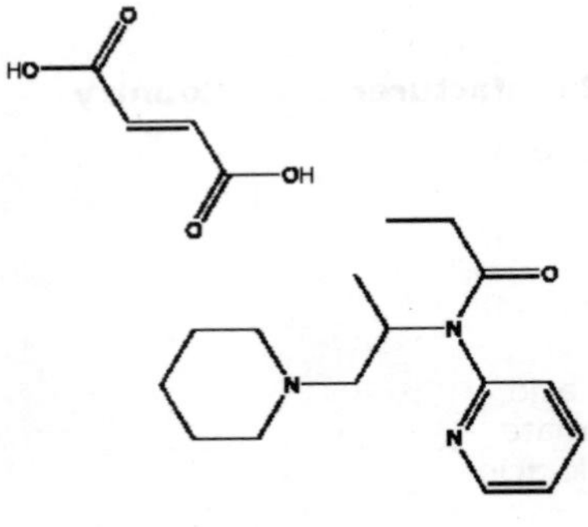

Chemical Abstracts Registry No.: 13717-04-9; 15686-91-6 (Base)

Trade Name	Manufacturer	Country	Year Introduced
Algeril	Bayropharm	Italy	1974
Algeril	Bayer	W. Germany	1974
Dirame	Schering	-	-

Raw Materials

Fumaric acid
2-(1-Piperidine-isopropyl)aminopyridine
Propionic anhydride

Manufacturing Process

20 g of 2-(1-piperidino-isopropyl)aminopyridine and 50 ml of propionic anhydride are heated to 120°C for 8 hours. The mixture is then evaporated under vacuum and the residue taken up in water. The base is precipitated from the solution with a caustic soda solution, taken up in ether and dried with potassium carbonate. After driving off the ether and distillation under vacuum, there are obtained 18 grams of N-propionyl-2-(1-piperidino-isopropyl)aminopyridine of BP 162°-163°C/0.5 mm Hg. The base is then reacted with fumaric acid to give the final product.

References

Merck Index 7733
Kleeman & Engel p. 772
DOT 10 (11) 309 (1974)
I.N. p. 815
Hiltmann, R., Wollweber, H., Hoffmeister, F., Wirth, W. and Kroneberg, H.-G.; US Patent 3,163,654; December 29, 1964; assigned to Farbenfabriken Bayer AG, Germany
Wollweber, H., Hiltmann, R., Hoffmeister, F. and Kroneberg, H.-G.; US Patent 3,594,477; July 20, 1971; assigned to Farbenfabriken Bayer AG, Germany

PROPOFOL

Therapeutic Function: Anesthetic

Chemical Name: Phenol-2,6-diisopropyl

Common Name: Disoprofol; Propofol

Structural Formula:

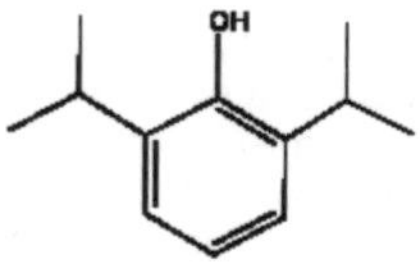

OH

Chemical Abstracts Registry No.: 2078-54-8

Trade Name	Manufacturer	Country	Year Introduced
Cleofol Inj.	Themis Pharmaceuticals Ltd.	India	-
Diprivan	AstraZeneca UK Limited	Italy	-
Diprivan	Zeneca	Italy	-
Diprivan	ICI India Limited	India	-

Trade Name	Manufacturer	Country	Year Introduced
Pofol	Dong Kook Pharmaceutical Co.	Korea	-
Propofol	Baxter	-	-
Propofol Abbott	Abbott Laboratories	USA	-
Propofol 1% Fresenius	Fresenius Kabi	Austria	-
Propofol Lipuro	B. Braun Melsungen AG	Germany	-
Propovan	Bharat Serum and Vaccines Pvt. Ltd.	India	-
Recofol	Leiras OY	Germany	-

Raw Materials

Phenol
Aluminum turnings
Propylene
Isopropyl (2-isopropylphenyl) ether
Fluorided alumina

Manufacturing Process

2 Methods of preparation of 2,6-diisopropylphenol

1. To vessel with flushed nitrogen at an elevated temperature to 165°C 490 parts of phenol was placed, then 4.5 parts of aluminum turnings were added in small increments.The reaction mixture was accompanied by evolution of hydrogen for 15 min, then the mixture was allowed to cool to about 60°C and agitation discontinued. Aluminum phenoxide catalyst mixture was ready.

The reaction vessel was heated to 150°C and pressurized with propylene. The temperature then increased slowly. The start of the reaction was evidenced by a drop in the propylene pressure at 190°C and at a pressure of 21-35 atm. The product was hydrolyzed and fractionated to yield 105 parts of 2,6-diisopropylphenol. Boiling point: 135.5-136.5°C.

2. To 15 g of isopropyl (2-isopropylphenyl) ether was added 8 g of 1% fluorided alumina. The mixture was placed in a 300 cc stirred autoclave and the system was flushed with nitrogen and left under a nitrogen atmosphere. The autoclave was heated to 150°C for 1 h with stirring during which time the pressure reached 200 psig. The cooled reaction mixture was taken up in acetone, filtered, and the solvent was removed on a rotary evaporator. The residue was analyzed by gas-liquid phase chromatography (glpc) which showed the presence of 2,6-diisopropylphenol (60%).

References

Ecke G.G., et al.; US Patent No. 2,831,898; April 22, 1958; Assigned: Ethyl Corporation, New York, N.Y., a corporation of Delaware

Firth B.E., Rosen T.J.; US Patent No. 4,447,657; May 8, 1984; Assigned: UOP Inc., Des Plaines, III

PROPOXYPHENE HYDROCHLORIDE

Therapeutic Function: Analgesic

Chemical Name: (S)-α-[2-(Dimethylamino)-1-methylethyl]-α-phenylbenzeneethanol propanoate hydrochloride

Common Name: Dextropropoxyphene hydrochloride

Structural Formula:

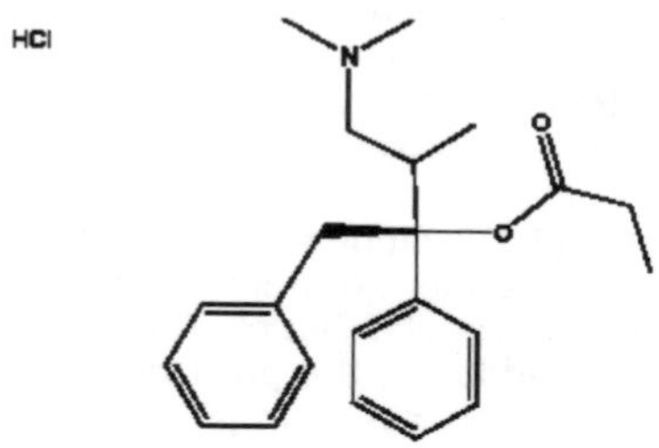

Chemical Abstracts Registry No.: 1639-60-7; 469-62-5 (Base)

Trade Name	Manufacturer	Country	Year Introduced
Darvon	Lilly	US	1957
Antalvic	Houde	France	1963
SK-65	SKF	US	1973
Propoxychel	Rachelle	US	1973
Dolene-65	Lederle	US	1973
Prophen 65	Halsey	US	1981
Darvocet-N	Lilly	US	-
Depronal SA	Warner	UK	-
Develin	Goedecke	W. Germany	-
Doloxene	Lilly	UK	-
Erantin	Boehringer Mannheim	W. Germany	-
Liberen	Lisapharma	Italy	-
Lorcet	U.A.D. Labs	US	-
Wygesic	Wyeth	US	-

Raw Materials

Benzyl chloride
Magnesium
Hydrogen chloride
Propionic anhydride
α-Methyl-β-dimethylaminopropiophenone

Manufacturing Process

A solution of benzylmagnesium chloride prepared from 63.3 grams (0.5 mol)

of benzyl chloride, 30.5 grams (1.25 mol) of magnesium and 750 cc of ether was added dropwise with stirring to a solution of 61.9 grams (0.35 mol) of α-methyl-β-dimethylaminopropiophenone (prepared by the method of Burchalter et al, JACS 70 page 4186, 1948), in 150 cc of ether. When all of the Grignard reagent had been added, the solution was refluxed for about 1 hour. The reaction mixture was then decomposed by the addition of saturated aqueous ammonium chloride solution. The ether solution containing the 1,2-diphenyl-2-hydroxy-3-methyl-4-dimethylaminobutane formed in the reaction was decanted from the granular precipitate and dried over anhydrous magnesium sulfate.

Dry hydrogen chloride gas was passed into the ether solution until precipitation was completed. The solid was removed by filtration and was recrystallized from a mixture of methanol and ethyl acetate. The α-dl-1,2-diphenyl-2-hydroxy-3-methyl-4-dimethylaminobutane hydrochloride thus obtained melted at about 231° to 232°C.

A mixture of 50 grams of α-dl-1,2-diphenyl-2-hydroxy-3-methyl-4-dimethylaminobutane hydrochloride, 50 grams of propionic anhydride and 50 cc of pyridine was refluxed for about 5 hours. The reaction mixture was cooled to 50°C and ethyl ether was added to the point of incipient precipitation. The hydrochloride salt of α-dl-1,2-diphenyl-2-propionoxy-3-methyl-4-dimethylaminobutane formed in the reaction precipitated upon cooling and was removed by filtration and washed with anhydrous ether. On recrystallization from a mixture of methanol and ethyl acetate, α-dl-1,2-diphenyl-2-propionoxy-3-methyl-4-dimethylaminobutane hydrochloride melted at 170°-171°C.

References

Merck Index 7739
Kleeman & Engel p. 285
PDR pp. 993, 1044, 1606, 1723, 1808, 1996, 1999
OCDS Vol. 1 pp. 50, 298 (1977) and 2, 57 (1980)
I.N. p. 816
REM p.1114
Pohland, A.; US Patent 2,728,779; December 27, 1955; assigned to Eli Lilly and Company

PROPRANOLOL HYDROCHLORIDE

Therapeutic Function: Beta-adrenergic blocker

Chemical Name: 1-(Isopropylamino)-3-(1-naphthyloxy)-2-propanol hydrochloride

Common Name: -

Chemical Abstracts Registry No.: 318-98-9; 525-66-6 (Base)

Structural Formula:

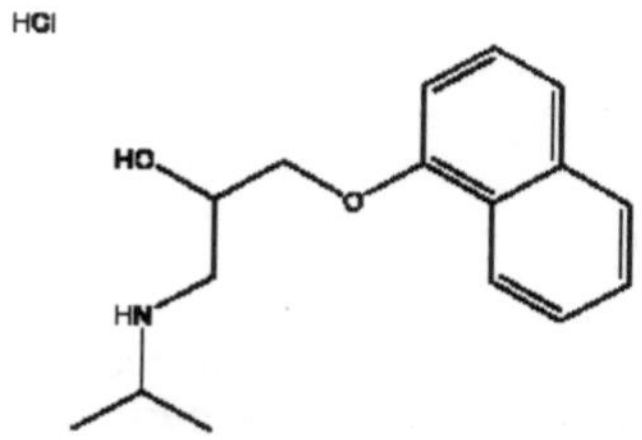

Trade Name	Manufacturer	Country	Year Introduced
Inderal	I.C.I.	UK	1965
Dociton	Rhein Pharma	W. Germany	1965
Avlocardyl	I.C.I.	France	1967
Inderal	Ayerst	US	1968
Angilol	D.D.S.A.	UK	-
Arcablock	Arcana	Austria	-
Bedranol	Berk	Switz.	-
Berkolol	Lagap	UK	-
Beta-Neg	Ellem	Italy	-
Beta-Tablinen	Sanorania	W. Germany	-
Cardinol	Protea	Australia	-
Caridolol	Sankyo	Japan	-
Corotrend	Siegfried	Switz.	-
Deralin	Abic	Israel	-
Detensol	Desbergers	Canada	-
Dideral	Dif-Dogu	Turkey	-
Frekven	Ferrosan	Denmark	-
Herzbase	Nichiiko	Japan	-
Herzul	Ono	Japan	-
Inderide	Ayerst	US	-
Indobloc	Homburg	W. Germany	-
Kemi	Otsuka	Japan	-
Nedis	Omega	Argentina	-
Noloten	Beta	Argentina	-
Novopranol	Novopharm	Canada	-
Obsidan	Iris-Chemie	E. Germany	-
Oposim	Richet	Argentina	-
Pranolol	A.L.	Norway	-
Pronovan	A.L.	Norway	-
Propranolol	Lederle	US	-
Propranur	Henning	W. Germany	-
Pur-Bloka	Lennon	S. Africa	-
Pylapron	Kyorin	Japan	-
Reducor	Leiras	Finland	-
Sawatal	Sawai	Japan	-
Tonum	Tubi Lux Pharma	Italy	-

Raw Materials

1-Naphthol
Isopropylamine
Epichlorohydrin
Hydrogen chloride

Manufacturing Process

In a first step, 1-naphthol was reacted with epichlorohydrin to give 1-chloro-3-(1-naphthoxy)-2propanol. A mixture of 4.4 parts of 1-chloro-3-(1-napntnoxy)-2-propano and 16 parts of isopropylamine is heated in a sealed vessel at 70°-80°C for 10 hours. The vessel is cooled and to the contents there are added 50 parts of water. The mixture is acidified with 2 N hydrochloric acid, and washed with 50 parts of ether. The aqueous phase is decolorized with carbon, and then added to 50 parts of 2 N sodium hydroxide solution at 0°C. The mixture is filtered, The solid residue is washed with water, dried, and crystallized from cyclohexane. There is thus obtained 1-isopropylamino-3-(1-naphthoxy)-2-propanol, MP 96°C.

The base may be converted into the hydrochloride as follows. 4.65 parts of the base are dissolved in 60 parts of warm acetone. To the warm solution there are added 2 parts of 10 N hydrochloric acid. The mixture is allowed to cool, and is then filtered. The solid residue is washed with acetone and then dried. The solid is crystallized from propanol, and there is thus obtained 1-isopropylamino-3-(1-naphthoxy)-2-propanol hydrochloride MP 163°C.

References

Merck Index 7740
Kleeman & Engel p. 773
PDR pp. 622, 993, 1999
OCDS Vol. 1 p. 117 (1977) and 2, 105, 107, 212 (1980)
DOT 19 (3) 172 (1983)
I.N. p. 816
REM p. 906
Crowther, A.F. and Smith, L.H.; US Patent 3,337,628; August 22, 1967; assigned to Imperial Chemical Industries Limited, England

PROPYLHEXEDRINE

Therapeutic Function: Nasal decongestant

Chemical Name: N,α-Dimethylcyclohexaneethanamine

Common Name: Hexahydrodesoxyephedrine

Chemical Abstracts Registry No.: 101-40-6; 6192-98-9 (Hydrochloride salt)

Structural Formula:

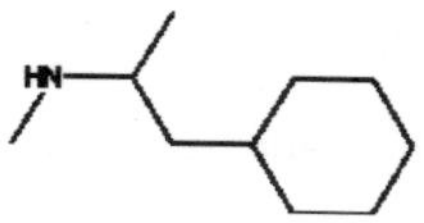

Trade Name	Manufacturer	Country	Year Introduced
Benzedrex	SKF	US	1949
Dristan	Whitehall	US	-
Eggobesin	Fahlberg-List	E. Germany	-
Eventin	Minden	W. Germany	-

Raw Materials

Cyclohexylacetone
N-Methylformamide
Sulfuric acid
Sodium hydroxide

Manufacturing Process

33.6 grams of cyclohexylacetone, a compound known to the art, dissolved in 13 grams of 85% formic acid is caused to interact with 72.0 grams of N-methyl formamide at 160°-180°C for 4 hours. This results in the formation of the formyl derivative of the amine, according to the following reaction:

C_6H_{11}-CH_2-C(O)-CH_3+ 2HCONHCH_3 => C_6H_{11}-CH_2-CH(NCH_3-CHO)-CH_3+ CH_3NH_2+ CO_2

The formyl derivative is then hydrolyzed by refluxing with 50% sulfuric acid for about 4 hours, after which the hydrolysate is extracted with ether to remove the acid-insoluble material and the aqueous solution made strongly alkaline with any suitable alkalizing agent, for example, sodium hydroxide, to liberate the amine.

The amine is then taken up in ether, dried over potassium hydroxide and purified by distillation, preferably under reduced pressure. β-cyclohexylisopropylmethylamine thus obtained boils at 90.0°-92°C at 22 mm Hg.

References

Merck Index 7761
Kleeman & Engel p. 774
OCDS Vol. 1 p. 37 (1977)
I.N. p. 817
REM p. 890
Ullyot, G.E.; US Patent 2,454,746; November 23, 1948; assigned to Smith, Kline and French Laboratories

PROPYLIODONE

Therapeutic Function: Diagnostic aid

Chemical Name: 1(4H)-Pyridineacetic acid, 3,5-diiodo-4-oxo-, propyl ester

Common Name: Propiliodon; Propyliodone

Structural Formula:

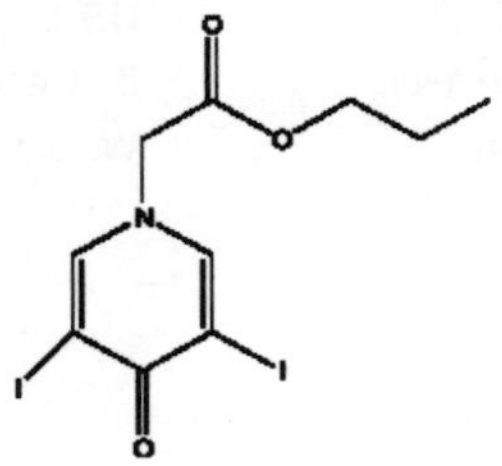

Chemical Abstracts Registry No.: 587-61-1

Trade Name	Manufacturer	Country	Year Introduced
Propyliodone	GlaxoSmithKline	-	-
Dionosil oily	GlaxoSmithKline	-	-

Raw Materials

Chloroacetic acid
3,5-Diiodo-4-1H-pyridone
Isopropanol

Manufacturing Process

15 parts of 3,5-diiodo-4-pyridone-N-acetic acid, (prepared from 3,5-diiodo-4-1H-pyridone and chloroacetic acid), 60 parts of isopropanol, and 1 part of concentrated sulfuric acid are boiled for an hour in a vessel fitted with a reflux condenser. The solution is then cooled, whereupon crystals of isopropyl 3,5-diiodo-4-pyridone-N-acetate separate out by filtration, washed with ethanol and dried. MP: 215°C.

References

Branscombe D.J.; G.B. Patent No. 517,382; July 25, 1938; Imperial Chemical Industries Limited, of Imperial Chemical House, Millbank, London

Pharmazeutishe Wirkstoffe von A. Kleemann und J. Engel. 2., neubearbeitete und erweiterte Auflage; Georg Thime Verlag Stuttgart New York 1982; p. 775

PROPYROMAZINE BROMIDE

Therapeutic Function: Spasmolytic, Anticholinergic

Chemical Name: 1-Methyl-1-(1-phenothiazin-10-ylcarbonylethyl) pyrrolidinium bromide

Common Name: Diaspasmyl; Propyromazine bromide

Structural Formula:

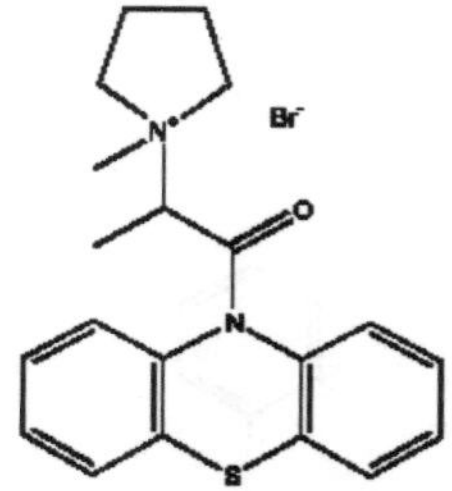

Chemical Abstracts Registry No.: 145-54-0

Trade Name	Manufacturer	Country	Year Introduced
Propyromazine bromide	Astra (AstraZeneca)	-	-

Raw Materials

Pyrrolidine
2-Bromo-1-phenothiazin-10-yl-propan-1
Methyl bromide

Manufacturing Process

185 g pyrrolidine and 334 g 2-bromo-1-phenothiazin-10-yl-propan-1 were refluxed in 2500 ml toluene for three hours. On cooling the solution the precipitated pyrrolidine hydrochloride was separated and collected, and the filtrate was evaporated to dryness. The crystalline residue 272 g was recrystallized from a mixture of five parts of light petroleum. MP of 1-phenothiazin-10-yl-2-pyrrolidin-1-yl-propan-1-one 94.5-95.5°C. Its brommethylate 1-methyl-1-(1-methyl-2-oxo-2-phenothiazin-10-yl-ethyl)pyrrolidinium, bromide was prepared by reaction with equivalent quantity of methyl bromide.

References

Dahlbom J.R., T.K.I. B. Ekstrand; US Patent No. 2,615,886; January 10, 1951; Assigned to Aktiebolaget Astra, Apotekarnes Kremiska Fabriker, Sodertalje, Sweden

Pharmazeutishe Wirkstoffe von A. Kleemann und J. Engel; 2., neubearbeitete und erweiterte Auflage, p. 776, 1982; Georg Thime Verlag Stuttgart New York

PROQUAZONE

Therapeutic Function: Antiinflammatory

Chemical Name: 1-Isopropyl-7-methyl-4-phenyl-2(1H)-quinazolinone

Common Name: -

Structural Formula:

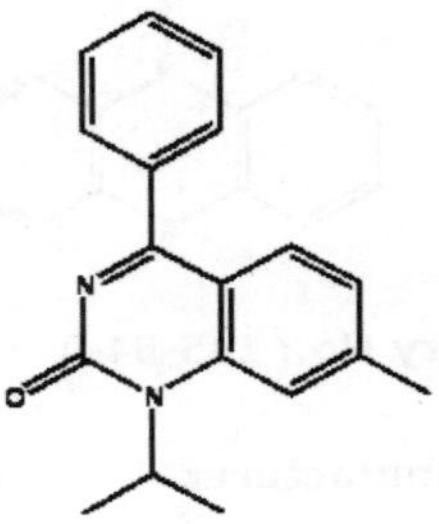

Chemical Abstracts Registry No.: 22760-18-5

Trade Name	Manufacturer	Country	Year Introduced
Biarison	Sandoz	Italy	1977
Biarison	Sandoz	Japan	1977
Biarison	Sandoz	France	1977
Biarison	Sandoz	Switz.	1977
Biarison	Wander	W. Germany	1979

Raw Materials

4-Methyl-2-isopropylaminobenzophenone
Urethane

Manufacturing Process

A mixture of 5.9 g of 4-methyl-2-isopropylaminobenzophenone, 13.9 g urethane and 500 mg of zinc chloride is heated at a temperature of 190°C for 1½ hours. There is then additionally added 7 g of urethane and 250 mg of zinc chloride, and the heating continued at a temperature of 190°C for an additional 2½ hours. The resulting mixture is cooled to about 100°C and diluted with chloroform. The resulting mixture is then filtered and the filtrate washed first with water and then with brine. The organic phase is separated, dried over anhydrous sodium sulfate and concentrated in vacuo to remove substantially all of the chloroform and obtain an oily residue which is dissolved in a small amount of about 20 ml of methylene chloride. The resulting solution is then diluted with about 40ml of ethyl acetate and concentrated in vacuo to crystallize 1-isopropyl-7-methyl-4-phenyl-2(1H)-quinazolinone; melting point 137°C to 138°C.

References

Merck Index 7775
DFU 1 (11) 540 (1976)
Kleeman and Engel p. 777
OCDS Vol. 2 p. 386 (1980)
DOT 8 (3) 116 (1972) and 13 (12) 534 (1977)
I.N. p. 818
Linder, J., Mattner, P.G. and Salmond, W.G.; US Patent 3,759,720; September 18, 1973; assigned to Sandoz-Wander Inc.
Denzer, M.; US Patent 3,793,324; February 19, 1974
Ott, H.; US Patent 3,925,548; December 9,1975; assigned to Sandoz, Inc.

PROSCILLARIDIN

Therapeutic Function: Cardiotonic

Chemical Name: 3-[(6-Deoxy-α-L-mannopyranosyl)oxy]-14-hydroxybufa-4,20,22-trienolide

Common Name: -

Structural Formula:

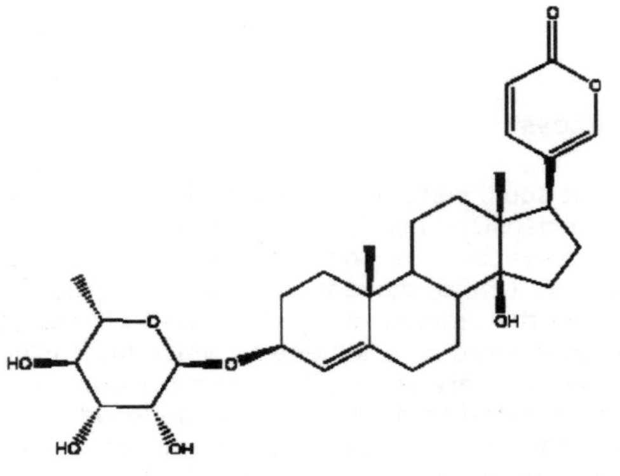

Chemical Abstracts Registry No.: 466-06-8

Trade Name	Manufacturer	Country	Year Introduced
Talusin	Knoll	W. Germany	1964
Talusin	Biosedra	France	1968
Apocerpin	Kotani	Japan	-
Bunosquin	Seiko	Japan	-
Caradrin	Kowa	Japan	-
Cardimarin	Santen	Japan	-
Cardiolidin	Nichiiko	Japan	-
Cardion	Nippon Chemiphar	Japan	-
Cardon	Kanto	Japan	-
Herzo	Toho	Japan	-
Mitredin	Nippon Shoji	Japan	-
Procardin	Mohan	Japan	-
Procillan	Hokuriku	Japan	-
Proherz	Shinshin	Japan	-
Proscillan	Streuli	Switz.	-
Proscillar	Toyo Jozo	Japan	-
Prosiladin	Sawai	Japan	-
Prostosin	Iwaki	Japan	-
Proszin	Teisan	Japan	-
Protasin	Bayropharm	W. Germany	-
Purosin-TC	Tatsumi	Japan	-
Sandoscill	Sandoz	W. Germany	-
Scillaridin	Morishita	Japan	-
Silamarin A	Wakamoto	Japan	-
Stellarid	Tobishi-Mochida	Japan	-
Talusin	Dainippon	Japan	-
Urgilan	Simes	Italy	-
Wirnesin	Inpharzam	W. Germany	-

Raw Materials

Squill

Manufacturing Process

350 g of dried and cut squill were fermented at 50°C for two hours in 1.1 liters of water. The suspension was then extracted three times with 1.1 liters of ethyl acetate. The extracts were united and evaporated to dryness, the residue was dissolved in 2 ml of dioxane and chromatographed in a twenty-fold quantity (based on the amount of dried residue) of silica gel. The proscillaridin was then eluated with toluene to which increasing quantities of a methanol-dioxane mixture were added. The main fraction, containing proscillaridin, was evaporated to dryness. The residue was crystallized out of methanol. Pure proscillaridin was obtained with a melting point of 227°C to 230°C; α_{20}^{D} = -93.5°C (in methanol).

The same result was obtained by fermentation on the aqueous suspension of the cut squill at room temperature for 24 hours and working up in the manner described.

References

Merck Index 7776
Kleeman & Engel p. 777
DOT 3 (3) 97 (1967)
I.N. p.819 Steidle, W. US Patent 3,361,630; January 2, 1968; assigned to Knoll A.G. (Germany)

PROTHIPENDYL HYDROCHLORIDE

Therapeutic Function: Sedative, Antihistaminic

Chemical Name: N,N-Dimethyl-10H-pyrido[3,2-b][1,4]benzothiazine-10-propanamine hydrochloride

Common Name: -

Structural Formula:

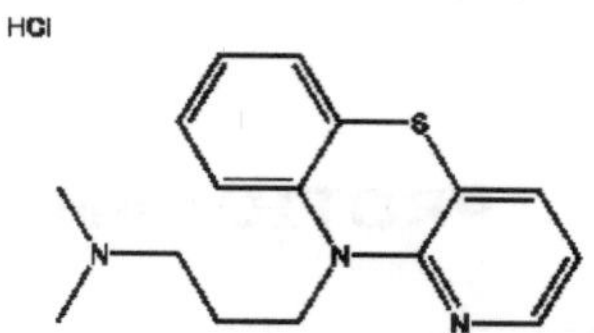

Chemical Abstracts Registry No.: 1225-65-6; 303-69-5 (Base)

Trade Name	Manufacturer	Country	Year Introduced
Timovan	Ayerst	US	1960
Dominal	Homburg	W. Germany	-
Prosyl	Kanto	Japan	-
Tolnate	SKF	UK	-

Raw Materials

1-Azaphenothiazine
Sodium amide
3-Dimethylaminopropyl chloride
Hydrogen chloride

Manufacturing Process

A mixture of 20 g (0.1 mol) of 1-azaphenothiazine, 4.3 g (0.11 mol) of sodamide and 300 ml of dry toluene is stirred and refluxed for eight hours. A slow stream of dry nitrogen gas is used to sweep out the ammonia as formed. The mixture is cooled and 110 ml of a 1 M solution of 3-dimethylaminopropyl chloride in toluene is added dropwise, with stirring. Subsequently, the mixture is stirred and refluxed for fifteen hours, cooled, and concentrated in vacuo. The viscous residue is refluxed with 500 ml of chloroform and filtered hot. The chloroform filtrate is treated with activated charcoal and again filtered. The filtrate is concentrated and the residue distilled to give about 19.8 g (69% yield) of product, an oil distilling at about 195°C to 198°C (under 0.5 mm pressure of mercury).

To a solution of 16.4 g (0.058 mol) of the free base in 75 ml of dry acetonitrile is added dropwlse while cooling (ice bath) and stirring 14.5 ml (0.053 mol) of 3.6 N ethereal hydrogen chloride. An equal volume of anhydrous ether is added and the product altered, dried and recrystallized from monochlorobenzene. The product melts at about 177°C to 178°C with sintering at about 176°C. The yield is about 11.0 g (60%).

References

Merck index 7789
Kleeman & Engel p. 779
OCDS Vol. 1 p. 430 (1977)
I.N. p. 821
Yale, H.L. and Bernstein, J.; US Patent 2,943,086; June 28, 1960; assigned to Olin Mathieson Chemical Corp.

PROTIONAMIDE

Therapeutic Function: Antitubercular

Chemical Name: 2-Propyl-4-pyridinecarbothioamide

Common Name: α-Propyl-isonicotinic thioamide

Structural Formula:

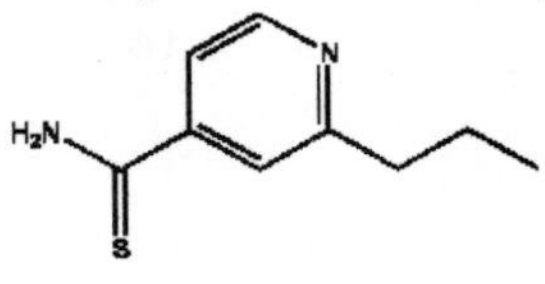

Chemical Abstracts Registry No.: 14222-60-7

Trade Name	Manufacturer	Country	Year Introduced
Ektebin	Bayer	W. Germany	1969
Protionizina	Farmitalia	Italy	1970
Entelohi	Kyowa	Japan	-
Peteha	Saarstickstoff-Fatol	W. Germany	-
Promid	Biofarma	Turkey	-

Trade Name	Manufacturer	Country	Year Introduced
Prothionamide	Toho	Japan	-
Trevintix	Theraplix	France	-
Tuberamin	Meiji	Japan	-
Tuberex	Shionogi	Japan	-
Tubermide	Sankyo	Japan	-

Raw Materials

Ethyl oxalate
Hydrogen chloride
Phosphoric anhydride
Cyanacetamide
Ammonia
Sodium ethylate
Hydrogen
Methyl-n-propyl ketone
Phosphorus oxychloride
Hydrogen sulfide

Manufacturing Process

(A) Ethyl Butyryl-Pyruvate: 146 grams of ethyl oxalate are condensed with 86 grams of methyl-(n)-propyl-ketone in the presence of sodium ethylate prepared from 25 grams of sodium. 135 grams of product, having a boiling point of 113°C/6 mm, are obtained.

(B) 3-Cyano-4-Carbethoxy-6-(n)-Propyl-2-Pyridone: The 135 grams of the product just obtained are condensed with 62 grams of cyanacetamide in the presence of 24 cc of piperidine in 1200 cc of 95% alcohol. 64 grams of a product, melting at 152°C, are obtained.

(C) 6-(n)-Propyl-2-Pyridone-4-Carboxylic Acid: The 64 grams of the product just obtained are treated with 500 cc of concentrated hydrochloric acid at boiling point. 40 grams of a product, having a melting point of 285°C, are obtained.

(D) Ethyl 2-Chloro-6-(n)-Propyl-Isonicotinate: The 40 grams of the acid just obtained are treated with 80 grams of phosphorus oxychloride and 95 grams of phosphorus pentachloride. The phosphorus oxychloride is distilled and the reaction mixture is treated with 400 grams of absolute alcohol. 40 grams of chlorinated ester, having a BP of 115°-116°C/2 mm, are obtained.

(E) Ethyl 2-(n)-Propyl-Isonicotinate: The product just obtained is dechlorinated by catalytically hydrogenating it in an alcoholic medium in the presence of palladium black and potassium acetate. 30 grams of ester, having a boiling point of 121°-125°C/7 mm, are obtained.

(F) 2-(n)-Propyl-Isonicotinamide: The 30 grams of the ester just obtained are treated with 40 cc of concentrated ammonia saturated with gaseous ammonia. 20 grams of product, having a melting point of 135°C, are obtained.

(G) 2-(n)-Propyl-Isonicotinic-Nitrile: The 20 grams of the amide just obtained are treated with 32 grams of phosphoric anhydride. 11 grams of nitrile, having a BP of 90°-95°C/4 mm, are obtained.

(H) 2-(n)-Propyl-Isonicotinic Thioamide: The 11 grams of nitrile just obtained, dissolved in 40 cc of ethanol containing 4 grams of triethanolamine, are treated with hydrogen sulfide. 8 grams of the desired product, having a melting point of 142°C, are obtained.

References

Merck Index 7791
Kleeman & Engel p. 780
DOT 3 (1) 24 (1967)
I.N. p. 821
Chimie et Atomistique, France; British Patent 800,250; August 20, 1958

PROTIZINIC ACID

Therapeutic Function: Antiinflammatory

Chemical Name: 7-Methoxy-α,10-dimethylphenothiazine-2-acetic acid

Common Name: -

Structural Formula:

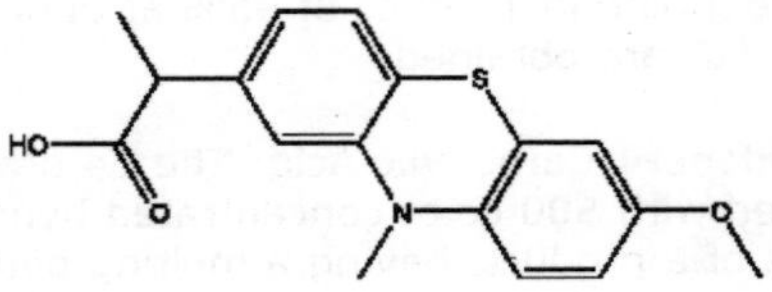

Chemical Abstracts Registry No.: 13799-03-6

Trade Name	Manufacturer	Country	Year Introduced
Pirocrid	Theraplix	France	1974
Pirocrid	Mochida	Japan	1979
P.R.T.	Mochida	Japan	-

Raw Materials

Sodium	Hydrogen chloride
thanol	Methyl iodide
Diethyl carbonate	Sodium hydroxide
Methyl (7-methoxy-10-methyl-3-phenthiazinyl)acetate	

Manufacturing Process

Methyl ethyl (7-methoxy-10-methyl-3-phenthiazinyl)malonate is prepared by reacting a solution of sodium (4.37 grams) in anhydrous ethanol (110 cc) with a solution of methyl (7-methoxy-10-methyl-3-phenthiazinyl)acetate (59 grams) in ethyl carbonate (180 cc). The reaction mixture is heated at about

105°-110°C for 3 hours and the ethanol formed is distilled off as it is formed.

The reaction mixture is acidified with N hydrochloric acid (200 cc) and the oil formed is extracted with methylene chloride (200 cc). The methylene chloride solution is washed with water (210 cc), treated with decolorizing charcoal (5 grams), dried over anhydrous sodium sulfate and concentrated to dryness under reduced pressure (20 mm Hg) giving an oil (77 grams) which is crystallized from methanol (300 cc) to yield methyl ethyl (7-methoxy-10-methyl-3-phenthiazinyl)-malonate (62.4 grams) melting at 80°-82°C.

Methyl ethyl (7-methoxy-10-methyl-3-phenthiazinyl)malonate (62.2 grams) followed by methyl iodide (45.7 grams) is added to a solution of sodium (4.45 grams) in anhydrous ethanol (500 cc). The reaction mixture is heated under reflux for 1 hour at 45°C, then for 6 hours at 55°C, and finally concentrated to dryness under reduced pressure (20 mm Hg). The residue is taken up in methylene chloride (300 cc) and water (250 cc), filtered in the presence of a filtration adjuvant, washed with methylene chloride (150 cc) and water (150 cc), and decanted. The aqueous solution is extracted once again with methylene chloride (100 cc), and the combined organic solutions washed with water (100 cc), aqueous 0.1 N sodium hyposulfite solution (200 cc) and finally with water (200 cc). After drying over anhydrous sodium sulfate and evaporation to dryness under reduced pressure (20 mm Hg), there is obtained an oil (64.8 grams) which is dissolved in methylene chloride (100 cc) and chromatographed over alumina (650 grams). After elution with methylene chloride, a fraction of 2.5 liters is recovered and concentrated to dryness under reduced pressure (20 mm Hg) to give methyl ethyl methyl-(7-methoxy-10-methyl-3-phenthiazinyl)malonate (59.7 grams) melting at 70°-72°C.

1 N sodium hydroxide solution (296 cc) is poured over a period of 3 hours into a solution of methyl ethyl methyl-(7-methoxy-10-methyl-3-phenthiazinyl)malonate (59.7 grams) in ethanol (600 cc) heated under reflux in an atmosphere of nitrogen. The reaction mixture is concentrated to dryness under reduced pressure (20 mm Hg), the residue obtained acidified with N hydrochloric acid (300 cc) and the gum formed extracted with methylene chloride (150 cc). The organic solution is washed with water (200 cc), treated with decolorizing charcoal (10 grams), dried over anhydrous sodium sulfate and concentrated to dryness under reduced pressure (20 mm Hg). The oil obtained (48 grams) is dissolved in N sodium hydroxide solution (200 cc) and the aqueous solution washed with diethyl ether (300 cc), treated with decolorizing charcoal (5 grams) and acidified with N hydrochloric acid (200 cc). The oil formed is dissolved in methylene chloride (350 cc), the solution washed with water (100 cc), treated with decolorizing charcoal (5 grams) and dried over anhydrous sodium sulfate. The solution is concentrated to dryness under reduced pressure (20 mm Hg) to give an oil (35.6 grams) which crystallizes slowly. On recrystallization from diisopropyl ether (180 cc) a product (19.5 grams), melting at 123°-124°C, is obtained. Further recrystallization from diisopropyl ether (290 cc) yields 2-(7-methoxy-10-methyl-3-phenthiazinyl)propionic acid (12.9 grams) melting at 124°-125°C.

References

Merck Index 7792
Kleeman & Engel p. 782
DOT 8 (12) 452 (1972)

I.N. p. 36
Farge, D., Jeanmart, C. and Messer, M.N.; US Patent 3,450,698; June 17, 1969; assigned to Rhone-Poulenc SA, France

PROTOKYLOL

Therapeutic Function: Bronchodilator

Chemical Name: 4-[2-[[2-(1,3-Benzodioxol-5-yl)-1-methylethyl]amino]-1-hydroxyethyl]-1,2-benzenediol

Common Name: -

Structural Formula:

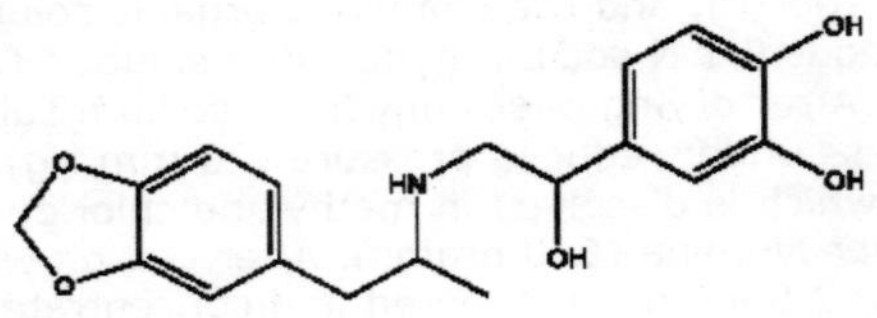

Chemical Abstracts Registry No.: 136-70-9; 136-69-6 (Hydrochloride salt)

Trade Name	Manufacturer	Country	Year Introduced
Caytine	Lakeside	US	1959
Ventaire	Marion	US	1974
Asmetil	Benvegna	Italy	-
Atma-Sanol	Sanol	W. Germany	-
Beres	Simes	Italy	-
Biturix	Nemi	Argentina	-
Palison	Farmasimes	Spain	-

Raw Materials

Chloroacetylcatechol
3,4-Methylenedioxyphenylisopropanolamine
Hydrogen

Manufacturing Process

3,4-Methylenedioxyphenylisopropanolamine is reacted with chloroacetylcatechol in a 3:1 mol ratio in 60% ethanol at reflux temperature with continuous stirring. Stirring and refluxing were continued for another five hours after which the reaction mixture was cooled and then acidified with 20 cc of concentrated aqueous HCl. The acid solution was concentrated in vacuo to a viscous consistency and the residue dissolved in acetone. On standing, the aminoketone precipitated and was filtered. The precipitate was dissolved in isopropyl alcohol and permitted to recrystallize. An alcoholic solution of this aminoketone precipitate was reduced with PtO_2 and hydrogen, clarified by filtration, concentrated to dryness in vacuo and the residue crystallized from acetone giving the desired product.

References

Merck Index 7798
Kleeman and Engel p. 783
I.N. p.821
Biel, J.H.; US Patent 2,900,415; August 18, 1959; assigned to Lakeside Laboratories, Inc.

PROTRIPTYLINE

Therapeutic Function: Psychostimulant

Chemical Name: N-Methyl-5H-dibenzo[a,d]cycloheptene-5-propylamine

Common Name: Amimetilina; 5-(3-Methylaminopropyl)-5H-dibenzo[a,d]cycloheptene

Structural Formula:

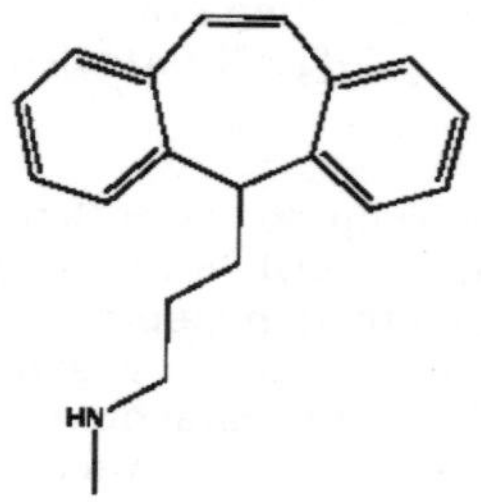

HN

Chemical Abstracts Registry No.: 438-60-8; 1225-55-4 (Hydrochloride salt)

Trade Name	Manufacturer	Country	Year Introduced
Vivactil	MSD	US	1967
Maximed	Sharp and Dohme	W. Germany	1968
Concordin	MSD	Italy	1972
Concordine	MSD	France	1973
Triptil	Merck-Frosst	Canada	-

Raw Materials

Formamide	3-Methylaminopropanol-1
Thionyl chloride	5H-Dibenzo[a,d]cycloheptene
Potassium amide	Potassium hydroxide

Manufacturing Process

Preparation of 3-(N-Formyl-N-Methyl)-Aminopropanol-1: A mixture of 40 grams of 3-methylaminopropanol-1 and 20 grams of formamide is heated while stirring for 4 hours at 165°C. The crude product is fractionated in vacuo using a Widmer column yielding substantially pure 3-(N-formyl-N-methyl)-aminopropanol-1.

Preparation of 3-(N-Formyl-N-Methyl)-Aminopropyl Chloride: 50 grams of 3-(N-formyl-N-methyl)-aminopropanol-1 obtained above is dissolved in a mixture of 100 ml of chloroform and 25 grams of pyridine. 40 grams of thionyl chloride is then slowly added while maintaining the temperature below 65°C. After 6 hours of refluxing, the mixture is washed with water, then with sodium bicarbonate solution and again with water and then dried over magnesium sulfate and the solvent distilled off in vacuo. Fractional distillation at 1 mm pressure yields substantially pure 3-(N-formyl-N-methyl)-aminopropyl chloride.

Preparation of 5-[3-(N-Formyl-N-Methyl)-Aminopropyl]-5H-Dibenzo[a,d]Cycloheptene: To a suspension of 3.9 grams of potassium amide is slowly added a solution of 19.2 grams (0.1 mol) of 5H-dibenzo[a,d]cycloheptene in 600 ml of ether with stirking. The suspension is refluxed with stirring for 3 hours, then cooled to room temperature and a solution of 0.1 mol of 3-(N-formyl-N-methyl)-aminopropyl chloride in 100 ml of ether added. The mixture is then refluxed with stirring for 5 hours and then 100 ml of water added. The ether layer is then washed with dilute hydrochloric acid, then water and then dried over magnesium sulfate and evaporated to dryness yielding 5-[3-(N-formyl-N-methyl)-aminopropyl]-5H-dibenzo[a,d]cycloheptene.

Preparation of 5-(3-Methylaminopropyl)-5H-Dibenzo[a,d]Cycloheptene from 5-[3-(N-Formyl-N-Methyl)-Aminopropyl]-5H-Dibenzo[a,d]Cycloheptene: 29.5 grams of 5-[3-(N-formyl-N-methyl)aminopropyl]-5H-dibenzo[a,d]cycloheptene is refluxed for 24 hours under nitrogen in a solution of 36.3 grams of potassium hydroxide in 378 ml of n-butanol. After cooling to room temperature, the solvent is evaporated in vacuo, the residue is stirred with 200 ml of water, 300 ml of n-hexane, the layers separated, the water layer extracted with 100 ml of n-hexane and the combined hexane layers washed with water (2 x 100 ml) and then with 0.5 N sulfuric acid (100, 80, 80 ml). The acid solution is then alkalized and extracted with ether (2 x 150 ml and 1 x 100 ml), dried over $MgSO_4$ and the solution evaporated to dryness yielding substantially pure 5-(3-methylarninopropyl)-5H-dibenzo[a,d]cycloheptene according to US Patent 3,244,748.

References

Merck Index 7804
Kleeman & Engel p. 783
PDR p. 1220
OCDS Vol. 1 p. 152 (1977)
I.N. p. 822
REM p. 1097
Tishler, M., Chemerda, J.M. and Kollonitsch, J.; US Patent 3,244,748; April 5, 1966; assigned to Merck & Co., Inc.

Tishler, M., Chemerda, J.M. and Kollonitsch, J.; US Patent 3,271,451; September 6, 1966; assigned to Merck & Co., Inc.

PROXAZOLE CITRATE

Therapeutic Function: Spasmolytic

Chemical Name: N,N-Diethyl-3-(1-phenylpropyl)-1,2,4-oxadiazole-5-ethanamine citrate

Common Name: Propaxoline citrate

Structural Formula:

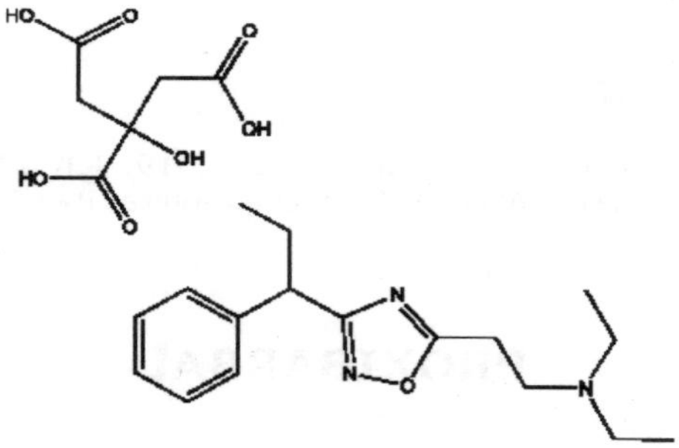

Chemical Abstracts Registry No.: 132-35-4; 5696-49-3 (Base)

Trade Name	Manufacturer	Country	Year Introduced
Recidol	Lampugnani	Italy	1967
Pirecin	Yoshitomi	Japan	1970
Mendozal	Beaufour	France	1976
Flou	Elea	Argentina	-
Solacil	Finadiet	Argentina	-
Toness	Angelini	Italy	-

Raw Materials

α-Ethylbenzamidoxime
Citric acid
β-Chloropropionyl chloride
Diethylamine

Manufacturing Process

α-Ethylbenzamidoxime and anhydrous potassium carbonate are suspended in chloroform. To this mixture, under continuous stirring and controlling of the reaction temperature to remain beyond 15°C, there is slowly added β-chloropropionyl chloride. After addition of the acid chloride, stirring is continued for a further hour. Then with cooling there is added portionwise a small amount of water. Further amounts of water are introduced into the reaction mixture and the chloroform solution containing the β-chloropropionyl α-ethylbenzamidoxime is separated.

To this solution there is added in about 20 minutes a solution of diethylamine in $CHCl_3$ while the temperature is kept below 35°C. The reacting mixture is heated to boiling, water formed during the reaction being distilled off thereby. After two hours the distillate contains no more water and the reaction is finished. Water is added to dissolve diethylamine hydrochloride formed during the reaction, and the chloroform layer containing the product is separated from the aqueous layer. The product may be purified by distillation; it boils at 132°C at 0.2 mm pressure. It is converted to the citrate by reaction with citric acid.

References

Merck Index 7805
Kleeman & Engel p. 784
OCDS Vol. 2 p. 271 (1980)
I.N. p. 822
Palazzo, G. and Silvestrini, B.; US Patent 3,141,019; July 14, 1964; assigned to Angelini Francesco, Aziende Chimiche Riunite, Italy

PROXIBARBAL

Therapeutic Function: Sedative

Chemical Name: 5-(2-Hydroxypropyl)-5-(2-propenyl)-2,4,6(1H,3H,5H) pyrimidinetrione

Common Name: Proxibarbital

Structural Formula:

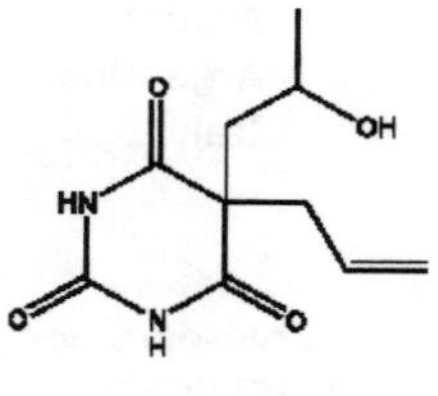

Chemical Abstracts Registry No.: 2537-29-3

Trade Name	Manufacturer	Country	Year Introduced
Axeen	Hommel	W. Germany	1962
Centralgol	Valpan	France	1965
Ipronal	Polfa	Poland	-
Vasalgin	Chinoin	Hungary	-

Raw Materials

Diallylbarbituric acid
Sulfuric acid
Water

Manufacturing Process

9 Parts of diallyl-barbituric acid are added to a precooled mixture of 15.5 parts of concentrated sulfuric acid and 0.5 part of water while stirring intensively, the mixture being cooled so that its temperature does not exceed 25°C. The honey-colored viscous solution is stirred vigorously and all at once into 45 parts of water, whereupon the mixture warms up to 35°C to 40°C and, after several seconds, solidifies into a thick pulp, which is then heated as quickly as possible to 95°C, at which temperature a clear solution is formed. This is cooled slowly until the 5-allyl-5-(β-hydroxypropyl)-barbituric acid begins to form coarse-grained crystals, after which the mass is cooled rapidly to 20%.

The crystallized 5-allyl-5-(β-hydroxypropyl)-barbituric acid is centrifuged off, 55 to 58 parts of mother liquor and 10 to 13 parts of crude product being obtained. The latter is dispersed in 20 parts of saturated aqueous sodium chloride solution and after two hours is again centrifuged off.

The thus-washed crude product is dissolved in a mixture of 12 parts of ethanol and 20 parts of benzene, with mild warming if necessary. 1 Part of sodium chloride and 1.5 parts of saturated aqueous sodium chloride solution are added to the obtained solution in ethanol-benzene, and whole thoroughly admixed. When the brine layer has settled, it is separated and the afore-described washing repeated. The clear solution is concentrated under reduced pressure until incipient formation of crystals and is then poured into 30 parts of benzene, whereupon a thick crystalline pulp is forthwith formed which, after being cooled to room temperature, is centrifuged off. The so-obtained 5-allyl-5-(β-hydroxypropyl)-barbituric acid is dried at 70°C under reduced pressure and can be used for therapeutic purposes without further purification. Melting point 164°C to 165°C. Yield: 5 parts.

References

Merck Index 7806
I.N. p. 822
Hommel A.G.; British Patent 953,387; March 25, 1964

PROXYMETACAINE

Therapeutic Function: Local anesthetic

Chemical Name: 3-Amino-4-propoxybenzoic acid 2-(diethylamino)ethyl ester

Common Name: Proparacaine; Proxymetacaine

Structural Formula:

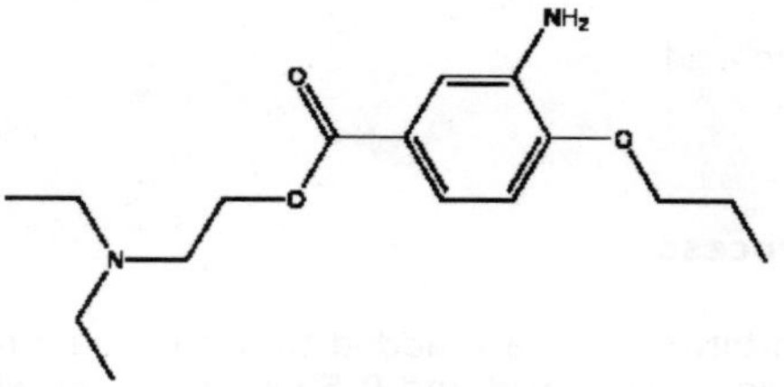

Chemical Abstracts Registry No.: 499-67-2

Trade Name	Manufacturer	Country	Year Introduced
Diocaine	Dioptic	-	-

Raw Materials

4-Propoxynitrobenzoyl chloride
Diethylaminoethanol
Granulated tin

Manufacturing Process

Equal molecular proportions of 4-propoxynitrobenzoyl chloride and diethylaminoethanol are mixed. They react forming hydrochloride of (diethylamino)ethyl 4-propoxy-3-nitrobenzoate as a pale yelllow leaflets; MP: 124.8-126.8°C. The reaction is completed when there is no further tendency to warm itself spontaneously. Ten parts by weight of the latter substance are dissolved in a mixture of twenty-five parts by weight of hydrochloric acid and twenty parts by weight of alcohol, and the solution treated with twelve parts by weight of granulated tin, keeping the temperature at about 35°C.

A colorless solution is obtained from which the tin is removed by precipitation with hydrogen sulfide. On addition of sodium carbonate solution, 3-Amino-4-propoxybenzoic acid 2-(diethylamino)ethyl ester separates as an oil. When treated with one equivalent of hydrochloric acid it forms a hydrochloride, which is readily soluble in water and crystallizes from a mixture of absolute alcohol and ethyl acetate in white prisms MP: 182.0-183.3°C.

References

Wildman E.A.; US Patent No. 1,317,250; September 30, 1919; Assigned to Parke Davis and Company, of Detroit,Michigan, a corporation
Climton R.O. et al.; J.A.C.S. v.74 p. 592-598, 1952

PROXYPHYLLINE

Therapeutic Function: Diuretic, Cardiac stimulant, Smooth muscle relaxant, Vasodilator

Chemical Name: 1H-Purine-2,6-dione, 3,7-dihydro-7-(2-hydroxypropyl)-1,3-dimethyl

Common Name: Hydroxypropyltheophylline; Proxifillina; Proxiphyllinum; Proxyphylline

Structural Formula:

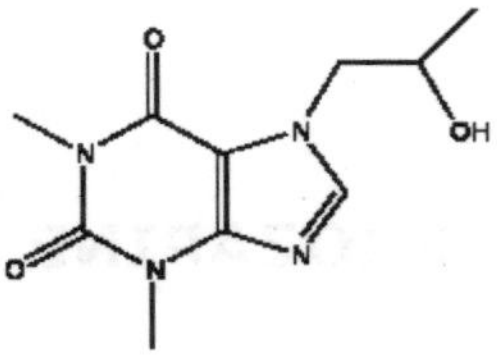

Chemical Abstracts Registry No.: 603-00-9

Trade Name	Manufacturer	Country	Year Introduced
Monophyllin	AFI	-	-
Monophyllin	Yoshitomi	-	-
Neofyllin	Abigo	-	-
Neofyllin	Pharmacia	-	-
Purophyllin	Siegfried	-	-
Spasmolysin	Kade	-	-
Theon	Draco	-	-

Raw Materials

Theophylline
1-Chloro-2-propanol
Sodium hydroxide

Manufacturing Process

A mixture of 270 g (1.5 moles) of anhydrous theophylline and 213 g (2.25 moles) of 1-chloro-2-propanol in 750 ml of water is heated to boiling in an apparatus equipped with a mechanical agitator, reflux condenser, thermometer, and dropping funnel. A 25 per cent solution of sodium hydroxide in water, containing 90 g (2.25 moles) of sodium hydroxide, is added to the refluxing mixture over a period of 2 hours. Refluxing is continued for 1 hour after all the sodium hydroxide has been added. The water is removed as completely as possible by distillation under reduced pressure, using a boiling water bath as the source of heat. The residue, consisting of a sticky, resinous mass or white solid, is treated with 700 ml of anhydrous ethyl alcohol and heated until the remaining insoluble solid is loose and granular. The solid is separated by filtering of the hot mixture. When the filtrate cools, a white, crystalline mass separates which is filtered off and washed with cold anhydrous ethyl alcohol. The material is purified by crystallization from anhydrous ethyl alcohol. The purified product is 7-β-hydroxypropyl theophylline, M. P. 135-136°C. The pH of a 5% per cent solution in distilled water falls within the range of 5.5 to 7.0. 1 g dissolves in approximately 1 ml of water at 20°C and in about 14 ml of anhydrous ethyl alcohol. It is considerably more soluble in boiling anhydrous ethyl alcohol.

References

Rice R. V.; US Patent No. 2,715,125; Aug. 9, 1955; Assigned to N. J., assignor to Gane's Chemical Works, Inc., Carlstadt, N. J., a corporation of New York

PROZAPINE

Therapeutic Function: Choleretic, Spasmolytic

Chemical Name: 1-(3,3-Diphenylpropyl)cyclohexamethyleneimine

Common Name: Prozapine; Hexadiphane

Structural Formula:

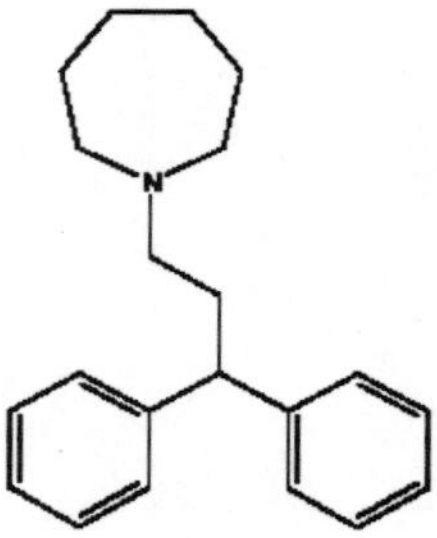

Chemical Abstracts Registry No.: 3426-08-2

Trade Name	Manufacturer	Country	Year Introduced
Prozapine	ZYF Pharm Chemical	-	-

Raw Materials

Hydrochloric acid	α,α-Diphenyl-7-hexamethyleneimino butyronitrile
Phenyl magnesium bromide	
Sodium amide	Sodium hydroxide
Hydrogen	Hexamethyleneiminoethyl phenyl ketone
Thionyl chloride	Palladium on charcoal

Manufacturing Process

The 1st method of preparation of the 1,1-diphenyl-3-hexamethyleneimino propane:

A vigorously stirred suspension of 0.2 to 1 mole of sodium amide in 200 ml of xylene, in which were dissolved 0.1 mole of α,α-diphenyl-7-hexamethyleneimino butyronitrile was boiled for 12 hours. Thereupon the excess of sodium amide was decomposed with water and the xylene layer was separated, washed with water and extracted with hydrochloric acid. This acidic extract was made strongly alkaline with concentrated lye and the separated base was extracted with ether. After drying, the ether was evaporated and the 1,1-diphenyl-3-hexamethyleneimino propane distilled in vacuo. The boiling point was 170-174°C/1 mm, the refractive index n_D^{20} = 1.5636, and the density d_4^{20} = 1.009. From the oil obtained several acid additions and quaternary ammonium salts can be obtained by reaction with acids containing a non-toxic anion or esters thereof. The hydrochloric acid salt, for instance, melts at 189-192°C, the methiodide at 174-177°C under decomposition.

Another method of preparation of the 1,1-diphenyl-3-hexamethyleneimino propane:

To a solution of 0.4 mole of phenyl magnesium bromide in ether were added 42.3 g (0.183 mole) of hexamethyleneiminoethyl phenyl ketone, dissolved in dry ether, followed by 250 ml of dry benzene. The temperature of the mixture was slowly raised until all the ether had been driven off, after which the solution was heated to boiling under reflux of the benzene for 6 hours. The reaction mixture was then cooled and a solution of ammonium chloride was added. The benzene layer was washed with water and dried on potassium carbonate, filtered from the potassium carbonate, and the solvent was then evaporated. Recrystallization of the residue from petroleum ether (B.P. 60-80°C) yielded 44 g of 1,1-diphenyl-3-hexamethyleneimino propanol-1 with a melting point of 81°C.

To 31 g of 1,1-diphenyl-3-hexamethyleneimino propanol-1, dissolved in chloroform, an excess of thionyl chloride was added and the mixture was heated under reflux for 3 hours. Thereupon the reaction mixture was evaporated to dryness under reduced pressure and the residue was recrystallized by dissolving in warm ethanol and diluting this solution with ethyl acetate. An aqueous solution of the 1,1-diphenyl-1-chloro-3-hexamethyleneimino propane hydrochloride thus obtained was hydrogenated with hydrogen gas in the presence of a buffered palladium-charcoal catalyst at a pressure of 3 atm. The 1,1-diphenyl-3-hexamethyleneimino propane obtained was purified by distillation under reduced pressure. The boiling point was 170-174°C/1 mm.

References

Paul A. J. Janssen, David K. De Jongh; US Patent No. 2,881165; Apr. 7, 1959; Assigned to N. V. Nederlandsche Cornbinatie voor Chemische Industrie, Amsterdam, Netherlands, a limited liability company of the Netherlands

PSEUDOEPHEDRINE SULFATE

Therapeutic Function: Bronchodilator

Chemical Name: Benzenemethanol, α-((1S)-1-(methylamino)ethyl)-, (αS)-, sulfate (2:1) (salt)

Common Name: Pseudoephedrine sulfate

Structural Formula:

Chemical Abstracts Registry No.: 7460-12-0; 90-82-4 (Base)

Trade Name	Manufacturer	Country	Year Introduced
Afrinol	Schering	-	-
Demazin sinus	Schering-Plough	-	-
Drixora	Magpharm Pharmaceuticals	S. Africa	-
Lertamine-D	White Pharma	-	-
Pseudoephedrine sulfate	BASF AG	Germany	-
Pseudoephedrine sulfate	Knoll AG	Germany	-

Raw Materials

Hydrobromic acid
Bromine
Methylamine
Propionaldehyde
Phenyl magnesium bromide

Manufacturing Process

Propionaldehyde was brominated by bromine and the propanoilbromide was obtained. Then propanoilbromide reacted with methanol and hydrobromic acid yielding 1,2-dibromo-1-methoxypropane.

1,2-Dibromo-1-methoxypropane in turn with phenylmagnesium-bromide gave the product, which after hydrolysis yielded 1-phenyl-1-methoxy-2-bromopropane.

To the solution of 1-phenyl-1-methoxy-2-bromopropane methylamine was added and as a result of the reaction 1-phenyl-1-methoxy-2-methylaminopropane was obtained. After that 1-phenyl-1-methoxy-2-methylaminopropane on hydrolysis with hydrobromic acid yielded 1-phenyl-1-hydroxy-2-methylaminopropane, i.e. racemic pseudoephedrine. The racemic base was resolved, by crystallization of their tartrates, into l- and d-pseudoephedrine. The base l-pseudoephedrine forms white rhombic crystals, melting point 118°C. The salt pseudoephedrine sulfate may be prepeared by treatment of pseudoephedrine base with sulfuric acid.

References

Manske and Holmes, The Alkaloids, Vol III, pp.343-344, 351-361, Academic Press (1953)

PSEUDONOREPHEDRINE

Therapeutic Function: Sympathomimetic, Anorexic

Chemical Name: Benzenemethanol, α-((1S)-1-aminoethyl)-, (αS)-

Common Name: Cathine, Katine; ψ-Norephedrine; Norisoephedrine; Norpseudoephedrine; Pseudonorephedrine

Structural Formula:

OH

NH_2

Chemical Abstracts Registry No.: 492-39-7

Trade Name	Manufacturer	Country	Year Introduced
Beloform	Neue Formulierung	-	-
Cathine	BlueRunners Trading Dev. Co.	-	-
Norpseudoephedrine	Shanghai Lansheng Corporation	-	-

Raw Materials

Thionyl chloride
Norephedrine hydrochloride, (+)-

Manufacturing Process

60 ml thionylchloride was added to 20 g (+)-norephedrine hydrochloride by cooling. Then the mixture was gradually heated to +35°C. Finally it was heated 20 minutes at 45°C. After that the excess thionylchloride was removed in vacuum, the residue was stirred with 100 ml acetone and 2-amino-1-chloro-1-phenylpropane hydrochloride was filtered off. It was dissolved in 100 ml 2 N hydrochloric acid, heated to reflux for 3 hours, and distilled to dryness in vacuum. The crystalline residue was mixed with acetone, filtered off and dried. Yield was 17.5 g crude product, which was cleaned by fractional

crystallization to give 13 g (65%) (+)-norpseudoephedrine hydrochloride 166°-168°C. The may be converted to the base by adding of equivalent of any base.

References

Pfanz H., Wieduwilt H.; D.D. Patent No. 13,785; Feb. 8, 1956

PYRANTEL PAMOATE

Therapeutic Function: Anthelmintic

Chemical Name: E-1,4,5,6-Tetrahydro-1-methyl-2-[2-(2-thienyl)vinyl] pyrimidine pamoate

Common Name: -

Structural Formula:

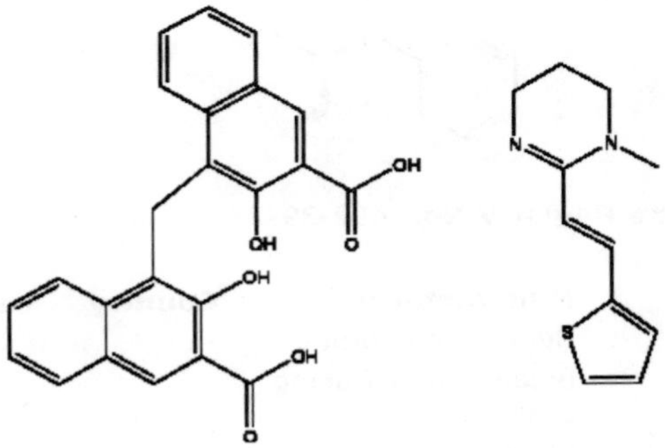

Chemical Abstracts Registry No.: 22204-24-6; 15686-83-6 (Base)

Trade Name	Manufacturer	Country	Year Introduced
Antiminth	Roerig	US	1972
Helmex	Roerig	W. Germany	1972
Cobantrin	Pfizer Taito	Japan	1973
Combantrin	Pfizer	France	1973
Combantrin	Pfizer	Italy	1975
Lombriareu	Areu	Spain	-
Piranver	ICN-Usafarma	Brazil	-

Raw Materials

Tartaric acid	Thiophene-2-carboxaldehyde
Pamoic acid	1,2-Dimethyl-1,4,5,6-tetrahydropyrimidine

Manufacturing Process

A solution of 0.1 mol of each of thiophene-2-carboxaldehyde and 1,2-dimethyl-1,4,5,6-tetrahydropyrimidine in dimethyl carbonate (0.2 mol) is held at 27°C for 48 hours. The reaction mixture is then stripped to give a 65% yield of product as the free base.

The base may be isolated as the tartrate as follows: A portion of reaction mixture is added to a well stirred solution of tartaric acid in ethanol at 27°C. The mixture is stirred for two hours and the product recovered by filtration. The filter cake is washed with cold ethanol followed by ether and air-dried. MP 144°-147°C.

The tartrate salt is recrystallized by dissolving in hot methanol, filtering, adding hot ethanol to the filtrate and cooling. The product is collected and air-dried. MP 148°-150°C. A second crop is obtained from the filtrate for a total yield of 59%. The tartrate is then metathesized with pamoic acid (Merck Index #6867) to give pyrantel pamoate as the product.

References

Merck Index 7856
Kleeman & Engel p. 786
PDR p. 1403
OCDS Vol. 1 p. 266 (1977) and 2, 303 (1980)
DOT 8 (11) 431 (1972); 17 (1) 41 (1981); and (6) 262 (1981)
I.N. p. 825
REM D. 1237
Kasubick, R.V. and McFarland, J.W.; US Patent 3,502,661; March 24,1970; assigned to Chas. Pfizer & Co., Inc.

PYRATHIAZINE

Therapeutic Function: Antihistaminic

Chemical Name: 10-[2-[1-(Pyrrolidinyl)ethyl]phenothiazine

Common Name: Parathiazine

Structural Formula:

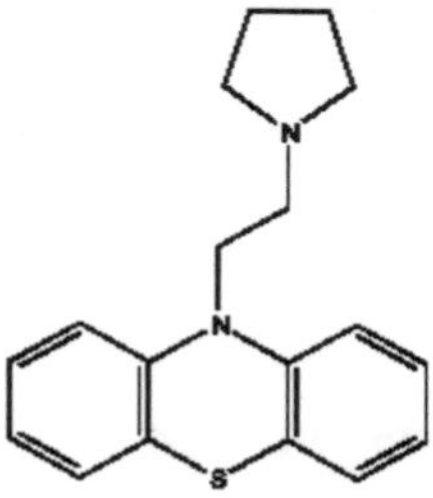

Chemical Abstracts Registry No.: 84-08-2

Trade Name	Manufacturer	Country	Year Introduced
Pyrrolazote	Upjohn	US	1949

Raw Materials

Phenothiazine
Sodium amide
β-Pyrrolidinoethyl chloride

Manufacturing Process

To a stirred suspension of 4.29 g (0.11 mol) of sodium amide in 100 ml of dry toluene was added 19.9 g (0.1 mol) of phenothiazine. The solution was heated at reflux for two hours, the sodium salt of phenothiazine precipitating from solution. The toluene suspension of the sodium salt of phenothiazine was cooled to room temperature, whereupon there was added dropwise with continued stirring 13.36 g (0.1 mol) of β-pyrrolidinoethyl chloride in 50 ml of dry toluene. After addition was complete, the solution was heated under reflux, with stirring, for an additional 15 hours. Upon cooling, the toluene was extracted with dilute hydrochloric acid and the toluene then discarded. The aqueous acid solution was made alkaline with dilute sodium hydroxide, the crude N-(β-pyrrolidinoethyl)-phenothiazine separating as a brownish oil.

The oil was extracted with ether, the ether solution dried with anhydrous magnesium sulfate, and then filtered. Dry hydrogen chloride was passed into the ether solution and a semisolid mass, which crystallized after scratching, separated therefrom. The crude N-(β-pyrrolidinoethyl)-phenothiazine was separated from the ether and, after two crystallizations from isopropanol, 17.0 g of desired product, melting at 196°C to 197°C (uncorr.), was obtained.

References

Merck Index 7857
OCDS Vol. 1 p. 373 (1977)
I.N. p. 731
Hunter, J.H. and Reid, W.B. Jr.; US Patent 2,483,999; October 4, 1949; assigned to The Upjohn Co.

PYRAZINAMIDE

Therapeutic Function: Antibacterial (tuberculostatic)

Chemical Name: Pyrazinecarboxamide

Common Name: -

Chemical Abstracts Registry No.: 98-96-4

Structural Formula:

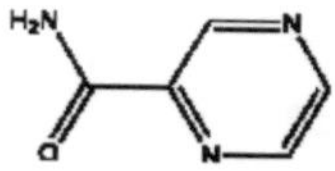

Trade Name	**Manufacturer**	**Country**	**Year Introduced**
Aldinamide	MSD	US	1955
Pirilene	Lepetit	France	1981
Eprazin	Krugmann	W. Germany	-
Isopyratsin	Leiras	Finland	-
Pezatamid	Hefa-Frenon	W. Germany	-
Piraldina	Bracco	Italy	-
Pirazimida	Madaus Cerafarm	Spain	-
Pyrafat	Saarstickstoff-Fatol	W. Germany	-
Pyrazide	SCS Pharmalab	S. Africa	-
P.Z.A.	Servipharm	Switz.	-
Tebrazid	Continental Pharma	Belgium	-
Tisamid	Orion	Finland	-
Zinamide	MSD	UK	-

Raw Materials

Pyrazine-2,3-dicarboxamide
Sodium hydroxide

Manufacturing Process

166 Parts of pyrazine-2,3-dicarboxamide (1 mol) is slurried in 1,000 parts of 1 N aqueous sodium hydroxide. The reaction mixture is heated at 95°C to 98°C until a clear solution results. Thereupon the mixture is cooled with ice to about 5°C and acidified to approximately a pH of 1. The cold reaction mixture is allowed to stand until precipitation of the pyrazine-2-carboxamide-3-carboxylic acid is substantially complete whereupon it is recovered by filtration and dried at 50°C to 60°C.

100 Parts of pyrazine-2-carboxamide-3-arboxylic acid is heated in a reaction vessel provided with an intake for inert gas. The reaction mixture is heated in a bath held at 220°C and nitrogen is introduced. The solid material melts and effervesces and sublimed pyrazinamide vapors are carried out of the reaction vessel in the nitrogen stream. They are introduced into a suitably cooled condenser, condensing in the form of a white sublimate. After the reaction is proceeding vigorously the bath temperature is raised to 255°C and then gradually and slowly allowed to drop to 190°C over a period of time sufficient to permit the reaction to go substantially to completion. The sublimed pyrazinamide, if desired, is further purified by recrystallization from water or alcohol.

References

Merck Index 7858
Kleeman & Engel p. 787
OCDS Vol. 1 p. 277 (1977)
I.N. p. 826
REM p. 1216
Webb, J.S. and Ark, H.G. Jr.; US Patent 2,780,624; February 5, 1957; assigned to American Cyanamid Co.

PYRIDINOL CARBAMATE

Therapeutic Function: Antiarteriosclerotic

Chemical Name: Bis[methylcarbamic acid]-2,6-pyridinediyldimethylene diester

Common Name: Pyricarbate

Structural Formula:

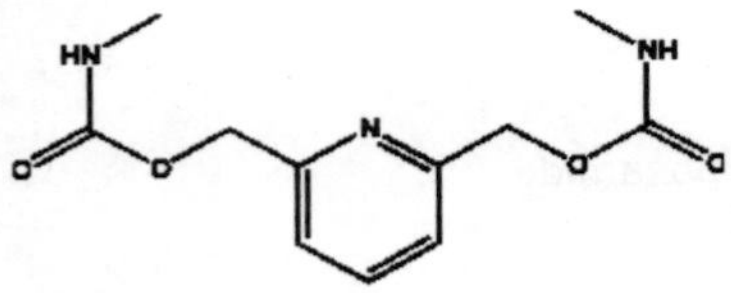

Chemical Abstracts Registry No.: 1882-26-4

Trade Name	Manufacturer	Country	Year Introduced
Movecil	Erba	Italy	1969
Angioxine	Roussel	France	1971
Anginin	Banyu	Japan	-
Angiovital	I.S.M.	Italy	-
Angioxil	Firma	Italy	-
Angiperl	Sawai	Japan	-
Arteriolangal	Lanzas	Spain	-
Aterin	Ilsan	Turkey	-
Aterofal	Nativelle	Italy	-
Atero-Flavin	Indelfar	Spain	-
Aterollano	Llano	Spain	-
Ateronova	Cheminova	Spain	-
Atover	Oti	Italy	-
Carbatona	Turro	Spain	-
Cicloven	A.G.I.P.S.	Italy	-
Colesterinex	Galenika	Switz.	-
Dual-Xol	Lifepharma	Spain	-

Trade Name	Manufacturer	Country	Year Introduced
Duaxol	Argentia	Argentina	-
Duvaline	Almirall	Spain	-
Gasparol	Castejon	Spain	-
Meduxal	Allard	France	-
Plavolex	Wolner	Spain	-
Prodectin	Kobanyai	Hungary	-
Ravenil	Caber	Italy	-
Sospitan	Kali-Chemie	W. Germany	-
Vasagin	Sidus	Italy	-
Vasapril	Cifa	Italy	-
Vasmol	Lifasa	Spain	-
Vasocil	Magis	Italy	-
Vasoverin	Biochimica	Switz.	-
Veranterol	Asla	Spain	-

Raw Materials

Methyl isocyanate
2,6-Dihydroxymethylpyridine hydrochloride

Manufacturing Process

(A) 15.7 g (0.1 mol) of 2,6-dihydroxymethylpyridine hydrochloride are suspended in 176 ml of acetonitrile, and 20.8 ml (0.15 mol) of triethylamine are added to the suspension. Thereafter 13 ml (0.22 mol) of methyl isocyanate are added dropwise to the reaction mixture at 20°C to 25°C. The reaction mixture is stirred at 20°C to 30°C for one hour, thereafter boiled for 3 hours, and finally the solvent is evaporated under reduced pressure. 35 to 40 g of a greyish, crystalline residue are obtained, which is a mixture of 2,6-dihydroxymethylpyridine-bis-(N-methylcarbamate) and triethylamine hydrochloride. The obtained residue is dissolved in 80 ml of hot water, decolorized with 2 g of activated carbon when hot, and filtered after 30 minutes of stirring. The filtrate is cooled, the resulting crystal suspension is stirred at 0°C to 5°C for 3 hours, the solids are filtered off, and dried at 50°C to 60°C.

23.3 g (94.4%) of 2,6-dihydroxymethylpyridine-bis(N-methylcarbamate) are obtained. The product melts at 134°C to 135°C; its purity is 99.8% (determined by UV spectrophotometry). When examined by thin layer chromatography, the product is uniform.

(B) 23.3 g of 2,6-dihydroxymethylpyridine-bis(N-methylcarbamate), prepared as described above, are dissolved in a boiling mixture of 46.6 ml of methanol and 46.6 ml of water. When the dissolution is complete, the solution is allowed to cool under slow stirring, without applying any external cooling means. The crystals start to separate at 48°C to 50°C. When the temperature of the mixture falls spontaneously below 35°C, it is cooled externally to 0°C to 5°C, and allowed to stand at this temperature for about 8 hours. The separated substance is filtered off and dried at 50°C to 100°C. 22.65 g of 2,6-dihydroxymethylpyridine-bis(N-methylcarbamate) are obtained. The quality of the product meets pharmaceutical requirements.

The yield of this crystallization procedure is 95.7%. The above process provides the γ_2 modification of 2,6-dihydroxymethylpyridine-bis(N-methylcarbamate), which can be tabletted directly. The substance melts at 134°C to 136°C, its purity is 99.9% (determined by UV spectrophotometry).

References

Merck Index 7874
Kleeman & Engel p. 787
DOT 5 (1) 16 (1969)
I.N. p. 826
Sprung, M., Toth, J., Kovatsits, M., Sztrokay, K., Szen, T., Gorgenyi, K., Boor, A., Forgacs, L.,Szabo, J. and Kruzics, A.; British Patent 1,548,334; July 11, 1979; assigned to Richter Gedeon Vegyeszeti Gyar R.T. (Hungary)

PYRIDOFYLLINE

Therapeutic Function: Coronary vasodilator

Chemical Name: 1H-Purine-2,6-dione, 3,7-dihydro-1,3-dimethyl-7-(2-(sulfooxy)ethyl)-, compd. with 5-hydroxy-6-methyl-3,4-pyridinedimethanol (1:1)

Common Name: Pyridofylline; Theodoxine

Structural Formula:

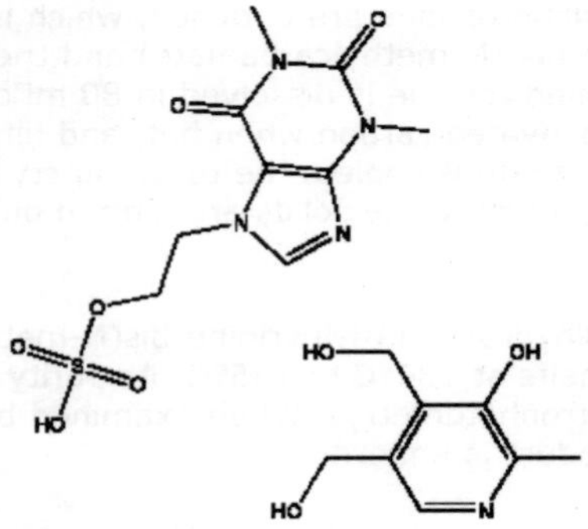

Chemical Abstracts Registry No.: 53403-97-7

Trade Name	Manufacturer	Country	Year Introduced
Pyridofylline	Debarge	-	-

Raw Materials

Theophylline
Chlorosulfonic acid
Pyridoxine
Potassium hydroxide
Monochlorhydrin ethylene glycol

Manufacturing Process

A solution of 100 g of theophylline in 500 ml 1 M solution of KOH was prepared potassium theophylline. To that potassium theophylline was added 120 ml monochlorhydrin ethylene glycol, a mixture was heated at 130°C for 4 hours. The product was dissolved in ethanol and filtered. After crystallization was obtained 7-(2-hydroxyethyl)theophylline.

100 g of 7-(2-hydroxyethyl)-theophylline was refluxed in 2.1 L of dry chloroform, then was added dropwise 47 g of HSO_3Cl and mixture was refluxed for 2 hours. After filtration the solid product [O-(7-theophyllinylethyl) sulfuric acid] was washed with chloroform and ether and dried in vacuum for 4 hours at 80°C.

Pyridofylline was obtained by mixing 110 g o-(7-theophyllinylethyl) sulfuric acid and 72 g pyridoxine in ethanol.

References

Albert M., Debarge E.J.J.; Brevet Special de Medicament; FR 828M, Dec. 23, 1960

PYRIDOSTIGMINE BROMIDE

Therapeutic Function: Cholinergic

Chemical Name: 3-[[(Dimethylamino)carbonyl]oxy]-1-methylpyridinium bromide

Common Name: -

Structural Formula:

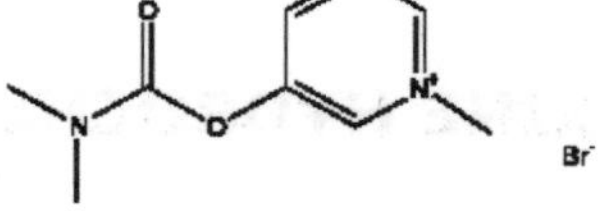

Chemical Abstracts Registry No.: 101-26-8

Trade Name	Manufacturer	Country	Year Introduced
Mestinon	Roche	US	1955
Mestinon	Roche	Japan	1970
Regonol	Organon	US	1973
Mestinon	Roche	France	1981
Kalymin	Arzneimittelwerk Dresden	E. Germany	-

Raw Materials

3-Hydroxypyridine
Dimethyl carbamic acid chloride
Methyl bromide

Manufacturing Process

12 parts by weight of dimethyl-carbamic acid chloride, dissolved in 20 parts by weight of xylol, are added dropwise to a boiling solution of 19 parts by weight of 3-hydroxypyridine in 120 parts by weight of xylol. Heating is continued under reflux for 3 hours. When the solution has cooled down, it is separated from the precipitated 3-hydroxypyridine hydro chloride and washed with water. After drying over sodium sulfate, the xylol is distilled off and the residue fractionated under reduced pressure. The N,N-dimethyl-carbamic acid ester of 3-hydroxypyridine distills at 148°C under a pressure of 15 mm.

A solution of 20 parts by weight of methyl bromide in 30 parts by weight of acetone is added to a solution of 35 parts by weight of N,N-dimethyl-carbamic acid ester of 3-hydroxypyridine in 70 parts by weight of acetone. After standing for a lengthy period (1 or 2 days), the N,N-dimethyl-carbamic acid ester of 3-hydroxy-1-methyl-pyridinium-bromide separates. It can be recrystallized from absolute alcohol. The colorless, strongly hygroscopic crystals melt at 151°-152°C.

References

Merck Index 7877
Kleeman and Engel p. 789
PDR pp. 1289, 1491
I.N. p. 826
REM p. 900
Urban, R.; US Patent 2,572,579; October 23,1951; assigned to Hoffmann-La Roche Inc.

PYRIDOXINE HYDROCHLORIDE

Therapeutic Function: Enzyme cofactor vitamin

Chemical Name: 3,4-Pyridinedimethanol, 5-hydroxy-6-methyl-, hydrochloride

Common Name: Adermine hydrochloride; Piridossina hydrochloride; Piridoxina hydrochloride; Pyridoxine hydrochloride; Pyridoxinium chloride; Pyridoxol hydrochloride; Vitamin B_6 hydrochloride

Structural Formula:

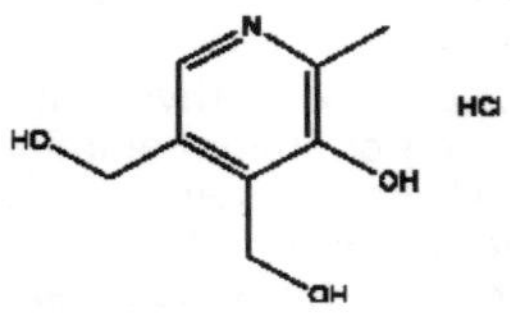

Chemical Abstracts Registry No.: 58-56-0; 65-23-6 (Base)

Trade Name	Manufacturer	Country	Year Introduced
Pyridoxine hydrochloride	Roche	-	-
Pyridoxine hydrochloride	Takeda Chemical Industry	Japan	-

Raw Materials

Formamide
Maleic anhydride
Sodium bicarbonate
Ethyl α-alaninate hydrochloride
Hydrogen chloride
Lithium aluminum hydride

Manufacturing Process

To 35 g of ethyl α-alaninate hydrochloride is added 10 g of formamide and the resulting mixture is heated slowly to 105°C over a period of 30 to 45 min. After heating at 105°C for 10 min, about 75 ml of toluene is added. After standing for about 1 h, the mixture is then refluxed for about 6 h. After cooling the ammonium chloride formed is removed and the resulting solution is evaporated to remove the solvent. The ethyl N-formyl-α-alaninate is obtained by distillation at 100°C.

To a moxture of 25 ml of alcohol free chloroform and 11.36 g of P_2O_5 is added over 20 min a solution of 5.81 g of ethyl N-formyl-α-alaninate in 15 ml of alcohol free chloroform at about 30°C the resulting reaction mixture is refluxed for 1 h, cooled and the solvent decanted. The hard mass remaining is broken up and a solution of 27 g of potassium hydroxide in 27 ml of water and 34 ml of methanol is gradually added keeping the temperature at 10-20°C. The resulting solution is refluxed for 1 h, cooled and extracted with 10 x 15 ml of methylene chloride. The 4-methyl-5-ethoxy oxazole is recovered, after removing the solvent, by distilling at 75-80°C at 10 mm pressure.

Upon mixing 1.27 g of 4-methyl-5-ethoxy oxazole (0.01 mole), 0.98 g of maleic anhydride (0.01 mole) and 2.5 ml of dry benzene, a yellow color appears and heat is evolved, requiring cooling. After 3-4 min the evolution of heat ceases and the color fades. The mixture is then refluxed for about 18 h, after which the solvent is decanted and the residue treated with a small quantity of water. To the residue is added 40 ml of ethanol and the solution is then saturated with gaseous HCl. The acidic solution is refluxed for 3.5 h. After cooling the solvent is evaporated and crystalline residue containing diethyl 2-methyl-3-hydroxy-pyridine-4,5-dicarboxylate hydrochloride is converted to the free base by reaction with aqueous sodium bicarbonate. The resulting solution is extracted with ether and the ether extacts dried.

The ether solution containing diethyl 2-methyl-3-hydroxy-pyridine-4,5-dicarboxylate is treated with 0.5 g of lithium aluminum hydride. The resulting mixture is stirred for 2 h and allowed to stand overnight. The ether layer is removed and the aqueous layer is saturated with carbon dioxide. The resulting residue is extracted three times with hot ethanol and gaseous HCl is passed into the ethanol extracts. After allowing the acidified ethanol solution to stand for 2-3 h, crystals of pyridoxine hydrochloride are deposited and recovered by filtration. Melting point 203.5°-205°C, dec.

References

Chase G.O.; US Patent No. 3,222,374; Dec. 7, 1965; Assigned: Hoffmann-La Roche Inc., Nutley, N. J.

Coffen D.L.; US Patent No. 4,026,901; May 31, 1977; Assigned: Hoffmann-La Roche Inc., Nutley, N. J.

PYRILAMINE

Therapeutic Function: Antihistaminic

Chemical Name: N-[(4-Methoxyphenyl)methyl]-N',N'-dimethyl-N-2-pyridinyl-1,2-ethanediamine (often used as the maleate)

Common Name: Mepyramine; Pyranisamine

Structural Formula:

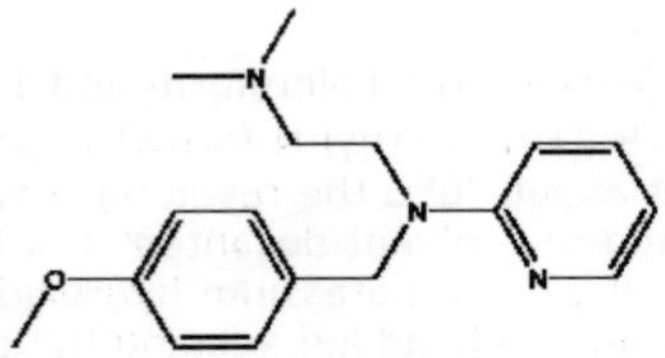

Chemical Abstracts Registry No.: 91-84-9; 6036-95-9 (Hydrochloride salt); 59-33-6 (Maleate salt)

Trade Name	Manufacturer	Country	Year Introduced
Neo-Antergan	MSD	US	1948
Thylogen	Rorer	US	1949
Statomin	Bowman	US	1950

Trade Name	Manufacturer	Country	Year Introduced
Pyra-Maleate	Mallinckrodt Inc.	US	1950
Copsmine	Durst	US	1950
Stamine	Tutag	US	1951
Albatussin	Bart	US	-
Allergan	Wiedenmann	Switz.	-
Amfeta	Bama-Geve	Spain	-
Anthisan	May and Baker	UK	-
Citra Forte	Boyce	US	-
Codimal	Central	US	-
Copsamine	Durst	US	-
Fiogesic	Sandoz	US	-
Histalet	Reid-Rowell	US	-
Histavet-P	Burns-Biotec	US	-
Kontristin	Eczacibasi	Turkey	-
Kriptin	Whitehall	US	-
Kronohist	Ferndale	US	-
Midol PMS	Glenbrook	US	-
Poly-Histine	Bock	US	-
Primatene	Whitehall	US	-
PV-Tussin	Reid-Rowell	US	-
Pyra	Mallinckrodt Inc.	US	-
Pyramal	Columbus	US	-
Statomin	Bowman	US	-
Triaminic	Dorsey	US	-

Raw Materials

4-Methoxybenzaldehyde
2-Aminopyridine
1-Dimethylamino-2-chloroethane
Sodium amide

Manufacturing Process

43 g of α-p-methoxybenzylaminopyridine (from 4-methoxybenzaldehyde reaction with 2-aminopyridine) are heated in 60 cc of toluene to 95°C to 100°C. 18 g of sodamide (85%) and 110 cc of a 40% toluene solution of 1-dimethylamino-2-chloroethane are added in small amounts alternately with shaking; the addition takes 1 hour. Toluene is distilled off, first at normal pressure, then under reduced pressure, until there remains a pasty mass. The mass is taken up with dilute hydrochloric acid and ether, neutralized to pH 7, and p-methoxybenzylaminopyridine separates. After making alkaline using excess of potash, it is extracted with benzene, dried and distilled. The product thereby obtained, N',N'-dimethylaminoethyl-N-p-methoxybenzyl-α-aminopyridine boils at 185°C to 190°C/2 mm. The monohydrochloride melts at 135°C (block Maquenne).

References

Merck Index 7883
Kleeman and Engel p. 561

PDR pp. 654, 674, 692, 784, 850, 875, 925, 1447, 1583, 1900
OCDS Vol. 1 p.51 (1977)
I.N. p. 597
REM p. 1129
Horclois, R.J.; US Patent 2,502,151; March 28, 1950; assigned to Societe des Usines Chimiques Rhone-Poulenc

PYRIMETHAMINE

Therapeutic Function: Antimalarial

Chemical Name: 5-(4-Chlorophenyl)-6-ethyl-2,4-pyrimidinediamine

Common Name: -

Structural Formula:

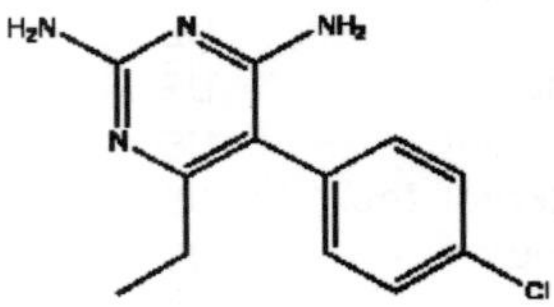

Chemical Abstracts Registry No.: 58-14-0

Trade Name	Manufacturer	Country	Year Introduced
Daraprim	Burroughs-Wellcome	US	1953
Daraprim	Burroughs-Wellcome	W. Germany	1969
Erbaprelina	Erba	Italy	-
Fansidar	Roche	France	-
Malocide	Specia	France	-
Pirimecidan	Cidan	Spain	-
Pyrimethamin-Heyl	Heyl	W. Germany	-
Tindurin	EGYT	Hungary	-

Raw Materials

p-Chlorophenylacetonitrile
Guanidine
Diazomethane
Sodium ethoxide
Ethyl propionate

Manufacturing Process

p-Chlorophenylacetonitrile (36.5 grams) and ethyl propionate (25.5 grams) were added to a solution of sodium ethoxide (from 5.75 grams sodium) in absolute ethanol (150ml). The solution was heated on a steam bath for 6 hours. After cooling, the whole was poured into water and the oil extracted well with ether, the ether solution was discarded and the aqueous solution neutralized with 1 N sulfuric acid. A heavy oil separated which was taken into ether, washed with water, bicarbonate solution and again with water. After drying, the ether was removed to give a thick oil which solidified on standing (34.6 grams). After recrystallization from an ether-petroleum ether mixture it formed needles, MP 108°-112°C.

The above keto-nitrile (15 grams) was methylated with a solution of diazomethane in ether. (The diazomethane solution was prepared using 20 grams of N-nitrosomethylurea.) The ether and excess diazomethane were evaporated on the steam bath and the oil dissolved in ethanol (50 ml). To this was added a solution of guanidine in ethanol (100 ml) (prepared from 8.1 grams of the hydrochloride). The solution was refluxed for 5 hours, the alcohol removed and the residue treated with 5 N sodium hydroxide. The insoluble material was then filtered. After purification by precipitation from dilute acetic acid with sodium hydroxide and by recrystallization from ethanol the product formed clear colorless needles (8.0 grams), MP 218°-220°C as described in US Patent 2,602,794.

References

Merck Index 7884
Kleeman & Engel p. 791
PDR pp. 741, 1484
OCDS Vol. 1 p. 262 (1977)
DOT 16 (5) 174 (1980)
I.N. p.827
REM p. 1219
Hitchings, G.H., Russell, P.B. and Falco, E.A.; US Patent 2,576,939; December 4, 1951; assigned to Burroughs Wellcome & Co. (USA.) Inc.
Hitchings, G.H. and Falco, E.A.; US Patent 2,579,259; December 18, 1951; assigned to Burroughs Wellcome & Co. (USA.) Inc.
Hitchings, G.H., Russell, P.B. and Falco, E.A.; US Patent 2,602,794; July 8, 1952; assigned to Burroughs Wellcome & Co. (USA.) Inc.
Jacob, R.M.: US Patent 2,680,740; June 8,1954; assigned to Societe des Usines Chimiques Rhone-Poulenc (France)

PYRISUCCIDEANOL DIMALEATE

Therapeutic Function: Cerebrotonic

Chemical Name: 2-(Dimethylamino)ethyl(5-hydroxy-4-(hydroxymethyl)-6-methyl-3-pyridyl)methyl succinate salt with maleic acid (1:2)

Common Name: Pirisuccideanoli maleas; Pirisudanol maleate; Pyrisuccideanol dimaleate

Chemical Abstracts Registry No.: 53659-00-0; 33605-94-6 (Base)

Structural Formula:

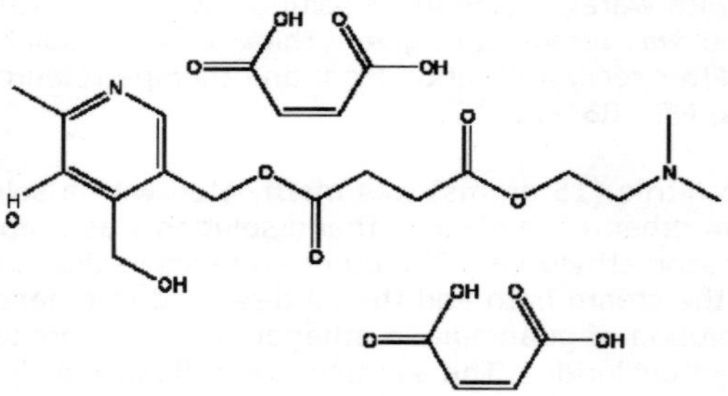

Trade Name	Manufacturer	Country	Year Introduced
Nadex Forte	Novartis Pharma	-	-

Raw Materials

Succinic anhydride
Thionyl chloride
Formic acid
Pyridoxine hydrochloride
Dimethylamino ethanol
Pyridine
Maleic acid

Manufacturing Process

Process of preparation of 2-(dimethylamino)ethyl(5-hydroxy-4-(hydroxymethyl)-6-methyl-3-pyridyl)methyl succinate includes four steps.

a. Into a 4 liter flask provided with a stirrer there were poured 2.1 liters of pure and anhydrous acetone, and 100 g of dried pyridoxine hydrochloride were added under stirring. The mixture was cooled whilst stirring at 0°C and then gaseous HCl was bubbled through the solution for 6 hours. After another hour the temperature was allowed to rise to room temperature and the stirring was maintained for a further hour. The mixture was then cooled to -15°C and the 3,4-isopropylidene pyridoxine was obtained in the form of its hydrochloride (100 g). The free base was obtained by treatment above hydrochloride with a solution of Na_2CO_3. 82 g of base was obtained melting at 111°C.

b. 100 g of succinic anhydride, 100 g of dimethylaminoethanol and 100 ml of anhydrous acetone were placed into a 2 liter flask and were refluxed for 3 hours. The solution was then concentrated to one third of its original volume by evaporation and cooled. A precipitate appeared, which was separated and recrystallized from acetone. 140 g of product was obtained, with a melting point of 78°C.

The chloride was prepared from this compound by treatment of 420 g of the compound with freshly distilled $SOCl_2$ 1.85 liter. After elimination of non-reacted $SOCl_2$, the product obtained was treated by benzene and dried, to yield 620 g of the hydrochloride of the chloride acid. 120 g of the compound of step (a) above (0.57 mole) were dissolved in 0.5 liter of pyridine. After cooling there were slowly added at about 5°C, during 90 minutes, 170 g (0.69 mole) of the hydrochloride previously obtained, dissolved in 0.2 liter of chloroform. The solution was stirred for 10 hours, then evaporated to dryness (350 g). The residue was dissolved in 0.3 liter of water and was neutralized by an aqueous solution of NH_3 saturated by K_2CO_3. There was obtained an oily substance which was extracted with chloroform. The extract was concentrated to dryness (170 g).

c. In this step the blocking group linking the OH in position 3 to the CH_2OH in position 4 of the pyridoxine ring was broken by hydrolysis with formic acid. 52 g of the product of step (b) above were treated with 1.650 liter of 1% solution of formic acid and 0.250 liter of ethyl alcohol. The mixture was boiled for 30 minutes, evaporated again treated with ethyl alcohol and evaporated. There were obtained 37 g of an oily substance.

d. Maleate was obtained by reacting 37 g (0.115 mole) of above product of step (c) dissolved in 120 ml of acetone with 27 g (0.230 mole) of maleic acid dissolved in 130 ml of acetone. Yield 51 g. MP: 134°C.

References

Esanu A.; US Patent No. 3,717,636; Feb. 20, 1973; Assigned to Societe d'Etudes Produits Chimiques Issy-Les-Moulineaux, France

PYRITHYLDIONE

Therapeutic Function: Hypnotic, Sedative

Chemical Name: 3,3-Diethyl-2,4-(1H,3H)pyridinedione

Common Name: -

Structural Formula:

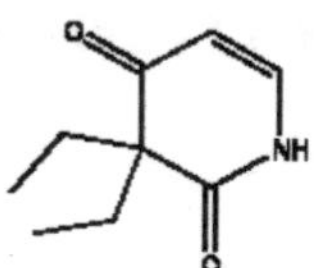

Chemical Abstracts Registry No.: 77-04-3

Trade Name	Manufacturer	Country	Year Introduced
Presidon	Roche	US	1948
Persedon	Roche	W. Germany	-

Raw Materials

Methyl formate
Diketene
Ethyl bromide
Sodium methylate
Ammonia

Manufacturing Process

108 g of sodium methylate were suspended in 500 ml of toluene. 120 g of methyl formate were dropped into the sodium methylate suspension thus formed at a rate so that temperature did not exceed 30°C. Thereafter a solution of 157 g of α,α-diethylacetoacetamide in 500 ml of toluene were added so that the temperature did not exceed 50°C. The mixture was stirred for one hour at 50°C and then overnight at room temperature. The reaction mixture was poured into 700 ml of ice water, permitted to stratify, the aqueous layer was separated, covered with a layer of 200 ml of toluene and then treated while stirring with 200 g of 50% sulfuric acid. Finally the reaction mixture, which was acid to congo red, was warmed at 50°C and the toluene containing layer was separated. The aqueous layer was extracted with four 200 ml portions of toluene at 50°C and then discarded. The toluene extracts were combined and then concentrated in vacuo at 60°C. There were obtained 135 g of crystalline residue which was recrystallized from 200 ml of toluene. The 3,3-diethyl-2,4-dioxo-1,2,3,4-tetrahydropyridine thus obtained melted at 96°C.

The α,α-diethylacetoacetamide used as starting material was obtained by converting diketene with aqueous ammonia to acetoacetamide and alkylating twice with ethyl bromide in the presence of sodium alcoholate.

References

Merck Index 7893
Kleeman & Engel p. 793
I.N. p. 828
Hinderling, R .,Lutz, A.H. and Schnider, O.; U.S. Patent 3,019,230; January 30, 1962; assigned to Hoffmann-La Roche Inc.

PYRITINOL

Therapeutic Function: Neurotropic

Chemical Name: 3,3'-(Dithiodimethyiene)bis[5-hydroxy-6-methyl-4-pyridine methanol]

Common Name: Pyrithioxin

Chemical Abstracts Registry No.: 1098-97-1; 10049-83-9 (Dihydrochloride salt)

Structural Formula:

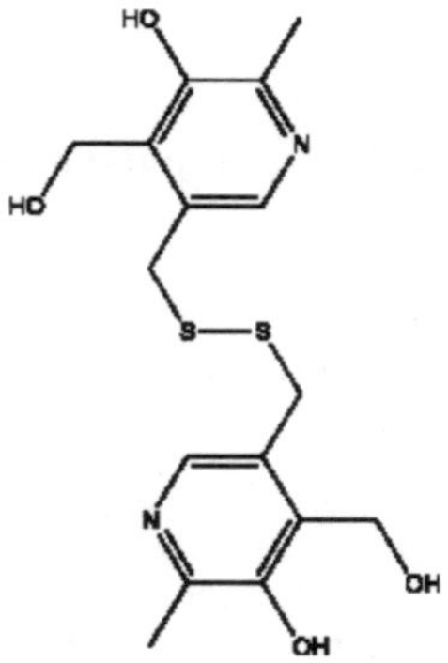

Trade Name	Manufacturer	Country	Year Introduced
Encephabol	Merck	W. Germany	1963
Enbol	Chugai	Japan	1971
Biocefalin	Benvegna	Italy	-
Bonol	Ikapharm	Israel	-
Cefalogen	Montefarmaco	Italy	-
Cerebropirina	Chemil	Italy	-
Cerebrotrofina	N.C.S.N.	Italy	-
Cervitalin	Savoma	Italy	-
Chioebon	Kyowa Yakuhin Osaka	Japan	-
Divalvon	Nippon Kayaku, Co.	Japan	-
Encebrovit	Sierochimica	Italy	-
Encefabol	Bracco	Italy	-
Encefort	Intersint	Italy	-
Encerebron	Pulitzer	Italy	-
Enerbol	Polfa	Poland	-
Evolubran	A.B.C.	Italy	-
Fulneurina	Fulton	Italy	-
Gladius	SKF	Italy	-
Leonar	Kalopharma	Italy	-
Life	S.I.T.	Italy	-
Maind	Also	Italy	-
Miriplex	Poli	Italy	-
Musa	Poli	Italy	-
Neurotin	Nakataki	Japan	-
Neuroxin	Yamanouchi	Japan	-
Piritinol	Magis	Italy	-
Piritiomin	Hishiyama	Japan	-
Sawaxin	Sawai	Japan	-
Scintidin	I.C.I.	Italy	-
Tonobrein	C.T.	Italy	-
Tonomentis	Ion	Italy	-

Raw Materials

Methanol
Potassium xanthogenate
Ammonia
3,4-Bisbromoethyl-4-hydroxy-5-methyl-pyridinium bromide

Manufacturing Process

To a solution of 60 g of potassium xanthogenate in 240 cc of water there is added dropwise, while being cooled with ice, a solution of 42 g of 3,4-bis-bromomethyl-4-hydroxy-5-methyl-pyridinium-bromide in 1 liter of water so that the temperature remains between 2°C and 5°C. After stirring for 1 hour at the same temperature, the water is decanted off and the residue is triturated with acetone. Yield: 25 g of 4-hydroxymethyl-5-hydroxy-6-methyl-pyridyl-(3)-methylxanthogenate; melting point: 170°C to 171°C (alcohol, decomposition).

40 g of 4-hydroxymethyl-5-hydroxy-6-methyl-pyridyl-(3)-methylxanthogenate are left standing at room temperature for 5 days in a mixture of 800 cc of alcohol and 400 cc of aqueous NH_3-solution, and subsequently concentrated under vacuum to about 50 cc. The precipitated bis(4-hydroxymethyl-5-hydroxy-6-methyl-3-pyridylmethyl) disulfide is sucked off. Yield: 20 g of the disulfide; melting point: 218°C to 220°C (butanol, decomposition).

References

Merck Index 7894
Kleeman & Engel p. 793
DOT 9 (6) 215 (1973)
I.N. p. 828
Zima, O. and Schorre, G.; US Patent 3,010,966; November 28, 1961; assigned to E. Merck A.G. (Germany)

PYROVALERONE HYDROCHLORIDE

Therapeutic Function: Psychostimulant

Chemical Name: 1-(4-Methylphenyl)-2-(1-pyrrolidinyl)-1-pentanone hydrochloride

Common Name: -

Chemical Abstracts Registry No.: 1147-62-2; 3563-49-3 (Base)

Trade Name	Manufacturer	Country	Year Introduced
Thymergix	Joullie	France	1973

Structural Formula:

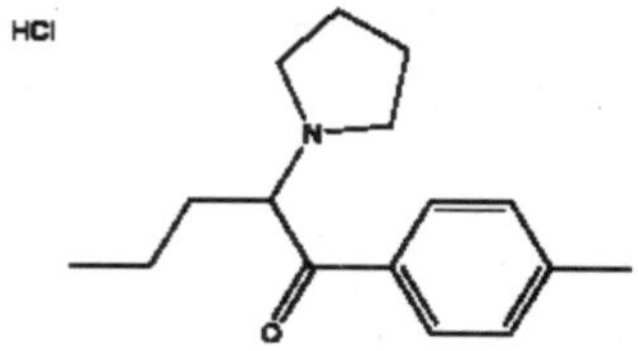

HCl

Raw Materials

p-Methylvalerophenone
Bromine
Pyrrolidine
Hydrogen chloride

Manufacturing Process

23.1 grams of α-bromo-p-methyl-valerophenone, obtained by bromination of p-methyl-valerophenone, are dissolved in 50 ml of benzene and 25 ml of pyrrolidine are added at 0°C. The whole is boiled for 20 minutes, cooled, washed twice with water, dried and acidified with about 50 ml of 2 N hydrochloric acid. After evaporation, it is recrystallized from methanol-acetone-ether. 22.6 grams of α-pyrrolidino-p-methyl-valerophenone hydrochloride, melting point 178°C, equivalent to a yield of 88.5% of the theoretical are obtained according to British Patent 927,475.

References

Merck Index 7914
Kleeman and Engel p. 794
OCDS Vol. 2 p. 124 (1980)
DOT 10 (5) 188 (1974)
I.N. p. 829
Dr. A. Wander SA, Switzerland; British Patent 927,475; May 29, 1963
Dr. Karl Thomae, GmbH, Germany; British Patent 933,507; August 8, 1963

PYRROBUTAMINE

Therapeutic Function: Antihistaminic

Chemical Name: 1-[4-(4-Chlorophenyl)-3-phenyl-2-butenyl]-pyrrolidine

Common Name: -

Chemical Abstracts Registry No.: 91-82-7

Structural Formula:

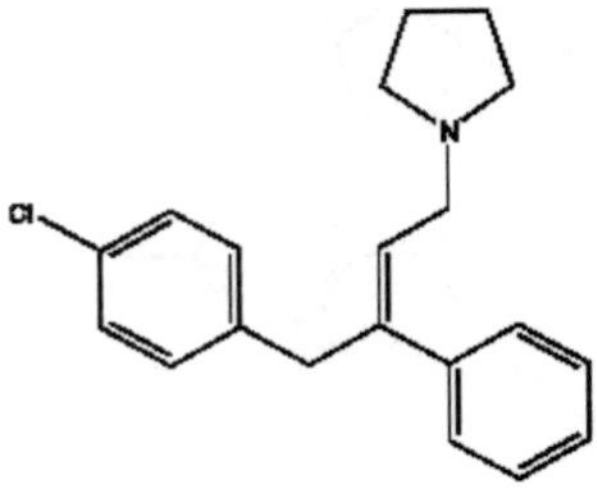

Trade Name	Manufacturer	Country	Year Introduced
Pyronil	Lilly	US	1952
Co-Pyronil	Lilly	UK	-
Proladyl	Lilly	-	-

Raw Materials

Pyrrolidine
Paraformaldehyde
Magnesium
Acetophenone
p-Chlorobenzyl chloride
Hydrogen chloride

Manufacturing Process

A mixture of 1,800 ml of absolute ethanol, 427 g (6 mols) of pyrrolidine, and a trace of methyl orange is cooled in an ice bath and gaseous hydrogen chloride is bubbled through the mixture until a red color develops, indicating that all of the amine has been converted to the hydrochloride. The addition of hydrogen chloride is stopped, the ice bath is removed and to the solution are added 720 g of acetophenone, 270 g of paraformaldehyde and 10 ml of concentrated hydrochloric acid. The mixture is stirred and refluxed vigorously for one hour. An additional 180 g of paraformaldehyde are then added, and refluxing is continued for about three hours. The hot solution is poured into 6 liters of acetone and the mixture is chilled overnight. A precipitate of ω-(N-pyrrolidino)-propiophenone hydrochloride separates. The precipitate is filtered off, washed with cold acetone, and dried in air.

ω-(N-pyrrolidino)-propiophenone hydrochloride thus prepared melted at about 163°C to 164°C after recrystallization from acetone.

To a suspension of 4 mols of ω-(N-pyrrolidino)-propiophenone hydrochloride in 1,500 ml of water and 100 g of ice in a separatory funnel are added a 50% aqueous solution containing 200 g of sodium hydroxide, and 2 liters of ether. The mixture is shaken vigorously until all of the suspended matter dissolves. The ether is then removed, washed with 1 liter of water and dried over anhydrous magnesium sulfate. The anhydrous ether solution of ω-(N-pyrrolidino)-propiophenone thus prepared is added to a Grignard reagent prepared from 6 mols of p-chlorobenzyl chloride and 6 mols of magnesium turnings in 3,000 ml of anhydrous ether. The ethereal solution of the ketone is added to the Grignard reagent at such a rate that rapid refluxing is maintained. After all of the ketone has been added, the reaction mixture is stirred for 2 hours and is decomposed by pouring it over a mixture of 500 g of ice and 6 mols of concentrated hydrochloric acid. The hydrochloric acid addition salt of 1-p-chlorophenyl-2-phenyl-4-N-(pyrrolidino)-butanol-2 formed in the reaction separates at the ether-water interface as a white crystalline material. The aqueous phase is removed and discarded, and the mixture of ether and hydrochloride salt is converted to 1-p-chlorophenyl-2-phenyl-4-(N-pyrrolidino)-butanol-2 by treatment with 10% sodium hydroxide solution. The base is removed by extraction with ether, and the ether extracts are dried over magnesium sulfate.

1-p-chlorophenyl-2-phenyl-4-(N-pyrrolidino)-butanol-2 melted at about 109°C to 110°C after recrystallization from petroleum ether.

A solution of 200 g of 1-p-chlorophenyl-2-phenyl-4-(N-pyrrolidino)-butanol-2 in 750 ml of concentrated hydrochloric acid is refluxed for 9 hours thereby causing a dehydration of the butanol compound. and the formation of the hydrochloric acid addition salt of a 1-p-chlorophenyl-2-phenyl-4-(N-pyrrolidino)-butene. The hydrochloride salt formed crystallizes in the oily lower layer of the two phase reaction mixture and is removed therefrom by filtration. The filtrate is again refluxed for 9 hours, cooled to 0°C, and a second crop of the hydrochloric acid addition salt of the dehydration product is obtained and filtered off. The filtrate containing residual amounts of 1-p-chlorophenyl-2-phenyl-4-(N-pyrrolidino)-butanol-2 is again refluxed for 9 hours to yield an additional crop of the salt of the dehydration product. The several fractions of the butene compound are combined and triturated with several small portions of hot acetone and recrystallized from alcohol-ether mixture. The hydrochloric acid addition salt of the dehydration product, 1-p-chlorophenyl-2-phenyl-4-(N-pyrrolidino)-butene hydrochloride, melts at about 227°C to 228°C.

References

Merck Index 7916
Kleeman & Engel p. 794
OCDS Vol. 1 p. 78 (1977)
I.N. p. 829 Mills, J.; US Patent 2,655,509; October 13, 1953; assigned to Eli Lilly & Co.

PYRROLNITRIN

Therapeutic Function: Antifungal

Chemical Name: 1H-Pyrrole, 3-chloro-4-(3-chloro-2-nitrophenyl)-

Common Name: Pyrrolnitrin

Chemical Abstracts Registry No.: 1018-71-9

Pyrrolnitrin

Structural Formula:

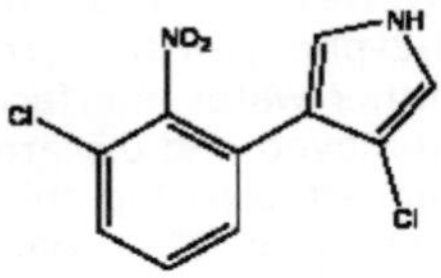

Trade Name	**Manufacturer**	**Country**	**Year Introduced**
Micutrin	Searle Farmaceutici	-	-
Lilly 52230	Eli Lilly and Company	-	-

Raw Materials

Piperidine
Sodium
Sulfuryl chloride
1-(2-Nitro-3-chlorophenyl)-1,3-butanedione
Diethyl aminomalonate

Manufacturing Process

A mixture of 2.0 g of 1-(2-nitro-3-chlorophenyl)-1,3-butanedione, 1.9 g of diethyl aminomalonate, 1.5 ml of absolute ethyl alcohol and two drops of piperidine was refluxed for 5 hours. After cooling, the reaction mixture was allowed to stand and then crystals were separated. The crystals were collected by filtration and then dried to obtain 2.5 g of colorless crystals. The crystals were recrystallized from a mixed solvent of benzene and ether to obtain diethyl N-[1-methyl-3-(2-nitro-3-chlorophenyl)-3-oxopropylidene]aminomalonate as colorless needles having MP: 134°-136°C.

A solution of 0.8 g of diethyl N-[1-methyl-3-(2-nitro-3-chlorophenyl)-3-oxopropylidene]aminomalonate in 4 ml of absolute tetrahydrofuran was added dropwise with stirring to a solution prepared with 8 ml of absolute ethanol and 100 mg of metallic sodium. After the reaction mixture was refluxed for 4.5 hours, the solvents were distilled off under reduced pressure. The residue was added with an ice-water and the solution was extracted with ether. The extract was washed with water, dried over anhydrous magnesium sulfate, after which ether was distilled off. The residue was recrystallized from benzene to obtain ethyl 3-(2-nitro-3-chlorophenyl)-5-methylpyrrole-2-carboxylate as colorless needles having MP: 220°-223°C.

1.2 g of ethyl 3-(2-nitro-3-chlorophenyl)-5-methyl-pyrrole-2-carboxylate is suspended in 12 ml of acetic acid. A solution of 2.1 g of sulfuryl chloride in 3 ml of acetic acid is added dropwise to the suspension with stirring at about 200°C. The reaction mixture stood overnight is stirred for one hour at 30°C, one hour at 40°C, and then two hours at 50°C. Thereafter, this mixture is poured into ice water. The mixture is extracted with ethyl acetate. The extract is washed with an aqueous solution of potassium hydrogen carbonate and dried over anhydrous magnesium sulfate, and then the solvent was distilled off. Ethyl 3-(2-nitro-3-chlorophenyl)-4-chloro- 5-trichloromethylpyrrole-2-carboxylate is obtained as pale-brown viscous oil.

A mixture of ethyl 3-(2-nitro-3-chlorophenyl)-4-chloro-5-

trichloromethylpyrrole-2-carboxylate, prepared from 40 mg of ethyl 3-(2-nitro-3-chlorophenyl)-4-chloro- 5-methylpyrrole-2-carboxylate and sulfuryl chloride, and 5 ml of 10% sodium hydroxide aqueous solution was heated for two hours on a water bath. The reaction mixture was acidified with 10% sulfuric acid and the resultant acidic solution was extracted with ethyl acetate and the extract was dried. The solvent was distilled off to yield 370 mg of 3-(2-nitro-3-chlorophenyl)-4-chloro-pyrrole-2,5-dicarboxylic acid having a melting point of 298°C. (decomp.).

Decarboxylation of this dicarboxylic acid gave almost quantitavily the disered 3-chloro-4-(2'-nitro-3'-chlorophenyl)pyrrole; MP: 125°C. (recrystallized from benzene).

References

Umio S. et al.; US Patent No. 3,428,648; Feb. 18, 1969; Assigned to Fujisawa Pharmaceutical Co., Ltd., Osaka, a comoany of Japan

Nakano H., Umio S. et al., Tetrahedron Letters No 7 pp. 737-740, 1966

PYRVINIUM PAMOATE

Therapeutic Function: Anthelmintic

Chemical Name: 6-(Dimethylamino)-2-[2-(2,5-dimethyl-1-phenyl-1H-pyrro-3-yl)ethenyl]-1-methylquinolium salt with pamoic acid (2:1)

Common Name: Pyrvinium embonate; Viprynium embonate

Structural Formula:

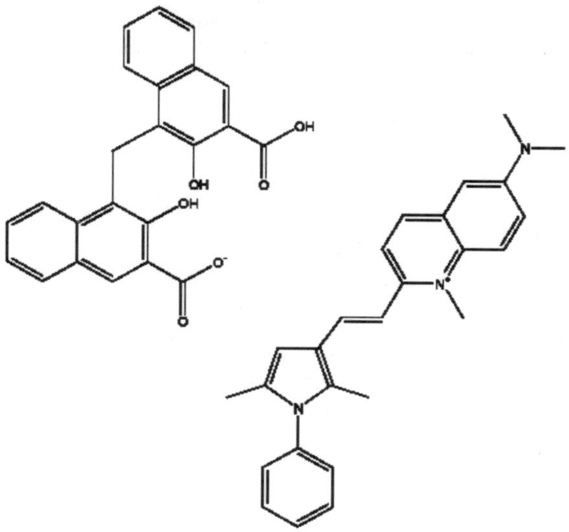

20212

Pyrvinium pamoate

Chemical Abstracts Registry No.: 3546-41-6

Trade Name	Manufacturer	Country	Year Introduced
Povan	Parke Davis	US	1959
Povanyl	Parke Davis	France	1981
Antioxur	Esteve	Spain	-
Molevac	Parke Davis	W. Germany	-
Neo-Oxypaat	Katwijk	Netherlands	-
Oxialum	Wolner	Spain	-
Pamovin	Merck-Frosst	Canada	-
Pamoxan	Uriach	Spain	-
Pirok	Bilim	Turkey	-
Poquil	Parke Davis Sankyo	Japan	-
Privonium	Rivopharm	Switz.	-
Pyrcon	Jenapharm	E. Germany	-
Pyrvin	Farmos	Finland	-
Tolapin	Taro	Israel	-
Tru	Elea	Argentina	-
Vanquin	Parke Davis	Italy	-
Vermitiber	Tiber	Italy	-

Raw Materials

Pyrvinium chloride
Sodium pamoate

Manufacturing Process

A hot, filtered solution of 2.27 grams of pyrvinium chloride dihydrate in 250 ml of water is added slowly to a solution of 2.25 grams of sodium pamoate monohydrate in 50 ml of water. A red precipitate immediately forms. The mixture is heated at about 90°-100°C for 5 minutes more and then filtered. The reaction product is washed with hot water and dried at about 75°C in a vacuum. This preparation melts at about 210°-215°C with prior softening from about 190°C.

References

Merck Index 7927
Kleeman & Engel p. 796
PDR p. 1384
I.N. p. 830
REM p. 1237
Van Lare, E. and Brooker, L.G.S.; US Patent 2,515,912; July 18, 1950; assigned to Eastman Kodak Company
Elslager, E.F. and Worth, D.F.; US Patent 2,925,417; February 16, 1960; assigned to Parke, Davis & Company